Studies in Surface Science and Catalysis 55

NEW DEVELOPMENTS IN SELECTIVE OXIDATION

Studies in Surface Science and Catalysis

Advisory Editors: B. Delmon and J.T. Yates

Vol. 55

NEW DEVELOPMENTS IN SELECTIVE OXIDATION

Proceedings of an International Symposium, Rimini, Italy, September 18–22, 1989

Editors

G. Centi and F. Trifiro'
Department of Industrial Chemistry and Materials, University of Bologna, V.le Risorgimento 4, 40136 Bologna, Italy

ELSEVIER Amsterdam — Oxford — New York — Tokyo 1990

ELSEVIER SCIENCE PUBLISHERS B.V.
Sara Burgerhartstraat 25
P.O. Box 211, 1000 AE Amsterdam, The Netherlands

Distributors for the United States and Canada:

ELSEVIER SCIENCE PUBLISHING COMPANY INC.
655, Avenue of the Americas
New York, NY 10010, U.S.A.

Library of Congress Cataloging-in-Publication Data

New developments in selective oxidation : proceedings of an international symposium, Rimini, Italy, September 18-22, 1989 / editors, G. Centi and F. Trifirò.
p. cm. -- (Studies in surface science and catalysis ; 55)
"Papers presented at the International Symposium on New Developments in Selective Oxidation ... organized by the Department of Industrial Chemistry and Materials of the University of Bologna in collaboration with the Catalysis Group of the Italian Chemical Society"--Pref.
Includes bibliographical references.
ISBN 0-444-88694-X
1. Oxidation--Congresses. I. Centi, G. (Gabriele), 1955- . II. Trifirò, F. (Ferruccio), 1938- . III. International Symposium on New Developments in Selective Oxidation (1989 : Rimini, Italy) IV. Università di Bologna. Dept. of Industrial Chemistry and Materials. V. Società chimica italiana. Catalysis Group. VI. Series.
TP156.09N48 1990
660'.2993--dc20 90-2988
CIP

ISBN 0-444-88694-X

This book is printed on acid-free paper.

Printed in The Netherlands

CONTENTS

Preface XIII

International Advisory Board, Organizing Committee and Financial Support XIV

Prospectives in Selective Liquid-Phase Oxidation with H_2O_2 and O_2

Catalytic oxidations in the manufacture of fine chemicals
Sheldon, R.A. 1

Selective oxidation with Ti-Silicalite
Romano, U., Esposito, A., Maspero, F., Neri, C., Clerici, M. 33

Cyclohexanone ammoximation: a break-through in the 6-caprolactam production process
Roffia, P., Leofanti, G., Cesana, A., Mantegazza, M., Padovan, M., Petrini, G., Tonti, S., Gervasutti,P. 43

Modified zeolites for oxidation reactions
Ferrini, C., Kouwenhoven, H.W. 53

Mn(III)-tetraarylporphyrins bearing a covalently anchored axial ligand: efficient catalysts in olefin epoxidations under two-phase conditions
Anelli, P.L., Banfi, S., Montanari, F., Quici, S. 63

Alkene epoxidation with hydrogen peroxide catalyzed by metalloporphyrins
Goor,G., Prescher,G., Schmidt, M. 71

Use of bimetallic systems for the selective oxidation of olefins with hydrogen peroxide
Strukul, G.,Zanardo, A., Pinna, F. 81

Platinum catalysts for cyclohexene epoxidation with an oxygen- hydrogen mixture
Kuznetsova, N.I., Lisitsyn, A.S., Boronin, A.I., Likholobov, V.A. 89

Selective oxidation of alcohols to carbonyl compounds by manganese (III) μ-oxo carboxylato complexes
Mimoun, H., Saussine, L., Menage, S., Girerd, J.-J. 97

Selective oxidations catalyzed by dioxo(porphyrinato) Ru(VI) species
Rajapakse N., James, B.R., Dolphin, D. 109

Ruthenium(II) catalysts for the homogeneous oxygenation of aliphatic hydrocarbons and ethers
Bressan, M., Morvillo, A. 119

Direct conversion of benzene to phenols under ambient conditions
Sasaki, K., Ito, S., Kunai, A. 125

A novel and effective oxygenation of 2,3,6-trimethylphenol to trimethyl-p-benzoquinone by dioxygen with copper(II) chloride/amine hydrochloride system catalyst
Takehira, K., Shimizu, M., Watanabe, Y., Orita H., Hayakawa, T. 133

Phenol oxidation with molecular O_2 in the presence of metallic copper
Ravasio, N., Gargano, M., Rossi, M. ... 139

Platinum catalyzed oxidation of 5-hydroxymethylfurfural
Vinke, P., van Dam, H.E., van Bekkum, H. ... 147

The use of precious metal catalysts supported on activated carbon in oxidation reactions for the synthesis of fine chemicals, especially for the selective oxidation of glucose to gluconic acid
Despeyroux, B.M., Deller, K., Peldszus, E. ... 159

Liquid-phase oxidation of hydrocarbons and alcohols catalyzed by heterogeneous palladium and platinum catalysts
Hronec, M., Cvengrosova, Z., Tuleja J., Ilavsky, J. ... 169

Catalytic oxidation of 1-alkenes with molecular oxygen and palladium nitro complexes
Kiers, N.H., Feringa, B.L., van Leeuwen, P.W.N.M. ... 177

Selective cyclohexane oxidation catalyzed by the GIF system
Schuchardt, U., Mano, V. ... 185

Metalloporphyrin-catalyzed oxidation of cyclohexane with dioxygen
Iwanejko, R., Mlodnicka, T., Poltowicz, J. ... 195

Oxometalates and dioxygen in catalytic oxidation
Bregeault, J.-M., El Ali, B., Mercier, J., Martin, J., Martin, C., Mohammedi, O. ... 205

^{1}H-NMR study of the mechanism of ethyleneglycol monoacetate formation in oxidative acetoxylation of ethylene catalyzed by Pd(II) complexes
Gusevskaya, E.V., Beck, I.E., Stepanov, A.G., Likholobov, V.A., Zamaraev, K.I. ... 213

Perspective in improvement of selectivity in liquid phase oxidation by dioxygen. New models of enzymatic oxidation
Skibida, I.P., Sakharov, A.M. ... 221

Nontraditional Approaches to Reactions of Heterogeneous Selective Oxidation

Catalytic nitroxidation of aliphatic and aromatic hydrocarbons by NO
Pajonk, G.M. ... 229

Selective oxidation of hydrocarbons by nitric oxide to nitriles
Belousov, V.M., Grinenko, S.B. ... 239

Selective oxidation of anthracene
Do, N.T., Kalthoff, R., Laaks, J., Trautmann, S., Baerns, M. ... 247

Vapour-phase oxidation of α-methylstyrene to phenylacrolein
Ai, M. ... 257

Partial oxidation of o-xylene to phthalic anhydride in a structured fixed bed containing a sequence of catalysts
Kotter, M., Li, D.X., Riekert, L. ... 267

$YBa_2Cu_3O_6$ - A selective ammoxidation catalyst
Otamiri, J.C., Andersson , A., Hansen S., Bovin, J.-O. 275

Catalytic properties of the high-temperature superconductor Y-Ba-Cu-Ag-O towards the oxidation of methanol
Klissurski, D., Pesheva, J., Dimitriev, Y., Abadjieva, N., Minchev, L. 287

Selective oxidation of propylene over rare earth-molybdate catalysts
Lopez Nieto, J.M., Bielsa, R., Kremenic, G., Fierro, J.L.G. 295

Oxyesterification of methanol to methylformate over V-Ti oxide catalysts
Elmi, A.S., Busca, G., Cristiani, C., Forzatti, P., Tronconi, E. 305

Oxidative transformation of methanol to higher aldehydes over zeolite-oxide catalysts
Jiru, P., Tvaruzkova, Z., Habersberger, K. 317

Tin-germanium phosphates as selective catalysts for the oxidative dehydrogenation of ethylbenzene to styrene
Turco, M., Bagnasco, G., Ciambelli, P., la Ginestra, A., Russo, G. 327

Advances in C_1-C_5 Alkane Selective Transformation

Methane Activation

Selective photochemical conversion of methane into worthier compounds
Ogura, K., Migita, C.T., Yamada,T., Chaki, S. 335

Working principle of Li-doped MgO applied for the oxidative coupling of methane
van Kasteren, J.M.N., Geerts, J.W.M.H., van der Wiele, K. 343

Investigations on the heterogeneously catalyzed oxidative coupling of methane over alkali doped metal oxides
Bartsch, S., Hofmann, H. 353

Oxidative coupling of methane over $LnLiO_2$ Catalysts. (Ln = Sm, Nd, La). Promoting effect of MgO and CaO
Kiennemann, A., Kieffer,R., Kaddouri, A., Poix, P., Rehspringer, J.L. 365

Behavior of metallic oxides supported on Li/MgO in the methane oligomerization
Baronetti, G.T., Lazzari, E.J., Castro, A.A., Scelza, O.A. 373

Oxidative coupling of methane, the effect of gas composition and process conditions
Roos, J.A., Korf, S.J., Biermann, J.J.P., van Ommen, J.G., Ross, J.R.H. 381

Role of some experimental parameters in the catalytic conversion of methane by oxidative coupling
Spinicci, R. 393

Models of the direct catalytic partial oxidation of light alkanes
McCarty, J.G., McEwen, A.B., Quinlan, M.A. 405

General mechanism for the oxidative coupling of methane
Forlani, O., Lupieri, M., Piccoli, V., Rossini, S., Sanfilippo, D., Dumesic, J.A., Aparicio, L.A., Rekoske, J.A., Trevino, A.A. 417

The mechanism of the oxidative coupling of methane: evidence that ethylene is a primary product
Keulks, G.W., Min Yu 427

Heterolytic mechanism of methane activation in oxidative dehydrodimerization
Sokolovskii, V.D., Buyevskaya, O.V., Aliev, S.M., Davydov, A.A. 437

SERS - In-situ study of the surface species formed in methane oxidative coupling
Kadushin, A.A., Krylov, O.V., Plate, S.E., Tulenin, Yu.P., Seleznev, V.A., Bobrov, A.V., Kimelfeld, Ya.M. 447

In-situ studies of the oxidative coupling of methane over Li-Ni-O catalysts
Pickering, I.J., Thomas, J.M., Maddox, P.J. 453

Selective oxidation of methane to formaldehyde at ambient pressure: the role of dopants in determining optimum carrier loading for the molybdena / silica system
MacGiolla Coda, E., Hodnett, B.K. 459

Oxidative dimerization of methane in high temperature eletrochemical devices
Belyaev, V.D., Bazhan O.V., Sobyanin, V.A., Parmon, V.N. 469

C_2-C_3 Alkane Selective Oxidation

Selective dehydrogenation of ethane by CO_2 over Fe-Mn oxide catalyst. An in-situ study of catalyst phase composition and structure
Mamedov, A.Kh., Shiryaev, P.A., Shashkin, D.P., Krylov, O.V. 477

Role of active oxygen forms and acidity in oxidative conversion of ethane on zeolites
Vereshchagin, S.N., Shishkina, N.N., Anshits, A.G. 483

Selective oxidation of propane to acrolein and ammoxidation to acrylonitrile over Ag-doped bismuth vanadomolybdate catalysts
Kim, Y.Ch., Ueda, W., Moro-oka, Y. 491

Oxidative dehydrogenation of alkanes over alkali and alkaline earth orthovanadates
Seshan, K., Swaan, H.M., Smits, R.H.H., van Ommen, J.G., Ross, J.R.H. 505

Propane ammoxidation on V-antimonate based mixed oxide
Centi, G., Grasselli, R.K., Patane', E.,Trifiro', F. 515

Selective catalytic oxidation of propane by sulphur dioxide
Osipova, Z.G., Ushkov, S.B., Sokolovskii, V.D., Kalinkin, A.V. 527

n-Butane and n-Pentane Oxidation

Fluidized bed oxidation of n-butane: a new commercial process for maleic anhydride
Stefani, G., Budi, F., Fumagalli, C., Suciu, G.D. 537

Butane oxidation in a transport bed reactor - Redox characteristics of the vanadium phosphorus oxide catalyst
Contractor, R., Ebner, J., Mummey, M.J. 553

Reactivity and structure of vanadyl pyrophosphate as a butane oxidation catalyst
Matsuura, I., Yamazaki, M. 563

Investigation of active and selective oxygen in V-P-O catalysts for n-butane conversion to maleic anhydride
Lashier, M.E., Moser, T.P., Schrader, G.L. 573

Comparative study of structure-sensitive oxidations of n-butane and 1-butene in maleic anhydride on two kinds of catalysts
Bordes, E. 585

Identification of a catalytically active copper oxychloride phase for the synthesis of maleic anhydride
Davies, M.J., Chadwick, D., Cairns, J.A. 595

Selective oxidation of n-butane to maleic anhydride over well-characterized vanadium-phosphorus mixed oxide
Misono, M., Miyamoto, K., Tsuji, K., Goto, T., Mizuno, N., Okuhara, T. 605

Product distribution in the V-P-O catalyst bed and an effect of reaction products on the catalyst in C_4 oxidation
Zazhigalov, V.A., Belousov, V.M., Pyatnitskaya, A.I., Komashko, G.A., Merkureva, Yu. N., Stoch, J. 617

A dynamic model of the oxidation of n-butane and 1-butene on various crystalline faces of $(VO)_2P_2O_7$
Ziolkowski, J., Bordes, E., Courtine, P. 625

Oxi-condensation of n-pentane to phthalic anhydride: a new type of heterogeneous oxidation reaction
Centi, G., Lopez Nieto, J.M., Pinelli, D., Trifiro', F., Ungarelli, F. 635

Electro- and Photo-Oxidation

Non-faradaic electrochemical modification of catalytic activity: partial oxidation of C_2H_4 on Ag and CH_3OH on Pt
Vayenas, C.G., Bebelis, S., Neophytides, S. 643

Electrogenerative oxidation of ethanol on platinized-carbon electrodes
Huang Zhong-Tao, Ye Daiqi, Pang Xianxing 653

Oxidation of L-phenylalanine and related reactions: preparation of L-DOPA
Blanchard, M., Bouchoule, C., Djaneye-Boundjou, G., Canesson, P. 661

Selective electrooxidation of 4-(di-n-propylsulfamyl)toluene to 4-(di-n-propylsulfamyl)benzoic acid
Michman, M., Weiss, M. 667

Photocatalytic oxidations at room temperature in various media
Herrmann, J.M., Courbon, H., Disdier, J., Mozzanega, M.N., Pichat, P. 675

Dynamic studies of photocatalytic oxidation of C_3H_6 with O_2 on vanadium oxide supported on porous Vycor glass
Anpo, M., Suzuki, T., Yamada, Y., Otsuji, Y., Giamello, E., Che,M. 683

New Aspects of the Mechanism and Surface Reactivity of Selective Oxidation Catalysts

On principles of catalyst choice for selective oxidation
Golodets, G.I. 693

Mutual orientation of oxygen and hydrocarbon molecules as factor determining the reaction pathway
Witko, M., Haber, J., Broclawik, E. 701

Characterizing oxygen adspecies for ethylene epoxidation over silver with the TAP reactor system
Rigas, N.C., Gleaves, J.T., Mills, P.L. 707

Ethylene oxidation over α-alumina supported silver-gold catalysts
Herrera, R., Varma, A., Martinez, E. 717

Ethylene epoxidation on silver-based alloy catalysts
Toreis, N., Verykios, X.E. 725

Ethylene oxidation over hydrophobic thin layer catalysts
Frusteri, F., Iannibello, A., Parmaliana, A., Cannizzaro, A., Giordano, N. 733

The mechanism of alkene epoxidation and epoxide conversion on single crystal silver surfaces
Hawker, S., Mukoid, C., Badyal, J.P.S., Lambert, R.M. 739

Active sites in propylene mild oxidation as studied on new [100] oriented MoO_3 catalyst
Abon, M., Mingot, B., Massardier, J., Volta, J.C. 747

Effects of added SnO_2, WO_3, MoO_3 and α-Sb_2O_4 on the catalytic properties of Bi_2MoO_6 in selective oxidation of isobutene to methacrolein
Weng, L.T., Sham, E., Doumain, B., Ruiz, P., Delmon, B. 757

MoO_3 on anatase and rutile low-surface-area TiO_2: characterization and oxidation of o-xylene and butene-1
Filipek, E., Grzybowska, B., Serwicka, E., Barbaux, Y., Bonnelle, J.P., Grimblot, J. 767

The role of the distribution of double bonded oxygen on surface of V-Mo-O catalysts in selective oxidation of benzene to maleic anhydride
Najbar, M., Wal, W., Chrzaszcz, J. 779

Effect of promoters on activity and selectivity of benzene oxidation on V_2O_5 catalysts: study by well-defined V_2O_5 catalysts
Satsuma, A., Nakata, M., Iwasaki, S., Hattori, T., Murakami, Y. 789

Cooperation between phases in mixed SnSbO selective oxidation catalysts
Weng, L.T., Patrono, P., Sham, E., Ruiz, P., Delmon, B. 797

IR spectroscopic identification of adsorbed surface species on oxidation catalysts exposed to propene and ethene in air
Pires, M.J., Do, N.T., Baerns, M., Portela, M.F. 807

Oxidation of propene and toluene on CoO-MgO solid solutions: a spectroscopic study
Giamello, E., Garrone, E., Coluccia, S., Spoto, G., Zecchina, A. 817

FT-IR study of selective oxidation intermediates of benzene on the surface of vanadia-titania "monolayer" catalysts
Busca, G., Ramis, G., Lorenzelli, V. 825

Proposal for active sites of iron phosphates in isobutyric oxidative dehydrogenation reaction
Millet, J.-M.M., Vedrine, J.C., Hecquet, G. 833

Heteropolyacid catalysts for oxydehydrogenation of isobutyric acid: kinetics and deactivation
Watzenberger, O., Haeberle, Th., Lynch, D.T., Emig, G. 843

A study of the catalytic oxydehydrogenation of ethylbenzene using nitrobenzene
Fiedorow, R., Beszterda, S., Przystajko, W., Dalla Lana, I.G. 853

The adsorption of aliphatic nitrocompounds on oxides investigated by FT-IR spectroscopy
Angevaare, P.A.J.M., Grootendorst, E.J., Zuur, A.P., Ponec, V. 861

The model of active centers for partial oxidation in the structure of heteropoly compounds
Kutyrev, M. Yu., Staroverova, I.N., Thiep, N.Z., Krylov, O.V 869

Author Index 881

Subject Index 885

Studies in Surface Science and Catalysis (other volumes in the series) 889

Preface

This Volume is a collection of the invited and research papers presented at the International Symposium on *New Developments in Selective Oxidation* held in Rimini, Italy, Sepember 18-22, 1989.

The Symposium was organized by the Department of Industrial Chemistry and Materials of the University of Bologna in collaboration with the Catalysis Group of the Italian Chemical Society and under the auspicies of the Italian National Research Council (CNR) and of the project "Chimica Fine II" of the CNR.

The objectives of the Symposium were to present new developments in fundamental research and in industrial applications of selective oxidation processes. At this meeting various trends were reflected:

- a wide interest in the selective oxidation of sophisticated substrates, both in the liquid and vapour phase for the synthesis of fine chemicals;
- growing possibilities offered by the use of light alkanes as feedstocks in selective oxidation processes;
- new opportunities for fundamental research created by new concepts in reactors;
- promising industrial prospects for the application of new zeolites in the liquid phase;
- a continuing search for new catalytic systems and nontraditional reactions for the functionalization of substrates by selective oxidation.

Further aims of the Symposium were to (i) bring together specialists of various origins and backgrounds working in the fields of homogeneous or heterogeneous catalysis and in photo-, electrochemical- or more traditional oxidation, to exchange ideas and experiences regarding the use of different oxidizing agents such as H_2O_2, O_2, NO, as well as of type of substrates (alkanes, alkenes, aromatics, etc.) , (ii) discuss and disseminate knowledge in specialized areas of selective oxidation and (iii) serve as a springboard for new ideas as well as to foster innovation and creativity.

The symposium was attended by over 300 researchers from 30 countries. More than 50% of the participants came from the major industries operating in the field, providing a further opportunity for interchange and cross-fertilization between academic and industrial points of view.

The Editors would like to thank the Authors for the quality of their presentations and for contributing to this Volume. Thanks also are extended to the International Advisory Board and to all referees for the time and effort spent to ensure the highly scientific level of this Volume.

The Editors also thank the Organizing Committee and all the Chairmen of the Sessions for willingly giving their time and expertise to the Symposium.

A special thank you is due to Professor Angelo Vaccari and Professor Alfredo Riva as well as to all researchers in the Department of Industrial Chemistry and Chemical Engineering (Politecnico Milano) and of the Institute of Chemistry (University of Genova), whose invaluable efforts made possible the concrete realization of the Symposium.

G. Centi and F. Trifiro', Editors

Bologna, December 1989

New Developments in Selective Oxidation

Rimini, Italy, September 18-22, 1989

ORGANIZED BY

Gruppo Interdivisionale di Catalisi della Societa' Chimica Italiana (GIC-SCI)
Consiglio Nazionale delle Ricerche (CNR)
Progetto Finalizzato "Chimica Fine II" del CNR
Universita' di Bologna, Dipartimento di Chimica Industriale e dei Materiali

INTERNATIONAL ADVISORY BOARD

B. Delmom (Belgium)
R.K. Grasselli (U.S.A.)
J. Haber (Poland)
W. Hölderich (Germany)
O. Krylov (U.S.S.R.)
H. Mimoun (France)
M. Misono (Japan)
I. Pasquon (Italy)
R.A. Sheldon (The Netherlands)
F. Trifiro' (Italy)

ORGANIZING COMMITTEE

G. Busca, University of Genova, Italy
G. Centi, University of Bologna, Italy
P. Forzatti, Politecnico of Milano, Italy
A. Riva, University of Bologna, Italy
P. Ruiz, University of Louvain-la-Neuve, Belgium
F. Trifiro', University of Bologna, Italy
A. Vaccari, University of Bologna, Italy
P. Villa, Politecnico di Milano, Italy

SPONSORING

The Organizing Committee gratefully acknowledges financial support from:

- **Air Liquide (France)**
- **Alusuisse Italia (Italy)**
- **BP America (USA)**
- **Carlo Erba Strumentazione (Italy)**
- **Degussa (BRD)**
- **Dutral (Italy)**
- **Enimont (Stabilimento di Ravenna) (Italy)**
- **Eniricerche (Italy)**
- **Hellma Italia Srl (Italy)**
- **IGI Italiana Gas Industriali (Italy)**
- **Interox (U.K.)**
- **Mitsubishi Kasei Corporation (Japan)**
- **Mobil R&D Corporation (USA)**
- **Monsanto (USA)**
- **Montedipe (Italy)**
- **National Research Council (CNR) (Italy)**
- **Norsolor - Groupe Orkem (France)**
- **Progetto Finalizzato "Chimica Fine II" of CNR (Italy)**
- **Repsol Petroleo (Spain)**
- **Rhône Poulenc (France)**

G. Centi and F. Trifiro' (Editors), *New Developments in Selective Oxidation* 1
1990 Elsevier Science Publishers B.V., Amsterdam — Printed in The Netherlands

CATALYTIC OXIDATIONS IN THE MANUFACTURE OF FINE CHEMICALS

Roger A. Sheldon

Andeno B.V., P.O. Box 81, 5900 AB VENLO, The Netherlands

SUMMARY

In recent years increasingly stringent environmental constraints have led to a burdgeoning interest in the application of catalytic oxidation methods to fine chemicals manufacture. Whereas with bulk chemicals the choice of oxidant is often limited to molecular oxygen the economics of fine chemicals manufacture allow for a broader choice of primary oxidant. For example, relatively inexpensive single oxygen donors, such as H_2O_2, RO_2H and NaOCl, in combination with a variety of metal catalysts provide a wide range of synthetically and economically useful oxidants. The various types of oxidation processes are outlined on the basis of type of catalyst, primary oxidant, mechanism and functional group transformation. Since fine chemicals are often relatively complex, multifunctional molecules the chemo- regio- and stereoselectivity of such processes are emphasized.

Recent developments in catalytic oxidations in the liquid phase are reviewed, such as (a) the use of phase transfer catalysts to facilitate catalytic oxidations with H_2O_2 and NaOCl in biphasic, aqueous-organic mixtures, (b) transition metal-substituted (redox) zeolites as heterogeneous catalysts for liquid phase oxidation, (c) heteropolyanions as novel, oxidatively-stable ligands, (d) new developments in catalytic asymmetric oxidation.

INTRODUCTION – WHY CATALYTIC OXIDATION?

As a result of increasingly stringent environmental constraints it is becoming more and more difficult to carry out industrial scale oxidations with traditional stoichiometric oxidants, such as dichromate, permanganate, etc. Consequently, there is a general trend towards the development of catalytic processes which do not generate aqueous effluents containing large quantities of inorganic salts.

An illustrative example is the industrial synthesis of hydroquinone (see scheme 1).

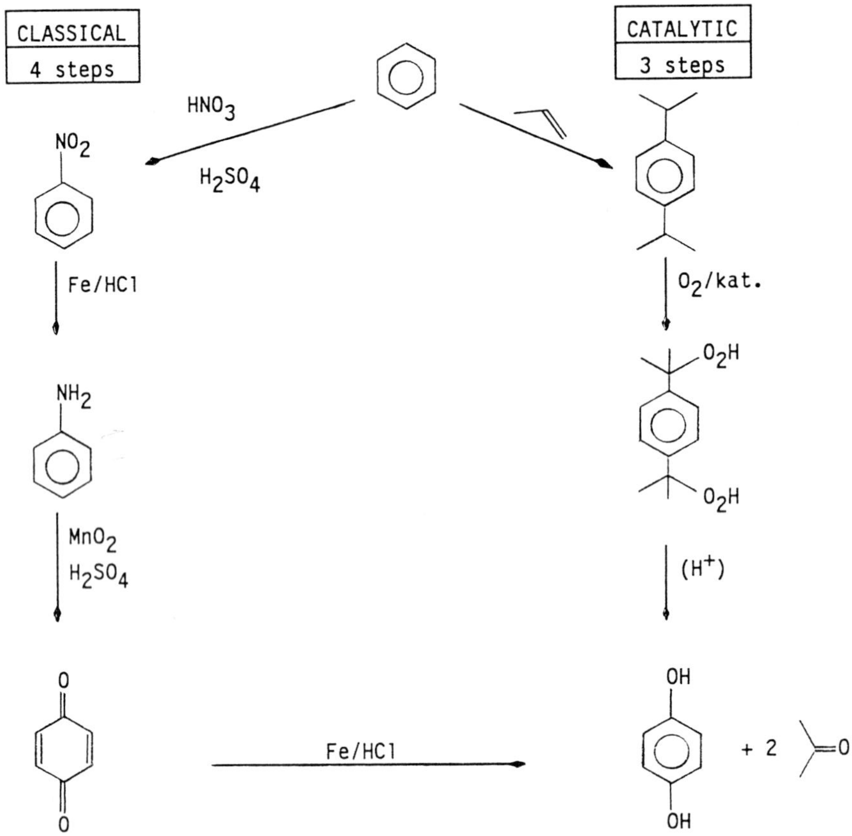

ROUTE	CLASSICAL	CATALYTIC
kg SALTS per kg PRODUCT	>10	<1 (Hydroquinone light)

SCHEME 1. Two routes to hydroquinone.

Traditionally hydroquinone was manufactured by oxidation of aniline with stoichiometric amounts of manganese dioxide to give p-benzoquinone, followed by reduction with iron and hydrochloric acid. The aniline was derived from benzene via nitration and reduction. In the overall process more than 10 kg of inorganic salts ($MnSO_4$, $FeCl_2$, Na_2SO_4, NaCl) are produced per kg of hydroquinone. In contrast, a more modern route to hydroquinone involves the catalytic oxidation of p-diisopropylbenzene followed by acid-catalyzed rearrangement of the bis-hydroperoxide, and produces <1 kg of inorganic salts per kg of hydroquinone.

In the bulk chemical industry traditional environmentally unacceptable processes have long been replaced with cleaner catalytic oxidations. Indeed, catalytic oxidation is the most widely applied technology for the conversion of base chemicals such as olefins and aromatics to commercially important oxygenated products (see table 1 for examples). Both heterogeneous, gas phase oxidations and homogeneous, liquid phase oxidations are applied (see table 1).

TABLE 1. Bulk Chemicals manufacture USA (1987).

CHEMICAL	VOLUME (10^6 tons)	REACTION	CATALYST
Terephthalic acid	4.0	Oxidation	Homogeneous
Styrene	4.0	Dehydrogenation	Heterogeneous
Methanol	3.7	$CO + H_2$	Heterogeneous
Formaldehyde	3.0	Oxidation	Heterogeneous
Ethylene oxide	2.8	Oxidation	Heterogeneous
Phenol	1.6	Oxidation	Homogeneous
Acetic acid	1.6	Carbonylation	Homogeneous
Propylene oxide	1.3	Oxidation	Homogeneous
Acrylonitrile	1.3	Ammoxidation	Heterogeneous
Vinyl acetate	1.2	Oxidation	Homogeneous

In the fine chemicals industry, on the other hand, much smaller volumes are involved and there has been much less pressure in the past to replace traditional stoichiometric oxidants. But times are rapidly changing. The fine chemicals industry is also under increasing pressure to develop cleaner, more efficient processes. This is not so surprising when one considers that although the absolute volumes are significantly less, the number of kilo's byproducts per kilo product are generally much higher.

This is partly due to the fact that fine chemicals are often produced via multi-step syntheses.

The simplest, cheapest and cleanest oxidant of all is molecular oxygen (dioxygen). However, the reaction of dioxygen with organic molecules is fraught with several difficulties :

- Dioxygen has a triplet ground state which means that its reaction with most organic molecules is a spin-forbidden process. Consequently, although its reactions with hydrocarbons are thermodynamically favored, they generally exhibit high activation energies. Once underway, however, they are difficult to control and the thermodynamically most favored products are carbon dioxide and water.

- Primary oxidation products (alcohols, aldehydes, epoxides, etc.) are generally more easily oxidized than the hydrocarbon substrate. Reactions are therefore, often carried out to low conversions necessitating recycling of large quantities of substrate.
- Dioxygen is largely indiscriminate, i.e. shows little chemo- or regioselectivity in its reactions with organic substrates.

Catalytic oxidations of organic molecules with dioxygen can, therefore, be typified by two extremes (see scheme 2). One extreme is complete oxidation which is of importance in the context of control of automobile exhaust emissions. The other extreme is exemplified by the chemo- regio- and enantioselective hydroxylation of progesterone mediated by the microorganism, Rhizopus nigricans.

1. COMPLETE OXIDATION

$$C_nH_{2n+2} + (n+1)O_2 \xrightarrow{[Pt]} nCO_2 + (n+1)H_2O$$

Catalytic conversion of exhaust gases

2. SELECTIVE (PARTIAL) OXIDATION

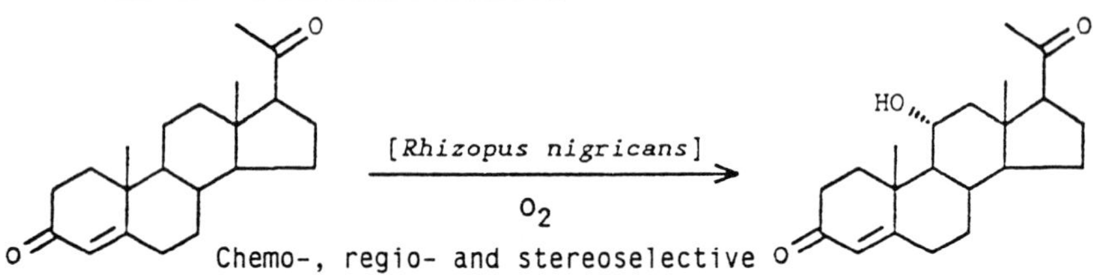

SCHEME 2. Two extremes.

CHARACTERISTICS OF FINE vs BULK CHEMICALS MANUFACTURE

Although they obviously have many things in common there are several basic differences between fine and bulk chemical processing which can influence process selection :

- Substrates are generally complex, multi-functional molecules with limited thermal stability thus necessitating reaction in the liquid phase at moderate temperatures.

- Chemo- regio- and stereoselectivity are often important requirements.
- Processing is multi-purpose and batch-wise in contrast to dedicated and continuous in bulk chemicals. This means that not only raw materials costs but also simplicity of operation and multi-purpose character of the installations are important economic considations (i.e. different ratio of variable to fixed costs).

TYPES OF OXIDANT

A consequence of the last point is the fact that hydrogen peroxide is, in principle, the oxidant of choice even though it is more expensive than dioxygen. Moreover, because of the higher price commanded by the products the choice of oxidant available (see table 2) to the fine chemist is obviously much larger than that to the bulk chemist who is largely limited to dioxygen. Next to price and ease of handling the two important economic considerations are the nature of the byproduct and the percentage available oxygen. The former is obviously important in the context of environmental considerations and the latter generally has a direct influence on the volume yield (kg product per unit reactor volume per unit time).

Hydrogen peroxide is obviously 'Mr. Clean', its by-product being water. We note, however, that the by-product from organic oxidants, such as tert-butylhydroperoxide (TBHP) and amine oxides, is readily recycled via reaction with hydrogen peroxide. The overall process produces water as the by-product, but requires one extra chemical step compared to the corresponding reactions with hydrogen peroxide.

TABEL 2. Oxygen donors.

DONOR	% ACTIVE OXYGEN	BYPRODUCT
H_2O_2	47.0	H_2O
$t\text{-}BuO_2H$	17.8	t-BuOH
NaClO	21.6	NaCl
$NaClO_2$	35.6	NaCl
NaBrO	13.4	NaBr
$C_5H_{11}NO_2$*	13.7	$C_5H_{11}NO$
$KHSO_5$	10.5	$KHSO_4$
$NaIO_4$	29.9**	NaI
PhIO	7.3	PhI

*N-Methylmorpholine-N-oxide

**Assuming all four oxygen atoms are utilized

With other inorganic oxygen donors environmental considerations are relative. Thus, sodium chloride (from NaClO or $NaClO_2$) and potassium sulfate (from $KHSO_5$) are obviously preferred above chromium, manganese or lead salts.

In addition to the standard examples compiled in table 2 other interesting oxygen donors have been described in the recent literature. For example, sodium perborate ($Na_2B_2[O_2]_2(OH)_4.nH_2O$) is an inexpensive bulk chemical (ca. 1 million tons per annum) which is used primarily in detergents, bleach and antiseptic mouthwash. Recently, McKillop and coworkers have reported its use as a selective oxidant in organic synthesis.[1,2] For example, anilines were oxidized to the corresponding nitrobenzenes, sulfides to sulfoxides or sulfones, ketones to esters and phenols or hydroquinones to the corresponding 1,4-benzoquinones. Similarly, the use of 'sodium percarbonate' ($Na_2CO_3.3/2H_2O_2$) as a selective oxidant has been described by Ando and coworkers.[3]

Two interesting classes of organic single oxygen donors are the dioxiranes (1)[4] and the oxoammonium salts (2).[5,6] The former are prepared from $KHSO_5$ and an appropriate ketone (reaction 1) and the latter from a dialkylhydroxylamine and aqueous NaOCl (reaction 2).

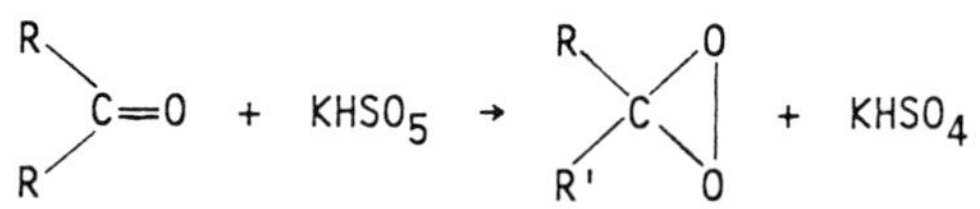

(1)

(1)

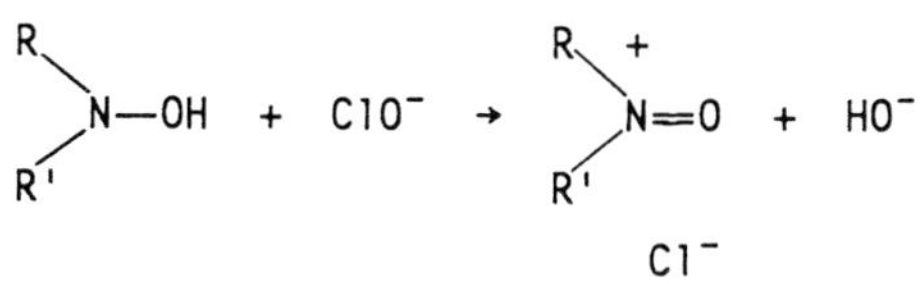

(2)

(2)

Since the oxidation of organic substrates with (1) and (2) leads to the formation of the corresponding ketone and dialkylhydroxylamine, respectively, the latter may be considered as organic catalysts for oxygen transfer processes with $KHSO_5$ and NaOCl, respectively.

TYPES OF CATALYTIC OXIDATION PROCESSES

Catalytic oxidations may be basically divided into three types based on the type of reaction involved in the key oxidation step.[7,8]

a. Free radical (aut)oxidation

Catalysis involves the metal ion–induced decomposition of H_2O_2 or RO_2H. In reactions of hydrocarbons with dioxygen this is followed by the classical autoxidation scheme :

$$RO_2^{\bullet} + RH \rightarrow RO_2H + R^{\bullet} \tag{3}$$

$$R^{\bullet} + O_2 \rightarrow RO_2^{\bullet} \tag{4}$$

Metal catalysis in these reactions results in rate acceleration but has little or no effect on the selectivity.

b. Oxygen transfer

This involves the reaction of an oxygen donor (see above) with an organic substrate in the presence of a metal (or an organic) catalyst according to scheme 3.

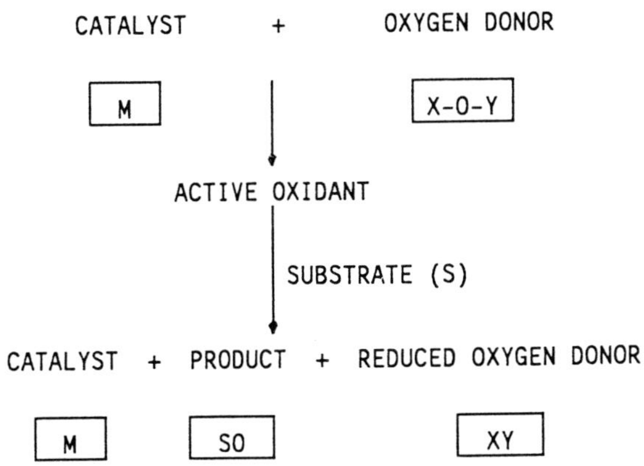

SCHEME 3. Catalytic oxygen transfer.

The active oxidant in these processes can be an oxometal or a peroxometal species (see scheme 4). Some metals (e.g. vanadium) can, depending on the substrate, operate via either mechanism.[7]

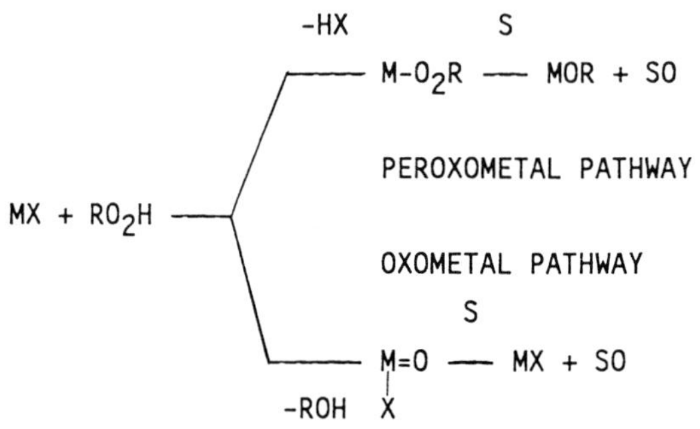

SCHEME 4.

c. Metal ion oxidations

In this class the key step involves the oxidation of the substrate by a metal ion. Examples include the palladium (II) catalyzed oxidation of olefins (Wacker process) and the oxidative dehydrogenation of alcohols where the key steps are reactions (5) and (6), respectively.

$$RCH{=}CH_2 + Pd^{II}X_2 + H_2O \longrightarrow RCOCH_3 + Pd^0 + 2HX \qquad (5)$$

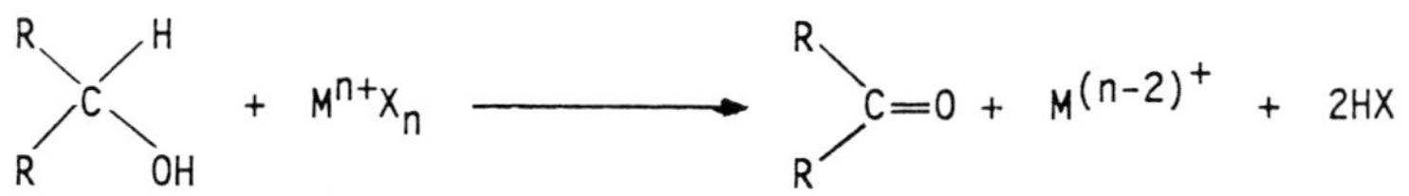

(6)

$$M^{n+} = Pd^{2+},\ Pt^{2+},\ Ru^{3+},\ etc.$$

The oxidized form of the metal ion is subsequently regenerated by reaction of the reduced form with the terminal oxidant which could, in principle, be dioxygen or an oxygen donor. In the latter case this is merely a third type of oxygen transfer process.

Although reactions of the first type (free radical autoxidation) are widely applied in bulk chemicals they are largely confined to relatively simple molecules with one reactive group (e.g. $ArCH_3 \rightarrow ArCO_2H$). The methods of choice in fine chemicals are, therefore, those involving catalytic oxygen transfer.

EXAMPLES OF CATALYTIC OXYGEN TRANSFER

Catalytic oxygen transfer is a reaction with tremendous scope.[7] In addition to the substantial number of relatively inexpensive oxygen donors which are available (see earlier) virtually all of the transition metals and several main group elements (e.g. Sn, As, Se) can be used as catalysts. Hence, the number of permutations and combinations is enormous. Probably the most well-known example is the catalytic epoxidation of olefins with alkyl hydroperoxides (reaction 7).[9-11]

$$CH_3CH{=}CH_2 + RO_2H \xrightarrow{\text{(Catalyst)}} CH_3\overset{\;O\;}{CH{-}CH_2} + ROH \qquad (7)$$

Catalyst : Homogeneous : Mo^{VI}, W^{VI}, V^{V}, Ti^{IV} (Arco)
Heterogeneous : Ti^{IV}/SiO_2 (Shell)

$$R = (CH_3)_3C\text{-} \text{ or } PhCH(CH_3)\text{-}$$

The reaction is catalyzed by compounds of high-valent metals such as Mo^{VI}, W^{VI}, V^{V} and Ti^{IV}. Molybdenum compounds are particularly effective as homogeneous catalysts. A heterogeneous Ti^{IV}/silica catalyst developed by Shell is also highly effective and can be used in continuous, fixed-bed operation. The economic importance of reaction 7 is underscored by the fact that it accounts for more than one million tons annual production of propylene oxide worldwide.

Analogous epoxidations of a wide variety of olefins are readily performed in hydrocarbon solvents at moderate temperatures (generally 80–120°). These RO_2H-metal catalyst reagents are particularly useful for chemo-, regio- and stereoselective epoxidations.[9] The reactions proceed via a peroxometal mechanism (see earlier) involving rate-limiting oxygen transfer from an electrophilic alkylperoxometal species to the olefin (reaction 8).

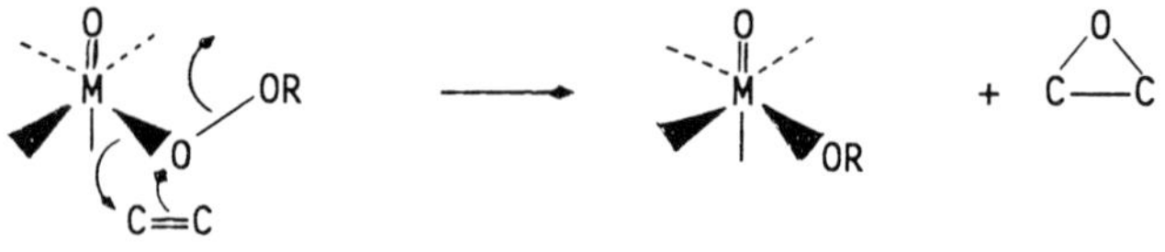

(8)

These reagents (metal catalyst–RO_2H or H_2O_2) have in recent years been widely applied to the chemoselective oxidation of alcohols and the regioselective oxidation of diols. They constitute environmentally attractive alternatives to the classical stoichiometric reagents based on chromium (VI). Thus, molybdenum–[12–17], vanadium[18] and titanium[19]–based catalysts in combination with TBHP mediate the selective oxidation of secondary alcohols. Zirconium, on the other hand, catalyzes the selective oxidation of primary alcohols to aldehydes (without further oxidation to carboxylic acids) and the chemoselective oxidation of allylic alcohols to the corresponding α,β-unsaturated aldehydes.[20] Chromium[21,22] and cerium[21,23] catalysts also mediate the selective oxidation of secondary alcohols using TBHP[21], peracetic acid[22] or $NaBrO_3$[21,23] as the oxygen donor. Another excellent catalyst for both primary and secondary alcohol oxidation is ruthenium[24] which has been used in conjunction with H_2O_2[25], RO_2H[25–27], NaOCl[28,29], $NaBrO_3$[30], $NaIO_4$[31] N-methylmorpholine-N-oxide[32,33] $K_2S_2O_8$[34], PhIO[35] and even dioxygen[36,37] as the terminal oxidant. A few illustrative examples are shown below (reactions 9–13).

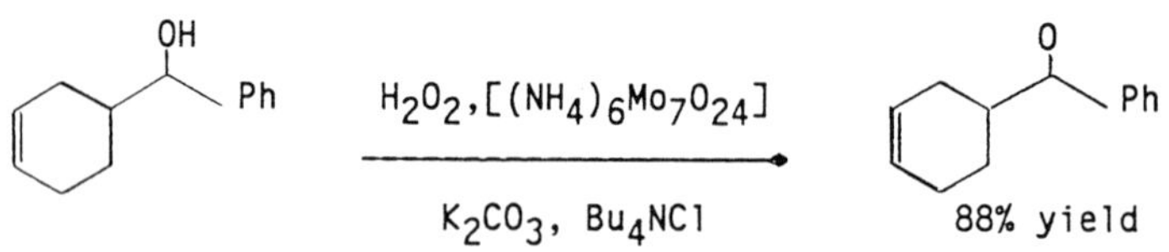

(9)[18]

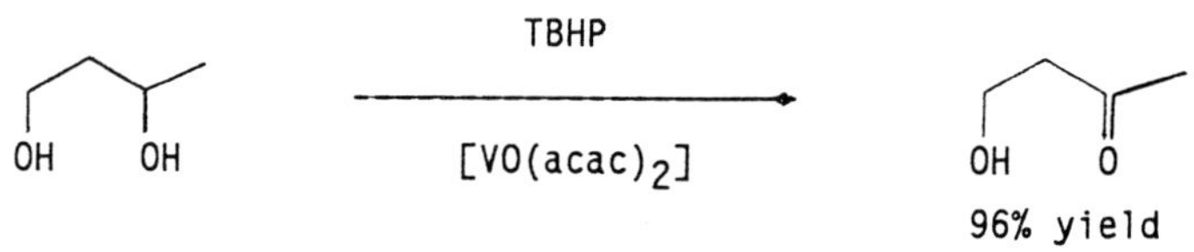

(10)[19]

$$RCH_2OH \xrightarrow[\text{[ZrO(acac)}_2\text{]}]{\text{TBHP}} RCHO$$

85-95% yield (11)[20]

(12)[21]

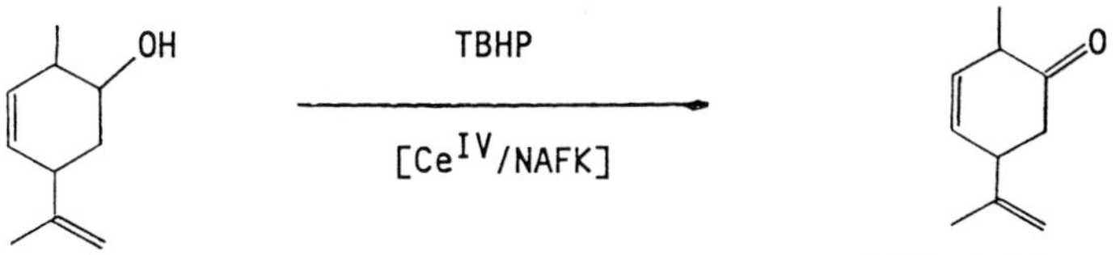

(13)[21]

98% yield

NAFK = Nafion ® 511 perfluorinated ion exchange resin.

$$\mathrm{RCH(OH)CN} \xrightarrow[\text{[RuCl}_2\text{(PPh}_3\text{)}_2\text{]}]{\text{TBHP}} \mathrm{RC(=O)CN} \qquad (14)^{26}$$

77-99% yield

In some instances the use of different oxygen donors with the same metal catalyst can lead to dramatic changes in chemoselectivity, e.g.,[38]

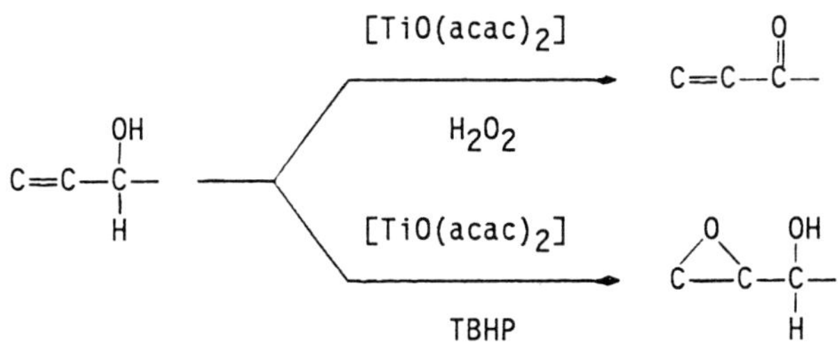

(15)

(16)

A possible explanation is that the water present in aq. H_2O_2 seriously inhibits epoxidation of the double bond.

The alcohol oxidations outlined above can proceed via peroxometal or oxometal pathways depending on the catalyst used (see scheme 5). Thus, metals which are strong oxidizing agents in their highest oxidation state (e.g. Cr^{VI}, V^{V}, Ce^{IV}, Ru^{VIII}) react via oxometal species whilst weakly oxidizing metal ions (e.g. Mo^{VI}, Zr^{IV}, Ti^{IV}) involve peroxometal species in the key oxidative dehydrogenation step. Which mechanism is operating can be easily demonstrated by carrying out the reaction stoichiometrically in the absence of terminal oxidant. Systems involving peroxometal species as the active oxidant will obviously give no reaction under these conditions.

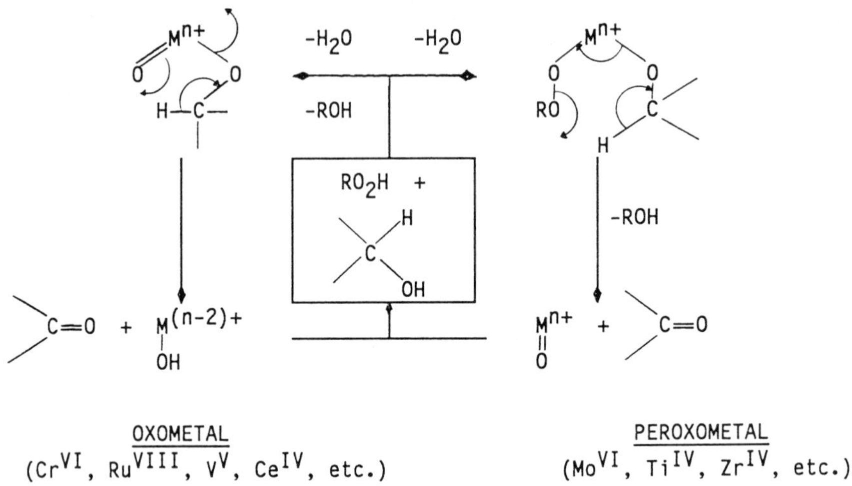

SCHEME 5.

A further variation on this theme is the use of organic oxygen transfer catalysts. For example, the oxoammonium salts (2) referred to earlier catalyze the oxidation of primary and secondary alcohols with NaOCl.[5] The reactions are carried out in a two-phase CH_2Cl_2–H_2O system at 0°C and primary alcohols afford the corresponding aldehydes in high yield. In the presence of a quaternary ammonium salt, as a phase transfer catalyst, the aldehyde undergoes rapid further oxidation to the corresponding carboxylic acid. The proposed mechanism[5] for the key oxidation step (scheme 6) in these alcohol oxidations is completely analogous to the oxometal mechanism outlined in scheme 5.

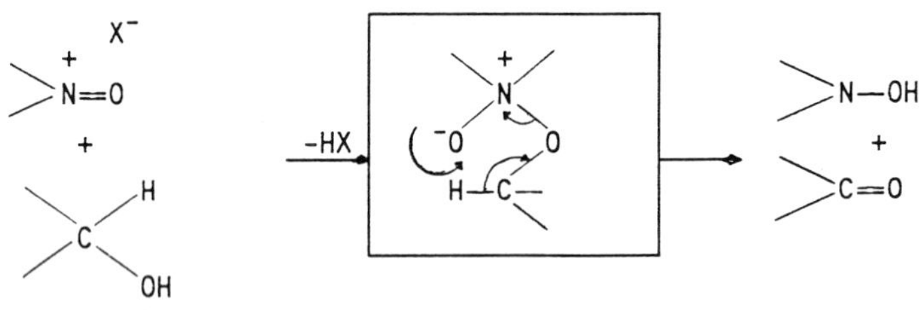

SCHEME 6.

Another reaction of practical interest is the oxidative cleavage of vinical diols (reaction 17) which is traditionally carried out using the stoichiometric reagents, periodate or lead tetra-acetate. Some of the catalytic oxygen transfer reagents described above, e.g. $VO(acac)_2$/TBHP[39], $WO_4^{2-}/PO_4^{3-}/H_2O_2$[40], $H_3PW_{12}O_{40}/H_2O_2$[16], $RuCl_3$/NaOCl[28] and $RuCl_3/H_2O_2$[25], have been successfully applied to this reaction, thus providing attractive alternatives to the classical stoichiometric reagents.

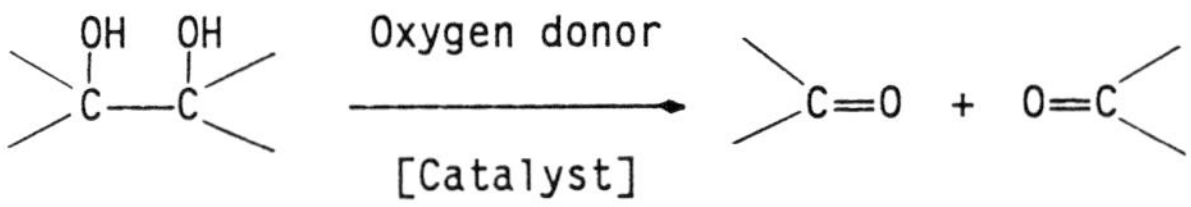

(17)

An interesting variation on this theme is the recently reported[41] use of ruthenium pyrochlore oxides ($A_{2+x}Ru_{2-x}O_{7-y}$ where A is Pb or Bi) as heterogeneous catalysts for the liquid phase oxidative cleavage of vinical diols with dioxygen, e.g. in the conversion of cyclohexane-1,2-diol to adipic acid :

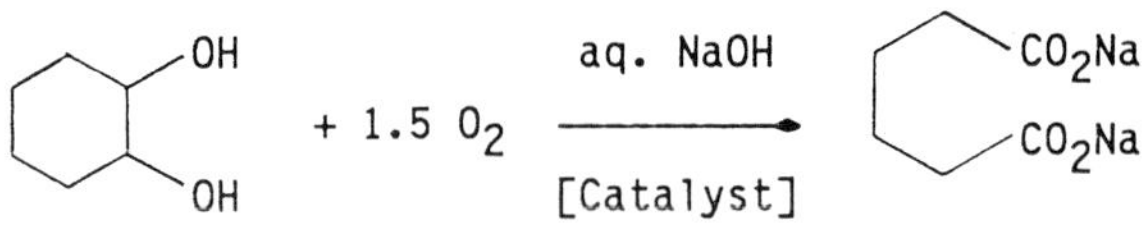

(18)

81-87% yield

Catalyst : $Pb_{2.63}Ru_{1.37}O_{6.5}$ or $Bi_{2.39}Ru_{1.61}O_{7-y}$

The reactions were carried out in batch autoclaves or in continuous trickle bed reactors.

In addition to the epoxidation of olefins mentioned earlier metal catalyst-oxygen donor reagents can effect a variety of potentially useful transformations of olefins (see scheme 7).[7,42]

Similarly, reactions of aromatics with oxygen donors can, in principle, afford products derived from nuclear hydroxylation, side-chain oxidation or oxidative cleavage of the aromatic ring. Catalytic oxidation of substituted toluenes to the corresponding carboxylic acids ($ArCH_3 \rightarrow ArCO_2H$) is relatively straightforward. There is still a need, however, for good methods (see later) for selective oxidation to the corresponding aldehydes ($ArCH_3 \rightarrow ArCHO$). The great remaining challenge in this area is the development of good methods for regioselective nuclear hydroxylation (but see later).

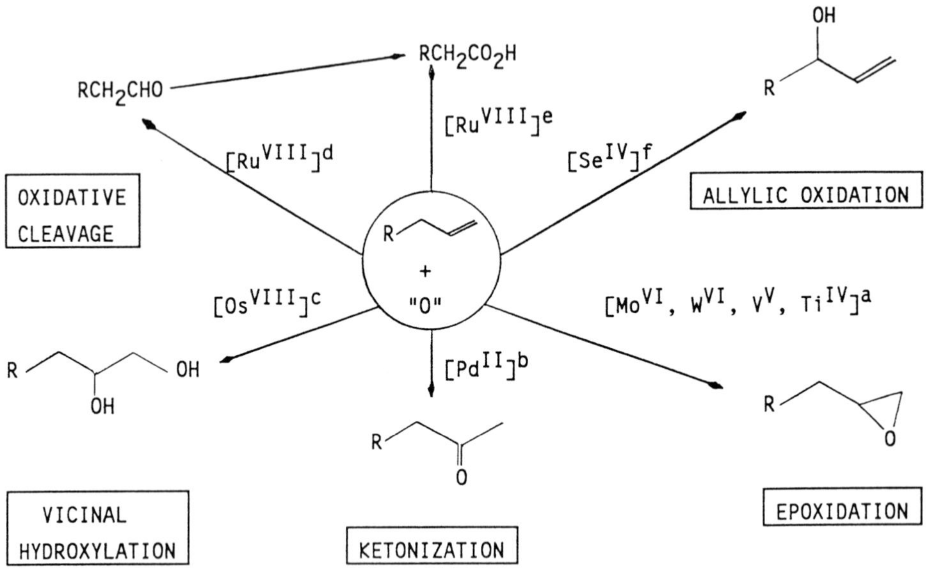

Oxygen donors of choice : (a) TBHP (b) O_2 or TBHP[43] (c) N-methylmorpholine-N-oxide[44] (d) H_2O_2[24] (e) NaOCl[24] (f) TBHP[45]

SCHEME 7. Oxidative transformations of olefins.

HYDROGEN PEROXIDE AS OXIDANT-PHASE TRANSFER CATALYSIS

As noted earlier the oxidant of choice in the fine chemicals industry is 30% H_2O_2. Unfortunately, H_2O_2 (in common with other useful oxygen donors such as NaOCl) is insoluble in many common organic solvents. This practical problem has been overcome by the application of phase transfer catalysis. This involves the transfer of a water-soluble anion to the organic phase as a quaternary ammonium salt. In catalytic oxidations this can involve the transfer of the terminal oxidant (e.g. ClO^-, $S_2O_8^{2-}$) or the catalyst (Scheme 8).

TRANSFER OF TERMINAL OXIDANT AS ANION

$$M + Q^+ClO^-_{(org)} \longrightarrow M{=}O + Q^+Cl^-_{(org)}$$

TRANSFER OF CATALYST AS ANION

$R_3NO_{(aq)}$ + $Q^+RuO_3^-$ ⟶ R_3N + Q^+RuO_4 (oxometal)

$H_2O_{2(aq)}$ + $Q^+HMoO_4^-$ ⟶ H_2O + $Q^+HMoO_6^-$ (peroxometal)

TRANSFER OF CATALYST/REAGENT AS NEUTRAL SPECIES

e.g. $R_4NX{\cdot}H_2O_2$; 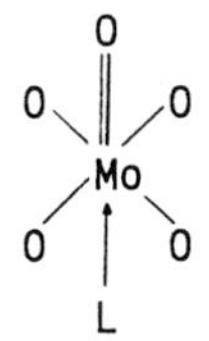 L = R_3PO, R_3NO

SCHEME 8. Phase transfer catalysis in catalytic oxidations.

The first example of the application of phase transfer catalysis in a catalytic oxidation is the ruthenium-catalyzed cleavage of olefins with NaOCl (reaction 19) reported by Foglia and coworkers.[46]

$$CH_3(CH_2)_{12}CH{=}CH_2 \xrightarrow[\text{NaOCl, NaOH, } CH_2Cl_2/H_2O]{[RuO_2/Bu_4NBr]} CH_3(CH_2)_{12}CO_2Na \quad \text{100\% yield} \tag{19}$$

$$ArCH_3 \xrightarrow[\text{NaOCl, NaOH, } ClCH_2CH_2Cl/H_2O,\ 25°C,\ pH = 9]{[RuCl_3/Bu_4NBr]} ArCO_2Na \quad \text{92-98\% yield} \tag{20}$$

More recently Sasson and coworkers[47] applied this technique to the selective oxidation of deactivated methylbenzenes to the corresponding carboxylic acids (reaction 20). The same group used a $H_2O_2/RuCl_3$ system under phase transfer conditions for the oxidation of alcohols[48], the side-chain oxidation of aromatics[49] and the oxidative cleavage of olefins.[50] Reaction of styrene with

$H_2O_2/RuCl_3$, for example, afforded benzaldehyde in 64% yield.[50] With $PdCl_2$ as catalyst, under the same conditions acetophenone was the major product (56%)[50] :

$$PhCH{=}CH_2 + H_2O_2 \xrightarrow[\text{[RuCl}_3\text{]}]{\text{[R}_4\text{NBr]}} PhCHO\ (64\%) \qquad (21)$$

$$PhCH{=}CH_2 + H_2O_2 \xrightarrow[\text{[PdCl}_2\text{]}]{\text{[R}_4\text{NBr]}} PhCOCH_3\ (56\%) \qquad (22)$$

The first example of a succesful catalytic epoxidation with aqueous H_2O_2 under phase transfer conditions was reported by Venturello and coworkers[51] :

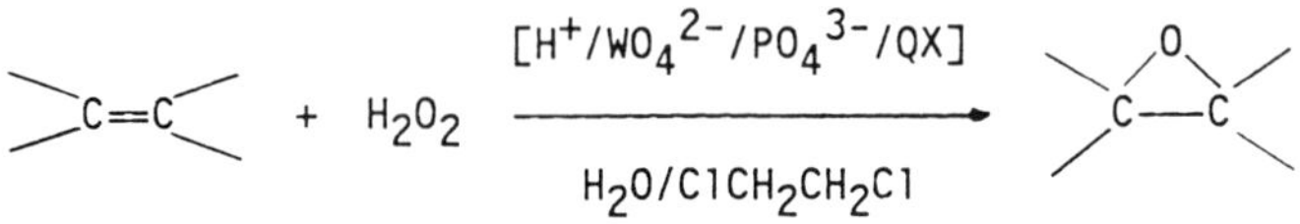

(23)

QX = onium salt

Subsequently this and analogous tungsten and molybdenum-based catalysts have been widely applied to the epoxidation of olefins[51–56], the oxidation of alcohols and the oxidative cleavage of diols[54,59,60] in aqueous/organic biphasic systems. Both simple molybdate and tungstate as well as Mo- and W-based heteropolyanions have been employed as catalysts. A typical example of the latter is the $H_3PM_{12}O_{40}$ (M=Mo or W)/cetylpyridinium chloride combination which catalyzes the efficient epoxidation of olefins and allylic alcohols under biphasic conditions.[54] The analogous oxidations of secondary alcohols to ketones and oxidative cleavage of 1,2-diols, on the other hand, gave the best results under homogeneous conditions in tert-butanol as solvent[54], e.g.

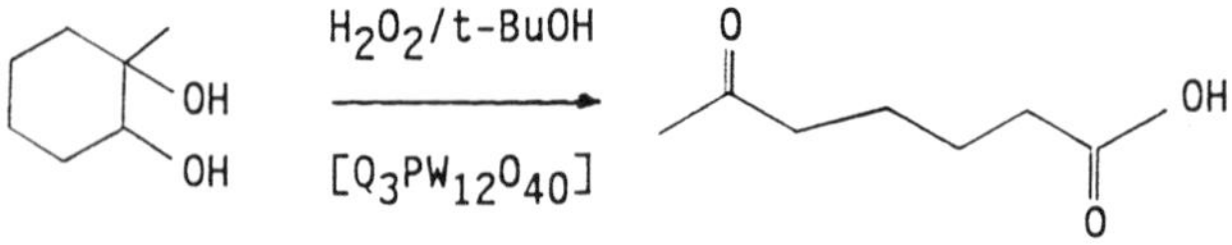

(24)

95% yield

LIGAND STABILITY, BIOMIMETIC OXIDATION AND HOMOGENEOUS vs HETEROGENEOUS CATALYSIS

As noted above two factors which have an important influence on the efficiency and selectivity of catalytic oxidations are the nature of the metal catalyst and the primary oxidant. A third important factor is the nature of the ligands surrounding the metal ion. In principle, the steric and electronic properties of catalysts can be finely tuned by an appropriate choice of ligand. This is particularly important in asymmetric oxidations (see later). Unfortunately, most organic ligands are unstable in strongly oxidizing media. This is nowhere more apparent than in the cytochrome P450-dependent monooxygenase enzymes which catalyze a wide variety of in vivo oxidative biotransformations.[61] The prosthetic group of these enzymes contains an iron (III) porphyrin complex and the active oxidant is generally accepted to be a high-valent oxoiron (V) porphyrin species.

However, this powerful oxidant is not only capable of oxidizing a wide variety of organic substrates it can also self-destruct by oxidative degradation of its own porphyrin ligand. Hence, cytochrome P450-dependent enzymes are not stable for any significant length of time outside the cell. Because this is a great disadvantage in the context of practical applications, there have been numerous studies[62] aimed at designing simple model systems capable of effecting the same, often highly regio- and stereoselective oxidations. Most of these model systems involve iron or manganese porphyrin catalysts in combination with single oxygen donors such as NaOCl[63], $KHSO_5$[64] and in a few cases H_2O_2.[65]

Unfortunately, virtually all of these systems suffer from the same disadvantage as the natural enzyme, i.e. they contain expensive, unstable ligands. There is a need, therefore, for oxidatively resistant ligands which can stabilize high-valent oxometal species in the same way that porphyrin ligands can. In principle, this can be achieved by 'fixing' the appropriate metal ion in an inorganic matrix such as a heteropolyacid(anion) or zeolite lattice.

HETEROPOLYACIDS AS OXIDATION CATALYSTS

Heteropolyacids (HPA's)[66] and their salts are polyoxocompounds incorporating anions (heteropolyanions) having metal-oxygen octahedra (MO_6) as the basic structural units. They contain one or more heteroatoms (Si, Ge, P, As, etc.) which are usually located at the centre of the anion. The MO_6 octahedra are linked together to form an extremely stable and compact structure for the heteropolyanion. One of the most common types of HPA comprises the so-called Keggin anions, $XM_n^1M^2_{12-n}O_{40}^{X-}$ (where M^1 = Mo^{VI}, W^{VI} and M^2 = V^V). Despite their rather complex formulae HPA's are very easy to synthesize by acidification of aqueous solutions containing the heteroelement and the alkali metal molybdate, tungstate or vanadate. They possess several rather unique properties which make them interesting in the context of (oxidation) catalysis :

- Strong Brønsted acids } Bifunctional
- Multi-electronxoxidants } catalysts
- Soluble in water and oxygenated organic solvents ('soluble oxides')
- Transition metal substituted HPA's can be considered as oxidatively resistant analogues of metalloporphyrins. The HPA anion functions as a multi-electron ligand and is able to stabilize reactive high-valent oxometal species.

Up till fairly recently applications have been largely limited to heterogeneous gas phase transformations[67] but it is becoming increasingly apparent that HPA's are very useful catalysts for heterogeneous and homogeneous liquid phase oxidations. In fact they may be considered as 'soluble oxides' and as such form a bridge between heterogeneous gas phase oxidations and liquid phase homogeneous oxidations.

As discussed earlier some Mo and W-based heteropolyacids have already been used, in combination with H_2O_2 as the primary oxidant under phase transfer conditions, for a variety of oxidative transformations. Some HPA's, e.g. $H_3PMo^{VI}_{12-n}V^V_nO_{40}$ (PMoV-n), are strong oxidants in their own right and can be used in combination with dioxygen as the primary oxidant. For example, $H_5PMo_{10}V_2O_{40}$ (PMoV-2) complexed with tetraglyme catalyzes the oxidative bromination of organic substrates with HBr/O_2 at ambient temperatures in chlorocarbon solvents.[68] This reagent was used for the regioselective para-bromination of phenol (reaction 25).

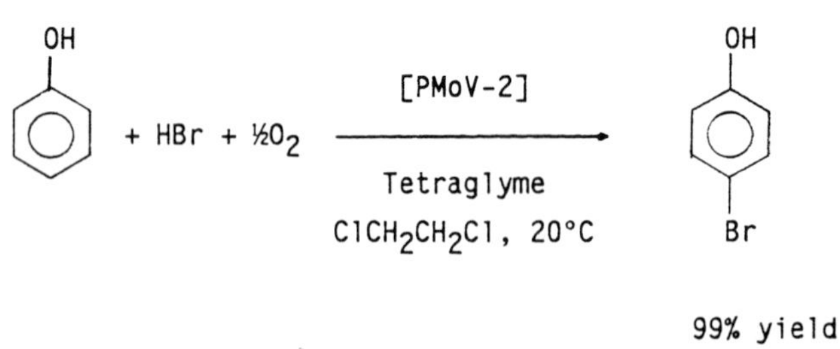

(25)

These reactions proceed via the following steps :

$$2HBr + PMoV\text{-}2_{(ox)} \rightarrow Br_2 + PMoV\text{-}2_{(red)} \quad (26)$$

$$Br_2 + RH \rightarrow RBr + HBr \quad (27)$$

$$PMoV\text{-}2_{(red)} + \tfrac{1}{2}O_2 \rightarrow H_2O + PMoV\text{-}2_{(ox)} \quad (28)$$

PMoV-2 also catalyzes the homogeneous liquid phase oxidation of organic sulfides to the corresponding sulfoxides and sulfones by dioxygen at 100–150° and 9–80 bar.[69]

The scope of HPA's as oxidation catalysts is further extended by incorporation of other redox metals. In these systems the HPA anion functions both as a (multi-electron) ligand and a co-oxidant. For example, Pd^{II}–HPA catalysts have been used for the Wacker oxidation of olefins[66,70] and the nuclear hydroxylation/ acetoxylation of aromatics with NaOAc and dioxygen in aqueous acetic acid.[71] Similarly, Mn^{II} and Co^{II}–substituted heteropolytungstates of general formula $(R_4N)_4HMP.W_{11}O_{39}$ ($M=Mn^{II}$, Co^{II}) catalyze the epoxidation of olefins with PhIO[72] and the hydroxylation of alkanes with TBHP.[73] These reactions bear a close resemblance to the cytochrome P450 model systems referred to earlier and the transition metal substituted polyoxometalates may be considered as oxidatively resistant inorganic analogues of metalloporphyrins.

On the basis of the above examples we conclude that heteropolyacid-based catalysts have a very promising future in the synthesis of fine chemicals via selective oxidation.

REDOX ZEOLITES AS SELECTIVE OXIDATION CATALYSTS

Another way of 'fixing' redox metal ions in stable inorganic matrices, thereby creating catalysts with interesting activities and selectivities, is to build them into a zeolite lattice. For example, the synthetic titanium(IV) zeolite, titanium silicalite (TS-1), developed by Enichem workers[74,75], catalyzes a variety of useful oxidations with 30% H_2O_2 such as olefin epoxidation[76,77], oxidation of primary alcohols to aldehydes[78], aromatic hydroxylation[79], and ammoxidation of cyclohexanone to cyclohexanone oxime[80] (see Scheme 9). The TS-1-catalyzed hydroxylation of phenol to a 1:1 mixture of catechol and hydroquinone has already been commercialized by Enichem.[74]

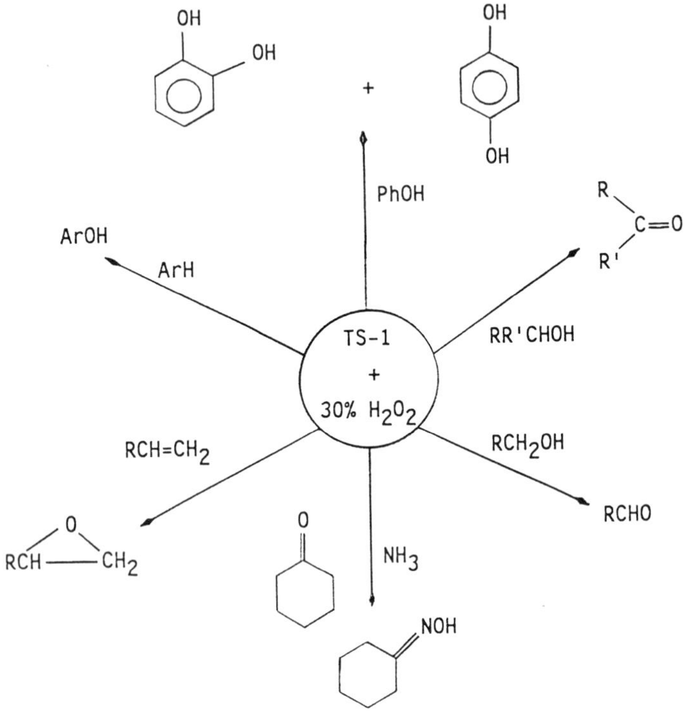

SCHEME 9. Oxidations catalyzed by titanium silicalite (TS–1).

The TS–1 catalyst exhibits some quite remarkable activities and selectivities. Thus, ethylene is epoxidized with 30% H_2O_2 in tert–butanol at ambient temperature, giving ethylene oxide in 96% selectivity at 97% H_2O_2 conversion.[76] Interestingly TS–1 also catalyzes the rearrangement of styrene oxides to the corresponding beta–phenyl–acetaldehydes (reaction 29).[81]

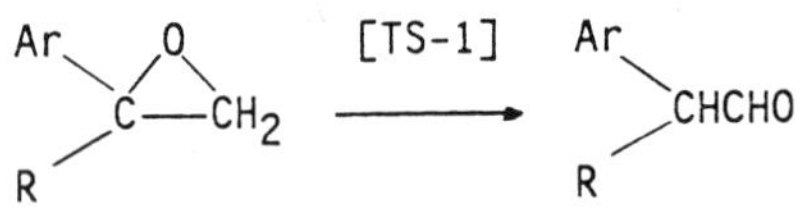

(29)

90-98% yield

From a mechanistic viewpoint it is noteworthy that the TS–1 catalyst contains the same chemical elements in roughly the same proportions as the Shell Ti^{IV}/SiO_2 catalyst referred to earlier. In the latter case we postulated[9] the formation of catalytically active, isolated titanyl (Ti=O) species to explain the unique activity of this catalyst :

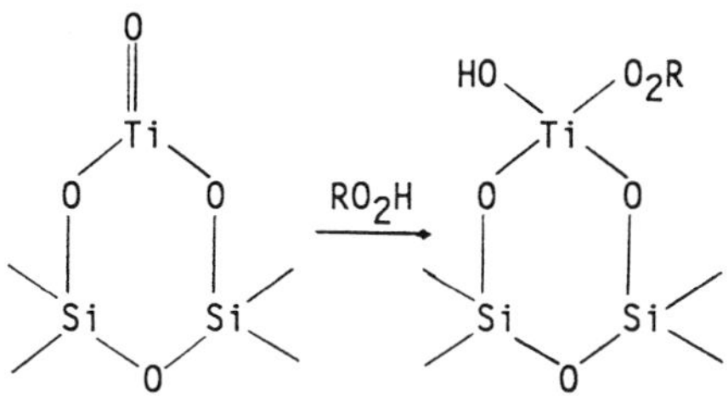

(30)

The formation of isolated titanyl groups is presumably an important prerequisite for catalytic performance since Ti^{IV} has a strong tendency to assume a high coordination number via the formation of Ti–O–Ti bonds. This presumably leads to the formation of titanium centres which are only capable of catalyzing the homolytic decomposition of hydrogen peroxide. Despite their similarities the TS–1 catalyst displays a broader range of activities than the Ti^{IV}/SiO_2 catalyst. The paramount question is then : what is the essential difference between these catalysts? A possible explanation is that the TS–1 contains more (or more active) isolated titanyl centres than the Ti^{IV}/SiO_2.

Based on the quite remarkable results obtained with TS–1 we predict a bright future for the use of redox zeolites, i.e. zeolites modified via isomorphous substitution of Si^{IV} with redox metals in the crystal lattice, as selective oxidation catalysts.

WHAT DETERMINES THE CHEMO– AND REGIOSELECTIVITY?

An understanding of the factors which determine chemo– and regioselectivities is of paramount importance in the context of designing selective oxidation catalysts. As discussed earlier chemoselectivities are influenced by the nature of the metal, its surrounding ligands and the primary oxidant.

Regioselectivity is of particular importance in oxidations of hydrocarbons (alkenes, arenes and alkanes). For example, what determines the extent of allylic vs vinylic (double bond) attack in olefin oxidations? In the first place, this is influenced by the nature of the metal (see Scheme 10). High–valent oxometal complexes such as Fe^{V}=O and Mn^{V}=O are very strong electrophiles and give predominantly attack at the (nucleophilic) double bond. High-valent oxometal complexes such as $Mo^{VI}=O$ and $Se^{IV}=O$, on the other hand, are weaker electrophiles and give predominantly attack at the allylic C–H bond. Examples are the gas-phase, bismuth molybdate-catalyzed oxidation of propylene to acrolein and the liquid-phase, SeO_2-catalyzed allylic hydroxylation of olefins with TBHP discussed earlier.

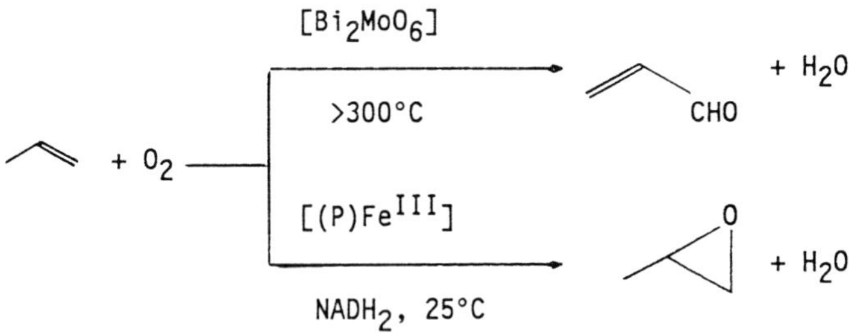

The key selectivity-determining step :

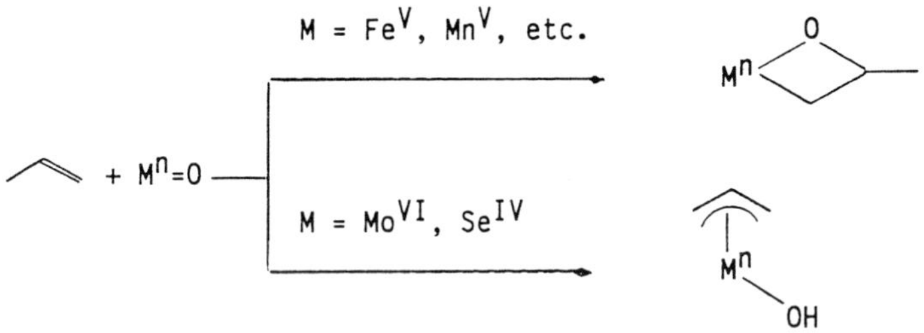

SCHEME 10. What determines allylic vs vinylic attack?

As has recently been pointed out by Lyons[71] both the oxidation state of the metal and the nature of the surrounding ligands are critical factors in determining allylic vs vinylic attack. Thus, the palladium(II)-catalyzed oxidation of olefins (Wacker process) to give aldehydes, ketones or vinyl esters, involves nucleophilic attack of water on a palladium(II)-olefin π complex. In these reactions carbon-hydrogen bond activation (β-hydrogen elimination) follows nucleophilic attack. In contrast, Pd^0 catalysts afford π-allylpalladium(II) species via oxidative addition of the allylic C–H bond to the coordinatively unsaturated palladium (O) centre. In other words, C–H activation precedes nucleophilic attack (see Scheme 11).

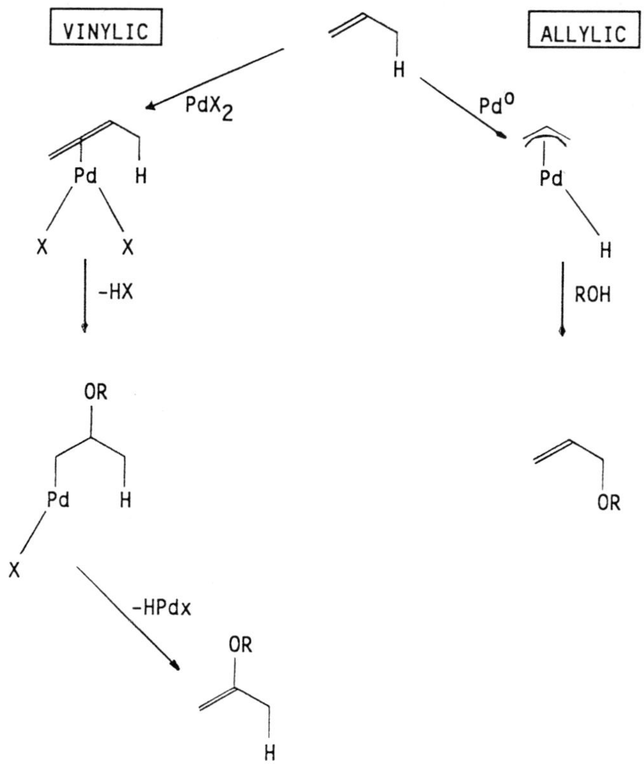

SCHEME 11. Allylic vs vinylic oxidation of olefins.

Such mechanistic insights led to the development of methods for the preparation of acrylic acid and allyl acetate via liquid phase, Pd/C-catalyzed oxidation of propylene, in water and acetic acid respectively, under mild conditions.

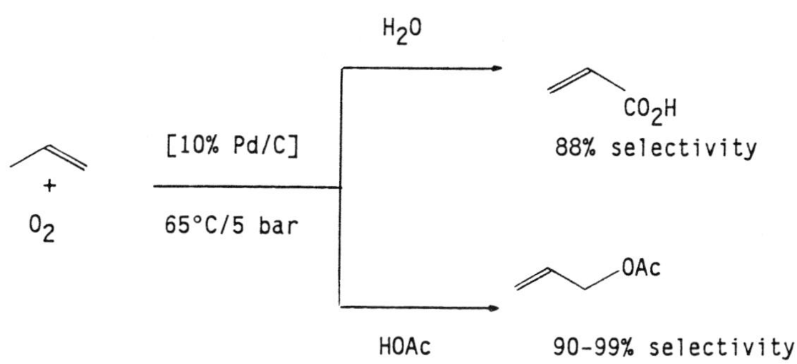

(31)

(32)

In order to achieve selective allylic oxidation it was necessary to preactivate the catalyst by treatment with propylene in the absence of oxygen, presumably to generate active Pd(O) centres.

Similarly, the oxidation state of the catalyst is a crucial factor in determining ring vs side-chain oxidation of aromatics.[71] Thus, palladium (II) catalyzes the nuclear oxidative acetoxylation of aromatics by strong oxidants ($Cr_2O_7^{2-}$, MnO_4^- or $S_2O_8^{2-}$) in acetic acid. In contrast 10% Pd/C is a very effective catalyst for the side-chain oxidation of aromatics under mild conditions :

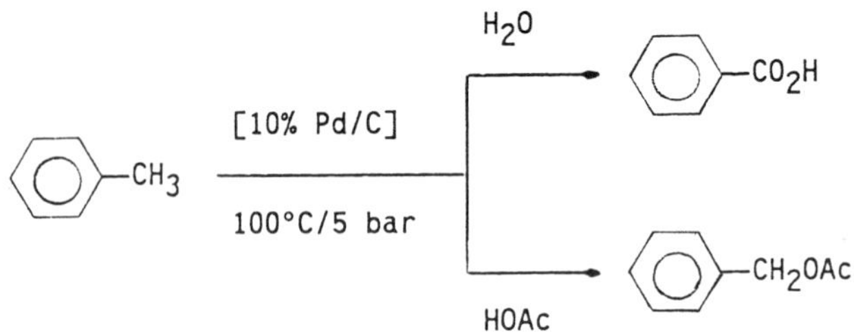

(33)

(34)

THE ULTIMATE CHALLENGE IN SELECTIVITY–ENANTIOSELECTIVE CATALYSIS

As noted earlier much effort has been devoted in recent years towards designing simple chemocatalysts which are able to emulate Nature's selective and versatile biocatalysts, the monooxygenases. This has led to the development of a variety of relatively simple metal catalyst/oxygen donor reagents which can mediate the same reactions, e.g. olefin epoxidation, alkane hydroxylation, etc., as the monooxygenases. The ultimate challenge in biomimetic oxidations is, however, the design of relatively simple chemocatalysts able to emulate the enantioselectivity characteristic of the monooxygenase-mediated transformations. In this context it is worth noting, however, that Nature is far from perfect. Thus, microbial expoxidation of propylene mediated by a Nocardia coralline species, for example, affords R-propylene oxide with 'only' 83% e.e.[83]

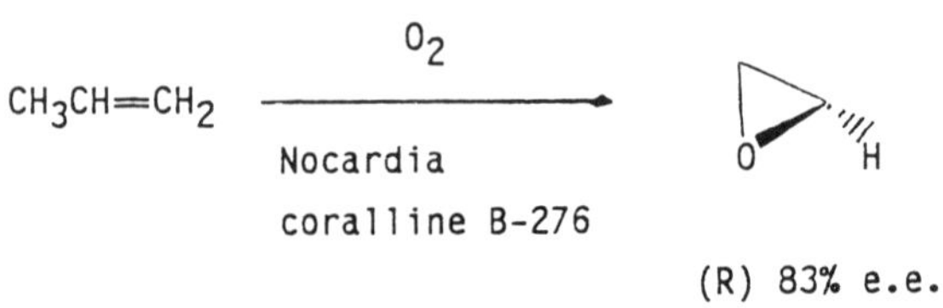

(35)

Although attempts to achieve biomimetic enantioselective epoxidation of non-functionalized olefins have so far met with little success, quite spectacular results have been obtained in two related catalytic asymmetric oxidations of olefinic substrates.

The first example of a highly enantioselective catalytic oxidation is the now well-known catalytic asymmetric epoxidation of allylic alcohols (Scheme 12) developed by Sharpless and coworkers.[84,85] The same titanium(IV)/TBHP reagent was subsequently applied, by Kagan and coworkers[86], to the enantioselective catalytic oxidation of sulfides to the corresponding sulfoxides.

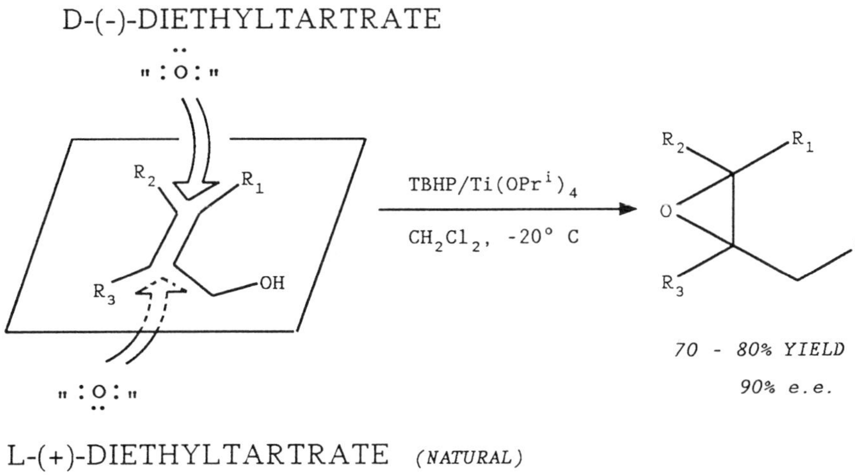

SCHEME 12. Catalytic asymmetric epoxidation.

The second important example, also developed by Sharpless and coworkers[87–91], is the asymmetric vicinal-dihydroxylation of olefins by N-methyl-morpholine-N-oxide (NMO) in the presence of an OsO_4 catalyst (0.2–0.4% m) and dihydroquinine or dihydroquinidine esters as chiral ligands (Scheme 13).

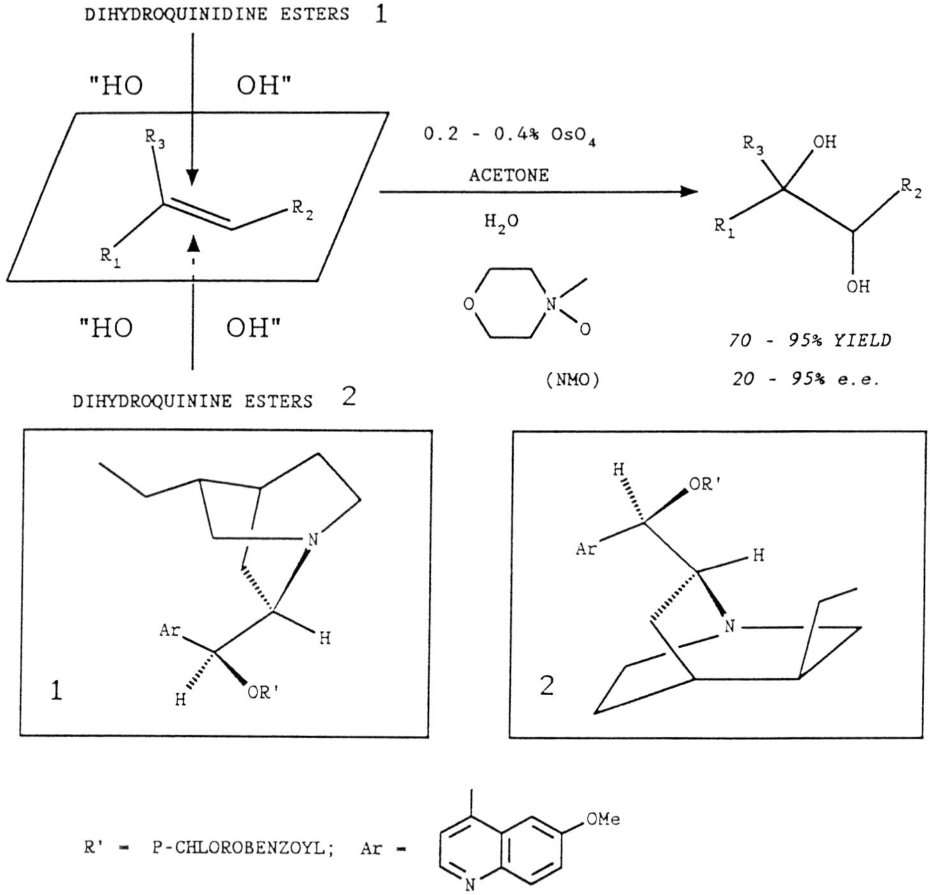

SCHEME 13. Catalytic asymmetric vicinal dihydroxylation.

An important prerequisite for high enantioselectivity in such processes is that coordination of the metal ion to the chiral ligand results in a substantial rate acceleration. Sharpless[87] coined the term ligand-accelerated catalysis to describe this phenomenon. Thus, if the metal-chiral ligand complex (M-L chiral) rapidly exchanges its ligands in solution then high enantioselectivities will be possible only when M-L is a much more active catalyst than M (see Scheme 14).

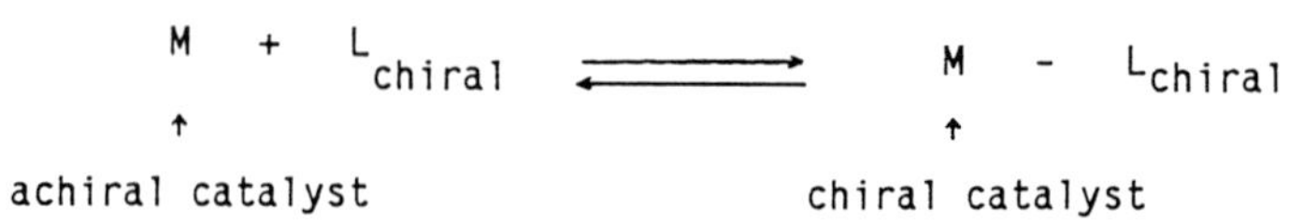

SCHEME 14. Ligand accelerated catalysis.

Application of this principle to other transition-metal catalyzed oxidations should lead to the development of other catalytic asymmetric oxidations in the future. Tartaric acid derivatives and cinchona alkaloids[92] appear to constitute attractive ligands in such processes. An interesting variation on this theme is the use of quaternary derivatives of cinchona alkaloids as chiral phase transfer catalysts in the base-catalyzed autoxidation of ketones to the corresponding α-hydroxyketones (reaction 36).[93]

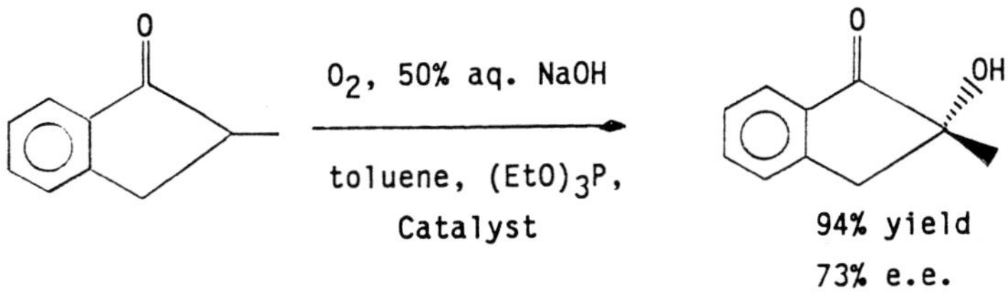

(36)

Catalyst :

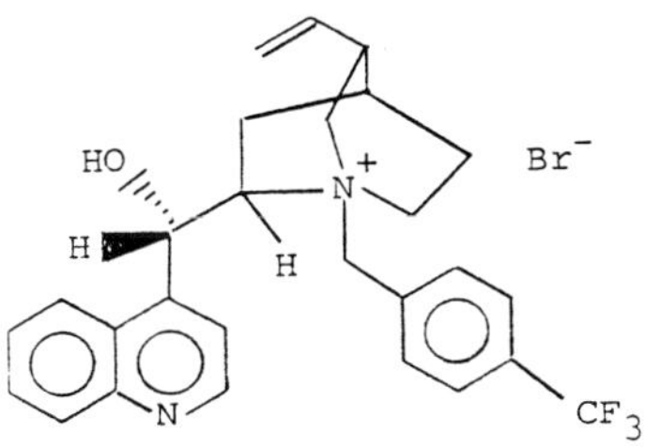

CONCLUDING REMARKS

Based on its wide choice of catalysts and oxygen donors, its diversity of mechanism and its scope in organic synthesis catalytic oxidation is surely the most fascinating and versatile of all catalytic processes. Moreover, with the added stimulation of the need to replace environmentally inefficient stoichiometric procedures, we expect that the application of catalytic oxidation techniques to the manufacture of fine chemicals will continue to be a very fruitful area of research in the future.

In particular, we have great expectations regarding the broader application of transition metal substituted zeolites and heteropoly acids as liquid phase oxidation catalysts and last but not least the development of more catalytic asymmetric oxidations.

REFERENCES

1. A. McKillop and J.A. Tarbin, Tetrahedron Lett., 24, 1505–1508 (1983).
2. A. McKillop and J.A. Tarbin, Tetrahedron, 43, 1753–1758 (1987).
3. T. Ando, D.G. Cork and T. Kimura, Chem. Lett., pp. 665–666 (1986).
4. For a recent review see W. Adam, R. Curci and J.O. Edwards, Acc. Chem. Res., 22, 205–211 (1989).
5. P.L. Anelli, C. Biffi, F. Montanari and S. Quici, J. Org. Chem., 52, 2559-2562 (1987).
6. T. Miyazawa and T. Endo, J. Org. Chem., 50, 3930–3931 (1985); T. Miyazawa, T. Endo, S. Shiihashi and M. Okawara, J. Org. Chem., 50, 1332–1334 (1985).
7. R.A. Sheldon, Bull. Soc. Chim. Belg., 94, 651–669 (1985).
8. For an excellent recent review see, H. Mimoun in Comprehensive Coordination Chemistry, Vol. 6, 317, G. Wilkinson, R.D. Gillard and J.A. McCleverty, Eds., Pergamon, New York, 1987.
9. R.A. Sheldon in "Aspects of Homogeneous Catalysis", Vol. 4 (R. Ugo, Ed.), D. Reidel, Dordrecht, 1981, p. 1.
10. P. Landau, G.A. Sullivan and D. Brown, Chem. Tech., p. 602 (1979).
11. For a recent review of transition metal-catalyzed epoxidations see : K.A. Jorgensen, Chem. Rev., 89, 431–458 (1989).
12. Y. Kurusu and Y. Masuyama, Polyhedron, 5, 289–296 (1986); J. Macromol. Sci. Chem., A24, 389–401 (1987).
13. Y. Masuyama, M. Takahashi and Y. Kurusu, Tetrahedron Lett., 25, 4417–4420 (1984).
14. Y. Kurusu, Y. Masuyama and M. Saita, Bull. Chem. Soc. Japan, 58, 1065–1066 (1985).
15. K. Yamawaki, T. Yoshida, H. Nishihara, Y. Ishii and M. Ogawa, Synth. Commun., 16, 537–541 (1986).
16. Y. Ishii, K. Yamawaki, T. Ura, H. Yamada, T. Yoshida and M. Ogawa, J. Org. Chem., 53, 3587–3593 (1988).
17. B.M. Trost and Y. Masuyama, Tetrahedron Lett., 25, 173–176 (1984).
18. K. Kaneda, Y. Kawanishi, K. Jitsukawa and S. Teranishi, Tetrahedron Lett., 24, 5009-5010 (1983).
19. K. Yamawaki, Y. Ishii and M. Ogawa, Chem. Express, 1, 95–98 (1986).
20. K. Kaneda, Y. Kawanishi and S. Teranishi, Chem. Lett., 1481–1482 (1984).
21. S. Kanemoto, H. Saimoto, K. Oshima and H. Nozaki, Tetrahedron Lett., 25, 3317–3320 (1984).
22. E.J. Corey, E.P. Barette and P.A. Magriotis, Tetrahedron Lett., 26, 5855–5858 (1985).
23. S. Kanemoto, H. Tomioka, K. Oshima and H. Nozaki, Bull. Chem. Soc., Japan, 59, 105–108 (1986).
24. For a review of Ru-catalyzed oxidations see : E.S. Gore, Plat. Metals Rev., 27, 111–125 (1983).
25. Japanese patent 80,102,527 (1980) to Mitsui Petrochemical Company, Chem. Abstr., 94, 46783b (1981).
26. S. Murahashi, T. Naota and N. Nakajima, Tetrahedron Lett., 26, 925–928 (1985).
27. M. Tanaka, T. Kobayashi and T. Sakakura, Angew. Chem., 96, 519–520 (1984).
28. S. Wolfe, S.K. Hasan and J.R. Campbell, Chem. Commun., p. 1420 (1970).
29. T. Sugiura, T. Saeki, S. Matsumoto and Y. Shizume, Japanese Patent, 86,289,053 (1986) to Osaka Yuki Kagaku Kogyo Company; Chem. Abstr., 106, 175781q (1987).
30. S. Giddings and A. Mills, J. Org. Chem., 53, 1103–1107 (1988); Y. Yamamoto, H. Suzuki and Y. Moro-oka, Tetrahedron Lett., 26, 2107–2108 (1985).
31. P.H.J. Carlsen, T. Katsuki, V.S. Martin and K.B. Sharpless, J. Org. Chem., 46, 3963–3938 (1981).
32. W.P. Griffith, S.V. Ley, G.P. Whitcombe and A.D. White, J. Chem. Soc. Chem. Commun., 1625–1627 (1987).
33. K. Vijayasri, J. Rajaram and J.C. Kuriacose, J. Mol. Catal., 39, 203–217 (1987).
34. G. Green, W.P. Griffith, D.M. Hollinshead, S.V. Ley and M. Schröder, J. Chem. Soc. Perkin Trans., I, 681–686 (1984); K.S. Kim, S.J. Kim, Y.H. Song and C.S. Hahn, Synthesis, pp. 1017–1018 (1987).

35. P. Müller and J. Godoy, Tetrahedron Lett., 22, 2361-2364 (1981).
36. C. Bilgrien, S. Davis and R. Drago, J. Am. Chem. Soc., 109, 3786-3787 (1987).
37. M. Matsumoto and S. Ito, J.C.S. Chem. Commun., 907-908 (1981).
38. K. Kaneda and S. Teranishi, in "Activation of Dioxygen Species and Homogeneous Catalytic Oxidation", Abstracts of the International Symposium held in Galzignano, Italy, June 24-29, 1984.
39. M. Zviely, A. Goldman, I. Kirson and E. Glotter, J. Chem. Soc. Perkin Trans, 229-231 (1986).
40. C. Venturello and M. Ricci, J. Org. Chem., 51, 1599-1602 (1986).
41. T.R. Felthouse, J. Am. Chem. Soc., 109, 7566-7568 (1987).
42. R.A. Sheldon in "The Chemistry of Functional Groups, Peroxides", S. Patai, Ed., Wiley, New York, 1982, pp. 161-200.
43. H. Mimoun, R. Charpentier, A. Mitschler, J. Fischer and R. Weiss, J. Am. Chem. Soc., 102, 1047 (1980).
44. M. Schröder, Chem. Rev., 80, 187 (1980).
45. M.A. Umbreit and K.B. Sharpless, J. Am. Chem. Soc., 99, 5526 (1977).
46. T.A. Foglia, P.A. Barr and A.J. Malloy, J. Am. Oil Chem. Soc., 54, 858A-861A (1977).
47. Y. Sasson, G.D. Zappi and R. Neumann, J. Org. Chem., 51, 2880-2883 (1986).
48. G. Barak, J. Dakka and Y. Sasson, J. Org. Chem., 53, 3553-3555 (1988).
49. G. Barak and Y. Sasson, J. Chem. Soc. Chem. Commun., 637-638 (1988).
50. G. Barak and Y. Sasson, J. Chem. Soc. Chem. Commun., 1266-1267 (1987).
51. C. Venturello, E. Alneri and M. Ricci, J. Org. Chem., 48, 3831-3833 (1983).
52. C. Venturallo and R. D'aloisio, J. Org. Chem., 53, 1553-1557 (1988).
53. J. Prandi, H.B. Kagan and H. Mimoun, Tetrahedron Lett., 27, 2617-2620 (1986) (quaternary phosphonium salts as phase transfer agents).
54. Y. Ishii, K. Yamawaki, T. Ura, H. Yamada, T. Yoshida and M. Ogawa, J. Org. Chem., 53, 3587-3593 (1988).
55. Y. Matoba, H. Inoue, J. Akagi, T. Okabayashi, Y. Ishii and M. Ogawa, Synth. Commun. 14, 865-873 (1984).
56. O. Bortolini, F. DiFuria, G. Modena and R. Seraglia, J. Org. Chem., 50, 2688-2690 (1985).
57. K. Yamawaki, H. Nishihara, T. Yoshida, T. Ura, H. Yamada, Y. Ishii and M. Ogawa, Synth. Commun., 18, 869-875 (1988).
58. O. Bortolini, V. Conte, F. DiFuria and G. Modena, J. Org. Chem., 51, 2661-2663 (1986).
59. C. Venturello and M. Ricci, J. Org. Chem., 51, 1599-1602 (1986).
60. M. Schwegler, M. Floor and H. van Bekkum, Tetrahedron Lett., 29, 823-826 (1988).
61. P. Ortiz de Montellano, Ed., "Cytochrome P450 : Structure, mechanism and biochemistry", Plenum, New York, 1986.
62. For leading references see : B. Meunier, Gazz. Chim. Ital., 118, 485-493 (1988).
63. For leading references see : J.P. Collman, J.I. Brauman, B. Meunier, T. Hayashi, T. Kodadek and S.A. Raybuck, J. Am. Chem. Soc., 107, 2000 (1985); B. Meunier, M.E. de Carvalho and A. Robert, J. Mol. Catal., 41, 185 (1987); A.W. van der Made, M.J.P. van Gerwen, W. Drenth and R.J.M. Nolte, J. Chem. Soc. Chem. Commun., 888 (1987).
64. B. de Poorter, M. Ricci and B. Meunier, Tetrahedron Lett., 26, 4459-4462 (1985).
65. P. Battioni, J.P. Renaud, J.F. Bartoli, M. Reine-Artiles, M. Fort and D. Mansuy, J. Am. Chem. Soc., 110, 8462-8470 (1988).
66. For recent reviews see : I.V. Kozhevnikov, Russ. Chem. Rev., 56, 811-825 (1987) and I.V. Kozhevnikov and K.I. Matveev, Appl. Catal., 5, 135-150 (1983).
67. M. Misono, Catal. Rev. Sci. Eng., 29, 269 (1987).
68. R. Neumann and I. Assael, J. Chem. Soc. Chem. Commun., 1285-1287 (1988); see also T.A. Gorodetskaya, I.V. Kozhevnikov and K.I. Matveev, Kinet. Katal. (Engl. Transl.), 23, 842-844 (1982).
69. I.V. Kozhevnikov, V.I. Simagina, G.V. Varnakova and K.I. Matveev, Kinet. Catal. (Engl. Transl.), 20 416-419 (1979).

70. H. Ogawa, H. Fujinami, K. Taya and S. Teratani, J.C.S. Chem. Commun., 1274–1275 (1981).
71. J.E. Lyons, Catalysis Today, 3, 245–258 (1988).
72. C.L. Hill and R.B. Brown, J. Am. Chem. Soc., 108, 536–538 (1986).
73. M. Faraj and C.L. Hill, J. Chem. Soc. Chem. Commun., 1487–1489 (1987).
74. B. Notari, Stud. Surf. Sci. Catal., 37, 413–425 (1988).
75. G. Perego, G. Bellussi, C. Corno, M. Taramasso, F. Buonomo and A. Esposito, Stud. Surf. Sci. Catal., 28, 129–136 (1986).
76. C. Neri, B. Anfossi, A. Esposito and F. Buonomo, Eur. Pat. Appl., 100,119 (1984) to ANIC; Chem. Abstr., 101, 38336f (1984).
77. M.G. Clerici and U. Romano, Eur. Pat. Appl., 100,625 (1987) to Enichem.
78. A. Esposito, C. Neri and F. Buonomo, Eur. Pat. Appl., 102,655 (1984) to ANIC; Chem. Abstr., 100, 209167n (1984).
79. A. Esposito, M. Taramasso, C. Neri and F. Buonomo, Br. Pat., 2,116,974 (1985) to ANIC; G. Belussi, M. Clerici, F. Buonomo, U. Romano, A. Esposito and B. Notari, Eur. Pat. Appl., 200,260 (1986) to Enichem.
80. P. Roffia, M. Padovan, E. Morelti and G. De Alberti, Eur. Pat. Appl., 208,311 (1985) te Montedipe.
81. C. Neri and F. Buonomo, Eur. Pat. Appl. 100,117 (1982) to ANIC; Chem. Abstr., 100, 209389 (1984).
82. J.E. Lyons, Homogeneous and Heterogeneous Catalysis", Y. Yermakov and V. Likholobov, Eds., VNU Science Press, Utrecht, 1986, pp. 117–138.
83. K. Furuhashi, Chem. Econ. Eng. Rev., 18 (7), 21–26 (1986).
84. T. Katsuki and K.B. Sharpless, J. Am. Chem. Soc., 102, 5976–5978 (1980).
85. For recent reviews see : K.B. Sharpless, Chem. Brit., 38–44 (1986); K.B. Sharpless, S.S. Woodward and M.G. Finn, Pure Appl. Chem., 55, 1823 (1983); M.G. Finn and K.B. Sharpless in 'Asymmetric Synthesis', J.D. Morrison, Ed., Academic Press, New York, 1985, Vol. 5, Chapter 8.
86. H.B. Kagan, Phosphorus and Sulfur, 27, 127–132 (1986); H.B. Kagan, E. Dunach, C. Nemecek, P. Pitchen, O. Samuel and S.H. Zhao, Pure Appl. Chem., 57, 1511–1517 (1985); P. Pitchen and H.B. Kagan, Tetrahedron Lett., 25, 1049–1052 (1984).
87. E.N. Jacobson, I. Markó, W.S. Mungall, G. Schröder and K.B. Sharpless, J. Am. Chem. Soc., 110, 1968–1970 (1988).
88. E.N. Jacobson, I. Markó, M.B. France, J.S. Svendson and K.B. Sharpless, J. Am. Chem. Soc., 111, 737–739 (1989).
89. J.S.M. Wai, I. Markó, J.S. Svendsen, M.G. Finn and K.B. Sharpless, J. Am. Chem. Soc., 111, 1123–1125 (1989).
90. I.E. Markó, lecture presented at the Chiral Synthesis Symposium and Workshop, organized by Spring Innovations Ltd., in Manchester, April 1989.
91. For further examples of stoichiometric asymmetric dihydroxylations with OsO_4 and chiral diamine ligands see : M. Tokles and J.K. Snyder, Tetrahedron Lett., 27, 3951–3954 (1986); K. Tomioka, M. Nakajima and K. Koga, J. Am. Chem. Soc., 109, 6213–6215 (1987); K. Tomioka, M. Nakajima, Y. Iitaka and K. Koga, Tetrahedron Lett., 29, 573–576 (1988).
92. For a review of asymmetric catalysis by cinchona alkaloids see H. Wijnberg in "Topics in Stereochemistry", E.L. Eliel and S. Wilen, Eds., Wiley, New York, 1986, Vol. 16.
93. M. Masui, A. Ando and T. Shioiri, Tetrahedron Lett., 29, 2835–2838 (1988).

For further background reading on the subject see the following excellent review articles :

F. di Furia and G. Modena, Pure Appl. Chem., 54, 1853 (1982); Rev. Chem. Interm. 6, 51 (1985).

H. Mimoun, Angew. Chem. Int. Ed. Engl., 21, 734 (1982).

J. Ruiz (Université Catholique de Louvain, Belgium) :
Grafting of organometallic compounds (porphyrins, phthalocyanines, etc.) into acid supports (zeolites, pillared clays) seems to be a promising approach to obtain active and selective catalysts for the preparation of fine chemicals. What is your opinion concerning these new materials?

R.A. Sheldon (Andeno B.V., Venlo, The Netherlands) :
As I mentioned in my lecture the problem associated with the use of metal porphyrins and related complexes is the limited stability of such ligands under oxidizing conditions. If their stability could be improved by immobilization in the pores of zeolites or pillared clays this could be a useful approach. However, my personal opinion is that 'fixing' redox metal ions in stable inorganic matrices such as zeolites or heteropolyacids represents a more promising approach. This allows for the creation of isolated metal sites which are potentially more active since deactivation of active intermediates (oxometal, peroxometal) via dimerization is precluded. Moreover, heteropolyanions (oxometalates) are multi-electron ligands and may be considered as stable, inorganic equivalents of porphyrins.

J. Kiwi (Fed. Inst. Technology, Lausanne, Switzerland) :
You have stated that active intermediates in oxidative transformations mediated by cytochrome P450-dependent mono-oxygenases are oxoiron (V) species. Are these stable, isolated iron (V) species characterized by physical techniques such as EPR? Or are they hypothetical cyt. P450–Fe^{V}=O species where porphyrin ligands mediate charge transfer? In such cases only iron (IV) species have been verified experimentally.

R.A. Sheldon (Andeno B.V., Venlo, The Netherlands) :
Although the active intermediate in cytochrome P450-mediated oxidations is often described as a formally oxoiron (V) porphyrin I agree that it is now generally accepted that this intermediate is more correctly described as an oxoiron (IV)-porphyrin radical cation complex ($P^{+\cdot}Fe^{IV}$=O). However, it is worth pointing out that from the point of view of electron counting in oxidation processes it is often convenient to regard the intermediate as being formally oxoiron (V). Moreover, an oxoiron (IV)-porphyrin radical cation complex is obviously different to a simple oxoiron (IV) species.

J. Haber (Institute of Catalysis and Surface Science, Polish Academy of Sciences, Krakow, Poland) :
When discussing the prospects of catalytic oxidations we should consider not only the development of novel catalytic systems, so spectacularly presented in your review, but also the possibilities of modifying the electron density at the metal centre via photostimulation or the application of electric potential. I think that the combination of organometallic chemistry with photo– and electrocatalysis may open exciting new fields in both research and technology.

R.A. Sheldon (Andeno B.V., Venlo, The Netherlands) :
I agree wholeheartedly. Although I did not mention these aspects in my talk photo– and electrocatalysis are potentially very useful techniques for generating active catalysts, particularly in the context of fine chemicals manufacture.

O. Krylov (Institute of Chemical Physics, Academy of Sciences of the USSR, Moscow) :
What do you think about the possible future application of intermediate systems between homogeneous and heterogeneous catalysts, such as catalysts in reverse micelles?

R.A. Sheldon (Andeno B.V., Venlo, The Neterlands) :
I think that there is potentially a great future for such systems in catalytic processes in general. The key advantage of heterogeneous over homogeneous catalysts is facile recovery and recycling of the catalyst. Much effort has been devoted in the last two decades, therefore, to immobilizing homogeneous catalysts on insoluble, solid supports. This approach has met with little success. There are other promising approaches, however, which up till recently have not received much attention. For example, the use of two–phase liquid systems in combination with a phase transfer catalyst or the use of catalysts attached to soluble polymers, catalysts in reverse micelles, catalytic membranes, etc. I think that we shall see much more effort devoted to these approaches in the future. Finally, I would like to note that we are always talking about immobilizing homogeneous catalysts; maybe we should think more about 'mobilizing' heterogeneous catalysts.

G. Centi and F. Trifiro' (Editors), *New Developments in Selective Oxidation* 33
1990 Elsevier Science Publishers B.V., Amsterdam — Printed in The Netherlands

SELECTIVE OXIDATION WITH TI-SILICALITE

by U.Romano, A.Esposito, F.Maspero, C.Neri,
ENICHEM SYNTHESIS, Milano (Italy)
and M.G.Clerici, ENIRICERCHE, Milano (Italy)

* * *

The synthesis and crystal structure of Titanium silicalite-1 (TS-1). a new synthetic zeolite of the ZSM family, has been reported (Ref.1). Characteristic features of this zeolite are: the total absence of Al atoms and strong acidic sites; the vicariant positions assumed by the Ti atoms in the silicalite framework; the uniform distribution of Ti random in the crystal, so that each tetrahedrally coordinated Ti atom is surrounded by four -O-Si units. This type of distribution has been suggested to be responsible for the typical reactivity of this material (Ref.2).

TS-1 is an efficient and selective catalyst for oxidations with H2O2. No oxidative reaction was detected using other peroxides (e.g. tert-butyl , cumene hydroperoxide).

The most important catalytic reactions observed are:

-Alcohol oxidation (Ref.3);

-Olefin epoxidation (Ref.4,5,6,7);

-Aromatic nuclear hydroxylation (Ref.8);

-Amine oxidation;

-Cyclohexanone ammoximation (Ref.9).

Moreover, a competitive pathway of H2O2 decomposition is normally present during the course of other catalytic reactions; its amount depends on the type of reaction and experimental conditions.

Further oxidation steps take place in more drastic conditions: aldehydes are oxidized to carboxylic acids; hydroxylated aromatic compounds may undergo ring opening cleavage.

REACTION OF TS-1 AND H2O2

After interaction with H2O2, TS-1 exhibits a yellow colour, a modified IR spectrum, with disappearing of the original band at 980 cm-1, and a new, intense EPR signal (Ref.10).

These modifications persist for some time, even in the separated zeolite; they cannot be reversed by addition of polar, donor or protic compounds (e.g. H2O), even in a large excess. The original material can be obtained upon interaction with oxidable species (propylene, phenol), or by prolonged staying, whereby a slow O2 development takes place.

The dependence of the H2O2 decomposition rates on catalyst concentration and temperatures has been found.

ALCOHOL OXIDATION

The oxidation of methanol at low conversions yields selectively CH_2O and dimethyl formale; no formic acid nor its ester were detected in these conditions.

The H_2O_2 decomposition to O_2 is normally about 20-30 % of the total H_2O_2 consumption.

The influence of kinetic parameters (temperature, catalyst concentration, H_2O_2 concentration) has been studied, obtaining an activation energy of about 20 Kcal/mole, a simple first order dependence on the catalyst concentration , and a kinetic order respect to the H_2O_2 concentration in the range 0 - 1. This kinetic behaviour is in accord with a 2-steps consecutive process:

(1) $Ti + H_2O_2 \rightleftharpoons [Ti,H_2O_2]$

(2) $[Ti,H_2O_2] + CH_3OH \rightarrow Ti + CH_2O + 2\ H_2O$

The formation of the acetale is a catalytic not oxidative reaction:

(3) $CH_2O + 2\ CH_3OH \rightarrow CH_2(OCH_3)_2 + H_2O$

Primary alcohols have been tested up to C8.

The selectivity to aldehyde is lower than for methanol, depending on the alcohol conversion. At low conversions, 1-propanol shows a high selectivity to aldehyde and acetale (up to 95 %):

(4) $EtCH_2OH + H_2O_2 \rightarrow EtCHO + 2\ H_2O$

(5) $EtCHO + 2\ EtCH_2OH \rightleftharpoons EtCH(OCH_2Et)_2 + H_2O$

As expected, the selectivity to aldehyde is lower at high conversions, with increasing contribution of the consecutive reaction:

(6) $EtCHO + H_2O_2 \rightarrow EtCOOH + 2\ H_2O$

The ester formation appears to be independent on the acid concentration; possibly, it could be due to a very fast oxidation of the hemiacetalic intermediate:

(7) $EtCHO + EtCH_2OH \rightarrow EtCH(OH)OCH_2Et \xrightarrow{(+H_2O_2)} EtCOOCH_2Et + 2\ H_2O$

The H_2O_2 decomposition to O_2 in primary alcohol solution is in general lower than for methanol (< 10 %).

The influences of kinetics parameters on the oxidation rates (temperature, catalyst and H_2O_2 concentrations) were similar to those found for methanol.

All the primary alcohols tested are oxidized faster than methanol. Ethanol is the most easily oxidized, and the rates decrease regularly with increasing chain length; while the isobutyl term shows a a lower oxidation rate due to the branched chain.

Secondary alcohols are very selectively oxidized to ketones. Except for some terms reacting very slowly (cyclohexanol), no other product can be detected, and also the H_2O_2 decomposition is negligible.

Runs at different alcohol concentrations in methanol solutions gave a first order dependence on the secondary alcohol.

Competition kinetics have been used to compare the rates of different alcohols. While the oxidation rates decrease regularly with increasing chain length (2-Butanol : 2-Pentanol = 1.2), there is a striking effect of the position of the OH group in the chain, (2-Pentanol : 3-Pentanol = 13), suggesting that the accessibility of the reactive groups to the catalytic sites plays a role in kinetically important step. Also the very low rate of cyclohexanol is probably due to its large size, which makes it difficult to approach a catalytic sites. All these facts are in accordance with an inner-channel catalysis.

OLEFIN EPOXIDATION

Propylene is selectively epoxidized, since all side-products found are consecutive ring opening products:

(8) C_3H_6 + H_2O_2 ---> CH_3CH—CH_2 (epoxide, bridged by O) + H_2O

(9) CH_3CH—CH_2 (epoxide, bridged by O) + CH_3OH ---> $CH_3CH(OH)$—$CH_2(OCH_3)$ + $CH_3CH(OCH_3)$—$CH_2(OH)$

(10) CH_3CH—CH_2 (epoxide, bridged by O) + H_2O ---> $CH_3CH(OH)$—CH_2OH

(11) CH_3CH—CH_2 (epoxide, bridged by O) + H_2O_2 ---> $CH_3CH(OOH)$—CH_2OH etc.

In particular, no allylic oxidation product can be detected. The final selectivities are normally about 98 % at high H_2O_2 conversions (95 %)

A solvent is needed to make the solutions homogeneous with H_2O_2. If methanol is used, its oxidation is negligible, as well as the H_2O_2 decomposition to O_2.

Similarly to the alcohol oxidation, rates are first order respect to the catalyst concentration, and of order < 1 respect to H_2O_2. Unlike in the alcohol oxidation, where no or little product inhibition effect is observed, added propylene oxide does retard the epoxidation process. The side-products have no significant inhibition effect.

Olefins higher than C3 exhibit the same reactivity feature with regard to selectivity and kinetic effects as propylene does. Moreover, no isomerization product is formed (i.e., cis epoxide only is formed from cis olefin):

(12) 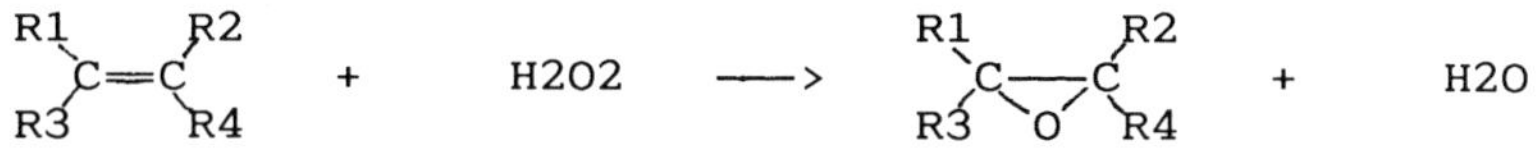

Competition kinetics were performed in order to compare oxidation rates of different olefins within a strictly homogeneous series, using methanol as solvent. The reactivity ratios, as obtained from the product ratios, are:

cis-2-Butene : 1-Butene : iso-Butene : trans-2-Butene =
= 13 : 4.5 : 3.6 : 1

This trend is different from what expected with homogeneous electrophylic catalysis, which is: iso-Butene > cis-2-Butene > trans-2-Butene > 1-Butene

Olefins having particularly hindered structures react more slowly than expected with homogeneous electrophylic catalysis: the reactivity ratio of 1-hexene : cyclohexene is about 30.

Unsaturated compounds bearing a second functional group, even if oxidable, are normally epoxidized. Allyl alcohol and esters give glycidyl derivatives. Allyl methacrylate is epoxidized in the allyl group .

A special regioselectivity is exhibited by several poliunsaturated compounds. In particular, while the selectivity in monoepoxidation in butadiene is not much higher than found with other catalytic systems, those exhibited by diallyl carbonate and diallyl ether are higher than expected with homogeneous catalysis.

AROMATIC COMPOUNDS

Aromatic substrates are generally oxidized to phenols or phenol derivatives.

Both regio and chemio selectivity of this reaction depend on the structure of the substrates and particularly on the presence of substituent of the aromatic ring.

In the case of benzene, phenol is obtained at low conversions, otherwise the hydroxylation process proceeds further to dihydroxybenzenes. On the other side, the hydroxylation in substituted activated aromatic compounds is selective towards mono substitution. Products derived from side chain reactions (toluene, ethyl benzene) have been detected.

A remarkable difference between TS1 and homogeneous catalysis (Refs.11,12) in the isomeric distribution has been found with a prevalence of para substitution which indicates the existence of a "restricted transition state selectivity".

In competition tests toluene and benzene have shown similar reactivities, while a reactivity ratio of ì 10 : 1 has been reported for homogeneous hydroxylation (Ref.13).

Particularly bulky substituents have a considerable retarding effect (isopropyl). All the deactivated aromatic substrates (e.g. benzonitrile, clorobenzene, benzoic acid, nitrobenzene) appear to be non reactive independently on the bulkness of the substituent.

Cresols show the expected prevalence of hydroxy group in orienting effect.

Aromatic compounds bringing a reactive substituent, hydroxyalkyl or unsaturated, react selectively at the side chain: styrene is oxidized to phenyl acetaldehyde probably via styrene oxide, which is fastly isomerized by TS 1 in same conditions (Refs.13,14). Benzyl alcohol follows the usual alcohols oxidation pathway as well as 1 phenyl-and 2 phenyl-ethanol.

PHENOL HYDROXYLATION PROCESS

Phenol hydroxylation has been studied in deptht due to the industrial interest for the hydroquinone-cathecol production process.

This reaction is performed in excess of phenol in the presence of an organic cosolvent and water which is in any case present in the H2O2 feed, as well as a reaction coproduct.

Beside hydroxylation, tars are produced in the course of this reaction together with minor amounts of O2, CO2 and organic acids, due to competitive coupling reactions and consecutive oxidation of products.

The dependences of selectivity and para/orto ratio on reaction conditions, catalyst concentration, solvent, temperature and phenol conversion are in accordance with a shape selective catalysis.

A new process for cathecol and hydroquinone production from phenol has been developed by ENICHEM SYNTHESIS and an industrial plant of 10.000 tons/year total capacity has been started up in 1986, in the Ravenna factory.

The new process displays a high selectivity at high phenol conversions, achieved in selected reaction conditions with a high stationary catalytic efficiency of TS1.

REFERENCES

(1) M.Taramasso, G.Perego, B.Notari, US.Pat.4,410,501 (1983); M.Taramaso, G.Manara, V.Fattore, B.Notari, U.S.Pat. 4,666,692 (1987); G.Perego, G.Bellussi, C.Corno, M.Taramasso, F.Buonuomo, A.Esposito, in Y.Murakami, A.Tijima, J.W.Ward (Eds.), Proc. Seventh Int. Conf. on Zeolites, Tokyo 1986, Tonk Kodansha Amsterdam Elsevier, p.129.

(2) B.Notari, Stud. Surf. Sci. and Catal., 37,(1988), 37.

(3) A.Esposito, C.Neri, F.Buonuomo, U.S.Pat. 4,480,135 (1984).

(4) C.Neri, A.Esposito, B.Anfossi, F.Buonuomo, Eur.Pat. 100.119 (1984).

(5) C.Neri, B.Anfossi, F.Buonuomo, Eur.Pat. 100.118 (1984).

(6) F.Maspero, U.Romano, Eur.Pat. 190.609 (1986).

(7) M.G.Clerici, U.Romano, Eur.Pat. 230949 (1987).

(8) A.Esposito, M.Taramasso, C.Neri, F.Buonuomo, U.K.Pat. 2.116.974 (1985); A.Esposito, M.Taramasso, C.Neri, U.S.Pat. 4,396,783.

(9) P.Roffia, M.Padovan, E.Moretti, G.De Alberti, Eur.Pat. 208.311 (1987).

(10) G.Bellussi, G.Perego, A.Esposito, C.Corno, F.Buonuomo, Proc. of Sixth Con. on Catal., Cagliari 1986, p.423.

(11) R.O.C.Norman, R.Taylor, Electrophylic Substitution in Benzenoid Compounds, ed. C.Eaborn, Elsevier, (1965).

(12) J.Varagnat, Ind.En.Chem.Prod.Res.Dev. 15-3,(1976),212.

(13) C.Neri, F.Buonuomo, U.S,Pat. 4,495,371 (1985).

(14) C.Neri, f.Buonuomo, U.S.Pat. 4,609,765 (1986)

J.HABER (Institute of Catalysis and Surace Chemistry, Krakow, Poland):

An obvious question which arises is as to what extent the TS-1 catalyst in oxidation of complex molecules behaves as a zeoliyte. In homogeneous oxidation you have the solvent, the oxidant and the oxidized molecule. Can the molecules of the solvent in a narrow pore of the zeolite be considered as being in the liquid phase? What happens when in this narrow channel you have a big molecule of the reactant in vicinity of the molecule of hydrogen peroxide and the molecule of the solvent. Certainly the degrees of freedom will be diferent from those in the liquid phase. Or it is only the outermost surface of the zeolite grains which is involved in the reaction?

F.MASPERO : Unlike in homogeneous catalysis, where both H2O and methanol exhibit a strong inhibitory effect due to the formation of stable metal complexes, the reactions catalized by TS-1 can be advantageously carried out in these solvents; this fact strongly suggests that H2O and CH3OH do not form stable complexes with the metal center, as they do in homogeneous catalysis.

On the other hand, some results are influenced by the nature of the solvent. An example is the selectivity in the phenol hydroxylation, which is higher in water/acetone than in the aqueous solution. Obviously the zeolitic catalysis is influenced by the different diffusion parameters for H2O, acetone and substrate, which depend on the organophylic character of TS-1, similarly to what happens with silicalite. Thus, the actual dilution of reactants within the pores of zeolite will depend very much on the nature of the solvent.

J.C.VEDRINE (Institut de Catalyze, Villeurbane, France):

A question already asked yesterday (A2 paper) is related to the reaction occurring with all reactants within the pores or on the external surface. If the reaction occurs within the pores the selectivity in para isomer in phenol hydroxylation should greatly depend on crystallite size. Did you check it and what was the result? Obviously one has to avoid isomerisation (thermodynamical equilibrium) on the surface site after reaction inside in a batch type reactor.

F.MASPERO (Enichem Synthesis, S.Donato Milanese, Italy):

Not only the para/orto ratio, but the products yield and the tars selectivity, and also the overall reaction rate, are greatly dependent on the crystal size. The change of the para/orto ratio is mainly due to the decomposition effect on the catechol to form tars. In the conditions of our experiments no isomerisation is observed.

Generally speaking, all evidences indicate that the reaction takes place mainly inside the zeolite pores; but we cannot rule out the possibility to have an outer surface reaction at a lower extent.

J.M.HERRMANN (Ecole Centrale de Lyon, Ecully Cedex, France):

I have found in your results several similarities with hetherogeneous photocatalytic reactions carried out on illuminated TiO_2 at room temperature with air. They concern the oxidation of alcohols, the epoxidation of propene and the hydroxylation of aromatic compounds. We proposed that the photocatalytic epoxidation of propene in the gas phase was due to a dissociated atomic species (see ref. J.M.Herrmann et al. in this book), whereas the hydroxylation in the aqueous phase was preferentially due to OH. radicals. Perhaps, the active species involved in your system are of the same nature.

Concerning the influencew of temperature, you mention a decrease of activity for T > 90°C. We have observed the same phenomenon for the rates of various photocatalytic oxidation reactions presented in an Arrhenius plot. We have attributed this decrease to te act that the reactions became limited by the adsorption of the substrate. Perhaps the same phenomenon occurs in your system.

F.MASPERO: No doubt that several indications of a radicalic reactivity can be found in the hydroxylation of aromatic hydrocarbons (see the side-chain oxidation), but the general pattern of aromatic nuclear hydroxylation is similar to the acid catalyzed reaction (no resorcine in the phenol hydroxylation), with the superimposed effect of shape-selectivity . The fact that some radicalic species of the aromatic ring are actually observed (epr spectra) are strongly suggestive, but not conclusive of a radicalic mechanism.

All the general question about mechanism is not completely defined at this stage.

The lowering of the selectivity at T > 100°C is due to a decomposition step, rather than to a slower reaction.

M.HADDAD (Amoco Chemical Company, Naperville, Illinois,USA):

To shed more light on whether the molecular sieve structure and the location of the Ti are important for the reported catalysis, have you tested the following catalysts:

1. Titanium impregnated on silicalite;
2. The hydrothermal reaction product of Titanium with amorphous silica under conditions which do not promote the formation of the sieve structure.

F.MASPERO : Catalysts containing supported TiO_2 both on amorphous silica and on silicalite give much more tars and H_2O_2 decomposition in the case of aromatic hydroxylation, and poor selectivites of epoxidation in the conditions of our experiments (protic solvents, H_2O_2 as oxidant).

J.F.BRAZDIL (BP Research, Cleveland, Ohio, USA):

Do Ti-Silicalites have catalytic activity for the oxidation of C2-C4 paraffins (ethane, propane, butane) with H2O2? If so, what are the major products and yields?

F.MASPERO : This reactivity has been studied, and will be published soon.

A.J.PAPA (Union Carbide Corp., Charleston WV, USA):

Were any catalyst spectral structural changes observed during product inhibition in propylene epoxidation?

F.MASPERO : Structural changes are totally absent; otherwise, these could be easily detected after catalyst recovering.

Spectral changes, other than those cited (i.e. after interaction of TS-1 with H2O2 and after subsequent reaction with the olefine) were not observed until now; the spectral properties of the catalyst are unchanged after recovering, except the powder X-ray spectra which reveal some pores occlusion, and turn back identical to the original material ater catalyst regeneration.

J.KIWI (EPFL, Lausanne, Switzerland):

Do you have any experimental evidence for the Ti(H2O2) initial reaction you postulate in your catalytic hydroxylation process? Have you used o-toluidine or infrared techniques to measure the amount and type of peroxytitanate formed in the initial step of hydroxylation?

F.MASPERO : We have no direct IR evidence of the intermediate Ti + H2O2, just the disappearing of the typical bands present in TS-1, followed by reappearing after H2O2 decomposition or after reaction with oxidable species.

Our hypothesis is based on the general kinetic pattern, which is consistent with the intermediate formation of an adduct, but we cannot define it more exactly at this stage.

G.BELLUSSI (Eniricerche, S.Donato Milanese, Italy):

I would like to add a comment about the question whether the reactivity takes place on external surface or inside the pores of TS-1. One of the major properties of TS-1 catalyst is to prevent secondary reactions. For instance in the phenol hydroxylation the amount of heavy polinuclear compounds produced by TS-1 is very low in comparison with Ti-supported silicalite or silica gel. This is a proof that reactions must take place inside the pores of TS-1.

G. Centi and F. Trifiro' (Editors), *New Developments in Selective Oxidation*

CYCLOHEXANONE AMMOXIMATION: A BREAK THROUGH IN THE 6-CAPROLACTAM PRODUCTION PROCESS

P. ROFFIA[1], G.LEOFANTI[1], A. CESANA[1], M. MANTEGAZZA[1], M. PADOVAN[1], G. PETRINI[1], S. TONTI[2], P. GERVASUTTI[2]

[1]Montedipe S.p.A., Research Unit of Bollate, Via S. Pietro n. 50, 20021 Bollate, Milano, Italy.

[2]Montedipe S.p.A., Research Center of P.to Marghera, Via Dell'Elettricitã 41, 30175 Porto Marghera, Venezia, Italy.

SUMMARY

As a part of a research devoted to selective oxidation of organic compounds, we have studied the synthesis of cyclohexanonoxime by direct ammonia oxidation in presence of cyclohexanone. By using titanium silicalite as catalyst the reaction took place in a smooth and selective way. The effect of the most important experimental conditions was investigated. The role of the catalyst and some possible reaction mechanisms were discussed. Our results provided a promising starting point for the development of a new technology for cyclohexanonoxime synthesis.

INTRODUCTION

The development of techniques for the selective oxidation of petroleum derivatives under mild reaction conditions is one of the most important tasks in Chemical Science both for its economic and environmental implications.

A number of commodity and specialty chemicals have been manufactured by oxidation using transition metal compounds as specific catalysts, which make selective reactions otherwise indiscriminate (refs. 1-5).

Part of our research in the selective oxidation was devoted to an once through synthesis of cyclohexanonoxime, an intermediate for caprolactam production, by causing cyclohexanone and ammonia to react in presence of an oxidizing agent. Our aim was to carry out a reaction, known as ammoximation (ref. 6), represented by the following equation:

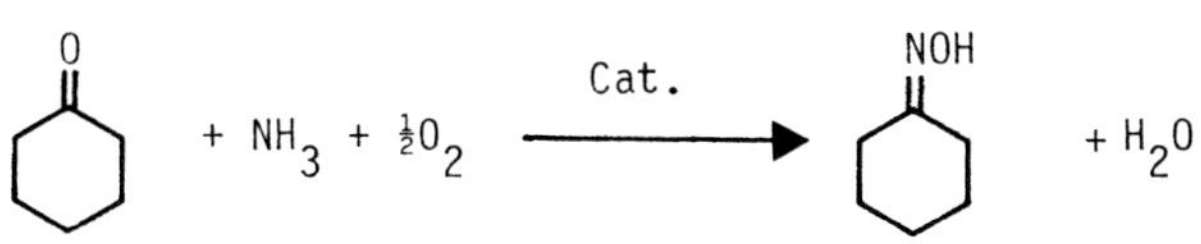

Only two main references regarding ammoximation processes are found in the literature. The first one concerns ammoximation in the liquid phase catalyzed by phosphotungstic acid. This process, developed by Toagosei,does however have a number of disadvantages, i.e. substantial loss of the hydrogen peroxide due to oxygen formation and rapid decomposition of the catalyst (ref. 7). The second process, claimed by Allied, concerns the gas phase ammoximation by oxygen on a silica gel catalyst (ref. 6). The low yield and selectivity calculated on cyclohexanone, as well as the rapid fouling and low productivity of the catalyst itself, exclude valid applications of the claimed results.

On choosing the oxidant, our preference was given to hydrogen peroxide because of its ability to perform oxidation under mild conditions and its cost, calculated for captive use,acceptable for the cyclohexanonoxime synthesis. By employing hydrogen peroxide as oxidant,the reaction model could be the peroxidic oxygen transfer to a nucleophilic substrate.Under this point of view the role of the catalyst should be the weakening or the polarization of the peroxidic bond conferring electrophilic properties to the oxidant (ref.8). Another possibility should be the conferring to the hydrogen peroxide the properties of oxene which is able to produce electrophilic insertion on the substrate to be oxidized (ref.9). In both cases the transfer of a peroxidic oxygen causes two-electrons oxidation as required by the ammoximation reaction.

Among the transition metals,used for transferring peroxidic oxygen to organic substrates (ref.10), titanium attracted our interest because titanium catalysts had been used for amines oxidation with hydroperoxides (refs. 11-12) and the oxygenated titanium compounds showed definite reactivity to hydrogen peroxide (ref.13). In particular we tried to use titanium silicalite,a zeolite derived from silicalite by isomorphous substitution of silicon by titanium atoms, because this catalyst has recently been successfully used for olefins epoxidation, aromatic hydrocarbons hydroxylation and alcohols oxidation using hydrogen peroxide, even diluted, as oxidant (refs. 14-15).

EXPERIMENTAL

Samples of titanium silicalite were synthesized according to patent literature (ref.16) and calcined at 500°C.The cristallinity,determined by X-ray diffraction was higher than 95%; the percentage of titanium determined by atomic absorption was in the 1.4÷1.6% (wt.) range and the IR spectra clearly showed the characteristic absorption band produced by the introduction of titanium atoms into the zeolite framework.Samples of H-ZSM-5 and silicalite were synthesized following the procedure reported in patent literature (refs.17-18). The silica supported titania was obtained by impregnating a commercial microspheroidal silica (420 m^2g^{-1}) with a diisopropyl-bistriethanoloamino-titanate solution and calcining the product at 200°C.

The ammoximation reaction was carried out by dispersing the catalyst in an ammonia-cyclohexanone aqueous-organic liquid phase and by feeding the hydrogen peroxide to the well-mixed slurry. More precisely a glass reactor, equipped by a stirrer and a heating jacket was preliminarily pressurized by an inert gas. After loading catalyst, water, ammonia and solvent, the whole was vigorously stirred, the temperature raised to the desired value and the cyclohexanone introduced by a syringe. Hydrogen peroxide was then added using a metering pump. At the end, after cooling, the liquid was analysed by gaschromatography.

RESULTS AND DISCUSSION

Catalysts evaluation

Results obtained in ammoximation experiments by different catalysts are reported in Table 1. As shown, titanium silicalite supplied excellent catalytic properties in the ammoximation reaction. Also titanium supported on amorphous silica showed good catalytic activity, even though the best performances were obtained by titanium silicalite. On the contrary, by using amorphous silica alone or non containing titanium zeolites,the ammoximation occurred only to a negligible extent, as without any catalyst, even if the ammonia-cyclohexanone--hydrogen peroxide system showed a high self reacting behaviour.

TABLE 1

Ammoximation in water/t-butanol by different catalysts: catalyst concentration 2% (wt.); temperature 80°C; NH_3/H_2O_2 molar ratio 2.0; reaction time 1.5 hours for all the runs except for the starred one effected in 5 hours

Catalyst	Ti %	H_2O_2/Cy-hexanone molar ratio	Cyclohexanone Conv. %	Cyclohexanone Oxime Select. %	Oxime yield based on H_2O_2 %
None	-	1.07	53.7	0.6	0.3
SiO_2 amorphous	0	1.03	55.7	1.3	0.7
Silicalite	0	1.09	59.4	0.5	0.3
H-ZSM-5	0	1.08	53.9	0.9	0.4
TiO_2/SiO_2	1.5	1.04	49.3	9.3	4.4
TiO_2/SiO_2*	9.8	1.06	66.8	85.9	54.0
Tit.Silicalite	1.5	1.05	99.9	98.2	93.2

Choice of solvent

A suitable reaction medium should be a good solvent for both the reagents and the reaction product and have a good stability to hydrogen peroxide.

The results obtained operating by different solvents considered suitable for our reaction are reported in Table 2. T-butanol proved to fit very well our requirements,even if other solvents were used with similar results. A water-alcohol mixture (weight ratio 1:1) showed a good solvent power for both cyclohexanone and cyclohexanonoxime and was stable during the reaction. By working in such a solution the conversion (about 90%) and the selectivity (96-99%) of the cyclohexanone to cyclohexanonoxime were very high for all tested catalysts, as well as the oxime yields based on the oxidant (89-95%). The reproducibility of the results was very good.

TABLE 2

Ammoximation in different solvents:temperature 80°C; NH_3/H_2O_2 molar ratio 2.

Solvent	Catalyst	H_2O_2/Cy-hexanone molar ratio	Cyclohexanone Conv. %	Cyclohexanone Oxime Select. %	Oxime yield based on H_2O_2 %
Benzene	A 28	1.03	99.7	95.0	91.7
Toluene	A 28	1.07	99.8	97.0	90.0
t-amyl alcohol	A 28/2	0.86	94.5	95.6	94.0
H_2O/t-butanol	A 28/2	0.95	88.8	99.5	93.4
"	A 30/1	0.94	89.7	99.5	94.9
"	A 30/2	0.95	87.8	99.5	92.1
"	A 30/3	0.97	89.6	96.4	89.4

Choice of temperature

The results obtained at temperatures between 60 and 95°C are recorded in Table 3. At 80 and 95°C the cyclohexanone selectivity to oxime and its yield based on hydrogen peroxide were similar and in both cases very high. When the reaction was performed at 70°C a reduction in the oxidant yield was observed, while at 60°C a considerable competition of non catalytic reactions decreased both selectivity and yield.

TABLE 3

Ammoximation in water/t-butanol at different temperatures: NH_3/H_2O_2 molar ratio 2.

Catalyst	H_2O_2/Cy-hexanone molar ratio	Temperature °C	Cyclohexanone Conv. %	Cyclohexanone Oxime Select. %	Oxime yield based on H_2O_2 %
A 28/2	0.93	60	81.5	87.0	76.4
A 28/2	0.97	70	90.2	96.4	89.4
A 28/2	0.83	80	80.1	98.8	95.0
A 28/2	0.88	95	83.0	99.9	94.0

Concentration of the catalyst

As in any catalyzed reaction, the catalyst concentration revealed a great importance in determining the ammoximation trend. The useful catalyst concentration is determined by the need to produce significant reaction rate with low reagents concentration to avoid side reactions.

Tests performed with different catalyst concentrations are recorded in Table 4.The results obtained using a 30 g/l concentration of titanium silica lite were excellent for both selectivity and yield,whereas the results obtained with a 10 g/l concentration indicated a considerable decrease in the cyclohexanone selectivity and in the oxidant yield to the oxime.

TABLE 4

Ammoximation in water/t-butanol with different catalyst concentrations: temperature 80°C; NH_3/H_2O_2 molar ratio 2.

Catalyst	g/l	H_2O_2/Cy-hexanone molar ratio	Cyclohexanone Conv. %	Cyclohexanone Oxime Select. %	Oxime yield based on H_2O_2 %
A 28	30	1.05	99.9	98.2	93.2
A 28	15	1.05	98.9	96.9	91.6
A 28	10	1.04	81.4	86.0	67.0
A 30	10	1.05	78.4	84.0	62.5
A 30	10	1.05	76.3	82.6	60.0

AMMOXIMATION MECHANISM

The use of titanium silicalite as catalyst, has been the specific and determining factor for performing excellent ammoximation yeld.In our opinion the activity and selectivity of this catalyst in the studied reaction are the result of a synergism between the presence of isolated titanium in a coordinative state different than usual and the zeolite framework of which titanium is a constituent component and a catalytically active site.

Tests performed with silicalite having the same structure as the titanium silicalite showed the catalytic activity in ammoximation was not caused by the zeolitic structure. A similar negative result was obtained using H-ZSM-5,a zeolite derived from silicalite by partial substitution of silicon atoms with aluminum atoms. On the other hand,samples prepared by supporting titanium dioxide on silica with a large surface area pointed out a good catalytic activity although lower than that of the titanium silicalite.

The experimental evidence (see Table 1) suggests that the titanium main role in ammoximation is to promote the selective oxidation of the ammonia nitrogen through activation of the hydrogen peroxide. The hydrogen peroxide itself is not able to produce ammoximation if not in trace quantities, because in the basic reaction environment it decomposes to oxygen and by-produces nitrites and nitrates (ref.8). It is possible the interaction between oxidant and catalyst is similar to that suggested for the reaction between porphyrinic titanium compounds and hydroperoxides in the oxidation of olefins or dialkyl sulfides (refs.19-20). It is possible therefore this interaction and activation of hydrogen peroxide can thus occur through substitution of the titanol groups.

The different activity among titanium containing catalysts can not be attributed to a different number of titanium atoms which are equal or greater in the silica supported titania than in titanium silicalite. However in the two types of compared catalysts titanium is mostly present in a completely different state of aggregation and coordination which can reasonably explain such a difference in catalytic activity. The technique of supporting or anchoring tetravalent titanium to an amorphous silica most probably cannot fix firmly in a tetrahedral structure isolated atoms of titanium to the surface of the silica itself, but instead produces or leads rapidly to the formation of surface clusters of titanium dioxide in the octahedral coordination even at low titanium concentration (ref.21). These aggregations of titanium atoms are not evidently as active as the isolated ones in the crystalline structure of the titanium silicalite (ref. 22).

In order to explain the high selectivity achieved with titanium silicalite, it must be assumed that the first reaction step, i.e. the interaction of hydrogen peroxide with the titanium atoms in the zeolite framework, takes place fairly rapidly to suppress all the reactions which occur in the absence of the catalyst.

The second step of the reaction is the transfer of the peroxidic oxygen added to the titanium to the substrate to be oxidized. The high selectivity obtained in the cyclohexanone ammoximation implies this reaction step occurs through a concerted mechanism involving the interaction of all three reagents on the active sites of the catalytic system and the direct formation of the cyclohexanonoxime. This hypothesis would appear the most realistic and preferable to a hypothesis of synthesis through a sequence of reaction stages which would require the formation of intermediates. It is unlikely that these would be able to coexist in solution with the very reactive system and this is not consistent with the high reaction selectivity.

CONCLUSIONS AND PERSPECTIVE

The results obtained using titanium silicalite were very promising. The cyclohexanone selectivity to oxime was generally higher than 99%,only traces of organic by-products were formed. Providing a H_2O_2/cyclohexanone near stoichiometric molar ratio, the hydrogen peroxide loss was very small and mainly determined by the formation of some inorganic by-products.

Moreover our new ammoximation route allows to widen the application field of this catalyst for activating diluted hydrogen peroxide, even in an alkaline solution, and for exploiting its oxidizing power through selective oxygen transfer to ammonia nitrogen atom.

REFERENCES

1 G.W. Parshall, Homogeneous Catalysts, Wiley, New York, 1980
2 R.A. Sheldon, J. Mol. Catal., 20 (1983) 1-26
3 J.E. Lyons, Hydr. Proces. (1980) 107-119
4 I.V. Spirina, V.P. Maslennikov, Yu. A. Aleksandrov, Russ. Chem. Rev., 56 (Engl. Trasl.) (1987), 670 - 681
5 F.Cavani, G.Centi, F.Trifirò, R.K.Grasselli,Catal.Today 3, (1988) 185-198
6 J.N. Armor, in J.R. Kosak (Ed.), Catalysis of Organic Reactions, Vol. 18, Marcel Dekker, New York and Basel, 1984
7 S. Tsuda, Chem. Econ. Eng. Rev., (1970) 39-41
8 J.P. Schirmann, S.Y. Delavarenne, Hydrogen Peroxide in Organic Chemistry, Edition et documentation Industrielle, Paris, 1979
9 D.T. Sawyer, Chem. Tech., (1988) 369-375
10 R. Sheldon, Bull. Soc. Chim. Belg., 94 (1985) 651-670
11 J.L. Russel, J. Kollar, US Pat. 1100672 (1965)
12 G.N. Koshel, M.I. Farberov, L.L. Zalygin, G.A. Krushinskaya, J.Appl.Chem. USSR, 44 (1971) 885
13 J.A. Connor, E.A.V. Ebsworth in N.J. Emeleus, A.G. Sharpe (Editors), Adv. Inorg. Chem. Radiochem., Vol. 6, Academic Press, N.Y. and London, 286
14 G. Perego, G. Bellussi, C. Corno, M. Taramasso, F. Buonomo, A. Esposito, Titanium silicalite: Proc. 7th Int. Zeolite Conference, Tokyo, August 17-22, 1986, Elsevier, Amsterdam, 129-136
15 B. Notari, Stud. Surf. Sci. Catal., (1988) 413-25
16 M.Taramasso, G. Perego, B. Notari, US Pat. 4410501 (1983)
17 R.J. Argauer, G. R. Landolt, US Pat. 3702886 (1972)
18 E. Moretti, M. Padovan, M. Solari, C. Marano, R. Covini, Ital. Pat. Appl. 19238 A/82 (1982)
19 H.J. Ledon, F. Varescon, Inorg. Chem. 23 (1984) 2735
20 O. Bortolini, F. Di Furia, G. Modena, J. Mol. Cat. 33 (1985) 241-244
21 J. Chen, Ph.D. Thesis, Carnegie-Mellon Univ. (1986) and refs. included
22 M.R.Boccuti, K.M.Rao, A.Zecchina, G.Leofanti, G.Petrini, Proc.Eur. Conf. on Struct. and React. of Surface, Trieste, September 13-16, 1988, in press

J. C. VEDRINE (Institut de Recherches sur la Catalyse)
69626 Villeurbanne (France):
You have clearly shown how Ti-silicalite gives high catalytic properties for oxime and ammoxime formation with respect to TiO_2/SiO_2 Catalyst. Zeolite type materials exhibit narrow channels (5.5 Å in Ti-silicalite) which should hinder the reactants to reach alltogether the inner active sites. Moreover in liquid solution there is no driving force for the reactant to enter the pores preferentially to the solvent molecules. In your opinion are active sites on the surface of the cristallites rather than inside the pores. If not, is the reaction diffusion limited.

PAOLO ROFFIA
Laboratory tests have shovn that all the reagents are rapilidy absorbed into the channels of the titanium silicalite, so in our opinion the ammoximation takes place inside the catalyst channels.
If so, I do not exclude, as you suggest, the existence of mass--transfer problem. At this time we did not make specific experiments aimed to evidence the presence of this problem but we have in program to do it.

G.M. PAJONK
69622 Villeurbanne (France):
According to one slide it seemed important to pour cyclohexanone all at once in your reactor, why?
When NH_3 reacts only with H_2O_2 over the titanium ion containing catalysts (TiO_2/SiO_2 as well Ti-Silicalite) what were the reaction products?
In particular is NO detected in this case?

PAOLO ROFFIA
The all at once addition of cyclohexanone is only required to semplify the way of performing the ammoximation.
The ketone can be advantageously fed in a continuous way as for the hydrogen peroxide.
To answer the second question, I can say that the main product in the ammonia oxidation is nitrogen.

L. ZULIANI - Chimica del Friuli - 33050 Torviscosa (Udine):
My question is: Do you know the role of excess ammonia to suppress undesirable side reactions.
Thank you.

PAOLO ROFFIA
In our ammoximation experiments we used an excess of ammonia (molar ratio NH_3/cyclohexanone >2). However we have observed at lower ratios the reaction begins to become less clean. This parameter is still under investigation.

P. JIRU - Dolejskova 3, 18223 Prague 8, Czechoslovakia:
The oxime yield based an H_2O_2 is always lower (~2-7%) in comparison with cyclohexanone selectivity. Also the selectivity of trasfirmation of H_2O_2 to oxime in your patents is always lower (80-90%). What are the reasons: decomposition of H_2O_2, formation of other products (organic peroxides, NH_2OH, NO, N_2...).

PAOLO ROFFIA
As I said in my report the H_2O_2 selectivity is a bit lower in comparison with that of cyclohexanone. The loss of hydrogen peroxide is mainly determined by inorganic products formation and in small amount nitrates and nitrites. The hydrogen peroxide decomposition to oxigen does not occurs at all.
As for as the claimed selectivity in our patent is concerned, the lower values are justified by using a titanium silicalite with lower catalytic performances.

G. Centi and F. Trifiro' (Editors), *New Developments in Selective Oxidation* 53

MODIFIED ZEOLITES FOR OXIDATION REACTIONS

Cristina Ferrini and Herman W. Kouwenhoven

Technisch Chemisches Laboratorium, ETH-Zentrum, Universitätsstrasse 6, 8092 Zürich, Switzerland

SUMMARY

Synthetic Ti-silicates with the MFI structure-type (Ti-silicalites or TS-1) have been reported to be selective catalysts in (ep) oxidation reactions of aqueous H_2O_2 with unsaturated compounds. We have applied secondary synthesis to introduce Ti into various zeolites by reaction with $TiCl_4$. It appears that Ti is incorporated into the zeolite framework. The catalytic activity of these modified zeolites was screened using the oxidation of phenol with H_2O_2 as a test reaction. Their performance depends on crystal structure, crystallite size, Ti content and preparation procedure and is compared with that of the non-modified acidic zeolites The product composition is solvent dependent. Deactivated catalysts may be regenerated by conventional techniques.

INTRODUCTION

TS-1, a titanium silicate with the MFI structure, is a selective catalyst for the (ep) oxidation of various aromatic and unsaturated compounds using aqueous H_2O_2 as an oxidant. This very interesting invention was reported by scientists from Enichem and affiliated companies (refs. 1,2,3,4). TS-1 is shape selective in phenol oxidation, favouring the formation of catechol and hydroquinone (ref. 5). In this reaction TS-1 is supposedly a better catalyst than H-mordenite and H-zeolite Y (ref. 6) or H-ZSM-5 (ref. 7) which are all strongly acidic solids. TS-1 is synthesized directly from SiO_2 and TiO_2 precursors by hydrothermal reaction. Secondary synthesis of Ti-modified ZSM-5 by reaction of gaseous $TiCl_4$ with acid-extracted ZSM-5 (Si/Al>2000) has been described and the properties of the product are equal to those characteristic for synthetic TS-1 (ref. 8). Moreover, as a catalyst in the hydroxylation of phenol with aqueous H_2O_2, its performance is comparable to that of TS-1 (ref. 9). It is assumed that Ti occupies tetrahedral lattice sites.

In the present contribution we report on the secondary synthesis of Ti-modified ZSM-5, zeolite Beta and zeolite Y, by reaction with $TiCl_4$ or Ti-tetraisopropylate with H-zeolite samples. Compared to those earlier reported, the present materials contain after secondary synthesis both Al and Ti and are therefore expected to be different catalysts. Catalyst performance was screened in the oxidation of phenol with aqueous H_2O_2 as a test reaction in comparison with TS-1, silicalite and the H zeolites. The effect of the addition of various solvents on product composition and conversion after 4 h was investigated.

EXPERIMENTAL

Materials

TS-1 was synthesized according to (ref. 2) example 1. Zeolite Beta was synthesized following the recipe given in (ref. 10). Samples of zeolite Y, (type FAU PY-32/F1), H-ZSM-5, (type FZ 21/G) and amorphous silica, (type C gel C-560) were obtained from the Chemische Fabrik Uetikon. Silicalite was synthesized according to (ref. 10), small crystals, 2 mm and (ref. 11), large crystals, 25 mm.

Activation and Secondary Synthesis

Zeolite Beta and Y were converted into the H-form according to ammonium exchange techniques usually used for the preparation of USY. H-ZSM-5 was exchanged further with 1 molar aqueous HCl for deep Na^+ removal. Reaction conditions for the T-site substitution reaction with $TiCl_4$ are given in Table 1.

TABLE 1
Secondary synthesis of Ti(Al)silicates by gas phase reaction with $TiCl_4$

Reaction step	Temp.K	Time.Hr
Drying	523	1
$TiCl_4$ Reaction	473-773	1-2
Stripping	623	1
Steaming	573	1
Calcination	813	1

Characterization

Materials were characterized by the following methods: Mid IR spectra, using a Perkin Elmer IR spectrometer 983. XRD, using a Guinier -de Wolff camera. Unit-cell size, using a Scintag Diffractometer PAD-X. N_2 adsorption, using a Micromeretics ASAP-2000. Elemental analysis by AAS or ICP after dissolution in HF. IR traces are reproduced in Fig 1, other analytical data are collected in Tables 2 and 3.

Test Reaction

Test reactions were performed in a stirred, closed glass reactor with and without added solvent at temperatures between 353 and 400 K in N_2 atmosphere, using 35%aq H_2O_2 and phenol, (Fluka purum). H_2O_2 was added dropwise over 10 min. The mixture was sampled 4 h after finishing H_2O_2 addition. The products were analysed by standard GC and HPLC techniques. Product analyses are given in Tables 4,5 and 6. The analysis is not quantitative, it was however verified that there is no appreciable phenol loss (>10%) in the experiments using solvents.

RESULTS AND DISCUSSION

Characterization

TS-1 and Ti modified ZSM-5 (Si/Al>2000) are usually identified by their mid IR spectra and XRD. IR shows an absorption at 960 cm^{-1}, the intensity of this absorption is dependent on the Ti content of the sample.The interpretation of spectroscopic data from TS-1 samples is discussed in (ref. 12). The IR spectra collected in Fig 1 show that the 960 cm^{-1} band is present in the samples after reaction with $TiCl_4$, indicating that indeed secondary synthesis may be applied to prepare Ti substituted zeolites. It is interesting to note that also in amorphous silica Ti appears to be incorporated in similar sites. The question remains however what type of Ti distribution is obtained by secondary synthesis.

Structural integrity was inspected by routine XRD using a Guinier-de Wolff camera. Crystallinity and changes in unit-cell size were measured by quantitative XRD methods. A linear increase with Ti content was reported (ref. 3) for Al-free samples with the MFI structure. Introduction of Ti into ZSM-5 with a Si/Al ratio of 25 by reaction with $TiCl_4$ results in an increase in unit-cell volume. The Si/Al ratio of the zeolite is not changed by the secondary synthesis (Table 2) which indicates that either substituted Al remains in the structure as extra framework alumina or that the reaction with $TiCl_4$ is not a substitution of Al in T-sites by Ti, but rather a reaction of $TiCl_4$ with functional OH groups. The change in unit-cell volume is, however, an indication that Ti is eventually incorporated into the framework, in tetrahedral sites. In Table 3 it is shown that a loss of 10-20% in N_2 adsorption capacity occurs, possibly due to a comparable loss in crystallinity.

TABLE 2
Unit Cell Parameters of H-ZSM-5 and Ti-H-ZSM-5

Properties	H-ZSM-5	Ti-H-ZSM-5
Si/Al	25	25
Si/Ti	2800	36
Symmetry	orth.	orth.
a (Å)	20.111(6)	20.16(1)
b (Å)	19.916(6)	19.96(1)
c (Å)	13.401(4)	13.447(8)
Unit Cell Vol.($Å^3$)	5367	5415

TABLE 3
Properties of catalyst samples

Material	$TiCl_4$ [K]	Surface area [m^2/g]	Si/Ti	Si/Al
H-ZSM-5		390	2800	25
Ti-H-ZSM-5	473	360	36	25
Ti-H-ZSM-5	673	360	42	25
Ti-H-ZSM-5	773	370	58	25
H-Beta		630		
Ti-H-Beta	523	530		
TS-1		400	46	

all samples are crystalline

Using our preparation method the Si/Ti ratio of the product increases with the temperature of the secondary synthesis. This result differs from those obtained with acid-extracted ZSM-5 having a Si/Al ratio of >2000 (ref. 8)

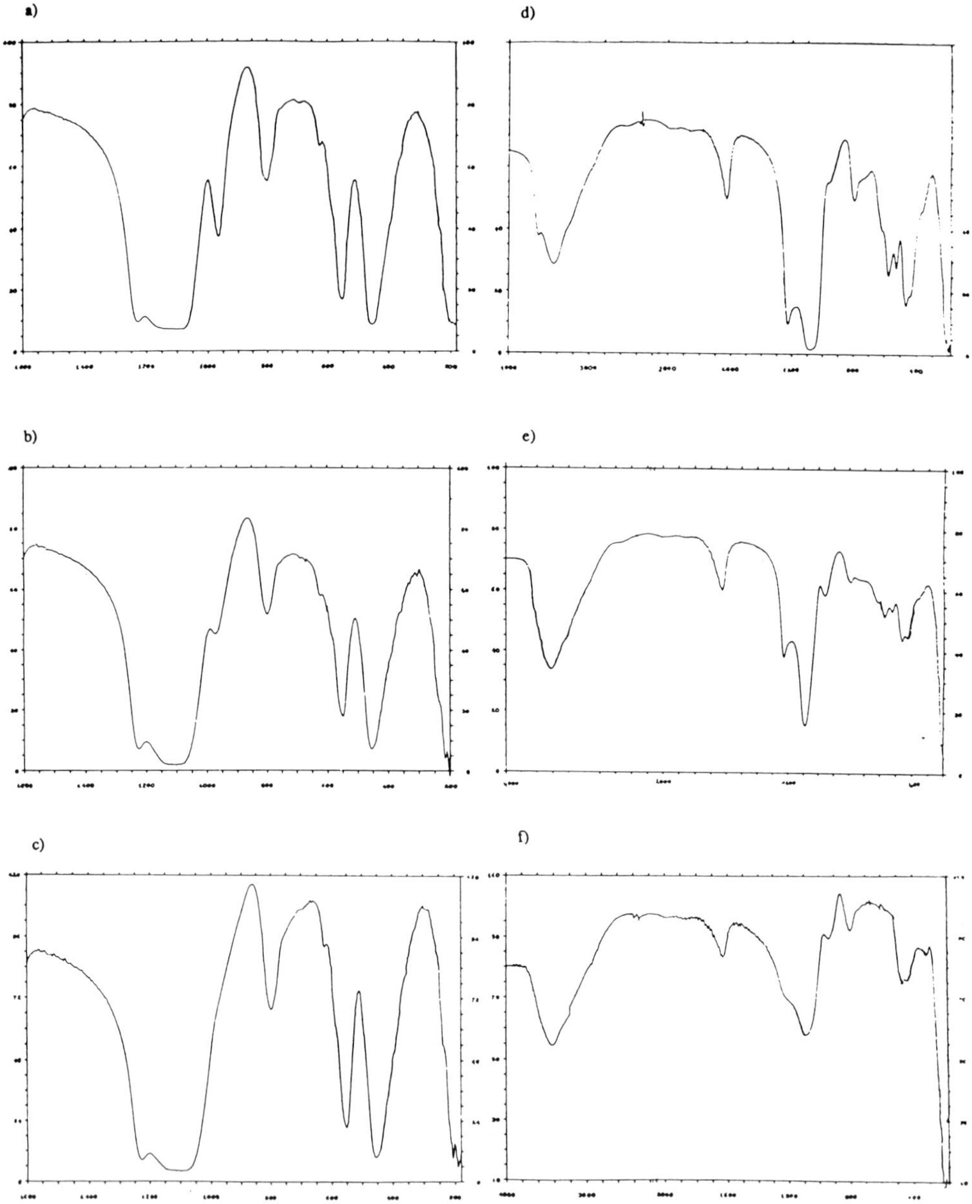

Fig. 1: IR Traces: a) TS-1, b) Ti-H-ZSM-5, c) H-ZSM-5, d) Beta, e) Ti-Beta, f) Ti-SiO_2 (amorph) Spectra a, b and c between 1600 - 200 cm^{-1}, spectra d, e and f between 4000 - 200 cm^{-1}

In the literature it is stressed that samples used in catalytic reactions with H_2O_2 should not contain free TiO_2 (ref. 9), since this causes in H_2O_2 decomposition and poor catalyst performance. In fact phenol hydroxylation may be applied as an elegant test for the presence of free TiO_2 in Al-free catalyst samples (ref. 9) and it is detected by a quick dark discoloration of the reaction mixture. This method does not apply to the Al-containing samples of solid acids such as H-ZSM-5, since these generally show a brown discoloration of the reaction mixture. Free TiO_2 is conveniently identified by its IR absorption at 380 cm^{-1}. TiO_2 has a poorly resolved IR spectrum in the range 400 - 800 cm^{-1}, the absorption at 380 cm^{-1} is however a relatively sharp peak, Fig 2, amorphous TiO_2 gel also shows this 380 cm^{-1} absorption. Small quantities of free TiO_2 (>0.5%w) can thus be detected in zeolites, provided that no zeolitic absorption occurs between 320 and 400 cm^{-1} (Fig. 2).

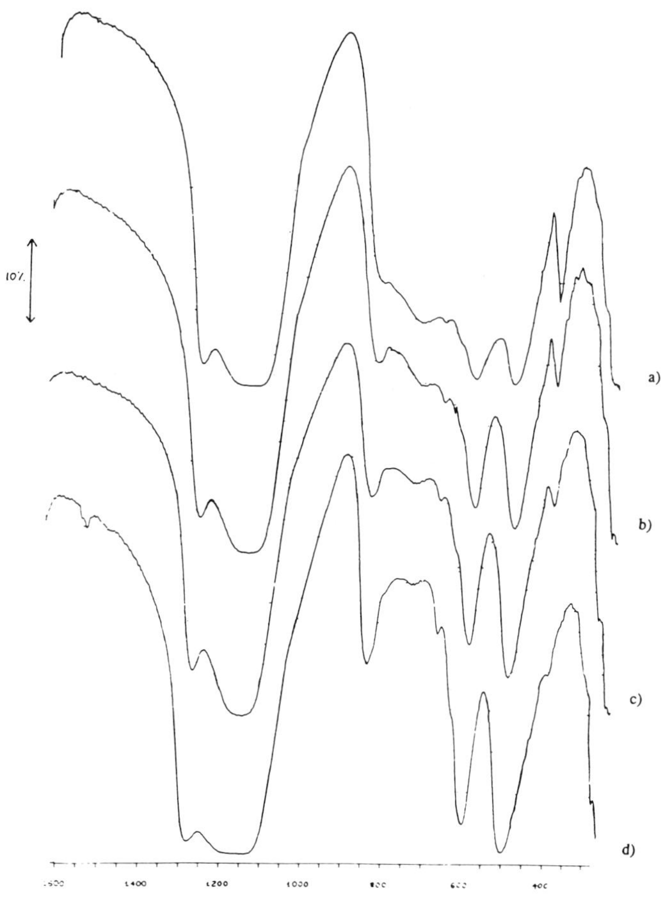

Fig 2: IR Traces: Mixtures of TiO_2 and ZSM-5 (wt.% TiO_2) a) 30%, b) 14%, c) 8%, d) 0.5%

<u>Catalytic tests</u>

Apart from the recent studies mentioned before, hydroxylation of aromatics using a zeolitic catalyst is hardly mentioned in the literature. Data on the application of acidic zeolites are conflicting, H-ZSM-5 is claimed to be an active and selective catalyst in (ref. 7), while (ref. 13) states that H-ZSM-5 is not a good catalyst for this reaction. Conditions mentioned in the references differ appreciably as far as the use of diluents/solvents is concerned, but high conversions and high selectivities are reported in all cases. It may be expected that the solvent will have an effect on the reaction depending on the nature of the surface: high Si/Al ratio zeolites are hydrophobic, low Si/Al ratio zeolites are hydrophilic. Accordingly we have tested the catalysts with and without addition of a solvent and we have also varied the type of solvent in a limited number of exploratory experiments.

1) Tests without additional solvent, Table 4. In these experiments water is always present since it is added with H_2O_2. Results confirm that H-ZSM-5 is a catalyst for the test reaction. Secondary synthesis with $TiCl_4$ improves its performance. Silicalite (Si/Al>500) has no catalytic activity, neither is it active after reaction with $TiCl_4$ unless it is given an additional thermal treatment at 1070K. Crystallite size appears to have a strong influence, the activity of a material with 20 mm particle size is very low. The selectivity of the catalysts for the para product is >5 indicating that Ti atoms are most probably located on the inner surface of the catalyst. Ti-USY did not show any activity in this test although it has an IR absorption at 960 cm^{-1} after treatment with $TiCl_4$.

TABLE 4
Reaction of H_2O_2 and Phenol using various catalyst in the absence of a solvent.

Catalyst	conv.after 4 hr.	para	ortho
H-ZSM-5	9	8	1
Ti-H-ZSM-5*	13	12	1
Silicalite	<1	<1	<1
Ti-Silicalite*	<1	<1	<1
Ti-Silicalite calc. 1073 K*	11	11	<1
Ti-HZSM-5 (25µm)*	<1	<1	<1
Ti-Y*	<1	<1	<1

*$TiCl_4$ at 473K, 2hrs.
strip at 350K, 1hrs.
steam at 573K, 1hrs.

Conversion: mol products/mol H_2O_2 in

Conditions: Temperature 353 K, 2 ml 35% H_2O_2, 7.36 g Phenol, 0.15 g Catalyst

2) Tests using a 1/1 methanol/water mixture as a solvent, Table 5. Both H-USY and H-Beta are active and selective catalysts, the product p/o ratio is 0.7 for USY and 0.8 for Beta, which might be due to the pore size of these materials (> 0.65 nm). TS-1 is the most active catalyst, the p/o ratio in the product is 1. Ti-modified H-ZSM-5 is less active than TS-1, its product p/o ratio is about 2. After modification with Ti zeolite Beta has a lower conversion than the parent material and its product p/o ratio is high and similar to that of the equally active $TiCl_4$ treated amorphous SiO_2. The last sample in this series is a catalyst made by reaction of Ti-tetraisopropylate in IPA at 348 K with a silicalite having

2 μm crystallites. The activity of this sample is very low and the initial conversion after 4 h is below detection limit. After 23 h however a high conversion is observed and the product p/o ratio is 0.5. The low initial activity of this sample might indicate that the Ti atoms are mainly located on the outer surface of the silicalite crystals (<1m^2/g). Because of its size Ti tetraisopropylate is not expected to enter the pores of silicalite

TABLE 5
Comparison of Catalysts in the reaction of phenol and H_2O_2
(Solvent CH_3OH/H_2O 1/1 w/w)

Catalyst	$TiCl_4$ 2h at K	Conv after 4hrs	product p/o ratio
Ti- H-ZSM-5	673	13	1.9
Ti-SiO_2	573	4	no ortho
H-USY		25	0.7
H-Beta		16	0.8
Ti-Beta	523	4	no ortho
TS-1		28	1
Ti-Silicalite	473	<1	-
Ti-Silicalite*		30**	0.5

*prepared by reaction of Ti-tetraisopropylate (in IPA) with acid-treated Silicalite,Temp. 348 K, time 21 hrs.
**after 23 hrs. reaction
Conversion: mol products/mol H_2O_2 in
Conditions: Temperature 353 K, 2 ml 35% H_2O_2, 5 g Phenol, 0.2 g catalyst, 4 g solvent

3) Comparison of solvents, Table 6. Using a sample of Ti-silicate prepared according to the recipe given for TS-1 in (ref. 2) example 1, acetone, methanol and water and methanol/water mixtures were used as solvents. It appeared that under our conditions the highest conversion was obtained using water even with this low-polarity catalyst and that activity does not vary much for water/methanol ratios higher than 1.

TABLE 6
Reaction of H_2O_2 and Phenol over TS-1*:
Solvent effect.

Solvent	Ratio w/w	Conv.(%) after 4h	product p/o
water		34	1
water/methanol	4/1	31	1
water/methanol	3/1	26	1
water/methanol	1/1	28	1
water/methanol	1/3	20	1
methanol		4	2.3
acetone		2	no ortho

Conditions: Temperatur 353 K, 2 ml 35% H_2O_2, 5 g Phenol, 0.2 g catalyst, 4 g solvent
Conversion: mol products/mol H_2O_2 in
*Ref. (2) example 1

CONCLUSIONS

Although the present results leave many questions open the following conclusions are drawn:

* Treatment of H-zeolites with $TiCl_4$ as described here is a method to insert Ti atoms into the zeolite framework.
* H-zeolite Y and H-zeolite Beta are active catalysts for the hydroxylation of phenol.
* The effect of the $TiCl_4$ treatment on product catalytic activity depends on the zeolite structure.
* Ti deposited on the outside of zeolite crystals is active in the hydroxylation of phenol although at a low activity level and shows a lower selectivity for p substitution than a material having the active sites located in the zeolite pores.

We would like to thank Ciba-Geigy for financial support for C.F.

REFERENCES

1 M.Taramasso, G.Perego and B.Notari, DEOS 3.047.798, 1981, Snamprogetti SpA.

2 M.Taramasso, G.Perego and B.Notari, USPat. 4.410.501, 1983, Snamprogetti SpA.

3 G.Perego, G.Bellusini, C.Corno, M.Taramasso, F.Buonomo and E.Esposito in Y.Murakami, A.Iijima, J.W.Ward (eds), Proc.Seventh Int.Conf. on Zeolites, Tokyo 1986, Tonk Kodanska Elsevier, Amsterdam, p 129.

4 B.Notari, in P.J.Grobet, W.J.Mortier, E.F.Vansant, G.Schulz-Ekloff (eds), Innovation in Zeolite Materials, Elsevier, Amsterdam, 1988, p 413.

5 A.Esposito, C.Neri and F.Buonomo, DEOS 3.309.669, 1983 Anic SpA.

6 H.S.Bloch, USPat. 3.580.956, 1971, UOP Corp.

7 C.D.Chang and D.Hellring, USPat. 4.578.956, 1986, Mobil Oil.

8 B.Kraushaar-Czarnetzki and J.C.vanHooff, Catal.Lett.1,(1988),81.

9 B.Kraushaar-Czarnetzki and J.C.vanHooff, Catal.Lett.2,(1989),43.

10 P.A.Jacobs and J.A.Martens, Synthesis of High Silica Aluminosilicate Zeolites, Elsevier, Amsterdam, 1987.

11 J.L.Guth, H.Kessler and R. Wey, in ref. 3, p. 121.

12 M.R.Boccuti, K.M.Rao, A.Zecchina, G.Leofanti and G.Petrini in C.Morterra, A.Zecchina and G.Costa (eds), Structure and Reactivity of Surfaces, Elsevier, Amsterdam 1989, p. 133.

13 G.Bellusi, M.G.Clerici, A.Giusto and F.Buonomo, EPA 226.258, 1986, Eniricerche.

G. BELLUSSI (ENIRICERCHE San Donato Milanese, Italy) : The dimensions of $TiCl_4$ are in the range between 5 and 6 Å depending whether one considers the molecule is rotating around its geometrical center or is in a fixed position. This dimension is close to the pore diameter of TS-1. Considering that $TiCl_4$ can easily react with the surface silanol group it seems unlikely that this molecules can penetrate inside the pore structure. More likely they can react on the surface as in the case of chemical vapor deposition of silicon compounds described by Prof. Murakami. Have you any direct evidence of Ti framework insertion as for example from IR spectra?

H.KOUWENHOVEN (ETH Zürich, Switzerland) : Reaction products were characterized by IR, XRD and BET surface area measurements (as reported in the paper) and the results are very similar to those published for TS-1. Additionally we applied the hydroxylation of phenol as a yardstick for the position of titanium. We have compared the performance of samples without titanium with that of a sample which contains titanium on the outer surface (sample treated with Ti-tetraisopropylate), a sample prepared by repeating the synthesis of TS-1 (Pat. Nr. 4.410.501, Example 1) and finally with that of titanium zeolites prepared by secondary synthesis. We measured the o/p ratio of the products and checked that the presence of tar products was less than 10% of the phenol reacted. The o/p ratio of the reaction product obtained using a sample containing Ti in the outer surface is higher than that obtained with "secondary synthesis Ti-zeolites" and over a TS-1 sample. The good agreement of the results obtained with TS-1 with those obtained over the "Ti-Silicates" prepared from H-ZSM-5 and the observation that the catalyst activity decreases with increasing crystal size indicate that the conclusions are most probably correct. We are aware that more data are required to establish the titanium distribution in the samples.

P. JACOBS (Lab. Oppervlaktechemie Leuven, Belgium) : From the results in your paper and from your presentation, it seems to me that your data on the preparation of Ti-zeolites with $TiCl_4$ are at variance with those of Kraushaar et al. (ref. 5 and 7 in your paper). Indeed, the latter author claims that it is essential to start with a Al-zeolite and to dislodge Al from the framework by steaming and/or acid treatment. Apparently, you start off with a pure H-form and do not use such treatments. Could you elaborate on this apparent contradiction.

H.KOUWENHOVEN (ETH Zürich, Switzerland): For our preparation of Ti-zeolites using $TiCl_4$ we always started with an acid treatment in order to introduce hydroxyl groups for the subsequent reaction with $TiCl_4$. The acid treatment was performed with 1N HCl during one hour at reflux temperature (50 ml pro gram). The product was subsequently filtered and washed with demi. water. This treatment was repeated three times. Elemental analysis by ICP of the pretreated H-ZSM-5 samples showed only little dealumination: Si/Al ratio increased from 21 to 25. This is in contrast with the results of ref. 5 and 7 where a higher degree of dealumination is observed (Si/Al 2000) and could be a reason for the apparent contradiction. During the subsequent reaction with $TiCl_4$ we did not observe a substitution of Al by Ti (after $TiCl_4$ reaction the Si/Al ratio is still 25), we prefer accordlingly a mechanism involving the reaction of surface -OH groups with $TiCl_4$.

J. VEDRINE (Institut de Catalyse, CNRS, France): You have answered to Bellussi that your conclusion for Ti incorporated in the lattice at framework position was based on Ti salt interaction effect on unit cell volume increase. I do not agree with the latter statement, IR spectroscopy of vibrational mode characterization would have been better. As a matter of fact unit cell volume is greatly dependent on water content and you should absolutely compare samples with the same water (or any adsorbate) content.

H.KOUWENHOVEN (ETH Zürich, Switzerland): The XRD measurements are only one of our characterization methods. The materials were characterized also by IR and BET surface area measurement and by the test reaction. These data are for our "secondary synthesis catalyst" in agreement with those published for TS-1. XRD data were obtained on samples having the same Si/Al ratio, which were carefully equilibrated under identical conditions. The differences which were observed in the diffraction patterns are significant for c and have an acceptable reliability for a and b (table 2).

J.KIWI (EPFL Lausanne, Switzerland). The high degree of catalysis observed when Ti-isopropylate is used on the zeolite surface has been explained by you on grounds of different Ti-deposits than when $TiCl_4$ is used. Is it not due to the fact that with $TiCl_4$ you affect profoundly the acid character of the zeolite surface which is not the case when Ti-isopropylate is used? Have you measured the acid character of the surface in both cases?

H.KOUWENHOVEN (ETH Zürich, Switzerland): The acidity of the various materials was not measured. We observed however that a $TiCl_4$ treated silicalite as a catalyst is comparable to TS-1 and $TiCl_4$ treated H-ZSM-5 and shows a relatively high p/o ratio in the reaction product. The silicalite treated with Ti-tetraisopropylate is different and has a much lower p/o ratio in its reaction product and we assume that this difference is caused by the fact that Ti in this case is deposited on the outside of the crystals. According to our IR and XRD results for samples with the MFI structure, the Ti is located in sites having a coordination comparable to that of Ti in TS-1.

G. Centi and F. Trifiro' (Editors), *New Developments in Selective Oxidation* 63

Mn(III)-TETRAARYLPORPHYRINS BEARING A COVALENTLY ANCHORED AXIAL LIGAND: EFFICIENT CATALYSTS IN OLEFIN EPOXIDATIONS UNDER TWO-PHASE CONDITIONS.

P.L. ANELLI, S. BANFI, F. MONTANARI*, and S. QUICI*

Dipartimento di Chimica Organica e Industriale dell'Università e Centro CNR, Via Golgi 19, 20133 Milano, Italy.

SUMMARY

Imidazole or pyridine "tailed" Mn-tetraarylporphyrins are efficient catalysts for olefin epoxidations carried out at 0°C with NaOCl or 30%-H_2O_2 under aqueous-organic two-phase conditions. Terms bearing Cl atoms in the o,o'-positions of the aromatic rings are also more resistant to oxidative demolition.

INTRODUCTION

Fe(III) and Mn(III) complexes of synthetic tetraarylporphyrins which mimic the biological activity of natural monoxygenases are efficient catalysts for the epoxidation of olefins and for the hydroxylation of saturated hydrocarbons (refs. 1-4). Reactions carried out under two-phase conditions with NaOCl or 30%-H_2O_2 as oxygen donors are particularly attractive (ref. 5). In both cases reaction rates are greatly enhanced by the presence of lipophilic heterocyclic bases (alkyl substituted imidazoles or pyridines) as electron donating axial ligands coordinated to the metal centre.

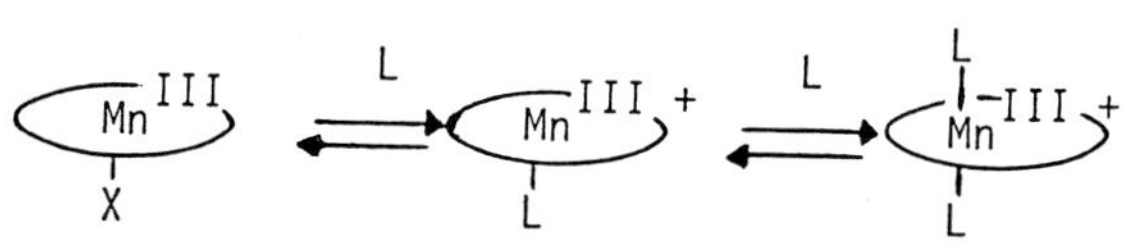

1 **2** **3**

Monoligated species **2** are highly reactive, whereas **1** and **3** are poorly reactive and completely inactive species, respectively (ref. 6).

The object of the present paper is the investigation of the catalytic activity and chemical stability of Mn(III)porphyrins bearing an imidazole or pyridine, as axial ligand, covalently bonded to the porphyrin ring through an appropriate spacer chain. Examples of these catalysts are Mn(III)porphyrins **4-7**.

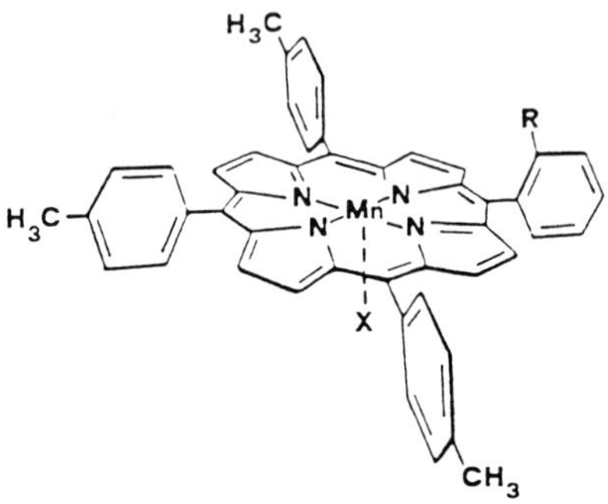

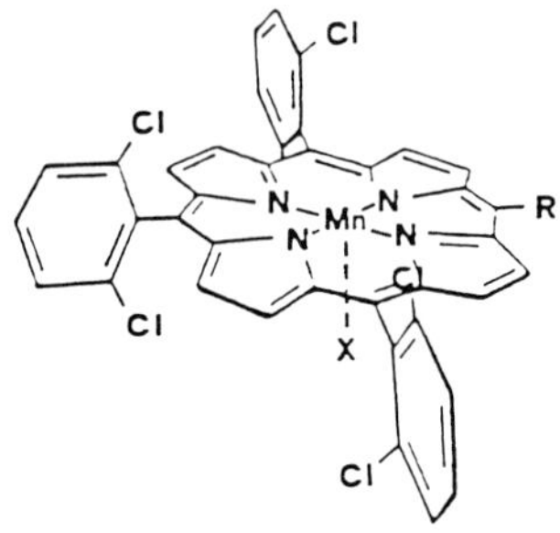

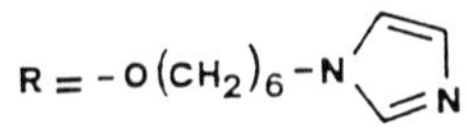

4

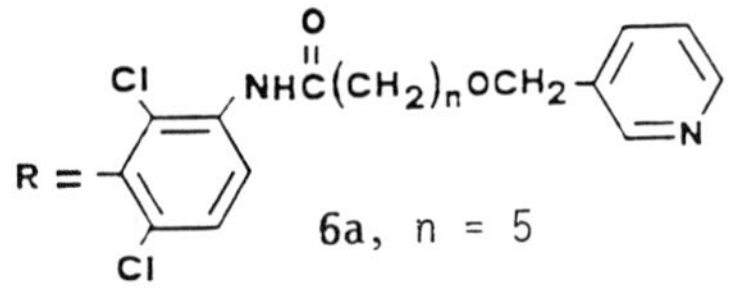

6a, n = 5

6b, n = 10

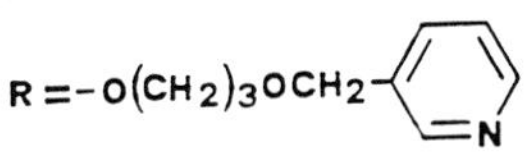

5

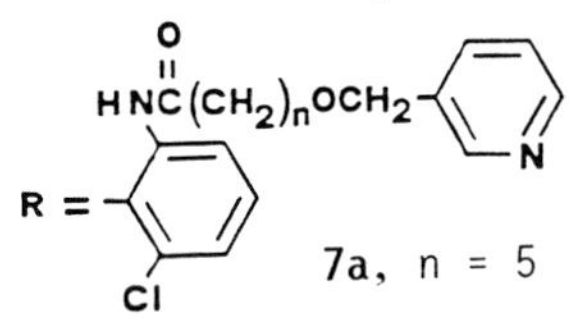

7a, n = 5

7b, n = 10

9 R = H

8 R = 2,6-$Cl_2C_6H_3$-

The catalysts, especially **6** and **7**, have been synthetized in the light of previous investigations on olefin epoxidation catalyzed by Mn(III)porphyrins and promoted by NaOCl and 30%-H_2O_2 (refs. 6-8).

We had found (ref. 6) that the effect of the ligand/porphyrin ratio (L/P) on the reaction rate depends of both the nature of the porphyrin and of the substrate. In the case of Mn(III)-tetra-(2,6-dichlorophenyl)porphyrin **8** and N-hexylimidazole as the axial ligand, the maximum reaction rate is observed for L/P = 1, which corresponds to the maximum concentration of the monoligated species **2** at the equilibrium. By increasing the L/P ratio the reaction rates progressively decrease; furthermore they are strongly slowed down in the absence of axial ligand.

The optimum conditions for oxidations carried out with NaOCl in a

CH_2Cl_2-H_2O two-phase system are achieved by buffering the pH of the aqueous phase at 10.5 (refs. 6,9-11). Catalytic amounts of quaternary onium salts as phase-transfer catalysts, currently used by other authors in epoxidations promoted by NaOCl at pH 12.7 (refs. 1,2), accelerate the reaction rate but favour the oxidative degradation of the porphyrin and of the axial ligand.

In the epoxidation reactions promoted by hydrogen peroxide (refs. 7,8), carried out under CH_2Cl_2-H_2O two-phase conditions, the pH of 30%-H_2O_2 commercial solution ($\dot{\sim}$ 2.5) must be adjusted in the 4.5-5.0 range. Lower pH values decrease the reaction rates whereas higher values promote H_2O_2 decomposition. Reaction rates are strongly accelerated by the presence of very small amounts of benzoic acid.

RESULTS AND DISCUSSION

In the synthesis of porphyrins **4-7** the following parameters have been considered: i) the linkage to the porphyrin ring and the length of the spacer chain; ii) the nature of the axial ligand; iii) the structure of the porphyrin. Porphyrins bearing covalently attached imidazole or pyridine ligands have been described by several authors (ref. 12) and the metal complexes of some have been used as catalysts in hydrocarbon oxygenation reactions. However, in this last case, the axial ligand was placed in the middle of a bridge connecting two opposite aromatic rings of tetraarylporphyrin (ref. 13). We chose the more easy to synthesize porphyrins **4-7**, confident that the very high complexation constants between imidazoles or pyridines with Mn(III)tetraarylporphyrins (refs. 6,14) would allow the spontaneous coordination of the metal and the axial ligand hung through the flexible chain (ref. 15). The chains have been linked through ether or amido bonds in the ortho or meta positions of a meso phenyl ring of the porphyrin. The number of the linearly disposed atoms in the chain was 6-14.

The porphyrin rings are those of meso-tris(p-tolyl) phenyl porphyrin **9** and tetra-(2,6-dichlorophenyl) porphyrin **8**; the latter being particularly stable under oxidative conditions (ref. 6). In catalysts **6** and **7** only pyridine has been attached to the spacer chain. Indeed, when imidazole or pyridine are used as extra bases they are both oxidized, together with the olefins; but while imidazoles are completely demolished, pyridines give, at least in the first step, the corresponding N-oxides which still behave as efficient axial ligands (ref. 6).

Porphyrins **4** and **5** have been used in a previous investigation (ref. 9). Epoxidations promoted by HOCl and catalysed by **4** and **5** are extremely fast and at 0°C are over in a few minutes. However porphyrins undergo oxidative degradation in the course of the epoxidation and at the end of the reaction are completely bleached. Present day knowledge led us to expect this behaviour, since porphyrins without bulky and/or electron withdrawing substituents in the o,o'-positions of the meso phenyl rings are extremely unstable under oxidative conditions (refs. 10,11).

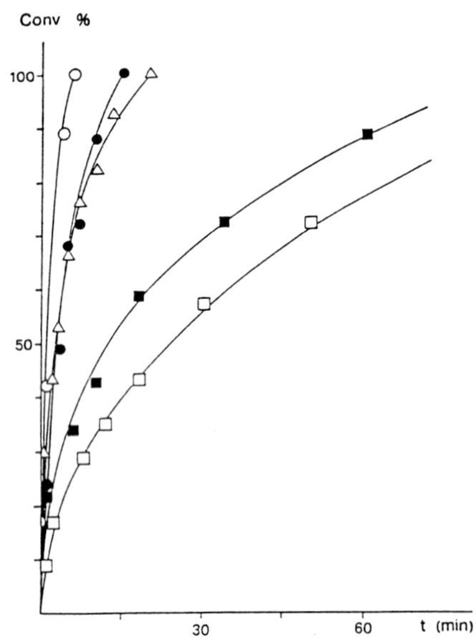

Fig.1. NaOCl cyclooctene epoxidation catalyzed by Mn(III)-porphyrins: **6a** (○), **6b** (●), **7a** (□), **7b** (■), and **8** (△). Reactions conditions: CH_2Cl_2--H_2O, 0°C, pH 10.5; Mn(III)-porphyrin: cyclooctene: 0.35M-NaOCl= 1:200: 700 molar ratios. With Mn(III)porphyrin **8** and N-hexyl imidazole, L/P = 1.

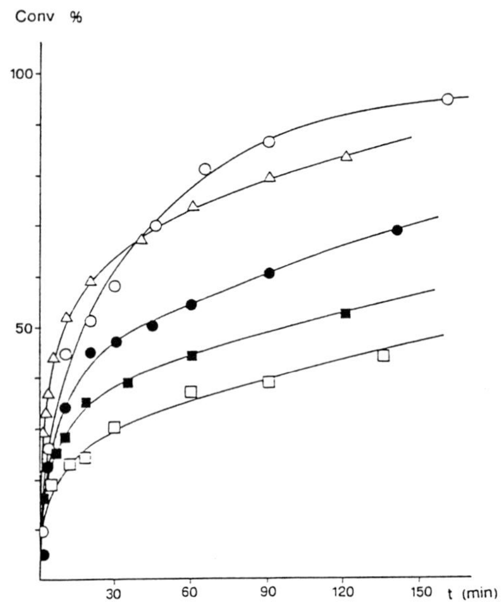

Fig.2. NaOCl 1-dodecene epoxidation catalyzed by Mn(III)-porphyrins: **6a** (○), **6b** (●), **7a** (□), **7b** (■), and **8** (△). Reaction conditions as in Fig. 1.

Epoxidation reactions promoted by HOCl or 30%-H_2O_2 and catalyzed by Mn(III) porphyrins **6** and **7** have been performed by using cyclooctene and 1-dodecene as models of reactive and poorly reactive substrates, respectively. The reaction conditions and the most significant results are reported in Fig. 1-4. Unexpectedly, porphyrins **7a** and **7b**, in which only one out of eight chlorine atoms of porphyrin **8** has been replaced by the amido group bearing the flexible chain, proved to be poorly stable under both HOCl and H_2O_2 oxidation conditions.

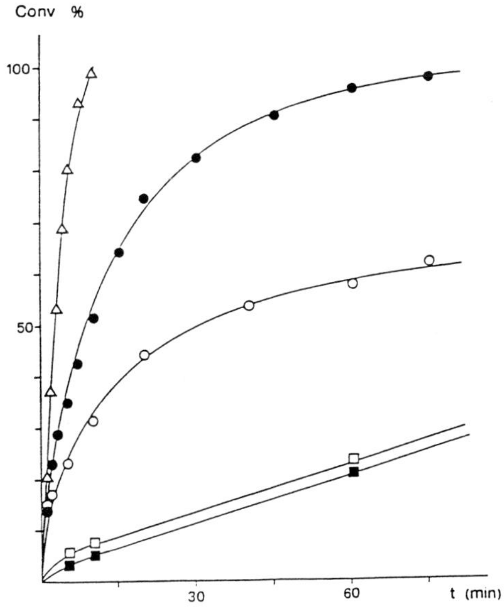

Fig.3. H_2O_2 cyclooctene epoxidation catalyzed by Mn(III)-porphyrins: **6a** (○), **6b** (●), **7a** (□), **7b** (■), and **8** (△). conditions: CH_2Cl_2-H_2O, 0°C, pH 4.5; Mn(III)-porphyrin: benzoic acid: cyclooctene: 30%-H_2O_2 = 1:1:200:400 molar ratios. With Mn(III)porphyrin **8** and N-hexyl imidazole, L/P = 1.

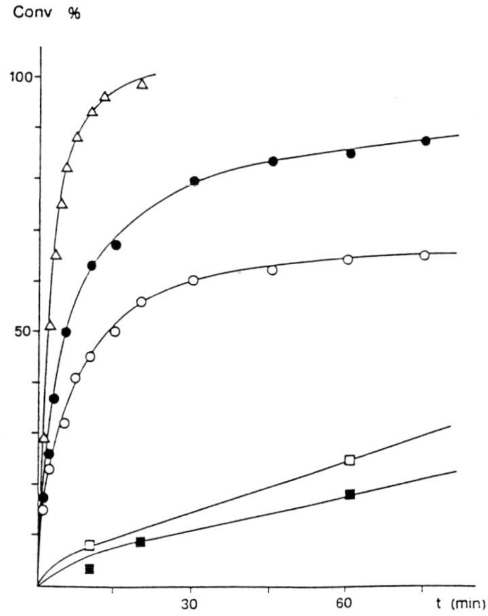

Fig.4. H_2O_2 1-dodecene epoxidation catalyzed by Mn(III)-porphyrins: **6a** (○), **6b** (●), **7a** (□), **7b** (■), and **8** (△). Reaction conditions: CH_2Cl_2-H_2O, 0°C, pH 5.0; Mn(III)-porphyrin: benzoic acid: 1-dodecene: 30%-H_2O_2 = 1:4:200:400 molar ratios. With Mn(III)porphyrin **8** and N-hexyl imidazole, L/P = 1.

Chemical stability is strongly enhanced in the case of porphyrins **6a** and **6b** in which the amido group is in the meta position of the phenyl rings, thus leaving unaffected , around the metal, the structure of Mn(III)-porphyrin **8**. Mn(III)-porphyrins **6** are more efficient catalysts then **7**, independently of the substrate and of the oxidant used. Reactivity of **6a** is equal or higher than that of Mn(III)-porphyrin **8** in the presence of N-hexylimidazole or 4-tert-butylpyridine (L/P = 1). It must be stressed that, in order to get complete conversion, in the HOCl epoxidations of 1-dodecene with **8** a higher amount of axial ligand is required, due to the oxidative degradation of the latter (ref. 6).

Catalytic activity of tailed porphyrins gets worse in the epoxidations promoted by 30%-H_2O_2 with both olefins (Fig. 3, 4). In this case olefin oxidation is accompanied by an extensive decomposition of the catalysts, mostly concerning the chain bearing the axial ligand.

CONCLUSIONS

The aim of obtaining Mn(III)porphyrins bearing a covalently anchored axial ligand, which are epoxidation catalysts at the same time efficient and sufficiently stable, has been reached in the case of porphyrins **6a** and **6b**. Unexpectedly, the less sterically demanding derivatives **7a** and **7b** are poorer catalysts. The reasons of this behaviour are still unknown: further efforts should be made in order to identify all the parameters involved to build up "tailed" Mn-porphyrins with higher catalytic activity.

REFERENCES.

1 B. Meunier, Bull. Soc. Chim. France, 578 (1986).

2 P.R. Ortiz de Montellano, Cytochrome P-450, Structure, Mechanism and Biochemistry, Plenum press, New York and London (1986).

3 I. Tabushi, Coordination Chem. Rev., 86, 1 (1988).

4 D. Mansuy, Pure Appl. Chem., 59, 759 (1987).

5 F. Montanari, S. Banfi, S. Quici, Pure Appl. Chem., 61, 1631 (1989).

6 S. Banfi, F. Montanari, S. Quici, J. Org. Chem., 54, 1850 (1989).

7 P.L. Anelli, S. Banfi, F. Montanari, S. Quici, J. Chem. Soc., Chem. Commun., 779 (1989).

8 S. Banfi, A. Maiocchi, F. Montanari, S. Quici, Gazz. Chim. Ital., in press.

9 F. Montanari, M. Penso, S. Quici, P. Viganò, J. Org. Chem., 50, 4888 (1985).

10 S. Banfi, F. Montanari, M. Penso, V. Sosnovskikh, P. Viganò, Gazz. Chim. Ital., 117, 689 (1987).

11 S. Banfi, F. Montanari, S. Quici, J. Org. Chem., 53, 2863 (1988).

12 a) J.E. Baldwin, P. Perlmutter, Top. Curr. Chem., 121, 181 (1984); b) B. Morgan, D. Dolphin, in "Structure and Bonding", J.W. Buckler (Ed.), 64, 115 (1987), and references therein.

13 a) M. Momenteau, Pure Appl. Chem., 58, 1493 (1986); b) B. Meunier, M.-E. De Carvalho, O. Bortolini, M. Momenteau, Inorg. Chem., 27, 161 (1988).

14 Unpublished results from this laboratory.

15 An extensive NMR investigation is currently under way in order to identify the possible presence of intermolecular complexations.

ALAN J. CHALK (Givaudan Corp., Clifton, N.J. 07014 USA): These catalysts have shown a reversal of the normal olefin reactivity towards epoxidation and this has been ascribed to steric bulk of the ligands about the manganese, what steric effects of this nature have you seen?

F. Montanari (Dipartimento Chimica organica Industriale, Università di Milano, Italy): It is not true that these catalysts have shown a reversal of the normal olefin reactivity towards oxidation, because the epoxidation rate of cyclooctene is always faster than that of 1-dodecene.

J. HABER (Institute of Catalysis and Surface Chemistry, Polish Academy of Sciences, Krakow, Poland): Our quantum chemical calculations of electron distribution in peroxide or peracid linked to a metal centre in the porphyrin have shown that the density of electrons on the terminal oxygen of the peroxide group, i.e. its electrophilic properties and hence reactivity towards olefins, strongly depends on the electron density on the metal centre, which obviously is a function of the axial ligand. Could your results of lesser activity of those porphyrins which have smaller steric hindrance be explained by different electron withdrawing properties of the ligands as the second important factor besides steric properties?

F. Montanari (Dipartimento Chimica Organica Industriale, Università di Milano, Italy): There is not doubt that electronic effects of substituents in the porphyrin ring affect the electron density on the metal centre and as a consequence electrophilic properties and reactivity of the intermediate metal oxene. Our data (**J. Org. Chem.**, <u>54</u>, 1850, 1989 and unpublished results) obtained from a series of Mn-tetraarylporphyrins showed a strong dependence of mono and bis-coordination constants K_1 and K_2 of axial ligands (alkyl substituted imidazoles and pyridines). These same effects should operate in "tailed" porphyrins **6** and **7**. However we feel that the main difference between **6** and **7** should be attributed to the easier cleavage of the amido bond of the orto substituted **7** with respect to the meta substituted **6** (this point is under current investigation). The more difficult intramolecular coordination in **6** is likely balanced by a easier intermolecular coordination in **7**.

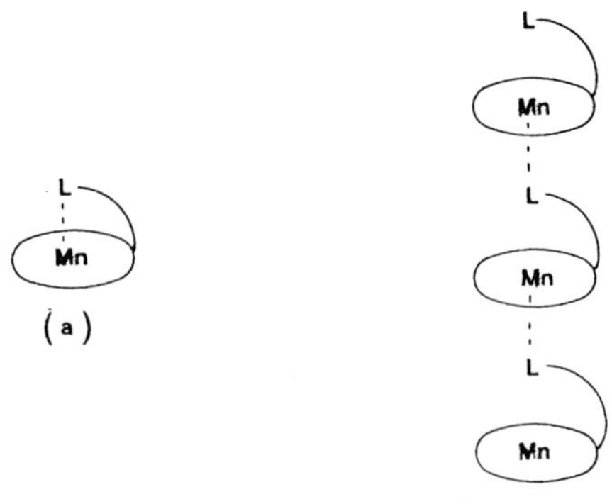

(a) (b)

B.R. JAMES (Department of Chemistry, University of British Columbia, Vancouver, BC, Canada, V6T 1Y6): Do you have data regarding stereoselectivity for these Mn systems; for example epoxidation of cis or trans stilbene?

F. Montanari (Dipartimento Chimica organica Industriale, Università di Milano, Italy): We do not have yet stereoselectivity data for Mn-porphyrins **6** and **7**. However in the epoxidation catalyzed by Mn-tetrakis(2,6-dichlorophenyl)-porphyrin and carried out with H_2O_2-benzoic acid, cis-stylbene afforded only cis-epoxide, whereas trans-stylbene did not react. The less rigid trans-oct-4-ene afforded the corresponding trans epoxide in fairly good yield (**J.C.S. Chem. Comm.**, 779, 1989).

G. Centi and F. Trifiro' (Editors), *New Developments in Selective Oxidation*
© 1990 Elsevier Science Publishers B.V., Amsterdam — Printed in The Netherlands

ALKENE EPOXIDATION WITH HYDROGEN PEROXIDE, CATALYZED BY METALLOPORPHYRINS

G. Goor, G. Prescher and M. Schmidt
Degussa AG, Forschung Organische Chemie
Rodenbacher Chaussee 4, D-6450 Hanau 1 (FRG)

SUMMARY

Metalloporphyrins of the type O-M(P)X, where M = Mo, W, (P) = tetraarylporphyrinato, octaethylporphyrinato, and X = halogene, OH, OR, OAc, SCN, $OClO_3$, are used as homogeneous catalysts in the epoxidation of alkenes by hydrogen peroxide. Selectivities and yields strongly depend on the nature of the alkene. In some cases considerable amounts of 1,2-diol are formed which strongly depends on the solvent used. 2-Hydroxyalkyl hydroperoxides are intermediates in the diol formation.

INTRODUCTION

Direct epoxidation of olefins by hydrogen peroxide has been a long-standing goal in oxidation chemistry. Contrary to oxidants such as organic peracids, hydroperoxides, chlorine via chlorohydrines (ref.1) as well as iodosylbenzene (ref.2) and hypochlorite (ref. 3) hydrogen peroxide is attractive from the standpoint of cost, and because water is its only reduction product.

Unfortunately the oxidizing power of hydrogen peroxide itself is rather low. Epoxidation reactions can be achieved only by means of suitable catalysts which are based mainly on group 5A,B and 6A,B metal oxides (ref. 4, 5). To date, however, valuable results in the case of simple olefins have been obtained only by working under virtually anhydrous conditions (ref. 6, 7) or by applying highly pretentious catalyst systems respectively (ref. 8).

The objective of our research is the development of selective homogeneous catalysts for epoxidation reactions with hydrogen peroxide as oxidant.

In this paper we present the results of studies on the catalytic activity of metal-ion porphyrin(V) complexes (Scheme 1) in the epoxidation of various alkenes. Both, aqueous and anhydrous solutions of hydrogen peroxide have been applied.

RESULTS AND DISCUSSION

Synthesis and Characterization of Metalloporphyrins

In general metalloprophyrins are synthesized by metal insertion (ref. 9). Oxomolybdenum(V) porphyrins, e.g. tetraphenylporphyrins derivative, MoO(TPP)X, are easily accessible by refluxing a metal-free porphyrin, e.g. H_2(TPP), with $Mo(CO)_6$ in decaline (ref. 10).

The homologous oxotungsten(V) complexes, WO(TPP)X are formed from $W(CO)_6$ only on prolonged refluxing in DMF (ref. 11). To prepare oxotungsten (V) octaethylporphyrin derivatives, $K_3W_2Cl_9$ in refluxing benzonitrile or H_2WO_4 in molten phenol have been used, yields however were low (ref. 12).

SCHEME 1

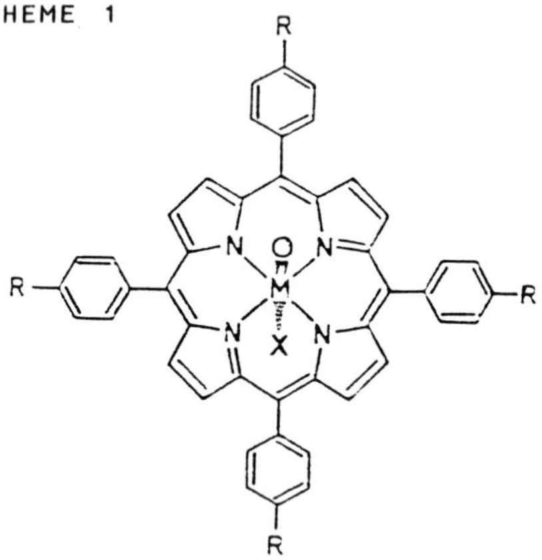

MO(TRP)X

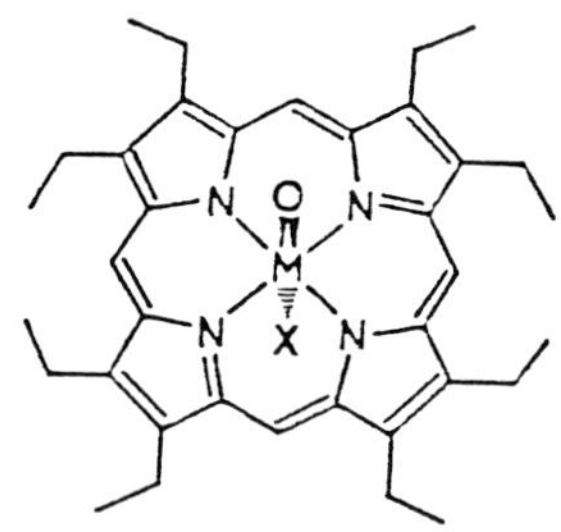

MO(OEP)X

Specification of

a) the porphyrins

R	(TRP)
OCH_3	(TAP)
Cl	(TClP)
H	(TPP)
CH_3	(TTP)

b) the metals M

M
Mo
W

c) the axial ligands X

X
O/2[a)]
OCH_3
$OClO_3$
Cl
Br
OAc
F
OH

a) μ-Oxo complex.

The metal insertion procedure could be improved for oxo-tungsten(V) complexes resulting in new tetra porphyrin derivatives, WO(TRP)X (ref. 13, Scheme 1).

The new porphyrin complexes are violet-green glittering, airstable crystals and form green solutions in organic solvents that display UV/VIS absorption spectra with three to four bands the wavelengths of which are very characteristic for the coordinated anion X. The composition are determined by elemental analysis and field ion desorption mass spectra. The axial ligands X and the terminal WO ("tungstyl") group produce specific vibrations in the IR spectra. The WO frequency is subject to a trans-effect of X (ref. 14): as the π-donor capacity of X increases, the WO frequency decreases.

SCREENING OF THE CATALYTIC ACTIVITY

Oxo-porphyrinato molybdenum(V) complexes as catalysts

Catalytical activity of these metalloporphyrins in epoxidation reactions was tested with different olefins. The results of epoxidation with hydrogen peroxide in the presence of (5,10,15,20)-tetraphenylporphyrinatooxomolybdenum(V) complexes (ref. 15, 16, 17) are summarized in Table 1.

The epoxide yields strongly depend upon the alkene applied. For 1,5-cyclooctadiene, both in aqueous and anhydrous hydrogen peroxide high yields of epoxide are obtained, even when using a substrate to catalyst ratio as high as 3000 (run 5). When cyclohexene or 2-methyl-2-butene is used as substrate, a relatively large amount of the respective 1,2-diol is formed (run 2, run 7), whereas in the epoxidation of 1,5-cyclooctadiene the formation of the corresponding 1,2-diol is not observed. Furthermore, in the epoxidation of cyclohexene a third compound was formed in addition to epoxide and 1,2-diol. With GC-MS this compound could be identified as 2-hydroxycyclohexyl hydroperoxide. As no 1,2-diol is formed before any 2-hydroxycyclohexyl hydroperoxide, it can be assumed that the latter compound is an intermediate in the 1,2-diol formation.

The oxoporphyrinato molybdenum(V) complexes are not destroyed by the oxidant and remain active after recycling (run 5 and 6). At higher initial hydrogen peroxide concentrations a larger part of the original catalyst is converted into the blue trans-diperoxo-complex $(O_2)_2Mo(VI)(TPP)$ (ref. 18). This is indicated by a strong absorption in the i.r. spectrum at 959 cm-1 which can be assigned to $\nu[Mo(O_2)]$. In the UV/VIS spectrum the Soret peak at 444 nm and weak absorptions at 573 nm and 613 nm belong to the transdiperoxo-compound which does not catalyse the epoxidation of alkenes anymore.

TABLE 1
Epoxidation of various alkenes with hydrogen peroxide in the presence of oxo(5,10,15,20-tetraphenylporphyrinato)molybdenum(V) complexes.[1,2]

Run	Catalyst (mmol)	Alkene (mmol)	H_2O_2 (mmol)	Time (h)	Yield[3] %	1,2-Diol[4] %
1	μ-O[OMo(TPP)]$_2$ (24×10^{-2})	1,5-COD[5] (72)	85% aq. (24)	18	90.2	0
2	μ-O[OMo(TPP)]$_2$ (24×10^{-2})	2-methyl-2-butene (72)	25.9% in nPAC[6] (20)	6	89.9	46.9
3	OMo(TPP)SCN (12×10^{-2})	1,5-COD (36)	25.9% in nPAC (10)	6	87.5	0
4	OMo(TPP)Cl (96×10^{-2})	1,5-COD (288)	25.9% in nPAC (79)	12	93.3	0
5	OMo(TPP)Cl[7a] (48×10^{-3})	1,5-COD (144)	25.9% in nPAC (48)	24	95.7	0
6	OMo(TPP)Cl[7b] (96×10^{-2})	1,5-COD (288)	25.9% in nPAC (79)	6	88.3	0
7	OMo(TPP)Me (12×10^{-2})	Cyclohexene (36)	85% aq. (12)	1	30.8	29.3

[1] Runs 1-6 in n-propyl acetate, run 7 in tert.-butanol; [2] T = 60°C;
[3] Yield of (epoxide + diol) relative to hydrogen peroxide;
[4] Diol to (epoxide + diol) ratio x 100 %; [5] 1,5-COD = 1,5-cycloctadiene;
[6] nPAC=n-propyl acetate; 7a Catalyst recycled once; 7b Catalyst recycled twice.

OXO-PORPHYRINATO TUNGSTEN(V) COMPLEXES AS CATALYSTS

On account of the marked catalytic activity of the molybdenum complexes, it was, therefore, of interest to investigate the analogous reactions catalyzed by tungsten porphyrins (Table 2), (ref. 19).

Table 2: Epoxidation of alkenes catalyzed by tungsten-porphyrin complexes[1)]

Run	Olefin (mmol)	Catalyst (mmol)	H_2O_2 (mmol)	time (h)	H_2O_2-conversion(%)	selectivity[2)] (%)	Epoxide/Diol
1	1,5-COD[3)] (72)	OW(TPP)Br (24×10^{-2})	25,9 %[3)] (24)	1	96.1	86.5	no diol
2	1,5-COD[3)] (72)	OW(TPP)Br (24×10^{-2})	85 % (24)	1	97.2	98.9	16.0
3	2-Methyl-2-butene[3)] (72)	OW(TPP)Br (24×10^{-2})	85 % (24)	1 6	83.6 92.8	51.1 75.3	1.13 0.65
4	1-Octene[3)] (72)	OW(TPP)Br (24×10^{-2})	85 % (24)	6	96.4	52.3	0.54
5	Allyl alcohol[4)] (36)	$OW(TPP)OCH_3$ (12×10^{-2})	85 % (12)	6	90.2	77.7[5)]	0.17
6	Cyclohexene[3)] (72)	OW(TPP)Br (24×10^{-2})	25,9 %[3)] (24)	1	83.2	81.6	0.32
7	Cyclohexene[4)] (72)	OW(TPP)Br (24×10^{-2})	25,9 % (24)	1,5	83.3	81.1	0.27
8	Cyclohexene[6)] (72)	OW(TPP)Br (12×10^{-2})	85 % (24)	6	84.8	67.7	only epoxide
9	Cyclohexene[7)] (36)	--	85 % (12)	18	69.6	55.3	1.6

1) At 60°C; 2) Selectivity to (epoxide + diol) rel. to converted H_2O_2; 3) in n-propylacetate; 4) in tert. butyl alcohol; 5) glycerol-2-glycidol ether included; 6) acetonitrile/dioxane = 2/1; 7) in acetonitrile.

High selectivities together with very short reaction times are reached with 1,5-cyclooctadiene, both with 85 % aqueous and waterfree hydrogen peroxide (runs 1, 2). With internal- and terminal olefins like 2-methyl-2-butene and 1-octene, respectively, the tungsten porphyrin catalysts seem to be less selective. Although the hydrogen peroxide conversion is high, the yield of epoxide is even lower than in the corresponding reactions catalyzed with molybdenum porphyrins. With allyl alcohol, no glycidol is observed. Glycerin-2-glycidol ether, formed in the consecutive reaction is the only product. Concerning the epoxidation of 1,5-cyclooctadiene, almost exclusively monoepoxide, 1,2-epoxycyclooctene-5, is formed.

Regarding the epoxide/diol ratio, a marked diol formation is obvious in all cases, except for 1,5-cyclooctadiene. Furthermore, as mentioned for the molybdenum porphyrin catalyzed epoxidation reaction with cyclohexene, 2-cyclohexyl hydroperoxide again was formed in addtion to epoxide and 1,2-diol. The formation of this intermediate and the dependence of the epoxide formation with time is shown in Figure 1.

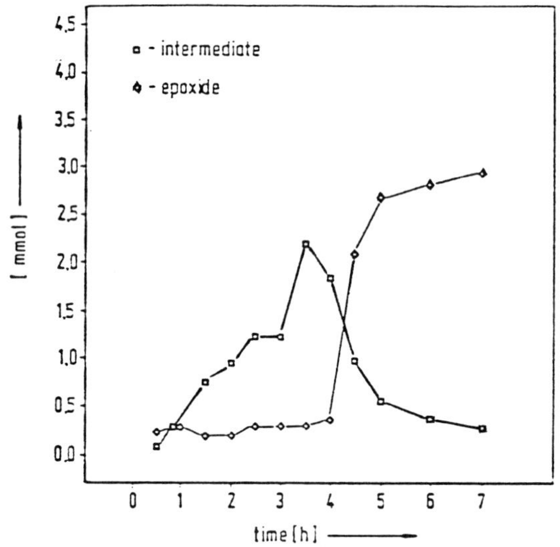

fig. 1

After a slight excess of epoxide at the beginning of the reaction, the formation of the intermediate incereases until 3 to 4 hours, whereas the epoxide stays at a very low concentration. After this time (maximum of intermediate) the yield of epoxide suddenly increases by a factor of 5 within one hour, with a concomitant decrease of the hydroxohydroperoxide intermediate. The course of this reaction is in very good agreement to a mechanism first proposed by Mattuci et al. (ref. 20) (Scheme 2) for the epoxidation of alkenes with hydrogen peroxide in the presence of inorganic peracids.

SCHEME 2

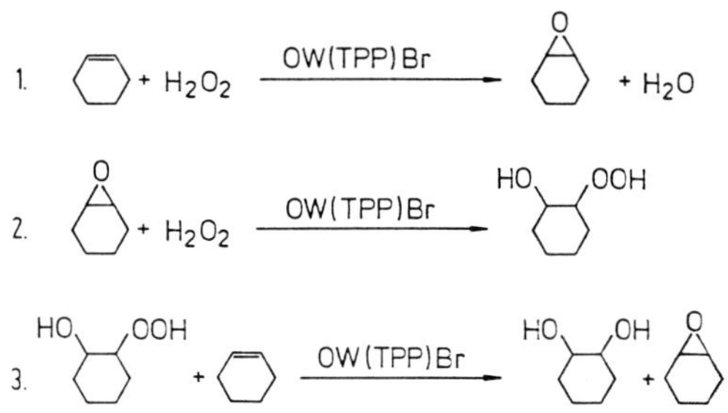

At first the expected epoxidation reaction takes place (eqn. 1). However, in the second step the epoxide reacts with further hydrogen peroxide to the hydroxo-cyclohexyl hydroperoxide intermediate (eqn. 2) which in turn further reacts with alkene to yield epoxide and diol (eqn. 3). When acetonitrile and methanol is used as a solvent mixture (1/1), 2-hydroxycycloheyl hydroperoxide can be obtained as main product.

Attempts were made to increase the low epoxide/diol-ratio:

a) aceotropic removal of water
b) epoxidation at different pH values
c) epoxidation at different H_2O_2 concentrations
d) epoxidation at different olefin : H_2O_2 ratios
e) variation of the solvent/-mixture.

However, none of the experiments a) to d) were suited to increase the epoxide portion.

The low epoxide/diol ratio was markedly increased when aceto-nitrile/dioxane (2/1) was used as solvent mixture (run 8). Although acetonitrile alone is able to react with alkenes (run 9), (ref. 21), the addition of the oxotungsten(V) porphyrin catalyst distinctly increases the reaction rate.

CONCLUSION

The observed activity of Mo(V) porphyrins in catalytic epoxidation with hydrogen peroxide has enormously stimulated the investigation of these complexes and has demonstrated that hydrogen peroxide can be a very selective oxidizing agent when proper conditions are used.

REFERENCES

1 (a) Inf. Chimie 247, March 1984, p. 131; (b) German Patent 2519287 to 301, Bayer-Degussa; (c) J. Kollar, Halcon International, US-Patent 3350442 (1967); (d) R.B. Stobaugh, V.A. Calarco, R.A. Morris and L.W. Stroud, Hydrocarbon Proc., 52 (1973) 99.

2 (a) J.T. Groves and T.E. Nemo, J. Am. Chem. Soc., 105 (1983) 5786; (b) J.R. Lindsay Smith and P.R. Sleath, J. Chem. Soc., Perkin Trans. 2 (1982) 1009; (c) A.W. van der Made, M.J.P. van Gerwen, W. Drenth and R.J.M. Nolte, J. Chem. Soc., Chem. Commun., (1987) 888 and references therein.

3 B. Meunier, E. Guilmet, M.E. De Carvalho and R. Poilblanc, J. Am. Chem. Soc., 106 (1984) 6668.

4 (a) G.B. Payne and P.H. Williams, J. Org. Chem., 24 (1959) 54; (b) H.J. Reich, F. Chow and S.L. Peake, Synthesis 1978, 299; (c) T. Hori and K.B. Sharpless, J. Org. Chem., 43 (1978) 1689.

5 S.E. Jacobsen, F. Mares and P.M. Zambri, J. Am. Chem. Soc., 101 (1979) 6946.

6 M. Pralus, J.C. Lecoq and J.P. Schirmann, Fundamental Research in Homogeneous Catalysis, in: M. Tsutsui (Ed.), Plenum, New York, Vol. 3, 1979, pp. 327-343 and references therein.

7 R.A. Sheldon and J.K. Kochi in Metal-catalysed Oxidations of Organic Compounds, Academic Press, New York, 1981, pp.275-288

8 J.-P. Renaud, P. Battioni, J.F. Bartoli and D. Mansuy, J. Chem. Soc., Chem. Commun., 1985, 888.

9 J.W. Buchler, in: D. Dolphin (Ed.), The Porphyrins, Vol. 1, Academic Press, New York, 1978, pp. 439-447 and references cited therein.

10 E.B. Fleischer and T.S. Srivastava, Inorg. Chim. Acta 5 (1971) 151.

11 E.B. Fleischer, R.D. Chapman and M. Krishnamurthy, Inorg. Chem., 18 (1979) 2156.

12 J.W. Buchler, L. Puppe, K. Rohbock and H.H. Schneehage, Chem. Ber., 106 (1973) 2710.

13 J. Buchler, G. Herget, M. Schmidt and G. Prescher, DE 3800973 A1, Degussa AG.

14 B. Plesnicar, in: W.S. Trahanovsky (Ed.), Oxidation in Organic Chemistry, part C, Academic Press, New York, 1978, p. 211.

15 M. Schmidt, G. Prescher and H. Müller, Eur. Pat. 0222 167, Degussa AG.

16 M. Schmidt and G. Prescher, Eur. Pat. 0282 708 A1, Degussa AG

17 G. Legemaat, W. Drenth, M. Schmidt, G. Prescher and G. Goor, submitted for publication.

18 B. Chevrier, T. Diebold and R. Weiss, Inorg. Chim. Acta, 19 (1976) L57.

19 M. Schmidt, G. Prescher, J. Buchler and A. Kleemann, DE 3800974 C1, Degussa AG.

20 A.M. Mattuci, E. Perrotti and A. Santambrogio, J. Chem. Soc., Chem. Commun., (1970) 1198.

21 J.-P. Schirmann and S.Y. Delavarenne, Hydrogen Peroxide in Organic Chemistry, Ed. Documentation Industrielle, Paris, 1980, p. 23.

B.R. JAMES (University of British Columbia, Vancouver, Canada): Could you provide details on the preparation of anhydrous H_2O_2-solutions?

G. GOOR (DEGUSSA AG; FRG): Preparation of anhydrous solutions of H_2O_2 in organic solvents is described in patent literature. Use of both solvents which form an azeotropic mixture with water, which azeotropic mixture boils below the boiling point of H_2O_2, e.g. n.propyl acetate, (ref. 1) and of high boiling solvents (ref. 2) have been described. However, in the course of an extensive safety examination on anhydrous solutions of H_2O_2 in low boiling solvents, i.e. those described in (ref. 1), it was observed at Degussa that if the solution catches fire the solvent will evaporate. As a consequence highly concentrated solutions of H_2O_2 in organic solvent result which can detonate. Therefore, use of H_2O_2-solutions in low boiling solvents was discontinued.

(1) Degussa AG, EP 98 427
(2) Degussa AG, EP 121 660.

R.A. SHELDON (Andeno BV, The Netherlands): Porphyrin ligands can be quite easily exchanged or oxidatively destroyed under these reaction conditions. How do you know for sure that the active catalyst is and remains the Mo- or W-porphyrin complex?

G. GOOR (DEGUSSA AG; FRG): During the course of the epoxidation reaction and after the reaction was finished samples were taken and analyzed by UV-VIS spectroscopy: neither demetallation nor oxidative degradation of the porphyrin skeleton could be observed.

R.K. GRASELLI (Mobile R&D Corp., USA): Could you comment on the enormous difference in reactivity which you observed between OW(P)X and OMo(P)X in epoxidation reactions using H_2O_2?

G. GOOR (DEGUSSA AG, FRG): We have no explanation for the much higher activity of OW(P)X as compound to OMo(P)X. It is interesting to note that OW(P)X and OMo(P)X also show different behavior against action of H_2O_2 : with OW(P)X the cis-oxo-peroxo complex of W is formed (ref. 1), whereas with OMo(P)X the trans-diperoxo complex of Mo is obtained (2).

(1) V.L. Goedken et al., J. Chem. Soc., Chem. Commun., (1985) 1425-26.
(2) B. Chevier, Th. Didold and R. Weiss, Inorg. Chim. Acta, 1976, 19, L57.

G. Centi and F. Trifiro' (Editors), *New Developments in Selective Oxidation* 81

USE OF BIMETALLIC SYSTEMS FOR THE SELECTIVE OXIDATION OF OLEFINS WITH HYDROGEN PEROXIDE.

Giorgio STRUKUL*, Andrea ZANARDO and Francesco PINNA.
Dipartimento di Chimica, Università di Venezia, Dorsoduro 2137, 30123 Venezia - ITALY

SUMMARY

The oxidation of olefins with hydrogen peroxide catalyzed by a Pt(II)/M bimetallic system is reported where M = Rh, Ir, Pd complexes. While with the former two metals no oxidation is obtained, with $(dppe)Pd(CF_3)(solv)^+$ the reaction selectively produces ketones. This appears to be a genuine example of bifunctional catalysis.

INTRODUCTION

Although hydrogen peroxide is a major chemical commodity, its direct use as oxidant in transition metal catalyzed reactions for industrial organic chemistry has found, to date, only limited success. This is mainly due to the unavoidable presence of the co-product water which is otherwise very appealing for environmental reasons.

We have recently reported (ref. 1) on the epoxidation of simple olefins with H_2O_2 catalyzed by a variety of Pt(II) complexes of the type: $P_2Pt(CF_3)X$ (P_2 = various diphosphines; X = solvent, -OH). These catalysts are very effective and versatile for a number of reasons: (i) they give high reaction rates, complete conversion of the oxidant into products and recovery of the catalyst; (ii) they can be easily modified with chiral diphosphines to give asymmetric epoxi=dation (ref. 2); (iii) they allow also the use of other hydroperoxidic oxidants like tert-butylhydroperoxide and potassium monoperoxysulfate (ref. 3) with which different selectivities in the oxidation of olefins are obtained.

In a kinetic study of this system (ref. 4) we have been able to demonstrate that the oxygen transfer step is a bimolecular reaction involving the nucleophilic attack of a PtOOH species onto an olefin activated on a different Platinum center (reaction 1).

$$Pt(ol)^+ + PtOOH \rightarrow epoxide + Pt^+ + PtOH \quad (1)$$

This observation illustrates an example of bifunctional catalysis, where the activation of the two reactants takes place on two distinct metal centers which combine together in the rate determining step. While this behavior is typical of most heterogeneous catalysts, to our knowledge it has very little counterpart in homogeneous catalysis, the only prior example in the field of oxidation being due to Mares and coworkers (ref. 5) who suggested that some Co-nitro complexes can transfer an oxygen atom to alkenes activated by Tl(III). The ability of Platinum to increase the nucleophilicity of hydrogen peroxide through formation of PtOOH complexes is rather uncommon and is shared only with some Pd species (refs. 6 and 7), on the other hand the role of olefin activator performed by Platinum in this system could be in principle carried out more efficiently by a different group VIII transition metal. In this work we report the results obtained in the oxidation of olefins with H_2O_2 as the oxidant, in the presence of a bimetallic catalyst system PtOOH/M(olefin) where M = Rh, Ir, Pd. Other very efficient olefin activators such as cyclopentadienyl Iron(II) complexes have not been considered because of the ease with which Fe(II) species undergo one-electron redox processes which lead to the decomposition of hydrogen peroxide.

RESULTS AND DISCUSSION

Rhodium

All reactions were carried out in a stoichiometric fashion. First, a very simple Rh-olefin complex has been considered:

$$[(COD)RhCl]_2 + (diphoe)Pt(CF_3)(OOH) \xrightarrow{CH_2Cl_2} (diphoe)Pt(CF_3)Cl + \text{"Rh" unknown} \quad \underline{\text{No Oxidation Product}} \qquad (2)$$

Since this reaction may indicate that the Pt oxidant is killed by exchange of the chloride present we have attempted the approach with a different Rh complex.

$$[(dppe)Rh(COD)]ClO_4 + (diphoe)Pt(CF_3)(OOH) \xrightarrow{THF} \text{No Reaction} \qquad (3)$$

The lack of reactivity might be due to the difficulty with which COD is oxidized by the Pt-OOH oxidant. In fact, because of the dominant role of steric effects in this class of platinum catalysts (ref. 8) even with $(diphoe)Pt(CF_3)(solv)^+$ as olefin activator no oxidation is observed with internal olefins (ref. 1). We have therefore devised a way to introduce 1-octene as the olefin which is very reactive in the case of the Pt^+/PtOOH system, according to the following procedure:

$$(dppe)Rh(COD)^+ + H_2 \xrightarrow{DCE} (dppe)Rh(solv)_2H_2^+ \xrightarrow[-H_2]{\text{vacuum}}$$

$$\xrightarrow{\text{1-octene}} (dppe)Rh(\text{1-octene})_2^+ \xrightarrow[24\ hr]{(diphoe)Pt(CF_3)(OOH)} (diphoe)Pt(CF_3)(OH) + \text{"Rh" brown} \quad \underline{\text{No Oxidation Product}}$$

However even in this case, no organic oxidation products were observed. All the Rhodium complexes used are described in the literature (ref. 9).

Iridium

The stoichiometric oxidation of 1,5-cylooctadiene coordinated to the $[(dppe)Ir(COD)]ClO_4$

complex (ref. 10) was tested:

$$[(dppe)Ir(COD)]ClO_4 + (diphoe)Pt(CF_3)(OOH) \xrightarrow{THF} (diphoe)Pt(CF_3)(OH) + \text{"Ir" dark-green} \quad (4)$$

GC analysis shows free COD and a small amount (corresponding to about 3% of the Pt introduced) of an unknown new product. This behavior is reminiscent of some previously observed reactions involving Iridium complexes and hydroperoxides (ref. 11):

$$(PPh_3)_2Ir(CO)(OH) + H_2O_2 \rightarrow \text{dark-green solution} \quad (5)$$

$$(PPh_3)_2Ir(CO)Cl + t\text{-}BuOOH \rightarrow \text{"} \quad (6)$$

$$[(COD)Ir(OH)]_2 + H_2O_2 \rightarrow \text{blue-green solution} \quad (7)$$

All the above reactions have been recognized to lead to decomposition of the hydroperoxide through Haber-Weiss mechanism promoted by the Ir(I)/Ir(II) redox couple.

The conclusion that emerges is that, albeit for opposite reasons, neither Rh nor Ir may be employed as cocatalysts in this oxidation system.

Palladium

Since Pd(II) centers are known to promote nucleophilic attack on coordinated olefins more efficiently than Pt(II) (ref. 12), we have tried to carry out catalytically the epoxidation reaction with H_2O_2 using a bimetallic catalyst system consisting of (diphoe)Pt(CF_3)(OH) and the homologous complex [(dppe)Pd(CF_3)(CH_2Cl_2)]BF_4.The latter complex was described by us some years ago (ref. 6) and is known to react both with H_2O and H_2O_2 and therefore is likely to be sufficiently stable under the catalysis conditions.

The system was tested in the oxidation of a variety of simple and substituted olefins and a summary of the results obtained is reported in Table 1. The reactions were carried out in a one phase THF/H_2O medium at 65°C. Attempts to work at room temperature either in THF/H_2O or in a two phase DCE/H_2O medium gave modest yields and selectivities. As shown in Table 1 in these reactions the selectivity is inverted with respect to the analogous system consisting only of Platinum, the ketone being the major oxidation product. Significant epoxidation is evident only in the case of 1-octene, while other oxidation products are formed in all cases, which include the iso-alcohols and species which may arise from further reaction of the epoxides formed, like glycols or benzaldehyde in the case of styrene. In the oxidation of butylvinyl ether partial hydrolysis of the ketonization product butylacetate leading to butanol and acetic acid is observed.

As a general trend, this system gives good amounts of oxidation products in the case of terminal olefins like styrene, butylvinyl ether, 1-octene and allylacetate, while very modest yields in ketones are observed with internal olefins like cyclohexene, cyclooctene and cis-4-methyl-2-pentene. With respect to the same substrates, Platinum alone is reactive only

TABLE 1

Oxidation of different olefins in the presence of a 1/1 $[(dppe)Pd)(CF_3)(CH_2Cl_2)^+/$ $(diphoe)Pt(CF_3)(OH)$ catalyst mixture[a].

Olefin	Time (h)	% Products[b]		
		ketone	epoxide	others[c]
Allyl Alcohol			No reaction	
Allyl Acetate	2	0.5	-	0.1
	6	1.5	-	2.4
	24	6.4	-	14.4
Styrene	2	5.6	benzaldehyde 2.6	0.6
	6	19.2	3.8	4.3
	24	26.2	4.8	5.8
Cyclohexene	2	0.3	0.3	2.4
	6	0.4	0.3	4.0
	24	0.8	-	5.2
Cis-4 Methyl-2-Pentene	2	0.2	0.6	3.3
	6	0.3	0.8	8.4
	24	0.4	1.0	9.0
Cycloctene	2	0.1	0.4	-
	6	0.2	0.6	-
	24	2.3	1.5	1.0
Butylvinyl ether	1	9.5	-	2.8
	2	26.0	-	3.2
	6	38.1	-	7.2
1-Octene	3	4.4	3.0	0.8
	7	12	3.6	6.4
	24	22.4	6 0	5.4

a Experimental Conditions: $[Pd^+]=[PtOH]$ $2x10^{-3}M$, [1-octene] 1.4 M, $[H_2O_2]$ 0.7 M, solvent THF, T 65°C.

b Yields calculated with respect to H_2O_2.

c Mainly iso-alcohols, glycols and C(n-1) aldehydes.

toward 1-octene and very slightly toward styrene, producing epoxides as the exclusive oxidation products (ref. 1). The increased reactivity of this bimetallic system seems to reflect the expected order for nucleophilic attack onto a coordinated olefin.

In order to get better insight into the role of Palladium on the activity and selectivity of this bimetallic system, we have studied the oxidation of 1-octene as a function of the catalyst composition by varying the Pd/Pt molar ratio. The results are summarized in Table 2. Again, in addition to 2-octanone and 1,2-epoxy-octane other oxidation products are formed including 2-octanol and 1,2-octandiol as the major components. Their total yield varies between 2-6.5% depending on the amount of Palladium. A representative reaction profile is shown in Fig. 1.

TABLE 2

Catalytic activity of the Pd^+/PtOH system in the oxidation of 1-octene as a function of the Pd/Pt ratio[a].

Pd/Pt	Time (h)	% Products[b]			10^5xRate (Ms^{-1})	Cat. lifetime[c] (h)
		ketone	epoxide	others		
0.5	3.5	9.2	3.2	1.8	1.91	6
	8	24	3.5	3.6		
	24	23	2.4	5.6		
1.0	3	4.4	3.0	0.8	0.54	> 8
	7	12	3.6	6.4		
	24	22.4	6.0	5.4		
3.0	0.5	1.8	4.2	0.2	1.24	4
	3	9.4	5.6	3.5		
	8	15.6	3.4	5.2		
	24	15.0	0.3	6.0		
7	2	11.8	3.8	3.5	2.06	3
	3	15.4	3.2	6.2		
	24	15.5	-	6.4		
Pd^+ alone[d]	3	5.3	0.3	2.6[e]	1.15	> 8
	6	18.2	0.8	6.0[e]		
	24	17.5	0.4	6.4[e]		
PtOH alone	24	-	2.4	-	0.07	> 24

a Experimental Conditions: [PtOH] $2x10^{-3}$M, [1-octene] 1.4 M, [H_2O_2] 0.7 M, solvent THF, T 65°C.

b Calculated with respect to H_2O_2.

c Taken when the colorless reaction mixture fades to brown.

d [Pd^+] $2x10^{-3}$M.

e about 60% heptanal

Typically, the formation of the ketone is initially slower than that of the epoxide, while the epoxide formed after reaching a maximum is partly destroyed. This behavior seems to suggest, at least in part, the occurrence of two consecutive reactions, where the epoxide is partly isomerized to ketone and partly converted into other oxidation products. Another important difference of this system with respect to the analogous system consisting only of Platinum is the catalyst lifetime that is rather limited due to the lower stability of the $[(dppe)Pd(CF_3)(solv)]^+$ complex compared to that of the homologous $[(dppe)Pt(CF_3)(solv)]^+$. When Pd(0) starts forming the reaction practically stops while the remaining hydrogen peroxide is very rapidly destroyed.

An analysis of the effect of the Pd/Pt ratio on the catalytic activity is not easy. Table 2 shows that the maximum amount of products does not significantly change when increasing Pd concentration, since an increase in the rate of conversion is balanced by the lower stability of

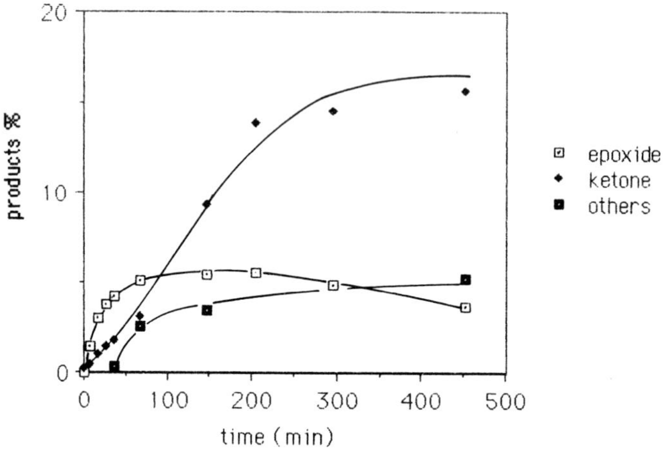

Fig. 1. Reaction profile for the oxidation of 1-octene with the Pd^+/PtOH catalytic system. Reaction conditions as in Table 2, Pd/Pt = 3.

the catalyst. Moreover the maximum rates of olefin conversion are observed both at the lowest and highest Pd concentrations.

In Table 2 the two blank reactions employing either Pd alone or Pt alone as catalysts are reported. While the latter is a very poor catalyst under the usual experimental conditions, Pd shows an activity similar to the bimetallic mixture but producing negligible amounts of epoxide.

The data obtained in the blank reactions seem to suggest that with the bimetallic system the two metal are actually working independently. So it seems important to determine whether there is actually a cooperative effect between the two metals. Indeed according to the mechanism determined for Pt only (ref. 4) reaction 1 in the case of the bimetallic system should read as

$$Pd(ol)^+ + PtOOH \rightarrow \text{oxidation products} + PdOH + Pt^+ \quad (8)$$

$$Pt(ol)^+ + PdOOH \rightarrow \text{oxidation products} + PtOH + Pd^+ \quad (9)$$

$$Pd(ol)^+ + PdOOH \rightarrow \text{oxidation products} + PdOH + Pd^+ \quad (10)$$

$$PdOH + Pt^+ \rightleftharpoons Pd^+ + PtOH \quad (11)$$

reaction 8. In principle in the next catalytic cycle the two roles would be inverted (reaction 10). However, together with the two heterobimetallic oxidation steps even the two homobimetallic oxidation steps are possible i.e. reaction 1 and reaction 10. Of course, the relative weight of these four possibilities on the overall catalytic activity will depend on the -OH exchange equilibrium i.e. reaction 11.

The position of equilibrium 11 was determined by ^{19}F NMR spectroscopy. The complexes (dppe)Pd(CF_3)(OH) (0.05 mmol; $\delta(CF_3)$: -29.71 ppm (dd), $^3J_{FPcis}$ 24.7 Hz, $^3J_{FPtrans}$ 63.4 Hz) and [(diphoe)Pt(CF_3)(CH_2Cl_2)]BF_4 (0.05 mmol; $\delta(CF_3)$: -28.58 ppm (dd), $^3J_{FPcis}$ 8.6

Hz, $^3J_{FPtrans}$ 56.7 Hz, $^2J_{FPt}$ 518 Hz) were dissolved in CD_2Cl_2 (1 mL) and a ^{19}F NMR spectrum of the mixture was run showing the presence of only $[(dppe)Pd(CF_3)(CH_2Cl_2)]BF_4$ ($\delta(CF_3)$: -26.97 ppm (dd), $^3J_{FPcis}$ 23.0 Hz, $^3J_{FPtrans}$ 63.7Hz) and (diphoe)Pt(CF_3)(OH) ($\delta(CF_3)$: -27.46 ppm (dd), $^3J_{FPcis}$ 9.7 Hz, $^3J_{FPtrans}$ 57.1 Hz, $^2J_{FPt}$ 565 Hz) indicating that equilibrium 11 is completely shifted to the right. Identical ^{19}F NMR results were obtained starting from $[(dppe)Pd(CF_3)(CH_2Cl_2)]BF_4$ and (diphoe)Pt(CF_3)(OH). These experiments suggest reaction 8 as the main oxidation pathway in this bimetallic system.

In conclusion the data obtained with Palladium as olefin activator prove that the bimetallic catalytic system in these oxidation reactions is indeed involved and that the activity and selectivity of the system can be modified. Although the search for the appropriate combination of metal complexes was not straightforward the results here reported show that bifunctional-bimetallic catalysis is indeed possible and this is in principle another possible way in which a homogeneous catalytic reaction can be "tuned" to the achievement of the desired properties.

EXPERIMENTAL SECTION.

Apparatus.

IR spectra were taken on a Perkin-Elmer 597 spectrophotometer either in Nujol mulls (CsI plates) or in solution (NaCl windows). ^{19}F NMR spectra in CD_2Cl_2 were recorded on a Varian FT 80 A spectrometer operating in FT mode, using as reference external $CFCl_3$. Negative chemical shifts are upfield from the reference. GLC measurements were taken on a Hewlett-Packard 5890 A gas chromatograph equipped with a Hewlett-Packard 3390 A integrator. Identification of products was made with GLC by comparison with authentic samples.

Materials.

Solvents were dried and purified according to standard methods. Olefins were purified by passing through neutral alumina, distilled and stored under N_2 in the dark. Hydrogen peroxide (34%) (Fluka) was a commercial product and used without purification. The preparation of the complexes was performed under dry N_2 by conventional Schlenk and syringe techniques.

The following compounds were prepared by literature methods: $[(COD)RhCl]_2$ (ref. 9), $[(dppe)Rh(COD)]ClO_4$ (ref. 9), $[(dppe)Rh(1\text{-octene})_2)]ClO_4$ (ref. 9), $[(dppe)Ir(COD)]ClO_4$ (ref. 10), (diphoe)Pt(CF_3)(OH) (ref. 13), $[(diphoe)Pt(CF_3)(CH_2Cl_2)]BF_4$ (ref. 13), (diphoe)Pt(CF_3)(OOH) (ref. 6), $[(dppe)Pd(CF_3)(CH_2Cl_2)]BF_4$ (ref. 6), (dppe)Pd(CF_3)(OH) (ref. 6).

Abbreviations: dppe = 1,2-diphenylphosphinoethane; diphoe = cis-1,2-diphenylphosphinoethylene; COD = 1,5-cyclooctadiene; ol = olefin; solv = solvent; THF = tetrahydrofuran; DCE = 1,2-dichloroethane.

Reactivity.

Catalytic Reactions were carried out in a 25 mL round-bottomed flask equipped with a stopcock for vacuum/N_2 operations, a reflux condenser and a side-arm fitted with a screw-capped silicone septum to allow sampling. Constant temperature (65°C) was maintained by an external oil bath equipped with heating coil and a Vertex thermometer for temperature control. Stirring was performed by a teflon-coated bar driven externally by a magnetic stirrer.

The general procedure here reported was followed in all cases. In a typical experiment the appropriate amounts of complexes were placed solid in the reactor which was evacuated and placed under N_2 atmosphere. Dry, N_2-saturated THF was added, followed by the olefin to be oxidized. After stirring the mixture up to the desired temperature the H_2O_2 solution was injected and the time was started. The conversion was monitored by sampling periodically the reaction mixture with a microsyringe.

ACKNOWLEDGEMENTS.

This work was supported jointly by the European Economic Community (Brussels) and Degussa AG (Frankfurt) through the special program BRITE. Special thanks are expressed to Drs G. Goor and M. Schmidt (Degussa AG) to Professor W. Drenth (University of Utrecht) and to Professor J. W. Buchler (University of Darmstadt) for stimulating discussions.

REFERENCES.

1 G. Strukul and R.A. Michelin, J. Chem. Soc. Chem. Commun., (1984) 1538; G. Strukul and R.A. Michelin, J. Am. Chem. Soc., 107 (1985) 7563.
2 R. Sinigalia, R.A. Michelin, F. Pinna and G. Strukul, Organometallics, 6 (1987) 728.
3 G. Strukul, R. Sinigalia, A. Zanardo, F. Pinna and R.A. Michelin, Inorg. Chem., 28 (1989) 554.
4 A. Zanardo, F. Pinna, R.A. Michelin and G. Strukul, Inorg. Chem., 27 (1988) 1966.
5 S.E. Diamond, F. Mares, A. Szalkiewicz, D.A. Muccigrosso and J.P. Solar, J. Am. Chem. Soc., 104 (1982) 4266.
6 G. Strukul, R. Ros and R.A. Michelin, Inorg. Chem., 21 (1982) 495.
7 M. Roussel and H. Mimoun, J. Org. Chem., 45 (1980) 5387.
8 A. Zanardo, R.A. Michelin, F. Pinna and G. Strukul, Inorg. Chem., 28 (1989) 1648.
9 J. Chatt and L.M. Venanzi, J. Chem. Soc. A, (1957) 4735; R.R. Schrock and J.A. Osborn, J. Am. Chem. Soc., 93 (1973) 2327.
10 M. Green, T.A. Kuc and S.H. Tailor, J. Chem. Soc. A, (1971) 2334.
11 G. Strukul, Unpublished results; B.L. Booth, R.N. Haszeldine and G.R.H. Neuss, J. Chem. Soc. Dalton Trans., 37 (1982).
12 R.A. Sheldon and J.K. Kochi, Metal Catalyzed Oxidations of Organic Compounds, Academic Press, New York, 1981, p. 191; F.R. Hartley, Chem. Rev., 69 (1969) 799; F.R. Hartley, J. Chem. Educ., 50 (1973) 263.
13 R.A. Michelin, M. Napoli and R. Ros, J. Organomet. Chem., 175 (1979) 239.

B. R. JAMES (University of British Columbia, Vancouver, Canada): It is possible that the Platinum/Palladium complexes catalyze, in solution, the isomerization of epoxides to ketones, a factor which could affect directly the observed selectivities. Have you checked this possibility?

G. STRUKUL (University of Venice, Italy): The Platinum complex used does not catalyze the isomerization of epoxides to ketones; however, as you suggest, this is a likely possibility in the case of Palladium that we have not checked. A complementary and/or alternative possibility is that the Palladium complex promotes further reaction of the epoxide with H_2O_2/H_2O as seems to be suggested by the reaction profile (Fig. 1) and by the nature of the other products reported in Tables 1 and 2.

G. Centi and F. Trifiro' (Editors), *New Developments in Selective Oxidation* 89

PLATINUM CATALYSTS FOR CYCLOHEXENE EPOXIDATION WITH AN OXYGEN-HYDROGEN MIXTURE

N.I. KUZNETSOVA, A.S. LISITSYN, A.I. BORONIN and V.A.LIKHOLOBOV
Institute of Catalysis, Novosibirsk 630090 (USSR)

SUMMARY

Upon simultaneous oxidation of c-hexene and hydrogen with molecular oxygen, epoxide of c-hexene is formed on Pt catalysts in absence of a porphyrin-like co-catalyst, provided that a part of Pt in the catalysts is in a metallic state and (i) the catalysts were reduced at moderate temperature or (ii) special combination of two Pt catalysts was used or (iii) hydrogen chloride was added into the reactor. The nature of the active sites responsible for the epoxide formation is discussed.

INTRODUCTION

Co-oxidation of two or several substrates sometimes allows one to obtain desirable product under milder conditions and/or with the use of a simpler oxidant. In particular, when olefins interact with O_2 in the presence of $NaBH_4$ (ref. 1), sodium ascorbate (ref. 2), Zn/CH_3COOH (ref. 3) or H_2 (refs. 4-6), such valuable compounds as epoxides are formed:

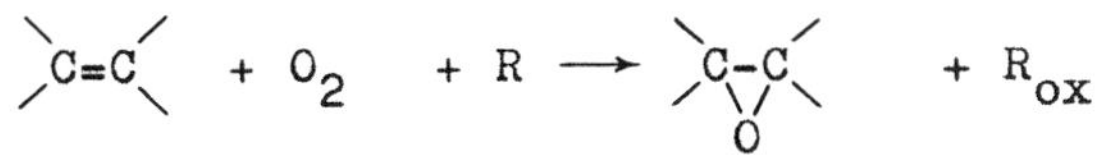 (1)

When R is hydrogen, metallic Pt together with the Mn or Fe porphyrins is used to catalyze the process (refs. 4-6). Recently it has been found, however, that the porphyrin complexes are not necessary component of the catalytic mixture and epoxide can form on Pt catalysts alone (ref. 7). Here we present new data on properties of the catalysts, with main attention to factors affecting the catalytic activity in epoxidation.

EXPERIMENTAL

i) Preparation of catalysts. Silica (200 m^2/g, BET) was impregnated with aqueous solutions of H_2PtCl_6, K_2PtCl_6 or K_2PtCl_4 (commercial substances) to provide a Pt content in the catalysts of 2.0 wt.%. The samples were dried in air for 20 h at ambient temperature and then treated in a flow of hydrogen (0.5-1 $cm^3/s/g$)

at a stepwise increase in temperature (0.5 h at 320, 370 and 420 K, 4h at 520 K and 7 h at 720 K). Before contact with air, they were cooled in H_2 and flushed with N_2. For designation of catalysts see Table 1. With the exception of I-720, studied by XPS were earlier prepared catalysts (ref. 7). Following impregnation, they were dried at 330 K and treated in H_2 at heating at once to desired temperature.

ii) XPS-study was performed on a VG ESCALAB electron spectrometer. Samples were supported on a nickel net and spectra calibrated relative to E_B Si_{2p} equal to 103.6 eV.

iii) Catalytic runs were carried out following the same technique as described earlier (ref. 7). A static two-neck reactor was supplied with a magnetic stirrer and connected with a volumetric burrete filled with dibutylphthalate. After loading a catalyst (usually 20 mg) into the reactor, acetonitrile (1 ml) was introduced to cover the catalyst, then the system was flushed with premixed O_2 and H_2 and experiment started by introducing c-hexene (10 μl). In special experiments, a freshly prepared solution of HCl in acetonitrile (via mixing acetonitrile and concentrated aqueous HCl) was added (several μl) before c-hexene. Products were analyzed chromatographically (sampling with a syringe through a silicon gasket in one of the reactor necks).

The laboratory system was placed behind a protective screen. (All other precautions necessary for operations with explosive mixtures should be followed if repeating the experiments).

RESULTS

Treatments of the samples with supported Pt chlorides at progressively higher temperatures led to decrease in intensity and then disappearance of Cl_{2p} lines in photoelectron spectra, which was accompanied by a shift of the Pt_{4f} lines to E_B value typical for metallic Pt (71.0 eV for Pt $4f_{7/2}$, (ref. 8)). Nevertheless, ion etching of the samples (Ar^+, $\sim$ 10-20 nm) resulted in a significant back shift of the Pt_{4f} lines to high binding energy (cf. spectra 5 and 4 in Fig. 1) and reappearance of the Cl_{2p} lines.

Catalytic properties of the samples prepared in different conditions and from different Pt compounds are given in Table 1. In this case, only on the Pt(IV)-derived catalysts reduced in H_2 in a narrow range of temperature ($320 \lesssim T \lesssim 420$ K) was formation of c-hexene epoxide observed. To some but minor extent, the up-

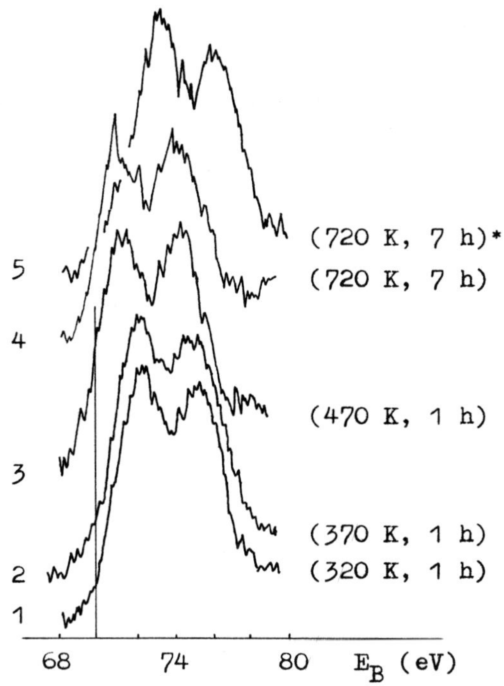

Fig. 1. X-ray photoelectron spectra of H_2PtCl_6-derived catalysts pretreated in hydrogen under different conditions (indicated in brackets near each spectrum; in case (5) the spectrum was recorded after ion etching the sample).

per and lower values of feasible temperature and catalytic properties were influenced by such parameters as temperature of drying the samples, duration of reduction, etc. (cf. data for newly and earlier prepared H_2PtCl_6-based catalysts in Table 1). The main alteration during catalytic runs was suppression of the side-hydrogenation of c-hexene, so that the epoxide to c-hexane ratio increased with time.

It has been found, however, that epoxide is formed when "over-reduced" catalysts are combined with those non-subjected to pre-reduction and contained Pt(II) or Pd(II) chlorides (Table 2). Although each component of the catalytic mixtures appeared nearly or entirely inactive if being tested separately, the velocity and selectivity of epoxide formation at the combined action were sometimes higher than even for best catalysts in Table 1.

Meanwhile, in case of $(CH_3CN)_2PdCl_2$ as second component, an elemental analysis showed rapid decomposition of the complex

TABLE 1

Conditions of preparation and catalytic properties of Pt catalysts in co-oxidation of c-hexene and hydrogen under standard conditions (H_2/O_2 = 1 v/v, 293 K, 1 atm, catalyst amount 20 mg, 0.1 mmol c-hexene in 1 ml acetonitrile)

Designation of catalyst	Pt compound used	Treatment in H_2, T_{max} (K)	Colour of sample	O_2/H_2 consumption (mmol)[a]	c-Hexene conversion (μmol)[a] total	into c-hexane	into epoxide
I-init	H_2PtCl_6	no	yellow	0.2	25	7	none
I-320	– " –	320	yellow	0.25	20	10	trace
I-370	– " –	370	grey	0.45	50	4	7
I-420	– " –	420	grey	1.1	15	12	trace
I-520	– " –	520	grey	1.25	17	15	none
I-320[b]	– " –	320	light-grey	-	60	10	10
I-370[b]	– " –	370	grey	-	65	9	12
I-420[b]	– " –	420	grey	-	40	25	3
II-init	K_2PtCl_6	no	yellow	0.15	13	10	none
II-320	– " –	320	yellow	0.15	8	6	none
II-370	– " –	370	grey	0.2	8	4	1
II-420	– " –	420	grey	0.3	10	4	1
III-init	K_2PtCl_4	no	pink	0.1	13	8	none
III-320	– " –	320	grey	0.3	13	11	none
III-370	– " –	370	grey	0.2	20	17	none
III-520	– " –	520	grey	0.9	10	5	none

[a] over 1 h; [b] prepared and tested earlier (ref. 7)

TABLE 2

Properties of some binary catalytic mixtures in c-hexene epoxidation (standard consitions)

No	Composition	c-Hexene conversion (%) over 1 h total	into epoxide
1	I-420 + I-297	69	18
2	I-420 + K_2PtCl_4/SiO_2	38	14
3	I-420 + H_2PdCl_4/SiO_2	63	24
4	I-420 + $(CH_3CN)_2PdCl_2$	63	17

[a] by 10 mg of each supported catalyst, with an equimolar Pt to Pd ratio in case (4)

under experimental conditions and absence of Pd in the solution

TABLE 3

Properties of Pt catalysts in co-oxidation of c-hexene and hydrogen when HCl (5 μmol) is added to the reaction mixture (0.1 mmol c-C_6H_{10} in 1 ml acetonitrile, 20 mg catalyst, 293 K)

Catalyst[a]	O_2/H_2 consumption (mmol)[b]	Total c-hexene conversion (μmol)[b]	c-Hexane formed (μmol)[b]	c-Hexene epoxide formed (μmol)[b]
I-320	0.15	25	3	trace
I-370	0.15	50	2	6
I-420	0.25	50	2	6
I-520	0.35	50	1	7
I-720	0.35	55	1	7
II-320	0.1	4	1	none
II-370	0.25	25	2	1
II-420	0.4	50	1	8
III-init	0.05	10	2	none
III-320	0.05	20	1	2
III-370	0.3	35	1	5
III-520	0.3	45	1	7

[a] as designated in Table 1; [b] over 1 h

after catalytic run. But when the solid was filtered off from the reaction mixture and used in new catalytic run without adding $(CH_3CN)_2PdCl_2$, epoxide of c-hexene was not formed.

An attempt has also been made to recreate the active sites responsible for epoxidation through a back treatment of the "over-reduced" samples with hydrogen chloride. When being pretreated with an acetonitrile solution of HCl, such samples remained inactive in epoxidation; however, the epoxidation took place if an appropriate quantity of HCl was added immediately into the reactor for catalytic testings. The main results obtained in this latter case are present in Table 3 and Figs. 2-5. As under standard conditions, other products were c-hexane, c-hexenol and presumably c-hexenon (mass-spectrometry data), with combined yield near 50% on a converted c-hexene basis. Also detected were c-hexane chlorhydrine and some heavy unidentified products. Addition of HBr turned out less effective and that of $HClO_4$ or CF_3COOH non-effective.

DISCUSSION

The present study reveals some new conditions under which one

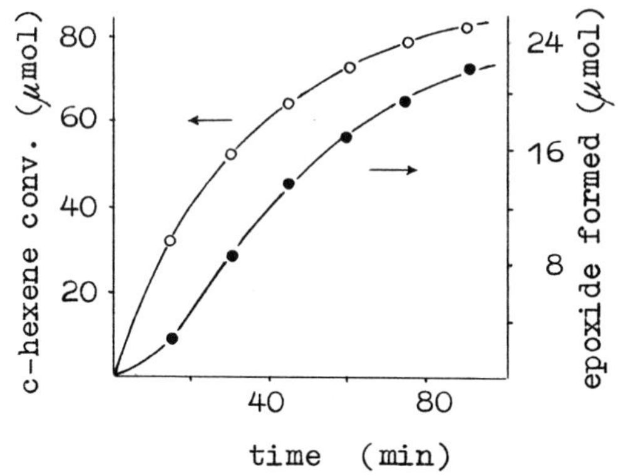

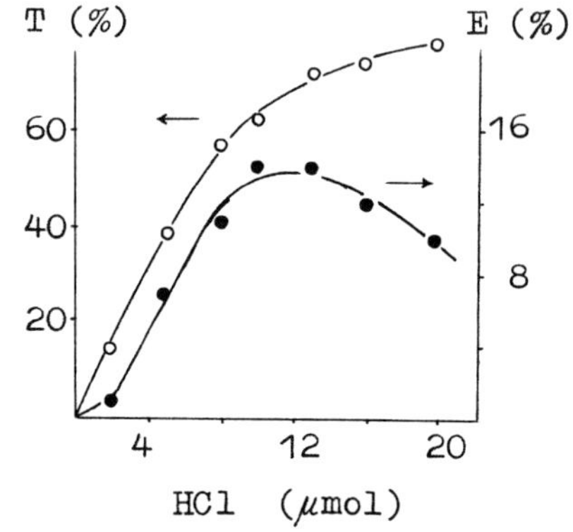

Fig. 2. Typical catalytic performance of Pt catalysts in c-hexene epoxidation with an oxygen-hydrogen mixture upon addition of HCl (12 μmol at 0.1 mmol c-C_6H_{10} in 1 ml acetonitrile, catalyst I-720).

Fig. 3. Fraction of c-hexene converted over 1 h into all products (T) and into epoxide (E) as a function of the HCl amount added to the reaction mixture (0.1 mmol c-C_6H_{10} in 1 ml acetonitrile, O_2/H_2 = 1, catalyst I-520).

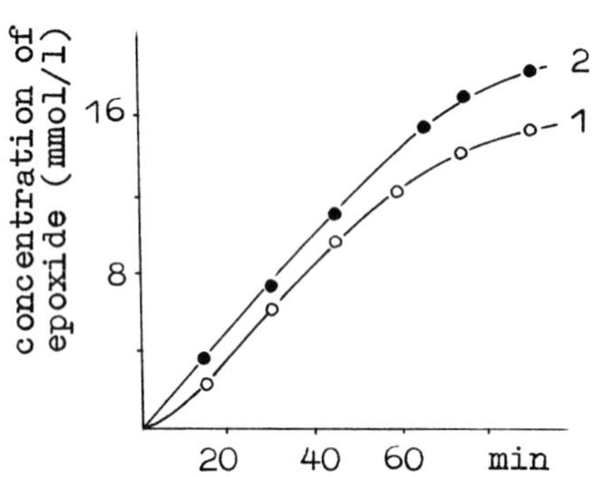

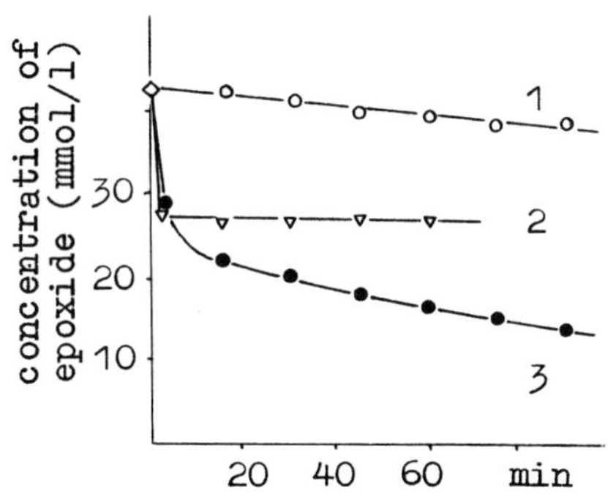

Fig. 4. Formation of c-hexene epoxide : 1 - in usual catalytic run with addition of HCl; 2 - when equimolar amount of c-hexene epoxide is also introduced to consume the HCl added. Other conditions: 10 μmol HCl, 0.1 mmol c-hexene, 1 ml acetonitrile, catalyst III-520.

Fig. 5. Decomposition of c-hexene epoxide in acetonitrile solution under O_2/H_2 gas phase: 1 - in presence of Pt catalyst (I-720, 20 mg); 2 - under the action of HCl (was added to provide $[HCl]_0$ = 12 mmol/l); 3 - in presence of both the catalyst and HCl.

should expect for the epoxide formation at interaction of c-hexene with O_2/H_2 mixture on Pt catalysts. It seems more reliable to use binary catalytic compositions (Table 2) or promote Pt^0-containing samples with HCl (Table 3). In opposite case (table 1) the Pt catalysts must be prepared by very specific method, and it makes understandable why their capability of catalyzing epoxidation has been revealed only recently (ref. 7).

Based on some literature data and inactivity (under standard conditions) of both initial and "overreduced" samples, a two-stage scheme of the catalytic process has been proposed (ref. 7) which assumes peroxide formation from O_2 and H_2 on metallic Pt and subsequent interaction of the peroxide and olefin in presence of Pt ions. It is supported by the results obtained in this work with binary catalytic mixtures (Table 2). Presence of metallic Pt in all the active catalysts is clearly indicated by their colour and consistent with the XPS data (Table 1, Fig. 1). Besides, tests with KI have been made, which did witness for the peroxide formation during catalytic runs.

Although bifunctional action of the catalysts can, in general, be accepted, there is uncertainty in exact nature of the necessary active sites. At the moment, it seems most probable that formation of both peroxide and epoxide takes place on the surface of metallic Pt, but with chlorine-containing species serving as a necessary modifier of the surface. These surface chloride adducts can form in presence of HCl or the metal chlorides which are specially added or present in the samples treated in H_2 under mild conditions. (The Pt chlorides which remain in the catalysts after a high-temperature reduction can hardly participate in the catalytic process. In this case they are probably incapsulated in micropores of the support or inside crystallites of metallic Pt (in accord with TEM data on the I-470 and I-720 catalysts there are Pt crystallites of 2-20 and more nm in size). It explains the inactivity of such "overreduced" catalysts in epoxidation and why the oxidized Pt is hardly developed in photoelectron spectra (3) and (4) in Fig. 1, recorded without ion etching the samples.) Perhaps, the role of metal chlorides is limited simply to supplying HCl, through their partial or complete reduction during catalytic runs, as it has been observed for the Pd complex as the second component (case 4 in Table 2).

It should be pointed out, however, that HCl reacts rapidly with c-hexene epoxide and promotes not only its formation but decomposition as well (Fig. 5). Obviously, it is responsible for an apparent induction period in c-hexene epoxide formation seen in Figs. 2,4 (note that it is absent in case (2) in Fig. 4) and a volcano-like plot of the dependence of epoxide yields vs. amount of HCl added (Fig. 3).

The unsuccessful attempts to activate the "overreduced" catalysts with HCl in other way than in situ show the surface chloride species to be of labile nature and exist in equilibrium with some chlorine-containing compounds in the reaction solution. Because HCl is rapidly consumed with epoxide, the role of such compounds during catalytic runs is played, presumably, by c-hexane chlorhydrine.

At last, it is not excluded that acetonitrile participates in the process not as a solvent only, and more detailed information on mechanisms of epoxide and by-products formation would be valuable for improving the rate and selectivity of epoxidation.

REFERENCES

1 H. Sakurai, Y. Hataya, T. Goromaru and H. Matsuura, J. Mol. Catal., 29 (1985) 153.
2 D. Mansuy, M. Fontecave and J.F. Bartoli, J. Chem. Soc. Chem. Commun. (1983) 253.
3 P. Battioni, J.F. Bartoli, P. Ledue, M. Fontecave and D. Mansuy, J. Chem. Soc. Chem. Commun. (1987) 791.
4 I. Tabushi and R. Morimitsu, J. Amer. Chem. Soc., 106 (1984) 6871.
5 I. Tabushi, M. Kodera and M. Yokoyama, J. Amer. Chem. Soc., 107 (1985) 4466.
6 Van Beijnum, A.I. van Dillen, I.W. Geus and W. Drenth, in: Yu.I. Yermakov and V.A. Likholobov (Eds.), Homogeneous and Heterogeneous Catalysis, Proc. 5th Intern. Symp. on Relat. between Homogeneous and Heterogeneous Catalysis, VNU Sci. Press, Utrecht, 1986, p. 293.
7 N.I. Kuznetsova, A.S. Lisitsyn and V.A. Likholobov, React. Kinet. Catal. Lett., 38 (1989) 205.
8 G.E. Moilenberg (Ed.), Handbook of X-Ray Photoelectron Spectroscopy, Perkin-Elmer, Minnesota, 1979.

J. KIWI (Federal Inst. of Techn., EPEL Chimie Phys., Switzerland): If you combine $O_2 + H_2$ (1:1) over Pt/SiO_2 catalyst as you have done, you have a high risk of explosion. To avoid this in which order do you mix, the catalyst, solvent and reactants gases with cyclohexene ?

N.I. KUZNETSOVA (Inst. of Catalysis, USSR) : To minimize the risk of explosion, the catalyst was always brought in contact with the O_2/H_2 mixture after it had already been covered with solvent (acetonitrile) and presence of dry catalyst on the reactor walls was thoroughly eliminated. All experiments (more than 100) have gone smoothly when we followed these precautions.

G. Centi and F. Trifiro' (Editors), *New Developments in Selective Oxidation* 97
1990 Elsevier Science Publishers B.V., Amsterdam — Printed in The Netherlands

SELECTIVE OXIDATION OF ALCOHOLS TO CARBONYL COMPOUNDS BY MANGANESE (III) μ-OXO CARBOXYLATO COMPLEXES

Hubert MIMOUN[1*], Lucien SAUSSINE[1], Stéphane MENAGE[2] and Jean-Jacques GIRERD[2]

[1]Institut Français du Pétrole, B.P. 311, 92506 Rueil Malmaison (France)

[2]Laboratoire de Chimie Inorganique, UA CNRS 420, Institut de Chimie moléculaire d'Orsay, Université Paris-Sud, Bat 420, 91405 Orsay (France)

Summary

Novel manganese (III) μ-oxo complexes with the formula $Mn_2^{III}O(C_3F_7CO_2)_4(bipy)_2$ (<u>1</u>) were synthesized from the oxidation of $Mn_2^{II}(C_3F_7CO_2)_2(bipy)_2$ (<u>2</u>) by potassium persulfate. The X-ray crystal structure of (<u>1</u>) revealed a $[Mn_2^{III}O(RCO_2)_2]^{2+}$ μ-oxo core, with one $C_3F_7CO_2$ group linked to each Mn(III) ion in a monodendate fashion. The X-ray structure of (<u>2</u>) indicated a cupric acetate type structure with four binucleating carboxylato groups. Compound (<u>1</u>) selectively oxidizes alcohols to carbonyl compounds, giving (<u>2</u>) in quantitative yields. Kinetic studies of this reaction, followed by visible spectroscopy in CH_2Cl_2 solution, suggested that the Mn^{III}-O-Mn^{III} core dissociates into Mn^{II} and active Mn^{IV}=O (or Mn^{III}-O°) species which react with alcohols in a way similar to highvalent metal oxo species.

Introduction

The active oxygenated species involved in catalytic oxidations have been to date almost exclusively limited to metal peroxides and highvalent metal oxo species (Scheme 1). [1]

Scheme 1

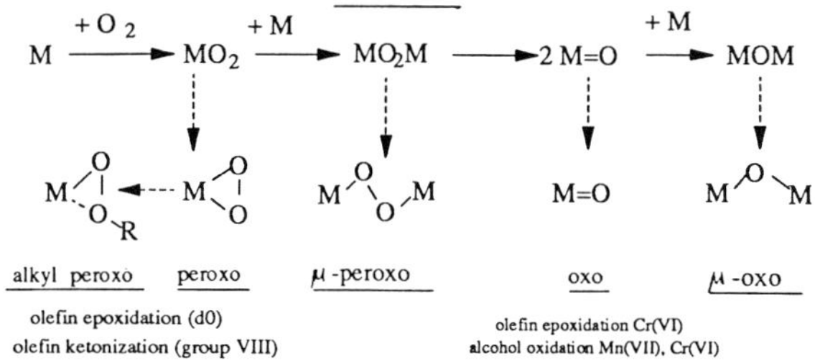

Reactive binuclear μ-oxo complexes are rare. To our knowledge, there only exists one reported example of allylic alcohol oxidation by μ-oxo-bis(chlorotriphenylbismuth(V) [2]. In fact, μ-oxo complexes are generally considered as dead-end species in various oxidative processes such as the catalytic oxidation by O_2 of phosphines in the presence of Mo(VI) *cis*-dioxo dialkyldithiocarbamates [3], and the catalytic epoxidation of olefins by O_2 in the presence of ruthenium tetramesityl porphyrins. [4]

Binuclear Mn(III) μ-oxo complexes bearing tridendate nitrogeneous bases, [5] or bipyridine (bipy) [6] as terminal ligands have recently been prepared, but their reactivity is not known.

We report here the synthesis and characterization of novel Mn(III) μ-oxo complex with the formula $Mn_2^{III}O(C_3F_7CO_2)_4(bipy)_2$ (<u>1</u>), and its reactivity with alcohols to give selectively $Mn_2^{II}(C_3F_7CO_2)_2(bipy)_2$ (<u>2</u>) and the corresponding carbonyl compound, according to scheme 2.

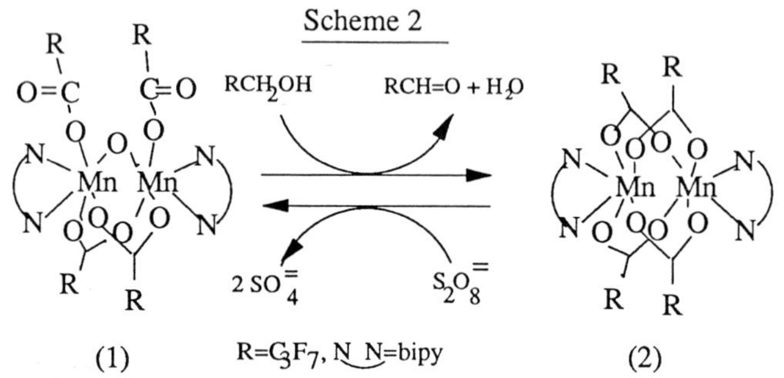

RESULTS AND DISCUSSION

<u>Oxidized Complex (1)</u>

This purple complex was prepared by precipitation from the reaction of manganese(II) sulfate with heptafluorobutyric acid, 2,2'-bipyridine, and potassium persulfate $K_2S_2O_8$ in water during 30 min at 85°C (see Experimental Section). Crystals were obtained by slow evaporation of a CH_2Cl_2 solution of (<u>1</u>).

The X-ray crystal structure of (<u>1</u>) consists of dimeric units (Fig 1) built by a diad axis: an oxygen atom lies on this axis and is shared by two manganese atoms (Mn-O1=1.75(3) Å, Mn-O1-Mn=130(4)°). The environment of each Mn atom is completed by 2 oxygen atoms of 2 bridging carboxylate groups, one oxygen atom of a monodendate carboxylate group and 2 nitrogen atoms of the bipy ligand. The coordination polyhedron is then an octahedron. Complex (<u>1</u>) contains the shorter Mn-O bond compared to other μ-oxo Mn(III) complexes (3), (4) and (5) having bridging acetates, and the larger Mn-O-Mn angle (see Table 1) . Details of the X-ray diffraction data and structure determination of (1) are described elsewhere.[7]

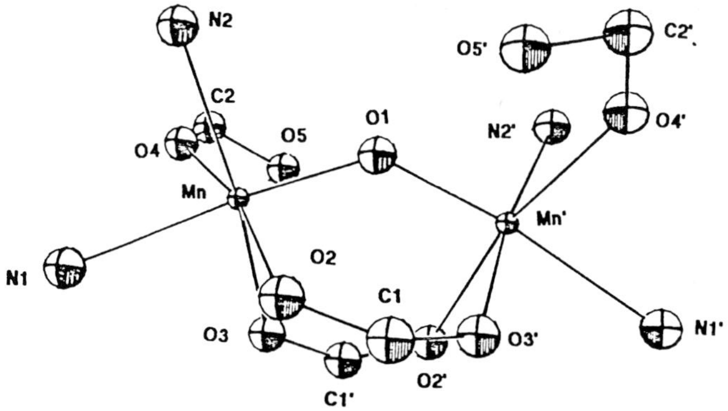

Figure 1. Structure of $Mn_2O(C_3F_7CO_2)_4(bipy)_2$ (1)

The infrared spectrum of (1) exhibits an intense absorption at 1700 cm^{-1} with a shoulder at 1670 cm^{-1}, which can be assigned to the asymmetric ν(C=O) vibrations of the bidendate and the monodendate heptafluorocarboxylate groups, respectively. A medium absorption at 730 cm^{-1} has been attributed to the ν_{as}(Mn-O-Mn) bridge vibration. Treatment of (1) with a CH_2Cl_2-H_2O^{18} mixture results in an isotopic exchange of the oxygen atom of the Mn-O-Mn bridgs, and the apparition of a new band at 690 cm^{-1}.

The UV-visible spectrum of (1) in acetone consists of a broad absorption at 710 nm (ε= 80), and three bands at 520 (ε=204), 500 (ε=210) and 480 nm (ε=220). The three absorptions around 500 nm are also found in the other μ-oxo Mn(III) complexes (3), (4), and (5) (Table 1), and confirm the μ-oxo μ-carboxylato structure of (1)

The magnetic susceptibility of (1) $\chi_M.T$ was found invariant at 5.86 cm^3mol^{-1} from 50 to 250 K, with no antiferromagnetic coupling between the two hign spin Mn(III) atoms. This value is close to that found for (5) (5.98 cm^3mol^{-1} at 300 K), for which no magnetic exchange was found (ref 4c), but contrasts with that of compounds (3) and (4) for which a an antiferromagnetic coupling was evidenced.

The NMR spectrum of (1) in deuterated acetone exhibits eight resonances at 56.5, 17.0, 1.2, -8.8, -28.2, -28.9, -79.5, and -103.0 ppm corresponding to the protons of bipyridine , shifted due to the S=2 electronic spin of Mn(III). The observation of the 8 resonances is in agreement with the C2 type structure found in the XRD structure, where one the nitrogen atoms of the bipy ligand is trans to the μ-oxo group.

Table 1. XRD (Distances and angles Mn-O-Mn), Infrared and UV-Visible properties, and magnetic susceptibilities of μ-oxo bis μ-carboxylato bridged dimanganese (III) species.

Compounds	(1)	(3)	(4)	(5)
d(Mn-O)/Å	1.75	1.783	1.81	1.780
Mn-O-Mn /°	130	122.9	120.9	125.1
Infrared ν(Mn-O-Mn)/cm^{-1}	730	-	730	712
UV-Vis $\lambda_{max}(\varepsilon)$/nm	480(220)[a] 500(210) 520(200) 710(80)	490(340)[a] 520(300) 535(280) 640(200)	486[b] 521	486(337)[c] 503(190) 524(175) 540(165) 582(95)
Magnetic susceptibility χ_M.T (287 K)/$cm^3mol^{-1}K$	5.86 not coupled	5.46 J=-6.8 cm^{-1}	6.69 J=+18 cm^{-1}	5.98 not coupled

Compound (3): $Mn_2O(OAc)_2(bipy)_2(PF_6)_2(H_2O)_{1.75}$ (ref 5); Compound (4): $[L_2Mn_2O(OAc)_2](ClO_4)_2.H_2O$ where L= N,N',N"-trimethyl-1,4,7-triazacyclononane (ref 4a); Compound (5): $Mn_2O(OAc)_2(HB(pz)_3)_2$ where $HB(pz)_3$= hydrotris(1-pyrazolyl)borate (ref 4c). [a]Solvent=acetone (this work and ref 5); [b]Solvent=CH_3CN (from ref 4b). [c]Solvent=CH_2Cl_2 (from ref 4c)

Reduced Complex (2)

This white complex was prepared from the reaction of manganese(II) carbonate with $C_3F_7CO_2H$ and bipy in water at 70°C for 1 hour, or as pure crystals from the reaction of (1) with ethanol (see Experimental Section).

The X-ray crystal structure of (2) consits of centrosymmetric dimeric units in which each manganese atom is bonded to four carboxylate oxygen atoms and two bipy nitrogen atoms in a skew trapezoidal bipyramidal configuration (see Fig.2). The four carboxylate groups are bidendate, bridging the two manganese atoms with a Mn-Mn distance of 3.679(3) Å. The four oxygen atoms arising from the carboxylate groups and adjoining the the manganese atoms are coplanar with maximum deviations from the least-squares plane of + 0.002(15) Å. The coordination of the Mn atom is completed by the two nitrogen atoms of the bipy molecule on the same side of the square constructed from the oxygen atoms of the carboxylate groups. The manganese atom lies at 0.737(2) Å of this plane and is directed towards the N atoms. The two pyridine rings of the bipy ligand are coplanar (maximum deviation from the best plane : + 0.05(2) Å) and almost perpendicular to the oxygen planes (87.8(4)°) with the N1-N2 line parallel to the O2-O3 and O1-O4 edges of the oxygen squares.

The tetra-μ-carboxylate bridged structure is novel for manganese, but rather common for other metal ions such as Cu(II), Cr(II), Rh(II), Mo(II) [8]. In (2), the Mn-Mn distance is particularly long (3.679 Å) compared to the other metal-metal bonds. Details of the X-ray diffraction data and structure determination of (2) are described in ref 7.

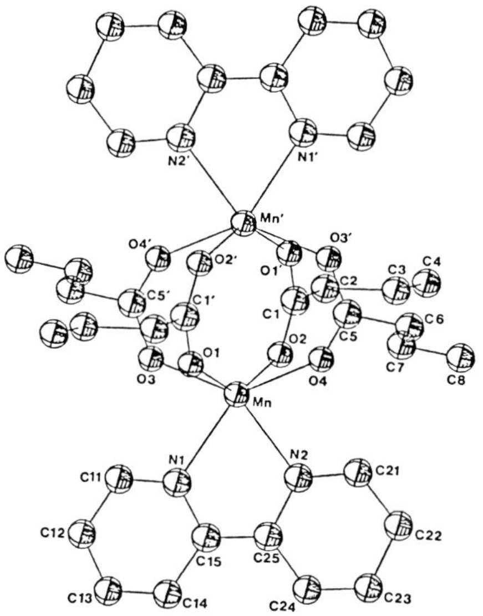

Figure 2. Structure of $Mn_2(C_3F_7CO_2)_4(bipy)_2$ (2)

Reactivity of (1)

Reaction with olefins.

Olefins such as cyclohexene, tetramethylethylene and isoprene were found unreactive towards (1) in dichloromethane and no oxidation products were detected. However, when the reaction of (1) with 1-pentene was carried out in acetone, an acetonylation reaction occurred, giving rise to the formation of 2-octanone in ca. 50% yield based on complex (1), according to equation (1).

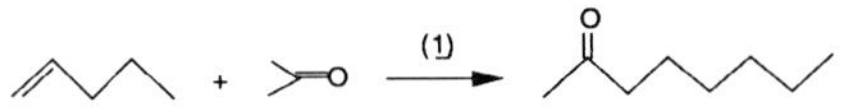

(1)

This homolytic addition of acetone to the double bond in the presence of Mn(III) species has already been previously described.[9]

Oxidation of alcohols to carbonyl compounds

Addition of alcohols such as ethanol to a solution of complex (1) in acetone or CH_2Cl_2 results in a progressive decoloration of the solution, and the almost quantitative formation of the corresponding carbonyl compound together with the reduced colorless complex (2), according to the stoichiometry of equation (2)

$$Mn_2O(C_3F_7CO_2)_4(bipy)_2 + RCH_2OH \longrightarrow Mn_2(C_3F_7CO_2)_4(bipy)_2 + RCHO + H_2O \quad (2)$$

Table 2 shows that primary alcohols are more reactive than secondary alcohols, and that amongst primary alcohols, benzylic and allylic alcohols are very easily oxidized by (**1**). This is probably due to the conjugative stabilisation between the developing carbonyl group and the olefin or the aromatic π-system. The same reactivity order of alcohols was previously observed in the reaction of chromic compounds and manganese dioxide.[10]

Table 2. Stoichiometric Oxidation of Alcohols by (1)[a]

Substrate	Product(s)[b]	Yield (%)[c]	Reaction time[d] (min)
C_2H_5OH	$CH_3CHO + CH_3CH(OEt)_2$[e]	100	340
C_6H_5-CH_2OH	C_6H_5-CHO	100	120
C_6H_5-CH=CH-CH_2OH	C_6H_5-CH=CH-CHO	95	10
CH_2=C(CH_3)-CH_2OH	CH_2=C(CH_3)-CHO	97	50
1-octanol	octanal	40	200
isopropanol	acetone	40	250
2-octanol	2-octanone	10	250

[a]*Reaction conditions:* Temperature= 30°C, solvent= CH_2Cl_2, alcohol concentration= 0.72 mol.l^{-1}, (**1**)= $1.22.10^{-3}$ mol.l^{-1}. [b]Products were identified by gc.ms coupling. [c]Yields are based on initial (**1**). [d]Reaction time for which the yield was obtained. [e]CH_3CHO=86%, acetal=14%.

Kinetic study of the reaction of (1) with EtOH in CH_2Cl_2.

The reaction was monitored by UV-Vis photometry at 485 nm in CH_2Cl_2 at 30°C. This wavelength corresponds to the maximum of absorption for (**1**), but to a weak absorption of (**2**). We used the formula :

$$[C] = \frac{(DO - \varepsilon_2[C]_0)}{(\varepsilon_1 - \varepsilon_2)} \quad (3)$$

where [C] is the concentration of binuclear species, DO the optical density of the solution, $[C]_0$ the initial concentration of (**1**), and ε_1 and ε_2 the extinction coefficients of (**1**) and (**2**), respectively.

Figure 3 shows the variation of concentration vs. time of (1). A striking result is obtained when the ethanol concentration is kept constant and when $[C]_0$ is varied (see Figure 4). The half reaction time increases when $[C]_0$ increases. This strongly suggests a dissociative step in the mechanism, as shown in equations 4-6 (X= $C_3F_7CO_2^-$, L=bipy).

$$xEtOH + Mn^{III}{}_2OX_4L_2 \underset{k'_{-1}}{\overset{k'_1}{\rightleftharpoons}} Mn^{II}X_2L(EtOH)_x + Mn^{IV}OX_2L \tag{4}$$

$$Mn^{IV}OX_2L + EtOH \xrightarrow{k'_2} CH_3CHO + Mn^{II}X_2L + H_2O \tag{5}$$

$$Mn^{II}X_2L(EtOH)_x + Mn^{II}X_2L \longrightarrow Mn^{II}{}_2X_4L_2 + xEtOH \tag{6}$$

If we assume that $Mn^{IV}O$ species react very quickly and are in a stationary concentration, and that the concentration of ethanol, used in large excess, remains constant, we can resolve the following simplified scheme:

$$A \underset{k_{-1}}{\overset{k_1}{\rightleftharpoons}} B + C \tag{7}$$

$$C \xrightarrow{k_2} D \tag{8}$$

where A stands for (1), B for the Mn(II) intermediate, C for the $Mn^{IV}O$ one, and D for the reduction product of C. The following kinetic expression has been derived for the relation : concentration of (**1**) versus time:

$$\left(\frac{1}{k_1}+\frac{[C]_0}{Kk_2}\right)\mathrm{Ln}\frac{[C]}{[C]_0}-\frac{[C]-[C]_0}{Kk_2}=-t \tag{9}$$

where $k_1 = k'_1[ROH]^x$, $k_{-1} = k'_{-1}$, $k_2 = k'_2[ROH]$ and $K = k_1/k'_1$
which gives an increasing half reaction time when $[C]_0$ increases. The theoretical curves are represented on figures 3 and 4 and are in good agreement with the experimental values. The corresponding parameters Kk_2 and k_1 are listed in Table 3.

The variation of k_1 and Kk_2 in function of the initial concentration of ethanol $[ROH]_0$ are represented in Figure 5 and give the following expressions :

$$k_1 = k'_1[ROH]_0^{2.5} \text{ and } Kk_2 = K'k'_2[ROH]_0^{3.5}$$

which are coherent with the value of x = 2.5. The origin of this non-integer value is not really known, but it might be due to the initial step involving the coordination of the alcohol to the Mn(III) μ-oxo complex (1), which was not included in our treatment. This might be responsible for the discrepancy existing between experimental and theoretical values at the beginning of the reaction.

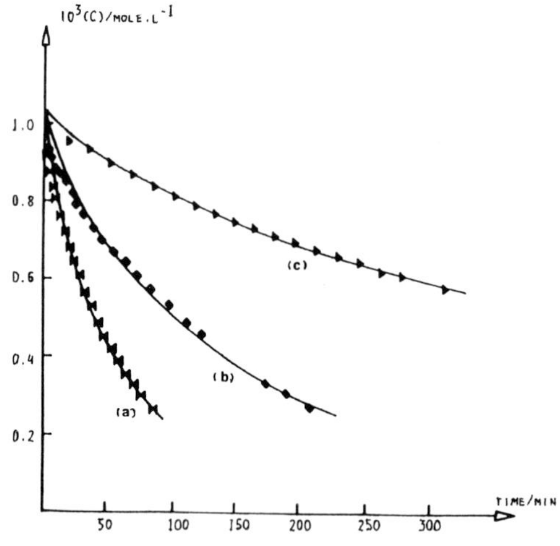

Fig.3. concentration vs. time dependence for the reaction of (**1**) with EtOH in CH_2Cl_2, followed by spectrometry at 485 nm and 30°C. The curves correspond to the best fit to the expression (9). $[C]_0$=1.04 M. $[EtOH]_0$= (a) 0.71 M; (b) 1.18 M; (c) 1.76M.

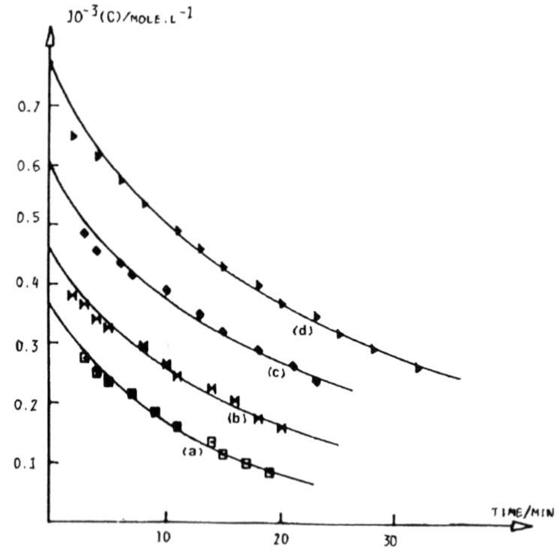

Fig.4. Concentration vs. time dependence for the reaction of (**1**) with EtOH in CH_2Cl_2 at 30°C. The curves correspond to the best fit to the expression (9). $[EtOH]_0$=2.17 M; $[C]_0$= (a) 0.37 mM; (b) 0.47 mM; (c) 0.61 mM; (d) 0.78 mM.

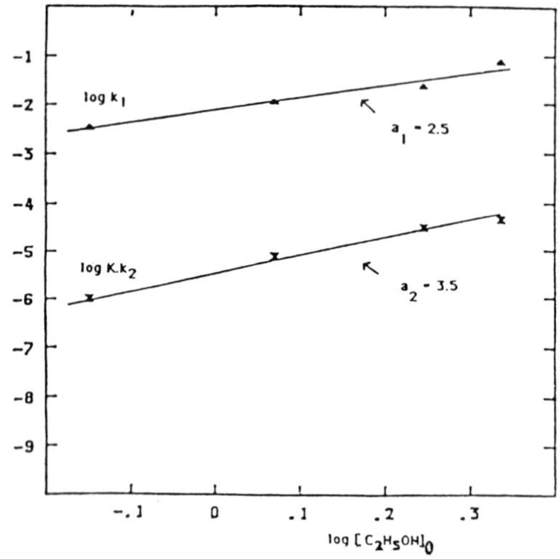

Figure 5. Variation of log k_1 and log Kk_2 in function of log $[EtOH]_0$. a_1 and a_2 refer to the slopes.

Table 3. Rate constants for the oxidation of ethanol by complex (1) at 30°C in CH_2Cl_2.

$[EtOH]_0$ $mol.l^{-1}$	$k_1.10^2$ min^{-1}	$Kk_2.10^5$ $mol.l^{-1}.min^{-1}$
2.17	7.4	2.0
1.76	2.2	2.95
1.18	1.0	7.35
0.71	0.35	0.1

Isotopic Effects.

Isotopic substitution of ethanol by deuterium slows down the oxidation by (**1**). A k_H/k_D isotopic ratio of 2.3 was obtained when C_2H_5OD is used instead of C_2H_5OH as substrate. This value is close to that observed in the oxidation of ethanol by acid permanganate (k_H/k_D=2.6).[11] This indicates that hydrogen abstraction by (**1**) from the hydroxyl group of the alcohol represents a rate-determining step in the reaction.

When C_2D_5OD was used, a different reaction occurred: a partial decoloration was first observed, but then the optical density of the solution raised again and no oxidation of the wholly deuterated alcohol took place.

CONCLUSION

The kinetic studies of the oxidation of alcohols by Mn(III) μ-oxo species strongly suggest that i) a dissociation of (**1**) into a reactive Mn^{IV}=O (or Mn^{III}-O°) and an unreactive Mn(II) complex occurs. ii) This dissociation is induced by the presence of alcohols. NMR studies of acetone solutions of (**1**) showed that the proton resonances are not affected by the presence of paramagnetic Mn(II) species. iii) The formation of Mn^{IV}=O species reactive towards alcohols accelerate the disproportionation.

Whether the active species is Mn^{IV}=O acting a two-electron heterolytic oxidant, or Mn^{III}-O° acting as an homolytic two consecutive one electron oxidant is not clear at the moment. Although the acetonylation of olefins suggest an homolytic mechanism, the reactivity order of alcohols and the isotopic effect are closer to the reactivity of MnO_2 or permanganates which are known to be heterolytic in nature. We therefore suggest the mechanism shown in Scheme 3, which involves the formation of an hydroxy-alkoxy Mn(IV) intermediate, which decomposes in a concerted two-electron transfer reaction to give the carbonyl compound; water and the reduced Mn(II) species.

Scheme 3

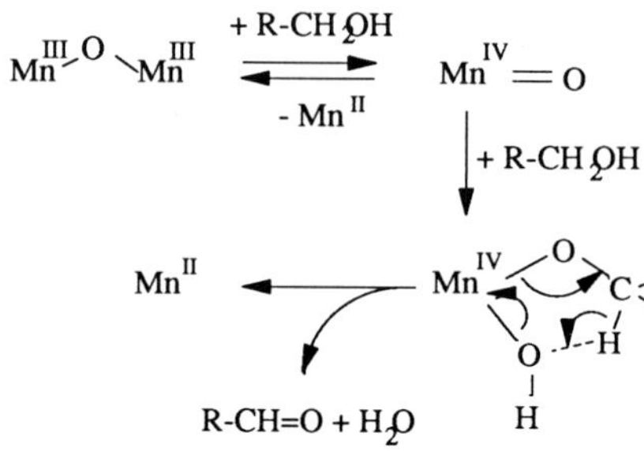

Experimental Section

Synthesis.

Synthesis of $Mn_2O(C_3F_7CO_2)_4(bipy)_2$ (1). To a solution of 1.1 g of $MnSO_4.H_2O$ (5 mmol) in water (25 ml) were added 1.3 ml (10 mmol) of $C_3F_7CO_2H$ (n-heptafluorobutyric acid) and 2.5 g of $K_2S_2O_8$. The mixture was heated for 3 min at 85°C, after which a dark brown color developed. 0.8 g of solid 2,2'-bipyridine were then added, and the resulting mixture was heated at 90°C during 30 min. During this time, a purple solid precipitated off, which was filtered, washed with distilled water and diethylether, and dried *in vacuo*. Yield = 1.36 g. *Anal.* Calcd for $C_{18}N_8N_2O_{4.5}F_{14}Mn$: C, 33.51; H, 1.25; N, 4.34; F, 41.23. Found: C, 33.61; H, 1.30; N, 4.20; F, 40.61. Crystals were obtained by slow evaporation of a dichloromethane solution of (1) under air.

Synthesis of $Mn_2(C_3F_7CO_2)_4(bipy)_2$ (2). *Method 1.* By refluxing (1) in ethanol, the reduced complex (2) was quantitatively obtained as a white cristalline powder. *Anal.* Calcd for $C_{18}H_8N_2O_4F_{14}Mn$: C, 33.91; H, 1.26; N, 4.40; F, 41.76. Found: C, 33.46; H, 1.25; N, 4.32, F, 39.74. *Method 2.* 1.2 g (10 mmol) of manganese (II) carbonate $MnCO_3.H_2O$ were dissolved in degassed water (25 ml); 2.6 ml (20 mmol) of $C_3F_7CO_2H$ were added to the solution, and the mixture was heated at 70°C for 15 min. After CO_2 gas has evolved, the brownish solution was filtered, and 1.57g of bipy were added to the filtrate. A yellow solution was obtained and heated for 1 hour at 70°C. Evaporation of water under vacuum gave complex (1) as a white solid.

Apparatus

UV-Visible spectra were recorded on a varian 2300 spectrometer. Infrared spectra were recorded on a Perkin-Elmer infrared spectrometer and NMR spectra on a Bruker AM 250 MHz instrument. Magnetic measurements in the 3-300 K temperature range were carried out with a Faraday type magnetometer equipped with a helium continuous-flow cryostat. $HgCo(NCS)_4$ was used as a susceptibility standard.

Reactivity experiments

The formation of carbonyl compounds was followed by gas chromatography (DEGS column 10% 4m) and by UV-Vis spectrometry of CH_2Cl_2 solution in thermostated cells. Alcohols and solvents were purified by standard procedures before use.

References

1. H. Mimoun in "*Comprehensive Coordination Chemistry*", 1987, **11**, 513. Pergamon Press, Oxford.

2. D.H.R. Barton, J.P. Kitchin, and W.P. Motherwell, *J. Chem. Soc. Chem. Commun.; 1978, 1099.*

3. R. Barral, C. Bocard, I. Seree de Roch, L. Sajus, *Tetrahedron Lett.*; 1972, 1693.

4. J.T. Groves and R. Quinn, *J. Am. Chem. Soc.;* 1985, **107,** 5790

5. a) K. Wieghardt, U. Bossec, D. Ventur and J. Weiss, *J. Chem. Soc. Chem. Commun.;* 1985, 347. b) K. Wieghardt, U. Bossec, B. Nuber, J. Weiss, J. Bonvoisin, M. Corbella, S.E. Vitols, J.J. Girerd, *J. Am. Chem. Soc.; in press.* c) J.E. Sheats, R.S. Czernuszewicz, R.S. Dismukes, A.L. Rheingold, V. Petrouleas, J. Stubbe, W.H. Armstrong, R.H. Beer, S.J. Lippard, *J. Am. Chem. Soc.;* 1987, **109**, 1435.

6. S. Ménage, J.J; Girerd, and A; Gleizes, *J. Chem. Soc. Chem. Commun.;* 1988, 431.

7. S. Ménage, S.E. Vitols, J.J. Girerd, C. Cartier, M. Verdaguer, H. Mimoun, L; Saussine, P. Charpin, M. Nierlich, and C. Merienne, *J. Am. Chem. Soc.; submitted for publication.*

8. R.C. Mehrotra and R. Bohra in "*Metal Carboxylates*" Academic Press, New York, 1983.

9. E.J. Corey, M.C. Kang, *J. Am. Chem. Soc.;* 1984, **106**, 5384. B.B. Snider, R. Mohan, S.A. Kates, *J. Org. Chem.;* 1985, **50**, 3659.

10. P. Muller in "*The chemistry of the functional groups, Ethers, Crown Ethers and Hydroxyl Groups*", 1980, 1, 469-538. S. Patai, ed. Wiley, Chichester (U.K) and references therein.

11. K.K. Banerji, *Bull. Chem. Soc. Japan,* 1973, 46, 3623.

Y. Moro-oka (Tokyo Inst. of Techn., Japan): I think your work is very important concerning understanding the mechanism of several kinds of monooxygenase, such as methane monooxygenase and tyrocynase.
(1) It is generally known that secondary alcohol is more easily oxidized than primary one. Do you have any explanation why primary alcohol is oxidized faster than secondary one by your complex ?
(2) If metaloxo complex is involved the reaction, do you have any idea why olefin is not oxidized to corresponding epoxide ?
H. Mimoun (IFP, Rue Malmaison, France): (1) There is a strong commitement of steric effects in this reaction. Further this reactivity is not unusual with respect to alcohol oxidation by MnO_2.
(2) Olefin epoxidation requires available coordination sites on the metal which are lacking here.

R.A. Sheldon (Andeno B.V., Venlo, The Netherland): What is the mechanicistic explanation for the dissociation of the μ-oxo complex (Mn(III)-O-Mn(III)) into Mn(II) and Mn(IV)=O in the presence of ethanol?
H. Mimoun: I believe that the oxidation of alcohol is the driving force for this dissociation.

G. Centi and F. Trifiro' (Editors), *New Developments in Selective Oxidation*
© 1990 Elsevier Science Publishers B.V., Amsterdam — Printed in The Netherlands

SELECTIVE OXIDATIONS CATALYZED BY DIOXO(PORPHYRINATO)RUTHENIUM(VI) SPECIES

NIMAL RAJAPAKSE, BRIAN R. JAMES* and DAVID DOLPHIN

Department of Chemistry, University of British Columbia, Vancouver, British Columbia, Canada V6T 1Y6

SUMMARY

The complexes *trans*-Ru(porp)$(O)_2$, where porp = the dianion of 5,10,15,20-tetramesitylporphyrin (TMP) or 5,10,15,20-tetra(2,6-dichlorophenyl)porphyrin (OCP), are readily formed in benzene by treatment of the Ru(II) bis(acetonitrile) precursors with O_2 or air. Such dioxo species in solution utilize both oxygen atoms for oxygenation of thioethers to sulfoxides, phenol to hydroquinone, and (as noted by other groups) olefins to epoxides; 2-propanol is also dehydrogenated to give acetone. Catalytic O_2-oxygenation has been demonstrated for the thioether and olefinic substrates.

INTRODUCTION

There remains extreme interest in selective, catalytic oxidation of organics, especially using O_2 or air, the cheapest oxidant (refs. 1-3). Some enzyme systems of the mono- and dioxygenase type, where one or both oxygen atoms of O_2, respectively, are incorporated into a substrate molecule, function via an iron-porphyrin centre. These monooxygenases, such as cytochrome P450, utilize reductive activation of O_2, where one O-atom is reduced to H_2O and the second O-atom becomes available within a high-valent metal oxo species ($O{=}Fe^{V}$(porp) $\leftrightarrow$ $O{=}Fe^{IV}$(porp$^{+\cdot}$)) for the oxygenation process; the dioxygenases are much less well-defined (refs. 1-8).

With the aim of mimicking the Fe-porphyrin systems and to learn more of the mechanisms, work on the second-row analogues (Ru-porphyrins) was initiated in this department (UBC) some 15 years ago, and this included both protein and non-protein work (refs. 5, 9-13). In attempts to attain higher oxidation state Ru-oxo species for emulating P450 systems, by using what has become a standard procedure, namely, addition of an O-atom donor to a suitable metalloporphyrin precursor, Groves and Quinn (ref. 14) and the UBC group (ref. 15) independently synthesized *trans*-Ru(TMP)$(O)_2$, 2, by reaction of meta-chloroperbenzoic acid with Ru(TMP)(CO). The presence of the ortho-methyl groups of the TMP ligand is advantageous in preventing sterically formation of μ-oxo species (Ru-O-Ru), which are inactive, thermodynamic sinks in Ru-oxygen chemistry (refs. 5, 15-17).

It soon became evident that **2** was readily formed *in situ* on exposing solutions of Ru(TMP)L_2 (L = THF, MeCN, or vacant) to air or O_2 (refs. 18-20), and that the system catalyzes stereospecific O_2-oxidation of olefins to epoxides (refs. 18, 21), oxidation of tertiary phosphines to the oxides (ref. 20) and thioethers to the sulfoxides (ref. 22). Both O-atoms of **2** are transferred to substrate, with formation of two mole-equivalents of monooxygenated substrate per mole of **2**; thus, in effect, **2** acts as a dioxygenase while the substrate undergoes monooxygenase conversion.

More generally, the use of porphyrin and nonporphyrin oxo complexes of metals in solution (with Ru dominating the more recent literature) for stoichiometric and catalytic oxidation reactions is extensively documented (refs. 23-30); the systems include O-atom transfer, and dehydrogenation with concomitant formation of H_2O, but the use of O_2 as the primary oxidant *in the absence of a sacrificial added reductant* is rare, and dioxygenase-type activity is thus far unique for the title *trans*-Ru(porp)$(O)_2$ complexes.

The present paper gives further details of our thioether oxidations reported recently (ref. 22), as well as results on the use of **2** for oxygen incorporation into phenol, and for conversion of 2-propanol to acetone; some data on another analogous, sterically hindered, octachloroporphyrin system, *trans*-Ru(OCP)$(O)_2$, are presented also.

EXPERIMENTAL METHODS

The carbonyl Ru(TMP)(CO) was prepared from $Ru_3(CO)_{12}$ and the free-base porphyrin H_2TMP (ref. 31) according to a literature procedure (ref. 32). The carbonyl was converted to the bis(acetonitrile) complex Ru(TMP)$(MeCN)_2$, **1**, by a standard photolysis method described elsewhere (refs. 9, 19) using a 450W Hanovia Hg vapour lamp; **1**, characterized previously by ^1H NMR and IR data (ref. 19), has now been subjected to X-ray crystallographic analysis, the structure (Fig. 1) being solved by conventional heavy-atom methods (ref. 33). Addition of O_2 to benzene or toluene solutions of **1** rapidly generates *in situ* solutions of *trans*-Ru(TMP)$(O)_2$, **2** (refs. 18, 19), and complete removal of solvent from such solutions gives quantitative yields of spectroscopically pure **2**. The corresponding *trans*-Ru(OCP)$(O)_2$, **3**, was synthesized by analogous routes via the $Ru_3(CO)_{12}$ and H_2OCP (ref. 34) precursors. The dioxo species are formed also by addition of meta-chloroperbenzoic acid to the carbonyl derivatives (refs. 14, 15).

The above synthetic reactions, and subsequent reactions of **2** and **3** with thioethers, phenol, and 2-propanol, are readily monitored by UV/vis and ^1H NMR spectroscopy, using specially designed optical cells (ref. 35) or NMR tubes that could be attached to a vacuum line, and/or fitted with serum caps via which the substrate could be added. Kinetic data were obtained using a thermostatted Perkin-Elmer 552A spectrophotometer, and ^1H NMR on C_6D_6 solutions using a Varian XL-300 instrument.

The Ru(porp)(O$\underline{S}$R$_2$)$_2$ complexes containing S-bonded sulfoxide ligands (R_2 = Me_2, Et_2, $^nPr_2$, $^nBu_2$, Me(n-decyl)), the final Ru(II) diamagnetic products formed from reaction of **2** or **3** with the R_2S thioethers, are isolated from this type of reaction (20°C, in C_6H_6 for 12h) as described previously for R_2=Et_2 (ref. 22), or by reaction of the Ru(porp)(MeCN)$_2$ species with the appropriate R_2SO sulfoxide in excess: the latter process is rapid in C_6H_6 at room temperature, and the bis(sulfoxide) either precipitates directly, or is obtained following chromatography on Activity II alumina using $CH_2C\ell_2$ as elutant following hexane to remove excess sulfoxide.

Reaction of **2** with excess phenol under O_2 at 20°C for 12h in C_6H_6 gives the bis(quinolato) complex Ru(TMP)(OC_6H_4OH)$_2$, **2c**, which is eluted with benzene as a brown band following chromatography of the reaction mixture on Activity I alumina.

All isolated complexes give satisfactory elemental analyses, and have been characterized especially using ^1H NMR and UV/vis spectroscopy (Table 1).

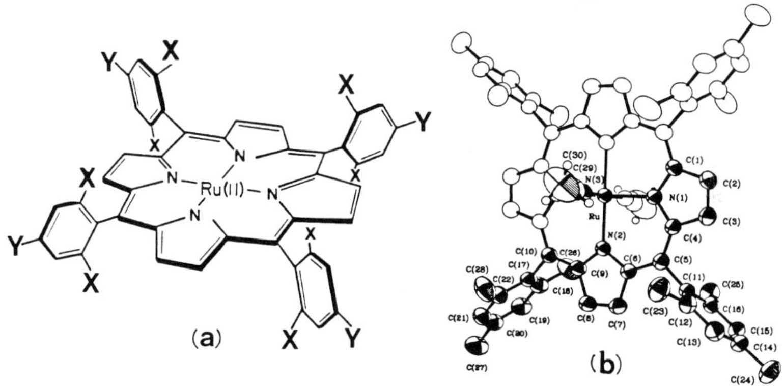

Fig. 1 (a) Diagram of Ru(porp) moiety; for TMP, X=Y=Me; for OCP, X=Cℓ, Y=H. (b) An ORTEP view of the Ru(TMP)(MeCN)$_2$ molecule. Some dimensions (A° or °) are: Ru-N(3) 2.027, Ru-N(2) 2.052, Ru-N(1) 2.051, N(3)-C(29) 1.126, C(29)-C(30) 1.457, Ru-N(3)-C(29) 172.5, N(3)-C(29)-C(30) 179.5, N(2)-Ru-N(3) 92.81, N(1)-Ru-N(3) 86.16.

RESULTS AND DISCUSSION

As noted in the Introduction, the presence of the ortho-substituents (Me or Cℓ) within the TMP and OCP systems (Fig. 1) appears to be critical for successful generation of easily handled *trans*-dioxo species such as **2** and 3 which have been shown to transfer both oxygens to substrates. In the presence of O_2, the systems become catalytic for the following processes: olefins → epoxides, PR_3 → OPR_3 and R_2S → R_2SO (refs. 18, 20-22).

Addition of thioethers to **2** in benzene, under Ar or O_2 from 10-30°C, results initially in production of the O-bonded Ru(TMP)($\underline{O}SR_2$), **2a**, via a process that is kinetically first-order in both Ru and R_2S (ref. 22); clean isosbestic points are observed in the UV/vis, and detailed ^1H NMR studies show unambiguously that **2a** is the product, eqn. (1) ($\underline{Ru}$ = Ru(TMP), ref. 22):

$$\mathbf{2}\ ,\ \underline{Ru}(O)_2 \xrightarrow[k]{SR_2} [\underline{Ru}(O)(\underline{O}SR_2)_2] \xrightarrow[fast]{SR_2} \underline{Ru}(\underline{O}SR_2)_2,\quad \mathbf{2a} \qquad (1)$$

Species 2a subsequently converts slowly to the mixed $\underline{Ru}(\underline{O}SR_2)(O\underline{S}R_2)$ species and then to the bis(S-bonded) derivative $\underline{Ru}(O\underline{S}R_2)_2$, (ref. 22); these latter species (R_2 = dialkyl), which have been isolated (see Experimental), are substitution-inert and are unreactive toward O_2. An observed, limited catalytic O_2-oxidation of Et_2S to the sulfoxide using **2** at 6 x 10^{-3}M (a maximum of 15 turn-overs over 15h at 65°C) is thus pictured as occurring via **2a** with O_2 replacing the more labile O-bonded sulfoxides (ref. 36) to regenerate **2**. At the end of this catalysis, the TMP ligand has clearly undergone degradation (perhaps via reactivity with a Ru-(di)oxo species) as indicated by loss of the Soret absorption maximum in the 410-420nm region and the ^1H NMR signal for the pyrrole protons.

Species **3**, containing the oxidant-resistant octachloroporphyrin (ref. 34), is a much more effective catalyst for the thioether oxidation; at 2 x 10^{-3} M in C_6D_6, **3** under O_2 (~1 atm at 20°C) in a sealed NMR tube effects close to complete conversion of 0.035M Et_2S to the sulfoxide and sulfone Et_2SO_2 (4:1 mixture) in a few hours at 100°C. There is no evidence for decomposition of the Ru(OCP) moiety which at the end of the catalysis is present as two Ru(OCP)$(OSEt_2)_2$ species (Table 1): further ^1H NMR selective decoupling experiments are required to elucidate whether the sulfoxide ligands are O- and/or S-bonded. Conditions are being sought for the selective O_2-oxidation of thioethers to the corresponding sulfoxide, this being a reaction of industrial importance (refs. 25, 37).

Some k^{20} values (eqn. (1)) together with the corresponding activation parameters are given in Table 2. The rates increase with increasing alkyl chain length within R_2S, the differences *perhaps* being reflected more in differences in $\Delta S^{\ddagger}$ than in $\Delta H^{\ddagger}$; the O-atom transfer, if induced by strong ν(Ru=O) vibrational motion (ref. 25), might be more efficient on encountering a bulkier substrate and this would be reflected in relatively higher $\Delta S^{\ddagger}$ values, although as expected (and seen) the coupling reaction is entropically unfavourable. Corresponding data for O-atom transfer from $[Ru(bipy)_2(py)O]^{2+}$ to Me_2S are $\Delta H^{\ddagger}$ ~34 kJ mol^{-1} and $\Delta S^{\ddagger}$ ~-110 $JK^{-1}mol^{-1}$ (ref. 25). Of interest, diphenylsulfide and methyl $\underline{p}$-tolylsulfide do not react with benzene solutions of **2** at 20°C, implying that π-acceptor aromatic groups on the thioether impede O-atom transfer via electronic effects. The selectivity for oxidation of dialkylsulfides contrasts with that found for an FeCℓ(TPP)/PhIO system that utilizes the iodosobenzene as the O-donor and effects catalytic formation of sulfoxides from dialkyl-, alkylaryl- and diarylsulfides (ref. 38); the intermediate proposed was "CℓFe(TPP)O". Consideration of this, and our data, which show that $(Et_2\underline{S}O)$Ru(TMP)O is a more effective O-atom donor than **2** (eqn. (1)), and those of Groves and Ahn (ref. 20), which show that **2** is a more potent oxo-transfer agent than 5-coordinate Ru(TMP)O (for oxidation of PPh_3), demonstrates the likely critical role of the ligand trans to the oxo ligand, a key factor in biologically important oxoiron(IV) porphyrin systems (refs. 4-6).

TABLE 1
^{1}H NMR data[a] (and some UV/vis absorption maxima[b]) for selected Ru(TMP)(=$\underline{Ru}$) and Ru(OCP)(=Ru') species.

	Complex	$H_{pyrrole}$	$\underline{m}$-H[c]	$\underline{p}$-Me	$\underline{o}$-Me[d]
	$\underline{Ru}(O\underline{S}{}^nPr_2)_2$[e]	8.60	7.23	2.50	2.10
	$\underline{Ru}(O\underline{S}{}^nBu_2)_2$[f]	8.64	7.23	2.50	2.22
2b,	$\underline{Ru}$(HO–⟨O⟩–OH)$_2$[g]	8.64	7.25	1.70	2.50
2c,	$\underline{Ru}$(O–⟨O⟩–OH)$_2$[h]	-30.45	7.63	3.00	2.90
	$\underline{Ru}(OH)_2$[i] ?	-12.50	7.52	2.85	2.90
				$\underline{p}$-H	
	Ru'$(MeCN)_2$[j]	8.54	7.42	6.90	
3,	Ru'$(O)_2$[k]	8.90	7.85	7.75	
	Ru'$(OSEt_2)_2$[ℓ]	8.66, 8.48	7.44m, 7.36d	7.24m, 6.92t	

[a] δ ppm from TMS in C_6D_6, unless stated otherwise; NMR data for $\underline{Ru}(MeCN)_2$, $\underline{Ru}(O)_2$ and the three $\underline{Ru}(OSEt_2)_2$ species are given in refs. 19, 22. All resonances integrate correctly and are singlets unless stated otherwise.
[b] In C_6H_6; UV/vis data for $\underline{Ru}(O)_2$ (refs. 18, 22) and $\underline{Ru}(O\underline{S}Et_2)_2$ (ref. 22) have been reported.
[c] Meta-proton; single peak shows presence of a porphyrin mirror plane.
[d] Ortho-methyl; single peak shows presence of a porphyrin mirror plane.
[e] In toluene-d_8. For axial ligands: δ -0.20 (t, CH_3), -0.43 and -0.70 (m, β-CH_2), -1.62 and -1.97 (m, α-CH_2).
[f] For axial ligands: δ 0.18 (m, CH_2CH_3), -0.40 and 0.60 (m, β-CH_2), -1.56 and -1.78 (m, α-CH_2).
[g] For axial ligands: δ 5.85 ($\underline{m}$-H), 5.76 ($\underline{o}$-H).
[h] In toluene-d_8. For axial ligands: δ 49.68, -68.19, -71.85 (unassigned as yet).
[i] For axial ligands: δ -0.35 (OH?).
[j] λ max at 408, 507nm. δ -1.44 (MeCN).
[k] λ max at 420, 510nm.
[ℓ] Mixture of 2 isomers with $\underline{O}$- and/or $\underline{S}$ bonded Et_2SO.

TABLE 2

Kinetic data for O-atom transfer from $Ru(TMP)(O)_2$ to alkyl thioethers, eqn. (1).

	Me(n-decyl)S	$^{n}Bu_2S$	Et_2S
$k^{20°}$, $M^{-1}s^{-1}$	0.11	0.012	0.0075
$\Delta H^{\ddagger}$, kJ mol^{-1}	56.5 ± 1.7	47.4 ± 7.1	58.3 ± 2.7
$\Delta S^{\ddagger}$, JK^{-1} mol^{-1}	-70 ± 6	-120 ± 30	-86 ± 9

The reaction of **3** with Et_2S (cf. eqn. (1)) is again cleanly first-order in **3** and in Et_2S, with k = 0.072 $M^{-1}s^{-1}$ at 20°C. This value is about 10 times that for the TMP system (Table 2), and shows that the electron-withdrawing chlorine substituents favour at least the first O-atom transfer, presumably by increasing the electrophilicity of the coordinated oxo ligands(s).

Preliminary data suggest that benzene solutions of $Ru(TMP)(O)_2$, **2**, react with phenol under O_2 according to the steps and stoichiometry shown in eqn. (2) (Ru = Ru(TMP)):

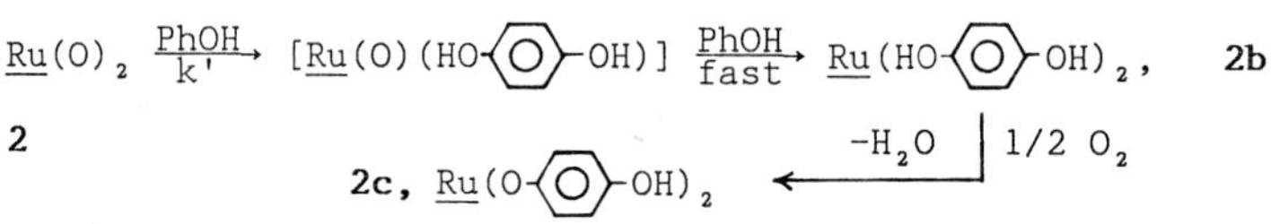

(2)

Complex **2c** has been isolated; a solution magnetic moment (ref. 39) of ~3.0μ_B is consistent with the Ru(IV) formulation with S=1, exactly analogous to dihalogenoruthenium(IV) porphyrin complexes; the ^1H NMR chemical shift data (Table 1), which vary almost linearly with inverse temperature, are also typical of such paramagnetic species (refs. 11, 40-42).

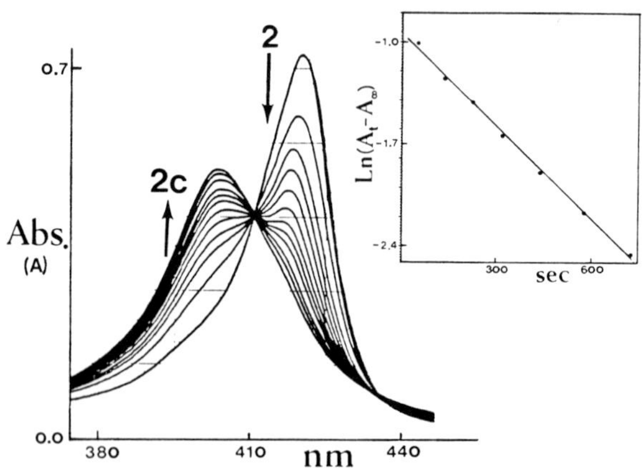

Fig. 2 UV/vis spectral changes observed in C_6H_6 at 20°C for reaction of **2** with phenol to give **2c**, eqn. (2); [Ru] = 2.5 x 10^{-6}M, [phenol] = 3.72 x 10^{-2}M. Inset shows pseudo-first order plot for the disappearance of 2 (A_t and A_∞ are absorbances at 420nm at times t and ∞, respectively).

Monitoring the reaction by UV/vis reveals a clean conversion of **2** → **2c** (Fig. 2), with a rate given by k'[2][PhOH]; k'(= 0.069 $M^{-1}s^{-1}$ at 20°C) is assigned to the step shown in eqn (2), which corresponds to that invoked by Meyer's group for attack of phenol on $[Ru(bipy)_2(py)O]^{2+}$ (ref. 26). The diamagnetic, bis(hydroquinone) Ru(II) intermediate **2b** is detected by ^1H NMR (Table 1), when the reaction of **2** with phenol is carried out under O_2-free conditions.

Conditions for effective catalytic hydroxylation of phenol to give hydroquinone using **2** under O_2 have yet to be realized, but can be envisioned to occur via **2b** if its axial ligands, the hydroquinone product, can be removed and isolated from the oxidizing medium (*cf.* catalytic oxidation of the thioethers via **2a**, eqn. (1)).

Benzene solutions of 2 under O_2 react with 2-propanol and UV/vis changes akin to those shown in Fig. 2 are observed at corresponding conditions, although the rates are ~500 times slower. ^1H NMR data for *in situ* reactions show that as **2** is consumed a paramagnetic Ru porphyrin product and acetone, as well as H_2O, are generated; coordinated propoxide ligands are not observed. The preliminary data suggest a non-catalytic conversion of 2-propanol to acetone and water, with concomitant loss of the Ru-dioxo complex, perhaps to give $Ru(TMP)(OH)_2$ (Table 1). Further studies are in progress to isolate and characterize the Ru product.

SUMMARY

In summary, dioxo(porphyrinato)ruthenium(VI) species are capable of transferring O-atoms to, or abstracting H-atoms from, several diverse types of substrates; as O_2 is the O-atom source, the systems represent a major advance in O_2-oxidation chemistry and offer an excellent opportunity for detailed mechanistic insight into oxidations of biological and industrial importance. The scene is comparable to that of catalytic hydrogenation in the early 1960's, when many transition metal hydrides were being synthesized using H_2, and their catalytic properties were being discovered (ref. 43).

ACKNOWLEDGEMENT

We thank the Natural Sciences and Engineering Research Council of Canada (B.R.J.) and the U.S. National Institute of Health (Grant AM17989 to D.D.) for financial support, and Johnson Matthey Ltd. for the loan of Ru.

REFERENCES

1 Report of the International Workshop on Activation of Dioxygen Species and Homogeneous Catalytic Oxidations, T.J. Collins (Ed.), Galzignano, Italy, 1984.

2 L.L. Ingraham and D.L. Meyer, Biochemistry of Dioxgyen, Plenum, New York, 1985.

3 Oxygen Complexes and Oxygen Activation by Transition Metals, A.E. Martell and D.T. Sawyer (Eds.), Plenum, New York, 1988.
4 J.T. Groves in P. Ortiz de Montellano (Ed.), Cytochrome P-450: Structure, Mechanism, and Biochemistry, Plenum, New York, 1986, Ch. 1.
5 B.R. James, in A.E. Shilov (Ed.), Fundamental Research in Homogeneous Catalysis Vol. 5, Gordon and Breach, New York, 1986, p. 309.
6 D. Mansuy, Pure Appl. Chem., 59 (1987) 759.
7 I. Tabushi, Coord. Chem. Rev., 86 (1988) 1.
8 T.C. Bruice, Aldrichimica Acta, 21 (1988) 87.
9 M.J. Camenzind, B.R. James, D. Dolphin, J.W. Sparapany and J.A. Ibers, Inorg. Chem., 27 (1988) 3054.
10 B.R. James, A. Pacheco, S.J. Rettig, I.S. Thorburn, R.G. Ball and J.A. Ibers, J. Mol. Catal., 41 (1987) 147.
11 M. Ke, S.J. Rettig, B.R. James and D. Dolphin, J. Chem. Soc. Chem. Commun., 1987, 1110.
12 D.R. Paulson, A.W. Addison, D. Dolphin and B.R. James, J. Biol. Chem., 254 (1979) 7002.
13 B.R. James, A.W. Addison, M. Cairns, D. Dolphin, N.P. Farrell, D.R. Paulson and S. Walker, in M. Tsutsui (Ed.), Fundamental Research in Homogeneous Catalysis Vol. 3, Plenum, New York, 1979, p. 751.
14 J.T. Groves and R. Quinn, Inorg. Chem., 23 (1984) 3844.
15 M.J. Camenzind, B.R. James and D. Dolphin, unpublished results, Oct. 1984.
16 J.P. Collman, C.E. Barnes, P.J. Brothers, T.J. Collins, T. Ozawa, J.C. Gallucci and J.A. Ibers, J. Am. Chem. Soc., 106 (1984) 5151.
17 H. Masuda, T. Taga, K. Osaki, H. Sugimoto, M. Mori and H. Ogoshi, J. Am. Chem. Soc., 103 (1981) 2199.
18 J.T. Groves and R. Quinn, J. Am. Chem. Soc., 107 (1985) 5790.
19 M.J. Camenzind, B.R. James and D. Dolphin, J. Chem. Soc. Chem. Commun., 1986, 1137.
20 J.T. Groves and K-H. Ahn, Inorg. Chem., 26 (1987) 3833.
21 J-C. Marchon and R. Ramasseul, J. Chem. Soc. Chem. Commun., 1988, 298.
22 N. Rajapakse, B.R. James and D. Dolphin, Catalysis Letters, 2 (1989) 219.
23 R.A. Sheldon and J.K. Kochi, Metal-Catalyzed Oxidations of Organic Compounds, Academic, New York, 1981, Chs. 3,4,6,8,9,12.
24 W.P. Griffith, S.V. Ley, G.P. Whitcombe and A.D. White, J. Chem. Soc. Chem. Commun., 1987, 1625.
25 L. Roecker, J.C. Dobson, W.J. Vining and T.J. Meyer, Inorg. Chem., 26 (1987) 779.
26 W.K. Seok, J.C. Dobson and T.J. Meyer, Inorg. Chem., 27 (1988) 3.
27 J.C. Dobson, J.H. Helms, P. Doppelt, B.P. Sullivan, W.E. Hatfield and T.J. Meyer, Inorg. Chem., 28 (1989) 2200.
28 G. Parkin and J.E. Bercaw, J. Am. Chem. Soc., 111 (1989) 391.
29 C-M. Che and C.K. Poon, Pure Appl. Chem., 60 (1988) 495.
30 M.M.T. Khan, H.C. Bajaj, R.S. Shukla, and S. Mirza, J. Mol. Catal., 45 (1988) 51.
31 J.T. Groves and T.E. Nemo, J. Am. Chem. Soc., 105 (1983) 6243.
32 D.P. Rillema, J.K. Nagle, L.F. Barringer and T.J. Meyer, J. Am. Chem. Soc., 103 (1981) 56.
33 M.J. Camenzind, S.J. Rettig, B.R. James and D. Dolphin, in preparation.
34 P.S. Traylor, D. Dolphin and T.G. Traylor, J. Chem. Soc. Chem. Commun., 1984, 279.
35 D.V. Stynes and B.R. James, J. Am. Chem. Soc., 96 (1974) 2733.
36 J.A. Davies, Adv. Inorg. Chem. Radiochem., 24 (1981) 115.
37 D.P. Riley, M.R. Smith and P.E. Correa, J. Am. Chem. Soc., 110 (1988) 177.
38 W. Ando, R. Tajima and T. Takata, Tetrahedron Lett., 23 (1982) 1685.
39 D.F. Evans, J. Chem. Soc. 1959, 2003.
40 C. Sishta, M. Ke, B.R. James and D. Dolphin, J. Chem. Soc. Chem. Commun., 1986, 787.
41 M. Ke, Ph.D. Dissertation, Univ. of B.C., 1988.
42 K. Rachlewicz and L. Latos-Grazynski, Inorg. Chim. Acta, 144 (1988) 213.
43 B.R. James in G. Wilkinson, F.G.A. Stone and E.W. Abel (Eds.), Comprehensive Organometallic Chemistry, Vol. 8, Pergamon, Oxford, Ch. 51, 1982.

U. SCHUCHARDT (University of Estadual de Campinas, Brasil): Have you tried to oxidize saturated hydrocarbons with your systems and could you say something about the turnover numbers?

B.R. JAMES (University of British Columbia, Canada): Reports on reactivity of the Ru(porp)$(O)_2$ species toward saturated hydrocarbons have not appeared. We find that benzene or toluene solutions of Ru(TMP)$(O)_2$ and Ru(OCP)$(O)_2$ are stable for long periods at room temperature, showing that the aromatic and activated methyl C-H bonds are not hydroxylated at the ambient conditions. At > 60°C, the Ru(TMP)$(O)_2$ solutions (but not those of Ru(OCP)$(O)_2$) are bleached, showing destruction of the TMP ligand, presumably due to oxidation by a Ru=O moiety. We plan to test a substrate such as cyclooctane under mild conditions (1 atm O_2 is sufficient to form the dioxo species), but are not hopeful for postive results.

F. MONTANARI (University of Milan, Italy): Sulphides are oxidized at room temperature. What are the conditions for the oxidation of alkenes catalyzed by the Ru-octachloroporphyrin dioxo species?

B.R. JAMES (University of British Columbia, Canada): We have not studied alkene oxidation by Ru(OCP)$(O)_2$. With the Ru(TMP)$(O)_2$ species, catalytic olefin epoxidation occurs at ambient conditions with 1 atm O_2; turnover numbers are in the range of 15-45 per day, depending on the olefinic substrate (see refs. 18, 21 in our paper). Based on commonly observed increased activity for halogenated porphyrins (e.g. ref. 1), somewhat higher activity is predicted for the octachloro system.

1 P.E. Ellis Jr., J.E. Lyons, J. Chem. Soc. Chem. Commun., 1989, 1189.

J.M. BREGEAULT (University of P. and M. Curie, France): How is the Ru$(O)_2$ species formed from the bis(acetonitrile) complex and O_2?

B.R. JAMES (University of British Columbia, Canada): We find that the reaction of Ru(TMP)$(MeCN)_2$ with 1 atm O_2 in toluene is fast, but probably amenable to study by the stopped-flow technique. In an NMR-tube, the reaction is possibly diffusion controlled, because Groves and Ahn (ref.20) were able to detect, via NMR studies (stated to be in C_6D_6, although the spectral data shown appear to be in CD_2Cl_2), Ru^{IV}(TMP)O enroute to the dioxo species, and they suggested formation of the latter by disproportionation of the Ru^{IV}=O intermediate. Thus, a plausible route for the reaction is as follows ($\underline{Ru}$ = Ru(TMP), MeCN and possible toluene ligands omitted):

$$\underline{Ru}^{II} \xrightarrow{O_2} [\underline{Ru}O_2] \xrightarrow{\underline{Ru}^{II}} \underline{Ru}^{III}O_2\underline{Ru}^{III} \longrightarrow 2\underline{Ru}^{IV}O \longrightarrow \underline{Ru}^{VI}(O)_2 + \underline{Ru}^{II}$$

It should be noted that we have isolated the 4-coordinate Ru(TMP) complex (ref. 19).

G. Centi and F. Trifiro' (Editors), *New Developments in Selective Oxidation* 119

RUTHENIUM (II) CATALYSTS FOR THE HOMOGENEOUS OXYGENATION OF ALIPHATIC HYDROCARBONS AND ETHERS.

M. BRESSAN and A. MORVILLO

Dipartimento di Chimica Inorganica and Centro C.N.R., Universita' di Padova, v. Marzolo 1, 35131 Padova (Italy)

ABSTRACT
Hydroxylation or ketonization of alkanes and selective conversion of ethers to esters or β-keto-ethers are achieved with hypochlorite and a choice of ruthenium(II) complexes as catalysts, in a biphasic water-dichloromethane system.

INTRODUCTION

There is a constant need to develop new mild and selective systems for the preparation of oxygenated compounds by the direct oxidation of hydrocarbons and various transition metal complexes have been tested as catalysts in conjunction with a number of oxidants (ref. 1). Ruthenium tetroxide was early introduced as useful stoichiometric reagent and ever since only scattered reports have appeared, dealing with ruthenium-catalyzed oxidations of aliphatic ethers (ref. 2) and saturated hydrocarbons (ref.3). We previously reported that a variety of ruthenium(II) complexes mediate the transfer of oxygen from hypochlorite and other single-oxygen donors to alkenes (ref.4a,b) and other organic substrates (ref.4c). In this study we describe their catalytic behavior in the oxygenation of alkanes and aliphatic ethers. In the presence of ruthenium(II) complexes, hypochlorite, inexpensive and easy to handle, shows high reactivity towards ethers (up to 2 turnovers per minute), thus making this procedure a possible synthetic method. The complexes examined are $[RuCl(DPP)_2]PF_6$, trans-$[RuCl_2(DPP)_2]$ (ref. 5a) {DPP=1,3-bis(diphenylphosphino) propane} and cis-$[RuCl_2(Me_2SO)_4]$ (ref.6a). $[OsCl(DPP)_2]PF_6$ (ref.5b) and cis-

$[RuCl_2(phen)_2]$ (ref.6b) (phen=1,10-phenanthroline) were also tested, but gave negligible yields.

RESULTS AND DISCUSSION

In a typical experiment, the substrate ($1x10^{-3}$ mol) and the catalyst ($2x10^{-6}$ mol) were dissolved in CH_2Cl_2 (1 mL) and stirred vigorously with 1 mL of aqueous LiClO (1 - 1.4 M, as determined by iodometric titration). Aliquots of the organic layer were periodically analyzed by glc, and the products were identified by mass spectrometry. The use of dilute solutions of LiClO (down to 0.2 M) results in a proportional decrease in the oxidation rate.

The ruthenium-complexes examined here promote the oxidation of saturated cyclic hydrocarbons (up to ca 3 turnovers per hour were achieved; see Table 1): each yield increases rapidly in the initial stage and shows a tendency to be saturated after more than 2 days reaction. Tertiary CH groups give the corresponding tertiary alcohols, whereas methylene groups are mainly converted to ketones, with minor amounts of secondary alcohols being detected. Since secondary alcohols are very effectively converted into ketones by the present catalytic system (ref.4c), it is reasonably to postulate a two-stage oxidation of CH_2 groups: first, and relatively slow, to alcohol and then, much faster, to ketone (eq.1).

$$-CH_2- \rightarrow -CH(OH)- \rightarrow -CO- + H_2O \qquad (1)$$

Oxidation rate ratios for cyclohexane/cyclohexane-d_{12} are in the 5÷6 range, in agreement with a significant C-H bond breaking in the transition state. The large ratio of oxidation of the tertiary position to the secondary, in adamantane and methylcyclohexane, also implies that the catalytic reaction follows a free-radical path, where oxidation happens in preference on the hydrogen carried by tertiary carbons. Also the methylene and methine groups of the alkyl-chains of alkyl-aromatics (ethylbenzene, cumene) are effectively transformed into ketones and ter-alcohols respectively. Aliphatic linear alkanes undergo selective oxidation at the secondary positions only, following the ω-1 rule (ref.7). In all examined cases little or no oxidation of methyl groups is observed.

Ethers, both cyclic and linear, were selectively (>99%) converted to ω-lactones and esters, respectively, and these latters were no further oxidized to anhydrides, thus indicating that the methylene group adjacent to the alkyl oxygen in esters, unlike the α-methylene groups of ethers, is strongly deactivated.

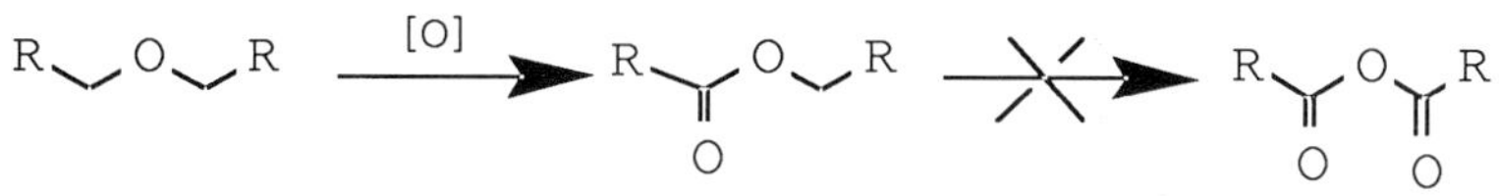

The most striking result consists in the fact that not only the α-methyl groups (see methyl-n-butyl-ether), but also the tertiary α-carbon atom in ethers (see 2,5-dimethyl- and 2-methyl-tetrahydrofuran), contrary to the secondary ones, remain unaffected by the oxidation. This strongly suggests that the reaction does not proceed by a simple hydrido or hydrogen atom transfer involving the α-carbon atom only. Indeed, in those cases, where α-methine and β-methylene groups are present, as in 2,5-dimethyl- and 2-methyl-tetrahydrofuran, unexpected β-ketonization occurs. Nevertheless, the sizeable kinetic isotope effect, resulting from complete deuteration of the substrates (for tetrahydrofuran, k_H/k_D = 5÷6), points to carbocationic or carbon radical intermediates.

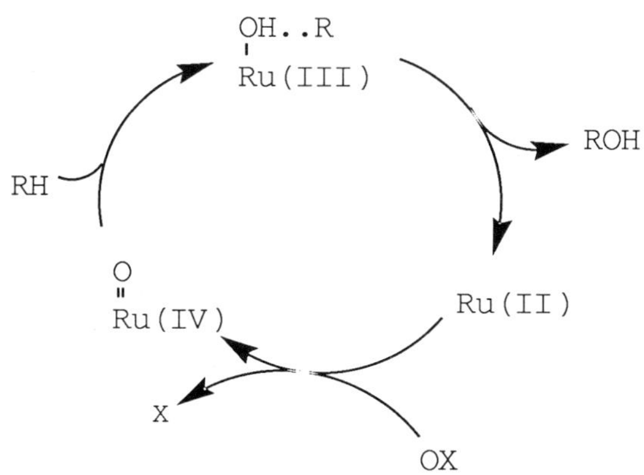

In the proposed mechanistic pathway, ruthenium(IV)-oxo intermediates are responsible for the oxygen transfer: a well-characterized oxo-derivative of ruthenium has been reported to convert tetrahydrofuran into γ-butyrolactone (ref.8) and we previously reported positive evidence of oxo-species during the PhIO-epoxidation of alkenes by $[RuCl(DPP)_2]PF_6$ (ref.4b). The oxidation of ethers might be somewhat different from that commonly proposed for the oxidation of the alkanes, although both reactions appear to be radical in nature, as shown by the sizeable deuterium kinetic effects observed for cyclohexane and tetrahydrofuran oxidation.

A first distinctive feature is the anomalous reactivity order of the α-carbon in ethers, i.e. sec >> tert, prim ≈ 0, whereas the tert > sec > prim reactivity pattern is conventionally associated to non-activated alkanes. A simple explanation, accounting the lack of reactivity of tertiary ethers to the presence of the heteroatom: however, hardly agrees with the remarkably high rates observed for methylene groups of ethers, either in α- or β-positions, which are some orders of magnitude larger than those in unactivated alkanes (see Table 1 and 2). It is therefore possible that the oxygenation of ethers proceeds in a different way, similar, for example, to that previously suggested by Meyer (ref.9) for the oxidation of alcohols by ruthenium-oxo complexes, and involving a simultaneous abstraction of two hydrogen atoms from both the α- and the β-position, with formation of an α-β unsaturated ether. The latter can undergo further oxidation (ref.10), but in the cases of α-tertiary ethers, β-keto-ethers, and not α-alcohol-ethers, are likely to be formed.

A final remark deals with the consumption of the oxidant during the reactions. Inspection of the aqueous phase revealed complete exhaustion of the oxidant, once ca 200 cycles have been completed. By addition of fresh hypochlorite, oxygenation of the substrates starts again and the conversions can be made practically quantitative, in the presence of a large excess of oxidant (about 10 times, as calculated from the stoichiometric requirements of eq.2 and 3).

$$R'CH_2OR + 2\ [O] \rightarrow R'COOR + H_2O \qquad (2)$$

$$R'CH_2R + 2\ [O] \rightarrow R'COR + H_2O \qquad (3)$$

TABLE 1

Ruthenium-catalyzed oxygenation of alkanes by LiClO.[a]

Substrates	Products	Catal.activity[b] I	II	III
adamantane[c]	1-adamantanol	1.5	1.8	3
	adamantanone	0.13	0.18	0.3
cyclo-octane[d]	cyclo-octanone	0.7	1.6	2.2
	cyclo-octanol	0.02	0.13	0.7
cyclohexane	cyclohexanone[e]	0.3	1.2	1.2
methylcyclohexane	1-methylcyclohexanol	0.3	0.1	0.6
	methylcyclohexanones	0.3	0.07	1.3
hexane[f]	2-hexanone	0.1	0.14	0.15
	3-hexanone	0.01	0.04	0.04
ethyl-benzene[d]	acetophenone	0.1	0.1	0.5
cumene[d]	2-phenylpropan-2-ol	0.04	0.1	0.4

[a] Catalyst, 2 mM, and substrate, 1 M, in CH_2Cl_2; LiClO, 1.4 M, in H_2O; 22°C.
[b] Mol of product per mol of catalyst, formed in 1 hour: I, cis-$[RuCl_2(DMSO)_4]$; II, trans-$[RuCl_2(DPP)_2]$; III, $[RuCl(DPP)_2]PF_6$.
[c] 0.5 M.
[d] LiClO 1 M.
[e] Together with traces of cyclohexanol
[f] 2 M.

TABLE 2

Ruthenium-catalyzed oxygenation of ethers by LiClO.[a]

Substrates	Products	Catal.activity[b] I	II	III
di-n-propyl-ether	propyl-propionate	0.8	1.1	0.4
methyl-n-butyl-ether	methyl-butyrate	0.4	0.4	0.3
tetrahydropyran	δ-valerolactone	0.3	0.2	0.2
tetrahydrofuran	γ-butyrolactone	0.7	0.9	0.4
2,5-diMe-tetrahydrofuran	2,5-diMe-dihydrofuran-3-one	2.9	1.6	0.5
2-Me-tetrahydrofuran	2-Me-dihydrofuran-3-one	0.2	0.2	0.1
	γ-valerolactone	0.2	0.1	0.1

[a] Catalyst, 2 mM, and substrate, 1 M, in CH_2Cl_2; LiClO, 1 M, in H_2O; 22°C.
[b] Mol of product per mol of catalyst, formed in 1 minute: I, cis-$[RuCl_2(DMSO)_4]$; II, trans-$[RuCl_2(DPP)_2]$; III, $[RuCl(DPP)_2]PF_6$.

The ruthenium-complexes clearly promote also the dismutation of hypochlorite (to chloride and oxygen), which is a well-established process, commonly triggered by metal ions. Independent experiments showed that 2 mM solutions of $[RuCl(DPP)_2]PF_6$ in dichloromethane catalyze the dismutation of 1 M aqueous solutions of LiClO, at a rate of ca 0.5 turnovers per min. The process is unaffected by the presence of alkanes, but usually accelerated by the presence of small amounts of ethers, irrespective of their reactivity. It is therefore likely that ethers act as phase transfer agents, by complexing the lithium cations and making the hypochlorite soluble in the organic phase.

REFERENCES

1 W.J. Mijs and C.R.H.I. deJonge, Organic Syntheses by Oxidation with Metal Compounds, Plenum Press, New York, 1986.

2 (a) A.B. Smith, and R.M. Scarborough, Synth. Commun., 1980, 10, 205; (b) P.H.J. Carlsen, T. Katsuki, V.S. Martin, and K.B. Sharpless, J. Org. Chem., 1981, 46, 3936.

3 (a) D.Dolphin, B.R. James, and T.Leung, Inorg. Chim. Acta, 1983, 79, 25 and 180; (b) M.M. Taqui Khan, H.C. Bajaj, R.S. Shukla, and S.A. Mirza, J. Mol. Catal., 1988, 45, 51; (c) T.C. Lau, C.M. Che, W.O. Lee, and C.K. Poon, J. Chem. Soc. Chem. Commun., 1988, 1406; (d) G. Barak, and Y. Sasson, J. Chem. Soc. Chem. Commun., 1988, 637.

4 (a) M. Bressan, and A. Morvillo, J. Chem. Soc. Chem. Commun, 1988, 650; (b) Inorg. Chem., 1989, 28, 950; (c) J. Chem. Soc. Chem. Commun., 1989, 421.

5 (a) M. Bressan, and P. Rigo, Inorg. Chem., 1975, 14, 2286; (b) M. Bressan, R. Ettorre, and P. Rigo, Inorg. Chim. Acta, 1977, 24, L57.

6 (a) I.P. Evans, A. Spencer, and G. Wilkinson, J. Chem. Soc. Dalton Trans., 1973, 204; (b) F.P. Dwyer, H.A. Goodwin, and E.C. Gyarfas, Aust.J. Chem, 1963, 16, 42.

7 J. March, Advanced Organic Chemistry, 3rd ed., Wiley Interscience, New York, 1986, p.621.

8 V.W.W. Yam, C.M. Che, and W.T. Tang, J. Chem. Soc. Chem. Commun., 1988, 100.

9 M.S. Thompson, and T.J. Meyer, J. Am. Chem. Soc., 1982, 104, 4106.

10 G. Piancatelli, A. Scettri, and M. D'Auria, Tetrahedron Lett., 1977 , 3483.

G. Centi and F. Trifiro' (Editors), *New Developments in Selective Oxidation* 125
© 1990 Elsevier Science Publishers B.V., Amsterdam — Printed in The Netherlands

DIRECT CONVERSION OF BENZENE TO PHENOLS UNDER AMBIENT CONDITIONS

KAZUO SASAKI, SOTARO ITO AND ATSUTAKA KUNAI

Department of Applied Chemistry, Hiroshima University
Saijo-cho, Higashi-Hiroshima 724 (Japan)

SUMMARY

A new catalytic reaction system for producing phenols from benzene is described. The system is basically composed of two gaseous reactants (H_2 and O_2) and silica supported Pd-Cu catalyst and operates in neat benzene under ambient conditions. Either phenol or hydroquinone can be obtained with a remarkably high selectivity. This system may be utilized as one of the general method of aromatic hydroxylation.

INTRODUCTION

There is a strong demand in chemical industry to convert benzene directly to phenol as an alternative to the most efficient process currently operating in the chemical industry, the Cumene Process. In fact the latter process necessitates multi-step operations throughout the whole process and gives by-production of equi-molar amount of acetone, which might become a serious drawback in the future.

During the last decade, we have engaged in a new process capable to replace the Cumene Process. Our method basically utilizes the charge transfer reaction between monovalent copper ions and gaseous oxygen, eq. (1):

$$O_2 + 2\,Cu^+ + 2\,H^+ \longrightarrow H_2O_2 + 2\,Cu^{2+} \quad (1)$$

$$H_2O_2 + Cu^+ + H^+ \longrightarrow OH + H_2O + Cu^{2+} \quad (2)$$

OH radicals thus produced attack benzene nucleus to yield either phenol (PhOH) or hydroquinone (HQ). The selectivity towards PhOH or HQ can readily be altered at will (eq. 3). The reaction mechanism has been studied in some detail and reported in our previous papers (refs. 1-3).

The reaction sequence expressed by eqs. (1) and (2) produces bivalent copper ion in the reaction system and it is necessary to return it to monovalent state in order to set up a continuous production system. We have firstly studied the electrolytic regeneration of Cu(I) species. This method has been proved to be quite attractive particularly for producing p-benzoquinone (BQ) (ref. 4).

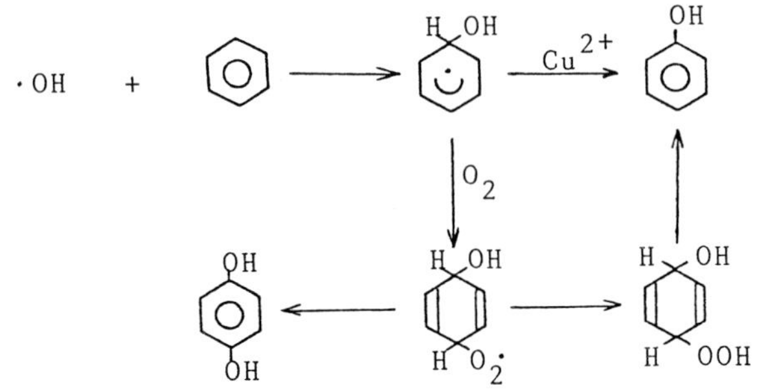

(3)

In particular, we were able to establish an ideal electrolysis system, which we named "duet electrolysis", where a single product, p-benzoquinone, is produced both at cathode and anode electrodes simultaneously from a single starting material, benzene (refs. 5-7). Electrochemical method seems, however, to be not always advantageous, if one aims at phenol as the final product. We thus attempted to develop another possibility that is the chemical reduction of bivalent copper to monovalent one. Hydrogen seemed to be appropriate for reducing chemicals because proton is a necessary reagent for hydrogen peroxide. In order to perform reduction of Cu(II) with hydrogen, however, we had to prepare some suitable heterogeneous catalyst and Pd loaded on silica surface was employed tentatively. This paper deals mainly with the use of palladized silica catalyst.

EXPERIMENTS

Catalyst

A proper amount of palladium chloride was precipitated on commercial silica gel (Merck No. 9385, 230-400 mesh ASTM) and dried by heating. In some of catalysts, to be used in neat benzene, several types of copper(II) salt were coprecipitated on the silica support.

Oxidation reaction

A desired amount of catalyst was put into an Erlenmeyer flask (50 ml) containing 20 ml of neat benzene and reaction was started by feeding the reacting gases. Two different modes of gas feeding were employed, i.e., alternate and simultaneous feedings. In the former mode, hydrogen and oxygen (normally air) were fed alternately in a programmed manner. In this case, the catalyst is activated during the hydrogen stage and promotes the oxidation of benzene during the oxygen stage. In the latter mode, no particular treatment for catalyst activation was made and both hydrogen and oxygen were fed simultaneously.

RESULTS

Reaction of benzene

In Table 1, are shown the results obtained with a catalyst composed of $CuSO_4$ and $PdCl_2$ supported on silica. The catalyst was first activated by streaming hydrogen for 2 h and then hydrogen was switched to air stream and continued for 1 h. The data show that a combined use of Pd and Cu is essential for realizing a high catalyst activity (runs 1 to 3). Silica support seems to play an important role (runs 8 and 9) and cannot be eliminated. Table 2 shows other data which were obtained with the simultaneous gas feeding. Except for the initial period of reaction, phenol accumulates linearly in benzene with increasing reaction time.

TABLE 1
Effect of catalyst composition on the phenol yield.[a]

Run	Catalyst composition			Additive	Product
	SiO_2/g	$CuSO_4$/mmol	$PdCl_2$/mmol	CH_3COOH/g	Phenol/μmol
1	2	2	0.1	1	83
2	2	2	0.1	0	51
3	2	2	0.2	0	50
4	2	2	0	1	0
5	2	2	0	0	0
6	2	0	0.1	1	3
7	2	0	0.1	0	trace
8	0	2	0.1	1	trace
9	0	2	0.1	0	trace
10	2	2	b	0	4
11	c	2	0.1	0	5

[a]The reactions were performed using the catalysts corresponding to type A by applying alternate feeding of H_2 for 2 h and air for 1 h. In runs 8 and 9, powdery mixture of $CuSO_4$ and $PdCl_2$ were used without support.
[b]H_2PtCl_6 (0.1 mmol) was used instead of $PdCl_2$.
[c]Alumina (2g) was used instead of silica.

TABLE 2
Catalytic oxidation of benzene by simultaneous feeding of hydrogen and oxygen.[a]

Reaction time/h	0.5	1	2	3	4	5	6	7	8	9
PhOH /μmol	4.7	15	48	77	114	141	167	183	205	229

[a]The reaction was carried out using the catalyst C by simultaneous feeding of H_2 and O_2.

Fig. 1 shows the effect of repetition of the alternating feeding. In this case, hydrogen and air feedings were alternated for every 30 min and the alternation was repeated up to six times. Three lines stand for three catalysts different in their process of preparation but have the same composition (see ref. 8). Although the slope of lines are different depending on the process of catalyst preparation, it is clearly indicated that the reaction continues steadily without any loss of catalyst activity.

It should be noted that not only the process of catalyst preparation but also the nature of counter ions of copper salt being supported on the silica surface affects the catalyst performance. Roughly speaking, the acid strength of corresponding free acid of counter ions determines the relative selectivity between benzoquinone and phenol (Q/P) : the stronger is the acid strength, the higher is the selectivity of phenol relative to quinones. The difference in counter ion in copper salt also affects the reaction rate. An example is shown in Fig. 2, where phenol yield is plotted as a function of time elapsed for activating the catalyst. The figure indicates that cupric acetate is more slowly activated than cupric sulfate does.

The selectivity is also affected by the oxygen partial pressure in the surrounding gas. This is shown in Table 3. The effect is appreciable when the catalyst is made from cupric acetate. As far as the oxygen source is ordinary air, the product ratio, Q/P, is only 0.2 even in the highest case. At 5% level of acetic acid added deliberately, the value increases from 0.16 to 1.53 corresponding to the change in oxygen pressure from 0.2 (air) to 3 atm, respectively. At 6 atm, the value tend to saturate. When the catalyst is made from cupric sulfate, the Q/P ratio never exceeds 0.15 even if pressurized oxygen is supplied. This is because, at lower pH, the saturation of pressure effect appears at lower pressure. A similar observation was obtained also in aqueous phase reaction (ref. 7).

TABLE 3

Effect of oxygen pressure on the product selectivity.[a]

Cu Salt	AcOH/vol%	Ratio of BQ/PhOH (Total products/μmol)			
		air, 1 atm	O_2, 1 atm	O_2, 3 atm[b]	O_2, 6 atm[b]
$Cu(OAc)_2$	2.5	-	0.79 (47.1)	1.50 (29.6)	-
$Cu(OAc)_2$	5.0	0.16 (24.6)	0.63 (44.2)	1.53 (57.6)	1.39 (29.6)
$Cu(OAc)_2$	10.0	0.20 (41.0)	0.61 (62.6)	0.84 (45.8)	-
$Cu(OAc)_2$	20.0	-	0.45 (43.4)	-	-
$CuSO_4$	0	0.10 (56.2)	0.11 (37.0)	0.12 (70.6)	-
$CuSO_4$	10.0	0.01 (31.2)	0.15 (39.0)	-	-

[a] The reactions were carried out in the same manner as Table 1.
[b] Hydrogen reduction was also done under the same pressure as the oxidation.

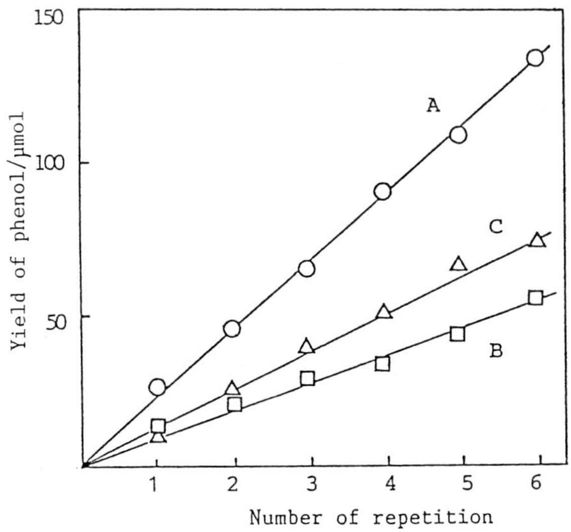

Fig. 1. Change in catalytic activity due to the method of catalyst preparation. Alternate gas feeding for each 30 min was applied.

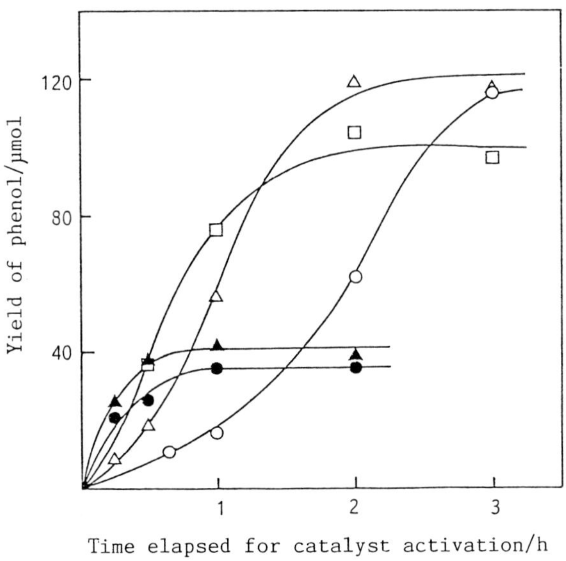

Fig. 2. Time elapsed for catalyst activation and its effect on the phenol yield. Coordinate axis represents the yield of phenol produced in 1h of oxidation reaction.

(●) $PdCl_2$ 0.1mmol, $CuSO_4$ 2mmol/SiO_2 2g.
(▲) $PdCl_2$ 0.2mmol, $CuSO_4$ 2mmol/SiO_2 2g.
(O) $PdCl_2$ 0.1mmol, $Cu(OAc)_2$ 2mmol/SiO_2 2g.
(△) $PdCl_2$ 0.2mmol, $Cu(OAc)_2$ 2mmol/SiO_2 2g - AcOH (1g) was added.
(□) $PdCl_2$ 0.3mmol, $Cu(OAc)_2$ 2mmol/SiO_2 2g.

Reaction of naphthalene

In principle, there is a possibility that our reaction system can be utilized as the general method of aromatic hydroxylation. We have thus studied tentatively the reaction of naphthalene in place of benzene. In this case, however, we have to find a suitable solvent which is inert to the attack of hydroxyl radicals. Several candidates were tested including cyclohexane, acetone, ethyl acetate as well as some aliphatic alcohols. Among these, cyclohexane was found most promising when benzene was reacted in it. Unfortunately, however, the reaction of naphthalene in this solvent was much slower than that of benzene. Probably, hydrophobic nature of cyclohexane retards the chance of encounter between solute molecule and the catalyst surface. Accordingly, the use of acetic acid was finally examined. Solid catalyst used in this system contained only palladium (0.1 mmol Pd /g SiO_2) and cupric acetate was dissolved in the solution phase (40 mM). Gases were supplied through a sintered glass-ball disperser at a rate of 7.5 ml/min for both hydrogen and oxygen. Both the alternate and simultaneous feeding were tested for comparison. Results obtained are listed in Table 4.

TABLE 4
Reactions in acetic acid.[a]

Reactant	Feeding mode	Catalyst[b]	Product/μmol			
			(Phenol)	(HQ)	(BQ)	(Total)
Benzene	Simultaneous	1 g	169	379	126	674
Benzene	Simultaneous	1 g	186	349	115	650
Benzene	Simultaneous	4 g	212	437	137	785
Benzene	Alternate	4 g	164	112	232	508
			(1-NpOH[c])	(2-NpOH)	(1,4-NQ)	(Total)
Naphthalene	Simultaneous	4 g	215	<121	649	985
Naphthalene	Alternate	4 g	172	< 88	320	580

[a] $Cu(OAc)_2$ was dissolved in solution phase (40mM).
[b] Composition was 0.1mmol Pd /g SiO_2.
[c] NpOH and NQ stand for naphthol and naphthoquinone, respectively.

In this experiment, total reaction time was fixed at 2 h irrespective of the feeding modes. Since in the alternate mode, the two gases were altered at every 15 min, effective reaction time elapsed was one half of that of the simultaneous feeding. If this is taken in mind, the data suggest that the simultaneous feeding is not always superior in the production rate and the final judgment is still a subject of debate. In any case, Table 4 clearly indicates that naphthalene is oxidized at a rate comparable with that of benzene.

REFERENCES

1 A.Kunai, S.Hata, S.Ito, and K.Sasaki, J. Org. Chem., 51 (1986) 3471-3474.
2 A.Kunai, S.Hata, S.Ito, and K.Sasaki, J. Am. Chem. Soc., 108 (1986) 6012-6016.
3 S.Ito, T.Yamasaki, H.Okada, S.Okino, and K.Sasaki, J. Chem. Soc., Perkin Trans. 2, (1988) 285-293.
4 S.Ito, H.Okada, R.Katayama, A.Kunai, and K.Sasaki, J. Electrochem. Soc., 135 (1988) 2996-3000.
5 S.Ito, R.Katayama, A.Kunai, and K.Sasaki, Tetrahedron Lett., 30 (1989) 205-206.
6 S.Ito, N.Fukumoto, A.Kunai, and K.Sasaki, Chem. Lett., (1989) 745-746.
7 S.Ito, A.Kunai, H.Okada, and K.Sasaki, J. Org. Chem., 53 (1988) 296-300.
8 A.Kunai, T.Wani, F.Iwasaki, Y.Kuroda, S.Ito, and K.Sasaki, Bull. Chem. Soc. Jpn., 62 (1989) 2613-2617.

D.Olivier (Institut de Recherche sur la Catalyse, France) :
What is the proof that the reaction is not performed in liquid phase ? Have you checked the supported metal loading after reaction ?

K.Sasaki (Dept. Applied Chem., Hiroshima University, Japan) :
No. We haven't studied the catalyst composition after use yet. However, we believe that no appreciable amount of Pd had been lost in the solution phase during a limited time of reaction.

B.R.James (University of Brit. Columbia, Canada) :
1. Use of H_2-O_2 mixtures for monooxygenase-type reaction was first demonstrated, to my knowledge, by my group in 1969 (Can.J.Chem.). Such systems suffer from competing direct hydrogenolysis of O_2 to water. Do your simultaneous feeding system suffer by competing water production and, if so, what is the ratio of phenol : water production at various conditions ?
2. In your simultaneous feeding experiments, what O_2:H_2 ratios were used and were explosion limits carefully avoided ?

K.Sasaki : 1. Yes. An appreciable part of hydrogen seem to yield water directly. Although no material balance has been studied in the reaction system here reported, we have made a separate experiment to study the percentage conversion of hydrogen to phenol. According to that, the conversion was ca.10 % at the highest. 2. In the present report, the two gases were flown through the space over benzene layer at an equal rate, 7.5 ml/min.

A NOVEL AND EFFECTIVE OXYGENATION OF 2,3,6-TRIMETHYLPHENOL TO TRIMETHYL-*p*-BENZOQUINONE BY DIOXYGEN WITH COPPER(II) CHLORIDE/ AMINE HYDROCHLORIDE SYSTEM CATALYST

Katsuomi Takehira,* Masao Shimizu, Yoshihito Watanabe, Hideo Orita and Takashi Hayakawa
National Chemical Laboratory for Industry, Tsukuba Research Center, Tsukuba, Ibaraki 305, Japan

SUMMARY

2,3,6-Trimethylphenol was selectively and quickly oxygenated to trimethyl-*p*-benzoquinone by molecular oxygen in the presence of catalytic amount of copper(II) chloride/amine hydrochloride system in alcoholic solvent at ambient temperature.

INTRODUCTION

Trimethyl-*p*-benzoquinone (TMQ) is a key compound for the synthesis of Vitamin E, etc., and the current method of its production on an industrial scale is *p*-sulfonation of 2,3,6-trimethyl-phenol (TMP) followed by oxidation with MnO_2. The methods for one step synthesis by the oxidation of TMP have been investigated so far using several oxidizing reagents, such as hydrogen peroxide,[1] etc. Use of molecular oxygen as the oxidizing reagent seems the most promising method where cobalt(II)-Shiff base[2] and copper (II) chloride[3] have been tested as the catalyst. Usually, the latter catalyst was used as the binary system combined with LiCl, resulting in high selectivity for TMQ formation.[4] The catalyst life of the former is very short though its activity is high, while the catalytic activity of the latter is so low even when coupled by LiCl that an almost stoichiometric amount of copper(II) salt is required to complete the oxidation.[4] We herein communicate a novel and effective oxygenation of TMP to TMQ by copper (II) chloride/amine hydrochloride system catalyst.

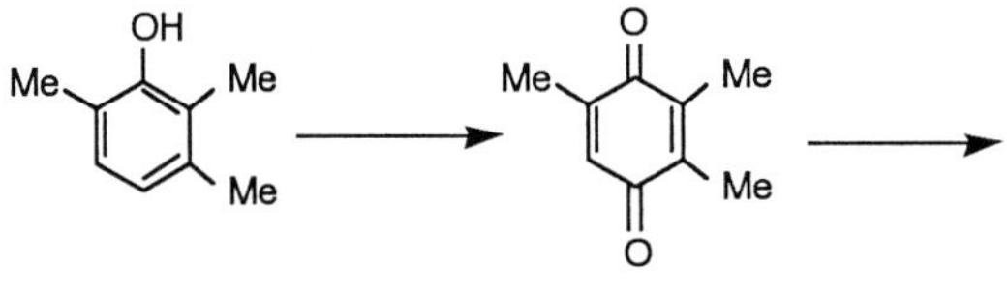

Vitamin E

EXPERIMENTAL

The oxidation of TMP (1, 2 mmol) was carried out using a copper compound (0.2 mmol) and an additive (0.2 mmol) as the catalyst in alcohol (2 ml) as the solvent at 40-60°C under 1atm. of oxygen atmosphere, where the amount of oxygen consumed was measured by a gas burette. TMQ (2), 4-chloro-2,3,6-trimethylphenol (3) and 4,4'-dihydroxy-2,2', 3,3', 5,5'-hexamethylbiphenyl (4) as the products were determined by GLC and HPLC method using a Thermon 3000 and an Inertsil ODS as columns, respectively.

RESULTS AND DISCUSSIONS

The results of the oxidation of **1** for 5 h with several copper-amine system catalysts are shown in Table 1. The maximum rate of O_2 consumption is shown as "dO_2/dt." $CuCl_2 \cdot 2H_2O$ alone showed a very low activity for the production of **2**, while $Cu(OAc)_2 \cdot H_2O$ afforded a substantial

TABLE 1
Oxidation of **1** with the copper - additive system catalyst.

Cu compound (0.2 mmol)	Additive (0.2 mmol)	dO_2/dt (mmol/h)	Conversion of **1** (%)	Yield (%) of **2**	**3**	**4**
$CuCl_2 \cdot 2H_2O$	-	0.062	24.9	8.9	8.9	7.5
$Cu(OAc)_2 \cdot H_2O$	-	0.225	79.4	0	0	36.6
$CuCl_2 \cdot 2H_2O$	LiCl	0.353	90.7	68.8	6.7	16.5
$Cu(OAc)_2 \cdot H_2O$	LiCl	0.910	100	6.5	0	14.6
$CuCl_2 \cdot 2H_2O$	NH_4Cl	0.162	54.3	28.4	7.0	12.0
$CuCl_2 \cdot 2H_2O$	$(CH_3)_4NCl$	0.038	12.6	1.5	5.1	2.9
$CuCl_2 \cdot 2H_2O$	$(C_2H_5)_4NCl$	0.527	100	81.8	1.6	14.7
CuCl	$(C_2H_5)_3N \cdot HCl$	1.54	100	29.3	0	0
$CuCl_2$	$(C_2H_5)_3N \cdot HCl$	1.44	100	83.7	0	12.3
$CuCl_2 \cdot 2H_2O$	$(C_2H_5)_3N \cdot HCl$	1.45	100	80.5	0	13.5

amount of the dimer **4**. An addition of LiCl to the $CuCl_2$ system caused an increase in the activity of 2 production as claimed in the several patents.[4] It took 7 - 8 h for completing the oxidation of **1** by the $CuCl_2 \cdot 2H_2O$ - LiCl system in the present conditions. When the amount of LiCl was changed, its use of 0.2 mmol resulted in the highest activity suggesting an 1/1 active complex formation between $CuCl_2$ and LiCl. A replacement of LiCl by NH_4Cl or $(CH_3)_4NCl$ caused a decrease in the activity, while $(C_2H_5)_4NCl$ was effective as the co-catalyst for the production of 2. It is noteworthy that $(C_2H_5)_3N \cdot HCl$ was the most effective among the additives tested and shortened the reaction time to 2 - 3 h for completing the oxidation of **1**. An use of CuCl caused generally rapid consumption of **1**, where the amount of **2** formed was however substantially lowered.

The results of the oxidation of **1** with $CuCl_2 \cdot 2H_2O$ - amine system catalyst are shown in Table 2. Many kinds of amines were effective when used as the salt form with inorganic proton acid, among which not only hydrochloric acid but also sulfuric and hydrobromic acids were included.

TABLE 2
Oxidation of **1** with the copper[a)] - amine system catalyst.

Amine (0.2 mmol)	dO_2/dt (mmol/h)	Conversion of **1** (%)	Yield (%) of **2**	**3**	**4**	React. time(h)
$CH_3NH_2 \cdot HCl$	0.715	100	79.5	0	21.6	5
$(CH_3NH_2)_2 \cdot H_2SO_4$[b)]	1.28	95.8	71.3	3.4	21.4	5
$(CH_3)_2NH \cdot HCl$	1.60	100	84.2	0	13.3	5
$(CH_3)_3N \cdot HCl$	1.13	100	82.6	0	20.4	5
$C_2H_5NH_2 \cdot HCl$	1.27	100	79.2	0	13.3	4
$C_2H_5NH_2 \cdot HBr$	1.21	100	83.3	0	15.4	4
$(C_2H_5)_2NH \cdot HCl$	1.55	100	84.2	0	13.3	4
$(C_2H_5)_3N \cdot HCl$	1.45	100	80.5	0	13.5	4
n-$C_3H_7NH_2 \cdot HCl$	2.05	100	80.5	0	15.5	4
$(n$-$C_3H_7)_2NH$	1.07	100	50.0	1.9	0	4
$(n$-$C_3H_7)_2NH \cdot HCl$	1.46	100	79.8	0	17.3	4
$(n$-$C_3H_7)_3N$	1.13	100	26.0	0	0.7	4
$(n$-$C_3H_7)_3N$ + HCl[c)]	0.990	100	77.5	0	19.1	4
$(i$-$C_3H_3)_2NH \cdot HCl$	0.720	100	84.4	0	16.8	5
n-$C_4H_9NH_2 \cdot HCl$	1.59	100	85.5	0	15.5	4

a)$CuCl_2 \cdot 2H_2O$ (0.2 mmol) was used. b)0.1 mmol was used. c)0.2 mmol of HCl was used in the form of its 36 % aqueous solution.

Secondary amine, such as $(CH_3)_2NH \cdot HCl$ or $(C_2H_5)_2NH \cdot HCl$ was preferable than primary or tertiary amine in the case with short alkyl chains, while the primary amine also caused a high activity in the case with long alkyl chains. When the amount of the amine $(C_2H_5)_2NH \cdot HCl$ was changed, the highest activity was obtained by its use of 0.2 mmol, suggesting formation of 1/1 active complex between $CuCl_2$ and $(C_2H_5)_2NH$. Cyclic amine, such as morpholine, aromatic amine, such as aniline, and aminoalcohol, such as ethanolamine, were effective as the co-catalyst in the presence of hydrochloric acid.

Table 3 shows the results obtained with the $CuCl_2 \cdot 2H_2O$ - amino acid system catalyst. Amino acid was also useful when the ability of bidentate coordination was weakened by the alkyl substitution at the amino group or by the esterification of the carboxyl group: glycine was not effective even when coupled by hydrochloric acid, while N,N-dimethylglycine or glycine ethyl coupled by hydrochloric acid showed high activity. D,L-Alanine was effective as the co-catalyst, though it has both free amino and carboxyl groups: this may be due to the steric hindrance which weakens the coordinating ability of the amino group to copper. Both of the two optical isomers, *i.e.*, D- and L-alanine, showed similar activity to that obtained by the racemic mixture, *i.e.*,

TABLE 3
Oxidation of **1** with the copper[a] - amino acid system catalyst.

Amino acid (0.2 mmol)	dO_2/dt (mmol/h)	Conversion of **1** (%)	Yield (%) of **2**	**3**	**4**	Reaction time (h)
NH_2CH_2COOH	0.078	24.9	7.2	4.8	7.0	5
$NH_2CH_2COOH{\cdot}HCl$	0.130	25.2	3.2	12.4	3.6	4
$(CH_3)_2NCH_2COOH{\cdot}HCl$	0.670	100	79.1	3.8	17.7	5
$NH_2CH_2COOC_2H_5{\cdot}HCl$	1.48	100	78.5	0	15.0	5
$NH_2(CH_2)_2COOH$	0.200	60.4	36.2	2.8	14.5	5
$NH_2(CH_2)_2COOH$ + HCl[b]	0.274	62.7	37.2	6.3	14.2	5
D,L-$CH_3CH(NH_2)COOH$	0.825	100	74.1	0	20.6	5
D,L-$CH_3CH(NH_2)COOH$ + HCl[b]	0.810	100	78.2	0	16.3	5

[a] $CuCl_2{\cdot}2H_2O$ (0.2 mmol) was used. [b] 0.2 mmol of HCl was used in the form of its 36% aqueous solution.

D,L-alanine. β-Alanine oppositely showed very low activity even when coupled by hydrochloric acid: this can be reasoned by its strong bidentate coordination because of the absence of steric hindrance. Neither diamine, such as ethylenediamine, nor dipyridyl was effective as the co- catalyst. It seems thus that the bidentate ligand occupies the active site of the copper complex resulting in the low activity.

All the amine compounds tested did not reveal the high activity in the form of free amine, *i.e*., in the absence of inorganic proton acid: the addition of the acid caused an increase in the activity of copper - amine system catalyst. When the amount of HCl added to $CuCl_2{\cdot}2H_2O$ (0.2 mmol) - $(C_2H_5)_3N$ (0.2 mmol) system was changed, its optimum amount was not observed at a fixed value: the addition of 0.2 mmol of HCl was sufficient to obtain a good yield of **2** (> 80 %) and the further addition did not affect substantially the rate and the yield of **2** production. This result and the effectiveness of HBr and H_2SO_4 as the additive as shown in Table 1 suggest that the acids works as the proton source during the oxidation of **1** to **2**.

The effect of the solvent on the yield of 2 in the oxidation with $CuCl_2{\cdot}2H_2O$ - $(C_2H_5)_2NH{\cdot}HCl$ or $(CH_3)_2NH{\cdot}HCl$ system is shown in Table 4. Use of alcohol of low molecular weight. *i.e.*, ethanol, propanol or isopropanol, as the solvent resulted in a decrease in the yield: this might be partly due to the low solubility of the amine in these solvents. When an aromatic compound, *i.e*., benzene or toluene, was mixed in the ratio of 1/1 into alcohol as the solvent, the rate of the oxidation became two times higher compared to the case without the aromatic solvent and the

TABLE 4
Oxidation of **1** with copper - amine system catalyst.

Amine	Solvent (ml)	dO_2/dt (mmol/h)	Conversion of **1** (%)	Yield (%) of **2**	**3**	**4**
A	EtOH(2)	0.452	83.0	60.2	5.4	4.1
A	Bz(1)+EtOH(1)	0.700	96.9	78.8	2.7	17.6
A	1-PrOH(2)	0.565	92.4	71.3	3.7	13.0
A	Tol(1)+1-PrOH(1)	1.01	100	86.4	0	10.5
A	*i*-PrOH(2)	0.575	81.6	64.9	5.4	6.8
A	Bz(1)+*i*-PrOH(1)	0.975	100	87.6	1.4	3.1
A	Tol(1)+*i*-PrOH(1)	0.955	100	89.4	0	1.8
B	EtOH(2)	0.555	96.8	73.9	2.9	18.5
B	Bz(1)+EtOH(1)	1.33	100	83.6	0	7.7
B	1-PrOH(2)	0.770	100	82.6	0	14.9
B	Tol(1) + 1-PrOH(1)	0.970	100	85.4	0	11.8
B	*i*-PrOH(2)	0.525	100	86.4	1.3	7.7
B	Bz(1) + *i*-PrOH(1)	1.02	100	90.0	0	2.5

a)**1**, 2 mmol; $CuCl_2{\cdot}2H_2O$ 0.2 mmol; $(C_2H_5)_2NH{\cdot}HCl$(A) or $(CH_3)_2NH{\cdot}HCl$(B), 0.2 mmol; Reaction temperature, 40°C; Reaction time, 5 h.

yield of **2** increased up to about 90 %. It seems that the mixed solvent can work as a good medium for the present oxidation by dissolving well each component of the reaction: **1** and the amines are well soluble in the aromatics, while $CuCl_2.2H_2O$ is in the alcohols.

The activity of the $CuCl_2{\cdot}2H_2O$ - $(C_2H_5)_2NH{\cdot}HCl$ system catalyst thus obtained was 5 - 6 times higher than that of the well known $CuCl_2{\cdot}2H_2O$ - LiCl system catalyst and the yield of **2** reached a

Scheme I

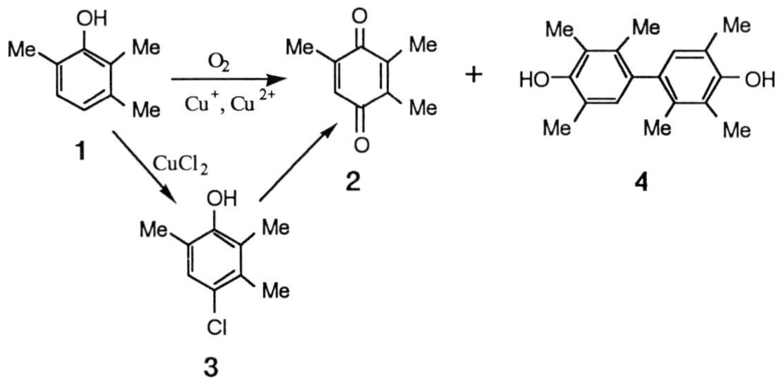

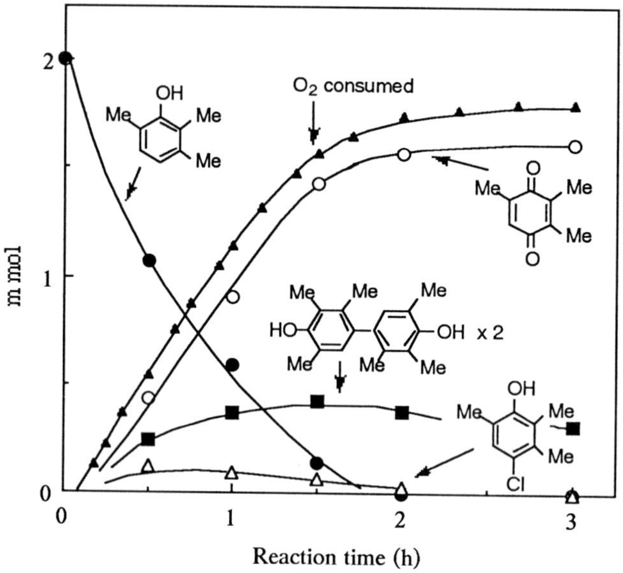

Fig. 1 Oxidation of **1** with $CuCl_2{\cdot}2H_2O$ - $(C_2H_5)_2NH{\cdot}HCl$ catalyst.

TMP, 2 mmol; $CuCl_2{\cdot}2H_2O$, 0.2 mmol; $(C_2H_5)_2NH{\cdot}HCl$, 0.2 mmol n-Hexanol, 2 ml; Reaction temperature, 60°C; pO_2 860 mmHg.

value of about 90 % in a few hours of the reaction with the former catalyst. Time course of the reaction with this catalyst system (Fig. 1) suggests a plausible oxidation scheme as follows: a main pathway may be a direct oxidation of **1** to **2** accompanied by a formation of small amount of **4**. A part of **2** can be formed *via* **3**, *i.e.*, by *p*-chlorination of **1**. (Scheme **I**) Attempts to increase the catalytic activity of the copper system catalyst and to establish a more complete view of the mechanism of the TMP oxidation are under way.

REFERENCES

1. M. Shimizu, H. Orita, T. Hayakawa and K. Takehira, *Tetrahedron Lett.*, **30**, 471 (1989) and references cited therein.
2. R. A. Sheldon and J. K. Kochi, "*Metal Catalyzed Oxidations of Organic Compounds*," p.373. Academic Press, New York, 1981.
3. *Ref.* 2, p.369.
4. Japan Patent 225,137(1984) to Mitsubishi Gas Chemical Co.

G. Centi and F. Trifiro' (Editors), *New Developments in Selective Oxidation*

PHENOL OXIDATION WITH MOLECULAR OXYGEN IN THE PRESENCE OF METALLIC COPPER

N. RAVASIO[1], M. GARGANO[1] and M. ROSSI[2]
[1]Centro C.N.R. sulle Metodologie Innovative di Sintesi Organiche, Dipartimento di Chimica dell'Università, via Amendola 173, I-70126 Bari

[2]Dipartimento di Chimica Inorganica e Metallorganica, Università di Milano, via Venezian 21, I-20133 Milano

SUMMARY

The reactivity of phenol, α- and β-naphtol with molecular O_2 in the presence of metallic copper in methanol/pyridine solution has been investigated and the results obtained compared with those already reported with Cu(I) and Cu(II) amine homogeneous systems.

Phenol gave 4,5-dimethoxy-1,2-benzoquinone **1** with fairly good yield (up to 46%) while the reaction of α-naphtol afforded up to 43% of 2-methoxy-1,4-naphtoquinone **2.**

This reaction represents a new and simple route for the synthesis of methoxyquinones, useful intermediates for the production of drug and photosensitive materials.

Moreover, the total selectivity exhibited towards ortho-hydroxylation in the reaction of phenol, mimics the monophenolase activity of copper enzyme tyrosinase.

β-naphtol gave mainly [1,1'-binaphtalene]-2,2'-diol.

INTRODUCTION

The activity of copper complexes in promoting oxygen activation in chemical and biological systems has been widely investigated[1,2].

Particular attention has been devoted to the oxidation of phenols due to either industrial applications of polyphenylene ethers and quinones or interest in modelling the action of oxygenases such as tyrosinase and pyrocathecase.

Thus the oxidative coupling of 2,6-dialkyl and 2,6-diaryl substituted phenols in the presence of homogeneous catalysts derived from copper salts and amines, is a well established method for the synthesis of polyphenylene ethers[3] which requires, however, continuous efforts to enhance the specificity[4].

Concerning the industrial production of hydroquinone, the direct air oxidation of phenol to p-benzoquinone in polar aprotic solvent, in the presence of copper ions and inorganic bases, followed by catalytic hydrogenation developed by Lyons and Hsu[5], represents a quite actractive alternative to the hydrogen peroxide oxidation of phenol.

An unusual high selectivity to p-benzoquinones has also been claimed with $CuCl_2$ and NEt_3 in acetonitrile as a catalytic system[6].

A more fundamental research concerns mechanistic studies on the copper mediated oxygenolysis of phenol to cis,cis-muconic acid monomethylester that mimics the action of pyrocathecase enzyme[7,8] and the selective oxidation of catechols to the corresponding o-quinones which represents a model for the diphenolase activity of the copper enzyme tyrosinase[9-11].

It is generally accepted that homogeneous systems containing the redox couple Cu(I)/Cu(II) are responsible for the oxygen activation and therefore catalytic systems are prepared by using either mono or divalent copper compounds.

The activation of O_2 towards phenols on metallic Cu has been scarcely investigated. An interesting example however, is reported by Capdevielle and Maumy which deals with the synthesis of copper catecholates from metallic Cu, phenols and oxygen, catalyzed by CuCl[12], while 1,2-dicarbonyls undergo oxidative C-C bond scission to give carboxilate Cu(II) complexes by reaction with Cu(0) and amines[13].

Our recent work on the activation of molecular oxygen on metallic Cu[14] lead to the conclusion that organic Bronsted acids such as methanol, nitromethane and benzoic acid easily react with Cu and O_2 according to the following scheme (X=-OMe,-OCOPh):

$$Cu + 1/2\ O_2 \rightarrow Cu \sim O \quad \text{(surface)}$$

$$Cu \sim O + HX \rightarrow Cu{<}^{OH}_{X}$$

$$Cu{<}^{OH}_{X} + HX \rightarrow CuX_2 + H_2O$$

Scheme 1

Thus, the interaction of molecular O_2 with Cu forms a layer of atomic oxygen, as supported by different surface analytical techniques (EELS,XPS,UPS)[15]; this oxygen covered surface can interact with proton donor molecules to give an hydroxylation reaction[16].

This latter surface species can evolve only in the presence of a base, to give Cu(II) complexes, this evolution being responsible of the bulk reaction of copper.

As an extension of this research, we have investigated the possibility to use metallic Cu as a catalyst for the oxidation of monohydroxilated aromatics as an alternative to the use of CuCl and $CuCl_2$.

As previously observed a Cu(II) complex should be expected in solution after interaction of Cu(0) with O_2 and the organic reagent.

However, a different catalytic behaviour can be suspected starting from metallic Cu, respect to classical Cu(I) and Cu(II) salts, on the basis of either a different mechanism for oxygen activation or stabilization of different oxidation states.

We here report preliminary results obtained by reaction of phenol, α- and β-naphtol with molecular oxygen in the presence of metallic copper in methanol/pyridine solution and compare them with those already reported with Cu(I) and Cu(II) homogeneous systems.

RESULTS

Phenol reacts smoothly with oxygen at atmospheric pressure and room temperature in the presence of metallic Cu, methanol and pyridine. From the resulting solution orange-yellow needles of 4,5-dimethoxy-1,2-benzoquinone **1** precipitate. Best yields were obtained with a phenol/Cu ratio of 6 and an O_2/phenol ratio of 1.5. Under these conditions 46% of **1** was collected.

When the solution was allowed to adsorb a higher amount of oxygen, products derived from C1-C2 bond cleavage began to form.

In every case, no para-products could be identified at any stage of the reaction.

We can assume that an o-hydroxylation reaction followed by oxidation to o-quinone takes place; nucleophilic addition to the activated 4 and 5 positions and reoxidation of the substituted catechols formed, give account of the observed products.

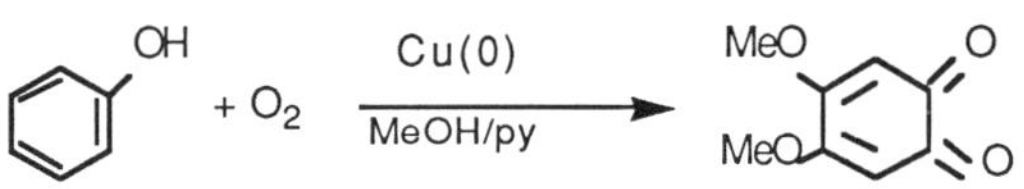

1

Starting from catechol, oxidation and dimethoxylation are also observed, under the conditions employed for phenol, and **1** can be obtained beside a catecholate Cu(II) pyridine complex.

The hydroxylation-oxidation of naphtols in the presence of Cu(0) shows a quite different behaviour in comparison with phenol. The α isomer readily consumes 1 mole of oxygen to give 2-methoxy-1,4-naphtoquinone **2** (43% at 90% conversion) besides unsubstituted 1,4-naphtoquinone (15%).

Therefore, a selective para-oxidation takes place in this case, whereas only monomethoxylation was observed.

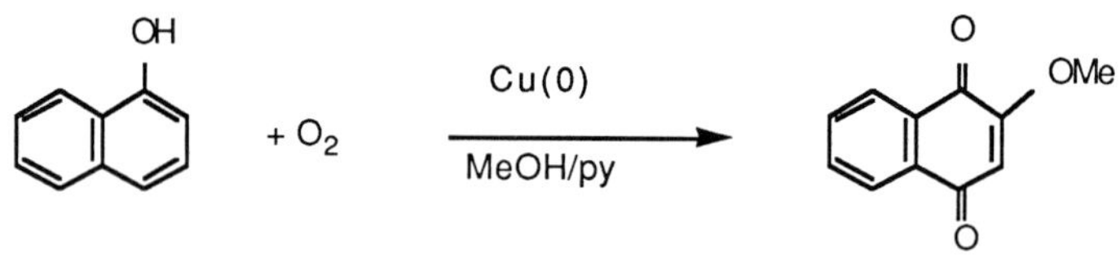

2

The reaction of β-naphtol shows a more different pathway. After adsorption of 1 equivalent of O_2 we could isolate [1,1'-binaphtalene]-2,2'-diol **3** (41%) and the fission product 3-(2-carboxyphenyl)-2-propenoic acid monomethylester **4** (9%), besides non reacted naphtol (19%).

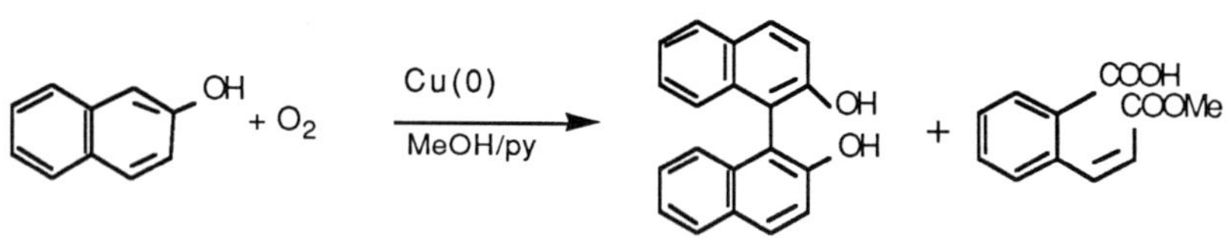

3 **4**

The presence of **3** and **4** suggests activation of o-position in β-naphtol and in particular the formation of an o-quinonic intermediate as precursor of **4**. The oxidation of phenol under similar conditions but using Cu(II) salts as catalysts has been already studied. In particular, according to Brackman and Havinga[17] phenol does not react in the presence of pyridine and methanol. The reaction occurs in the presence of a secondary amine as morpholine (M): o-hydroxylation followed by oxidation takes place and 4,5-dimorpholino-1,2-benzoquinone can be obtained.

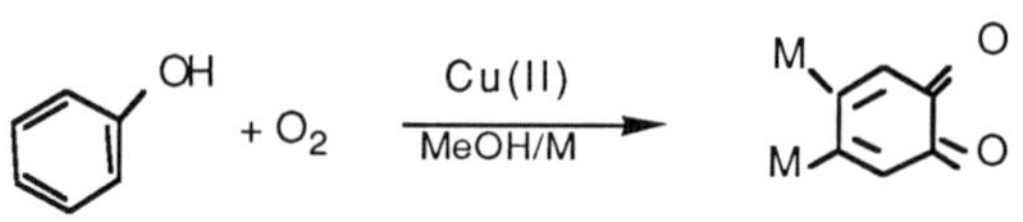

On the other hand, **1** can be obtained according to Rogic and Demmin in a similar way but starting from the already o-functionalized catechol by using Cl_2Cupy_2[7].

A few reports deal with the oxidation of naphtols in the presence of Cu salts. The Brackman-Havinga system transforms both α- and β-naphtol in 4-morpholino-1,2-naphtoquinone; however dinaphtol derivatives are also formed, particularly during reaction of the α isomer[17].

It is worth noting that alcohols do not normally add to quinones, metal ions catalysis being always needed[18,19] and only a few examples of direct conversion of phenols and α-naphtol to alkoxyquinones are known[20,21].

In particular, the reaction of phenol here reported represents a novel and simple route for the synthesis of **1**.

o-Benzoquinone derivatives are conventionally prepared by oxidation of the corresponding catechols. The preparation of **1** , used in the synthesis of redox polymers, has been achieved by oxidation of catechol with PbO_2 in methanol and in the presence of CH_3ONa [22,23] or with $NaIO_3$ in methanol[24], yields never exceeding 60%.

The easy hydroxylation observed when phenol reacts with oxygen in the presence of metallic Cu can be rationalized on the basis of the following model. The acidic phenol interacts with the oxygen-covered Cu surface according to Scheme 1 to produce a copper phenolate intermediate.

According to EELS determinations, the oxygen covered Cu surface produces different species[15] and in particular the existence of peroxo groups can be inferred by the presence of a stretching vibration band around 880 cm^{-1}.

Therefore, the coordinated phenolate can react with the peroxo species owing to partial electrophilic character of one oxygen atom of the peroxide unit as suggested by Solomon[25] to explain the monophenolase activity of tyrosinase.

This model agrees with the well accepted existence of a μ-1,2-peroxo species as active intermediate of oxy-tyrosinase.

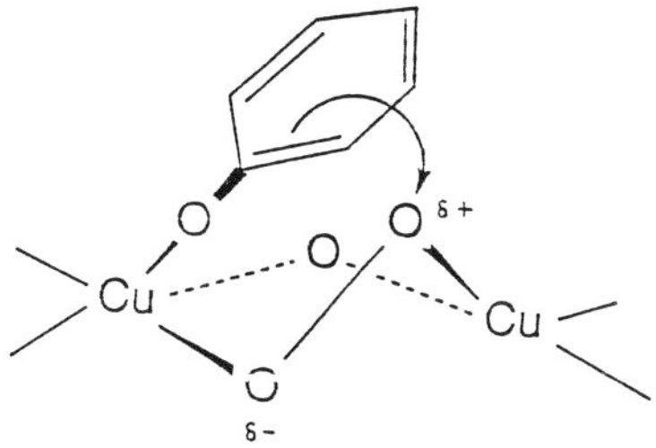

REFERENCES

1 - Houben-Weyl, "Methoden der Organische Chemie", IV/1b, Georg Thieme Verlag, Stuttgart, 1976, p.55-67

2 - K.D.Karlin and J.Zubieta (eds.), "Copper Coordination Chemistry: Biochemical and Inorganic Perspectives", Adenine Press, Guilderland, N.Y., 1983

3 - a) General Electric Co., Neth. Appl. 295,748, C.A. **64**:9843a; b) Dynamit Nobel A.G., Neth. Appl.6,610,017, C.A. **66**:116152t; c) Hay,A.S., U.S.Patent 3,432,466, C.A. **70**:97406t; d) Toyo Rayon Co. Ltd., Fr.1,523,821, C.A. **71**:13525r; e) Hori,R., Kataoka,T., Kodama,H., Japan 7001,633, C.A. **72**:112034g; f) Hay,A.S., Polym.Eng.Sci. **16** (1976)1.

4 - C.E.Koning, G.Challa, F.B.Hulsbergen and J.Reedjik, J.Mol.Catal. **34** (1986) 355 and references therein

5 - C.Y.Hsu and J.E.Lyons, Eur. Pat. Appl.EP 93,540, C.A. **100**:67996f; EP 104,937, C.A. **101**:54719p; EP 105,067, C.A. **101**:54721h; EP 107,427, C.A. **101**:170888s; U.S.Patent 4,442,036, C.A. 101:6829c; U.S. 4,482,756, C.A. **102**:113013z.

6 - Showa Denko K.K., Jpn. Kokai Tokkyo Koho JP 60,123,440 (85,123,440), C.A. **104**:5640j.

7 - T.R.Demmin and M.Rogic, J.Org.Chem. **45** (1980) 4210

8 - J.Tsuji and H.Takayanagi, Tetrahedron **34** (1978) 641

9 - J.S.Thompson and J.C.Calabrese, J.Am.Chem.Soc. **108** (1986) 1903

10 - G.Speier, J.Mol.Catal. **37** (1986) 259

11 - S.Tsuruya, H.Kuwahara and M.Masai, J.Catalysis **108** (1987) 369 and references therein

12 - P.Capdevielle and M.Maumy, Tetrahedron Lett. **23** (1982) 1577

13 - G.Speier and Z.Tyeklar, J.C.S.Dalton Trans. (1988) 2663

14 - M.Gargano, N.Ravasio, M.Rossi, A.Tiripicchio and M.Tiripicchio Camellini, J.C.S. Dalton Trans. (1989) 921

15 - K.Prabhakaran, P.Sen and C.N.R.Rao, Surface Sci. **177** (1986) L971

16 - K.Prabhakaran, P.Sen and C.N.R.Rao, Surface Sci. **169** (1986) L301

17 - W.Brackman and E.Havinga, Rcl. Trav.Chim. Pays-Bas, **74** (1955) 937, 1021, 1070, 1100, 1107

18 - Y.Kitayama and T.Sato, Nippon Kagaku Kaishi **9** (1980) 1309, C.A. **94**:14755c

19 - A.Takuwa, O.Soga, H.Iwamoto and K.Maruyama, Bull.Chem.Soc.Jpn. **59** (1986) 2959

20 - Showa Denko K.K., Jpn. Kokai Tokkyo Koho JP 60,123,441 (85,123,441), C.A. **104**:5641k.

21 - O.Reinaud, P.Capdevielle and M.Maumy, Tetrahedron Lett. **26** (1985) 3993

22 - H.W.Wanzlich and U.Jahnke, Chem.Ber. **101** (1968) 3744; Ger. Pat. No. 1294969, C.A. 71:030237

23 - A.I.Zvonok, P.Matusevich, N.M.Kuz'menok, A.I.Kumachev, USSR Pat. 638,537, C.A. **90**:87050W

24 - Y.Itoh, T.Karuta, M.Hirano and T.Marimoto, Bull. Chem. Soc. Jpn. **52** (1979) 2169

25 - D.E.Wilcox, A.G.Porras, Y.T.Hwang, K.Lerch, M.E.Winkler and E.I.Solomon, J.Am.Chem.Soc. **107** (1985) 4015

U. TAKAKI (Mitsui Toatsu Chem. Inc., Japan): 1) You get binaphthyl compounds from β-naphtol. Why you do not get p,p'-biphenol from phenol, since you use the same experimental conditions?
2) Why don't you synthesize chiral Binap using chiral ligands by your coupling technique?

N. RAVASIO (C.N.R. MISO, Università di Bari, Italy): 1) Minor amounts of diphenoquinones and other polymeric products are in fact present in the residue of phenol reaction mixture. The products distribution found in the described reactions depends on the substrates oxidation potential, the presence of different copper oxidation states and reciprocal orientation of the intermediates coordinated to the metal.
2) Work is in progress on the synthesis of chiral binaphtol.

G. Centi and F. Trifiro' (Editors), *New Developments in Selective Oxidation*

PLATINUM CATALYZED OXIDATION OF 5-HYDROXYMETHYLFURFURAL

P. VINKE, H.E. van DAM[a] and H. van BEKKUM

Laboratory for Organic Chemistry, Delft University of Technology, P.O. Box 5045, 2600 GA Delft, The Netherlands.

[a]Present address: Norit N.V., P.O. Box 105, 3800 AC Amersfoort, The Netherlands.

SUMMARY

The liquid phase oxidation of 5-hydroxymethylfurfural (HMF) over platinum on alumina catalysts is described. The main intermediate is 5-formyl-2-furancarboxylic acid (FFCA), indicating that the hydroxymethyl group of HMF is oxidized first in the presence of the aldehyde group. The catalyst is not deactivated by oxygen, due to a strong metal/substrate interaction via the π-electrons of the furan nucleus.

INTRODUCTION

In recent years the interest in the use of carbohydrates as chemical feedstock is growing considerably (1, 2, 3, 4). Attention is given both to the use of renewables as starting material for existing products as well as to the replacement of oil-derived chemicals by new products made from renewables. One of the options of the second category is the acid catalyzed dehydration of carbohydrates (e.g. fructose) yielding 5-hydroxymethylfurfural (HMF) (5). HMF may serve as a starting material in several industrial applications (6). For example, the oxidation products of HMF (see scheme 1) are potential building units for polymers.

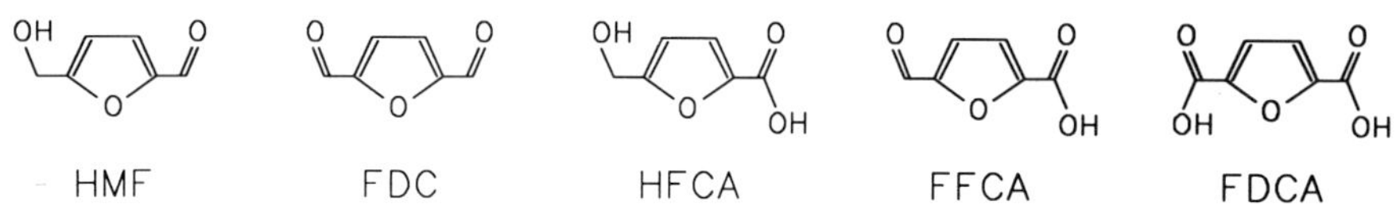

Scheme 1. Oxidation products derived from HMF: 2,5-furandicarboxaldehyde (FDC), 5-hydroxymethyl-2-furancarboxylic acid (HFCA), 5-formyl-2-furancarboxylic acid (FFCA) and 2,5-furandicarboxylic acid (FDCA).

Several methods have been described to synthesize FDC, HFCA and FDCA. Thus, FDC can be prepared by oxidation of HMF with manganese dioxide (7) or barium permanganate (8). HFCA is formed by oxidation of HMF with molecular oxygen over a combined silver oxide/copper oxide catalyst (9). FDCA can be obtained by oxidation with oxygen over a palladium on carbon catalyst (9). Up to now the preparation of FFCA from HMF has not been described.

Noble metals (e.g. platinum and palladium) on a carrier are widely used as catalyst in oxidation reactions. Very little is known about the detailed mechanism of the oxidation reaction which takes place at the noble metal surface. Some mechanistic considerations on the noble metal catalyzed oxidation sequence, primary alcohol → aldehyde → carboxylate, are given below.

SUPPORTED NOBLE METAL OXIDATION CATALYSTS

The platinum catalyzed oxidation of primary alcohol groups can be described as a stepwise oxidative dehydrogenation as shown in Scheme 2. During the oxidation the metal surface is largely covered by hydrogen which is oxidized by adsorbed atomic oxygen.

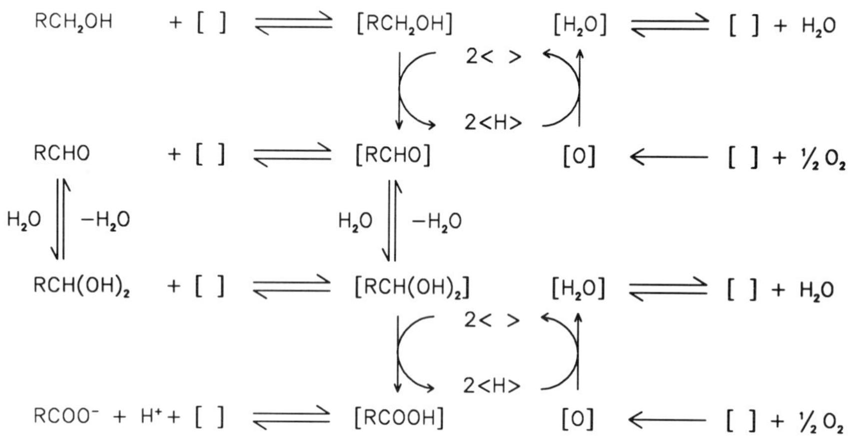

Scheme 2. Mechanism of the oxidative dehydrogenation of alcohols over a platinum catalyst in aqueous media. (From (10), with permission.)
[ ] metal surface site, < > (sub-surface) hydrogen site.

In non-aqueous media the reaction stops in the aldehyde stage. If water is present the aldehyde is hydrated to a geminal diol and further dehydrogenated yielding a carboxylate group.

When the rate of dehydrogenation of the substrate is lower than the rate of oxidation of adsorbed hydrogen the catalyst is deactivated due to the excessive adsorption of oxygen. Then, the metal is covered by chemisorbed oxygen, probably as hydroxyl species, and sub-surface hydrogen can not be formed anymore. Upon longer exposure to oxygen, the metal surface is covered with an amorphous oxide layer and all catalytic activity is lost. Thus, the catalyst surface can appear in three forms, (i) active catalyst covered with sub-surface hydrogen, (ii) deactivated catalyst with low activity covered with chemisorbed hydroxyl species, and (iii) poisoned catalyst with an amorphous oxide layer. If necessary, this deactivation by oxygen can be prevented by applying low oxygen partial pressures or by using 'diffusion stabilized' catalysts (11).

In principle, all noble metals which are able to perform (i) the dehydrogenation of the substrate and (ii) the oxidation of adsorbed hydrogen at the same time, are suitable catalysts for the oxidation of HMF. In practice only the platinum group metals (Pt, Pd, Rh, Ru and Ir) can be used. However, large differences in turnover number (TON) and sensitivity for oxygen deactivation have been found (12). In the case of methanol oxidation, platinum is the least sensitive for oxygen and has the highest TON of the metals mentioned before.

The present paper describes the platinum-catalyzed oxidation of HMF with a focus on the selective formation of FFCA. A model for the oxidation mechanism is proposed. The influences of reaction conditions, such as temperature and pH, on the selectivity and rate of the oxidation were studied.

MATERIALS AND PROCEDURES

Oxidation equipment

Experiments were performed in a thermostatted glass batch reactor of 300 ml, equipped with a glass gastight stirrer (1500 rpm). The pH was kept constant using a pH meter (Metrohm 654) coupled to a pH control unit (Metrohm 614) and an automatic burette (Metrohm 655, 10 ml piston) containing 2.00 M potassium hydroxide. The oxygen partial pressure of the gas phase could be adjusted to any desired value between 0.05 and 1 and was kept constant during the reaction using an automatic oxygen supply system. This system consisted of a motor burette filled with water as displacing liquid, a thermostatted (30 °C) gas burette filled with oxygen, and a differential pressure sensor, which operated the motor burette. The oxygen and hydroxyde uptakes were recorded during the reactions. The oxygen concentration in the liquid phase could be monitored too, by using an Orion 970899 oxygen electrode. The oxidation set-up is shown in Figure 1.

Oxidation procedure

(i) Reduction of the standard catalyst. 1 g of dry powdered catalyst (5% platinum on alumina, platinum dispersion 0.30, Janssen Chimica, Beerse, Belgium) was introduced in the reactor, 50 ml of water was added and the system was flushed with nitrogen (ca. 500 ml/min) to remove oxygen from the reactor. Then, hydrogen was conducted through the reactor for 5 min at high flow and low stirring speed, followed by an additional 25 min at low flow and high stirring speed. Finally the hydrogen was removed from the gas phase by flushing with nitrogen for 5 min.

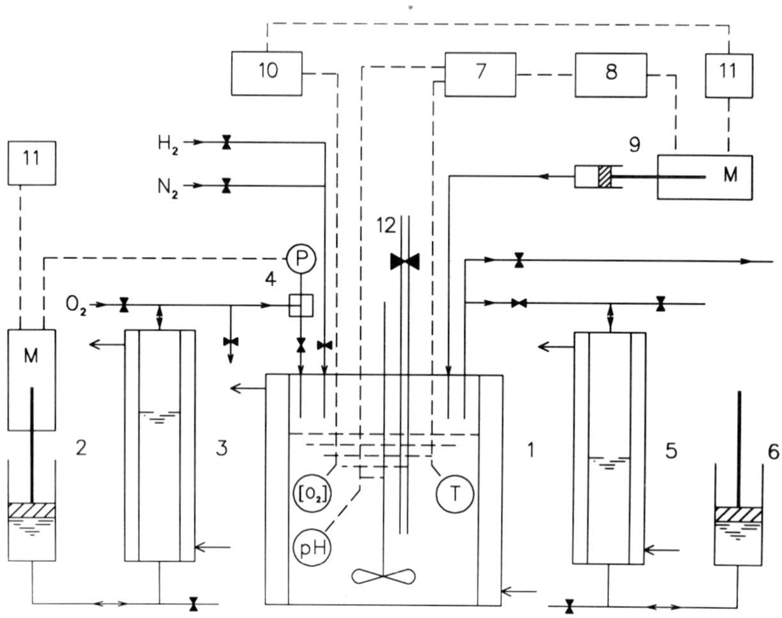

Figure 1. Batch oxidation equipment. 1 thermostatted batch reactor, 2 motor burette, 3 gas burette, 4 differential pressure sensor, 5 gas burette, 6 manually operated piston, 7 pH meter, 8 pH control unit, 9 automatic motor burette with storage vessel, 10 oxygen sensor, 11 recorders, 12 sample tube.

(ii) Starting the reaction. 8 mmol of substrate in 30 ml of water was added to the reduced catalyst under a low nitrogen flow, to prevent introduction of oxygen. After the system was equilibrated at the preset temperature, the desired oxygen partial pressure was set by sucking a calculated amount of gas out of the reactor with piston 6 (see Figure 1), which was automatically replaced by pure oxygen. The reaction started after the pH was adjusted by activating the pH control system. After 3 minutes the first sample was drawn.

(iii) Sample preparation for HPLC. Samples of ca. 0.4 ml were spinned in a small tube (V= 2 ml) for 1 min to allow the catalyst to settle down. The clear solution was collected and stored at -20 °C. Just before HPLC analysis 200 μl of solution was added to 200 μl of 1,5-pentanediol solution (20 mg/ml), which was used as internal standard.

HPLC analysis

The system consisted of a Waters 590 chromatography pump, a Waters R401 differential refractometer and a Perkin-Elmer ISS-100 autosampler. A Biorad

HPX87H column (strong cation exchange resin in the H^+ form) was used with $3*10^{-3}$ M trifluoroacetic acid (TFA) as mobile phase at 60 °C.

For HPLC-MS measurements a similar system with a Waters 510 chromatography pump was used, connected to a VG 70-SE mass spectrometer. The ionisation was accomplished by plasma spray.

NMR measurements

^{13}C NMR spectra were recorded on a Varian VXR-400S spectrometer. Sample concentrations were ca. 0.3 M and at a pH of 9. By applying long relaxation times the spectra could be interpreted quantitatively. Deuterium oxide was added to lock the signal. Proton spectra were recorded on a Nicolet NT-200 WB apparatus. Sample concentrations were 0.2 M in deuterium oxide. No internal standard was applied.

RESULTS AND DISCUSSION

Selective oxidation of HMF to FFCA

In Figure 2 the reaction mixture composition of a typical oxidation experiment of HMF is shown as a function of time. In every oxidation experiment the final oxidation product was FDCA, which was formed in quantitative yields.

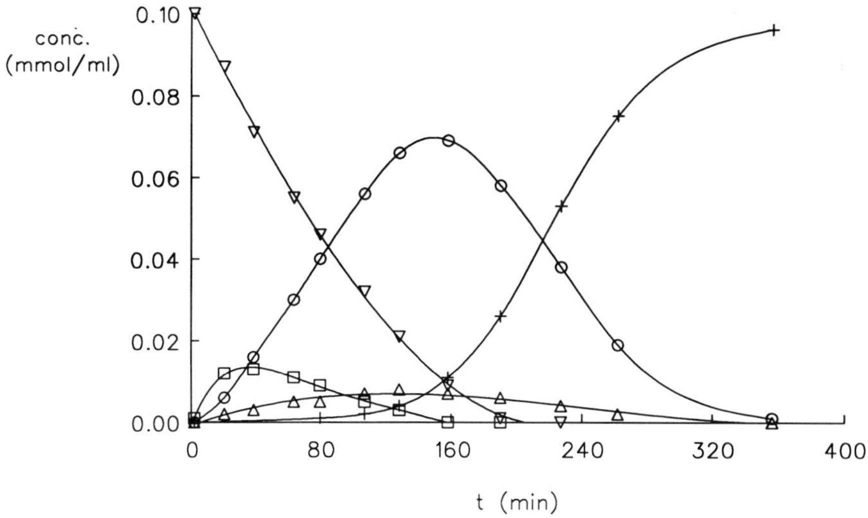

Figure 2. Oxidation of HMF over a platinum catalyst. Reaction conditions: T 60 °C, pH 9.0, $p(O_2)$ 0.2 atm, p(total) 1.0 atm, C_0(HMF) 0.1 M, 1.00 g 5% platinum on alumina powdered catalyst, $V(H_2O)$ 80 ml. ∇= HMF, □= FDC, Δ= HFCA, ○= FFCA and += FDCA.

At this stage, the catalyst was deactivated by oxygen chemisorption because of the low reactivity of FDCA. This paper will focus on the selective formation of the intermediate FFCA and the factors determining the selectivity.

At the present conditions, aldehydes usually are more reactive than primary alcohols. So, the intermediate formed with a maximum yield at t= 160 min, was expected to be HFCA. However, the product formed in high yields proved to be FFCA. Its structure was determined by HPLC coupled to a mass spectrometer (HPLC-MS). The molecular mass peak of the intermediate was 141 (M_{FFCA}+ 1). The fragmentation pattern of the intermediate was in accordance with that of FFCA. The ^{13}C NMR spectrum of the reaction mixture at t= 160 min confirmed that FFCA was the intermediate (main peaks at δ= 183.1 ppm (-C=O), δ= 166.5 ppm ($-CO_2^-$), δ= 155.5 ppm (C5 furan), δ= 153.3 (C2), δ= 126.5 ppm (C4), δ= 117.5 ppm (C3). Comparison of the experimental and calculated oxygen uptake also showed that FFCA is the main intermediate.

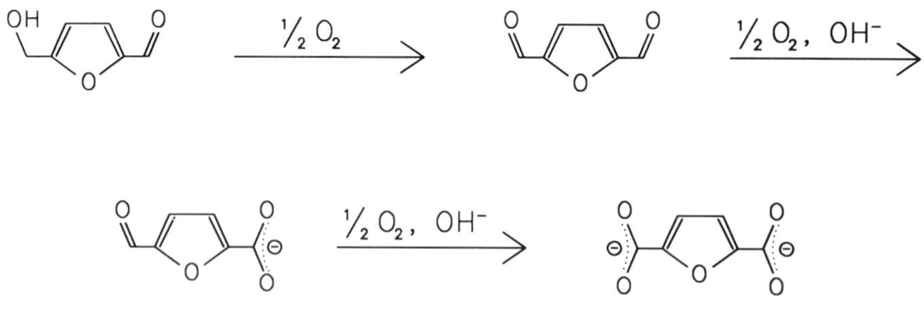

Scheme 3. Main oxidation sequence of HMF upon platinum catalyzed oxidation.

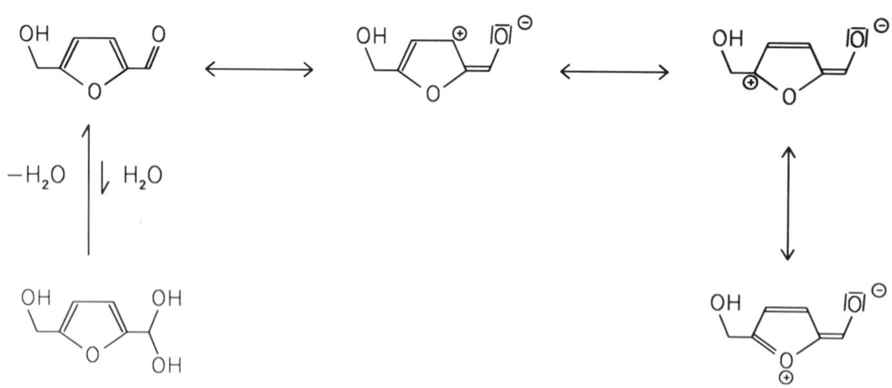

Scheme 4. Resonance structures and equilibrium hydration of HMF.

In Scheme 3 the main oxidation reaction sequence is shown. The unusual initial selectivity for alcohol instead of aldehyde oxidation may be caused by the conjugation of the aldehyde group with the aromatic nucleus (cf. Scheme 4). This could prevent hydration of the aldehyde to the reactive geminal diol. Indeed, ^1H NMR measurements indicated that the aldehyde group is hydrated for less than 1% in aqueous solutions at a temperature range of 30-70 °C and a pH of 8-11.

The effect of varying reaction conditions on the rate of reaction

In Figure 3 the rates of reaction at low conversion and the maximum yields of FFCA are shown for several different reaction conditions. The reaction rates are determined at low conversion (ca. 5%) and differ slightly from the true initial rates.

(i) Variation of oxygen partial pressure (not shown in Figure 3). The rate of reaction is essentially first order in oxygen partial pressure in the gas phase.

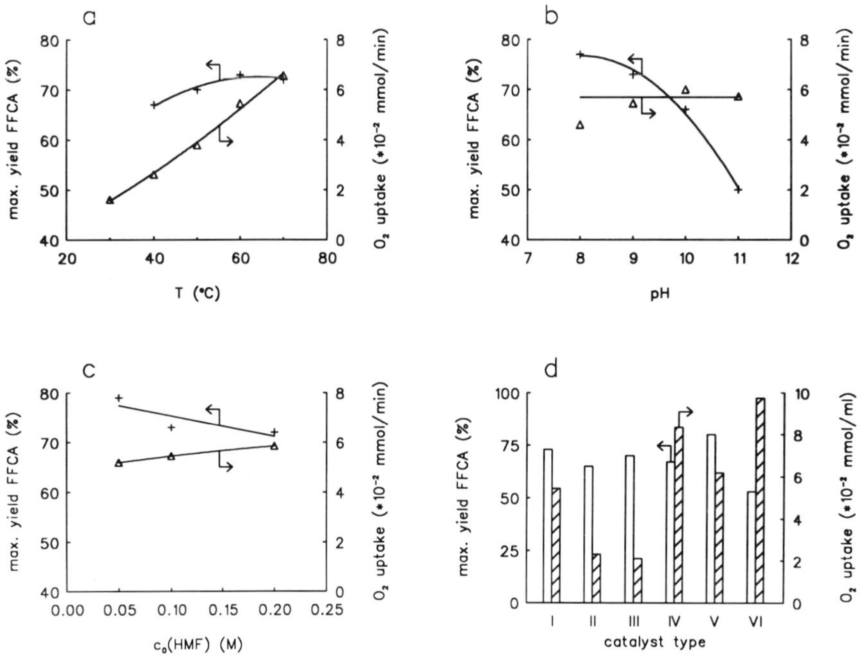

Figure 3. The rate of reaction at low conversion and the maximum concentration of FFCA versus (a) temperature, (b) pH, (c) initial HMF concentration, and (d) catalyst type. I 5% Pt/alumina, II 1% Pt/alumina, III 5% Pt/alumina extrudates, IV 5% Pt/carbon, V Adams' catalyst, and VI 5% Pd/alumina. Standard conditions: T 60 °C, pH 9.0, $p(O_2)$ 0.2 atm, p(total) 1.0 atm, C_0(HMF) 0.1 M, 1.00 g 5% Pt/Al_2O_3 powdered catalyst, $V(H_2O)$ 80 ml.

At 30 °C and a $p(O_2)$ of 0.2 atm the oxygen concentration in the liquid phase during reaction is 6-7 ppm (saturation 7.5 ppm). Therefore the gas/liquid mass transfer of oxygen is not rate limiting, at least not at low temperatures. The diffusion into the catalyst particles is unlikely to be rate limiting as can be seen for the oxidation with Adams' catalyst (powdered pure platinum) which has a comparable rate of reaction. Other studies also indicate that oxygen diffusion into the catalyst particle is not rate limiting (13). Thus, the chemisorption of oxygen on to the platinum surface has to be the rate determining step.

(ii) Variation of temperature. In the temperature range studied, the rate of oxidation follows an exponential curve as a function of temperature. Assuming (see above) the reaction to be first order in $[O_2]_L$, rate constants can be calculated and the Arrhenius parameters can be obtained. It has to be taken into account that the solubility of oxygen is temperature dependent, so the rate of reaction has to be corrected for these differences. The activation energy for this reaction appears to be 37.2 kJ/mol which is in good accordance with other platinum catalyzed oxidations (e.g. glucose oxidation, $E_A \approx 40$ kJ/mol) (14).

(iii) Variation of pH. The rate of reaction is independent of the pH in the range of pH 8-11, which result is in contrast to other studies in this field, reporting increasing reaction rates at higher pH values (14). These authors explained the increase in reaction rate by assuming a higher degree of ionisation of the hydroxyl (or hydrated aldehyde) group of the substrate. Apparently, this process is of no importance for the rate of the present oxidation reaction.

At low pH values (pH$\leq$ 8) the reaction rate decreases somewhat because carbon dioxide, which is formed in small amounts by oxidative cleavage of the substrate, evolves from solution, thus lowering the oxygen partial pressure in the gas phase. At higher pH values the carbon dioxide is kept in solution as (bi)carbonate.

(iv) Variation of initial substrate concentration. The initial substrate concentration did not influence the reaction rate significantly (zero order), except for very low values of the concentration. In that case the dehydration of the substrate is slower than the oxidation of chemisorbed hydrogen and the catalyst is deactivated.

(v) Variation of catalyst type. The reaction rate is strongly dependant on the type of catalyst used. This confirms the fact that the gas/liquid mass transfer of oxygen is not rate limiting. The reaction rate for palladium is much higher than that for platinum. Even at a high oxygen concentrations in the liquid phase ($[O_2]_L$= 6-7 ppm) the palladium catalyst remains active, in contrast to oxidation of methanol where the catalyst is poisoned at $[O_2]_L \approx 1$ ppm (12).

In general, noble metal catalyst are often deactivated when the oxygen concentration in the liquid phase is too high. In the case of HMF, however, the catalyst remains active and stable, even at very high oxygen concentrations in solution. This can be explained by assuming a strong metal/substrate interaction, in which the substrate is adsorbed strongly onto the metal surface. The interaction of a hydroxyl or aldehyde group with the metal is probably not strong enough to prevent oxygen chemisorption, as can be concluded from the oxidation of methanol or glucose where catalyst deactivation occurs. Therefore the interaction of the π-electron system of the aromatic furan ring is believed to be responsible for this strong adsorption. This model is supported by the fact that the rates of reaction are zero order in initial substrate concentration. So, at any time during reaction the platinum surface is largely covered with substrate molecules. In this way the oxygen coverage is kept low and the catalyst remains active.

The rates of reaction of FDC and HFCA are different from HMF (O_2 uptake at standard conditions $2.71*10^{-2}$ mmol/min for FDC, and $3.87*10^{-2}$ mmol/min for HFCA). This could be caused by differences in rates of dehydrogenation, leading to a difference in hydrogen occupation of the platinum. Due to the difference in hydrogen coverage, the chemisorption of oxygen is affected.

At low degree of conversion (5 to 25%) the rate of reaction is decreased, compared to the initial rate. At the start of the oxidation the platinum surface is covered with HMF. Even at very low conversions, part of the metal surface will be occupied by the first intermediate FDC, due to the very strong interaction of FDC with the metal. Because FDC has a lower rate of oxidation, this causes a decrease in overall reaction rate, which is experimentally found. At higher conversions, when the amount of FDC is almost zero, the rate of reaction is increasing again. The adsorption of the second intermediate FFDC on to the platinum surface is less strong and the oxidation reaction proceeds faster.

The selectivity towards FFCA

In Figure 3 the maximum yields of FFCA are shown for several different reaction conditions.

(i) Variation of oxygen partial pressure (not shown in Figure 3). The variation of oxygen partial pressure does not have any effect on the selectivity of the reaction, which is in accordance with the model presented above.

(ii) Variation of temperature. A change in temperature has little effect on the selectivity. At lower temperatures the maximum yield of FFCA is somewhat less. This may be caused by a change in adsorption characteristics of HMF and FDC at the metal surface.

(iii) Variation of initial substrate concentration. The substrate concentration has little influence on the selectivity of the oxidation reaction, which is in accordance with the model of the reaction.

(iv) Variation of the pH. At high pH values the selectivity decreases significantly. The loss of selectivity is caused by the concurrent formation of HFCA as intermediate and not by direct oxidation of FCD to FDCA. Apparently, at high pH values (≥10) the oxidation of the aldehyde group proceeds more easily. Possibly, the hydrated aldehyde is stabilized at the platinum surface by ionization of the geminal diol. Because a geminal diol is more reactive in oxidation reactions than an alcohol, a large amount of HFCA is formed.

In principle, the oxidation of HFCA can give FFCA in high yields, but experimentally this is only 40%. This low selectivity for FFCA upon oxidizing HFCA can be explained by assuming that the adsorption of HFCA and FFCA on to the metal surface will not differ significantly. Consequently, the two substrates will be oxidized simultaneously.

The high selectivity for FFCA upon oxidation of HMF at moderate pH values can be explained too. The hydrated aldehyde is not 'stabilized' by ionization and therefore the alcohol group will be oxidized selectively, yielding FDC, which is oxidized to FFCA. The ionized carboxylate group of FFCA will decrease the metal/substrate interaction, so FDC is adsorbed predominantly, even in the presence of FFCA. Experimentally, oxidation of FDC is giving FFCA in 95% yield.

(v) Variation of the catalyst type. The selectivity is dependant on the type of platinum catalyst used. Possibly the dispersion of the catalyst influences the interaction of the substrate with the platinum. At high dispersions the larger density of steps and edges on the noble metal crystallite surface could decrease the interaction of the substrate with the metal, thus lowering the selectivity.

CONCLUSIONS

The oxidation of HMF over platinum on alumina catalysts proceeds with high selectivity towards the intermediate FFCA. This is believed to be caused by the conjugation of the carbonyl bond with the aromatic furan nucleus. Thus, the aldehyde is only slightly hydrated to a geminal diol, which is the reactive species in the oxidative dehydrogenation to the corresponding carboxylic acid.

The oxygen concentration in the liquid phase is rate determining for this oxidation, although no diffusion limitation is observed. This can be explained by assuming a strong metal/substrate interaction, which also prevents oxygen to deactivate the catalyst. The aromatic nuclei are believed to be responsible for this interaction, which is supported by the fact that the reaction is zero order in substrate concentration but dependant on type of substrate.

ACKNOWLEDGEMENTS

We wish to thank Süddeutsche Zucker A.G. for generously providing a sample of HMF, and Dr. Jan Douwstra of TNO, Division of Technology for Society, The Netherlands for providing FDC. The investigation was supported by the Netherlands Organization for Scientific Research (NWO).

REFERENCES

1 A. Fuchs, Starch/Stärke, 10, (1987), 335-43.
2 A.J.J. Straathof, A.P.G. Kieboom, and H. van Bekkum, Carbohydr. Res., 146, (1986), 154-9; A.J.J. Straathof, A.P.G. Kieboom, and H. van Bekkum, Starch/Stärke, 40, (1988), 229-34; A.J.J. Straathof, Alkyl glucoside surfactants from starch and sucrose, Thesis Delft University of Technology, The Netherlands, (1988).
3 H. Schiweck, K. Rapp, and M. Vogel, Chem. Ind., 4, (1988), 228-34.
4 J.L. Hickson (editor), Sucrochemistry, ACS Symposium Series 41, American Chemical Society, Washington D.C., (1977).
5 H.E. van Dam, A.P.G. Kieboom, and H. van Bekkum, Starch/Stärke, 3, (1986), 95-101.
6 A. Faury, A. Gaset, and J.P Gorrichon, Inf. Chim., 214, (1981), 203-9.
7 A.F. Oleinik, and K.Y. Novitskii, J. Org. Chem. USSR, 6, (1971), 2643.
8 T. El-Hajj, J.-C. Martin, and G. Descotes, J. Heterocyclic Chem., 20, (1983), 233-235.
9 B.W. Lew, US Patent 3.326.944, (1967).
10 H.E. van Dam, A.P.G. Kieboom, and H. van Bekkum, Appl. Catal., 33, (1987), 361-72.
11 H.E. van Dam, P. Duijverman, A.P.G. Kieboom, and H. van Bekkum, Appl. Catal., 33, (1987), 373-82.
12 H.E. van Dam, Carbon supported noble metal catalysts in the oxidation of glucose-1-phosphate and related alcohols, Thesis Delft University of Technology, The Netherlands, (1989).
13 P.J.M. Dijkgraaf, H.A.M. Duisters, B.F.M. Kuster, and K. van der Wiele, J. Catal., 112, (1988), 337-44.
14 P.J.M. Dijkgraaf, Oxidation of glucose to glucaric acid by Pt/C catalysts, Thesis Eindhoven University of Technology, The Netherlands, (1989).J

B. DELMON (Université Catholique de Louvain, Belgium): At non-perfect conditions (inadequate support or O_2 pressure) you observe a deactivation of your catalyst. There are, in principle, two reasons at least why such a deactivation could occur:
- oxidation of the Pt surface
- polymerization of the aldehyde group

You nicely solved the problem. Nevertheless, it would be interesting to identify the origin of deactivation (and, thus, the real role of the favorable modifications you make).
One can notice that both possible causes of deactivation can be affected by O_2, (*i*) the oxidation of the Pt surface, which is obvious, and (*ii*) the condensation of the aldehyde through the effect on acidity of the support by spill-over oxygen. Do you have physico-chemical information on the possible cause of deactivation, and did you try other supports?

P. VINKE (Delft University of Technology, The Netherlands): First of all I have to emphasize that in the case of oxidation of aromatic compounds such as 5-hydroxymethylfurfural (HMF) this deactivation does not occur untill the oxidation is completed. This is probably caused by a protective interaction of the π-electrons of the aromatic nucleus with the noble metal surface. However, in many other cases deactivation of the catalyst is a serious problem. In our laboratory the deactivation of the catalyst is studied using activated carbon as carrier and methanol as substrate (1). We found that the electrochemical potential of the catalyst particles is changing during deactivation, indicating a change in chemical structure of the noble metal. These potential measurements lead to the conclusion that the metal is changing from the reduced state into the oxidized state during deactivation. This clearly shows that direct oxidation of the noble metal surface causes the catalyst poisoning.

1. H.E. van Dam and H. van Bekkum, Recl. Trav. Chim. Pays-Bas, in press.
H.E. van Dam, Carbon supported noble metal catalysts in the oxidation of glucose-1-phosphate and related alcohols, Thesis Delft University of Technology, The Netherlands, (1989).

D. ARNTZ (Degussa A.G., Hanau, BRD): The comparison on activity was made only in view of precious metal content. Because of the different dispersions due to different preparation methods a better characterization would be a correlation between activity and number of active centers. Are there measurements on the dispersions of the active phase and did you correlate them to the activity?

P. VINKE (Delft University of Technology, The Netherlands): Indeed, it is interesting to relate the activity with the amount of exposed noble metal. Therefore, I will give you the TON's (turnover numbers) as mol O_2/mol metal exposed/min for the different catalysts as described in Figure 3.

Table. TON's for the six catalysts tested (see Figure 3).

catalyst code	I	II	III	IV	V	VI
catalyst type	5% Pt/ Al_2O_3	1% Pt/ Al_2O_3	5% Pt/ Al_2O_3 [1]	5% Pt/ C	Pt black	5% Pd/ Al_2O_3
dispersion[2]	0.30	0.15	0.07	0.51	0.02	0.07
TON (/min)	0.71	3.02	1.31	0.61	2.46	3.28

[1] extrudates, [2] measured as mol CO adsorbed per mol noble metal

As can be seen from these results, the TON's differ significantly. Not only the two noble metals show different values, but the platinum catalysts used are not comparable either. A trend can be observed towards higher TON's at lower dispersions. Therefore it can be concluded that the dispersions also influence the TON for this oxidation reaction.

G. Centi and F. Trifiro' (Editors), *New Developments in Selective Oxidation*

THE USE OF PRECIOUS METAL CATALYSTS SUPPORTED ON ACTIVATED CARBON IN OXIDATION REACTIONS FOR THE SYNTHESIS OF FINE CHEMICALS, ESPECIALLY FOR THE SELECTIVE OXIDATION OF GLUCOSE TO GLUCONIC ACID

B. M. Despeyroux, K. Deller, E. Peldszus
Degussa AG, Geschäftsbereich Anorganische Chemieprodukte
Abt. AC-AT 3-CK, Postfach 13 45, D-6450 Hanau 1

ABSTRACT

For the oxidation of glucose to gluconic acid in the liquid phase a trimetallic catalyst Pd-Pt-Bi on activated carbon has been developed. Activities of more than 4000 g [gluconic acid]/g [precious metal] x h were found. The selectivity obtained is higher than 96 mol%.

A detailed investigation of this catalyst as well as Pd-Bi on activated carbon, Pt on activated carbon and Pt-Bi on activated carbon has been carried out showing the effect of each metal component or the combination of them.

The role of Pt as a booster for activity and Bi as a booster for selectivity is shown.

There is no correlation between the results obtained by carrying out usual physico-chemical characterization techniques and the catalytic behaviour of this Pd-Pt-Bi on activated carbon system.

INTRODUCTION

The use of precious metal containing supported catalysts for the synthesis of fine chemicals under oxidative conditions in the liquid phase is mainly reported in the literature for the oxidation of alcohols, the oxidation of alkenes and the oxidation of sugars.

Especially for the catalytic oxidation of glucose, the main products are gluconic acid (industrially used as chelating agent for cleaning applications) and/or glucaric acid.

Another reaction product is 2-keto-gluconic acid (used in the manufacture of Vitamin C). By-products are fructose (isomerization of glucose) and other carboxylic acid products (cleavage oxidation reaction).

Catalysts with high activity, high selectivity and high stability are desirable to enable such a process to compete economically with the industrial fermentative synthesis of gluconic acid. Different catalysts are known already, most of them based on Pd on activated carbon (refs. 1 - 5). The use of Bi resp. Pb, leads to the selective formation of gluconic acid from glucose especially under alcaline reaction conditions.

We have investigated how the different possible catalyst systems Pd-Bi on activated carbon, Pt on activated carbon and Pt-Bi on activated carbon influence activity and selectivity for the oxidation of glucose to gluconic acid. Based on these results we have finally developed a trimetallic catalyst Pd-Pt-Bi on activated carbon.

We will also report results on the influence of reaction and catalyst parameters both on activity and selectivity.

EXPERIMENTALS

Materials

Glucose as glucose monohydrate from Fluka ("p.a." quality) resp. Riedel-de Haen ("rein" quality) was used as received. Pure O_2 was used.

The following Degussa catalysts were used:

CEF	196 RA/W	4 % Pd, 1 % Pt, 5 % Bi	(Pd-Pt-Bi on activated carbon)
CF	196 RA/W	5 % Pt, 5 % Bi	(Pt-Bi on activated carbon)
CE	196 RA/W	5 % Pd, 5 % Bi	(Pd-Bi on activated carbon)
F	196 RA/W	5 % Pt	(Pt on activated carbon)
F	196 B/W	5 % Pt	(Pt on activated carbon)

Catalyst preparation

An activated carbon powder (BET surface area: 1000 m^2/g) with high macropore content was used as support. The metallic phase on the catalyst was prepared using a solution of Bi_2O_3 (dissolved in HCl conc.) and/or hexachloroplatinic acid and/or palladium(II)chloride which was added to an aqueous suspension of activated carbon.

Co-precipitation by treatment with NaOH and final reduction leads to the desired metallic phase. The catalyst suspension was filtered and washed. The catalyst was used in the glucose oxidation without any further treatment.

Reaction conditions of glucose oxidation

All reactions were carried out in a 150 ml stirred glas vessel in batch type at a constant pressure and temperature including pH-control. A stirrer (type: BR1 from Buddeberg GmbH) assured a thorough gas mixing during the reaction.

The reaction products were neutralized with NaOH to the corresponding sodium salts during the reaction.

After a certain reaction time samples were taken and the catalyst was separated from the product containing solution. The filtrate was analyzed by HPLC, ion chromatography and thin layer chromatography. The stability of the catalyst was determined by recycling the catalyst and measuring the level in activity.

Standard reaction conditions

16 g glucose (~ 17,6 g glucose monohydrate), dissolved in 100 ml water

temperature:	55 °C
O_2-pressure:	10 mbar
pH:	10,0
NaOH-solution:	10 weight% for neutralization
stirrer rate:	1800 Rpm
catalyst conc.:	0,24 g (1.5 weight% based on glucose)

RESULTS AND DISCUSSION

Catalyst system Pd-Bi on activated carbon

The results obtained with different Pd-Bi on activated carbon catalysts are listed in Table 1.

TABLE 1
5 % Pd-5 % Bi on activated carbon catalyst for the glucose oxidation under standard reaction conditions: Influence of the catalyst preparation method on glucose conversion, gluconic acid selectivity and catalyst activity.

	catalyst A		catalyst B		catalyst C	
reaction time (min)	35	60	35	60	35	60
conversion of glucose (mol%)	69	100	79	100	100	100
selectivity to gluconic acid (mol%)	95	93	95	96	96	90
catalyst activity g[gluconic acid]/ g[palladium] x h	1700	1300	2000	1400	2400	1300

catalyst A: Prepared from the corresponding Pd on activated carbon catalyst followed by Bi-impregnation.

catalyst B: Prepared from the corresponding Bi impregnated activated carbon followed by Pd-impregnation and reduction.

catalyst C: Prepared by co-precipitation of the metal phase followed by reduction. Degussa catalyst CE 196 RA/W 5 % Pd, 5 % Bi

The use of Bi-Pd on activated carbon catalyst where first the Bi-salt is fixed on the activated carbon followed by the palladium impregnation shows high selectivities of gluconic acid, Na-salt but the activity of the catalyst is low.

The use of Pd-Bi on activated carbon catalyst where the fresh prepared Pd on activated carbon catalyst has been impregnated with the Bi-salt shows an activity less or comparable to the one described above (1500 g [gluconic acid/ g[palladium] × h).

Pd-Bi on activated carbon catalysts prepared by co-precipitation of the Bi-salt and the palladium(II)chloride acid with NaOH showed an increase of the activity to 2400 g [gluconic acid]/g [palladium] × h without affecting the selectivity to gluconic acid.

<u>Catalyst system Pt on activated carbon</u>

Pd an Pt behave quite different during the oxidation of glucose.

Under low catalyst concentration (standard reaction conditions) the oxidation of glucose with Pt-containing catalysts leads to yields of gluconic acid obtained less than 70 %.

At high concentration of Pt on activated carbon catalyst (standard reaction conditions but with catalyst to glucose ratio ~ 20 weight%), the oxidation of glucose produces glucaric acid at a longer reaction time (refs. 5 - 8). High selectivity values of more than 80 mol% of glucaric acid can be obtained (see Fig. 1). The formation of by-products results from the oxidative degradation of the gluconic acid resp. the glucaric acid formed.

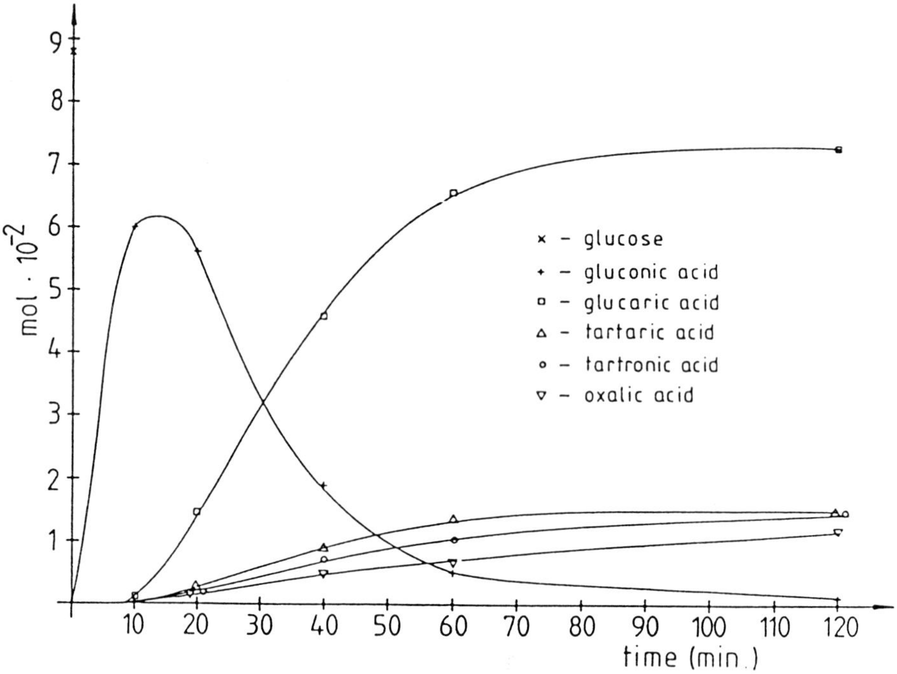

Fig. 1. Pt on activated carbon catalyst: glucose oxidation under standard reaction conditions but under catalyst to substrate ratio 20 : 100 with Degussa catalyst F 196 B/W 5 % Pt.

Catalyst system Pt-Bi on activated carbon

Pt-Bi on activated carbon catalysts improve the yield of gluconic acid obtained under standard reaction conditions (in comparison to Pt on activated carbon catalysts) but the yield of glucose is limited to 80 - 90 mol%.

At higher catalyst concentration (catalyst to glucose ratio: 20 weight%) more cleavage products and very unselective formation of glucaric acid were obtained.

The oxidation of the α-position of glucose is described in the literature (refs. 9 - 10). This oxidation of the α-position was also reported for other reaction types (alcohol oxidation) (refs. 11 - 13). Under the reaction conditions used in the present study this behaviour could not be observed.

Catalyst system Pd-Pt-Bi on activated carbon

Fig. 2 shows the rate of formation of gluconic acid as a function of time using a trimetallic Pd-Pt-Bi on activated carbon catalyst under standard reaction conditions. It is evident that the use of Pt can boost the activity of the Pd-Bi on activated carbon catalyst without influence on the selectivity. Activity values higher than 4000 g [gluconic acid]/g [precious metal] x h can be obtained.

The above mentioned catalysts systems are plotted on Fig. 2 for comparison.

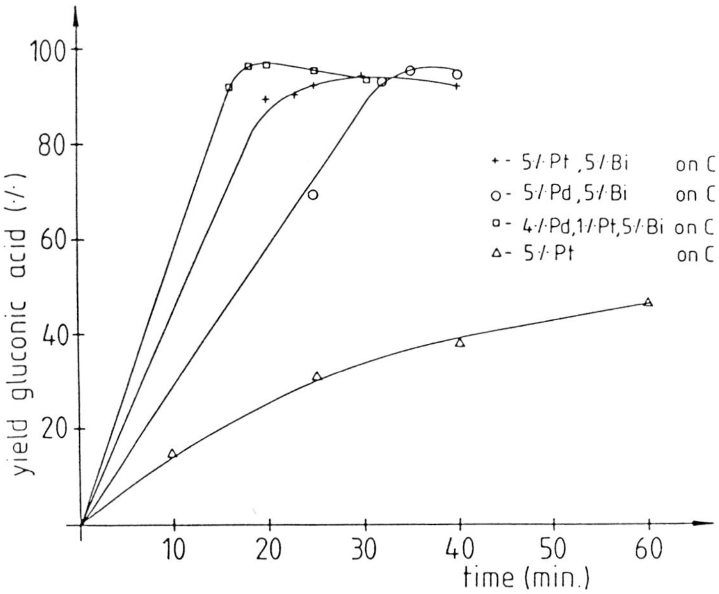

Fig. 2. Yield of gluconic acid obtained for the different catalyst systems used in the oxidation of glucose under standard reaction conditions.

In Table 2 the formation of all reaction products as a function of time for the trimetallic catalyst is shown.

The influence of the temperature and the pH of the reaction as well as the metal concentrations used in the trimetallic catalyst was investigated.

It was found out that a metal content of 4 % Pd, 1 % Pt and 5 % Bi gives the best results. The optimized reaction conditions already given in this study were used as standard to compare the different catalyst systems.

The catalyst stability has been investigated. The recycling of the catalyst more than 50 times is possible with regeneration of the catalyst. Further investigations are planned.

TABLE 2
Pd-Pt-Bi on activated carbon catalyst: Glucose oxidation under standard reaction conditions with Degussa catalyst CEF 196 RA/W 4 % Pd, 1 % Pt, 5 % Bi.

	reaction time [min]			
amount of substances [mol × 10^{-2}]	18	20	25	30
glucose	0,15	< 0,01	< 0,01	< 0,01
gluconic acid	8,50	8,58	8,44	8,13
fructose	0,08	0,13	0,13	0,13
glucaric acid	0,03	0,05	0,09	0,32
tartaric acid	< 0,01	< 0,01	0,05	0,07
tartronic acid	< 0,01	< 0,01	0,09	0,17
oxalic acid	< 0,01	0,01	0,06	0,14

conversion (20'): 100 %

selectivity (20'): 98 %

activity (20'): 4200 g[gluconic acid]/g[precious metal] × h

CATALYST CHARACTERIZATION OF DEGUSSA CATALYST CEF 196 RA/W
4 % Pd, 1 % Pt, 5 % Bi

The analysis of the metallic phase by energy dispersive analysis X-ray (EDX) shows that the catalyst particles are totally impregnated and the metal very well homogeneously dispersed throughout the catalyst particles.

The metal dispersion on the catalyst surface is low (measured by CO-adsorption) and comparable with values of other Pd/Pt bimetallic catalysts without Bi.

The cristallite size was measured by TEM and revealed well cristallized Bi in rod shape besides Pt-Pd agglomerates of about 2 - 5 nm size (comparable to cristallite size of Pd/Pt bimetallic catalysts without Bi).

ESCA/SIMS investigations demonstrate that under optimized preparation conditions the Pd-phase is still mainly oxidized, whereas the Pt-phase is mainly reduced. The Bi-phase was found to be in the oxide form as Bi_2O_3 and $Bi_2O_2CO_3$. The last compound could be interpreted as an interaction of Bi with the support leading to the carbonate formation.

The predominant role of Bi in the very selective oxidation of glucose to gluconic acid still remains undisclosed. No interaction of Bi with the precious metals Pd and Pt could be detected. No alloy formation could be seen in ESCA.

Pure Bi on activated carbon catalyst (without precious metal) is totally inactive in this reaction. The presence of precious metal (Pd or Pt) is necessary.

SUMMARY

The use of a trimetallic catalyst Pt-Pd-Bi on activated carbon proved to be superior in activity, selectivity and stability in comparison to other bimetallic Pd-Bi on activated carbon or Pt-Bi on activated carbon catalyst systems for the gluconic acid formation from glucose. The enhancement in activity by the addition of Pt to Pd-Bi on activated carbon catalyst is surprising and could not be explained by the expected behaviour of both Pd (selective oxidation of the aldehyde function of the glucose) and Pt (selective oxidation of the position 6 resp. the position 2 of the glucose chain) alone.

Also the preponderant role of Bi as a selectivity booster in the gluconic acid formation remains undisclosed and could so far not be clarified by usual physical characterization methods. Only the formation of $Bi_2O_2CO_3$ could be observed showing a chemical interaction between the Bi-phase and the support.

REFERENCES

1 Kao Corporation
Eur. Pat. EP 142725 on May 29, 1985; CA 103(23): 196366m
Jap. Pat. JP 60/92240 on May 23, 1985; CA 103(II): 88175q
Jap. Pat. JP 59/205343 on Nov. 20, 1984; CA 102(17): 149721t
Jap. Pat. JP 58/72538 on Apr. 30, 1983; CA 99(II): 88543g
Eur. Pat. EP 48974 on Apr. 7, 1982; CA 97(5): 39311e
Jap. Pat JP 55/7230 on Jan. 19,1980; CA 93(7): 72211n
Belg. Pat. BE 851804 on June 16, 1977; CA 88(23): 170441d
2 Roquette Frères
Eur. Pat. EP 233816 on Jan. 30, 1986; CA 108(21):187206k
Eur. Pat. EP 232202 on Jan. 30, 1986; CA 108(21): 187205j
3 Towa Kasei Kogyo Co., Ltd.
Jap. Pat. JP 59/225140 on Dec. 18, 1984; CA 102(21): 185439r
4 Asahi Chemical Industry Co. Ltd.
Jap. Pat JP 55/47672 on Apr. 4, 1980; CA 94(I): 4223p
Jap. Pat JP 55/40606 on March 22, 1980; CA 93(23): 221020d
5 Johnson Matthey and Co., Ltd.
Brit. Pat. GB 1208101 on Oct. 7, 1970; CA 74(4): 14347h
6 P.J.M. Dijkgraaf, H.A.M. Duisters, B.F.M. Kuster, K. van der Wiele
Journal of Catalysis 112, 329 - 336 (1988)
Journal of Catalysis 112, 337 - 344 (1988)
7 P.J.M. Dijkgraaf, oxidation of glucose to glucaric acid by Pt/C catalysts, PhD, TU Eindhoven, Netherlands (1989)
8 H.E. van Dam, A.P.G. Kieboom, H. van Bekkum
Appl. Cat. 33, 373 (1987)
9 Akzo N.V.
Eur. Pat. EP 151498 on Aug. 14, 1985; CA 103(19): 160805q
10 Mitsui Toatsu Chemicals Inc.
U.S. Pat. US 4599446 on Jul. 8, 1986; CA 105(23): 209345x
Jap. Pat. JP 60/54338 on March 28, 1985; CA 103(13): 105264n
Jap. Pat. JP 57/163340 on Oct. 7, 1982; CA 98(13): 107688f
11 Bayer AG
Ger. Offen. DE 2836327 on Feb. 28, 1980; CA 93(5): 46193p
Ger, Offen. DE 2824407 on Dec. 13, 1979; CA 92(21): 180833e
12 Ube Industries, Ltd.
Jap. Pat. JP 55/22615 on Feb. 18, 1980; CA 93(3): 26425n
13 Mitsui Toatsu Chemicals, Inc.
Jap. Pat. JP 56/158733 on Dec. 7, 1981; CA 96(21): 180978p
Brit. Pat. GB 2018773 on Oct. 24 1979; CA 93(I): 7662w

H. Hoffmann (Univ. of Erlangen, West-germany): Can you indicate how the pH value changed during the reaction ?
How did you stabilize an alkaline pH ?
B.M. Despeyroux (Degussa AG, West-Germany): The pH was maintained constant during the reaction by the use of a pH-regler and adding NaOH. A pH value of 10 ± 0.1 could be achieved.

R. Chunk (Lonza AG, Switzerland): The role of the bismuth promotor in improving the selectivity remains unclear. Since this observation is not restricted to this reaction, but is an often observed phenomenon in heterogeneous catalysis, it seems to me important to understand the role of promoters in oxidation catalysis. An understanding of the mechanism/structures involved would allow us to "tailor make" catalysts for specific oxidation reactions.

G. Centi and F. Trifiro' (Editors), *New Developments in Selective Oxidation*

LIQUID-PHASE OXIDATION OF HYDROCARBONS AND ALCOHOLS CATALYZED BY HETEROGENEOUS PALLADIUM AND PLATINUM CATALYSTS

M. HRONEC, Z. CVENGROSOVA, J. TULEJA and J. ILAVSKY
Faculty of Chemistry, Slovak Technical University
812 37 Bratislava (Czechoslovakia)

SUMMARY

Activity and selectivity of supported Pd and Pt catalysts have been studied in the liquid-phase oxidation of hydrocarbons and alcohols to ketones and carboxylic acids. It was found that the rate of these reactions is mostly controlled by mass transfer effects. At higher partial pressure of oxygen the catalysts are reversibly deactivated by oxygen. Higher resistance against deactivation and higher catalytic activity of Pd and Pt catalysts is achieved by doping them with some metals.

INTRODUCTION

Palladium and platinum supported on charcoal are known as selective catalysts for the oxidation of hydrocarbons, alcohols and other organic compounds [1-4]. The reactive metal centers of these catalysts are capable to activate oxygen and C-H bonds, but during the processes a deactivation by oxygen often occurs. An important influence on the oxidation reaction catalyzed by these catalysts has the nature of a solvent. Thus, in n-heptane solution, primary alcohols are oxidized to aldehydes, but in water at alkaline pH, the corresponding acids are produced.

Much still needs to be done to explore the effect to other metals on the activity and selectivity of Pd and Pt catalysts. In the literature only a few such data are available. There is also a lack of data in the influence of the structure of the oxidized substrate on the catalyst activity and the deactivation process.

METHODS

Materials

α-Pinene, 1-methoxy-2-propanol (MPOL) and derivatives of phenoxyethanol were purified by distillation. 2,3;4,6-Di-isopropylidene-α-L-sorbose (DIS) was purified by double crystallization from methanol. Other reagents were of analytical purity.

Apparatus

Oxidation experiments were performed in two types of reactors. A 150 ml stainless steel reactor was fitted with a magnetic stirrer (≃ 1000 rpm), air inlet at the bottom and outlet through a condenser. The second reactor was a 80 cm high bubble column (i. d. 3.1 cm) equipped with an air introduction through a porous sparger (mean pore size less than 0.2 mm). During the reaction the outlet gases from the reactors were monitored continuously for oxygen.

Catalysts

The palladium and platinum catalysts were prepared by impregnation of charcoal (surface area 1265 $m^2.g^{-1}$, particle size < 0.12 mm) and $CaCO_3$ (2.9 $m^2.g^{-1}$, particle size <0.08 mm) with $PdCl_2$ or H_2PtCl_6 (60 °C; 8 h), followed by a reduction with formaldehyde [5]. Some part of each catalyst was re-impregnated (80 °C; 5 h) with Co, Bi, Cd, Zn, Mn water soluble salts (nitrates, chlorides, sulfates) which were subsequently transformed to hydroxides, adding a solution of NaOH. The catalysts thus obtained were thoroughly washed with water and stored moist under nitrogen. The metal content of the catalysts was determined by polarography (after their transformation to soluble salts).

Analysis

Samples of the reaction mixtures from MPOL and α-pinene oxidation were analyzed by GC (Hewlett Packard 5830) after separation of the catalyst and doping them with internal standards. The reaction mixtures from DIS and phenoxyethanols oxidation were neutralized with HCl to pH ≃ 3 after separation of the catalyst, and the formed acids extracted (3×), esterified and analyzed by GC (using an internal standard). The products were confirmed by GC-MS and NMR spectroscopy.

RESULTS AND DISCUSSION

A series of Pt and Pd catalysts were tested during the oxidation of following compounds:

(i) 1-methoxy-2-propanol to 1-methoxy-2-propanon

$$CH_3\text{-O-}CH_2\text{-}\underset{\substack{|\\OH}}{CH}\text{-}CH_3 + 1/2\ O_2 \rightarrow CH_3\text{-O-}CH_2\text{-}\underset{\substack{\|\\O}}{C}\text{-}CH_3 + H_2O$$

(ii) α-pinene to verbenol and verbenon

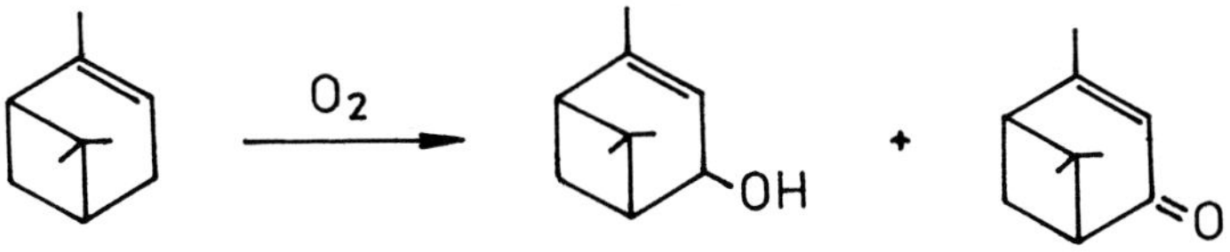

(iii) DIS to 2,3;4,6-diisopropylidene-2-keto-L-gulonic acid

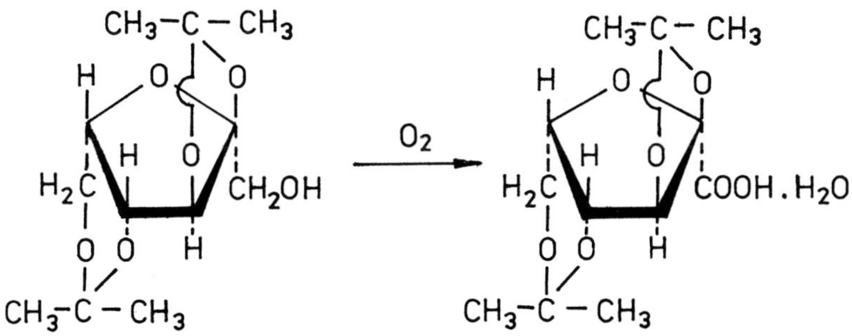

(iv) derivatives of phenoxyethanol to corresponding phenoxyacetic acids

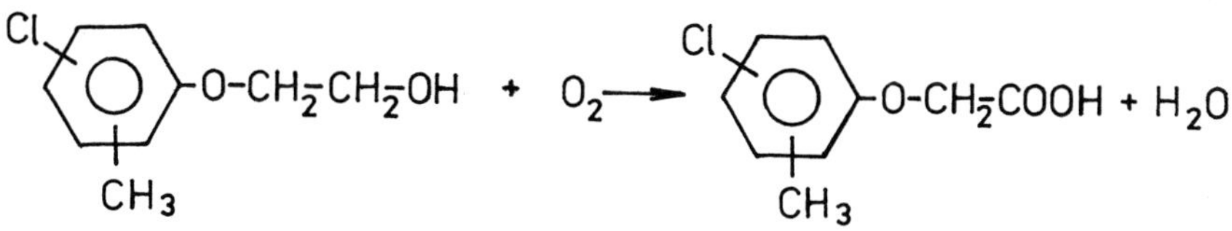

These compounds are used in the preparation of pesticides, pheromones and Vitamine C.

The results in Table 1 show effect of the support and its surface area on the activity of palladium catalyst during the oxidation of 2-methoxyphenoxyethanol in an aqueous solution of NaOH.

TABLE 1

Effect of the support on the activity of palladium catalyst for the oxidation of 2-MPE.

Support	Surface area $m^2.g^{-1}$	Reaction time min	Conversion %	Yield, mol % 2-MPA	2-MP
Active carbon	1 265	200	97.8	95.2	0.90
	970	210	97.2	89.4	0.70
	443	200	95.4	94.6	0.75
$CaCO_3$	2.9	250	98.3	94.6	0.60

2-MPE = 2-methylphenoxyethanol; 2-MP = 2-methylphenol; 2-MPA = 2-methylphenoxyacetic acid

4.5 g 2-MPE; 66 g H_2O; 1.3 g NaOH; 0.71 g catalyst (5 % Pd/support); 99 ^oC; 0.2 MPa; gas flow (67 vol % O_2 in N_2-O_2 mixture) = 20 cm^3 min^{-1}; batch reactor

From the results it is seen that the nature of the support and its surface area affect the catalytic properties of supported palladium. However, a very high yield of 2-MPA is obtained with Pd/$CaCO_3$ catalyst having a veru low surface area. Moreover, this catalyst remains active and selective upon reuse (see Table 2).

TABLE 2

Change of activity and Pd content of the catalyst upon reuse for the oxidation of 2-MPE.

Number of runs	Reaction time min	Conversion %	Yield, mol % 2-MPA	2-MP	% Pd/$CaCO_3$ wt %
1	270	98.1	94.7	0.50	4.31
2	440	98.2	95.2	0.60	-
12	450	97.9	95.0	0.45	1.56
16	450	98.4	94.8	0.70	1.11

Catalyst: 4.31 % Pd/$CaCO_3$ (1.3 g); after each oxidation 0.13 g fresh catalyst was added to compensate losses during the filtration

The drop of the oxidation rates is observed only after the

first run, and then it remains unchanged. Surprising is that despite more than 73 % loss of palladium from the catalyst, the activity remains unchanged.

Kinetic measurements in the bubble column and the stirred reactor show that the reaction is always controlled by mass transfer phenomena (see Table 3).

TABLE 3

Kinetic data of oxidation of phenoxyethanol derivatives in two types of reactors.

Reactor	Substrate	Mass transfer resistance; s	
		$1/k_La$	$1/k_sa_s + 1/k_r$
Batch reactor	2-MPE	0.39	2.71
	3,4-MCPE	0.42	2.09
	2,4-MCPE	0.59	1.86
Bubble column	2-MPE	3.51	1.72

3,4-MCPE = 3-methyl-4-chlorophenoxyethanol; 2,4-MCPE = 2-methyl-4-chlorophenoxyethanol; k_La = volumetric gas-liquid mass transfer coefficient; k_sa_s = liquid-solid mass transfer coefficient; k_r = reaction rate constant

The kinetic regime of the oxidation cannot be reached at any conditions. The problem is that at higher partial pressures of oxygen (above 0.25 MPa), the catalyst reversibly deactivates. However, after decreasing the oxygen pressure, the original activity of the catalyst is reached again.

Hydrocarbons having secondary or tertiary C-H bond are in the presence of Pd and Pt catalysts oxidized to alcohols and ketones. As it is evident from Table 4, a strong influence on the activity of these catalysts have some metals deposited on the catalyst surface in the form of hydroxides and oxides.

Their presence on the catalyst surface does not influence the distribution of formed alcohols and ketone (ratio 2 : 1) and the ratio of cis/trans isomers of verbenols. The highest promoting effect on both, Pd and Pt catalysts has a mixture of cobalt and cadmium.

TABLE 4

Effect of the catalyst composition on α-pinene oxidation

Catalyst wt % metal	Conversion %	Selectivity, % verbenon	Selectivity, % verbenol	cis/trans verbenol
5% Pd/C	18.1	26.9	54.8	1.2
5%Pd-7.1%Bi-0.4%Zn/C	21.3	27.1	57.9	1.2
5%Pd-7.1%Bi-0.4%Cd/C	20.1	26.1	55.5	1.3
5%Pd-2%Co-0.7%Cd/C	44.4[a]	29.7	58.5	1.3
5%Pd-2%Mn-0.7%Cd/C	25.4	16.6	33.7	1.2
5%Co/C	6.8	28.7	58.5	1.0
5%Pt/C	1			
5%Pt-2.09%Co-0.7%Cd/C	35.6	33.9	49.7	1.1

[a] 9.6 wt % hydroperoxides in the reaction mixture

80 ^{o}C; 0.2 MPa; oxygen flow 300 cm^{3} min^{-1}; catalyst 1.5 g; α-pinene 250 ml; reaction time: 5 h; bubble column reactor

It is expectable that the activity and selectivity of palladium and platinum catalysts will be different during the oxidation of various organic compounds. However, as we have found, the oxidized substrate plays also an important role during the reactivation of the catalyst which was deactivated by molecular oxygen. Thus, the monometallic catalyst, 5 % Pd/C stored moist, is highly active for the oxidation of DIS, but when this catalyst is dried and exposed to air before the reaction, its activity sharply decreases (see Table 5).

On the other hand, the same Pd/C catalyst deactivated in this manner has the same activity during the oxidation of phenoxyethanol derivates, MPOL and another alcohols. As it is shown in Table 5, the resistance of Pd/C catalyst against irreversible deactivation by oxygen is achieved by doping it with some metals, e.g. Co and Cd.

The palladium and platinum catalyzed oxidation of alcohols in aqueous solution proceeds via a dehydrogenation mechanism. This reaction proceeds on the catalyst surface and obbeys the Langmuir-Hinshelwood kinetics. It means that Pt and Pd surface can be covered with oxygen, oxidized substrate, hydrogen atoms and products formed in the dehydrogenation process. The fraction of the surface covered by each component depends on experimental conditions and the type of organic substrates.

TABLE 5

Effect of the catalyst history on its activity for the oxidation of DIS

Catalyst	Reaction time	Conversion	Selectivity
	h	%	%
4.9%Pd-2%Co-0.7%Cd/C	5.5	100	99.8
4.9%Pd-2%Co-0.7%Cd/C[b]	5.5	100	99.7
5% Pd/C	6	96.5	99.5
5% Pd/C[a]	7	52.3	99.1
5% Pd/C[b]	7	7.3	99.4

[a] 1 week exposed to air; [b] 1 month exposed to air

130 ^{o}C; 0.35 MPa; 15 g DIS; 150 ml H_2O; 3.02 g NaOH; air flow 20 cm^3 min^{-1}; batch reactor

According to the L-H mechanism, an alcohol adsorption and its rate of oxidation are influenced by oxygen concentration and adsorptive properties of alcohol. Thus, during the oxidation of phenoxyethanols at partial pressure above 0.25 MPa, oxygen completely covers the catalyst surface and totaly deactivates it. When the pressure decreases, the catalyst is again active. However, in the oxidation of DIS, a higher pressure of oxygen (above 0.4 MPa) is needed to deactivate only partly the catalyst [6].

The dehydrogenation of alcohol is a reversible reaction and the hydrogen on the catalyst surface is continuously oxidized. In some cases it can also hydrogenate the formed product. This is suggested by the results of 1-methoxy-2-propanol oxidation to ketone MPON which proceeds only to a ca. 50 % conversion with various catalysts at any experimental conditions. The added MPON retards the rate of oxidation and supresses the conversion of alcohol.

On the basis of the above mentioned results and the literature data [6-8], we assume that the equilibrium concentrations of reactants adsorbed on Pd and Pt surfaces are responsible for the activity and the deactivation of these catalysts. The observed promoting effect of some metals deposited on the catalyst is probably connected with their ability to change the fractional concentration of the surface oxygen and oxidized substrate. In order to prove this assumption, additional physicochemical investigation is continued.

REFERENCES

1 R.A. Sheldon and J.K. Kochi, Metal Catalyzed Oxidation of Organic Compounds, Academic Press, 1981

2 US Patent 4 599 446; C.A. 105, 209345

3 German Offen 3 135 946; C.A. 99, 70217

4 US Patent 4 579 689; C.A.105, 116991

5 Belg. Patent 851 804; C.A. 88, 170471

6 M. Hronec, Z. Cvengrosova and M. Stolcova, React. Kinet. Catal. Lett., 20 (1982) 207

7 H.E. von Dam, P. Duijverman, A.P.G. Kieboom and H. van Bekkum, Appl. Catal. 33 (1987) 373

8 P.J.M. Dijkgraaf, H.A.M. Duisters, B.F.M. Kuster and K. van Wiele, J. Catal. 112 (1988) 337

H.Mimoun (Inst. Francais du Petrole, France): In the case of oxidation of cis-pinene, is your reaction a radical chain one?

M.Hronec: Oxidation of cis-pinene proceeds via a free radical mechanism. Palladium and platinum catalysts probably activate the C-H bond in the hydrocarbon. Since the influence of these catalysts on the hydroperoxide decomposition is very low, hydroperoxides formed as the primary products are accumulated in the reaction mixture (see Table 4).

S.Coluccia (Dipartimento di Chimica, Torino): You show that the activity does not change significantly during several runs, in spite of a substantial decrease of the metal concentration. Does this observation suggest any hypothesis on the actual extent of active sites and possibly on their structure?

M.Hronec: We suggest that only a part of the metal loaded on the carrier is actually catalytically active. It is based on the measurements of the activity of the palladium catalysts having a different amount of the loaded metal. For example, the Pd/C catalyst containing only 1.11 % Pd after sixteen reuses (see Table 2) was still several times more active than the freshly prepared Pd/C catalysts with a 1.2 - 2.5 % content of palladium.

The palladium catalysts before and after reaction have been studied by ESCA and electrochemical methods. From the ESCA measurement followed that the surface of the catalyst always contains the Pd-phase and PdO. No correlation was found between catalysts differing in composition and their redox properties measured by electrochemical method.

G. Centi and F. Trifiro' (Editors), *New Developments in Selective Oxidation*

CATALYTIC OXIDATION OF 1-ALKENES WITH MOLECULAR OXYGEN AND PALLADIUM NITRO COMPLEXES

N.H. KIERS[1], B.L. FERINGA*[1] and P.W.N.M. van LEEUWEN[2]

[1]University of Groningen, Department of Organic Chemistry, Nyenborgh 16, 9747 AG Groningen (The Netherlands)

[2]Koninklijke/Shell-Laboratorium, Amsterdam (Shell Research B.V.), Badhuisweg 3, 1031 CM Amsterdam (The Netherlands)

SUMMARY

$(CH_3CN)_2PdClNO_2$ is capable of catalysing the oxidation of 1-alkenes to methyl ketones, epoxides (refs. 1-9) and aldehydes (ref. 6) using molecular oxygen. In this paper we report the influence of solvent, co-catalyst and additional ligands on the reactivity and selectivity in the oxidation of 1-alkenes to aldehydes by $(CH_3CN)_2PdClNO_2$.

INTRODUCTION

Selective catalytic oxidations of alkenes with molecular oxygen are commercially important and synthetically useful processes (ref. 10). It is well-known that 1-alkenes can be selectively oxidized to methyl ketones (ref. 11). Based on this oxidation reaction alkenes can be regarded as masked ketones.

Oxidation reactions of alkenes with molecular oxygen mediated by $(CH_3CN)_2Pd$-$ClNO_2$ have been described (refs. 1-9). Alkenes are generally oxidized to the corresponding ketones (refs. 1,2,4,6,7). With specific alkenes epoxides were formed (refs. 3,4,7,8,9). However, we observed aldehyde formation in a good yield using $(CH_3CN)_2PdClNO_2$ as catalyst with t-butyl alcohol as solvent and $CuCl_2$ as co-catalyst (ref. 6). It is assumed that the oxidation of alkenes to ketones goes by an intramolecular nucleophilic attack of the nitro group on the palladium bonded alkene followed by a hydride shift (refs. 1-9,12,13). We assume that formation of aldehydes goes by a comparable mechanism (scheme 1).

We now report the influence of solvent, co-catalyst and additional ligands on the catalytic oxidation reaction.

RESULTS AND DISCUSSION

In a typical oxidation reaction 1-octene was converted using a catalyst (5 mol %) prepared from $(CH_3CN)_2PdClNO_2$, $CuCl_2$ in an oxygen-saturated solution of t-butyl alcohol. After a reaction time of 16 hours under an oxygen atmosphere octanal and 2-octanone (ratio 60:40) were obtained in a 970 % combined yield based on $(CH_3CN)_2PdClNO_2$. A low isomerization activity was observed resulting in the formation of 80 % (based on Pd) octene isomers. The proposed mechanism for the catalytic oxidation of 1-alkenes to aldehydes is given in scheme 1.

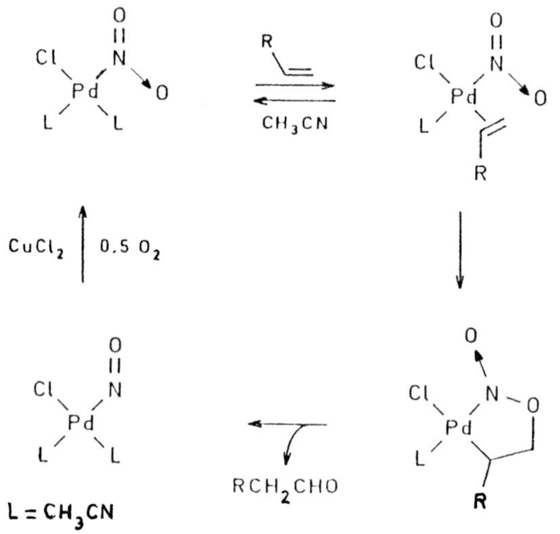

Scheme 1. The proposed mechanism for the catalytic oxidation of 1-alkenes to aldehydes by molecular oxygen mediated by $(CH_3CN)_2PdClNO_2$.

We propose that the regioselectivity in the cycloaddition of the alkene coordinated to the palladium nitro catalyst, determines the aldehyde to ketone ratio. It may be anticipated that the constitution of the palladium nitro complex and the nature of the ligands strongly influence the stereoselectivity in the cyclisation step. Subtle effects on the stereochemical results of 1,3-dipolar cycloaddition to alkenes are well precedented and the mechanistic pathways described above certainly show similarities with 1,3-dipolar cycloaddition reactions (ref. 14).

Preliminary experiments showed that several factors like metal salts, solvent and ligands influence the reactivity and selectivity of the oxidation reaction. In order to assess these factors we have undertaken a systematic investigation. The results on variation in solvent, co-catalyst and ligands are described herewith.

The influence of solvent on the oxidation reaction is summarized in table 1. Entry 2,3,4 and 7 show the strong tendency of Pd(II) complexes to catalyse the selective oxidation of 1-alkenes to methyl ketones. Coordinating solvents almost completely inhibits the oxidation reaction and results in isomerization of the starting 1-alkene. Possibly this effect is due to blocking of the necessary coordination places at palladium or is the result of a fast substitution of the coordinated alkene.

TABLE 1

Oxidation of 1-octene. 0.2 mmol $(CH_3CN)_2PdClNO_2$ + 0.8 mmol $CuCl_2$ + 4 mmol 1-octene, 25 ml solvent, 50°C. (Product determination (GC) after 16 hours, amounts in % based on Pd).

Entry	Solvent	Octanal	2-Octanone	Octene isomers
1	t-butyl alcohol	580	390	80
2*	t-amyl alcohol	170	1490	460
3	isopropyl alcohol	<30	500	130
4	hydroxyacetone	<30	1000	<30
5	2-hydroxypropionitrile	-	-	-
6	acetonitrile	-	-	-
7	toluene	<30	300	100
8	nitromethane	40	30	50
9	acetone	-	<30	500
10	HMPA	-	<30	500
11	DMF	<30	<30	<100

*Reaction time of 8 hours.

In apolar solvents we only observed a very slow reaction, partially due to the low solubility of the catalyst in these solvents. So far t-butyl alcohol is the only solvent in which aldehyde than ketone can be obtained.

The role of $CuCl_2$, the co-catalyst, in the classical Wacker oxidation is to oxidize Pd(0) to Pd(II). The effect of co-catalyst on the oxidation of 1-octene with $(CH_3CN)_2PdClNO_2$ in t-butyl alcohol is summarized in table 2.

TABLE 2

The influence of co-catalyst on the oxidation of 1-alkene. Reaction in t-butyl alcohol, at 30°C with 20 equivalents (based on Pd) of 1-octene. The amount of co-catalyst and of products (GC) are based on Pd.

Entry	Co-catalyst (eq)	Reaction time (h)	Octanal (eq)	2-Octanone (eq)	Octene isomers (eq)
1*	4 $CuCl_2$	16	6	4	1
12	4 $CuCl_2$	4	6	3	1
13	1 $CuCl_2$	19	1	4	9
14	10 $CuCl_2$	5	2	0.5	1
15	15 $CuCl_2$	5	2	0.5	1
16	4 $CuCl_2$	0.5	-	1	19
17	4 CuCN	2	0.2	0.7	10

TABLE 2 (continued)

Entry	Co-catalyst (eq)	Reaction time (h)	Octanal (eq)	2-Octanone (eq)	Octene isomers (eq)
18**	4 $Cu(ClO_4)_2.6H_2O$	1	5	19	56
19	4 $Cu(CO_3)Cu(OH)_2$	2	-	-	8
20	4 $CuCl_2$ + 1 LiCl	7	5	6	2
21	4 $CuCl_2$ + 1 LiF	4	4	3	6
22	2 $CuCl_2$ + 2 $SnCl_2$	0.5	1	3	15
23	2 $CuCl_2$ + 2 $CoCl_2$	1	-	-	15
24	2 $CuCl_2$ + 2 $NiCl_2$	1	-	-	14

*Reaction temperature of 50°C.
**100 Equivalents of 1-octene.

Increasing the amount of $CuCl_2$ shows a maximum in the reactivity and the selectivity to aldehyde formation using four equivalents of $CuCl_2$. Decreasing the amount of $CuCl_2$ gives an increase in isomerization rate and a decrease in oxidation rate and selectivity. Increasing the amount of $CuCl_2$ beyond four equivalents shows only a small influence on the oxidation reaction, but increases the isomerization rate. Substituting two equivalents of $CuCl_2$ for other metal salts like $SnCl_2$, $FeCl_3$, $ZnCl_2$, $NiCl_2$, $HgCl_2$, $CoCl_2$ or $PdCl_2$ results in the acceleration of isomerization reaction and gives only small amounts of ketones and aldehydes. The enhanced isomerization using other metal halides than $CuCl_2$ might be attributed to a lewis acid effect on the palladium catalysed reactions.

The role of $CuCl_2$ in the Wacker oxidation is well established. The role of $CuCl_2$ in the oxidation of 1-alkenes by $(CH_3CN)_2PdClNO_2$ is however not completely clear. In table 2 it is shown that not only the oxidation rate but also the selectivity of the oxidation reaction strongly depends on the amount of $CuCl_2$ used. We therefore assume that the active species in the formation of aldehydes is not just a palladium nitro complex, but a species that also contains $CuCl_2$, presumably a chloride bridged binuclear complex. Unfortunately we were not yet able to isolate such a species to prove our assumption.

In table 1 it is shown that the oxidation reaction was almost completely inhibited using coordinating solvents (entry 5,6,10,11). The effects of additional ligands on the oxidation reaction of 1-octene by $(CH_3CN)_2PdClNO_2$-4$CuCl_2$ in t-butyl alcohol under an oxygen atmosphere is summarized in table 3. It was shown (ref. 12) that NO_2^- easily dissociates from palladium. However, we found that increasing the amount of NO_2^- in solution by adding KNO_2 (entry 26,27) leads to an increase in the isomerization rate and in a decrease of the oxidation reaction rate. This effect was even much stronger using an NO_2 atmosphere above the reaction medium.

TABLE 3

Ligand effect on the oxidation of 1-octene. Reaction of $(CH_3CN)_2PdClNO_2$-$4CuCl_2$ in t-butyl alcohol at 50°C with 20 equivalents (based on Pd) of 1-octene, amount of ligand and product (GC) based on Pd.

Entry	ligand	Amount (eq)	Reaction time (h)	Octanal (%)	2-Octanone (%)	Octene isom.(%)
12	-	-	4	600	300	100
25	NO_2 (atmosphere)		5	<20	180	1350
26	KNO_2	2	4	125	125	375
27	KNO_2	4	4	50	50	300
28	acetonitrile	4	8	200	250	850
29	2-hydroxypropionitrile	2	3	80	50	<10
30	trichloroacetonitrile	2	0.1	<50	<50	>500
31	HMPA	2	3	<50	<50	700
32	$(CH_3CN)_2PdBrNO_2$	-	19	390	690	540

Addition of nitriles increases the isomerization rate and decreases the oxidation rate, probably by blocking the necessary coordination places on palladium or by a fast substitution of the coordinated alkene by the additioned ligands.

CONCLUSIONS

The reactivity and selectivity of the oxidation reaction of 1-octene to octanal using molecular oxygen and $(CH_3CN)_2PdClNO_2$ strongly depends on the reaction conditions. Reasonable amounts of aldehyde are only observed in t-butyl alcohol using four equivalents of $CuCl_2$ as the co-catalyst. The oxidation of other alkenes is now under investigation and will be reported later.

REFERENCES

1 B.S. Tovrog, S.E. Diamond and F. Mares, J. Am. Chem. Soc., 1979, 101, 270.
2 M.A. Andrews and K.P. Kelly, J. Am. Chem. Soc., 1981, 103, 2894.
3 A. Heumann, F. Chauvet and B. Waegell, Tetrahedron Lett., 1982, 23, 2767.
4 M.A. Andrews, T.C.-T. Chang, C.-W.F. Cheng and K.P. Kelly, Organometallics, 1984, 12, 1777.
5 M.A. Andrews, T.C.-T. Chang, C.-W.F. Cheng and K.P. Kelly, J. Am. Chem. Soc., 1984, 106, 5913.
6 B.L. Feringa, J. Chem. Soc., Chem. Commun., 1986, 909.
7 J.P. Solar, F. Mares and S.E. Diamond, Catal. Rev.-Sci. Eng., 1985, 27(1), 1.

8 M.A. Andrews and C.-W.F. Cheng, J. Am. Chem. Soc., 1982, 104, 4268.
9 P.K. Wong, M.K. Dickson and L.L. Sterna, J. Chem. Soc., Chem. Commun., 1985, 1565.
10 G.W. Parshall, Homogeneous Catalysis, Wiley, New York, 1980.
11 P. Henri, Palladium-Catalysed Oxidation of Hydrocarbons, Riedel, Dordrecht, 1980.
12 M.A. Andrews, T.C.-T. Chang and C.-W.F. Cheng, Organometallics, 1985, 4, 268.
13 B.S. Tovrog, F. Mares and S.E. Diamond, J. Am. Chem. Soc., 1984, 102, 6618.
14 A. Padwa (Ed.), 1,3-Dipolar Cycloaddition Chemistry, Vol. 1, 2, Wiley, New York, 1984.

ACKNOWLEDGEMENTS

This investigation was supported by the Netherlands Foundation for Chemical Research (SON) with financial aid from the Netherlands Organization for Scientific Research (NWO).

H. MIMOUN (Lab. d'Oxydation Inst. Franc. Petrole, France): Did you observe Pd(0) precipitation? How can you explain the antimarkovnikov oxidation induced by t-BuOH? Did you try t-BuO^- ligands on Pd?

N.H. KIERS (University of Groningen, The Netherlands): In some cases we observed a small amount of Pd(0) precipitation after about 4 turnovers, resulting in an increase of the isomerisation rate and a decrease in the selectivity towards aldehyde formation. We were not able to make a complex containing both NO_2^- and t-BuO^- ligands. Oxidation reactions using $Cu(O^tBu)$ or $Cu(O^tBu)_2$ instead of $CuCl_2$ were not succesful.

G. FRANZ (Fa. Huls AG, BRD): Comment on Wacker Process: Scientists of some institutes of Sibirian Branch of Academie of Sciences of USSR have some years ago succeeded in replacing $CuCl_2$ by heteropolicacids in the Wacker Process. Therefore Cu is not essential in this system. Pd is a very effective decarbonylation catalyst of aldehydes. Did you ever pay attention to decarbonylation products during your experimental studies?

N.H. KIERS (University of Groningen, The Netherlands): The main difference between the Wacker oxidation and our oxidation process is the use of a Pd(II)-NO_2 / Pd(II)-NO instead of a Pd(II) / Pd(0) couple. In our system Pd(II) is not reduced. We followed the oxidation reactions with GC/MS and we did not observe the formation of decarbonylation products.

J.M. BRÉGEAULT (Univ. P et M. Curie, France): Did you observe the formation of chlorinated products when you work with an excess of $CuCl_2$? What about the decay period of your system?

N.H. KIERS (University of Groningen, The Netherlands): We did not observe chlorinated products. The decay period stongly depends upon the conditions under which the catalytic system is pre-prepared. Whithout pre-prepareti on we observed a slow reaction and a decay of the reaction after about 4 hours. With pre-preparetion of the catalyst by stirring the mixture of $(CH_3CN)_2PdClNO_2$ and $CuCl_2$ under an oxygen atmosphere for about 5 hours we observed a fast oxidation reaction but also a rapid decay of the reaction after about 20 minutes.

G. STRUKUL (Dip. di Chimica Univ. Venezia, Italy): The formation of aldehydes in your system requires oxygen transfer at the C_1 instead of C_2 as normally happens. Also, it is clear that this unusual behavior depends on t-BuOH and the Cu co-catalyst. Do you have any suggestions for this unusual behavior?

N.H. KIERS (University of Groningen, The Netherlands): What we propose as the catalyst is a dinuclear complex, containing both Pd and Cu in which t-BuOH is acting as a bridging ligand. Steric effects are probably the main reasons for the coordination of the alkene in such a way that oxygen transfer to C_1 becomes more facile.

A.J. CHALK (Givaudan Corp., U.S.A.): Is it possible that terminal epoxides are intermediates in this reaction and that they rearrange to the mixture of ketone and aldehyde? Have you tried adding terminal epoxides to the reaction mixture to see if they isomerise to test this possibility?

N.H. KIERS (University of Groningen, The Netherlands): Epoxides do react under the reaction conditions used, but they give a mixture of t-butyl ethers which are not observed in the oxidation reaction of 1-alkenes. We did not observe the corresponding aldehydes or ketones in the reaction with terminal epoxides.

B.R. JAMES (Dept. of Chem. Univ. of British Columbia, Canada): You state that "the role of $CuCl_2$ in the Wacker oxidation is well established". I am not convinced that this is so. I believe that mixed, metal species (Pd_2Cu?), but ill-characterised, have been isolated from Wacker systems. In your system you imply that $CuCl_2$ is a ligand, and so this part of the catalytic cycle may not be so different to that in the Wacker cycle. It would be of value to continue the attempt to isolate such bimetallic species.

N.H. KIERS (University of Groningen, The Netherlands): The Wacker oxidation process does not work without a co-catalyst like $CuCl_2$ to oxidize Pd(0) to Pd(II). However, $CuCl_2$ is not essential for the reoxidation of Pd-NO to Pd-NO_2, but has an enormeous influence on the selectivity of the oxidation reaction. This implies a different role for $CuCl_2$ in our system. We will continue our efforts to isolate a (bimetallic) complex which is active in the oxidation of 1-alkenes to aldehydes.

G. Centi and F. Trifiro' (Editors), *New Developments in Selective Oxidation*

SELECTIVE CYCLOHEXANE OXIDATION CATALYZED BY THE GIF SYSTEM

ULF SCHUCHARDT and VALDIR MANO

Instituto de Química, Universidade Estadual de Campinas, Caixa Postal 6154, 13081 - Campinas, SP (Brasil)

SUMMARY

The turnover number and selectivity of the cyclohexane oxidation by the Gif system were studied as functions of the quantities of cyclohexane and catalyst and of the reaction temperature and time. It was found that the cyclohexanone-producing catalytic species is only formed after the reaction has started and that the turnover number decreases after the first 60 min. Under an atmosphere of pure oxygen turnover numbers as high as 100 h^{-1} were observed, but the Gif system loses its selectivity and reacts with cyclohexanone to produce further compounds.

INTRODUCTION

In a series of publications, Barton et al. (refs. 1-6) described a new system for the selective oxidation of saturated hydrocarbons at ambient conditions. This Gif system (ref. 4) consists of an iron catalyst in the presence of a reducing agent (normally metallic zinc) and pyridine with a carboxylic acid (normally acetic acid) as solvent and proton source, plus molecular oxygen. The main features of this system are its high selectivity for the oxidation of secondary carbon atoms, forming ketones as the major products, and the high yields obtained compared with those of analogous systems (ref. 7). The Gif system was successfully used for the selective oxidation of steroids (ref. 8) and other natural products (ref. 9), but most of the zinc is lost in an useless side reaction with acetic acid to form zinc acetate (ref. 6). This prompted Barton to develop in collaboration with Balavoine et al. the Gif-Orsay system (refs. 10,11) in which the oxygen is reduced electrochemically. This system oxidizes saturated hydrocarbons with the same specificity as the original Gif system, but with a much better efficiency, giving coulombic yields of up to 59% (ref. 12).

Barton et al. studied mostly adamantane in these oxidation reactions, obtaining adamantanone as the principal product with a selectivity of up to 88% (ref. 12). The turnover number (mmol of oxidized products per mmol of catalyst per hour) is normally small, but for very low catalyst concentration it is reported to be higher than 100 h^{-1} (ref. 5). The selectivity for the products depends very much on the flow rate of air and the stirring velocity (ref. 5).

In the oxidation of cyclohexane, Barton found a selectivity for cyclohexanone (one/ol) as high as 22.1 with a turnover number of 30.3 h^{-1} (ref. 5). We have reexaminated the oxidation of cyclohexane with the Gif system in order to determine which quantities of catalyst and cyclohexane permit the highest turnover number. We have, furthermore, studied the influence of the reaction temperature and time on the selectivity and turnover number. Kinetic studies have been performed on the oxidation under an atmosphere of pure oxygen.

EXPERIMENTAL

All reagents and solvents used were analytical grade. Cyclohexane was purified by washing with conc. sulfuric acid, water, 5% sodium hydroxide solution and water and then distilled. The iron catalysts $Fe^{II}Fe_2^{III}O(CH_3CO_2)_6(C_5H_5N)_3$ (ref. 13) and $Fe(bipy)_3Cl_2$ (ref. 14) were prepared according to the literature.

The oxidation reactions in open air were performed in a 125 ml erlenmeyer, using 28 ml of pyridine, 2.3 ml of acetic acid, 1.8 ml of water, 1.31 g (20 mmol) of finely powdered zinc, normally 1.1 ml (10 mmol) of cyclohexane and 8 μmol of the catalyst. The reaction temperature was adjusted with a thermostated water bath at the value indicated. The reaction mixture was magnetically stirred for 4 h at 1000 rpm, which proved to be appropriate for having all the zinc in suspension. In the kinetic measurements, 1 ml of solution was taken out of the reaction mixture every 30 min. In the other experiments, the reaction mixture was filtered after 4 h. The liquid products were analyzed with a CG 37 gas chromatograph equipped with a 4 m packed column of 5% Carbowax 20 M on Chromosorb WHP coupled to a flame ionization detector and temperature programmed at $8^{o}C\ min^{-1}$ from 80 to $170^{o}C$. Cyclooctane was added as an internal standard and the observed retention times were: cyclohexane (1.5 min), cyclooctane (3.5 min), cyclohexanone (7.8 min) and cyclohexanol (9.3 min).

The reactions carried out under an atmosphere of pure oxygen (99.5%) were performed in a 125 ml round bottom Schlenk flask. After introduction of the same amounts of solvents and cyclohexane as used in the previous experiments, pure oxygen was passed through the flask for approximately 5 min. The catalyst and 1.31 g (20 mmol) of zinc were then added and the flask was sealed with a septum already connected to silicon tubing whose other end was inserted into a 500 ml graduated cylinder filled with oxygen and immersed up-side-down in a water reservoir. The reaction mixture was magnetically stirred at 1000 rpm. The oxygen consumption was measured every 3 min after equalizing the water level inside the cylinder with the level of the water reservoir. Every 30 min, 1 ml of the solution was taken out of the reaction mixture with a syringe, inserted into the flask through the septum. The reaction products were analyzed as described before.

RESULTS AND DISCUSSION

Reactions in open air

The catalyst $Fe(bipy)_3Cl_2$ shows exactly the same reactivity as $Fe^{II}Fe_2^{III}O(CH_3CO_2)_6(C_5H_5N)_3$ in the oxidation of cyclohexane. This confirms the result of Barton *et al.*, who believe that $Fe(bipy)_n^{2+}$ is the active species in the oxidation reaction (ref. 6). As the first complex is much easier to prepare, it was used in all experiments described in this paper.

The values obtained for the turnover number and the selectivity for cyclohexanone (one/ol) depend strongly on the stirring velocity. If it is too low, some of the zinc adheres to the wall of the reaction flask and the turnover number lowers considerably. If it is too high, the selectivity is strongly reduced. These effects were already observed by Barton *et al.* (ref. 6). In our experiments the stirring velocity of 1000 rpm was a good compromise and was carefully maintained in all experiments in order to make the results comparable.

The influence of the quantity of cyclohexane in the reaction mixture on the turnover number and the selectivity is shown in Fig. 1. Both values increase up to a quantity of 10 mmol of cyclohexane and then stay approximately constant, showing that the cyclohexane concentration is sufficiently high and does not control the kinetics of the reaction anymore. Under these conditions, 0.969 mmol of cyclohexanone and 0.075 mmol of cyclohexanol are obtained after 4 h of reaction.

The quantity of the catalyst exhibits a strong influence on the turnover number and selectivity (Fig. 2). As observed by Barton *et al.* in the oxidation

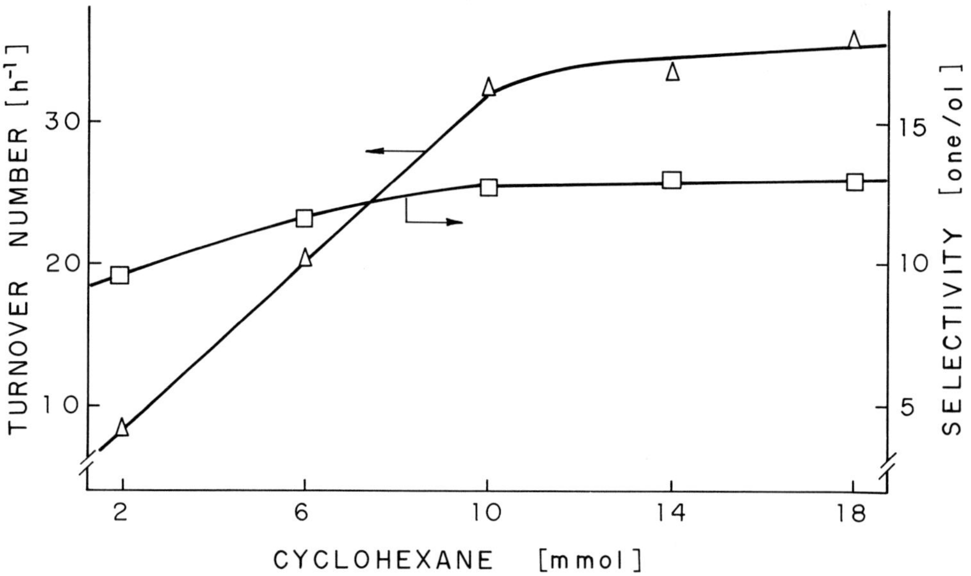

Fig. 1. Turnover number and selectivity as a function of the quantity of cyclohexane in the reaction mixture (8 μmol of $Fe(bipy)_3Cl_2$, 20°C, 4 h).

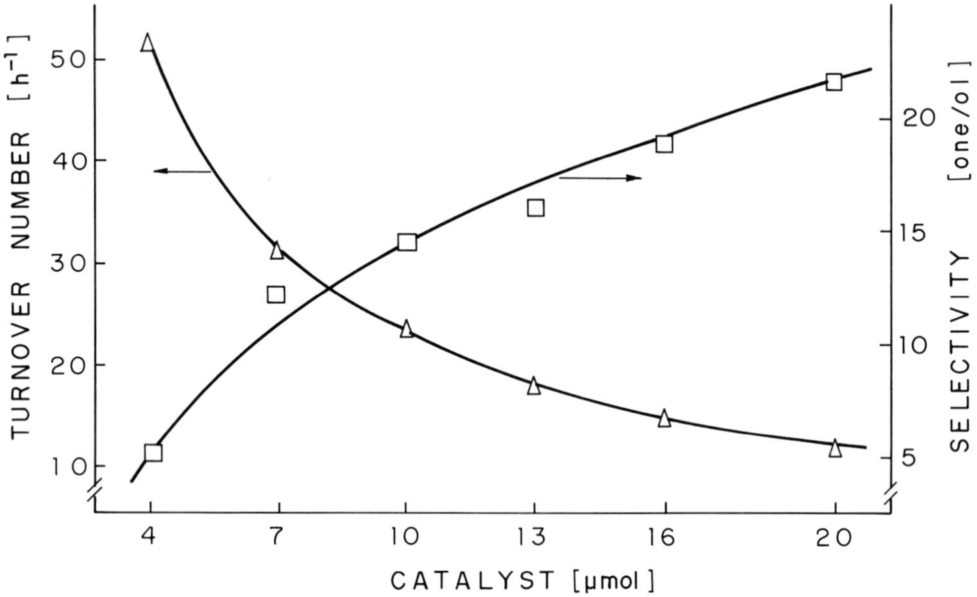

Fig. 2. Turnover number and selectivity as a function of the quantity of catalyst in the reaction mixture (10 mmol of cyclohexane, 20°C, 4 h).

of adamantane (ref. 5), the turnover number increases strongly with the reduction of the catalyst quantity. With 4 μmol of catalyst, 0.831 mmol of oxidized products are formed after 4 h, which only increases to 0.966 mmol if five times the quantity of catalyst (20 μmol) is used. On the other hand, the selectivity is much better for the higher catalyst concentrations, reaching a value of 22 for 20 μmol of catalyst.

The best turnover number of 35.6 h^{-1} is obtained at 20°C. At higher temperatures, the turnover number decreases but the selectivity increases slightly (Fig. 3). This was also observed by Barton et al. for the oxidation of adamantane (ref. 6). The reduction of the turnover number at 10°C, we believe, cannot be attributed to a loss of activity of the catalyst, but to the slower reaction of the zinc with molecular oxygen, which now controls the reaction rate.

As can be seen in Fig. 4, the turnover number is not constant during the reaction course. In the begining of the reaction it increases to a value close to 50 h^{-1} after approximately 60 min and then decreases steadily. On the other hand, more than 90% of cyclohexanol is formed in the first 30 min. After this the catalyst becomes very selective, producing nearly exclusively cyclohexanone, which makes the overall selectivity increase from 1.4 to 10.6. This shows clearly, that the mechanism of the oxidation in the first minutes is different. The cyclohexanone-producing catalytic species is formed only after the reaction has started. The decrease in the turnover number at longer reaction times

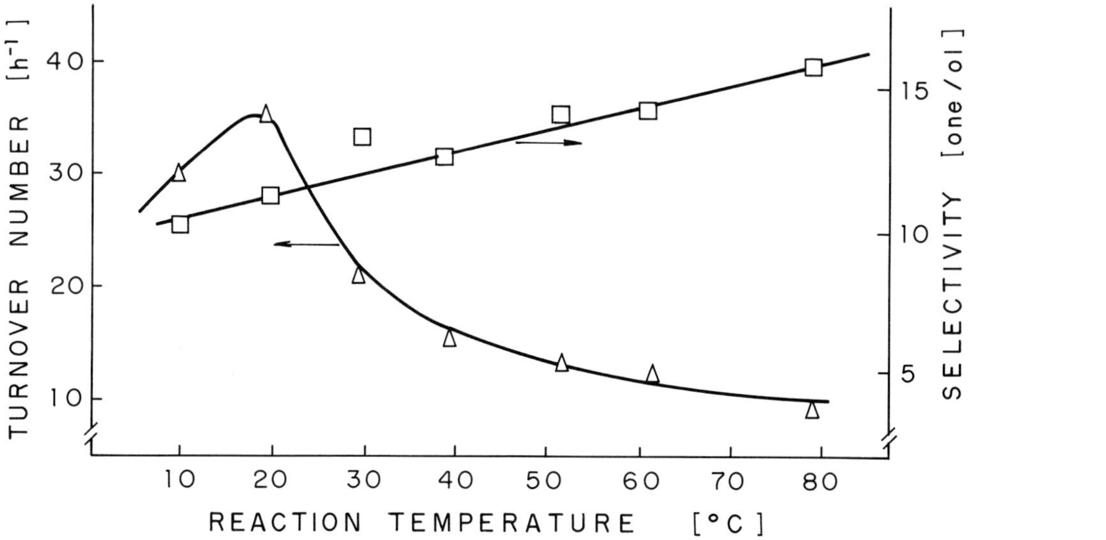

Fig. 3. Turnover number and selectivity as a function of the reaction temperature (7.4 μmol of $Fe(bipy)_3Cl_2$, 10 mmol of cyclohexane, 4 h).

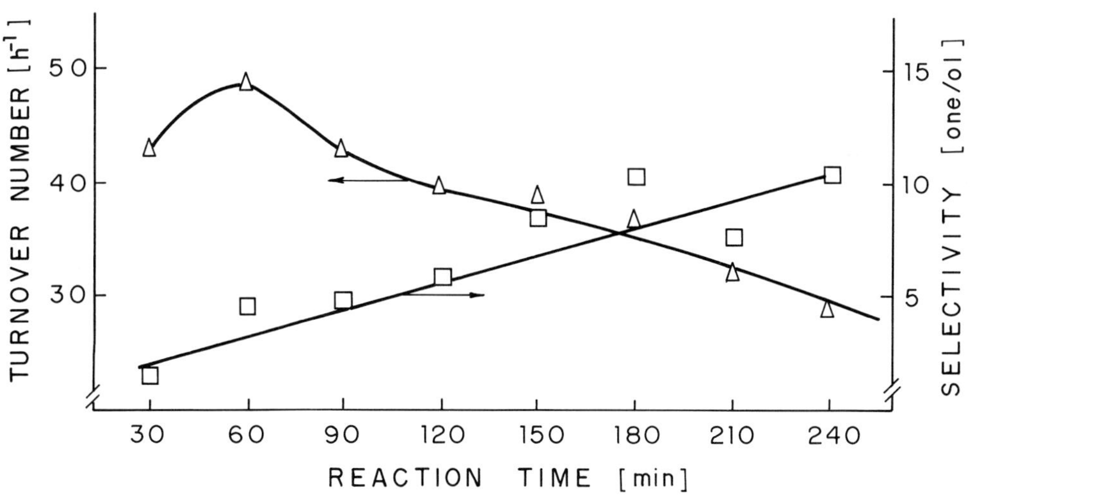

Fig. 4. Turnover number and selectivity as a function of the reaction time (7.9 μmol of $Fe(bipy)_3Cl_2$, 10 mmol of cyclohexane, 20°C).

cannot be attributed to the reduction of the cyclohexane concentration, as approximately 6 mmol of cyclohexane and 7 mmol of zinc can be recovered after the reaction. Presently we believe that the catalytically active species decomposes slowly under the reaction conditions.

Under the same reaction conditions cobalt bis(dimethylglyoximate), manganese diacetate and manganese diacetylacetonate show very poor activity for cyclohexane oxidation (turnover number < 1 h^{-1}) and produce only cyclohexanol. Cobalt tetraphenylporphyrin gives a better turnover number (4.2 h^{-1}) but the selectivity for cyclohexanone (0.5) is poor. Zinc can be replaced by ascorbic acid (turnover number 7 h^{-1}, selectivity 2.1), powdered iron (turnover number 5.2 h^{-1}, selectivity 13.8) or powdered copper (turnover number 9.6 h^{-1}, selectivity 8.5), but the quantity of oxidized products formed is smaller. Substitution of the acetic acid by oxalic, malonic or adipic acid reduces both the catalytic activity and the selectivity of the catalyst. Substitution of half of the pyridine by acetone reduces the turnover number (24.6 h^{-1}) and the selectivity (7.8). In pure acetone nearly no oxidation of cyclohexane is observed (turnover number 0.8 h^{-1}).

We have also varied the pH of the reaction mixture by substituting the 1.8 ml of water by the same amount of 1 M solutions of perchloric acid, sulfuric acid or potassium hidroxide. A reduction of the turnover number from 35.6 h^{-1} to 11.2, 20.7 and 22.5 h^{-1}, respectively, is observed and only for sulfuric acid does the selectivity increase slightly from 12.9 to 14.2, while it is reduced to values between 5 and 6 for the other systems. Addition of 1 mmol of the electron transfer reagents hydroquinone, antraquinone or phenantroquinone does not improve the oxidation reaction of cyclohexane. Under the conditions used, the turnover number drops to values between 13 and 16 h^{-1} and the selectivity to values between 5 and 7.

Reactions under an atmosphere of pure oxygen

The reactions performed under an atmosphere of pure oxygen show a very steady consumption of oxygen, which stops after all the zinc is used up. The total amount of oxygen consumed (15.8 mmol) is not much smaller than the quantity of zinc used in the reactions (20 mmol). The rate of oxygen consumption cannot be used for a kinetic analysis of the cyclohexane oxidation. Therefore, we have determined the turnover number and selectivity after every 30 min of reaction. The results are shown in Fig. 5. In the begining of the reaction the turnover number increases slightly, reaching a maximum of more than 100 h^{-1} at 60 min. Then it decreases linearly until 210 min, when all the zinc has reacted. After this, the decrease is even more accentuated. The selectivity increases up to 120 min, reaching a moderate value of 8.6 and then decreases. The total quantity of cyclohexanone and cyclohexanol produced is 0.79 mmol after 60 min and 1.35 mmol after 120 min, when more than 5 mmol of cyclohexane is still present in the reaction mixture. Although the consumption of cyclohexane continues, the total quantity of the products only increases slightly to 1.42 mmol after 180 min and 1.44 mmol after 210 min. After this, when all zinc has already reacted, the total quantity of products drops to 1.02 mmol after 240

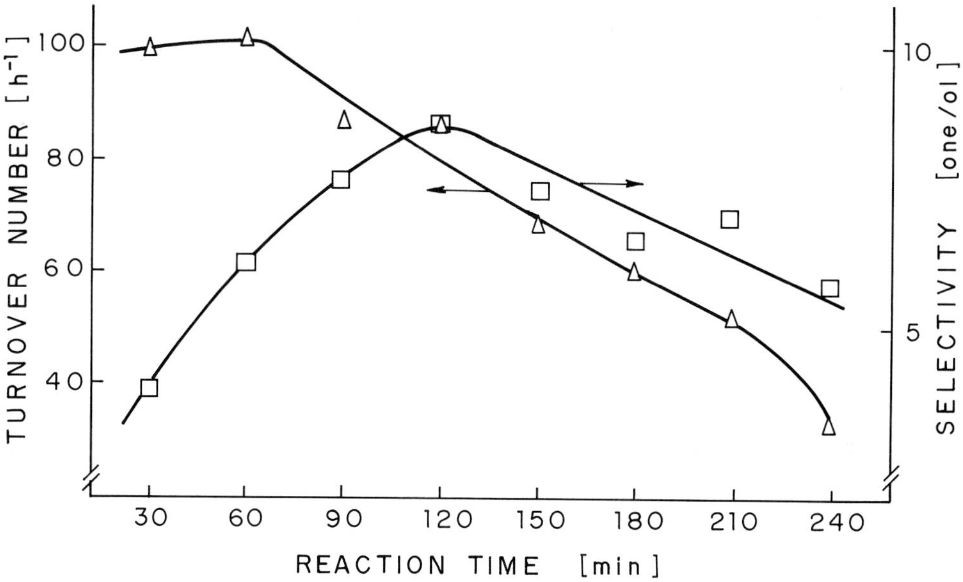

Fig. 5. Turnover number and selectivity as a function of the reaction time under an atmosphere of a pure oxygen (7.85 µmol of Fe(bipy)$_3$Cl$_2$, 10 mmol of cyclohexane, 20°C).

min, when approximately 3 mmol of cyclohexane is still left in the reaction mixture. These results show clearly that the oxidation products suffer further reactions under an atmosphere of pure oxygen.

Similar to the results in open air, the selectivity increases sharply during the first 120 min. Approximately half of the cyclohexanol is formed during the initial 30 min, then the catalyst becomes more selective for the cyclohexanone production. From 120 min on, the quantity of cyclohexanol formed continues to increase slightly, while the quantity of cyclohexanone stays approximately constant. This, we believe, is due to the fact that cyclohexanone is further oxidized at the same rate as it is produced. In agreement, the selectivity decreases during this part of the reaction. After 210 min, when the cyclohexanone is only consumed, the turnover number decreases sharply and the selectivity drops to 5.8.

When 1 mmol of antraquinone is added to the reaction mixture, the initial turnover number is only 69.3 h^{-1} and reduces during the reaction to 22.0 h^{-1}. The selectivity is poor with values between 3.6 and 5.5. This shows that antraquinone is also uneffective as an electron transfer reagent in a pure oxygen atmosphere under the reaction conditions employed.

CONCLUSIONS

The results show that the reaction system used by Barton *et al.* is the best and that any substitutions result in lower turnover numbers and/or selectivities in the cyclohexane oxidation reaction. On the other hand, the catalytically active species is not known. The selectivity change in the beginning of the reaction and the reduction of the turnover number after 60 min of reaction time show that this species suffers modifications during the reaction course, which are at present not explained. Under an atmosphere of pure oxygen the Gif system loses its selectivity, resulting in further reactions of the oxidation products. More research is required to identify the catalytically active species, to explain the loss of selectivity and identify the products formed under an atmosphere of pure oxygen.

ACKNOWLEDGEMENTS

This work was financed by Nitrocarbono S.A.. Fellowships from the Conselho Nacional de Desenvolvimento Científico e Tecnológico (CNPq) are acknowledged. The authors thank Prof. Keith U. Ingold, National Research Council of Canada, for helpful discussions and Sir Derek H.R. Barton, Texas A&M University, for his interest in our work.

REFERENCES

1 D.H.R. Barton, M.J. Gastiger and W.B. Motherwell, J. Chem. Soc. Chem. Commun., (1983) 41-43 and 731-733.
2 D.H.R. Barton, R.S. Hay-Motherwell and W.B. Motherwell, Tetrahedron Lett., 24 (1983) 1979-1982.
3 D.H.R. Barton, J. Boivin, N. Ozbalik and K.M. Schwartzentruber, Tetrahedron Lett., 25 (1984) 4129-4222.
4 D.H.R. Barton, J. Boivin, N. Ozbalik, K.M. Schwartzentruber and K. Jankowski, Tetrahedron Lett., 26 (1985) 447-450.
5 D.H.R. Barton, J. Boivin, M. Gastiger, J. Morzycki, R.S. Hay-Motherwell, W.B. Motherwell, N. Ozbalik and K.M. Schwartzentruber, J. Chem. Soc. Perkin Trans. I, (1986) 947-955.
6 D.H.R. Barton, J. Boivin, W.B. Motherwell, N. Ozbalik and K.M. Schwartzentruber, Nouv. J. Chim., 10 (1986) 387-398.
7 R.A. Sheldon and J.K. Kochi, Metal-Catalyzed Oxidations of Organic Compounds, Academic Press, New York, 1981, Ch. 8, p. 215.
8 D.H.R. Barton, J. Boivin and C.H. Hill, J. Chem. Soc. Perkin Trans. I, (1986) 1797-1804.
9 D.H.R. Barton, J.-C. Beloeil, A. Billion, J. Boivin, J.-Y. Lallemand and S. Mergui, Helv. Chim. Acta, 70 (1987) 273-280.
10 G. Balavoine, D.H.R. Barton, J. Boivin, A. Gref, N. Ozbalik and H. Rivière, Tetrahedron Lett., 27 (1986) 2849-2852.
11 G. Balavoine, D.H.R. Barton, J. Boivin, A. Gref, N. Ozbalik and H. Rivière, J. Chem. Soc. Chem. Commun., (1986) 1727-1729.
12 G. Balavoine, D.H.R. Barton, J. Boivin, A. Gref, P. Le Coupanec, N. Ozbalik, J.A.X. Pestana and H. Rivière, Tetrahedron, 44 (1988) 1091-1106.
13 C.T. Dziobkowski, J.T. Wrobleski and D.B. Brown, Inorg. Chem., 20 (1981) 679-684.
14 J.E. Fergusson and G.M. Harris, J. Chem. Soc. (A), (1966) 1293-1296.

B.R. James (Dept. of Chemistry, Univ. of British Coulumbia, Vancouver, Canada): You observe a maximum turnover at about 20°C and attribute this a slow reaction of Zn with O_2 at the lower temperature (10°C). You also describe the temperature reaction profile as "enzyme-like". Considering that your reactions are diffusion controlled, I think the analogy to enzymas tenuous at best. Have the effects of stirring rate been measured at all temperatures, and do you have direct evidence for a slower Zn + O_2 reaction at 10°C ? It is not obvious how this could give rise to a maximum value within the given temperature range (what is the Zn + O_2 reaction ?).

U. Shuchardt (State Univ. of Campinas, Brazil): Zinc reacts with oxygen to form the superoxide anion, which is the real oxidant. The effects of stirring rate have only been measured at 20°C, whose the best turnover numbers are obtained. We have no direct evidence for a slower Zn + O_2 reaction at 10°C. Turnover numbers are slightly better with a higher stirring rate at this temperature, while they decrease with increase of the stirring rate at higher temperatures. On the other hand, the stirring found best is quite vigorous and the reduction of the turnover numbers at higher temperatures cannot be explained by diffusion phenomena.

H. Mimoun, IFP, Ruel Malmaison, France): What is the electronic yield of the reaction ? i.e. how much Zn is used for every mole of oxygenated product produced ?

U. Shuchardt:Considering that more or less 10 mmol of zinc are used up in the reactions, that two atoms of zinc are needed to produce one molecule of cyclohexanone and that slightly more than 1 mmol of oxidized products is formed, the electronic yield is only around 20%. The other zinc reacts with the acetic acid to form zinc acetate.

M. Baerns (Ruhr Univ. Bochum, BRD): You have mentioned that selectivity depends on stirring speed without giving any explanation for this observation. Could you imagine that diffusional effects in the liquid play a role; this would be the case if any segregation effects prevail ?

U. Shuchardt: At low stirring velocities the selectivity is good but the turnover numbers are small. At very high stirring velocities the selectivity is strongly reduced, as the amount of oxygen dissolved in the reaction mixture is large, favoring radical reactions which produce cyclohexanol. As the stirring used is very effective, I do not believe that diffusional effects in the liquid paly a role nor that any segregation effects are encountered.

J. Kiwi (EPFL, IPC II, Lousanne, Switzerland): (a) Could you attribute the decrease in turnover and selectivity observed in your reaction to the disappearence of some species produced by $Fe(bpy)_3Cl_2$ added at the beginning of the reaction ?

(b) Could you find other metal$(bpy)_x$ compound , e.g. Co, Ru-complexes to avoid many problems you have encountered when you use $Fe(BPY)_3$ systems ?

U. Shuchardt: (a) As evidenced by UV-visable spectroscopy the $Fe(BPY)_3^{2+}$ cation dissociates immediately on dissolution in the reaction mixture. The species produced by the dissociation suffers further modifications during the reaction course, probably by oxidation, which reduce the turnover number and under an atmosphere of pure oxygen also the selectivity. The true nature of the catalytically active species is not yet known.

(b) $Co(bpy)_3^{n+}$ and $Mn(bpy)_3^{n+}$ cations show only very poor catalytic activity under the conditions of the Gif system. $Ru(bpy)_x$ complexes have not been tried yet, but we hope that their use will avoid some of the problems encountered with the $Fe(bpy)_3$ system.

G. Centi and F. Trifiro' (Editors), *New Developments in Selective Oxidation*

METALLOPORPHYRIN-CATALYZED OXIDATION OF CYCLOHEXANE WITH DIOXYGEN

R. IWANEJKO, T. MŁODNICKA and J. POŁTOWICZ

Institute of Catalysis and Surface Chemistry, Polish Academy of Sciences, 30-239 Krakow, Poland

SUMMARY

Oxidation of cyclohexane with dioxygen in the presence of propionaldehyde and some metalloporphyrins as catalysts have been investigated. The main products of the hydrocarbon oxidation are cyclohexanol and cyclohexanone. Also, the products of propionaldehyde oxidation such as peroxypropionic acid and carbon dioxide are present. The yields and the product distribution appeared to be dependent on the character of the metal centre. These differences in the catalytic behaviour shed light on the character of the active forms of the catalyst and on the mechanism of the investigated reaction.

INTRODUCTION

Oxygen containing complexes of some metalloporphyrins have recently become the subject of great interest and intensive investigations. This interest was raised by efficiency of biological oxidation processes in which iron protoporphyrin IX effects important functions. This species constitutes the essential part of cytochrome P-450 which mediates a broad range of biological oxygenations. The modelling of this system has led to the development in the field of liquid phase oxidation of hydrocarbons i.e. epoxidation of olefins and hydroxylation of alkanes. Chemical models of cytochrome P-450 consist of a metalloporphyrin, dioxygen and reducing agent such as H_2/Pt (ref. 1), borohydride (refs. 2, 3), ascorbate (ref. 4) or metalloporphyrin and single atom oxygen donor such as iodosylbenzene (refs. 5,6), peroxy acid (ref. 7), hypochlorite (ref. 8), potassium hydrogenpersulphate (ref. 9) and hydrogen peroxide (refs. 10, 11).

Our investigations focused first on epoxidation of olefins

with dioxygen in the presence of aldehyde as reducing agent and some metalloporphyrins as catalysts (ref. 12) and then on oxidation of cyclohexane under similar conditions.

EXPERIMENTAL

Tetra-p-tolylporphyrins (TTP) with metal cntres such as Cr^{3+}, Mn^{3+}, Fe^{3+}, Co^{2+}, Ni^{2+}, Cu^{2+} and Zn^{2+} have been prepared according to the procedure described in (ref. 13) and purified by column chromatography on alumina. Chloro-tetrakis(2,6-dichlorophenyl) porphyrinatomanganese(III) Mn(TDCPP)Cl has been prepared according to the procedure described in (ref. 14). Propionaldehyde produced by Fluka was redistilled before each series of measurements under argon. Benzene used as solvent produced by POCH-Gliwice was of purity grade. Cyclohexane of purity grade was also produced by POCH-Gliwice.

Reactions were carried out in a thermostated glass reactor equipped with magnetic stirrer (ref. 15) at 30°C.

In a standard experiment the reactor was filled with molecular oxygen under normal pressure and catalyst was introduced into it. Then 10 ml of cyclohexane and propionaldehyde solution in benzene was added. The amounts of products quoted in the text correspond to this volume. The reagent mixture was vigorously stirred and the reaction was carried out for 90 minutes. The progress of the reaction was followed by measuring the oxygen uptake.

The yields of cyclohexanol and cyclohexanone were determined using GC Chrom 5 with columns filled with Tenax GC. The yield of carbon dioxide was determined using TCD with columns filled with Porapak QS. The amount of peroxy acid was determined by iodometric titration.

RESULTS

Cr(III), Mn(III), Fe(III), Co(II), Ni(II), Zn(II) and Cu(II)-tetra-p-tolylporphyrins (TTP) have been used as catalysts in the reaction of oxidation of cyclohexane with dioxygen in the presence of propionaldehyde as reducing agent in benzene solutions. Some kinetic experiments have also been carried out with chloro - tetrakis (2,6-dichlorophenyl) porphyrinatomanganese (III) (Mn(TDCPP)Cl). The main reaction products of the hydrocarbon oxidation were cyclohexanol and cyclohexanone. Also, the products derived from aldehyde oxidation such as peroxypropionic acid, carbon dioxide and propionic acid were found. No oxidation of cyclohexane takes place when one of the reagents is missing.

The induction time , oxygen consumption as well as the yields of products and their distribution appeared to be dependent on the character of the metal centre of the porphyrin molecule. The experimental data are listed in Table 1.

Fig. 1 shows the oxygen uptake curves for the reactions carried out in the presence of the investigated metalloporphyrins while Fig. 2 illustrates the changes in the product concentrations during the reaction course. The latter reaction was carried out in the presence of Mn(TDCPP)Cl which in comparison with other metalloporphyrins is more stable in oxidizing media. As it follows from Fig. 2 the amounts of peroxy acid as well as cyclohexanol and cyclohexanone grow parallelly. When the conversion of the aldehyde reaches 100% the concentration of peroxy acid starts to diminish while the concentrations of cyclohexanol and cyclohexanone remain practically at the same level. CO_2 evolution is still observed but at much lower rate.

TABLE 1

Oxidation of cyclohexane.

catalyst	induction time (min.)	oxygen consumption (mmoles)	yield of products (mmoles) alcohol	ketone	peroxy acid	CO_2	turnover frequency (cycles/min) 1	2
Cr(TTP)J	52	6.6	0.01	0.02	4.0	0.7	2.6	0.02
Mn(TTP)Cl	10	5.8	0.03	0.05	0.1	2.2	8.7	0.3
Fe(TTP)Cl	0	7.8	0.03	0.06	5.7	2.4	68	0.7
Co(TTP)	10	8.4	0.075	0.07	0.6	4.0	16	0.5
Ni(TTP)	33	6.9	0.05	0.03	6.5	0.3	8	0.1

[cyclohexane]=1.8×10^{-1}M, [propionaldehyde]=1 M, [catalyst]=2×10^{-3}M 1 - corresponds to the total amounts of products, 2 - to the amount of cyclohexanol and cyclohexanone, t=30 C.

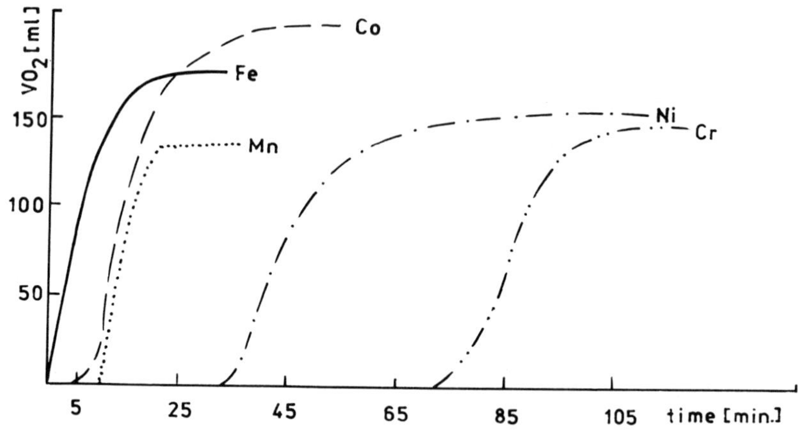

Fig.1 Oxygen absorption during the course of the reactions.

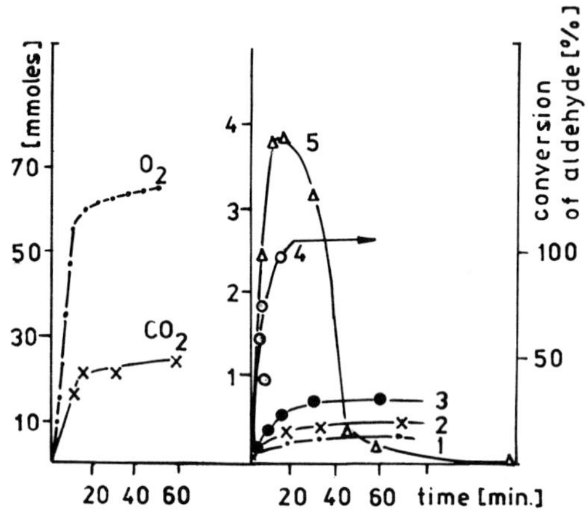

Fig.2 The distribution of products during the course of the reaction catalyzed by Mn(TDCPP)Cl. 1-cyclohexanol, 2-cyclohexanone, 3-total(1+2), 4-conversion of aldehyde, 5-peracid.

DISCUSSION

It was shown in our earlier paper (ref. 12) that the reduction step plays an essential role in the investigated process. The metalloporphyrins are transferred to lower oxidation states and simultaneously acyl radicals are generated:

$$M^{n}TTP^{+} + RCHO \longrightarrow M^{n-1}TTP + RCO^{\cdot} + H^{+} \tag{1}$$

Acyl radicals are then free to react with dioxygen which has biradical character and this process gives rise to the oxygen uptake :

$$RCO^{\cdot} + O_2 \xrightarrow{RCHO} RCOOOH + RCO^{\cdot} \qquad (2)$$

Hence the induction time may be related to the ease of reduction of the metalloporphyrins used as catalysts. Thus Fe(III) and Mn(III)-porphyrins which have low reduction potentials give short induction times while for Ni(II) and Cr(III)-porphyrins which are more difficult to reduce the induction time is much longer. Reduced forms of some metalloporphyrins as for instance Fe(II)TTP and Mn(II)TTP are reoxidized in the process of binding dioxygen but the oxygenated adducts have different forms, character and ability to react with organic molecules (refs. 16, 17). At variance with Mn(III) and Fe(III)-porphyrins, Co(II)- porphyrin is difficult to reduce, however, this latter binds readily molecular oxygen and the odd electron is transferred from the cobalt centre to the antibonding oxygen orbitals. The oxygenated adduct reacts with aldehyde molecule in the rate determining step in which acyl radicals are released (ref. 17):

$$Co(III)TTPO_2^- \xrightarrow{RCHO} Co(III)TTPO_2^-HC(=O)R \longrightarrow Co(III)TTP\text{-}O_2H + RCO^{\cdot} \qquad (3)$$

It might also be expected that under reducing conditions dominating at the beginning of the reaction so called reductive oxygen activation may happen, however, as yet no experimental support has been found in favour of this assumption.

Zn(II) and Cu(II)-porphyrins are difficult to reduce and may undergo only ring reduction which means that the accepted electron resides on the porphyrin ligand. Such an anion radical is difficult to oxidize by oxygen since the latter can only be bound to the metal centre when the conditions for the orbital overlap exist and thus the reaction cannot be recycled. For these reasons these porphyrins do not show any catalytic activity.

Peroxy acid is a more efficient oxidant than molecular oxygen and on one hand may oxidize metalloporphyrins to higher oxidation states and on the other hand interacts with the porphyrin species to form intermediate complexes responsible for the insertion of an oxygen atom to the hydrocarbon molecule.

Peroxy acid may react with the metalloporphyrin molecule in a homolytic (one electron) and heterolytic (two electron) pathway. The homolytic process is accompanied by decomposition of peroxy acid resulting in formation of carbon dioxide (Eqn. 4) (ref. 19).

$$M^{n}TTP + RCOOOH \longrightarrow M^{n+1}TTP^{+} + OH^{-} + R^{\cdot} + CO_2 \qquad (4)$$

$$M^{n}TTP + RCOOOH \longrightarrow M^{n+2}(TTP) = O + RCOOH \qquad (5)$$

The heterolytic pathway results in formation of high-valent metal-oxo species (Eqn. 5). The latter are believed to be able to incorporate an oxygen atom to the hydrocarbon molecule. However, it is reported that the efficiency of the process depends on the character of the oxidizing agent used to produce these active oxo species (ref. 9). In such a case the structure and character of an intermediate complex composed of a metalloporphyrin molecule, oxidant and hydrocarbon molecule would play the decisive role in the catalytic step.

As seen from the data in the Table and oxygen uptake diagram, Fe(TTP)Cl shows the highest activity which is manifested by the highest total yield of products and absence of induction time. However, the absorbed oxygen is predominantly involved in formation of peroxy acid. The explanation of this phenomenon lies in the fact that Fe(II)TTP, resulting from reduction of the initial porphyrin at the first stage of the reaction reacts rapidly with dioxygen in the following sequence of reactions (ref. 19):

$$Fe(II)TTP \xrightarrow{O_2} Fe(II)TTPO_2 \xrightarrow{Fe(II)TTP} TTPFe(III){-}O{-}O{-}(III)FeTTP \longrightarrow$$

$$Fe(IV)(TTP){=}O \xrightarrow{Fe(II)TTP} TTPFe(III){-}O{-}Fe(III)TTP \qquad (6)$$

Fast reaction with dioxygen gives priority to the formation of five coordinate μ-oxo dimer which is an inactive form of the metalloporphyrin. The access of both aldehyde and peroxy acid molecules to the iron centres involved in dimeric structure is difficult and the compound is resistant to reduction (reduction potential = -0.9 V) as well as to oxidation. Therefore large amount of peroxy acid is found at the end of the reaction. Under the described conditions the active forms i.e. O=Fe(IV)TTP and

$O=Fe(V)TTP^+$ which could effect oxidation of hydrocarbon cannot be generated in significant concentrations in the course of the reaction.

Longer than for Fe(TTP)Cl induction time at higher oxygen consumption and more significant amounts of cyclohexanol and cyclohexanone are observed for CoTTP. Simultaneously, large part of peroxy acid is decomposed to CO_2. These results are consistent with our previous investigations on epoxidation of propylene under similar conditions which showed that Co-porphyrin exhibited the highest activity in both oxygen consumption and epoxide production (refs. 12, 20). It has been shown that the active form of the cobalt porphyrin is its Π cation radical obtained in the process of homolytic oxidation of the initial porphyrin with the generated peroxy acid which is simultaneous with CO_2 evolution starting before the epoxide is detected. The induction time is necessary to generate peroxy acid and Π cation radical which together with another peracid molecule forms a precursor of an active intermediate complex capable to incorporate an oxygen atom to the hydrocarbon molecule :

$$[Co(III)(TTP)X^{+}\cdots RCOOOH] + RH \longrightarrow ROH + RCOOH \qquad (7)$$

An alternative proposal given in (ref. 21) is formation of a cobalt(V)- oxo species.

Still less active appeared Mn(TTP)Cl which shows the lowest oxygen uptake and smallest yield of products. However, the amount of cyclohexanol and cyclohexanone consists more important part of the total yield of products than in the case of other metalloporphyrins. It means that the system is more selective with respect to these products. Also, the amount of CO_2 found is comparatively higher which indicates that homolytic decomposition of peroxy acid is more important here. This is justified by high number of oxidation states (II-V) accessible for Mn-porphyrins. In a heterolytic reaction with peroxy acid such catalytically active species as $O=Mn(V)TTP^+$ and O=Mn(IV)TTP are likely to be generated. The former has already been recognized as responsible for oxygen atom transfer to hydrocarbon molecules (refs. 6, 8). Recently Groves et al. have shown that TMPMn(IV)=O species are also active in epoxidation of olefins (ref. 22).

NiTTP and Cr(TTP)Cl show the lowest activity and the latter yields predominantly peroxy acid and negligible amounts of cyclohexanol and cyclohexanone. The former porphyrin also yields large amounts of peroxy acid but the amounts of cyclohexanol and cyclohexanone are comparable to those found for Mn-porphyrin. It means that the complex is more effective in heterolytic reaction with peroxy acid than in its homolytic decomposition. According to Kochi et al. a putative oxo-nickel(IV) and/or μ-oxo-nickel(III) intermediates are engaged in oxidation of hydrocarbons (ref. 23).

The investigated system employs two oxidizing agents: molecular oxygen and peroxy acid. Both oxidants are involved in generation of oxygen containing metalloporphyrin species. However, the structure and activity of the oxygenated forms depend on the character of the metal centre and on kinetics of their formation.

ACKNOWLEDGMENT

The authors wish to express their gratitude to Dr P.Battioni from Université René Descartes Paris VI for chemicals and scientific guidance in preparation of Mn(TDCPP)Cl.

REFERENCES

1. I.Tabushi, A. Yazaki, J.Am.Chem.Soc., 103 (1981) 7371-7375.
2. I.Tabushi, N. Koga, J.Am.Chem.Soc., 101 (1979) 6456-6458.
3. M.Perree-Fauvet, A.Gaudemer, J.Chem.Soc.Chem.Commun.,(1981) 874.
4. M.Fontecave, D.Mansuy, Tetrahedron, 40 (1984) 4297-4311.
5. C.L.Hill, B.C.Schardt, J.Am.Chem.Soc., 102 (1980) 6374-6375.
6. J.T.Groves, W.J.Kruper and R.C.Haushalter, J.Am.Chem.Soc., 102 (1980) 6375-6377.
7. J.T.Groves,Y.Watanabe,T.J.McMurry, J.Am.Chem.Soc., 105 (1983) 4489-4490.
8. B.Meunier, Bull.Soc.Chim.Fr., (1986) 578-594.
9. A.Robert, B.Meunier, New.J.Chem., 12 (1988) 885-896.
10. J.P.Renaud, P.Battioni, J.F.Bartoli, D.Mansuy, J.Chem.Soc. Chem.Commun., (1985) 888-889.
11. P.Battioni, J.P.Renaud, J.F.Bartoli, D.Mansuy, J.Chem.Soc. Chem.Commun., (1986) 341-343.
12. J.Haber, T.Młodnicka, J.Połtowicz, J.Mol.Catal. in press.

13. A.D.Adler, F.R.Longo, F.Kampas, J.Kim, J.Inorg.Nucl.Chem., 32 (1970) 2443-2446.
14. A.W.van der Made, E.J.H.Hoppenbrouwer, R.J.M.Nolte, W.Drenth, Rec.Trav.Chim.Pays Bas., 107 (1988) 15-16.
15. J.Haber, A.Marchut, T.Młodnicka, J.Połtowicz, J.J.Ziołkowski, React.Kinet.Catal.Lett., 8 (1977) 281-286.
16. R.D.Jones, D.A.Summerville, F.Basolo, Chem.Rev. ,79 (1979) 139-179.
17. T.Młodnicka, J.Mol.Catal., 36 (1986) 205-242.
18. R.A.Sheldon, J.K.Kochi, Metal-Catalyzed Oxidation of Organic Compounds, Academic Press, New York, 1981, pp.43-45.
19. A.B.Hoffman, D.M.Collins, W.V.Day, E.B.Fleischer, T.S.Srivastava, J.L.Hoard, J.Am.Chem.Soc., 94 (1972) 3620-3626.
20. J.Haber, T.Młodnicka, M.Witko, J.Mol.Catal., 52 (1989) 85-97.
21. W.A.Lee, T.C.Bruice, Inorg.Chem., 25 (1986) 131-135.
22. J.T.Groves, M.K.Stern, J.Am.Chem.Soc., 109 (1987) 3812-3814.
23. J.D.Koola, J.K.Kochi, Inorg.Chem., 26 (1987) 908-916.

B.R. James (Dept. of Chemistry, Vancouver, Canada): In principle, you are using an aldehyde as a co-reductant in reductive-activation of O_2, and in one step invoke reduction of the metalloporphirin by RCHO to generate $RCO^{\cdot}$. Such acyl radicals may lose CO rapidly and lead to decarboxylation of aldehydes (which we have demonstrated with Ru and Fe prphyrins, Can. J. Chem. (1988)). Do you see reaction of M(TTP) with aldehydes (in the absence of O_2) to give any decarboxylation products ? For example, C_2H_6 from propionaldehyde ?
T. Mlodnicka: The reaction was tested in the absence of O_2 and no products at approciable concentrations were detected. However, it would be interesting to check it once more paying special attention to this problem.

J. Kiwi (EPFL, Lousanne, Switzerland): How stable are the peracids you postulate formed in your system ? If they are stable have you measured by a specific method or a general titration method ?

T. Mlodnicka: At the given concentrations of the solution components and the temperature of the reaction mixture, peroxy acid is a comparatively stable species and its concentrations were determined by iodometric titration.

U. Shuchard(Brazil): Could you say anything about the mass balance with respect to the cyclohexane ?
T. Mlodnicka: We worked at a significant excess of cyclohexane. In such a case the determination and evaluation of mass balance is a difficult task with a large experimental error. To my knowledge many investigators have the same problem.

G. Centi and F. Trifiro' (Editors), *New Developments in Selective Oxidation*

OXOMETALATES AND DIOXYGEN IN CATALYTIC OXIDATION

J.-M. BRÉGEAULT[1], B. EL ALI[1], J. MERCIER[2], J. MARTIN[1], C. MARTIN[1] and O. MOHAMMEDI[1]

[1]Département de Chimie, Université P. et M. Curie, Catalyse et Chimie des Surfaces; Tour 44 - 2e ; 4, Place Jussieu; 75252 Paris Cedex 05 (France)

[2]Département de Chimie, Université P. et M. Curie, Laboratoire de Chimie Organique Structurale, Bât. 74

SUMMARY

The application of oxovanadium (IV) or (V) complexes $|VO(acac)_2|$, $|VO\{OCH(CH_3)_2\}_3|$ and heteropolyacids $H_{3+n}|PMo_{12-n}V_nO_{40}|$ "HPA-n", as catalyst precursors is examined for the oxidative cleavage of ketones. In the presence of dioxygen, at room temperature or at 60°C, they react with benzylic ketones, $Ar\text{-}CH_2\text{-}C(O)R$, to produce the corresponding carboxylic acids, R-COOH, and benzaldehyde and/or benzoic acid in high yield. Substituted cycloalkanones, α-ketol, α-diketones and 1-phenylalkanones are also oxidatively cleaved by HPA-2 and dioxygen under very mild conditions. The experiments show that the efficiency of HPA-2 is related to the key role of Va(V) and to the large solubility of this "acidic complex" in organic media.

INTRODUCTION

As part of our continuing investigation into oxidation processes, we have been interested in the use of vanadium (V) precursors in homogeneous catalysis (ref. 1). The oxidation of organic compounds by quinquevalent vanadium has, besides its own interest, implications concerning the behavior of vanadium oxide catalysts (ref. 2). The subject was first examined by Morette *et al.* essentially from an analytical viewpoint (ref. 3) and later by Littler *et al.* (ref. 4); *molecular oxygen has no significant effect* on the rates of oxidation and all these oxidation processes are stoichiometric. In this context, it was interesting to test Matveev's systems as oxidizing agents : palladium (II) complexes and salts of heteropolyacids abbreviated to "Pd(II)/HPA-*n*" (ref. 5),

$$\text{HPA-}n + \text{Red} + iH^+ \rightarrow H_i|\text{HPA-}n| + \text{Ox} \quad (1)$$

Thus, HPA-*n* in eqn. 1 is a heteropolyacid with the Keggin structure : $H_{3+n}|PM_{12-n}V_nO_{40}|$; M = Mo, W; Red is a reducing agent involving *i* electrons : a reduced form of a catalyst or a substrate. The reduced form, $H_i|\text{HPA-}n| \equiv H_i|H_{3+n}\{PM^{VI}_{12-n}Va^{V}_{n-i}V^{IV}_{i}O_{40}\}|$ or the *"heteropoly-blue" can be reoxidized by dioxygen under very mild conditions :*

$$H_i\,|HPA\text{-}n| \;+\; (i/4)O_2 \;\rightarrow\; HPA\text{-}n \;+\; (i/2)H_2O \qquad (2)$$

Since polyoxometalates have no sophisticated oxidizable ancillary ligands, they have considerable potential as long-lived catalysts. According to eqn. (1) and (2), HPA-*n* can be used as cocatalysts, for example, in association with precious metal complexes (ref. 5-6) or as direct catalysts (ref. 7).

While testing recently-described systems : "Pd(II) or Rh(III) complexes with or without a cocatalyst and in the presence of dioxygen" (ref. 5-6, 8) for the oxidation of several olefinic substrates, we obtained clear evidence for secondary catalytic processes; one of these is *the oxidation of ketones*. These results on the vanadium(V) catalytic oxidative cleavage of ketones with dioxygen are discussed.

RESULTS AND DISCUSSION

Oxidation of 1-phenylpropan-2-one, **I** ; evidence for a precursor effect

In order to compare several precursors we took 1-phenylpropan-2-one, **I** , as a model substrate.

Ketone **I** is oxidized without an organic solvent. Results are summarized in Table 1.

TABLE 1

Oxidation of 1-phenyl-2-propanone, **I** , by heteropolyacids or vanadium oxo-complexes and dioxygen[a]

Run	Precursors $(0.20\ mol\ l^{-1})$	T (°C)	time (h)	Conversion[b] (%)	Products (% Yields[c]) IX	X	XI	XII	XIII
1	$H_3\,|PMo_{12}O_{40}|\cdot 30\ H_2O$	20	6	6.5	1.5	–	6.5	–	5
2	$H_4\,|PMo_{11}V_1O_{40}|\cdot 33\ H_2O$	20	6	70	37	20	62	3.5	9.5
3	$H_5\,|PMo_{10}V_2O_{40}|\cdot 36\ H_2O$	20	6	99	78	10	88	10.5	0.5
4	$H_6\,|PMo_9V_3O_{40}|\cdot 34\ H_2O$	20	6	99	77	10	88	10.5	0.5
5	$H_3\,|PW_{12}O_{40}|\cdot 29\ H_2O$	20	6	3	traces	traces	traces	–	traces
6	$H_4\,|PW_{11}V_1O_{40}|\cdot 30\ H_2O$	20	6	46	28	7	38	4.5	7
7	$|VO\{OCH(Me)_2\}_3|$	60	24	82	10	68	78	3	–
8	$|VO(acac)_2|$	60	24	75	10	56	66	6	–

[a] Reaction conditions . runs 1-6 : substrate = 3 cm³ (22.4 mmol) ; runs 7-8 : substrate = 1.5 cm³; $p\,(O_2) = 10^5$ Pa; without organic solvent; [b] % of substrate consumed; [c] Products analysed by coupled GC-MS (OV 17 and OV 105 columns); internal standard anisole.

In all experiments (runs 1-8), oxidation leads to benzaldehyde, **IX** and acetic acid, **XI**, as major products, and to benzoic acid, **X**, resulting from benzaldehyde cooxidation. Two minor products, 1-phenylpropan-1,2 dione, XII and *trans*-stilbene, XIII, are formed. The latter was identified in a reaction between benzaldehyde and I (ref. 9). Comparison of runs 1 and 5 with runs 2-4, 6-8 shows that vanadium (V) is a key-element in the catalytic systems. Vanadium (V) oxoalkoxide, $|VO\{OCH(Me)_2\}_3|$ or vanadium (IV) acetylacetonate can have also a catalytic effect with dioxygen, but the substrate/vanadium ratio, the conversion and the reaction times indicate that they are less efficient than HPA-1 with Mo (run 2).

These results and the easy synthesis (ref. 10) of the acid "$H_5|PMo_{10}V_2O_{40}|$. 30-36 H_2O", or of the acid salts "$H_{5-x}Na_x|PMo_{10}V_2O_{40}|$, y H_2O" led us to choose these precursors for the oxidative cleavage of other ketones.

Oxidation of benzylic ketones

HPA-2 can be used in the oxidation of benzylic ketones (ref. 11). All data are compatible with scheme 1 :

$$X\text{-}C_6H_4\text{-}CH_2\text{-}C(O)\text{-}R \xrightarrow[20°C;\ MeCN]{HPA\text{-}2;\ O_2} X\text{-}C_6H_4\text{-}CHO + R\text{-}COOH$$

major products
(yields : 70 - 90 %)

$$+ \quad X\text{-}C_6H_4\text{-}COOH + X\text{-}C_6H_4\text{-}C(O)\text{-}C(O)R$$

minor products
(yields ≤ 10%)

X : H R : Me; Et; Ar or Ar-CH_2
X : *p*-OMe R : Me

Scheme 1

The oxidative cleavage of phenylacetones gives high yields of acetic acid and of benzaldehyde (or of functionalized benzaldehyde). The reaction can be applied to different benzylic ketones.

Oxidation of α-ketol, α-diketones and 2-methyl cyclohexanone; evidence for a solvent effect

In MeCN the conversions and the selectivities are usually close to the values obtained with the other pure liquid substrates (ref. 12). Radical chain reactions can be involved in the mechanism of oxidative cleavage of ketones (*vide infra*); these reactions are less subject to interference from polar effects. Nevertheless, for some substrates we observed a dramatic change on going from MeCN to ethanol. For an α-ketol, 2-hydroxy-2-phenylacetophenone, II, the conversion is low in acetonitrile (run 9 - Table 2) but reaches 95% in ethanol (run 10) even at room temperature; other non-alcoholic solvents cause inhibition.

TABLE 2

Oxidation of a ketol and α-diketones by HPA-2 and O_2[a]

Run	Substrate (mmol)	HPA-2 (mmol)	Solvent	Time (h)	Conversion[b] (%)	Products (% Yields[c]) IX	X or X'	XI
9	II (4.8)	0.05	MeCN	24	25	24	24	
10	II (4.8)	0.05	EtOH	6	95	93	91	
11	III (7.4)	0.03	EtOH	3	≃ 100	–	≃ 100	≃ 100
12	IV (2.45)	0.05	EtOH	24	52		102	

[a] Reaction conditions . HPA-2 = $H_5[PMo_{10}V_2O_{40}]$. 30-36 H_2O ; solvent : 6 cm³ ; T : 20°C ; $p(O_2) \simeq 10^5$ Pa

[b, c] see Table 1

Benzaldehyde, IX, and ethyl benzoate, X', are the only isolable products. α-Diketones such as Ph-C(O)-C(O)-Me III (or XII) and Ph-C(O)-C(O)-Ph, IV, are not cleaved in MeCN, while they react with high selectivity in ethanol at room temperature (runs 11-12). These surprising results led us to examine the solvent effect on the oxidative cleavage of 2-methylcyclohexanone by HPA-2 (Table 3). Some solvents (benzene, dichloro-1,2 ethane,...) inhibit the reaction or reduce the rate of oxidation of 2-methylcyclohexanone, V. The novel catalyst systems studied can operate in alcoholic media (methanol, ethanol, *tert*-butanol,...) but the conversion is lower than with MeCN or acetic acid and nitromethane. The main product is the 6-oxoheptanoic methylester XIV' (run 15) or the corresponding keto-acid XIV (run 16), with a ring-contraction product (cyclopentanone, XV) corresponding to ethyl group elimination. Nitromethane favors the formation of this minor product, although the conversion is equal to that obtained with MeCN (run 14). Mixed solvents (runs 17-18) such as MeCN/MeOH give also XIV′, the methyl ester. With ethylene glycol dimethyl ether (run 19), moderate yields of the esters, XIV', are obtained, due to the partial methylation of the ketoacid.

Oxidation of cycloalkanones by HPA-2 and O_2

The results for the reaction of some cycloalkanones are presented in Table 4. Each ketone was subjected to nearly identical oxidation conditions with HPA-2. Treatment of 2-methylcyclopentanone, VI, for 2h (run 20) gives 5-oxohexanoic acid, XVI, in high yield (94%). 2-methylcyclohexanone, V, gives 6-oxoheptanoic acid, XIV, (run 21); 2,6-dimethylcyclohexanone, VII, produces mainly 6-oxo-2-methylheptanoic acid, XVII, also in good yield (89%) (run 22). Other cyclic ketones can be cleaved, but the selectivity depends on the substrate and on the catalytic system; for example, under these conditions,

TABLE 3

Oxidation of 2-methylcyclohexanone, V, by HPA-2 and O_2 : solvent effect[a]

Run	Solvent (cm^3)	Conversion[b] (%)	Time (h)	Products (% Yields[c]) XIV	XIV'	XV
13	MeCN (6)	98	4	85	2	9
14	$MeNO_2$ (6)	98	4	81.5	1.5	14.5
15	MeOH (6)	54	6	0	49	2.5
16	MeCOOH (6)	81	6	69	1	10.5
17	MeCN/MeOH (5) (1)	96	6	4	86	3
18	$MeNO_2$/MeOH (5) (1)	96	6	6.5	81.5	5
19	$MeO-(CH_2)_2OMe$ (6)	96	24	74	16	4.5

[a] Reaction conditions . HPA-2 : 0.075 mmol ; substrate : 12.4 mmol ; T : 60°C ; $p(O_2) = 10^5$ Pa ;
[b, c] see Table 1 ; internal standard : heptanoic acid

TABLE 4

Oxidation of cycloalkanones by HPA-2 and O_2[a]

Run	Substrate (mmol)	HPA-2 (mmol)	Solvent (cm^3)	Time (h)	Conversion[b] (%)	Products (% Yields[c])
20	VI (4.4)	0.025	MeCN (6)	2	96	XVI (94)
21	V (12.4)	0.075	MeCN (6)	4	98	XIV (85)
22	VII (11.0)	0.075	MeCN (6)	6	91	XVII (89)
23	VIII (4.7)	0.05	MeCN/MeOH (5) (1)	24	100	XVIII (40) XIX (17)

[a] Reaction conditions . HPA-2 = $H_5[PMo_{10}V_2O_{40}]$. 30-36 H_2O ; T : 60°C ; $p(O_2) \simeq 10^5$ Pa
[b, c] see Table 1 ; internal standard : heptanoic acid

cyclohexa-1,3 dione, which exists mainly in the monoenolic form, can be cleaved with loss of one or two carbons; it gives two major products in MeCN-MeOH :

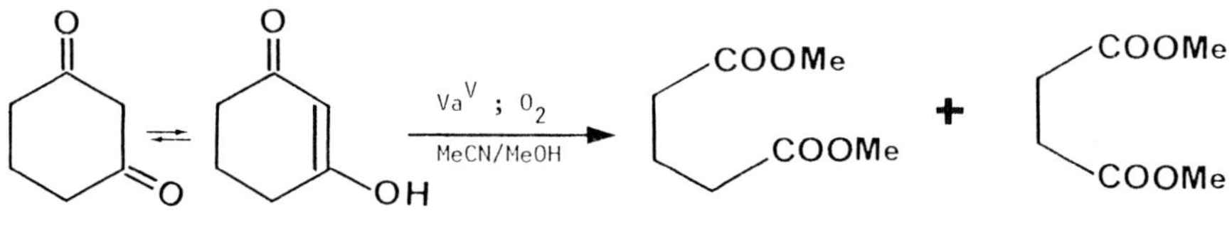

VIII XVIII XIX

The species undergoing cleavage has not been isolated, but it could be different from the triketone formed with the sodium periodate system (ref. 13) which gives mainly glutaric acid.

Mechanism of oxidation and search for intermediates

EPR results give clear evidence only of isolated V^{IV} species. To determine whether one mechanism is more plausible than another, we carried out a labeling experiment using $^{18}O_2$ but without clear-cut results. Considering the experimental facts, we can propose a plausible mechanism as shown in scheme 2 for cycloalkanones in which the formation of a vanadium enolate is a key step; it

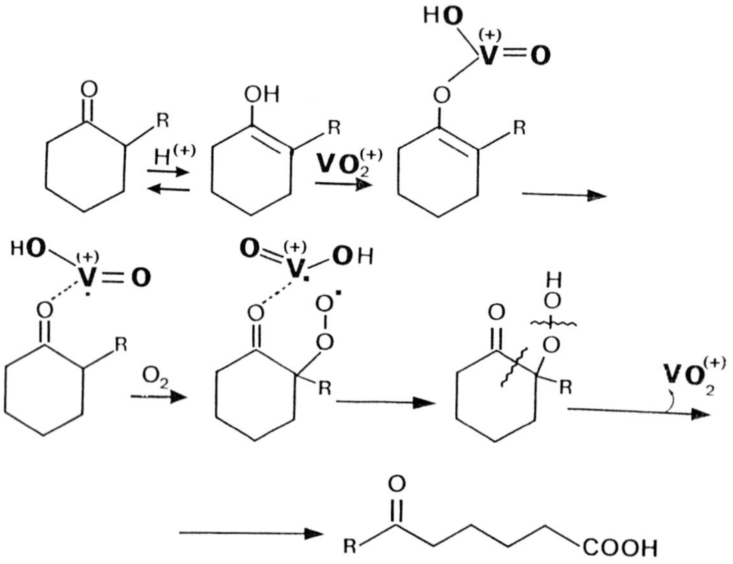

Scheme 2

could generate short-lived radical species which could interact with O_2 in a vanadium-assisted pathway. The intermediate peroxide could undergo direct or vanadium-assisted decomposition to yield the keto-acid. The most intriguing step is the homolytic cleavage of V-O bond to give the short-lived radical species. A more thorough study of the system is now in progress.

CONCLUSION

A new catalytic method for the oxidative cleavage of some open-chain ketones or of substituted cycloalkanones has been found. It employs a rather inexpensive "PMoV" association as the catalyst in a homogeneous phase in combination with dioxygen as the primary oxidant. Some of the reactions reported here may have synthetic potential : some keto-acids have been utilized in the synthesis of macrocyclic lactones. Other applications include the preparation of catecholamine conjugates and natural-product total synthesis. The reaction can be extended to other carbon-carbon bond cleavages using dioxygen; for example α -diols have been smoothly cleaved (ref. 14) by a catalytic amount of $H_5[PMo_{10}V_2O_{40}]$. 30-36 H_2O or of $[VO\{OCH(CH_3)_2\}_3]$ under molecular oxygen and very mild conditions.

REFERENCES

1 J.-M. Brégeault, F. Derdar, J. Martin, C. Martin et J. Mercier, Proc. 6th Int. Symp. Homogeneous Catalysis, Vancouver, August 21-26, 1988, p. 34; J.-M. Brégeault, B. El Ali, J. Mercier, J. Martin and C. Martin, C.R. Acad. Sci. Paris, 307 (1988) série II, 2011-2014.

2 G. Centi, J. Lopez Nieto, C. Iapalucci, K. Brückman and E.M. Serwicka, Appl. Catal., 46 (1989) 197-212; J.G. Highfield and J.B. Moffat, J. Catal., 98 (1986) 245-258; M. Misono, Catal. Rev.-Sci. Eng., 29 (1987) 269-321

3 A. Morette et G. Gaudefroy, Bull. Soc. Chim. France, (1954) 956-964.

4 J.S. Littler, J. Chem. Soc., (1962) 832-837; J.S. Littler and W.A. Waters, J. Chem. Soc., (1959) 3014-3019.

5 I.V. Kozhevnikov and K.I. Matveev, Russ. Chem. Rev., 51 (1982) 1075-1088; Appl. Catal. , 5 (1983) 135-150; I.V. Kozhevnikov, Uspeckhi Khimii, 56(1987) 1417-1443; E.G. Zhizhina, L.I. Kuznetsova and K.I. Matveev, React. Kinet. Catal. Lett., 31 (1986) 113-120.

6 B. El Ali, J.-M. Brégeault and J. Martin, J. Organometal. Chem., 327 (1987) C9-C14.

7 I.V. Kozhevnikov, V.I. Simagina, G.V. Varnakova and K.I. Matveev, Kinet. i Katal., 20 (1979) 506-510.

8 O. Mohammedi, Ph. D., Université P. et M. Curie, may 11, 1987.

9 W.V. Miller and G. Rohde, Berichte, 23 (1890) 1070-1079.

10 G. Canneri, Gazz. Chim. Ital., 56 (1926) 871-889; P. Courtin, Rev. Chim. Min., 8 (1971) 75-85; G.A.T. Tsigdinos and C.J. Hallada, Inorg. Chem., 7 (1968) 437-441; J.-M. Brégeault *et al.*, unpublished results.

11 B. El Ali, J.-M. Brégeault, J. Martin, C. Martin and J. Mercier, New J. Chem., 13 (1989) 173-175.

12 B. El Ali, Ph. D., Université P. et M. Curie, june 21, 1989.

13 M.L. Wolfrom and J.M. Bobbit, J. Amer. Chem. Soc., 78 (1956) 2489-2493.

14 J.-M. Brégeault, B. El Ali, J. Mercier, J. Martin et C. Martin, C.R. Acad. Sci. Paris, 309 (1989) série II, 459-462.

DISCUSSION CONTRIBUTION

B.R. JAMES (University of British Columbia, Vancouver, Canada) : I am confused regarding your implications for the mechanism. You note that the reduced heteropoly-blue H_i(HPA-*n*) is reoxidized by O_2 to give the oxidized form HPA-*n*, yet in Scheme 2 you make a mechanism showing unchanged oxidation state in the HPA moiety (VO_2^+) with O_2 attacking the coordinated cycloalkanone to give a peroxo radical. Do you favour O_2 playing a role within the HPA moiety or within the organic moiety ?

J.-M. BREGEAULT (Université P. et M. Curie, Paris, France) : The equations of the introductory part are not those of a mechanism (i.e. elementary processes) but only those of the presumed overall process. Scheme 2 shows the formation of intermediate vanadium (IV) species, but omit the formation of an alkylperoxidic complex which would involve free radical-dioxygen-vanadium interaction. This intermediate species could assist reoxidation of vanadium (IV). At present, we have no experimental results which show the preferred interaction of dioxygen with the organic rest, so the third and fourth step of the mechanism (Scheme 2) indeed, has the character of a proposal.

R.A. SHELDON (ANDENO, The Netherlands) : Do any of the oxometalate or vanadium catalyzed oxidative cleavage of 1,2-diols also work under neutral or basic conditions ?

J.-M. BREGEAULT (Université P. et M. Curie, Paris, France) : Vicinal diols can also be cleaved by some HPA-salts, but the rate of reaction, the conversion and the yields are lower than those obtained with HPA-2. It should be mentioned that reoxidation of the reduced form(s) is controlled by the acidity function of the medium.

H. MIMOUN (Institut Français du Pétrole, Rueil-Malmaison, France) : What is the stability of the HPA during the cleavage reaction ?

J.-M. BREGEAULT (Université P. et M. Curie, Paris, France) : The stability of HPA-*n* under our reaction conditions has not yet been studied in detail, but till now we have no experimental evidence of its instability. Work on this is in progress.

G. Centi and F. Trifiro' (Editors), *New Developments in Selective Oxidation* 213

^{1}H NMR STUDY OF THE MECHANISM OF ETHYLENEGLYCOL MONOACETATE FORMATION IN OXIDATIVE ACETOXYLATION OF ETHYLENE CATALYZED BY Pd(II) COMPLEXES

E.V. GUSEVSKAYA, I.E. BECK, A.G. STEPANOV, V.A. LIKHOLOBOV and K.I. ZAMARAEV

Institute of Catalysis, Novosibirsk 630090, USSR

SUMMARY

A detailed mechanism of ethylene oxidation by $Pd(NO_n)ClL_2$ complexes (n=2,3; $L=CD_3CN$) in a chloroform-acetic acid mixture is studied by ^{1}H NMR spectroscopy. The end reaction products are ethyleneglycol monoacetate (EGMA), acetaldehyde, nitroethylene and compounds with the general formula CH_3-CHXY (X,Y = OH, OAc, Cl, NO_2), whose ratio depends upon the solvent composition. Kinetic and spectral data obtained indicate the formation of a number of intermediates. The structure and routes of decomposition of the intermediates to EGMA and other reaction products are suggested.

INTRODUCTION

Oxidation of olefins catalyzed by Pd(II) complexes is a rapidly developing trend in selective synthesis of oxygen-containing organic compounds. The main product of α-olefins oxidation in acetic acid solutions containing salts of nitric acid and palladium(II) is glycol monoacetate, while in the absence of nitrate ions carbonyl compounds and vinyl ethers are formed (ref. 1).

The mechanism of formation of glycol monoacetates from α-olefins has been studied for $Pd(OAc)_2/LiNO_3/HOAc$ (ref. 2) and $Pd(NO_2)Cl(CH_3CN)_2/HOAc$ (ref. 3) systems using lithium nitrate or nitro ligands in Pd(II) complexes labelled by heavy isotopes of oxygen. It has been established that the resulting glycol monoacetate contains labelled oxygen in the carbonyl position of the acetate group. Based on data obtained on distribution of labelled oxygen in reaction products, the authors have suggested the mechanism of glycol monoacetate formation; however, the structure of intermediates has not been confirmed by spectroscopy.

The objective of this work was to study the mechanism of ethylene oxidation by $Pd(NO_n)ClL_2$ complexes in chloroform-acetic acid solution by ^{1}H NMR spectroscopy.

METHODS

$Pd(NO_n)ClL_2$ complexes were prepared as in (ref. 4). ^1H NMR spectra were recorded using a Bruker CXP-300 spectrometer with a magnetic field induction of 7 T. Chemical shifts of signals were measured with respect to the internal reference hexamethyldisiloxane. The temperature of samples was continuously monitored with a precision of 1° by a VT-1000 thermocouple. In all experiments, the concentration of palladium in solution was 2×10^{-2} Ml^{-1}; $5\div10$ mol of ethylene per palladium ion being introduced into the solution of complex. $CDCl_3$ and CD_3COOD(DOAc) were used as solvents.

RESULTS AND DISCUSSION

Addition of ethylene (I) to solutions of $Pd(NO_n)ClL_2$ complexes in chloroform-acetic acid medium (content of DOAc is 0-100%) gives rise to the appearance of several lines in the NMR spectra. Analysis of the changes in the line intensities with reaction time permitted us to isolate groups of lines, whose intensities varied in the same manner and that could, for this reason, be attributed to the same compounds. For this purpose the parameters of J(H-H) of multiplet lines were also used.

Reaction products

Seven groups of lines that do not disappear for a long period of time can be assigned to end reaction products whose ratio depends upon concentration of DOAc in solution.

Acetaldehyde (II) (δ = 2.17 ppm (d), δ = 9.73 ppm (qd)) is the main product (95-97% per reacted olefin) of the reaction in chloroform; its yield tends to decrease with increasing concentration of DOAc in solution.

During ethylene oxidation in chloroform nitroethylene (III) (δ = 5.91 ppm (dd), δ = 6.65 ppm (dd), δ = 7.14 ppm (dd)) is accumulated (up to 5%) with a long induction period; in the presence of DOAc nitroethylene is formed in trace amounts.

In solutions containing DOAc one of the products of ethylene oxidation is EGMA (IV) (δ = 3.77 ppm (m), δ = 4.14 ppm (m)), in glacial DOAc the yield of EGMA is 95-97%.

In the range of DOAc concentrations 2-20 vol.% ethylene oxidation gives rise to the formation of compounds V-VIII (total yield up to 45%), whose NMR spectra are similar in line structures and positions (δ = 1.35-1.71 ppm (d) and δ = 6.46-6.97 ppm (qd) with intensity ratio 3:1). Analysis of NMR spectra of compounds V-VIII and peculiarities of their accumulation in solution permitted us to suggest the following composition for these products: $CH_3CH(OAc)_2$ (V), $CH_3CH(OAc)(OH)$ (VI), $CH_3CH(OAc)(Cl)$ (VII) and $CH_3CH(OAc)(NO_2)$ (VIII).

Intermediates

Based on the initial increase and subsequent decrease of their intensities with time the groups of lines IX-XVI (see table 1) seem to belong to intermediates formed during the reaction. As a result of ^1H NMR studies of the kinetics of ethylene oxidation by $Pd(NO_n)ClL_2$ complexes at various concentrations of DOAc in chloroform, we have registered intermediates that may be responsible for the formation of observed reaction products. The maximum observed line intensities of the intermediates formed during ethylene oxidation in $CDCl_3$ solutions with different concentrations of DOAc are shown in Fig. 1 for $Pd(NO_3)ClL_2$ and in Fig. 2 for $Pd(NO_2)ClL_2$. An analysis of ^1H NMR spectra of intermediates and kinetic curves of accumulation-decomposition of the intermediates and end products at various concentrations of DOAc allowed us to suggest the structures of compounds IX-XVI (table 1) as well as the possible routes of their formation and decomposition.

Mechanism of ethylene oxidation

Palladium complexes with NO_2 ligands in chloroform solutions exist as two isomers: $Pd(ONO)ClL_2$ (complex A) and $Pd(NO_2)ClL_2$ (complex B) (ref. 5); in the presence of DOAc $Pd(OAc)ClL_2$ complex C) may be also formed. Then it is reasonable to suggest that in the first step of ethylene oxidation displacement of the neutral ligands from complexes A,B,C and $Pd(NO_3)ClL_2$ (complex D) and formation of the corresponding π-olefin complexes of palladium $\underline{A}$, $\underline{B}$, $\underline{C}$ and $\underline{D}$ take place. Due to insertion of coordinated ethylene into Pd-O bonds in complexes $\underline{A}$, $\underline{C}$ and $\underline{D}$ and into the Pd-N bond in complex $\underline{B}$, organopalladium intermediates XII, XI, IX and XIII, respectively, are formed (table 1).

A wide variety of ethylene oxidation products is determined by the step of decomposition of organometallic compounds IX, XI-XIII $Pd-CH_2CH_2Z$. The transformation of these key intermediates depend on the nature of substituent Z, ligands in the palladium complex and solvent composition. Based on the results of IR and NMR spectroscopy studies on the mechanism of ethylene oxidation by Pd(II) complexes (for chloroform solutions the results have been

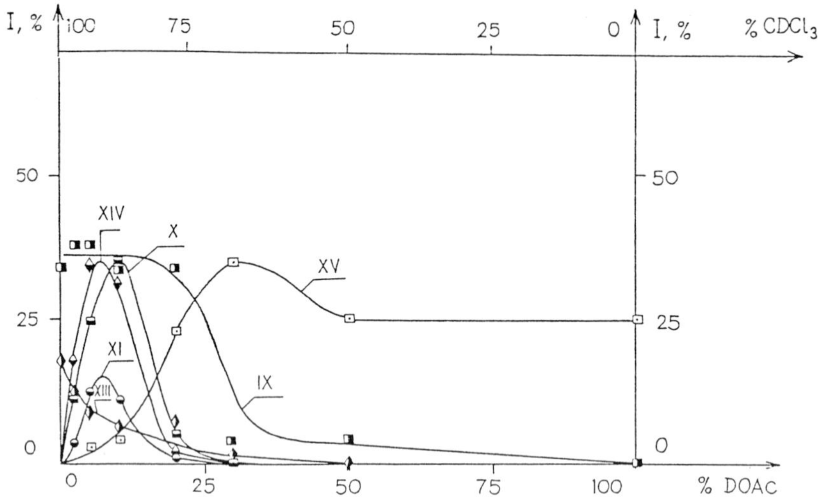

Fig. 1. Maximum observed intensities* of NMR lines of intermediates registered during ethylene interaction with $Pd(NO_3)ClL_2$ vs. solvent composition at 295 K.

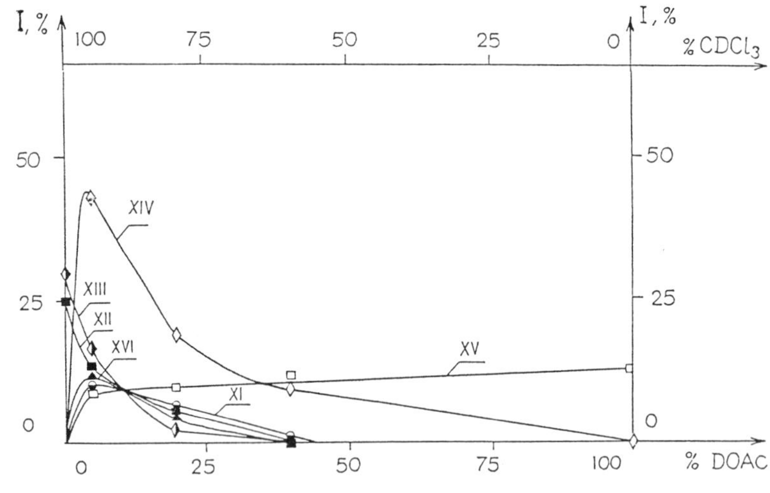

Fig. 2. Maximum observed intensities* of NMR lines of intermediates registered during ethylene interaction with $Pd(NO_2)ClL_2$ vs. solvent composition at 295 K.

* Line intensities in spectra are given per one proton; the total quantity of reacted (during observation time) ethylene is taken as 100%. **The yield of products is also given per reacted ethylene.**

TABLE 1

Characteristics of ^{1}H NMR spectra lines attributed to reaction intermediates and their proposed structures

Compound	Line structure	δ(ppm)	J(H-H) (Hz)	Intensity ratio	Proposed structure
IX	a triplet	1.61	7.3	I	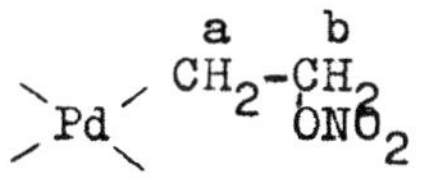
	b triplet	4.30	7.3	I	
X	a triplet	1.56	6.2	I	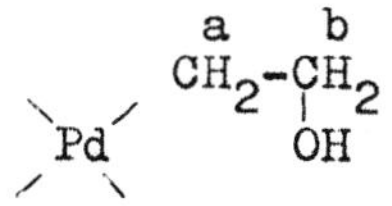
	b triplet	3.72	6.2	I	
XI	a triplet	1.67	6.8	I	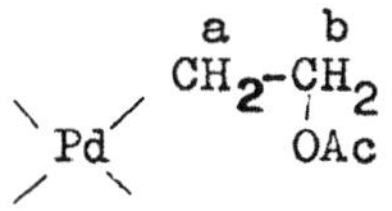
	b triplet	3.92	6.8	I	
XII	a broad line	2.37-3.25	-	I	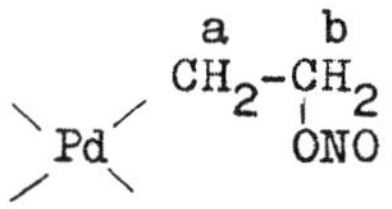
	b triplet	4.05-4.3	6.3	I	
XIII	a triplet	1.88	7.2	I	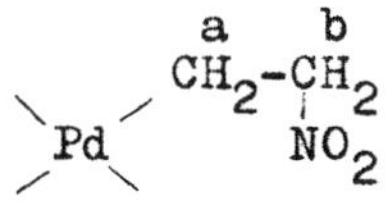
	b triplet	4.22	7.2	I	
XIV	a doublet	2.34	3.5	2	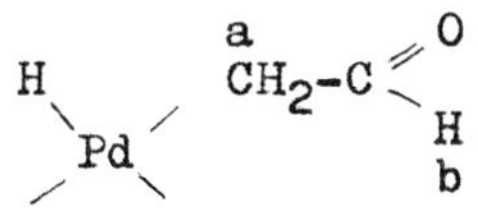
	b triplet	9.56	3.5	I	
XV *	a multiplet	4.29		I	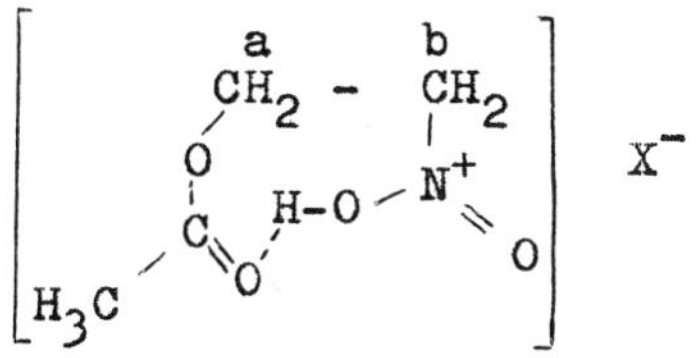
	b broad multiplet	4.89		I	
XVI	a triplet	2.67	5.6	I	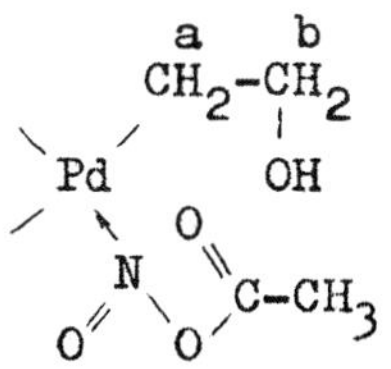
	b triplet	3.54	5.6	I	

* The spectrum typical for a four-spin system with $J_{AA'}$ = 1.4 Hz; $J_{BB'}$ = 3.1 Hz; J_{AB} = $J_{AB'}$ = $J_{A'B}$ = $J_{A'B'}$ = 4.55 Hz.

published in (refs. 5,6)) we propose the following possible mechanism of the formation of 1,1- (i.e. containing an ethylidene fragment) and 1,2-addition products.

<u>1,2-Addition products</u>. During ethylene oxidation by $Pd(NO_2)ClL_2$ EGMA seems to form at least by three parallel routes via key intermediate E:

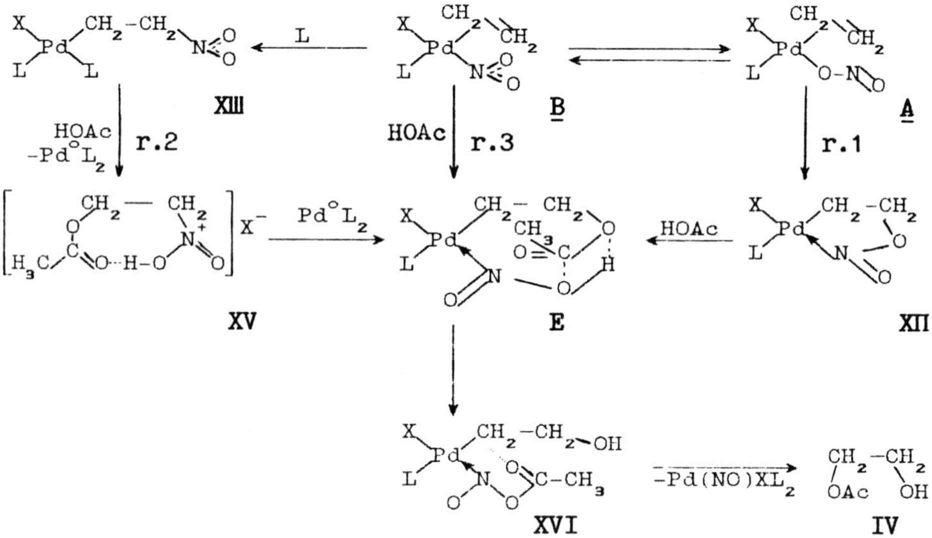

Organometallic intermediate E may be formed: from β-nitritoethylpalladium complex XII via reesterification by acetic acid (route 1); by heterolysis of the Pd-C bond in β-nitroethylpalladium complex XIII under the influence of DOAc resulting in 1-nitro-2-acetoxyethane XV followed by oxidative addition of the Pd(O) complex to the C-N bond in intermediate XV (route 2) and by direct acetoxypalladation of ethylene in palladium complex with nitro ligand (route 3). Then intramolecular rearrangement of intermediate E leads to the Pd(II) complex with a hydroxyalkyl ligand and acetylnitrite XVI.

Decomposition of complex XVI to form EGMA and nitrosyl complex of Pd(II) by heterolysis of the Pd-C bond under the action of the coordinated molecule of acetylnitrite.

It should be noted, that the mechanism proposed here is consistent with stereochemical data, labeling studies and the regiochemistry observed in (refs. 2,3,7).

It has been established that during ethylene oxidation by $Pd(NO_3)ClL_2$ EGMA is formed directly from the ethylene nitrate complex of Pd(II). The mechanism of interaction of the ethylene nitratopalladation product IX with DOAc seems to be similar to r. 2 for the nitrite system. Although in the nitrate system the EGMA formation intermediate analogous to XV was not found probably due to its high reactivity, intermediate X (analogous to XVI), which might contain acetylnitrate as a possible ligand, was registered. Heterolysis of the Pd-C bond in the complex X under the action of the coordinated acetylnitrate molecule yields EGMA. Unlike acetylnitrite, acetylnitrate can easily be displaced from the palladium complex followed by decomposition of the ethylene oxypalladation product into 1,1-addition products (mainly acetaldehyde) by β-hydride elimination.

<u>1,1-addition products</u>. The products of 1,1-addition (acetaldehyde and CH_3CHXY) seem to form during the decomposition of organopalladium intermediates IX-XIII via the following scheme: (a) reversible σ-π-rearrangement of complexes IX-XIII; (b) π-σ-transformation of hydridepalladiumolefin complexes via the attack of coordinated vinyl ether by the nucleophile X leading to the formation of either registered intermediate XIV (acetaldehyde precursor) or complex G (precursor of CH_3CHXY products); (c) decomposition of hydride complexes XIV and G via reductive elimination producing acetaldehyde and compounds V-VIII:

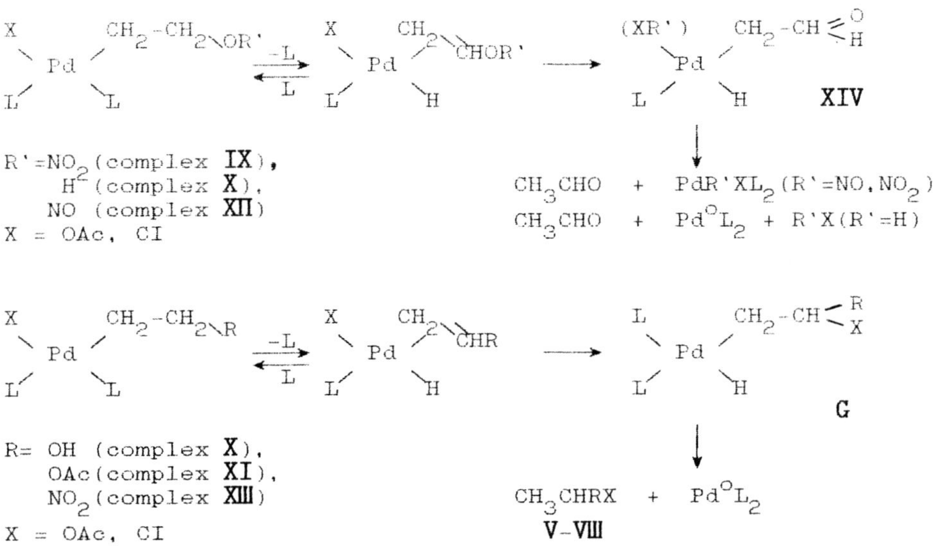

During the formation of acetaldehyde the Pd(O) complexes are oxidized by nitroxyl or nitrosyl chloride (or by the corresponding acetates); in the other cases the palladium black is formed, along with the products of 1,1-addition.

In chloroform, during the decomposition of intermediate XIII nitroethylene is formed, as has been described by us in (ref. 5).

REFERENCES

1 P.M. Henry, Palladium-catalyzed oxidation of hydrocarbons, Reidel, Dordrecht, 1980, p. 99.

2 V.A. Likholobov, N.I. Kuznetsova, M.A. Fedotov, Yu.A. Lokhov and Yu.I. Yermakov, Interaction between oxidants and olefins in solutions containing palladium complexes, in: 6th Nat. Symp. Recent Advances in Catalysis and Catalytic Reaction Engineering, Pune, India, 1983, pp. 217-228.

3 F. Mares, S.E. Diamond, F.J. Regina and J.P. Solar, Formation of glycol monoacetates in the oxidation of olefins catalyzed by metal nitro complexes: mono- vs. bimetallic system, J. Am. Chem. Soc., 107 (1985) 3545-3552.

4 I.E. Beck, E.V. Gusevskaya, V.A. Likholobov and Yu.I. Yermakov, Synthesis of Pd(II) nitro and nitrate complexes and studies of their reactivity towards oxidation of olefins in organic solvents, React. Kinet. Catal. Lett., 33 (1987) 209-214.

5 E.V. Gusevskaya, I.E. Beck, A.G. Stepanov, V.A. Likholobov, V.M. Nekipelov, Yu.I. Yermakov and K.I. Zamaraev, Study on the mechanism of ethylene oxidation by a nitrite complex of palladium in chloroform medium, J. Molec. Catal., 37 (1986) 177-188.

6 I.E. Beck, E.V. Gusevskaya, A.G. Stepanov, V.A. Likholobov, V.M. Nekipelov, Yu.I. Yermakov and K.I. Zamaraev, Study of the mechanism of ethylene oxidation by palladium(II) complexes containing nitro and/or nitrato ligands in chloroform, J. Molec. Catal., 50 (1989) 167-179.

7 Jan-E. Bäckvall and A. Heumann, A comment on the recently proposed mechanism for the oxidation of olefins with $PdCl(NO_2)(CH_3CN)_2$, J. Am. Chem. Soc., 108 (1986) 7107-7108.

G. Centi and F. Trifiro' (Editors), *New Developments in Selective Oxidation*
© 1990 Elsevier Science Publishers B.V., Amsterdam — Printed in The Netherlands

PERSPECTIVES IN IMPROVEMENT OF SELECTIVITY IN LIQUID PHASE OXIDATION BY DIOXYGEN. NEW MODELS OF ENZYMATIC OXIDATION

I.P.SKIBIDA and A.M.SAKHAROV

Institute of Chemical Physics, Academy of Sciences of the USSR, 117334 Moscow, Kosygin street, 4 (USSR)

SUMMARY

$[Cu(I)(o\text{-}phen)_2]^+$ complexes effectively catalyze the oxidation of primary alcohols to aldehydes by dioxygen in non aqueous solutions at pH > 7 and moderate temperatures and can be considered as an adequate model of galactose-oxidase. The anion coordination on Cu(I)-centers results in a strong increase in anion reactivity towards O_2. The use of D_4-labeled methanol confirms that dioxygen reduction in the course of oxidation occurs by a concerted two-electron mechanism.

INTRODUCTION

Many different catalytic systems involving metal complexes and phase transfer catalysts were proposed recently as essentially good models of enzymatic oxidation [1]. Most of them uses hydrogen peroxide or other single-oxygen donors as the oxidants. As before the idea of using dioxygen as the cheepest and most ecologically pure oxidant remains attractive. However despite the important successes in this field, the problem is still far from solving.

At the same time in the presence of enzymes dioxygen is known to oxidize various substrates at high rates and with good selectivity even at room temperature. In this connection the interest in modelling enzymatic oxidations greatly increased in recent years.

Of special interest is to mimic such enzymes as tyrosinase catalyzing the oxidation of o-phenols to quinones [2], galactose-oxidase in which presence the $-CH_2OH$ groups react with O_2 to yield aldehydes [3,4], dioxygenase catalizing oxygenation of catechols [5]. One specific mechanistic feature of the mechanism of catalysis by these enzymes is the participation of the substrate coordinated to enzyme active center in dioxygen activation.

Certain copper complexes having a N or O donors set catalyze alcohols oxidation at pH > 7 and can be considered as possible chemical models of galactose-oxidase [6-8].

It was demonstrated in our previous reports [7,9] ,that copper/o-phen complexes can be used as very active catalysts for primary alcohols one-step oxidation in aprotic solvents and pH > 7 giving aldehydes (cuprous catalysis) or acids (cupric catalysis) .

The present communication submits the additional proves for the mechanism of one-step oxidation of alcohols to aldehydes and demonstrates that the proposed catalytic system represents an adequate model of galactose-oxidase.

RESULTS AND DISCUSSION

Methanol was oxidized by dioxygen in the presence of copper complexes and bases at 30-50°. Both cuprous and cupric complexes with 1,10-phenanthroline (o-phen) ; 5,6-dimethyl-1,10-phenanthroline (CH_3-phen) ; 5-nitro-1,10-phenanthroline (NO_2-phen) and bipyridile (bipy) were used as catalysts. The complexes were prepared by dissolving $CuCl_2$ or CuCl and the ligands in alcohol. The reaction began after addition of alkali.

In neutral media methanol virtually is not oxidized at moderate temperature even in the presence of a catalyst. It seems that just as in many enzymatic systems [7], the formation of an anionic form of the substrate is an obligatory condition for an increase of its activity towards oxygen. The use of $Ba(OH)_2$, $Ca(OH)_2$ or triethylamine as bases does not provoke effective deprotonation of methanol in non agueous solutions , and the oxidation rate in the presence of these bases and the catalyst is still low. High rates of oxygen consumption are observed only in catalytic oxidation of methanol in the presence of strohg bases such as NaOH , KOH or $NaOCH_3$.

The semiconvertion time of primary alcohols under these conditions is about ~10 min , $n = [RCH_2OH]/[Cat]_o = 10^3$, turnover time ~1 sec. Fig.1 shows the rate of oxygen consumption as a function of alkali concentration in the oxidation of 20 % mol. methanol solutions in acetonitrile in the presence of $[Cu(o\text{-}phen)_2]^+$ at 30°. The increase in the oxidation rate with base concentration seems to be connected with increasing of the methylate anions concentration. Very high medium alkalinity results in a decline of the oxidation rate caused by hydrolysis of copper complexes as observed earlier for the catalytic oxidation of ketones in the presence of copper/o-phen complexes at pH > 7 [10].

The methanol oxidation rate is dependent not only on concentration but also on the donor capacity of alcoholate ions in solution. The dilution of methanol with aprotic solvents (benzene , acetonitrile, DMFA , etc.) increases essentially the rate of CH_3OH oxidation. In the presence of small amounts of water the reaction becames complitely passivated due to much lower electron donating activity of anions in water-containing solutions as compared to that in aprotic solvents.

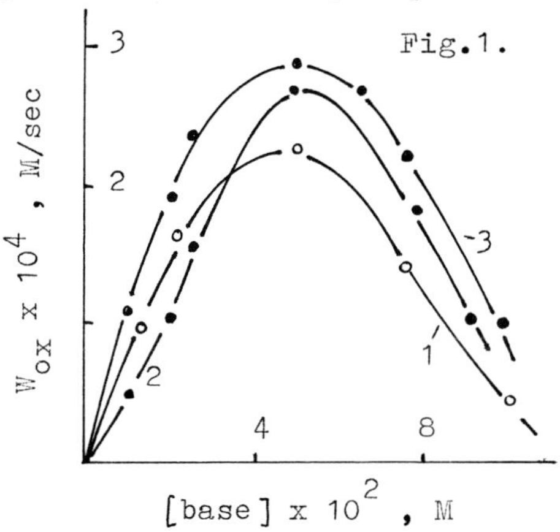

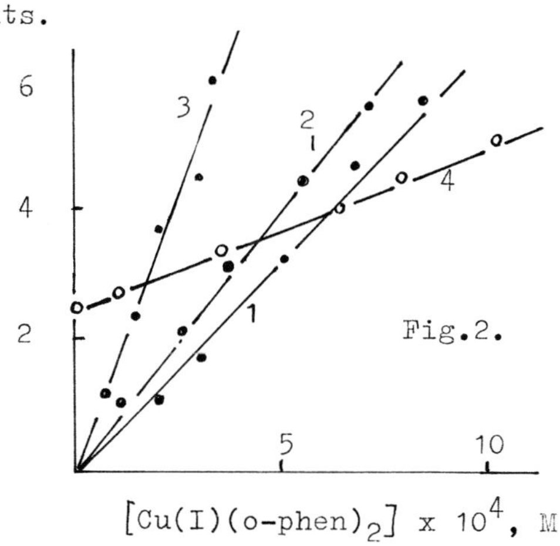

Fig.1. The oxygen consumption rate as a function of KOH (1) ; NaOH (2) ; $NaOCH_3$ (3) concentration in methanol oxidation. $[CuCl_2]$ = 2.5 x 10^{-3} M ; [o-phen] = 1.0 x 10^{-2} M ; $[CH_3OH]$ =20% vol., acetonitrile as solvent , 30°.

Fig.2. The rate of oxygen consumption as a function of $Cu(I)L_2$ concentration in methanol oxidation : L = o-phen (1,4), bipy (2) , CH_3-phen (3).
$[Cu^{n+}]_o$ = 1.0 x 10^{-3} M , [NaOH] = 4.0 x 10^{-2} M , $[CH_3OH]$= 20% mol. acetinitrile (1-3) and DMFA (4) as solvents

Cu(I) complexes act as active catalytic species in the oxidation of alcoholate ions in acetonitrile or in bulk. Thus ,when $Cu(II)(o\text{-}phen)_2^{\ 2+}$ are used as catalyst CH_3OH oxidation occurs with some induction period. During this period Cu(I) is accumulated in solution. No induction period is observed when using $Cu(I)(o\text{-}phen)_2^{\ +}$ as catalyst.

The Cu(I) concentration and the rate of methanol oxidation after induction period is finished are independent on the copper valency state in the initial catalyst and are a function of the experimental conditions (solvent , alkali and ligand concentration etc.)

The rate of methanol oxidation linearly increases with Cu(I) concentration*. One of the most important factors responsible for the catalysis by [Cu(I)(o-phen)$_2$]$^+$ complexes seems to be the increasing reactivity of A^- ions in LCu(I).A^- adducts towards dioxygen. This is supported by the fact that the complexes catalytic activity increases with the electron donating activity of ligand (L). In the range L = o-phen , bipy , CH_3-phen the highest oxidation rates were observed for [Cu(I)(CH_3-phen)$_2$]$^+$ (curve 3) and it is just for CH_3-phen the donor activity is the highest in the above mentioned range.

Low activity of Cu(II) complexes in the catalytic oxidation of primary alcohols in the presence of bases seems to be due to the fact that cupric complexes act usually as electron acceptors and the coordination of alcoholate ions to Cu(II)-centers results in lowering of their donor activity and thus in decreasing of the ions reactivity towards dioxygen.

It would have been expected that electron withdrawing substituents (such as-NO_2) in o-phen molecule make the Cu(I) complexes activity lower. It however appears that the activity of copper/NO_2-phen complexes is rather high: the rate of methanol oxidation in the presence of copper/NO_2-phen complexes is twice that in the presence of copper/o-phen complexes, other conditions being equal.

When the [Cu(II)(NO_2-phen)$_2$]$^{2+}$ complex is used as catalyst, a short induction period is observed. However [Cu(I)(NO_2-phen)$_2$]$^+$ accumulation is not detected in the system in contrast with methanol oxidation in the presence of copper complexes with other ligands.The spectrum of cuprous complexes disappears completely in the course of reaction also when [Cu(I)(NO_2-phen)$_2$]$^+$ is used as a catalyst. It seems that the introduction of a strong electron withdrawing NO_2-group into the ligand sharply increases the rate of one-electron reaction of methylate ions with catalyst yielding to free radicals. The ligand interaction with free radicals results in fast irreversible consumption of NO_2-phen.Similarly when catalytic oxidation of ketones at pH > 7 occurs in the presence of [Cu(II)(o-phen)$_2$]$^{2+}$ and redox-active additives the reaction results in irreversible consumption of the ligand and increasing of the role of free radical reactions.

* The value of Cu(I) was varied by changing o-phenanthroline concentration in solution.

As found in [9] aldehydes are the main products of Cu(I)-catalyzed oxidation of primary alcohols (propanol , benzyl alcohol , etc.) in aprotic solutions. With a low medium basicity ([NaOH]/ /[Cat] = 10 ÷ 20) the selectivity of aldehyde formation attains 90%. In accordance with these results formaldehyde must be the primary reaction product of methanol oxidation in the presence of Cu(I)-complexes.

However at the first stages of methanol oxidation only very small amounts of formaldehyde can be detected in solution due to high rate of its condensation at high pH. The condensation products are further oxidized to form acids with very high rate.

When the medium basicity is lowered due to acids formation and the concentration of Cu(I)-complexes is rather high CH_2O is accumulated in considerable amounts.

The ratio of oxygen and alkali, consumed in methanol oxidation ($\Delta[O_2]/\Delta[NaOH]$) in the presence of Cu(I)-complexes is always higher than 1, and in low alkaline solutions it can attain 5 ÷ 10 because of formation of nonacidic products, such as formaldehyde or its condensation products ; the rate of oxygen consumption depends to a great extent on O_2 partial pressure. A decrease in p_{O_2} from 1 to 0.4 atm results in an almost ten-fold decrease in the oxygen consumption rate. Simultaneously the $[Cu(I)(o\text{-phen})_2]^+$ concentration drops.

The relative rate of oxygen uptake ($W_{O_2}/[Cu(I)]$) increases linearly with p_{O_2} within the oxygen pressure variation from 0 to 1 atm. This seems to be an evidence of dioxygen participation in the rate determing step of the oxidation process.

An very important specific feature of the catalytic system under investigation is its activity only in the oxidation of primary alcohols. Secondary alcohols are not oxidized under our conditions and can be used as inert solvents in the oxidation of primary alcohols.

DISCUSSION

The Mechanism of Primary Alcohols Oxidation

(i) Cu(I) catalysis. It is apparent from the above discussion that the oxidation of primary alcohols takes place due to oxygen interaction with the $Cu(I).A^-$ adducts whose reactivity towards dioxygen is higher than that of non-coordinated anions.

It should be suggested that the oxidation of alcoholate

ions to aldehydes in our system occurs by one-electron mechanism according to reaction :

$$RCH_2O^- \ldots Cu(I) \xrightarrow{\bar{e}\ \curvearrowright\ O=O} RCH_2O^{\bullet} \ldots Cu(I) \xrightarrow{O_2^{\dot{-}}} RCHO + Cu(I) + HO_2^- \quad (1)$$

However the one-electron reduction of dioxygen in reaction (1) can not explaine the chemio selectivity of the oxidation reaction : as mentioned above in the presence of $[Cu(I)(o\text{-}phen)_2]^+$ at pH > 7 only primary alcohols can be oxidized.

The alternative two-electron mechanism of oxygen reduction has been suggested by us in [9]:

$$Cu(I) \ldots R\text{-}\overset{\curvearrowright O=O}{C}\overset{H}{\underset{OH}{<}} \longrightarrow Cu(II) \ldots R\text{-}\dot{C}\text{—}H \xrightarrow{HO_2^-} RCHO + Cu(I) + HO_2^- \quad (2)$$

It should be mentioned that the HO_2^- formation suggested in both (1) and (2) reactions does not lead to any change in the reaction rate and does not favour the one-electron oxidation since under the given conditions hydrogen peroxide decomposes quickly to form H_2O and O_2 and does not contribute to CH_3OH oxidation.

Oxygen protonation is the most important step in the proposed two-electron oxygen reduction. The same was suggested also for O_2 reduction over Cu(I)-centers of dopamine-ß-monooxygenase [12]. Such an assumption permits to explain why i-propanol and other sec.alcohols are not oxidized under the reaction conditions: hydrogen transfer from substrate to oxygen to form HO_2^- ,by the reaction similar (2) is obviously impossible for sec. alcohols.

The kinetic regularities of deuterated methanol (CD_3OD) oxidation were studied in order to provide evidence for the possibility of simultaneous transfer of two electrons and a proton (formally corresponding to transfer of a hydride ion to dioxygen)from the anion coordinated to Cu(I) center to the dioxygen molecule.

Fig.4 shows the kinetic curves for oxygen consumption and Cu(I) accumulation in the oxidation of methanol (curves 1,3) and deuterated methanol (curves 2,4). It can be seen that the rate of CD_3OD oxidation is almost by one order of magnitude lower than that of CH_3OH oxidation. This is in part due to the lesser concentration of $[Cu(I)(o\text{-}phen)_2]^+$. The variations in $Cu(I)(o\text{-}phen)_2{}^+$ concentration must necessarly be taken into account in calculation of the kinetic isotope effect. The ratio of the effective first oder rate constants, $k_{eff} = W\ /\ Cu(I)$, can be taken as measure of the kinetic isotope effect.

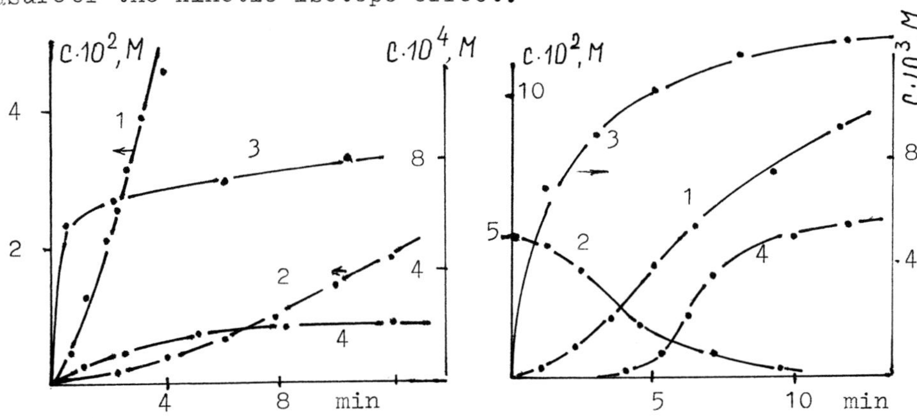

Fig.4 Fig.3

Fig.3. The kinetic curves of O_2 (1) and $NaOCH_3$ (2) consumption and $[Cu(I)(o\text{-}phen)_2]^+$(3) and formaldehyde (4) accumulation in methanol oxidation.

$[CuCl_2] = 2.5 \times 10^{-3}$M ; [o-phen] = 1,0 x 10^{-2} M ; acetonitrile as solvent , 30°.

Fig.4. The kinetic curves of oxygen consumption (1,2) and Cu(I)/ /o-phen complexes accumulation in oxidation of 20% vol. solutions of CH_3OH (1,3) and CD_3OD (2,4).

$[CuCl_2] = 1.0 \times 10^{-3}$ M; [o-phen] = 2.0 x 10^{-2} M ; [NaOH] = 0.05 M; acetonitrile as solvent , 30° .

For the methanol oxidation the kinetic isotope effect calculated from the data represented on Fig.4 is $k_H/k_D = k^H_{eff}/k^D_{eff} = 2.7$. When both methanol and CD_3OD are oxidized using Cu(I)/o-phen as catalyst in the presence of o-phen excess the concentrations of cuprous complexes during the oxidation coincide and are about 80% of $[Cu(I)]_o$. The ratio of the rates of oxygen consumption in this conditions is 2.6 .

The obtained values of kinetic isotope effect are close to that for some hydride transfer reactions occuring via a non-linear activated complex [13].

Thus, the measured isotope effect values for methanol oxidation catalysed by Cu(I) complexes provide convincing evidence for the importance of hydride ion transfer by interaction of coordinated methylate ions with O_2 , i.e. these values are in favour of the two-electron mechanism of alcohol oxidation in the catalytic system under investigation.

<u>(ii) Cu(II)-catalysis.</u> It appeared that when DMFA is used as a solvent not only cuprous but also cupric complexes are active catalyst for methanol oxidation at pH > 7 (fig.2, curve 4) that

seems to be accounted for by the higher DMFA donor activity compared to that of acetonitrile [11]. The donor capacity of methylate ions coordinated to Cu(II)-centers seems to be sufficient in this case to ensure the high rate of their interaction with dioxygen. Formic acid is the main product of methanol oxidation in the presence of Cu(II) complexes.

The one-step oxidation of alcohols to acids catalysed by Cu(II) complexes occurs by two-electron mechanism [9]:

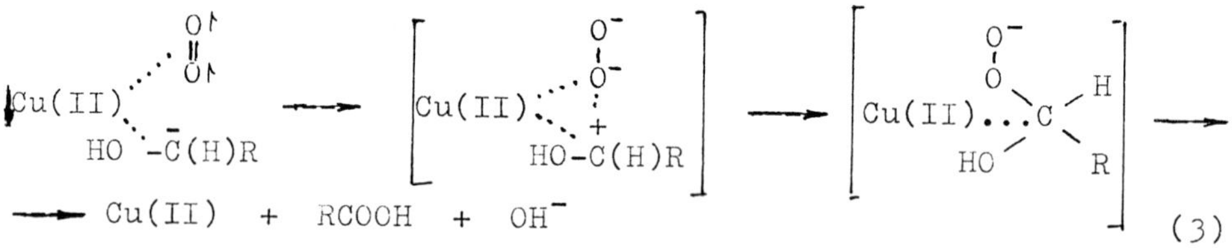

(3)

As expected, the rates of CH_3OH and CD_3OD oxidation in DMFA (Cu(II)-catalysis) virtually coincide in agreement with (3).

The participation of Cu(III) ions in the mechanism of primary alcohols oxidation to aldehydes in the presence of galactose-oxidase [2] or some Cu(I) complexes [14] in neutral media was proposed. However for the system under investigation the Cu(III) ions catalysis is not very probable becose this oxidant cannt act as a chemioselective one (the system is quite inactive in oxidation of secondary alcohols) [10].

REFERENCES

1. B.Meunier, Bull.Soc.Chim.France, 1986 (4) 578-584.
2. E.L.Solomon, in T.G.Spiro (ed.) Metal Ions in Biology, v.2, Wiley, N.-Y., 1981, 41-1o2.
3. G.A.Hamilton, P.K.Adolf , J.de Jersey, G.S.Du Bois, J.Amer. Chem.Soc., 100 (1978) 1899-1901.
4. A.M.Klibanov, R.N.Alberty, M.A.Marletta, Biochim.Biophys.Res. Commun., 108 (1982) 804.
5. L.Que, Jr., Coord.Chem.Rev., 50 (1983) 73-78.
6. W.Brackman, C.L.Gaasbeek. Rec.trav.chim.Pay-Bas , 85(2) (1966) 242-256.
7. I.P.Skibida, A.M.Sakharov, in: Itogi nauki i tekhniki, ser. Kinetika i kataliz , v.15 (1986) 110-234
8. N.Kitajama, K.Wang, Y.Moro-oka, A.Uchida, Y.Sasada, J.Chem.Soc. Chem.Commun., 1986(2) 1504-1506.
9. A.M.Sakharov, I.P.Skibida , Izv.AN SSSR , ser.khim., 1980 (2) 523-528.
10. A.M.Sakharov, I.P.Skibida,J.Molec.cat.,48(2-3)(1988) 157-174
11. V.Guttman, Coord.Chem.Rev., 18 (1976) 225-228.
12. S.M.Miller, L.R.Klinman, Biochemistry , 24 (1985) 2114-2116.
13. W.P.Jenks, Catalysis in Chemistry and Enzymology, Mc Graw Hill 1969.
14. P.Capdevielle, P.Audebert,M.Maumy, Tetr.Lett., 25 (1984) 4397.

G. Centi and F. Trifiro' (Editors), *New Developments in Selective Oxidation*
© 1990 Elsevier Science Publishers B.V., Amsterdam — Printed in The Netherlands

CATALYTIC NITROXIDATION OF ALIPHATIC AND AROMATIC HYDROCARBONS BY NO (review)

G.M. PAJONK

Université Claude Bernard Lyon I, ISM, Laboratoire de Thermodynamique et Cinétique Chimiques, 43 bd du 11 novembre 1918, 69622 Villeurbanne Cedex, France.

SUMMARY

Catalysts containing supported NiO or PbO are very active and selective for the transformation of paraffins, olefins, toluene and the three xylenes in their corresponding nitriles when NO is reacted with them.

Paraffins and olefins give unsaturated nitriles while aromatics lead to aromatic mono and/or di-nitriles.

The reaction mechanism disclosed is of the redox type for all the hydrocarbons studied.

INTRODUCTION

The industrial way of making nitriles, especially unsaturated ones uses the ammoxidation process (refs. 1, 2) i.e. the reaction between the hydrocarbon and a mixture of ammonia and air or oxygen in the presence of mixed oxides (BiMoO, SbSnO) as catalysts. This process suffers from several drawbacks such as : very high exothermicity, necessity to neutralize the ammonia in excess by sulfuric acid leading to commercial products of very low values, necessity to obey severe safety conditions because of the production of considerable amounts of HCN and also of the use mixture of O_2 and hydrocarbons as reactants.

In this laboratory a new process was developed to synthesize the nitriles in order to escape these drawbacks by reacting NO instead of the conventional ammoxidation mixture (refs. 3, 4). This type of making nitriles has been called nitroxidation and it exhibits a much lower exothermicity than the ammoxidation, it does not release HCN and operates at reaction temperatures lower by 100 K than the industrial ones (683 up to 723 K for nitroxidation reactions). The nitroxidation catalysts are very specific, they are based upon NiO and PbO oxides supported by alumina, silica or magnesia (ref. 5). For instance they are able to nitroxidize aliphatic as well as aromatic hydrocarbons (which is not the case for the ammoxidation process), they are also active and selective in ammoxidation conditions which is not true in the reverse situation.

Schematically nitroxidation consists in substituting 3 H atoms bonded to a same carbon atom of the hydrocarbon by a N one as in equation (1) :

$$\begin{matrix} A \\ A' \end{matrix}\!\!>CH_3 + \frac{3}{2}NO \longrightarrow \begin{matrix} A \\ A' \end{matrix}\!\!>CN + \frac{3}{2}H_2O + \frac{1}{4}N_2 \qquad (1)$$

A : aliphatic group

A' : aromatic group

EXPERIMENTAL AND RESULTS

All the tests described were performed in U pyrex microreactors under flowing conditions and in differential conversion (from 1 to 10 %). The reactant feed was diluted by He and its total pressure was the atmospheric one. All data were recorded at steady state of activity unless otherwise stated and in the chemical regime. The catalysts were activated in situ at 713 K in O_2 for 24 hours before any run.

GC chromatography was used online to analyze the reaction partners with two detectors : katarometer and flame ionization.

Beside the main products (nitriles, CO_2, H_2O, N_2) it was always detected traces of NH_3 in the outlet stream. After catalysis the active oxide was always, partially, in a reduced state.

Preparation of the catalysts

The catalysts were prepared according to the sol-gel procedure and dried as xero- or aero-gel (refs. 6, 7). Aerogels containing NiO on alumina with Ni/Al ratios of 1 and 0.5 were labelled respectively NA VIII and NA V, the corresponding xerogels were noted X NA VIII and X NA V. The catalysts containing PbO on alumina were denominated respectively PA VIII and PA V for aerogels, and X PA VIII and X PA V for xerogels the ratios Pb/Al being again respectively 1 and 0.5.

The pure supports were totally inactive in the reaction conditions while unsupported NiO was very active and selective towards nitroxidation but very unstable with time on stream while pure PbO activity was close to nil.

Precursors of the catalysts were respectively nickel and lead acetate concerning the active phase and respectively tetramethoxisilane and aluminium secondary butylate.Mixtures of adequate active and support precursors in alcool were cohydrolysed and dried as aerogels (in an autoclave) or as xerogels (ref. 8). Xerogels were prepared in water, and dried in an oven (in air).

Surface areas were measured with N_2 using the BET method while XRD analysis was performed to determine the structures of the catalysts.

Table 1 gives these data for the catalysts described in the forthcoming nitroxidation reactions.

TABLE 1

Properties of the catalysts and supports

Catalyst	S in m^2/g	X RD analysis
XNA VIII	127	NiO
NA VIII	208	NiO
XNA V	193	NiO, $NiAl_2O_4$
NA V	350	NiO, $NiAl_2O_4$
NiO (xerogel)	23	very well cristallized
Al_2O_3 (aerogel)	254	amorphous
X PA VIII	8	$Pb(OH)_2$
PA VIII	84	Pb_3O_4, $PbO_{1.37}$
X PA V	36	badly cristallized
PA V	132	Pb_3O_4, $PbO_{1.37}$
Pb_3O_4 (xerogel)	1	very well cristallized

After each run Ni^o and Pb^{2+} were detected in all corresponding catalysts.

Nitroxidation of paraffins

Propane and isobutane were selectively converted by NiO based catalysts containing also chromia or Fe_2O_3 as shown in Tables 2 and 3 respectively (ref. 4).

TABLE 2

Nitroxydation of propane in acrylonitrile and acetonitrile at 753 K.

Catalysts	% in Cr^{3+}	Selectivities in %			Activities in 10^{-8} moles/g/s.	
		C_3H_3N	C_2H_3N	CO_2	C_3H_3N	C_2H_3N
NAC25	25	20	27	52	17	23
X NAC30	30	30	25	45	14	11

TABLE 3

Nitroxidation of isobutane in metacrylonitrile, acrylonitrile and acetonitrile at 753 K.

Catalysts	% in Cr^{3+} or Fe^{3+}	Selectivities in %			Activities in 10^{-8} moles/g/s.		
		C_4H_5N	C_3H_3N	C_2H_3N	C_4H_5N	C_3H_3N	C_2H_3N
NA V	0	39	8	20	16	3	8
NA VIII	0	32	6	20	17	5	11
NAF5	5	45	11	14	21	5	7
X NAC50	50	43	9	20	18	3	5

One can easily remark that a double functionalisation of each paraffin has been obtained (insertion of the nitrile group and creation of a double bond) even on a pure NiO-alumina aerogel.

Nitroxidation of olefins

For both catalysts, with isobutylene or with propylene the selectivities in metacrylonitrile or acrylonitrile were comprised between 80 and 87 % as shown respectively in Tables 4 and 5 (refs. 3, 4).

TABLE 4

Nitroxidation of isobutylene in metacrylonitrile and acetonitrile at 683 K

Catalysts	Selectivities in % C_4H_5N	C_2H_3N	CO_2
NA V	82	8	7
NA VIII	85	8	5
X NA VIII	82	1	13

The activity of formation of metacrylonitrile was of the order of 30.10^{-8} moles/g/s.

TABLE 5

Nitroxidation of propylene in acrylonitrile and propionitrile at 683 K.

Catalysts	Selectivities in % C_3H_3N	C_2H_3N	CO_2
NA VIII	77	9	11
X NA VIII	79	7	14
NA V	75	13	8
PA V	87	-	14

The activity of the NiO based catalysts was of 150.10^{-8} moles/g/s towards acrylonitrile formation and it was even greater with the PbO based catalyst 260.10^{-8} moles/g/s.

All the feeds containing an aliphatic hydrocarbon were characterized by a NO/hydrocarbon ratio 1:9 excepted in the case of isobutylene where it was as low as 1:2.

Selectivities are very large and in the case of isobutylene higher than those reported in the literature (refs. 1, 2) concerning the ammoxidation process.

Nitroxidation of toluene

For both types of catalysts the selectivity towards benzonitrile formation was superior to 87 % as shown in Table 6 and again the PbO catalysts exhibited higher activities than the NiO catalysts (ref. 7).

TABLE 6

Nitroxidation of toluene in benzonitrile at 723 K.

Catalysts	Selectivities in % C_6H_5N	CO_2	Activities in 10^{-8} moles/g/s. in C_6H_5N
NA V	91	7	20
X NA V	91	8	12
NA VIII (a)	86	11	16
X NA VIII (b)	86	8	5
PA V	93	3	79
X PA V	95	2	57
PA VIII	94	-	78
X PA VIII	88	-	22

(a) no steady state achieved. Values recorder after 5 hrs of reaction
(b) same as in (a) but values recorded after 10 min of reaction.

From the data collected in Table 6 it is clear that the nickel oxide catalysts are much less stable with time on stream (when their ratio Ni/Al is equal to unity) than the catalysts containing lead oxide. Another difference is registered between PA V and NA V concerning the transient state before reaching the steady one : PA V is much more rapidly at steady state than NA V.

These facts emphasize once more that lead oxide supported by alumina develops very good catalytic properties contrary to its usual reputation.

Nitroxidation of o, p and m xylene

Orthoxylene is transformed at 673 K into the mononitrile (orthotolunitrile) with a selectivity of more than 90 % on NiO as well as on PbO catalysts.

But the dinitrile (phtalonitrile) is obtained selectively (S = 20 %) only at 713 K and only with the PbO type catalysts. The other products being benzonitrile (S = 20 %) and orthotolunitrile (S = 44 %). Thus the total selectivity into nitriles is of the order of 84 % nevertheless.

Again the metaxylene is selectively (S > 90 %) converted into metatolunitrile at 673 K on both kinds of catalysts. But once more only PbO catalysts are able to convert the hydrocarbon into the dinitrile (is ophtalonitrile (S = 13 %) at 713 K. The other products being metatolunitrile (S = 68 %) and benzonitrile (S = 11 %). The total selectivity in nitriles is of 92 % for the conversion of metaxylene (ref. 9).

Finally paraxylene is equally well converted into paratolunitrile (S = 42 %) on both types of catalysts at 673 K, but again only PbO is able to convert this xylene into terephtalonitrile at 713 K with in this case a good selectivity in the dinitrile (S = 43 %) which is here the major product of the reaction (benzonitrile S = 25 % and paratolunitrile S = 18 %).

Therefore it seems that PbO is even better than chromium oxide based catalysts for the same conversion which give only a selectivity of 5 % in terephtalonitrile, the major products being paratolunitrile (S = 65 %) and benzonitrile (S = 16 %)(ref. 10).

Nitroxidation of o, m and p tolunitrile

It was of interest from a mechanistic point of view to convert the three mononitriles in the corresponding dinitriles in order to assess if a one or a two step(s) reaction mechanism (ref. 11) is working during the catalysis of xylenes conversions.

Orthotolunitrile was not transformed into phtalonitrile but into benzonitrile, benzene and CO_2 whatever the catalysts or the reaction conditions.

However PAV catalyst was able to give isophtanonitrile from metatolunitrile with a selectivity of 16 % at 713 K but NA V was incapable to give the dinitrile.

Finally paratolunitrile gave terephtalonitrile only in the presence of PA V catalysts with a selectivity of 87 %, the other nitrile product as benzonitrile (ref. 11).

In summary it can be said that the three xylenes can be selectively converted in the mononitriles on both catalysts (NiO and PbO) but only lead catalysts give selectively the dinitriles.

DISCUSSION

Proposed reaction mechanism

Independently of the nature of the hydrocarbons tested in this work, a redox mechanism is able to explain the whole kinetic results as follows :

- Hydrocarbon + Oxidized Cat $\xrightarrow{k_r}$ Reduced Cat + Adsorbed dehydrogenated hydrocarbon adsorbed (releasing up to 3 hydrogen atoms).

- NO + Reduced Cat $\xrightarrow{k_o}$ Oxidized Cat + N(adsorbed).

The next step, a fast one, is the combination of the dehydrogenated hydrocarbon species with the N atoms giving the corresponding nitrile.

When aliphatic hydrocarbons are involved the adsorbed species is of a dehydrogenated π-allyl type and when aromatics are the reactants then the adsorbed species is of a dehydrogenated benzylic type.

The formation of traces of NH_3 in the effluent gas was considered as an indirect proof of the dissociative adsorption of NO giving N as adatoms (and finally the nitrile) and O adatoms (which oxidize the reduced catalysts). A check of this idea was performed by trying to convert NO by H_2 into NH_3 on both types of catalysts which was indeed observed, while it was impossible to transform the classical ammonia syngas on the same catalysts.

The ascertain the possibility of the redox mechanism, values of k_o and k_r were measured for a series of four nitroxidations and collected in Table 7.

TABLE 7

Values of k_o, k_r for nitroxidation reactions.

Arbitrary units	Nitroxidation of Isobutane	Propylene	Isobutylene	Toluene
k_o	0.66	7.5	3.33	4.7
k_r	0.19	3.5	2.44	1.4
References	(11)	(3)	(12)	(6)

The values of k_o, k_r for each converted hydrocarbon are very close to each other which is a good fit of the model. Moreover it is clear that in every case $k_o > k_r$ which means that reoxidation is easier than reduction of the catalysts and this remark is in good agreement with the composition of the reactant feeds always richer in hydrocarbons (with respect to NO) than the stoechiometric ones.

COMPARISON BETWEEN AMMOXIDATION AND NITROXIDATION

It was checked that the nitroxidation catalysts were able to give nitriles in the ammoxidation conditions (no NO). The selectivities exhibited in nitriles were of the order of 30-40 % only.

The conventional ammoxidation catalysts such as Sb-Sn-O, Bi-Mo-O, V_2O_5/Al_2O_3 were unable to give nitriles in the nitroxidation conditions. They were also inactive in the conversion of NO by H_2 into NH_3. Therefore it is possible to claim that a necessary (but not sufficient) condition for a catalyst to be selective in nitroxidation is its ability to dissociate NO into N and O species.

CONCLUSIONS

To convert aliphatics into unsaturated nitriles is possible on NiO as well as on PbO based aerogels or xerogels. Generally speaking the aerogels are more active than the corresponding xerogels.

Aromatics like toluene is easily transformed into benzonitrile by both types of catalysts while PbO catalysts are more efficient and stable with time on stream in order to convert selectively the xylenes or the monotolunitriles.

It is worth to mention the particuliar good catalytic nitroxidation properties exhibited by catalysts containing PbO.

REFERENCES

1 T. Dumas, W. Bulani, Oxidation of Petrochemicals : Chemistry and Technology, Applied Science, Londres, 1974.

2 D.J. Hucknall, Selective Oxidation of Hydrocarbons, Academic Press, Londres, 1974.

3 F. Zidan, G. Pajonk, J.E. Germain and S.J. Teichner, J. Catalysis, 52 (1978) 133-143.

4 F. Zidan, G. Pajonk, J.E. Germain and S.J. Teichner, Z. Phys. Chem., 111 (1978) 91-103.

5 V.M. Belousov, V.V. Korovina, M. Ya. Rubanik, Kataliz i Katalizatory, Vol. 6, Naukova Dumka, Kiev, 1970, 89-100.

S.B. Grinenko, V.M. Belousov, Kinetika i Kataliz, Vol. 15 (1974) n° 2, 522-524.

V.M. Belousov, S.B. Grinenko, Kataliz i Katalizatory, Vol. 14, Naukova Dumka, Kiev, 1976, 27-31.

6 G.E.E. Gardes, G. Pajonk, S.J. Teichner, Bull. Soc. Chim. France, 1976, 1321-1326.

7 S. Abouarnadasse, G.M. Pajonk, J.E. Germain and S.J. Teichner, Appl. Catal;., 9 (1984) 119-128 ; J. Chem. Eng., 62 (1984) 521-525.

8 G.M. Pajonk in Proceed. 2nd Int. Symp. on Aerogels in press. Les Editions de Physique, Paris 1989.

9 S. Abouarnadasse, G.M. Pajonk and S.J. Teichner, Appl. Catal., 16 (1985) 237-247 ; Proceed. 9th ICC Calgary, M.J. Philips and M. Ternan Eds, 4, p. 1936-1943. The Chemical Institute of Canada, Ottawa, 1988.

10 S. Zine, A. Sayari and A. Ghorbel, Can. J. Chem. Eng. 65 (1987) 127.

11 S. Abouarnadasse, G.M. Pajonk and S.J. Teichner in "Heterogeneous Catalysis and Fine Chemicals", M. Guisnet et al. Eds, Elsevier, Amsterdam, 1988, p. 371-378.

B. DELMON (Université Catholique de Louvain, Belgique). I have some reservation with respect to your emphasis on a redox mechanism. Ni and Pb are extremely different with respect to oxido-reduction behaviour. On the other hand, both metals could interact with alumina, adjusting adequately the acidity of the latter, thus explaining the similitude of the catalytic behaviour. I suggest the role of acidity could be investigated.

G.M. PAJONK (Université Claude Bernard, France). It has been shown as reported in (ref. 1) that the acidity did not play a role at steady state at least in the case of the synthesis of methacrylonitrile (from isobutene and NO) upon the selectivities into the nitrile.

The acidity seemed to intervene only during the transient period before reaching the steady state, the greater the acidity of the catalyst the shorter the transient period and the lower the simultaneous degradation activity during this regime.

Moreover the presence of traces of NH_3 in the outlet stream allows to assume that the acidity is probably neutralized at steady state.

1 A. Sayari, A. Ghorbel, G.M. Pajonk and S.J. Teichner, Bull. Soc. Chim., 16, 1981 (see also reply to G. Golodets).

R. CHUCK (Lonza A.G., Switzerland). 1. Is nitroxidation limited to $-CH_3$ side-chains, or can longer-chain alkyl groups be oxidized ?

2. What is the % of NO in the exhaust gas ?

3. Is there a possible loss of Pb/Ni in the environment ?

Compare to ammoxidation : (with e.g. V/Ti catalysts)

No NO in atmosphere (recycling of NH_3 necessary)

No heavy metal problems

Is not limited to methyl side-chains.

G.M. PAJONK (Université Claude Bernard, France). No experiment was performed on other aromatics than the xylenes. Ethylbenzene is under study at present.

No NO is detected in the exhaust gas, only N_2, N_2O are present, probably resulting from the disproportion of NO (which is observed with pure NO over the catalysts).

As the same steady state is observed at least for tens of days it is likely that no loss of Pb or Ni occurs during catalysis.

Compared to ammoxidation I agree with the comments which can also be made for nitroxidation with the exception of the last point (in progress now).

J. OTAMIRI (University of Lund, Sweden). In your paper you stated that a necessary condition for a catalyst to be selective in nitroxidation is the ability to dissociate NO into N and O.

V_2O_5 is a known catalyst for NO_x reduction and hence should meet your requirement, however nitroxidation does not occur on it.

Will it not be more appropriate to suggest that the necessary condition will be for the catalyst to be able to form NH_3, or at least NH_3-precursors at the surface ?

G.M. PAJONK (Université Claude Bernard, France). The study presented here involved only NO (and not the NO_x as a whole). For example N_2O (instead of NO) resulted in a total oxidation of the hydrocarbons and it was checked directly that the catalysts were unable to synthesize NH_3 from a N_2 + H_2 mixture whereas NH_3 was obtained quantitatively with a NO + H_2 feed.

No attempt at reacting NO + H_2 on a V_2O_5/Al_2O_3 aerogel catalyst was carried out. This type of catalyst was indeed not selective towards the nitroxidation reaction (ref. 1).

1 S. Abouarnadasse, Ph.D. Doctoral Dissertation Lyon 1986, n° 86-46 (France).

F. VAN DEN BRINK (DSM Research BV, Netherlands). 1. Experimental results presented were obtained in a differential reator, so presumably conversions of the hydrocarbon were low (< 10 % ?). Could you indicate the dependance of the selectivity upon conversion ?

2. What was the ratio NO/hydrocarbon used and how does this influence selectivity and yield ?

3. Comment : HCN is a valuable by product from the production of acrylonitrile ; toxicity of acrylonitrile is also very high, although not as high as of HCN. The fact that HCN is not a by product of the nitroxidation is therefore hardly an advantage.

G.M. PAJONK (Université Claude Bernard, France). The conversion used in this work varied between 1 and 10 - 15 % (at most). No systematic study was performed at higher conversions.

The NO-hydrocarbon ratio was of the order of 1:8 for aliphatics and 1:3 for aromatics. Only under conditions where the hydrocarbon was in a fairly large excess with respect to stoechiometry were the selectivities as high as reported even in the case of conversions reaching a value of 10 % (yields were of the order of 9 %). This is also true when one considers the stability with time on stream exhibited by both types of catalysts.

Now considering ammoxidation, if HCN is produced only under the form of traces for instance then the severe safety conditions necessitated by the process are very expensive for a poor yield and on the contrary if HCN is obtained in large amounts then it is at the expense of the desired product and therefore it competes with the well known Andrussow's process.

G. GOLODETS (Institute of Physical Chemistry, URSS). 1. Have you any idea on the reasons why PbO is a better catalyst for the nitroxidation ?

2. What are the experimental evidences in favour of the proposed mechanism ?

G.M. PAJONK (Université Claude Bernard, France). The reasons why PbO based catalysts are better than the NiO ones are not yet known.

The arguments of favoring a redox mechanism are based on the observation that during catalysis Ni^{2+} is at least partially, reduced in Ni° (ferromagnetic properties) and Pb^{4+} is also reduced in Pb^{2+} cations as seen from XRD analysis.

By flowing the hydrocarbons (without NO) over our catalysts reduction was always recorded and subsequently shifting to NO (without hydrocarbon) resulted in a reoxidation of the reduced form of the catalyst, see ref. 1 for instance for chromia-alumina aerogel using EPR spectrometry.

1 H. Zarrouk, A. Ghorbel, G.M. Pajonk and S.J. Teichner, Procedings, IXth Ibero American Symp. on Catalysis, Lisbon, 1984, 339.

G. Centi and F. Trifiro' (Editors), *New Developments in Selective Oxidation* 239

SELECTIVE OXIDATION OF HYDROCARBONS BY NITRIC OXIDE TO NITRILES

V.M.BELOUSOV and S.B.GRINENKO

The L.V.Pisarzhevski Institute of Physical Chemistry, Academy of sciences of Ukrainian SSR, 252028, Kiev (USSR)

SUMMARY

The oxidation of 20 hydrocarbons, having different structure, by nitric oxide has been investigated on composite lead oxide catalysts. The CH_3-group, conjugated with a double bond or the aromatic ring, is attacked by NO to form cyano group. Nitric oxide is reduced to N_2O and N_2. Pb-Ti-O, Pb-Sn-O and Pb-Zr-O systems proved to be the most effective catalysts. Pb^{2+} acts as an active component.

INTRODUCTION

The Heterogeneous catalytic interaction between nitric oxide and hydrocarbons leads to the formation of the nitriles of carboxylic acids. For example, acrylonitrile is formed from propene and benzonitrile from toluene:

$$CH_2{=}CH{-}CH_3 + 1.5NO = CH_2{=}CH{-}CN + 1.5H_2O + 0.25N_2$$

$$C_6H_5{-}CH_3 + 1.5NO = C_6H_5{-}CN + 1.5H_2O + 0.25N_2$$

The reaction of nitroxidation of hydrocarbons is more selective in nitriles than the oxidative ammonolisis of hydrocarbons.

RESULTS AND DISCUSSION

Catalysts

The reaction is catalyzed by silver (refs. 1,2) and oxides of several metals. In the nitroxidation of propene we have investigated 17 metal oxides as catalysts, which, by total consumption of NO at 673 K, are arranged in the order (ref. 3):

$CuO > MnO_2 > Co_2O_3 > V_2O_5 > Cr_2O_3 > Fe_2O_3 > NiO > Bi_2O_3 > ZnO > U_3O_8 > PbO > SnO_2 > TiO_2 > ZrO_2 > WO_3 > Sb_2O_4 > MoO_3$.

The yield of acrylonitrile at 673 K decreases in the series:

$Co_2O_3 > Bi_2O_3 > V_2O_5 > MnO_2 > NiO > PbO > Fe_2O_3 > CuO > Cr_2O_3 > ZnO > SnO_2 > WO_3 > U_3O_8 > Sb_2O_4 > TiO_2 > ZrO_2$.

However at higher temperatures the lead, nickel and zinc oxides are the most active in acrylonitrile formation. Our results correlate with correspondens data on the oxidation of propene by oxy-

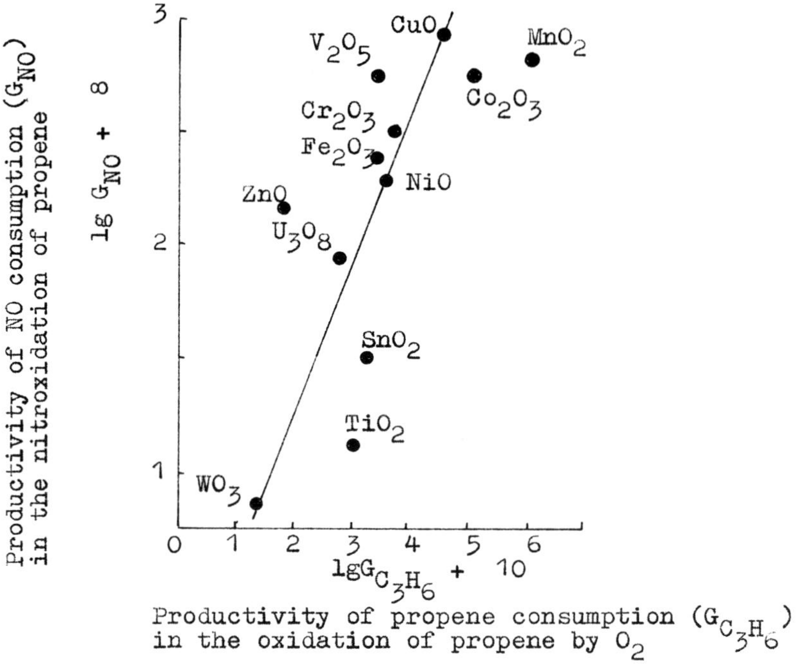

Fig. 1. The correlation between the catalytic activity of different oxides in propene nitroxidation at 673 K and in propene oxidation by oxygen at 573 K.

gen (Fig. 1). However, there are some differences:

1 The rate of propene nitroxidation is lower, then of its oxidation by oxygen.

2 The reduction of nitric oxide proceeds by a parallel-consecutive scheme:

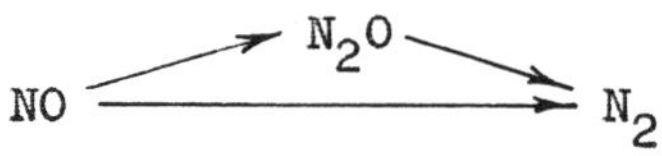

3 The most selective catalysts for the one reaction have poor selectivity in the other and vice versa.

Since lead oxide proved to be the most selective among individual oxides its catalytic properties were investigated in some detail. The dependence of the activity and selectivity of propene nitroxidation on the valent state of lead in oxides was investigated by the nonstationary catalysis method (ref. 4). It has been shown, that the selectivity of propene conversion to acrylonitrile on PbO is higher than that on PbO_2. On the other hand, PbO_2 is more active, than PbO by an order of magnitude. The pro-

TABLE 1

Phase composition and catalytic activity of lead-titanium oxide catalysts

No	Composition		Acrylonitrile productivity* mol/m^2.s x10^8	Selectivity calculated for NO consumpted %
	Chemical	Phase		
1	PbO	Phase PbO - yellow, rhombic modification (I).	6.0	58
2	$9PbO.TiO_2$	(I) - main phase. Admixed phase - $PbTiO_3$ perovskite structure (II)	5.5	60
3	$3PbO.TiO_2$	(II) - main phase. Small amount of unidentified phase.	5.0	65
4	$PbO.TiO_2$	(II) - clean	3.8	35
5	$PbO.3TiO_2$	(II) - main phase. TiO_2 in a smaller quantity.	2.0	30
6	$PbO.9TiO_2$	(II) and TiO_2 - in comparable amounts.	0.8	28
7	TiO_2	TiO_2 - rutile structure	0.2	12

*T = 713 K, reaction mixture: 30 vol.% C_3H_6, 10 vol.% NO, N_2 is the rest.

perties of the initial oxides are equalized as the number of reaction mixture pulses fed to catalysts is increased. After 3 pulses PbO_2 markedly reduces to PbO. In this case the selectivity of acrylonitrile formation increases sharply and, hence, the conversion of propene decreases. On PbO, during the first 3 - 5 pulses, the selectivity also somewhat increases due to the removal of the chemisorbed oxidant. Moreover, the activity of PbO drops because in the course of catalysis the active surface area decreases (ref. 3).

Composite lead-titanium, lead-tin and lead-zirconium oxide catalysts are more stable and as efficient as PbO (refs. 5,6). Tables 1 and 2 represent phase composition of these catalysts. The comparison of catalytic properties and phase composition indicates that the catalytic activity is in agreement with the amount of PbO, $PbTiO_3$ and Pb_2SnO_4 phases. Thus, Pb^{2+} cationes are responsible for catalysis, while the second component in the composite catalysts ensures the stabilisation of the lead cation in the valent state of two. Moreover, the refractory tin and titanium oxide phases prevent the catalysts from sintering.

TABLE 2

Phase composition and catalytic activity of lead-tin oxide catalysts

No	Composition: Chemical	Composition: Phase	Acrylonitrile productivity* mol/m^2.s x10^8	Selectivity calculated for NO consumpted %
1	PbO	Phase PbO - yellow, rhombic modification (I).	6.0	58
2	$9PbO.SnO_2$	(I) - main phase. Smal amount of Pb_2SnO_4 isostructural to red lead (III).	6.5	30
3	$3PbO.SnO_2$	(III) - main phase. (I) - admixed phase. There is smal amount of unidentified compound (IV).	8.0	30
4	$PbO.SnO_2$	(III) - main phase. SnO_2 - admixed phase	8.0	30
5	$PbO.2SnO_2$	SnO_2 - tetragonal, isostructural to rutile, main phase. Admixture of an unidentified compound.	5.6	30
6	$PbO.3SnO_2$	The same	4.0	30
7	$PbO.9SnO_2$	Solid solution based on phase SnO_2	3.2	30
8	SnO_2	Phase SnO_2	0.6	5

*T = 713 K, reaction mixture: 30 vol.% C_3H_6, 10 vol.% NO, N_2 is the rest.

Reactivity of hydrocarbons

We have studied the interaction of nitric oxide with 20 hydrocarbons of different structure on silver and composite lead oxide catalysts (refs. 7,8). Some results are given in Table 3 and Figure 2. The regularities in the influence the structure of hydrocarbons on their reactivity in nitroxidation are essentially similar to those observed in their heterogeneous catalytic oxidation by oxygen. In the both cases the two reactions were found to proceed: the complete oxidation to CO_2 and H_2O and the selective oxidation of CH_3-group, conjugated with a double bond or the aromatic ring, to either a cyano or carboxylic group.

The observed regularities may be formulated as follws:

1 The unsaturated aliphatic hydrocarbons are oxidized faster than paraffins; the rate of oxidation increases in the order: paraffins < monoolefines < acetilenes.

TABLE 3

The products of the oxidation of hydrocarbons by nitric oxide on silver and on composite lead oxide catalysts

Hydrocarbon	Products	Selectivity in the sum of nitriles mol %
Ethene	HCN, CO_2, H_2O	5
Propene	acrylonitrile	80
i-Butene	acrylonitrile, acetonitrile	50
n-Butene	the same	30
Pentene-1	the same	20
Hexene-1	the same	20
Isoprene	the same	10
Hexane	CO_2, H_2O	0
Cyclohexane	the same	0
Pentane	the same	0
Benzene	the same	0
Toluene	benzonitrile	98
p-Xylene	p-tolunitrile, terephtalodinitrile, benzonitrile	89
m-Xylene	m-tolunitrile, isophtalodinitrile, benzonitrile	94
o-Xylene	o-tolunitrile, phtalodinitrile, benzonitrile	85
p-Chlorotoluene	p-chlorobenzonitrile, benzonitrile	20
o-Chlorotoluene	o-chlorobenzonitrile, benzonitrile	30
p-Tolunitrile	terephtalodinitrile, benzonitrile	80
m-Tolunitrile	isophtalodinitrile, benzonitrile	80
o-Tolunitrile	phtalodinitrile, benzonitrile	70

2 In the aliphatic hydrocarbon homologous row the rate of oxidation increases with increasing of the number of carbon atoms, for example: ethene < propene < butene < pentene-1.

3 The rate of oxidation of olefins increases on branching: n-butene < i-butene.

4 The substitution of the hydrogen atom in the aromatic ring by a chlorine atom, a CN- or CH_3-group increases the reactivity of the molecule and the conversion of hydrocarbon increases: benzene < toluene < o-, m-, p-xylenes < o-, m-, p-tolunitriles;

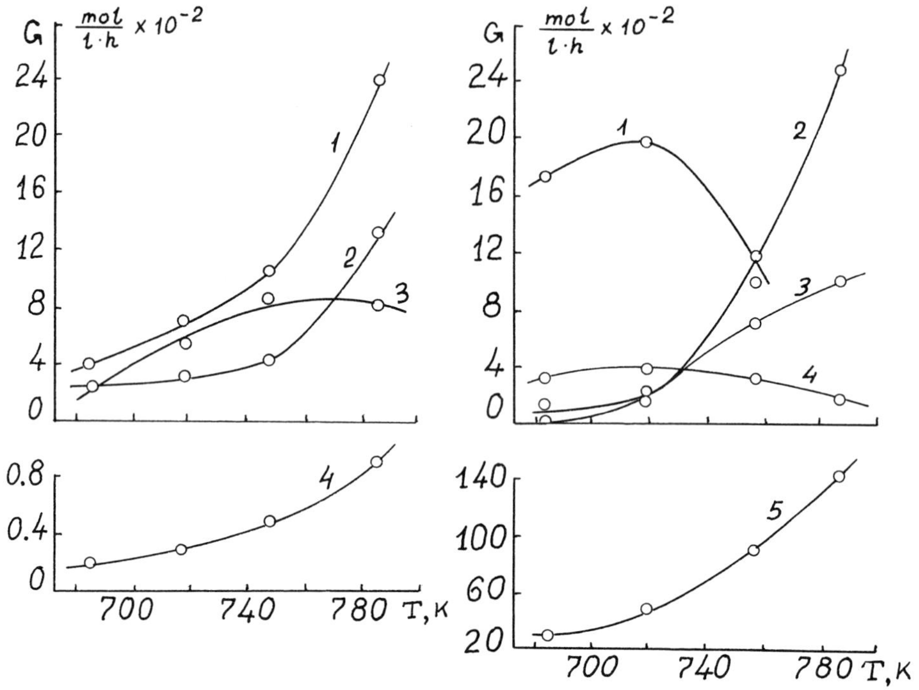

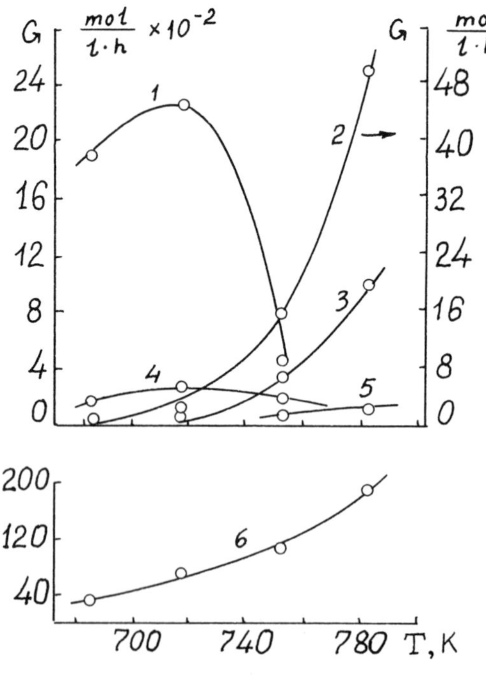

Fig. 2. The nitroxidation of

A toluene: 1 - benzonitrile, 2 - CO_2, 3 - N_2O, 4 - benzene;

B o-chlorotoluene: 1 - N_2O, 2 - toluene, 3 - benzonitrile, 4 - o-chlorobenzonitrile, 5 - CO_2;

C p-chlorotoluene: 1 - N_2O, 2 - toluene, 3 - benzonitrile, 4 - p-chlorobenzonitrile, 5 - o-chlorobenzonitrile, 6 - CO_2.

Catalyst $2PbO.SnO_2$ 7 cm^3, [hydrocarbon] = 5 vol.%, [NO] = 35 vol.%, space velocity = 55 cm^3/min.

toluene < o-chlorotoluene < p-chlorotoluene.

The reactivity of hydrocarbons correlates with their ionisation potentials or the π-electron charge in the reacting CH_2-groups of the corresponding radicals (refs. 7,8).

Mechanisms and kinetics

The stage mechanisms of the formation of the products in the nitroxidation of propene (ref. 9), toluene (ref. 10), xylenes (refs. 11,12) and other hydrocarbons are described by parallel-consecutive schemes. However, the kinetics equations may be derived on the basis of a simplified schemes. For example, in the case of propene nitroxidation on the lead-titanium oxide catalyst the parallel formation of acrylonitrile (AN) and CO_2 predominates. The experimental data are expressed by the equations:

$$\frac{d[AN]}{dt} = k_{AN}[C_3H_6][NO]^{-0.1}$$

$$\frac{d[NO]}{dt} = k_{NO}[C_3H_6]^{0.8}.$$

On the same catalyst the nitroxidation of toluene is described by the system of equations based on the consecutive scheme:

$$C_6H_5CH_3 \xrightarrow{k_1} C_6H_5CN \xrightarrow{k_2} CO_2 + H_2O \;;$$

$$-\frac{d[C_7H_8]}{dt} = k_1[C_7H_8]^{0.9}[NO]^{0.1},$$

$$\frac{d[C_7H_5N]}{dt} = k_1[C_7H_8]^{0.9}[NO]^{0.1} - k_2\frac{[C_7H_5N]^{0.8}}{[C_7H_8]}[NO]^{1.3},$$

$$\frac{d[CO_2]}{dt} = 7k_2\frac{[C_7H_5N]^{0.8}}{[C_7H_8]}[NO]^{1.3},$$

$$\frac{d[N_2O]}{dt} = k_3[NO]^{1.7}$$

$$-\frac{d[NO]}{dt} = 1.5\frac{d[C_7H_5N]}{dt} + 2.57\frac{d[CO_2]}{dt} + \frac{d[N_2O]}{dt}.$$

Thus, the application of nitric oxide as a reactant is promising, because it allows to obtain useful products and to develop new technological processes.

REFERENCES

1 US patent 2736729, RZhKhim, (1960), 108808P.

2 I.Ya.Mulik, M.Ya.Rubanik, V.M.Belousov, Kataliz i katalizatory, Vol. 3, Naukova dumka, Kiev, 1967, pp. 121 - 128.

3 V.M.Belousov, V.V.Korovina, M.Ya.Rubanik, Kataliz i katalizatory, Vol. 6, Naukova dumka, Kiev, 1970, pp. 89-96.

4 A.S.Plachinda, V.M.Belousov, Ukr. Khim. Zhurn., 39, (1973), 975 -978.

5 V.M.Belousov, D.A.Dulin, A.I.Gelbschtein, S.S.Stroeva, V.V. Korovina, V.S.Roginskaya, Kataliz i katalizatory, Vol. 10, Naukova dumka, Kiev, 1973, pp. 37 - 42.

6 V.M.Belousov, D.A.Dulin, A.I.Gelbschtein, S.S.Stroeva, V.V. Korovina, V.S.Roginskaya, Kataliz i katalizatory, Vol. 11, Naukova dumka, Kiev, 1974, pp. 123 - 128.

7 V.M.Belousov, M.Ya.Rubanik, I.Ya.Mulik, Kinet. i Katal., 10, (1969), 841 - 846.

8 V.M.Belousov, S.B.Grinenko, Kataliz i katalizatory, Vol. 14, Naukova dumka, Kiev, 1976, pp. 27 - 32.

9 I.Ya.Mulik, V.M.Belousov, V.V.Korovina, A.V.Gerschingorina, M.Ya.Rubanik, Kataliz i katalizatory, Vol. 5, Naukova dumka, Kiev, 1969, pp. 46 - 51.

10 V.M.Belousov, Kataliz i katalizatory, Vol. 26, Naukova dumka, Kiev, 1989, pp. 8 - 18.

11 S.B.Grinenko, V.M.Belousov, Dokl. AN Ukr. SSR, ser. B, (1973), 1028 - 1031.

12 V.P.Bodrov, V.M.Belousov, S.B.Grinenko, Dokl. AN Ukr. SSR, ser. B, (1975), 317 - 320.

G. Centi and F. Trifiro' (Editors), *New Developments in Selective Oxidation*
© 1990 Elsevier Science Publishers B.V., Amsterdam — Printed in The Netherlands

SELECTIVE OXIDATION OF ANTHRACENE

N.T. Do, R. Kalthoff, J. Laacks, S. Trautmann and M. Baerns

Ruhr-University Bochum, POB 10 21 48, D-4630 Bochum

SUMMARY

The oxidation of anthracene was studied on different unsupported V/Mo/P oxides catalysts as well as on alumina- and silica-supported catalysts at 673 and 723 K. Selectivity of the reaction to anthraquinone and phthalic anhydride was affected by catalyst composition, support material and temperature. A kinetic reaction scheme was set up and the kinetic parameters were determined. Adsorbate structures of anthracene, anthraquinone and phthalic anhydride on both the supported catalysts and on the pure support materials were derived from in-situ FTIR transmission spectroscopic measurements.

INTRODUCTION

Polycyclic aromatic hydrocarbons such as anthracene, phenanthrene and fluorene being formed in coal pyrolysis may be used as feedstocks for producing quinones and dicarboxylic anhydrides. Conversion of some of the compounds by heterogeneous catalytic gas-phase oxidation or in the liquid phase with chromic acid are well known technologies (ref. 1). For the gas-phase reaction, mixtures of vanadium oxide and the oxides of molybdenum, manganese (ref. 2), tungsten (ref. 3), iron, alkali (ref. 4) and phosphorus are used as catalytic active components; alumina, silica and titania are often applied as support materials (refs. 5-7). In the present work anthracene has been subjected to catalytic gas-phase oxidation to study the effect of catalyst composition and of the support material on selectivity; furthermore a kinetic reaction scheme is proposed for characterization of the catalysts by kinetic parameters. Finally, adsorbate structures of anthracene, anthraquinone and phthalic anhydride were determined from in-situ IR spectroscopic measurements. The investigations aim at a better understanding of the fundamentals of the anthracene oxidation.

EXPERIMENTAL

Preparation of catalysts

Vanadium oxide was used as base component for the catalysts; it was modified by adding molybdenum oxide and phosphoric acid; in some instances silica and alumina were applied as support materials. Vanadium oxide was dissolved in concentrated hydrochloric acid as reducing agent at 80°C for 2 h. Subsequently, molybdenum oxide and phosphoric acid were added. When preparing supported catalysts the carrier material was dispersed in the afore mentioned solution. After solvent evaporization the solid material was dried at 120°C and subsequently calcined at 500°C for 16 h. The compositions and surface areas of the unsupported catalysts and the supported ones used in the oxidation of anthracene are given in Table 1.

Apparatus

A schematic diagram of the apparatus used for catalytic testing is given in Fig. 1. Anthracene was oxidized in an electrically heated fixed-bed quartz reactor (length 300 mm, I.D. 8 mm). Axial temperature profiles in the catalyst bed were measured by a movable thermocouple. Anthracene and some of the oxygenates were analyzed by on-line GC. All condensable products of the effluent from the reactor were collected at room temperature and analyzed by off-line GC and HPLC (ref. 8). The carbon oxides CO and CO_2 were determined by on-line GC.

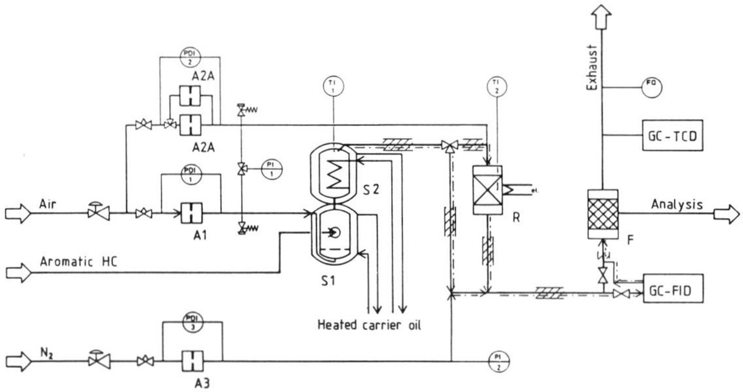

Fig. 1. Schematic diagram of the apparatus for catalytic testing (A: capillary flow meter, R: fixed bed reactor, F: separator, S: saturator for anthracene).

For measuring IR transmission spectra a FTIR spectrometer (Perkin-Elmer model 1710) was used. Two identical IR cells made of quartz were incorporated into the spectrometer. A schematic diagram of the IR cell which could be used as a reactor when the catalyst was inserted is shown in Fig. 2.

The IR cell was composed of a cylindrical quartz tube (length 100 mm, E.D. 35 mm) which was sealed on both ends by NaCl windows. The catalyst sample, hold in a quartz frame, was kept in a fixed position by guide ledges containing heated filaments for direct heating of the catalyst up to 773 K.

The catalyst powder was pressed at 32 bar to a 10x30 mm specimen of about 10-30 mg/cm^2. A continuous gas stream of about 30 l/h loaded with anthracene (0.1 vol.%) was passed through both cells one containing the catalyst sample. Both cells were alternatively moved into the IR beam; hereby the spectrum of the gas phase surrounding the catalyst could be eliminated. The adsorbate spectrum was then obtained by dividing the transmittances obtained for the catalyst plus the adsorbate by the respective transmittances of the clean catalyst, i.e. without adsorbate as measured before adsorption (refs. 9, 10).

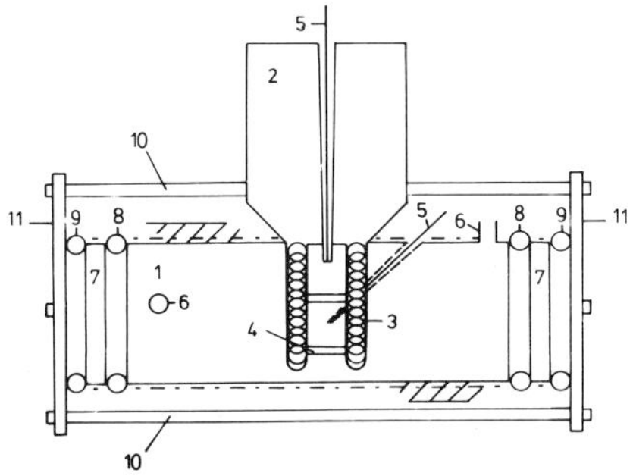

1: quartz cylinder
2: sample holder
3: heated filament
4: guide ledges
5: thermocouple
6: gas inlets/outlets
7: NaCl window
8: graphite washer
9: viton ring
10: pole
11: Al-ring

Fig. 2. In-situ IR cell.

RESULTS AND DISCUSSION

Catalytic testing

Oxidation of anthracene (0.25 vol.% in air) was carried out using the catalysts listed in Table 1 (grain size: 0.5 to 0.7 mm) at 673 and 723 K. The concentrations of anthracene and of the products were measured as a function of contact time ($m_{cat}/\dot{V}$). Therefrom, the dependence of the selectivities on anthracene conversion was derived. Moreover, a reaction scheme was set up. Assuming a

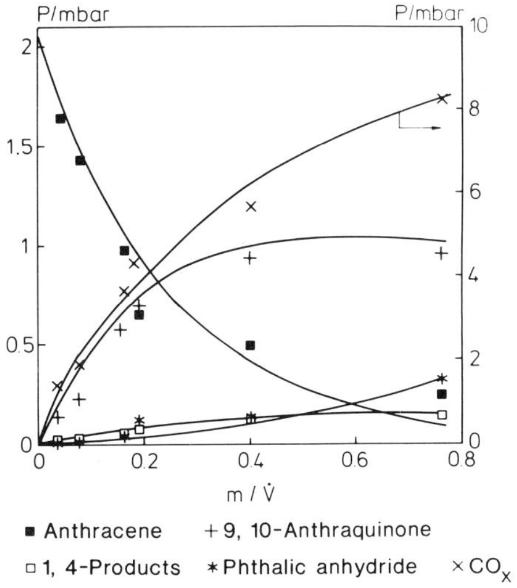

Fig. 3. Dependence of the partial pressures on contact time at 723 K. Catalyst: V:Mo = 4.17; P:V = 0.11.

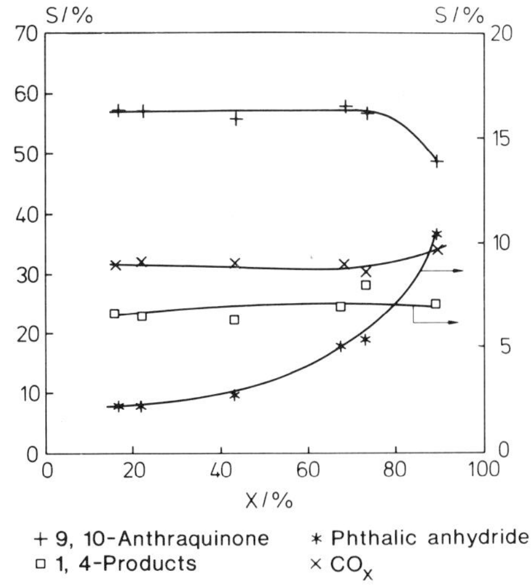

Fig. 4. Dependence of the selectivities on anthracene conversion at 723 K. Catalyst: V:Mo = 4.17; P:V = 0.11.

first-order reaction with respect to the hydrocarbon compounds the kinetic parameters were determined characterizing catalyst performance.

For illustration, a typical dependence of the partial pressures on contact time is given in Fig. 3 for a selected catalyst (symbols: measured data, lines: calculated according to the kinetic data, which are reported further below); the corresponding dependence of the selectivities on anthracene conversion is presented in Fig. 4.

The pattern of the relationships shown in Fig. 3 and 4 indicate that 9,10- and 1,4-anthraquinone as well as the carbon oxides can be considered as primary products. With increasing anthracene conversion 9,10-anthraquinone is further oxidized to phthalic anhydride under simultaneous formation of carbon oxides.

Reaction scheme

The oxidation of anthracene can occur in the 9,10- and/or 1,4-positions. An attack of the oxygen in the 9,10-positions leads to 9,10-anthraquinone while an 1,4-attack results in 1,4-anthraquinone. The 9,10-anthraquinone reacts further to phthalic anhydride while the 1,4-anthraquinone is oxidized further to 2,3-naphthalic anhydride and finally to pyromellitic anhydride. A simplified reaction scheme for the anthracene oxidation as derived from the kinetic relationships is presented in Fig. 5.

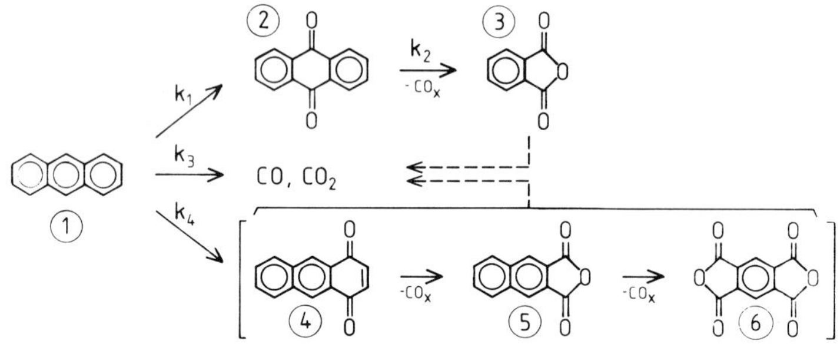

1: Anthracene
2: 9,10-Anthraquinone
3: Phthalic anhydride
4: 1,4-Anthraquinone
5: 2,3-Naphthalic anhydride
6: Pyromellitic anhydride

Fig. 5. Reaction scheme of the anthracene oxidation.

Catalyst performance

The effect of the V/Mo ratio on the selectivity of different catalysts with a constant P/V ratio of 0.11 was studied at 673 and 723 K; the selectivities are compared at an anthracene conversion of about 80%. An increase in temperature from 673 to 723 K results in a higher 9,10-anthraquinone selectivity. The results presented in Fig. 6 show that the selectivity of 9,10-anthraquinone decreases with increasing V/Mo ratio.

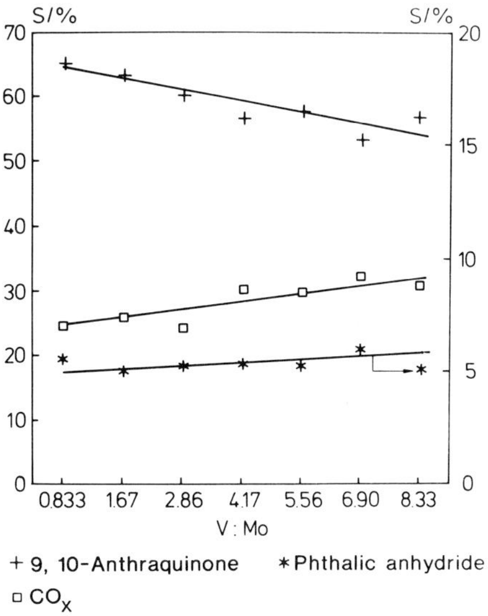

Fig. 6. Effect of the ratio V:Mo on the selectivities at 723 K at an anthracene conversion of about 80%.

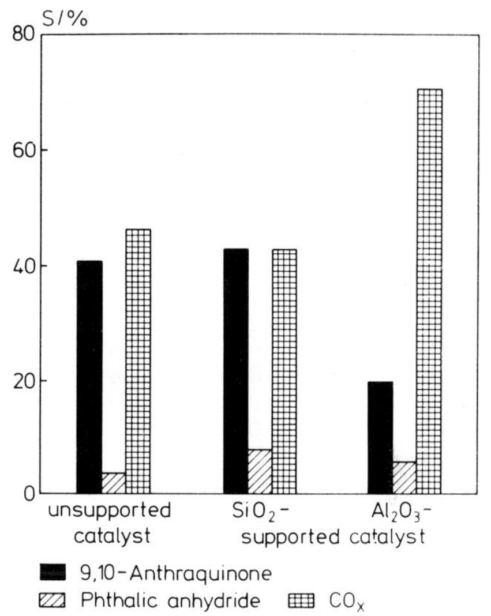

Fig. 7. Effect of support material on the selectivities at 673 K at an anthracene conversion of about 80%.

When using a support material for the catalytic compounds, catalyst activity increases for the oxidation of anthracene. The selectivity is differently affected depending on the support applied. The effect of a silica and an alumina support on the selectivity loaded with catalytic material (V/Mo = 0.83 and P/V = 0.11) at 673 K and at an anthracene conversion of about 80% is shown in Fig. 7. The unsupported and the SiO_2-supported catalyst show almost the same selectivity behaviour; 9,10-anthraquinone selectivity decreased, however, markedly when using alumina as support material.

Kinetic characterization of the catalysts

A statistical discrimination between different kinetic models based on different reaction schemes showed that the total oxidation of the oxygenates, i.e., 9,10-anthraquinone, phthalic anhydride and the other 1,4-products could be neglected as a first approximation up to anthracene conversions of about 90%; for simplification all the products formed by the 1,4-attack of anthracene were lumped as a pseudo-component (1,4-products; cp. Fig. 5). All the reaction steps were assumed to be first-order with respect to anthracene and to the various oxygenates; this appeared justified because of the low concentration of these compounds. For catalyst characterization various ratios of rate constants were defined. The dependence of these values on the V/Mo/P atomic ratio and on the support material used for the catalytic materials are shown in Tab. 1.

TABLE 1.

Ratio of the rate constants for the oxidation of anthracene

a) P:V = 0.11

Temperature		673 K			723 K		
V:Mo	S_{BET} m^2/g	$\frac{k_2}{k_1}$	$\frac{k_1}{k_1+k_3+k_4}$	$\frac{k_3}{k_1+k_3+k_4}$	$\frac{k_2}{k_1}$	$\frac{k_1}{k_1+k_3+k_4}$	$\frac{k_3}{k_1+k_3+k_4}$
0.83	1.4	0.2	0.5	0.4	0.3	0.6	0.3
1.67	1.2	0.3	0.5	0.4	0.3	0.7	0.2
2.86	0.8	0.4	0.5	0.4	0.2	0.7	0.2
4.17	0.5	0.4	0.4	0.4	0.2	0.7	0.2
5.56	0.7	0.6	0.4	0.4	0.4	0.6	0.3
6.90	1.4	0.5	0.4	0.5	0.2	0.5	0.4
8.33	0.6	0.8	0.4	0.5	0.2	0.6	0.3
0.83	74	Al_2O_3-supported[1,2]			0.5	0.2	0.8
0.83	128	SiO_2-supported[1,3]			0.5	0.6	0.4

b) P:V = 0.4

V:Mo	S_{BET} m^2/g	$\frac{k_2}{k_1}$ (673 K)	$\frac{k_1}{k_1+k_3+k_4}$ (673 K)	$\frac{k_3}{k_1+k_3+k_4}$ (673 K)	$\frac{k_2}{k_1}$ (723 K)	$\frac{k_1}{k_1+k_3+k_4}$ (723 K)	$\frac{k_3}{k_1+k_3+k_4}$ (723 K)
1.67	0.3	0.6	0.4	0.5	0.2	0.6	0.3
8.33	0.8	0.6	0.4	0.5	0.4	0.5	0.4

[1] 10 mass-% of catalytic material, 90 mass-% support material
[2] Alon-C/Degussa; S_{BET} = 95 m^2/g
[3] Aerosil-200/Degussa; S_{BET} = 140 m^2/g

The ratio k_2/k_1 is a measure for the consecutive oxidation of the primary product 9,10-anthraquinone to phthalic anhydride while the ratios $k_1/(k_1+k_3+k_4)$ and $k_3/(k_1+k_3+k_4)$ represent the extent of the selective reaction route to 9,10-anthraquinone and of the total oxidation to CO_x respectively (cp. Fig. 5). The following results can be derived by a comparison of the numerical values:

i) The consecutive oxidation of 9,10-anthraquinone to phthalic anhydride increases with an increase of the V/Mo ratio at the low reaction tempe rature of 673 K and at the low P/V ratio of 0,11.

ii) Total oxidation is more marked at 673 K than at 723 K.

iii) Silica as support material results in better selectivities than alumina.

iv) No significant effect of the P/V ratio was observed in the range from 0.11 to 0.40.

These results are in agreement with the qualitative data described above.

In-situ IR spectroscopic identification of adsorbate structures

When oxidizing anthracene CO_x formation was increased by the use of the alumina support material for the V/Mo/P oxides catalysts while by the use of silica no significant change in selectivity was observed. To elucidate this behaviour the following experiments were conducted. Anthracene was adsorbed on both the supported catalysts between 573 and 723 K in the presence of air. IR transmis-

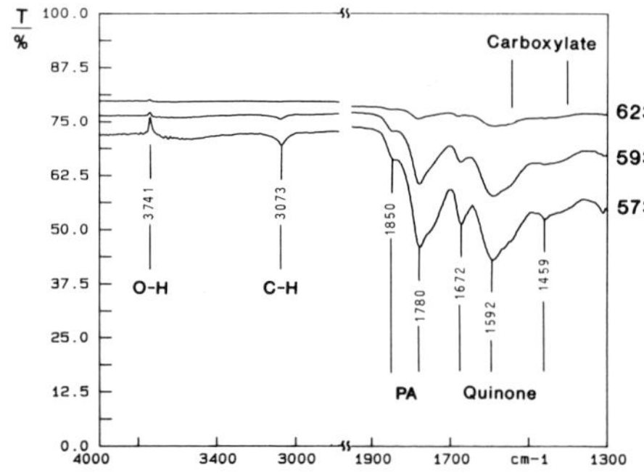

Fig. 8. IR transmission spectra of the anthracene adsorbates on the SiO_2-supported catalyst.

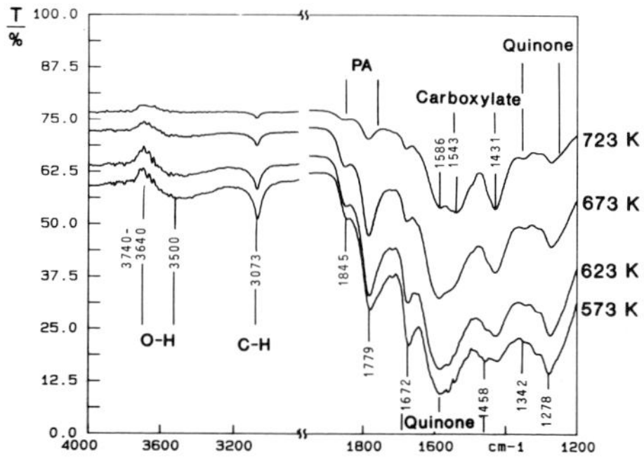

Fig. 9. IR transmission spectra of the anthracene adsorbates on the Al_2O_3-supported catalyst.

sion spectra of anthracene adsorbates are shown in Figs. 8 and 9.

Anthraquinone (νC=O: 1672 cm^{-1}), phthalic anhydride (νC=O: c. 1850 and 1780 cm^{-1}) and carboxylate complexes ($\nu_{as}COO^-$: 1543 and $\nu_s COO^-$: 1431 cm^{-1}) (ref. 11) were observed as adsorbate structures on the surface of both supported catalysts.

The intensity ratios of the carboxylate bands to those of anthraquinone and phthalic anhydride bands respectively are larger on the Al_2O_3-supported catalyst than on the SiO_2-supported one. Furthermore, a strong product adsorption was observed on the Al_2O_3-supported catalyst up to 723 K while on the SiO_2-supported catalyst no adsoption was observed any more above 623 K. The adsorbate spectra of anthracene on the Al_2O_3-supported catalyst showed additional strong negative OH bands of Al_2O_3 at 3640 - 3740 cm^{-1} (ref. 12) as well as bridged OH bands at 3500 cm^{-1} and a strong CH band of adsorbates at 3073 cm^{-1} while on the SiO_2-supported catalyst the negative OH band of SiO_2 at 3741 cm^{-1} and the CH band at 3073 cm^{-1} were very weak; the negative bands are ascribed to an interaction of OH groups with the reactants. From these results it can be derived that the interaction between the catalyst and the reactants, i.e. intermediates and products was stronger on the Al_2O_3-supported catalyst than on the SiO_2-supported catalyst. This could be confirmed by desorption measurements at 723 K: complete desorption was observed within less than 1 minute on the SiO_2-supported catalyst while desorption on the Al_2O_3-supported catalyst took more than 30 minutes. The IR spectroscopic results indicate that non-selective CO_x formation is favored on the Al_2O_3-supported catalyst due to the formation of carboxylate structures which are considered as precursors to oxidative degradation of phthalic anhydride being a consecutive oxidation product of anthraquinone. Furthermore, it was shown that anthracene adsorbed between 573 and 723 K only on pure Al_2O_3; no adsorption was observed on pure SiO_2. When pure SiO_2 and the silica supported catalyst were exposed to gaseous anthraquinone and phthalic anhydride no adsorption was observed while on pure Al_2O_3 and on the alumina-supported catalyst strong carboxylate formation occured on the solid surface.

CONCLUSIONS

Catalysts composed of V/Mo/P oxides are suitable for the oxidation of anthracene to anthraquinones. For 9,10-anthraquinone a maximum selectivity of 65% was obtained (T = 723 K, X = 85%); summing-up all the valuable products, i.e., 1,4-anthraquinone, 2,3-naphthalic anhydride, pyromellitic anhydride and phthalic anhydride a total selectivity of about 85% was achieved. The catalytic performance of the various solids used as a catalyst could be quantitatively described by first-order rate constants. IR spectroscopic studies showed that adsorbate structures of different concentrations were formed on the catalyst surface when using alumina or silica as support materials. The alumina support having higher surface acidity when compared to silica resulted in an extensive formation of surface carboxylates which are considered to be precursors to oxidative degradation of the valuable oxygenates.

ACKNOWLEDGEMENTS

Financial support by Deutsche Forschungsgemeinschaft (grant SFB-0218/B3) is gratefully acknowledged.

REFERENCES

1 Ullmanns Encyclopädie der Technischen Chemie, Vol. 7, 4th edn., Verlag Chemie, Weinheim-New York, 1974, pp.578.
2 J. Vymetal and J. Norek, Czech., CS Pat. 205981 (1983).
3 J.E. Germain, Catalytic Conversion of Hydrocarbons, Academic Press, New York, 1969, pp.256.
4 W. Wettstein and L. Valpiana, Swiss Pat. 407079 (1966).
5 Ullmanns Encyclopädie der Technischen Chemie, Vol. 17, 4th edn., Verlag Chemie, Weinheim-New York, 1974, pp.510.
6 J. Vymetal, J. Norek and V. Cermak, Chem. Prum., 34(9) (1984) 467.
7 H. Yasi and K. Ota, Jpn. Kokai Tokkyo Koho JP, 75, 108254 (1975).
8 A. Zein and M. Baerns, J. Chromat. Science, 27 (1989) 249.
9 A. Ramstetter and M. Baerns, J. Catal., 109 (1988) 303.
10 N.T. Do and M. Baerns, Appl. Catal., 45 (1988) 9.
11 L.J. Bellamy, The Infrared Spectra of Complex Molecules, 3rd edn., Chapman and Hall Ltd., London, 1975.
12 A.V. Kiselev and V.I. Lygin, Infrared Spectra of Surface Compounds, John Wiley & Sons, New York-Toronto, 1975.

B. Delmon (Université Catholique de Louvain, Belgium): You have obtained a very low selectivity when your V/Mo/P catalyst was supported on alumina. It is known that Mo, in its oxide form, has such a strong affinity for Al_2O_3 that it forms strongly adhering monolayers. A very likely reason for the low activity of the Al_2O_3-supported catalyst is that MoO_3 segregates out of the V/Mo/P compound for reacting with Al_2O_3 (phosphorous, to a certain extent, could do the same). I now refer to your IR-spectra of Fig. 9. Did you take, for comparison, similar spectra for MoO_3/Al_2O_3 (and P_2O_5/Al_2O_3)? The formation of a MoO_3 monolayer could explain the presence of the species you detect.

M. Baerns (Ruhr-University Bochum, W.-Germany): We have no IR-spectra of MoO_3/Al_2O_3 or P_2O_5/Al_2O_3 but we studied the pure carrier materials under the same reaction conditions. The adsorbates on Al_2O_3 showed similar IR-spectra as the Al_2O_3-supported catalyst. Our conclusion was that the low selectivity of the Al_2O_3-supported catalyst is mainly affected by the support material.

S. L. Kiperman (N. D. Zelinskii Institute of Organic Chemistry, Academy of Scienes, USSR): First question: In this work the authors have proposed that all the reaction steps are of first order with respect to hydrocarbon compounds. This was assumed to be possible as concentrations of reagents and their products were small; but low gas phase concentrations do not mean that surface coverings were also small. Therefore, the first order of all the reactions was not proved. What do the authors think about it?
Second question: Do you have a possibility to measure the surface concentrations of the reaction components?

M. Baerns (Ruhr-University Bochum, W.-Germany): (1) We have described our kinetic results within the range of conditions studied by the first-order reactions; this was possible since this type of rate equation was applicable and described our experimental data sufficiently. If, however, the reactant concentrations are varied over a wider range this simplification cannot any longer be used; more complex kinetics of the Hougen-Watson type are required.
(2) We have no possibility to measure the absolute surface concentrations of the reaction compounds. We only can estimate the relative surface concentrations of the adsorbed species from the IR-spectra by comparison of the areas below the bands.

G. Centi and F. Trifiro' (Editors), *New Developments in Selective Oxidation*

VAPOR-PHASE OXIDATION OF ALPHA-METHYLSTYRENE TO PHENYLACROLEIN

M. AI

Research Laboratory of Resources Utilization, Tokyo Institute of Technology, 4259 Nagatsuta, Midori-ku, Yokohama 227 (Japan)

ABSTRACT

Various Mo-Te-based ternary oxides were tested as catalysts for the vapor-phase air oxidation of α-methylstyrene to form phenylacrolein (atropoaldehyde). The best catalyst performances were found with Mo/Te/Ti, Mo/Te/W, and Mo/Te/Fe oxides: the one-pass yield of phenylacrolein attained 63 mol% at the α-methylstyrene conversion of 96.5 %. The catalytic functions obtained with these combinations of oxides were also investigated in the light of both acid-base and oxidizing functions obtained from the catalytic activity for dehydration-dehydrogenation of 2-propanol and oxidation of 1-butene.

INTRODUCTION

Propylene is oxidized to acrolein with a very high selectivity over Bi-Mo- and Sb-based mixed-oxide catalysts. Further, methyl-substituted propylene, i.e., n-butene and isobutene, are oxidized to butadiene and methacrolein, respectively, over similar type of mixed-oxide catalysts. However, in the oxidation of aromatic compounds, these catalysts cannot usually exhibit an excellent performance. Indeed, in the oxidation of ethylbenzene to styrene, the catalysts proposed to be effective are different from those used in the oxidation of propylene and butenes (ref. 1). In the oxidation of toluene to benzaldehyde, V-P oxides (ref. 2), Mo-P oxides (ref. 3), Mo-P-based ternary oxides (ref. 4), Mo-based oxides (ref. 5), and V-Ti oxides (ref. 6) have been proposed to be effective.

As for the oxidation of phenyl-substituted propylene, i.e., α-methylstyrene [2-phenylpropene] to form α-phenylacrolein [atropoaldehyde, 2-phenylpropenal] (abbreviated hereafter as PhA), there have been very few studies. Adams (ref. 7) reported that Bi-Mo oxide is not effective; the yield of PhA is about 3 mol% at the α-methylstyrene conversion of 45 %. Recently, Grasselli et al. (ref. 8) reported a 30 mol% yield of PhA at the conversion of 50 % over Nb-promoted Sb-U oxides.

In the preceding study (ref. 9), it was found that the best performance for the production of PhA is obtained with MoO_3 among the various single-oxides tested and that Mo/Te atomic ratio = 10/4 oxide exhibits the best performance among the various Mo-based binary oxides tested. The yield of PhA attained

48.5 mol% at the conversion of 96.3 %. It was also found that α-methylstyrene is much more reactive than propylene and butenes.

In this study, for purpose of exploring more effective catalysts for this partial oxidation, various oxides were combined with the Mo/Te atomic ratio = 10/4 oxide and their catalytic properties were tested. Then, the functions of oxide required for catalyzing this oxidation reaction were investigated.

EXPERIMENTAL

Catalysts

The catalysts used in this study were Mo/Te/X (X is the third component) atomic ratio = 10/4/x (x = 0 to 16) ternary oxides. They were supported on 8- to 20-mesh size pumice originating from volcanic stone consisting of macropores (packing density = ca. 0.4 g/ml and specific surface area = 0.3 to 0.6 m^2/g).

For example, the Mo/Te/W = 10/4/8 catalyst was prepared as follows. $(NH_4)_6$-$Mo_7O_{24}\cdot 4H_2O$ (35.3 g) was dissolved in hot water and 41.8 g of $(NH_4)_{10}W_{12}O_{41}\cdot 5H_2O$ was also dissolved in another hot water using oxalic acid. The two solutions were mixed and 18.0 g of H_6TeO_6 was dissolved to the solution. Excess water was evaporated off with stirring with the aid of hot air current, yielding a sticky syrup. Thereafter, 100 ml of the pumice was added to the sticky syrup and the mixture was evaporated to dryness with stirring with the aid of hot air current. The obtained solid was evaporated again in an oven at 200°C for 4 h and then it was calcined at 450°C for 6 h in a stream of air.

Reaction procedures

The vapor-phase contact oxidation of α-methylstyrene was conducted with a conventional continuous-flow system. The reactor was made of a steel tube, 50 cm long and 1.8 cm i.d., mounted vertically and immersed in a lead bath. Air was introduced from the top of the reactor, with α-methylstyrene being injected into the preheating section of the reactor by means of a syringe pump.

Unless otherwise indicated, the feed rates were fixed as follows: air, 400 ml (at 20°C)/min (ca. 1.0 mol/h); α-methylstyrene, 11.9 mmol/h (ca. 1.19 mol% in air). The effluent gas from the reactor was led successively into four chilled scrubbers containing 2-propanol to recover the 2-propanol-soluble compounds. Afetr 1 h time-on-stream, the content of the scrubbers was collected (120 ml). The reaction products and unreacted α-methylstyrene were analysed by gas chromatograph: a 1-m column of Molecular sieve 13X for CO; a 6-m column of propylene carbonate for CO_2; a 2-m column of PEG 20M at 160°C for α-methylstyrene, benzaldehyde, and 2-propanol; a 1-m column of AT-1200 + H_3PO_4 at 160°C for PhA, benzaldehyde, maleic anhydride, and benzoic acid. The amount of total acid was also checked by titration with 0.1 N NaOH using a pH meter. The amount was usually in accord with the sum of maleic anhydride and benzoic acid measured by gas chromatograph.

Since PhA is not available as a chemical agent, the identification was performed by means of GC-MS (Hitachi H-80) and the quantity was determined on the basis of the approximation that the peak area of PhA recorded in gas chromatograph (FID) is equal to that of cinnamaldehyde [β-phenylacrolein, 3-phenylpropenal].

The yield and selectivity of a particular product were defined as mole percentage yield and selectivity on a carbon-account-for basis. As for the yield of carbon oxides [CO_x], the CO_x accompanied with the formation of benzoic acid, benzaldehyde, and maleic anhydride was excluded.

RESULTS AND DISCUSSION

Performances of the Mo-Te-based ternary oxide catalysts

The results obtained over 10 g-portions of Mo-Te-based ternary oxide catalysts at the optimal reaction temperatures are listed in Table 1, according to the classification of oxide in view of both acid-base and oxidizing functions (ref. 9,10). The results may be summarized as follows.

(1) The addition of WO_3, TiO_2, and Fe_2O_3 to the Mo/Te = 10/4 oxide enhances markedly both the oxidation activity and selectivity to PhA. The presence of an optimum amount was observed for each third component.

(2) The best results are obtained with the Mo/Te/Ti = 10/4/4 oxide; the one-pass yield of PhA attains 63.0 mol% at the α-methylstyrene conversion of 96.5 %.

(3) The second best results are obtained with the Mo/Te/W = 10/4/8 oxide; the PhA yield attains 59.5 mol% at the conversion of 98.4 %.

(4) The third best results are obtained with the Mo/Te/Fe = 10/4/4 oxide; the PhA yield attains 58.0 mol% at the conversion of 96.7 %.

(5) The addition of ZrO_2, Bi_2O_3, and Co_3O_4 enhances the oxidation activity, but it enhances the selectivity only a little.

(6) The effect of V_2O_5 is small.

(7) The addition of U_3O_8, SnO_2, ZnO, NiO, and MnO_2 enhances the oxidation activity, but it decreases the selectivity.

(8) The addition of an acidic oxide such as P_2O_5, B_2O_3, and Sb_2O_5 decreases markedly the oxidation activity and it does not improve the selectivity.

Performances of the binary oxide catalysts

For understanding the function of each component in the Mo/Te/W, Mo/Te/Ti, and Mo/Te/Fe ternary oxides, the catalyst performance for the oxidation of α-methylstyrene obtained over each binary oxide consisting of the ternary

TABLE 1

Performances of Mo–Te-based ternary oxide catalysts*

Catalyst	atomic ratio	T (°C)	Conv (%)	Yield (mol%) PhA	Baci	Bald	MA	CO_x	other	S_{PhA} (mol%)
Mo		450	81.5	18.5	12.5	15.9	5.6	4.1	24.9	23.0
		460	90.0	20.5	13.8	17.1	7.0	6.4	15.2	23.0
Mo/Te	10/4	430	92.5	45.0	11.3	2.5	2.1	4.4	27.2	48.5
		440	96.3	48.5	13.7	2.0	2.0	5.0	25.1	50.5
Mo/Te/P	10/4/4	468	95.7	44.0	12.4	4.3	4.5	8.1	22.4	46.0
Mo/Te/B	10/4/4	500	58.5	26.5	3.9	2.5	2.0	3.3	20.3	45.0
Mo/Te/Sb	10/4/4	490	78.0	42.0	8.2	7.7	3.0	4.8	12.7	54.0
Mo/Te/Zr	10/4/4	415	91.4	51.5	11.5	5.0	4.5	7.0	11.9	56.0
Mo/Te/W	10/4/2	395	88.8	52.5	11.9	3.0	3.4	4.7	13.3	59.1
		400	95.9	55.5	14.6	5.2	3.3	3.7	13.6	57.9
	10/4/4	390	54.0	37.0	6.0	2.4	1.5	1.8	5.3	69.0
		410	93.5	53.5	11.3	4.5	1.6	5.0	17.6	57.0
	10/4/8	385	89.0	58.5	12.3	3.0	2.6	3.8	8.8	66.0
		390	96.2	58.5	13,2	3.6	2.5	4.4	14.0	61.0
		395	98.4	59.5	16.9	4.2	2.9	4.9	10.0	60.5
	10/4/16	375	51.0	21.8	8.7	3.6	2.9	3.0	11.0	43.0
		385	85.5	28.2	16.3	8.4	5.0	7.8	19.8	33.0
Mo/Te/V	10/4/4	435	88.5	46.5	15.8	6.7	2.5	2.9	14.1	52.5
		440	95.0	45.5	14.1	7.1	2.5	6.0	19.8	48.0
Mo/Te/U	10/4/4	400	94.7	44.1	14.1	4.8	4.7	4.8	22.2	46.5
Mo/Te/Ti	10/4/2	440	83.7	47.8	8.1	3.6	3.2	2.6	18.4	57.0
		450	88.3	49.3	9.3	4.8	2.8	4.1	18.0	56.0
	10/4/4	390	83.5	59.3	13.7	2.4	3.5	2.3	2.3	71.0
		400	96.5	63.0	15.0	3.6	3.2	3.8	7.4	65.0
	10/4/8	400	90.6	50.5	12.8	2.4	3.2	5.5	16.2	56.0
		410	96.7	56.8	15.7	4.8	3.2	5.6	10.6	59.0
Mo/Te/Sn	10/4/4	390	90.0	27.0	12.3	3.6	7.2	24.7	14.7	30.3
Mo/Te/Fe	10/4/4	420	89.0	56.8	11.4	4.0	3.4	0.6	12.8	64.0
		430	96.7	58.0	13.3	3.6	3.5	0.6	17.2	60.0
	10/4/8	375	86.5	40.7	22.0	9.6	4.5	1.4	8.3	47.0
		385	90.0	44.1	18.5	8.4	3.0	3.7	12.3	49.0
Mo/Te/Bi	10/4/4	395	83.2	47.6	9.6	3.0	2.1	7.5	13.3	57.2
Mo/Te/Zn	10/4/4	410	96.2	47.2	16.0	4.8	5.0	4.7	18.5	49.0
Mo/Te/Ni	10/4/4	390	96.4	40.5	15.8	4.2	4.6	10.5	20.8	41.8
Mo/Te/Co	10/4/4	390	82.0	46.1	14.7	2.4	0.7	6.0	12.1	56.2
		400	94.8	49.3	14.7	3.6	3.6	7.7	15.9	52.0
Mo/Te/Cr	10/4/4	405	89.6	40.2	17.4	4.8	5.0	7.0	15.0	44.6
Mo/Te/Mn	10/4/4	410	92.7	36.3	15.3	5.4	4.1	10.7	20.9	39.1

* T = temperature, PhA = phenylacrolein, Baci = benzoic acid, Bald = benzaldehyde, MA = maleic anhydride, CO_x = carbon oxides, other = [(overall conversion of α-methylstyrene) – (sum of the yields of PhA + Baci + Bald + CO_x)], S_{PhA} = selectivity to PhA, amount of catalyst used = 10 g.

TABLE 2

Comparison of the performances of the ternary oxides with those of binary oxides

Catalyst atomic ratio		T (°C)	Conv (%)	Yield (mol%) PhA	Baci	Bald	MA	CO_x	other	S_{PhA} (mol%)
Mo/Te	10/4	440	96.3	48.5	13.7	2.0	2.0	5.0	25.1	50.5
W/Te	10/4	400	88.0	32.5	11.5	4.8	4.1	12.9	22.2	37.0
Mo/W	10/4	410	95.8	46.2	18.6	9.6	4.0	4.0	13.4	48.3
Mo/Te/W	10/4/8	395	98.4	59.5	16.9	4.2	2.9	4.9	10.0	60.5
Mo/Te	10/4	440	96.3	48.5	13.7	2.0	2.0	5.0	25.1	50.5
Ti/Te	10/4	410	52.0	23.3	4.9	2.4	2.1	14.7	4.6	45.0
Mo/Ti	10/4	350	89.0	6.4	9.4	5,4	15.3	31.4	21.1	7.2
Mo/Te/Ti	10/4/4	400	96.5	63.0	15.0	3.6	3.2	3.8	7.4	65.0
Mo/Te	10/4	440	96.3	48.5	13.7	2.0	2.0	5.0	25.1	50.5
Fe/Te	10/4	345	33.0	1.7	0.	4.8	0.	17.5	9.0	5.1
Mo/Fe	10/4	370	79.4	18.1	14.7	7.8	6.5	4.9	27.4	22.8
Mo/Te/Fe	10/4/4	430	96.7	58.0	13.3	3.6	3.5	0.6	17.2	60.0

Abbreviations are the same as for Table 1.
The amount of catalyst used is 10 g.

oxides were compared with those obtained over the ternary oxides. The tests were performed using 10 g-portion of the catalysts. The results are shown in Table 2.

The results may be summarized as follows.

(1) WO_3 by itself is not effective as a catalyst for this oxidation (ref. 9) and TeO_2 has no oxidation activity. However, the combination of the two oxides generates a high oxidation activity, but the selectivity to PhA is lower and the formation of CO_x is much greater than those obtained with the Mo/Te oxide. On the other hand, the addition of WO_3 to MoO_3 enhances markedly both the oxidation activity and selectivity: the performance of the Mo/W = 10/4 oxide is comparable with that of the Mo/Te = 10/4 oxide. Therefore. it is considered that the addition of TeO_2 improves the Mo/W oxide much as it improves the MoO_3 alone catalyst.

(2) The performances of the Ti/Te and Fe/Te oxides are much lower than that of the W/Te oxide. Further, the addition of TiO_2 and Fe_2O_3 to MoO_3 increases the oxidation activity, but it decreases the selectivity to PhA. However, an excellent catalytic performance is obtained by the addition of TeO_2 to the Mo/Ti and Mo/Fe oxides, suggesting that the presence of MoO_3 and TeO_2 is essential for a catalyst to be effective for this oxidation.

Acid-base properties of the catalysts

The acid-base properties of the ternary and binary oxide catalysts were studied. Simce the catalysts are colored, the indicater/titration method is

TABLE 3

Catalytic activity for dehydration and dehydrogenation of 2-propanol*

Catalyst atomic ratio		S (m^2/g)	($\times 10^4$ mol/h m^2) r_p	r_a	r_a/r_p
Mo/Te	10/4	0.61	7.0	15.7	2.3
W/Te	10/4	2.4	2.3	2.0	0.9
Mo/W	10/4	1.1	15.4	26.5	1.7
Mo/Te/W	10/4/4	1.05	8.6	20.0	2.4
Ti/Te	10/4	3.7	0.18	1.0	5.7
Mo/Ti	10/4	8.6	28.7	7.6	0.26
Mo/Te/Ti	10/4/4	0.9	13.3	18.6	1.4
Fe/Te	10/4	24.6	0.11	0.96	8.6
Mo/Fe	10/4	1.6	15.7	15.6	1.0
Mo/Te/Fe	10/4/4	0.45	4.6	8.7	1.9

* S, surface area; r_p, rate of dehydration; r_a, rate of dehydrogenation.

not applicable. Therefore, the properties were estimated indirectly from the catalytic activities for acid- and base-catalyzed test-reactions. As a measure of the acidic property, the activity for dehydration of 2-propanol to propylene, and as a measure of the basic property, the (activity for oxidative dehydrogenation of 2-propanol to acetone)/(activity for dehydration of 2-propanol) ratio, were employed (refs. 11-14). The activities were measured under the following conditions: temperature, 220°C; 2-propanol concentration, 1.3 mol% in air; feed rate of air, 400 ml/min.

The results are listed together with the specific surface area in Table 3. They may be summarized as follows.

(1) The oxides which are poor in the acidic property are not effective as catalysts for the formation of PhA; for example, the Ti/Te and Fe/Te oxides.

(2) The oxides which are poor in the basic property are not effective in the oxidation; for example, the Mo/Ti, W/Te, and Mo/Fe oxides.

(3) The addition of TeO_2 suppresses the acidic property and enhances the basic property, to a certain extent.

(4) The possession of a certain level in both the acidic and basic properties seems to be required to achieve a good performance in the oxidation.

Performances in the oxidation of 1-butene.

To know the characteristic features of the ternary oxides which show a good performance in the oxidation of α-methylstyrene, the performances of these oxides in the oxidation of 1-butene were studied. The reaction was conducted under the following conditions; 1-butene concentration, 2.03 mol% in air; feed

TABLE 4

Performances in the oxidation of 1-butene*

Catalyst Atomic ratio		T (°C)	Conv (%)	Yield (mol%) C_4H_6	Yield (mol%) Acid	$S_{C_4H_6}$ (mol%)
Mo/W	10/4	440	27.7	20.2		74
		460	39.0	22.6		58
Mo/Te/W	10/4/4	440	66.3	60.3		91.
		460	79.4	65.9		83
		480	88.8	64.7	14.4	73.
Mo/Ti	10/4	360	72.0	5.4	38.0	7.5
Mo/Te/Ti	10/4/4	420	60.0	55.0		92.
		440	75.4	65.5		87.
		460	92.1	66.5		72.
		480	94.5	55.0	25.5	58.
Mo/Fe	10/4	440	47.5	11.3		24.
		460	57.0	12.0		21.
Mo/Te/Fe	10/4/4	460	34.0	33.0		97.
		480	41.0	39.5		96.5

*$S_{C_4H_6}$, selectivity to butadiene; amount of catalyst used, 20 g.
Acid was measured by the titration and the amount was calculated as acetic acid or maleic anhydride.

rate of air, 280 ml/min; amount of catalyst used, 20 g. The yields of butadiene and acid (mainly acetic acid and maleic anhydride) and the selectivity to butadiene are listed in Table 4.

The results may be summarized as follows.

(1) The ternary oxides which show a high selectivity in the oxidation of α-methylstyrene to PhA, show a very high selectivity in the oxidation of 1-butene to butadiene, too. At a high conversion, a fair amount of acetic acid and maleic anhydride is formed. Possibly, they may be formed by the consecutive oxidation of butadiene.

(2) The Mo/Ti and Mo/Fe oxides are not effective in the oxidation of 1-butene to butadiene much as in the oxidation of α-methylstyrene to PhA. The selectivity to butadiene decreases in the order of Mo/W > Mo/Fe > Mo/Ti. This order is in conformity with that of the selectivity to PhA.

(3) The addition of TeO_2 to the Mo/W oxide enhances the catalytic activity in oxidation of both α-methylstyrene and 1-butene. Whereas, the addition of TeO_2 to the Mo/Ti and Mo/Fe oxides strongly decreases the activity in the both oxidation reactions. The TeO_2 enhances the basic property of the Mo/W oxide, whereas it suppresses the acidic property of the Mo/Ti and Mo/Fe oxides (Table 3).

Discussion

The addition of TeO_2 to the Mo/Ti and Mo/Fe oxides decreases markedly the oxidation activity. This may be ascribed to the decrease in the surface area.

Since both α-methylstyrene and PhA are basic compounds, the oxidation of α-methylstyrene to PhA is a "base → base type reaction" (ref. 10). Therefore, the possession of both acidic and basic properties in a proper level is required as a catalyst for this type of partial oxidation (refs. 10,11,13). The Mo/Te/Ti, Mo/Te/W, and Mo/Te/Fe ternary oxides may best fit the required balance and/or level of the two opposite properties.

The presence of MoO_3 in the catalyst may be essential to have acidic and redox properties. The WO_3, TiO_2, and Fe_2O_3 play a role in enhancing the acidic property, but the property may be too strong to suppress the side-reactions; for example, consecutive oxidation of basic products and C–C bond fission. The addition of TeO_2 to the Mo-based binary oxides suppresses the acidic property to a proper level and also enhances the basic property.

It should be noted that the oxides which show a good performance in the oxidation of 1-butene to butadiene, do not always show a good performance also in the oxidation of α-methylstyrene to PhA. For example, Bi-Mo- and Sb-based oxides are effective for oxidation to butadiene, but are not effective for the oxidation of α-methylstyrene to PhA.

At present, it is still hard to explain the reason. We feel that a more strict level of acid-base properties, which we cannot measure now, is required to achieve a good performance in the oxidation of α-methylstyrene.

REFERENCES

1 G. Emig and H. Hofman, J. Catal., 84 (1983) 15-26.
2 M. Ai, Kogyo Kagaku Zasshi, 73 (1970) 946-950: Chem. Abst., 73 (1970) 76790k.
3 M. Ai, Kogyo Kagaku Zasshi, 74 (1971) 1636-1639: Chem. Abst. 75 (1971) 109992z.
4 M. Ai, Nippon Kagaku Kaishi, (1972) 1151-1156: Chem. Abst., 77 (1972) 66559k.
5 N.K. Nag, T. Frasen and P. Mars, J. Catal., 68 (1981) 77-85.
6 A.J. Van Hengstum, J.E. Ommen, H. Bosch and P.J. Gellings, Appl. Catal., 8 (1983) 369-382.
7 C.R. Adams, J. Catal., 10 (1968) 355-361.
8 R.K. Grasselli, J.D. Burrington, D.D. Suresh, M.S. Friedrich and M.A.S. Hazle, J. Catal., 68 (1981) 109-120.
9 M. Ai, J. Catal., (in press).
10 M. Ai, in T. Seiyama and K. Tanabe (Eds.), Proc. 7th Int. Congr. Catal., Tokyo, June 30 - July 4, 1980, Kodansha, Tokyo/Elsevier, Amsterdam, 1981, pp. 1060-1069.
11 M. Ai, J. Catal., 40 (1975) 318-326 and 327-333.
12 M. Ai, Bull. Japan Petrol. Inst., 18 (1976) 50-54.
13 M. Ai, Bull. Chem. Soc. Japan, 49 (1976) 1328-1334.
14 M. Ai, J. Catal., 52 (1978) 16-24.

V. CORTÉS CORBERÁN (Inst. Catalisis y Petroquimica, Spain): (1) which have been your initial criteria for the selection of metallic cations, and atomic ratios between them, for the catalyst compositions you have used. (2) The systems you have used are well known by their properties in the catalytic selective oxidation of olefins and the overall tendencies finally found for this reaction (for example, addition of tellurium) parallel those previously known for selective oxidations of olefins. Would you have expect, a priori, different tendencies, and if so, why?

M. AI (Tokyo Inst. Tech., Japan): (1) I had no initial criterion; we tested various kinds of single and binary oxide systems and, then, we selected some ternary systems basing on the information obtained from the tests of binary oxides. (2) Many kinds of mixed oxide systems have been known to be effective as catalyst for oxidation of olefins. This study indicates at least that all of them are not effective for oxidation of α-methylstyrene. I expected, a priori, different tendencies, because Bi-Mo-type catalysts are not effective for this oxidation.

R.K. GRASSELLI (Mobil Research and Develop., USA): You state in your conclusions that Sb-based catalyst while effective for the oxidation of 1-butene to butadiene are not effective for the oxidation of α-methylstyrene to atropoaldehyde, apparently in contrast to Mo-Te-Ti, Mo-Te-W, and Mo-Te-Fe based catalysts whichyou studied. I should like to remind you that our work which you kindly quoted as reference 8, clearly showed that α-methylstyrene is effectively converted to atropoaldehyde with Nb-U-Sb based catalysts, in fact the selectivities which we reported with our antimony based catalysts rival your best systems, while I agree that catalytic systems must be optimized for each given reaction. There is no a priori reason to exclude antimony for the oxidation ofα-methylstyrene.

M.AI (Tokyo Inst. Tech., Japan): I could not get a good performance with Sb, Mo-Sb, and Mo-Te-Sb oxides, but I did not try to test with U-Sb oxides because you had already tested with them. Therefore, I think that your are reason.

J. KIJENSKI (Warsaw Politechnika, Poland): What parameter is for you a measure of acidity or basicity of molecule and allow you to consider a particular reaction as e.g., "acid-base" process?

M. AI (Tokyo Inst. Tech., Japan): I have not special opinion about the definition of acid-base. Indeed, ordinary indicator/titration method is not applicable for oxidation catalysts because of their dark color. On the other hand, the gas phase adsorption method contains some problems. Therefore, at present the measurement of catalytic activity for acid-base catalyzed model reactions seems to be the most combinient, though there remains arguments about the definition and strength of acid-base.

B. GRZYBOWSKA (Inst. Catal. Surface Chem., Poland): (1) In the Me-Mo-Te ternary oxide systems there exist well defined compounds (e.g., telluromolybdates of Co, Mn, Ni, Cd) synthesized and characterized by Sloczynski (ref. 1) and by Forzatti, Trifiró, and Villa (ref. 2). They have shown to be active also in aromatic hydrocarbon oxidation (ref. 3). There exist also well defined phases in Mo-Te-O system. Both acid-base and oxidizing properties will depend strongly on phase composition of your catalysts and mode of mutual arrangement of the phases and not only on the presence of particular ions in predefined environment. I believe that these facts should be taken into account when considering the oxidation mechanism. (2) Adding some comments to the question of Prof. Kijenski: let me remind that some authors consider the ionization potential of a molecule as a measure of acidity, basicity [see reviews by Ruckenstein et al (ref. 4)]. (3) The discussion on relation between one-electron and two-electron (acid-base) properties is still open. In the case of solid for instance there are some attempts to reconcile the both properties (ref. 5) - showing that surface one-electron acceptor states can be identical and involve the same orbitals as Lewis acidic sites, provided the acceptor level lies below the Fermi energy level.

1 J. Sloczynski, Z. Anorg. Allg. Chem., 438 (1978) 287.
2 P. Forzatti, F. Trifirò, P.L. Villa, J. Catal., 55 (1978) 52.
3 B. Grzybowska, M. Czerwenka, J. Sloczynski, Catal., Today, 1 (1987) 157.
4 D.B. Dadyburjor, S.S. Jewur, E. Ruckenstein, Catal. Rev., 19 (1979) 293.
5 S.R. Morrison, Surf. Sci., 50 (1975) 329.

M.AI (Tokyo Inst. Tech., Japan): Thank You for your comments.

J. KIJENSKI (Warsaw Politechnika, Poland): Comment to the remark of Prof. Grzybowska: Inonization energy cannot be considered as a measure of acidity or basicity which are ionic, i.e., two electron properties. There is no general parallelism between the basicity and electron donor properties.

J. HABER (Inst. Catal. Surface Chem., Poland): (1) What is the reproductivity of your results. The data seem to be considerably spread which may indicate their dependence on uncontrolled factors. (2) One of the important side reactions of methylstyrene is certainly cracking which will occur on more acidic catalysts and will rivaluate the mechanistic conclusions. Did you try to determine the contribution from cracking by carring test experiments in the absence of oxygen.

M. AI (Tokyo Inst. Tech., Japan): (1) I did not find that the reproductivity of this oxidation is specially low. However, phenylacrolein is not stable, i.e., it tends to polymerize to dimer and trimer. Indeed, I cannot measure the amount of these polymers. Therefore, "other" may consit mainly of polymers. (2) In the predecing work (ref. 9 in the text), the effect of oxygen concentration was studied. The consumption of α-methylstyrene increases almost in proportional to the oxygen concentration. Therefore, the cracking may be small at least over the Mo-Te-based catalysts.

G. Centi and F. Trifiro' (Editors), *New Developments in Selective Oxidation*
1990 Elsevier Science Publishers B.V., Amsterdam — Printed in The Netherlands

Partial Oxidation of o-Xylene to Phthalic Anhydride in a Structured Fixed Bed Containing a Sequence of Catalysts

M. Kotter, D.X. Li, L. Riekert

Institut für Chemische Verfahrenstechnik, Universität Karlsruhe,
Kaiserstr.12, Postfach 6980, 7500 Karlsruhe 1, FRG

Abstract

It is shown that an elevated yield of phthalic anhydride (PAA) can be obtained in a fixed bed consisting of a sequence of different catalysts. A suitable sequence of catalysts can be determined by a computerized search, following the strategy of biological evolution on the basis of experimentally obtained kinetic data.

Introduction

Overall selectivity is the most important objective in the design of catalytic processes, the desired product in general being unstable relative to other possible species which can be generated from the starting material. Selectivity will depend on the relative extent of several parallel and consecutive reactions. The catalyst can be considered as a guide leading the material through composition space on a path which avoids everywhere descent to thermodynamically stable but undesired products. Composition and possibly also temperature change along the length of a fixed bed, using a sequence of different catalysts might therefore be more suitable to obtain high selectivity than using any single catalyst from the set in the sequence alone. This proposition was investigated experimentally. Partial oxidation of o–xylene to phthalic anhydride (PAA) in a fixed bed of catalysts containing oxides of V and Ti as active components was chosen as a typical example. The literature on this subject has been reviewed by Wainwright at al [1] and Saleh at al [2]. We found that the network of parallel and consecutive reactions taking place in this system can be described by the simplified scheme shown in Fig.1. O–tolualdehyde, o–toluylic acid and phthalide are generated in only small amounts and can be lumped into one pseudospecies ("intermediates"). The effect of different reactions in this scheme on overall integral selectivity will depend on the local composition of the gas phase (conversion of o–xylene), which means on the relative length into the bed. For example it will be expedient to prevent reaction $1 \rightarrow 4$ near the inlet where the concentration of o–xylene is high, whereas reaction $3 \rightarrow 4$ is not of much concern where almost

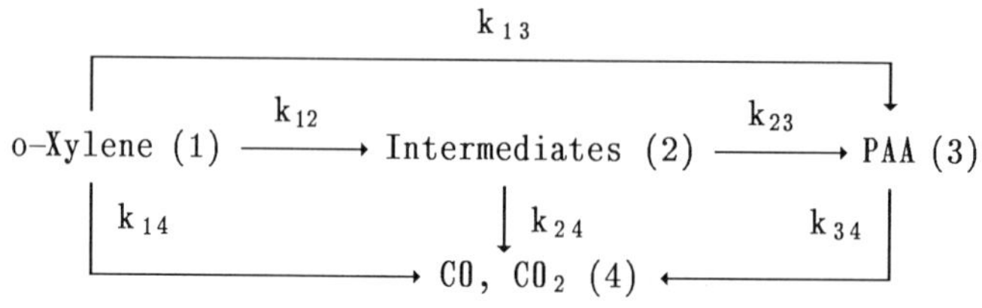

Fig. 1 Simplified scheme of reaction in catalytic oxidation of o–xylene

no PAA is present. The opposite must be true towards the outlet where almost no o–xylene is present and total oxidation of PAA has to be prevented. It appears unlikely that both requirements can be met by the same single catalyst in an optimal way, whereas it seems feasible that a structured bed consisting of several different catalysts in series (Fig. 2) can possibly fulfill both conditions and analogous requirements concerning suppression or allowance of remaining reaction pathways at different locations.

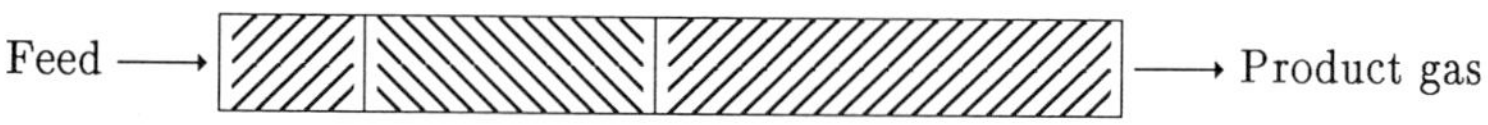

Fig. 2 Structured fixed bed reactor (SFBR)

Physical implementation of this concept and experimental verification of its merits proceeded through the following four steps in sequence:

(1) Preparation of a set of catalysts with different properties, containing the same metals (Ti,V,Cs,Li) as oxides in the active component in different relative amounts. The experimental investigation was thus restricted to a set ("family") of similar but different catalysts.

(2) Mathematical modelling of the catalytic properties for all catalysts: quantitative determination of the kinetics of all reactions in the scheme shown in Fig. 1 for each single catalyst prepared.

(3) Computer calculation of the expected selectivity behaviors of various sequences consisting of several catalysts of different amounts from the set prepared, based on the data from (2); search for an optimal sequence. The number S of possible sequences of N given catalysts is $S(N)=\sum_{i=1}^{N}\binom{N}{i}\cdot i!$; we have already $S\approx10^7$ for N=10. It is therefore impracticable to compute and compare the behavior of all possible sequences. The search for an optimal sequence was based on the strategy of biological evolution as a known shortcut to a relative optimum in a multi–parameter system [3,4].

(4) Experimental verification of the selectivity of the structured bed consisting of the sequence of catalysts found in step (3).

Experimental

Catalyst preparation

Two groups of supported catalysts were prepared and investigated. The first group comprises catalysts in the form of monoliths, the support being cordierite of low porosity with parallel channels of 1x1 mm^2 cross section. This carrier was impregnated with solutions of $Ti(OC_3H_7)_4$ and $VO(OC_3H_7)_3$ in isopropanol with some water added [5]. After impregnation the monoliths were exposed to air saturated with water vapor for 14 h at 70°C, then calcined at 450°C for 4 h.

The second group of catalysts was prepared by coating nonporous spheres of steatite of 2–3 mm diameter with a porous layer consisting of TiO_2 (anatase), V_2O_5 and small amounts of Cs_2O or Li_2O as modifiers. The carrier was immersed in a suspension of anatase in a solution of $VO(OC_3H_7)_3$ in isopropanol and water, Li or Cs being added in the form of nitrates to the solution. After immersion for 10 min the spheres are removed and dried at 60°C for 20 h, then calcined at 450°C for 3 h. The procedure was repeated to increase the thickness of the porous layer.

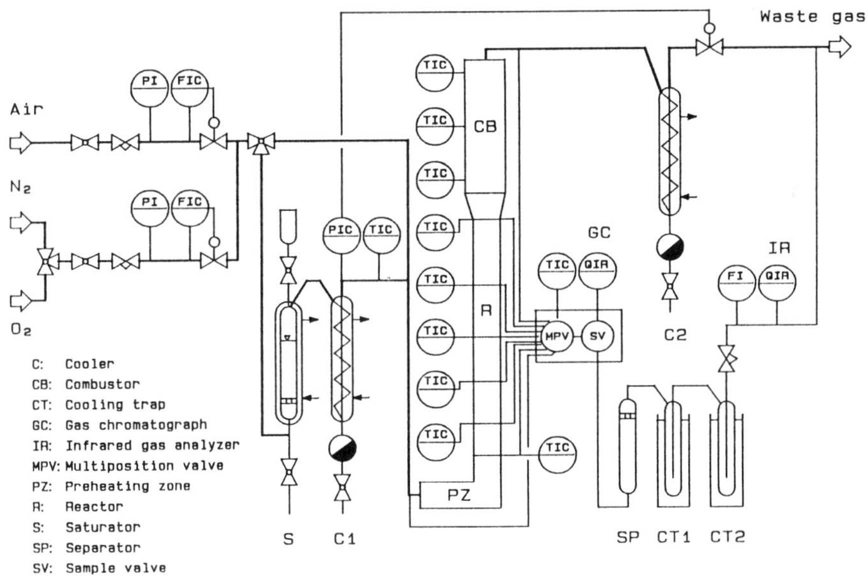

Fig. 3 Apparatus for the catalytic oxidation of o–xylene

Kinetic measurements

Kinetic measurements were performed by monitoring composition along the length of a fixed bed of catalyst 1500 mm long and 15 mm in diameter. The reactor was made of stainless steel, consisting of 5 segments with individual temperature control in each segment (Fig. 3). Heated capillaries were located at the inlet, outlet and between segments, leading to a multiposition valve which fed gas samples to the analysis train, consisting of nondispersive IR–analyzers for CO and CO_2 and a chromatograph with FID for separation and determination of organics. The feed was prepared by saturating an air stream with o–xylene at controlled temperature. Unattended continuous operation of the unit was possible as all functions were actuated, controlled and recorded by computer. The reactor with its high length/diameter–ratio was treated as an isothermal plug–flow system. The following range of reaction conditions was investigated:

Temperature:	$370^{\circ}C$ to $410^{\circ}C$
Total pressure:	1.6 bar
Mole fraction of o–xylene in feed:	$0.3 \cdot 10^{-2}$ to $0.8 \cdot 10^{-2}$
Mole fraction of O_2 in feed:	0.15 to 0.3
Volumetric flow rate measured:	20 to 240 ml/s ($0^{\circ}C$, 1.013 bar)

Results and Discussion

The rate of individual reactions in the system shown in Fig. 1 can be represented by the rate equation

$$r_{ij} = \frac{k_{ij} \cdot c_i}{1 + b \cdot c_1 / c_{1,0}} \qquad (1)$$

where r_{ij} is in $mol \cdot g^{-1} \cdot s^{-1}$, c_1 and $c_{1,0}$ representing the local and initial concentration of o–xylene, respectively. In order to obtain a set of coefficients k_{ij} and b for a given catalyst the experimental results were first represented by polynomial series. The coefficients in the set of 6 simultaneous differential equations of type (1) were then found by linear regression. Fig. 4 shows as an example composition as function of conversion of o–xylene and of space time as represented by the mathematical model in this way together with experimentally observed points. It is clearly evident from Fig. 4 that parallel and consecutive reactions are occurring simultaneously in this system. The temperature coefficients (activation energies) of different reactions in the network were found to be different, reaction path and selectivity are therefore sensitive to temperature. An increase in temperature can be beneficial with respect to initial differential selectivity of certain catalysts. Addition of Li or Cs to the active component reduces the activity of the catalysts and leads to an increase in differential selectivity at low conversion.

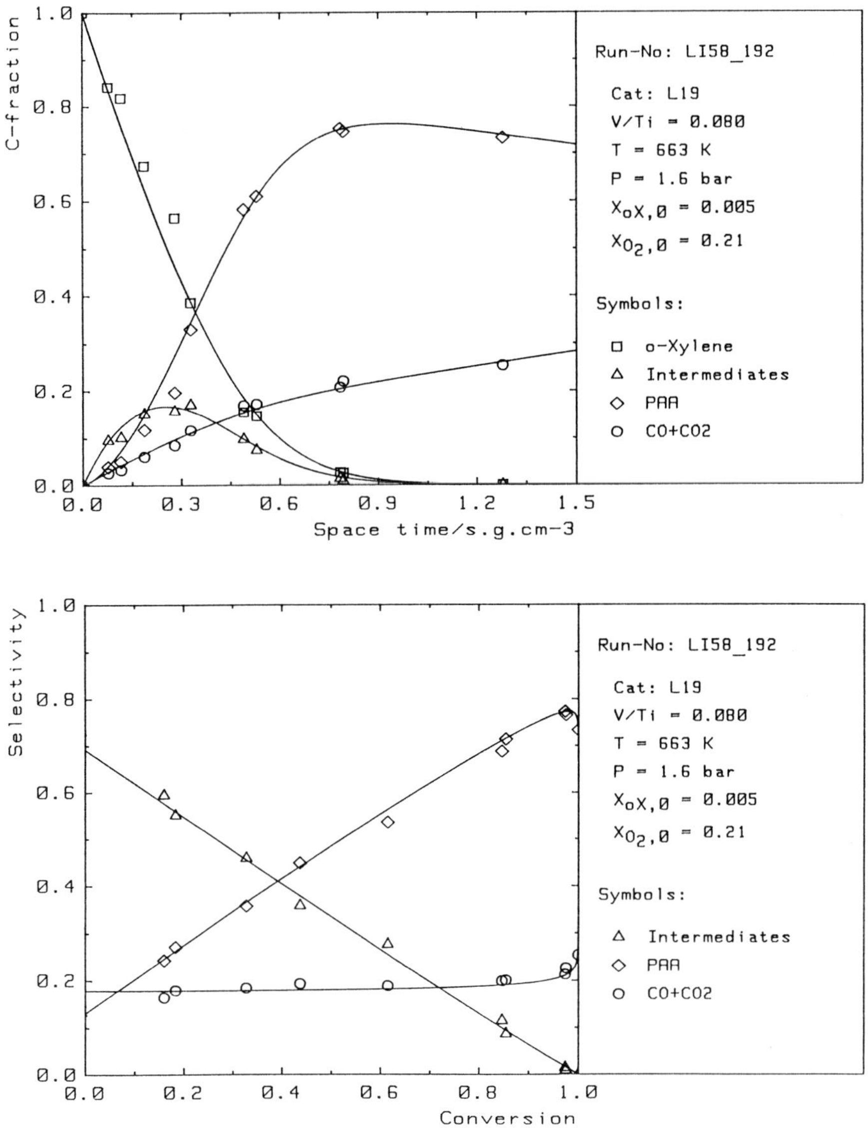

Fig. 4 Product distribution

Evolution results from endless repetition of change by various principles, such as mutation, substitution, selection etc. As an example the principle of mutation in the computerized search–procedure for an optimal sequence of different catalysts on the basis of mathematical models describing the behavior of individual catalysts is depicted in Fig. 5. From a set of μ different sequences of 9 catalysts (numbered 1 to 9) one sequence is chosen at random. In this sequence the position of two catalysts chosen at random is interchanged, thereby a new member in the set is generated which now contains μ+1 members. For all members the attainable maximum yield is computed numerically by *Runge–Kutta* method and considered to be a criterion of quality. The member with the lowest quality is then dropped, so that a new set of μ sequences results. This procedure is repeated until no improvement in the quality of the best member of the set results when the cycle is repeated a certain number of times. Yield of PAA (integral selectivity times conversion of o–xylene) is thus the objective function in the optimization procedure.

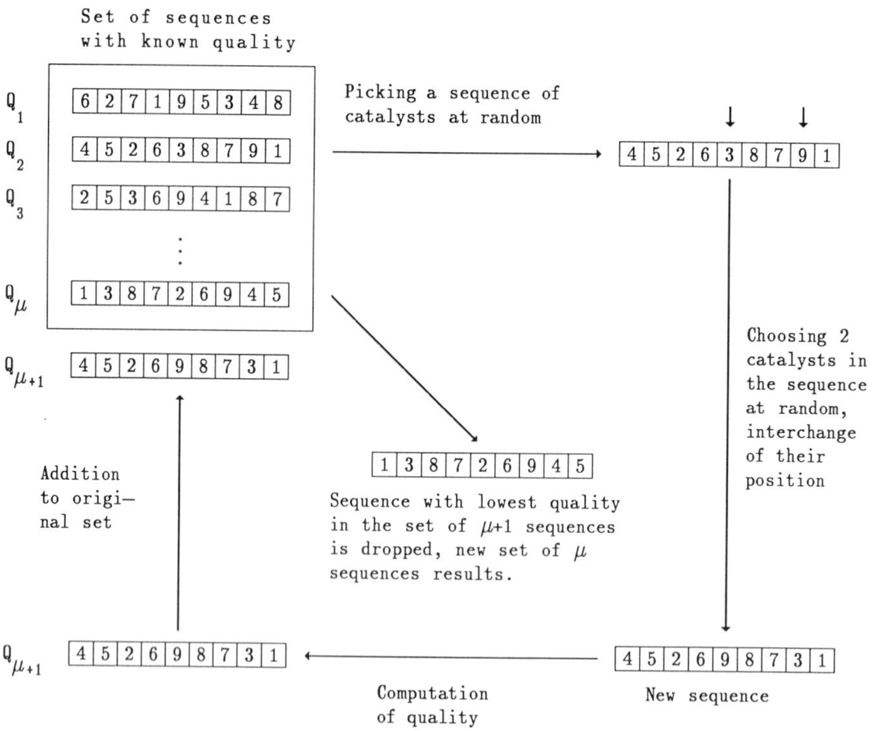

Fig. 5 Mutation

Table 1 shows results obtained in this way for 3 sets of individual catalysts containing each 9 members. The resulting optimal sequence, arrived at after a few hundred cycles of evolution contains only between 2 and 4 different catalysts. In all these optimized sequences the ratio of V/Ti in the catalyst increases from the inlet towards the outlet of the reactor. The computed maximum yield of PAA which can be achieved in a structured

Table 1 Results of the optimization

<table>
<tr><td colspan="2">Catalyst set</td><td>1</td><td>2</td><td>3</td></tr>
<tr><td rowspan="2">Number of catalysts</td><td>Available</td><td>9</td><td>9</td><td>9</td></tr>
<tr><td>Selected</td><td>2</td><td>4</td><td>3</td></tr>
<tr><td colspan="2">Evolution steps</td><td>560</td><td>600</td><td>600</td></tr>
<tr><td colspan="2" rowspan="2">Composition of the structured fixed bed</td><td>Cat %</td><td>Cat %</td><td>Cat %</td></tr>
<tr><td>A01 15.6
A04 84.4</td><td>B10 4.2
B08 6.9
B03 52.8
B05 36.1</td><td>L21 14.9
L18 14.5
L23 70.6</td></tr>
<tr><td colspan="2">Temperature</td><td>683 K</td><td>643 K</td><td>663 K</td></tr>
<tr><td rowspan="3">Attainable maximum yield of PAA</td><td>With 1 cat</td><td>0.643</td><td>0.692</td><td>0.780</td></tr>
<tr><td>SFBR, cal</td><td>0.709</td><td>0.724</td><td>0.794</td></tr>
<tr><td>SFBR, exp</td><td>——</td><td>0.73</td><td>0.79</td></tr>
</table>

Table 2 Composition of the catalysts

<table>
<tr><td rowspan="2">Active component</td><td>Cat V/Ti</td><td>Cat V/Ti</td><td>Cat V/Ti Prom</td></tr>
<tr><td>A01 0.20
A04 0.40</td><td>B10 0.115
B08 0.13
B03 0.20
B05 0.50</td><td>L21 0.02 Cs
L18 0.06 —
L23 0.06 Cs</td></tr>
<tr><td>Support, Form</td><td colspan="2">Cordierite, Monolith</td><td>Steatite, Sphere</td></tr>
<tr><td>Prep. method</td><td colspan="2">1</td><td>2</td></tr>
</table>

fixed bed containing several catalysts lies between 1.4 and 6.6 percentage points above the yield which could be achieved with a single catalyst from the set under consideration. Two such optimal sequences resulting from evolution in the computer were filled into the reactor in order to verify the result experimentally. The observed yields corresponded to expectation, as shown in the last line of table 1.

References

[1] *M. S. Wainwright and N. R. Forster*, Catal. Rev.–Sci. Eng., **19** (1979) 211–292

[2] *R. Y. Saleh and I. E. Wachs*, Appl. Catal., **31** (1987) 87–98

[3] *I. Rechenberg*, Evolutionsstrategie – Optimierung technischer Systeme nach Prinzip der biologischen Evolution, F. Fromman Verlag, Stuttgart (1973)

[4] *G.L. Stebbins*, Evolutionsprozesse, Fischer Verlag, Stuttgart (1968)

[5] *M. Kotter and L. Riekert*, Chem.–Ing.–Tech., **59** (1987) 733–734

Keywords

Fixed bed reactor, oxidation of o–xylene, phthalic anhydride, V/Ti–catalyst

G. Centi and F. Trifiro' (Editors), *New Developments in Selective Oxidation* 275

$YBa_2Cu_3O_6$ - A SELECTIVE AMMOXIDATION CATALYST

J.C. OTAMIRI[1], A. ANDERSSON[1], S. HANSEN[2] and J.-O. BOVIN[2]

[1]Department of Chemical Technology, Chemical Center, University of Lund, P.O. Box 124, S-221 00 Lund (Sweden)

[2]Department of Inorganic Chemistry 2, Chemical Center, University of Lund, P.O. Box 124, S-221 00 Lund (Sweden)

SUMMARY

$YBa_2Cu_3O_{6+x}$, $0 \leq x \leq 1$, was used as a catalyst for the oxidation of toluene in presence of oxygen and ammonia. The partial pressures of reactants were varied. It was observed that when x is above zero, the material is active for total combustion. The activity is highly dependent on the value of x. After reductive treatment of sample having x>0 in reactant stream without molecular oxygen producing $YBa_2Cu_3O_6$, the partial pressure of oxygen was increased from low to high. At low oxygen pressure, the material was active and selective for formation of benzonitrile. A dramatic transition from selective to non-selective region was observed to occur at a distinct pressure, which according to X-ray diffraction analysis is due to incorporation of oxygen species into the lattice. Under selective conditions, $YBa_2Cu_3O_6$ is more active for nitrile formation in comparison with vanadium oxide catalysts. Catalytic behaviours are discussed considering possible surface structures.

INTRODUCTION

Earlier it was discovered that CuO can be used as a catalyst for oxidation of propene to acrolein. However, its usefulness is limited due to the difficulty of maintaining the surface coverage of oxygen in a suitable range. It was demonstrated that when clusters of more than 5 adjacent oxygens are present at the surface, total combustion is predominant. A hypothesis was advanced that for selective reaction to occur, it is necessary to have structurally isolated sites of appropriate metal-oxygen bond strength [1-3]. Considering the recently discovered superconductor material $YBa_2Cu_3O_7$ [4], it is of potential interest for use in catalytic oxidation since it has structurally isolated Cu-layers containing mobile oxygen species, which can easily be abstracted [5]. When they are all removed, $YBa_2Cu_3O_6$ is formed having characteristic layers of Cu^{1+} [6], which hypothetically can adsorb oxygen giving Cu^{3+} and nucleophilic oxygen species. The latter are believed to be involved in selective oxidation and ammoxidation mechanisms [7-9]. Therefore, in order to gain some insight into the catalytic behaviour of these new materials, they are in the present investigation used as catalysts for the ammoxidation of toluene to produce benzonitrile, which is used e.g. in the synthesis of benzoguanamine [10].

METHODS

$YBa_2Cu_3O_{6+x}$, with x equal to 1 and 0, Y_2BaCuO_5 and $SmBa_2Cu_3O_7$ were prepared from stoichiometric mixtures of appropriate and pure, >99 %, chemicals of Y_2O_3, Sm_2O_3, $BaCO_3$ and CuO according to the procedure described elsewhere [11].

Catalytic activity investigations of prepared phases were carried out in a differential and isothermal plug flow reactor made of Pyrex glass and maintained at 400 °C. Reactants, oxygen, ammonia, and toluene, mixed with inert nitrogen, were introduced and controlled by HI-TEC mass flow controllers. Products, benzonitrile, benzaldehyde, CO_2, and CO, were analyzed on a Varian Vista 6000 gas chromatograph. Reactor and analysis setup has earlier been described in detail [9].

Freshly prepared $YBa_2Cu_3O_{6+x}$, with x=1, and reduced sample, with x=0, were studied at a constant high and low partial pressure of oxygen, respectively, while varying the partial pressures of ammonia and toluene, and also at constant pressures of ammonia and toluene, while varying pressure of oxygen.

Powder X-ray data were recorded at room temperature using a Guinier-Hägg camera with quartz monochromator, $CuK\alpha_1$ radiation and Si as internal standard. Accurate lattice constants were obtained by least-squares refinement. The smooth variation of lattice constants with lattice oxygen content observed for $YBa_2Cu_3O_{6+x}$ phases [12] was used to estimate the value of x.

RESULTS

Catalysis on $YBa_2Cu_3O_7$

Figure 1 shows the dependency of rates over freshly charged $YBa_2Cu_3O_7$ as partial pressure of oxygen (P_O) is varied from high region towards low at constant pressures of ammonia (P_A) and toluene (P_T). The rates show partial order dependency, however, it is worthy to note that the rates hardly increase at high pressures of oxygen. At zero pressure of oxygen the rate for nitrile formation decreases slowly with time, whereas other rates rapidly go to zero. The x-value is dependent on the partial pressure of oxygen. After use at P_O = 17.30 kPa, the composition is $YBa_2Cu_3O_{6.4}$. When the partial pressure of oxygen is set to zero, the x-value gradually approaches zero.

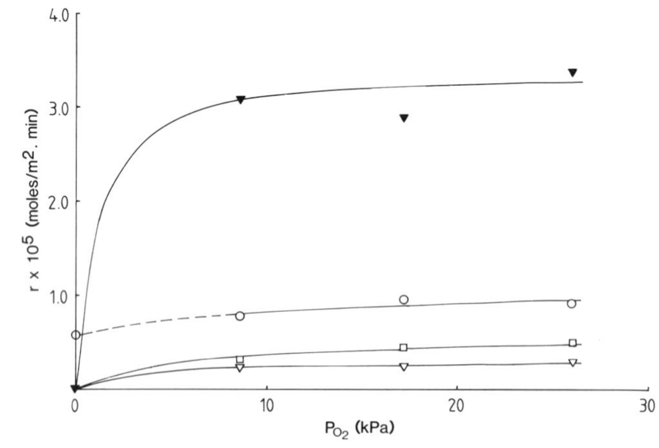

Fig. 1. Rates for formation of nitrile ○, aldehyde □, CO_2 ▼, and CO ▽ over $YBa_2Cu_3O_{6+x}$, x>0, as a function of partial pressure of oxygen. P_A = 2.58 kPa and P_T = 0.77 kPa.

In Fig. 2 is given the dependency of rates on partial pressure of ammonia with pressures of oxygen (high) and toluene maintained constant. The rates for formation of benzaldehyde, CO_2, and CO decline with increase in pressure of ammonia while that of nitrile increases and remains constant at higher pressures.

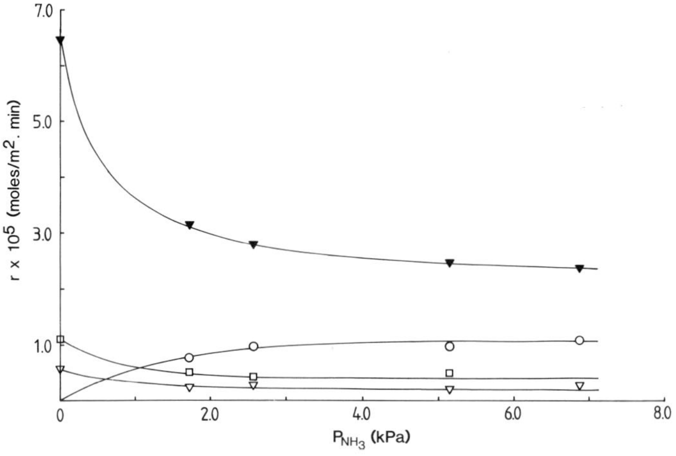

Fig. 2. Rates for formation of products over $YBa_2Cu_3O_{6+x}$, x>0, versus partial pressure of ammonia. P_O = 17.30 kPa and P_T = 0.77 kPa. Notations: cf. Fig. 1.

The variation of rates with partial pressure of toluene, Fig. 3, shows also partial order dependency. In this figure, it could be seen, that the dependencies for benzaldehyde and CO_2 are strong, whereas for nitrile and CO, the rates are virtually constant at high pressures of toluene.

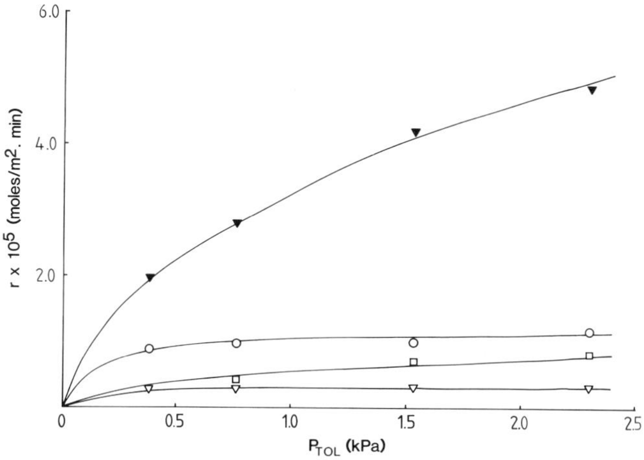

Fig. 3. Influence of partial pressure of toluene on rates for formation of products over $YBa_2Cu_3O_{6+x}$, x>0. P_O = 17.30 kPa and P_A = 2.58 kPa. Notations cf. Fig. 1.

Reductive treatment

Fresh samples, $YBa_2Cu_3O_7$, $SmBa_2Cu_3O_7$ and Y_2BaCuO_5 which is a wellknown contaminant in superconductor materials [13] were heated to reaction temperature (400 °C) in presence of oxygen. Reduction of samples was carried out by performing the experiments in the absence of molecular oxygen for one hour at fixed conditions of temperature (400 °C) and pressures of toluene (0.77 kPa) and ammonia (2.58 kPa). Then, the pressure of oxygen was increased to the level of selective conditions for nitrile formation, and activities were measured as a function of time. The results are given in Table 1. For comparison data are also included for a sample freshly prepared as $YBa_2Cu_3O_6$ and heated to reaction temperature in nitrogen.

TABLE 1

Reaction rates[a] at 400 °C as a function of time-on-stream for various samples after reductive treatment.

Sample	Rate x 10^5 (moles m^{-2} min^{-1})			Time (min)
	Nitrile	CO_2	CO	
$YBa_2Cu_3O_6$	1.97	0.36	0.02	10
	2.05	0.27	0.02	25
	2.14	0.23	0.01	40
$YBa_2Cu_3O_7$	1.65	0.33	0.02	10
	1.61	0.32	0.02	25
	1.59	0.32	0.02	40
$SmBa_2Cu_3O_7$	1.77	0.33	0.02	10
	2.08	0.34	0.03	25
	2.10	0.33	0.03	40
Y_2BaCuO_5	0.27	0.89	0.03	10
	0.42	0.63	0.03	25
	0.51	0.52	0.03	40

[a] P_O = 2.16 kPa, P_A = 2.58 kPa, and P_T = 0.77 kPa.

From the table it could be observed, that the behaviour of reduced $YBa_2Cu_3O_7$, and $SmBa_2Cu_3O_7$ is similar to that of $YBa_2Cu_3O_6$, which is active and selective for toluene ammoxidation under the conditions used in the experiments. The Y_2BaCuO_5 compound is found to be less active and less selective.

Catalysis on $YBa_2Cu_3O_6$

Reduced $YBa_2Cu_3O_7$ sample, with a composition close to $YBa_2Cu_3O_6$, was then used for experiments in which the partial pressures of reactants were varied. The results are given in Figs. 4-6. In series where the partial pressure of ammonia or toluene was varied, the partial pressure of oxygen was kept low.

In Fig. 4 are the rates obtained when the partial pressure of oxygen was varied gradually from low region towards high. This figure shows some features worth noting: i) There is a clear region of sharp transition in selectivity, ii) At low partial pressure of oxygen, the catalyst is selective for nitrile formation, iii) At higher pressures, the activity towards total combustion dramatically increases and is about ten times higher than before reduction, and iv) The passing of the rate for CO formation through a maximum.

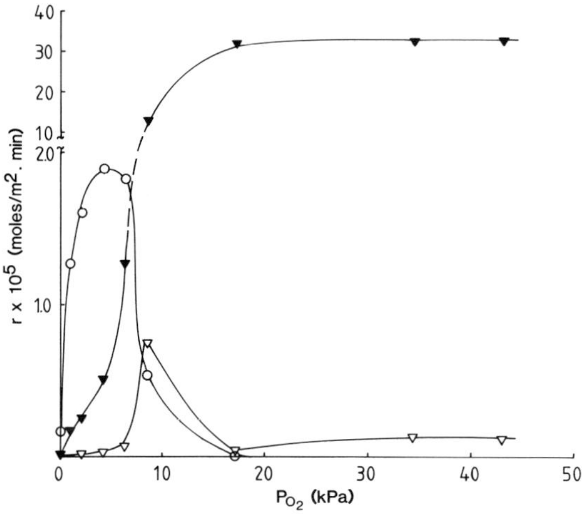

Fig. 4. Effect of partial pressure of oxygen on rates for formation of products over $YBa_2Cu_3O_{6+x}$, x = 0. P_A = 2.58 kPa and P_T = 0.77 kPa. Notations: cf. Fig. 1.

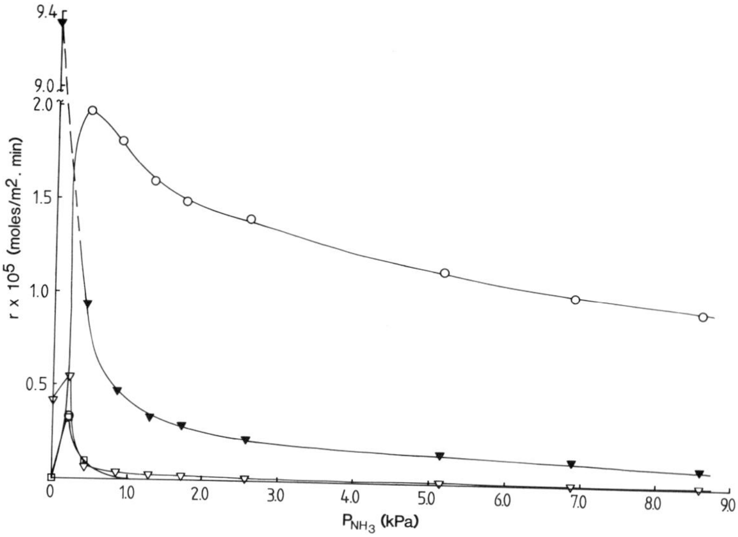

Fig. 5. Rates for formation of products on $YBa_2Cu_3O_{6+x}$, x = 0, versus partial pressure of ammonia. P_O = 2.16 kPa and P_T = 0.77 kPa. Notations: cf. Fig. 1.

Figure 5 (above) shows how the rates vary as the partial pressure of ammonia is varied from high region towards low at fixed partial pressures of oxygen and toluene. The rate for nitrile formation passes through a maximum and is higher than before reduction, cf. Fig. 2. Benzaldehyde is formed at low partial pressure of ammonia but not in its absence. A sharp increase in the rate of CO_2 formation occurs as low partial pressures are approached. The rate for CO formation also increases but declines at zero pressure of ammonia.

The dependency of rates on partial pressure of toluene is given in Fig. 6. There is an almost first order dependency of rates on pressure of toluene. Comparison with Fig. 3 shows that the rates for formation of nitrile and CO_2 have reversed places. The rate for formation of CO_2 before reduction was higher than after reduction, whereas for nitrile formation the opposite is the case. Another feature is the fact that after reduction aldehyde is not formed when the partial pressure of oxygen is maintained low.

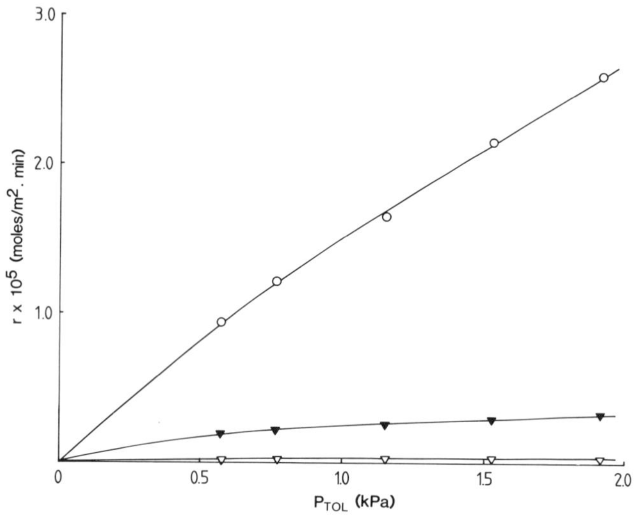

Fig. 6. Influence of partial pressure of toluene on rates for formation of products over $YBa_2Cu_3O_{6+x}$, x = 0. P_O = 2.16 kPa and P_A = 2.58 kPa. Notations: cf. Fig. 1.

In Fig. 7 are reaction rates plotted as a function of reaction time for a $YBa_2Cu_3O_6$ sample which before use had been stored under ambient conditions for 10 days. Initially, the material though active was non-selective. After use for few hours, the rate for CO_2 formation dropped to a very low value, while that for nitrile formation increased more than twice. This behaviour was always observed when using $YBa_2Cu_3O_6$ samples which had been stored in an air atmosphere for several days. It is probably due to removal of some oxygen species which have been incorporated into the lattice during storage.

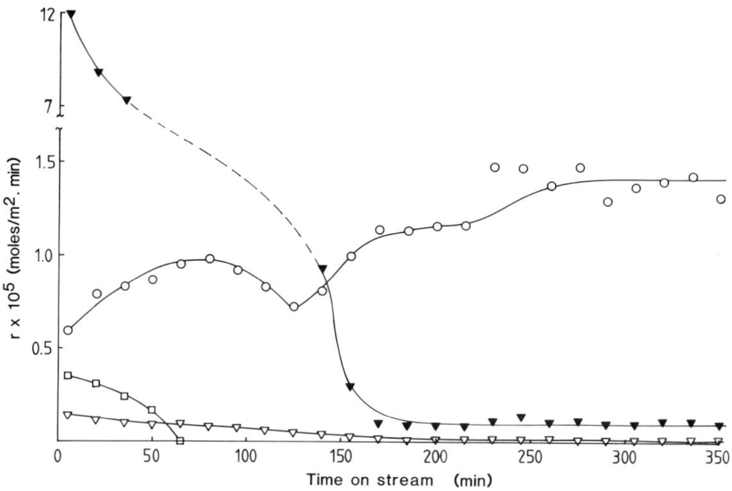

Fig. 7. Reaction rates over $YBa_2Cu_3O_6$ as a function of time. P_O = 2.16 kPa, P_A = 2.58 kPa, and P_T = 0.77 kPa. Notations: cf Fig. 1.

X-ray diffraction analysis

TABLE 2

Lattice constants (Å) and oxygen content (x) of catalysts.

Catalyst	$YBa_2Cu_3O_{6+x}$	$YBa_2Cu_3O_{6+x}$	$SmBa_2Cu_3O_{6+x}$	Y_2BaCuO_5
Fresh sample	orthorhombic a=3.8203(8) b=3.8853(7) c=11.679(2) x=1	tetragonal a=b=3.8582(2) c=11.830(1) x=0	orthorhombic a=3.855(1) b=3.892(1) c=11.712(2) x=1	orthorhombic a=12.177(2) b=5.6571(9) c=7.130(1)
At high P_O, before reductive treatment[a]	tetragonal a=b=3.8581(5) c=11.764(1) x=0.4			
At low P_O, after reductive treatment[b]	tetragonal a=b=3.8572(4) c=11.830(1) x=0	tetragonal a=b=3.8569(2) c=11.834(1) x=0	tetragonal a=b=3.8841(5) c=11.829(2) x=0	orthorhombic a=12.172(1) b=5.6590(5) c=7.1294(7)
At high P_O, after reductive treatment[a]	tetragonal a=b=3.8572(4) c=11.818(2) x=0	tetragonal a=b=3.8571(4) c=11.819(2) x=0		

[a] P_O = 17.30 kPa, P_A = 2.58 kPa, and P_T = 0.77 kPa
[b] P_O = 2.16 kPa, P_A = 2.58 kPa, and P_T = 0.77 kPa.

Lattice constants determined by X-ray diffraction are given in Table 2 for various samples. Also included are x-values as estimated using the published relationship between lattice constants and oxygen content of $YBa_2Cu_3O_{6+x}$ phases [12]. The composition of Sm-substituted samples was estimated by comparing cell parameters and catalytic activity with corresponding values for $YBa_2Cu_3O_{6+x}$ phases. From the table, it can be concluded that the x-value of used $YBa_2Cu_3O_{6+x}$ sample not being subjected to reductive treatment is well above zero. After reductive treatment and further use at low and high partial pressure of oxygen, respectively, the oxygen content of catalysts is close to 6 oxygen atoms/unit cell. However, the c axis repeat of catalysts used at high oxygen pressure, non-selective conditions, was always found to be slightly shorter than that measured after use at low oxygen pressure, selective conditions. This implies that the x-value for catalysts run under non-selective conditions is slightly above zero. The lattice constants determined for Y_2BaCuO_5 are identical for freshly prepared and used samples, and they also agree with those reported in the original structure determination [14].

DISCUSSION

A drawing of the $YBa_2Cu_3O_7$ structure is shown in Fig. 8. There are two structurally different Cu positions, noted Cu(1) and Cu(2) [15]. The formers are connected via O(4), thus, forming chains in the [010] direction between Ba-layers. Cu(1)-chains are connected to Cu(2)-layers by O(1). In the Cu-layers, Cu(2) is coordinated to five oxygen species, 2 x O(2), 2 x O(3) and 1 x O(1). It has been shown that there are no distinct Cu^{2+} and Cu^{3+} sites. The valence of Cu in both sites is intermediate between +2 and +3 [6]. Oxygen O(4) in the chains have been found to be mobile and can be totally abstracted [5]. When this occurs, the structure changes from orthorhombic $YBa_2Cu_3O_7$ to tetragonal $YBa_2Cu_3O_6$. The latter structure can simply be derived from the former by removal of O(4) so that the coordination of Cu(1) is changed from square planar to linear twofold [16]. As a result, distinct Cu^{1+} at Cu(1) sites, and Cu^{2+} at Cu(2) sites are formed [6].

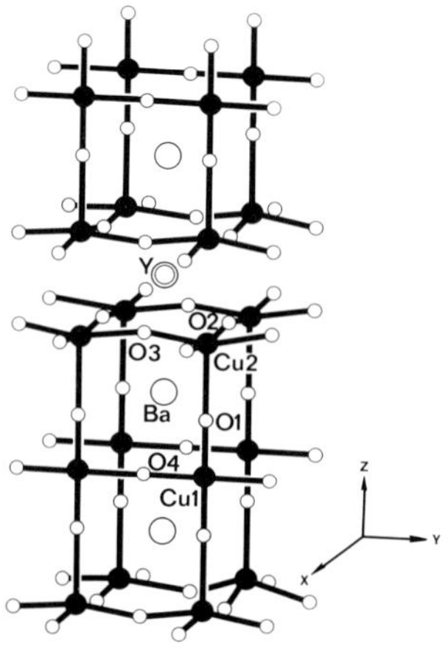

Fig. 8. Drawing of the $YBa_2Cu_3O_7$ structure.

From the fact that the main difference between the structures of the orthorhombic and tetragonal phases is connected to the coordination of Cu(1), it follows that it is reasonable to compare their catalytic behaviours in terms of possible surface coordinations of Cu(1).

At the surface of $YBa_2Cu_3O_7$, undercoordinated Cu(1) and Cu(2) can exist, serving as possible adsorption sites for toluene and ammonia. The number of undercoordinated species depends on the partial pressure of oxygen. Molecular oxygen can adsorb in the form of diatomic species. As a consecutive step, when dissociation is possible, monoatomic oxygen species can also be formed. However, dissociation is probably not facile due to lack of oxygen vacancies in the bulk. A common feature of oxygen species projecting from the surface is that they are undercoordinated, which renders them electrophilic in character. It has been established that electrophilic oxygen participates in the degradation of hydrocarbons leading to total combustion [7,9,17]. Indeed, $YBa_2Cu_3O_{6+x}$, with x well above zero, was found to be non-selective in catalytic (amm)oxidation, cf. Figs. 1-3.

In $YBa_2Cu_3O_6$, Cu(1) is two-coordinated due to that O(4) positions are vacant. After adsorption of molecular oxygen, two options are possible depending on the partial pressure of oxygen. At low pressure of oxygen, adsorbed diatomic oxygen can react with co-adsorbed ammonia to give water under simultaneous oxidation of low valent Cu(1) to Cu^{3+} and formation of nucleophilic Cu=O and Cu=NH species. Substantial evidence exist for nucleophilic oxygen species and imido species to be involved in selective oxidation and ammoxidation mechanisms, respectively [7-9], which is verified by the present investigation. Figures 4-6 show that $YBa_2Cu_3O_6$ is selective for nitrile formation at low partial pressure of oxygen. Furthermore, the finding that the rate for formation of benzaldehyde is zero in absence of ammonia, and passes through a maximum as the partial pressure of ammonia is increased suggests that co-adsorption of ammonia is a prerequisite for formation of nucleophilic oxygen species. On the contrary, when the partial pressure of oxygen is high, the catalyst is non-selective, cf. Fig. 4. This can be seen as a result of the facile dissociation of adsorbed diatomic oxygen at $YBa_2Cu_3O_6$. One of the oxygen species can migrate into a neighbouring oxygen vacancy situated between two Cu(1) sites. Consequently, the remaining monoatomic surface species will have electrophilic character due to that Cu has to share its available valence electrons between both oxygen species.

The rate for formation of CO_2 over $YBa_2Cu_3O_{6+x}$ at high pressure of oxygen depends on the value of x. When the value is small, the rate is much higher compared to when x is high, cf. Figs. 1 and 4. Several explanations are possible for this behaviour, of which a few will be mentioned briefly. One is that the electronic properties of surfaces must be influenced by the occupancy frequency of exterior O(4) positions, cf. Fig. 7, consequently affecting adsorption and reactivity properties. Another factor of importance is that the number of active sites increase when the value of x decrease. In case of $YBa_2Cu_3O_7$, if extending the bulk structure to the surface, Cu(1) at (100) faces cannot adsorb projecting single coordinated oxygen species. When the composition is close to $YBa_2Cu_3O_6$, such an adsorption is possible producing electrophilic oxygen species on the condition that neighbouring O(4) positions are only partly filled.

At low pressure of oxygen, 2.5-5 kPa, the rates for formation of nitrile and CO_2 over $YBa_2Cu_3O_6$ at 400 °C are 16-19 and 2-4 μmole m^{-2} min^{-1}, respectively. Over V_2O_5, under the same conditions, the corresponding rates are 2-4 and 0.3-0.5 μmole m^{-2} min^{-1}, respectively [18,19]. In conclusion, it has been shown that $YBa_2Cu_3O_6$ is an active and selective catalyst for ammoxidation of toluene at low partial pressures of oxygen.

ACKNOWLEDGMENT

Financial support from the National Swedish Board for Technical Development (STU) and the Swedish Natural Science Research Council (NFR) is gratefully acknowledged.

REFERENCES

1 J.L. Callahan and R.K. Grasselli, *AIChE J.*, 9 (1963) 755.
2 R.K. Grasselli and J.D. Burrington, in D.D. Eley, H. Pines and P.B. Weisz (Eds.), *Advances in Catalysis*, Vol. 30, Academic Press, New York, 1981, pp. 133-163.
3 F. Cavani, G. Centi, F. Trifiro and R.K. Grasselli, *Catal. Today*, 3 (1988) 185.
4 M.K. Wu, J.R. Ashburn, C.J. Torng, P.H. Hor, R.L. Meng, L. Gao, Z.J. Huang, Y.Q Wang and C.W. Chu, *Phys. Rev. Lett.*, 58 (1987) 908-911.
5 A. Manthiram, J.S. Swinnea, Z.T. Sui, H. Steinfink and J.B. Goodenough, *J. Am. Chem. Soc.*, 109 (1987) 6667-6669.
6 M. O'Keeffe and S. Hansen, *J. Am. Chem. Soc.*, 110 (1988) 1506-1510.
7 J. Haber, in J.P. Bonnelle, B. Delmon and E. Derouane (Eds.), *Surface Properties and Catalysis by Non-Metals*, Reidel, Dordrecht, 1983, Ch. 1, pp. 1-45.
8 R.K. Grasselli, J.F. Brazdil, and J.D. Burrington, *Proc. 8th Int. Congr. Catalysis, Berlin(West), July 2-6, 1984*, Verlag Chemie, Weinheim, 1984, Vol. V, pp. 369-380.
9 A. Andersson and S. Hansen, *J. Catal.*, 114 (1988) 332-346.
10 *Kirk-Othmer, Encyclopedia of Chemical Technology*, 3rd edn., Vol. 15, Wiley, New York, 1981, p. 906.
11 S. Hansen, J. Otamiri, J.-O. Bovin and A. Andersson, *Nature*, 334 (1988) 143-145.
12 C.N.R. Rao, *J. Solid State Chem.*, 74 (1988) 147-162.
13 H. Steinfink, J.S. Swinnea, Z.T. Sui, H.M. Hsu and J.B. Goodenough, *J. Am. Chem. Soc.*, 109 (1987) 3348-3353.
14 C. Michel and B. Raveau, *J. Solid State Chem.*, 43 (1982) 73-80.
15 F. Beech, S. Miraglia, A. Santoro and R.S. Roth, *Phys. Rev.*, B35 (1987) 8778-8781.
16 J.S. Swinnea and H. Steinfink, *J. Mater. Res.*, 2 (1987) 424-426.
17 A.M. Gasymov, V.A. Shvets and V.B. Kazansky, *Kinet. Katal.*, 23 (1982) 951-954.
18 J.C. Otamiri and A. Andersson, *Catal. Today*, 3 (1988) 211-222.
19 J.C. Otamiri and A. Andersson, *Catal. Today*, 3 (1988) 223-234.

B. DELMON (Univ. Catholique de Louvain, Belgium): Due to the fact that catalyst surfaces are usually reduced in their steady state during catalytic oxidation, it might seem doubtful that copper remains in the Cu^{1+} or Cu^{2+} oxidation state, with no Cu^0 and, consequently, Cu crystallites being formed. The absence of new lines in X-ray diffraction cannot be a fully convincing proof since small crystallites might not be detectable.

Did you find a change in the intensity ratio of Cu/Y or Cu/Ba XPS lines, or changes of ISS signals after use of the catalyst?

If really no Cu^0 were formed, this would indicate a really exceptional strength of the chemical bonds involving Cu. In ammoxidation, ammonia is a very strong reducing agent, even in the presence of O_2. If this is so, this could give a clue to the very special electronic structure of superconductors.

A. ANDERSSON (University of Lund, Sweden): For both freshly prepared samples and used samples only X-ray diffraction lines belonging to $YBa_2Cu_3O_{6+x}$ could be detected. Use in catalytic reaction did not cause any change in the intensity of the X-ray lines that cannot be explained as due to change in oxygen content. Also, XPS analysis did not show formation of Cu^0. However, the ratio of Cu/Y and Cu/Ba XPS lines showed some dependence on reaction conditions. In this regard, it should be noted that $YBa_2Cu_3O_{6+x}$ faces can expose both Cu-, Y-, and Ba-layers and that their distribution possibly depends on the composition of the reactant stream.

O.V. KRYLOV (Acad. of Sciences, Moscow, USSR): In connection with an interesting observation of Dr. Andersson and his collaborators I should like to comment about many similarities between high temperature semiconductors and oxide catalysts of partial oxidation. Both of them have oxygen-deficient lattice. In the case of high temperature semiconductors, oxygen vacancies in the lattice must be stable and only motion of electron pairs must be observed. On the contrary, in oxide catalysts of partial oxidation such vacancies must move. It is very possible now to search new high temperature semiconductors from oxidative catalysis.

A. ANDERSSON: Thank you for your comment, we believe that such an approach may yield fruitful results.

M. MISONO (The University of Tokyo, Japan): Very interesting results. I would like to know more about the chemical reactivity and the composition of the surface of $YBa_2Cu_3O_{6+x}$.

Is it stable at high temperatures against CO_2, H_2O, etc.? Is the surface composition the same as in the bulk? Segregation of certain elements (Ba, etc.) has often been indicated in the reported papers of electric conductivity.

A. ANDERSSON: Our XPS results, that will be published elsewhere, clearly show the existence of carbonate species both in freshly prepared samples and in used samples. In fresh samples, the amount is highly dependent on the preparation method used. After use in catalytic reaction, only a minor variation of the amount of carbonate species in comparison with fresh samples was observed.

Examination of the catalyst before and after use in the reactor by high-resolution transmission electron microscopy, reveals an increase in the number of crystal structure defects on the (001) plane (ref. 1). Such defects are possibly formed under the influence of H_2O (ref. 2). However, once a steady state has been reached, no change with time in the formation of products was detected for the period it was used, which was up to 3 days.

1 S. Hansen, J. Otamiri, J.-O. Bovin and A. Andersson, Nature, 334 (1988) 143.
2 B.G. Hyde et al., Nature, 327 (1987) 402.

G.M. PAJONK (Univ. Claude Bernard Lyon I, France): I would like to know if your catalyst is stable with time on stream.

Due to the mobility of oxygen inside the structure of your high T_c superconductor, why did you not try to oxidize, e.g., propylene which could have been less complex to interpret with respect to the reaction mechanism?

A. ANDERSSON: Referring to the answers given to professors Delmon and Misono, some structural changes were observed as a result of the catalytic reaction. Once a steady state was reached, the performance of the catalyst was stable as long as it was used (up to 3 days).

In comparison with toluene oxidation, we do not think that the mechanism of propylene oxidation is less complex.

G. Centi and F. Trifiro' (Editors), *New Developments in Selective Oxidation* 287

CATALYTIC PROPERTIES OF THE HIGH - TEMPERATURE SUPERCONDUCTOR Y-Ba-Cu-Ag-O TOWARDS THE OXIDATION OF METHANOL

D. KLISSURSKI[1], J. PESHEVA[1], Y. DIMITRIEV[2], N. ABADJIEVA[1] and L. MINCHEV[3]

[1]Institute of General and Inorganic Chemistry, Bulgarian Academy of Sciences, Sofia 1040 (Bulgaria)

[2]Higher Institute of Chemical Technology, Sofia 1756 (Bulgaria)

[3]Institute of Kinetic and Catalysis, Bulgarian Academy of Sciences, Sofia 1040 (Bulgaria)

SUMMARY

The behaviour of oxygen in superconducting ceramic materials of the systems Y-Ba-Cu-O and Y-Ba-Cu-Ag-O and their catalytic activity and selectivity with respect to the oxidation of methanol have been studied simultaneously. The catalytic properties of the high-temperature superconductors are compared with their structure and phase composition and the reactivity of surface and bulk oxygen. It has been shown that at least two different forms of oxygen are presented in Y-Ba-Cu-Ag-O catalysts. It is found that this class of compounds catalyzes mainly complete oxidation. On the contrary the Y-Ba-Cu-O catalyst is selective towards the oxidation of methanol to formaldehyde. Comparative studies of the two classes of compounds have shown that the selectivity with respect to mild oxidation of methanol depends strongly on the structure and phase purity of the superconducting materials.

INTRODUCTION

Extensive studies of the physico-chemical properties of various compositions of the Y-Ba-Cu-O system have been carried out (refs. 1, 2) in association with the high-temperature superconductivity (above 90 K) established for the $YBa_2Cu_3O_{7-x}$ (1:2:3) phase. Since the oxygen content in this phase can vary within definite limits ($6 < x < 7$) and depends strongly on the additional treatment (ref. 3), it can be assumed that materials of this kind would be of both practical and scientific interest.

The investigations carried out up to now show that the structure and the electric properties of the compound $YBa_2Cu_3O_{7-x}$ strongly depend on the values of "x" (ref. 2). The changes in the transition temperature T_c are ascribed to the transformation of the crystalline phase from an ortho-rhombic to a tetragonal structure. It is shown that T_c decreases monotonically when the values of "x" change in the range 7-6,4, which is accompanied by destruction of the ortho-rhombic structure (ref. 3). It is evident that the changes in oxygen content of the $YBa_2Cu_3O_{7-x}$ phase and the treatment of the materials are the main factors affecting the superconducting properties.On the other hand, it is known that a number of physicochemical properties of the complex oxide systems and their catalytic activity during oxidation processes depend strongly on their stoichiometry. This provoked a parallel study of the behaviour of oxygen in supercon - ducting ceramic materials of the systems Y-Ba-Cu-O and Y-Ba-Cu-Ag-O (refs. 4-6) and the catalytic activity and selectivity of these systems. The test reaction used was oxidation of methanol.

METHODS

Samples of the system Y-Ba-Cu-O were synthesized by a classical ceramic technology. The mixture consisting of Y_2O_3, $BaCO_3$ and CuO with a Y:Ba:Cu ratio of 1:2:3 was baked up to 930°C for 12h. Samples of the system Y-Ba-Cu-Ag-O were prepared from the same initial materials with 1-10 wt% Ag_2O (ref. 7). X-ray phase analysis was performed with a DRON 2M diffractometer. The specific surface areas were determined by the BET method using krypton adsorption. The mobility of oxygen in the samples with the composition Y-Ba-Cu-O and Y-Ba-Cu-Ag-O (5 wt% Ag_2O) was estimated by quantitative determination of unstable oxygen. For that purpose, a direct thermal desorption method developed by one of the authors (ref.8) was used for studying the stoichiometric deviations in oxide catalysts and other oxide materials.

Temperature-programmed desorption of oxygen from the same compositions was carried out at 25-800°C and a heating rate of 25°C/min. A flow apparatus and a He flow (99,9%) with a rate of 60 ml/min were utilized.

The photoelectron spectra of the samples Y-Ba-Cu-O and Y-Ba-Cu-Ag-O (5 wt% Ag_2O) were registrated with an ESCALAB-1 apparatus at room temperature. The spectra were calibrated along the C1s (284,8eV)line.

The main kinetic parameters of methanol oxidation were investigated on the compositions Y-Ba-Cu-O and Y-Ba-Cu-Ag-O (5 wt% Ag_2O) and a flow-circulation apparatus was used (refs. 9-11). The measurments were made within the temperature range 300-425°C at methanol con - centrations of 3,36-3,40 vol% and catalyst grains of 0,3 - 0,6 mm. The rate of the initial methanol-air mixture varied between 6,0 - 8,0 ml/m^2 s.

RESULTS AND DISCUSSION

The X-ray analysis of sample Y-Ba-Cu-O shows that the product obtained has an ortho-rhombic structure with characteristic "d" values (Å) ranging from 20-60° (d = 1,58; 1,94; 2,33; 2,72; 2,74; 3,89 Å). It was established that the sample Y-Ba-Cu-Ag-O (5 wt% Ag_2O) is polyphase and contains, in addition to the main 1:2:3 phase, a certain amount of CuO and a 2:1:1 phase (see Fig. 1). Earlier studies on the microstructure showed silver to appear mainly at the crystallite interfaces as metal inclusions or a copper-containing alloy (ref. 7). It was shown that part of the silver might be dissolved in the superconducting (1:2:3) phase (0,23 - 1,15 wt%) (ref. 7). The crystal size of Ag varies in a relatively large range: 5-20 μ (ref. 7). No change in the silver crystal size in the sample after the catalytic test was found by the electron-micro analysis (Philips SEM 505 EDEX). The following superconductivity parameters werefound for the same composition: $T_c(0,5R_n)$=90,7 K; $T_{c\ end}$= 87,0 K; ΔT_c= 3,0 K.

The main results from a study on the catalytic properties of the composition Y-Ba-Cu-O are presented in Fig. 2. Formaldehyde was the main reaction product in the temperature range 310-360°C.

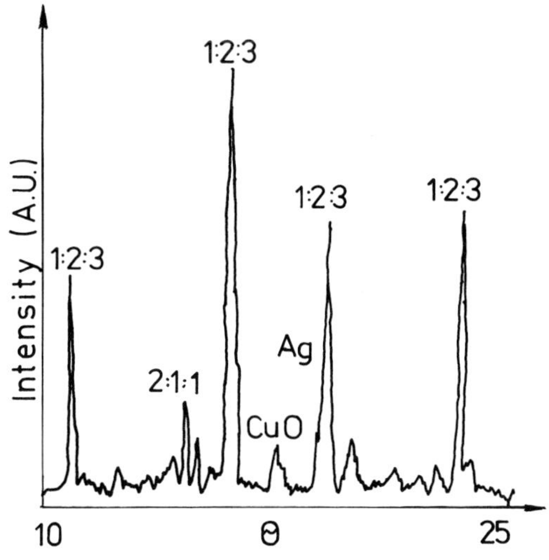

Fig. 1. X-ray Diffraction of Y-Ba-Cu-Ag-O (5 wt% Ag_2O).

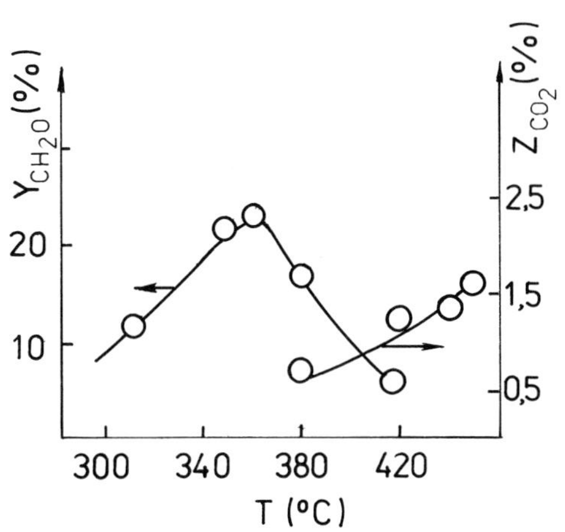

Fig. 2. Temperature dependence of the conversion of methanol to formaldehyde and CO_2 on a Y-Ba-Cu-O catalyst.

The results on the catalytic properties of a sample of the Y-Ba-Cu-Ag-O system are presented in Fig. 3. It was established that for the Ag-containing composition the main reaction product was CO_2. Carbon monoxide and hydrogen were not found under the experimental conditions used. The yield of the formaldehyde over the whole temperature range was relatively low (6,4-7,9%). This was also observed with investigations of the catalytic properties of some oxides- Co_3O, NiO, Mn_2O_3 during the oxidation of methanol (ref. 12).The selectivity towards the complete oxidation depending on temperature ranged from 41-91%, whereas the total conversion degree of methanol reached 86%.

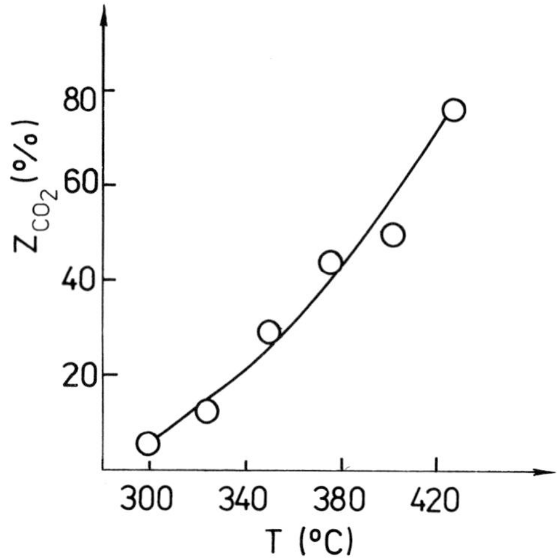

Fig. 3. Temperature dependence of the conversion of methanol to CO_2 on a Y-Ba-Cu-Ag-O catalyst.

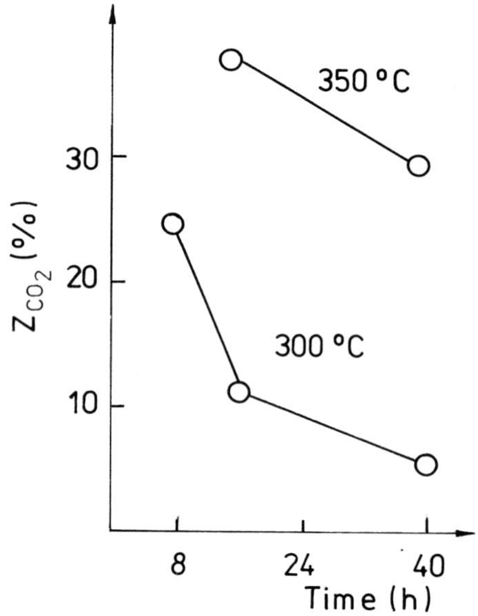

Fig. 4. Dependence on time of the conversion degree of methanol to CO_2 on a Y-Ba-Cu-Ag-O catalyst at 300 and 350°C.

The parameters of the catalytic process (conversion degrees of methanol to formaldehyde and carbon dioxide,%) were determined after establishing a regime corresponding to the steady state of the catalyst. Fig. 4 shows the dependence of the conversion degree of methanol to carbon dioxide on the time of contact of the sample with the methanol-air mixture at two different temperatures. At 300°C the amount of formed dioxide decreases two times for the first eight hours. With a rise of temperature,up to 400°C and subsequent cooling to 300°C, after the 40th hour the amount of carbon dioxide in the reaction mixture remains constant. The same trend to a decrease in activity of the sample was observed at 350°C. These results permit the assumption that during the catalytic process a steady state composition of the sample is attained, which significantly differs from the composition of the fresh (initial) samples. Thedecrease of the catalytic activity could be attributed to the gradual evolution of unstable (weakly bound) oxygen.The specific surface areas of the fresh and used Y-Ba-Cu-Ag-O catalyst are in the range 0,42 - 0,5 m^2/g.The similarity of these values shows that there is not a process of sintering during the catalytic test.

Fig. 5 shows the temperature dependence of the amounts of evolved and uptaken oxygen for fresh and used Ag-containing catalysts. Measurable amounts of oxygen are evolved from the fresh sample already at 380°C, while the total amount of evolved oxygen at 20 - 700°C is about

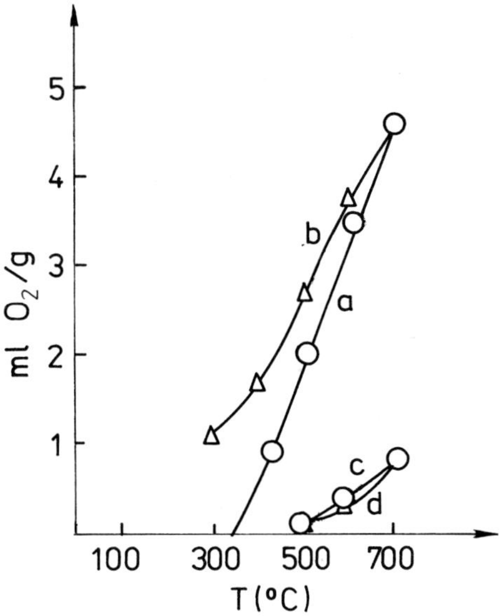

Fig. 5. Temperature dependence of the amounts of evolved (a,c) and uptaken (b,d) oxygen for fresh (a,b) and used (c,d) Y-Ba-Cu-Ag-O catalysts.

4,6 ml/g (curve a). Obviously, the total amount of oxygen cannot be uptaken for the time of the experiments.(curve b). These results are in agreement with the studies of Tripathi et al performed by other methods (ref.13). Curves c and d show the thermal desorption and subsequent adsorption of oxygen for the used catalyst. Oxygen desorption is observed only at 500°C, the total amount of evolved oxygen being considerably smaller. The drop of temperature results in adsorption of practically the whole amount of desorbed oxygen (curve d).

Fig. 6 presents the TPD curves of the samples Y-Ba-Cu-O and Y-Ba-Cu-Ag-O. Evidently, the oxygen chemisorbed on the catalyst surface of the sample Y-Ba-Cu-Ag-O exists in at least two forms, to which desorption maxima at 530 and 740°C correspond. The TPD curve of the used catalyst indicates practically no oxygen desorption at 400-600°C, and the evolved amount of oxygen is two times lower. The TPD curve of a fresh Y-Ba-Cu-O sample has a different shape. Only one desorption peak at 600°C is observed.

The chemical analysis of the surface of the investigated catalysts confirms the TPD results. The photoelectron spectra of a Y-Ba-Cu-O sample contain single peaks corresponding to the binding energies of the 3d electrons of Y (156,4 eV) and Ba (779,8 eV) and 2p electrons of Cu (933,9 eV). The spectra of a Y-Ba-Cu-Ag-O (5 wt% Ag_2O) sample exhibit double peaks with the following 3d electron binding energies: Y - E_B = 155,7; 157,4 eV; Ba - E_B = 779,7; 782,6 eV; Ag - E_B = 367,8; 369,9 eV. It can be assumed that on the surface of the sample with 5 wt% Ag_2O the oxygen is bonded to the separate elements in two different ways which probably correspond to two adsorption forms of oxygen.

The presence of two adsorption forms of oxygen is usually observed with simple and complex oxides (Cr_2O_3, MnO_2, Co_3O_4) in the presence of which deep oxidation of organic substances is achieved (ref. 14).

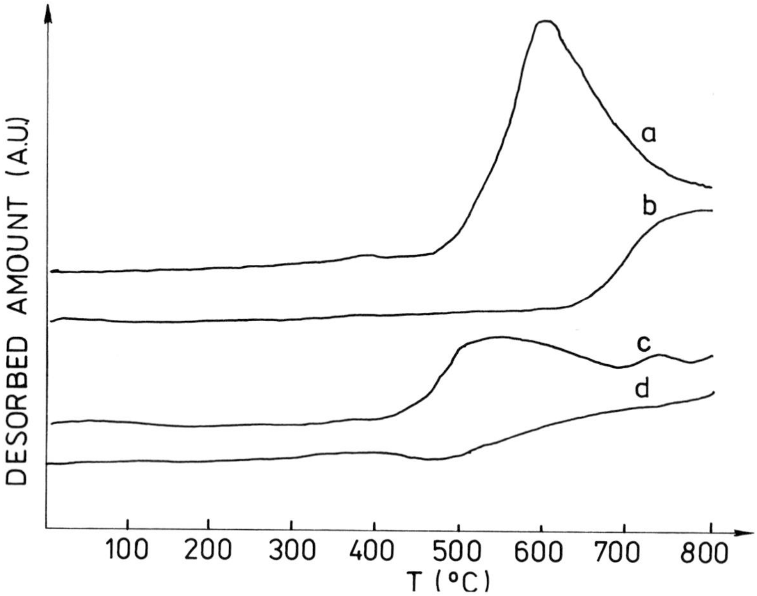

Fig. 6. TPD curves of oxygen from Y-Ba-Cu-O (a- fresh, b- used) and Y-Ba-Cu-Ag-O (c- fresh, d- used) catalysts.

The low- temperature form of oxygen adsorption is defined as "weak bonded". It has values close to these of the bond energy of chemisorbed surface oxygen possessing a high reactivity.

It can be assumed that during the catalytic process, weakly bound oxygen forms from the Y-Ba-Cu-Ag-O surface are the first to react with substance being oxidized with rising temperature. This determines a high initial catalytic activity of the fresh catalyst with respect to the deep oxidation of methanol at relatively low temperatures. According to current concepts (ref. 15) complete, i.e. destructive oxidation of methanol to CO and CO_2 prevails at low bond energies of surface oxygen.

The selectivity with respect to the partial oxidation of methanol undoubtedly depends strongly on the phase purity of the superconducting materials. In the case under consideration the presence of free Cu(II) oxide and Ag favours the destructive oxidation of methanol.

The high- temperature superconductors are of great interest as catalysts for oxidation processes. Extensive studies on their structure and physical properties allow looking for new correlations between these parameters and the catalytic activity and selectivity. The conclusions on the catalysts selectivity towards partial oxidation presuppose a very precise study of their phase composition, structure and the behaviour of the oxygen in them.

REFERENCES

1 M.K. Wu, J.R. Ashburn, C.J. Torng, P.H. Hor, R.L. Meng, L. Gao, Z.J. Huang, Y.Q. Wang and C.W. Chu, Superconductivity at 93 K in a new mixed- phase Y-Ba-Cu-O compound system at ambient pressure, Phys. Rev. Lett., 58 (9) (1987) 908 - 911.

2 R.J. Cava, B. Batlogg, R.B. van Dover, D.W. Murphy, S. Sunshine, T. Siegrist, J.P. Remeika, E.A. Reitman, S. Zahurak and G. Espinosa, Bulk superconductivity at 91 K in single-phase oxygen-deficient perovskite $Ba_2YCu_3O_{9-x}$, Phys. Rev. Lett.,58 (16) (1987) 1676-1679.

3 W.E. Farneth, R.K. Bordia, E.M. McCarron III, M.K. Crawford and R.B. Flippen, Influence of oxygen stoichiometry on the structure and superconducting transition temperature of $YBa_2Cu_3O_x$, Solid State Commun., 66 (9) (1988) 953 - 959.

4 E. Gattev, E. Vlakhov, V. Kovachev, S. Djambasov, S. Tinchev and M. Taslacov, Anomalous superconductivity in the system Y-Ba-Cu-Ag-O, High Temperature Superconductivity, in: R.M. Metzger (Ed), Proc. Int. Conf. High Temperature Superconductivity, Tuscaloosa, USA, April 11-13, 1988, Gordon and Breach Science Publishers, N.Y.,p. 141.

5 V. Kovachev, E. Vlakhov, K.A. Nenkov, V.A. Zovchinov, D.P. Lepkova and S. Djambasov, Superconductivity of Y-Ba-Cu-Ag-O system, in: R.G. Scurlock and C.A. Bailey (Ed), Proc. ICEC,12, Southampton, UK, July 12 - 15, 1988, pp. 1026 - 1029.

6 Pat. Bulg. 80263, 1987.

7 Y. Dimitriev, B. Samuneva, Y. Pirov, E. Gattev, Y. Ivanova, V. Dimitrov, E. Kashchieva, V. Kovachev and E. Vlakhov, Phase-formation and superconductivity in the Y-Ba-Cu-Ag-O system, in: R.G. Scurlock and C.A. Bailey (Ed), Proc. ICEC,12, Southampton, UK, July 12 - 15, 1988, pp. 982 - 986.

8 D.G. Klissurski, A new method of determination of non-stoichiometric oxygen in oxide catalysts, in: Proc. 8th Int. Congr. on Catalysis, vol. III, Berlin (West), July 2-6, 1984, Verlag Chemie, Weinheim, 1984, pp. III-165 - 174.

9 V.N. Bibin, B.I. Popov, Formaldehyde oxidation on iron-molybdenum catalyst, Kinet. Catal., 9 (3) (1968) 618 - 622.

10 V.N.Bibin, B.I. Popov, Kinetic of methanol oxidation in air on iron- molybdenum oxide catalyst, Kinet. Catal., 10 (6) (1969) 1326 - 1335.

11 G. Bliznakov, M. Marinov, D. Klissurski, V. Kozhukharov, J. Pesheva, Oxidation of methanol to formaldehyde on V_2O_5 - TeO_2 catalysts, Commun. Chemystry, Bulg. Acad. Sci.,15 (3) (1982) 261 - 266.

12 G.K. Boreskov, B.I. Popov, V.N. Bibin, E.S. Kosishnikova, Catalytic properties of the IVth periode oxides in methanol oxidation, Kinet. Catal., 9 (1968) 796 - 803.

13 R.B. Tripathi, R.K. Kotnala, S.M. Khullar, B.S. Khurana, Satbir Singh, K. Jain, B.V. Reddi, R.C. Goel, K.C. Nagpal, S. Singal and B.K. Das, Oxidation studies of Y-Ba-Cu superconducting oxides, Solid State Commun., 68 (3) (1988) 319 - 322.

14 D. Klissurski, A. Licourghiotis, N. Abadjieva, L. Guyrova, Studies of stoichiometric deviations in -Cr_2O_3 with different dispersities,in:Proc.Int.Symp. of Solid State Chem., Carlovy Varie, CSSR, October 27-30, 1986, pp. 145 - 149.

15 D. Klissurski, Regularities in the selection of oxide catalysts for reactions of the type: methanol oxidation to formaldehyde, in:Proc. IV Int. Congr. on Catalysis, Moscow, USSR, 1968, Academiai Kiado, Budapest, 1971, vol. I, pp. 477 - 488.

E.MICHELI (Snamprogetti, Milano, Italy): Do you correlate the different selectivity of the two classes of compounds with their structure ?

J.PESHEVA (Institute of General and Inorganic Chemistry, Bulg. Acad.Sci., Sofia, Bulgaria): Obviously, the different chemical and phase composition as well as the surface properties determine the different catalytic behaviours of these types of compounds. The both initial X-ray diffraction patterns show the presence of an orthorhombic structure and a transition to a tetragonal structure appears after the catalytic test. On the other hand, the X-ray analysis show that Y-Ba-Cu-Ag-O samples are multiphase and contain in addition to the main 1:2:3 phase, some amount of CuO and a 2:1:1 phase. This can be related with the lower selectivity of these materials with respect to the partial oxidation of methanol.

E.MICHELI : How have you determined the amounts of evolved and uptaken oxygen by increasing and subsequantly drop of the temperature ?

J.PESHEVA : This is a new direct and sensitive thermodesorption method for determination of non- stoichiometric oxygen in oxide catalysts. The method is developed by Klissurski D. and is reported in the references of the paper (ref. 8).

G. Centi and F. Trifiro' (Editors), *New Developments in Selective Oxidation*

SELECTIVE OXIDATION OF PROPYLENE OVER RARE EARTH-MOLYBDATE CATALYSTS

J.M. LOPEZ NIETO, R. BIELSA*, G. KREMENIC'[1] and J.L.G. FIERRO

Instituto de Catálisis y Petroleoquímica, C.S.I.C., Serrano 119, 28006 Madrid (Spain)
*On leave from INTEC-CONICET, 3000 Santa Fe (Argentina)

ABSTRACT

Catalytic activity for the selective oxidation of propylene over Mo-RE-O (RE=Pr,Sm,Tb,Yb) catalyst series, with Mo/(Mo+RE) atomic ratios ranging between 0 and 1, has been studied. For all catalyst series, both activity and selectivity to partial oxidation products exhibited a relative maximum in the Mo-rich compositions region. These data are interpreted in terms of surface and bulk characteristic of the catalysts as revealed by X-ray diffraction, temperature-programmed reduction, laser-Raman and X-ray photoelectron spectroscopic methods.

INTRODUCTION

Molybdenum-based catalysts are commonly used in many industrial processes which involve selective oxidation of olefins **[1]**. Rare earth (RE) oxides catalyse a great variety of reactions and promote the partial oxidation of light hydrocarbons **[2]**. With the only exception of Ce-containing catalysts **[3]**, the role of rare earth oxide on the selective oxidation process is not well understood **[4]**. Recent studies carried out in our laboratory **[5-8]** revealed that catalytic behaviour markedly depends on the composition and type of phases present in the catalysts. This work is part of a broad study to investigate the effects of the rare earth promoters on the structure and reactivity of Mo-based catalysts. For this purpose, the information revealed by several bulk and surface sensitive techniques is compared with activity and selectivity of the binary Mo-RE-O preparations.

EXPERIMENTAL

Catalyst preparation. The catalysts were prepared by mixing ammonium heptamolybdate and/or RE nitrate solutions of selected

---

[1] Deceased

concentration and volume to obtain fixed Mo/(Mo+RE) ratios. The solutions were evaporated until dryness and then the remaining solids calcined in a forced flow of air at 823 K for 14 h [5].

Catalyst testing. Details of the experimental technique used for catalytic activity experiments have been given elsewhere [5-8]. In short, 1.0 g-samples (particle size between 0.42 and 0.59 mm) were mixed with SiC (in a volume ratio, catalyst:SiC= 1:4). The molar ratio of the components in the reactant mixture was $C_3H_6:O_2:He:H_2O$ = 20:30:30:20 and the contact time W/F= 30-90 $g.h(mol\ C_3H_6)^{-1}$. Experiments were carried out over the temperature range 623-723 K, at atmospheric pressure. The efluents of the reactor were analyzed by gas chromatography.

Catalyst characterization. Specific surface areas of catalysts were calculated by the BET method from the Kr adsorption isotherms at 77 K. X-ray diffraction (XRD) patterns were obtained using a Phillips PW-1100 diffractometer using Ni-filtered CuK_α radiation (λ = 0.15406 nm). Raman spectra (LRS) were recorded using a Jarrell-Ash 25-300 spectrometer equipped with halographic gratings. X-ray photoelectron spectra (XPS) were acquired with a Leybold Heraeus LHS10 electron spectrometer equipped with a magnesium anode (MgK_α = 1253.6 eV) and a hemispherical electron analyzer. The binding energies were referenced to the C1s line at 284.6 eV. Details of all these techniques are given elsewhere [5-8]. TPR experiments were made in a Cahn microbalance.

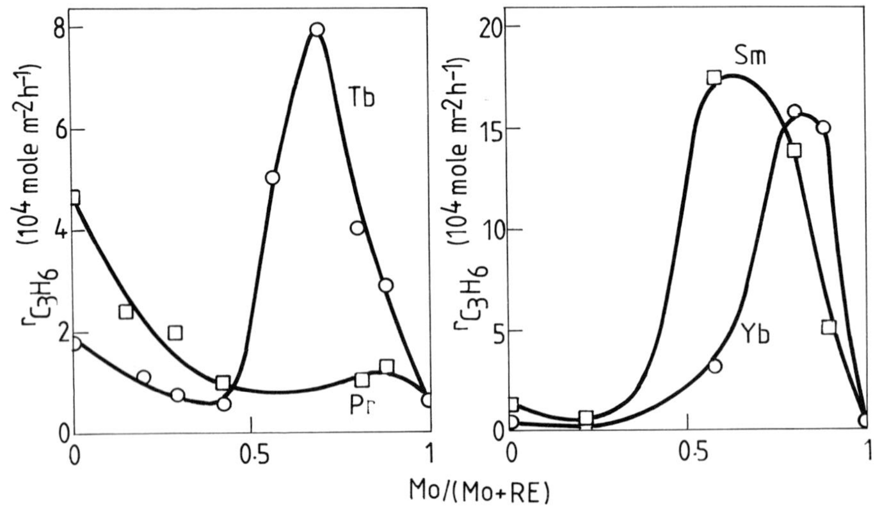

Figure 1. Reaction rate for C_3H_6 at 673 K over Mo-RE-O (RE= Pr, Tb, Sm, Yb) catalyst series. Contact time W/F= 30 $g.h.mol^{-1}$.

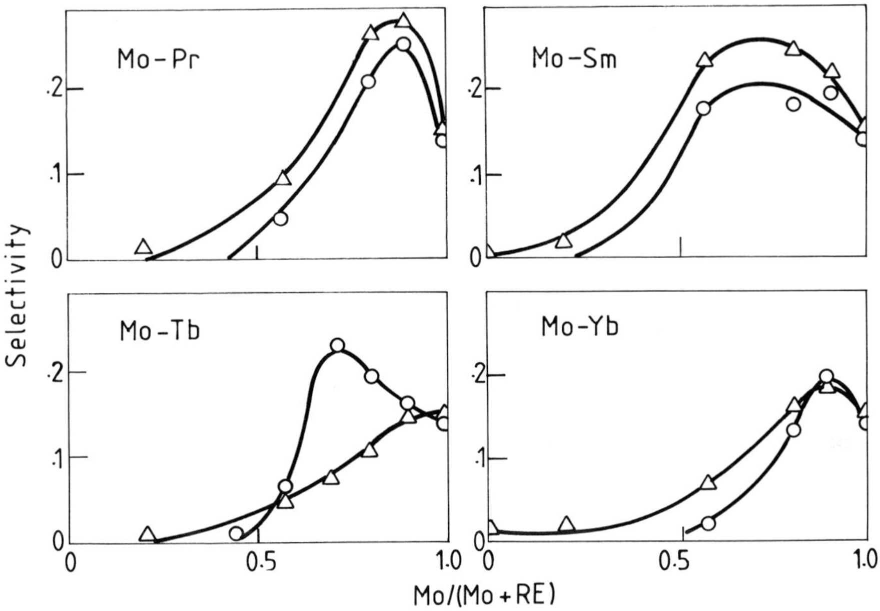

Figure 2. Selectivity to acrolein (○) and acetaldehyde + acetic acid (△) at 673 K for a propylene conversion = 5 mole %.

Samples (0.2-0.3 mg) were first heated to 773 K in helium flow (7.2 dm^3 h^{-1}), and the cooled to ambient temperature. After this, they were contacted with hydrogen (7.2 dm^3 h^{-1}) and heated at a rate of 240 K h^{-1} to a final temperature of 793 K. This temperature was maintained about 0.5 h.

RESULTS AND DISCUSSION

The selective oxidation of propylene over Mo-RE-O (RE= Pr, Sm, Tb, Yb) catalysts has been shown to depend strongly on the catalyst composition. As Fig.1 shows, all four catalyst series display a clear maximun for the rate of propylene conversion in the Mo-rich composition region. However, the compositions (expressed by the Mo/(Mo+RE) atomic ratios) at which the maximum appears, varies according to sequence Mo-Yb-O(0.89), Mo-Pr-O(0.89), Mo-Tb-O(0.70) and Mo-Sm-O(0.60). With the only exception of the Mo-Pr-O catalyst series, a further decrease of the Mo/(Mo+RE) ratios, e.g. increasing the fraction of RE oxide added, induces a marked decrease of the specific catalytic activity. Beside that, from the data of Fig.1 the activity sequence for the pure RE oxides ((Mo/(Mo+RE)= 0.0) follows the order, Pr_6O_{11} > Tb_4O_7 >

Sm_2O_3 > Yb_2O_3, which agrees with the one reported by Minachev et al. **[9]** for the same reaction.

Selectivity values to acrolein and acetyl (acetic acid + acetaldehyde) (Fig. 2) also show a similar maximun to the one found on the activity profiles in the Mo-rich composition region (Mo/(Mo+RE) between 0.60 and 0.89), while carbon oxides are almost the unique C-containing molecules.

As already shown by the XRD patterns, formation of quite disimilar crystalline phases occurs as catalyst composition is varied (Table 1). In agreement with literature findings **[10,11]**, the Mo-rich composition range exhibits the MoO_3 phase as the major crystalline entity, in parallel with small amounts of stoichiometric molybdates, and probably some type of tetra- and hexamolybdates **[11]**, whose abundance decreases for the less Mo-rich preparations. One important point to be considered is that catalysts with maximun in activity profiles are those having the largest proportion of molybdates among the overall crystalline phases. Of course, the Mo-Pr-O is the exception as no crystalline phases were detected along all compositions range.

TABLE 1
Crystalline Phases as Identified from X-Ray Difrattion Patterns.

$\frac{Mo}{(Mo+RE)}$	Pr	Sm	Tb	Yb
0	Pr_6O_{11}	Sm_2O_3 **(a)**	Tb_4O_7	Yb_2O_3 **(a)**
<0.30	Pr_6O_{11}	Sm_2O_3 **(b)**	Tb_6MoO_{12} **(c)**	Yb_2O_3 **(d)**
0.57	Pr_6O_{11}	$Sm_2Mo_2O_9$ **(e)**	$Tb_2(MoO_4)_3$ **(f)**	$Yb_2(MoO_4)_3$ **(e)**
0.70	MoO_3 **(e)**	ns	$Tb_2Mo_4O_{15}$ **(g)**	ns
0.80	MoO_3 **(e)**	MoO_3 **(h)**	MoO_3 **(h)**	$Yb_2Mo_4O_{15}$ **(h)**
0.89	MoO_3	MoO_3 **(k)**	MoO_3 **(k)**	MoO_3 **(h)**

a= cubic; ns= not studied.
Minor phases: b= $9Sm_2O_3.4MoO_3$; c= Tb_4O_7; d=$Yb_2(MoO_4)_3$; e= RE_2O_3; f=$Tb_2Mo_2O_9$; g= MoO_3; h=$RE_2O_3.4MoO_3$ (likely); k= $RE_2O_3.6MoO_3$ (likely).

Laser Raman spectra of the Mo/(Mo+RE)= 0.8 (RE= Pr, Sm, Tb, Yb) catalyst samples were also recorded to monitor the presence of molybdate structures. As shown in Fig. 3, all spectra show the bands at 998 and 820 cm^{-1} characteristic of Mo=O stretch and antisymmetric Mo-O-Mo stretching, respectively in MoO_3 isolate

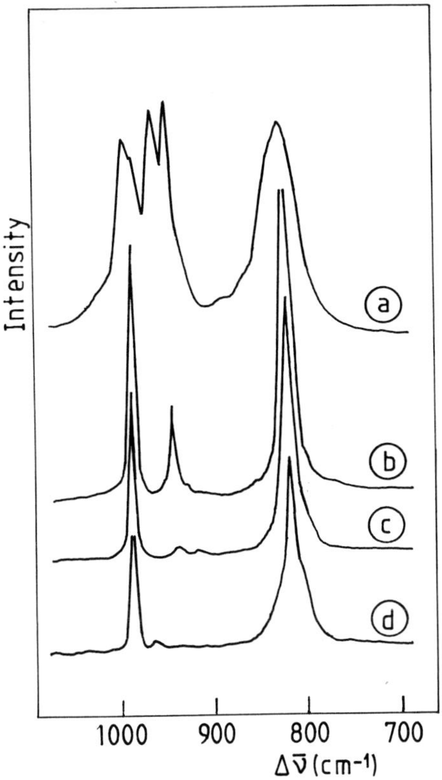

Figure 3. Laser Raman Spectra of Mo-RE-O catalysts (atomic ratio Mo/(Mo+RE)= 0.8): **a)** Mo-Yb-O; **b)** Mo-Tb-O; **c)** Mo-Sm-O; **d)** Mo-Pr-O catalysts.

phase. Other bands in the region 800-960 cm^{-1}, very intense for Mo-Yb-O, moderately intense for Mo-Tb-O and very low for Mo-Sm-O catalysts have been assigned, in agreement with XRD patterns, to that vibrations in RE molybdates **[12]** as its intensity increased with decreasing Mo-loading. The exception is Mo-Pr-O catalyst in which small bands in the same region seem to be due to polymolybdates in a separate phase **[6]**.

To obtain an estimate of the metal-oxygen strength as well as to explain activity and selectivity changes as a function of catalyst composition. TPR profiles were obtained for all preparations. Table 2 summarizes the reduction degree of catalysts obtained at 793 K. One important point to be considered is the strong dependence of TPR profiles upon catalyst composition. For example in the RE-rich preparations, mostly Mo-Pr-O **[5]** and Mo-Tb-O catalysts series, the reduction degree is larger than in Mo-rich preparations, and also the kinetics of reduction decreases continuously with time indicating that this process takes place

TABLE 2

Reduction Degree (α) at 793 K Derived from TPR Profiles.

$\frac{Mo}{Mo+RE}$	$\alpha^{(a)}$			
	Mo-Pr-O (b)	Mo-Tb-O (b)	Mo-Sm-O	Mo-Yb-O
0.00	1.04	1.20	0.00	0.00
0.20	0.80	0.70	1.04	1.27
0,57	0.70	0.75	0.68	1.27
0,80	0.40	1.00	0.75	1.28
0,89	0.13	0.20	0.31	0.91

(a) Calculated by the ratio between the experimental weight loss and the theoretical one espected for the quantitative reduction of MoO_3 to MoO_2 (α=1). **(b)** Reducible oxides such as Pr_6O_{11} and Tb_4O_7 present in the catalysts were considered to be reduced to Pr_2O_3 and Tb_2O_3, respectively.

according to the contracting sphere model. However, Mo-rich catalysts begin to reduce at higher temperatures and present S-shaped TPR profiles, i.e., they reduce according to a nucleation model.

Photoelectron spectroscopy (XPS) has also been used from a quantitative point of view to reveal the surface composition of catalysts. The dependence between the Mo/(Mo+RE) XPS ratios and those corresponding to the chemical analysis are given in Fig.4. As can be observed, for the Mo-RE-O (RE= Pr, Sm, Yb) catalyst series there is, in general, a good correlation between surface XPS and chemical compositions, while for Mo-Tb-O series an important RE surface enrichment is clearly observed throughout the explored compositions. In this latter case a Tb molybdate-phase a few layers thick seems to be formed over MoO_3 nuclei, as also suggested by the well resolved LRS spectra of Tb-molybdates (Fig.3).

When comparing activity and selectivity data for oxidation of propylene with those of catalyst characterization it results that partial oxidation products are more likely to occur on catalysts with lattice oxygen of a lower reactivity, viz., more difficult to be reduced. Moro-oka et al.**[15]** found the more active oxides for total oxidation of hydrocarbons to be those with lower heat of formation of the oxide (ΔH_{M-O}). Pr_6O_{11} and Tb_4O_7 have low ΔH_{M-O} values and an important part of unstable lattice oxygen of a high mobility, thus explaining their tendency to form deep oxidation products when present as separate phases in RE rich Mo-RE-O (RE= Pr, Tb) preparations (Figs.1 and 2). As already

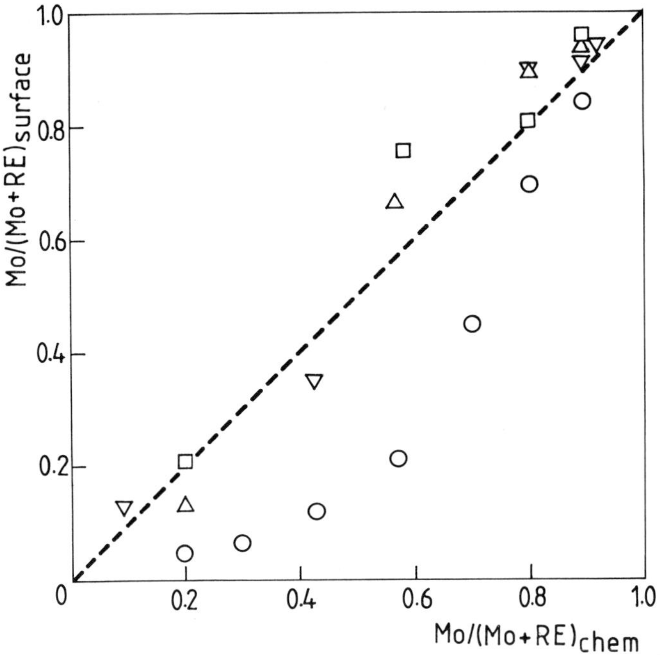

Figure 4. Dependence between the surface XPS and chemical Mo/(Mo+RE) atomic ratios: RE= Pr (▽); Sm (□); Tb(○); Yb (△). In this calculation, the integrated Mo3d and RE4d intensities and published sensitivity factors **[14]** were considered.

shown by TPR, ΔH_{M-O} tends to be larger for catalyst which are more difficult to reduce. The reduction degree (α) at 793 K in the region Mo/(Mo+RE)= 0.7-0.8 is the lowest but simultaneously selectivity to partial oxidation products is the highest (Fig.2). A similar correlation among catalyst reduction and conversion and selectivity were found by Sachtler and de Boer **[16]** in the propylene oxidation over metallic molybdate catalysts.

These results are closely related to those reported by Trifiro' et al. **[17]**, who found that the most selective catalysts (within a series of molybdates) for the same reaction are those exhibiting the lowest diffusion rate of lattice oxygen. Oxygen may be removed by diffusion of lattice oxygen to the interface reduced phase in all the ternary catalyst systems employed in this study. Thus, the diffusion rate of oxygen ions will be lower and the selectivity will be higher for catalysts with lower reducibility, as it effectively occurs. The fact that maximun selectivity to partial oxidation products occurs for Mo/(Mo+RE) ratios in the region 0.7-0.8, where XRD patterns and LRS spectra revealed excess of MoO_3 and several kinds of molybdates, indicates that nucleophilic oxygen species, which then would lead to allylic oxidation, are optimized.

AKNOWLEDGEMENTS

The authors are indebted to CSIC and CAICYT for sponsorship of this work (Project No. 120).

REFERENCES

[1] a) R.K. Grasselli, J.D. Burrington, **Adv. Catal.**, 30 (1981) 133.
b) C.F. Cullis, D.J. Hucknall, in G. Bond & G. Webb (Eds.), "Catalysis", Vol. 5, Specialist Periodical Reports The Chemical Society, London, (1982) ch. 7, p. 273.

[2] a) M.P. Rosynek, **Catal. Rev.- Sci. Eng.**, 16 (1977) 111.
b) P. Pomonis, **React. Kinet. Catal. Rev.**, 18 (1981) 247.

[3] a) J.C.J. Bart, N. Giordano, **J. Catal.**, 75 (1982) 134.
b) J.F. Brazdil, R.K. Graselli, **J. Catal.**, 79 (1983) 104.

[4] a) J.J. Kim, S.W. Weller, **Appl. Catal.**, 33 (1987) 15.
b) V.M. Khiteeva, Sh.M. Rzakulieva, **Russ. J. Phys. Chem.**, 55 (1981) 1202.

[5] J.M. López Nieto, J.L.G. Fierro, L. Gonzalez Tejuca, G. Kremenic', **J. Catal.**,107 (1987) 325.

[6] J.M. López Nieto, G.Kremenic', A. Martinez Alonso, J.M.D. Tascón, **J. Mater. Sci.**, (in press).

[7] G. Kremenic',J.M. López Nieto, J. Soria, J. Martí, **Proc. Inter. Conf. Rare Earth Dev.& Appl.**, Beijing, China, September 1985, Vol. 1, p. 614.

[8] G. Kremenic', J.M. López Nieto, J.L.G. Fierro, L.G. Tejuca, **J. Less-Common Met.**, 136 (1987) 95.

[9] K.M. Minachev, D.A. Kontratev, G.N. Antoshin, **Kinet. Kata.**, 8 (1967) 131.

[10] a) K. Nassau, J.W. Shiever, E.T. Keve, **J. Solid State Chem.**, 3 (1971) 411.
b) L.H. Brixner, P.E. Biersted, A.W. Sleight, M.S. Lisic, **Mat. Res. Bull.**, 6 (1971) 545.

[11] a) F.P. Alekseev, E.I. Get'man, G.G. Koshchoev, M.V. Mokhosoev, **Russ. J. Inorg. Chem.**, 14 (1969) 1558.
b) E. Ya Rode, G.V. Lysanova, L.Z. Gokhman, **Inorg. Mater.**, 7 (1971) 1875.

[12] H. Jeziorowski, H. Knözinger, **J. Phys. Chem.**, 83 (1979) 1166.

[13] J.M. López Nieto, A.G. Valdenebro, J.L.G. Fierro, in preparation.

[14] C.D. Wagner, L.E. Davis, M.V. Zeller, J.A. Taylor, R.H. Raymond, L.H. Gale, **Surf. Interface Anal.**, 3 (1981) 211.

[15] Y. Moro-oka, Y. Morikawa, A. Ozaki, **J. Catal.**, 7 (1967) 23.

[16] W.M.H. Sachtler and N.H. de Boer, **Proc. 3rd. Int. Congr. Catal.**, Amsterdam, 1964 (W.M.H. Sachtler, G.C.A. Schuit and P. Zwietering, Eds), Wiley, New York, 1965, Vol.1, p.252.

[17] F. Trifiro', P. Centola, I. Pasquon and P. Jiru, **Proc. 4th. Int. Congr. Catal.**, Moscow, 1968 (B.A. Kazansky, Ed.), Adler, New York, 1968. Vol.I, p.252.

J.C. VEDRINE (I. de Recherche sur la Catalyse, Villeurbanne, France): I was surprised that you concluded that selective molybdate catalyst exhibit lower diffusion rate of lattice oxygen. Using ^{18}O labelled CO_2 as a probe we have observed that lattice O of bismuth molybdates (α or β phases, kown to be very selective in propene oxidation to acrolein) are exceptionally labile involving both surface and bulk lattice oxygen. How did you determine the lattice oxygen lability of your samples?

J.M. LOPEZ NIETO (I. Catálisis y Petroleoquímica, Madrid, Spain): The term diffusion rate of oxygen in the rare earth molybdates refers here to the relative ease with which oxygen can be released from the catalyst. We found that the catalyst whose Mo/(Mo+RE) ratio is 0 and 1 are poorly selective to partial oxidation products, viz. carbon oxides and water were the major oxidation products.To explain this behavior, it was assumed that catalysts with these extreme compositions have highly reactive oxygen species, such as oxygen adsorbed. On the contrary, in the region of intermediate Mo/(Mo+RE) ratios, where molybdates were found to ocour, the bulk lattice oxygen seems to be involved in the selective oxidation of adsorbed hydrocarbon. The mobility of the latter oxygen species must be high as confirmed by the observation that the surface prereduction of the different molybdates at temperatures close to 600 K is faster than the subsequent oxygen adsorption on the partially reduced surface. This particular behaviour has been explained as due to partial restoration of the original surface, upon surface reduction, by diffusion of bulk lattice oxygen to the surface which then adsorbs oxygen slowly until initial state recovery.

J.C. VEDRINE (Ins. de Recherche sur la Catalyse, Villeurbanne, France): You also found high selectivity in acetic acid and acetaldehyde which was interpreted as electrophilic attack of propene rather than allylic. The last is giving acrolein. In a recent paper by us on MoO_3/SiO_2 (ref.1) much allylic attack was detected at low Mo coverage but yielded propanal. Did not you observed any propanal in your products? Acetic acid results from a more complex reaction mechanism with C-C cleavage as for acetaldehyde.

J.M. LOPEZ NIETO (I. Catálisis y Petroleoquímica, Madrid, Spain): For the MoO_3 and MoO_3/SiO_2 systems, Vedrine et al. (ref.1) found high selectivity toward propanal at conversions levels below 1%. For the MoO_3/SiO_2 catalysts studied early in our laboratory, we did not detect propanal at conversion levels as high as 15-20% (ref.2). In this study working at conversion levels around 10% on Mo-RE-O systems, no propanal was detected in any case. Only acrolein, acetic acid and acetaldehyde were observed. Acetaldehyde is mainly a primary product (from propene degradation), but it also forms by decomposition of acrolein (ref.3). However, acetic acid is formed by oxidation of C_2- and C_3-oxigenated products.

R. LARSSON (I. Inorganic Chemistry, Lund, Sweden): Have you determined the activation energies of these reaction?

J.M. LOPEZ NIETO (I. Catálisis y Petroleoquímica, Madrid, Spain): Temperature coefficients for propene oxidation on the various rare earth molybdates have been calculated. They have been not summarized for practical reasons. In general, the values obtained summarized for practical reasons. In general, the values obtained do not vary significantly along the explored compositions with the exception of the Mo/(Mo+RE) ratios with maxima in activity and selectivity which led to values substantially higher. To illustrate this, the temperature coefficients for the Mo-Pr-O catalyst series were 106-119 kJ/mole for compositions Mo/(Mo+RE) < 0.88, while a value of 143 kJ/mole was obtained for the most active Mo/(Mo+RE)= 0.91 catalyst.

R. LARSSON (I. Inorganic Chemistry, Lund, Sweden): Have you IR spectra of the catalysts?

J.M. LOPEZ NIETO (I. Catálisis y Petroleoquímica, Madrid, Spain): Exploratory experiments using IR technique revealed the appearance of several M-O-M lattice vibrations, however the unambiguous assignment of that bands to specific compounds was not straighforward. Very recently, an in-depth analysis of these molybdate series was carried out by Laser Raman Spectroscopy in our laboratory. This study will constitute the next step of the research of the bulk and surface properties of the rare earth molybdates.

1) T. Liu, M. Forissier, G. Coudurier, J. C. Vedrine, **J. Chem. Soc., Faraday Trans.** 1, 85 (1989) 1607.
2) J. M. López Nieto, G. Kremenic', A. Martinez-Alonso, J. M. D. Tascón, **J. Mater. Sci.**, 24 (1989) (in press).
3) J. M. López Nieto, J. M. D. Tascón, G. Kremenic', **Bull. Chem. Soc. Jpn.**, 61 (1988) 1383.

G. Centi and F. Trifiro' (Editors), *New Developments in Selective Oxidation*

OXYESTERIFICATION OF METHANOL TO METHYLFORMATE OVER V-Ti OXIDE CATALYSTS

A.S. ELMI[1], G. BUSCA[2], C. CRISTIANI[1], P. FORZATTI[1] and E. TRONCONI[1]

[1] Dipartimento di Chimica Industriale e Ingegneria Chimica del Politecnico, P.zza L. Da Vinci 32, 20133 Milano (Italy)

[2] Dipartimento di Chimica, Facoltà di Ingegneria dell' Università, 16129 Genova (Italy)

SUMMARY

Based on previous work and on new data for various V-Ti oxide systems, generalized results are presented concerning their physico-chemical characterization, their catalytic behavior in the oxyesterification of methanol to methyl formate, and the related reaction mechanism. The feasibility of industrial process configurations for the the production of methyl formate, possibly combined with formaldehyde, are discussed.

INTRODUCTION

Methyl formate is regarded as a convenient intermediate in the synthesis of several chemicals. The current technology for its production involves carbonylation of methanol in the liquid phase in the presence of basic catalysts, typically sodium methylate, at low temperatures and under moderate CO pressures (ref.1),

$$CH_3OH + CO \rightarrow HCOOCH_3$$

A route using the gas-phase dehydrogenation of methanol over Cu-based catalysts has been recently proposed (ref.2),

$$2\ CH_3OH \rightarrow HCOOCH_3 + 2\ H_2$$

for which attractive yields in methylformate have been claimed. An alternative oxidative route, based on the reaction

$$2\ CH_3OH + O_2 \rightarrow HCOOCH_3 + 2\ H_2O$$

was studied by Ai (ref.3) over various Mo- and W- based catalysts. We have reported that this reaction occurs also over V-Ti oxide catalysts prepared either by coprecipitation (refs.4-5) or by impregnation (ref.6) techniques. Encouraging preliminary results concerning global selectivities and yields of methyl formate (ref.4) prompted us to perform a complete process variable study (ref.5), and to address the mechanistic features by an FT-IR study on the interaction of methanol and its oxidation products with the V-TiO_2 surface (ref.6). The characterization of the catalyst samples was also fully investigated (refs.4-9). Recently, our understanding of the reaction mechanism has been completed and refined by the results of a series of flow reactor experiments where reaction products and intermediates were used as reagents, which have confirmed the oxidative nature of the reaction step leading to methyl formate ("oxyesterification") as compared to the disproportionation mechanism previously suggested (ref.7).

Based on our previous work as well as on new data for various coprecipitated V-Ti systems, in this paper we present generalized results concerning their physico-chemical characteristics, their catalytic behavior and the related mechanistic features. The effects of the catalyst preparation parameters (V/Ti a.r. and calcination temperature) and of the operating conditions is discussed in connection with the selection and the feasibility of alternative process configurations (production of methyl formate only versus coproduction of formaldehyde and methyl formate).

EXPERIMENTAL

V-Ti oxide samples with V/Ti atomic ratios (a.r.) 0 - 0.5 were prepared by coprecipitation from $VOCl_3$ and $TiCl_4$ at r. t., followed by drying and calcination. Different samples were obtained varying the calcination temperatures between 500 and 700 °C. The procedures and the equipment used in catalyst characterization and in flow reactor experiments have been described elsewhere (refs. 4-9).

RESULTS AND DISCUSSION

Catalyst Characterization

Coprecipitated V-Ti oxide catalysts have been characterized with respect to the influence of both calcination temperature and V/Ti a.r. Samples with low V/Ti a.r. and low activation temperature are constituted by the anatase phase only. XRD, UV-visible diffuse reflectance, ESR, FT-IR and chemical analysis provide evidence for the presence of a solid state solution characterized by the incorporation of V^{4+} in the bulk (ref.8). For the samples with V/Ti a.r.=0.0375, on increasing the calcination temperature the rutile phase becomes predominant (T_c= 650 °C), and a sudden drop in surface area is observed. On the other hand, on increasing the

V/Ti a.r. at a calcination temperature of 600 °C, V_2O_5 appears in the samples with V/Ti a.r.≥ 0.0625, again causing a drop in surface area. Appearance of a rutile phase of TiO_2 is detected in the sample with V/Ti a.r.= 0.5. For a calcination temperature of 700 °C, the rutile phase is first detected at V/Ti a.r. ≥0.0125, while V_2O_5 is observed at higher V loadings. Table 1 presents the specific surface area of the coprecipitated V-TiO_2 catalysts and the detected phases other than the anatase phase as functions of the sample calcination temperature and of the V/Ti a.r..

TABLE 1

Effects of calcination temperature and V/Ti a.r. on the surface areas (m^2/g) and on the phase composition of V-Ti oxide samples. V = V_2O_5 detected. R = rutile phase detected.

V/Ti (a.r.)	CALCINATION TEMPERATURE (°C)						
	500°	550°	600° (ref.5)	625°	650°	675°	700° (ref.4)
0	70	-	53	-	-	-	20
0.01	-	-	-	-	-	-	16
0.0125	82	-	48	-	-	-	9 R
0.025	83	59	44	39	28	9 R	9 R
0.0375	78	54	37	34	8 R	6 R	6 R
0.05	-	64	28	-	-	-	-
0.0625	80	62	5 V	-	-	-	1 R
0.125	30	-	4 V	-	-	-	-
0.25	-	-	6 V	-	-	-	3 R+V
0.50	-	-	7 R+V	-	-	-	-

The boundary between the samples constituted by the anatase phase only, and those where also the rutile phase and/or V_2O_5 are detected is seen to correspond to a dramatic reduction of the surface area. The results of elemental chemical analysis further indicate that V interacts with the support in the form of V^{4+} and that it is also present at the surface as V^{5+} (ref.7). The characterization by adsorption of probe molecules and a combined FT-IR and Laser Raman microscopy study demonstrate that both V and Ti centers, and specifically mono-oxo vanadyl centers with a coordinative unsaturation, are present at the catalyst surface (refs. 5,9).

Catalytic behavior of V-TiO_2 samples: effects of V/Ti a.r. and of calcination temperature

During the flow reactor experiments, the observed reaction products included HCHO, methyl formate, water, dimethylformal (DMFL), CO, CO_2 and formic acid (refs. 4-5). Varying the V content of the catalysts was found to affect significantly both the conversion of methanol and the distribution of products. For the samples calcined at 600 °C, the overall conversion is seen to go through a maximum (V/Ti a.r.= 0.0625-0.125), which can be attributed to the contrasting effects resulting from increasing the V loading: while this enhances the oxidizing capacity of the

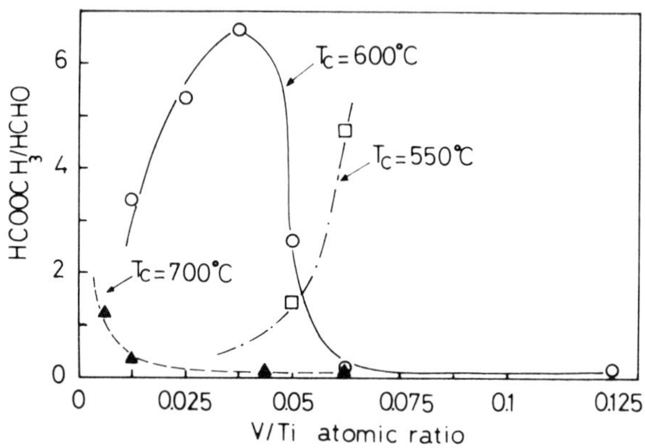

Fig.1 - Effects of the calcination temperature T_c and of the V/Ti a.r. on the $HCOOCH_3$/HCHO molar ratio in the oxidation of methanol over V/Ti oxide catalysts.

catalyst, it also reduces its surface area (see Table 1). For the same calcination temperature, Fig. 1 shows that a maximum is evident also in the $HCOOCH_3$/HCHO molar ratio. As discussed in a later section, this ratio is of specific interest for the implementation of an industrial process for the production of methyl formate: depending on its value, different process designs have to be considered. As compared to the optimal V content in Fig. 1, the low selectivities to $HCOOCH_3$ corresponding to low and high V contents appear to be associated with poorly active systems due to deficiency of oxidizing capacity, and to deficiency of surface area and excess of V, respectively. This interpretation is consistent with a reaction mechanism where formation of $HCOOCH_3$ occurs consecutively to the formation of HCHO, requiring V-related catalytic centers and adequate surface areas.

In this work the influence of changes in the catalyst calcination temperature has also been studied, as shown in Fig. 1. Higher calcination temperatures correspond to lower surface areas for the same V/Ti a.r., as indicated in Table 1, and also to greater amounts of V at the surface: accordingly, the $HCOOCH_3/HCHO$ ratio is seen only to decrease in the case of the catalysts calcined at 700 °C, where the optimal $HCOOCH_3/HCHO$ ratio is shifted to lower V loadings; on the other hand, for the catalysts calcined at 550 °C, associated with higher surface areas and lower V contents at the surface, only the rising branch of the curve is apparent, the maximum being shifted to greater V/Ti a.r..

Catalytic behavior of $V\text{-}TiO_2$ samples: effects of the process variables

The effects of the main process variables, including methanol and water feed concentrations, space velocity, temperature and pressure have been investigated over various catalysts. The O_2 feed molar concentration was fixed at ≈ 10% in all runs in order to remain below the flammability limits of methanol/oxygen mixtures.

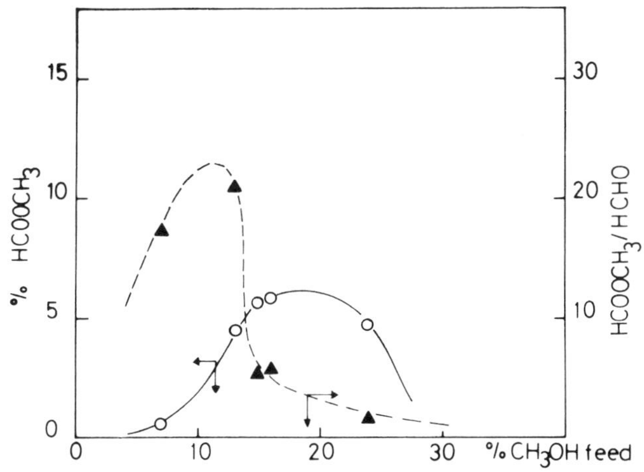

Fig.2 - Effect of methanol feed content on % $HCOOCH_3$ and on the $HCOOCH_3/HCHO$ molar ratio. Catalyst: sample with V/Ti a.r.= 0.0625 calcined at 550°C. Reaction conditions: T= 165 °C, 10% O_2 feed, F/W_C= 10 cc/g min.

For a catalyst with V/Ti a.r. = 0.0625, T_c= 550 °C, Fig. 2 illustrates the effects of the methanol feed level on the concentration of methylformate in the products and on the $HCOOCH_3/HCHO$ molar ratio. Distinct optimal values of the methanol feed content exist for the output concentration of methylformate and for the $HCOOCH_3/HCHO$ ratio. Selectivities to valuable products ($HCOOCH_3$+ HCHO+ DMFL) in excess of 90% were achieved with methanol concentrations greater than 15%. An excess of methanol enhanced DMFL with respect to HCHO, and almost suppressed the formation of CO and CO_2.

The effect of reaction temperature at two space velocities on the $HCOOCH_3/HCHO$ is presented in Fig. 3 for a catalyst with V/Ti a.r.= 0.0375 calcined at 600 °C. Both high temperatures and long contact times are seen to favor methyl formate with respect to formaldehyde, in line with the consecutive nature of the reaction scheme. At temperatures higher than 180 °C, however, a sudden drop in the selectivity to methyl formate has been observed for prolonged contact times. The addition of H_2O to the feed was found to depress the overall conversion of methanol, and also reduced the ratio $HCOOCH_3/HCHO$.

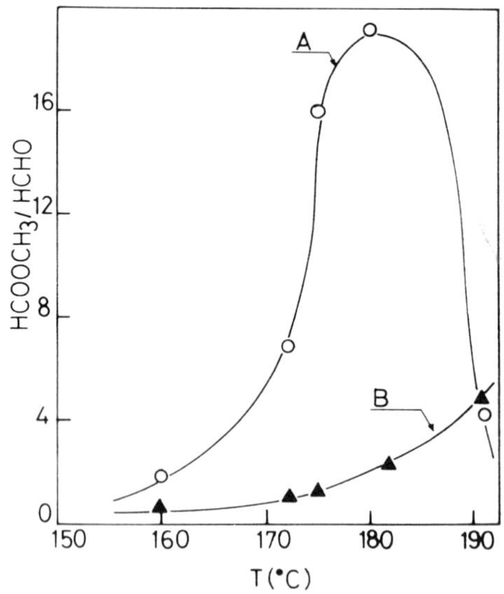

Fig.3 - Effect of reaction temperature and of contact time on the $HCOOCH_3/HCHO$ molar ratio. Catalyst: sample with V/Ti a.r.= 0.0375 calcined at 600 °C. Reaction conditions: 10% CH_3OH, 10% O_2 feed; F/W_c = 12 cc/g min (curve A) and 24 cc/g min (curve B).

Similar effects of the process variables had been previously observed with other V-Ti catalysts (refs. 4,5), so that they seem to be representative of the general catalytic behavior of such systems. They are interpreted in the following section in the light of our findings on the reaction mechanism.

Mechanism of the oxidation of methanol over V/Ti oxide catalysts

The mechanistic features of the oxidative route to methyl formate over V-Ti oxide catalysts have been studied by FT-IR techniques, investigating the interaction of methanol and its oxidation products with the catalyst surface (ref.6), and by running a series of flow reactor experiments where intermediates and reaction products were used as reactants (ref.7). The results are supportive of the reaction scheme presented in Fig. 4, consisting essentially of successive oxidation steps. Each of these steps has received experimental validation by FT-IR and/or specifically designed flow reactor runs. Thus, in the case of the route leading from formaldehyde to methyl formate, IR spectroscopy has provided evidence for a Cannizzaro-type disproportionation of dioxymethylene (step 11), and the occurrence of this reaction has been confirmed by flow reactor experiments with a HCHO + He feed, where $HCOOCH_3$ was produced. However, the results of similar experiments with a HCHO + O_2 + He feed show that the oxidation route (step 6) is considerably faster under typical, oxidizing reaction conditions.

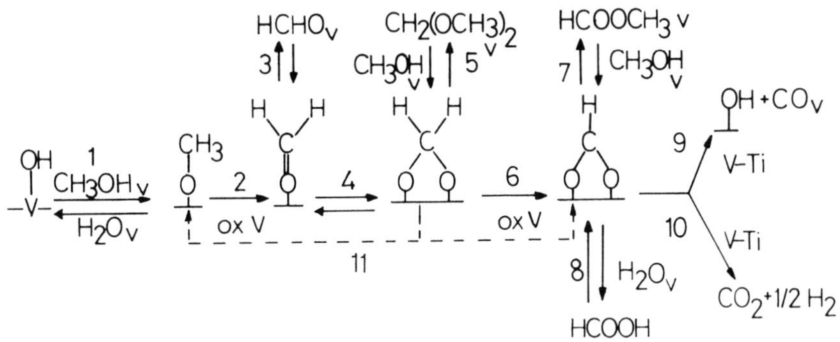

Fig.4 - Reaction mechanism for the oxidation of methanol over V/Ti oxide catalysts.

Flow reactor runs with HCOOH in the feed have proved that CO and CO_2 originate through decomposition of formate groups. The esterification of formate groups to methyl formate appears however to be faster than their decomposition, provided that methanol is available in the reaction mixture.

All of the effects of the operating variables can be interpreted on the basis of the scheme in Fig.4. Thus, step 1 is consistent with the observed inhibiting effect of water on the conversion of methanol. The effect of the CH_3OH feed concentration can be rationalized by observing that, for $CH_3OH < 10\%$, the excess of oxygen favors the oxidation steps, leading preferentially to the terminal products $HCOOCH_3$, CO and HCOOH. On increasing the methanol feed content, the steps involving gaseous methanol are beneficially affected, resulting first in a decreased selectivity to CO and HCOOH, corresponding to an increased selectivity to $HCOOCH_3$, and finally in enhanced selectivities to HCHO and particularly to DMFL. The data in Fig. 3 are explained considering that an increase in temperature and contact time results in enhanced methanol conversions, and reduces the concentration of gaseous methanol. Accordingly, first the selectivity to $HCOOCH_3$ grows at the expense of HCHO + DMFL, then the selectivity to CO is favored at the expense of methyl formate.

Catalytic tests for the oxidation of methanol over pure TiO_2 (ref. 5) have confirmed the fundamental role of Vanadium in the oxidative steps of the mechanism (steps 2 and 6 in Fig. 4).

Process considerations

The general results of the flow reactor experiments indicate that the ratio $HCOOCH_3/HCHO$ in the products can be adjusted within a wide range of values by appropriate choices of both the catalyst preparation parameters and of the reaction conditions, depending on the desired features of the reaction product. One possible goal is to design a process aimed at the production of methyl formate only. Fig. 1, 2 and 3 illustrate a few examples where the production of methyl formate can be optimized by a suitable selection of either V/Ti a.r. and calcination temperature, or of the methanol feed concentration or of the reaction temperature. Along these lines we have achieved values of $HCOOCH_3/HCHO$ as high as 20, corresponding to weight ratios =40/1, with productivities to $HCOOCH_3$ exceeding 200 g/Lh.

Alternatively, the V-TiO_2 catalysts appear suitable also for the industrial coproduction of $HCOOCH_3$ and HCHO by the mild gas-phase oxidation of methanol. In this case, the ratio $HCOOCH_3$/HCHO is expected to have a strong impact on the design of the separation section of such a process, for which a tentative schematic diagram is given in Fig.5.

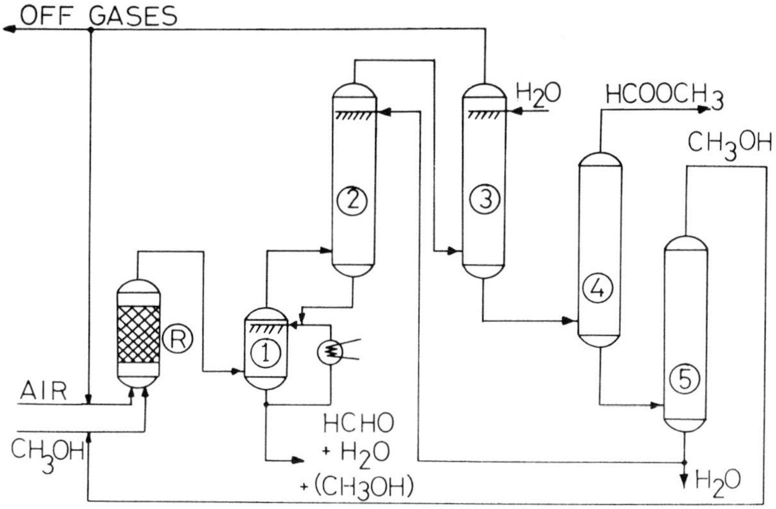

Fig.5 - Tentative process scheme for the coproduction of methyl formate and formaldehyde by mild gas-phase oxidation of methanol.

Units 1 and 2 are devoted to the separation of HCHO, which is dissolved in water, and to the concentration of the resulting aqueous solution. The remaining separation units effect removal from the gaseous stream of the inert gases (unit 3), which may be in part recycled to dilute the oxygen in the air feed stream, of methyl formate (unit 4), and eventually of unreacted methanol (unit 5), which is recirculated to the synthesis reactor. In this scheme, the trickiest section is that effecting separation of formaldehyde, its target being the production of a commercial aqueous solution of HCHO. If concentration of the solution is required, a lower bound exists on the acceptable content of formaldehyde in the reaction products. This implies that it may be desirable to design operation of the reactor without necessarily maximizing the $HCOOCH_3$/HCHO ratio.

Preliminary calculations of the separation section were performed assuming a reactor outlet stream containing 10.5% H_2O

and 1.5% HCHO. Results demonstrate that concentrations of HCHO of 20% w/w and more in the final solution are feasible by autothermal operation of units 1 and 2 under slight pressure. Polymerization of HCHO can be prevented by allowing a small concentration of residual methanol in the final solution. The final choice of the reactor working conditions, however, is controlled by a balance between the increase in revenues expected from maximization of the $HCOOCH_3$ production and the increased costs resulting from concentration of more dilute aqueous solutions of HCHO, for which a detailed economic analysis is required.

ACKNOWLEDGEMENTS
This work was supported by Ministero Pubblica Istruzione (Roma).

REFERENCES

1 The Leonard Process Co. - Kemira OY, Formic Acid, Hyd. Process., November (1983).

2 S.P. Tonner, D.L. Trimm, M.S. Wainwright and N.W. Cant, Dehydrogenation of Methanol to Methyl Formate over Copper Catalysts, I&EC Prod.Res.Dev., 23 (1984) 384.

3 M. Ai, The Production of Methyl Formate by the Vapor Phase Oxidation of Methanol, J. Cat., 77 (1982) 279.

4 P. Forzatti, E. Tronconi, G. Busca and P. Tittarelli, Oxidation of Methanol to Methyl Formate over V-Ti Oxide Catalysts, Cat. Today, 1 (1987) 209.

5 E. Tronconi, A.S. Elmi, N. Ferlazzo, P. Forzatti, G. Busca and P. Tittarelli, Methyl Formate from Methanol Oxidation over Coprecipitated V-Ti-O Catalysts, I&EC Res., 26 (1987) 1269.

6 G. Busca, A.S. Elmi and P. Forzatti, Mechanism of Selective Methanol Oxidation over Vanadium Oxide - Titanium Oxide Catalysts: A FT-IR and Flow reactor Study, J. Phys. Chem., 91 (1987) 5263.

7 A.S. Elmi, E. Tronconi, C. Cristiani, J.P. Gomez Martin, P. Forzatti and G. Busca, Mechanism and Active Sites for Methanol Oxidation to Methyl Formate over Coprecipitated Vanadium-Titanium Oxide Catalysts, I&EC Res., 28 (1989) 387.

8 G. Busca, P. Tittarelli, E. Tronconi and P. Forzatti, Evidence for the Formation of an Anatase-Type V-Ti Oxide Solid State Solution, J. Solid State Chem., 67 (1987) 91.

9 C. Cristiani, P. Forzatti and G. Busca, On the Surface Structure of Vanadia-Titania Catalysts: Combined Laser-Raman and FT-IR Investigation, J. Cat., 116 (1989) 586.

V. CORTES CORBERAN (Instituto de Catalisis y Petroleoquimica, CSIC, Madrid, Spain): Concerning the composition of the different V-Ti-O catalysts, have the authors experimental evidence of the surface composition ? And, if so, does the bulk composition represent the actual surface composition of samples having different V/Ti atomic ratios ?

G. BUSCA (University of Genova, Italy): A qualitative analysis of the catalyst surface composition has been performed for one of the most active catalysts (V/Ti a.r. 0.0375, calcined at 600 °C (1)) using FT-IR spectroscopy. It has been shown that vanadyl species (responsible for a well evident surface-sensitive IR band at 2050 cm^{-1}, first overtone of the V=O stretching) and of coordinatively unsaturated Ti^{4+} ions (responsible for the formation of carbonyl species absorbing at 2195 cm^{-1} when the catalyst is put into contact with CO gas) are both present on the surface. By measuring the intensities of these bands and by compearing them with those are measured on pure TiO_2 and on V-TiO_2 catalysts prepared by impregnation with measured amounts of vanadium compounds, a quantitative evaluation can be obtained (2). In this case, a surface enrichment of vanadium seems evident.

1. E. Tronconi, A.S. Elmi, N. Ferlazzo, P. Forzatti, G. Busca and P. Tittarelli, **Ind. Eng. Chem. Res.**, 26 (1987) 1269.

2. G. Ramis, G. Busca and V. Lorenzelli, **Z. Phys. Chem., Neue Folge**, 153 (1987) 189.

G. Centi and F. Trifiro' (Editors), *New Developments in Selective Oxidation*
© 1990 Elsevier Science Publishers B.V., Amsterdam — Printed in The Netherlands

OXIDATIVE TRANSFORMATION OF METHANOL TO HIGHER ALDEHYDES OVER ZEOLITE - OXIDE CATALYSTS

P. Jiru, Z. Tvaruzkova and K. Habersberger
J. Heyrovsky Institute of Physical Chemistry and Electrochemistry,
Dolejskova 3, 182 23 Prague 8, Czechoslovakia

SUMMARY

Methanol may be in the temperature range 350-500°C catalytically oxidized in one step to acetaldehyde and benzaldehyde over bifunctional catalysts containing the redox active component Bi-Mo-O together with the HZSM-5 zeolite. In comparison with HZSM-5 zeolite alone (without the redox active component Bi-Mo-O) a 16 times higher selectivity to C_2+ aldehydes was attained. The analysis of the infrared spectra of adsorbed d_3-acetonitrile and of the surface complexes of methanol formed in its interaction with the proton donor sites of the bifunctional catalyst at 400°C led to the suggestion of the probable mechanism of the oxidative transformation of methanol to higher aldehydes.

INTRODUCTION

Together with the development of C_1 chemistry, methanol becomes a possible source for C_{2+} oxygen derivatives (acroleine, acetaldehyde, benzaldehyde etc.). According to the present state of art such a production would be performed in two steps, namely the transformation of methylalcohol to C_{2+} hydrocarbons, followed by the selective oxidation of hydrocarbons to the desires products. The present study investigates the possibility of a single step process of oxidative methanol transformation over a bifunctional catalyst and represents a continuation of the previously published short communication (ref. 1).

EXPERIMENTAL

The following types of bifunctional catalysts were prepared: i) Catalyst A - - HZSM-5, containing 3wt% of Bi-Mo-O component, prepared by hydrotermal synthesis. Tetrapropylammonium ions were used, as structure directing agent. The starting mixture for the hydrothermal synthesis (performed in an autoclave for 6 days at 150°C) contained 6.67% SiO_2 (as silica sol), 0.41% $AlCl_3.6H_2O$, 1.86% NaOH, 3.15% $(C_3H_7)_4NBr$, 0.39% $Bi(NO_3)_3.5H_2O$ and 0.10% $(NH_4)_6Mo_7O_{24}.4H_2O$. ii) Catalyst B - a mechanical mixture of HZSM-5 and 5wt% of $Bi_2O_3.MoO_3$ was prepared according to (ref. 2). iii) Catalyst C - $Bi_2O_3.MoO_3$ (23wt%) supported on HZSM-5, was prepared by impregnation, first with a solution of $(NH_4)_6Mo_7O_{24}.4H_2O$ and, after calcination at 500°C in a flow of dry oxygen, with a solution of $Bi(NO_3)_3$. The Si/Al ratio of the original HZSM-5 was 19. All samples were before the catalytic test activated in a flow of dry oxygen for 2 hours at 350°C.

Both the activity and the selectivity of the catalysts were investigated in an integral flow microreactor, the products were determined by gas chromatography and (after adsorption in water) by polarography. The measurements were performed in the reaction temperature range 350-500°C, with WHSV = 1 to 2 h^{-1} and the molar ratio CH_3OH/O_2 = 3-12 in the feed. The grain size of the catalysts was 0.3-0.6 mm.

FT-IR spectroscopy was used both for the determination of the crystallinity of the catalysts (which was found to correspond in all cases to HZSM-5 structure and was preserved also after the catalytic test) in the skeletal vibration range and for the investigation of the structure and reactivity of the proton donor sites of the catalysts after their interaction with d_3-acetonitrile and methanol, performed in vacuum cuvettes on selfsupporting catalyst pellets of 10 mg/cm^2 thickness. The experimental details have been published previously (refs. 3,4)

RESULTS AND DISCUSSION

Catalytic activity and selectivity

In the investigated range of reaction conditions (WHSV, temperature, feed, ratio $MeOH/O_2$) higher hydrocarbons, carbon dioxide, formic acid, formaldehyde, acetaldehyde and benzaldehyde were found as reaction products of the oxidative transformation of methanol over all investigated catalysts. Practically no acroleine (yield $<$ 0.02%) was present in the products. The relative selectivity $S^{ox}_{C_{2+}}$ to C_{2+} oxygen containing products (here acetaldehyde and benzaldehyde) was defined as the percentage represented by the sum of the respective selectivities to acetaldehyde and benzaldehyde in the total sum of the respective selectivities to all oxygen containing products, i.e. $\sum^{ox}_{C_1} + \sum^{ox}_{C_{2+}}$. A survey of the catalytic tests (reaction conditions and selectivities to the individual reaction products) with the respective highest relative selectivities $S^{ox}_{C_{2+}}$ obtained over the individual catalysts A,B and C is given in Table 1. According to this criterion, the highest relative selectivity $S^{ox}_{C_{2+}}$ was obtained over the catalyst A, catalyst C being the second in order. Over the catalyst B (consisting of a mechanical mixture of both components, zeolite and oxide) the lowest value of this criterion was attained. The lower value of $S^{ox}_{C_{2+}}$ (see Tab. 1) attained over the catalysts B and C is connected with the fact that these two catalysts convert methanol into formaldehyde to a much higher extent than catalyst A: the respective conversion degrees to formaldehyde are X^{Fd}_B = 19.6%, X^{Fd}_C = 4.2%, whereas X^{Fd}_A = 0.16% only. The catalysts B,C may therefore be used in a single-step preparation of a mixture of formaldehyde with higher aldehydes (acetaldehyde, benzaldehyde). It may be assumed that over the catalyst A the reaction rate of the methanol transformation to higher aldehydes is either approximately the same or higher than the rate of the methanol oxidation to formaldehyde. Although the differences in the deep oxidation of methanol to CO_2 over the individual catalysts are not so

TABLE 1

Catalytic oxidation of CH_3OH to higher aldehydes

Cat.	^oC	WHSV	$\frac{MeOH}{O_2}$	conv. MeOH	selectivity CO_2	HCCOH	Fd	ΣC_1^{ox}	Ac	Bz	ΣC_{2+}^{ox}	$S_{C_{2+}}^{ox} = \frac{\Sigma C_{2+}^{ox}.100}{\Sigma C_1^{ox}+\Sigma C_{2+}^{ox}}$
A	450	1.3	12	99	2.07	0.05	0.17	2.29	0.35	0.4	0.76	24.9
B	450	2	3	70	4.39	0.00	28.13	32.59	1.53	0.21	1.74	5.08
C	450	2	6	82	7.91	0.00	5.16	13.07	2.15	0.20	2.35	15.24

Fd - formaldehyde, Ac - acetaldehyde, Bz - benzaldehyde

significant as in the case of the formaldehyde formation, a higher formation of CO_2 contributes also to a decrease in the relative selectivity $S_{C_{2+}}^{OX}$

Our further discussion of these results in this paper will concentrate on the catalyst A and its function at reaction conditions leading to the lowest values in methanol oxidation to both formaldehyde and CO_2. By this choice we obtain the reaction conditions and the type of catalyst where most probably both the transformation rate of methanol to higher hydrocarbons (the total methanol conversion attains 99%) on the acidobasic sites of the zeolitic component of the catalyst and the subsequent oxidative transformation rate of these higher hydrocarbons to higher aldehydes on the redox sites of the Bi-Mo-O component ($S_{C_{2+}}^{OX}$ = 24.9%) are comparable.

The total conversion of methanol and the respective selectivities for its transformation to the individual products over catalyst A as function of the reaction temperature at constant values of WHSV and $MeOH/O_2$ are given in Figs. 1 and 2. In Fig. 1a also the total conversions obtained under the same conditions over the hybridic catalyst B (mechanical mixture of zeolitic and oxidic component) are given for comparison. Similarly in Fig. 2c comparative selectivities obtained over the original HZSM-5 zeolite (Si/Al = 19) without Bi-Mo-O redox component are given.

From the data shown both in Figs. 1 and 2 and in Table 1 the following conclusions about the action of the investigated catalysts may be drawn:

1) The respective values of the total conversion of methanol in the temperature range of 400-500^oC over the original HZSM-5 and over the hybridic catalyst B (mechanical mixture) are very near to each other whereas they differ from the values obtained over the catalyst A, where the redox component has been added in the hydrothermal synthesis. From this fact it follows that on the catalyst A methanol is transformed on a different type of active sites.

2) Both the hybride catalyst B and the original zeolite HZSM-5 exihibit in the reaction temperature range of 300-500^oC in the methanol transformation significantly higher selectivities to both formaldehyde and CO_2 than the catalyst A (Figs. 1b, 1c). On the different type of active sites, specific for catalyst A,

Fig. 1. Dependence of the total conversion of methanol - x_{MeOH} (a), of the selectivity of methanol transformation to CO_2 - S_{CO_2} (b) and of the selectivity of CH_3OH transformation to formaldehyde - S_{Fd} (c) on the temperature of the catalytic reaction over the catalysts A ⊙, B ● and the original HZSM-5 (without Bi-Mo-O component) ⊡, at WHSV = 1.33 and $MeOH/O_2$ = 12.

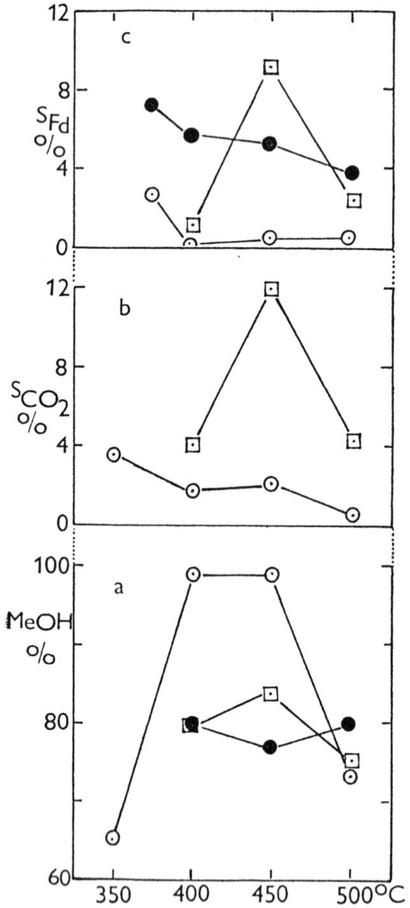

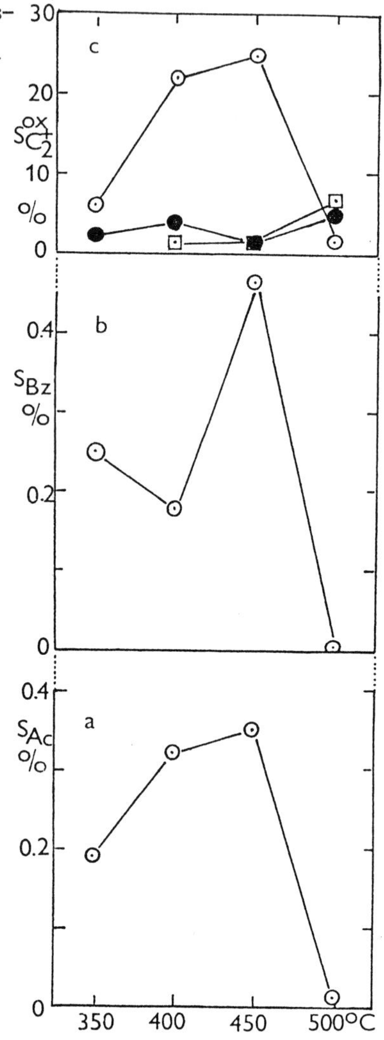

Fig. 2. Dependence of the selectivity of methanol transformation to acetaldehyde - S_{Ac} (a), of the selectivity of methanol transformation to benzaldehyde - S_{Bz} (b) and of the relative selectivity to C_{2+} oxygen containing products - $S^{ox}_{C_2}$ (c) on the reaction temperature over the catalysts A ⊙, B ● and the original HZSM-5 (without Bi-Mo-O component) ⊡, at WHSV = 1.33 and $MeOH/O_2$ = 12.

the oxidation of methanol to formaldehyde and CO_2 is reduced. Both these reactions are therefore probably connected with the active sites on the original zeolite HZSM-5 and not with the presence of the Bi-Mo-O component.
3) With the specific type of active sites on catalyst A a higher relative selectivity $S^{ox}_{C_{2+}}$ may be achieved, resulting from the formation of acetaldehyde and benzaldehyde (Figs. 2a, b, c). This type of sites is not at disposal on the catalyst B, where therefore substantially lower values of $S^{ox}_{C_{2+}}$ are obtained.
4) The results in Table 1 (the values of $S^{ox}_{C_{2+}}$ and S^{Fd}) show that the catalytic properties and therefore also the active sites of catalyst C (prepared by impregnation with the Bi-Mo-O component) are more similar to those of catalyst A (hydrothermal addition of Bi-Mo-O) than to those of catalyst B (mechanical mixture).

Surface properties

In Fig. 3 the IR spectra of the catalysts A,C in the wavenumber range of the structural OH groups are presented. With regard to the analogy with the respective band in the spectrum of the original HZSM-5 zeolite, the band at 3610 cm^{-1} may be ascribed to the proton-donor centres. The band at 3740 cm^{-1} corresponds to the terminal OH groups which do not exhibit acidobasic properties. The lower values of absorbancies of the band at 3610 cm^{-1}, found with both catalysts, indicate that a part of the Bi^{3+} ions exchanged into the cationic sites of the zeolite, and an interaction of the Bi-Mo-O redox component with the OH groups (their "neutralization") takes place. Such a neutralizing effect has been observed already before (ref. 5). The concentration of OH groups is therefore reduced, when compared to the original HZSM-5 zeolite.

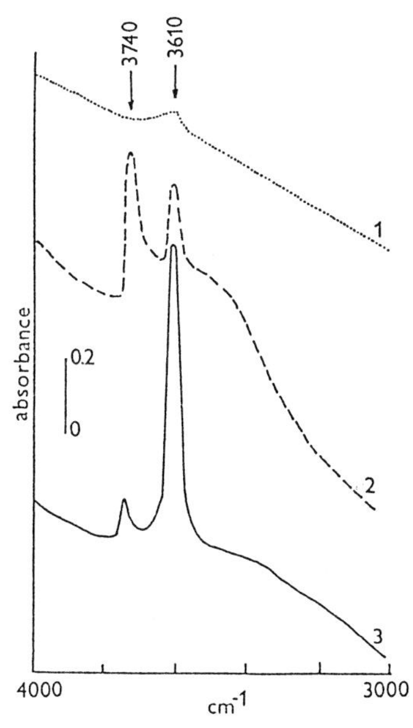

Fig. 3
IR spectra of the structural OH groups after the evacuation at 350°C overnight: 1- catalyst C, 2- catalyst A, 3- HZSM-5 without Bi-Mo-O component.

The presence of acidobasic sites in the catalysts A and C is indicated also by the IR spectra of adsorbed d_3-acetonitrile. In order to give a more precise characterization of the acidobasic properties of catalyst A, Fig. 4 presents the IR spectra of adsorbed d_3-acetonitrile at the equilibrium adsorption pressure of 2 mm Hg, adsorption temperature of 20°C and subsequent desorptions for 30 minutes at 20 and at 100°C. For comparison, also the IR spectra of d_3-acetonitrile sorption on the original HZSM-5 zeolite, registrated under the same conditions, are given. The spectrum exhibits three bands which are characteristic for d_3--acetonitrile adsorbed on different centres of the zeolitic catalysts, which are characterized by their different thermal stability in the course of the desorption (see the spectrum after desorption at 100°C). With regard to the data given in the literature (refs. 6, 7), the band at 2320 - 2330 cm^{-1} may be ascribed to the interaction of d_3-acetonitrile with the electron-acceptor sites of the catalyst A, whereas the band at 2294 - 2275 cm^{-1} to the interaction with either the proton-donor centres or the electrostatic field of the cations present (an interaction with the Bi^{3+} cations cannot be excluded). The third band at 2266 cm^{-1} has with both samples the same intensity and is thermally unstable. This band corresponds probably to a weak unspecific physical sorption of d_3-acetonitrile on the zeolite. The shift of the band at 2321 cm^{-1} (HZSM-5) to 2300 cm^{-1} (catalyst A) indicates that the electron-acceptor sites of the catalyst A are less

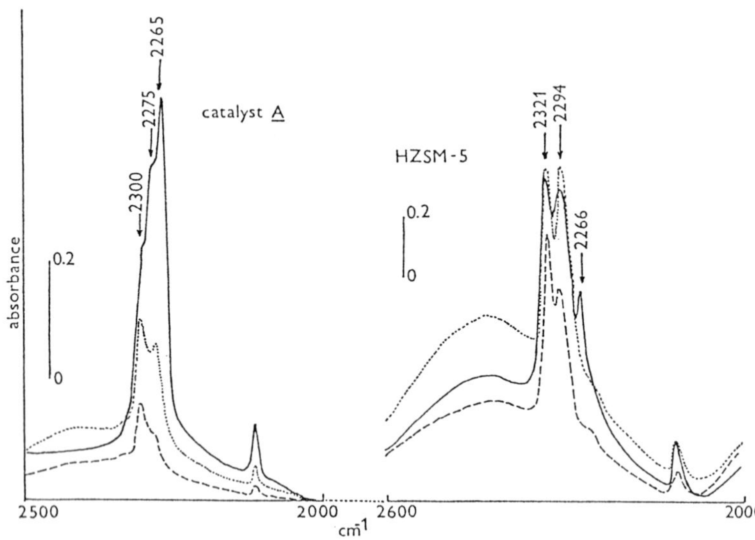

Fig. 4. IR spectra of adsorbed d_3-acetonitrile on catalyst A and on original HZSM-5 without Bi-Mo-O component after preceding evacuation at 350°C for 4 hours: sorption at 25°C ———, subsequent desorption at 25°C for 30 minutes ..... and desorption at 100°C for 30 minutes ----- .

acidic than those of the original HZSM-5. This conclusion is sustained by the lower values of both absorbance and thermal stability of this band in the case of catalyst A; this indicates also a lower concentration of electron-acceptor centres. The interpretation of the IR spectra of adsorbed d_3-acetonitrile on catalyst A represents thus another independent information on the acidobasic properties of this catalyst which is in agreement with the views given above.

Fig. 5a presents the IR spectra of the catalyst A after its interaction with methanol in the temperature range of 20 - 400°C in the wavenumber range of the vibrations of structural OH groups, Fig. 5b shows the IR spectra of this catalyst in the range of the ν (C-H) vibrations in the same temperature range.

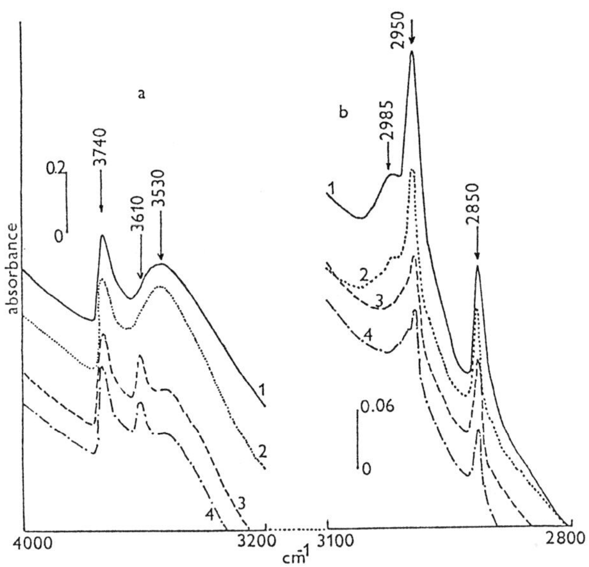

Fig. 5
IR spectra of catalyst A after its interaction with CH_3OH after preceding evacuation at 350°C for 4 hours: interaction at 25°C (1), at 100°C (2), at 200°C (3) and at 400°C (4).

After the adsorption of methanol at 20°C, followed by a short desorption of the gaseous phase at the same temperature, the spectrum in the region of vibrations of structural OH groups exihibits a broad band at 3530 cm^{-1}, shifted with respect to the band 3610 cm^{-1} to a lower wavenumber indicating thus the strong interaction of CH_3OH with proton-donor centres of the catalyst. In the region of the ν (C-H) vibrations two strong bands at 2950 cm^{-1} and 2850 cm^{-1} respectively, are formed, accompanied by a shoulder at 2985 cm^{-1}.

When the temperature of the interaction is raised to 100, 200 and 400°C, respectively, both strong bands appear in the spectrum again, but with a lower intensity when compared to the interaction at 20°C (see Fig. 5b). These stepwise changes in intensity with the increase of the temperature of interaction indicate that a part of methanol has either desorbed or reacted to its transformation products; in the course of this process a partial regeneration of the original proton-donor sites took place. Similar, although substantially less significant changes were observed in the spectrum of the catalyst C (prepared by impregnation).

The interpretation of the IR spectra in this region, indicating the formation of reactive methoxy groups, has been described previously (refs. 3, 8).

The catalyst A represents a typical bifunctional catalyst with two types of active centres, namely: i) acidobasic sites represented by proton-donor structural OH groups as well as (as it follows from the interpretation of the IR spectra of d_3-acetonitrile) electron-acceptor centres, ii) redox sites of the Bi-Mo-O component.

Presently there is no indication whether the latter component is finely dispersed on the HZSM-5 zeolite or at least partially inserted in the zeolite network.

The methoxy groups formed in the interaction of CH_3OH with an OH group seem to act as precursors in the formation of ethylene and higher hydrocarbons by participation of both types of acidobasic sites according to a mechanism described previously (8). These products can subsequently interact with the surface oxygen of the Bi-Mo-O redox component of the bifunctional catalyst under formation of the respective aldehydes. The participation of the surface oxygen in this process seems to be indicated by the change of the colour of the catalyst after its interaction with CH_3OH in the IR cuvette at higher temperatures from yellow to bluishgreen. Such a cycle of interactions (alternatively with methanol and with oxygen) may be performed repeatedly.

The priority in the formation of acetaldehyde and benzaldehyde over these catalysts is probably caused i) by the lower reactivity of these aldehydes in comparison with e.g. acroleine, ii) by the low reactivity of ethylene and toluene when compared to the olefins C_3-C_4, the latter being more readily oligomerized to surface polyenes than oxidized so that the possibility of the formation of other aldehydes is reduced.

REFERENCES

1 P. Jiru, Z. Tvaruzkova and Habersberger, React. Kinet. Catal. Lett. in press.
2 A. Batist, J.F.H. Bouwens and G.C.A. Schuit, J. Catal., 25 (1972) 1.
3 L. Kubelkova, J. Novakova and P. Jiru, Structure and Reactivity of Modified Zeolites, (P.A. Jacobs et al., Eds.) Elsevier, Amsterdam 1984, Stud. Surf. Sci. Catal. Vol. 18 p. 217.
4 Z. Tvaruzkova, M. Tupa, P. Jiru, A. Nastro, G. Giordano and F. Trifirò, Zeolites as Catalysts, Sorbents and Detergent Builders, (H.G. Karge and J. Weitkamp Eds.), Elsevier, Amsterdam 1989, Stud. Surf. Sci. Catal. in press.
5 Z. Tvaruzkova, G. Centi, P. Jiru and F. Trifirò, Appl. Catal. 19 (1985) 307.
6 C.L. Angell and M.V. Howell, J. Phys. Chem. 73 (1969) 2551.
7 H. Knözinger and H. Krietenbrinck, J. Chem. Soc. Faraday I, 71 (1976) 184.
8 E.R. Derouane, J.B. Nagy, P. Dejaifve, J.H.C. van Hoff, B.P. Speahman, J.C. Vedrine and C. Naccache, J. Catal., 53, (1978) 40.

J. HABER (Institute of Catalysis and Surface Chemistry, Polish Academy of Sciences, Krakow, Poland): In the Mobil process methanol is passed over the ZSM catalyst in the absence of oxygen, and hydrogen transfer reactions are important steps in the chain leading from methanol to higher olefins and aromatics. One may wonder whether the failure to obtain aldehydes in your experiments was not due to the fact that the presence of oxygen in the feed has strongly perturbed the hydrogen transfer steps?

P. JIRŮ (J. Heyrovsky Institute of Physical Chemistry and Electrochemistry, Czechoslovak Academy of Sciences, Prague, Czechoslovakia): In the oxidative transformation of CH_3OH over the investigated catalysts A,B and C the total conversion of CH_3OH (see Tab. 1) was in the range of 70-90%, whereas the conversion of CH_3OH to oxygenated products was in the range of 2-24%, the remaining methanol transformation products being higher hydrocarbons. From this fact it follows that the presence of oxygen in the feed has no significant effect on the hydrogen transfer step.

P. JACOBS (Katholieke Universiteit Leuven, Laboratory of Surface Chemistry, Leuven, Belgium): When a ZSM-5 zeolite is synthesized in presence of Mo and Bi, a very complex material may result. Mo and/or Bi may be occluded in the zeolite crystal or even in minor amounts be substituted in the lattice. Furthermore, Bi-Mo-O phases external to the zeolite may also be present. Your IR spectra seem to suggest that at least part of the negative framework charge is neutralized by Bi or Mo (oxide). My question, therefore, is how you do visualize in terms of chemical composition and morphology your catalyst A?

P. JIRŮ (J. Heyrovsky Institute of Physical Chemistry and Electrochemistry, Czechoslovak Academy of Sciences, Prague, Czechoslovakia): The synthesized catalyst represents in effect a very complex system. By electrone microscopy it was found that the Bi-Mo-O component in this case most probably forms a thin layer covering the surface of the ZSM-5 zeolite. Our experiments give no evidence whether a part of Bi and/or Mo has entered the framework of the zeolite or not. The fact that, in comparison with the original zeolite, the intensity of the absorption bands of the OH groups of the ZSM-5 zeolite with the Bi-Mo-O component is decreased (see Fig. 3) indicates either that the hydrogen ions are partly substituted by Bi ions or that the Bi-Mo-O redox component interacts with the OH groups of the ZSM-5 zeolite. These results indicate that a part of the redox component is really located in the pores of the zeolite.

G. Centi and F. Trifiro' (Editors), *New Developments in Selective Oxidation* 327
1990 Elsevier Science Publishers B.V., Amsterdam — Printed in The Netherlands

TIN-GERMANIUM PHOSPHATES AS SELECTIVE CATALYSTS FOR THE OXIDATIVE DEHYDROGENATION OF ETHYLBENZENE TO STYRENE

Turco M.[1], Bagnasco G.[1], Ciambelli P.[2], La Ginestra A.[3], Russo G.[1]

[1]Dipartimento di Ingegneria Chimica, Università di Napoli, Italy

[2]Dipartimento di Chimica, Università di Napoli, Italy

[3]Dipartimento di Chimica, Università "La Sapienza", Roma, Italy

ABSTRACT

Tin-Germanium phosphates, with formula $Sn_xGe_{1-x}(HPO_4)_2.H_2O$ $(0\leq x\leq 1)$ were synthesized, characterized and tested as catalysts for the oxydehydrogenation of ethylbenzene to styrene. X-ray analysis showed that mixed compounds form solid solutions, in agreement with thermal analysis. Surface acid sites concentration increased with Ge content. Mixed compounds were more active than pure phosphates and were highly selective to ST (up to 97%), giving mainly CO_x as byproducts. The role of acidity is discussed.

INTRODUCTION

Styrene (ST) is produced on industrial scale by catalytic dehydrogenation of ethylbenzene (EB). Such a process requires high temperatures (600-700°C) and high amounts of steam giving rise to conversions of abt. 50% with selectivity to ST higher than 90%. ST production could be performed through EB oxidative dehydrogenation at lower temperatures with ST yields not limited by thermodynamics.

Among the catalysts described in literature for this reaction metal phosphates showed high selectivity and activity (1,2).

Recent works have indicated that mixed Zr-Ti, Zr-Sn phosphates present improved catalytic properties in respect to pure compounds (3, 4), although no correlation between catalytic activity and structural properties was found. On the base of these results we have studied the Sn-Ge mixed phosphates, with formula $Sn_xGe_{1-x}(HPO_4)_2\cdot H_2O$ $(0\leq x\leq 1)$, as catalysts for the oxydehydrogenation of EB to ST.

EXPERIMENTAL

Pure Sn and Ge phosphates were synthesized as described in (5). The mixed Sn-Ge phosphates were obtained by coprecipitation from the corresponding chlorides in different molar ratios (3:1, 1:1, 1:3) by adding H_3PO_4 and HNO_3.

The mixtures were refluxed for 100 h and the white precipitates were filtered and washed with ethanol.

X-ray analysis was carried out by a Philips diffractometer using filtered CuK_a radiation. Simultaneous DTA and TG thermal analysis was performed by a Stanton mod. 781 thermoanalyzer (Pt-Pt/Rh thermocouples) at 2-5°C/min heating rate. SEM micrographs were obtained by Hitachi 2300 apparatus. Surface areas were measured by N_2 adsorption at -196°C using a Carlo Erba Sorptomatic 1800. Acidity measurements were effected by NH_3 TPD as described in (6). The NH_3 adsorption was effected at room temperature on samples pretreated at 450 and 600°C.

The measurements of catalytic activity were performed in a fixed bed flow reactor (i. d.=12 mm) loaded with 0.7-4.0 g of catalyst. The samples were pretreated in a 10% O_2/N_2 mixture at the same temperature of the reaction overnight. EB was fed by a metering pump. The reaction products were analyzed by gaschromatography. The catalytic tests were effected in following conditions: reaction temperature from 450 to 550°C, gaseous flow rate of 10 Nl h^{-1}, contact time from 0.16 to 0.9 g cat/g EB/h, EB molar fraction=0.1.

RESULTS AND DISCUSSION

Table 1 lists the chemical composition of the materials. In Fig. 1 are reported the simultaneous TG and DTA curves of the five samples. The dehydration process occurs for SnP between 40 and 200°C while for GeP it occurs at higher temperatures (200-350°C). For the mixed phases we can observe an intermediate behaviour. Moreover GeP shows a well evident reversible phase transition at 400°C, absent in pure SnP. Also the mixed phases with high Ge content show the same phase transition between 350 and 400°C. All the samples are condensed to pyrophosphates between 400 and 650°C. Between 900 and 1000°C the mixed phases show an exothermic effect due to the transformation to cubic pyrophosphates.

In Table 1 the d_{002} values obtained from X-ray analysis on the hydrogen phosphates phases are reported. They indicate that the mixed Sn-Ge phopsphates give rise to solid solutions in all the range of compositions examined. After treating at 650°C for 12 h pure GeP and SnP samples are completely transformed to layered pyrophosphates phases while the condensation process of mixed compounds is not complete. After treatment at 450°C such process is incomplete for all the samples. After treatment at 1000°C the samples show the signals of cubic pyrophosphates: their gradual shift with composition indicates that solid solutions are present.

TABLE 1. Chemical composition and d_{002} values of Sn-Ge phosphates.

SAMPLE	CHEMICAL FORMULA	d_{002} Å
GeP	$Ge(HPO_4)_2 \cdot H_2O$	7.70
SnGe13	$Sn_{0.20}Ge_{0.80}(HPO_4)_2 \cdot H_2O$	7.78
SnGe11	$Sn_{0.41}Ge_{0.59}(HPO_4)_2 \cdot H_2O$	7.82
SnGe31	$Sn_{0.52}Ge_{0.48}(HPO_4)_2 \cdot 1.1H_2O$	7.85
SnP	$Sn(HPO_4)_2 \cdot 1.5H_2O$	7.89

In Fig. 2 SEM micrographs are reported. GeP sample shows aggregates of laminar shaped particles, 1 to 2μm large and 0.1μm thick. In micrograph of SnP sample we can observe complex structure formed by lamellar shaped elements (diameter=2μm, thickness=0.05μm), linked together as a regular succession. SnGe13 and SnGe31 show aggregates of particles with dimension smaller than those of pure compounds (0.5μm); the morphology of particles varies gradually with composition; moreover the tendency to give aggregates increases with tin content.

After treatment at 450 and 600°C surface areas of mixed compounds are higher than those of pure compounds (Table 2), achieving a maximum for SnGe11 sample. However the treatment at 600°C leads to a marked increas of surface area of pure SnP and GeP samples.

The surface concentration of acid sites, evaluated as in (6), and temperature of peak maxima in TPD spectra are also reported in Table 2. We can observe an increase of acid sites concentration with Ge content. The acid strength increases with thermal treatment at 600°C, particularly for pure compounds. The large acid site concentration of GeP treated at 600°C could be releated to the hydrolysis of Ge-O-P bonds leading to formation of Ge-O-H groups.

In Fig. 3 the EB total conversion, selectivity to ST and BA, and surface area as function of chemical composition are shown . Mixed compounds give higher EB conversion in respect to pure phosphates. Moreover these values are

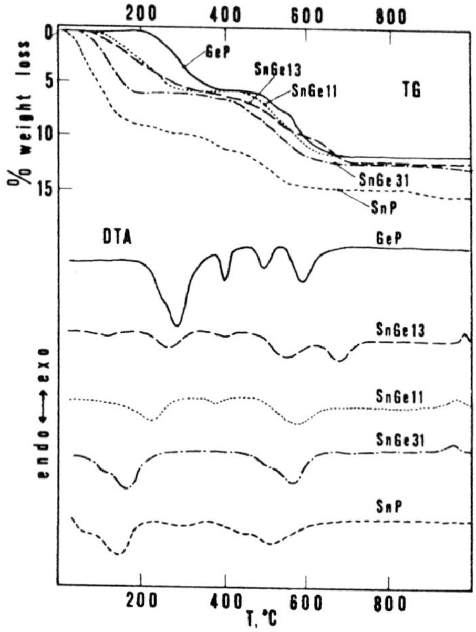

Fig. 1. TG and DTA curves of Sn-Ge phosphates

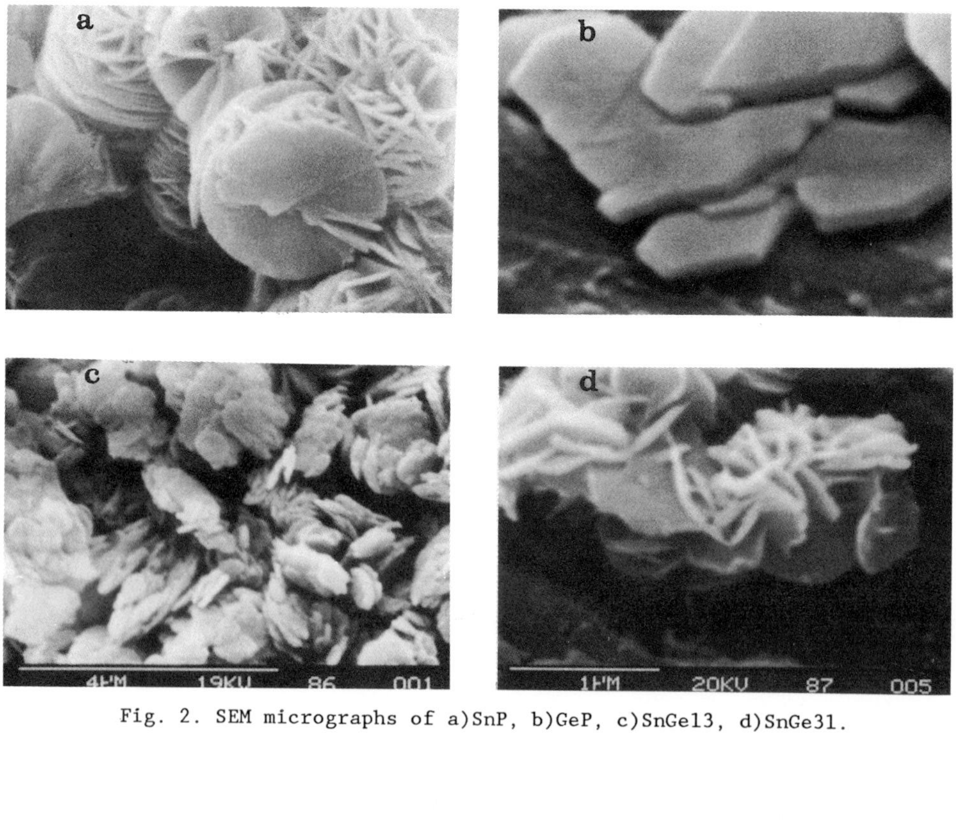

Fig. 2. SEM micrographs of a)SnP, b)GeP, c)SnGe13, d)SnGe31.

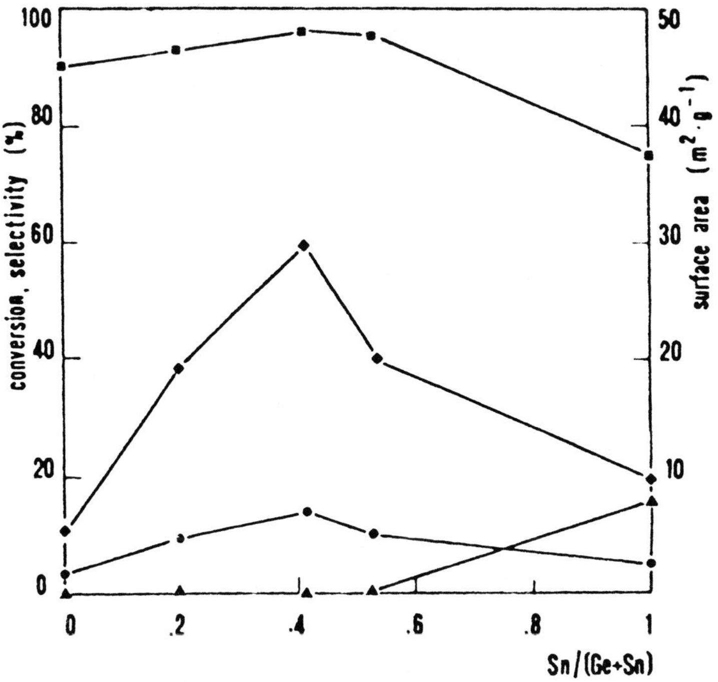

Fig. 3. (●) EB conversion, (■) selectivity to ST, (▲) selectivity to BA, (◆) surface area as a function of chemical composition of Sn-Ge phosphates. Contact time-0.4 g cat./EB/h; T-450°C; other conditions in the text.

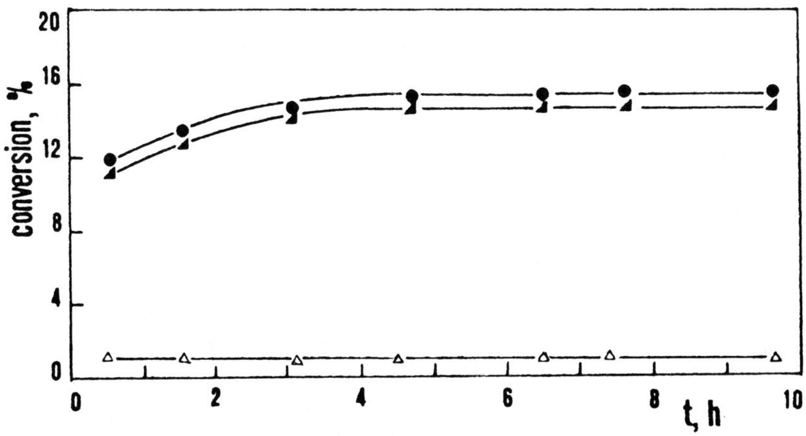

Fig. 4. (●) EB conversion, conversions to (◢) ST and to (△) CO_x as a function of time on stream for SnGeII sample. Contact time-0.42 g cat/g EB/h, T-450°C; other conditions in the text.

TABLE 2. Surface areas and surface acidity of Sn-Ge phosphates.

SAMPLE	SURFACE AREA ($m^2\ g^{-1}$)		ACID SITES CONCENTRATION ($cm^{-2}x10^{-14}$)		T_{max} (°C)	
	a)	b)	a)	b)	a)	b)
GeP	5.2	15.3	5.5	7.0	126	154
SnGe13	18.7	18.4	4.6	4.1	140	140
SnGe11	29.7	31.0	5.0	4.6	142	154
SnGe31	20.0	20.0	4.2	3.6	140	148
SnP	10.0	19.7	2.2	2.2	140	170

a) after treatment at 450°C, b) after treatment at 600°C.

TABLE 3. Catalytic activity of Sn-Ge phosphates.
T=450°C, other conditions as reported in the text.

SAMPLE	CONTACT TIME h	EB CONVERSION %	SELECTIVITY %		
			ST	BA	CO_x
GeP	0.16	1.4	80	0	20
	0.42	4.0	90	0	10
	0.90	12.0	96	0	4
SnGe13	0.16	4.2	93	0	7
	0.42	9.5	93	1	6
	0.90	16.8	93	1	6
SnGe11	0.16	7.7	97	0	3
	0.42	14.5	96	1	3
	0.90	26.0	95	1	4
SnGe31	0.16	3.7	96	0	4
	0.42	7.0	96	0	4
	0.90	16.0	97	1	2
SnP	0.16	3.4	56	35	9
	0.42	5.6	75	16	9
	0.90	7.5	92	0	8

higher than those found over different mixed systems such as Sn-Zr phosphates (4) because of the neglectable conversion to BA.

As shown in Table 3 EB conversion is strongly dependent on contact time in the range 0.16-0.9 g cat./g EB/h. Selectivity to ST over Sn-Ge systems is maintained at higher EB conversion, whereas it is enhanced over pure compounds. It must be remarked that SnP improved selectivity to ST as a result of the progressive decreasing of conversion to BA.

The effect of temperature on conversion and selectivity over Sn-Ge catalysts is shown in Table 4. For all the systems the large increase of EB conversion at 500 and 540°C resulted in a decrease of selectivity to ST due to improved oxidation to CO_x. Therefore the performance of Sn-Ge catalysts at 450°C is comparable to that reported for Zr phosphate (1), but at higher temperature selectivity to ST is lower than that reported in (2) for different phosphates.

TABLE 4. Catalytic activity of Sn-Ge phosphates at a) T=500°C and b) T=540°C. Other conditions in the text.

SAMPLE	EB CONVERSION %		SELECTIVITY %					
			ST		BA		CO_x	
	a)	b)	a)	b)	a)	b)	a)	b)
SnGe13	21.0	32.6	85	69	0	0	15	31
SnGe11	31.7	54.0	88	64	1	0	11	36
SnGe31	20.0	28.1	84	67	1	0	15	33

In all experimental conditions investigated catalytic activity increased during the first 2-3 hours of the reaction up to constant conversions, as shown for SnGe11 sample in Fig. 4. Coke formation was observed for both pure and mixed compounds without loss of activity after 10 hours on stream. This behaviour is typical of many different acid catalysts reported in the literature for EB oxydehydrogenation (1, 2, 7, 8). Tagawa et al. (7) correlated the oxydehydrogenation activity of various acidic catalysts with their moderate acid strength, whereas Fiedorow (8) suggested that coke formed on the sites of moderate acid strength was the effective catalyst for ST formation on alumina. More recently Hattori et al. (10) showed that over SnO_2/SiO_2 catalysts prepared by chemical vapor deposition, which were very selective for ST formation, deposition of coke was not observed. With reference to metal phosphates Schraut et al. (11) proposed that an active coke deposited on zirconium phosphate during the first hours of reaction is the actual catalyst, but surface acidity does not play any role, as it disappears at the reaction temperature. Vrieland (2) found very high conversion and selectivity over many different metal pyrophosphates and proposed that a carbonaceous overlayer formed on pyrophosphate groups of moderately acid strength is the actual catalyst surface. The results obtained by us can be interpreted by this "active coke model", but we think that the role of acidity should be carefully taken into account. In fact, we have shown by NH_3 TPD (6) that after treatment at 600°C surface acidity of Zr phosphate is still present and it has been confirmed by IR measurements (12). The same conclusion can be extended to Sn-Ge mixed systems. For all these phosphates the surface sites are not pyrophosphate groups, but P-OH sites of medium strength with lower amount of stronger sites. These medium strength sites should be responsible for the formation of a selective carbonaceous surface layer, as proposed by Vrieland (2). On the other hand high selectivity to BA seems to be specific of SnP. Therefore the eventual role of metal ion on its formation should be not excluded.

REFERENCES.

1) Emig, G., and Hofmann, H., J. Catal. 84, 15 (1983).
2) Vrieland, G. E., J. Catal. 111, 1 (1988).
3) Frianeza, T. N., and Clearfield, A., J. Catal. 85, 398 (1984).
4) Galli, P., La Ginestra, A., Patrono, P., Massucci, M. A., Ferragina, C., Ciambelli, P., and Bagnasco, G., Italian Patent 21587 A/86.
5) La Ginestra, A., Patrono, P., Berardelli, M. L., Galli, P., Ferragina, C., and Massucci, M. A., J. Catal. 103, 346 (1987).
6) Turco, M., Ciambelli, P., Bagnasco, G., La Ginestra, A., Galli, P., and Ferragina, C., J. Catal. 117, 355 (1989).
7) Tagawa, T., Hattori, T., Murakami, Y., J. Catal. 75, 56 (1982)
8) Alkhazov, T. G., Lisovskij, A. E., Safarov, M. G., and Dadasheva, A. M., Kinet. Catal. 13, 509 (1972).
9) Fiedorow, R., Przystajko, W., Sopa, M., and Dalla Lana, I. G., J. Catal. 68, 33 (1981).
10) Hattori, T., Itoh, S., Tagawa, T., and Murakami, Y., Studies Surf. Sci. Catal. 31, 113 (1987).
11) Schraut, A., Emig, G., and Hofmann, H., J. Catal. 112, 221 (1988).
12) Ramis, G., Busca, G., Lorenzelli, V., La Ginestra, A., Galli, P., and Massucci, M. A., J. Chem. Soc. Dalton Trans., 881 (1988).

G. Centi and F. Trifiro' (Editors), *New Developments in Selective Oxidation*
1990 Elsevier Science Publishers B.V., Amsterdam — Printed in The Netherlands

SELECTIVE PHOTOCHEMICAL CONVERSION OF METHANE INTO WORTHIER COMPOUNDS

K. OGURA*, C. T. MIGITA, T. YAMADA and S. CHAKI

Department of Applied Chemistry, Yamaguchi University, Tokiwadai Ube 755 (Japan)

SUMMARY

Methane and ammonia has been photochemically converted with no catalyst to methylamine and ethylenediamine. Methylamine is produced by the reaction of CH_3 and NH_2 radicals which are provided by the hydrogen-abstraction from methane and ammonia, respectively. The reaction of $CH_3\cdot$ with $NH_2\cdot$ competed with the formation reactions of methanol and ethane, but the former reaction became predominant in the presence of higher concentration of ammonia. Three radical species, CH_3O, CH_2NH_2, and CH=NH were detected as the reaction intermediates by the spin trap-ESR method, and it was suggested that the generation of ethylenediamine is attributed to the coupling of CH_2NH_2 radicals.

INTRODUCTION

Methane is the main constituent of natural gas which is not so unevenly distributed as crude oil. Prospects of future carbon resource after the depletion of petroleum are good, provided that natural gas can be used as raw material for the chemical industry. However, direct chemical utilization of methane is very difficult, and methane "activation" by some means is essential. For instance, Solymosi et al. (ref. 1) have reported the conversion of methane to formaldehyde with N_2O over a Bi_2O_3-SnO_2 catalyst at 550 °C. Kitayama and Schwartz (ref. 2) employ silica-supported rhodium complexes in the catalytic conversion of methane to methylchloride and hydrogen chloride with lesser amounts of other chlorinated methane. In our previous works (refs. 3, 4), methane was activated with hydroxyl radicals which are formed by the photolysis of water. In the process, the initial activation of methane is caused by hydrogen-abstraction from methane. The reaction of methyl radical with $OH\cdot$ and $CH_3\cdot$ leads to methanol and ethane, respectively.

$$H_2O \xrightarrow{h\nu} \cdot H + \cdot OH \qquad (1)$$

$$CH_4 + \cdot OH \longrightarrow \cdot CH_3 + H_2O \qquad (2)$$

$$\cdot CH_3 + \cdot OH \longrightarrow CH_3OH \quad (3)$$

$$\cdot CH_3 + \cdot CH_3 \longrightarrow C_2H_6 \quad (4)$$

In the present work, methane was converted to nitrogen- and oxygen-containing compounds in the photolysis of the gas mixtures of methane, ammonia, and water. The major products were methylamine and ethylenediamine except methanol, ethane, and hydrogen, and the mechanism for the formation of these products was investigated.

EXPERIMENTAL

The photochemical reaction apparatus consists of the reaction chamber, water-cooled condenser, solution reservoir, and pressure-adjusting tank. A 50 W low-pressure mercury lamp made of synthetic quartz, which was used as the light source, was put in the reaction chamber (2.5 dm^3). The radiation was polychromatic, and the wave lengths of 185 and 254 nm were main. A flask (1 dm^3) containing 0.4 dm^3 of water was connected to the reaction chamber, and the total volume of the whole system was about 6.6 dm^3. Various initial concentrations of methane were adjusted by changing the volume ratio of methane to nitrogen. After the whole system was filled with methane, a given amount of aqueous ammonia was admitted.

The reaction products were found in the condensed solution and gas phase for the most part. The quantitative analyses of the products were performed with a gas chromatograph (Shimadzu GC-8A, JEOL JGC-1100), a steam chromatograph (Ohkura Model SSC-1), and a high-performance liquid chromatograph (HPLC, Hitachi 655A). The gas chromatograph was used at 100 °C with a flame ionization detector (FID) and a Porapak Q column or at 30 °C with a thermal conductivity detector (TCD) and a molecular sieve 5A column. The steam chromatograph was employed at 130 °C with a FID and a Porapak R column, and the high-performance liquid chromatograph was employed at 60 °C with an UV monitor and a GL-C610H column. Formaldehyde was determined by a colorimetric analysis using chromotropic acid. The absorption spectrum of the solution was obtained with a Hitachi 100-50 type double-beam spectrometer.

Reaction intermediates were detected by the spin trap-ESR method. The spin trap was the silica gel-PBN (α-phenyl N-tert-butylnitrone) mixture. The sample for the purpose of ESR measurements was prepared by flowing the photolyzed gases onto the spin trap and dissolving the PBN-adducts in benzene. ESR spectra were recorded on a JES-ME-1X spectrometer with 100 kHz field modulation and 1 mW x-band microwave power.

RESULTS AND DISCUSSION

The photolysis of the CH_4-NH_3-H_2O mixture led to the formation of methylamine and ethylenediamine with small amounts of acetonitril, monoethanolamine, diethanolamine, and nitromethane in addition to oxygen-containing species, ethane, and hydrogen found in the photochemical reaction of CH_4 with H_2O. The results obtained are shown in Table 1 where aqueous ammonia was added by two different methods: (A) before beginning of the photolysis and (B) every 30 min during the photolysis. It is seen that the yield of ethane is smaller in method B than in method A, but methylamine

TABLE 1

Products (μmol) obtained in the photolysis of the CH_4-NH_3-H_2O mixture.[a]

	Ammonia (mmol)			
	28.8		57.6	
	A[b]	B	A	B
CH_3NH_2	1809	2313	2247	2761
CH_3NO_2	4	5	-	9
CH_3CN	62	103	75	67
$NH_2C_2H_4NH_2$	730	195	651	651
$NH_2C_2H_4OH$	-	9	4	15
$NH(C_2H_4OH)_2$	3	1	-	9
CH_3OH	1067	1686	1578	1658
C_2H_5OH	115	87	59	58
HCHO	86	98	72	163
C_2H_6	4620	3080	2920	1810
H_2	18940	19160	19040	19540

[a]Amount of methane, 275 mmol; reaction time, 5h; temperature, 100 °C.
[b]A, ammonia was added before beginning of the electrolysis; B, 5 ml of aqueous ammonia was added every 30 min during the photolysis.

gives the opposite results.

The hydrogen atom of NH_3 is abstracted by OH radical formed in reaction 1:

$$NH_3 + \cdot OH \longrightarrow \cdot NH_2 + H_2O \quad (5)$$

Methylamine is produced by the reaction of $NH_2\cdot$ with $CH_3\cdot$:

$$\cdot CH_3 + \cdot NH_2 \longrightarrow CH_3NH_2 \quad (6)$$

Reactions 3, 4 and 6 are competitive, and high ammonia concentration during the photolysis may expedite reaction 6, leading to higher yield of methylamine.

As shown in Table 1, considerable amount of hydrogen is formed regardless of the presence or absence of ammonia. Two possible routes for the formation of hydrogen are considered:

$$\cdot H + \cdot H \longrightarrow H_2 \quad (7)$$

$$CH_4 + \cdot H \longrightarrow H_2 + \cdot CH_3 \quad (8)$$

However, the formation rate of hydrogen via reaction 8 is probably much smaller than that via reaction 7, which bases on the

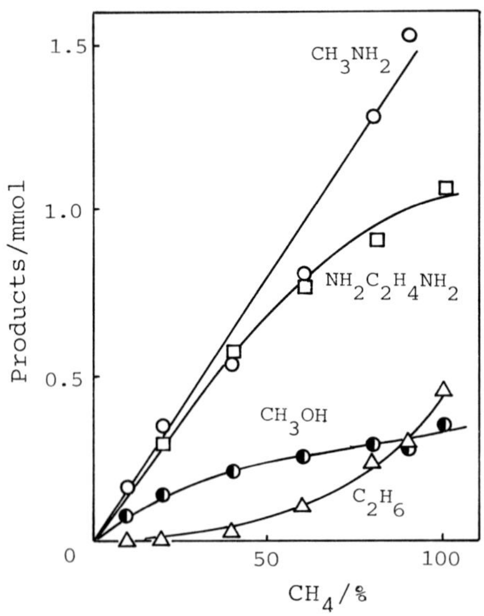

Fig. 1. Products versus initial volume % of methane. Reaction time, 5h; temperature, 100 °C; [NH_3], 432 mmol.

large activation energy (11.9 kcal mol^{-1}) for reaction 8 (ref. 5). Moreover, the major source of CH_3 radicals may be ascribed not to reaction 8 but to reaction 2, because the activation energy for this reaction is about one-half (4.8 kcal mol^{-1}) of that for reaction 8 (ref. 5).

The four products, which were most abundant, are displayed versus the volume percentage of methane in Fig. 1. In this photochemical system, 100 volume % of methane was equivalent to 275 mmol. The added quantity of ammonia was kept to 432 mmol, and hence there was an excess of ammonia. The yield of methylamine is approximately proportional to the initial concentration of methane. The yields of ethylenediamine and methanol are both inclined to be saturated at higher concentration of reaction gas. On the other hand, the yield of ethane shows the exponential increase from about 40 volume % of methane. These results indicate that in the presence of the excess ammonia, $CH_3\cdot$ first reacts with $NH_2\cdot$ and $OH\cdot$ to form methylamine and methanol, ethylenediamine is the secondary product from methylamine as described below, and the formation

TABLE 2

Products (μmol) obtained in the photolysis of CH_3NH_2-H_2O and C_2H_6-NH_3-H_2O mixtures.[a]

	CH_3NH_2[b]	C_2H_6 + NH_3[c]
CH_3NH_2	-	1081
CH_3NO_2	0	22
CH_3CN	13	322
$NH_2C_2H_4NH_2$	211	57
$NH_2C_2H_4OH$	26	1
$NH(C_2H_4OH)_2$	5	86
CH_3OH	138	1434
C_2H_5OH	3	840
HCHO	35	113
C_2H_6	310	-
CH_4	1810	3890
H_2	17250	19920

[a]Reaction time, 5h; temperature, 100 °C.
[b]3.86 mmol.
[c]C_2H_6, 41.6 mmol; NH_3, 144 mmol.

of ethane becomes conspicuous at higher concentration of methane.

Methylamine and ethane plus ammonia both could be a candidate as the starting material for the formation of ethylenediamine. To examine this, the photolysis of methylamine and ethane plus ammonia was made in the presence of water, and the product distribution is shown in Table 2. As seen from this table, methylamine gives about four times as much ethylenediamine as ethane does though the initial concentration of ethane is 10 times that of methylamine. The photolysis of ethane plus ammonia leads to large amounts of methanol and methane, and in this photolysis the splitting of the C-C bond rather than the hydrogen-abstraction from ethane seems to be predominant. Hence, the reaction route via methylamine is reasonable for the formation of ethylenediamine, and the diamine observed in the photolysis of C_2H_6 + NH_3 (see Table 2) is probably originated from the methylamine produced in the photochemical reaction of ethane with ammonia.

Hyperfine coupling constants of the PBN-adducts obtained from CH_4-NH_3-H_2O system are shown in Table 3 along with the data which have been reported previously by the liquid-phase trapping method. From these values, three radicals were assigned: $CH_3O\cdot$, $\cdot CH_2NH_2$, and $\cdot$CH=NH. The assignment for methylene iminoyl radical bases on the data reported by Janzen et al. (fer. 7). The finding of CH_2NH_2 radical leads to following mechanism for the formation of ethylenediamine.

TABLE 3
Hyperfine coupling constants of the PBN-adducts obtained from CH_4-NH_3-H_2O system.

Radicals	a(N)/mT	a(β-H)/mT	
CH_3O	1.35	0.20	a
	1.36	0.20	b
CH_2NH_2	1.48	0.35	a
	1.47	0.35	b
CH=NH	1.49	0.64	a

[a]Solid-phase trapping (this work).
[b]Liquid-phase trapping (ref. 6).

$$CH_3NH_2 \quad + \quad \cdot OH \longrightarrow \cdot CH_2NH_2 \qquad (9)$$

$$\cdot CH_2NH_2 \quad + \quad \cdot CH_2NH_2 \longrightarrow NH_2C_2H_4NH_2 \qquad (10)$$

CH_3O and CH=NH radicals are formed in the photolysis of methanol (ref. 8) and by hydrogen-abstraction from CH_2NH_2 radicals (ref. 9), respectively.

The formation rate (R_m) of methylamine was 39.2 and 63.0 μmol dm^{-3} h^{-1} at 70 and 90 °C, respectively, and the Arrhenius plot of R_m is shown in Fig. 2a. The corresponding activation energies were 3.1 and 6.9 kcal mol^{-1} at the NH_3 concentrations of 86.4 and 28.9 mmol, respectively. The NH_3 concentration-dependency of the activation energy indicates apparently the existence of the secondary reaction of methylamine. As described above, ethylenediamine is formed through the coupling of CH_2NH_2 radicals, and the total formation rate of methylamine may be considered as the sum of the formation rates of methylamine and ethylenediamine:

$$R_{m+e} = R(CH_3NH_2) \quad + \quad R(NH_2C_2H_4NH_2) \qquad (11)$$

In Fig. 2b, R_{m+e} is plotted against reciprocal temperature, and the activation energy was 4.8 kcal mol^{-1} indipendent of the added

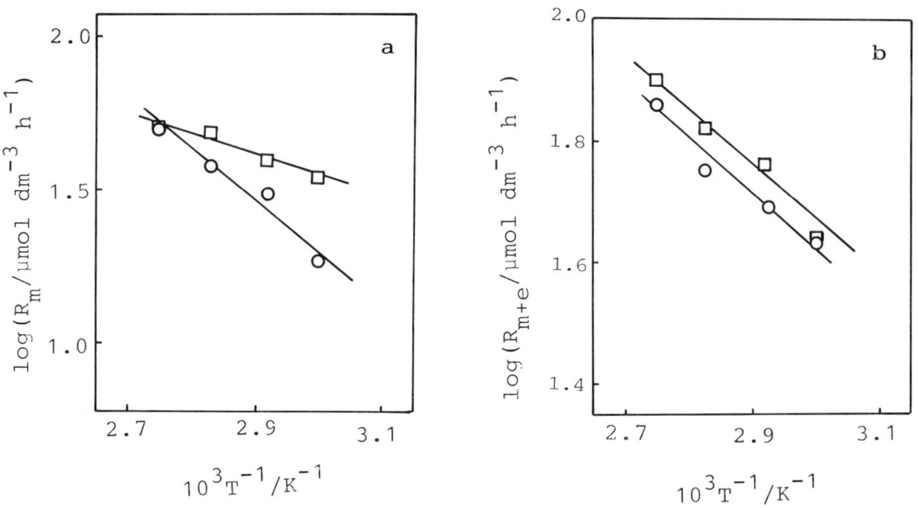

Fig. 2. Arrhenius plot of the formation rates on the presence of ammonia: (□) 86.4 and (○) 28.8 mmol; [CH_4], 80%. (a) Methyamine (R_m). (b) Methylamine plus ethylenediamine (R_{m+e}).

concentration of ammonia. Accordingly, this result also supports the formation route of ethylenediamine via methylamine.

Thus, the gas mixtures of methane, ammonia, and water can be converted to nitrogen-containing compounds such as methylamine and ethylenediamine by the photochemical reactions, and the dimerization of $\cdot CH_2NH_2$ leads to the formation of ethylenediamine. This process is noble as a scientific idea, but further works to improve the yields of the products are required for practical application.

REFERENCIES

1 F. Solymosi, I. Tombácz and G. Kutsan, Partial oxidation of methane by nitrous oxide over Bi_2O_3-SnO_2, J. Chem. Soc., Chem. Commun., (1985) 1455-1456.
2 N. Kitajima and J. Schwartz, Activation of methane by supported rhodium complexes, J. Am. Chem. Soc., 106 (1984) 2220-2222.
3 K. Ogura and M. Kataoka, Photochemical conversion of methane, J. Mol. Catal., 43 (1988) 371-379.
4 K. Ogura, C. T. Migita and M. Fujita, Conversion of methane to oxygen-containing compounds by the photochemical reaction, Ind. Eng. Chem. Res., 27 (8) (1988) 1387-1390.
5 M. Imoto, Free radicals (in Japanese), Kagaku Dojin, Kyoto, 1975, pp. 111-116.
6 K. Ogura, C. T. Migita and T. Yamada, Photochemical formation of methylamine and ethylenediamine from gas mixtures of methane, ammonia and water, Chem. Lett., (1988) 1563-1566.
7 E. G. Janzen and H. J. Stronks, Assignment of the ESR spectrum of the cyanyl radical spin adduct of phenyl tert-butyl nitrone, J. Phys. Chem., 85 (1981) 3952-3954.
8 C. T. Migita, S. Chaki and K. Ogura, ESR spectroscopic detection of methoxyl radicals formed in the photochemical gas-phase reaction of methane and water, J. Phys. Chem., in press.
9 C. T. Migita, S. Chaki and K. Ogura, Application of the spin-trap-ESR method for the detection of radical intermediates produced in the photochemical reaction gases, Nippon Kagaku Kaishi, (1989) 1233-1239.

G. Centi and F. Trifiro' (Editors), *New Developments in Selective Oxidation*

WORKING PRINCIPLE OF Li DOPED MgO APPLIED FOR THE OXIDATIVE COUPLING OF METHANE

J.M.N. van Kasteren, J.W.M.H. Geerts and K. van der Wiele

Department of Chemical Technology, University of Technology,
P.O. Box 513, 5600 MB Eindhoven (The Netherlands)

SUMMARY

The nature of the active compound in Li doped MgO was investigated by comparing the activity and deactivation of Li/MgO catalysts with that of Li_2CO_3 supported on an inert carrier (ZrO_2). The conclusion is that Li_2CO_3 itself is a very active catalyst (or a catalyst precursor). Also the role of the catalyst in the oxidative coupling of methane was determined: The selectivity of the active catalyst is mainly due to a very high production rate of methyl radicals.

INTRODUCTION

According to Driscoll et al. (ref. 1), the catalytic activity of Li/MgO is due to the presence of Li^+O^- centres, stabilized in the MgO matrix. He has shown that the formation of methyl radicals is proportional to the amount of the centres as detected by EPR. However, the reported amount of lithium dopant needed for an optimal activity is extremely high. On the other hand, deactivation occurs due to loss of lithium, in contrast with the assumed stability of Li^+ centres in the MgO matrix.

Therefore, the lithium content and catalytic activity of Li/MgO catalysts during deactivation were examined in more detail. Moreover, the performance of Li/MgO was compared to that of Li_2CO_3 on ZrO_2, a carrier which does not interact with lithium compounds at the temperatures used.

EXPERIMENTAL

Lithium doped magnesia was prepared by wet impregnation of MgO with Li_2CO_3. The lithium content was varied from 1 to 7wt% Li/Li+MgO. After drying of the paste at 120°C for 16 hours the catalyst was calcined for 4 hours at 900°C. The catalyst was ground and sieved to the desired particle size of 0.3 - 0.5 mm.

The methane oxidation experiments were carried out in a micro fixed bed flow set-up described elsewhere (ref. 2). The process conditions used were: Temperature 800°C, atmospheric pressure,

CH_4/O_2=5, CH_4/He=1.25, pseudo contact time (W/F) varied between 0.3-0.6 g.s/cm^3(s.t.p.).

RESULTS AND DISCUSSION

Figure 1 shows the normal behaviour of our "home made" 7 wt% Li/MgO catalyst during the first 50 hours on stream. The oxygen conversion increases during the first 16 hours to a maximum which can only be maintained for a few hours. After this, deactivation sets in and the oxygen conversion decreases continuously. Also in the same figure, the total lithium content of the catalyst is plotted as function of the time on stream showing a continuous loss of lithium. Although the lithium content decreases from the beginning of the reaction, the activity for oxidative coupling initially increases. This means that only part of the total lithium present in the catalyst is responsible for the activity.

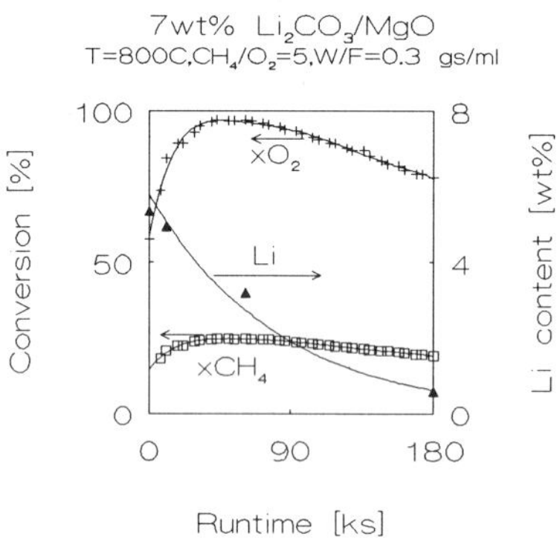

Fig. 1. CH_4, O_2 conversion and lithium content of a 7 wt% Li/MgO catalyst versus runtime.

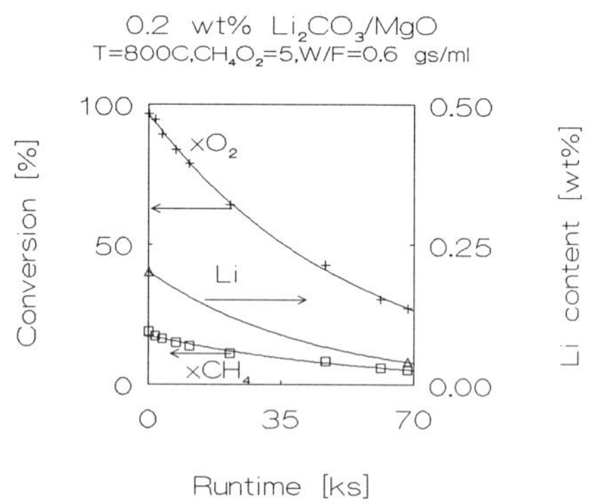

Fig. 2. CH_4, O_2 conversion and lithium content of a 0.2 wt% Li/MgO catalyst versus runtime.

Only very small amounts of lithium are responsible for an active and selective catalyst. This is proven by Figure 2 in which the oxygen conversion is plotted against time on stream for a 0.2wt% Li/MgO catalyst. At pseudo contact times of 0.6 g.s/cm^3(s.t.p.) almost complete oxygen conversion can be reached when starting with a fresh catalyst. Because of the low lithium content the catalyst deactivates from the beginning of the reaction and the deactivation rate is proportional to the rate of lithium loss.

The reason for the lithium loss is twofold: 1) Lithium is lost by evaporation of LiOH which is formed by reaction of water and lithium carbonate. 2) The lithium carbonate, which is a liquid at reaction conditions, reacts with quartz (reactor wall material) to form lithium silicates which are almost inert for the oxidative coupling of methane. Dilution of the catalyst bed with quartz particles accelerates the loss of lithium (ref. 2).

Driscoll et al. (ref. 1) stated that Li^+O^- centres stabilized in the MgO matrix were the active centres for the generation of methyl radicals from methane. Especially the role of MgO is essential in his theory because of the substitution of Mg^{2+} ions ($r_{Mg}2+=0.66$ Å) in the MgO lattice by Li^+ ions ($r_{Li}+=0.68$Å) from the Li_2CO_3 phase. Korf et al. (ref. 3) have shown that carbon dioxide, continuously added to the gas phase, reduces the activity of the Li/MgO catalyst, while a short treatment of a deactivated Li/MgO catalyst with carbon dioxide restores the initial activity for some time. From these experiments it appears that the presence of Li_2CO_3 is essential for an active catalyst.

To prove that Li_2CO_3 can generate an active catalyst, Li_2CO_3 was impregnated on an inert carrier: ZrO_2. Figure 3 shows the activity of Li_2CO_3/ZrO_2 as function of time on stream. Clearly the oxygen conversion increases to a maximum followed by a decrease to almost no activity.

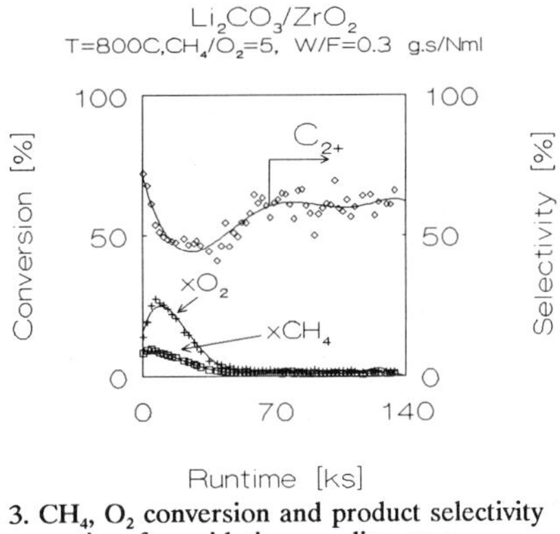

Fig. 3. CH_4, O_2 conversion and product selectivity versus runtime for oxidative coupling over Li_2CO_3/ZrO_2.

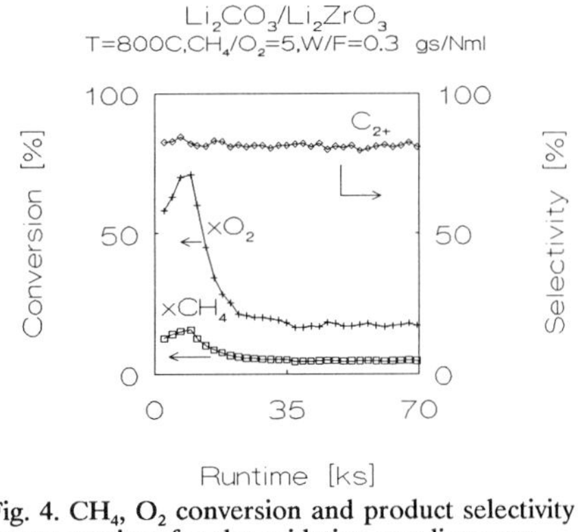

Fig. 4. CH_4, O_2 conversion and product selectivity versus runtime for the oxidative coupling over Li_2CO_3/Li_2ZrO_3.

Indeed the performance of this catalyst is identical to Li/MgO, except for a lower activity due to a lower surface area. The activity lasts as long as lithium carbonate is present. The interaction of ZrO_2 ($r_{Zr^{4+}}$=0.79 Å) with Li_2CO_3 (r_{Li^+}=0.68Å) at 800°C is far less than that of Li_2CO_3 with MgO. Only at very high temperatures (>1000°C) detectable amounts of lithium zirconate (Li_2ZrO_3) are formed. This lithium zirconate is itself a catalyst for the oxidative coupling of methane with a reasonable activity and a high C_{2+} selectivity. However Li_2ZrO_3 is not the active phase in the Li_2CO_3/ZrO_2 catalyst, because also the activity of Li_2ZrO_3 can be increased temporarily by doping it with Li_2CO_3 (Figure 4). Also this catalyst loses its activity more rapidly than Li/MgO. Due to the interaction of Li_2CO_3 with MgO the loss of the lithium phase is retarded. In that respect the carrier plays an essential role: stabilization of the lithium phase. These results clearly show that Li_2CO_3 is essential for an active lithium catalyst. Combining of all this leads to a possible working principle of the Li/MgO catalyst shown in Figure 5.

Li_2CO_3 decomposes in the presence of oxygen to an active centre and CO_2. This active centre reacts with methane to form a methyl radical. Deactivation of the catalyst occurs due to reaction of Li_2CO_3 with water to LiOH which evaporates or with quartz to lithium silicates which are almost inert.

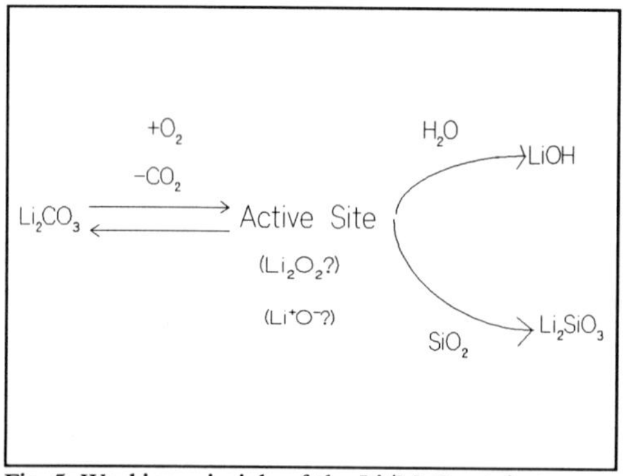

Fig. 5. Working principle of the Li/MgO catalyst.

Knowing the working principle of the catalyst, the role of the catalyst in the oxidative coupling of methane can be understood as well. As shown by Geerts et al.(ref. 4) the catalyst plays an important role in the generation of methyl radicals and in the oxidation of CO to CO_2. Ethane is formed by coupling of methyl radicals in the gas phase. Ethylene is formed by dehydrogenation of ethane and in turn is oxidized to CO. Figure 6 shows a comparison of selectivities for the oxidative coupling of methane between a totally deactivated catalyst and no catalyst.

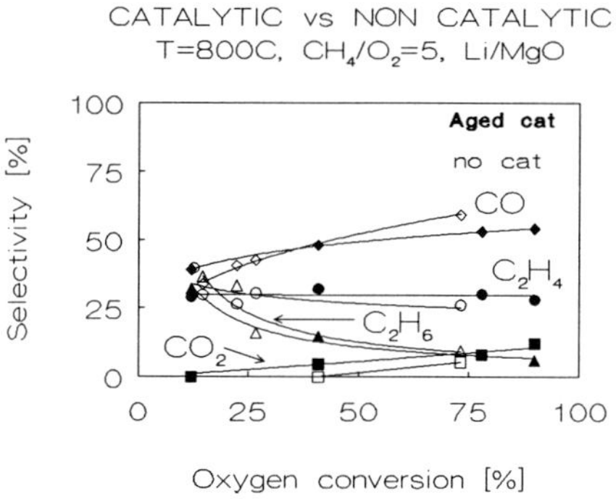

Fig. 6. Comparison between catalytic (Li/MgO) (solid symbols) and non catalytic (open symbols) oxidative coupling.

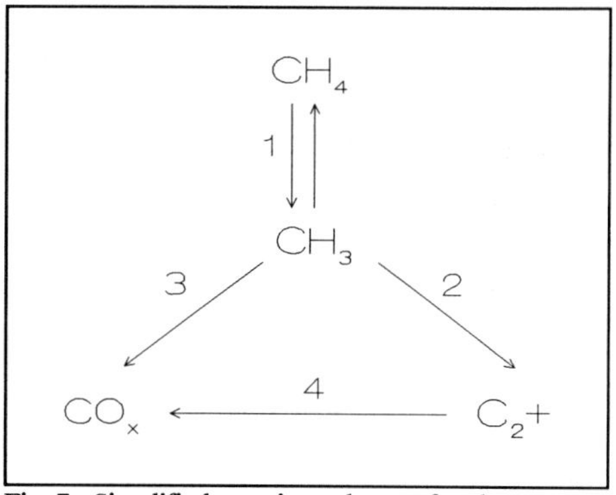

Fig. 7. Simplified reaction scheme for the oxidative coupling of methane.

Remarkably the selectivity of the deactivated catalyst is identical to the selectivity of the homogeneous gas phase reaction, at the same oxygen conversion. However, the reaction rates are much higher with the aged catalyst than in the empty tube. This means that the surface acts as a radical initiator, which releases radicals into the gas phase, without further interfering with the course of the reaction. A fresh catalyst obviously releases (methyl) radicals at a much higher rate. This fact alone might be responsible for the selectivity of a fresh catalyst: high methyl radical concentrations may cause relatively high reaction rates to ethane, as the coupling is second order in methyl radicals, while the oxidation is probably first order. Thus a catalyst may improve selectivity, if it produces methyl radicals at (locally) high concentrations. This reasoning is visualised by the simplified reaction scheme in Figure 7. Reaction 1 is the abstraction of a hydrogen atom from methane. Reaction 2 is the coupling reaction of the methyl radicals to C_2+ components. Reaction 3 and 4 are the total oxidation reactions in which C-O bonds are formed irreversibly. Thus the methyl radical is the key to selectivity. This hypothesis is supported by calculations with a computer model that simulates the gas phase oxidation by taking account of practically all elementary radical reactions (ref. 5). The effect of the catalyst is simulated by adding an extra equation that increases the rate of formation of methyl radicals. The results are shown in Table 1. This table shows a comparison between two simulations: with and without increased methyl radical production.

TABLE 1.
Computer simulations of gas phase oxidative coupling of methane with increased methyl radical production.

Methyl radical production	Normal	Increased	
Contact time (s)	2.7	2.7	0.13
CH_4 conversion (%)	0.06	3.4	0.09
O_2 conversion (%)	0.1	8.2	0.1
C_{2+} selectivity (%)	79	70	97
CO_x selectivity (%)	21	30	3

At the same contact time, the higher methyl radical production rate increases both the methane and oxygen conversion, as expected, while the C_{2+} selectivity is somewhat lower. However, when results at the same conversion level are compared (first and last column of Table 1) it is clear that much higher selectivities are achieved at increased methyl radical production rate, in accordance with the hypothesis proposed.

CONCLUSIONS

The presence of the lithium carbonate phase in the Li/MgO catalyst is essential for the activity. Lithium carbonate itself can generate an active catalyst if supported on an inert carrier like ZrO_2. Very small amounts of lithium are sufficient to create an active and selective Li/MgO catalyst. The main function of the Li-catalyst in the oxidative coupling of methane is the activation of methane. This results in high local methyl radical concentrations which favour the coupling reaction to ethane.

ACKNOWLEDGEMENT

The financial support for this research, which was provided by the European Communities under contract number EN3C-0038-NL and the Netherlands Organization for Scientific Research (NWO), is gratefully acknowledged.

REFERENCES

1 D.J. Driscoll, W. Martir, J-X. Wang, J.H. Lunsford, J.Am.Chem.Soc., 107 (1985) 58-63.
2 J.M.N. van Kasteren, J.W.M.H. Geerts and K.van der Wiele, Proceedings 9th ICC Calgary, Alberta, Canada, Vol 2 (1988) 930-936.
3 S.J. Korf J.A. Roos, N.A. de Bruijn, J.G. van Ommen, J.R.H. Ross, J.Chem.Soc.,Chem.Commun. (1987) 1433-34.
4 J.W.M.H. Geerts, J.M.N. van Kasteren and K. van der Wiele, Catal. Today, 4 (1989) 453-461.
5 J.W.M.H. Geerts, Q. Chen, J.M.N. van Kasteren and K. van der Wiele, Catal. Today, submitted for publication.

WORKING PRINCIPLE OF Li DOPED MgO APPLIED FOR THE OXIDATIVE COUPLING OF METHANE

J.M.N. van Kasteren, J.W.M.H. Geerts and K. van der Wiele

Department of Chemical Technology, University of Technology, P.O. Box 513, 5600 MB Eindhoven (The Netherlands)

1. **J.C. Volta** (Institut de Catalyse, Vileurbanne, France) :
   1) What is the loss of Li in the case of your Li_2CO_3/Li_2ZrO_3 catalyst?
   2) Is there some isomorphity between Li_2CO_3 and Li_2ZrO_3?

   J.M.N. van Kasteren (University of Technology, Eindhoven, The Netherlands): We did not measure the lithium loss for our Li_2CO_3/Li_2ZrO_3 catalyst but it is very probable that the lithium carbonate which we impregnated on this catalyst is almost completely lost during reaction. The performance of the catalyst after a few hours is exact that of pure Li_2ZrO_3, which we tested separately. The cristal structure of the two phases is not similar.

2. **J.R.H. Ross** (University of Twente,The Netherlands) : I am glad to see that your conclusions are similar to those which we have reached (S.J. Korf, N.A. de Bruijn,J.G. van Ommen and J.R.H. Ross, Catalysis Today 2(1988) 535) in relation to the nature of the active sites. However, we differ with you about the importance of surface reactions in steps such as the reaction $C_2H_6 + 0.5O_2$ ----- $C_2H_4 + H_2O$, $C_2H_x + O_2$ ---- CO_x, etc. (j.A. Roos, S.J. Korf, R.H.J. Veehof, J.G. van Ommen and J.R.H. Ross, Catal. Today, 4(1988) 471; Appl. Catal., 52 (1989) 147). Do you think that your conclusions are still applicable at lower temperatures?

   J.M.N. van Kasteren (University of Technology, Eindhoven, The Netherlands): We agree with you that at lower temperatures gas phase reactions lose importance compared to surface reactions. However, the best results with respect to C_{2+} yield have been reached at conditions where gas phase reactions play a very important role. Our conclusions are in this way applicable at lower temperatures that the lower C_{2+} yield can be explained by the occurrence of more total oxidation reactions which occur mainly at the catalyst surface. A low temperature means a lower activity and thus lowering of methyl radical concentrations especially in the gas phase. This will favour the total oxidation reactions thus a lowering of the C_{2+} selectivity. Also at low temperatures gas phase reactions can not be excluded totally because radical coupling reactions have no activation energy and are possible even at room temperature (6).

3. **J.F. Brazdil** (BP Research, Ohio, USA): My question relates to your computer simulation of the methane coupling reaction.

Increasing local concentration ot methyl radicals should increase the yield of C_{2+} products. Have you looked for conditions with the model that will give a maximum yield of C_{2+}? If so, what is the maximum yield predicted by your model?

J.M.N. van Kasteren (University of Technology, Eindhoven, The Netherlands): We did not look for process conditions which give the optimum C_{2+} yield, but this is an objective for the future.

4. **J. Kiwi** (EPFL, Lausanne, Switserland) : You report CO and CO_2 formation in your processes leading to C_{2+} at 800°C on Li/MgO catalysts. How does in time scale the CO and CO_2 evolve? Does the CO_2 with time form at the expense of CO? How do you account for this in your model?

J.M.N. van Kasteren (University of Technology, Eindhoven, The Netherlands): Indeed does the CO_2 form at the expense of CO. it can be shown that a fresh catalyst converts almost all CO to CO_2, while a deactivated catalyst produces mostly CO (3). Also the gas phase oxidative coupling of methane produces only CO and this can be well described with our computer model (4). We are at the beginning of our catalytic modelling and our object is to add a dummy reaction set analogous to the methane activation steps to simulate the CO to CO_2 catalytic reaction.

5. **V. Cortes Corberan** (Institute of catalysis, CsIc, Madrid, Spain) :

1) Have you experimental evidence of the presence of Li_2CO_3 phase in the working catalyst?
2) In the initial stage of catalytic test of Li/MgO catalyst (Fig. 1) activity increases while Li content is decreasing: I wonder if the important factor for activity is not the Li content but the nature of the Li containing phase, taking also into account that carbonate could be decomposed at lower temperatures with a reducing atmosphere (such as reaction conditions) than with an oxidant atmosphere (such as in the calcination step). Then, possibly the active phase could be a lithium oxide (or peroxide) instead of carbonate, being the latter a precursor of the active phase.
3) Have you tried the catalytic activity of pure Li_2CO_3?

J.M.N. van Kasteren (University of Technology, Eindhoven, The Netherlands): We know that the decomposition of the Li_2CO_3 phase plays an important role during the oxidative coupling of methane over Li/MgO. Addition of CO_2 to the feed gas lowers immediately the activity of the catalyst. Also the initial activity of a deactivated Li/MgO catalyst can be temporarily restored by a treatment with CO_2 as shown by Korf et al. (2). At 800°C in a stream of oxygen the Li_2CO_3 is decomposing: the loss of CO_2 can be measured. However, when the reaction is started CO_2 is formed immediately and this reacts with the active site and with the Li_2O to form Li_2CO_3 again. Under reaction conditions an equilibrium exists between Li_2CO_3 , Li_2O and CO_2. We have tested pure Li_2CO_3 itself although this is not easy at 800°C, because it melts at 723°C. We constructed a bubble reactor with which it is possible to bubble CH_4 and O_2 through the liquid Li_2CO_3 (5). The C_{2+} selectivity is much lower compared to the Li/MgO catalyst. The addition of MgO to the melt of Li_2CO_3 leads to improvement of the catalytic performance. The conclusion from this work is that MgO is needed to give a good coupling catalyst.

REFERENCES

1 J.M.N. van Kasteren, J.W.M.H. Geerts and K.van der Wiele, Proceedings 9th ICC Calgary, Alberta, Canada, Vol 2 (1988) 930-936.
2 S.J. Korf J.A. Roos, N.A. de Bruijn, J.G. van Ommen, J.R.H. Ross, J.Chem.Soc.,Chem.Commun. (1987) 1433-34.
3 J.W.M.H. Geerts, J.M.N. van Kasteren and K. van der Wiele, Catal. Today, 4 (1989) 453-461.
4 J.W.M.H. Geerts, Q. Chen, J.M.N. van Kasteren and K. van der Wiele, Catal. Today, submitted for publication.
5 J.W.M.H. Geerts, J.M.N. van Kasteren and K. van der Wiele, to be published.
6 J.M.N. van Kasteren, P. de Been, J.G.A. Holscher, IXth European Sectional Conference on Atomic and Molecular Physics of Ionized Gases, Lissbon, Portugal, 351-352 (1988).

G. Centi and F. Trifiro' (Editors), *New Developments in Selective Oxidation* 353

INVESTIGATIONS ON THE HETEROGENEOUSLY CATALYZED OXIDATIVE COUPLING OF METHANE OVER ALKALI DOPED METAL OXIDES

S. BARTSCH, H. HOFMANN
Institut für Technische Chemie I der Universität Erlangen-Nürnberg
Egerlandstr. 3, 8520 Erlangen

ABSTRACT

The heterogeneously catalyzed oxidative coupling of methane into ethane and ethene using alkali doped cerium oxide has been investigated. Among all promotors tested the activity towards C_2 products decreased in the order Li>Na>Cs>K. An optimal lithium loading of 5 wt% Li_2CO_3 in CeO_2 was found. The effect of the BET surface area for this catalyst was negligible. For a cerium doped Li/MgO catalyst the influence of the operating conditions has been studied to find conditions which produce an optimal yield of C_2 hydrocarbons.

INTRODUCTION

Because of its large deposits natural gas will become the most important primary energy source in the future (ref.1). Consequently there is great interest in using methane, which is the predominant component of natural gas, as a chemical feedstock. Different approaches have been proposed to provide new routes for producing useful chemicals by partial oxidation of methane into methanol, formaldehyde or C_2-hydrocarbons (ref.2). Although a great deal of work has been done in this area the economic barrier has not been overcome. This circumstance leads to the situation that at present $7 \cdot 10^9$ Nm^3/a of natural gas are burned off in the oil fields or at remote natural gas deposits (ref.3).

Since the pioneering work of Keller and Bhasin in 1982 (ref.4) the heterogeneously catalyzed oxidative coupling of methane into ethane and ethene by means of metal oxides has attracted increasing attention. In some recent publications on the use of alkali doped metal oxides as catalysts good chances for the above reaction have been opened up (refs.5-17).

It is reported (refs.5,18) that an O^- species on the catalyst surface is an active site which is able to abstract a hydrogen atom from methane. This process is considered as an initiating step in the methane coupling reaction. As to the Li/MgO system, it was shown that such O^- species can be produced by solid diffusion of Li^+ cations into the MgO lattice (refs. 5, 19). A similar creation of active sites can be assumed for other alkali doped metal oxides. Figure 1 illustrates schematically how the reaction is thought to occur: If effects of adsorption and desorption are neglected, there are two principal steps involved: on the one hand, methane molecules react with lattice oxygen forming hydroxy groups and methyl radicals which can produce ethane by recombination. On the other hand, the regeneration of the active sites has to take place with oxygen from the gas phase. Water is formed as a by-product.

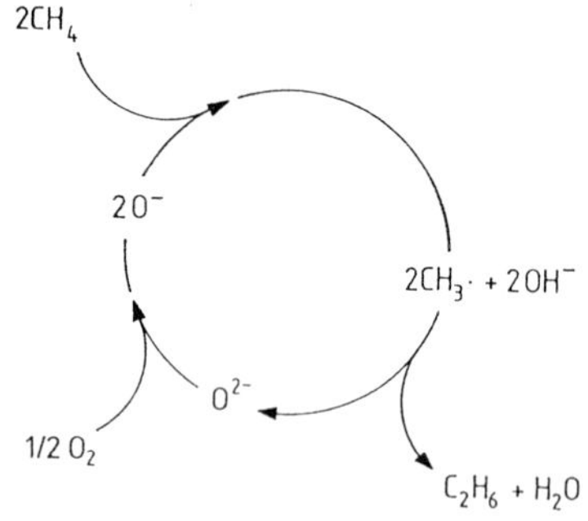

Fig.1:
Radical generating mechanism

In a previous paper (ref.20) we reported that the addition of cerium oxide as a third compound to a Li/MgO catalyst improves the C_2 yield. The Ce/Li/MgO system was proposed for the first time by Y. Bi et al (ref.15). We have investigated the role of cerium in this system by means of pulse technique (ref.21). It could be demonstrated that the Ce/Li/MgO catalyst produces ethane as well as ethylene from methane even if no gas phase oxygen is available. The amount of C_2 products decreases with respect to the number of pulses (see Figure 2). After reduction with methane, the catalyst was reoxidized with oxygen pulses. This treatment enabled the catalyst to show its initial behaviour. As these results could not be achieved by the use of pure Li/MgO in place of the cerium doped catalyst, it is assumed that lattice oxygen acts as a highly selective oxidizing agent and cerium changes its oxidation state during the methane coupling reaction. We therefore concluded that the rate-controlling step of the coupling reaction is not the formation of methyl

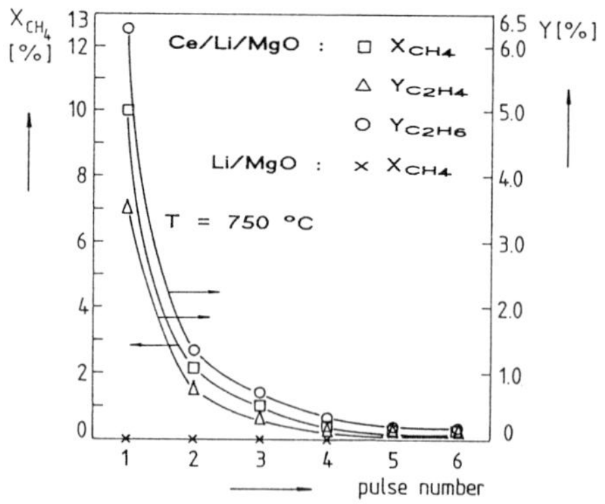

Fig.2: CH_4 conversion (X) and C_2 yield (Y) with respect to the number of CH_4 pulses over Li/MgO and Ce/Li/MgO.

radicals but the regeneration of the active sites on the catalyst surface. This reaction step might be enhanced by the use of a multivalent metal oxide as a charge carrier as depicted in Figure 3. A number of other metal oxides have been tested as substitutes for cerium but without success (ref. 21).

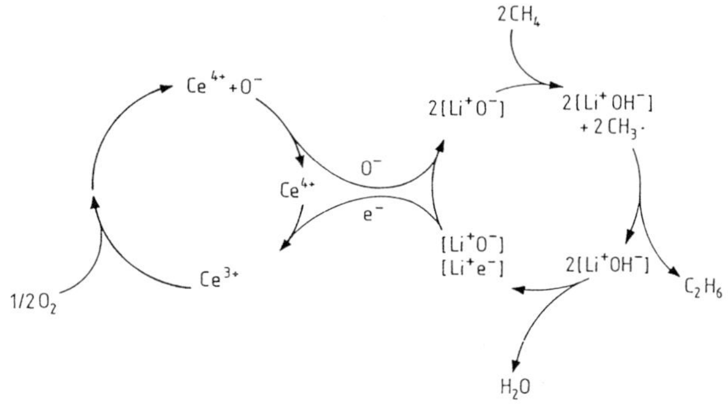

Fig.3:
Speculative radical generating mechanism

This paper presents results of our investigations into the effect of alkali dopands on the activity and selectivity of CeO_2. The influence of the operating conditions on the C_2 yield over a Ce/Li/MgO catalyst is discussed.

EXPERIMENTAL

The catalyst testing experiments were carried out with a standard flow system. In Figure 4 the catalytic section of the reactor tube as well as the operating conditions are shown. The axial temperature

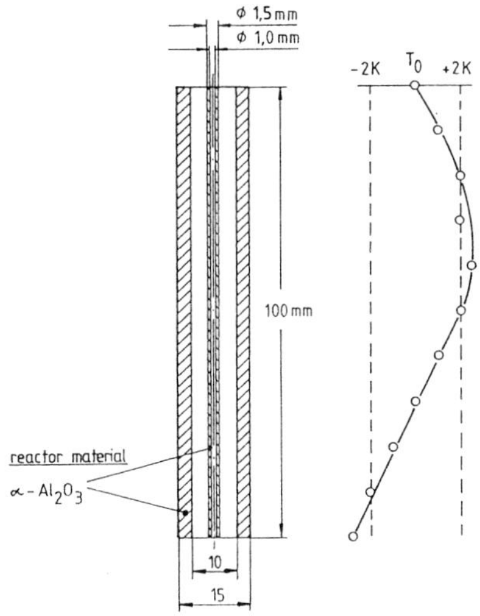

OPERATING CONDITIONS

T_o	=	650	-	800	°C
p_o	=	1.2	-	7.0	bar
F	=	1.6	-	25	Nml/s
W_{CAT}	=	1.0			g
W_{INERT}	=	16.3			g
$x(N_2)$	=	0.9			
$x(CH_4)$	=	0.06	-	0.1	
$x(O_2)$	=	0.01	-	0.03	
d_p	=	0.8	-	1.0	mm
ε	=	0.39			

Fig.4: Catalytic section of the reactor tube and operating conditions. x(i): mol fraction of component i. d_p: particle diameter. ε: void fraction of the fixed-bed. The temperature profile was measured under reaction conditions.

profile is measured by means of a thermocouple which can be moved inside a capillary tube made of α-alumina. In order to avoid undesired temperature gradients in the catalytic fixed-bed, the feed gas consists of 90 mol% N_2 and 10 mol% of the desired CH_4/O_2 ratio. Thus, a typical temperature profile with respect to the reactor length as depicted in Figure 4 was achieved under reaction conditions.

For the examination of the selective oxidation of hydrocarbons at temperatures above 500 °C, the construction material of the reactor becomes important. Stainless steel was shown to enhance the unselective oxidation of light hydrocarbons (ref.22). In order to eliminate any catalytic influence of the reactor itself, a special fixed-bed reactor constructed from α-alumina was developed. The tube has a length of 110 cm and is filled with ceramic particles. The catalyst particles are spread over a length of 10 cm in the middle of the tube (catalytic section). It can be heated electrically up to 800 °C.

Catalyst preparation has been reported on previously (ref.20) and was carried out according to prescriptions given in the literature (ref.5). Table 1 lists all catalysts which were studied in the investigations covered in this paper.

Using CeO_2 as the main compound, the amount of lithium loading (Catalysts 1-5) as well as the effect of alkali dopands (Catalysts 3,6,7,8) were investigated. These catalysts were prepared in such a way that the molar ratio of alkali/CeO_2 was constant.

TABLE 1

1st compound	2nd compound	wt%	Catalyst
CeO_2	Li_2CO_3	2	1
CeO_2	Li_2CO_3	3.5	2
CeO_2	Li_2CO_3	5	3
CeO_2	Li_2CO_3	7.5	4
CeO_2	Li_2CO_3	10	5
CeO_2	Na_2CO_3	7.1	6
CeO_2	K_2CO_3	9	7
CeO_2	Cs_2CO_3	18.9	8

In order to find conditions which produce an optimal yield of C_2 hydrocarbons, the influence of the operating conditions on conversion and selectivity over the Ce/Li/MgO catalyst (ref.20) were studied in detail. The values of T,p,W/F and $p^o_{CH_4}/p^o_{O_2}$ were varied corresponding to those given in Figure 4. For an interpretation of these results it was necessary to eluminate the reaction scheme. Therefore mixtures of $C_2H_6/O_2/N_2$ and $C_2H_4/O_2/N_2$ were also used as feed gases. The operating conditions in these runs were the same as stated above, but temperature (750 °C) and pressure (atmospheric) were kept constant. In order to distinguish between homogeneous and heterogeneously catalyzed reaction steps each run was carried out with and without catalyst.

Blank runs with $CH_4/O_2/N_2$ (molar ratio 0.67/0.33/9) showed that homogeneous oxidation is negligible up to a temperature of 770°C.

RESULTS AND DISCUSSION

Alkali doped Cerium oxide

Alkali doped CeO_2 was found to be a selective catalyst for the methane coupling reaction. This must be due to the presence of dopands, since pure CeO_2 yields only total oxidation products, a fact that is observed by several researchers (refs.23,24). Figure 5 presents a comparison of the C_2 yield obtained over CeO_2 doped with different alkali metals, such as lithium, sodium, potassium and cesium. The influence of such dopands on the catalytic behaviour of MgO (ref.25) as well as of ZnO (ref.26) was studied by Matsuura and co-workers. They found that Lithium was the most attractive alkali dopand for both systems, ZnO and MgO. In agreement with these results it is shown that the amount of C_2 hydrocarbons decreases in the order Li>Na>Cs>K. Li/CeO_2 is an effective catalyst yielding 12 % C_2 hydrocarbons under the following operating conditions: T= 750 °C, W/F= 0.16 g·s/ml, $p^o_{CH_4}/p^o_{O_2}$ = 2, atmospheric pressure.

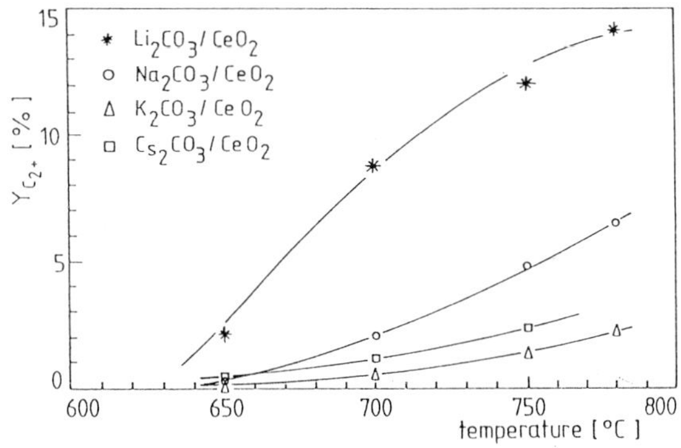

Fig.5: Effect of different alkali metals as dopands in CeO_2.

In literature it is suggested that the similarity in the ionic radii of the main catalyst compound and the respective dopand might have a decisive influence on the formation of the active sites (ref.5). The alkali cation should fit into the cation vacancies of a higher valent metal oxide matrix in order to create O^- centres. The catalytic properties of Li/MgO, Li/ZnO and Na/CaO can be explained in this way.

The radii of some interesting cationes are summarized in Table 2. As to CeO_2 one would expect the Na^+ cation as the most attractive alkali dopand but the obtained C_2 yields over the alkali/CeO_2 system are not coherent with this theorie. There is only a marginal difference in the activities of Na/CeO_2, K/CeO_2 and Cs/CeO_2, but Li/CeO_2 is a much more active and selective catalyst for the methane coupling reaction.

TABLE 2

Cation	Li^+	Na^+	K^+	Cs^+	Mg^{2+}	Ca^{2+}	Zn^{2+}	Ce^{3+}	Ce^{4+}
Radius	0.68	0.97	1.33	1.67	0.66	0.99	0.74	1.07	0.94

radii values in 10^{-10}m (ref.28).

Another important fact that governs the catalytic properties is the amount of alkali loading. Figure 6 shows the C_2 yield obtained at different reaction temperatures with respect to the wt% Li_2CO_3 used in catalyst preparation. At each temperature the yield of C_2 hydrocarbons goes through a maximum at a value of 5 wt% Li_2CO_3. Similar behaviour is reported by Matsuura et al (ref.25) for Li/MgO, by Iwamatsu et al (ref.13) for Na/MgO and Rb/MgO and by Otsuka et al (ref.6) for Li/Sm_2O_3. There has been a discussion in the literature on the fact, that alkali doping causes the formation of active sites and therefore an increase in activity, but a simultaneous reduction in surface area. Since these two properties have opposite effects on the C_2 yield, there should exist an optimal alkali loading with respect to the C_2 yield (ref.27). We cannot confirm this consideration in the case of lithium doped CeO_2, because the BET surface area of pure cerium oxide is slightly reduced by the lithium loading and no correlation with respect to the C_2 yield is observed as depicted in Figure 6.

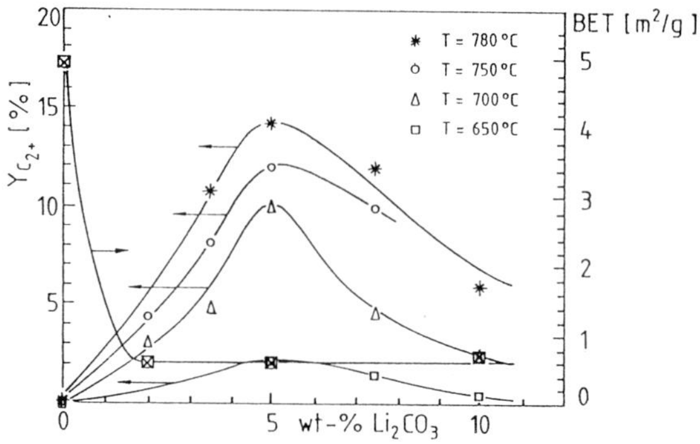

Fig.6: Effect of the amount of Li_2CO_3 as dopand in CeO_2

Effect of the operating conditions

In Figure 7 the influences of temperature (650 - 800 °C) as well as of the CH_4/O_2 ratio at the reactor inlet (0.2 - 10) over Ce/Li/MgO are shown. With increasing temperature, the conversion and the selectivity also increase. A large excess of oxygen increases the methane conversion, but the selectivity declines simultaneously so that an optimal yield exists at $p^o_{CH_4}/p^o_{O_2} = 2$.

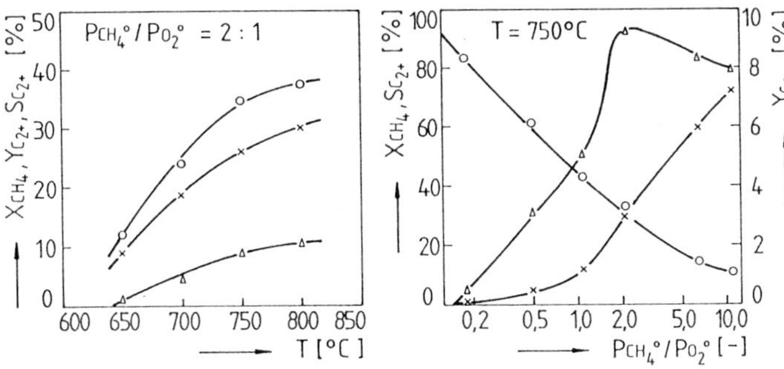

Fig. 7: Effect of temperature and $p^o_{CH_4}/p^o_{O_2}$ on CH_4 conversion (o), C_2 selectivity (×) and C_2 yield (Δ) over a Ce/Li/MgO catalyst. p_{abs} = 1.2 bar, W/F = 0.16 g·s/Nml.

A temperature of 750 °C and a CH_4/O_2 ratio of 2 were used for the examination of W/F and the total pressure on the C_2 yield. Because of the dilution of the reactants (see Figure 4) a total pressure of e.g. 6 bar is equivalent to partial pressures of 400 mbar CH_4 and 200 mbar O_2 respectively. With increasing residence time the C_2 yield also increases. At each value of W/F the C_2 yield moves through a maximum with respect to pressure. At lower values of W/F this maximum is shifted to higher values of pressure.

In order to achieve a high yield of hydrocarbons over this catalyst, one must operate at a high value of W/F and at a low pressure.

These results must be due to the complexity of the reaction system consisting of homogeneous as well as heterogeneously catalyzed reactions, some of which are parallel or consecutive steps.

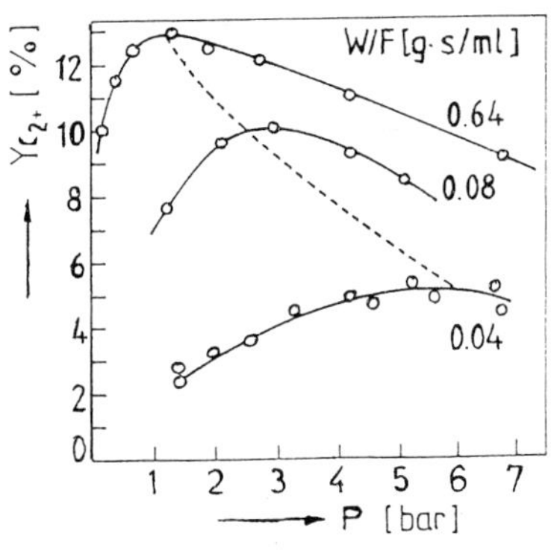

Fig. 8: Influence of W/F and total pressure on the C_2 yield over a Ce/Li/MgO catalyst at T = 750 °C and $p^o_{CH_4}/p^o_{O_2} = 2$.

Figure 9 summarizes the main reaction pathways that were found in the catalytic (left hand scheme) and non-catalytic (right hand scheme) experiments. Methane is oxidized into CO_2 in both cases, but only in the presence of the catalyst ethane is produced. Ethane is oxidatively dehydrogenated into ethylene with and without catalyst. However dehydrogenation of ethane also takes place to some extent in the absence of oxygen. Ethylene is homogeneously converted into CO which undergoes further oxidation into CO_2 by a catalytic reaction step.

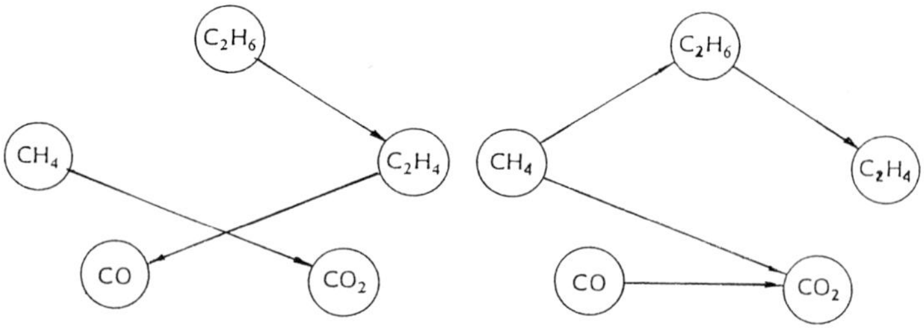

Fig.9:
Catalytic (left hand) and non-catalytic (right hand) reactions

In order to gain a clear understanding of the results shown in Figures 7 and 8 extensive kinetic studies on each reaction step pointed out in Figure 9 have to be done. This is the main prospect which our future work will encompass.

REFERENCES

1 W. Häfele, W. Terhost, Chem. Ind., 37 (1985) 10
2 R. Pitchai, K. Klier, Catal. Rev. - Sci. Eng., 28 (1986) 1
3 S. Maier, F.-J. Müller, Chem. Ing. Tech., 58 (1986) 287
4 G.E. Keller, M.M. Bhasin, J. Catal., 73 (1982) 9
5 T. Ito, J.X. Wang, C.H. Lin, J.H. Lunsford, J. Am. Chem. Soc., 107 (1985) 5062
6 K. Otsuka, Q. Liu, M. Hatano, A. Morikawa, Chem. Lett., (1986) 467
7 C.A. Jones, J.J. Leonard, J.A. Sofranko, J. Catal., 103 (1987) 311
8 K. Otsuka, Sekiyu Gakkaishi, 30 (1987) 385
9 J.B. Kimble, J.H. Kolts, Chemtech, August (1987) 501
10 C.H. Lin, J.X. Wang, J.H. Lunsford, J. Catal, 111 (1988) 302
11 M.Y. Lo, S.K. Agarwal, G. Marcelin, J. Catal, 112 (1988) 168
12 H.S. Zhang, J.X. Wang, D.J. Discroll, J.H. Lunsford, J. Catal., 112 (1988) 366
13 E. Iwamatsu, T. Moriyama, N. Takasaki, K. Aika, J. Catal., 113 (1988) 25
14 G.J. Hutchings, M.S. Scurell, J.R. Woodhouse, J. Chem. Soc., Chem. Commun., (1987) 1862
15 Y. Bi, K. Zhen, Y. Jiang, C. Teng, X. Yang, Appl. Catal., 39 (1988) 185
16 R. Burch, G.D. Squire, S.C. Tsang, Appl. Catal., 43 (1988) 105
17 J.M. DeBoy, R.F. Hicks, Ind. Eng. Chem. Res., 27 (1988) 1577
18 K. Aika, J.H. Lunsford, J. Phys. Chem., 81 (1977) 1393

19 Y. Chen, H.T. Tohver, J. Narayan, M.M. Abraham, Phys. Rev., 16 (1977) 5535
20 S. Bartsch, J. Falkowski, H. Hofmann, Catalysis Today, 4 (1989) 421
21 S. Bartsch, H. Hofmann, submitted for publication
22 S. Mahajan, W.R. Menzies, L.F. Albright, Ind. Eng. Chem., Proc. Des. Dev., 16 (1977) 271
23 K.D. Campbell, H.Zhang, J.H. Lunsford, J. Phys. Chem., 92 (1988) 750
24 K. Otsuka, K. Jinno, A. Morikawa, Chem. Lett., (1985) 499
25 I. Matsuura, Y. Utsumi, T. Doi, Y. Yoshida, Appl. Catal., 47 (1988) 299
26 I.Matsuura, Y. Utsumi, M. Nakai, T. Doi, Chem. Lett., (1986) 1981
27 E. Iwamatsu, T. Moriyama, N. Takasaki, K. Aika, J. Chem. Soc., Chem. Commun., (1987) 19
28 D'Ans·Lax, Taschenbuch für Chemiker und Physiker, Springer-Verlag, Berlin/Heidelberg 1967

R.K. Graselli (Mobil Res.& Dev. Corp., USA): Just a brief comment on your interesting paper. You conclude from your work that Li is the most effective alkali dopant for your CeO_x system, while the ionic radii of Li^+ (0.68) and Ce^{3+} (1.03), Ce^{4+} (0.92) are rather poorly matched; other other alkalies e.g. Na^+ (0.97) [or K^+ (1.33)] would be a better size match for Ce^{3+}/Ce^{4+}. I should line to offer the following possible explanation for your finding that the effectiveness of alkali dopants dereases in the order Li>Na>Cs≈K. Not only should an ionic fit between dopant and base catalyst elements be considered, but also the elektronic factors of them. Considering the electronegativities of the alkali series, Cs is the most basic while Li is the least basic in this series. Thus Cs will force Ce more readily into the Ce^{4+} state than Li. This results in a stronger Ce-O bond and thus a more difficult release of lattice oxygen to the hydrocarbon (CH_4) in the case of Cs and/or K doping, while doping with Li will allow a more facile yet selective release of lattice oxygen; resulting in an optimum for the system Li/Ce among alkali dopants.

J.G. van Ommen (University of Twente, The Netherlands): In your reaction scheme of heterogeneously catalyzed reactions, you exclude the oxidation of C_2 (C_2H_6 + C_2H_4) products to CO_2. Do you have experimental evidence to support this hypothesis?

S. Bartsch (University of Erlangen, Germany): We have carried out experiments under catalytic and non-catalytic conditions using $CH_4/O_2/N_2$ or $C_2H_6/O_2/N_2$ or $C_2H_4/O_2/N_2$ as feed gas. Up to now we did not use mixtures of those hydrocarbons in our investigations. In each case we obtained profiles of partial pressures of the reactants with respect to the reactor length. All these data were obtained at atmospheric pressure, T = 750 °C, F = 6 Nml/s, W = 0.1 $g_{catalyst}/cm_{reactor\ length}$, $p_{hydrocarbon}/p_{oxygen}$ = 2...10.

Comparing these profiles we found the main reaction pathways as depicted in Figure 9: Ethylene is mainly homogeneously oxidized into CO which undergoes further oxidation into CO_2 in the presence of a catalyst (ref. 1). Ethane reacts into CO and CO_2 to some extent with and without catalyst, but the main product is ethylene. The catalyst enhances ethylene production, while the CO_x formation remains almost unchanged. The small amounts of CO_x are due to the consecutive oxidation of ethylene as already mentioned. Only in the case of catalytic CH_4 oxidation remarkable amounts of CO_2 were found. The amount of CO_2 cannot be explained by consecutive oxidation of C_2 hydrocarbons, indicating that methane is directly converted into CO_2,

1 S. Bartsch, H. Hofmann, submitted for publication in Catalysis Today

J. Kiwi (EPFL Lausanne, Switzerland): 1) You have not elaborated on the stability in time and reuse of your ceric catalyst doped with Li at 780 °C; What is the situation as shown by your experimental results?

2) CeO_2 as support for Li is used in your work. What is the loss of Li at 700 - 800 °C during experiments?

S. Bartsch (University of Erlangen, Germany): 1) We have investigated the conversion of methane, the C_2 selectivity and the C_2 yield as a function of time on stream for a Li/CeO_2 catalyst. The results are shown in the following Figure:

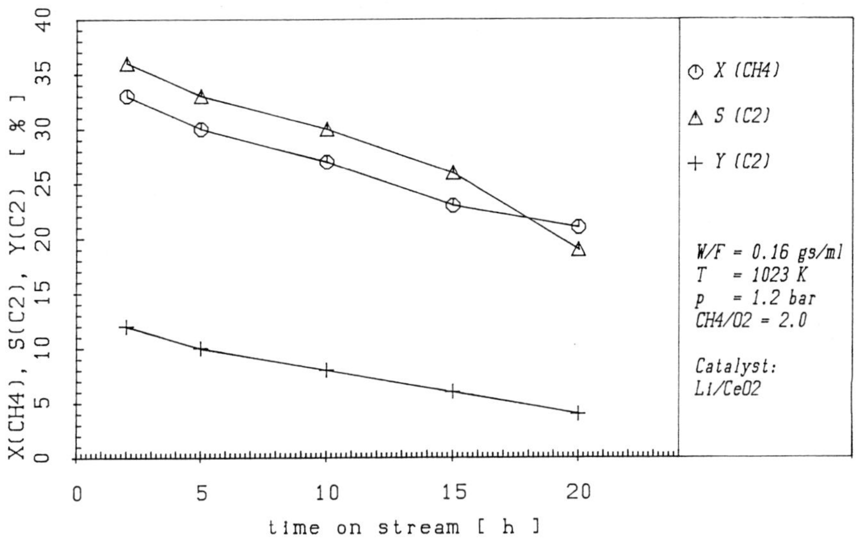

All of the experimental results presented in our paper have been obtained with fresh catalyst in order to avoid the influence of deactivation effects.

2) It is known from the literature that lithium loss is the main reason for the deactivation of lithium containing catalysts (ref.1). We have not studied the effect of lithium loss of a Li/CeO_2 catalyst, but we assume similar behaviour as we found for a Ce/Li/MgO catalyst which has been investigated in our laboratory (ref.2). In summarizing these results it was found, that lithium loss is mainly caused by the influence of temperature, time on stream and turnover rate. The lithium loading decreases nearly to zero under severe reaction conditions.

1 S.J. Korf, J.A. Roos, N.A. de Bruijn, J.G. van Ommen, J.H.R. Ross, Catalysis Today, 2 (1988) 535

2 S. Bartsch, H. Hofmann, submitted for publication in Catalysis Today

G.I. Golodets (Ukrainian Academy of Science, USSR): What is the degree of oxidation of lattice oxygen which is, as you told us, "the oxidizing agent in the oxidative coupling of methane" ?

S. Bartsch (University of Erlangen, Germany): Up to now we have no detailed information about the nature of the active oxygen species (e.g. oxidation state) in our catalyst. This will be one of the main aspects of our future work on cerium containing catalysts.

W.J. Vermeiren (K.U. Leuven, Belgium): There is now enough evidence that gas phase reactions between methane and oxygen are possible, especially at high partial pressure of oxygen in the feed. I think that the optimum, you obtained in Figure 8 is due to a combination of gas phase and catalytic reactions. In these conditions of CH_4/O_2 = 2 the gas phase reactions produce low amounts of C_2 products. This is the reason for the decrease of C_2 yield at higher partial pessures of methane and oxygen.

Did you perform experiments with the same conditions as shown in Figure 8, but with an empty reactor to investigate the contribution of gas phase reactions?

S. Bartsch (University of Erlangen, Germany): We did not perform the same set of experimental runs as depicted in Figure 8 without catalyst, but we checked the influence of gas phase reactions under the following operating conditions: $p^o_{CH_4}/p^o_{O_2}$ = 2.0, F = 6 Nml/s, no catalyst. The results are summarized in the following Table:

p (bar)	T (°C)	X_{CH_4} (%)	Y_{CO} (%)	Y_{CO_2} (%)	$Y_{C_2H_6}$ (%)	$Y_{C_2H_4}$ (%)
1.2	708	1.13	n.e.	n.e.	n.e.	n.e.
1.2	755	1.85	0.0	1.49	0.18	0.18
1.2	780	2.17	n.e.	n.e.	n.e.	n.e.
1.2	800	3.11	n.e.	n.e.	n.e.	n.e.
10.0	750	11.10	3.27	6.46	0.80	0.57

n.e. not evaluated

From these data it is clear, that our results (see Figure 8) cannot be simply explained by the influence of homogeneous gas phase reactions, because of the very low conversion of methane. However, we agree that the interaction between homogeneous and heterogeneously catalysed reactions, especially the consecutive reactions of ethane and ethylene, must be clarified to gain a clear understanding of our results.

G. Centi and F. Trifiro' (Editors), *New Developments in Selective Oxidation* 365
© 1990 Elsevier Science Publishers B.V., Amsterdam — Printed in The Netherlands

OXIDATIVE COUPLING OF METHANE OVER $LnLiO_2$ CATALYSTS. (Ln = Sm, Nd, La). PROMOTING EFFECT OF MgO AND CaO.

A. KIENNEMANN, R. KIEFFER, A. KADDOURI
Laboratoire Chimie Organique Appliquée U.A. CNRS 469
E.H.I.C.S. 1, Rue Blaise Pascal 67008 Strasbourg (France)

P. POIX, J.L. REHSPRINGER
Groupe de Chimie des Matériaux Inorganiques I.P.C.M.S.
E.H.I.C.S. 1, Rue Blaise Pascal 67008 Strasbourg (France)

SUMMARY

The catalytic activity in methane coupling oxidation on definite $LnLiO_2$ (Ln = Sm, Nd, La) compounds where the alkali is enclosed in the crystal lattice is reported. $LnLiO_2$ structures have the advantage to be a host structure for promoting cations (Mg, Ca). The doped systems provide significant increases in activity and selectivity towards C_2 hydrocarbons.

INTRODUCTION

The proved world reserve of methane is widely superior to that of higher hydrocarbons. Furthermore, the different methane production fields are often located wide away from its use spot (ref. 1). That's why many countries have undertaken studies on the valorization of natural gas. Today, the most common transformation way (except the manufacturing of halogenated derivatives, of CS_2 or of acetylene) passes through synthesis gas (CO, CO_2 , H_2) as intermediate. An important energy loss takes place upon conversion of natural gas to a CO, CO_2, H_2 mixture. Therefore, any direct transformation of methane is interesting. In heterogeneous catalysis the direct production of aromatics, of methanol and the oxidative coupling of methane to ethylene and ethane are worth mentioning (ref. 2-5). The first catalysts, in methane oxidative coupling of methane worked in an sequential way (e.g. with alternate addition of N_2O and methane to the gas flow), but now the simultaneous addition of the oxidating agent and of methane is possible. The catalysts working on a sequential mode are rather easily reducible metallic oxides (e.g. Pb, Mn, Bi, Sb...) whereas the catalysts operating on a simultaneous mode are non - or hardly reducible oxides like MgO, CaO or alkali (Li, Na, K) doped rare earth oxides.

The operating conditions of these catalysts (temperature, constituents partial pressure, CH_4/O_2 ratios) vary in a large range. The volatilization of either the active metal (Pb) or of the promoter (Li) is the main cause of catalyst deactivation which is very often reported in the literature. Thus, lithium when deposited on the catalytic surface migrates and reacts with the quartz reactor in the reaction conditions.

Magnesium, samarium or lanthanum oxide catalysts doped by alkalis (Li, Na) or alkaline earth (BaO, CaO, SrO) oxides (refs 6-8) are mentioned to lead to the highest C_2 hydrocarbons yields. The amount of lithium added to surface varies strongly depending on the authors : (from a few % up to 25% or even more) (refs 7,11).

Our catalytic systems are based on rare earth oxides and alkalis. They were prepared with the following aims :

- good distribution and controlled localization of alkali atoms
- raised amount of lithium in the catalytic formulation (50 mole %)
- decreasing of the lithium loss during the reaction.

In order to achieve this goal definite compound of general formula $LnLiO_2$ (Ln = Sm,Nd,La) were prepared by reaction of the rare earth oxide with the alkalis instead simply depositing the latter on the rare earth oxides. Taking into account the ionic radius of Ln^{3+} in a six fold coordination, the partial substitution of Ln and Li by suitable cations is possible. If the oxide of these cations is active in oxidative coupling of methane, one can expect the obtention of active, selective and long lived catalysts. Definite compounds were prepared by this method. They correspond to $(LnLiO_2)_{(1-x)}(MgO,CaO)_x$ or $(LnLiO_2)_{(1-x)}(MgO,SrO)_x$ compounds in which Mg^{2+} substitutes for Li^+ and Ca^{2+} or Sr^{2+} for Ln^{3+} ($0.1 \leq x \leq 0.33$).

EXPERIMENTAL PART

Activity and selectivity of the different samples were determined in a fixed bed quartz reactor (6.6 mm ID) in the following conditions : unlet temperature : 600-750°C; feed gas partial pressures : 0.133 atm. CH_4, 0.0665 atm. O_2 and 0.8 atm. He; gas flow : 4.5 $l.h^{-1}$ $g.cat^{-1}$ (N.T.P.); catalyst weight : 0.67 g; CH_4/O_2 ratio : 2 ($2CH_4 + O_2 \longrightarrow C_2H_4 + 2H_2O$).

Methane conversion is expressed as : moles of transformed CH_4 X 100/moles of introduced CH_4.

Selectivity in product (i) is defined as : moles of CH_4 transformed into product (i) X 100/moles of transformed CH_4.

Yield in product (i) is given as conversion X selectivity X 100.

The catalysts were prepared from aqueous solutions of the given rare earth nitrate and the lithium carbonate or hydroxide. The solids are obtained by evaporation to dryness at 100-120°C of the solution or of the suspension in which the rare earth has been precipitated as oxalate by oxalic acid (pH = 2.2). These solids were then calcined at 750°C during 24 hours. The following definite $LnLiO_2$ compounds characterized by then XRD mesh parameters (ref. 12) were obtained by this method : $SmLiO_2$, $NdLiO_2$ and $LaLiO_2$. Compounds having a $(LnLiO_2)_{1-x}$ $(MgO,CaO)_x$ or $(LnLiO_2)_{1-x}$ $(MgO,SrO)_x$ structure were prepared as follows :

i) an ethanol solution of lanthanide (Sm, Nd or La), magnesium, calcium or strontium nitrates was precipitated by oxalic acid (pH = 2.2). The solids were obtained by evaporation of the suspension to dryness at 110-120°C and then heated between 550 and 650°C under argon during 24 hours to decompose the oxalates.

ii) the obtained solid was then suspended in an ethanolic solution containing lithium hydroxide or carbonate. The mixture was stirred during one hour and the solvent was eliminated by evaporation at 110-120°C. The solid residue was heated under argon (24 h.) at 850°C. The obtained catalyst was only taken out of the furnace after cooling to room temperature under argon.

RESULTS

Although the obtained $LnLiO_2$ compounds have a definite structure, the precursors used in the preparation play an important role. Thus the selectivity results are markedly altered when the starting rare earth compound changes from hydroxide to oxalate or nitrate and lithium hydroxide to lithium carbonate.

The results obtained for $NdLiO_2$ are given in table I.

TABLE 1

Precursors effects on $NdLiO_2$ activity.

$NdLiO_2$	Conversion CH_4	O_2	Selectivity C_2H_4	C_2H_6	C_2	CO_2	CO	ratio C_2 sat./ C_2 unsat.	yield C_2
A	26.3	51.6	5.9	8.3	14.2	81.6	4.2	1.4	3.7
B	30.4	56.1	14.0	24.0	38.0	60.0	2.0	1.7	11.6
C	31.2	46.7	9.6	17.4	27.1	70.9	2.0	1.8	8.5

A : neodynium oxalate and lithium carbonate
B : neodymium oxalate and lithium hydroxide
C : neodymium nitrate and lithium carbonate
(T = 700°C, ratio CH_4/O_2 = 2, gas flow 4.5 $l.h^{-1}$ g^{-1} catalyst; weight catalyst : 0.7 g, P = 1 atm : 0.133 atm CH_4; 0.0665 atm. O_2; 0.8 atm He).

As for $SmLiO_2$ (ref. 13) one can notice that the CH_4 conversion remains more or less constant with a slight increase for the preparation based on neodymium nitrate and lithium carbonate. The selectivity into C_2 hydrocarbons is most favoured for the catalyst obtained from neodymium oxalate and lithium hydroxide (B). In a previous work (ref. 13) an attempt of explanation based on samarium oxalate and nitrate calcination temperature and on compared basicity of $LiCO_3$ and LiOH was given. No difference for the three preparations (A,B,C) is apparent in the XRD spectra of the $NdLiO_2$ samples. The BET specific area of the catalysts after calcination at 750°C are : 0.6, 5.75 and 4.00 m^2/g for A,B and C preparations respectively. The surface independant CH_4 conversion suggests that other structural or homogeneity factors may play an important role.

Insertion of other oxides into $LnLiO_2$ structures.

The substitution of samarium and lithium atoms from a $LnLiO_2$ structure can be achieved by cations having close metal-oxygen distances (coordination number : six, monoclinic structure). Our choice went to Mg^{2+} and Ca^{2+} for which the literature reports excellent properties in oxidative coupling of methane. For electric balance reasons samarium and lithium substitution must take place simultaneously. However the following substitution schemes :

$$Sm^{3+} + Li^{+} \longrightarrow 2Ca^{2+} \quad \text{or} \quad Sm^{3+} + Li^{+} \longrightarrow 2Mg^{2+}$$

must, a priori, be discussed because of the incompatibility of dimensional factors between calcium and lithium, magnesium and samarium. The metal-oxygen distances computed by the invariant method (ref. 14-16) are respectively 2.405A (Ca^{2+}); 2.135A (Li^{+}); 2.473A (Sm^{3+}) and 2.106A (Mg^{2+}).

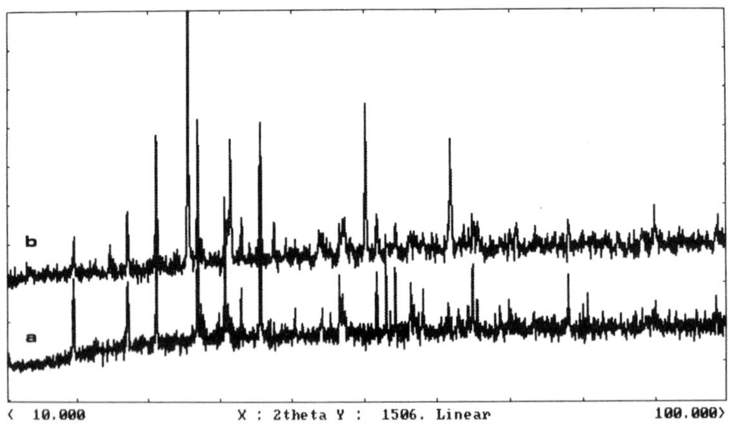

Fig. 1 : X.R.D. data of $SmLiO_2$ and substitued $SmLiO_2$.
a : $SmLiO_2$ b : $SmLiO_{2\ 1-x}(MgO,CaO)_x$

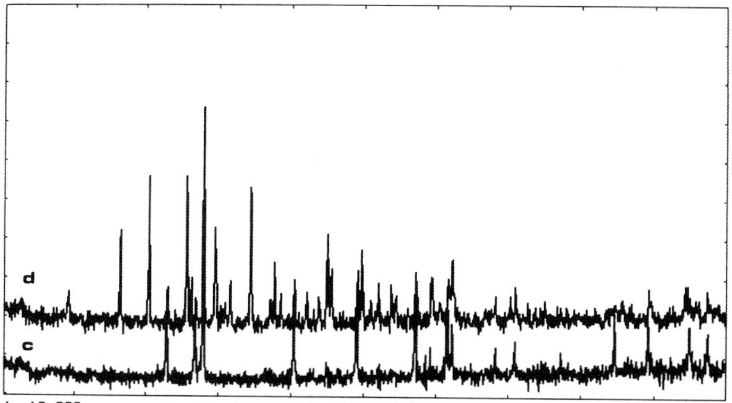

Fig. 2 : X.R.D. data of $LaLiO_2$ and substitued $LaLiO_2$.
c : $LaLiO_{2(1-x)}(MgO-SrO)_x$ d : $LaLiO_{2(1-x)}n(MgO-CaO)_x$

Thus dimensional factors are consistent between Ca^{2+} and Sm^{3+} and between Li^+ and Mg^{2+}, but the difference is too large between Ca^{2+} and Li^+ or Mg^{2+} and Sm^{3+}. Therefore the substitution by MgO and CaO must be undertaken simultaneously. The substitution by Mg^{2+} and Ca^{2+} taken in equal amounts fits fairly well since the compatibility is realized between Sm^{3+}(2.473A) and Ca^{2+}(2.405A) and between Li^+(2.135A) and Mg^{2+}(2.106A). Definite systems of $(SmLiO_2)_{1-x}(MgO,CaO)_x$ composition can be obtained as confirmed by XRD analysis (Fig. 1 and 2).

Table 2 summarizes the results obtained at 700°C after substitution by MgO and CaO, x being equal to 0.33.

An increased C_2 hydrocarbon (60%) selectivity as well as a changed C_2 saturated/C_2 unsaturated hydrocarbon ratio when compared to $LnLiO_2$ or Ln_2O_3 are observed. Conversion is slightly lower. When MgO is used alone in the substitution, the XRD spectra shows the presence of free MgO and the catalytic system works as if MgO was deposited on the $SmLiO_2$ surface (higher activity but similar selectivity).

The value of x can vary in a large range. For $0.1 \leqslant x \leqslant 0.33$, fig. 3 represents the evolution of selectivity and conversion. For $SmLiO_2$ an enhanced selectivity is reached for a x value as low as 0.1. A decreased conversion appears too for a low substitution by MgO and CaO. Additional amounts of MgO and CaO seem to influence neither selectivity nor activity further.

Table 2 : Activity and selectivity of pure Sm and Nd oxides, $SmLiO_2$ and $NdLiO_2$ and substitued by Mg and Ca of $SmLiO_2$ and $NdLiO_2$.

catalysts	Conversion CH_4	O_2	Selectivity C_2H_4	C_2H_6	C_2	CO_2	CO	Ratio C_2 sat./ C_2 unsat.	Yield C_2
Sm_2O_3	25.5	45.7	14.7	10.4	25.1	67.9	7.0	0.7	6.4
Nd_2O_3	28.6	65.7	14.5	16.5	31.0	63.5	5.5	1.1	8.9
$SmLiO_2$	31.9	50.3	4.6	24.1	28.7	65.5	5.8	5.2	9.2
$NdLiO_2$	30.4	56.1	14.0	24.0	38.0	60.0	2.0	1.7	11.6
$SmLiO_2$(1-x) xMgO-CaO*	26.5	51.3	30.8	28.2	59.0	38.6	2.4	0.9	18.2
$NdLiO_2$ (1-x) xMgO-CaO*	21.9	47.7	28.4	21.8	50.2	46.8	2.9	0.8	11.0
$SmLiO_2$ (1-x) xMgO*	35.0	63.1	26.2	15.8	41.9	54.8	3.3	0.6	14.7

*x = 0.33. Same conditions as in Table 1.

The evolution is slightly different for $(NdLiO_2)_{1-x}$ $(MgO,CaO)_x$. The fact that the Ca^{2+} substitution for Nd^{3+} is less favourable (Nd^{3+} = 2.513A; Ca^{2+} = 2.405A) than for Sm^{3+} must be underlined here. The substitution in $LaLiO_2$ is even less likely (La^{3+} = 2.596A) and the obtained results are more haphazardous. Except in one case, the selectivity is close to that obtained with $LaLiO_2$ alone but the conversion is increased.
That's why for lanthanum oxide, the chosen alkaline earth is strontium (Sr^{2+} = 2.580A; La^{3+} = 2.596A. Results for x = 0.33 are given in table 3. The reactivity of $(LaLiO_2)_{1-x}(MgO, SrO)_x$ versus x is given in Fig. 3. Here too, the addition of even weak amounts of MgO and SrO is sufficient to increase significantly the selectivity.

CONCLUSION

The present work shows the possibility to work with compounds of definite structure in the oxidative coupling of methane instead with catalysts obtained by impregnation. Structures such as $LnLiO_2$ can be used alone or as host structure for other cations (Mg,Ca,Sr...) which are active in the oxidative coupling. Introduction of MgO and CaO into the crystal frame of $SmLiO_2$ increases the selectivity up to 60% in C_2 hydrocarbons compare to 25% and

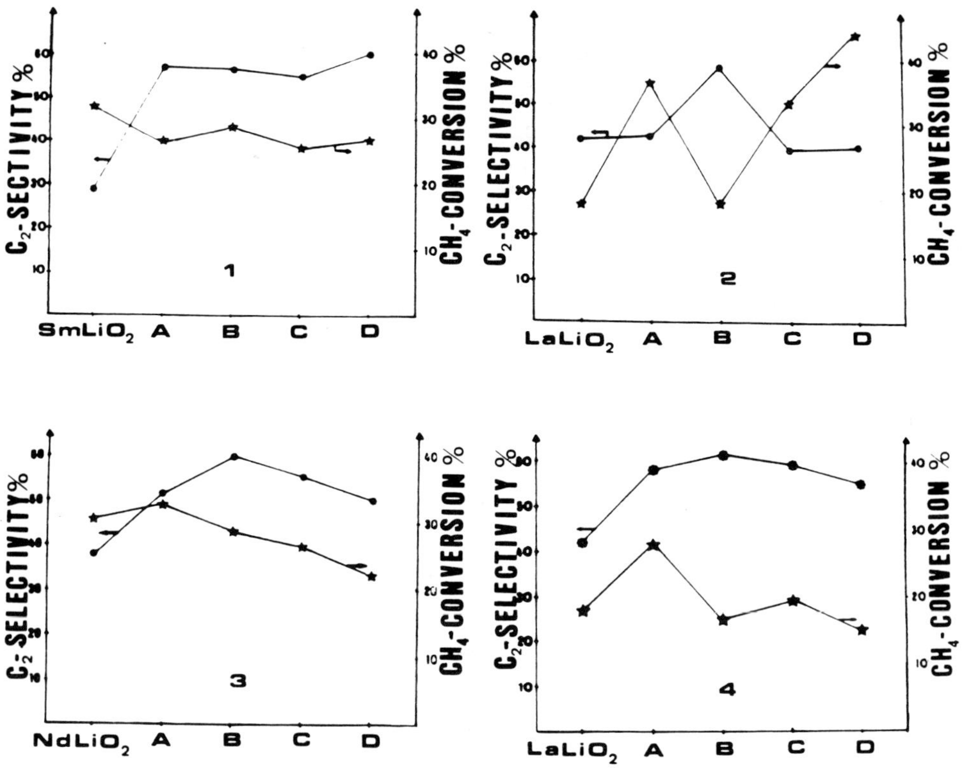

Fig. 3 : Evolution of C_2 selectivity with cation substitution.(Mg and Ca)
1 : $SmLiO_2$ 2 : $LaLiO_2$ 3 : $NdLiO_2$ 4 : $LaLiO_2$: substitution by Mg and Sr
A : x = 0.10 ; B : x = 0.16 ; C : x = 0.22 ; D : x = 0.33

Table 3 : Activity and selectivity of La_2O_3, $LaLiO_2$ and substitued by Mg and Ca, Mg and Sr of $LaLiO_2$ (x = 0.33). Same conditions as in Table 1

Catalyst	Conversion		Selectivity					Ratio	Yield
	CH_4	O_2	C_2H_4	C_2H_6	C_2	CO_2	CO	C_2 sat./ C_2 unsat.	C_2
La_2O_3	26.9	46.9	12.3	11.2	23.5	61.2	15.3	0.9	6.2
$LaLiO_2$	17.7	39.0	13.1	29.8	42.9	56.0	1.1	2.3	7.6
$LaLiO_2$(1-x) xCaO-MgO	44.8	84.3	23.0	16.8	39.8	48.5	11.7	0.7	17.8
$LaLiO_2$(1-x) xSrO-MgO	14.9	51.9	18.6	37.1	55.7	40.3	3.9	2.0	8.3

29% for Sm_2O_3 and $SmLiO_2$ respectively. Methane conversion, although slightly diminished remains higher than 25%.

LITERATURE

1 H. Mimoun, New Journal Chem. 11 (1987) 513-525
2 "Kirk-Othmer Encyclopedia of Chemical Technology" Wiley, New-York Vol. 1, p. 193 (2nd Edition)
3 British Petroleum European Patent 93 543 (1983)
4 M. Ito and J.H. Lundsford, Nature 314 (1985) 721-722
5 W. Hinsen, W. Bytyn and M. Baerns, Proc. 8th Int. Congr. Catal., Berlin, 2-6 July, 1984, Springer Verlag, 1984, Vol. III, pp. 581-592
6 T. Ito, J.X. Wang, C.H. Lin and J.H. Lundsford, J. Am. Chem. Soc. 107 (1985) 5062-5068
7 K. Otsuka, Q. Liu, M. Hatano and A. Morikawa, Chem. Lett. (1986) 467-468
8 T. Moriyama, N. Takasaki, E. Iwamatsu and K. Aika, Chem. Lett. (1986) 1165-1168
9 N. Yamagata, K. Tanaka, S. Sasaki and S. Okazoki, Chem. Lett. (1987) 81-82
10 J.M. De Boy and R.F. Hicks, Ind. Eng. Chem. Res. 27 (1988) 1577-1582
11 S.J. Korf, J.A. Ross, N.A. de Bruijn, J.G. Van Ommen and J.R.H. Ross, Chem. Comm. (1987), 1433-1434
12 M. Gondrand, Bull. Soc. Fr. Mineral. Cristallog. (1967) XC 107-108
13 A. Kaddouri, R. Kieffer, A. Kiennemann, P. POIX and J.L. Rehspringer, Appl. Catal. 51 L1-L6 (1989)
14 P. Poix, "Liaison Interatomique et propriétés physiques des composés minéraux " 1. SEDES (1966) 82-120
15 P. Poix, C.R. Acad. Sci. Paris C 270 (1970) 1852-1853
16 P. Poix, C.R. Acad. Sci. Paris C 268 (1969) 1139-1140

G. Centi and F. Trifiro' (Editors), *New Developments in Selective Oxidation*
© 1990 Elsevier Science Publishers B.V., Amsterdam — Printed in The Netherlands

BEHAVIOR OF METALLIC OXIDES SUPORTED ON Li/MgO IN THE METHANE OLIGOMERIZATION

G.T. BARONETTI, E.J. LAZZARI, A.A. CASTRO and O.A. SCELZA

Instituto de Investigaciones en Catálisis y Petroquímica -INCAPE-
Santiago del Estero 2654 - 3000 Santa Fe, Argentina

SUMMARY

The effect of Li addition to MgO on the performance in the methane oligomerization was determined. Besides, Li/MgO doped with different lanthanide oxides (Pr, La, Ce and Sm oxides) were also studied in the same reaction. It was found a beneficial effect of Li addition to MgO. Moreover, the addition of Pr or Ce oxide enhances the performance of Li/MgO catalyst.

INTRODUCTION

Methane selective oxidation into C_2 hydrocarbons is a very interesting process since that more valuable chemicals could be obtained from natural gas.

However, methane conversion into higher hydrocarbons presents a thermodynamic barrier. In fact, homogeneous oligomerization in gas phase is only feasible at temperature higher than 1000°C but with low C_2 yield (ref. 1). This thermodynamic barrier could be eliminated by using a substance able to provide the necessary oxygen to react with the hydrogen producing water. Thus, reducible metal oxides could be used as oxygen source, as Keller and Bhasin showed in their pioneer paper (ref. 2). Since then, a great effort has been made in order to find active and selective catalysts to produce higher hydrocarbons from methane (refs. 3-6).

Several recent papers reported a good performance of basic solids promoted with alkaline-metal ions (refs. 7-9). Thus, it has been reported that the activity of MgO is due to its capacity to produce methyl radicals, since that this material posses intrinsic cationic vacancies which can react with O_2 to produce O^- centers (active centers for the methane activation) (refs. 10, 11). Alkaline-metal addition to MgO enhances the capacity to abstract H• from methane producing CH_3• which could dimerize in gas phase. Other catalysts based on MgO have been reported in the literature (refs. 12-14).

The influence of the Li content in Li/MgO catalysts on the activity and selectivity in methane oligomerization is studied in this paper. Likewise, the behavior of different lanthanide oxide/Li/MgO catalysts (Pr, Sm, Ce or La oxide) is reported.

Besides, in order to elucidate the effect of Pr and Sm addition to Li/MgO, a more detailed study on these catalysts was carried out.

EXPERIMENTAL

MgO-based catalysts (MgO, Li/MgO, lanthanide oxide/MgO, lanthanide oxide/Li/MgO) were obtained by using the slurry method proposed by Ito et al. (ref. 15). Magnesium oxide p.a. from Merck was used for the catalysts preparation.

Li/MgO catalysts with different Li content (0.5, 1, 4, 7 and 15wt%) were obtained by using Li_2CO_3 as lithium precursor.

Pr_6O_{11}/Li(1wt%)/MgO catalysts with different Pr loading (2, 5 and 10 mol lanthanide per 100 mol Li) were prepared by simultaneous addition of $Pr(NO_3)_3 \cdot 6H_2O$ and Li_2CO_3 to the slurry which contained MgO and water. Besides, Li(1wt%)/MgO doped with Ce, La and Sm oxides were prepared in the same way by using nitrates as lanthanide precursors. In these catalysts the lanthanide content was 2 mol lanthanide per 100 mol Li.

After obtaining the MgO slurry with the different components, the residual water was evaporated, then the catalysts were dried at 120°C and finally calcined in a flowing air stream at 750°C during 5 h.

Several intimate mechanical mixtures (MM) between bulk samarium or praseodymium oxide and MgO or Li(1wt%)/MgO were also tested in the methane oligomerization. For these cases the catalytic bed contained the same lanthanide oxide amount as catalysts prepared by the slurry method. One sample of bulk samarium oxide was impregnated with Li_2CO_3 and calcined in air at 750°C.

Praseodymium and samarium bulk oxides were prepared by decomposition of the corresponding nitrate at 750°C.

Methane oligomerization reaction was carried out at atmospheric pressure in a fixed bed quartz reactor by using a CH_4-air mixture as a feed. The residence time was 3.35 g cat, s/ml CH_4 STP, the CH_4/O_2 molar ratio in the feed was either 2 or 5 and the reaction temperature ranged between 650 and 800°C. Reaction products were analyzed by G.C. The selectivity to C_2 was defined as the percentage of the methane converted into ethane and ethylene.

RESULTS AND DISCUSSION

Table 1 shows the values of the CH_4 conversion (X) and selectivities to C_2 (S_{C_2}) and CO_2 (S_{CO_2}) at 750°C for MgO promoted with different lithium loadings and for different CH_4/O_2 molar ratios in the feed. It can be observed that unpromoted MgO is highly selective to deep oxidation products, but the addition of a small quantity of Li (0.5wt%) enhances the CH_4 conversion and produces a drastic change in the selectivity, increasing the C_2 hydrocarbons formation. The effect of the lithium addition on the activity and selectivity is more noticeable when higher CH_4/O_2 in the feed is used. Besides, Table 1 shows that both the selectivity to C_2 and methane conversion present a broad maximum between 0.5 and 4wt% Li.

The role of the lithium in MgO doped catalyst has been discussed by Lunsford

TABLE 1

Values of methane conversion (X), selectivity to ethane + ethylene (S_{C_2}) and selectivity to CO_2 (S_{CO_2}) for MgO catalysts doped with different contents of lithium. T = 750°C, CH_4/O_2 molar ratio (R) = 2 and 5

Catalyst	R = 2			R = 5		
	X, %	S_{C_2}, %	S_{CO_2}, %	X, %	S_{C_2}, %	S_{CO_2}, %
MgO	26.8	7.8	92.2	10.2	6.1	93.9
0.5wt% Li/MgO	32.1	30.9	69.1	18.9	54.8	45.2
1.0wt% Li/MgO	32.9	29.9	70.1	18.5	56.3	43.7
4.0wt% Li/MgO	22.7	38.3	61.7	16.0	59.4	40.6
7.0wt% Li/MgO	30.9	23.2	76.8	9.5	56.8	43.2
15.0wt% Li/MgO	24.8	33.7	66.3	12.1	48.8	51.2

et al. (ref. 15), who show a relationship between the production of $CH_3\bullet$ radicals and the density of paramagnetic O^- species on the Li/MgO surface. It is accepted that the active sites of the H• abstraction from CH_4 would be (Li^+O^-) surface species. These active sites would be produced from a Mg^{+2} substitution by a lithium cation. Moreover, it has been reported that the addition of higher amounts of alkaline metals to MgO leads to a great diminution of the surface area (ref. 16).

Considering the following reaction scheme (ref. 17):

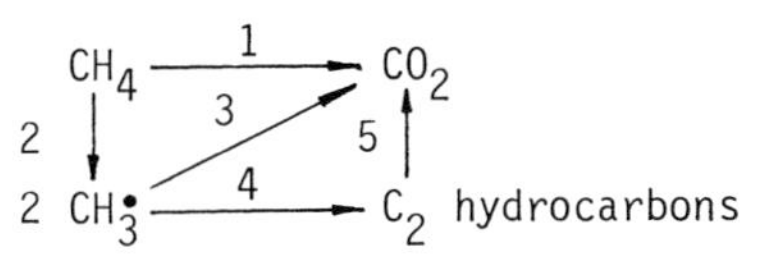

the steps 1, 2, 3 and 5 take place on the catalyst surface meanwhile step 4 occurs in gas phase. With a high surface area of the catalyst, steps 1, 2, 3 and 5 are favored. When a small quantity of lithium is added to MgO not only the active sites concentration increases but also the surface area decreases. Hence, the step 2 is enhanced, meanwhile the steps which lead to the hydrocarbon oxidation products are negatively affected. For high lithium content the active sites (Li^+O^-) concentration is drastically reduced (ref. 11) and the formation of the oxidation products are favored in such conditions. Hence a maximum value of the selectivity to C_2 hydrocarbons must be expected for a given Li content such as the results of Table 1 display.

It must be indicated that a drastic drop in the MgO surface area after lithium addition was observed in our experiments. In fact, areas values of 59.7, 3.4 and 1 m^2/g for 0, 4 and 15wt% Li/MgO were obtained.

The effect of doping Li(1wt%)/MgO with Pr, Ce, Sm and La oxides (2 mol% lanthanide with respect to the Li content) was also analyzed. X, S_{C_2} and the yield to C_2 (Y_{C_2} = X S_{C_2}) values for CH_4/O_2 molar ratio (R) = 2 and different reaction temperatures are shown in Table 2. It can be seen that the methane conversion increases when the reaction temperature increases, meanwhile S_{C_2} and Y_{C_2} present maxima values at 750°C for all the catalysts. By comparing the Y_{C_2} values at 750°C, it can be concluded that the addition of Pr and Ce oxides to Li/MgO leads to a better performance of the catalyst, meanwhile the addition of Sm and La oxides appears to have a negative effect.

The effect of the different Pr loading on Li(1wt%)/MgO was also studied and the results are shown in Table 3. It can be observed a maximum in the methane conversion and in the selectivity to C_2 for a Pr loading of 2% (mol Pr/mol Li).

In order to explain the beneficial effect of the Pr addition to Li(1wt%)/MgO, Pr_6O_{11}/MgO, Pr_6O_{11}/Li(1wt%)/MgO, and the intimate mechanical mixtures (MM) between praseodymium oxide and MgO or Li(1wt%)/MgO were tested in methane oligomerization at 750°C. Figure 1 shows S_{C_2} and Y_{C_2} values for the above mentioned catalysts. By comparing the results of Pr_6O_{11}/MgO with those of the MM

TABLE 2

Values of methane conversion (X), selectivity to C_2 hydrocarbons (S_{C_2}) and yield to C_2 hydrocarbons (Y_{C_2}) at different reaction temperatures for Li(1wt%)/MgO catalysts doped with Pr, Ce, Sm and La oxides (2 mol% lanthanide respect to the Li content) and for R = 2

Temperature, °C	Catalyst	X, %	S_{C_2}, %	Y_{C_2}, %
	Pr_6O_{11}/Li/MgO	8.4	14.2	1.2
	CeO_2/Li/MgO	6.9	13.2	0.9
650	Sm_2O_3/Li/MgO	6.7	12.7	0.8
	La_2O_3/Li/MgO	6.5	26.9	1.7
	Li/MgO	8.0	18.9	1.5
	Pr_6O_{11}/Li/MgO	34.8	33.7	11.7
	CeO_2/Li/MgO	36.0	35.6	12.8
750	Sm_2O_3/Li/MgO	29.8	22.7	6.8
	La_2O_3/Li/MgO	31.4	27.4	8.6
	Li/MgO	32.9	29.9	9.8
	Pr_6O_{11}/Li/MgO	35.7	18.4	6.5
	CeO_2/Li/MgO	36.8	13.2	4.8
800	Sm_2O_3/Li/MgO	35.0	15.3	5.3
	La_2O_3/Li/MgO	35.0	20.4	7.1
	Li/MgO	33.2	16.6	5.5

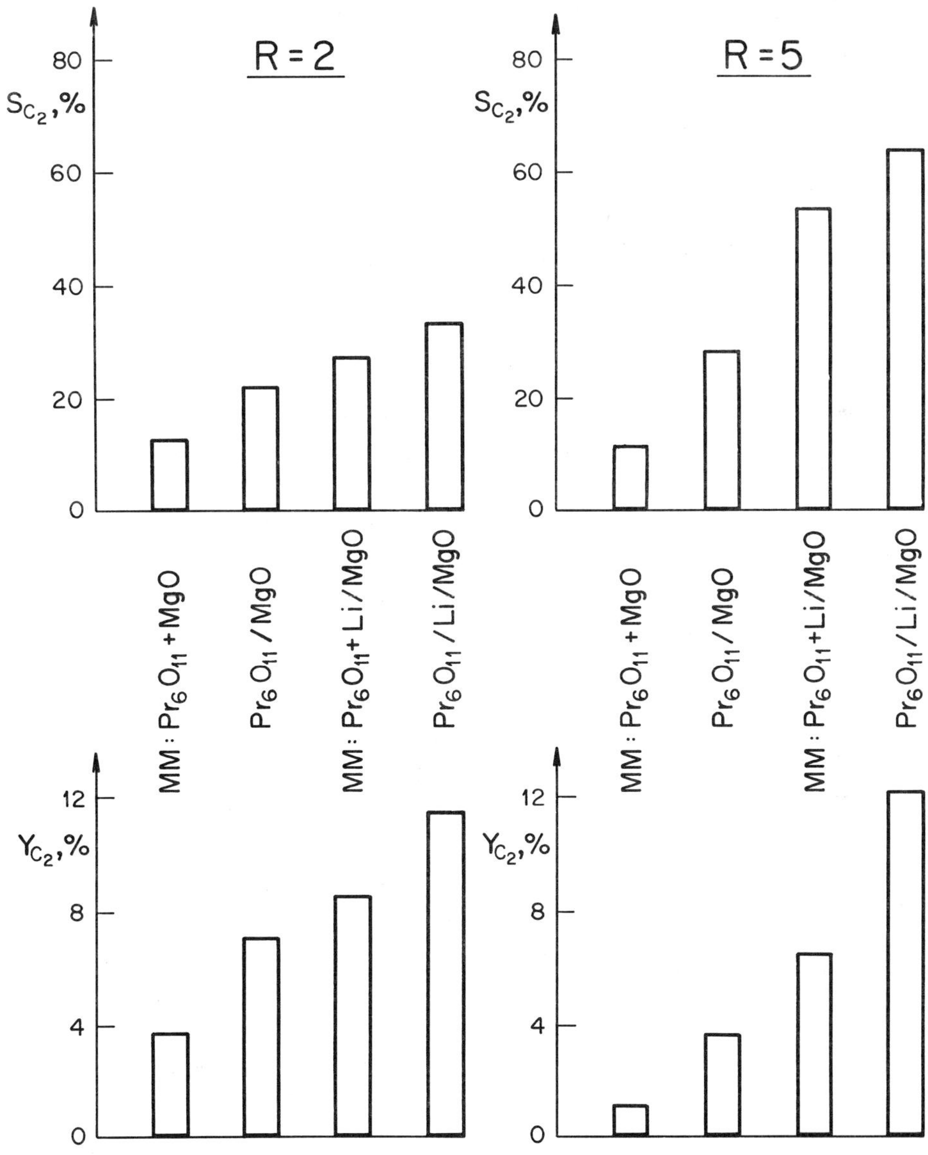

Fig. 1. Selectivity to C_2 and yield to C_2 at 750°C for different catalysts. MM: mechanical mixture. The lithium content in $Pr_6O_{11}/Li/MgO$ was 1wt%. Pr loading in Pr_6O_{11}/MgO was the same that in $Pr_6O_{11}/Li/MgO$ (2 mol Pr/100 mol Li). The proportion of the components in mechanical mixtures was the same that in catalysts prepared by the slurry method

TABLE 3

Performance of Pr_6O_{11}/Li(1wt%)/MgO catalysts with different Pr loading in the methane oligomerization at 750°C and different CH_4/O_2 molar ratios (R)

Catalyst	Pr/Li molar ratio, %	R = 2		R = 5	
		X, %	S_{C_2}, %	X, %	S_{C_2}, %
Li/MgO	0	32.9	29.9	18.5	56.3
Pr_6O_{11}/Li/MgO	2	34.8	33.7	19.6	64.3
Pr_6O_{11}/Li/MgO	5	31.9	22.8	17.2	51.7
Pr_6O_{11}/Li/MgO	10	33.6	28.9	14.4	50.7

(Pr_6O_{11} + MgO), it can be observed that the first catalyst is more selective to C_2 than the mechanical mixture. Hence, these results indicate that the catalysts obtained by the slurry method can not be considered as the sum of the isolated components. There appears to take place a certain interaction between the lanthanide oxide and MgO.

Considering the results of the MM composed by Pr_6O_{11} + Li(1wt%)/MgO and those of the Pr_6O_{11}/Li(1wt%)/MgO catalyst (Figure 1), it must be noted again that when this catalyst is obtained by the slurry method there is a certain interaction between the dopants and the support. In fact, the characteristic praseodymium oxide lines were not found in the XRD patterns of the Pr_6O_{11}/Li(1wt%)/MgO catalyst. However, when the Pr_6O_{11}/Li(1wt%)/MgO catalyst is prepared by impregnation of Li(1wt%)/MgO with $Pr(NO_3)_3.6H_2O$ the XRD patterns clearly show the characteristic peaks of Pr_6O_{11}.

It could be explained the beneficial effect of Pr addition to Li/MgO considering that the praseodymium oxide could act as a charge carrier for the regeneration of the active sites of lithium-doped MgO catalysts. Similar effects for alkali promoted-Pr_6O_{11} and CeO_2/Li/MgO were reported in the literature (refs. 18, 19).

Previously, it has been shown that Li(1wt%)/MgO catalysts promoted with Sm displays lower S_{C_2} and Y_{C_2} values than those of Li(1wt%)/MgO catalysts (Table 2). In order to study this system, bulk samarium oxide (Sm_2O_3), Sm_2O_3/Li(1wt%)/MgO, and a mechanical mixture (MM) between bulk Sm_2O_3 and Li(1wt%)/MgO were tested at 750°C and R = 2 (Table 4).

The results show that the yield to C_2 of the catalyst prepared by simultaneous deposition of Sm, Li and MgO is lower than that obtained for the isolated components and for the mechanical mixture. Hence, when Sm, Li and MgO are in an intimate contact, an unfavorable effect would take place.

One additional experiment was carried out by using bulk samarium oxide doped

TABLE 4

X, S_{C_2} and Y_{C_2} values at 750°C and R = 2 for Sm catalysts

Catalyst	Sm/Li molar ratio, %	X, %	S_{C_2}, %	Y_{C_2}, %
Sm_2O_3	-	12.8	61.2	7.8
Sm_2O_3 + Li(1wt%)/MgO (mechanical mixture)	2	27.3	27.4	7.5
Sm_2O_3/Li(1wt%)/MgO	2	29.8	22.7	6.8
Li(1wt%)/MgO	-	32.9	29.9	9.8

with Li_2CO_3 (Sm/Li molar ratio = 2%). It showed S_{C_2} values 10% lower than that of Sm_2O_3, meanwhile the methane conversion was similar to that of Sm_2O_3. These results indicate a negative effect of Li on the performance of bulk samarium oxide. Korf et al. (ref. 20) reported that the addition of lithium to Sm_2O_3 produces a structural modification from cubic to monoclinic structure with a simultaneous diminution in the selectivity to C_2. On the other hand, the results reported by Otsuka et al. (ref. 21) were opposite to those of Korf et al. (ref. 20) and to our results. In fact, Otsuka found that the Li-doped Sm_2O_3 catalyst was more active and selective for C_2 hydrocarbons production than Sm_2O_3. However, the negative effect of samarium oxide addition to Li/MgO can not be clearly explained and much effort will be needed in order to understand this behavior.

REFERENCES

1 D.R. Ardiles, O.A. Scelza and A.A. Castro, Rev. Fac. Ing. Quím. Santa Fe, 46 (1984) 7-16.
2 G.E. Keller and M.M. Bhasin, J. Catal., 73 (1982) 9-19.
3 W. Hinsen, W. Bytyn and M. Baerns, Proc. 8th. Int. Congr. Catal., 3 (1984) 581-92.
4 K. Otsuka, K. Jinno and A. Morikawa, Chem. Lett., (1985) 499-500.
5 J.A. Sofranko, J.J. Leonard and C.A. Jones, J. Catal., 103 (1987) 302-10.
6 K. Asami, S. Hashimoto, K. Fujimoto and H. Tominaga, in D.M. Bibby, C.D. Chang, R.F. Howe and S. Yurchak (Eds.), Methane Conversion, (1988) 403-07.
7 F.P. Larkins and M.R. Nordin, in D.M. Bibby, C.D. Chang, R.F. Howe and S. Yurchak (Eds.), Methane Conversion, (1987) 127-31.
8 J.A. Roos, A.G. Bakker, H. Bosch, J.G. van Ommen and J.R.H. Ross, Catalysis Today, 1 (1987) 133-45.
9 Y.L. Bi, K.J. Zhen, Y.T. Jiang, C.W. Teng and X.G. Yang, Appl. Catal., 39 (1-2) (1988) 185-90.
10 D.J. Driscoll and J.H. Lunsford, J. Phys. Chem., 89 (1985) 4415-18.
11 D.J. Driscoll, W. Martir, J.-X. Wang and J.H. Lunsford, J. Am. Chem. Soc., 107 (1985) 58-63.
12 I.T. Ali Emesh and Y. Amenomiya, J. Phys. Chem., 90 (1986) 4785-89.
13 E. Iwamatsu, T. Moriyama, N. Takasaki and K. Aika, J. Chem. Soc. Chem. Comm., (1987) 19-20.
14 E.J. Ereckson and A.L. Lee, Preprints, Div. Petrol. Chem., ACS, 33 (3) (1988) 443-44.
15 T. Ito, J.-X. Wang, C. Liu and J.H. Lunsford, J. Am. Chem. Soc., 107 (1985)

5062-68.
16 T. Moriyama, N. Takasaki, E. Iwamatsu and K. Aika, Chem. Lett., (1986) 1165-68.
17 E. Iwamatsu, T. Moriyama, N. Takasaki and K. Aika, J. Catal., 113 (1988) 25-35.
18 A.M. Gaffney, C.A. Jones, J.J. Leonard and J.A. Sofranko, Preprints, Div. Petrol. Chem., ACS, 33 (3) (1988) 445-52.
19 S. Bartsch, J. Falkowski and H. Hofmann, Catalysis Today, 4 (1989) 421-31.
20 S.J. Korf, J.A. Roos, J.M. Diphoorn, R.H.J. Veehof, J.G. van Ommen and J.R.H. Ross, European Workshop on Methane Conversion, May 1988, FRG.
21 K. Otsuka, Q. Liu and A. Morikawa, J. Chem. Soc. Commun., (1986) 586-7.

J.M. VAN KASTEREN (University of Technology, The Netherlands): Did you observe any deactivation or lithium loss during your experiment? Did you measure your Li content after calcination and after reaction?

O.A. SCELZA (INCAPE, Argentina): It was not observed deactivation during the reaction. The lithium contents on the catalysts after reaction were the same as those of the fresh catalysts (calcined catalysts).

J. KIWI (EPFL Chimie Physique, Switzerland): You use Pr_6O_{11} on MgO at 750°C when you prepare your catalyst. It is possible that Pr_yMgO_x (Li-doped) epitaxial growth is produced during the preparation of the catalyst or during the reaction at this relative high temperature. Do you have any evidence for this?

O.A. SCELZA (INCAPE, Argentina): Pr_6O_{11}/Li/MgO prepared by the "slurry" method and by impregnation were examined by XRD after the calcination step. The XRD patterns showed the characteristic peak of Pr_6O_{11} only for the second catalyst type. For the first catalyst type, no Pr compound (including Pr oxides) was detected.

J.R.H. ROSS (University of Twente, The Netherlands): How reproducible are your results if you prepare more than one sample of the same composition? Our experience is that the behavior over the first few hours of the reaction depends very much on the pre-history of the material even though the final behavior after some hours is the same or very similar.

O.A. SCELZA (INCAPE, Argentina): Several samples of different catalysts were prepared and tested in the methane oligomerization (the reaction time was about 1 h). Results showed a good reproducibility. However, I believe that the catalytic behavior strongly depends on the preparation and calcination conditions.

G. Centi and F. Trifiro' (Editors), *New Developments in Selective Oxidation*
1990 Elsevier Science Publishers B.V., Amsterdam — Printed in The Netherlands

Oxidative Coupling of Methane, the Effect of Gas Composition and Process Conditions.

J.A. Roos, S.J. Korf, J.J.P. Biermann, J.G. van Ommen and J.R.H. Ross
Faculty of Chemical Technology, University of Twente, PO Box 217,
7500AE, Enschede, The Netherlands

ABSTRACT

The effect of process conditions on the oxidative coupling of methane has been studied; factors examined included the effect on the product composition of the oxygen concentration in the reactor feed and of backmixing of the gas in the reactor. It appears that the desired C_2 products are susceptible to degradation reactions and that this has consequences for the method of reactor operation.

INTRODUCTION

A topic which has recently attracted interest is that of the direct oxidative conversion of methane into products such as methanol, ethane and ethylene. A large number of publications (see for example [1-14]) have appeared over the last few years which show that several types of catalyst can be used to give reasonably high conversions and selectivities for the formation of the latter two products according to the all-over reactions (1) and (2):

$$2\ CH_4 + \tfrac{1}{2}\ O_2 \ \text{--->}\ C_2H_6 + H_2O \quad (1)$$

$$2\ CH_4 + O_2 \ \text{--->}\ C_2H_4 + 2\ H_2O \quad (2)$$

In all cases, CO and CO_2 are also formed:

$$CH_4 + 1\tfrac{1}{2}\ O_2 \ \text{--->}\ CO + 2\ H_2O \quad (3)$$

$$CH_4 + 2\ O_2 \ \text{--->}\ CO_2 + 2\ H_2O \quad (4)$$

Ito et al. [1,2] have shown that lithium-doped magnesium oxide is an active and selective catalyst for the oxidative coupling reaction; they suggested that the lithium ions just fit into cation-vacancies of the MgO matrix, the ionic radii of the Li^+ and Mg^{2+} ions being almost the same. Using EPR spectroscopy they showed that $[Li^+O^-]$ centres are formed when Li^+ is added to MgO. Based on investigations of the formation of methyl radicals from methane over the same type of material, they proposed a radical mechanism in which methyl radicals are produced on the $[Li^+O^-]$ sites; they assumed that the methyl radicals then recombine to yield ethane or react with oxygen to form methoxy radicals which ultimately give rise to the formation of CO and CO_2. Other successful catalyst systems which have been reported are PbO supported on γ-Al_2O_3, various lanthanide oxides (i.e. Sm_2O_3 [4]) or promoted transition metal oxides e.g. NiO or MnO_2 [5, 6]. It appears that doping the lanthanides or transition metal oxides with alkali metal ions results in an enhancement of the activities of the metal oxides for the formation of C_2-compounds. When alkali chlorides were used, a high C_2H_4/C_2H_6 ratio was found. This is thought to be a result of the dehydrogenation of C_2H_6 by Cl•-radicals liberated by the catalyst [5, 6]. According to Otsuka and his coworkers, the methane activation step takes place on a peroxide anion species (O_2^{2-}) present at the surface of the catalyst [7]. Alkaline earth metal oxides (or carbonates), other than MgO, such as CaO, SrO or BaO, have also been found to be active for methane coupling, especially if they are promoted with alkali metal ions [8]. Most authors appear to accept that the oxidative coupling reaction occurs by the radical mechanism proposed by Lunsford et al. [1,2]; however, some authors favour a redox mechanism [3,9].

The aim of our work on selective methane oxidation has been to gain a greater understanding of the factors which govern the activity and selectivity in the coupling reaction with a view to developing more active, selective and stable catalysts. We therefore first compared the supported lead oxide system with the Li/MgO catalyst under the same circumstances [11]. We next studied the effects of pretreatment of Li/MgO catalysts [12,13] and the addition of promoters to Sm_2O_3 catalysts [14]. In the course of these investigations, we also found that process conditions, such as linear gas velocity, gas composition, reactor geometry and temperature can have a profound influence on the formation of C_2 products. We have therefore studied in some detail,in the work reported in this paper, the effects of these variables using a standard Li/MgO catalyst. The influence on the composition of the product mixture of the oxygen partial pressure in the reactor feed under normal process conditions is presented first. This is followed by a discussion of experiments showing the influence on selectivity of back-mixing in the reactor. In most of our work published to date, these effects have been taken fully into account.

EXPERIMENTAL

The catalysts used for these experiments were samples of Lithium-doped Magnesium oxide [1,2] prepared by wet impregnation of MgO powder by an aqueous solution of LiOH (both chemicals obtained from B.D.H. Ltd.). Two types of Li/MgO catalysts were prepared: in the case of type B, CO_2 was passed through the evaporating solution during impregnation [13], while in the case of type A the CO_2 treatment was omitted.(Passing CO_2 through the evaporating solution during impregnation results in a more active catalyst [13].) After impregnation, the samples were dried in air overnight at 140°C and then calcined in air in quartz or fused alumina crucibles at 850°C for 6 h. After calcination, the catalysts were crushed and sieved to a grain size of 0.3-0.6 mm or

0.1-0.3 mm. The particular catalyst used for any specific experiment is given in the appropriate Figures and the Table.

The catalytic experiments were carried out in vertically-placed quartz fixed-bed reactors (5 mm i.d., 40 cm heated length) operated at a pressure of one atmosphere. The temperature of the reactor, heated in an electric oven, was measured by a thermocouple (shielded by a quartz capillary), placed on top of the catalyst bed. Unless otherwise stated, the bed consisted of a mixture of catalyst particles with the same weight of quartz particles of the same size. The gases were analysed by gas chromatography using a Carbosieve B column (2.25 m length; 2 mm i.d.). The reactor could be used in two modes: single-pass or recycle. In the first case, the gas passed through the reactor and was directly fed to the gas chromatograph; in the second case, the greater part of the gas flow was recycled. The recycle ratio (R) and the gas flow are given in the appropriate table; R is defined as the ratio of the recycle flow to the net flow through the system.

The conversion of the reactants (α_{CH4} and α_{O2}) is defined as the number of moles of the reactant that have been converted, divided by the number of moles of the reactant in the feed. The selectivity of a product is defined by the number of moles of CH_4 that have reacted to give this product, divided by the total mumber of moles of CH_4 that have been converted. The yield of a product is given by the product of CH_4 conversion and the selectivity to this specific product.

In the experiments carried out to examine the effect of the variation of the oxygen concentration in the reactor feed, 93 mg of Li/MgO catalyst was used (0.3-0.6 mm). The catalyst patricles were diluted with the same weight of quartz paticles of the same size. The gas feed contained 50 kPa CH_4 and a variable amount of O_2 (the balance being He), with a total pressure of 1 atm (101 kPa). The gas flow rate was 0.42 cm^3s^{-1} (STP).

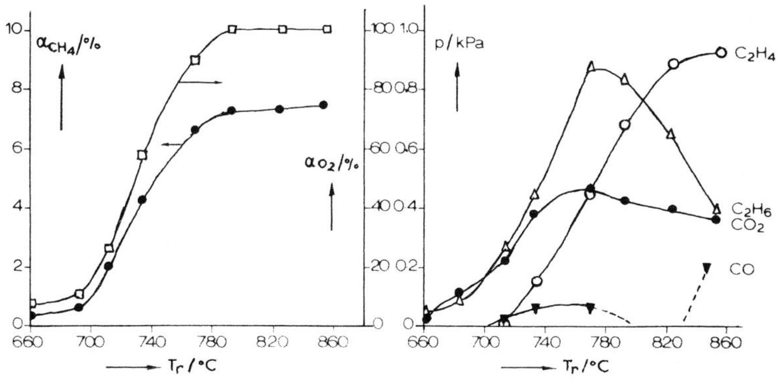

Figure 1. Methane and oxygen conversions and product concentrations as a function of reactor temperature, T_r; 93 mg of Li/MgO catalyst (type A, 7 wt% Li, 0.3-0.6 mm.) with quartz dilution and a gas feed containing 50 kPa CH_4, 2 kPa O_2 and 49 kPa He, with a flow rate of 0.42 cm^3s^{-1} (STP).

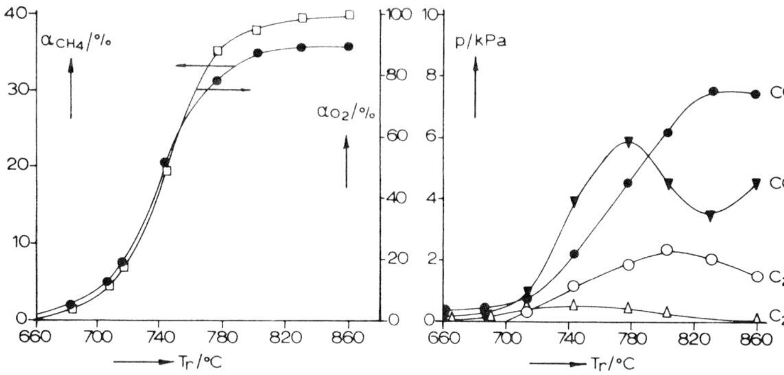

Figure 2. Methane and oxygen conversions and product concentrations as a function of reactor temperature, T_r; 93 mg of Li/MgO catalyst (type A, 7 wt% Li, 0.3-0.6 mm.) with quartz dilution and a gas feed containing 50 kPa CH_4, 21 kPa O_2 and 30 kPa He, with a flow rate of 0.42 cm^3s^{-1} (STP) .

RESULTS and DISCUSSION

The Effect of the Concentration of Oxygen in the Gas Feed.

A number of experiments were carried out, using a Li/MgO catalyst, to show the effect of the partial pressure of oxygen in the feed on the product composition. Figures 1 and 2 give the results (as a function of temperature) obtained at the extremes of the range of oxygen partial pressure studied: 2.0 kPa in the experiment of Figure 1 and 20.7 kPa in that of Figure 2. In both cases, the partial pressure of CH_4 in the feed was kept at 50 kPa; He was used to keep the total pressure at 101 kPa. When a low oxygen concentration was present in the feed (Figure 1), ethane and ethylene were the major products, with the consequence that the selectivity for the formation of C_2 products was high; a value of 88% was found at a reaction temperature of 800°C. When a high concentration of oxygen was present in the feed (Figure 2), CO and CO_2 were the the major products, with the consequence that the C_2 selectivity was much lower (32% at 800°C). In the latter case, the ratio of C_2H_4 to C_2H_6 was higher than in the former, indicating that there was a higher rate of oxidative dehydrogenation of C_2H_6 to C_2H_4 at higher oxygen concentrations. The dip in the partial pressure of CO as function of temperature observed in both Figures 1 and 2 at temperatures around 830°C was probably caused by a competition for the available oxygen between catalytic reactions (producing CO_2) and gas-phase reactions (producing CO) [16]; at temperatures around 790°C and higher, the available oxygen was (almost) completely consumed in both experiments. The catalytic reactions have relatively higher rates at temperatures of ca. 790-820°C than do the gas-phase reactions [16]; this has the consequence that the amount of oxygen available for CO formation in the gas-phase is reduced. At temperatures well above 800°C, however, the rates of the gas-phase reactions are much increased [16] and gas-phase reactions compete successfully with catalytic reactions for the available oxygen, resulting in the higher CO production observed in the experiments shown in Figures 1 and 2.

Figure 3 shows the partial pressures of the C_2 products and total oxidation products (at the exit of the reactor) as a function of the partial pressure of oxygen in the feed at reactor temperatures of 700, 750 and 800°C; the results of Figures 1 and 2 are included, together with those from experiments with partial pressures in the feed of 8.5 and 14.5 kPa. The following general conclusions can be drawn from these results:

(i) the C_2H_4/C_2H_6 ratio increases with increasing temperature and P_{O2} ;

(ii) the production of CO is low compared to the CO_2 production at lower temperatures and low values of P_{O2} ;

(iii) the partial pressures of CO and CO_2 are more affected by the partial pressure of oxygen in the reactor feed than are the partial pressures of the C_2 products, the sum of which remains almost constant with increasing P_{O2}.

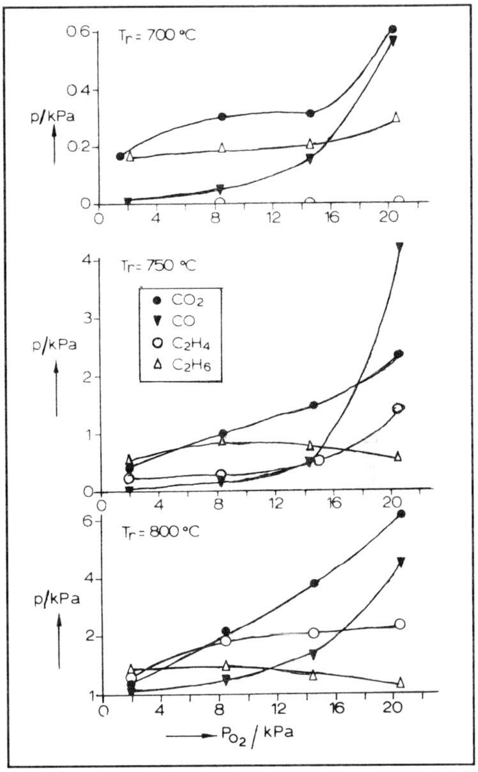

Figure 3. Product concentrations as a function of the partial pressure of oxygen in the reactor feed containing 50 kPa CH_4 (balance He, total pressure 101 kPa), with a flow rate of 0.42 cm^3s^{-1}.

It is generally accepted in the literature concerning the reaction network for this reaction that the C_2H_4 is formed from the C_2H_6 [2,3,4,11]. The increase in the C_2H_4/C_2H_6 ratio may thus be explained by an increase in the (oxidative) dehydrogenation of C_2H_6 with increase in oxygen concentration in the reactor and with increase in temperature.

The low values of the partial pressure of CO in the reactor effluent at lower P_{O2} and lower temperatures may be explained in the same way as done above in relation to the interplay of gas-phase oxidation reactions (producing CO) and catalytic oxidation reactions (producing CO_2) [16]. The CO production will be relatively low compared to the production of CO_2 at lower temperatures (gas-phase reactions are favoured at higher reaction temperatures) and at lower partial pressure of oxygen in the empty volume of the reactor between the catalyst bed and the

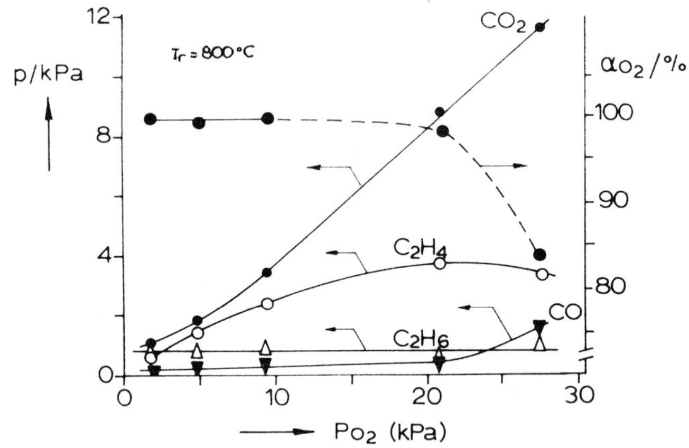

Figure 4. Product concentrations and oxygen conversion as a function of the partial pressure of oxygen, P_{O2}, in the reactor feed containing 50 kPa CH_4 (balance He, total pressure 101 kPa, flow rate = 0.42 cm^3s^{-1}), using 1.5 g of the Li/MgO catalyst (type A, 7 wt% Li, 0.3-0.6 mm.).

reactor exit (the "post-catalytic volume"). The oxygen partial pressure in the post-catalytic volume will be low if the oxygen conversion by the catalyst bed is high, which will be the case at low oxygen partial pressures in the reactor feed; compare Figures 1 and 2. We can thus explain the relatively high rate of CO production (compared to CO_2) at a reaction temperature of 750°C when there is a partial pressure of oxygen of 20.7 kPa in the feed; see Figure 3. At this temperature, the conversion of oxygen is relatively low (58%, see Figure 2), and this means that there is much oxygen available for gas-phase oxidation to produce the CO. At a reaction temperature of 700°C, the gas-phase reactions are probably too slow to produce much CO. At 800°C, the oxygen consumption in the catalyst bed is much higher: relatively more CO_2 is produced on the catalyst and less oxygen remains available for gas-phase oxidation (oxygen conversion: 95%). This explanation of the observed ratio of CO and CO_2 was also used to explain the dip in the CO production as a function of temperature as shown in Figures 1 and 2, see above.

The results of the experiments shown in Figure 4 further illustrate the importance of the interplay between gas-phase and catalytic reactions. In this experiment, 1.5 g of catalyst was used instead of the 0.093 g used for the experiments of Figures 1-3. Because of the high catalyst weight, the oxygen conversion in the catalyst bed is high and it is thus to be expected that little gas-phase oxidation, resulting in CO production, will take place after the catalyst bed. Figure 4 shows that this is indeed the case. Only if the oxygen conversion is not complete does CO production start to increase.

The data of both Figure 3 and 4 show that the sum of the partial pressures of the C_2 products (and hence the rate of their production) increases less with an increase in the partial pressure of oxygen than do the partial pressures of the total oxidation products. We conclude elsewhere [16,

21] that this increase in the degree of total oxidation (in the presence of a Li/MgO catalyst) is due to a high rate of further oxidation of the C_2 compounds.

If the possibility of the sequential total oxidation of the C_2 products is taken into account, some observations which have been reported in the literature can be explained. Otsuka et al. [5] showed in their experiments with a LiCl-Sm_2O_3 catalyst (T=750°C) that when they increased the sum of the partial pressures of the reactants in the reactor feed (CH_4/O_2 constant at 5/2) by lowering the partial pressure of the helium diluent (at a constant total pressure of 101 kPa), there was a decrease in the C_2 selectivity and yield. Calculations based on their data show that the sum of the C_2H_6 and C_2H_4 partial pressures levelled off at a value of approximately 1.9 kPa when the sum of the partial pressure of the reactants was 43 kPa; the maximum of the sum of the partial pressures of the C_2 products achieved was 2.7 kPa when the sum of the partial pressures of the reactant was 99 kPa. The corresponding partial pressures of C_2H_4 for the two situations were 1.7 and 2.1 kPa respectively. An increase in the absolute C_2 concentration brought about by increasing the absolute concentration of the reactants thus leads to a decrease of the (relative) C_2 selectivity and C_2 yield and this is probably caused by an increased oxidation of the C_2 products at higher partial pressures of these products.

A rather general observation which may be explained in the light of our conclusions is that approximately the same absolute optimum C_2 partial pressure has been reported by different authors using different types of catalyst. For example, Otsuka et al. [6] obtained a total partial pressure of C_2 products of 5 kPa over a lithium doped NiO catalyst; the same total C_2 partial pressure has been reported by Matsuura et al. [17] for a lithium-doped ZnO catalyst and by ourselves [18] for a lithium doped MgO catalyst (under conditions which were better optimised than those shown above). We have presented direct evidence for the occurrence of total oxidation of the desired C_2 products in the case of a Li/MgO and a Ca/Sm_2O_3 catalyst [16, 21]; this limits the maximum partial pressures of C_2 products that may be reached with these catalysts. These total oxidation reactions of the C_2 hydrocarbons may take place not only in the gas-phase, but also on the catalyst: experiments with a Li/MgO catalyst at 720°C (when the rates of gas-phase oxidation reactions of ethane and ethylene are negligible) show that all CO and CO_2 are formed from C_2H_4 by a sequential reaction scheme [21]:

CH_4 ---> C_2H_6 ---> C_2H_4 ---> CO_x

We have no direct evidence for the occurrence of total oxidation of C_2 products on the Li/NiO and Li/ZnO catalysts mentioned above; however, we believe that the the striking similarity in the partial pressures of C_2 products which can be achieved with these systems shows that the total partial pressure of C_2 products is limited by sequential oxidation reactions. The same conclusions were reached by Labinger and Ott for a Na-MnO_x/MgO catalyst, used in the cyclic mode of operation [22].

The Effect of Back-Mixing

The importance of the occurrence of sequential oxidation reactions of the C_2 hydrocarbons is demonstrated by the effect on the product distribution of back-mixing in the reactor. From a reactor engineering standpoint, it is to be expected that the selectivity of a reaction system in which there is the possibility of sequential reactions will be highest in a plug-flow reactor (In an ideal plug-flow reactor, back-mixing is absent [19]). In the oxidative coupling of methane, back-mixing of the (relatively unstable) intermediate C_2 products will lead to transport of these products to regions (nearer to the entrance) of the reactor where a higher partial pressure of oxygen exists, and this will lead to an increased degree of sequential oxidation of the C_2 products and consequently a lower C_2 selectivity [18]. In order to investigate this effect, two experiments involving different conditions of back-mixing (but with the same amount of catalyst and the same residence time in the reactor) were performed and the results of these are shown Table 1. In the first experiment of Table 1, the reactor was used as above in the single-pass mode (approaching plug flow); in the second experiment, it was used in the recycle mode, with a recycle ratio of 10 (under conditions approaching those of an ideally mixed reaction system). As the other process conditions were exactly the same in both experiments, any differences between the two experiments can only be caused by the different degree of back-mixing. Table 1 shows that there are distinct differences between the results of the two experiments. For the single-pass experiment, both the conversion and selectivity are superior. The lower conversion obtained by the well-mixed reactor is normal for this type of reactor: for reactions with an order greater than zero, a well mixed reactor always gives lower conversions than does a plug-flow reactor (if the residence time and the amount of catalyst is the same), due to the effect of back-mixing [19]. The difference in selectivities in the two reactors provides further evidence for the suggestion that sequential oxidation reactions take place.

Table 1 The effect of backmixing: a gas mixture was contacted with 500 mg of Li/MgO catalyst (group B, 2.8 wt% Li, 0.1-0.3 mm., no quartz dilution) with a recycle ratio R. Reactor feed: P_{CH4} = 50 kPa, P_{O2} = 10 kPa, flow rate: 0.83 cm^3s^{-1}, T_r = 800°C

R	CH_4 conversion /%	O_2 conversion /%	C_2 selectivity /%	C_2 yield /%
0	25	95	67	16.4
10	17	82	54	9.0

With respect to the effect of back mixing on selectivity, it is also important to know the degree of back-mixing in the experiments of Figures 1, 2 and 3 (showing the effect on product composition of variation of the oxygen concentration in the reactor feed). Both measurements of the residence time distribution in the reactor (using a N_2 pulse in a He flow [18]) and calculations based on the method of Hsiung and Thodos [20] showed that the flow pattern in the reactor in these experiments lay in between that of plug flow and of ideal mixing. Experiments carried out with higher flow rates in the same reactor with the same value of W/F of 0.22 $g.s.cm^{-3}$ (i.e., at the same residence time in the catalyst bed), in order to improve the plug-flow character of the gas flow through the reactor (and also to decrease the residence time in the post-catalytic volume) resulted in an increase in the selectivity with increasing flow rate [18]. With a flow rate of 3.36 cm^3s^{-1}, the selectivity reached a maximum and further increase gave no improvement. We attempted to repeat the experiments of Figures 1, 2 and 3 with these optimal process conditions. However, due apparently to the higher throughput of gas under these conditions, temperature instabilities (i.e., the occurrence of severe hot-spots) occurred if the oxygen concentration in the reactor feed was increased above 10 kPa, even in the narrow-bore reactor used. It thus follows from our results that the use in the reactor feed of partial pressures of oxygen above ca 10 kPa results in unnecessary loss of selectivity and also to temperature-instabilities when higher flow rates are used.

CONCLUSIONS

1. An increase of the oxygen concentration in the reactor feed gives an adverse effect on the selectivity. The total C_2 concentration levels off at higher values of the oxygen inlet concentration, this being caused by total oxidation of the C_2 products.
2. As ethane and ethylene are susceptible to further oxidation, optimum selectivity is reached under plug-flow conditions.
3. To obtain optimum C_2 yields, non-selective gas-phase reactions must be minimised by minimising the residence time (and oxygen concentration) in the post-catalytic volume.

REFERENCES

1. T. Ito and J.H. Lunsford, Nature vol. 314 (1985) 721.
2. T. Ito, J.-X. Wang, C.-H. Lin and J.H. Lunsford, J. Am. Chem. Soc., 107 (1985) 5062.
3. W. Hinsen, W. Bytyn and M. Baerns, Proc. 8th Int. Congr. Catal. Berlin (1984) III, 581.
4. K. Otsuka, K. Jinno and A. Morikawa, J. Catal, 100 (1986) 353.
5. K. Otsuka, Q. Liu and A. Morikawa, J. Chem. Soc., Chem. Commun., (1986) 586.
6. K. Otsuka, Q. Liu, M. Hatano and A. Morikawa, Chem. Lett., (1986) 903.
7. K. Otsuka, A.A. Said, K. Jinno and T. Komatsu, Chem. Lett., (1987) 77.
8. K.-I. Aika, T. Moriyama, N. Takasaki and E. Iwamatsu J. Chem. Soc., Chem. Commun., (1986) 1210.
9. I.T. Ali Emesh and Y. Amenomiya, J. Phys. Chem., 90 (1986) 4785.
10. G.-A. Martin and C. Miradotos, J. Chem. Soc., Chem. Commun., (1987) 1393.
11. J.A. Roos, A.G. Bakker, H. Bosch, J.G. van Ommen and J.R.H. Ross, Catalysis Today, 1 (1987) 133.
12. S.J. Korf, J.A. Roos, N.A. de Bruijn, J.G. van Ommen and J.R.H. Ross, J. Chem. Soc., Chem. Commun., (1987) 1433.
13. S.J. Korf, J.A. Roos, N.A. de Bruijn, J.G. van Ommen and J.R.H. Ross, Catalysis Today, 2 (1988) 535.
14. S.J. Korf, J.A. Roos, J.M. Diphoorn, R.H.J. Veehof, J.G. van Ommen and J.R.H. Ross, Catal. Today, 4 (1988) 279.

15. H.M.N. van Kasteren, J.W.M.H. Geerts and K van der Wiele, Proc. 9th Int. Congr. Catal. Calgary, (1988) II 930.
16. J.A. Roos, S.J. Korf, R.H.J. Veehof, J.G. van Ommen and J.R.H. Ross, Catal. Today, 4 (1988) 471
17. I. Matsuura, Y. Utsumi, M. Nakai and T. Doi, Chem Lett. (1986) 1981.
18. J.A. Roos, S.J. Korf, A.G. Bakker, N.A. de Bruijn, J.G. van Ommen and J.R.H. Ross, "Methane Conversion", Eds. D.M. Bibby, C.D. Chang, R.F. Howe and S. Yurchak, Elsevier Sci. Publ., Amsterdam, (1988) 427.
19. K.R. Westerterp, W.P.M. van Swaaij and A.A.C.M. Beenackers, Chemical Reactor Design and Operation, John Wiley & Sons, Chichester (1984).
20. T.H. Hsiung and G. Thodos, Int. J. Heat Mass Transfer, 20 (1977) 331.
21. J.A. Roos, S.J. Korf, R.H.J. Veehof, J.G. van Ommen and J.R.H. Ross, Paper presented at the 196th National Meeting of the American Chemical Society, Division Colloid & Surface Science, Los Angeles, 1988; Appl. Catal. 52 (1989) 147.
22. J.A. Labinger and K.C. Ott, J. Phys. Chem. 91 (1987) 2682.

M. BAERNS (University of Bochum, FRG): Considering homogeneous gas-phase reactions in oxidative methane coupling; the post-catalytic gas volumes of the plug-flow and the recycle reactor should be accounted for when comparing the results obtained in these two types of reactors (When high recycle ratios are applied the reactants pass frequently through the post-catalytic volume).

J.R.H. ROSS (University of Twente, The Netherlands): In principle, you are correct. However, in practice, the post-catalytic volume was expressly made as small as possible so that such effects would be negligible in the types of experiments which we have reported here. We have shown that the rate of oxidation of the C_2 products in the gas phase is negligible for residence times and temperatures used here.

K. VAN DER WIELE (Technical University of Eindhoven, The Netherlands): Your suggestion to improve the selectivity of the methane oxidative coupling is to develop more active "methane activation" catalysts and to work at lower temperatures, thus avoiding consecutive homogeneous reactions (oxidation of C_2 's to CO_x). In my opinion possibilities are very limited because any methane activation catalyst will presumably activate ethane and ethylene as well, the latter requiring a lower activation energy. So lower temperatures will soon favour catalytic oxidation of C_2 's. Moreover the product composition unfavourably shifts from ethene to ethane at lower temperatures. Do you agree with my opinion ?

J.R.H. ROSS (University of Twente, The Netherlands): Your comment is correct <u>if</u> the pre-exponential factor is the same for both reactions. Although the same (number of) active sites are likely to be involved, the entropy of activation is not necessarily the same as it will depend on factors such as the geometries of the reacting molecules. Hence, it should be possible to find catalysts favouring the oxidation of methane over that of ethane even if the activation energy for the latter reaction is lower. Experience has indeed shown that the selectivity to ethane increases at lower operation temperatures. However, this need not be a problem if the process in which the C_2 hydrocarbons are used either includes a dehydrogenation step or allows reaction of C_2H_4 and C_2H_6 simultaneously.

J.G. MC CARTY (SRI International, USA): I appreciate your comments about the need to avoid post-reactor homogeneous reactions (with O_2) and internal mixing. However, even with plug flow reactors, product ethane and ethene must rise (for higher yields) to levels where product oxidation takes place. Does this not place limits on the C_{2+} yields that can be achieved with oxidative coupling ?

J.R.H. ROSS (University of Twente, The Netherlands): We agree. Our aim was to show under which circumstances the various factors which we have described are of importance. Our point is that there is no sense in talking about chemical limits if limits placed by method are not fully recognized.

W.J. VERMEIREN (University of Leuven, Belgium): I am not convinced of your conclusion that the main part of the CO_x is coming from C_2 oxidation. I have two comments:
1. The adverse effect on the C_2 selectivity upon increasing the oxygen concentration in the feed, is caused by an increasing contribution of gas phase reactions. The selectivity for CO with increasing oxygen concentration in the feed, is typical for gas phase reaction in methane-oxygen mixture. I believe that the oxidation of methyl radicals in the gas phase has a greater probability than the oxidation of C_2 products.
2. You compare a single-pass operation with an operation, approaching ideally mixing. However, the decrease of C_2 selectivity is not that high in the ideal mixed reactor and results still in a C_2 selectivity of 54%. According to me, this is an indication that the oxidation of C_2 products is not so important as you think.

J.R.H. ROSS (University of Twente, The Netherlands):
1. Our conclusions are based not only on the results given in our paper but also on other results presented elsewhere [1,2,3]. We have shown that the gas-phase reaction is not of significant importance if the residence time in the reactor is short and the temperature is relatively low. We quite agree that high temperatures and residence times favour the gas-phase process, in which undoubtedly methane can react with oxygen. When we see CO as product, we accept that it comes mainly from gas-phase oxidation but we think that gas-phase oxidation of C_2H_6 and C_2H_4 also contribute.
2. The drop in selectivity reported here is well outside experimental error so we cannot accept your remarks. The results reported here are only supportive of our arguments, reported fully elsewhere, that C_2 oxidation is the predominant route to CO_x [1,2,3].

1. J.A. Roos, S.J. Korf, A.G. Bakker, N.A. de Bruijn, J.G. van Ommen and J.R.H. Ross, "Methane Conversion", Eds. D.M. Bibby, C.D. Chang, R.F. Howe and S. Yurchak, Elsevier Sci. Publ., Amsterdam, (1988) 427.
2. J.A. Roos, S.J. Korf, R.H.J. Veehof, J.G. van Ommen and J.R.H. Ross, Catal. Today, 4 (1988) 471.
3. J.A. Roos, S.J. Korf, R.H.J. Veehof, J.G. van Ommen and J.R.H. Ross, Applied Catalysis, 52 (1989) 147.

ANONYMOUS: From the results you presented on the competitive reaction of ethane or ethylene with methane, can you quantify the relative rates for methane versus ethane versus ethylene oxidation ?

J.R.H. ROSS (University of Twente, The Netherlands): In order to give a semi-quantitative answer to your question, we have to refer to results given in [1]. It is possible to calculate the data of the following Figure from the results given in reference [1]. The Figure shows the ratio $Z = R_{COx}/R_{CH4}$ (where R_{COx} is the rate of formation of CO + CO_2 and R_{CH4} is the rate of consumption of methane) plotted against the partial pressures of C_2H_4 and C_2H_6 added to the feed of a plug-flow reactor.

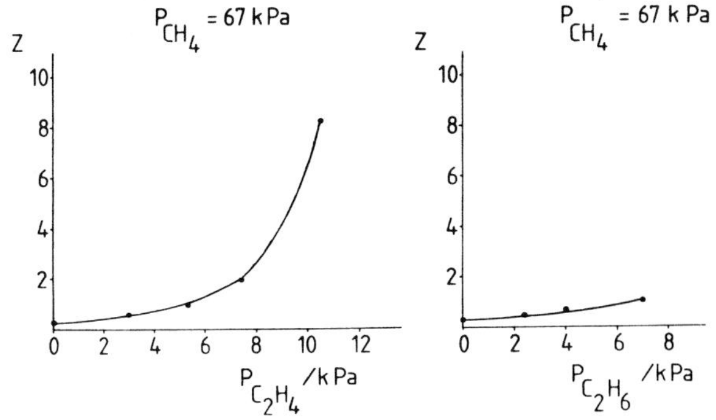

Figure Ratio $Z = R_{COx}/R_{CH4}$ as a function of the partial pressures of C_2H_4 and C_2H_6 added to the gas feed of a plug-flow reactor.

A value of Z > 1 means that at least some of the CO_x most be formed from the C_2 hydrocarbon; the higher the value, the more that the C_2 must be involved. A value of Z < 1 could imply that all the CO_x comes from the methane, but this is not essential. It is thus clear that the Z values are upper limits to the relative rates of reaction of the C_2 hydrocarbons to those of methane. It can be seen from the Figure that ethylene is much more reactive than methane and that the rate of reaction of ethylene at least at high ethylene concentrations, is much higher than that of methane. For example, with a gas-phase concentration of P_{CH4} = 67 kPa, P_{C2H4} = 10.5 kPa and P_{O2} = 7.0 kPa (balance He), the value of Z is 8.3.

1. J.A. Roos, S.J. Korf, R.H.J. Veehof, J.G. van Ommen and J.R.H. Ross, Catal. Today 4 (1988) 471.

G. Centi and F. Trifiro' (Editors), *New Developments in Selective Oxidation*

ROLE OF SOME EXPERIMENTAL PARAMETERS IN THE CATALYTIC CONVERSION OF METHANE BY OXIDATIVE COUPLING.

R.SPINICCI

Dipartimento di Energetica, Università di Firenze, Via S.Marta 3, Firenze (Italy).

SUMMARY

A study on the oxidative coupling of methane has been carried out over CaO and over K_2O/CaO in order to define the effect of the dilution of the reactants in the carrier gas, expecially on the selectivity; subsequently the attention has been addressed to the study of the effect caused by the presence of the main products in the reactant mixture, with the aim to obtain informations from the presence of species which can favour or hinder the reaction. Moreover a kinetic study of the reaction has been performed in differential conditions in order to gain a deeper insight in the development of the reaction: the whole of the experimental results have been employed to propose a reaction mechanism, which takes into account the differences between the two types of catalysts.

INTRODUCTION

The increasing interest in the activation of methane is born by the attempts to exploit at the best the large reserves of this substance and in this context, the interest in the oxidative coupling of methane is surely due to the interest in the synthesis of higher hydrocarbons and expecially ethylene. The reaction occurs at high temperatures on a number of catalysts (1,2) but it has been undoubtedly established (3,4) that the reaction develops homogeneously also, in the gas phase. In any case the presence of a proper catalyst appears to be determining, on the basis of the experimental result that there is no reaction with no catalyst: therefore it appears very interesting to clear out the role of the catalysts in this reaction.

Following a previous work on the methane coupling over calcium oxide or over other oxides supported on calcium oxide (5), it appeared interesting to focuse the attention mainly on CaO and K_2O/CaO, that is on typically basic oxides, in order to try to clear out the role of the basic centers related to the presence of adsorbed oxygen species. Indeed the results previously obtained and many literature reports (6,7) lead to the conclusion that the observed selectivities of many catalysts towards the formation of C_2-hydrocarbons in methane coupling can be mainly attributed to the presence of proper surface oxygen species.

The investigation was therefore addressed towards the study of the influence of the dilution of reactants, and to the investigation on the competitive presence of the products in the reactant mixture. As a logical development, it appeared necessary to get informations about the reaction mechanism by means of a kinetic analysis of the reaction, and therefore studies have been then carried out, either about the overall oxidation or about the oxidative coupling, in order to define the intervention of the catalysts in the reactions.

METHODS

Calcium oxide was obtained by the decomposition for 5 h at 950 °C of powdered white marble MICROCAL FM-600, provided by Ingram: its surface area, measured by means of a Perkin Elmer Sorptometer 512-D, was found to be 2.5 m^2g^{-1}.
The X-ray analysis showed the typical cubic structure of calcium oxide, while the semiquantitative analysis, performed in a scanning electron microscope, revealed the presence of 0.5 - 1% SrO. The supported samples were prepared by impregnating the calcium oxide with solutions of KNO_3 in such a way to obtain catalysts containing 7% K_2O: after evaporation of the excess solution, the samples were dried at 110 °C for 3 h and then calcined in air at 550 °C for 3 h. The experiments were carried out in a quartz tubular reactor (1 cm o.d.) at the desired temperature, using 0.1 - 0.3 g, after a pretreatment in a flow of helium plus oxygen (ratio 7.5/1) at 650 °C for 2 h. Methane and oxygen were fed with a helium carrier over the catalysts and in some scheduled runs a chosen amount of ethane, or ethylene, or carbon dioxide was added to the reactant mixture. The composition of the effluent gases was analyzed by means of a C. Erba 4200 gas-chromatograph, equipped with a hot wire detector and two 1/8" o.d. 10 ft columns packed with Carbosieve S-II.

RESULTS AND DISCUSSION

Above all it must be specified that experiments have been carried out with empty reactor in order to determine the activity of the reactor walls, which cannot surely be neglected above 750 °C; indeed in the range 750 - 850 °C the conversion of methane ranges from 0.5 to 0.9 moles% into carbon monoxide; from 1.4 to 2.2 moles% into carbon dioxide; from 1.1 to 1.9 moles% into ethylene and from 0.1 to 2.2 moles% into ethane. The values reported in the following results take always into account this reactivity of methane.

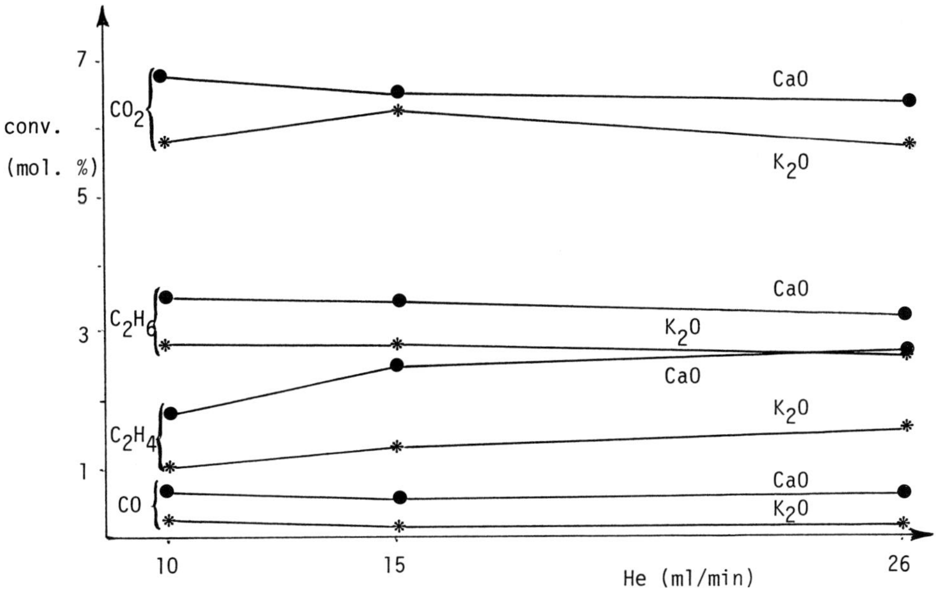

Fig.1 Conversion of methane (mol %) into the main products at 780 °C over samples of 0.25 g, employing a mixture of methane and oxygen (10 and 2 ml/min) with different amounts of carrier gas (He) and the two types of catalysts

In Fig. 1 the results are reported about the catalytic activity and the related selectivity, when varying the dilution of the reactant mixture. It can be seen immediately that there is an appreciable amount of carbon monoxide only on calcium oxide, while the presence of an active phase with monovalent ions seems to inhibit the formation of carbon monoxide: it must be underlined at this point that the formation of carbon monoxide is accompanied by a parallel production of hydrogen, testifying the common pathway of the formation of these two products. (It must be said that the amount of hydrogen has not been included in Fig. 1 because, being little, it is difficult to be evaluated from the chromatograms). Small amounts of C_2H_2 have been also revealed.

But two other results can be seen immediately, that is: 1) the formation of CO_2 also decreases with catalysts containing potassium oxide, and 2) that of C_2-hydrocarbons, and expecially C_2H_4, is generally favoured by diluting reactants in the carrier gas, that is by decreasing their partial pressures.

On the basis of these results it appears reasonable to suppose that the formation of hydrogen and the enhanced formation of carbon monoxide on CaO can be related to the enhanced development of the total and of the partial oxidation on this catalyst: since among the products of partial oxidation aldeydes can also be included it appeared logical to work about the hypothesis (8) that HCHO can be included among the products and that it can dissociate into hydrogen and carbon monoxide. Indeed T.P.D. runs performed on CaO samples with adsorbed HCHO show a peak with maximum at 510 °C which is due to desorption of formaldeyde with dissociation into hydrogen and carbon monoxide: the experimental result that the maximum of this peak is nearly coincident with that of carbon dioxide and with those of methane and other C_2-hydrocarbons (9) seems to lead to the conclusion that there is a common surface intermediate which can give oxygenated compounds. This surface intermediate has been supposed (9) to have its origin on the less basic centers of the catalyst, while the more basic centers seem to develop the formation of the C_2-hydrocarbons.

Subsequently runs have been carried out, over CaO and over 7% K_2O/CaO, introducing in the reactant mixture a known amount (1.5 ml/min) of ethylene or ethane or carbon dioxide, with the aim to check if their presence simply lowers the formation of the same product or enhances the formation of some others: the results obtained are collected in Fig. 2.

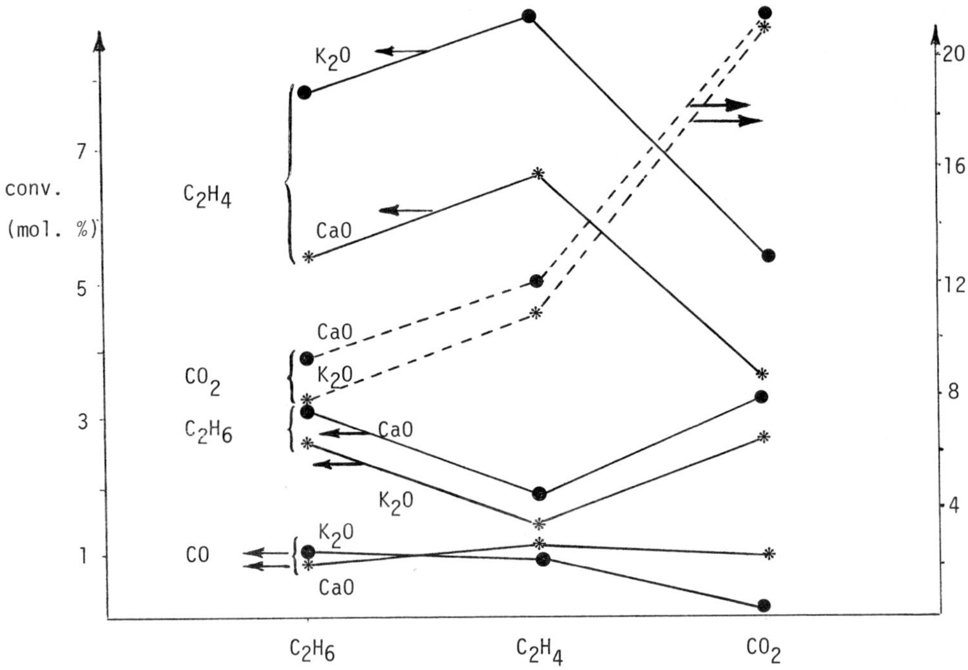

Fig. 2 Conversion of methane (moles %) into the main products at 800 °C over samples of CaO and 7% K_2O/CaO (0.25 g), employing a mixture of methane (10 ml/min) and oxygen (2 ml/min), with the addition of ethylene or ethane or carbon dioxide (1.5 ml/min) .

Some evidences can be immediately underlined, since the presence of ethane in the gas phase stimulates undoubtedly the formation of ethylene, expecially over K_2O/CaO. The hypothesis which seems more probable is that not only the C-H bond, but the C-C bond also can be activated, giving intermediate species which behave similarly to those formed by methane, and therefore (in addition to methane as a cracking product) can produce ethylene.

Particular interest can be now attributed to some parallel runs, concerning the execution of runs based on the oxidation of ethane, by means of oxygen: in these runs, where methane have been substituted in the same ratios by ethane, no dimerization products have been formed, but only cracking products, and above all ethylene. It must be stressed, however, that the contribution of the homogeneous reaction in the gas phase cannot be neglected and is important, either in the formation of methane or in the formation of ethylene:indeed the conversion of methane reach the value of about 0.70 at 800 °C.

The absolute amounts of C_2-hydrocarbons formed, on the contrary, seems not to be affected by the presence of the carbon dioxide in the feed: in these experiments their percentages appear slightly increased (expecially on K_2O/CaO) but this effect seems to be due to the coincidence of the adsorption sites for carbon dioxide and for methane, as shown in (9), which dcreases the absolute conversions of methane, as it is possible to check experimentally.

It was then assumed that a kinetic analysis of the reaction could give decisive informations about its mechanism and the features of the catalysts investigated: a kinetic analysis was therefore performed by means of isothermal experiments, which were carried out in such conditions that the partial pressure of the products can be neglected; these experiences were indeed performed either at constant methane pressure (11.7 kPa), or at constant oxygen pressure (2.8 kPa) in order to determine the effect of $P(O_2)$ and respectively of $P(CH_4)$ on the rates of formation of the products. The temperatures investigated were in the range 740 - 780 °C.

It was presumed that the data obtained in the experiments for examining separately either the pressure effects of methane or those of oxygen could be therefore described by one of the following basic kinetic equation, that is:

i) $r = k$

ii) $r = k \cdot K \cdot p$

iii) $$r = k \cdot \frac{K \cdot p}{1 + K \cdot p}$$

iv) $$r = k \cdot \frac{K^2 \cdot p^2}{1 + K^2 \cdot p^2}$$

It appeared indeed interesting to gain a deeper detail, by studying separately for the two catalysts the progression of the rate of conversion of methane into C_2-hydrocarbons, and respectively into carbon dioxide either in function of the methane pressure (at constant oxygen pressure), or in function of the oxygen pressure (at constant methane pressure). This follows the hypothesis that the formation of these two types of products proceeds on different types of active centers. In Fig. 3 these diagrams are reported for CaO and in Fig. 4 the corresponding plots are reported for 7% K_2O/CaO.

By examining these two figures it can be seen that the progress of the reaction rates in function of the partial pressures is similar to that of a Langmuir isotherm, and this suggest that the rate can described by the relation:

$$r = k \cdot \Theta(CH_4) \cdot \Theta(O_2)$$

where $\Theta(CH_4)$ and $\Theta(O_2)$ are the surface coverages of methane and respectively of oxygen: the application of this general relationship is based obviously on the hypothesis that either the rate of formation of C_2-hydrocarbons or that of carbon oxides is determined by the formation of methyl radical $CH_3\cdot$ during the

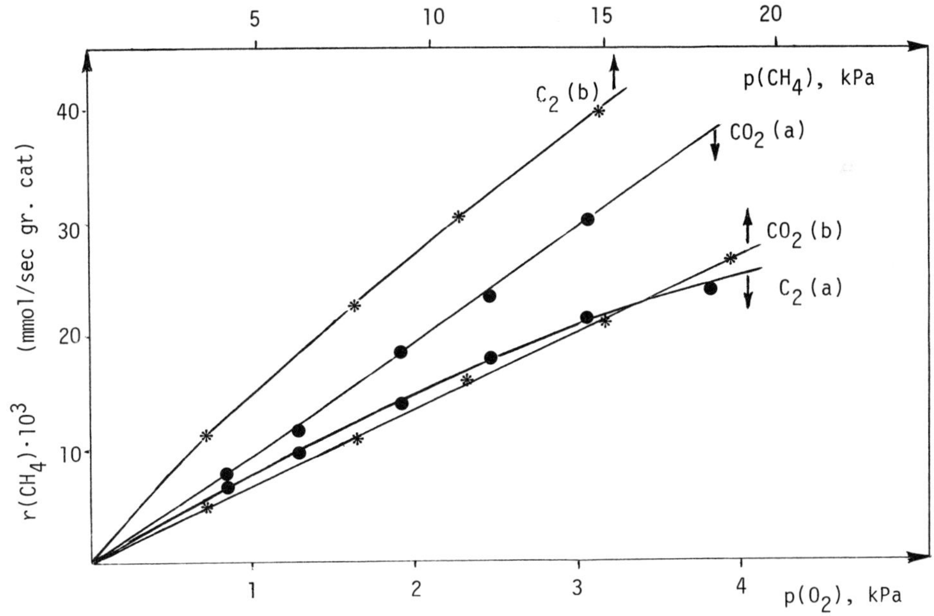

Fig. 3 Variation of the rate of C_2-hydrocarbons and of CO_2 production vs. the oxygen pressure (at constant methane pressure), (a), and versus methane pressure (at const. oxygen pressure), (b), at 780 °C, over CaO.

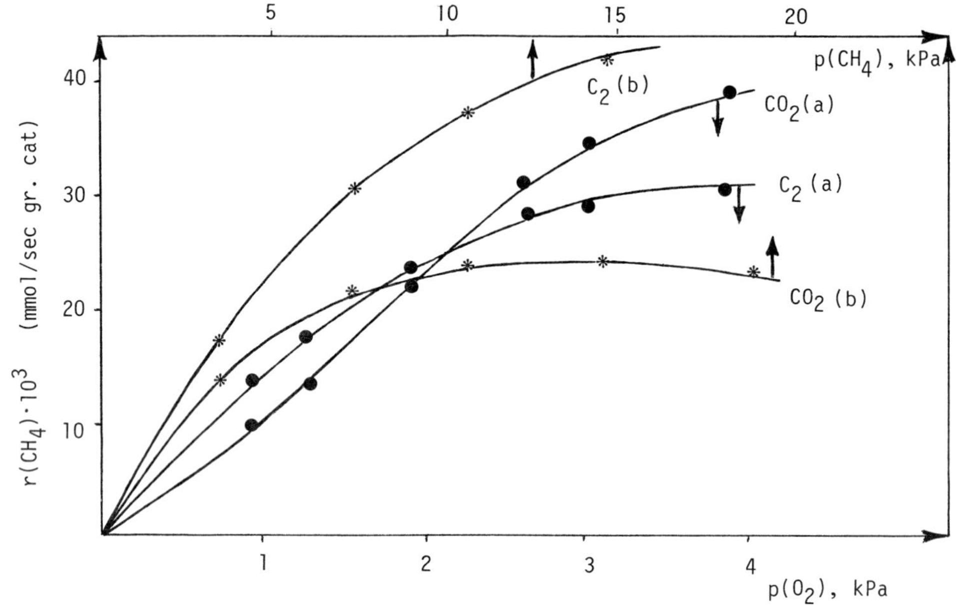

Fig. 4 Variation of the rate of C_2-hydrocarbons and of CO_2 formation versus the oxygen pressure (at constant methane pressure), (a), and versus the methane pressure (at constant oxygen pressure), (b), at 780 °C, over 7% K_2O/CaO.

surface reaction of methane and an oxygen surface species.

An interesting feature, which emerges from the examination of Figg. 3 and 4, is that for CaO there is a wide range of the partial pressure of oxygen, and to a less extent of methane, where the rate of production of carbon dioxide and of C_2-hydrocarbons varies linearly in function of $p(O_2)$ and respectively of p(CH4).

This suggests immediately that the reaction follows a first order kinetics, and that the adsorption constants could be therefore sufficiently small and could be neglected in the kinetic equation of type iii). This confirms the results obtained by the T.P.D. runs (9) with adsorbed oxygen or with adsorbed methane, which demostrate that at the reaction temperatures there is no more any stable adsorbed species on the catalyst surface.

The kinetic analysis has been performed by linearizing the equations i) - iv) and checking if the experimental data could agree with one of these linearized equations: intercept and slope allow us to calculate k and K. Fig. 5 shows the fitting of the experimental data with some of these equations.

As a result of this kinetic analysis it can be said that the formation of carbon oxides and of C_2-hydrocarbons at constant methane pressure follow a rate equation of type iii) over 7% K_2O/CaO, because of the good linear correlation observed by plotting 1/r values versus $1/p(O_2)$ values ; indeed, in the case of CaO, either for the production of C_2-hydrocarbons or for the production of carbon dioxide a nearly satisfactory correlation is obtained by plotting r versus p , testifying that in its formation the adsorptive term $K \cdot p$ for oxygen could be neglected.

As far as the formation of C_2-hydrocarbons and carbon dioxide at constant oxygen pressure is concerned, the good linear correlation observed by plotting 1/r versus $1/p(CH_4)$ for K_2O/CaO, shows that either the production of ethylene and ethane or that of carbon dioxide follows again a rate equation of the type

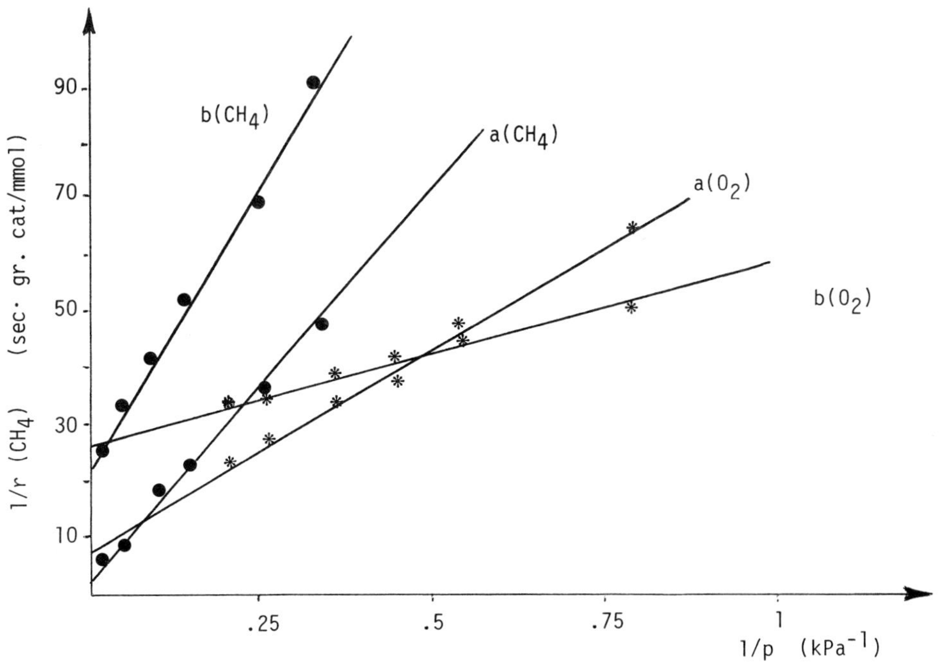

Fig. 5 Linearization of eq. iii) for the production of CO_2 (a), and for the production of C_2-hydrocarbons (b), on K_2O/CaO.

Table I

Kinetic parameters related to CaO (a) and 7% K_2O/CaO (b), as calculated from the formation of ethylene, ethane and carbon dioxide.

Temperature (°C)		K_m (kPa^{-1})		k (mmol sec^{-1} g^{-1})		K_o (kPa^{-1})		k (mmol sec^{-1} g^{-1})	
		a	b	a	b	a	b	a	b
740	C_2	0.011	0.15	0.018	0.012	0.015	0.18	0.022	0.011
	CO_2	0.008	0.09	0.1	0.09	0.022	0.092	0.11	0.12
760	C_2	0.005	0.11	0.025	0.02	0.009	0.15	0.029	0.018
	CO_2	0.007	0.05	0.17	0.14	0.013	0.08	0.19	0.16
780	C_2	0.001	0.06	0.041	0.035	0.003	0.11	0.045	0.031
	CO_2	0.003	0.02	0.23	0.18	0.006	0.06	0.25	0.19

iii); on CaO indeed it is more difficult to discriminate between the model ii) and the model iii) and this confirms the weak adsorption of methane on this catalyst, which however is not as weak as oxygen and therefore does not provide a sure criterion, for establishing the kinetic equation.

The value of the kinetic constants, as determined from the intercepts of the diagrams 1/r versus 1/p (or from the slope of the diagrams r versus p), and the values of the adsorption constants, as determined from the slopes of these diagrams are reported in Table I.

Therefore, in the hypothesis of a non-competitive adsorption of methane and oxygen, the overall rate equations for the formation of carbon dioxide and respectively of C_2-hydrocarbons on CaO should take the form:

a) for CO_2: $$r = k \cdot K_o \cdot p_o \cdot \frac{K_m \cdot p_m}{1 + K \cdot p_m} \quad \text{or} \quad r = k \cdot K_o \cdot p_o \cdot K_m \cdot p_m$$

b) for C_2-hydr. $r = k \cdot K_o \cdot p_o \cdot \dfrac{K_m \cdot p_m}{1 + K_m \cdot p_m}$ or $r = k \cdot K_o \cdot p_o \cdot K_m \cdot p_m$

while those for the formation of C_2-hydrocarbons and of carbon dioxide on K_2O/CaO should take the form:

a′) for CO_2: $r = k \cdot \dfrac{K_o \cdot p_o}{1 + K_o \cdot p_o} \cdot \dfrac{K_m \cdot p_m}{1 + K_m \cdot p_m}$

b′) for C_2-hydr. $r = k \cdot \dfrac{K_o \cdot p_o}{1 + K_m \cdot p_m} \cdot \dfrac{K_m \cdot p_m}{1 + K_m \cdot p_m}$

In the hypothesis that the activation energy is determined by the step concerning the activation of CH_4 to give $CH_3\cdot$ radicals it appears reasonable to consider the values of k for the reactions on 7% K_2O/CaO and on CaO, and to consider the possibility of determining the corresponding activation energy: by means of the appropriate Arrhenius plot it has been possible to find an average value of the activation energy of 152 kJ/mol for the formation of CO_2 and an average value of 164 kJ/mol for the formation of C_2-hydrocarbons on both catalysts.

CONCLUSIONS

From the whole of the experiments it is possible to check that at the reaction temperature oxygen and methane are more strongly adsorbed on 7% K_2O/CaO than on CaO and this confirms the results reported in (9) and indeed obtained with 7% K_2O/CaO also, from the TPD experiments with oxygen or methane adsorbed.

Indeed the kinetic results seem to indicate that on 7% K_2O/CaO a greater number of surface centers for oxygen and methane adsorption is occupied than on CaO: a wide pressure range, indeed, is found with this catalyst where the reaction rate depends linearly on the pressures of oxygen and, respectively, of methane.

The temperature dependence of the reaction rates is nearly equal in the two catalysts and therefore it is possible to suppose that on both catalysts the $CH_3\cdot$ radicals (formed dissociatively on the catalysts surface and recognized as the intermediate species, which lead probably to C_2-hydrocarbons, and to CO and CO_2, through distinct pathways according to the basicity of the surface sites) can dimerize on the surface or in the gas phase. In this perspective the result reported in Fig. 2, concerning the increased formation of ethylene, when ethane is present in the feed seems to support the hypothesis that ethane is an intermediate step in the production of ethylene.

The differences found with the two catalysts, as far as selectivity is concerned, can be therefore ascribed to the differences in the basicity of the surface centers of the two catalysts: on 7% K_2O/CaO a larger amount of more basic centers and therefore of more charged surface oxygen species addresses the reaction preferably towards the formation of C_2-hydrocarbons. Moreover the fact that on CaO the adsorption of methane and oxygen proceeds to a lesser extent allows us to suppose that with this catalyst the contribution of the homogeneous gas phase reaction is greater.

REFERENCES

1 M.S. Scurrel, Appl. Catal. 34 (1987) 1
2 J.S. Lee, S.T.Oyama, Catal. Rev.-Sci. Eng. 30 (1988) 249
3 O.V. Krilov, React. Kinet. Catal. Letters 35 (1987) 315
4 W. Martir, J.H. Lunsford, J. Phys. Chem. 84 (1980) 3079
5 R. Spinicci, Catal. Today 4 (1989) 311
6 T. Ito, J.X. Wang, J.H. Lunsford, J. Amer. Chem. Soc. 107 (1985) 5062
7 J. X. Wang, J.H. Lunsord, J.Phys. Chem. 90 (1986) 3890
8 J.A. Roos, A.G. Bakker, H. Bosch, J.G. van Ommen, J.R.H. Ross, Catal. Today 1 (1987) 133
9 R. Spinicci, A. Tofanari, Communication presented at 2nd Europ. Workshop on Methane Activation, Enschede, May 1989

G. Centi and F. Trifiro' (Editors), *New Developments in Selective Oxidation*
1990 Elsevier Science Publishers B.V., Amsterdam — Printed in The Netherlands

MODELS OF THE DIRECT CATALYTIC PARTIAL OXIDATION OF LIGHT ALKANES

J. G. McCARTY, A. B. McEWEN, and M. A. QUINLAN
SRI International, 333 Ravenswood Avenue, Menlo Park, California, USA 94025

SUMMARY

Application of homogeneous kinetic models to methane activation indicates that the higher hydrocarbon yield may be limited by homogeneous oxidation of methyl radical intermediates. In this paper, we discuss the development of a model that describes the homogeneous and heterogeneous chemistry involved in the selective oxidation of methane and light alkanes and the impact of this chemistry on alkene and higher alkane yields. We also present experimental results for methane activation and ethane dehydrogenation using stable non-volatile catalysts composed of alkaline and rare earth carbonates supported by refractory complex oxides.

INTRODUCTION

There is ample evidence that homogeneous reactions substantially contribute to the catalytic oxidative dimerization of methane and the catalytic oxidative dehydro-genation of ethane to ethene. The product distribution has often been cited[1-3] as being consistent with homogeneous free radical chemistry, but the definitive experiments are those of Lunsford, et al.,[4-6] who used matrix isolated electron paramagnetic resonance (MIEPR) measurements to determine the distribution of methyl radicals downstream of a Li/MgO catalyst bed. The MIEPR results of Campbell and Lunsford[6] show that product ethane forms downstream of the catalyst bed by homogeneous methyl radical recombination and verify, within a factor of two, the bimolecular recombination rate constant. Isotopic exchange experiments[7-10] also support the view that the methyl radical is the primary intermediate in the production of ethane. Various recent laboratory results indicate that direct catalytic conversion of premixed oxygen and methane into higher hydrocarbons approaches a single-pass limit of about 25% yield (on a C atom basis) regardless of catalyst and reaction conditions (Figure 1). Finally, observations that empty microreactors[11-14] can selectively produce ethane and ethylene affirms the significant role of homogeneous kinetics in all aspects of the reaction. These findings indicate that homogeneous oxidation kinetics play an important role in the selectivity of alkane partial oxidation reactions

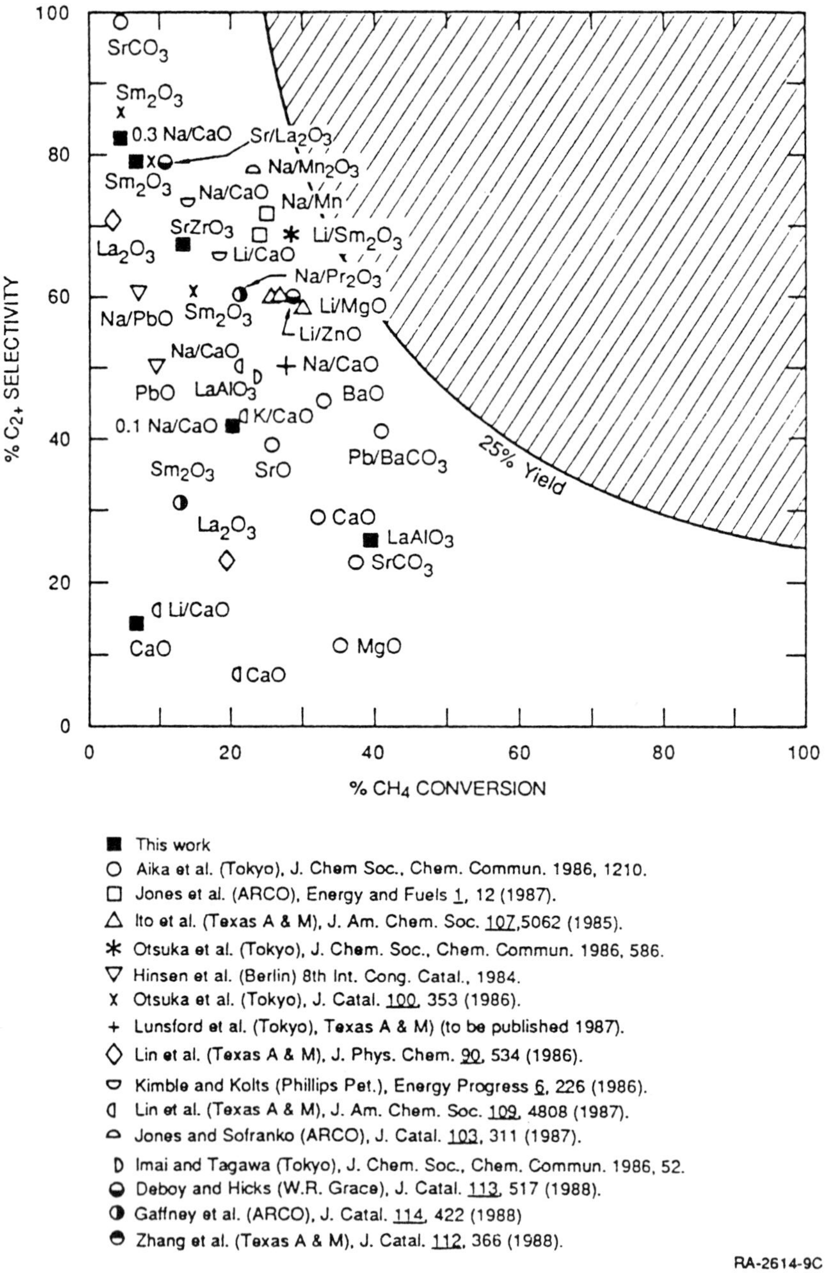

Fig.1. Laboratory fixed-bed catalytic oxidative coupling of methane with premixed oxygen

In this paper, we describe a model of catalytic light alkane partial oxidation used to evaluate the relative importance of individual homogeneous and heterogeneous reactions over a wide range of reaction conditions. The model incorporates key heterogeneous reaction steps into a network of known gas phase alkane free radical oxidation reactions. We also report the activity and selectivity of stable non-volatile strontium-based complex oxide catalysts for the direct oxidative conversion of methane into higher hydrocarbons and the direct oxidative dehydrogenation of ethane to ethene. Comparison of the experimental results and model calculations shows that the catalysts selectively oxidize intermediates such as methanol and carbon monoxide at rates higher than expected for heterogeneous hydrogen abstraction reactions.

METHODS

The premise of our model is that most of the reaction chemistry including by product formation occurs by homogeneous reactions in the catalyst pores, catalyst bed void space, and post-reactor volume. Our complete model[15] consists of 144 reactions, 134 reversible homogeneous reactions and 10 reactions which involve catalyst surface sites. Most of the gas phase reaction parameters were obtained from the review compilations of Frenklach[16], Warnatz[17], or Tsang[18]. The primary source of ethane in our mechanism of methane co-oxidative coupling is from the gas phase recombination of methyl radicals,

$$\bullet CH_3 + \bullet CH_3 \Longrightarrow C_2H_6 \qquad (1)$$

while ethene is produced from ethane by thermal (2) and oxidative (3) dehydrogenation.

$$\bullet C_2H_5 + M \Longrightarrow C_2H_4 + \bullet H \qquad (2)$$

$$\bullet C_2H_5 + O_2 \Longrightarrow C_2H_4 + \bullet O_2H \qquad (3)$$

A fundamental question is to what degree deep oxidation results from gas phase or surface chemistry.[19] The presence of premixed oxygen, although necessary to provide a sink for hydrogen and the thermodynamic driving force for the coupling process, unfortunately leads to undesired oxygenated by-products, e.g. CO, CO_2, and formaldehyde.

The first step in the catalytic cycle involves the activation of methane by a surface oxygen atom. The heterogeneous H abstraction step can be generalized to include scission of any C-H bond by an Eley-Rideal reaction with surface oxygen (O_S) to form a gas phase alkyl radical and a hydroxyl surface site (HO_S),

$$RH + O_s \Longrightarrow \bullet R + HO_s \qquad (4n)$$

where RH = (4a) CH_4; (4b) C_2H_6; (4c) C_2H_4; (4d) CH_3OH; and (4e) CH_2O. The activation energies used for the reaction of O_s with other C-H bonds (e.g. C_2H_6) reflect their bond strengths relative to methane. The rate determining step in the oxidative coupling of methane over Li/MgO was shown by Cant et al.[20] to be methane C-H bond scission,

$$CH_4 + O_s \Longrightarrow \bullet CH_3 + HO_s \qquad (4a)$$

based on a large, positive (1.5) deuterium isotope effect. Amorebieta and Colussi[21] showed at low pressure (10^{-6} to 10^{-4} atm) that methane conversion over Li/MgO is half order in oxygen and first order in methane. This result suggests that methane reacts with atomic surface oxygen. Labinger et al.,[22,23] report that with the Na/Mg/Mn catalyst, C_2H_6 converts 1.9 times faster than CH_4 . We have adjusted the rate constants for C_2H_6 to give this ratio (4b/4a = 1.9) at 1000 K. Rate constants for the other reactants (H-C_2H_3, H-CH_2OH, and H-CHO) were determined by fixing their frequency factors to that of ethane and adjusting their activation energies relative to methane in proportion to the differences in reaction enthalpy.

The reaction of the methyl radical with surface oxygen can significantly alter the selectivity predicted by the model. Labinger and Ott[22] analyzed their results and concluded that the oxidation of methyl radicals with the Na/Mg/Mn catalyst (without feed gas oxygen) was 2700 times the rate of methane activation. The ratio of heterogeneous oxidation to homogeneous coupling of methyl radicals is the essential factor governing the selectivity at low conversion. Therefore, the second key premise of our model is that alkyl radicals irreversibly react in a non-activated step with surface oxygen to form adsorbed complexes that are precursors to oxygenates.

$$\bullet R + O_s \Longrightarrow RO_s \qquad (5)$$

The heterogeneous rate parameters involving surface sites were optimized to fit experimental results for Na/CaO at 1073 K. The variable, ϕ_s, represents the initial active site density (essentially the sum of O_s and α_s). For a specific site density, the activation energy for C-H bond activation was then adjusted to obtain experimentally observed conversion rates. Reaction rate and product ratios were investigated as a function of the active site density at 1073 K with a methane to oxygen ratio of 10 (Fig. 2). When the O_s concentration and is high (i.e. $\phi_S = 10^{-2}$ to 10^{-4}) methane conversion is high and oxidation to CO is the dominant processes. At lower surface site concentrations ($\phi_s < 10^{-6}$), the conversion rate is lower and the C_2 selectivity is higher. The highest C_2 yield was found to be $\phi_s = 10^{-5}$. These optimized independent parameters that affect the product selectivity were used for all subsequent calculations ($\phi_s = 10^{-5}$ and an activation energy (E_a) for the rate determining abstraction step (reaction 4a) of 63.6 kJ mol^{-1}).

We used the Chemkin kinetic modeling program to solve the set of non-linear differential equations. In linking the heterogeneous reactions to the homogeneous reaction network, we used a constant surface to volume ratio and calculated the surface site concentration. The fraction of active centers on the surface (ϕ_s) was normalized to the amount of surface oxide cations. The reaction is not surface transport limited in our model calculations.

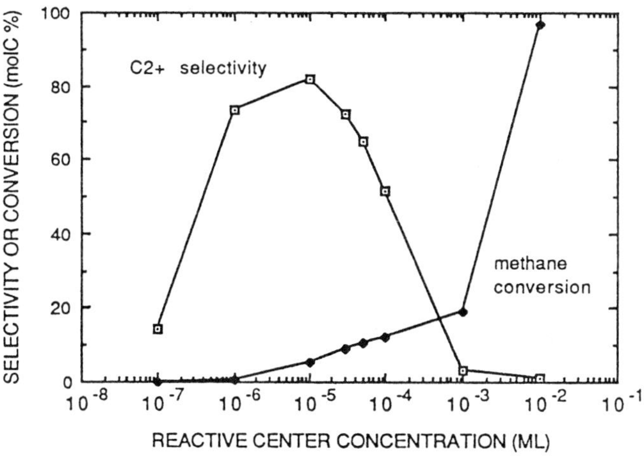

Fig. 2. Effect of reactive oxygen center surface concentration on methane conversion and higher hydrocarbon selectivity at 1073 K with 0.3 atm methane and 0.03 atm oxygen.

RESULTS

Model Calculations

Once the heterogeneous parameters were established, the temporal concentrations of co-oxidation products were determined for various reaction conditions . The results obtained at 1073 K with CH_4 and O_2 concentrations of 0.3 and 0.03 atm., respectively (Fig 3), show that ethane is the major carbon containing product, followed by CO and ethylene. Other significant products are methanol and formaldehyde, which decrease in relative importance with increased reaction time. The relative importance of several gas phase reactions at varies with initial partial pressures and reaction time (Table 1). At low pressure the main source of methyl radicals is the heterogeneous activation step, while at high pressure two additional gas phase sources of methyl radicals are reactions involving the hydrogen and hydroxyl radicals.

Higher hydrocarbon selectivity in the methane coupling process is very dependent on oxygen partial pressure. The effect oxygen partial pressure on methane conversion and product selectivity was systematically examined (Fig. 4) for fixed methane

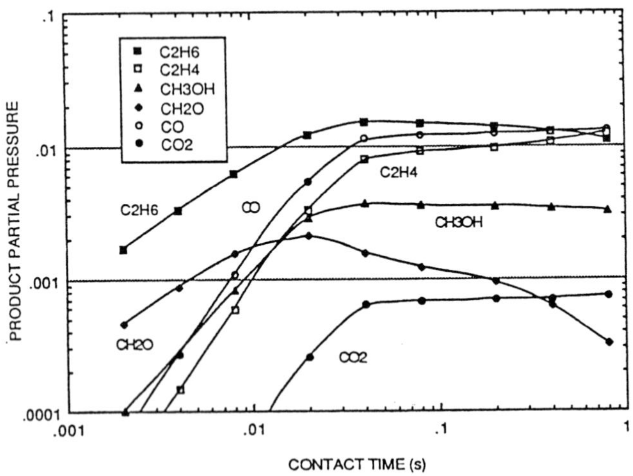

Fig. 3. Calculated product distribution vs. contact time for methane coupling at 1073 K with 0.3 atm methane and 0.03 atm oxygen.

TABLE 1

Reaction Rates for T = 1073 K, PCH4 = 0.3 atm, PO2 = 0.03, Phi = 1.0e-5

Reaction at 2ms	Rate	Reaction at 40 ms	Rate
CH3+O2=CH3O2	2.175E-05	CH4+H=CH3+H2	3.80E-06
CH3O2=CH3+O2	2.155E-05	CH4+MEOS-CH3+MEOHS	3.04E-06
CH4+MEOS-CH3+MEOHS	1.964E-05	MEOHS+MEOHS-H2O+MES+MEOS	2.05E-06
MEOHS+MEOHS-H2O+MES+MEOS	1.048E-05	C2H5=C2H4+H	2.01E-06
CH3+CH3=C2H6	9.74E-06	CH3+H2=CH4+H	1.98E-06
MES+MES+O2-MEOS+MEOS	5.67E-06	CH3+CH3=C2H6	1.94E-06
CH4+OH=CH3+H2O	2.84E-06	HCO+M=H+CO+M	1.41E-06
CH3+O2=CH2O+OH	1.47E-06	C2H6+H=C2H5+H2	1.13E-06
CH3+HO2=CH3O+OH	1.46E-06	MES+MES+O2-MEOS+MEOS	9.6E-07
CH3O+O2=CH2O+HO2	8.4E-07	C2H6+CH3=C2H5+CH4	5.8E-07
CH3O+CH4=CH3+CH3OH	7.6E-07	CH2OH+O2=CH2O+HO2	5.7E-07
CH4+H=CH3+H2	7.3E-07	CH3+HO2=CH3O+OH	5.5E-07
CH3+MEOS-CH3MOS	6.7E-07	CH2O+H=HCO+H2	5.3E-07
CH3MOS+MEOS-CH2O+MEOHS+MES	6.7E-07	CH2O+CH3=HCO+CH4	4.4E-07
HCO+O2=HO2+CO	5.0E-07	CH4+OH=CH3+H2O	4.2E-07
CH2O+MEOS-HCO+MEOHS	3.9E-07	C2H6+MEOS-C2H5+MEOHS	4.0E-07
CH4+HO2=CH3+H2O2	3.6E-07	C2H3+CH4=C2H4+CH3	3.4E-07
CH3+H2O2=HO2+CH4	3.2E-07	C2H4+CH3=C2H3+CH4	3.1E-07
C2H5=C2H4+H	3.0E-07	C2H4+H=C2H3+H2	2.9E-07
CH2O+CH3=HCO+CH4	2.9E-07	CH2O+MEOS-HCO+MEOHS	2.7E-07
HCO+M=H+CO+M	2.6E-07	CH4+HO2=CH3+H2O2	2.4E-07
C2H6+MEOS-C2H5+MEOHS	2.2E-07	CH3+H2O2=CH4+HO2	1.9E-07
CH3O2+CH3=CH3O+CH3O	2.0E-07	CH3+O2=CH3O2	1.9E-07
CH3O+M=CH2O+H+M	1.8E-07	CH3O2=CH3+O2	1.9E-07
C2H6+CH3=C2H5+CH4	1.4E-07	C2H3+O2=CH2O+HCO	1.8E-07
C2H5+O2=C2H4+HO2	1.2E-07	C2H3+H2=C2H4+H	1.7E-07
MEOS+MEOS-MES+MES+O2	1.1E-07	C2H5+H2=C2H6+H	1.5E-07

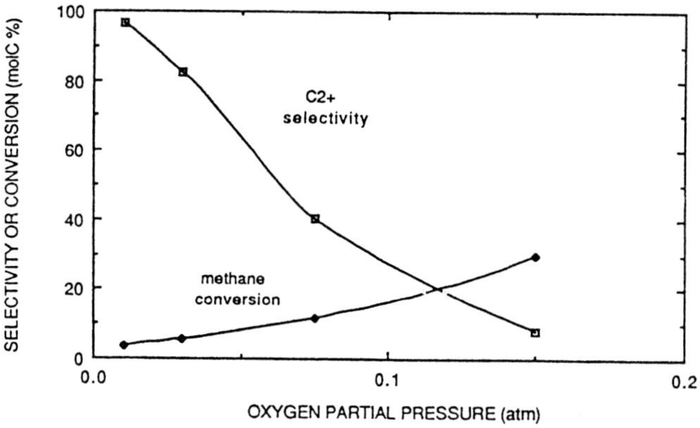

Fig. 4. Methane coupling conversion and higher hydrocarbon selectivity vs. oxygen partial pressure at 1073 K with 0.3 atm methane.

partial pressure (0.3 atm) and fixed active site concentration ($\phi_s = 10^{-5}$). The methane conversion increased, the C_{2+} selectivity decreased, while the ethylene to ethane ratio increased to a constant level with increasing oxygen partial pressure. Several homogeneous methyl radical oxidation pathways are important. Under high temperature and low pressure conditions the reaction of methyl radicals with hydrogen peroxy radicals is the prominent oxidation pathway,

$$\bullet CH_3 \ + \ \bullet O_2H \ ====> \ \bullet OCH_3 \ + \ \bullet OH \qquad (6)$$

while at high pressure and low temperature the reaction of methyl radicals with methyl peroxy radicals is prominent. The major sources of $\bullet O_2H$ are hydrogen abstraction reactions of unstable radicals such as •CHO and $\bullet C_2H_5$ with diatomic oxygen.

Experimental Alkane Conversion with $SrZrO_3$

Although alkali promoted catalysts have greater selectivity than unpromoted alkaline earth and rare earth oxides, there is some concern about stability of these catalysts given the high vapor pressures of alkali under reaction conditions. The volatility of alkali under typical alkane activation conditions is due to the high vapor pressure of the alkali hydroxide molecules in the presence of steam and oxygen, although the solid phase is likely to be alkali carbonate. Thermochemically stable carbonate salts with the low volatility in steam are SrCO3 and BaCO3.

Perovskite-supported alkaline earth carbonates, $Sr/SrZrO_3$ and $Ba/SrZrO_3$ are selective and stable methane activation catalysts (Figure 5), comparable or superior in this respect to Li-promoted MgO and Na-promoted CaO. Good ethene selectivity was also shown for co-oxidative dehydrogenation of ethane (Figure 6). Unlike the alkali-promoted alkaline earth catalysts, $SrZrO_3$ operated 20 hours at 1173 K with no evidence of evaporation or corrosive attack on our quartz reactors. These catalysts appear to achieve hydrocarbon selectivity approaching theoretical yields based on laboratory reaction conditions and predicted homogeneous oxidation rates.

DISCUSSION

As a result of our analysis with the heterogeneous-homogeneous model, we conclude that methane co-ox coupling processes with premixed oxygen and methane may be limited to a

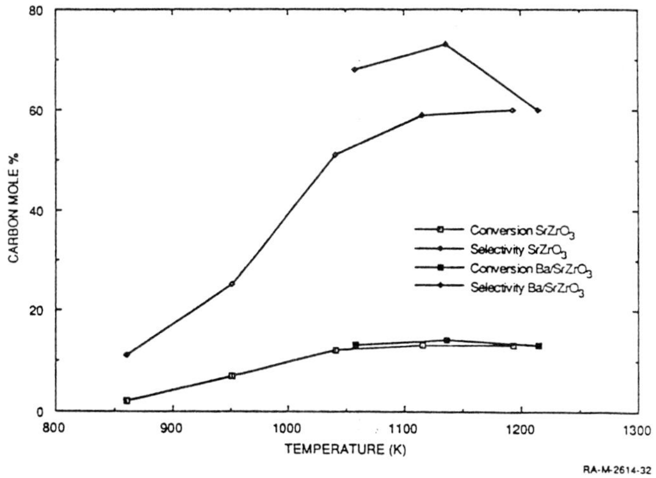

Fig. 5. Methane conversion and C_2 selectivity for $SrZrO_3$ versus reaction temperature. Conditions: 0.29-atm CH_4, 0.029-atm O_2, 3.3 mL s^{-1} (NTP) flow at 1-atm pressure.

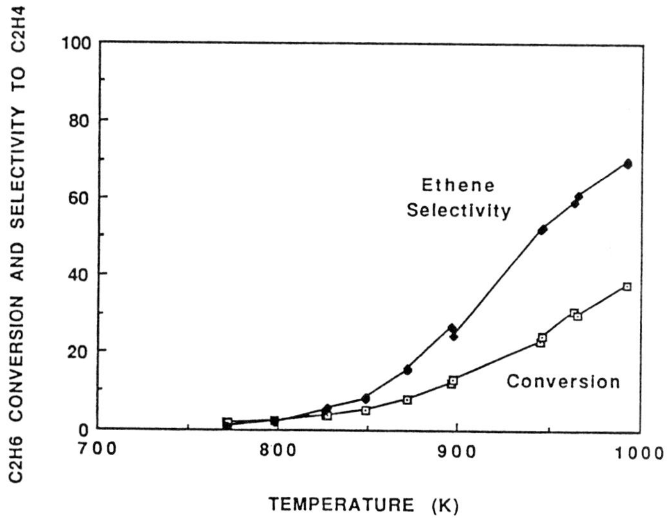

Fig. 6. Ethane partial oxidation on $SrZrO_3$ with 0.03 atm oxygen.

maximum higher hydrocarbon yield of about 30 mol% (carbon basis) and a maximum ethene/ethane ratio of about 2. Catalysts that can favorably influence the selectivity by combining high turnover rates for alkyl radical generation per reactive site with a very low surface concentration of reactive centers (optimally one part per 10^5 sites). Formation of a surface or bulk base metal carbonate layers may be one way of reducing the density of reactive oxygen centers to levels that avoid the rapid oxidation of alkyl radicals at the catalyst surface, and indeed our temperature programmed desorption experiments show that the surfaces of those solid-state basic oxide catalysts highly selective for light alkane dehydrogenation and methane coupling are consistently covered with at least one monolayer of carbonate.

Homogeneous reactions can fully account for the production of hydrocarbon products and the selectivity between coupling products and CO_x, but predict higher yields of CH_3OH, CH_2O, and H_2, and greater CO/CO_2 product ratios than observed experimentally. Heterogeneous oxidation reactions are evidently responsible for significant oxidation of oxygenate products, the substantial conversion of product hydrocarbons or radicals, and possibly direct oxidation of ethylene to CO_2 and H_2O in co-oxidation processes by catalysts with poor inherent selectivity.

ACKNOWLEDGEMENT

The authors gratefully acknowledge the support of the Methane Reaction Science Program directed by Dr. Daniel A. Scarpiello for the Gas Research Institute and associated Industrial Affiliate co-sponsors.

REFERENCES

1. Ito, T., Wang, J.-X., Lin, C.-H, and Lunsford, J. H., J. Am. Chem. Soc. (1985) **107**, 5062.
2. Sofranko, J. A., Leonard, J. L.., and Jones, C. A., J. Catal. (1987) **103**, 302.
3. Kimble, J. B.,and Kolts, J. H., Energy Progress (1986) **6**(4), 226.
4. Driscoll, D. J., Martir, W., Wang, J.-X., and Lunsford, J. H., J. Am. Chem. Soc. (1985) **107**, 58.
5. Campbell, K. D., Morales, E., and Lunsford, J. H., J. Amer. Chem. Soc. (1987) **109**, 7900.
6. Campbell, K.,and Lunsford, J. H., J. Phys. Chem. (1988) **92**, 5792.
7. Nelson, P. F., Lukey, C. A., and Cant, N.W., J. Phys. Chem. (1988) **92**, 6176.

8. Mims, C., (to be published).
9. Mirodatos, C., (to be published).
10. (a) Ekstrom, A. and Lapszewicz, J. A., J. Am. Chem. Soc. (1988) **110**, 5226. (b) Ekstrom, A. and Lapszewicz, J. A., J. Chem. Soc., Chem. Commun. (1988) 797.
11. van Kasteren, W. M. H., Geerts, W.M.H., and van der Wiele, K., Proceedings of the Ninth International Congress on Catalysis, **2**, p. 930, Phillips, M. J., and Ternan, M., eds., The Chemical Institute of Canada (1988).
12. Lo, M.-Y., Agarwal, S. K., and Marcelin, G., J. Catal. (1988) **112**, 168.
13. Lane, G. S., and Wolf, E. E., J. Catal. (1988) **113**, 144.
14. Asami, K., Omata, K., Fujimoto, K., and Tominaga, H., Energy and Fuels (1988) 2, 574.
15. McEwen, A.,B., Quinlan, M. A., and McCarty, J. G., (to be published, 1989).
16. Frenklach, M., (private communication).
17. Warnatz, J., in Combustion Chemistry, Gardiner, W. C., Jr., ed., Chpt. 5, p. 197,Springer-Verlag, (1984).
18. Tsang, W., J. Phys. and Chem. Ref. Data (1986) **15**, 1087.
19. Driscoll, D. J., Campbell, K. D., and Lunsford, J. H., Adv. Catal. (1987) **35**, 139.
20. Cant, N.W., Lukey, C. A., Nelson, P. F., and Tyler, R. J., J. Chem. Soc. Chem. Commun. (1988) 766.
21. Amorebieta, V. T., and Colussi, A. J., J. Phys. Chem. (1988), **92**, 4576.
22. Labinger, J. A., and Ott, K. C., J. Phys. Chem. (1987) **91**, 2682.
23. Labinger, J. A., Mehta, S., Ott, K. C., Rockstad, H. K., and Zoumalan, S., in Catalysis 1987, Ward, J. W., ed., p. 513, Elsevier (1988).

G. Centi and F. Trifiro' (Editors), *New Developments in Selective Oxidation*
1990 Elsevier Science Publishers B.V., Amsterdam — Printed in The Netherlands

General Mechanism for the oxidative coupling of methane

O.Forlani, M.Lupieri, V.Piccoli, S.Rossini* and D.Sanfilippo

Snamprogetti, S.Donato Milanese, 20097 Italy

J.A.Dumesic

Dep.Chem.Engineering, University of Wisconsin, Madison WI 53706

L.A.Aparicio, J.A.Rekoske and A.A.Trevino

Shanahan Valley Associates, Madison, WI 53711 USA

SUMMARY

A detailed mechanism, composed of 27 reactions, for the oxidative coupling of methane is described. The main products may derive either from a surface route or from a gas phase pathway. The kinetic parameters of the model, handled by a computer program, have been calculated from general chemical laws. The proposed mechanism has been calibrated on the data of the Li/MgO catalyst, studied by Lunsford and co-workers. The key steps are discussed in details and the fair good agreement between calculated and experimental data is given.

1. INTRODUCTION.

The huge availability and the low price of methane have led researchers in recent years to look for a route more direct than present technologies for the conversion of methane to more valuable chemicals . The promising advances, mainly obtained with oxide catalysts, have been reviewed by several authors (1)-(8). In general, the extensive literature on the subject deals with two main processes:

a) Direct methanol synthesis, catalyzed by oxides of altervalent metals;
b) Oxidative coupling (C2 hydrocarbon synthesis), carried out in one of the two ways:
   1.) with methane and oxygen co-fed to the reactor, catalysed by basic oxides(alkaline earth usually doped with alkali metals) and by rare earth oxides ;
   2.) with methane and oxygen alternatively fed to the reactor, catalyzed by low melting metal oxides.

Recognising the importance of such studies, we have begun a research program to investigate processes of type b.1).
We are using two complementary approaches in parallel:

i) The synthesis and testing of new catalysts;
ii) A semi-theoretical approach which combines experimental catalytic behaviour with surface science and general knowledge to guide new preparations.

This paper deals with point ii).

2. AIM OF THE WORK

The aim of our effort is to describe a catalytic reaction with a detailed kinetic mechanism that is consistent with intermediates identified spectroscopically and other in-house or literature observations. Every step is characterised by its rates, forward and reverse, given by the general equation (2.1):

$$r_i\ (x) = k(x)\ \pi_j\ C_j^{\mu}\ ; \qquad (2.1)$$

x = for, rev; π = product; μ = reaction order
C_j = gas phase concentration or surface coverage.

Starting from two basic equations, Arrhenius Law (2.2), allowing the calculation of the rate constant k, and Polanyi

$$k = A\ \exp(-DE_{att}/RT) \qquad (2.2)$$

Law (2.3), giving the activation energy DE_{att}, we have evaluated **"a priori "** all the kinetic constants, except a few experimentally available [(#1.14)- (#1.18)],

$$DE_{att} = DE_0 + \alpha\ DH_{reac} \qquad (2.3)$$

DE_0 = Constant term; α = Polanyi constant.
DH_{reac} = Gas phase or surface reaction heat.

trough heats of formation of every fragment involved in the mechanism, its bond strength with the surface and well known theories for preexponential factors A, such as the collision theory or the transition state theory. In this way we may discuss the rate determining steps in terms of surface chemistry and consequently tailor the catalyst that fulfills the model suggestions at best. This is actually our ultimate goal.

Naturally, the parameters cannot be calculated perfectly as a first temptative value, but usually they require a little adjustment (few Kcal/mol). So we have calibrated such parameters with literature data measured on a single Li/MgO catalyst over a wide range of conditions by Lunsford and co-workers(9). From this point on, we are developing the same approach to our own catalysts trying to correlate intimate properties and reactivity.

The whole set of reactions is managed by a computer program.

3. MECHANISM

It is generally accepted that C2 hydrocarbons are formed by a coupling of $CH_3(.)$ radicals; in particular C_2H_6 is obtained as primary product and C_2H_4(9.a) derives from a dehydrogenative step of the saturated hydrocarbons.

The presence of $CH_3(.)$ radicals has been demonstrated with Li/MgO catalyst by Lunsford et al.(9.b), while it is still questionable if the coupling takes place in the gas phase or on the surface.

We are convinced that a particular form of oxygen (O_{act}) is responsible for the homolytic cleavage of the C-H bond: following the suggestion of some authors, O_{act} may be defined as O^-, in Li/MgO and similar catalysts due to the formation of $[M^+O^-]$ centers(9) as well as in other completely different catalysts(10,11), or O_2^- as proposed for lanthanum oxide(12).

A bit more ambiguous is the genesis of carbon oxides; three main paths can be envisioned[9.a]:

a) 2 $CH_3(.)$ ———> C_2H_6 ———> C_2H_4 ———> CO_x (3.1)

b) $CH_3(.)$ + $O^{(2-)}surf$ ———> OCH_3- ———> CO_x (3.2)

c) $CH_3(.)$ + O_2 ———> $CH_3O_2(.)$ ———> CO_x (3.3)

We have tried to consider all these routes and we aim to differentiate them on the basis of the reaction rate values.

We have set up the mechanism described in Scheme #1 from this whole of considerations.

4. RESULTS

The simulations of the data of Lunsford et al.[9.a] are shown in Figures n.1-4. The agreement is satisfactory; the model fits the behaviour of the catalyst in a fair good way, except when considering the partial pressure of oxygen (see figure n. 4). This will be discussed in the next point.

O_2 + 2 M	———>	2 MO		(#1.1)
O_2 + 2 M	———>	2 MO_{act}		(#1.2)
CH_4 + MO + MO_{act}	———>	CH_3MO	+ MOH	(#1.3)
2 CH_3MO	———>	C_2H_6	+ 2 MO	(#1.4)
C_2H_6 + MO + MO_{act}	———>	CH_3CH_2OM	+ MOH	(#1.5)
CH_3CH_2OM + MO	———>	C_2H_4	+ MO + MOH	(#1.6)
CH_3CH_2OM + 2 MO	———>	$MOCH_2OM$	+ CH_3MO	(#1.7)
CH_3OM + MO + MO_{act}	———>	$MOCH_2OM$	+ MOH	(#1.8)
$MOCH_2OM$ + MO	———>	MOCHO	+ MOH + M	(#1.9)
MOCHO + MO	———>	CO	+ MOH + MO	(#1.10)
MOCHO + MO	———>	CO_2	+ MOH + M	(#1.11)
2 MOH	———>	H_2O	+ MO + M	(#1.12)
CH_3MO	———>	$CH_3(.)$	+ MO	(#1.13)
2 $CH_3(.)$	———>	C_2H_6		(#1.14)
$CH_3(.)$ + O_2	———>	$CH_3O_2(.)$		(#1.15)
$CH_3O_2(.)$	———>	CH_2O	+ OH(.)	(#1.16)

CH_2O	+ OH(.)	———>	CHO(.)	+ H_2O	(#1.17)
CHO(.)	+ O_2	———>	CO	+ HO_2(.)	(#1.18)
CH_3O_2(.)	+ M	———>	MO_2CH_3		(#1.19)
MO_2CH_3	+ M	———>	MO	+ CH_3MO	(#1.20)
CH_2O	+ M	———>	$MOCH_2$		(#1.21)
$MOCH_2$	+ MO	———>	$MOCH_2OM$		(#1.22)
HO(.)	+ M	———>	MOH		(#1.23)
CHO(.)	+ MO	———>	CO	+ MOH	(#1.24)
HO_2(.)	+ M	———>	MO_2H		(#1.25)
MO_2H	+ M	———>	MO	+ MOH	(#1.26)
CO	+ MO	———>	CO_2	+ M	(#1.27)

Scheme #1

5. DISCUSSION AND COMMENT

The reaction mechanism may be divided into two main sections:

1) Surface Reactions; [Reactions from (#1.1) to (#1.12)]
2) Gas Phase Reactions; [Reactions from (#1.13) to (#1.27)]

The main reaction worth discussing is reaction (#1.2), in which O_{act} species are formed. Molecular orbital calculation have shown that the edge of the valence band in MgO is composed by orbitals that are mainly 2p-oxygen in character(13).Hence, one can view the formation of O^- species as a process in which holes are generated in the valence band of MgO. The creation of a hole is an endothermic process, and this would be in good agreement with the fact that O^- species are usually observed experimentally only at high temperature. The creation of holes in the valence band of MgO can be envisioned in two ways:

1) through the transfer of an electron from the valence band to an acceptor level within the band gap. The acceptor level could be due to the presence of a doping agent or it could be a Schottky defect (a cation vacancy);
2) through the generation of ionized Schottky defects by

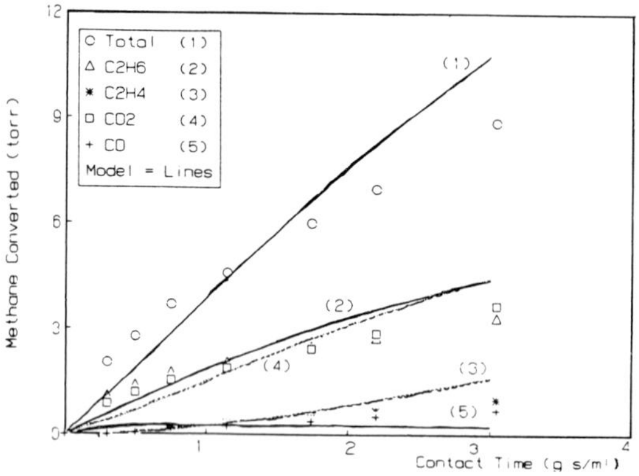

Figure n.1 Methane conversion vs. Contact Time; T=620°C;Inlet Press (Torr): CH4=300 ,O2=60, He=40. Catalyst = 1g

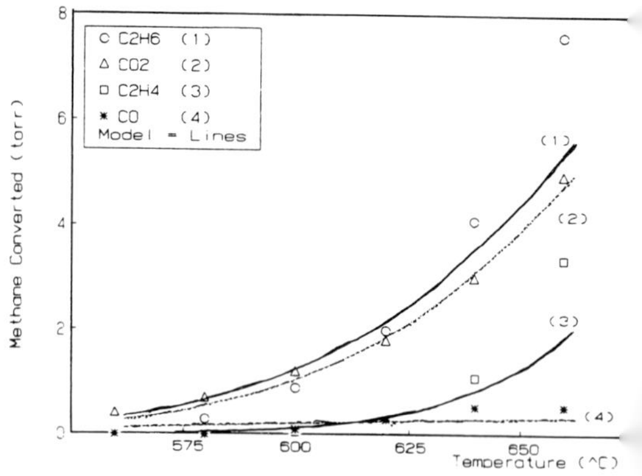

Figure n.2 Methane conversion vs. temperature; Flow=.83ml/s; Inlet Press (Torr): CH4=300, O2=60, He=40. Catalyst= 1g

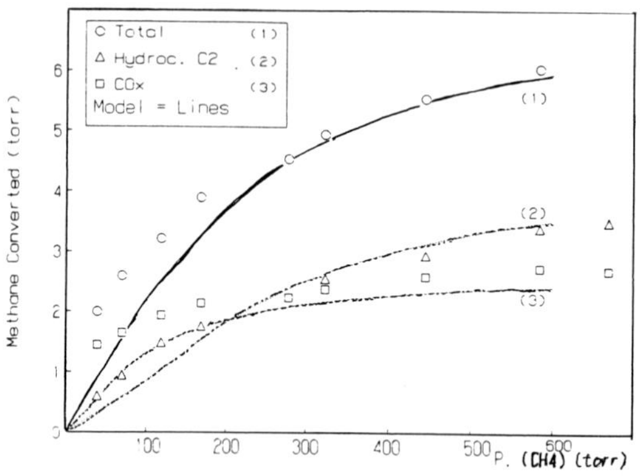

Figure n.3 Methane conversion vs. Methane pressure; T= 620°C Flow=.83 ml/s; Inlet Pressure: (torr) O2=60, total=760 Catalyst = 1g

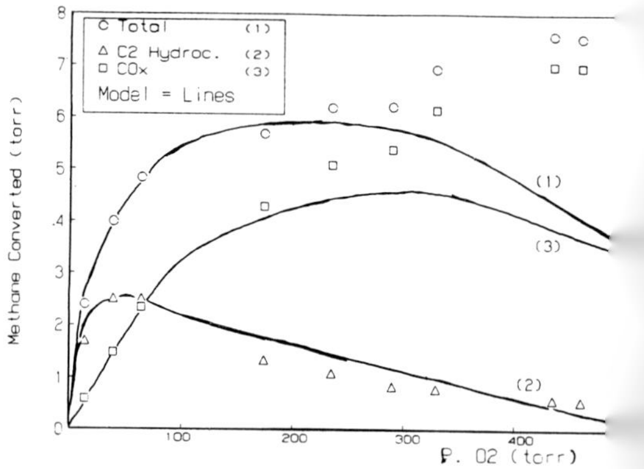

Figure n.4 Methane conversion vs. Oxygen Pressure; T=620°C Flow= .83ml/s; Inlet Pressure (Torr): CH4=300, total=760 Catalyst = 1g

oxygen from the gas phase.

A little of mathematics and some hypotheses allowed us to predict the formations of O_{act} according to these ideas.

Steps (#1.3) and (#1.5) are similar hydrocarbon adsorption reactions that form surface OR species through the cleavage of a C-H bond by O_{act}. O_{act} is involved also in step (#1.8) where another C-H bond is broken to form the surface CO_x precursor.

Ethane is produced by either the coupling of two surface OCH_3 species (#1.4) or by the coupling of two gas phase $CH_3(.)$ radicals (#1.14). Ethylene derives only from a dehydrogenation (#1.6) of a readsorbed ethane molecule, this way being in competition with the cleavage of a C-C bond leading to total combustion.

A gas phase $CH_3(.)$ in the mechanism can combine with O_2 to eventually yield CO [from step (#1.15) to (#1.18)] and CO_2 (#1.27) through some possible interaction with the solid [Step (#1.19) to (#1.26)].

The discrepancies between experimental and calculated data as function of oxygen partial pressure (v.Figure n.4) are probably due to an overestimation of oxygen-surface bond so that, at relatively high oxygen pressure, the model predicts an high oxygen coverage. The OCH_3 coverage is forced to almost zero and the total activity declines although the product distribution is preserved. We have been able to better follow the trend at high oxygen pressure, but doing so we were missing the quite characteristic and significant maximum in C2 products at about 50 torr of oxygen.

Under the low conversion conditions studied by Lunsford and co-workers, the model predicts that:

a) the coupling of gas phase $CH_3(.)$ radicals is negligible if compared to the coupling of OCH_3 species on the surface;
b) the dehydrogenation of ethane is orders of magnitude faster than the C-C bond cleavage, i.e. the way (3.1) is completely unsignificant at these levels of conversion;

c) the main route (ca.70%) to CO_x is the gas phase radical chain via methylperoxide while the surface combustion contributes for the remaining 30%.

The quite satisfactory agreement may be expressed in these following points:

1) At 620^C and low conversion, the main products are C_2H_6 and CO_2; the selectivities and yields of C_2 hydrocarbons reach a maximum at low oxygen pressure; both the production of C_2 and CO_x have reaction orders with respect to methane pressure less than one, the latter being the minor. This makes C_2 selectivity always increase with methane pressure although C_2 yield reaches a maximum.
2) At 720^C,the main C_2 product becomes C_2H_4 instead of C_2H_6
3) In the temperature range 550-675^C and low conversion conditions, the selectivity and yield of C_2 products increase with temperature.
4) Under none of the conditions studied by Lunsford et al. formaldehyde is a significant product.

6. CONCLUSIONS

The mechanism set up to simulate the methane oxidative coupling is quite satisfactory considering the many experimental values employed in the calculation. We believe that our model, now that it has been calibrated with literature data, will become a powerful tool in tailoring new catalyst formulation. Preliminary results are confirming this feeling.

7. REFERENCES

(1) Grigoryan E.A.; Russ.Chem.Rew.; 53(2) 210-220 (1984)
(2) Foster N.R.; Appl.Catal., 19, 1-11 (1985)
(3) Gesser H.D. and Hunter N.R.; Chem.Rew., 85(4) 235-244 (1985)
(4) Pitchain R. and Klier K.; Catal.Rev.-Sci.Eng., 28(1), 13-88 (1986)
(5) Mimoun H.; New Jour.Chem., 11(7), 513-525 (1987)

(6) Scurrell M.S.; Appl.Catal., 32, 1-22 (1987)
(7) Lee J.S. and Oyama T.S.; Catal.Rev.-Sci.Eng., 30(2), 249-280 (1988)
(8) Baerns M., van der Wiele K. and Ross J.R; Cat.Today, 4, 471-494 (1989)
(9) a. Ito T., Wang J., Lin C. and Lunsford J.H.; J.Am. Chem.Soc., 107, 5062-5068 (1985)
b. Driscoll D.J., Martin W., Wang J. and Lunsford J.H.; J.Am.Chem.Soc., 107, 58-63 (1985)
c. Lin C., Ito T., Wang J. and Lunsford J.H.; J.Am. Chem.Soc., 109, 4808-4810 (1987)
(10) a. Hill W., Shelimov B.N. and Kazansky V.B; J.Chem.Soc. Faraday Trans., 1, 83, 2381-2389 (1987)
(11) b. Kaliaguine S.L., Shelimov B.N. and Kazansky V.B; J.Catal., 55, 384-393 (1978)
(12) Wang J. and Lunsford J.H.; J.Phys.Chem., 90, 3890-3891 (1986)
(13) Mehandru S.P., Anderson A.B. and Bradzil J.F.; J.Am. Chem.Soc., 110, 1715 (1988)

G. Centi and F. Trifiro' (Editors), *New Developments in Selective Oxidation*

THE MECHANISM OF THE OXIDATIVE COUPLING OF METHANE: EVIDENCE THAT ETHYLENE IS A PRIMARY PRODUCT

GEORGE W. KEULKS and MIN YU
Laboratory for Surface Studies, University of Wisconsin-Milwaukee, P.O. Box 340, Milwaukee, Wisconsin, 53201, USA

SUMMARY

The oxidative coupling of methane was studied over a MgO supported Bi-P-K oxide catalyst. When CD_4 was substituted for CH_4, the reaction exhibited a kinetic isotope effect as expected, but the D-distribution in ethylene could not be explained by assuming the sole pathway to ethylene was via ethane. The results suggest that ethylene can be produced as a primary product as well as a secondary product.

INTRODUCTION

The oxidative coupling of methane to ethylene and ethane has generated considerable interest, following the published report by Keller and Bhasin [ref. 1] in 1982. Most of the work has been directed toward the identification of selective catalysts. A wide variety of oxides now have been reported to be active and selective for this reaction. Alkali, alkaline earth, and rare earth oxides have shown good activity-selectivity behavior for the formation of C_2 hydrocarbons [ref. 2].

The mechanistic details of the reaction, on the other hand, have received considerably less attention. Ito et.al. [ref. 3] detected CH_3 radicals in the gas phase over lithium-promoted magnesium oxide. They suggested that the formation of ethane involves the coupling of 2 CH_3 radicals in the gas phase.

ARCO workers [ref. 4,5] also suggested that ethane is formed via the gas phase coupling of CH_3 radicals. In addition, they suggested that ethylene is formed from ethane via an oxidative dehydrogenation process on the surface of the catalyst.

The rate limiting step of the reaction appears to be hydrogen atom abstraction from CH_4 to form a CH_3 radical. This is supported by the observation of a kinetic isotope effect [ref. 6]. The use of CD_4 also allows one to examine whether or not the isotopic composition of ethane is consistent with the coupling of two CH_3 radicals. Nelson et.al. [ref. 7] reported that only C_2H_6, CD_3CH_3, and C_2D_6 were formed when a mixture of CD_4 and CH_4 was passed over lithium-promoted magnesium oxide. They also found only C_2H_4, CD_2CH_2, and C_2D_4 in the ethylene product. This is consistent with the suggestion that ethylene forms from ethane and not as a primary product of the reaction.

We have been intriqued about the possibility that ethylene can form as a primary product in the reaction, not solely as a secondary product. During the surface initiation of CH_4 to form CH_3 radicals, it is conceivable that some CH_3 radicals react further with the surface. In fact, workers [ref. 8,9] have suggested that such a pathway may lead complete combustion. The formation of surface species such as CH_2, CH, C, however, would provide the opportunity for a direct pathway to ethylene. Carbene ($:CH_2$) species have been proposed to be involved in the formation of higher hydrocarbons [ref. 10] as well as the improvement in ethylene selectivity at higher temperatures [ref. 11].

In this work, we have further examined the possibility that ethylene is formed as a primary product in the oxidative coupling of methane.

EXPERIMENTAL

Catalyst

We previously reported [ref. 12] that a MgO supported Bi-P-K oxide catalyst was active and selective for the oxidative coupling of methane. The details of the catalyst preparation were described earlier.

Reaction Studies

All experiments were conducted in a single-pass flow reactor at atmospheric pressure. The feed gases, methane (Matheson, CP), oxygen (Amerigas, 99%), helium (Amerigas, 99.9%), were used without further purification. The individual gas feed rates were controlled by Tylan mass flow controllers (Model FC-260), except for C_2D_4 and C_2D_6 feed rates, which were controlled by a motor-driven, syringe pump (Sage Instruments, Model 341), equipped with a 10 cm^3 Hamilton gas-tight syringe.

The isotopically labelled gases used in this study were: CD_4 (99.6 at % D), C_2D_6 (99.5 at % D), C_2D_4 (99.2 at % D), and D_2 (99.5 at % D). All were purchased from Merck, Sharp, and Dohme, Canada Ltd., Montreal, Canada.

The reaction gas mixture of methane and oxygen, diluted with helium to achieve a total pressure of 1 atm., was passed over the catalyst while heating to the desired temperature. On-line analysis of the effluent gas was achieved by gas chromatography, using Porapak Q and molecular sieve 5A columns.

For analysis of the isotopic distribution in the products, the effulent gas was first allowed to pass through a 5 cm^3 sampling trap for 2 min. The 5 cm^3 sample was then injected into the gas chromatograph. Separation of methane, ethane, and ethylene was achieved by using a 4' x 0.25" (O.D.) Porapak Q (80-100 mesh) column at room temperature. The separated products were collected in individual traps that could be used for subsequent mass spectrometric analysis.

Before introducing the trapped product into the mass spectrometer, each trap was evacuated at -196 °C for 5 min. Ethylene and methane were analyzed at 13 e.v. and 15 e.v., respectively, in order to minimize fragmentation. In the case of ethane, however, the ionization potential of the molecular ion does not differ significantly from the appearance potential of the ethylene ion. Hence, elimination of fragment peaks was not possible and mass spectra were obtained at 70 e.v., utilizing calibrations with C_2H_6 and C_2D_6. The fragmentation patterns for C_2H_5D, $C_2H_4D_2$, $C_2H_3D_3$, $C_2H_2D_4$, and C_2HD_5 were calculated from the calibration data using the method suggested by Amenomiya and Pottie [ref. 13].

RESULTS

If the rate-determining step is the breaking of a C-H bond in methane, then a kinetic isotope effect should be observed when CD_4 is substituted for CH_4. The conversion of both CD_4 and CH_4 were determined under the following conditions; T=650 °C and 700 °C; CH_4/O_2 = 8/1; W/F= 0.06 gm-sec/ml. Under these conditions, the C_2 selectivity was 39.1% (650 °C, CH_4), 56.8% (650 °C, CD_4), 28.6% (700 °C, CD_4), 50.6% (700 °C, CD_4).

Because the experimental conditions are the same for both reactions, a kinetic isotope effect (k_H/k_D) can be calculated for the formation of ethane, ethylene, and CO_2 by using the relative product yields. The results are summarized in Table 1. The rate of formation of ethane exhibits a kinetic isotope effect, but no isotope effect is observed for the formation rates of ethylene and CO_2. The observed isotope effect is in good agreement with molecular data calculations [ref. 14].

TABLE 1. Kinetic Isotope Effect

Temp., °C	k_H/k_D Ethylene	Ethane	CO_2
650	0.96	1.75	1.00
700	1.04	1.22	0.95

The C_2 products, ethane and ethylene, as well as the recovered, unreacted methane, were analyzed by mass spectrometry. The concentration for each isotopic component was obtained by normalizing each compound to itself, e.g., % C_2H_4 = $C_2H_4/(C_2H_4+C_2H_3D+C_2H_2D_2+C_2H_3D+C_2D_4)$. The results are summarized in Table 2.

TABLE 2. Isotopic Distribution for Reaction of CD_4+O_2

Temp., °C	Methane d0	d1	d2	d3	d4	Ethylene d0	d1	d2	d3	d4	Ethane d0	d1	d2	d3	d4	d5	d6
650	-	-	-	1	99	35	35	-	2	28	-	-	-	-	-	-	95
700	-	-	-	1	99	17	17	-	2	64	-	-	-	-	-	-	95

The surprisingly high amount of H in ethylene caused concern that ethylene was undergoing an exchange reaction with a H-source. We examined several possible exchange reactions. The reactions studied are summarized in Table 3.

TABLE 3. Possible Exchange Reactions

Mixture*	Temp. °C	Ethylene d0	d1	d2	d3	d4	Ethane d0	d1	d2	d3	d4	d5	d6
C_2D_4/O_2	650	-	-	-	3	97	-	-	-	-	-	-	-
C_2D_6/O_2	650	-	-	-	2	98	-	-	-	-	-	-	99
$C_2D_4/CH_4/O_2$	650	2	-	-	3	95	99	-	-	-	-	-	-
$C_2D_6/CH_4/O_2$	650	4	-	-	2	94	2	-	-	-	-	-	96

*C_2D_4 starting material was 98% d_4, 2% d_3; total D = 99.5 at %.
C_2D_6 starting material was 99% d_6.

We passed a C_2D_4/O_2 mixture over the catalyst at 650 °C. Gas chromatographic and mass spectrometric analyses indicated that the products were CO_2, H_2O, and a trace of C_2D_3H. A mixture of

C_2D_6/O_2 produced traces of ethylene (98% C_2D_4) and the unreacted ethane showed no evidence of exchange.

We also studied the possibility of the exchange of C_2D_4 and C_2D_6 with CH_4. The ratio of CH_4 to C_2D_4 or C_2D_6 was 11 to 1. The formation of C_2H_6 and C_2H_4 (experiments 3 and 4 in Table 3) indicate that the CH_4 reacted, as expected, but the C_2D_4 and C_2D_6 passed through the catalyst unchanged.

Having convinced ourselves that exchange reactions could not explain the large amount of H-incorporation into ethylene, we examined the D distribution obtained when a mixture of CH_4 and CD_4 was oxidized at 650 °C. The results are shown in Fig. 1. The

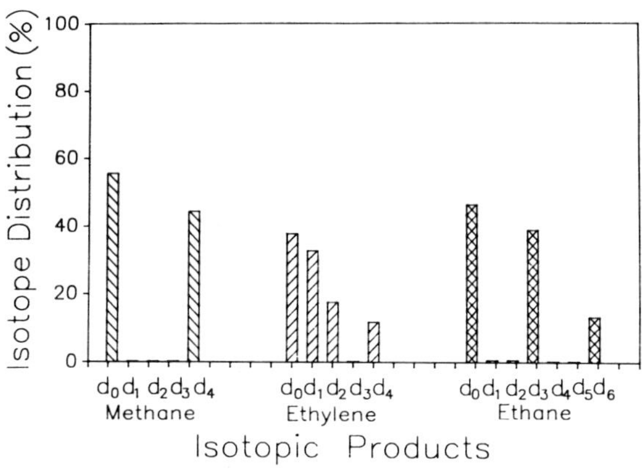

Fig. 1. Isotopic distribution of ethane, ethylene, and unreacted methane for the reaction of $CH_4/CD_4/O_2$ at 650 °C.

detection of C_2D_6, CH_3CD_3, and C_2H_4 is consistent with the suggested mechanism for the production of ethane via the coupling of the methyl radicals. The isotopically labelled ethylenes, C_2D_4, $C_2H_2D_2$, and C_2H_4, are the expected products resulting from the consecutive oxidative dehydrogenation of the labelled ethane species. However, the preponderance of C_2H_4 and C_2H_3D cannot be explained by the consecutive reaction sequence, $CH_4 \rightarrow C_2H_6 \rightarrow C_2H_4$. The most probable explanation is a parallel pathway involving the reaction of a hydrogen-deficient, surface intermediate with a hydrogen source on the catalyst.

Two additional experiments were conducted in an attempt to

gain additional evidence for the formation of ethylene as a primary product: (1) a study of the H-distribution in ethylene as a function of methane conversion and (2) a study of H-distribution in ethylene as a function of surface dehydroxylation.

The production of ethylene via ethane by a consecutive reaction pathway is obviously operative, as evidenced by the results obtained from the reaction of the mixture of CD_4 and CH_4. If the consecutive reaction rate for ethylene production is faster than the parallel reaction rate for ethylene production, then the D-distribution in ethylene should be a function of methane conversion. The reaction of CD_4/O_2 at 650 °C was studied at varying CD_4 conversion levels. The results are shown in Fig. 2.

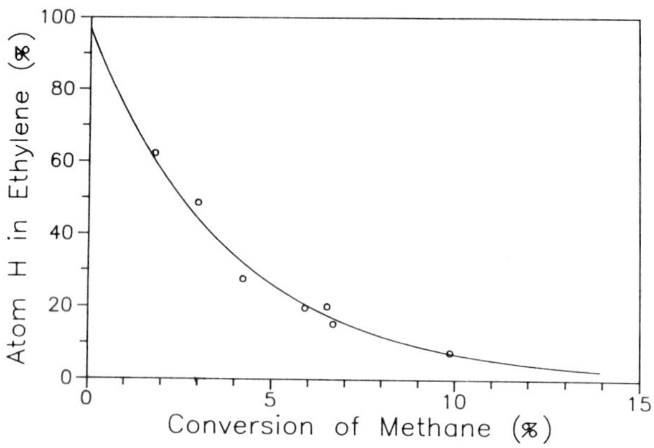

Fig. 2. Atom % H in ethylene for the reaction of CD_4+O_2 at 650 °C as a function of CD_4 conversion.

The at % H in ethylene decreases with increasing methane conversion. Thus, if C_2H_4 and C_2H_3D are indicators of ethylene production via a parallel pathway, and if C_2D_4 is an indicator of ethylene production via a consecutive pathway, then, as shown in Fig. 2, the at % H in ethylene should decrease as methane conversion increases.

Because pure CD_4 is used as the reactant, the H-source must be finite. This finite H-source is most likely to be surface hydroxyl groups on the catalyst. In an effort to dehydroxylate the catalyst surface, we pretreated the catalyst at increasingly higher temperatures with a flow of He/O_2. The results are presented in Table 4. The decrease in the H-concentration with

TABLE 4. D-Distribution from CD_4+O_2 at 650 °C as a Function of Catalyst Pretreatment

Pretreat	Methane					Ethylene					Ethane						
T (C)	d0	d1	d2	d3	d4	d0	d1	d2	d3	d4	d0	d1	d2	d3	d4	d5	d6
-	-	-	-	-	99	45	45	2	-	9	-	-	-	-	-	-	95
650	-	-	-	-	99	35	35	2	-	27	-	-	-	-	-	-	95
750	-	-	-	-	99	32	32	3	-	32	-	-	-	-	-	-	95
850	-	-	-	-	99	26	27	2	-	43	-	-	-	-	-	-	95

higher catalyst pretreatment temperatures is consistent with the assumption that the H-source is the catalyst surface hydroxyl groups, which are removed at elevated temperatures.

We have tried rehydroxylating the surface with D_2O and subsequently reacting CH_4/O_2, but little D is detected in the products. Apparently, once the surface hydroxyl groups are removed, they are difficult to replace or the reaction of OD with the surface intermediate is considerably slower than the reaction with OH.

DISCUSSION

The D-distribution in ethylene is best explained by a mechanism that allows for ethylene to be produced by both a consecutive pathway via ethane and a parallel pathway directly from methane. The parallel pathway involves a H-deficient surface intermediate.

The consecutive pathway is the same as that proposed by other workers [ref. 3-5]. The rate-limiting step in the activation of methane is the homolytic cleavage of a C-H bond to form CH_3 radicals. While our experiments provide no direct evidence as to the nature of the active site responsible for the activation of methane, an increasing body of evidence [ref. 15] suggests that surface O^- species may be the active site for methane activation.

Two CH_3 radicals combine in the gas phase to produce ethane as a primary product. The ethane undergoes a subsequent oxidative dehydrogenation by either reacting with active oxygen species on the surface or in the gas phase.

The parallel pathway leads to ethylene as a primary product, not a secondary product produced by further reaction of ethane. A site other than surface O^- is responsible for producing a

H-deficient surface intermediate. The exact nature of the H-deficient intermediate, responsible for the production of ethylene, is still unclear. A carbene ($:CH_2$) species would yield C_2D_4 when CD_4 is used as the reactant, unless the carbene intermediate undergoes rapid H-D exchange with the surface hydroxyls. Other possibilities include CH and C that could react with the surface hydroxyls to produce ethylene.

The results in Fig. 2 suggest that the formation of the H-deficient surface intermediate is fast compared to the formation of CH_3 radicals. At low methane conversions, the ethylene produced has a high at % of H and the ethylene/ethane ratio is large. As the methane conversion increases, the ethylene/ethane ratio and the at % of H in the ethylene decrease. This indicates that a larger fraction of the ethylene is formed by the consecutive pathway via ethane (C_2D_6).

CONCLUSIONS

The Bi-P system, which has been shown to be active and selective for the oxidative dimerization of propylene, also is active and selective for the oxidative coupling of methane. The isotopic tracer results suggest that two sites exist on the catalyst surface. One site is responsible for the formation of methyl radicals from gas-phase methane. The methyl radicals quickly dimerize to form ethane. Another site is responsible for producing hydrogen deficient surface intermediates. These surface intermediates produce ethylene as a primary product by reacting with surface hydroxyl groups.

REFERENCES

1 G.W. Keller and M.M. Bhasin, J. Catal., 73 (1982) 9.
2 J.M. DeBoy and R.F. Hicks, J. Chem. Soc., Chem. Commun., (1988) 982, and references therein.
3 T. Ito, J.X. Wang, C.H. Lin, and J.H. Lunsford, J. Phys. Chem., 90 (1985) 534.
4 J.A. Sofranko, J.J. Leonard, and C.A. Jones, J. Catal., 103 (1987) 302.
5 C.A. Jones, J.J. Leonard, and J.A. Sofranko, J. Catal., 103 (1987) 311.
6 N.W. Cant, C.A. Lukey, P.F. Nelson, and R.J. Tyler, J. Chem. Soc., Chem. Commun., (1988) 766.
7 P.F. Nelson, C.A. Lukey, and N.W. Cant, J. Phys. Chem., 92, (1988) 6176.
8 T. Ito and J.H. Lunsford, Nature (London), (1985) 314.
9 M.Y. Lo, S.K. Agarwal, and G. Marcelin, J. Catal., 112 (1988) 168.

10 G.A. Martin and C. Mirodatos, J. Chem. Soc., Chem. Commun., (1987) 1393.
11 C.H. Lin, K.D. Campbell, J.X. Wang, and J.H. Lunsford, J. Phys. Chem., 90 (1986) 534.
12 G.W. Keulks and M. Yu, React. Kinet. Catal. Lett., 35, (1987) 361.
13 Y. Amenomiya and R.F. Pottie, Can. J. Chem., 46 (1968) 1741.
14 Z. Melander, Isotope Effects on Reaction Rates, Ronald Press, New York, 1960, pp. 7-22.
15 J.H. Lunsford, Methane Conversion, Elsevier, Amsterdam, 1988, pp. 359-371.

G. Centi and F. Trifiro' (Editors), *New Developments in Selective Oxidation*

HETEROLYTIC MECHANISM OF METHANE ACTIVATION IN OXIDATIVE DEHYDRODIMERIZATION

V.D. SOKOLOVSKII, O.V. BUYEVSKAYA, S.M. ALIEV and A.A. DAVYDOV
Institute of Catalysis, Novosibirsk 630090, USSR

SUMMARY

The dependence of the rate of methane oxidative dehydrodimerization on oxides of alkaline earth metals on the concentration of base sites has been studied. The isotope CH_4-CD_4 exchange in conditions of methane oxidative dimerization has been examined. FIR spectroscopy data suggest the formation of metal-methyl groups during methane adsorption on MgO. A heterolytic mechanism of methane activation involving low-coordination surface sites is proposed.

INTRODUCTION

Oxidative dehydrodimerization of methane attracts attention of many researchers as one of the most promising ways of production of ethylene from non-oil raw material. In recent years, considerable advances have been achieved in the development of catalysts for this reaction (refs. 1-3); however, the mechanism and, primarily, the nature of methane activation on catalytic surface remain open to discussion. Most popular is the hypothesis of radical activation of methane on surface radical-ions of oxygen put forward by Lansford and co-workers who studied the reaction with Li/MgO catalysts (ref. 4).

Recent studies of selective oxidative transformations of saturated and unsaturated hydrocarbons via the C-H bond on oxide catalysts have allowed us to propose an alternative mechanism implying a heterolytic activation of the C-H bond (ref. 5). Such a mechanism does not require large concentrations of sites capable of producing oxygen radical forms, which may lead to complete oxidation (ref. 5).

The possibility of the heterolytic activation of methane in oxidative dehydrodimerization has been mentioned in (refs. 6,7).

In this work an attempt has been made to substantiate the heterolytic mechanism of methane activation in oxidative dehydrodimerization with base catalysts.

EXPERIMENTAL

Samples of alkaline earth metal oxides were obtained by calcination of nitrates (pure for analysis grade) in air at 1173 K. The catalytic activity was measured in a flow reactor, as described in (ref. 6). The reaction mixture composition was 80% CH_4 and 20% O_2. The concentration of base sites was determined by the adsorption of benzoic acid (ref. 6). IR spectra were registered at 293 K on a Bruker-113 V FIR spectrometer (ref. 8). The isotope exchange was examined directly in the course of oxidative dehydrodimerization, the reaction mixture composition being 45% CH_4, 45% CD_4 and 10% O_2.

RESULTS AND DISCUSSION

A study of the dependence of rates of methane conversion and C_2^+ products formation on concentration of base sites (see Fig. 1) has indicated that both the total reaction rate and the rate of oxidative dehydrodimerization of methane tend to increase with increasing concentration of base sites on catalyst surface.

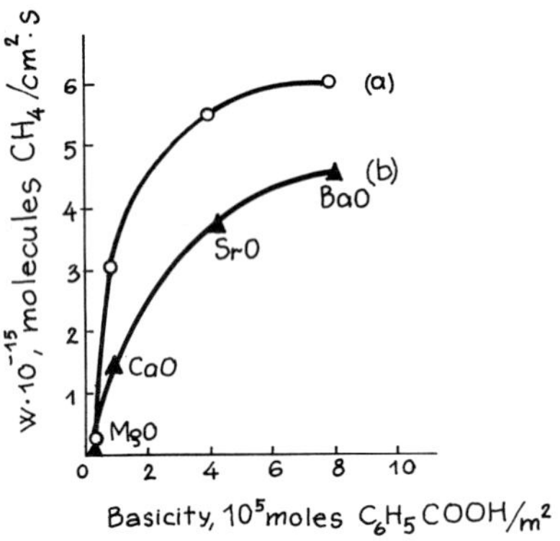

Fig. 1. Total rate of methane conversion (a) and rate of formation of C_2^+ hydrocarbons (b) vs. concentrations of base sites on the catalyst surface (T = 1153 K, GHSV = 18000 h^{-1}). Pretreatment: 1173 K, 4 h, air.

However, for magnesium oxide the rate of oxidative conversion is found to be lower than one might anticipate proceeding from the concentration of base sites on this catalyst.

A similar dependence has been found earlier for oxidative ammonolysis of propane on catalysts containing base sites (ref. 9). This result has made it possible to suggest a heterolytic dissociation of the C-H bond on base catalysts. Support for this conclusion comes also from data on deuterium-hydrogen exchange in molecules of lower paraffins on solid bases (refs. 10,11). The literature reports on some attempts to detect methane activation on solid bases by direct physical methods. The authors of (ref. 12) have succeeded in detecting the propyle heterolytic dissociation on MgO at room temperature. However, they have failed to observe methane chemisorption under the conditions employed (ref. 12).

To detect methane chemisorption on solid bases, we used magnesium oxide with a large surface area (200 m^2/g) and employed FIR spectroscopy which allows a drastic increase in sensitivity of experiment. The oxidative dehydrodimerization reaction is carried out at high temperatures with an excess reductant which should lead to the appearance of low-coordination sites on the surface of the oxide. With this in mind, prior to experiments magnesium oxide was outgassed at 1000 K when, according to UV diffuse reflectance spectra (ref. 12), the surface contains five-, four- and three-coordinated sites (Mg^{2+}_{5c} O^{2-}_{5c}, Mg^{2+}_{4c} O^{2-}_{4c} and $Mg^{2+}_{3c}O^{2-}_{3c}$, respectively). Propylene adsorption on such a sample at room temperature produced IR absorption bands (a.b.) corresponding to OH groups (3650 cm^{-1}) and σ- (1620 and 950 cm^{-1}) and π-allyl complexes (1550 and 1250 cm^{-1}) (see Fig. 2a). π-Allyl complexes of this type were first found by Kokes (ref. 13) and attributed to the anion type. Taking into account data on variations of UV diffuse reflectance spectra after propylene adsorption on magnesium oxide (ref. 14) we may conclude that in our case the adsorption occurs on low-coordination (4- and 3-coordinated) sites producing OH groups and an organometallic species of magnesium.

We have failed to observe dissociative adsorption of methane on this magnesium oxide sample at room temperature, which seems to be due to a stronger and less polar C-H bond in methane than in propylene. However, as the temperature of adsorption was raised to 573 K a.b. corresponding to OH groups (3600 cm^{-1}) and new a.b. in the region of stretching vibrations of the C-H bond (2940 and 2980 cm^{-1}) (see Fig. 2b) appeared in the IR spectrum. Note that beginning with these temperatures the H-D isotope exchange in methane molecules on magnesium oxide is typically observed (ref. 10). A comparison of the spectrum obtained with a.b. ascribed to metal-methyl groups (2920, 2990 cm^{-1} for adsorption of $(CH_3)_2SnCl_2$ on MgO) and oximethyl groups (2800, 2860 and

2920 cm^{-1} for adsorption of CH_3OH on MgO) suggests that in our case the heterolytic dissociative adsorption of methane producing hydroxyl and magnesium-methyl groups takes place (ref. 8).

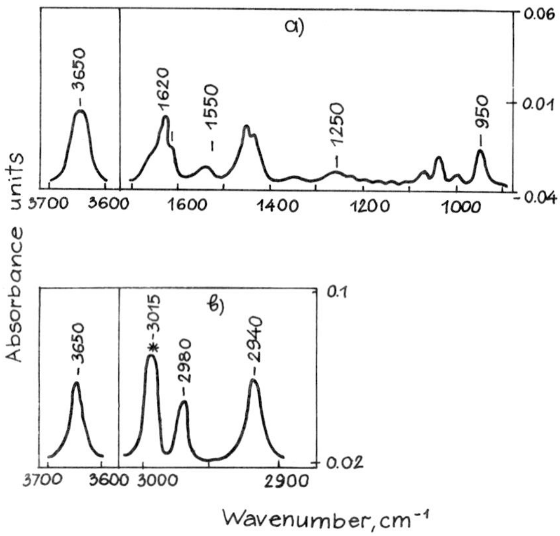

Fig. 2. IR spectra of hydrocarbons adsorbed on MgO. a - propylene adsorption at 300 K (subtracted background of MgO); b - methane adsorption at 573 K (subtracted background of MgO and partially compensated gas phase); * the band corresponding to the gas phase.

Probably, the adsorption occurs on low-coordination sites. As shown in (ref. 15), after methane adsorption on magnesium oxide containing low-coordination sites (treated under vacuum at 1123 K) the oxygen adsorption leads to the formation of radical ions O_2^-.

The authors of (ref. 15) (like those of refs. 12,14 who observed such an effect after propylene preadsorption) have made a conclusion about the presence on the surface of an anion form of a hydrocarbon residue from which an electron is transferred into the oxygen molecule.

It should be noted that the exposure of the sample in methane at 573 K gives rise to a 3085 cm^{-1} band characteristic of C-H vibrations at a double bond (see Fig. 2b). This may be taken as evidence for the formation of dimerization products (ethylene) in conditions of methane adsorption on this sample.

Thus, we have observed experimentally the heterolytic activation of methane on magnesium oxide containing low-coordination sites and suggested a possible role of this process in oxidative dehydrodimerization of methane.

As shown in (ref. 10), due to a high temperature treatment of magnesium oxide under vacuum, which results in the appearance of low-coordination sites, the catalyst reveals activity toward H-D isotope exchange of methane. However, oxygen adsorption leads to complete deactivation of the sample, most probably, by destroying low coordination sites. However, as has already been mentioned, heterolytic activation of methane occurs only on low coordination sites. To verify whether low coordination sites which can activate methane are retained during the course of oxidative dehydrodimerization in a methane-oxygen mixture, we have studied the CH_4-CD_4 isotope exchange directly in the process.

The results obtained are listed in Table 1. As can be seen in the table, on magnesium oxide a rather fast isotope exchange occurs, which may evidence for high concentration of low coordination sites on its surface.

TABLE 1
Comparison of rates of oxidative dehydrodimerization (W_d) and CH_4-CD_4 isotope exchange (W_o) in catalysis conditions on alkaline earth metal oxides (T = 1073 K, GHSV = 22500 h^{-1})

Catalyst	$W \cdot 10^{-14}$ $\frac{molec.}{cm^2 \cdot s}$		W_o+W_d	Basicity 10^6 mol C_6H_5COOH m^{-2}	$\frac{W_d}{W_o+W_d}$
	W_o	W_d			
MgO	1.9	0.09	1.99	2.14	0.045
CaO	12	6.5	18.5	7.94	0.35

As has already been mentioned, the isotope exchange in methane on magnesium oxide begins at fairly low temperatures (ref. 10). From this fact it follows that the primary heterolytic activation of methane is relatively fast. The total rate of methane activation (defined as a sum of observed rates of exchange and dimerization) increases with increasing basicity of the oxide, which is in agreement with our conclusion about heterolytic activation of methane at catalyst base sites. At the same time, the rate ratio of exchange and dimerization is different for different oxides (see Table 1).

A simultaneous occurrence of exchange and dimerization indicates that the rate of the primary activation of methane is high enough to provide the both reactions.

The general scheme of the activation process may be as follows:

$$\begin{array}{ccccccc} CH_4 & + & Me^{2+}O^{2-} & \underset{2}{\overset{1}{\rightleftharpoons}} & Me^{2+}\text{-}CH_3^- & + & O^{2-}\text{-}H^+ \\ & & & & \downarrow 3 & & \\ & & & & CH_3^{\bullet} & & \\ & & & & \downarrow & & \\ & & & & \text{dimer} & & \end{array}$$

It can be supposed that part of metal-methyl groups transforms to methyl radical producing the dimer and the remainder part is reversibly desorbed which leads eventually to the isotope exchange.

A slow step of the dehydrodimerization reaction may be the step of rupture of the metal-methyl bond. The energies of $Mg\text{-}CH_3$ binding for magnesium ions with different coordination numbers abstracted from (ref. 16) are listed in Table 2. The energies of the metal-methyl bond on low-coordination (3- and 4-coordinated) magnesium ions are close to activation energies of dimerization on magnesium oxide-based catalysts (ca. 200 kJ/mol) (ref. 4).

TABLE 2
Energies of the bond rupture in the $Mg\text{-}CH_3$ group vs. coordination number for magnesium (abstracted from ref. 16)

Coordination number	3	4	5
Binding energy, kJ/mol	288.0	192.7	153.4

The homolytic rupture of the metal-methyl bond is accompanied by an electron transfer from the methyl group into the catalyst. This transfer can be facilitated by acceptor sites of the catalyst (refs. 5,17). Calcium oxide is known to possess hole conductivity even at very low oxygen pressures (ca. 10^{-2} Torr) (ref. 18). Due to this property a large proportion of metal-methyl groups will be consumed at step 3, which may result in a higher ratio of the dimerization rate to the total rate of methane activation on CaO in comparison with magnesium oxide (see Table 1).

CONCLUSIONS

The data obtained allow us to conclude that the heterolytic mechanism of methane activation during the oxidative dehydrodimerization process on base catalysts is more probable than a homolytic one involving surface radical ions of oxygen.

1. The mechanism of methane activation with participation of radical ions O^- should lead to a considerable enhancement of complete oxidation processes, which has been recently observed at temperatures of oxidative dehydrodimerization on a series of magnesium-containing catalysts (ref. 19).

2. The homolytic mechanism of activation involving a radical abstraction of the hydrogen atom from methane by O^- should have a low activation energy. The high observed activation energy of methane dimerization on Li/MgO was explained (ref. 4) by that the rate-determining step is that of regeneration of radical sites Li^+O^-. However, as has been shown recently in (ref. 20), this explanation is in conflict with kinetic data on isotope effect for this reaction.

3. Most catalysts of oxidative dehydrodimerization of methane are solid bases. On the oxides containing low coordination sites the heterolytic activation of methane occurs very effectively. (It has been reported recently (ref. 21) that doping of magnesium oxide with lithium increases the number of low coordination sites). Thus, there is a correlation between the number of base sites and the rate of dimerization. By assuming that the rate-determining step is the decomposition of the metal-methyl species found in this work it is possible to explain the observed high energy of activation of oxidative dehydrodimerization.

REFERENCES

1 J.A.S.P. Carreiro, M. Baerns, Catalytic conversion of methane by oxidative coupling to C_{2+} hydrocarbons, React. Kinet. Catal. Lett., 35 (1987) 349.
2 K. Otsuka, K. Jinno, A. Morikawa, Active and selective catalysts for the synthesis of C_2H_4 and C_2H_6 vis oxidative coupling of methane, J. Catal., 100 (1986) 353.
3 T. Moriyama, N. Takasaki, E. Iwamatsu, K. Aika, Oxidative dimerization of methane over promoted magnesium oxide catalysts, Chem. Lett., (1986) 1165.
4 O.J. Driscoll, W. Martir, J.-X. Wang, J.H. Lunsford, Formation of gas-phase methyl radicals over MgO, J. Am. Chem.Soc., 107

(1985) 58.

5 V.D. Sokolovskii, Some principles of choosing catalysts for selective conversions of organic compounds at C-H bonds, React. Kinet. Catal. Lett., 35 (1987) 337.

6 O.V. Buevskaya, A.I. Suleimanov, S.M. Aliev, V.D. Sokolovskii, Activation of hydrocarbon in the oxidative dimerization of methane over alkaline earth oxides, React. Kinet. Catal. Lett. 33 (1987) 223.

7 G.A. Martin, C. Mirodatos, Evidence of carbene formation in oxidative coupling of methane over lithiumpromoted magnesium oxide, J. Chem. Soc. Chem. Commun., 1393 (1987).

8 A.A. Davydov, A.A. Budneva, S.M. Aliev, V.D. Sokolovskii, IR-spectra of methane adsorbed on MgO, React. Kinet. Catal. Lett. 36 (1988) 491.

9 S.Yu. Burylin, Z.G. Osipova, V.D. Sokolovskii, Kinet. Katal., Effect of C_2H_5Br on the catalytic oxidative ammonolysis of propane, Kinet. Katal., 24 (1983) 639.

10 M. Utiyama, H. Hattori, K. Tanabe, Exchange reaction of methane with deuterium over solid base catalysts, J. Catal., 53 (1978) 237.

11 R. Bird, C. Kemball, H.F. Leach, Reactions of alkanes with deuterium on lanthan in the temperature range 570 to 720 K, J. Catal., 107 (1987) 424.

12 E. Garrone, F.S. Stone, The behaviour of MgO as a Brönsted base in chemisorption and surface processes, Proc. 8th Int. Congr.on Catalysis, Verlag Chemie, Weinheim, 1984, v. 3, p. 441.

13 R.J. Kokes, Anionic intermediates in surface processes leading to O_2 formation on magnesium oxide, in: Catalysis. Progress in Research, Plenum Press, London, New York, 1973, p.75.

14 E. Garrone, A. Zecchina, F.S. Stone, The nature of active sites, J. Catal., 62 (1980) 396.

15 T. Ito, T. Tashiro, T. Watanabe, K. Toi, I. Ikemoto, Actovation of methane on the MgO surface at low temperatures, Chem. Lett., (1987) 1723.

16 N.U. Zhanpeisov, A.G. Pelmentschikov, G.M. Zhidomirov, Cluster quantum-chemical study of the interaction of molecules with MgO surface. Dissociative chemisorption of H_2, CH_4, C_2H_6, Kinet. Katal. (in press).

17 A.I. Suleimanov, E.G. Ismailov, S.M. Aliev, V.D. Sokolovskii, Contribution of one-electron acceptor centers to oxidative dimerization of methane, React. Kinet. Catal. Lett., 34 (1987) 51.

18 K. Hauffe, Reaktionen in und an Festen Staffen, Springer Verlag, Berlin, 1955, v. 1.

19 G.J. Hutchings, M.S. Scurrell, J.R. Woodhouse, The role of surface O^- in the selective oxidation of methane, J. Chem. Soc., Chem. Commun., (1987) 1388.

20 N.W. Cant, C.A. Lukey, P.F. Nelson, R.J. Tyler, The rate controlling step in the oxidative coupling of methane over a lithium-promoted magnesium oxide catalyst, J. Chem. Soc., Chem. Commun., (1988) 766.

21 M. Anpo, M. Sunamoto, T. Doi, I. Matsuura, Oxidative coupling of methane over ultrafine crystalline MgO doped with Li. Role of lower coordinative surface sites produced by Li-doping, Chem. Lett., (1988) 701.

CORTES CORBERAN, V. (Institut Catalisis y Petroleoquimica, Spain): As the basicity/acidity properties of metal oxides depend on the temperature, does it make sense to compare catalytic activity measurements at high temperature with number of basic sites as determined at room temperature? Would you expect that the observed overall tendency (activity increases as basic sites number increases) can be extrapolated?

SOKOLOVSKII V.D. (Institute of Catalysis, USSR): The observed concentration of basic centers depend not so much on the temperature of measurement (if adsorption of acid is quick and irreversible), as on the temperature of preparative treatment of specimens. The treatment of specimens during our experiments before measuring a catalytic activity and basicity was identical.

CORTES CORBERAN,V. (Institut Catalisis y Petroleoquimica, Spain): You have used only one probe molecule to correlate activity with basicity. Does it mean that centers with any basic strength are equally active and must be taken into account or would you expect that only very strong basic centers will be active?

SOKOLOVSKII V.D. (Institute of Catalysis, USSR): In these methodics we determined the basic centers concentration only, but not their strength. Benzoic acid, used as a test molecule, is rather weak (= 4.2) and so strong and medium strong centers are determined with its help. We suppose that activation of such an inert molecule as the methane, must be conducted on the strong basic centers.

BUSCO GUIDO (Istituto di Chimica, Italy): The Mg-CH_3 groups for cite must be responsible also for bands in the region 1500-1300 cm^{-1}, as well as for rocking modes at lower frequencies. However, if the CH bands you detect are due to oxygen-containing species, several characteristic bands would be observable in the region below 1800 cm^{-1}. Have you also investigated the low-frequency region to confirm your argument?

SOKOLOVSKII V.D. (Institute of Catalysis, USSR): Absorption bands were observed by us in the region of deformation bands of C-H bonds 1800 cm^{-1}. However, as these bands are not characteristic (ref. 1) for identification of Me-CH_3 groups, the region of changes in valence.

1 D.C. McKeam, G. McQuillau, I. Torto, A.R. Morrison, J. Molec. Struct., 141 (1986) 457.

OMMEN G.V. (University of Turent , The Netherlands): By a special treatment of the MgO surface (more or less reduceit) you measure Mg-CH_3 bands at 300°C. These organometallic type cannot be present during methane coupling reactions at 700°C and in the presence of oxygen. In the IR spectra of the same material treated under real coupling conditions we measure only oxygenates, possible precursors of the total oxidation. These species are more stable and can therefore be measured under these conditions. My question is, do you suppose that the measured Mg-CH_3 bands play an important role during oxidative coupling of CH_4 at 700°C in the presence of O_2?

SOKOLOVSKII V.D. (Institute of Catalysis, USSR): We also think that there is reason to believe that at 700°C under reaction conditions, the groups Me-CH_3 must be unstable. If not, they could not have acted as intermediates providing the reaction proceeding. Such intermediates must be formed and decay quickly in conditions of reaction. To fix their availability we have used a lower temperature (pre-catalysis conditions), at which these forms are stable enough to be recorded.

G. Centi and F. Trifiro' (Editors), *New Developments in Selective Oxidation*

SERS-IN SITU STUDY OF THE SURFACE SPECIES FORMED IN METHANE OXIDATIVE COUPLING

A.A.KADUSHIN, O.V.KRYLOV, S.E.PLATE, YU.P.TULENIN, V.A.SELEZNEV[1], A.V.BOBROV and YA.M.KIMELFELD[2]

[1]Institute of Chemical Physics of the U.S.S.R. Academy of Sciences, Kosygin str.4, 117334, Moscow, U.S.S.R.

[2]Institute of Spectroscopy of the U.S.S.R. Academy of Sciences, 142092, Troizk, Moscow Region

SUMMARY
The existence of stationary concentrations of CH_3- and probably CH_2-surface species bonded with carbon and noncarbon surface sites was shown on Sm_2O_3/MgO catalyst of methane oxidative coupling using SERS-in situ method in 670-970 K temperature interval. The CH_3- and CH_2-species could be intermediates in the formation of C_2-hydrocarbons (ethane and ethylene).

INTRODUCTION

Methane oxidative coupling to form higher hydrocarbons is gaining an increasing interest after pioneering work of Keller and Bhasin [1]. Recently, a large number of catalysts manifesting a high activity and selectivity in the reaction were developed.

However, a lack of information about the mechanism of the reaction and the nature of intermediate species still exists in the literature.

Lunsford et al. detected the formation of CH_3 radicals in the gas phase over Li/MgO [2] and La_2O_3 [3] using ESR matrix isolation method. Martin and Mirodatos [4] have postulated carben (CH_2-) intermediate species formation over Li/MgO on the basis of chemical evidence of cyclopropane formation by introduction of C_2H_4 into the reaction mixture. Nelson et al. [5] studied the oxidation of equimolar CH_4+CD_4 mixture over Li/MgO at 750°C and did not find any notable isotopic H-D exchange in methane molecules. The main C_2-products were C_2H_6, CH_3-CD_3 and C_2D_6. On the basis of these data the authors [5] concluded, that the ethane formation takes place as a result of CH_3 and CD_3-radicals recombination in the gas phase.

Surface enhanced Raman scattering (SERS) was used in our previous papers [6,7] for the investigation of CH_4 and O_2 interaction with Ni,Cd,Pb and Mg films at 77 K.

In the present work Raman spectra of the surface species were measured at 300-970 K during the interaction of CH_4 and O_2 on MgO and Sm_2O_3/MgO (1.0 wt. % Sm_2O_3) which are active in the CH_4 oxidative coupling [8].

METHODS

Granulated catalysts samples of fraction 0.25-0.5 mm were packed in a quartz reactor with small windows. Raman spectra were excited by a beam of an argon-ion laser at 4880 A and 150 mw and registered by a double monochromator in photon counting regime.

RESULTS AND DISCUSSION

Spectroscopic measurements revealed that the interaction of the CH_4+O_2 mixture with catalyst samples in the temperature range 670-970 K leads to appearance in the Raman spectra of several bands at 1190, 1290, 1380 and 1480 cm^{-1} (Fig.1, spectrum 1)

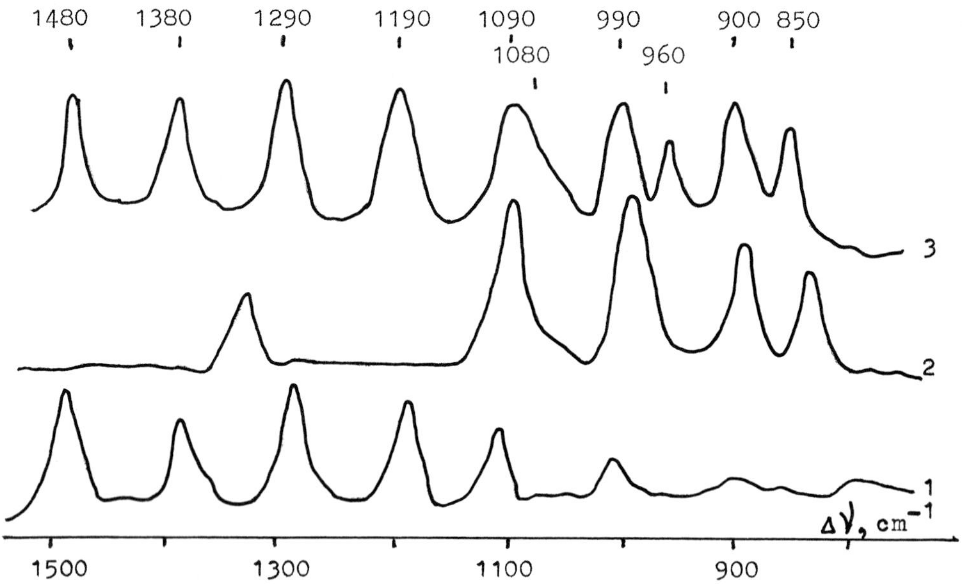

Fig.1. Raman spectra of the surface species formed during methane oxidative coupling over Sm_2O_3/MgO at 970° K. Feed mixtures: 1 - CH_4+ O_2 +He; 2 - CD_4+ O_2+ He and 3 - CH_4+ CD_4+ O_2 +He. Contact time (τ) 0.7 S.

This region is well characterised in the literature and these bands can be related to bending vibrations of CH_2 (1180 and 1290 cm^{-1}) and CH_3 (1380 and 1480 cm^{-1}) surface species. The quartet of these bands appears in the spectrum at the temperatures higher than 760 K when oxidative coupling begins to proceed with a noticeable rate. This was also confirmed by simultaneous analysis of the reaction products (Table 1):

TABLE 1
Catalytic properties of the 1% wt. Sm_2O_3/MgO
Feed mixture (% vol): CH_4-10.0; O_2-2.9; He-87.1; τ=0.7s

Temperature K	Conversion, %		Selectivity, %			
	CH_4	O_2	C_2H_6	C_2H_4	CO_2	CO
770	2.0	12.2	-	-	50.5	49.5
870	11.7	65.7	3.7	-	54.9	41.4
970	15.9	80.0	12.0	6.7	48.2	33.2

It should be noted that these bands can exist only in the presence of reaction mixture and disappear in an argon flow. This evidences in favour of an equilibrium between the gas phase and hydrocarbon species on the catalyst surface. This means that the bands observed in the Raman spectra during the reaction prove the existence of surface species stationary concentration. In connection with this it is interesting to note the work of Ekstrom and Lapszewicz [9] where unusually high adsorption of isotopically labelled methane molecules was observed on Sm_2O_3 in the same conditions.

Similar measurements have been made with CD_4 (spectrum 2). The isotopic band shift (Δ=300-370cm^{-1}) is in agreement with the interpretation of the hydrocarbon spectrum.

Spectrum 3 was obtained when an equimolar mixture of CH_4+CD_4 with O_2 was passed through the catalyst. A new band at 960 cm^{-1} and an unresolved low frequency shoulder of 1080 cm^{-1} at 1090cm^{-1} band appears in the spectrum due to H-D exchange between the hydrocarbon surface species. The Raman-spectra of ethane-O_2 and ethylene-O_2 mixtures, measured in the same conditions, differ from the spectrum of methane-O_2 mixture.

INTERPRETATION OF SPECTRA

The 1480 and 1380 cm^{-1} bands are ascribed to antisymmetric and symmetric bending vibrations of CH_3-groups, respectively. The positions of these bands indicate that the CH_3-groups are bonded with carbon atoms of the surface.

As to the interpretation of 1290 and 1190 cm^{-1} bands at least three possibilities can be proposed:

1. These bands can be assigned to internal and external bending vibrations of CH_2-groups in bridge structures

$$-M \diagup^{CH_2} \diagdown M-$$

, where M - is a noncarbon surface site. The positions of these bands in the spectrum are considerably lower, than those in hydrocarbon spectra. Similar spectra were observed by Chang et al. [10], who studied $FeCH_2$ and N_2FeCH_2 in argon and nitrogen matrixes by FTIR matrix isolation spectroscopy.

2. These bands can be assigned to the same vibrations of CH_2 -groups (as in point 1) in ethylene-like structure.

3. These bands may be ascribed to antisymmetric and symmetric bending vibration of CH_3-groups, bonded with noncarbon surface sites. Similar IR-spectra were observed by Billups et al. [11] during activation of methane with photoexcited atoms of some transition metals (matrix isolation method).

Unfortunately, experiments with CH_4+CD_4 mixture do not exclude any mentioned possibilities. We could not measure the C-H stretching vibrations due to large emission by the sample at high temperatures.

The 1080 and 960 cm^{-1} bands can be related to internal and external bending vibrations in -CHD, $-CH_2D$ or $-CD_2H$-groups bonded with noncarbon surface sites. For more detailed interpretation a further study is needed using other isotopically labelled methane molecules. Therefore our data do not allow at the present time to conclude finally that CH_2 species exist during the oxidative coupling of methane.

Some band intensities in the "oxygen" region change considerably, however, their detailed interpretation needs further research. We compared Raman spectra of surface species measured at 970 K for Sm_2O_3/MgO and MgO. Spectra for both of the samples are similar but in the case of MgO after some hours of the reaction a new 1570 cm^{-1} band appeared typical for C=C double bond. This fact points to deeper surface transformations, for instance, to dimerization of CH_2-fragments.

The SERS (enhance coefficient ~100) is untrivial for nonconducting oxide systems. Usually the SERS having an enhance coefficient up to 10^5-10^6 is observed on metal surfaces [12], although recently this effect was registered for colloidal α-Fe_2O_3 [13].

The above mentioned results are obtained for the first time and future systematic work is needed for detailed interpretation of the Raman spectra and their connection with the mechanism of methane oxidative coupling.

However, even now it is possible to make some conclusions:

1. The possibility is shown of Raman spectra measurements for granulated oxide catalysts (MgO, Sm_2O_3/MgO) in conditions of catalysis process at temperatures up to 1000 K. The theory has to explain the nature of the enhancement effect in the case of oxides and at high temperatures.
2. The existence is shown of stationary concentration of CH_3-fragments bonded with carbon atoms of catalyst surface and CH_3- or CH_2-fragments bonded with noncarbon sites of catalyst surface in conditions of methane oxidative coupling at 970 K.
3. A notable isotopic H-D exchange between surface hydrocarbon species is shown at 16% methane conversion.
4. The obtained results indicate a principally new level of Raman spectroscopy for the use in study of high temperature catalytic processes in situ.

REFERENCES

1 G.E.Keller and M.M. Bhasin, J.Catal., 73(1982) 9-19.
2 T. Ito, J.-X. Wang, C.-H. Lin, J.H. Lunsford, J.Am.Chem.Soc., 107(1985) 5062-68.
3 C.-H. Lin, K.D. Campbell, J.-X. Wang, J.H. Lunsford, J.Phys. Chem., 90(1986) 534-537.
4 G.-A. Martin, C. Mirodatos, J.Chem.Soc.,Chem.Commun., 18 (1987) 1393-94.
5 P.F. Nelson, C.A. Lukey, N.W. Cant, J.Phys.Chem., 92 (1988) 6176-79.
6 S.E. Plate, A.V. Bobrov, A.A. Kadushin, Ya.M. Kimelfeld, Kinetika i Kataliz, XXVII (1986) 495-497 (RUss).
7 A.B.Bobrov, S.E. Plate, Ya.M. Kimelfeld, A.A.Kadushin, XVIIIth European Congress on Molecular Spectroscopy, Amsterdam, August 30-September 4, 1987, Abstracts, P.275.
8 V.H. Korchak, A.A. Kadushin, Yu.P. Tulenin, V.A. Seleznev, Tezisy dokladov 6 konferenzii po okislitelnomu geterogennomu katalizu, Baku, November 15-17, 1988, pp.264-265 (Russ).
9 A.Ekstrom, J.A.Lapszewicz, J.Chem.Soc.,Chem.Commun.,12 (1988) 747-749.
10 S.C. Chang, R.H. Hauge, Z.H. Kafafi, J.L. Margrave, W.E. Billups, J.Am.Chem.Soc., 110(1988) 7975-80.
11 W.E. Billups, M.M. Kanarski, R.H. Hauge, J.L. Margrave, J.Am. Chem.Soc., 102(1980) 7393-94.

12 M. Fleischmann, P.J.Hendra, A.J. McQuillan, Chem.Phys.Letters, 26(1974) 163-166.
13 P. Zhang, Y. Wang, T. He, B. Zhang, X. Wang, H. Xen, F Liu, Chem.Phys.Letters, 153(1988), 215-218.

C.G. VAYENAS' (University of Patras, Greece) : It is surprising that SERS spectra have been obtained at temperatures up to 970 K on an oxide surface. Did you study the temperature dependence of the intensity of the SERS bands?

A.A. KADUSHIN' (Institute of Chemical Physics of the USSR Academy of Sciences, Moscow, USSR): The SERS is really untrivial for non-conducting oxide systems. Nevertheless we quite definitely established the presence of this effect in a wide temperature range. It may be, for example, supposed that the microclusters of carbon produced upon partial methane oxidation can be just those species which possess the metallic conductivity responsible for the appearance of bands in the spectra. These bands appear in the spectrum at 670 K and their intensity increases up to 970 K. No special investigation of the temperature dependence of these bands has been carried out.

J.R.H. ROSS (University of Twente, The Netherlands) : Is there any chance that gas-phase methane species can contribute to the spectra under your reaction conditions? Under similar conditions using Li/MgO catalysts, we can find no evidence with FTIR for anything other than oxygen-containing species on the catalyst surface (J.G. van Ommer et al., unpublished results).

A.A.KADUSHIN (Institute of Chemical Physics of the USSR Academy of Sciences, Moscow, USSR) : We did not observe in the spectrum any bands of gaseous methane, In the temperature interval 670-870 K we observed a series of bands in the range 700-1700 cm^{-1} which could be assigned to oxygen-containing surface species. But at higher temperatures these bands disappear.

In Situ Studies of the Oxidative Coupling of Methane Over Li-Ni-O Catalysts

I. J. Pickering, J. M. Thomas and P. J. Maddox

Davy Faraday Research Laboratory,
The Royal Institution of Great Britain,
21 Albemarle Street, London W1X 4BS, UK

Abstract

We describe a comparative *in situ* X-ray diffraction study of the lithium nickel oxide catalyst in the presence and absence of added gaseous oxygen, The results reveal several interesting features and in particular pinpoint the involvement of crystallographic phases implicated in the conversion of methane to various distinct gaseous products.

Introduction

Given that there is an abundance of naturally occurring methane, the question arises as how best to convert it to liquid fuel or other feedstocks, such as ethylene, that are more readily usable in the chemical industry. Two well-known methods already exist: (i) partial oxidation to synthesis gas (CO + H_2) followed by Fischer-Tropsch conversion, and (ii) steam-cracking to yield more reactive hydrocarbons and other useful by-products. These may well turn out to be the most practical way forward industrially, but they may not be as economically attractive as other, more subtle conversions to which methane can be subjected. One such conversion, the oxidative coupling of methane, is the focus of our attention here.

We report below on laboratory *in situ* X-ray diffraction studies of the conversion of methane to C_2 hydrocarbons over a lithium nickel oxide catalyst, both in the presence and absence of added gaseous oxygen (catalytic and non-catalytic reactions respectively). The latter proceeds by the extraction of the structural oxygen (of the Li-Ni-O system) which is also known (ref. 1) to be implicated in the catalysed oxidative coupling. In parallel with the gas chromatographic analysis of reaction products we recorded, under *in situ* conditions, the X-ray powder diffraction patterns of the solid oxide in both cases.

Experimental

The lithium nickel oxide catalyst used in the experiments was prepared by a solid-state reaction of an intimate mixture of Li_2CO_3 and NiO at 800°C in air. Atomic absorption spectroscopy showed (ref. 2) the composition to be $Li_{0.45}Ni_{0.55}O$. X-ray powder diffraction patterns showed the structure to be rhombohedral, R-3m, with some ordering of the lithium and nickel ions on alternate close-packed layers.

The *in situ* catalysis was carried out in a specially constructed reaction cell (ref. 3) which facilitates simultaneous monitoring of both diffraction patterns and gaseous products. The catalyst was placed in a sintered quartz sample holder inside this cell, and the reactive gases were passed over the sample in a tubular fashion, the outlet gases being analysed by gas chromatography. The cell was attached to a Siemens D500 diffractometer fitted with a rotating anode source and scintillation counter, permitting rapid characterization by X-ray diffraction. The conditions for the two experiments are shown in table 1:

Table 1. Experimental conditions for *in situ* catalytic and non-catalytic reactions.

	Mass of catalyst (grams)	Catalyst temperature (degrees C)	Gas composition (%)			Flowrate (ml/min)
			Methane	Oxygen	Nitrogen	
Catalytic reaction	1.01	700	20	3	77	50
Gas-solid reaction	1.04	700	20	0	80	50

Results

The various regimes of catalysis determined during the catalytic run have already been described (ref. 1). Briefly, there is an initial regime of near-constant conversion of methane, with selectivity for C_2 hydrocarbons decreasing. Conversions and selectivities during the initial regime are summarised in table 2:

Table 2. Conversions and selectivities recorded during the initial regime of the catalytic reaction.

Time after start of reaction (min)	20	100	200	300
Percentage conversion of methane	7.9	8.2	8.5	8.9
Percentage selectivity for C2 hydrocarbons	62	54	45	36

During the second and third regime there is rapid change as the catalyst breaks down. The second regime yields CO_2 as the dominant product, and the major nickel-containing phase is NiO. During the third regime CO is produced and the catalyst has been reduced to nickel metal. We concentrate here on the initial catalytic regime and its comparison with the non-catalytic results.

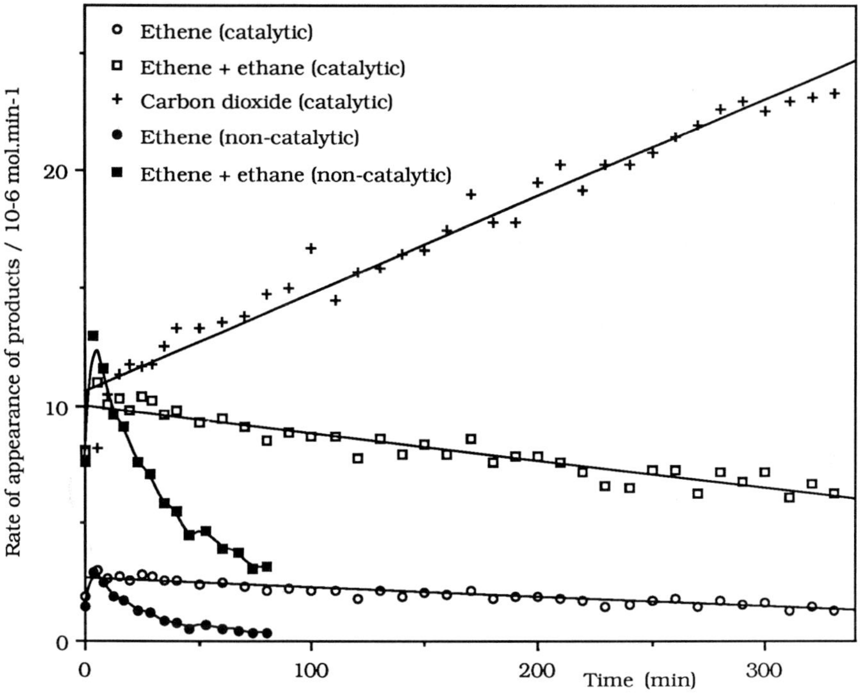

Figure 1. Rate of appearance of products for the catalytic reaction (open symbols and crosses) and the non-catalytic reaction (solid symbols). Carbon dioxide is negligible in the latter case.

To compare the reactions in the presence and absence of gaseous oxygen, it is instructive to examine the rate of appearance of products with time (figure 1). It is apparent that, in the presence of O_2, the rates of production of total C_2 and of ethene in particular are fairly constant, decreasing slowly over a period of some hours, whereas in the absence of added gaseous oxygen the rate of production of C_2s decays rapidly, dropping to less than $5x10^{-6}$ $mol.min^{-1}$ during the first 50 min. This decay is to be expected since, in this case, there is no replenishment of oxygen in the system. It is, however, noteworthy that the initial rates of production are very similar for both cases. Conversely the rate of production of CO_2 is very different, being significant in the case of the catalytic reaction, and essentially absent during the initial period of the gas-solid reaction.

The X-ray diffraction patterns for the entirety of the experiments are shown in figure 2. It is evident that, in the presence of oxygen, the initial phase remains essentially unchanged for 330 min; in the absence of oxygen the initial, ordered phase breaks down rapidly, as can be seen by the disappearance of the superlattice peaks at two-theta values of 18.5^o (003) and 36.2^o (101). This phase is replaced by one in which the lithium and nickel cations are randomly distributed in a rock-salt structure (ref. 4). Other phases may be identified during the experiment: for example, the peak at 25.7° two-theta is due to the strongest line of Li_2NiO_2 (ref. 5), and those appearing at the end are due to an orientated form of Li_2CO_3.

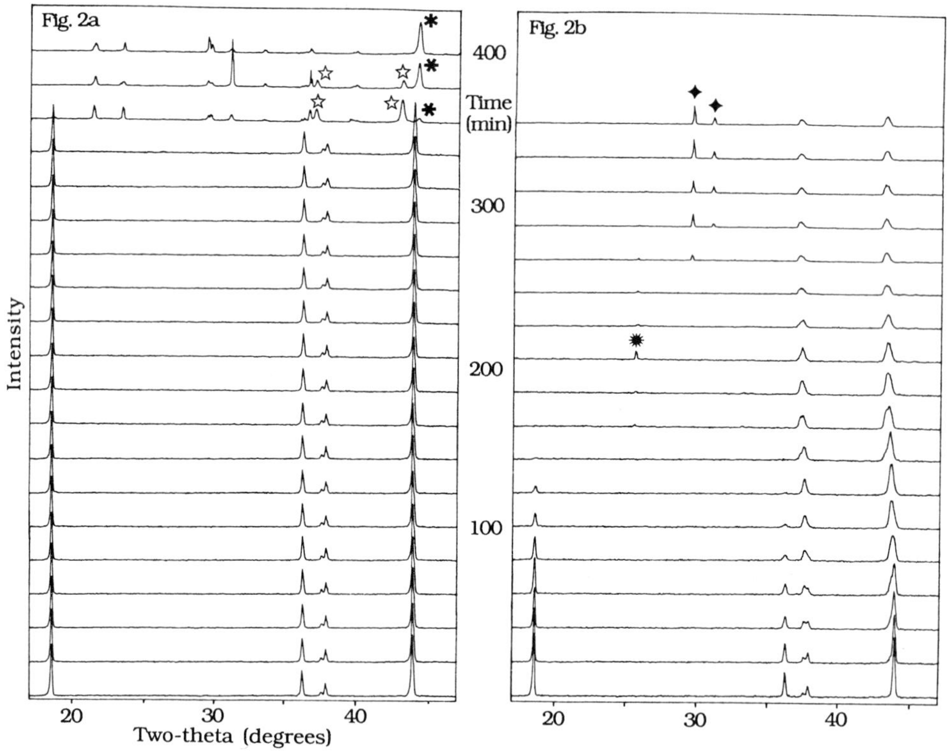

Figure 2. X-ray diffraction patterns for the duration of the experiments. Figure 2a shows those for the catalytic run: they remain essentially unchanged from the initial diffraction pattern, that of ordered Li-Ni-O, throughout the initial regime (about 330 minutes). There is subsequent rapid change, to yield first NiO (☆) and then nickel metal (✱), together with Li_2CO_3. Figure 2b depicts the diffraction patterns for the non-catalytic run: here the oxide decomposes much more quickly. ✸ is the strongest peak of Li_2NiO_2; ✦ are peaks due to a highly orientated form of Li_2CO_3.

Figure 3a shows how the unit-cell volume varies with time for both experiments. For the catalytic reaction the unit-cell volume is essentially invariant throughout the duration of the initial regime; in the gas-solid reaction it can be seen that the trend is to larger unit-cell volumes as time progresses. This trend suggests (ref. 6) phases with smaller lithium to nickel ratios. These unit-cell volumes reveal four Li-Ni-O phases: the initial one ordered, the other three random.

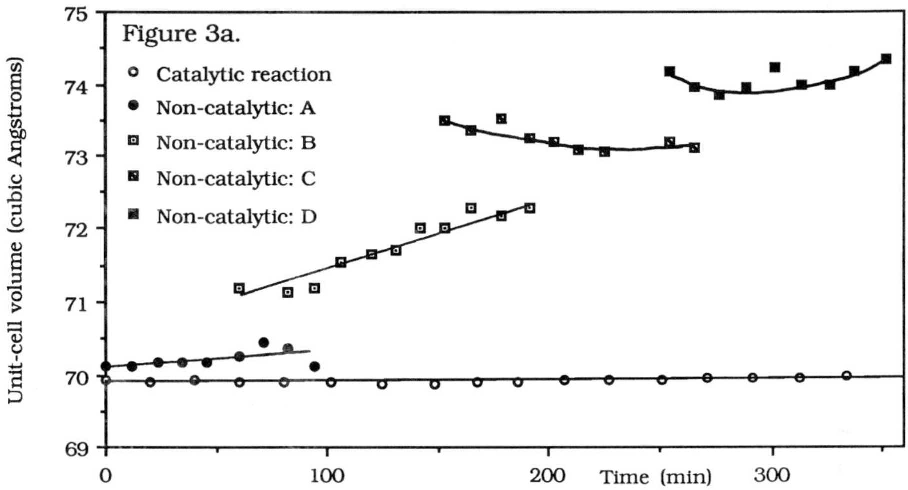

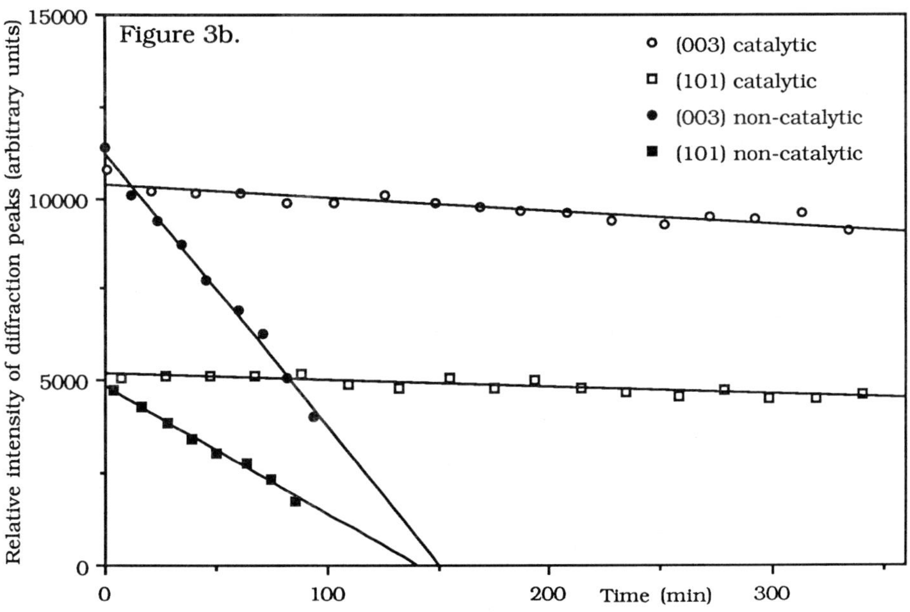

Figure 3a. The variation of unit-cell volume with time. Values are obtained from a lattice parameter refinement of the X-ray diffraction data,and are adjusted to be equivalent to the volume of 4 lattice motifs, *i.e.* the volume of the f.c.c. unit cell of the random phases or $^2/_3$ of the volume of the rhombohedral unit cell (with hexagonal setting).
Figure 3b. The variation with time of the intensity of two peaks of the initial phase of Li-Ni-O. The intensity is calculated as the integral area of a Pseudo-Voigt profile fitted to the diffraction data.

Figure 3b shows the intensity of two of the superlattice peaks as a function of time. Again we see the now familiar pattern that in the presence of gaseous oxygen the intensities of these peaks are largely unchanged, whereas in the absence of oxygen they decrease, this time in a linear fashion. This can be qualitatively linked to the amount of phase A present at a given time.

Discussion

From this comparison of catalytic and gas-solid reactions some interesting conclusions may be drawn about the catalytic reaction itself. In the initial stages of the reaction the rates of production of C_2 products are very similar for both conditions. This suggests that structural oxygen species is indeed responsible for the oxidation, as this is the only oxygen supply available in the gas-solid reaction. By a similar argument, the CO_2 which appears initially for the catalysis reaction is absent for the gas-solid reaction, and so this suggests that the CO_2 is due to oxidation by gaseous or adsorbed oxygen. These observations are in agreement with those of Otsuka (ref. 7-8), who also suggests that there are two distinct types of oxygen involved, just as there are in other selective oxidations of hydrocarbons (ref. 9).

If the graphs of figures 1, 3a and 3b and the diffraction patterns of figure 2 are examined, an interesting trend can be observed. The values of C_2 production, of unit cell volumes and of peak intensities, and the appearance of the diffraction patterns observed at the beginning of the two experiments are very similar, after which the values for the gas-solid reaction change rapidly, the catalytic values changing much more slowly. It is observed that the conditions near the end of the initial regime of the catalysis run correspond with those in the gas-solid run at about 20-30 minutes, and thus the presence of oxygen is stabilising the initial high-lithium content phase and, thereby prolonging the initial high rate of C_2 production.

References

1 I. J. Pickering, P. J. Maddox and J. M. Thomas, 'Probing changes in the structure and performance of a lithium nickel oxide catalyst during the high-temperature oxidative coupling of methane by *in situ* X-ray diffraction', Angew. Chem., Adv. Mat., (1989) (in press).
2 L. D. Dyer, B. S. Borie, Jr. and G. P. Smith, 'Alkali metal-nickel oxides of the type $MNiO_2$', J. Am. Chem. Soc., **76** (1954) 1499-1503.
3 P. J. Maddox, J. Stachurski and J. M. Thomas, 'Probing structural changes during the onset of catalytic activity by *in situ* X-ray diffractometry', Cat. Lett., **1** (1988) 191-4.
4 J. Deren and M. Rekas, 'Physico-chemical studies of NiO-Li_2O system', Roczniki Chemii, Ann. Soc. Chim. Polonorum, **46** (1972) 1411-9.
5 V. H. Rieck and R. Hoppe, 'Ein neuses Oxoniccolat: Li_2NiO_2', Z. Anorg. Allg. Chem., **392(3)** (1972) 193-6.
6 J. B. Goodenough, D. G. Wickham and W. J. Croft, 'Some magnetic and crystallographic properties of the system $Li^+{}_xNi^{++}{}_{1-2x}Ni^{+++}{}_xO$', J. Phys. Chem. Solids, **5** (1958) 107-116.
7 M. Hatano and K. Otsuka, 'Alkali metal-doped transition metal oxides active for oxidative coupling of methane', Inorg. Chim. Acta, **146** (1988) 243-7.
8 M. Hatano and K. Otsuka, 'The oxidative coupling of methane on lithium nickelate(III)', J. Chem. Soc., Faraday Trans. 1, **85(2)** (1989) 199-206.
9 L. M. Kaliberdo, M. I. Tselyutuia, A. S. Vaabel, V. M. Kalikhman and B. N. Shvetsov, 'The role of the catalyst lattice oxygen and the gas-phase oxygen in the oxidative dehydrodimerisation of propene', Russ. J. Phys. Chem., **53(6)** (1979) 843-5.

G. Centi and F. Trifiro' (Editors), *New Developments in Selective Oxidation*

SELECTIVE OXIDATION OF METHANE TO FORMALDEHYDE AT AMBIENT PRESSURE: THE ROLE OF DOPANTS IN DETERMINING OPTIMUM CARRIER LOADING FOR THE MOLYBDENA/ SILICA SYSTEM

E. MacGiolla Coda and B.K. Hodnett
Department of Materials Engineering & Industrial Chemistry,
University of Limerick, Limerick, Ireland.

ABSTRACT

Conversion of methane to formaldehyde from a 5:1 mixture of methane and nitrous oxide was investigated between 500 and 600C for a range of molybdena catalysts supported on silica. This support was also modified by treatment with a range of additives (sodium, phosphorus or copper) prior to impregnation with ammonium heptamolybdate. An optimal molybdena loading could be defined for each support; modification of the support generally increased the optimal molybdena loading so that higher conversions could be achieved without loss of selectivity. These changes are rationalised in terms of a modification of the redox properties of the supported molybdena.

INTRODUCTION

Large reserves of natural gas have led in recent years to renewed interest in C_1 chemistry (ref.1). To date this subject has been dominated by production processes involving synthesis gas. Since the mid-1980's a great deal of interest has been aroused in methane coupling to ethane and ethylene, and such has been the interest and progress made in this area that many commercial plants are already being envisaged (refs 2,3).

Alternatives to these two processes are few in number but there has always been an interest in direct oxidation of methane to methanol and/or formaldehyde (ref.4). Studies have appeared in the literature on this topic in recent years and these include homogeneous and heterogeneous systems. In general, the former (ref 4,5) have tended to be carried out at high methane pressures (>30 bar), in the temperature range 200-400C; reasonable yields of methanol have been attained in these systems with smaller amounts of formaldehyde reported depending on the exact experimental conditions used; the addition of heterogeneous catalysts have little effect and radical species are commonly cited in proposed reaction mechanisms.

By contrast studies of heterogeneous systems (refs 6-12) are usually carried out between 450 and 650C at ambient pressure with high CH_4:Oxidant ratios; in general formaldehyde is the usual selective oxidation product observed, but selectivities usually decrease dramatically as conversion exceeds ca. 1% of available CH_4. This is often related to the decomposition of formaldehyde within the catalytic reactor and for this reason most of these studies have attempted to achieve kinetic isolation of the selective oxidation product by operating with minimal contact times.

A primary indication of the stability of any hydrocarbon molecule in oxidising conditions can be got from the strength of its weakest C-H bond. These values are presented in table 1 (ref. 13) for a number of feed stocks and products which feature in some commercial selective oxidation processes.

Table 1

Comparison of C-H Bond Strengths

Feed	Product	C-H bond strength kJ $mole^{-1}$
$n\text{-}C_4H_{10}$		405
	$C_4H_2O_3$	412
C_3H_6		412
	CH_2CHCHO	366
CH_3OH		393
	CH_2O	366
CH_4		440

It is clear from this crude comparison that CH_4 is the most difficult hydrocarbon to activate and CH_2O is amoung the least stable of the selective oxidation products which implies that the selective oxidation route would be difficult to perfect.

It follows therefore that reasonable selectivities in formaldehyde by selective oxidation of methane can only be achieved through very careful control of all parameters involved in the process. Here we report efforts to optimise the support - supported phase ratio for catalysts based on the molybdena/silica system and used in the ambient pressure oxidation of methane to formaldehyde.

EXPERIMENTAL

Catalyst Preparation

Three sources of silica were used in his study i.e, fumed silicas Cab-o-sil M-5, and Aerosil supplied by the Cabot Corporation and Degussa, respectively, and spherosil, a porous silica. Sodium, phosphorus, lead or copper were added to these supports by impregnation in the way already described and molybdenum in the form of $(NH_4)_6Mo_7O_{24}$ was then added by further impregnation (ref 11,12).

Below catalysts will be cited as, for example 5Mo-2Na-Cabosil. This refers to a Cab-o-sil support impregnated with $NaCO_3$ so as to achieve 2 wt% sodium, followed by impregnation with $(NH_4)_6\ Mo_7O_{24}$ to achieve 5wt% MoO_3.

Testing

Catalysts were tested by passing a 5:1 ratio $CH_4:N_2O$ mixture at 0.4 ml s^{-1} over 0.1g of catalyst held between 400 and 600C in a 10mm i.d quartz reactor. Analysis was by on time G.C. Full details have already been presented (ref.12).

Characterisation

Catalysts were analysed before and after use by X-ray diffraction with a Philips diffractometer using Cu Kα radiation filtered through nickel.

In addition, samples were subjected to analysis by temperature programmed reduction. This was carried out by placing the equivalent of ca.5 mg of MoO_3 in a quartz reactor and passing a flow of 5% H_2 in N_2 over the catalyst at 20 ml min^{-1}. The temperature was linearly increased from room temperature to 800C at 10 C min^{-1} while hydrogen consumption was monitored using a thermal conductivity detector.

RESULTS

Table 2 presents the conversion of methane and selectivity to formaldehyde achieved at 500 and 600C over a range of supported molybdena catalysts in standard reaction conditions. Essentially, formaldehyde decomposition was small at 500C but appreciable at 600C for most systems studied (ref.12). Good selectivity was observed only over silica supported molybdena catalysts whereas other combinations of support and supported phase were not selective. In addition, the porous silica used (spherosil) exhibited good performance at 500C, but its selectivity diminished drastically at 600C, indicating that formaldehyde could not survive within the pores of this support at the higher reaction temperature.

Table 2

Conversion of Methane and Selectivity to Formaldehyde Over a Range of Catalysts.

		500c		600C	
	W/F	Conv%	Sel%	Conv%	Sel%
	g s ml^{-1}	CH_4	CH_2O	CH_4	CH_2O
Empty Reactor	0.25	0.01	0	0.01	0
Cabosil	0.25	0.01	0	0.03	0
2Na-Cabosil	0.25	0.01	0	0.02	0
2Mo-Cabosil	0.25	0.03	72	0.05	67
2Mo-Spherosil	1.25	0.57	58	0.90	3
2Mo-TiO_2	0.25	0.54	0	2.20	0
2Mo-2Na-TiO_2	0.25	-	-	0.12	0
2Mo-MgO	0.25	0.13	0	0.67	0
		400C		500C	
3Cu-Cabosil	0.25	0.09	0	0.31	0
0.1Pt-Cabosil	0.25	1.23	0	-	-

Table 3

Nature of the Support Additives for Fumed Silicas

	500C				600C			
	Conv%	Sel%			Conv%	Sel%		
	CH_4	CH_2O	CO	CO_2	CH_4	CH_2O	CO	CO_2
5Mo-Cabosil	0.04	0	15	85	-	-	-	-
5Mo-2Na-Cabosil	0.08	60	9	31	0.3	38	22	38
5Mo-3Pb-Cabosil	0.24	0	0	100	-	-	-	-
5Mo-2Pb-2Na-Cabosil	0.04	66	0	34	0.23	31	53	12
10Mo-Aerosil	0.04	71	2	27	0.23	0	35	64
10Mo-2Na-Aerosil	0.05	85	5	10	0.22	32	12	57
10Mo-5Cu-Aerosil	0.05	72	0	28	0.55	38	27	35
10Mo-2P-Aerosil	0.05	67	11	22	0.28	20	65	15

The influence of a number of support additives is presented in table 3 Sodium, phosphorus or copper, each impregnated onto the support prior to addition of the molybdenum component enhanced the selectivity towards formaldehyde particularly, during operating at 600C.

This effect is further elaborated upon in figures 1-3 which show the influence of nominal MoO_3 loading on the conversion of CH_4, the selectivity to formaldehyde and the rate of formaldehyde formation at 500 and 600C for the Cabosil, 2Na-Cabosil, 5Na-Cabosil series. For each temperature studied and for each support material an optimal nominal MoO_3 loading in terms of selectivity and rate of formaldehyde production can be identified. This optimal loading depends upon the additive loading of the support, but generally allows catalysts with vastly increased MoO_3 loadings to be made up without the severe loss in selectivity observed without the additive.

A final point to note is the inhibiting effect of added sodium at low MoO_3 contents. Essentially, the MoO_3 loading had to exceed a certain minimal value (2-3 wt % MoO_3 in the case of 2Na-Cabosil) before any catalytic activity set in.

A further point of interest for all catalysts studied is the production of large amounts of CO when formaldehyde selectivity diminished (Table 3). This finding has been reported elsewhere and associated with formaldehyde decomposition(ref.11).

Peaks due to Na_2MoO_4 appeared during X-ray diffraction analysis of most of the Na-Cabosil based catalysts used in this study before and after testing, with smaller amounts of MoO_3 detected. For the 5Na-Cabosil series it was possible to establish a correlation between the XRD phase composition and the rate of formaldehyde formation (figure 4), which demonstrate a clear link between formaldehyde production and the presence of crystalline Na_2MoO_4.

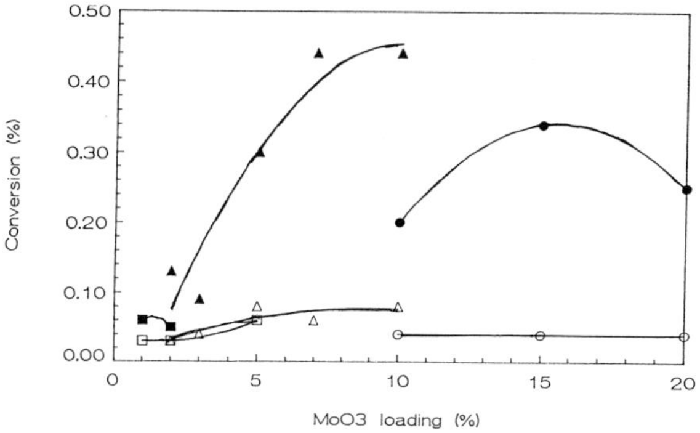

Figure 1: Influence of MoO_3 loading on the conversion of CH_4.
(■)xMo-Cabosil
(▲)xMo-2Na-Cabosil
(●)xMo-5Na-Cabosil
Open symbols 500C
Full symbols 600C.

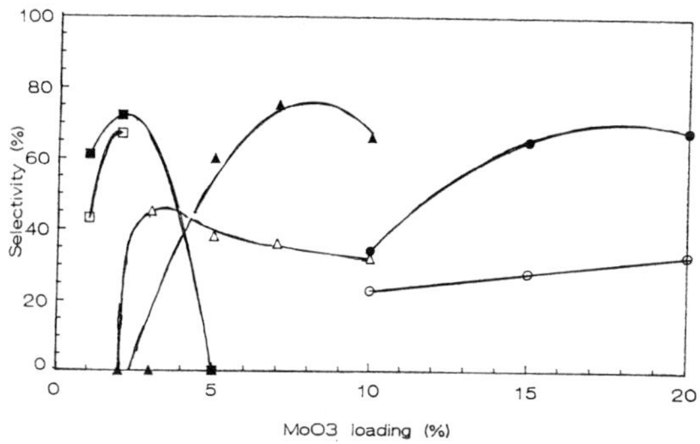

Figure 2 Influence of MoO_3 loading on the selectivity to formaldehyde.
(■)xMo-Cabosil
(▲)xMo-2Na-Cabosil
(●)xMo-5Na-Cabosil
Solid symbols 500C
Open symbols 600C

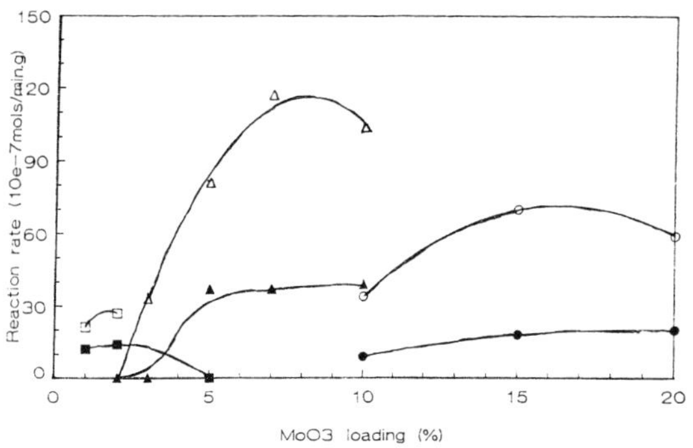

Figure 3 Influence of MoO_3 loading on the rate of formaldehyde production.
(■)xMo-Cabosil
(▲)xMo-2Na-Cabosil
(●)xMo-5Na-Cabosil
Full symbols 500C
Open symbols 600C.

The T.P.R. patterns of a representative selection of the catalysts tested here are shown in figure 5. These demonstrate that the best catalysts in terms of formaldehyde production and selectivity are all more readily reduced than unsupported MoO_3 and the sodium free molybdena/Cabosil catalysts.

DISCUSSION

Methane conversions achieved in this work were low (<1%) and at first sight the yields of formaldehyde are low when compared with other selective oxidation reactions. However, most selective oxidation processes operate with a hydrocarbon: air ratio below the lower explosion limit. In practice the partial pressure of hydrocarbon used can be as low as 0.015 atm in, for example the case of n-butane oxidation and the partial pressure of product generated (ca 0.01 atm of maleic anhydride) then compares with gas phase pressures of formaldehyde (ca 0.005 atm) achieved in this work (ref 14). To date however attempts at operating methane oxidation with low methane: oxidant ratios have failed due to the poor methane activation properties of the molybdena catalysts in these conditions.

We have already proposed that formaldehyde rather than methanol is the predominant selective oxidation product observed at ambient pressure because the following reactions occur either on the surface of the catalyst or in the vapour phase (refs 11,12):

$$CH_3\cdot + O \longrightarrow CH_3O\cdot \qquad [1]$$

$$CH_3O\cdot \longrightarrow CH_2O + H\cdot \qquad [2]$$

$$CH_3O\cdot + CH_4 \longrightarrow CH_3OH + CH_3\cdot \qquad [3]$$

In conditions of low methane partial pressure (ambient pressure) direct decomposition of the CH_3O radical (reaction 2) should be favoured. At high methane pressures collisions between radical species and methane molecules could occur more readily, so that methanol would be the predominant selective oxidation product in these conditions.

A recurring feature of the poor selectivities observed above particularly at high temperatures is the appearance of CO in the reaction products, presumably from the decomposition of CH_2O (ref.11). Therefore, for the reaction conditions used in this study i.e. high CH_4:N_2O ratios the molybdena silica catalysts can be classified as sufficiently active but lacking the selectivity necessary to permit the formaldehyde to exit the reactor without decomposition. In this regard examination of Table 3 reveals that catalysts based on pure silica and additive - silica achieved similar conversions of methane for a given nominal MoO_3 loading. However, improved selectivities were observed particularly at 600C when support additives were incorporated into these catalysts. It may be concluded therefore that the support additives somehow reduce the amount of CH_2O decomposition which occurs, thereby increasing selectivity.

A general finding when additives, almost irrespective of their nature, are

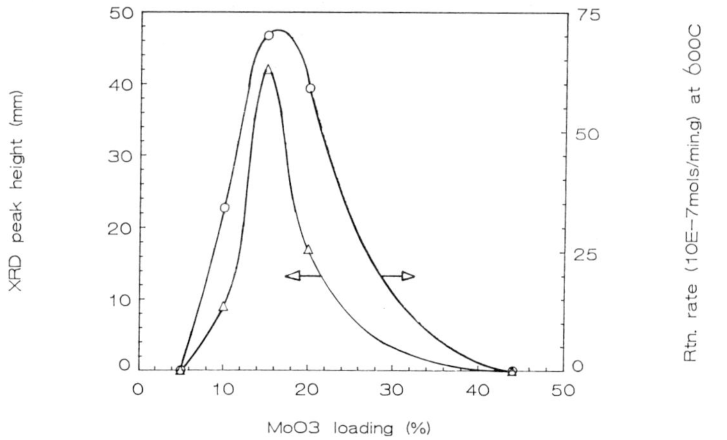

Figure 4: Rate of formaldehyde production over the 5Na-Cabosil series (O) and the relative intensity of the X.R.D. peak at d=5.24A for Na_2MoO_4 (Δ).

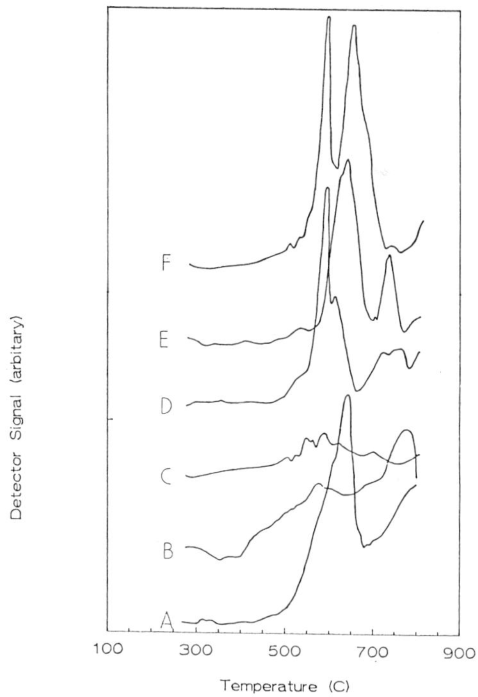

Figure 5 T.P.R. profiles of a) MoO_3
b) 1Mo-Cabosil,
c) 2Mo-2Na-Cabosil,
d) 7Mo-2Na-Cabosil,
e) 10Mo-5Na-Cabosil,
f) 20Mo-5Na-Cabosil.

incorporated into the molybdena/silica system is the very high molybdena loadings achievable (ref. 8) without the corresponding losses in activity and selectivity normally observed with the additive free systems. Several recent studies of the molybdena/silica system have attempted to relate selectivity to the presence of specific compounds on the silica surface. These include particularly heteropoly compounds (refs 8,10). Based on the T.P.R. data presented in figure 5 and the correlation observed in figure 4 it is proposed here that the additives bring about a change in the redox properties of the catalyst surface, making it easier to extract lattice oxygen at the reaction temperature. This in turn helps to establish an appropriate supply of lattice oxygen at the surface so as to achieve a balance between the activation of methane and the decomposition of formaldehyde.

Sodium has a beneficial effect in our test conditions provided the Mo:Na ratio exceeds a minimal value. It may be argued therefore that the sodium modulates the redox properties of the molybdena through the formation of a non-stoichiometric, hence defect rich, phase. In operation this phase is probably in a somewhat reduced state.

ACKNOWLEDGEMENTS

We gratefully acknowledge the support of the European Community non-nuclear energy programme for this work (contract no: EN3C-0034-IRL)

REFERENCES

1. N.R. Foster, Appl. Catal., 19 (1985) 1.
2 G.E. Keller and M.M. Bhasin, J. Catal., 73 (1982) 9.
3 Methane Activation, Proc. 1st European Workshop, Bochum, May, 1988 Catal Today, 4 (1989) nos 3-4.
4 H.D. Gesser and N. R. Hunter, Chem. Revs, 85 (1985) 235.
5 N.R. Hunter, H. D Gesser, J.A. Morton, P.S. Yarlagodda and D.P.C. Fung, Symp. on Hydrocarbon Oxidation, New Orleans, Sept, (1987).
6 H.F. Lui, R.S. Lui, K.Y. Liew, R.E. Johnson and J.H. Lunsford, J. Amer. Chem. So., 106 (1984) 4117.
7 M.M. Khan and G.A. Somorjai, J. Catal., 91(1985) 263.
8 S. Kasztelan and J.B. Moffat, J. Catal., 106 (1987) 512
9 N.D. Spencer, J. Catal, 109 (1988) 187.
10 Y. Barbaux, A.R. Elamrani, E. Payan, L. Gengembre, J.B. Bonnelle and B. Grazbowska, Appl. Catal., 44 (1988) 117.
11 E. MacGiolla Coda, E. Mulhall, R. Van Hoek and B.K. Hodnett, Catal. Today, 4 (1989) 383.
12 E. MacGiolla Coda, R. Van Hoek, E. Mulhall and B. K. Hodnett, Hydrocarbons, Lyons, Sept 1988.
13 Handbook of Chemistry and Physics, 68th Edition 1987-88, CRC Press.
14 J.C. Burnett, R.A. Keppel and W.D. Robinson, Catal. Today, 1 (1987) 537.

Prof.E.Bordes (Universite de Technologie de Compiegne): Have you tried adding Na, Cu, P, or Pb after loading molybdenum on silica to see how this influences the performance of your catalysts? You mentioned that additives bring about a change in the redox properties of the surface and I am in accordance with that. In the case of sodium and copper you can form Na_2MoO_4 (which you have seen)and $CuMoO_4$, whereas with phosphorus or silicon you can could form heteropolyanions. Do you see differences in reactivity for these two kinds of additives?

Dr B.K.Hodnett (University of Limerick, Ireland): We tried to reverse the order of impregnation with the sodium system, i.e., adding the dopant after the molybdenum component. In such conditions no appreciable beneficial effect was observed. We have detected the presence of Na_2MoO_4 on our sodium treated catalysts by XRD, but we have not detected any other complex oxide by XRD when other additives were used. Generally we have found that the presence of Na, Cu or P results in less combustion of the selective oxidation products at elevated temperatures.

Prof O. Krylov (Institute of Chemical Physics, Moscow): In connection with the interesting results reported by Dr Hodnett I would like to mention an interesting observation made by us. When we used the reversed catalyst, i.e. silica supported on molybdena, we have observed 100% selectivity in the oxidation of methane to formaldehyde in similar experimental conditions.

Dr B.K.Hodnett (University of Limerick, Ireland): It is gratifying to see a system which achieves activation of methane and a reasonable conversion without combustion of the selective oxidation product.

Dr Sinevmy (Institute of Chemical Physics, Moscow): What can you tell us about the efficiency of your catalysts in the presence of O_2 as oxidant? If they do not produce formaldehyde in these conditions does it mean that O_2 cannot reoxidize the active sites or that there are some other problems?

Dr B.K.Hodnett (University of Limerick, Ireland): We have observed selective oxidation with this system using O_2 as oxidizing agent.

Prof. Baerns (Ruhr-Universitat Bochum): In your presentation you defined the rate of formaldehyde production as yield which is contradictory to its usual definition. To make comparison with other data easier, please, indicate the degree of methane conversion (X) besides the selectivity data (S) to calculate the yield (Y) as commonly described:

$$Y\% = (S\% * X\%) / 100$$

Dr B.K.Hodnett (University of Limerick, Ireland): We have presented sufficient imformation in our paper which allows our yields to be calculated. However, we find this single index of catalytic performance to be misleading because it fails to take into account the partial pressure of hydrocarbon in the feed stream. In many conventional selective oxidation processes a hydrocarbon lean feed is used. In these conditions high conversion of hydrocarbon can be achieved. In our case we use a hydrocarbon rich feed, so that a lower conversion can still result in partial pressures of selective oxidation product being produced which compare with those produced in many conventional oxidation processes. It is for this reason we expressed formaldehyde production in terms of a reaction rate as this also takes into account the reactor loading and feed gas flow rate.

Dr. Grzybowska (Institute of Catalysis, Krakow): Your TPR data suggests that you have severl types of M-O_x species dispersed on silica which is in accord with the literature data on this subject quoted by you (refs 8-10) in your paper. On the other hand you show the correlation between the content of Na_2MoO_4 phase and the rate of aldehyde formation, stating that redox properties may play a role in CH_4 oxidation. Could you ascribe any of your TPR peaks then to the reduction of Na_2MoO_4?

Dr B.K.Hodnett (University of Limerick, Ireland): No.

G. Centi and F. Trifiro' (Editors), *New Developments in Selective Oxidation* 469

OXIDATIVE DIMERIZATION OF METHANE IN HIGH TEMPERATURE ELECTROCHEMICAL DEVICES

V.D. BELYAEV, O.V. BAZHAN, V.A. SOBYANIN and V.N. PARMON

Institute of Catalysis, Novosibirsk 630090, USSR

SUMMARY

Studied is oxidative dimerization of methane in electrocatalytic and standard catalytic conditions on electrode-catalysts made from Cu, Ag, Ni, Pt, Ag (20 at.%) - Pd (80 at.%) alloy and on lanthanum chromite-based oxide systems in high temperature electrochemical devices with solid oxides or molten carbonate oxygen-conducting electrolytes. Similarities and differences in reaction occurrence under the above conditions have been elucidated. As shown, the electrolytic regime may be more advantageous than the standard catalytic regime for the purpose of increasing the selectivity of production of C_2-hydrocarbons.

INTRODUCTION

The development of efficient methods of CH_4 conversion into valuable chemicals, in particular C_2 hydrocarbons (refs. 1-3) is one of the most serious problems of applied interest faced by modern catalysis science (refs. 1-3). In this connection of great interest are recent data on gas-phase electrocatalytic oxidative dimerization of CH_4 in high temperature electrochemical devices (fuel cells or oxygen pumps) (refs. 4-7). It is essential that when operated under definite conditions, such devices make it possible to produce simultaneously electricity and purpose products (operation with fuel cells). Also, these devices can be applied for production of only purpose products with the aid of external energy supply (operation with electrolizer or electrochemical pump). These two regimes of operation will be called hereinbelow as electrocatalytic.

The oxidation of CH_4 in electrocatalytic regimes has been studied (refs. 4-6) on Ag, Ag-Bi_2O_3, Ag-Li/MgO and $LiNiO_2$ electrode-catalysts which were in contact with a yttria-stabilized zirconia electrolyte (YSZ) providing oxygen transport to the reaction zone.

The objective of this work was to study oxidative dimerization of methane in high temperature fuel cells (FC) using a YSZ electrolyte (SOFC) on electrode-catalysts from Cu, Ni, Pt, Ag, Ag(80%)

Pd(20%) alloy and lanthanum chromite-based oxides as well as electrolyte in the form of molten carbonates of alkaline metals (MC) on Ni and Ag electrode-catalysts (MCFC).

METHODS

SOFC were test-tubes made from a YSZ electrolyte with composition 0.9 ZrO_2+ 0.1 Y_2O_3 on the internal side of which a working electrode (anode) was supported and on the external side - a counter electrode (cathode). The geometrical area of the electrodes was 10 cm^2.

MCFC were prepared on the basis of a porous $LiAlO_2$ disc matrix impregnated with a mixture of molten carbonate of lithium and potassium. The geometrical area of electrodes which were placed on the opposite sides of the matrix was 4 cm^2.

The methane oxidation reaction was studied under atmospheric pressure in electrocatalytic and, for comparison, in standard catalytic regimes at 650-680°C for MCFC and 700-890°C for SOFC. In the both regimes a flow of methane or a helium-methane mixture was fed into the anodic space of a fuel cell with a velocity of 1 cm^3/s. Simultaneously, the cathode was blown by air for SOFC and by an air-CO_2 mixture for MCFC.

In the electrocatalytic regime oxygen was fed directly into the reaction zone by passing the electric current through FC. The initial methane or helium-methane mixture flow had no oxygen. The operation of FC in this regime is shown in a schematic fashion in Fig. 1. For SOFC the passing of the electric current leads to the

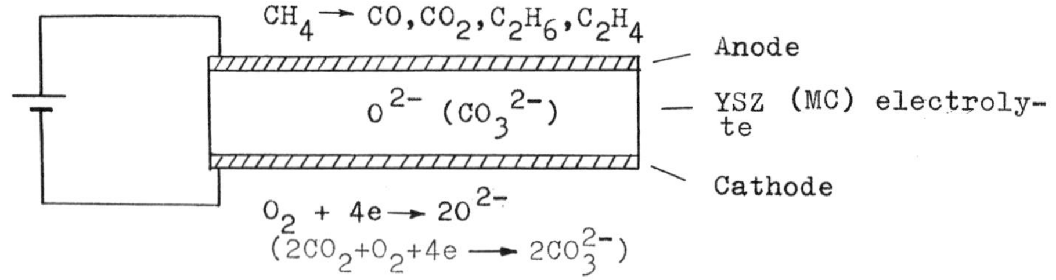

Fig. 1. Schematic diagram of electrocatalytic performance of methane oxidative dimerization.

reduction of molecular oxygen to O^{2-} ions on the cathode. The ions are then transferred through the YSZ into the anode on which they can either discharge producing O_2 or oxidize methane. Actually, the similar situation occurs during the electrocatalytic oxidation of methane in MCFC. The only difference is that oxygen is transferred from the cathode into the anode by CO_3^{2-} anions.

The relation between electric current (I) and the oxygen transfer flux (Q) from cathode to anode can be written as follows:

$$I = 4FQ,$$

where F is the Faraday constant.

In the standard catalytic regime oxygen was fed by portions into flows of methane or a helium-methane mixture prior to their feeding into the FC anodic space. The electric circuit of FC was disconnected and the anode served as conventional heterogeneous catalyst.

When the reaction was carried out in the electrocatalytic and catalytic regimes the oxygen flux into the reaction zone was expressed in the same units (amperes, A).

The gas mixture composition before and after the FC was analysed chromatographically. The experimental apparatus has been described in detail elsewhere (ref. 8).

RESULTS AND DISCUSSION

It has been found for all systems studied that methane oxidation in the electrocatalytic regime yields ethane and ethylene, along with CO, CO_2 and H_2O. Oxygen-containing compounds and higher hydrocarbons were in negligible amounts.

In the SOFC with Ni and Pt electrodes the rate of oxidative conversion of methane sharply and irreversibly decreased with time and current-voltage characteristics of working electrodes worsened. These phenomena result from delamination of Ni and Pt from the YSZ surface due to carbon which is formed during the reaction and coked at the metal electrolyte interface.

As distinct from the SOFC with Ni and Pt working electrodes, other FCs as a catalytic reactor were characterized by stable operation for a long period of time (up to 70 h).

Consider now the data obtained for these FCs in more detail. Figures 2 and 3 show dependences of steady-state rates of ethane and ethylene formation, methane conversion and selectivity toward C_2 hydrocarbons on oxygen flow in methane oxidation on Ag and Ag-Pd alloy-based electrode-catalysts supported on YSZ in electrocatalytic and catalytic regimes. As can be seen in Fig. 2a, when the reaction is performed on the Ag electrode in both regimes

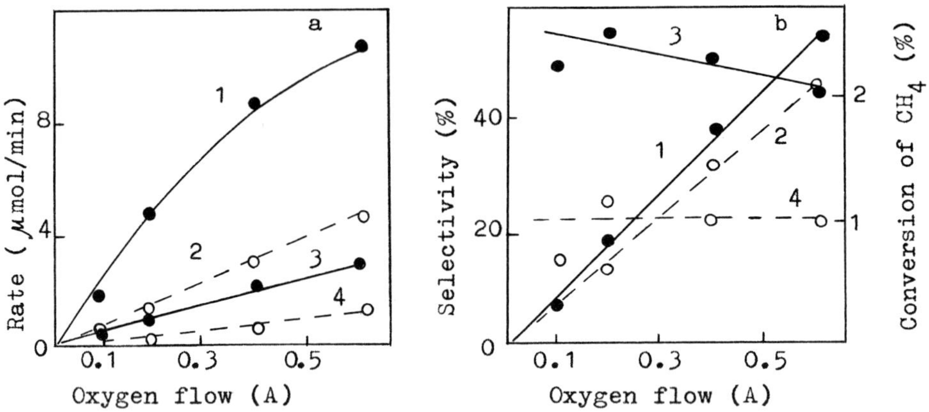

Fig. 2. Rates of ethane (1a,2a) and ethylene (3a,4a) formation, methane conversion (1b,2b) and selectivities toward C_2 hydrocarbons (3b,4b) vs. oxygen flow rate in methane oxidation on Ag contacting with YSZ in electrocatalytic (1a,1b,3a,3b) and catalytic (2a,2b,4a,4b) regimes. Reaction conditions: 795°C; CH_4 concentration 100 vol.%; methane flow rate 1 cm^3/s.

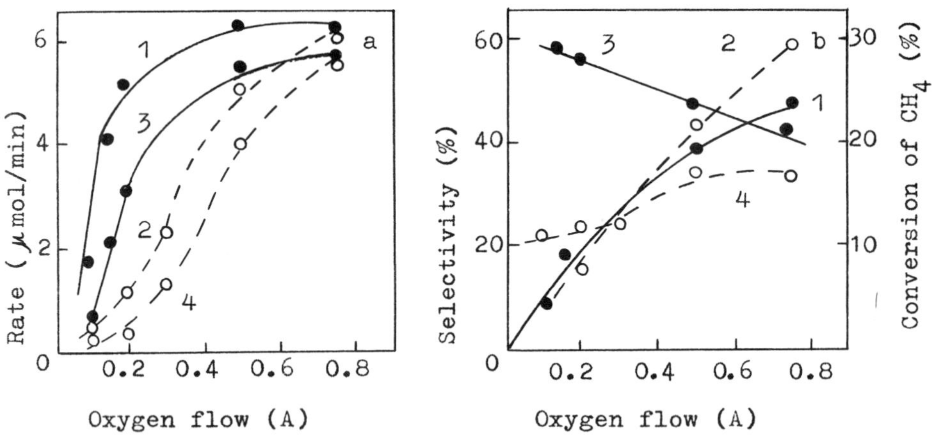

Fig. 3. Rates of ethane (1a,2a) and ethylene (3a,4a) formation, methane conversion (1b,2b) and selectivities toward C_2 hydrocarbons (3b,4b) vs. oxygen flow rate in methane oxidation on the Ag (80 at.%)-Pd(20 at.%) alloy contacting with YSZ in electrocatalytic (1a,1b,3a,3b) and catalytic (2a,2b,4a,4b) regimes. Reaction conditions: 840°C; methane concentration 10 vol.%; methane-helium flow rate 1 cm^3/s.

the rates of formation of C_2 hydrocarbons increase with increasing oxygen flow from 0.1 to 0.6 A (or the O_2/CH_4 flow rate ratio from 8 10^{-3} to 4 10^{-2}). According to data in Fig, 2b in the electrocatalytic regime the selectivity changed from 50 to 40% and in the catalytic regime remained almost unchanged (ca. 20%). In both regimes oxygen conversion was close to 100% and the conversion of methane depended linearly on oxygen flow being no higher than 3% (Fig. 2b).

A comparison of the results obtained indicates that in the electrocatalytic regime the yield, rate and selectivity towards C_2 hydrocarbons are 3-4 times higher than those achieved in the standard catalytic regime.

As is seen in Fig. 3a, when the reaction is carried out on the Ag-Pd alloy electrode, in both regimes the rates of formation of C_2 hydrocarbons first increase and then reach a plateau with increasing oxygen flow from 0.1 to 0.75 A (or the O_2/CH_4 flow rate ratio from 0.06 to 0.46). According to data in Fig. 3b the conversion of methane and the selectivity tend to change from 4-24% and 58-43% and from 4-29% and 21-34%, respectively, for the electrocatalytic and catalytic regimes.

A comparison of results obtained for oxygen flows smaller than 0.4-0.5 A (O_2/CH_4 = 0.25-0.31) evidences that in the electrocatalytic regime the yield of ethane is 3-5 times higher and of ethylene is 4-10 times higher than their yields in the standard catalytic regime. Note that under these conditions oxygen is completely converted in both regimes.

On increasing oxygen flow (> 0.55 A) the rates of formation of C_2 hydrocarbons in both regimes are comparable (Fig. 3a) and the selectivities are nearly the same (Fig. 3b). Interestingly, such an effect has also been observed for the Ag electrode-catalyst upon feeding methane diluted with helium into the anodic space. This effect is due to that in the electrocatalytic regime at high rates of oxygen flow the reaction of oxygen evolution into the gas phase is predominant

$$2\ O^{2-} \longrightarrow O_2 + 4\ e^-$$

and the oxidation of methane occurs as for standard catalysis. Thus, the results obtained show that performance of oxidative dimerization of methane on Ag and Ag-Pd alloy electrodes in the electrocatalytic regime is, under definite conditions, more advantageous than in the standard catalytic regime. As has been found for Ag electrode-catalyst (Fig. 4), the electrocatalytic regime becomes more advantageous than the catalytic one with increasing temperature. The differences observed between electrocatalytic and catalytic regimes seem to be due to an important role of electrochemical steps in the former regime which affect the state of oxygen on the electrode surface (ref. 9).

Results similar to those of Figs. 2 and 3 have been obtained for other electrode-catalysts too. Below we summarize the main results.

On the copper electrode which was in contact with YSZ at 800-890°C the steady-state rate of formation of C_2 hydrocarbons in the electrocatalytic regime of methane oxidation was more than 1.5 times higher than that in the catalytic regime. Different from Ag and Ag-Pd, in the electrocatalytic regime on Cu electrodes the rate of C_2-hydrocarbons formation slowly achieved its steady state. As can be seen in Fig. 5, more than a two times decrease in rate was observed. However, as current was switched off for only one minute and then switched on, the rate tended to increase to its initial value.

This observation suggests the possibility to enhance reaction efficiency by applying the electrocatalytic method in unsteady-state conditions. In particular, it has been found that upon periodic switching on a switching off the current passing through the cell (oxygen flow through the electrolyte) a period-average rate of formation of C_2 hydrocarbons is ca. 2 times higher than the steady-state reaction rate. This fact seems to be due to partial copper oxidation upon feeding oxygen through the electrolyte and its reduction with methane at switching off the current.

In SOFC on the electrode based on lanthanum chromite the oxidative dimerization of methane was studied at 815°C. As found experimentally, in both regimes the selectivity toward C_2 hydrocarbons did not exceed 4%, and the main reaction product was CO_2. However, addition of lithium chloride significantly enhanced the selectivity toward C_2 hydrocarbons. E.g., in the electrocatalytic regime upon varying the O_2/CH_4 flow rate ratio from 0.06 to 0.24 the selectivity achieved 60-45% at methane conversions 8-23%. In this instance the yield of ethylene was ca. 4 times higher than the yield of ethane being, under optimal conditions, 8.5%.

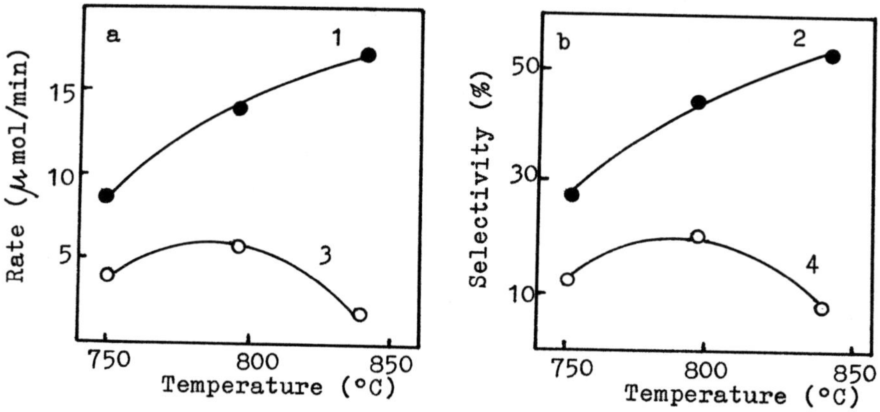

Fig. 4. Rates (a) and selectivities (b) of formation C_2 hydrocarbons vs. temperature in methane oxidation on Ag contacting with YSZ in electrocatalytic (1,2) and catalytic (3,4) regimes. Reaction conditions: methane concentration 100%, methane flow rate 1 cm^3/s, oxygen flow 0.6 A.

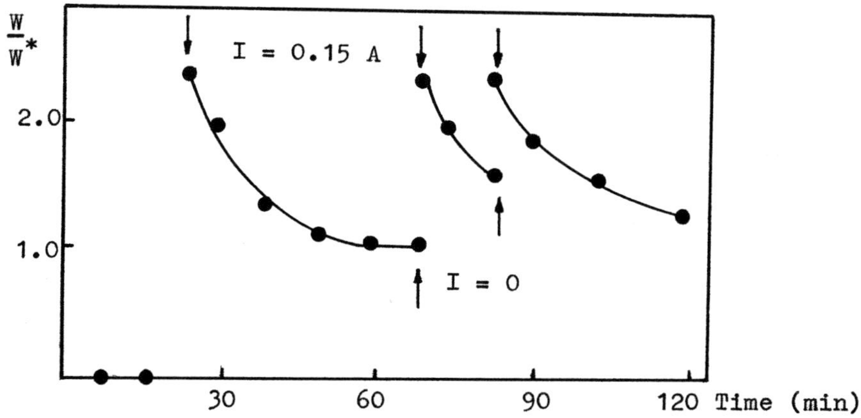

Fig. 5. Rates of formation C_2 hydrocarbons (W/W*) vs. time in methane oxidation on Cu contacting with YSZ in electrocatalytic regime. ↓↑- moments of switching on and off the current (I) through the cell (oxygen flow through the electrolyte). W and W* - current and steady-state rates. Reaction conditions: 860°C, methane concentration 100%, methane flow rate 1 cm^3/s.

In the catalytic regime the yield and selectivity toward C_2 hydrocarbons were 3 times lower than those in the electrocatalytic regime. Thus, modification of the electrode by additives that increase the efficiency of the standard catalytic reaction improve even to a larger extent the reaction parameters in the electrocatalytic regime.

In MCFC the oxidative conversion of methane was studied on Ag and Ni electrode-catalysts at 650-680°C. The rate of formation of C_2 hydrocarbons on these electrode-catalysts was found to be ca. 2.5 times higher in the electrocatalytic regime than in the catalytic one. The difference in selectivity toward C_2 hydrocarbons was the same; note that in any experiment the selectivity was low and did not exceed 10% even at low conversions of methane (1-2%). Such a low selectivity at methane oxidation in MCFC seems to be due to that (i) electrode-catalysts had the composition far from optimum and (ii) the temperature of operation of MCFC was by ca. 100° lower than the typical temperature of oxidative dimerization of methane.

CONCLUSIONS

Thus, with some electrode-catalysts the electrocatalytic regime for oxidative dimerization of methane is more advantageous than the catalytic one. The nature of this phenomena is not quite clear and much work remains to be done in this direction. However, we believe that the electrocatalytic method opens up a new way for oxidation reaction performance.

REFERENCES

1 T. Ito and J.H. Lunsford, Nature, 314 (1985) 721.
2 R. Pitcha and K. Klier, Catal. Rev., 28 (1986) 13.
3 Kh.M. Minachev, N.Ya. Usachev, V.N. Udut and Yu.S. Khodakov, Usp. Khim., 57 (1988) 385 (in Russ.).
4 K. Otsuka, S. Yokoyama and A. Morikawa, Chem. Lett., (1985) 319.
5 S. Seimanides and M. Stoukides, Electrochem. Soc., 133 (1986) 1535.
6 K. Otsuka, K. Suga and I. Yamanaka, Chem. Lett., (1988) 317.
7 V.D. Belyaev, O.V. Bazhan, V.A. Sobyanin and V.N. Parmon, Proc. 7th All-Union Conference on Electrochemistry, Chernovtsy, USSR, 1988, vol. 3, p. 134 (in Russ.).
8 V.D. Belyaev, V.A. Sobyanin, V.A. Arzhannikov and A.D. Neuimin, Dokl. AN SSSR, 305 (1989) 1389 (in Russ.).
9 V.D. Belyaev, V.A. Sobyanin and O.A. Mar'ina, Izv. SO AN SSSR Ser. Khim. Nauk (in Russ.) (accepted for publication).

G. Centi and F. Trifiro' (Editors), *New Developments in Selective Oxidation*
© 1990 Elsevier Science Publishers B.V., Amsterdam — Printed in The Netherlands

SELECTIVE DEHYDROGENATION OF ETHANE BY CARBON DIOXIDE OVER Fe-Mn OXIDE CATALYST. AN IN SITU STUDY OF CATALYST PHASE COMPOSITION AND STRUCTURE.

A.Kh.MAMEDOV, P.A.SHIRYAEV, D.P.SHASHKIN, O.V.KRYLOV
Institute of Chemical Physics of the USSR Academy of Sciences, 117334 Moscow, Kosygin st.4, USSR.

SUMMARY

The results on activity, phase composition and crystalline structure reconstruction studies concerning the Mn catalyst and Fe-Mn catalyst in the course of ethane dehydrogenation by carbon dioxide are presented.

It has been shown that manganese oxide systems modified by small quantities of iron are effective catalysts for ethane dehydrogenation by carbon dioxide.

INTRODUCTION

Dehydrogenation of methane and ethane by the unconventional oxidant-carbon dioxide is an interesting process from both scientific and practical point of view.

Carbon dioxide in equimolecular mixtures with methane converts the latter into syngas with $CO+H_2$ stoichiometric relationship (refs. 1-2):

$$CO_2 + CH_2 \rightarrow 2CO + 2H_2 \quad (1)$$

Conversion of ethane takes place by the reaction:

$$C_2H_6 + CO_2 \rightarrow C_2H_4 + CO + H_2O \quad (2)$$

The reaction (2) assumed to be selective is accompanied by the by-reaction of deep conversion:

$$C_2H_6 + 2CO_2 \rightarrow 4CO + 3H_2 \quad (3)$$

Under these conditions at 780°-850°C the following reactions are proceeding:

$$2C_2H_6 \rightarrow C_2H_4 + 2CH_4 \quad (4)$$

$$C_2H_6 + CO_2 \rightarrow CH_4 + 2CO + H_2 \quad (5)$$

The composition of the reaction products depends on the reactions (1)-(5) rates relationship.

EXPERIMENTAL

The reactivity experiments were carried out in pulse and flow reactors with the vibroliquified bed of the catalyst at 780°-850°. The X-Ray spectra in situ were taken using a diffractometer DRON-2.0 with $Fe_{K\alpha}$ radiation.

The reaction mixture C_2H_6 + CO_2 was passed at the rate of 3 cc/min through the X-Ray chamber-reactor which allows to analyse simultaneously the changes in the phase composition as well as the parameters of the catalytic reaction (ref.3). The studies in situ were carried out at 600°C, the catalyst volume was 1 cc.

The following systems have been investigated in the capacity of catalysts:

Mn - O / SiO_2 (I)

Fe - Mn - O / SiO_2 (II)

The catalysts were prepared by impregnation of the silicagel KSK with nitrates of Mn and Fe. The content of Mn and Fe in the samples amounted to 17% and 4% respectively.

RESULTS AND DISCUSSION

The catalyst Mn-O/SiO_2 is effective in the 50% C_2H_6 + 50% CO_2 mixture. Under the steady-state conditions the accumulation of coke deposits on the surface is not observed in the presence of this catalyst.

The ethane conversion on the Mn-O/SiO_2 catalyst increases when it is modified by Fe (table 1). This catalyst when calcined at 450°C showed the presence of only one phase of manganese: β-MnO_2. The SiO_2 carrier is highly dispersed. After the calcination at high temperatures of 600°-700°C partial reduction of β-MnO_2 into the Mn_2O_3 phase is observed.

TABLE 1

Data on dehydrogenation of ethane by carbon dioxide on Mn catalyst and Fe-Mn catalyst. C_2H_6/CO_2 = 1:1.6-1.8

Catalyst	T°C	Conversion,%		S, C_2H_4	Yield
		C_2H_6	CO_2	%	C_2H_4,%
Mn-O/SiO_2	770	49.8	57.8	51.5	25.5
	800	73.1	79.0	63.0	46.0
Fe-Mn/SiO_2	770	65.9	40.7	62.5	41.2
	810	81.0	52.8	72.4	58.7

In Fe-Mn-SiO_2 catalyst another phase α-Fe_2O_3 was detected. With increase of the iron oxide content at the preparation stage, only the change in the Fe_2O_3-MnO_2 phases relatioship occurs.

When preparing the binary Fe-Mn catalytic system partial screening of the Mn surface by the α-Fe_2O_3 phase is supposed to occur. The dimensions of the crystallites of the α-Fe_2O_3 are less (~200A) than the dimensions of the crystallites of the β-MnO_2 phase (~1000A). In the course of the reaction the phase composition of the Fe-Mn catalyst changes: Mn oxide is reduced to the full and the α-Fe_2O_3 phase is partially reduced.

In this case ethane convesion proceeds unsteadily and the rates of the products formation depend on time. Soon after the charge of the initial sample II ($V = 1800\ hr^{-1}$ $C_2H_6/CO_2 = 1$) the selectivity at 780°C is low (35%) as the result of C_2H_6 partial conversion into CO_2 on the more oxidized phase of Mn. As the reaction proceeds and Mn oxide is reduced the rate of CO_2 formation decreases; in this case the selectivity for C_2H_4 increases up to 60%.

Since the reduction of the catalyst by the reaction mixture proceeds under steady-state conditions, we have studied the interaction of ethane with the catalyst and reoxidation of the catalyst with CO_2 in order to model the processes accompanying the catalysis.

Figure 1 demonstrates the curves presenting the intensities changes of Mn_2O_3 and Fe_2O_3 while reducing the catalyst by ethane. The Mn_2O_3 is the main Mn phase available bafore the feeding of C_2H_6 at 600°C. At first the polymorphous transition of Mn_2O_3 to Mn_3O_4 proceeds; the most intensive period of it is at the beginning of the reduction. After the 70 minute exposure to C_2H_6 Mn_2O_3 disappears completely, the intensity of the MnO phase increases and Mn_3O_4 passes through a maximum.

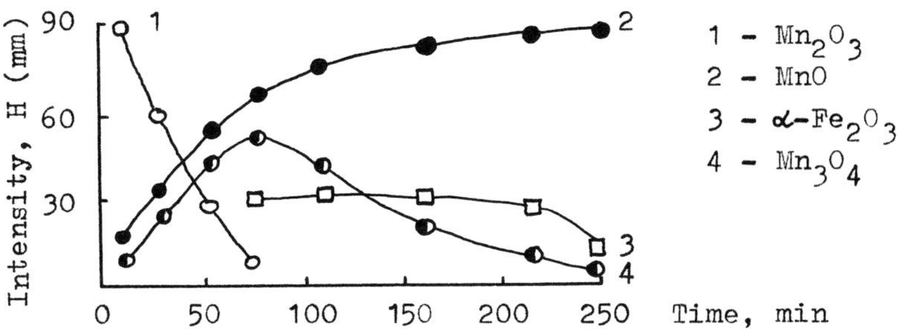

Fig. 1. Change of the catalyst phase composition as a function of time under the reaction conditions.

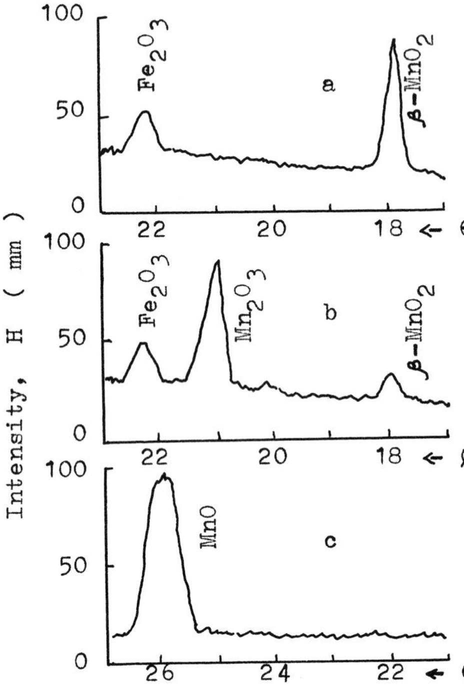

Fig. 2. XRD patterns of the catalyst Fe-Mn/SiO_2: (a) before heating; (b) after heating at 600°C; (c) after the reaction C_2H_6+CO_2.

The intensity of the Fe_2O_3 phase at the initial stage of the reduction remains unchanged and after the completion of the phase transition from Mn_3O_4 to MnO the intensity decreases. Absence of precise diffractional maxima for the Fe_3O_4 phase is likely to be due to the formation of this phase in highly dispersed state.

Figure 2 demonstrates parts of the Fe-Mn catalyst diffractograms at the preparation stage, after heating in a He stream and after reduction (reaction).

After feeding of ethane the level of its conversion decreases gradually. At this stage of reduction the selectivity for C_2H_4 increases (see Fig. 3).

Mainly, it occurs at the initial stage of reduction. When the curve of selectivity reaches the plateau, a maximum is observed on the curve of the MnO phase change. Its diffrac-

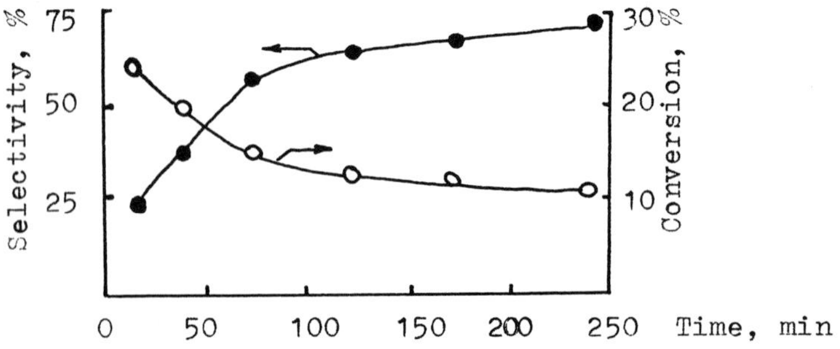

Fig. 3. Change of ethane conversion and selectivity for C_2H_4 as a function of the catalyst reduction in situ.

tion lines gradually widen in the process of reduction indicating high dispersity of this phase. Maximum selectivity is attained when the catalyst composition becomes steady-state.

A series of experiments on the dynamics of the catalyst phase composition change in the course of the regeneration with CO_2 has been carried out.

In this case the increase of the Fe_2O_3 content relative to the manganese-containing phase is likely to be due to the oxidation of the Fe_3O_4 phase. The phase transition is accompanied by CO evolution into the gas phase involving O_2 transfer from CO_2 to the reduced Fe phase. Modification of the manganese system by iron results in the formation of the multiphase system which is likely to promote facilitation of the exchange processes, involving oxygen removal from CO_2 on the Mn oxide - Fe oxide interface.

Under the steady-state conditions Mn is available in the MnO phase and Fe is available in the Fe_2O_3 and Fe_3O_4 phases.

C_2H_6 conversion can be presented by the following scheme:

		N(1)	N(2)	N(3)
C_2H_6	$\rightarrow C_2H_4 + H_2$	1	0	0
$Z''O + CO_2$	$\rightarrow Z''CO_3$	1	1	1
$C_2H_6 + Z''CO_3$	$\rightarrow C_2H_4 + H_2O + Z''O + CO$	0	1	1
$C_2H_6 + Z'O$	$\rightarrow C_2H_4 + H_2O + Z'$	0	1	0
$Z' + Z''CO_3$	$\rightarrow Z'O + Z''O + CO$	1	1	0
$Z'O + H_2$	$\rightarrow Z' + H_2O$	1	0	0

According to the scheme C_2H_6 conversion by CO_2 takes place mainly by two processes: oxidative dehydrogenation and ethane pyrolysis followed by the hydrogen oxidation.

Oxidative dehydrogenation of ethane takes place with the participation of the Fe_2O_3 (Z'O) phase and Mn-containing fragment ($Z''CO_3$).

Activation of the Mn-system by iron is likely to be due to the fact that the Mn-containing phase (MnO) in the course of reoxidation participates in the O_2 transfer from CO_2 contributing thus to the phase conversion $Fe_3O_4 \rightarrow Fe_2O_3$:

$$MnO...Fe_3O_4 \xrightarrow{CO_2} MnCO_3...2Fe_3O_4 \longrightarrow MnO...3Fe_2O_3 + CO$$

The MnO-Fe_3O_4 interface in this case makes the CO_2 reduction easy. This scheme agrees with the data of (ref. 3) where the transfer of O_2 from Mn-oxide to the reduced phase of Fe-oxide is shown. With increase of Fe in the catalyst up to 10% the improvement of the total ethane conversion is accompanied by a dramatic decrease of selectivity for ethylene. It seems to be related to the accumulation on the Fe-containing phase surface of carbon fragments and their removal upon interaction with CO_2. In this case the number of the sites of CO_2 activation decreases and the number of the active sites for the rupture of C-H, C-C bonds increases involving -(C)- or CO formation.

Therefore, dehydrogenation of ethane by carbon dioxide on the Mn-catalyst, modified by small quantities of Fe (2-3%), leads to the formation of ethylene, CO and H_2.

REFERENCES

1 Sh.A.Nuriev, I.A.Guliev, Ak.H.Mamedov, Proceedings 2 Republic Conference of Joung scientists-chemists, Baku, 1986.

2 K.I.Aika, T.Nishiyama, Proceedings 9 International Congress on catalysis, 1988, v.2, p.907.

3 P.A.Shiryaev, D.P.Shashkin, M.Sh.Zurmuhtashvili, L.Ya.Margolis, O.V.Krylov, Kinetika i Kataliz, 1984, v.25, N5, p.1164.

G. Centi and F. Trifiro' (Editors), *New Developments in Selective Oxidation*
1990 Elsevier Science Publishers B.V., Amsterdam — Printed in The Netherlands

ROLE OF ACTIVE OXYGEN FORMS AND ACIDITY IN OXIDATIVE CONVERSION OF ETHANE ON ZEOLITES

S. N. VERESHCHAGIN, N. N. SHISHKINA, A. G. ANSHITS

Institute of Chemistry and Chemical Technology, Siberian Branch of the USSR Academy of Science, Karl Marks st., 42, Krasnoyarsk, 660049, USSR

SUMMARY

Catalytic oxidation of ethane with nitrous oxide and oxygen has been investigated over ZSM-5 and mordenite. Selective conversion of ethane to ethylene and propylene was believed to involve surface oxygen atomic species which were formed by the interaction of nitrous oxide with zeolites. Catalytic activity was proportional to the number of N_2O-decomposition sites and did not depend upon the total number of strongly acidic sites measured by ammonia chemisorption.

INTRODUCTION

Zeolites have attracted considerable attention as catalysts for more than two decades as a result of their high activity and unusual selectivity for acid-catalyzed reactions. The lack of interest in zeolites as catalysts for selective oxidation may be a result of poor selectivities and high reaction temperatures. However it has recently been reported that the catalytic conversion of methane to hydrocarbons, including aromatics over HZSM-5-type zeolite in the presence of N_2O or O_2 was observed (refs. 1,2). Differences in the product distribution and catalytic activity with the two oxidants indicate that the mechanism is not the same for N_2O and O_2. Although the mechanism is unknown it is suggested that initial step of the catalytic reaction is the protonation of methane by superacid sites (ref. 3). It has been reported that O_2 and N_2O have extremely different activities and also the oxide radical ion O^-, which can be generated primarily in the decomposition of N_2O, shows higher reactivity than O_2^-, O_3^-, O^{2-} ions (refs. 4-6). We have therefore studied the oxidation of ethane by N_2O and O_2 over HZSM-5 and mordenite which will be of significant interest in the chemistry of active oxygen species and acid-base properties of the surface.

EXPERIMENTAL

Materials. High purity grade (99.8+%), ethane, oxygen and nitrous oxide were used without further purification. Helium was purified by passing through CaA-liquid nitrogen trap.

The acidic form of the ZSM-5 and mordenite was obtained by exchanging the Na cations with NH_4Cl 1.0 N at 90°C and further by calcining in air at 550°C. 1.5% Na-HZSM-5(41) was obtained by wet impregnation of HZSM-5(41) with aqueous solution of NaOH. Experiments have been performed on samples with SiO_2/Al_2O_3 ratios equal to 38, 41, 60, 90, 148 for HZSM-5, and 11 for HM, Na content was as low as 0.1%.

Equipment and Catalyst Evaluation. Zeolites were sieved to 60-80 mesh for use in the catalytic run. A mixture of C_2H_6, N_2O, (O_2) was passed through a fixed bed in a tubular flow reactor at atmospheric pressure. The catalytic run was carried out under the following conditions: catalyst weight 0.3 g, partial pressures of ethane 370 kPa, of N_2O (O_2) 37 kPa and of He 606 kPa respectively, reaction temperature 387°C, the extent of ethane conversion was upto 5%, and that of N_2O (O_2) upto 20%.

Products analyses were carried out by on-line gas liquid chromatography using flame-ionization detector and catharometer and two columns: Porapak Q and 5A molecular sieve.

Acidity measurements. After activation at 550°C under helium ammonia was adsorbed on to the catalyst at 100°C. TPD (heating rate 17°/min) indicated the number of acid sites, given as the millimoles of ammonia chemisorbed per gram of catalyst.

N_2O decomposition. Catalyst samples were heated (550°C) in the reactor under He (40 sccm at 1 atmosphere) for 3 hours. Pulses of pure N_2O (0.2 sccm) in He were introduced at 347°C.

The amount of oxygen held by the surface N_o was determined as

$$N_o = N_{N_2} - 2\,N_{O_2}$$

where N_{N_2} and N_{O_2} were the amount of nitrogen and oxygen released respectively.

RESULTS AND DISCUSSION

Upon passing the reactants over acidic form of zeolites the products CO, CO_2, C_2H_4 and H_2O were detected.

The nature of the oxidant used had a significant effect on the ethane conversion rate and product formation selectivity (Table 1).

With O_2 as the oxidant the main part of ethane undergoes the

Table 1

Catalytic oxidation of ethane by nitrous oxide over ZSM-5 and mordenite at 387°C

Catalyst	Oxidant	Rate of ethane conversion, 10^{18} molec/(g·s)	Selectivity to,%			Acidity, mmol/g	
			CO_2	C_2H_4	C_3H_6	total	h-peak*
HZSM-5(38)	O_2	0.5	42	58	-	0.72	0.39
	N_2O	6.3	2	88	8		
HZSM-5(60)	O_2	0.3	50	50	-	0.30	0.18
	N_2O	4.4	2	86	10		
HZSM-5(90)	O_2	0.2	60	40	-	0.24	0.14
	N_2O	3.6	1	90	7		
HZSM-5(148)	O_2	0.2	75	25	-	0.21	0.13
	N_2O	0.4	3	89	6		
HM-11	O_2	0.3	59	39	-	0.86	0.34
	N_2O	2.8	13	85	1		
HZSM-5(41)	N_2O	9.7	1	91	7	0.74	0.38
1.5% Na-HZSM-5(41)	N_2O	13.0	1	91	6	0.98	0.08

*h-peak - the amount of ammonia which is desorbed above 300°C.

deep oxidation and no trace of propylene and butenes was formed. By changing SiO_2/Al_2O_3 ratio from 38 to 148 catalytic activity of zeolites was reduced by three times.

The ethane conversion rate for reaction $N_2O-C_2H_6$ over HZSM-5(38) was by an order of magnitude greater than that for $O_2-C_2H_6$. Upon increasing SiO_2/Al_2O_3 ratio catalytic activity was reduced by a factor of 10. Ethylene and propylene were found to be the major products with CO_2, methane and C_4-hydrocarbons in minor amounts.

The different level of catalytic activity for $O_2-C_2H_6$ and $N_2O-C_2H_6$ can indicate, that there are two reaction mechanisms, caused by the different contributions of acidic and oxidative pathways of ethane conversion.

To elucidate the role of the catalyst acid-base properties catalytic activity was compared with acidity, evaluated by TPD spectra of ammonia.

Upon comparing the rates of C_2H_6 conversion for reaction of $O_2-C_2H_6$ it is evident, that the decrease of catalytic activity and the amount of strong acidic sites takes place concurrently. The correlation indicates that the activity of zeolites for $O_2-C_2H_6$ conversion can be caused by the acidity of the surface.

With N_2O as the oxidant the variation of zeolites activity is ranging from $0.2 \cdot 10^{18}$ to $13.0 \cdot 10^{18}$ molec.C_2H_6/(g·s) and does not correlate with the amount of ammonia adsorbed. The rate of $N_2O-C_2H_6$ conversion over HZSM-5(90) was 10 times greater than the rate observed over HZSM-5(148), these samples having the equal amount of strong acid sites. The sodium form of ZSM-5(41) was not active in oxidative conversion of ethane in $N_2O-C_2H_6$ mixture. Decationization of samples led to the appearence of strong acid sites. These samples exhibited also high activity in the reaction. Introduction of 1.5% Na into HZSM-5(41) suppresed high temperature ammonia adsorption form, but the rate of ethane conversion over 1.5% Na-HZSM-5(41) was even 1.3 times greater, than over acidic form HZSM-5(41).

Therefore, the high level of catalytic activity for $N_2O-C_2H_6$ can not be explained by the acidity of samples and could be connected with specific N_2O activation. This is consistent with the earlier study concerning the decomposition of nitrous oxide at 450°C over H-mordenite (ref. 7).

To check the possibility of N_2O activation the interaction between nitrous oxide and zeolites was studied at 347°C.

Experiments demonstrated that pure N_2O decomposed under studied conditions to give gaseous nitrogen and no appreciable evolution of oxygen was observed. The N_2O conversion was highest in the first pulse (Fig. 1), but after 10-15 pulses decomposition did not occur. Evolution of oxygen also did not occur at 347°C in flowing helium during an hour period. Hence, the amount of oxygen species held by the surface is equal to the amount of N_2O decomposed and the nitrogen released. This fact allows the calculation of the number of N_2O decomposition sites for each zeolites.

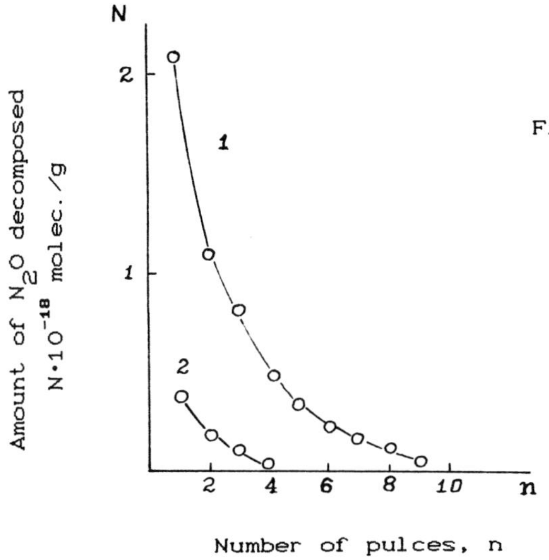

Fig 1. The amount of nitrous oxide decomposed on
1 - HZSM-5(60),
2 - HZSM-5(148)
as a function of the pulce number n.
Pulce volume 0.2 ccm, T=347°C.

The rate of C_2H_6 conversion as a function of N_2O decomposition sites is shown in Fig. 2. It is evident that for all examined samples linear correlation exists, that is the case even for HM-11 , which has different selectivities to CO_2, C_2H_4 and C_3H_6 formation.

The determining role of surface oxygen species in C_2H_6 conversion is also confirmed by pulse experiments. The surface oxygen produced by N_2O pretreatment of HZSM-5(38) reacted with ethane to form ethylene as a major product with minor amount of CO_2. Upon increasing the number of C_2H_6 pulses ethane conversion was sharply reduced to zero after 5-6 pulses. The total amount of ethane converted was $5.0\cdot10^{18}$ molec.C_2H_6/g. Taking into account the selectivity of ethylene and CO_2 formation it is possible to calculate the amount of surface oxygen consumed. This value is equal to $5.3\cdot10^{18}$ atom O/g that corresponds to 80% of initial

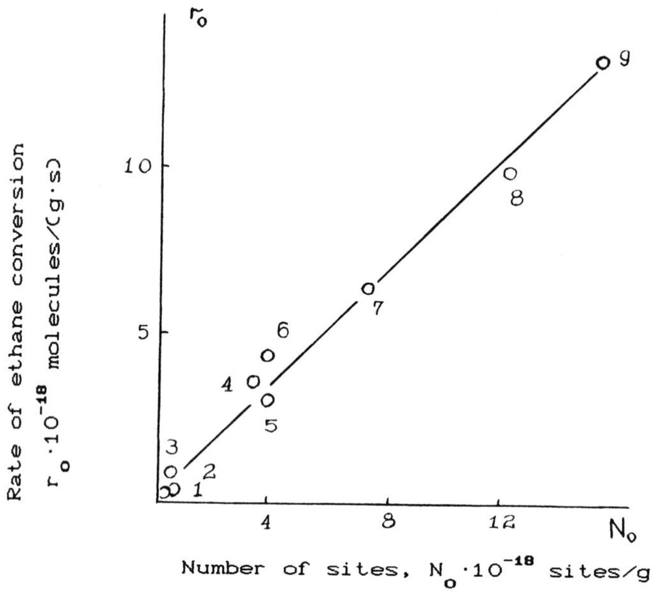

Fig. 2. Rate of C_2H_6 conversion r_o versus number of surface N_2O decomposition sites N_o in N_2O-C_2H_6 reaction at 387°C on zeolites: 1-NaZSM-5(41), 2-NaZSM-5(38), 3-HZSM-5(148), 4-HZSM-5(90), 5-HM-11, 6-HZSM-5(60), 7-HZSM-5(38), 8-HZSM-5(41), 9-1.5% Na-HZSM-5(41).

oxygen coverage ($6.8 \cdot 10^{18}$ atom O/g for HZSM-5(38)). Hence, a good agreement between the amount of oxygen held by the surface and the consumed oxygen , as well as the similarity of product composition obtained during pulse and flow experiments was observed. These results clearly show that the oxygen surface coverage determines high degree of catalytic activity and excellent selectivity to C_2-C_3 olefines for N_2O-C_2H_6 reaction.

Electron paramagnetic resonance studies have demonstrated that there are strong redox sites on HZSM-5 and mordenite capable to form ion-radical species. The appearence of organic cation-radicals was observed during the interaction of olefines with HZSM-5 (ref. 8), the anion-radical SO_2^- was detected by EPR on HM during adsorption of sulfur dioxide (ref. 9). Decomposition of nitrous oxide can be a result of N_2O interaction with structural defects on the zeolite framework (ref. 7) or impure metal ions, for example Fe^{3+}.

As to possible oxygen species responsible for the reaction the following ones may be mentioned: O^- and atomic oxygen. According to work (ref. 10) $O^-..^-O$-type sites can be formed by a high temperature treatment of HZSM-5 (SiO_2/Al_2O_3=70-140), the number being equal to 10^{17} spin/g as measured by EPR spectroscopy. Within an order of magnitude it corresponds to the number of N_2O decomposition sites which was found to be $5 \cdot 10^{17}$ sites/g for HZSM-5(148). The surface oxygen species formed by N_2O decomposition does not exhibit an EPR signal. It may be concluded that they are likely to be uncharged forms having an atomic character as proposed for oxidative dehydrogenation of ethane by nitrous oxide over cobalt-doped magnesium oxide (ref.11).

REFERENCES

1. S.S.Shepelev, K.G.Ione, React. Kinet. Catal. Lett., 23 (1983) 319.
2. S.S. Shepelev, K.G. Ione, React. Kinet. Catal. Lett., 23 (1983) 323.
3. S.Kowalak, J.B.Moffat, Applied Catalysis, 36(1-2) (1988) 139.
4. K.Aika, J.H.Lunsford, J. Phys. Chem., 81 (1977) 1393.
5. M.Iwamoto, J.H.Lunsford, J. Phys. Chem., 84 (1980) 3079.
6. M.Iwamoto, T.Taga, S.Kagawa, Chem. Lett., (1982) 1469.
7. A.A.Slinkin, T.K.Lavrovskaya, I.V.Mishin, Kinet. Katal., 19(4), (1978) 922.
8. S.J.Shih, J. Catal., 79 (1983) 390.
9 A.A,Slinkin, A.V.Kucherov, D.A.Kondratjev, T.N.Bondarenko, A.M.Rubinstein, Kh.M.Minachev, Kinet. Katal., 22 (1986) 156.
10. V.A.Poluboyarov, V.F.Anufrienko, N.G.Kalinina, S.N.Vosel, Kinet. Katal., 26 (1985) 751.
11. K.Aika, M.Isobe, K.Kido, T.Moriyama, T.Onishi, J. Chem. Soc. Faraday Trans. -1, 83 (1987) 3139.

G. Centi and F. Trifiro' (Editors), *New Developments in Selective Oxidation*
1990 Elsevier Science Publishers B.V., Amsterdam — Printed in The Netherlands

SELECTIVE OXIDATION OF PROPANE TO ACROLEIN AND AMMOXIDATION TO ACRYLONITRILE OVER Ag-DOPED BISMUTH VANADOMOLYBDATE CATALYSTS

Young-Chul KIM, Wataru UEDA and Yoshihiko MORO-OKA*

Research Laboratory of Resources Utilization, Tokyo Institute of Technology, Nagatsuta-cho 4259, Midori-ku, Yokohama, 227 Japan

SUMMARY

Acrolein was formed as a major product in the partial oxidation of propane using molecular oxygen over Ag-doped bismuth vanadomolybdate catalysts having a scheelite structure. The catalyst system was also effective for the selective ammoxidation of propane to acrylonitrile. The reaction seems to proceed via propylene involving a kind of autoxidation of propane at temperatures about 500°C.

INTRODUCTION

The selective oxidations of lower alkanes to other chemicals are becoming increasingly important both in fundamental and in industrial chemistry. However, selective oxidations of alkanes to partially oxidized products are accompanied by many difficulties due to their low reactivities compared to alkenes and dienes. Nevertheless, the selective oxidation of n-butane to maleic anhydride has been successfully established using V-P-O catalysts[1,2]. Now, much attention is being denoted to the catalytic oxidative dimerization of methane by many investigators.[3,4] On the other hand, little has been reported for the partial oxidation of propane. Giordano et al reported the oxidation of propane to acrolein over supported Te_2MoO_7 catalysts.[5] The catalysts such as $Ba_{5.55}Bi_{2.3}Te_3O_{18}$[6] and U-Sb-O[7] were clamed to be active for the oxidation to form acrolein in some patents. Direct formation of acrylic acid from propane was also reported on V_2O_5-P_2O_5-TeO_2 catalysts.[8] However, their catalytic performances are not yet sufficient for the practical applications. Compared to the simple oxidation of propane, more predominant results have been obtained in the ammoxidation to form acrylonitrile. A series of patents have been published by using multicomponent metal oxide catalysts containing Mo, Bi, Sb and many other elements.[9] The space time yield of the target product has been improved remarkably but a long distance still remains to the final goal. Here, we report our discovery that Ag-doped bismuth vanadomolybdates are promising catalysts for the conversions of propane both to acrolein and acrylonitrile.

EXPERIMENTAL METHODS

Ag-doped bismuth vanadomolybdate catalysts were prepared by an aqueous slurry reaction of ammonium metavanadate, ammonium heptamolybdate, bismuth nitrate and silver nitrate at pH = 10. After the evaporation of water, the obtained precursors were dried at 110°C and calcined at 520°C for 6 hr in an air stream. All other mixed metal oxides used in the survey to find effective catalyst were prepared according to the literature. All catalysts were used in the form of powder with 100-200 mesh after diluted five times by quartz tips.

The reaction at atmospheric pressure was carried out in a conventional flow system equipped with a quartz tube reactor of 18 mm inner diameter and a tubular furnace. Mixtures of propane, oxygen and nitrogen were fed in from the top of the reactor. The feed and products were analyzed by an on-line gas chromatograph operating with two sequential columns (Gaskuropack 54, 3m and Molecular sieve 5A, 2m).

RESULTS AND DISCUSSION

Survey of the effective catalyst

Various types of the catalysts reported for the selective partial oxidations of lower alkanes and alkenes were examined for the oxidation of propane to form acrolein. Table 1 shows some typical results obtained in the oxidation of propane over the various mixed oxide catalysts. It is well-known that V_2O_5-P_2O_5 systems give the best result for the selective formation of maleic anhydride in the oxidation of C_4-alkane, n-butane.[1)] However, the catalysts were extreamly inactive for the acrolein formation in the oxidation of propane. Although small amounts of acrylic acid were formed as reported,[8)] the major products were carbon oxides. Mixed oxides containing vanadium and magnesium with or without phosphorous pentoxide were reported active for the oxidative dehydrogenation of propane.[10)] Certainly, some amounts of propene were produced under the present reaction conditions but acrolein was not yielded at all. On the other hand, mixed oxide catalysts containing both vanadium and molybdenum showed better activity to form acrolein. Thus, catalysts based on molybdenum seem to have an ability for the formation of acrolein. Hence, various molybdate-based catalysts which are widely used for the selective oxidations of propene and 2-methyl propene to the corresponding unsaturated aldehydes were tested for the reaction. All these catalysts showed improved results for the acrolein formation. Amoung them, the scheelite type catalysts were found to show higher catalytic activity and selectivity to acrolein. The catalyst system having the following composition, $Bi_{1-x/3}V_{1-x}Mo_xO_4$[11)] was found to give the highest catalytic activity and selectivity to acrolein amoung the scheelite type catalysts tested.

TABLE 1

Conversion and selectivity to acrolein in the partial oxidation of propane over various mixed oxide catalysts.[a)]

Catalyst	C_3H_8/O_2 ratio	Reaction temperature (°C)	Propane conversion (%)	Acrolein selectivity (%)
V_2O_5-P_2O_5(P/V = 1/1)	0.8	469	30	tr
V_2O_5-P_2O_5(P/V = 3/2)	0.8	450	43	0
V-Mg-P-O(V/Mg/P = 1/1/0.1)	0.7	430	65	tr
V-Mg-O(V/Mg = 1/1)	0.7	425	70	tr
V-Sb-Mo-O(V/Sb/Mo = 1/1/4)	0.6	496	16	2
V_2O_5-MoO_3(V/Mo = 1/1)	0.6	440	55	10
V_2O_5-MoO_3(V/Mo = 1/2)	0.6	439	56	15
Bi_2MoO_6(γ-type)	1.0	475	36	18
$Bi_2Mo_3O_{12}$(α-type)	0.9	484	37	24
$Bi_2Mo_2O_9$(β-type)	0.6	500	20	34
$Bi_1Mo_{12}Fe_3Co_8O_x$	0.6	500	10	8
$Bi_{0.77}V_{0.3}Mo_{0.7}O_4$	0.7	478	16	34
$Bi_{0.85}V_{0.55}Mo_{0.45}O_4$	0.8	488	35	31
$Bi_{0.85}V_{0.55}Mo_{0.45}O_4$/purmice	0.7	473	19	34
$Bi_{0.85}V_{0.35}P_{0.2}Mo_{0.45}O_4$	0.9	486	40	18
$Bi_{0.97}V_{0.91}Mo_{0.09}O_4$	0.8	476	13	31

a) Space velocity; 1800 cm^3/g-cat·h, Feed gas; (C_3H_8 + O_2) = 91%, N_2 = 9%.

Improvement of bismuth vanadomolybdate catalyst by the doping of monovalent metal cations

Tricomponent metal oxide catalysts having the scheelite structure, $Bi_{1-x/3}V_{1-x}Mo_xO_4$ were first reported by Sleight et al[11)] as effective catalysts for the selective partial oxidation of propene to acrolein. Their catalytic behaviours under reaction conditions were extensively investigated by Moro-oka et al using an $^{18}O_2$ tracer.[12,13)] The product distribution in the propane oxidation was examined changing the catalyst composition by changing x value in the catalyst system. The results are shown in Fig. 1. The catalyst including no molybdenum, $BiVO_4$, showed very low catalytic activity to form acrolein. Dehydrogenation of propane to propene was mainly observed on this catalyst. Increasing yield of acrolein was obtained by increasing concentration of molybdenum in the catalyst system. The highest yield of acrolein was attained on the catalyst having the composition, $Bi_{0.85}V_{0.55}Mo_{0.45}O_4$.

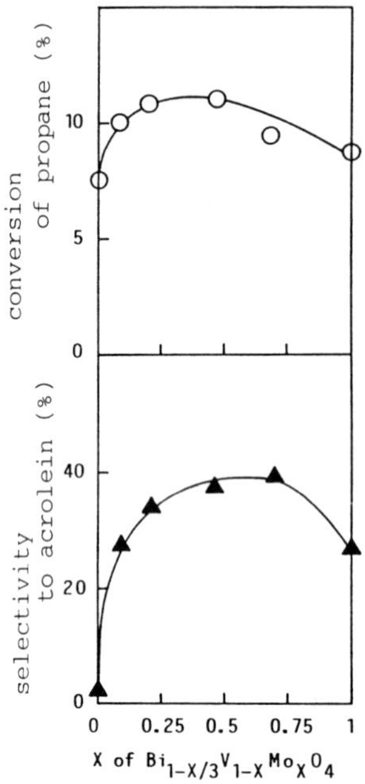

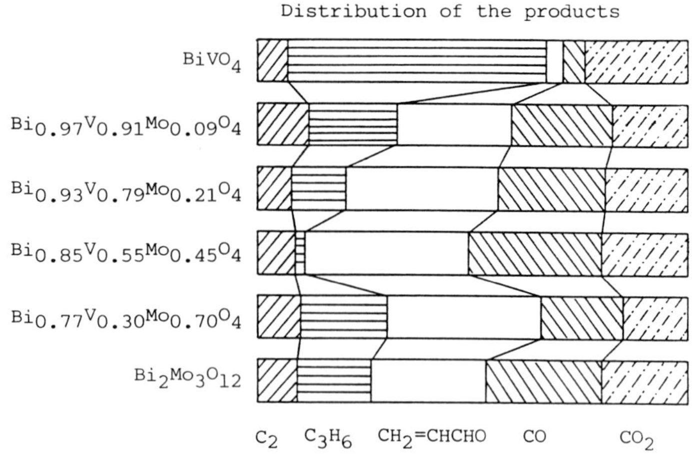

Fig. 1 Catalytic activity and selectivity of $Bi_{1-x/3}V_{1-x}Mo_xO_4$ for the oxidation of propane.
a) Dependencies of the catalytic activity and selectivity on the catalyst composition.
b) Products distribution at 10% conversion of propane.
Reaction temp.; 500°C, Space velocity; 3000 cm^3/g-cat·h, C_3H_8/O_2 ratio; 0.55.

In order to improve its catalytic performance, several kinds of monovalent metal ion-doped bismuth vanadomolybdates were prepared and employed for the oxidation of propane. The results are summarized in Table 2. It was found that dopings of K, Rb, and Tl rather decreased catalytic activity to form acrolein giving mainly propene by the dehydrogenation. Some improvements were obtained in the additions of Li and Na but the most prominent effect was observed in the Ag doping, where both catalytic activity and selectivity to acrolein were improved remarkably. Fig. 2 shows the conversion of propane and selectivity of products on $Ag_xBi_{0.85}V_{0.55-x}Mo_{0.54}O_4$ with variation of Ag amounts.[14] The selectivity was increased gradually by the silver addition at first, then decreased by the further addition, showing a maximum at x = 0.01. Thus, the following composite metal oxide, $Ag_{0.01}Bi_{0.85}V_{0.54}Mo_{0.45}O_4$, was obtained as the most effective catalyst for the oxidation of propane to acrolein.

Oxidation of propane over Ag-doped bismuth vanadomolybdate catalyst

(i) Effect of the reaction temperature. The oxidation of propane to acrolein on the $Ag_{0.01}Bi_{0.85}V_{0.54}Mo_{0.45}O_4$ catalyst was examined changing the reaction temperature from 380 to 540°C.[15] As shown in Fig. 3, observed changes in the conversion of propane and selectivity to acrolein with the reaction temperature

TABLE 2

Conversion and selectivity for the partial oxidation of propane over monovalent metal ion-doped bismuth vanadomolybdate catalyst.[a)]

Monobalent metal (M^I)[b)]	Conversion (%)	Selectivity[c)]				
		CH_2CHCHO	CO	CO_2	C_2	C_3H_6
none	11.0	38.5	31.5	19.5	10.5	tr
Li	7.8	41.5	28.0	21.6	8.8	0
Na	6.8	38.3	29.7	23.5	8.5	0
K	7.6	16.2	18.2	11.4	10.0	44.2
Rb	9.6	4.7	11.1	9.5	8.2	66.5
Ag	13.1	63.5	15.6	11.8	9.1	0
Tl	5.1	10.7	17.9	12.9	14.2	44.2

a) Reaction temp.; 500°C, Space velocity; 3000 cm^3/g-cat·h, C_3H_8/O_2 ratio; 0.55.
b) $M^I_{0.01}Bi_{0.85}V_{0.54}Mo_{0.45}O_4$. c) Normalized by carbon number in each product.

were quite unusual. Oxidation of propane started suddenly at about 400°C and showed the highest conversion at this temperature. The conversion of propane then decreased slightly with increasing the reaction temperature. On the other hand, the selectivity to acrolein increased markedly with increasing the reaction temperature. The products besides acrolein were propene at lower temperatures, CO, CO_2 and C_2-hydrocarbons. The phenomena shown in Fig. 3 are quite different from those observed in the ordinary catalytic oxidations and suggest strongly that the reaction involves a kind of autoxidation in the process and propene is an intermediate to the main product.

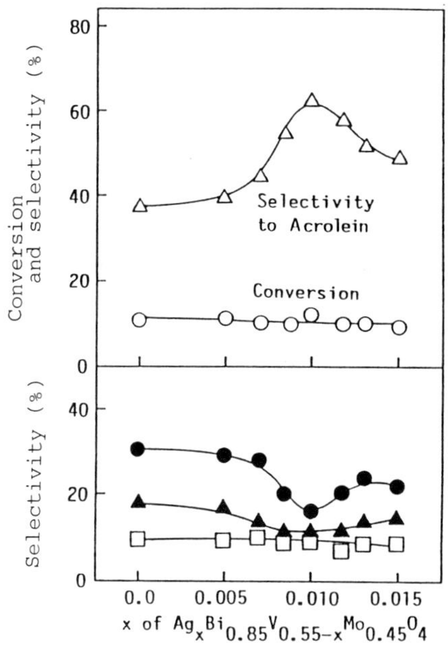

Fig. 2. Conversion of propene (○) and selectivity of products on $Ag_xBi_{0.85}V_{0.55-x}Mo_{0.45}O_4$ with variation of Ag amounts.
(△) acrolein, (▲) CO_2, (●) CO, (□) C_2, Space velocity; 3000 cm^3/g·h, C_3H_8/O_2 ratio; 0.55.

(ii) Effect of the space velocity

Effect of the space velocity on the reaction was examined at 500°C by varying the flow rate of the feed gas using a constant amount of the

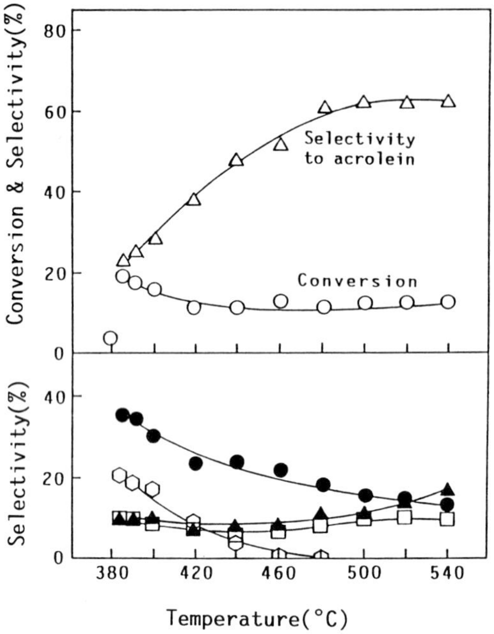

Fig. 3. Conversion of propane and selectivity of products on $Ag_{0.01}Bi_{0.85}V_{0.54}Mo_{0.45}O_4$ catalyst with variation of the reaction temperature. Space velocity; 3000 cm^3/g-cat·h.

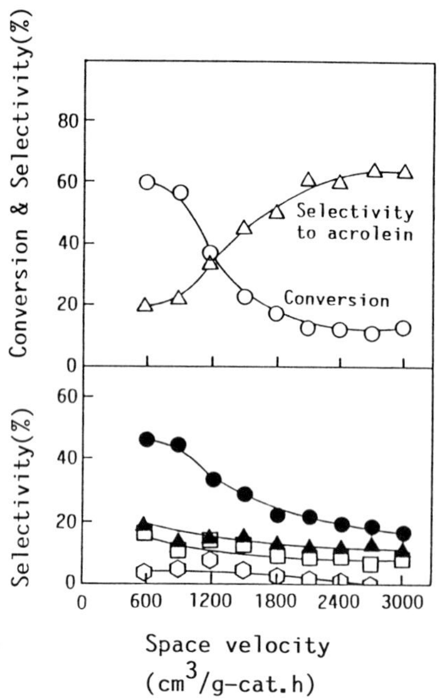

Fig. 4. Effect of the space velocity on the partial oxidation of propane. Reaction temp.; 500°C.

(△) CH_2CHCHO, (□) C_2, (⬡) C_3H_6, (●) CO, (▲) CO_2.
Feed gas; C_3H_8 32%, O_2 59%, N_2 9%.

$Ag_{0.01}Bi_{0.85}V_{0.54}Mo_{0.45}O_4$ catalyst. The results are shown in Fig. 4. The general tendency on the space velocity of the catalytic oxidation of hydrocarbon is also valid for this reaction. The conversion of propane rose and the selectivity to acrolein fell gradually with decreasing the space velocity and increasing the contact time. It is clear that some carbon oxides are formed in the consecutive oxidation of acrolein. Stable acrolein selectivity higher than 60% was observed only in the range of the space velocity higher than 2000 cm^3/g-cat·h.

(iii) Effect of the feed gas composition on the oxidation of propane Dependency of the reaction on the reactant gas composition was also examined by varying the C_3H_8/O_2 molar ratio in the feed gas. All runs were carried out under the steady reaction conditions (space velocity; 3000 cm^3/g-cat·h, reaction temp.; 500°C) where total concentration of propane and oxygen in the feed gas was fixed at 91%. As shown in Fig. 5, the conversion of propane increases with increasing the C_3H_8/O_2 ratio in the feed gas keeping a stable

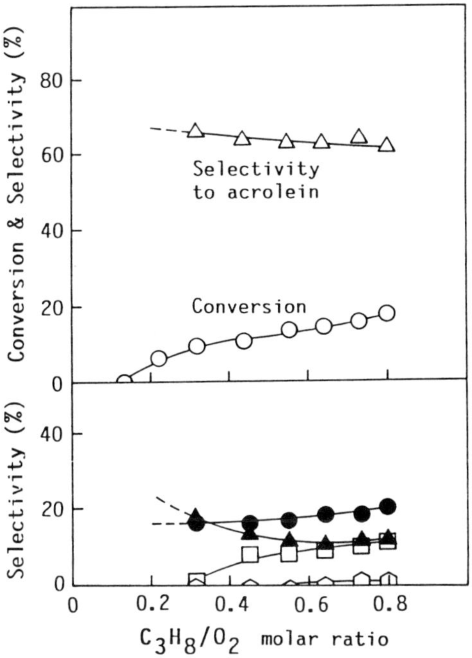

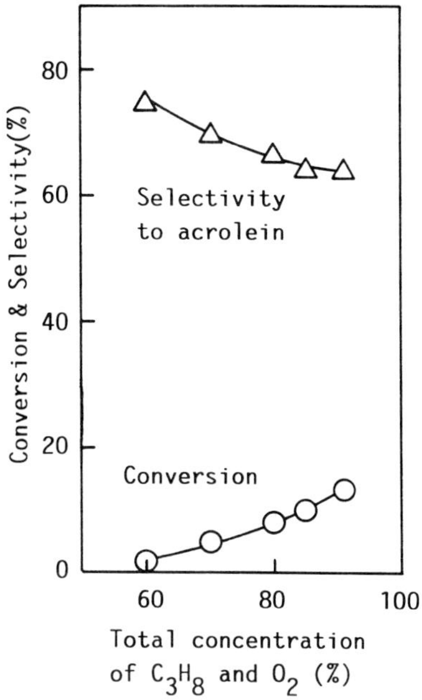

Fig. 5. Dependency of the reaction on the C_3H_8/O_2 molar ratio in the feed gas. (△) CH_2CHCHO, (□) C_2, (⬡) C_3H_6, (●) CO, (▲) CO_2.
Feed gas; $(C_3H_8 + O_2)$ 91%, N_2 9%.

Fig. 6. Effect of the total concentlation of propane and oxygen on the rate of the reaction.
Feed gas; C_3H_8/O_2 molar ratio 0.55, N_2 balance.

Conditions: Catalyst; $Ag_{0.01}Bi_{0.85}V_{0.54}Mo_{0.45}O_4$, Reaction temp.; 500°C, Space velocity; 3000 cm^3/g-cat·h.

selectivity to acrolein. It is noteworthy that the reaction stopped completely under the lower C_3H_8/O_2 ratio than 0.14. This does not mean that only the C_3H_8/O_2 molar ratio is important factor to promote the reaction. The rate of the propane oxidation was further determined at various total concentrations of propane and oxygen. In the measurement, the C_3H_8/O_2 molar ratio was fixed at 0.55 by replacing the reactant gas by nitrogen to keep balance. The results are shown in Fig. 6. The most striking is that the oxidation of propane does not proceed at all when the total concentration of propane and oxygen is below 60% in the feed gas. Thus, it is clear that fairly high concentration of propane is required to promote the reaction. All these results suggest strongly that the oxidation of propane involves a kind of autoxidation and the catalyst does not participate seriously in the activation of propane. Actually, considerable amounts of propene were formed without any catalyst in the oxidation of propane under the reaction conditions employed in this investigation. However, no acrolein was detected in the homogeneous gaseous oxidation of propane using no catalysts. It is concluded that propene is the intermediate to acrolein and mixed oxide catalysts mainly promote the oxidation of propene to acrolein in this reaction. It should be referred that excellent catalysts for the oxidation of propene to acrolein such as multicomponent bismuth molybdates are susally used at lower temperatures below 400°C. It was found that these catalysts are not so effective for the oxidation of propane to acrolein. As shown in Table 1, one of the best catalyst for the propene oxidation, $Bi_1Mo_{12}Fe_3Co_8O_x$, showed very poor selectivity to acrolein in the oxidation of propane. Excellent catalysts for the propane oxidation are required to act for the propene oxidation at more higher temperature where the homogeneous oxidative dehydrogenation of propane to propene proceeds efficiently.

Ammoxidation of propane to acrylonitrile

It has been well known that selective catalysts to form acrolein from propene are also active for the ammoxidation of propene to acrylonitrile. Most catalysts employed in this investigation were also examined for the ammoxidation of propane. It was found that considerable number of the catalysts showed very high activity and selectivity to form acrylonitrile. The results are partly shown in Table 3. The selectivities to acrolein obtained on the same catalysts are shown in the last column of the table for comparison. It is noteworthy that most of the catalysts effective for the ammoxidation of propane are not so to form acrolein. It seems that the difference mainly arises from the different

TABLE 3

Conversion and selectivity for the ammoxidation of propane.[a)]

Catalyst	Ammoxidation to acrylonitrile		Oxidation to acrolein	
	Conv. (%)	Selec. (%)	Conv. (%)	Selec. (%)
$Bi_3Fe_1Mo_2O_{12}$	12.8	51.5	14.3	17.0
$Bi_3Ga_1Mo_2O_{12}$	10.1	65.3	11.7	34.5
$Bi_3Fe_{0.3}Ga_{0.7}Mo_2O_{12}$	11.3	60.0	10.2	12.4
$Bi_4Nb_1Mo_{20}O_x$	9.0	55.4	7.8	42.9
$Pb_{0.88}Bi_{0.08}MoO_4$	8.4	52.5	9.7	40.1
$Bi_{0.85}Nb_{0.55}Mo_{0.45}O_4$	5.9	64.5	11.2	58.1
$Ag_{0.01}Bi_{0.85}V_{0.54}Mo_{0.45}O_4$	13.1	67.0	13.4	63.0

a) Reaction temp.; 500°C, Space velocity; 3000 cm^3/g-cat·h, Feed gas; NH_3 20%, C_3H_8 50%, O_2 30%.

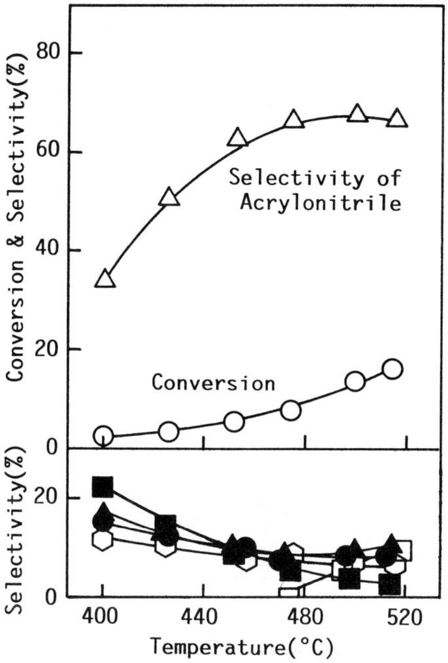

Fig. 7. Conversion of propane and selectivity to products in the ammoxidation of propane on $Ag_{0.01}Bi_{0.85}V_{0.54}Mo_{0.45}O_4$ catalyst with variation of the reaction temperature.

(△) CH_2CHCN, (□) C_2-hydrocarbon, (■) CH_3CN, (⬡) C_3H_6, (●) CO, (▲) CO_2.
Space velocity; 3000 cm^3/g-cat·h,
Feed gas; C_3H_8 34%, NH_3 20%, O_2 46%.

stabilities of two products at the reaction conditions. In fact, a fair number of effective catalysts have been insisted in patents for the ammoxidation of propane.[9] Amoung the catalysts tested, the best catalyst for the acrolein formation, $Ag_{0.01}Bi_{0.85}V_{0.54}Mo_{0.45}O_4$, also gave the most excellent results for the acrylonitrile formation. The conversion of propane and the selectivity to products with variation of the reaction temperature are shown in Fig. 7.

CONCLUSION

We have shown that propane can be converted selectively to acrolein and acrylonitrile in the oxidation and ammoxidation over mixed oxide catalysts.
At this stage, the conversion of propane is not satisfactory even on the best catalyst. Selectivities to the main products still remain some room to be improved. However, compaired to the oxidation or ammoxidation of propene, the concentration of propane in the feed gas in this investigation is 5 to 10 times higher than that of propene. Therefore, the concentrations of the main products in the effluent gas and space time yields reach almost the same values with those of the propene reactions. We think that this will stimulate further investigations for the selective oxidation of propane in the near future.

REFERENCES

1 R. L. Varma and D. N. Saraf, Ind. Eng. Chem. Prod. Res. Dev., 18 (1979) 7.
2 F. Cavani, G. Centi, A. Riva, and F. Trifiro, Catal. Today, 1 (1987) 17.
3 T. Ito and J. H. Lunsford, Nature (London), 314 (1987) 721.
4 K. Otsuka and T. Nakajima, J. Chem. Soc. Faraday Trans. I, 83 (1987) 1315.
5 N. Giordano, J. C. J. Bart, P. Vitarelli, and S. Cavallaro, Oxid. Commun., 7 (1984) 99.

6 W. C. Conner Jr., S. L. Soled, A. J. Signorelli, and B. A. DeRites, U.S. Patent 4472314.
7 J. Dewing, C. Barnett, and J. J. Rooney, Ger. Offen, 1903617.
8 M. Ai, J. Catal., 101, (1986) 389.
9 U.S. Patents, 4609502, 4736054, 4746641, 4760159, 4767739, 4769355, 4783545, 4784979, 4788173, 4788317.
10 M. A. Chaar, D. Patel, and H. H. Kung, J. Catal., 109 (1988) 463.
11 A. W. Sleight, K. Aykan, and D. B. Rogers, J. Solid. State Chem., 13 (1975) 231; A. W. Sleight, Advanced Materials in Catalysis, Academic Press, New York, 1977 p.181.
12 W. Ueda, K. Asakawa, C. L. Chen, Y. Moro-oka, and T. Ikawa, J. Catal., 101 ((1986) 360.
13 W. Ueda, C. L. Chen, K. Asakawa, Y. Moro-oka, and T. Ikawa, J. Catal., 101 (1986) 369.
14 Young-Chul Kim, W. Ueda, and Y. Moro-oka, J. Chem. Soc. Chem. Commun., in press.

P. F. Ruiz (Universite Catholique de Louvain, Belgium)

The figure 3 is typical of a decomposition of the catalysts as function of the temperature giving a two phase catalysts. It is possible to explain the increase of the selectivity by a cooperative effects between these phases (Remote control mechanisum), namely the control by a donor phase (acceptor), via oxygen spill over. I would like to know your opinion about these hypothesis.

Y. Moro-oka (Tokyo Institute of Technology, Japan)

I don't agree with your hypothesis that the results shown in figure 3 came from the decomposition of the catalyst. Used catalyst gave the same results as shown in figure 3 and the catalytic activity was quite stable for a long time at any reaction temperature.
The catalyst gave the same XRD pattern before and after the reaction.

P. Courtine (Université de Technologie de Compiègne, France)

Could you identify the reduced phase of the catalyst corresponding to the composition $Bi_{0.85}V_{0.55}Mo_{0.45}O_4$?

Y. Moro-oka (Tokyo Institute of Technology, Japan)

We reported the results on the reduction of $Bi_{0.85}V_{0.55}Mo_{0.45}O_4$ catalyst previously (ref. 13). It is noteworthy that the reduction of this catalyst does not take place in the vicinity of the surface layer of oxide because rapid migration of lattice oxide ions prevents the local reduction of the catalyst. It was found that the reduction spread over the whole oxide particles. Although numbers of oxide ion vacancies were formed, the catalyst kept its original scheelite type structure at least until the reduction to 6%. Thus, we found no new XRD peaks during the reduction of this catalyst.

Ashok Padia (Scientific Design, USA)

Your research is very interesting. My comments are
1) to explore regions of commercial interest and
2) to explore $i\text{-}C_4 \longrightarrow$ methacrolein?

Y. Moro-oka (Tokyo Institute of Technology, Japan)

1) We have checked the possibility to develop the reaction to the commercial scale by asking industrial specialists to evaluate its economical value. If unreacted propane is effectively recycled, the process may be comparable to the oxidation of propylene (SOHIO process). Several companies are now following the reaction.

2) We have examined to extend the reaction to $i\text{-}C_4$ oxidation to form methacrolein. Methacrolein was surely detected as one of the main products but selectivity of $i\text{-}C_4$ to it was fairly lower than that of propane to acrolein. We still continue to improve the $i\text{-}C_4$ oxidation.

Z. Osipova (Institute of Catalysis, USSR)

1) Because of different stability of acrolein and acrylonitrile in the reaction conditions there is a different dependence of selectivity on conversion for these compounds. Do you present the optimal yield of acrolein and acrylonitrile on your catalysts as you compare your results with those for oxidation and ammoxidation of propylene?

2) Comment. The activity of molybdenum containing catalysts in ammonia oxidation to nitrogen and nitrogen oxide is rather high. In your conditions the activity of these catalysts in ammonia oxidation may become higher than those in propane activation. In this case the antimony containing catalysts may be better than vanadium containing because of their low activity in ammonia oxidation.

Y. Moro-oka (Tokyo Institute of Technology, Japan)

1) We have tried to find optimum conditions to form acrolein and acrylonitrile as far as possible. As you know, the processes for acrolein and acrylonitrile production from propylene have been established on the continuous modification for 30 years. The reactions presented by us stay in far unpolished state compared to those for propylene. Thus, we expect that the reactions have great room for improvement in the future.

2) Thank you very much for your comment. I agree with your suggestion that antimony is one of the best candidate for the catalyst for this reaction.

G. M. Pajonk (Univ. C. Bernard Lyon 1, France)

You explained the large differences of selectivities between ammoxidation and partial oxidation by saying that acrolein is more unstable in your reaction conditions. From mechanistic point of view it is generally accepted that both reactions proceed through the same intermediates. So my question concerns the conversion of acrolein in your ammoxidation conditions, did you such an experiment?
Second, assuming your hetero-homogeneous reactions what is the species initiating the ammoxidation in the gas phase?

Y. Moro-oka (Tokyo Institute of Technology, Japan)

1) No, we didn't.
Recently, we obtained improved selectivities to acrolein using the same catalysts listed in Table III. We have written them in our revised manuscript. However, they are still lower than those for acrylonitrile. At this stage, I have no evidence to explain the difference.

2) I am sorry that I can not reply to your question about initiator of ammoxidation. Our estimation for the hetero-homo reaction mechanism is based on the following experimental results.

i) Considerable amounts of propylene were formed without any catalyst under the reaction conditions.

ii) Propylene was selectively converted to acrolein or acrylonitrile on every catalyst adopted in this reaction.

We have no direct informations about unstable intermediates of the reaction byond mentioned above.

O. Watzenberger (Institut für Technische Chemie I, Universitat Erlangen-Nürnberg, BRD)

1) Does lattice-oxygen "migration" proceed only on the surface, or is there oxygen ion transport in the bulk, too?

2) How did you measure oxygen "migration" (or didn't you)?

3) How can you confirm that it is really oxygen transport?

4) Do you have any values or estimation for the rate of oxygen migration?

Y. Moro-oka (Tokyo Institute of Technology, Japan)

We have long studied lattice oxide ion migration under the working state of the catalysts using $^{18}O_2$ tracer technique.
For example,

$$\text{S (hydrocarbon)} + {}^{18}O_2 \xrightarrow[{}^{16}\text{O-cat}]{} S^{16}O$$

The above reaction was clearly observed in the $Bi_{0.85}V_{0.55}Mo_{0.45}O_4$ catalyst system. Lattice oxide ions not only in the vicinity of the surface but also in the bulk of the oxide particles were involved into the oxidation reaction. Since the oxide ion incorporation to the reaction product was observed under the steady state catalysis, it is clear that the migration proceeds not only from the bulk to the surface but also from the surface to the bulk of the catalyst (ref. 12). Migration of bulk oxide ions was also confirmed by the XRD studies during the reduction of the catalyst (ref. 13). This involvement of the lattice oxide ion into the reaction does not depend on the simple exchange reaction between adsorbed oxygen species and the lattice oxide ion. Indeed, migration of oxide ions is not so rapid under the completely oxidized state of the catalysts. It takes place only under the partially reduced state in the presence of reductant such as hydrocarbon and this is the reason why we could not determine the absolute rate of oxide ion migration. Thus, we estimated the migration rate by measuring the degree of involvement of lattice oxide ions in the reaction using $^{18}O_2$ tracer (ref. 12). On the basis of the results, we proposed the following model of the catalyst.
Catalytic activity of the mixed oxide catalyst adopted in our investigation (most of them have scheelite structures) was parallel with the estimated value of the lattice oxide migration (ref. 12).

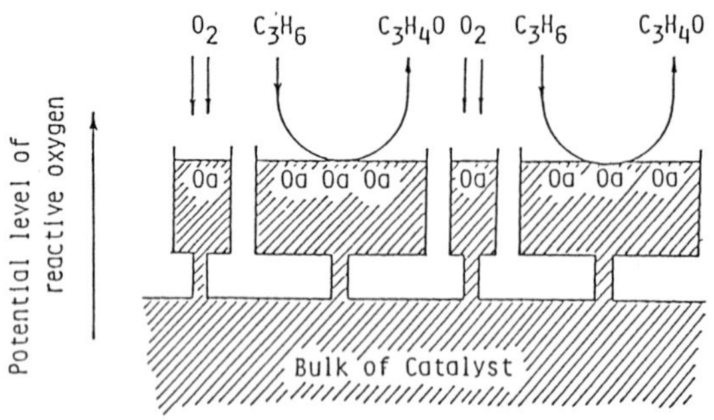

Water tank model of rapid equalization of chemical potential of active oxygen through bulk diffusion of oxide ions in the oxidation of propylene to acrolein.

J. Volta (Inst. de Catalyse, France)

1) Did you test your catalysts in the presence of water?

2) What do you think about the role of silver in your catalysis of propane

oxidation?

Y. Moro-oka (Tokyo Institute of Technology, Japan)

1) No, we didn't but I think that we should test it because I also know importance of the effect of water vapour in the oxidation of hydrocarbon.

2) At this stage, we have no direct evidence for the role of silver in this reaction but I think that it may serve in the second step in the process (oxidation of intermediary propylene to acrolein) by activation of molecular oxygen.

G. Emig (University of Karlsruhe, BRD)

1) Diameter of your tube reactor is relatively large. At the same time you leave higher concentrations of propane and oxygen. Didn't you get problem in keeping the bed temperature in radial and axial direction canstant?

2) How can you explain the different conversion vs temperature behavior for acrolein (fig. 3) and acrylonitrile (fig. 7) formation?

Y. Moro-oka (Tokyo Institute of Technology, Japan)

1) I should refer the effect of the reactor on the reaction. Since the reaction involves homogeneous steps, the results depend seriously on the type of the reactor. As you pointed out, temperature of the catalyst bed was no homogeneous under the reaction but we had no problem to control it.

2) We confirmed that the results shown in figures 3 and 7 are reproducible and did not come from the decomposition of the catalysts or some experimental faults. I think that conversion of propane to acrolein or acrylonitrile is controlled by the homogeneous steps. It has been well known that the homogeneous oxidation including radical reaction does not obey to the usual conversion vs temperature behavior and often shows a negative activation energy. Partial difference between the results shown in figures 3 and 7 may come from the presence and absence of ammonia in the reaction system.

M. Misono (The University of Tokyo, Japan)

1) Did you observe any XRD lines due to Ag for used catalysts?

2) The presence of Ag is necessary for oxidation to acrolein but it is not necessary for ammoxidation. Is this correct?

Y. Moro-oka (Tokyo Institute of Technology, Japan)

1) No, we didn't.

2) Addition of Ag component is effective for both oxidation and ammoxidation of propane but is more effective for the acrolein formation than acrylonitrile.

B. Delmon (Universite Catholique de Louvain, Belgium)

1) I accept your conclusion that homogeneous steps may be involved in your reaction. However, one could remark that Ag can produce electrophilic species which after migration on the oxidic pox, could attack, or cooperate in the attack of propane.

I notice that the selectivity increases when temperature increases. It is known that oxide surfaces become progressively depleted in electrophilic species, for the benefit of nucleophilic species, when temperature increases. One could interpret your results by saying that attack of the propane molecule needs a certain balance between electrophilic and nucleophilic oxygens. Would this interpretation correspond to your conclusion?

2) Ag might be an effective producer of oxygen mobile species (spill over oxygen). In our experiments, we observe that spill-over oxygen protects oxide catalysts from reductions. Did you compare the reduction state after catalytic work of two catalysts of the same composition except for Ag, which would be present in only one of them?

<u>Y. Moro-oka (Tokyo Institute of Technology, Japan)</u>

1) I agree with that nature of active oxygen species based on their negative charge is very important to determine the conversion and selectivity of the oxidation reaction. However, I don't think that Ag component plays an important role in the C-H activation of propane by controlling electrophilic or nucleophilic nature of active oxygen in our reaction. The reason is that oxidation of propane gave considerable amounts of propylene under the reaction conditions without any catalyst and propylene was easily converted to acrolein or acrylonitrile in the presence of catalysts adopted in this investigation.

2) I agree with your suggestion that Ag might be an effective producer of active oxygen. Most important feature of the catalyst system used in this reaction was demonstrated by the rapid migration of oxide ion through bulk diffusion (refs. 12 and 13). We think that the total reaction rate is controlled by the rapid migration of oxide ion through bulk diffusion (refs. 12 and 13). We think that the total reaction rate is controlled by the homogeneous steps but the positive effect of Ag is realized by its activation of oxygen and distribution of active species through bulk diffusion in the step of oxidation of intermediary propylene.

G. Centi and F. Trifiro' (Editors), *New Developments in Selective Oxidation* 505
© 1990 Elsevier Science Publishers B.V., Amsterdam — Printed in The Netherlands

OXIDATIVE DEHYDROGENATION OF ALKANES OVER ALKALI AND ALKALINE EARTH ORTHOVANADATES

K. SESHAN, H.M. SWAAN, R.H.H. SMITS, J.G. VAN OMMEN and J.R.H. ROSS,
Laboratory for Inorganic Chemistry, Materials Science and Catalysis, Faculty of Chemical Technology, University of Twente, P.O. Box 217, 7500 AE, Enschede, The Netherlands.

SUMMARY

The orthovanadates of the alkaline and alkaline earth metals and the un-modified alkaline earth oxides have been compared with vanadium pentoxide as catalysts for the oxidative dehydrogenation of ethane and propane. The orthovanadates do not contain the V=O species known to cause deeper oxidation; the presence of alkaline and alkaline earth metals also reduces the acidity of the vanadium pentoxide component. It was found that pure vanadium pentoxide gave substantial combustion. The orthovanadates gave an increase in selectivity towards olefin production in the order Ba<Sr<Ca<Mg and this was coupled with an increase in activity in the same order. The unmodified alkaline earth oxides also exhibited some activity and selectivity, the selectivity increasing in the same order; however, the activity increased from BaO to MgO. For the orthovanadate series, there was a shift in the V-O infra-red absorption band to higher frequencies. Selective oxidation over the orthovanadates thus seems to depend mainly on the strength of the V-O bond.

INTRODUCTION

The global abundance of liquefied petroleum gas (LPG), consisting mainly of propane and butanes, makes attractive the development of a new route for the production of olefins, important building blocks of the chemical industry. Conventional dehydrogenation of LPG is besieged with problems such as the equilibrium limitation of the reaction, and the consequent high temperatures required to obtain reasonable alkane conversions, as well as rapid coking and hence short catalyst life.

A possible solution to these problems is to carry out oxidative dehydrogenation (ODEH), i.e. dehydrogenation in the presence of oxygen. The possibility that total oxidation may occur is one of the major disadvantages associated with ODEH. However, this can probably be overcome if a way can be found to avoid strong interaction of the product olefins with the catalyst surface and to limit the oxidation activity to the abstraction of hydrogen from the alkanes.

It has been reported by Chaar and his colleagues that the V_2O_5/MgO catalyst system fulfills these conditions, showing appreciable activity and selectivity for the ODEH of propane (1) and butane (2). Their choice of magnesia as a support was based on the idea that it was necessary to prevent a strong interaction between the product olefins and the catalyst surface: a basic support such as MgO repels the nucleophilic olefin. The formation of surface compounds between the V_2O_5 and the MgO was found to suppress the activity of the vanadia component for deep oxidation. Chaar et al. have suggested that this is due to the absence of the V=O bonds found in pure V_2O_5 which cause the formation of oxygen containing products (2). Patent literature also cites vanadium and magnesium oxides containing systems as useful catalysts for ODEH of lower hydrocarbons (3). Of the various compounds which can be formed in the V-Mg-O system, magnesium orthovanadate has been reported to be the most effective as a catalyst, this compound being structurally most different from pure V_2O_5 (4).

Preliminary work in our laboratory on the use of various other V_2O_5-containing materials as catalysts for the ODEH reaction confirmed the importance of the absence of V=O bonds in the catalyst surface. It thus appeared that further work on other orthovanadates of the alkaline earth and alkaline metals for catalysis of the above reaction might provide greater insight into the requirements of a good ODEH catalyst: the relationships between the activity and selectivity for the reaction and parameters such as the nature of the V-O bonding, oxidation-reduction behaviour, the oxidation state and concentration of the components on the surface and the basicity of surface groupings. This paper describes the behaviour of a number of alkaline and alkaline earth orthovanadates for the oxidative dehydrogenation of propane and ethane and attempts to make a correlation between these results and infrared spectra obtained with the orthovanadates. We show that the criteria for a good ODEH catalyst put forward by Chaar et al. cannot be the only factors that govern the catalytic behaviour.

EXPERIMENTAL

The orthovanadates were prepared by coprecipitation using ammonia from solutions containing stoichiometric mixtures of ammonium metavanadate and the nitrate of one of a series of alkaline or alkaline earth metals. The precipitates were dried at 300 K, and then calcined in air at 1025 K for 2 hours. This procedure yielded monophasic orthovanadates. The pure alkaline earth oxides were prepared by decomposition of the corresponding hydroxides in air at 1025 K for 2 hours.

The phase structure of each of the orthovanadates was confirmed using x-ray powder diffraction. All the materials were found by N_2 adsorption to have low surface areas (<0.5 m^2/g). XPS measurements were made using a KRATOS XSAM 800. Details of the equipment used for TPR measurements are given elsewhere (5). The V^{4+} and V^{5+} contents of the catalysts after calcination were obtained by wet analytical methods (6). The FTIR measurements of the samples, in the form of thin discs of the powders mixed with KBr, were obtained with a Nicolet 20SXB spectrometer.

The catalysts were tested for the oxidative dehydrogenation of propane using a flow system with a tubular quartz reactor (30 cm long, 0.5 cm internal diameter). Powdered samples of the catalysts (normally 600 mg of 0.3 - 0.6 mm diameter particles) were supported in the reactor by quartz wool. A total gas flow of 137 cm^3 min^{-1} of composition $He:C_3H_8:O_2$ = 13:6:1. was used. The products were analysed using a VARIAN 3700 gas chromatograph having a Heysep Q column (total product analysis) and a MS 5A column (CO,O_2 analysis). A limited number of experiments were carried out with ethane in place of propane, the same gas composition being used.

In the results which are given below, the activity (or conversion) is defined as the percentage of the number of moles of propane (or ethane) fed which react; the selectivity is defined as the percentage conversion to a particular product relative to the total products, taking into account the number of carbon atoms in each product molecule.

RESULTS

Catalytic Results

Oxidative Dehydrogenation of Propane. Table 1 gives typical results for the oxidative dehydrogenation of propane at 873 K over the various oxides and mixed oxides tested. The choice of the temperature used was based on the fact that above 900 K thermal cracking began to take place and that a number of the catalysts showed very little activity below 873 K. The catalysts

Table 1 Results for the oxidative dehydrogenation of propane over various compounds

Catalyst	Carbon balance (%)	Conversion C_3H_8 (%)	Conversion O_2	Selectivity CO_x (%)	CH_4	C_2H_4	C_3H_6
MgO	99	12.6	45.6	44.1	8.5	24.1	23.2
CaO	101	2.1	13.0	49.1	8.3	19.4	23.2
SrO	100	3.2	8.5	8.5	18.9	25.7	47.0
BaO	100	0.9	8.4	25.8	18.1	0.0	56.1
V_2O_5	100	17.3	55.2	56.4	0.0	0.0	43.6
$Mg_3(VO_4)_2$	101	7.6	17.1	30.8	10.6	0.0	58.6
$Ca_3(VO_4)_2$	101	4.7	22.0	59.9	6.3	14.0	20.0
$Sr_3(VO_4)_2$	99	3.2	13.7	47.4	10.4	17.7	24.5
$Ba_3(VO_4)_2$	100	0.3	5.6	100.0	0.0	0.0	0.0
Li_3VO_4	101	3.0	18.4	40.0	7.9	17.9	28.9*
Cs_3VO_4	101	0.2	3.8	100.0	0.0	0.0	0.0

* rest C_4 ** An empty reactor under these experimental conditions did not show any reaction of propane below 900K

showed no change in activity or selectivity during typical experiments of four hours' duration. The carbon balance in all the experiments was 100 $\pm$ 1 %.

With the pure alkaline earth oxides, the propane conversion decreased in the order MgO>CaO>SrO>BaO. The selectivity to propylene was in the reverse order, BaO>SrO>>CaO=MgO, and some cracking to methane and ethylene was also found; SrO gave a remarkably low degree of total oxidation. Pure V_2O_5 gave a relatively high conversion but the only products were now propylene and the oxides of carbon.

With the alkaline earth orthovanadates, the propane conversion decreased in the same order as found for the corresponding pure alkaline earths although there were some small differences in the absolute values; for example, the magnesium vanadate was less active than the MgO while the calcium vanadate was more active than the CaO. The magnesium vanadate was now by far the most selective vanadate, being more selective than MgO, while the other vanadates were less selective than the corresponding oxides; the barium vanadate was totally unselective. The degree of cracking exhibited by the orthovanadates was lower than for the pure oxides.

Results for the pure lithium and cesium orthovanadates are also shown in Table 1. The former shows some selectivity; it also gives some cracking (compare pure vanadia). The cesium compound gives only combustion.

Oxidative Dehydrogenation of Ethane. The alkaline earth orthovanadates were found to have very low activities (conversion <1%) for the oxidative dehydrogenation of ethane even around 950K. This was to be expected as the activation of the C-H bond becomes more difficult as the chain length of the alkane decreases. Due to the low activities, a regular variation in selectivity could not be observed. However, the overall behaviour seems to be similar to that for propane. For example, magnesium orthovanadate gave a selectivity to ethylene of 97% and to carbon oxides of 3% at a reaction temperature of 950 K; the corresponding figures for barium orthovanadate were 40% and 60%, respectively. In both of these cases, the conversion was only about 0.5%.

Catalyst Characterisation

XPS Measurements. Table 2 gives the binding energies of the vanadium electrons for the various vanadium-containing compounds of Table 1. The table also shows the V/M and O/V ratios calculated for each sample from the appropriate peak intensities; the theoretical values are also given for comparison purposes. From the positions of the peaks, it can be concluded that the

Table 2 Details of XPS measurements on the catalysts

Catalyst	Binding energy (eV)			Surface concentration (mole ratios)	
	$V_{2p1/2}$	$V_{2p3/2}$	O_{1s}	V/M	O/V
V_2O_5	526.0	518.3	531.5	--	4.1
"	theoretical				2.5
$Mg_3(VO_4)_2$	524.9	517.3	530.0	0.8	4.9
$Ca_3(VO_4)_2$	524.4	516.8	529.7	0.4	5.5
$Sr_3(VO_4)_2$	524.2	516.6	529.5	0.4	5.1
$Ba_3(VO_4)_2$	524.0	516.0	529.5	0.4	5.6
$M_3(VO_4)_2^*$	theoretical			0.7	4.0
Li_3VO_4	526.5	519.2	531.9	0.3	6.5
"	theoretical			0.3	4.0

*M= Mg,Ca,Sr or Ba

vanadium was in the V^{5+} state in all the catalysts (7). However, there was a clearly distinguishable chemical shift to lower binding energies for the electrons of both the vanadium and oxygen ions of the alkaline earth orthovanadates; for lithium orthovanadate, there was a slight increase in the binding energies for both V and O. All the orthovanadate samples, with the probable exception of magnesium orthovanadate, showed lower amounts of vanadium on the surface than the stoichiometric requirement. All the catalysts had amounts of oxygen on the surface which were above the stoichiometric requirements.

Temperature Programmed Reduction. Temperature programmed reduction experiments carried out using 5% H_2 in argon did not show any reduction of the pure alkaline earth oxides at temperatures up to about 1275 K. Pure V_2O_5 underwent reduction in five stages to V_2O_3(via 1/3 V_6O_{13}, 2 VO_2, and V_6O_{11}), as has previously been reported (5). As is shown in Figure 1, the alkaline earth orthovanadates underwent reduction rather sluggishly; the first reduction maximum occurred well above the temperature used for the catalyst testing (873 K).

Chemical Analysis. The results for the analysis of the vanadium species in a number of the samples are given in Table 3. It can be seen that the proportion of V^{4+} in the different samples is relatively low and that this proportion does not appear to change to any appreciable extent after use.

Table 3. The amounts of V^{5+} and V^{4+} (wt%) in the fresh and used catalysts

Catalyst	Fresh catalyst		Used catalyst	
	V^{4+}	V^{5+}	V^{4+}	V^{5+}
V_2O_5	0.9	58.0	--	--
$Mg_3(VO_4)_2$	0.2	33.8	0.2	34.2
$Ca_3(VO_4)_2$	0.1	28.3	--	--
$Sr_3(VO_4)_2$	0.4	22.3	0.2	20.3
$Ba_3(VO_4)_2$	0.3	15.5	0.3	15.3

IR Measurements. The IR spectrum of the pure V_2O_5 sample (Fig.2) was characterised by absorption bands at 835 and 1020 cm^{-1}, corresponding, respectively, to vibrations of V-O-V groups in the (010) plane and V=O projecting perpendicularly from this plane. The IR spectrum of the orthovanadates, as expected, did not show the V=O bond but showed a gradual shift in the V-O band (there are no V-O-V bands because in orthovanadates the $(VO_4)^{3-}$ units are isolated by MgO_6 units) to lower frequencies in the order magnesium to barium orthovanadate. The spectrum for strontium and barium orthovanadates also showed increased formation of carbonate species (bands at 1700 cm^{-1}, not shown).

DISCUSSION

The basicity of alkaline earth hydroxides increase from $Mg(OH)_2$ to $Ba(OH)_2$, the latter having a basicity approaching that of alkali hydroxides. As the alkaline earth oxides easily form the corresponding hydroxides it is probable that this is the form present under reaction conditions. The results of Table 1 for the pure oxides showed an improved selectivity in the same order. It

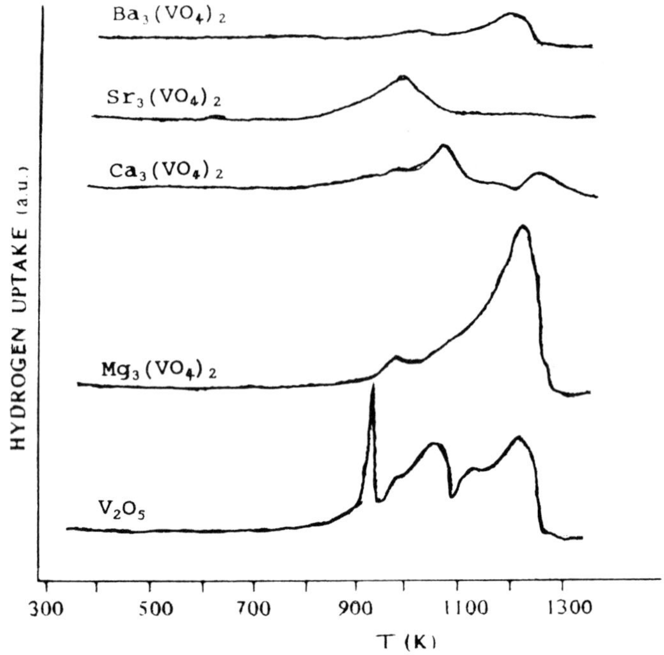

Fig. 1 TPR recordings of the catalysts

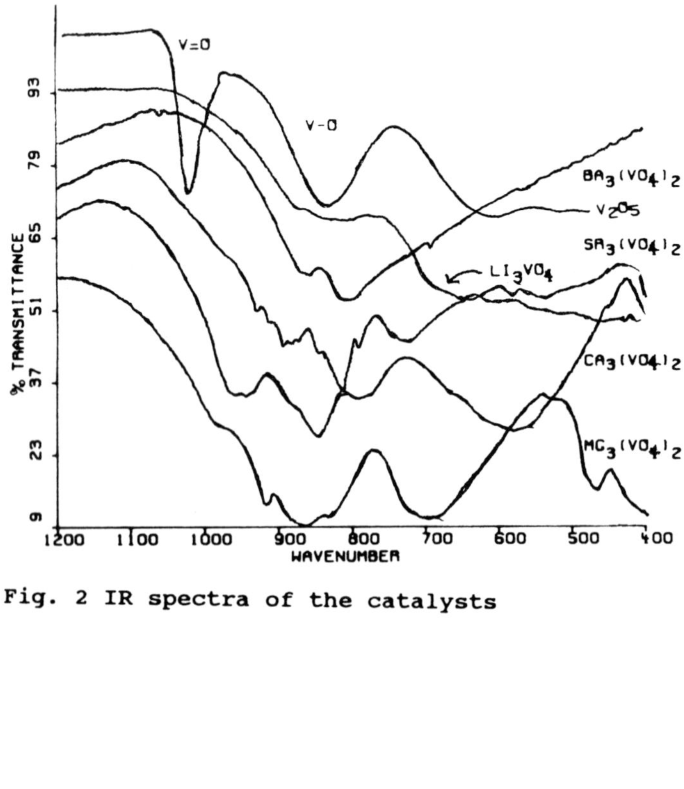

Fig. 2 IR spectra of the catalysts

thus seems that an increase in the basicity had the expected effect on the selectivity, ie. yielding more olefin as a result of a lower interaction between the catalyst and the olefin. However, when the alkali and alkaline earth oxides were combined with vanadium to form the orthovanadates, they showed a very different trend: ie. while the activity decreased in the same order, the selectivity to the formation of the olefin also decreased. These differences may be due to a change in the nature of the V-O bonding of these compounds.

XPS results show that the oxidation state of the surface vanadium is similar in all the fresh orthovanadates (Table 2). The chemical shift to lower binding energies from pure V_2O_5 to $Ba_3(VO_4)_2$ probably reflects a gradual weakening of the V-O bonding in this series; if anything, the bonding of the oxygen in the Li compound is slightly stronger than in the V_2O_5. The TPR results show that the orthovanadates are not reduced to any appreciable extent at reaction temperatures; any slight reduction of the catalyst under reaction conditions is probably offset by fast reoxidation from the gas phase. This conclusion is confirmed by the analysis of the catalysts which showed that the composition appeared to be almost the same before and after the reaction (Table 3).

The presence of reactive surface V=O species in the case of pure vanadium pentoxide, shown by the absorption band at about 1020 cm^{-1} in the FTIR spectrum (Fig. 2), seems to be responsible for the high selectivity to carbon oxides exhibited by this material. In contrast, there appear to be no V=O groups in the magnesium orthovanadate or the other orthovanadataes; the magnesium compound gave the best selectivity for the formation of propane and fewer combustion products (Table 1). However, on going along the series from the magnesium to barium orthovanadate or to the alkali orthovanadates, the amounts of carbon oxides formed increased in spite of the low alkane conversions that were achieved. This is surprising in view of the fact that the surface vanadium concentrations were lower in the Ca, Sr, and Ba orthovanadates than in the magnesium compound and also of the fact that one would expect the basicity of the materials to increase in the same order.

The FTIR spectra show that there are no V=O species in the orthovanadate catalysts but that there is a gradual shift in the V-O band to lower frequencies. Such shifts in V-O frequencies in orthovanadates have been attributed to changes in the effective nuclear charge on the cation in the structure (8). Patel et al. (4) found that the magnesium orthovanadate (IR band for V-O at 859 cm^{-1}) had superior catalytic properties to those of magnesium metavanadate (V-O-V band at 910 cm^{-1}) and magnesium pyrovanadate (V-O-V band at 975 cm^{-1}). They suggested that in the latter two samples the V-O-V band had shifted nearer to that for the V=O (band at 1020 cm^{-1}) and that these materials therefore has some of the character associated with that grouping. We might thus have expected that a move of the V-O band to lower frequencies, as observed for the other orthovanadates (Fig. 2), would lead to improved catalytic behaviour. However, a shift to lower frequencies from magnesium orthovanadate to the other orthovanadates also seems to be detrimental to the selectivity to olefin production (see Table 1). This loss of selectivity may be associated with a gradual weakening of the V-O bond, also evidenced by the chemical shifts in the XPS spectra (Table 2); this weakening will have the consequence that oxygen from the lattice is now available to give total oxidation.

It is difficult to explain why the ability to activate propane decreases for the alkaline earth oxides from MgO to BaO and in the same order for the orthovanadates. This observation may perhaps be associated with the increasing tendency of the higher molecular weight oxides to form stable surface carbonates. The presence of the stable carbonate species on the surface may mean that it is less easy to generate the active site under reaction conditions. The implication is that the active site on these catalysts may have some similarity to that required for a good methane coupling catalyst; results from our laboratory have shown that the active site on Li/MgO catalysts used for this reaction may be created by the decomposition in situ of Li_2CO_3 species (9). Further work to attempt to clarify these ideas is in progress.

ACKNOWLEDGEMENTS

We wish to thank W. Lengton and G.L. van Assen for performing the V^{5+} and V^{4+} analyses and Ing. H.J.M. Weierink for obtaining the IR spectra.

REFERENCES

1. M.A. Chaar, D. Patel, and H.H. Kung, J. Catal., 109, (1988) 463.
2. M.A. Chaar, D. Patel, M.C. Kung, and H.H. Kung, J. Catal., 105, (1987) 483.
3. M. Lee Fu , U.S. Patent 44607129, Philips Petroleum Co., 1986.
4. D. Patel, M.C. Kung and H.H. Kung, Eds. M.J. Phillips and M. Ternan, Proc. 9th Int. Cong. Catal., Calgary, 4,(1988), 1554.
5. H. Bosch, B.J. Kip, J.G. van Ommen and P.J. Gellings, J. Chem. Soc., Faraday Trans., 80, (1984) 2479
6. H.R. Grady, Treatise on Analytical Chemistry, Part II, (Eds) I.M. Kolthoff and P.J. Elving, Interscience Publishers, New York Vol.8, 1963, 224pp
7. Handbook of X-ray Photoelectron Spectroscopy, (Ed) G.E. Muilenberg, Perkin Elmer Corporation, Minnesota, 1979, p71
8. E.J. Baran and P.J. Aymonino, Z. Anorg. Allge. Chemie, 365, (1969) 211
9. S.J. Korf, J.A. Roos, N.A. de Bruijn, J.G. van Ommen and J.R.H. Ross, J. Chem. Soc., Chem. Comm. (1987) 1433.

J.C. VEDRINE (C.N.R.S., France): You have observed a shift in the binding energy values of V by XPS as a function of the alkaline earth cations and in parallel some shift in the vibrational bands in IR data. As you know all these parameters are related to the strength (ie. ionic character) of V-O bonds. I suggest you to think more about such a condition. However this may be ruled out if the samples are modified under catalytic conditions. Did you check that such modification did not occur by analysing the catalysts after catalytic reaction ?

K. SESHAN (University of Twente, The Netherlands): The catalysts did not undergo any modification during catalytic testing. The shift in the IR vibrational bands of V-O group in the presence of different alkaline earth cations is indicative of change in the V-O bond strength and hence the correlations drawn with catalytic activity takes care of your suggestion.

M. MICHMAN (Hebrew University of Jerusalem, Israel): You have shown the varying reactivities of catalysts containing Mg,Ca,Sr and Ba. Are the selectivities shown by the cations consistent with varying temperatures ?. Why have you chosen a specific temperature for the comparison of selectivities ?.

K. SESHAN (University of Twente, The Netherlands): The catalytic measurements have been carried out at 600°C only. This is because, above 625°C gas phase reactions dominate and below 590°C $Ba_3(VO_4)_2$ does not show any activity at all. This puts a restriction on the measurement of catalytic activities at different temperatures.

G. BUSCA (University of Genova, Italy): The vibrational spectra of solids are very complex and their interpretation is not straightforward. Have you Raman spectra that support the relation you observe between IR frequencies and catalytic activity of orthovanadates ?.

K. SESHAN (University of Twente, The Netherlands): The Raman spectra of these catalysts have not been recorded. The IR spectra of alkaline earth orthovanadates have been studied extensively (see references in the paper) and rather well understood. In my opinion, only the relationship between IR frequencies for the V-O grouping and the catalytic activity need further confirmation.

G. EMIG (University of Karlsruhe, FRG): In your table on product distribution, CH_4 appears always together with C_2H_4. Only in the case of $Mg_3(VO_4)_2$, there is no C_2H_4 at all formed !. How do you explain this ?. Is there a change in the mechanism on this type of catalyst ?.

K. SESHAN (University of Twente, The Netherlands): On the basis of the present set of data it is difficult to say if there is a different mechanism of action on $Mg_3(VO_4)_2$. However, the higher activity of this compound may be causing total oxidation of any ethylene that is formed. This probably may by be the reason for the absence of any C_2H_4.

J.C. VOLTA (C.N.R.S., France) : You compare the oxidative dehydrogenation activity of propane on different orthovanadates with the assumption that magnesium orthovanadate is the effective phase for the V-Mg-Osystem. We have recently investigated catalytic behaviour of Mg ortho, meta and pyro vanadates for the same reaction. For higher oxygen concentrations (propane/oxygen=1/90) the pyro phase was more selective to propylene formation than orthovanadate. This was confirmed also from the fact that on using propylene as the reactant, while the pyro vanadate partially transformed propylene to acrolein, the orthovanadate combusted propylene under identical conditions. Consequently we do not agree with the choice of orthovanadate. It appears to depend on propane to oxygen ratio. Why don't you compare selectivity values at iso conversions ?. Why do you use a ratio of propane to oxygen of 6 ?.

K. SESHAN (University of Twente, The Netherlands): To get iso conversions the parameters that needed variation were temperature, contact time or propane/oxygen ratio. Variation of temperature was rather difficult(see question by M. Michman). Variation of contact time was not attempted. Increasing oxygen content led to more combustion in the case of V_2O_5/MgO catalysts. The propane/oxygen ratio of 6 was chosen because at this composition highest propylene selectivities at reasonable propane conversions were obtained.

G. Centi and F. Trifiro' (Editors), *New Developments in Selective Oxidation*

Synthesis of Acrylonitrile from Propane on V-Sb-based Mixed Oxides

G. Centi[1], R.K. Grasselli[2], E. Patane'[1], F. Trifiro[1]

[1] *Dep. of Industrial Chemistry and Materials, V.le Risorgimento 4, 40136 Bologna, Italy*
[2] *Mobil Research and Development Corp., Paulsboro, NJ 08066, USA*

The catalytic behavior of propane ammoxidation to acrylonitrile of pure $VSbO_4$ + Sb_2O_4 (VSbO) samples (V:Sb atomic ratio in the range 0.2-1), of modified VSbO samples with Al_2O_3, TiO_2 or SiO_2 and of further promoted systems with W or Mo is analyzed. V-Sb-Al_2O_3 based systems show an interesting potential for the development of a new process of acrylonitrile formation directly from propane. Al-oxide does not act merely as a support for the VSbO phase, but has a specific structural effect and greatly promotes the formation of propylene from propane. Catalytic and characterization results suggest that the active phase in propane ammoxidation is not a pure $VSbO_4$ rutile phase plus Sb_2O_4, but rather an amorphous $VSbO_4/Sb_2O_4$ system epitaxially intergrown with $AlSbO_4$ rutile phase whose formation is catalyzed by the presence of V. This system may be further promoted by the addition of W or Mo. Kinetic analysis of the propane ammoxidation shows that acrylonitrile forms by two parallel pathways, one directly from propane and the second through the intermediate formation of propylene, the latter being the prevalent path. In all cases propylene is the main reaction product observed, especially at low conversion, with selectivities up to about 80%.

Introduction

There is notable interest in the development of a viable catalytic process for the conversion of propane to acrylonitrile because of the considerable difference in price between propylene and propane [1]. However, only limited information is available in the literature about this reaction. Vanadyl pyrophosphate, active phase in n-butane oxidation to maleic anhydride [2], shows selectivities of about 20-30% to acrylonitrile from propane [3], but only at high concentrations of ammonia (around 20%). Most of the ammonia, furthermore, is oxidized to nitrogen oxides making this catalyst unattractive.

V-antimonate catalysts also show a certain selectivity to acrylonitrile from propane [3]. In this system the vanadium is stabilized in a valence state lower than five by the formation of the rutile structure ($V^{III}SbO_4$ or V_2O_4) [4-8]. Recently, the V-Sb based system was patented for the ammoxidation of propane and isobutane [9], but no further information is available on its catalytic behavior. Our results on the direct synthesis of acrylonitrile from propane on V-Sb based mixed oxides are summarized in this paper in order to show how these mixed oxides represent a very interesting potential catalytic system for the development of a new process of acrylonitrile formation from propane.

Experimental

The catalysts were prepared by coprecipitation from an aqueous hydrochloric acid solution (1 N) containing soluble salts of V, Sb, W or Mo. The compounds were precipitated by the addition of ammonia or of an ammoniacal solution containing Al-, Si- or Ti-hydroxide as specified in the text.

The samples were filtered, washed, dried overnight at 150 C and then calcined in air or in a flow of 2% H_2 in Helium, first at 350 C for 24 h and then as specified in the text. A typical V:Sb ratio used was 1:5 and a typical amount of support [Al_2O_3] was 70% by weight. Further details on the preparation of unsupported VSbO samples have been reported previously ref. [3,7], as well as details on the preparation of supported and multicomponent samples [1,8]. The surface areas of the samples were in the 10- 20 m^2/g range for the $VSbO_4$-Sb_2O_4 systems and around 90-100 m^2/g for the V(1):Sb(5):[W (1) or Mo(7)]- Al_2O_3 [70% wt] systems calcined an air at 620 C. The surface area of the sample with SiO_2 is around 70 m^2/g and that of the sample with TiO_2 around 10 m^2/g. No great changes in the surface areas were observed for any of the samples after the catalytic tests.

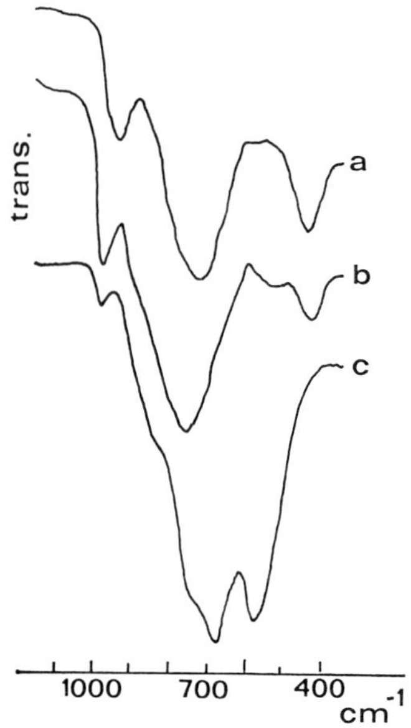

Figure 1 Infrared spectra of VSbO samples (V:Sb = 1) after drying at 70 C overnight (**a**) and after calcination at 300 C (12 h) (**b**) and at 400 C (12 h) (**c**).

The catalytic tests were made in a fixed bed tubular flow reactor at atmospheric pressure, with on-line gas chromatographic analyses of reagents and products [3,8]. Reaction conditions are specified in the legends of the Figures. The catalytic data refer to steady-state behavior observed after about 4 hours of time-on-stream.

The catalysts were characterized, before and after the catalytic tests, mainly by infrared and X-ray diffraction analyses; the mean oxidation state of vanadium was determined by chemical analyses after ammoniacal extraction of V^V and after consecutive complete dissolution of the samples in concentred sulphuric acid solution.

Results and Discussion

Development of the Catalytic System

System $VSbO_4$+Sb_2O_4. The synthesis of $VSbO_4$ reported in the literature [5,6,10] usually involves a solid state reaction at high temperature (around 600-700 C) between Sb_2O_3 and V_2O_5 or a redox reaction between NH_4VO_3 and Sb_2O_3 in an aqueous slurry medium [9]. In this study we preferred to utilize an alternative method of coprecipitation from an acid solution of all the components in order to obtain more homogeneous samples and higher surface areas.

Infrared (IR) analysis of dried samples shows main bands at 970 cm^{-1} due to V^{IV}=O and at 730, 590 and 450 cm^{-1} due to νSb-O in Sb^V-hydroxide (fig. 1 a). This indicates that vanadium is reduced by antimony. In acid solution [11] V^V gives rise to a two electron redox reaction with Sb^{III} forming V^{III} and Sb^V. However, V^{III} is not stable and tends to be reoxidized in solution. Chemical analyses

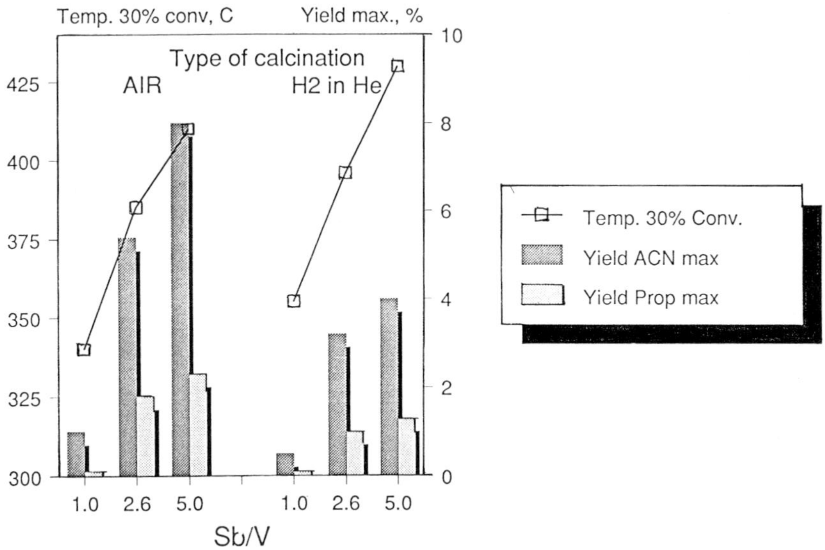

Figure 2 Catalytic behavior (temperature of 30% conversion, maximum yields of ACN and propylene) of the $VSbO_4$+Sb_2O_4 system in propane ammoxidation: effect of different of Sb/V atomic ratios and of the atmosphere of calcination, air at 450 C for 4 h or 2% H_2 in helium at 450 C for 4 h. Exp. conditions.: 1.6% propane, 10% oxygen, 2.8% NH_3, W/F = 2036 g.h/moles C3.

of the dried sample confirm that after precipitation vanadium is mainly present as V^{IV} and antimony as a mixture of Sb^{III} and Sb^{V}, whose proportion depends on the initial Sb:V ratio.

After calcination at 300 C (12 h), the IR spectrum is modified, indicating the oxidation of V^{IV} to V^{V} (shift of the band at 970 cm^{-1} to 995 cm^{-1}) and the formation of Sb_6O_{13} (fig. 1 b). At higher calcination temperatures (400 C, 1 h) (fig. 1 c) the IR spectrum is further modified with the formation of typical bands at 670 and 550 cm^{-1} due to νSb-O in the rutile $V^{III}Sb^{V}O_4$ structure [12]. Additional bands at 770, 745, 685, 610, 530 and 460 cm^{-1}, due to νSbO in distorted tetrahedra and octahedra of Sb^{V} and Sb^{III} in Sb_2O_4 are found when the starting Sb:V ratio is higher than 1.0. At calcination temperatures higher than 900 C, decomposition of the rutile structure with the formation of V_2O_5 and Sb_2O_4 occurs with consequent changes in the IR spectrum.

These data thus show that the formation of the rutile phase occurs at temperatures between 300 and 400 C and that this phase is stable up to temperatures of about 900 C. The band at 995 cm^{-1} shows that V^{V} is present in the sample. The intensity of this band decreases with increasing Sb:V ratio, indicating that excess antimony favours the formation of the rutile structure.

X-ray diffraction analysis confirms the presence of the rutile $VSbO_4$ phase in samples calcined at T > 400 C and of α-Sb_2O_4 for Sb:V ratios higher than 1.0. A heat treatment in reducing atmosphere (2% H_2 in helium) enhances the crystallinity of the samples with respect to comparable calcination in air; in particular, increasing the intensity and sharpening the XRD reflections of $VSbO_4$.

The catalytic results are summarized in Figure 2 where the temperature of 30% of conversion and the maximum yields found in acrylonitrile (**ACN**) and propylene (**Prop**) are reported as a function of the Sb/V ratio in the catalyst and of the calcination atmosphere [450 C, 4 hours]. It can be seen that (**i**) increasing the Sb:V ratio causes a considerable decrease in the activity, but increases the selectivity and maximum yield both of acrylonitrile and propylene, (**ii**) the increased crystallinity of the samples, due to the heat treatment in a flow of H_2/He, decreases both the activity and the

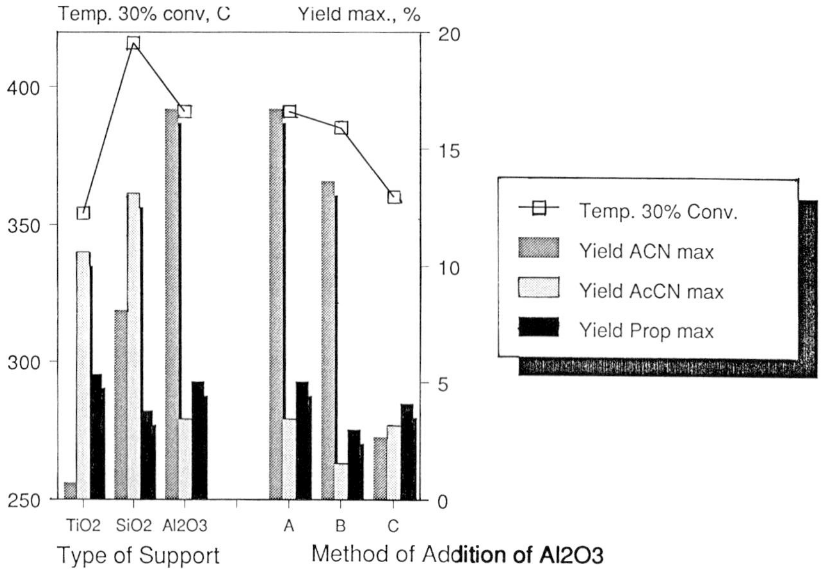

Figure 3 Catalytic behavior (temperature of 30% conversion, maximum yields of propylene, ACN and AcCN) in propane ammoxidation of the VSbO(V:Sb=1:5)-TiO_2, - SiO_2 or -Al_2O_3 systems [30% wt. of VSbO phase]. Samples calcined in air at 620 C. Methods of preparation: (**A**) precipitation of mixed oxide by addition af an ammoniacal suspension containing the Ti-, Si- or Al-hydroxide to an hydrochloric acid solution containing V and Sb. (**B**) precipitation of mixed oxide by addition of an ammoniacal solution to an hydrochloric acid solution containing V, Sb and Al. (**C**) impregnation with hydrochloric acid solution of Al_2O_3 support and the in-situ hydrolisis with ammoniacal vapours. Exp. conditions as in Fig. 1.

selectivity and thus is not favorable to obtain selective catalysts.

Since an Sb:V atomic ratio of 5.0 is preferable for good catalytic behavior, we refer in the following sections to catalysts having only this Sb:V ratio.

Role of the Support. The effect of the nature of the support and of the method of addition is summarized in Figure 3 which reports, as key indicative parameters, the temperature of 30% conversion and the maximum yields of acrylonitrile (**ACN**), acetonitrile (**AcCN**) and propylene (**Prop**) using standard reaction conditions and variable reaction temperatures.

For the comparison of the nature of the support the same method of preparation was used (addition of an ammoniacal suspension containing the support to the acid solution containing V and Sb) (fig. 3). It is shown that TiO_2 and SiO_2 are not a good support. The former enhances the activity, but gives rise to a considerable formation of acetonitrile, as found also in the case of the VSbO-SiO_2 system, which is less active. Al_2O_3, on the contrary, allows a significant increase in the yield of acrylonitrile and propylene. It should be noted that, the increase in activity is relatively limited notwithstanding the higher surface area (around 50 m^2/g) in comparison to the same sample without aluminium (16 m^2/g). In general, no correlation was observed between surface area and catalytic activity, indicating that aluminium does not act merely as a support.

The method of addition of the Al component has a considerable effect on the catalytic behavior (Fig. 3). Best performances are obtained when an intimate contact between the V-Sb-O phase and the Al-phase is realized (Method A: addition of an ammoniacal suspension containing the support to the acid solution with V and Sb ; Method B: precipitation with ammonia from an acid solution containing V, Sb and Al). In contrast, poor performances were found when the VSbO phase was

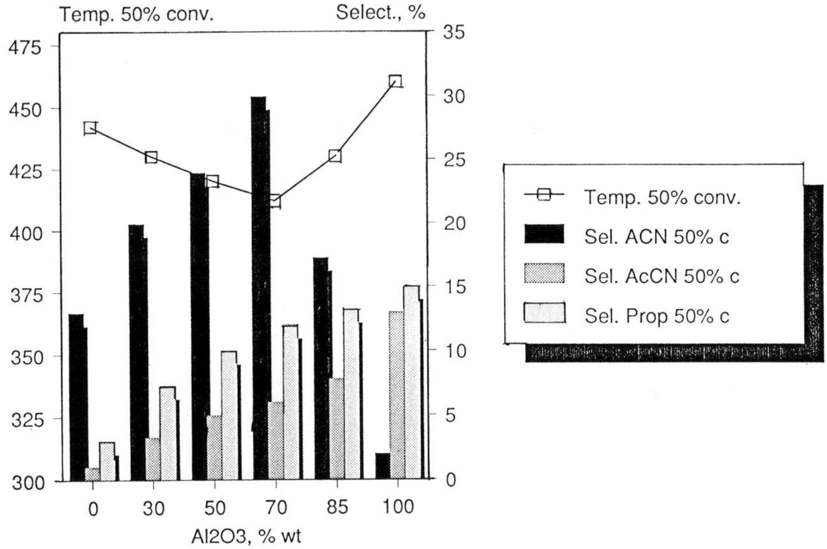

Figure 4 Effect of changing in the relative amounts by weight of the VSbO phase and the Al_2O_3 phase on the catalytic behavior in propane ammoxidation (temperature of 50% conversion and selectivities to ACN, AcCN and propylene at 50% conversion). Samples calcined in air at 620 C. Experimental conditions as in Fig. 1.

supported on preformed Al_2O_3 (Method C). This also suggests a specific structural effect of Al in the enhancement of the catalytic behavior of the VSbO system toward propane selective ammoxidation.

Figure 4 summarizes the effect of the relative amounts of the VSbO phase and Al_2O_3; the temperature of 50% propane conversion and the relative selectivities to acrylonitrile, acetonitrile and propylene are reported. As the relative amount of Al_2O_3 increases, the selectivity to acrylonitrile passes through a maximum at a value of around 70% Al_2O_3, in correspondence also to the higher activity of the catalyst. On the contrary, the selectivity to both propylene and acetonitrile increases in going from the pure V-Sb-O system to pure Al_2O_3 system. The surface areas increase continuously from 16 m^2/g (0% Al_2O_3) to around 50 m^2/g (70% Al_2O_3) up to values around 100 m^2/g (100% Al_2O_3). These data suggest that (**i**) a mixed V-Sb-Al-O system is prefer for the selective synthesis of acrylonitrile from propane, (**ii**) an increase in the surface area and the amount of Al_2O_3 increases the oxidative dehydrogenation of propane to propylene, but also the cracking activity connected, probably, to exposed Al^{3+} sites, (**iii**) a bifunctional catalyst is necessary to perform effectively both stages (propane to propylene and propylene to acrylonitrile).

Structural analysis of these catalysts by X-ray diffraction evidences that in the studied catalytic system (70% wt. of Al_2O_3, calcined at 620 C in air) only an amorphous phase is present with very broad diffraction lines centered at *d*= 2.40, 2.14 and 1.40 Å due to the presence of Al_2O_3. As the amount of alumina decreases [50% wt Al_2O_3], new diffraction lines appear at d= 3.20, 2.46, 2.15, 1.67 and 1.60 Å corresponding to the presence of the rutile $AlSbO_4$ phase. With further decreases in the amount of Al_2O_3, the *d* values of the rutile phase progressively shift up to that corresponding to rutile $VSbO_4$ [0% of Al_2O_3]. Weak lines of α-Sb_2O_4 are also detected in the sample without alumina. The crystallization of the $AlSbO_4$ rutile phase is induced by the presence of V and is not

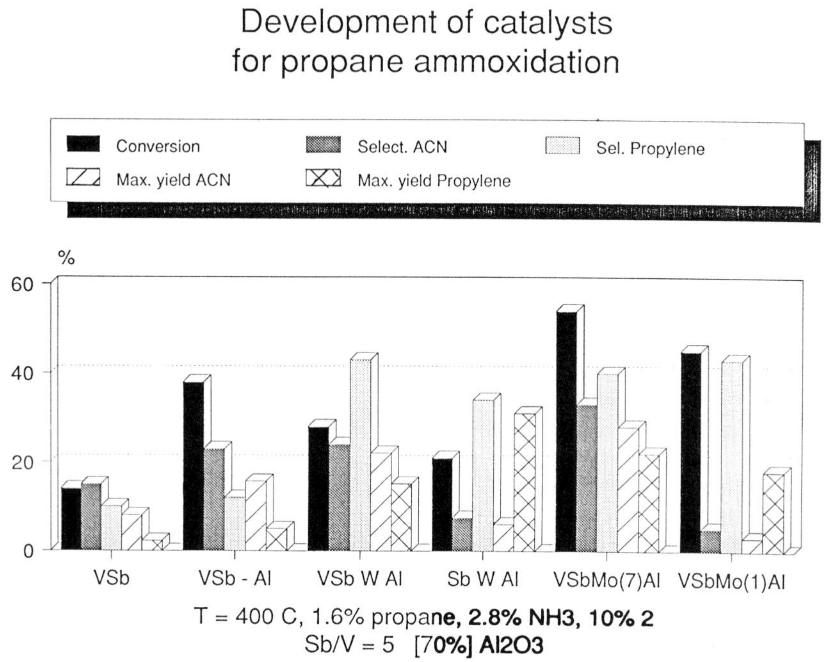

Figure 5 Comparison of the catalytic behavior at 400 C (conversion and selectivities to ACN and propylene) and maximum yields to ACN and propylene of V-Sb based mixed oxides. Exp. conditions as in Fig. 1.

observed at higher temperatures of calcination [up to 900 C] when vanadium is absent in the sample.

In conclusion, the X-ray diffraction characterization of these samples suggests that the active phase for propane ammoxidation is not a pure $VSbO_4$ rutile phase plus Sb_2O_4, but rather an amorphous $VSbO_4/Sb_2O_4$ system epitaxially intergrown with $AlSbO_4$ rutile phase, in which new modified geometries are formed at the interface.

Multicomponent Systems. The V-(Al,Sb)-SbO_4-Al_2O_3 system is active for propane conversion, but further improvments can be achieved by introducing a group 6b M^{6+} transition metal having typical properties of allylic oxidation and of oxygen insertion [13]. Cr was found not to be an effective promoter in propane ammoxidation, but W and Mo show good performances. Figure 5 summarizes the principal results; the conversion and selectivities at 400 C under standard conditions, and the maximum yield of acrylonitrile and propylene found in tests with variable reaction temperature are reported. It is shown that **(i)** the addition of Al effectively promotes the activity and selectivity of the catalyst as discussed before, **(ii)** the further addition of W decreases the activity, but drastically improves the selectivity to propylene as well as increases the maximum yields of acrylonitrile and propylene, **(iii)** when V is absent, the activity decreases, but especially the selectivity and yield to acrylonitrile decrease, whereas the selectivity to propylene is less affected, **(iv)** Mo produces a substantial increase in the activity of the catalyst and also selectivity to propylene, but the catalyst becomes selective in acrylonitrile synthesis only when the Mo:V ratio is 7 or higher, in contrast to a W:V ratio of 1.0. Molybdenum-based systems, furthermore, are very sensitive to reducing conditions, and become much less selective when the O_2 concentration is decreased or the NH_3 or propane concentrations are increased.

In all cases relevant amounts of propylene are obtained, with very high selectivities at low

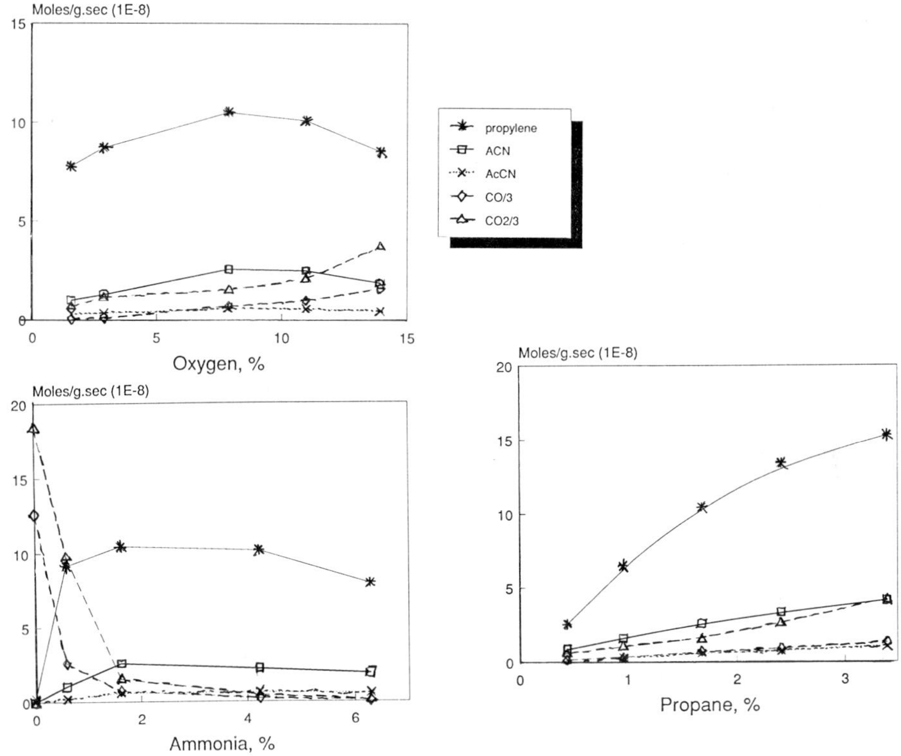

Figure 6 Effect of propane (**a**), of oxygen (**b**) and of ammonia (**c**) concentrations on the rates of formation of propylene, ACN, AcCN and carbon oxides at 433 C on VSbW(1:5:1)-Al_2O_3[70% wt] catalysts. Exp. conditions: (**a**) 7.9% O_2, 1.6% NH_3; (**b**) 1.6% propane, 1.6% NH_3; (**c**) 1.6% propane, 7.5% O_2. 0.5 g of catalyst, 3.29 L/h total flow at STP, conversion lower than 5-10 %.

conversion (75-80% up to about 20% conversion). Relatively small differences are noted in the selectivity to propylene for the various samples, whereas much more drastic differences in the selectivity to acrylonitrile are noted.

The data indicate that these catalysts selective activate propane to form propylene, but an additional critical step is necessery for the rapid transformation of propylene to acrylonitrile before desorption and possible re-adsorption on un-selective sites leading to waste reactions such as carbon oxides. This is partially accomplished by the addition of oxygen insertion promoters such as Mo or W.

We may thus conclude based on the analysis of the catalytic behavior of V-Sb-Al based mixed oxides and their spectroscopic characterization that the $AlSbO_4$ rutile phase modified with V is the active phase for the oxidative dehydrogenation of propane to propylene and V or other redox elements having M=O double bonds [13] (such as W or Mo) are the sites necessary for the consecutive transformation of propylene to acrylonitrile.

Kinetic Analysis of the Reaction

In order to gain more information on the reaction mechanism and key factors governing a potential improvement of the catalytic performances in acrylonitrile synthesis from propane, a kinetic study was undertaken on the V- Sb-W-(1:5:1)-Al_2O_3 [70%] system calcined at 620 C in air

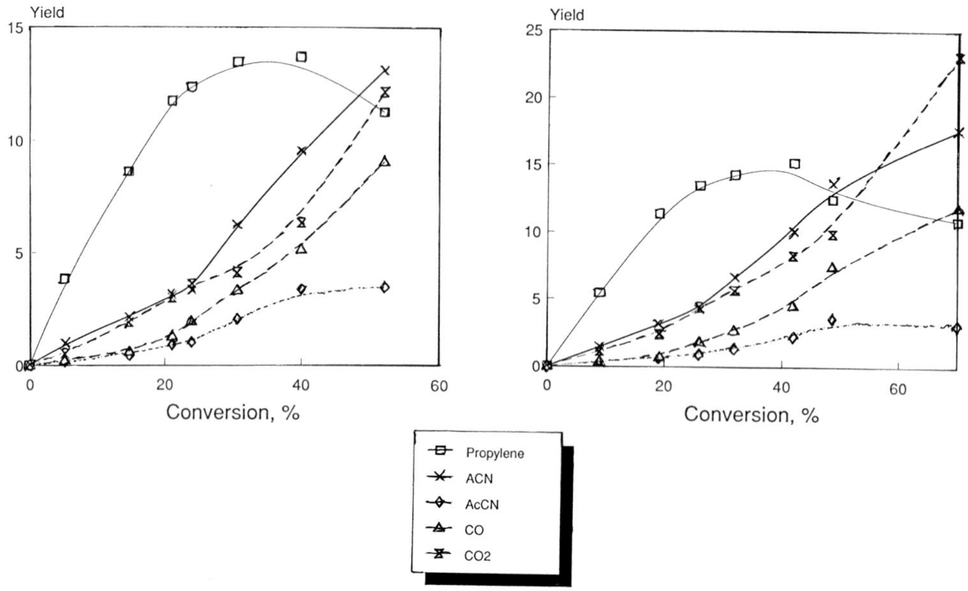

Figure 7 Dependence of the yields of propylene, ACN, AcCN and carbon oxides on the propane conversion at 432 C (**a**) and at 452 C (**b**) on VSbW(1:5:1)-Al_2O_3 [70% wt] catalyst calcined at 620 C in air. Exp. conditions: 1.62% propane, 7.42% O_2, 1.96% NH_3.

(surface area 88.6 m^2/g and 94.3 m^2/g before and after the catalytic tests, respectively), which was thought to be a representative sample for this analysis.

Figure 6 reports the effect of propane (Fig. 6a), oxygen (Fig. 6b) and ammonia (Fig. 6c) concentrations on the rates of formation of propylene, ACN, AcCN and carbon oxides measured in a differential-type flow reactor at conversions lower than 5-10%. The usual Langmuir-Hinshelwood dependence on hydrocarbon concentration with saturation of active sites is found for propylene, ACN and AcCN rates of formation, whereas that of carbon oxides formation depends linearly on the propane concentration. More complex dependences are observed for the variation in oxygen or ammonia concentrations. In both cases a maximum in the rates of formation of propylene, ACN and AcCN is noted, along with a different trend for the rates of carbon oxides formation. The latter increases considerably at the higher oxygen or lower ammonia concentrations. Higher selectivities to propylene, ACN and AcCN are thus observed at low O_2 or high NH_3 concentrations, but in the latter case the increase in ammonia causes an inhibition of the rate of propane transformation explaining the maximum observed in the rates of formation of propylene, ACN and AcCN. It should be noted that in all cases the principal product is propylene, which is formed with selectivities up to about 80% in these tests at low conversion in a differential-type flow reactor. A further important observation is the considerable lowering of the selectivity to propylene when ammonia is absent in the feed (Fig. 6c).

The effect of the conversion on the yield of products is shown in Figure 7 for two reaction temperatures [432 C (Fig. 7a) and 452 C (Fig. 7b)]. Clearly, acrylonitrile (ACN) and acetonitrile (AcCN) form by two parallel reactions, one directly from propane and the other by consecutive transformation of propylene. At the higher conversions, propylene, ACN and AcCN yields start to decrease due to consecutive transformations to carbon oxides. The effect is more evident at the higher reaction temperature (Fig. 7b). The reaction network thus may be summarized as follows:

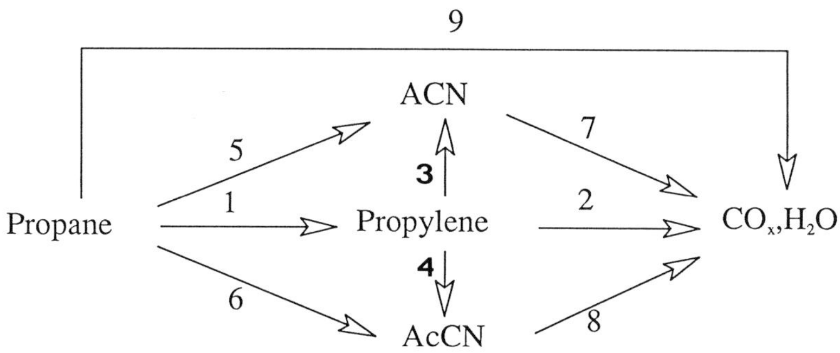

The rate constant **1** of propylene formation from propane is about four times higher than the rate constant **5** of direct formation of ACN from propane. The rate constant **9** of carbon oxides formation directly from propane is lower than the rate constants **3**, **4**, **7** and **8** of propylene consecutive transformation to carbon oxides and also lower than rate constants **1** and **5** of selective activation of propane. A slightly higher rate constant **2** of carbon oxides formation from propylene with respect to the rate constant **7** of ACN consecutive transformation is also observed. No drastic changes in the relative behavior are observed when the reaction temperature is increased from 410 to 450 C.

The kinetic analysis thus evidences that:

(**i**) there exist defined oxygen and ammonia concentrations for which the rates of propylene, ACN and AcCN are higher;

(**ii**) a very selective activation of propane to propylene and ACN occurs, with initial selectivities up to 80%, and the lowering of the selectivity is due to the consecutive oxidation of these compounds to carbon oxides;

(**iii**) acrylonitrile forms by two parallel reaction pathways, one directly from propane and the second through the intermediate formation of propylene;

(**iv**) due to parallel reaction of carbon oxides from propylene and the relatively slow rate of ACN formation from propylene, an improvment in the catalytic behavior can be realized by (**a**) adding a promoter for the direct ACN formation from propane or (**b**) more easily adding a co-catalyst that is more active in the transformation of propylene to ACN.

Conclusions

V-Sb-Al based mixed oxides are promising potential candidates as catalysts for the development of a new process of direct synthesis of acrylonitrile from propane. Actual yields to ACN are occurrently still relatively low, however, taking into account the large difference in the price of feedstocks, permit the development of an economical process with selectivities to ACN from propane lower than that to ACN from propylene. However, the high selectivity shown by these systems in the activation of propane to ACN and propylene indicates the possibility of improving performances by proper tuning of the catalytic functions and surface properties. In this respect, kinetic analysis shows that the main problem is the increase in the rate of consecutive ammoxidation of intermediate propylene, which requires the use of suitable promoters. We should also like to point out that the characteristics of these catalytic systems are very different from those of the usual catalysts for ACN synthesis from propylene. Starting with surface area about 6-10 times higher than propylene-based catalysts, new types of promoters and processing approaches are required for this new class of catalysts to be perfected to commercial reality.

References

[1] G. Centi, F. Trifiro', presented at the AIChE Spring Nat. Meeting, Symp. on Natural Gas Conversion, Houston TX, April 1989.

[2] G. Centi, F. Trifiro', J.R. Ebner, V. Franchetti, *Chem. Rev.*, 88 (1988) 55.

[3] G. Centi, D. Pesheva, F. Trifiro', *Appl. Catal.*, 33 (1987) 343.

[4] G. Centi, F. Trifiro', *Catal. Rev.-Sci. Eng.*, 28 (1986) 165.

[5] J. Birchall, A.W. Sleight, *Inorg. Chem.*, 15 (1976) 868.

[6] J. Berry, M.E. Brett, W.R. Patterson, *J. Chem. Soc. Dalton*, (1983) 9 and 13.

[7] N. Abadjieva, D. Pesheva, D. Klissurski, G. Centi, F. Trifiro', in Proceedings, 6th Int. Symp. Heterogeneous Catalysis, Varna (Bulgarien), Jun. 1987, Vol. I, p. 44.

[8] E. Patane', Thesis, Univ. Bologna, June 1988.

[9] A.T. Guttmann, R.K. Grasselli, J.F. Brazdil, U.S. Patent 4,746,641 May (1988), assigned to Standard Oil Co.

[10] G. Duquenoy, F. Josien, J. Livage, M. Michaud, G. Duquenoy, *Rev. Chim. Min.*, 18 (1981) 344 and 19 (1982) 211.

[11] B. Pal, K.K. Gupta, *Inorg. Chem.*, 14 (1975) 226.

[12] C. Rocchiccioli-Deltcheff, T. Depuis, R. Frank, M. Harmelin, C. Wadier, *C.R. Acad. Sci.*, B-541 (1970) 270.

[13] F. Trifiro', I. Pasquon, *J. Catal.*, 12 (1968) 412.

J.R. EBNER (Monsanto, St. Louis, USA): It is interesting to compare butane oxidation to propane oxidation. In butane oxidation we believe high selectivity and yields are possible because the initial heteroatom insertion is faster than desorption of olefins. In propane oxidation on this catalyst system, it appears that olefin desorption occurs readily. This presents the problem that the catalyst, which is designed to attack an alkane C-H bond, now can attack the highly reactive alkene. Thus, it would seem that major strides forward in this field will require catalysts that can directly insert heteroatoms as in the butane case. Your comments on this would be appreciated.
G. CENTI (Univ. Bologna, Italy): Our first approach for the design of catalysts for propane conversion was to utilize the same active phase for n-butane conversion to maleic anhydride (vanadyl pyrophosphate). Unfortunately, vanadyl pyrophosphate was found to be unselective in propane oxidation. Using an high propane concentration and a low oxygen concentration it was possible to increase the selectivity founding, however, only propylene as product of reaction. In an analogous way, adding ammonia to the propane/air mixture the main product on vanadyl pyrophosphate was propylene. It seems therefore that the sites of (allylic) oxygen insertion into propylene intermediate are not present and thus the nature of this sites is in someway different from that required for the selective conversion of butane to maleic anhydride. For this reason, we have used V-antimonate based catalysts. This systems is more selective in the stage of propane transformation to propylene, but again fail in the stage of consecutive quickly transformation of adsorbed propylene. However, it is evident that an increase of the rate of this step using suitable promoters is the major stride to improve the overall selectivity.

J.F. BRAZDIL (BP Research, Cleveland, USA): What is the evidence that the vanadium antimonate stoichiometry is $VSbO_4$ with V in the 3+ oxidation ? The work in the literature shows vanadium antimonate has the rutile structure with V in the 4+ oxidation state. The results in a nonstoichiometric defect structure containing cation vacancies.
G. CENTI: We have indicated the formulation $VSbO_4$ with V in the 3+ oxidation on the basis of the Mössbauer results of Sleight (ref. 5), but we do not have direct evidences about the real presence of V^{III} due to the difficulty in the chemical analysis of V^{III} (it oxidize to V^{IV} dissolving the sample) and due to the unclear evidences provide by physico-chemical techniques. However, in the analysis of the hydroxide precursor of vanadium antimonate we have observed the presence of mainly V(IV). The V(IV) oxidizes to V(V) during the calcination, at temperatures lower of that of formation of rutile phase. It is known that a two electron redox reaction may occurs between V^V and Sb^{III} giving V^{III} and Sb^V like in $V^{III}Sb^VO_4$ [B.B. Pal, K.K. Sen Gupta, Inorg. Chem., 11 (1975) 2268]. This suggests that at least partially V is in the 3+ valence state in the $VSbO_4$ rutile structure, but we cannot exclude the presence of a nonstoichiometric defect structure or of a solid solution with V_2O_4.

J. VEDRINE (CNRS, Inst. Catalyse, Villeurbanne, France): The best catalyst you have obtained involved both a badly crystallized material as detected by XRD and a large amount of Al_2O_3 (70% wt.). Does that mean that the active catalyst exhibit defects or noncrystalline phases ? Moreover, alumina may exhibit acidic properties. Do you think that Al_2O_3 as such or $AlSbO_4$ not yet in crystalline form are playing an important role in ammoxidation reaction ?
G. CENTI: I agree that in our system the best catalytic performances were obtained on amorphous systems and possibly on defective structures. When the same catalyst is calcined at higher temperatures and become crystalline, the catalytic performances worsen. However, it should be noted that also the surface area decreases and we believe that an high surface area in important.
I agree also that our results suggest that non crystalline $AlSbO_4$ (plus antimony oxide) may be the active phase for propane activation and the role of vanadium is more related to the further stage of propylene transformation and in the epitaxial crystallization of the $AlSbO_4$ phase, as shown by XRD results. However, these are only preliminary conclusions and more information are necessary to assess these statements.

P. RUIZ (Unitè de Catalyse, Univ. Catholique du Louvain, Belgium): We have evidences that when antimony is coprecipitated with other metals and the precursor is calcined, Sb_2O_4 is formed at the surface of the catalyst. Mechanical mixtures and impregnated supports with antimony show that this Sb_2O_4 is formed like a separate phase. This separate Sb_2O_4 exerts a powerful promoting action via emission of spill-over oxygen and remote control of catalysts. What are your arguments to affirm that in your catalysts the phases are formed epitaxially ?

G. CENTI: We have tested what is the effect of adding Sb_2O_4 by mechanical mixing to our catalyst, but the worsening of the catalytic performances indicate that in our case a separate Sb_2O_4 crystalline phase lead mainly to a surface coverage of the active centers. In addition, when we have discussed the formation of epitaxially intergrowth phases we refer to $VSbO_4$ and $AlSbO_4$ phases and not to antimony-oxide. In addition, it should be noted that in our best catalytic systems we do not have presence of crystalline Sb_2O_4.

J.M. HERRMANN (Ecole Centrale de Lyon, Ecully-Cedex, France): You mentioned in your last transparent that perovskites could be used as possible potential catalysts for mild oxidation reaction. Some years ago, we studied the oxidation of propene on tin-antimony mixed perovskites ($BaSn_{1-x}Sb_xO_3$). These solids appeared as total oxidation catalysts (only small traces of acrolein). This means that the structure of perovskite completely altered the mild oxidation properties of tin and antimony in mixed compounds. Have you some ideas about possible selective and active perovskites for mild oxidation reactions.
G. CENTI: I agree that due to general high ionic oxygen mobility perovskites behaves as total oxidation catalysts. However, the ionic oxygen mobility may be controlled choosing suitable doping elements. Also during the Symposium we have observed that Cu-based perovskites offer interesting possibilities as selective oxidation catalysts.

K. SESHAN (Univ. of Twente, Enschede, The Netherlands): You mentioned about gas phase promotor. On what basis do you select them ? Do you want to make a guess on the type of promoters ?
G. CENTI: We mentioned of gas phase promoters referring to some papers published on propane ammoxidation [Giordano et al., Oxid. Comm., 1-2 (1984) 99; Osipova et al., Kinet. Katal., 20 (1979) 510 and 910] that use halogen-based promoters in order to induce a radical-like pathways of propane transformation to propylene enhancing the selectivity in this step. A further type of gas phase promotor is SO_2 which has a beneficial effect on the control of selectivity.

G. Centi and F. Trifiro' (Editors), *New Developments in Selective Oxidation*

SELECTIVE CATALYTIC OXIDATION OF PROPANE BY SULPHUR DIOXIDE

Z.G. OSIPOVA, S.B. USHKOV, V.D. SOKOLOVSKII and A.V. KALINKIN
Institute of Catalysis, Novosibirsk 630090, U.S.S.R.

SUMMARY

Catalytic properties of Al, Ga, Si, Bi, Pb and Fe oxides and graphitized carbon towards propane oxidation by SO_2 were studied in a continuous flow reactor at 873-913 K. Oxidative condensation products (OCP) containing C,H,O and S which form during reaction over Ga,Al,Si oxides and carbon, were shown to be active towards propylene formation. Sulphidization of Fe,Bi and Pb oxides in the reaction did not increase the activity. As shown by XPS, OCP contain sulphur in S° state which participates in the limiting reaction step. A scheme of the mechanism of propylene formation over OCP is proposed.

INTRODUCTION

Selective oxidation of hydrocarbons by SO_2 has been the subject of many researches (refs. 1-7) but all the attempts to dehydrogenate propane have been failures (refs. 1-3). Most comprehensively studied is the reaction of oxidative dehydrogenation of ethylbenzene to styrene (refs. 4-7). T.G. Alkhazov, A.E. Lisovskii et al. have shown that when the reaction occurs over - Al_2O_3, oxidative condensation products (OCP) which contain C,H,S and O and selectively accelerate the interaction of ethylbenzene with SO_2 are deposited on the catalyst surface (refs. 4,5). A study of independent interactions of reagents with the catalyst allowed the authors to propose a concerted mechanism of styrene formation on OCP with participation of the oxygen of the oxidant (refs. 6,7). According to the authors sulphur in OCP is not involved in the interaction.

This paper reports the results of studies on the selective oxidation of propane over Bi,Pb,Fe,Ga,Si,Al oxides as well as carbon and also some ideas on the reaction mechanism. These compounds are recommended as catalysts and supports for propane dehydrogenation with oxygen in the presence of SO_2 and H_2S (refs. 9,10).

EXPERIMENTAL

Experiments were performed in continuous flow and pulse reactors with a steady catalyst bed at atmospheric pressure; the gas phase composition was analyzed chromatographically. Reaction mixtures containing propane, sulphur dioxide and helium were prepared in preliminarily outgassed steel balloons. Gases were preliminarily purified from admixtures.

Catalysts - Bi, Fe, γ-Al_2O_3 oxides were of analysis grade purity. Ga oxide was obtained via decomposition of Ga nitrate in air for 4 hours at 973 K. SiO_2 was of P.E.A. grade, carbon was sibunite obtained by granulating carbon black. Catalyst grains of 0.25-0.50 mm were placed in a quartz tube of 3 mm inner diameter. Vacant space in the reactor was filled with quartz granules of 0.25-0.50 mm. Without catalyst, filling-up of the reactor only with quartz does not lead to any noticeable propane oxidation under the experimental conditions employed. For the standard treatment the catalyst was aged in a helium flow for 1 h at the experimental temperature prior to experiments. The selectivity towards a product (%) was calculated as a proportion of this product in the products registrated without taking into account the OCP. The catalyst activity was defined as either an observable propane flow rate or a propylene formation rate.

The composition and amount of OCP were determined by DTA, DTG along with chemical analysis. The phase composition of catalysts was checked by X-ray and DTA methods. Concentration and electronic state of sulphur on carbon surfaces were determined by XPS on an ESCA-3 spectrometer with Al-K_α radiation. To calculate the binding energy of sulphur, C_{1s} graphite line was used as an internal reference.

RESULTS AND DISCUSSION

Oxides

Propylene, ethylene, methane, oxides of carbon and sulphur were the main products during propane oxidation over the studied catalysts. In the initial period of reaction CS_2 and H_2S were formed on a number of catalysts. For Ga, Al and Si oxides, propane conversion and selectivity towards propylene increase with time of stream (Fig. 1a), while for Fe, Bi, Pb oxides propane conversion decreases (Fig. 1b). The catalytic properties of investigated oxides are presented in Table 1.

As stated by the data of chemical analysis, DTA and X-ray, Ga, Al and Si oxides do not change their phase composition during the reaction, but OCP containing C, H, O and S are deposited on the surface. Depending on time, Fe, Bi and Pb oxides undergo phase changes, i.e., become sulphidized. No formation of OCP is observed on them.

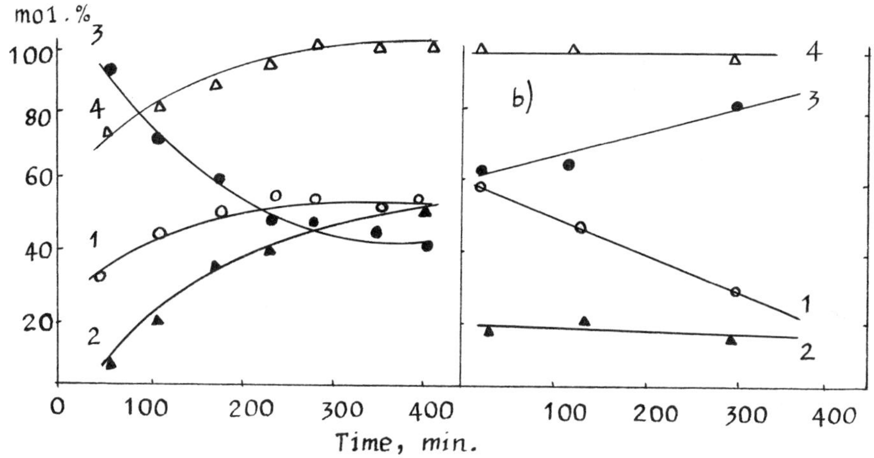

Fig. 1. Propane conversion (1), selectivity towards propylene (2), carbon oxides (3) and carbon balance (4) vs. time for γ-Al_2O_3 (a) and Fe_2O_3 (b) catalysts. Contact time for γ-Al_2O_3 1 s, for Fe_2O_3 25 s.

TABLE 1

Catalytic properties of oxides 3 h after the reaction start-up 913 K, mixture composition, mol.%, 10 C_3H_8, 10 SO_2, 80 He

Catalyst	S_{BET}[a)], m^2/g	Contact time[b)] s	Conversion, mol.% SO_2	C_3H_8	Selectivity, mol.% C_3H_6	C_2H_4	CH_4	CO	CO_2	Balance, C %
γ-Al_2O_3	229	1	100	50.6	35.9	5.9	6.6	20.4	30.5	85.9
SiO_2	40	10	90.4	71.7	56.8	16.2	6.0	4.9	10.7	90.0
Ga_2O_3[c)]	60	5	50.2	50.4	69.3	4.5	1.0	1.2	1.9	90.2
Fe_2O_3	11	25	100	23.3	14.3	2.9	-	-	82.8	99.5
PbO	0.2	20	9.1	2.5	12.0	-	-	-	88.0	100
Bi_2O_3	0.2	20	9.2	3.0	11.3		-	-	88.7	100

[a]Initial surface area of the catalysts (before work).
[b]Contact time (τ) = Catalyst volume (ml)/Mixture flow rate (ml/s)
[c]Data for 873 K.

For identical reaction mixture compositions, the composition of OCP formed is also similar (Table 2). As the surface of catalysts is packed with OCP, their activities become similar. Sulphidization of catalysts does not lead to an increase in propylene formation rate.

Thus, the active catalyst for selective propane oxidation with SO_2 is an oxidative condensation product occuring on the mineral substrate. The formation of surface Pb, Bi or Fe sulphides is unfavourable for the reaction.

The result obtained is in agreement with (ref. 3) indicating that for Pd/Al_2O_3 activity and selectivity towards propylene decrease with catalyst sulphidization.

TABLE 2
Effect of the reaction mixture on composition, surface area and catalytic properties of oxides. Reaction mixture composition mol.%, 10 C_3H_8, 10 SO_2, 80 He, 913 K

Catalyst	S_{BET}, m^2/g before work	S_{BET}, m^2/g after 6 h	OCP wt.%	Atomic ratio in OCP C : H : O : S	Rate $\frac{M \cdot 10^8}{m^2 \cdot s}$ 0.5h	Rate $\frac{M \cdot 10^8}{m^2 \cdot s}$ 6 h	Balance, C, %
γ- Al_2O_3	229	139	17.5	1:0.6:0.5:0.06	0.20	1.1	95.8
SiO_2	40	38	17.0	1:0.4:0.4:0.07	0.60	1.9	93.5
Ga_2O_3	60	30	10.0	1:0.5:0.5:0.06	0.25	1.2	94.2
PbO	0.2	0.2	0	0.8 wt.%(S^{2-})	0.1*	0.1*	100
Bi_2O_3	0.2	0.2	0	0.2 wt.%(S^{2-})	0.09*	0.09*	100
Fe_2O_3	11.2	1.2	0	20 wt.% (S^{2-})	0.4	0.3	100

*Reaction rate at ~ 2% propane conversion.

Since carbon was the main component of OCP, it was of interest to study catalytic properties of a carbon system. For this purpose carbon was used, and an active and selective catalyst for oxidative propane dehydrogenation was obtained by supporting Fe_2O_3 on carbon (ref. 8).

<u>Carbon</u>

Compared to OCP on γ-Al_2O_3, SiO_2 and Ga_2O_3, the rate of propylene formation is approximately 4 times higher at the initial moment of time.

TABLE 3

Catalytic properties of carbon, 913 K, τ = 3 s, 20% C_3H_8, 20% SO_2, 60% He (mol.%)

t, min	Conversion, mol.% C_3H_8	Selectivity, mol.% C_3H_6	C_2H_4	$CO+CO_2$	Balance, C,%	Rate C_3H_6 $\frac{M\ 10^6}{g\ s}$	$\frac{M\ 10^8}{m^2\ s}$	S_{BET}, m^2/g
3	58.0	78.4	14.7	6.9	61.7	1.88	4.7	400
5	62.9	78.1	17.8	6.3	72.1	2.04		
15	44.8	84.8	11.5	3.8	90.7	2.27		
100	26.2	85.9	11.9	2.3	94.2	1.42	18.7	76

The catalyst surface area decreases with time and disbalance with respect to carbon is observed, which indicates the formation of oxidative condensation products.

In order to reveal the reaction mechanism and the role of SO_2 as an oxidant, experiments were run in a pulse reactor with a 1.75 cm^3 pulse and 1.75 s time of pulse pass. Reaction mixtures of the following composition (mole %) were used: R (10 C_3H_8, 90

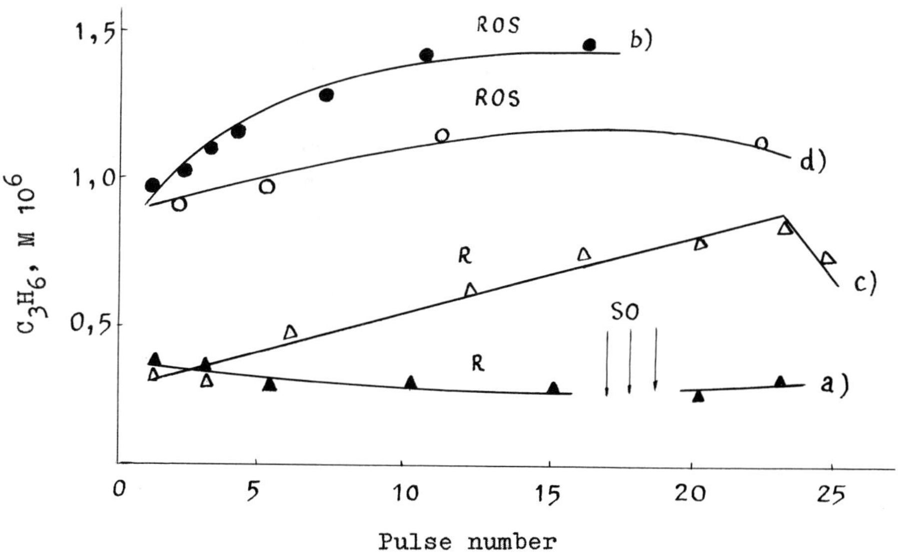

Fig. 2. Amount of propylene formed in R (▲ , △) and ROS (● , ○) pulses. 913 K, g = 0.132 g. a,b - only R (a) or ROS (b) pulses were fed to the catalyst; c,d - R (c) and ROS (d) pulses were fed in turn.

He), S (0.24 S in He), ROS (10 C_3H_8, 10 SO_2, 80 He), SO (10 SO_2, 90 He). Prior to experiments the catalyst was aged for 1 h in He flow at 913 K. The catalyst activity was characterized by the amount of propylene formed in one propane pulse. As propane passes through carbon, propylene is formed, the amount of which changes slightly from pulse to pulse (Fig. 2a). By further passing SO and after R pulses no propylene is observed in the reaction products. After 50 SO pulses, a slight (0.30 to 0.33 10^6 mol) increase of propylene formation is observed during R pulses. This indicates a weak interaction of both propane and SO_2 with the carbon surface. At ROS pulse through pure carbon the amount of formed propylene is higher compared to R pulse and increases from pulse to pulse (Fig. 2b). At passing ROS and R pulses in turn through the catalyst the amount of propylene formed in R pulses increases, thus becoming closer to that for ROS pulses.

Thus, accumulation of OCP results in greater contribution to the mechanism of propylene formation during propane interaction with the catalyst. This may be due to accumulation of oxidants participating in formation of propylene from propane.

It is known that elementary sulphur (refs. 10,11) can selectively dehydrogenate hydrocarbons; for this reason the effect of catalyst treatment with sulphur vapours on catalytic activity towards propylene formation has been tested (Fig. 3). The rate of propylene formation over the catalyst treated with sulphur vapours increases considerably. In all cases, treatment of the catalyst with the reaction mixture, sulphur vapours or sulphur dioxide resulted in an increase in the yield of propylene in pro-

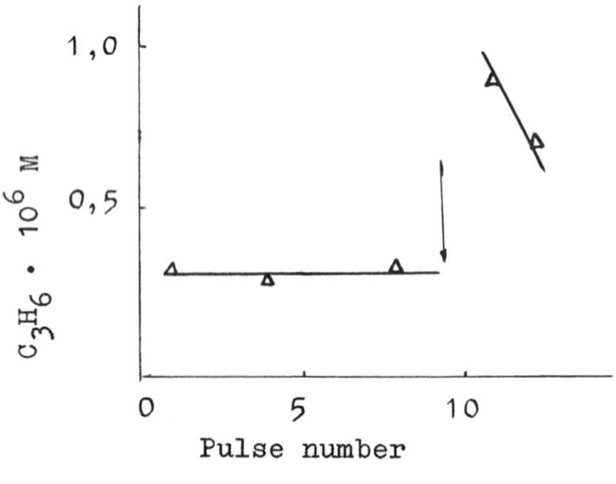

Fig. 3. Effect of carbon treatment with sulphur vapours ($3 \cdot 10^{-6}$ mol S_8) on propylene formation in propane pulses (R); 913 K, g = 0.132 g (g is catalyst weight).

pane pulses, with increasing sulphur content on the catalyst surface (Fig. 4).

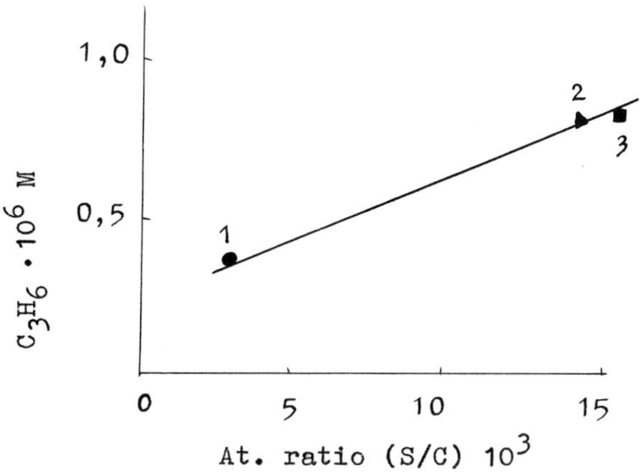

Fig. 4. Propylene yield in propane pulses vs. atomic S/C ratio on the carbon surface (from ESCA data).
1- treated with SO mixture (SO_2+He),
2- treated with ROS mixture (SO_2+C_3H_8+He),
3- treated with sulphur vapours (S_8+He),
g = 0.132, 913 K.

Independent of treatment conditions, only one state of sulphur corresponding to S° elementary sulphur was detected by ESCA (see Table 4).

TABLE 4

Values of binding energies for S_{2p} and C_{1s} on carbon vs. treatment conditions

Treatment conditions	C_{1s}	S_{2p}
Sulphur vapours	284.5	164.3
ROS mixture	284.5	164.5
SO mixture	284.5	164.4

The data obtained suggest that in selective propane oxidation with SO_2, the products of oxidative condensation containing C, H, O and elementary sulphur are the active species.

Taking into account the observed composition of reaction products and rate dependence, the scheme for propylene formation can be presented as follows:

1. $SO_2 + 3\ Z \longrightarrow ZS + 2\ ZO$

2. $ZS + RH_2 \longrightarrow R + SH_2Z$

3. $SH_2Z + ZO \longrightarrow ZS + H_2O + Z$

4. $ZS \rightleftharpoons Z + S$,

where Z are the active sites of the condensation product.

REFERENCES

1 C.R. Adams, Catalytic oxidation with sulphur dioxide, J. Catal., 11 (1968) 96-112.

2 F.M. Ashmavy, Catalytic oxidative dehydrogenation of propane to propylene, J. Catal., 46 (1977) 424-425.

3 F.M. Ashmavy, Kinetic investigations of the reaction of propane with sulphur dioxide on a palladium-alumina catalyst, J. Chem. Tech. Biotechnol., 34A (1984) 183-186.

4 T.G. Alkhazov, A.E. Lisovskii and Z.A. Talybova, Oxidative dehydrogenation of ethylbenzene to styrene with sulphur dioxide, Neftekhimiya, 17 (5) (1977) 687-689.

5 A.E. Lisovskii, Z.A. Talybove, A.E. Portyanskii, A.M. Musayev and T.G. Alkhazov, Condensation reactions in the process of oxidative dehydrogenation of ethylbenzene with oxygen and with sulphur dioxide, Neftekhimiya, 23 (5) (1983) 622-627.

6 Z.A. Talybova, A.A. Davydov, A.E. Lisovskii, T.G. Alkhazov, Investigation of the ethylbenzene interaction with oxygen and sulphur dioxide on the surface of -Al_2O_3, J. Phys. Chem., 58 (2) (1984) 453-457.

7 A.E. Lisovskii, Z.A. Talybova, T.G. Alkhazov, The regularities of oxidative dehydrogenation of ethylbenzene with sulphur dioxide on alumina oxide catalyst, Kinet. Katal., 25 (4) (1984) 862-867.

8 A.L.J. Solon and R.M.B. Northfield, Dehydrogenation of hydrocarbons employing a catalyst of iron oxide containing activated carbon, US Patent, 3647910 (1972).

9 S. Pasternak, M. Vadekar, Förfarande för dehydrering av ett ramaterial bestaende av kolväten med 2-8 kolatomer, Sweden Patent, 340451 (1971).

10 J.S. Pasternak, M. Vadekar, A.D. Cohen and N.J. Caspar, Sulphur promoted dehydrogenation of organic compounds, US Patent, 3585250 (1971).

11 P.H. Cortez, J.S. Land and L.J. Van Nice, New technology for selective production of propylene, AIChE Symp. Ser., 69 127 (1973) 134-135.

G. Halter (Yale University, USA): Could you elaborate on the second step of your mechanism? Specifically, do you intend for this step to be elementary or the summary of several steps that result in the reaction of gas phase propane and adsorbed H_2S? It appears to be too complicated to be an elementary step and one might expect H_2S to be adsorbed in dissociative manner.

Z. Osipova (Institute of Catalysis, USSR): We do not believe this step to be elementary. Certainly, it is a summary of several elementary steps that result in the formation of propylene. We have no sufficient results now to detail this step. This will be the subject of our future investigations.

M. Baerns (Ruhr Univ., Bochum, BRD): Would you please comment, how this reaction would be applicated from an applied point of view. How would you separate S, which is obviously a reaction product, from the catalyst and the reactants?

Z. Osipova (Institute of Catalysis, USSR): The following can corroborate that sulphur adsorbed on a surface takes part in the reaction:

1. Considerable increase in rate of formation of propylene during the treatment of a catalyst surface with sulphur.
2. No H_2S (hydrogen sulphide) in reaction products. That corroborates that reaction in a gaseous phase $S + C_3H_8 \rightarrow C_3H_6 + H_2S$ does not take place.
3. Regardless of the way we treat a catalyst, whether with fumes of sulphur, reaction mixture of propane and sulphur dioxide, or with sulphur dioxide, XPS method shows that there is one state of sulphur correspondent to S^o on a surface and that the speed of propylene formation is proportional to the quantity of S^o on a surface.
4. No gas-phase reactions: $C_3H_8 + 1/2\ SO_2 \rightarrow C_3H_6 + 1/2\ S + H_2O$ under experimental conditions.

Thus, active sulphur is sulphur stabilized on a surface by the oxidative condensation products. The reaction product is the gas-phase sulphur which can dehydrate propane at 700-740°C (Oil & Gas J., 1972, 70, No 23, 62-63), but is not active under conditions of our experiments.

G. Centi and F. Trifiro' (Editors), *New Developments in Selective Oxidation*

FLUIDIZED BED OXIDATION OF n-BUTANE: A NEW COMMERCIAL PROCESS FOR MALEIC ANHYDRIDE

G. Stefani[1], F. Budi[1], C. Fumagalli[1] and G.D. Suciu[2]

[1]Alusuisse Italia S.p.A. - Scanzorosciate, Bergamo (Italy)

[2]Lummus Crest Inc., Bloomfield, N.J. (USA)

SUMMARY

A new technology (ALMA) for producing maleic anhydride (MA) starting from n-butane has been developed jointly by Alusuisse Italia and Lummus Crest. It includes a fluid-bed oxidation system and a non-aqueous recovery system.

Extensive development work was carried out to successfully combine good catalytic activity and selectivity with high attrition resistance of the catalyst. A semi-commercial demonstration unit was used to confirm both the catalytic performance and the mechanical resistance of the catalyst and to optimize the hydrodynamic aspects of the fluid bed.

Several maleic anhydride producers have now chosen the ALMA technology which offers a remarkable economic advantage and makes available a low-cost maleic anhydride for new possible applications. First startup of an ALMA commercial plant was January 1989 at Shin-Daikyowa in Yokkaichi, Japan.

INTRODUCTION

MA is an important chemical intermediate. With its two functional groups (double bond and anhydride), it is very versatile and can be used to prepare many derivatives. The present end uses in Europe, USA and Japan are reported in Table 1.

TABLE 1

END USES OF MALEIC ANHYDRIDE (%)

	EUROPE	USA	JAPAN
Unsatured Polyester resins	55	57	35
Fumaric and Malic acids	12	10	25
Lube Oil Additives	8	9	10
Agricultural Chemicals	3	10	3
Paper sizing	2	4	5
Others	20	10	22

The demand is still growing and an annual increase of at least 5% is foreseen in the next few years. The present world capacity is in excess of 600,000 t/y. Table 2 presents the geographic distribution of the installed capacity in 1988 and the forecast for the next three years.

TABLE 2

MALEIC ANHYDRIDE INSTALLED CAPACITY (*)
(Thousand metric tons)

	1988	1989	1990	1991
West Europe	176	231	248	398
North America	185	203	227	227
Japan	86	97	108	108
Asia/Mid East (excl.Japan)	19	23	62	87
South America/Africa	38	41	56	63
East Europe	76	79	79	79
Total	580	674	780	962

(*) Alusuisse Italia information.

Until the early '70's practically all the maleic was produced from benzene. In 1974-75, at the conclusion of long research work, first Monsanto in the USA, and a few months later Alusuisse Italia in Italy, converted some commercial reactors using benzene feed to butane feed.

In subsequent years, several companies converted existing plants from benzene to butane and others constructed new plants specifically for butane feedstock. This happened mostly in the US where butane was easily available and environmental regulations for benzene emissions became very strict. At the present time, all the production in the US is based on butane.

Worldwide, half of the production is still based on benzene, but this technology is considered obsolete and not of interest for new plants, except for special isolated cases. The capacity increases shown in Table 2 derive almost exclusively from C_4 feedstocks and some of the older capacity based on benzene may be retired (future shutdowns not reflected in the table).

In recent years, scientific interest in selective oxidation of butane has increased significantly and many fundamental studies have been carried out on the catalytic system vanadium-phosphorus-oxygen (V-P-O) which is generally used in commercial processes. By the end of 1986, the number of patents and papers was reported to be 562 "(ref. 1)". In addition in the last years, some companies have undertaken research to apply fluid-bed technology to butane oxidation. At the same time, new recovery and purification systems have been studied which are based on organic solvent instead of aqueous systems. Two fluid-bed commercial plants have been started up recently in Japan by Mitsui Toatsu/BP (BP-UCB technology) and Shin-Daikyowa (ALMA technology). The latter also incorporates a non-aqueous recovery system based on a high-boiling organic solvent.

ALMA PROCESS DESCRIPTION

The schematic flow diagram (Fig. 1) depicts the standard inside battery limit (ISBL) configuration for the ALMA Process plus incineration. This consists of the following five areas:

Reaction
Maleic Anhydride Recovery
Maleic Anhydride Purification
Solvent Purification
Incineration

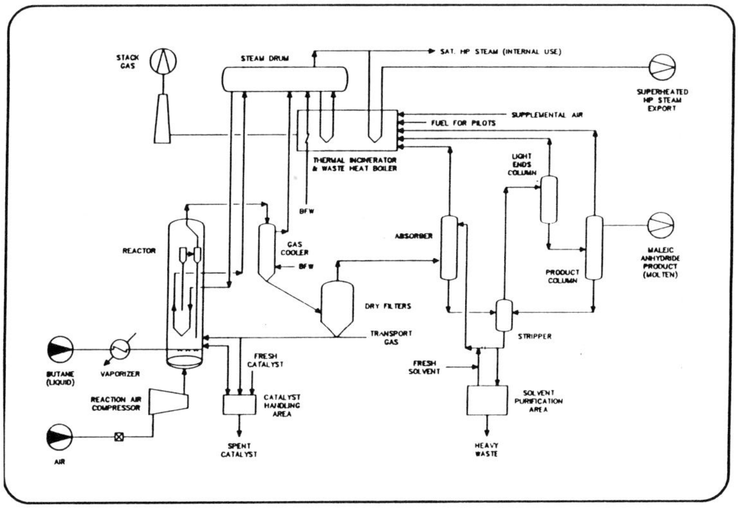

Fig. 1. Schematic flow diagram ALMA Process.

Reaction Area

The fluidized-bed reactor is fed separately with vaporized butane and compressed air. A start-up heater is used for initial commissioning and certain other nonsteady-state operations. Exothermic heat of reaction is removed from the reactor via production of HP steam in coil bundles located within the fluid bed, which is maintained at isothermal condition. The HP steam is used for internal requirements and for export after superheating in the incinerator.

Reactor effluent gases - after separation from elutriated bed solids in a cyclone separator system - are cooled, filtered to remove fines, and delivered to the maleic anhydride recovery area. The cyclone solids are returned directly to the reactor and the filter fines can be sent to either the reactor or the catalyst handling system for discharge, as preferred.

Maleic Anhydride Recovery Area

Reaction offgas is scrubbed with a proprietary solvent "(ref. 2)" to remove maleic anhydride before the offgas is exhausted to the incinerator. The rich solvent stream is heated and vacuum stripped to release the maleic anhydride. Crude maleic anhydride (stripper condensate) is sent to purification. Stripped solvent is cooled and returned to the maleic anhydride absorber. A solvent slipstream is withdrawn for purification.

Maleic Anhydride Purification Area

Crude maleic anhydride is fractionated to remove light ends. The small quantity of by-product light ends is delivered to the incinerator for destruction and waste heat recovery. The maleic anhydride is further fractionated to separate the small amount of solvent which accompanied it in the stripper overhead. The high-purity maleic anhydride product is delivered molten to the battery limits. The bottoms stream is returned to the maleic anhydride stripper.

Solvent Purification Area

A special step is included to remove any possible by-products from the reactor and solvent degradation in order to prevent the build-up of impurities in the solvent recycle loop.

Incineration Area

The absorber offgas is fed to the incinerator, where unreacted butane and reaction by-products (carbon monoxide, acetic and acrylic acids) are combusted and heat is recovered as superheated HP steam. The steam generated in the reaction area is also superheated in the incinerator. Because of the low air/butane ratio for the fluidized-bed reaction system, the combustibles in the offgas are concentrated enough that a thermal incinerator can be used without the need for auxiliary fuel (except for minimum control requirements) or feed preheat.

STEPS IN DEVELOPMENT PROGRAM

The development work for the ALMA fluidized bed system started in 1981 in cooperation between Alusuisse Italia and Lummus Crest. This combined the strength of Lummus Crest in fluid-bed technology and catalysts and Alusuisse Italia in maleic anhydride production and catalysts.

The main steps in the development program are summarized in Table 3.

TABLE 3

HIGHLIGHTS OF ALMA DEVELOPMENT PROGRAM

1981	- Concept development
1982	- Catalyst development in laboratory
	- Engineering studies of full-scale plant options
1983-84	- Extended laboratory test of selected catalyst
	- Design and construction of semi-commercial plant
	- Catalyst production for semi-commercial plant
	- Cold flow model simulations
1984	- Start of semi-commercial plant operation
1987	- First ALMA license to Shin-Daikyowa
Jan.1989	- Startup of Shin-Daikyowa plant

In 1982, hundred of catalysts were prepared and tested on a laboratory scale until the techniques to control the attrition resistance were identified.

From that point, optimization was begun to find the best combination between catalytic and mechanical properties, which culminated in the development of one or more catalysts possessing the optimum characteristics for a commercial fluid bed process. Concurrently, some patent applications were filed, which have subsequently been granted "(refs. 3-4-5-6)".

Extended laboratory tests with the selected catalysts demonstrated an excellent stability in both catalytic performance and attrition resistance. In parallel, kinetic simulations based on the projection of three different models were carried out for scaling up the fluid-bed reactor "(ref. 7)". At this point it was decided to build a semi-commercial plant to confirm the results of previous tests and studies, in order to ensure a successful scale-up. In the same period the production of the catalyst with commercial equipment was developed. The first batch of catalyst was used both for tests in a transparent cold flow model unit and for loading the semi-commercial plant, which started up in July '84. The size of the semi-commercial reactor (one meter diameter and height comparable to that for full-scale plants) was chosen in order to avoid any further scaling up risk to the full commercial size.

In the first period, several reactor configurations and particle size distributions were tested and optimized, together with other parameters like superficial velocity, bed height, pressure, temperature, butane concentration. At the end, the results previously anticipated have been confirmed and data to design the full-scale plant have been achieved. The butane concentration in the feed (at least 4% molar) and the conversion level (over 80%) make it possible to have an oxygen concentration in the off-gas low enough to avoid any possible flammable mixture. The MA molar yield is in excess of 50%.

In 1987, a license was sold to Shin-Daikyowa for the first

commercial plant, which started up in January '89. The yields obtained in the Shin-Daikyowa plant are consistent with those of the semi-commercial plant and no problems of reactor scale-up have been found.

CATALYST CHARACTERISTICS

A fluidized-bed partial oxidation catalyst must possess good properties in two important areas. The first, which is also common to fixed-bed catalysts, consists of the catalytic characteristics (activity, selectivity, stability): the second area, which is specific for the fluid-bed use, consists of the physical-mechanical properties like shape, particle size distribution and, which is most difficult to achieve, attrition resistance.

Improved resistance to attrition is frequently accompanied by decreased chemical performance because of the use of binders, supports, coatings or other additives. The success in developing the ALMA catalysts consisted of combining outstanding mechanical resistance with good catalytic properties.

Table 4 lists a few types of catalytic systems used in commercial plants and the corresponding typical Attrition Index (AI): the smaller the AI, the stronger is the catalyst.

TABLE 4

ATTRITION INDEX OF COMMERCIAL CATALYSTS

Type of Fluidized-bed System	Attrition Index (AI)
Zeolite-based catalysts for FCC	0.5 - 2
Older non-zeolite catalysts for FCC	15 - 25
Circulating reactor/regenerator dense bed system	4 - 8
Non-circulating dense bed system	8 - 15
ALMA catalyst	less than 8

The AI in Table 4 is measured in an accelerated test, using the apparatus of Fig. 2.

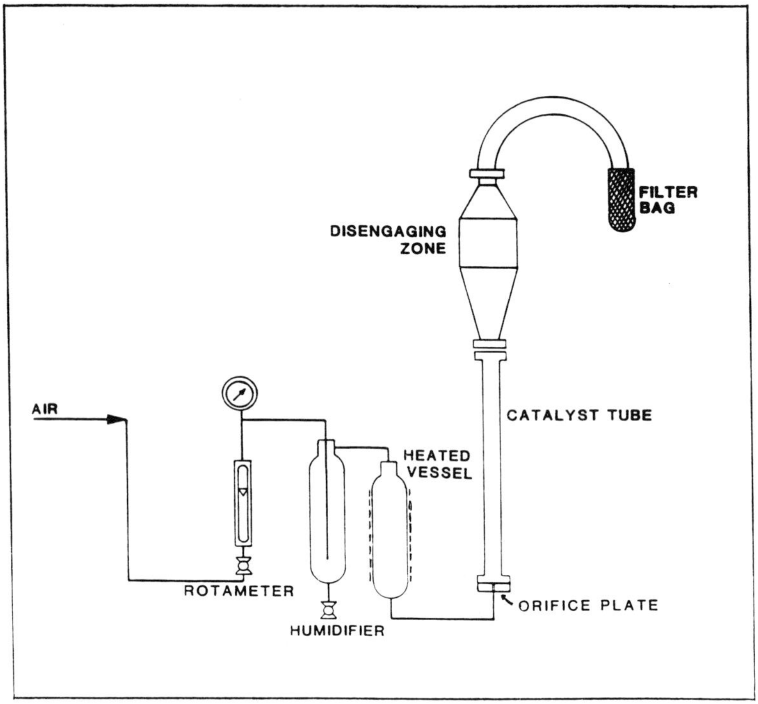

Fig. 2. Apparatus for attrition test.

The sample is loaded in the bottom of the tube and an air jet of sonic velocity is fed from the bottom through an orifice; the fines produced by attrition are collected in the filter.

The AI is the wt% of the fines, referred to the initial sample, collected in a 1 hour test.

Semi-commercial experience has confirmed that the attrition of the ALMA catalyst is almost negligible even in the most severe conditions. This has been further confirmed in the SDPC commercial plant.

Fig. 3 obtained with a scanning electron microscope (SEM) shows the typical spherical shape of a fluid-bed catalyst.

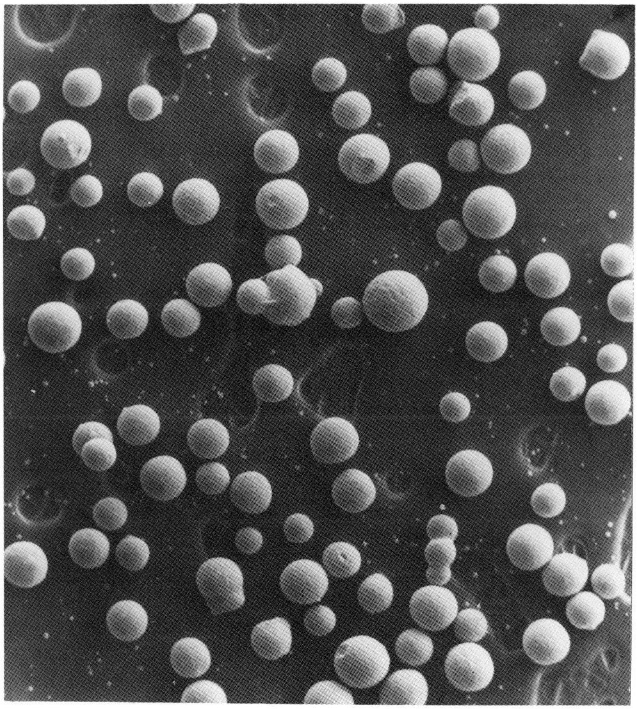

Fig. 3. Fluid-bed catalyst.

With regard to the chemical composition, the catalyst is basically composed of vanadium phosphorus oxide (VPO) complex. As mentioned above, many fundamental papers have been presented in recent years concerning VPO catalysts for selective oxidation of n-butane to maleic anhydride. An extensive review on the mechanistic aspects of the reactions has been made by G. Centi et al. "(ref. 8)".

E. Bordes "(ref. 9)" has supplied many details regarding the crystal structure of the various VPO complexes and the conditions for the transformation from one structure to another. The importance of the P/V ratio in the precursor preparation and the difference in P/V ratio between surface and bulk have been reported by B.K. Hodnett "(ref. 1)", N. Yamazoe "(ref. 10)" and others.

Fig. 4 shows a typical lamellar structure of a VPO precursor for a fluid bed catalyst. Fig. 5 is the FTIR spectrum of a fluid bed catalyst after about 1 month on-stream. It shows the presence of $(VO)_2P_2O_7$ at high purity level. Table 5 reports XPS (ESCA) data for a fluid bed catalyst after different times of operation.

Fig. 4. Crystal structure of a VPO precursor.

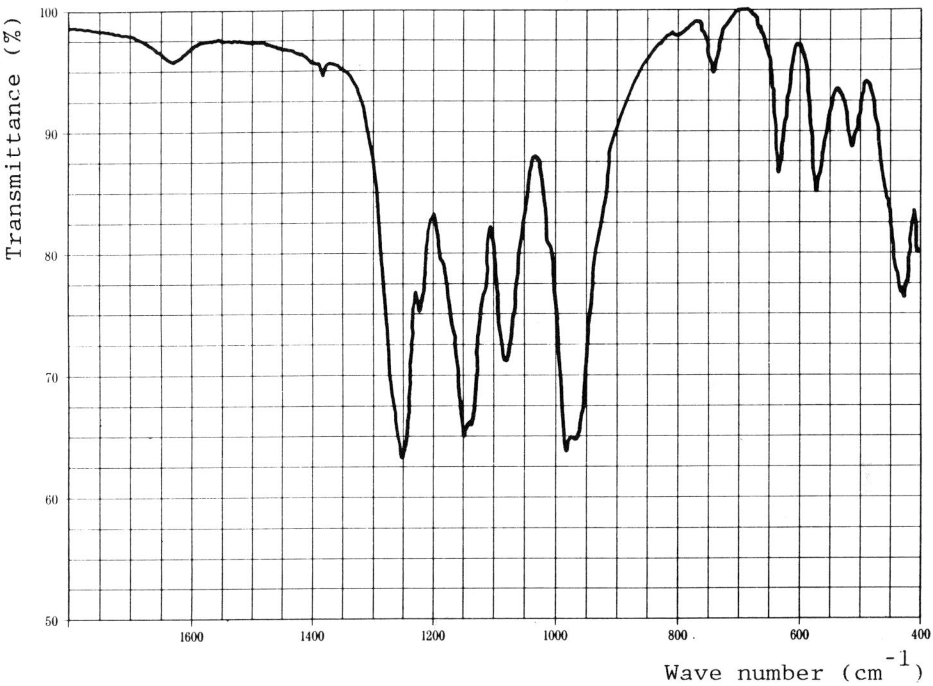

Fig. 5. FTIR of a fluid bed catalyst

TABLE 5

XPS DATA OF A FLUID-BED CATALYST

Days	P/V (atomic)	EBE (eV)
1	2.09	516.3
15	1.82	516.4
30	1.95	516.3
60	2.07	516.3
90	2.07	516.3

The XPS data show the following:

- electron beam energy (EBE) typical of vanadium + 4
- higher P/V ratio at the surface than in the bulk (ca. 1)
- stability of the surface P/V ratio and vanadium valence.

MODERN COMMERCIAL TECHNOLOGIES

Maleic anhydride production technology is currently undergoing a significant evolution. A list of the major new projects, all based on butane feedstock, is reported in Table 6.

TABLE 6

MAJOR NEW MALEIC ANHYDRIDE PROJECTS

Owner	Technology	Country	Capacity MTA	Startup
Mitsui Toatsu/BP	BP/UCB*	Japan	10,000	Oct. 1988
SDPC	ALMA	Japan	15,000	Jan. 1989
UPC/GP/Monsanto	Monsanto*	Taiwan	20,000	late 1990
TASCO	ALMA	Taiwan	14,500	late 1990
Nan Ya Plastics	ALMA	Taiwan	20,000	
CIEK	Mitsubishi*	Brazil	20,000	
Alusuisse Italia	ALMA*	Italy	50,000	
CEPSA/BP	BP/UCB*	Spain	30,000	
ORKEM/Monsanto	Monsanto*	France	54,000	
DSM/Alusuisse Italia	ALMA*	Northern Europe	40,000	
Dong Sung	ALMA	Korea	25,000	

*Licensor is part owner of plant.

There are also several recent or current minor projects (debottlenecking, modest expansion, small new plants up to 10,000 MTA capacity) which make use of fixed bed reactors, aqueous recovery systems and even, occasionally, benzene feedstock for localized special conditions.

The availability of maleic anhydride at lower cost, thanks to the modern technologies, can open new avenues to different important end uses.

The Dong Sung project, to produce 1.4-butanediol and related derivatives, is an example of such a new end use, which will further expand the maleic anhydride product market.

In addition to butane feedstock, all of these major projects incorporate modern technology in terms of fluidized-bed reactor and/or anhydrous recovery systems (Table 7). The reason for the selection of fluidized-bed reactors and anhydrous recovery systems is their widely recognized advantages "(ref. 11)".

Quantitative comparative information has been presented at the 1987 AIChE Summer National Meeting "(ref. 12)". A summary of the advantages of the most advanced current processing approaches is given in Tables 8 and 9.

TABLE 7

MODERN TECHNOLOGIES

Technology	Reactor	Recovery
ALMA	Fluidized bed	Anhydrous
BP/UCB	Fluidized bed	Aqueous
Mitsubishi	Fluidized bed	Aqueous
Monsanto	Fixed bed	Anhydrous

TABLE 8

REACTOR AREA (FLUIDIZED BED VERSUS FIXED BED)

- Increased steam production
- Shift of steam production to high-value superheated high-pressure steam suitable for conversion to power
- Reduced power consumption for air compressor
- Investment savings, especially for large plants
- Improvements in catalyst handling and downtime
- Large single-train capacity
- New technology, with high potential for future advances.

TABLE 9

PRODUCT RECOVERY AREA (CONTINUOUS ANHYDROUS VERSUS AQUEOUS)

- Large savings in steam consumption
- Substantial reduction in maleic anhydride losses, by-product formation, liquid wastes, and operational problems
- Operational, consumption, product quality benefits relative to batch recovery systems
- Investment savings, especially for large plants
- Newest technology, with greatest potential for future advances.

The current evolution of maleic anhydride production technology also includes the new "Recirculating Solids Reactor" being developed by Dupont and Monsanto. This technology, the evaluation of which is outside the scope of this paper, has been tested on a bench scale, but is not yet known to have been demonstrated in a pilot or semi-commercial plant.

CONCLUSIONS

As mentioned above, fluid-bed technology for maleic anhydride is relatively new and therefore possesses high potential for future improvement. Even now, several fluidized-bed technologies are surpassing previous processes based on fixed-bed reactors. The ALMA process is the only one which combines the fluid-bed with a modern non-aqueous recovery system.

REFERENCES

1 B.K. Hodnett, Catalysis Today, 1 (1987), pp. 475-498
2 A. Neri and S. Sanchioni, US Patent 4,314,946 (1982)
3 G.D. Suciu, G. Stefani and C. Fumagalli, US Patent 4,510,258 (1985)
4 G.D. Suciu, G. Stefani and C. Fumagalli, US Patent 4,511,670 (1985)
5 G.D. Suciu, G. Stefani and C. Fumagalli, US Patent 4,594,433 (1986)
6 G.D. Suciu, G. Stefani and C. Fumagalli, US Patent 4,654,425 (1987)
7 S. Dutta, S.C. Arnold, G.D. Suciu and L. Verde, Scale-up of a catalytic fluid-bed reactor involving complex kinetics, I. Chem. E. Symposium Series N. 87, pp. 517-526 (1984)
8 G. Centi, F. Trifirò, J.R. Ebner and V.M. Franchetti, Mechanistic Aspects of Maleic Anhydride Synthesis from C_4 Hydrocarbons over Phosphorus Vanadium Oxides, Chem. Reviews 88, 55-80 (1988)
9 E. Bordes, Crystallochemistry of V-P-O phases and application to catalysis, Catalysis Today 1 (5) 499-526 (1987)
10 N. Yamazoe, Successful design of Catalysts, pp. 15-24, Elsevier (1988)
11 S.D. Cooley and Bahrat Doshi, Maleic Anhydride from Normal Butane, AIChE 1987, Summer National Meeting
12 S.C. Arnold, J.W. Stanecki, D. Pedretti and M. Komeya, Development of ALMA Process Advances Maleic Anhydride Production Technology, AIChE 1987, Summer National Meeting.

DISCUSSION

M. HADDAD (Amoco Chem. Co.): You reported that the surface P/V ratio to be stable at about 2 between day 1 and days 90. How about the change in bulk P/V ratio ? Do you observe absolute P loss from the bulk ?

G. STEFANI (Alusuisse Italia): P/V ratio in the bulk is about 1 and it does not significantly change during the plant operation, even after longer times than 90 days.

J.R. EBNER (Monsanto Co.): In addition to the ALMA process, you mentioned the BP process for manufacture of maleic anhydride. There is an obvious difference in the back end purification method used. I would ask for you to compare and contrast the reactor/ catalyst with respect to several points.
1/ The catalyst method of preparation and attrition rates.
2/ Process conditions of butane concentration, pressure of operation and contact time.
3/ The use of special reactor design features such as autozones or special baffeling.
4/ Methods of minimizing back mixing.
As you may not be knowledgeable on all these areas for the BP process, please still comment on your process.

G. STEFANI (Alusuisse Italia): Of course I can answer only for our process:
1/ Attrition rate of the catalyst is less than 8, as measured in the apparatus of Fig. 2 and in the conditions described.
2/ Butane concentration is at least 4%, contact time is in the range 5-15 sec.
3-4/ Internal configuration of the reactor is designed to minimize bubble growth and back mixing.

P. RUIZ (Univ. Catholique du Louvain): Would you please give some data concerning the conversion and the yield of your commercial process ? Are there some differences between the conversion and the yield in commercial and pilot plant ?

G. STEFANI (Alusuisse Italia): Conversion is higher than 80% and molar yield is in excess of 50%. No problems of reactor scale-up were observed from semi-commercial (not pilot) scale to full scale.

J. VEDRINE (Inst. de Catalyse, CNRS - France): You have made a nice adverstisement to your problem but you did not speak about the particle size neither the binder used to get high attrition resistance. Could you tell us more details about particle size and binder and about the effect of the binder on VPO phase under catalytic conditions including presence of H_2O.

G. STEFANI (Alusuisse Italia): Particle size distribution is similar to type A powders according to Geldart classification (ref. 13). A good attrition resistance was obtained also without foreign binders and therefore without negative effect on VPO phase.

13 D. Geldart, Gas fluidization technology, p. 3, J. Wiley and Sons (1986)

G. Centi and F. Trifiro' (Editors), *New Developments in Selective Oxidation*
1990 Elsevier Science Publishers B.V., Amsterdam — Printed in The Netherlands

BUTANE OXIDATION IN A TRANSPORT BED REACTOR - REDOX CHARACTERISTICS OF THE VANADIUM PHOSPHORUS OXIDE CATALYST

RASHMI CONTRACTOR[1], JERRY EBNER[2], AND MICHAEL J. MUMMEY[2]

[1] E. I. du Pont de Nemours & Company, Wilmington, Delaware (USA)

[2] Monsanto Chemical Company, St. Louis, Missouri (USA)

ABSTRACT

Redox properties of vanadium phosphorous oxide (VPO) catalysts have been measured using butane and hydrogen as reductants and molecular oxygen as an oxidizing agent. These redox measurements indicate vanadium phosphorous oxide catalysts can vary markedly in oxygen capacity and redox rates. The VPO systems selected for this study range in oxygen carrying capacity from surface oxygen participation only to involvement of subsurface oxygen. The stability of the operative redox chemistry of a VPO catalyst is illustrated by a 30000 cycle butane reduction test. Butane oxidation results in a transport bed reactor are explained by fundamental redox properties of vanadium phosphorous oxide catalysts.

INTRODUCTION

The conversion of butane to maleic anhydride on vanadium phosphorus oxide catalysts represents an important commercial process (ref. 1). The Recirculating Solids Reactor (RSR), in which the hydrocarbon-catalyst reduction reaction is separated from the molecular oxygen-catalyst oxidation reaction, represents a recent innovation in the field (refs. 2-3). The key characteristics of a successful RSR butane to maleic catalyst are the catalyst's redox properties and its stability during continuous oxidation reduction cycles. We have developed methods of measuring these redox properties and have related fundamental redox properties to catalytic performance in a RSR. We have found the vanadium phosphorus oxide catalyst has a sufficient physical and chemical stability toward oxidation/reduction cycles, and this stability is explained in terms of the fundamental interactions between the catalyst surface and the reactant gases.

BACKGROUND

The best catalysts for the reaction of n-butane and oxygen to maleic anhydride are based on vanadium phosphorus oxides, and there is general agreement in the literature for the $(VO)_2P_2O_7$ solid state, crystalline phase predominating in these best catalyst systems (ref. 4). This selective oxidation reaction, which involves the removal of eight hydrogens as water and the insertion of three oxygen atoms, has a 3.5:1 oxygen to butane molar stoichiometry. The oxygen demand of this reaction is quite high. Earlier TAP (Temporal Analysis of Products) reactor studies of C4 hydrocarbon oxidation using the $(VO)_2P_2O_7$ catalyst indicate a primary source of selective oxygen is the metal oxide lattice (ref. 5). Isotopic labeling studies show lattice oxygen is involved in oxydehydrogenation and oxygen insertion reactions for both selective and non-selective reaction products. These studies also point to the importance of oxygen availability at the active surface to facilitate high selectivity by minimizing the desorption of reaction intermediates. The extent of lattice participation measured with VPO catalysts extracted from fixed bed reaction environments is limited to the surface lattice (refs. 5-7). This mechanism of supplying reactive oxygen, referred to as the Mars van Krevelen mechanism, is best described as a redox process in which oxidation of the hydrocarbon is achieved using surface lattice oxygen and reoxidation of the catalyst active sites is accomplished by reduction of oxygen at separate sites and the movement of oxide ions through the lattice framework. Recirculating Solids Reactor (RSR) technology allows decoupling and optimization of these redox reaction.

The application of a RSR to the heterogeneously catalyzed selective oxidation of butane offers several process advantages (refs. 8-9). Catalyst is oxidized in the catalyst regenerator and contacted with n-butane in the riser reactor where the oxidized catalyst converts n-butane to maleic and COx and is consequently reduced. The reduced catalyst is separated from the unconverted reactants and products and is returned to the regenerator for reoxidation.

EXPERIMENTAL

Catalyst Preparation

The catalysts used in this study were prepared in aqueous or organic media according to literature methods. The standard catalyst precursor used in the recirculating solids reactor (RSR) studies is prepared by partial reduction of V_2O_5 with isobutyl alcohol (IBA)/benzyl alcohol (BA), followed by precipitation of the blue $[VOHPO_4]_2{\cdot}H_2O$ precursor with addition of 85% phosphoric acid, tetraethyl orthosilicate, and long reflux (ref. 2). The final precipitate is filtered to produce a P/V ratio near the required 1:1 precursor stoichiometry. This preparation is identified as VPO-org1. In a second comparative organic preparation, the V_2O_5 is refluxed in isobutyl alcohol (IBA) saturated with SO_2 to reduce and dissolve the vanadium, and the precipitated precursor complex is obtained by addition of excess 105% H_3PO_4, reflux, and evaporation to remove solvent (ref. 10). This preparation, abbreviated VPO-org2, yields a precursor with excess phosphorus. The phosphorus to vanadium ratio by chemical analysis is $\approx$ 1.1. The aqueous preparation involves synthesis of the precursor by reduction of V_2O_5 to vanadium(IV) with aqueous HCl and precipitation by addition of 85% H_3PO_4 followed by distillation of solvent and filtration (ref. 11). This preparation is referred to as VPO-aq. For the organic based VPO catalysts used in RSR studies, the precursors are converted into fluidizable 45-150 μm microspheres by spray drying with silicic acid (ref. 12). These VPO microspheres are referred to as RSR catalysts in this paper. Both fixed bed and RSR "active" catalysts are obtained by air calcination at ≈ 673°K followed by running the butane oxidation reaction. Analysis of the activated catalyst samples reveals average vanadium oxidation states of 4.0. XRD analysis is consistent with the presence of a single crystalline phase, $(VO)_2P_2O_7$. However, the aqueous preparation yields larger crystals of $(VO)_2P_2O_7$ characterized by a smaller degree of disorder in the layering direction than the corresponding organic preparations. This is indicated in the XRD by the greater sharpness and intensity of the reflection at 3.87Å (020), and is consistent with published results (refs. 1, 4, 13). The surface areas for the activated catalysts are 30, 15, 20 and 12 m^2/g for the VPO-org1(RSR), VPO-org2(RSR), VPO-org2(Fixed bed) and VPO-aq, respectively.

TGA Redox Studies

Continuous Flow Reduction (CFR) studies are conducted using a Mettler TA-1 thermoanalyzer equipped with a quartz furnace with capabilities of controlled addition of reactive atmospheres (ref. 14). Reductions are carried out with hydrogen or butane. In a typical experiment, 50 mg of catalyst sample is pre-treated at 523°K in argon or helium to remove surface moisture. A weight loss of <1% is typically seen in this clean-up step. The sample is then cooled to room temperature and CFR data are obtained by ramping the temperature to 773°K at 10°K/min in a stream containing ≈ 30% reducing agent and holding at this temperature for ≈ 120 minutes. When the rate of weight loss reaches a constant value of < .005 mg/min, the sample is cooled to room temperature and the total weight loss is measured. This procedure eliminates the need for buoyancy corrections. When hydrogen is used as the reducing agent, the reaction product is water. Butane reaction products were not determined.

Pulse Flow Reduction (PFR) and Pulse Flow Oxidation (PFO) studies are conducted using a Du Pont 915 Thermogravimetric Analyzer, modified to permit pulsing the reducing or oxidizing agents. A large purge stream of nitrogen, and a short duration (15 seconds) reacting gas pulse alternating with a longer (75 seconds) nitrogen pulse are used to maintain the baseline of weight scale and near isothermal conditions. The gas stream contains 20% butane during the reduction pulse, and 20% oxygen during the oxidation pulse. In a typical experiment, about 60 mg of oxygen equilibrated catalyst sample is heated in TGA to 673°K in nitrogen purge and held for 10 minutes. Several oxygen pulses are introduced to insure that the catalyst is oxygen equilibrated. Then five butane pulses are introduced, and the weight loss due to oxygen extraction from the catalyst sample is recorded. This step is followed by oxygen pulses and the number of oxygen pulses needed to recover the catalyst weight loss is determined.

Reactor Redox Studies

A fluidized bed cyclic reactor apparatus was developed to test the catalyst's stability to repeated oxidation/reduction cycling (ref. 15). The cycle times and concentrations used were 30 seconds of 7.5 mole % butane in nitrogen followed by 120 seconds of 16 mole % oxygen in nitrogen. The reactor operating temperature was 693°K. The cycle times and reactor temperature were carefully chosen to expose the catalyst to a significant oxidation/reduction cycle without deliberate over-reduction of the catalyst by excessive exposures to the butane pulse.

RESULTS & DISCUSSION

TGA Redox Studies

Using a thermogravimetric balance, it is possible to separately study the reduction and oxidation properties of various VPO catalysts. In this paper, temperature programmed, continuous flow reduction/oxidation reactions and isothermal, pulsed flow reduction/oxidation reactions have been conducted using butane or hydrogen as the reducing agent. Reaction of the reducing agent with the catalyst results in a loss of weight associated with the reactive extraction of lattice oxygen. Hydrogen extracts lattice oxygen and forms water, and the surface remains clean. However, hydrogen is not a discriminating reducing agent and can be expected to remove lattice oxygen not available to butane. Presumably n-butane selectively removes oxygen through the active sites, but small amounts of oxidized hydrocarbon residues can remain on the surface (refs 5, 7). As a result of these surface residues, the amount of oxygen removed by n-butane in the TGA method could be slightly underestimated. Of interest to this study is the capacity of VPO catalysts to provide a reservoir of active oxygen (oxo-capacity), and the rates associated with its removal and restoration to the lattice.

The results of the CFR studies in hydrogen or butane are given in Table 1 for various catalysts. In the experiments in which hydrogen is the reducing agent, the onset temperature for removal of lattice oxygen for all catalysts studied is ≈ 573°K. Also, for all catalysts the rate of weight loss in mg/min increases nearly linearly with increase in temperature up to 773°K hold temperature. During the 120-minute time at the hold temperature, the rate of weight loss decreases gradually from as high as .02 mg/min to a constant value less than .001 mg/min. This experiment is a measure of the amount of O^{-2} available with deep reduction of the lattice. The difference between the VPO-aq and VPO-org2 systems in amount of oxygen extracted with hydrogen is quite large and cannot be explained on the basis of surface areas. There are ≈ 29237 μm of oxygen in a gram of $(VO)_2P_2O_7$. For one gram of $(VO)_2P_2O_7$, the estimated amount of oxygen in the surface lattice structure is ~50 μmoles/m^2 assuming a 5.5Å average surface depth. The 12 m^2/g VPO-aq catalyst can only provide ~1.1% of its total structural oxygen, which would not require subsurface lattice oxygen extraction. A ten-fold larger amount of oxygen is removed from VPO-org2, and some participation of subsurface layers is required. The larger number of defect sites and the greater exposure of (010) surface in catalysts derived from organic media (refs. 1, 16) may be the reason for these differences.

TABLE 1

TGA Studies of VPO Reduction [a]

Catalyst [b]	H_2 Reduction		n-Butane Reduction	
	mg O lost	micromoles lattice O	mg O lost	micromoles lattice O
VPO-aq (fixed bed)	5	312		
VPO-org1 (RSR)			34.3	2144
VPO-org2 (fixed bed)	51.9	3244		
VPO-org2 (RSR)	49.4	3088	10	625

[a] Values are expressed per gram of catalyst; [b] RSR catalysts have silica shells and fixed bed catalysts do not.

When butane is the reducing agent for VPO-org2, the amount of oxygen extracted is about five times less than for hydrogen. The VPO-org1(RSR) catalyst has an oxygen reservoir about four times greater than VPO-org2(RSR). Even assuming all the surface area of VPO-org1(RSR) is due to VPO (the surface area for RSR catalysts is actually the sum of the silica and VPO components), subsurface lattice oxygen accessibility is indicated. The reduction onset temperature for VPO-org1(RSR) is $\approx$ 573°K vs $\approx$ 633°K for VPO-org2(RSR), indicating greater oxygen lability in the former catalyst. These differences in oxygen availability may arise from the factor of two difference in surface area (30 versus 15 m^2/g) and the higher P/V ratio of VPO-org2(RSR). The literature is consistent in reporting reduced activity with high P/V catalyst systems (ref. 4). All the reduced catalysts were completely reoxidized when treated with oxygen at 773°K for 120 minutes.

The results of the PFR/PFO studies with butane and oxygen are given in Table 2 for the catalysts VPO-org1(RSR) and VPO-org2(RSR). The amount of oxygen extracted from VPO-org1 catalyst by five butane pulses is more than three times greater than that from VPO-org2; and the oxygen pulses needed to recover this weight loss is less than one-fourth for VPO-org1 than for VPO-org2. These higher redox rates and oxygen availability of VPO-org1 may arise from its higher surface area and lower P/V ratio.

TABLE 2

Pulse Flow TGA Studies of VPO Reduction/Oxidation

Catalyst	5 butane pulses mg O lost/gm cat	Number of oxygen pulses to regain wt loss
VPO-org1 (RSR)	3.13	48
VPO-org2 (RSR)	0.92	206

Reactor Redox Studies

The oxidation/reduction cycling stability test was run for a total of 31380 cycles which represent approximately 3-4 months of continuous operation in a RSR. The catalyst was characterized by determination of bulk density, surface area, pore volume, average pore diameter, X-ray diffraction powder pattern and phosphorous to vanadium ratio; tested for attrition resistance in a submerged jet attrition test mill according to DuPont's internal test procedure and performance tested in a bench scale RSR at the start of the test and after approximately every 8,000 oxidation/ reduction cycles for a total of 4 test sequences. Table 3 summarizes the results of the catalyst characterization measurements and attrition rate for the starting catalyst and after the 4 test periods during the 31,000 oxidation/reduction test. Figure 1 compares the X-ray diffraction powder patterns of these catalyst samples. No statistically significant trends are apparent in the data from Table 3 and Fig. 1 except for surface area. There appears to be a slight negative trend in surface area with increasing number of oxidation/reduction cycles; however, this trend did not have an impact on catalyst activity as shown below. There may be a slight improvement in attrition resistance with increasing oxidation/reduction cycles presumably due to a loss of weak catalyst particles from the cycling fluid bed as a function of time.

TABLE 3

Catalyst Characterization and Attrition Data

Oxid/red cycles	Fresh	8692	16216	25073	31380
Bulk Den g/ml	0.840	0.800	0.800	0.820	0.811
Surface Area m2/g	32.700	31.200	31.900	29.300	27.600
Pore Volume ml/g	0.110	0.120	0.110	0.110	0.100
Avg Pore Dia. Å	135.200	153.000	211.700	146.000	144.500
P/V Ratio	0.980	0.960	0.940	0.927	0.992
% Silica	13.400	12.100	13.050	13.300	13.230
Attrition Rate g/h	0.018	0.013	0.016	0.012	0.011

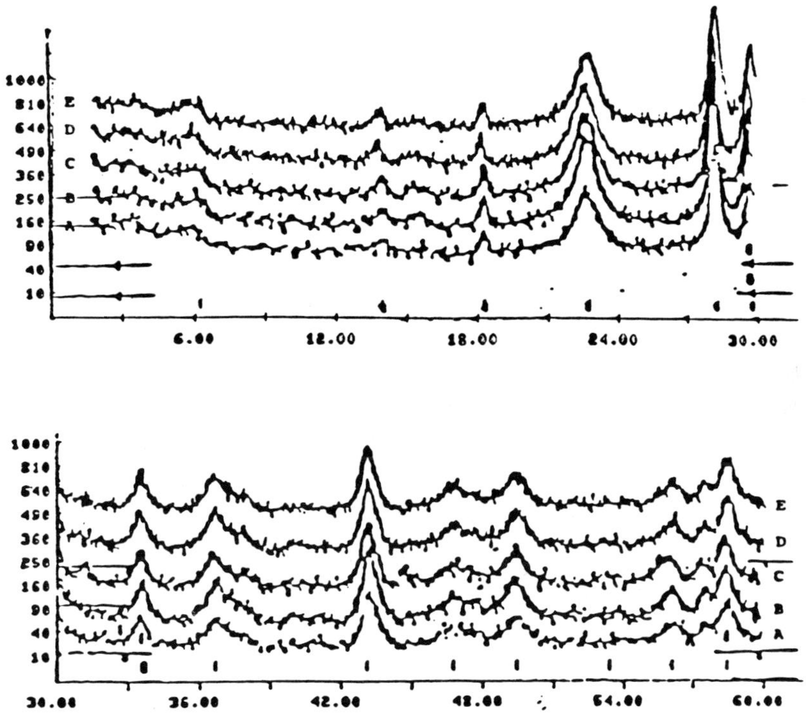

Fig. 1 TGA studies of VPO Reduction [a]
XRD of catalysts after various oxid/red cycles
(a) Fresh Catalyst, (b) After 8692 cycles, (c) after 16216 cycles, (d) After 25073 cycles, (e) After 31380 cycles

Performance of the catalyst for different numbers of cycles is summarized in Figs. 2 and 3. Figure 2 is a plot of the conversion selectivity relationship at constant process conditions for the fresh catalyst and catalyst exposed to different numbers of oxidation/reduction cycles. Fig. 3 is a plot of conversion vs. mole % butane at constant process conditions for various number of cycles. These data from Figs. 2 and 3 indicate that there were no changes in either the conversion/selectivity relationship of the catalyst or the activity of the catalyst as a function of the number of oxidation reduction cycles.

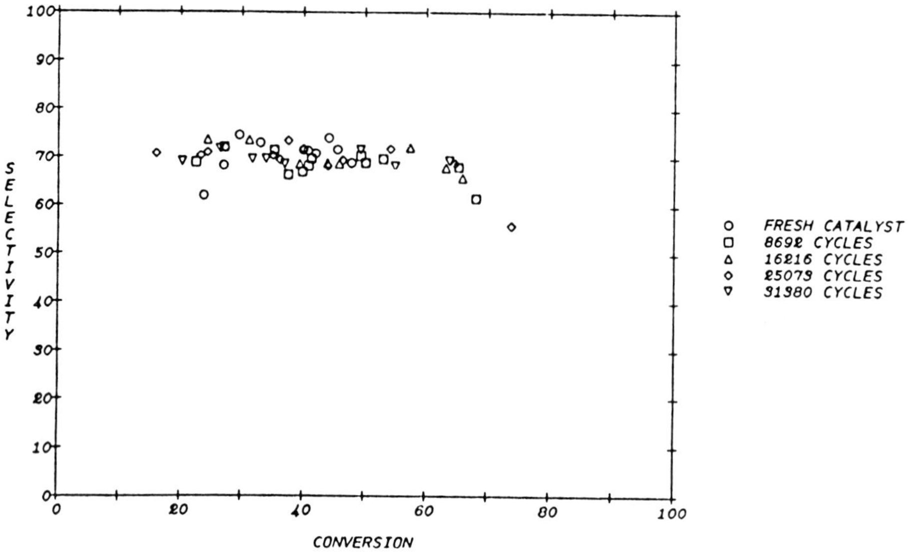

Fig. 2 Selectivity to MA as a function of butane conversion in RSR for catalyst after various oxid/red cycles.

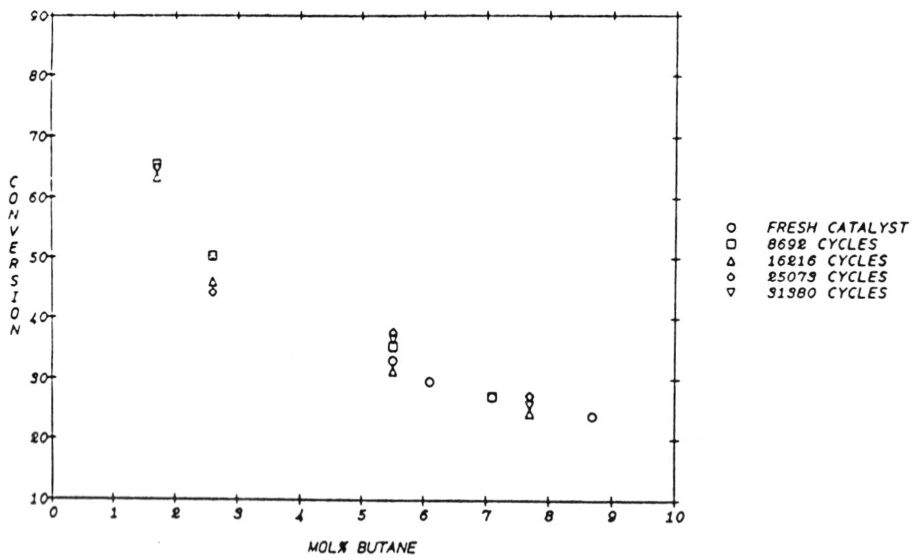

Fig. 3 Butane conversion as a function of butane concentration in RSR feed gas for catalysts after various oxid/red cycles.

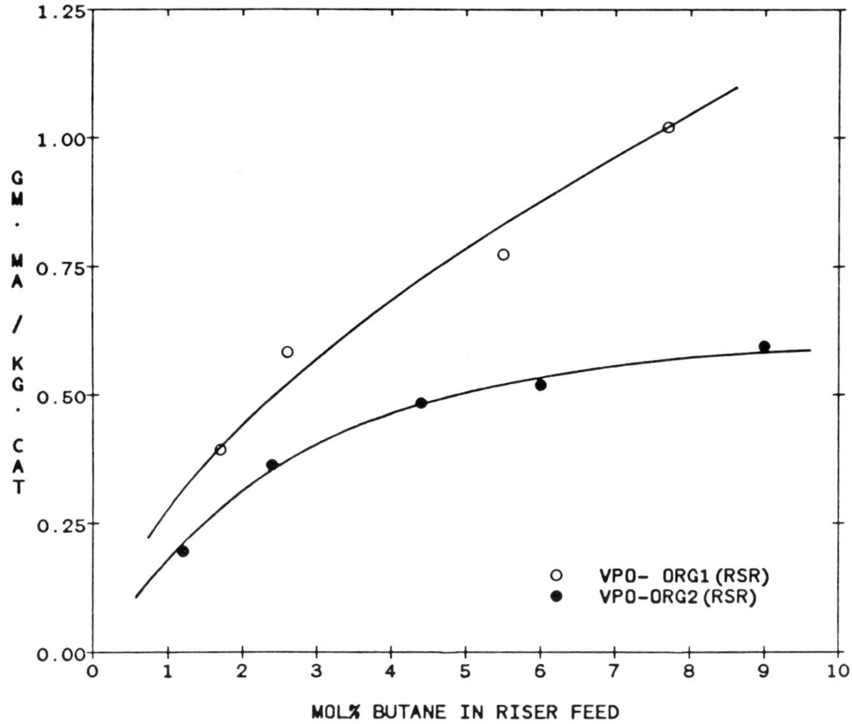

Fig. 4 Grams of maleic anhydride produced per kilogram of recirculating catalyst as a function of butane concentration in the RSR feed gas for two catalyst samples of different redox properties.

SUMMARY AND CONCLUSIONS

The oxidation/reduction properties of the VPO catalyst were earlier described as key to the successful application of the RSR to the selective oxidation of butane to maleic anhydride. The results reported in Tables 1 and 2 for VPO-org1 and VPO-org2 indicate significant differences in the oxo-capacity and the relative rates of removal and restoration of oxygen to the VPO lattice. If these properties are indeed key to the successful application of the RSR to the butane to maleic anhydride process, then VPO-org1 and VPO-org2 should produce dramatically different results in the RSR bench scale reactor. Figure 4 is a plot of the RSR data from both VPO-org1 and VPO-org2, showing grams maleic anhydride produced per kilogram of circulating catalyst as a function of mole % butane in the riser reactor feed gas. The two sets of experiments were made under identical reaction temperature (400 °C), gas residence time (2 seconds) and other test conditions. Increasing mole % butane in the riser feed extracts increasing amounts of oxygen from the catalyst and produces increasing amount of maleic anhydride per kilogram of catalyst. However, for VPO-org2, the catalyst deficient in oxo-capacity and with much lower rate of removal and restoration of oxygen to the catalyst lattice, the rate of increase in maleic anhydride production tapers off rapidly until essentially no increase in maleic anhydride production occurs with further increase in butane partial pressure. The catalyst VPO-org1 gives higher maleic anhydride production and continues to show increased production with increasing butane partial pressure.

REFERENCES

1 G. Centi, F. Trifiro', G. Busca, J. Ebner, and J. Gleaves, Faraday Discussion No. 87, Catalysis by Well Characterized Materials, University of Liverpool, April 11-13, 1989.
2. R. M. Contractor, H. E. Bergna, H. S. Horowitz, C. M. Blackstone, B. Malone, C. C. Torardi, B. Griffiths, U. Chowdhry, and A. W. Sleight, *Catalysis Today* 1 (1987) 49-48.
3. R. M. Contractor, U.S. Patent 4668802 issued May 26, 1987 to E. I. du Pont de Nemours and Company.
4. G. Centi, F. Trifiro', J. Ebner, and V. Franchetti, *Chem. Rev.* 88 (1988) 55-80.
5 J. R. Ebner and J. T. Gleaves in *Proceedings of the 5th IUCCP Symposium, "Oxygen Complexes and Oxygen Activation by Metal Complexes,* " eds. A. E. Martell and D. T. Sawyer, Plenum Press, New York (1988) 273.
6. J. S. Buchanan and S. Sundareson, *Appl. Catal.* 26 (1986) 211.
7. M. A. Pepera, J. L. Callahan, M. J. Desmond, E. C. Milberger, P. R. Blum, and N. J. Bremer, *J. Am. Chem. Soc.* 107 (1985) 4883.
8. R. M. Contractor, H. E. Bergna, H. S. Horowitz, C. M. Blackstone, U. Chowdhry, and A. W. Sleight in *Proceedings of the 10th North American Meeting of the Catalysis Society*, ed. J. W. Ward, Elsevier Science Publishers B V, Amsterdam (1988) 645-654.
9. R. M. Contractor and A. W. Sleight, *Catalysis Today*,3 (1988) 175-184.
10. J. T. Wrobleski, J. W. Edwards, C. R. Graham, R. A. Keppel, and H. Raffelson, U.S. Patent 4562268 issued Dec. 31, 1985 to Monsanto Company.
11. G. Busca, F. Cavani, G. Centi and F. Trifiro' *J. Catal.* 90 (1986) 400.
12. H. E. Bergna, U.S. Patent 4679477 issued Sept. 6, 1988 to E. I. du Pont de Nemours and Company.
13. F. Cavani, G. Centi, and F. Trifiro', *J. Chem. Soc. Chem. Commun.* , (1985) 492.
14. H. K. Yuen, G. W. Mappes and W. A. Grote, *Thermochemica Acta* 52 (1982) 143.
15. R. M. Contractor, A paper presented at CHEMECA 88, Syndey (1988).
16. H. S. Horowitz, C. M. Blackstone, A. W. Sleight and G. Teufer, Applied Catalysis, 38 (1988) 193-210.

G. Centi and F. Trifiro' (Editors), *New Developments in Selective Oxidation* 563

REACTIVITY AND STRUCTURE OF VANADYL PYROPHOSPHATE AS A BUTANE OXIDATION CATALYST

I. MATSUURA and M. YAMAZAKI

Faculty of Science, Toyama University, Toyama 930, Japan.

SUMMARY

Investigations were made on the structure and catalytic activity of vanadyl pyrophosphate. The structure of α, β, and γ-$(VO)_2P_2O_7$ were determined means of IR spectroscopic method. The catalytic activity decreases in the following order $\beta > \gamma > \alpha$, while the selectivity to MAA increases $\beta < \gamma < \alpha$. The catalytic properties of $(VO)_2P_2O_7$ depend on the arrangement of pyrophosphate ions in their crystals. Addition of phosphorus to β-$(VO)_2P_2O_7$ prevent the complete oxidation of butane.

INTRODUCTION

Vanadium phosphorus oxides are active and selective catalysts for the oxidation of butane to maleic anhydride. Many patents (1) and other studies indicate the best catalysts to have vanadium in the 4+ oxidation state and a P/V ratio of approximately 1. Vanadium pyrophosphate, $(VO)_2P_2O_7$, is the only well-characterized oxide with this oxidation state and P/V ratio. Recently, Misono et al. (2) prepared pure $(VO)_2P_2O_7$ and reported it to be an active and selective catalyst for the oxidation of butane to maleic anhydride. Bordes and Courtine (3) suggest that there are three phases, α, β, and γ-type of $(VO)_2P_2O_7$. The XRD patterns of α, β, and γ-$(VO)_2P_2O_7$ are similar, but their diffractograms show differences in the relative intensities I/I(200) of the main lines. They propose that the γ-$(VO)_2P_2O_7$ is the active and selective catalyst in butane oxidation to maleic anhydride whereas β-$(VO)_2P_2O_7$ is selective butane oxidation only.

The aim of this paper was to study the oxidation behavior of three phases and their catalytic activities for the oxidation of butane into maleic anhydride.

EXPERIMENTAL

Three types of $(VO)_2P_2O_7$ were prepared as follows. α-$(VO)_2P_2O_7$ was prepared according to the literature (4). V_2O_5 was added to an aqueous solution of $NH_2OH{\cdot}HCl$ and H_3PO_4. The mixture was stirred at 75°C until V_2O_5 was completely reduced. After the solution was evaporated to dryness at 170°C, the dried solid was washed by hot water until hydrochloride was removed. This material was calcined at 500°C for 2h. γ-$(VO)_2P_2O_7$ was prepared by the dehydration of $VOHPO_4 \cdot 0.5H_2O$ in vacuum at 600°C. To obtain $VOHPO_4 \cdot 0.5H_2O$, α-$VOPO_4 \cdot 2H_2O$ was reduced with n-butyl alcohol according to the literatures (5). β-$(VO)_2P_2O_7$ was prepared as follows: V_2O_5 was added to benzyl alcohol at 120°C. After the V_2O_5 was dissolved, non-aqueous H_3PO_4 was added. The precipitation was filtrated, washed with ethyl alcohol, and dried at 100°C. The dried material was evacuated at 600°C for 8h. The XRD and IR spectra of these solids are shown in Fig. 1-a and b.

The oxidation of butane was carried out in a conventional fixed-bed reactor from 360 - 480°C. The feed gas consisting of 2 vol% butane and artificial air was allowed to flow at a space velocity of 3000 ml/h·g-catalyst. The products were analyzed by gas chromatographs using a silica column for butane, CO, CO_2, and O_2 and a Porapak QS for maleic anhydride.

High temperature XRD measurements of the samples were carried out in a dried artificial air flow at 560°C on SHIMADZU XD-3A.

RESULTS AND DISCUSSION

First, it should be pointed out that the X-ray patterns of the three $(VO)_2P_2O_7$ were similar to those noted by Bordes and Courtine (5). There were no lines other than those belonging to the rhombic cell with space group of $C_{2v}{}^2$-P_{nc21}. Table 1 shows the inter-planar spacings d and relative intensities for α, β, and γ-$(VO)_2P_2O_7$.

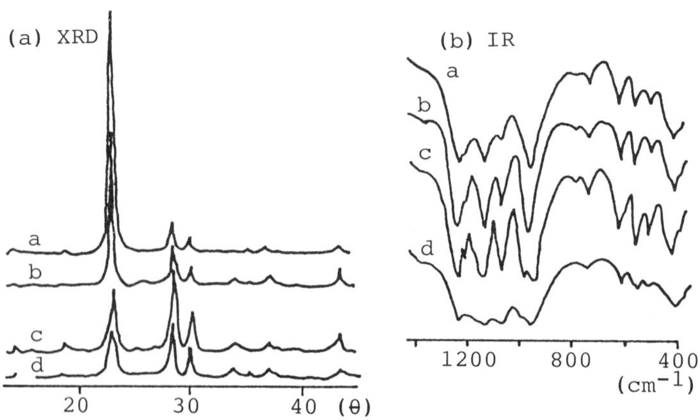

Fig. 1. XRD and IR spectrum of vanadium pyrophosphates.
(a) α-$(VO)_2P_2O_7$; (b) γ-$(VO)_2P_2O_7$; (c) β-$(VO)_2P_2O_7$;
(d) β-$(VO)_2P_2O_7$ with excess phosphorus (P/V = 1.1).

The diffratograms show that the relative intensities I(042)/I(200) of the main lines decreased in the following order $\alpha < \gamma < \beta$. It thus appears that α-$(VO)_2P_2O_7$ has a thin lamellar morphology along the plane (200).

As shown in Fig. 1-b, the IR spectra of the three $(VO)_2P_2O_7$ show a number of important differences. Some bands possess different contours differing intensity. In Table 2, we suggest the normal frequencies of IR for $P_2O_7^{4-}$ in the lattice norde D_{3h}, based on the spectra of $Ca_2P_2O_7$ (6). The V = O stretching and V-O-V bending frequencies are also indicated.

The structure of $(VO)_2P_2O_7$ has been solved by Gorbunova and Linde (7). The framework of the parallel plane to (100) is composed of pairs of pseudooctahedra sharing edges, where the vanadyl bonds are in a trans-position, linked by pyrophosphate. The layers are connected together through V - O = V and P - O - P bonds resulting in chains of vanadium octahedra sharing opposit corners and pyrophosphates as shown in Fig. 2-a. Crystallographic data are given in Table 3.

A comparison of the cryatallographic data and vibrational assignments for the pyrophosphate ion and V = O stretching in α, β, and γ-$(VO)_2P_2O_7$ confirmed the β-$(VO)_2P_2O_7$ to be in agreement with two kinds of pyrophosphate ions and trans-type vanadyl groups, as shown by Gorbunova and Linde (7). The α-$(VO)_2P_2O_7$ has only one kind of pyrophosphate ion and one kind of trans-type vanadyl group in the crystal. The proposal (100) plane is shown in Fig. 2-b. The arrangement of pyrophosphate ion and vanadyl group in γ-$(VO)_2P_2O_7$ crystal is not yet solved from the result of IR spectrum.

TABLE 1
X-RAY DATA OF VANADYL PYROPHOSPHATE

hkl	d(Å)	α-$(VO)_2P_2O_7$ I/I(200)	γ-$(VO)_2P_2O_7$	β-$(VO)_2P_2O_7$	β-$(VO)_2P_2O_7$(+P)
021	6.281		4	10	13
111	5.680		4	7	8
002	4.822	2	5	15	14
200	3.833	100	100	100	100
042	3.143	13	46	135	132
202	2.988	7	23	65	63
232	2.660	2	9	23	18
242	2.444	3	10	19	18
004	2.408	1	6	11	10
063	2.094	6	23	35	36
400	1.920	6	10	10	12
263	1.841	2	6	19	18
442	1.642		3	12	10

TABLE 2

OBSERVED FREQUENCIES OF PYROPHOSPHATE ION IN VANADYL PYROPHOSPHATE

Description	Class	α-$(VO)_2P_2O_7$	γ-$(VO)_2P_2O_7$	β-$(VO)_2P_2O_7$	Calc.Pyro-Phosphate ion
Terminal stretch	A_1'	1246	1248	1246 1266	1212
	A_2''	1222	1222 1186	1222 1187 1166	1165
	E'	1144	1146 1120	1148 1132 1116	1124
	E''	1087	1082	1088	999
Bridge stretch	A_1'	924	924	930	940
Terminal bend	E'	744	744	746 670	707
	E'	637	636	636 620	615
	E''	578	578	570	573
	A_2''	514	516	516 544	553
	A_1'	428	428 440	406 400	432
V = O		974	978	990 956	
V - O - V		800	800	800	

TABLE 3

CRYSTALLOGRAPHIC DATA OF VANADYL PYROPHOSPHATE (7)

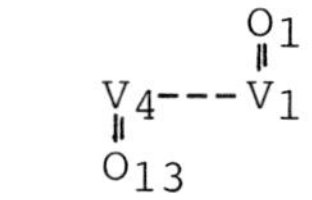

Bond	Length
V_1--O_1	1.730Å
V_4--O_{13}	1.551

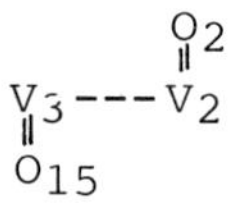

Bond	Length
V_2--O_2	1.704Å
V_3--O_{15}	1.531

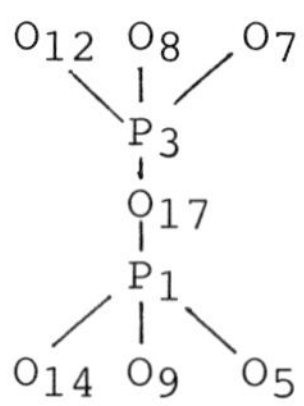

Bond	Length
P_3--O_7	1.568Å
P_3--O_8	1.514
P_3--O_{12}	1.482
P_3--O_{17}	1.561
P_1--O_{17}	1.569
P_1--O_5	1.566
P_1--O_9	1.489
P_1--O_{14}	1.521

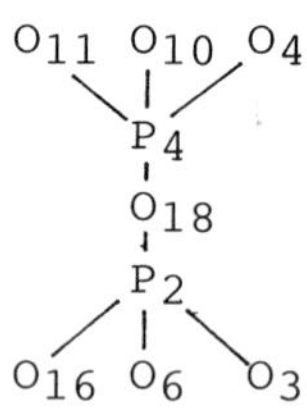

Bond	Length
P_4--O_4	1.475Å
P_4--O_{10}	1.523
P_4--O_{11}	1.558
P_4--O_{18}	1.582
P_2--O_{18}	1.577
P_2--O_3	1.482
P_2--O_6	1.559
P_2--O_{16}	1.520

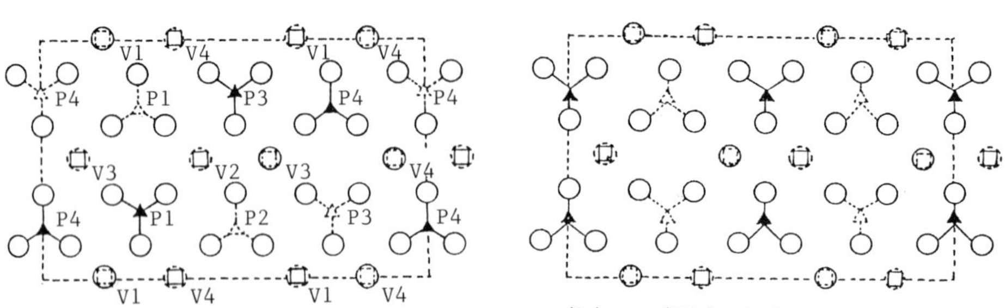

(a) β-$(VO)_2P_2O_7$ (b) α-$(VO)_2P_2O_7$

Fig. 2. Framework of (100) plane:
(a) β-$(VO)_2P_2O_7$; (b) proposal α-$(VO)_2P_2O_7$.

Table 4 shows the optimum yield of maleic anhydride from butane oxidation between 360°C to 480°C. α-$(VO)_2P_2O_7$ was not so active but had high selectivity to maleic anhydride. β-$(VO)_2P_2O_7$ was very active but possessed less selectivity compared to the α- and γ-$(VO)_2P_2O_7$. However, when excess phosphorus was present in the β-$(VO)_2P_2O_7$, the selectivity greatly increased.

TABLE 4
CATALYTIC ACTIVITY FOR BUTANE OXIDATION TO MALEIC ANHYDRIDE

Catalyst	React. Temp. (°C)	C_4H_{10}-Conv. (%)	MAA-Select. (%)	MAA-Yield (wt. %)
α-$(VO)_2P_2O_7$	460	65	77	85.1
γ-	400	75	58	74.0
β-	360	82	42	58.5
β- (P/V= 1.1)	420	72	78	95.1

Fig. 3-a, b, c, and d show the results of the oxidation of these materials tested at 560°C in an oxygen gas atmosphere using in-situ XRD apparatus. In the case of α-$(VO)_2P_2O_7$, the peak for $(VO)_2P_2O_7$ (d = 3.88, 3.14, 3.00 Å) attenuated and the peak for β'-phase (d = 4.62, 4.23, 3.19 Å) similar to that of B'-phase reported by Hodnett (8), by which the general formula of $(VO)_2^{2+}(VO)_{2n-1}^{3+}(P_2O_7)^{4-}(PO_4)_{2n-1}^{3-}$ (where n is 1,2,4,---,∞) (9) was proposed, increased with time. By further oxidation, β-$VOPO_4$ was observed. In the case of γ-$(VO)_2P_2O_7$, the peak for $(VO)_2P_2O_7$ attenuated with time, and β'-phase peaks and the peak for β-$VOPO_4$ (d = 4.62, 4.23, 3.19 Å) appeared. By further oxidation, the peak for $(VO)_2P_2O_7$ and β'-phase disappeared, leaving only those corresponding to β-$VOPO_4$. In the case of β-$(VO)_2P_2O_7$, the sample was oxidized directly to β-$VOPO_4$. However, β-$(VO)_2P_2O_7$ with excess phosphorus was oxidized to the β'-phase. No peaks to β-$VOPO_4$ appeared in this case. A probable oxidation scheme of vanadyl pyrophosphate is as follows;

$(VO)_2P_2O_7$ → β'-phase, $(VO)_2^{2+}(VO)_{2n-1}^{3+}(P_2O_7)^{4-}(PO_4)_{2n-1}^{3-}$ (n = 1,2,4,---,∞)
$(VO)_2P_2O_7$ → β-$VOPO_4$
β'-phase ↓ β-$VOPO_4$

The oxidation of α-$(VO)_2P_2O_7$, the selective oxidation catalyst, proceeds to the β'-phase. On the other hand, the non-selective oxidation catalyst of β-$(VO)_2P_2O_7$ was oxidized directly to β-$VOPO_4$. The different oxidation properties of α, β, and γ-$(VO)_2P_2O_7$ depend on the different arrangement of pyrophosphate ions in their crystals. The view of the structure looking down onto the (100) plane of α- and β-$(VO)_2P_2O_7$, β'-phase (n = ∞) and β-$VOPO_4$ is shown in Fig. 4.

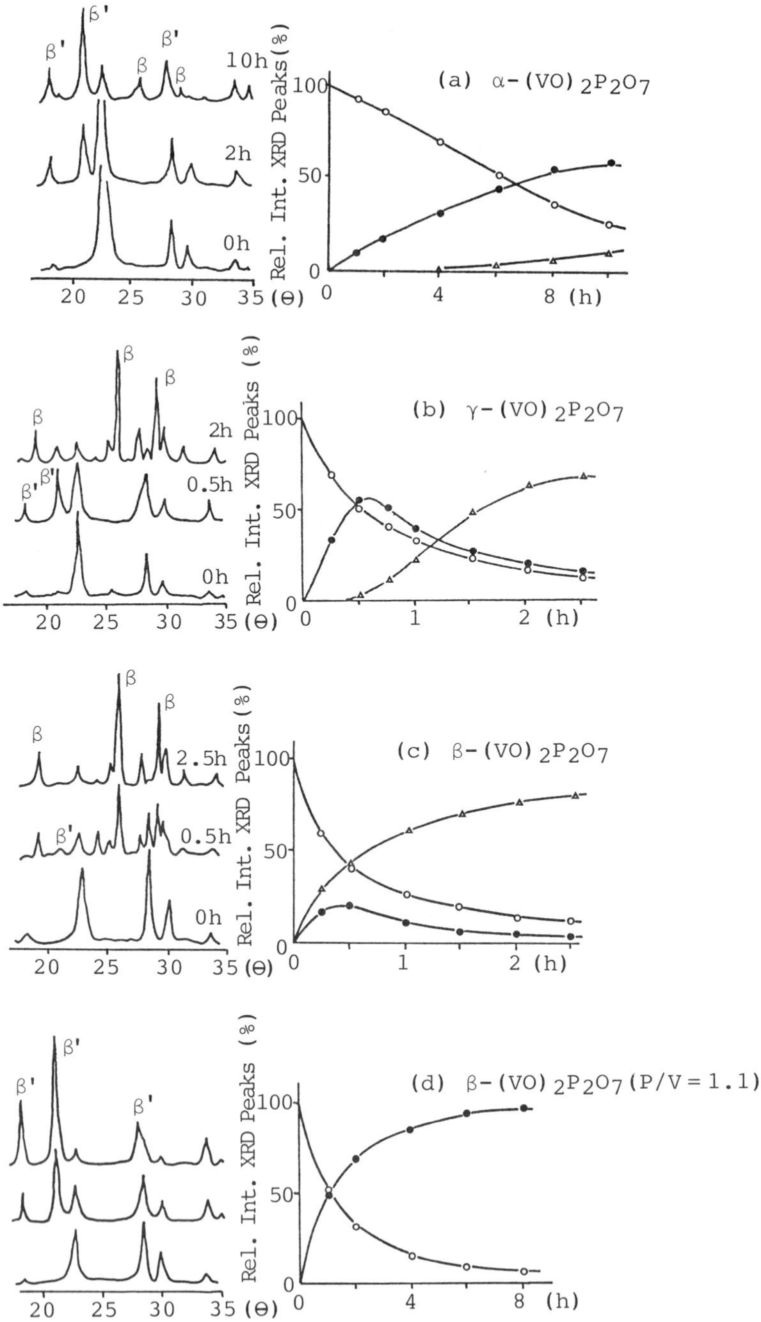

Fig. 3. Oxidation of vanadyl pyrophosphates.
(a) α-$(VO)_2P_2O_7$; (b) γ-$(VO)_2P_2O_7$; (c) β-$(VO)_2P_2O_7$;
(d) β-$(VO)_2P_2O_7$ with excess phosphorus (P/V = 1.1).
o; $(VO)_2P_2O_7$, ●; β'-phase, ▲; β-$VOPO_4$

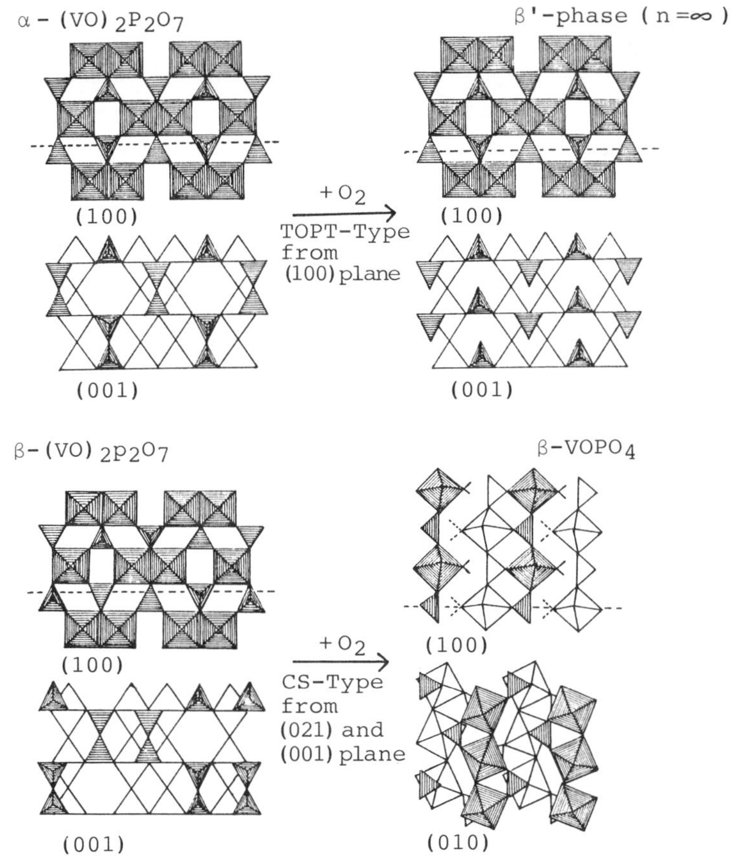

Fig. 4. Comparison of α-$(VO)_2P_2O_7$, β-$(VO)_2P_2O_7$, β'-phase (n = ∞) and β-$VOPO_4$ viewed in the (100) plane.

The oxidation of vanadyl pyrophosphate (V^{4+}) occured through two routes, oxidation to β-$VOPO_4$ through the (021) or (001) plane with crystallographical share and oxidation to the β'-phase topotactically from the (100) plane. Assuming that the redox reaction between vanadyl pyrophosphate (V^{4+}) and vanadium phosphate (V^{5+}) is introduced for butane oxidation, the formation of maleic anhydride may proceed with the topotactical redox reaction over the base of a catalyst (over the (100) plane) and the combustion of butane may proceed with crystallographical share redox reaction over the pillar face of catalyst (over the (021) and (001) planes).

Trifiro et al. (10) have proposed that the pair of vanadium ions coordinated with VO_6 edge-sharing octahedra is important to activate butane for the formation of maleic anhydride through oxidation. We consider their theory acceptable in consideration of the redox reaction between α-$(VO)_2P_2O_7$ and β'-phase over topotactically similar patterns of (100) plane during butane oxidation to maleic anhydride. The addition of extra phosphate while preparing β-$(VO)_2P_2O_7$,the non-selective oxidation catalyst, results in the formation of vanadium bis-metaphosphate $(VO)(PO_3)_2$ with the reaction of β-$VOPO_4$ on the pillar face ((021) and (001) planes) of β-$(VO)_2P_2O_7$. Therefore, the pillar face of β-$(VO)_2P_2O_7$ covered with $(VO)(PO_3)_2$ prevents the oxidation reaction with crystallographical share to β-$VOPO_4$ which controls the complete oxidation of butane.

REFERENCES

1 Chevron, U.S.Patent, 3864280 (1975), 4043943 (1977); Monsanto, U.S.Patent, 330354 (1977); Lonza, D.Patent, 2505844 (1975); I.C.I., Berg.Patent, 867189 (1978).

2 T.Shimoda, T.Okuhara, and M.Misono, Bull. Chem. Soc. Jpn., 58 (1985) 2163.

3 E.Bordes and P.Courtine, J. Chem. Soc., Chem. Commun., (1985) 294.

4 Mitsubishi, Japan Patent, S56-45815 (1981).

5 E.Bordes and P.Courtine, J. Solid Chem., 55 (1984) 270; G.Bussa, F.Cavani, G.Centi, and F.Trifiro, J. Catal., 99 (1986) 400.

6 A.Hezel and S.D.Ross, Spectrochim. Acta, 23A (1967) 1583.

7 Yu.Y.Gorbunova and S.A.Linde, Sov. Phys. Dokl., 24 (1979) 138.

8 B.K.Hodnett and B.Delmon, Appl. Catal., 9 (1984) 203.

9 I.Matsuura, K.Yoshida, and A.Mori, Chem. Lett., (1987) 535; I.Matsuura, A.mori, ans M.Yamazaki, Chem. Lett., (1987) 1897.

10 F.Cavani, G.Centi, and F.Trifiro, J.Chem. Soc., Chem. Commun., (1985) 492.

J.C. VOLTA (Inst. de Catalyse, France):
1. I am not convinced on the existence of three different $(VO)_2P_2O_7$ phases. Indeed IR spectra are similar and the X-ray lines of the three phases appear at the same position.
The differences observed can be explained by different morphologics of the pyrophosphate which should explain the difference in the catalytic redox, on the basis of a structure sensitivity of the butane mild oxidation. Can you comment on this?
2. What evidence do you have of the formation of vanadium bis metaphosphate on $(VO)_2P_2O_7$ doped with phosphorus?

I. MATSUURA (Toyama University, Japan): From the XRD data (see Table 1) we concluded that the V^{4+} positions in the unit cells of the three $(VO)_2P_2O_7$ phases are the same. However there are two kinds of pyrophosphate ions in the β-$(VO)_2P_2O_7$ phase. But there is only one pyrophosphate ion in the α-$(VO)_2P_2O_7$ phase, as shown in Table 2. The arrangement of the pyrophosphate in α-$(VO)_2P_2O_7$ and β-$(VO)_2P_2O_7$ are shown in figure 2. The difference in arrangement of the pyrophosphate is deciding for the catalytic properties. The oxidation reaction of butane proceeds on two planes of $(VO)_2P_2O_7$. The selective oxidation of butane to maleic anhydride occurs on the (100) plane with the topotactical redox reaction. The non selective oxidation of butane occurs on the (021) and (001) plane with the crystallographical share redox reaction. The addition of extra phosphate to $(VO)_2P_2O_7$ results in the formation of vanadium bis-metaphosphate $(VO)(PO_3)_2$ on the pillar face of $(VO)_2P_2O_7$ (021,001). This prevents the non selective oxidation taking place on the pillar face of $(VO)_2P_2O_7$. This maki-sushi (rolled sushi) type catalyst model is shown below.

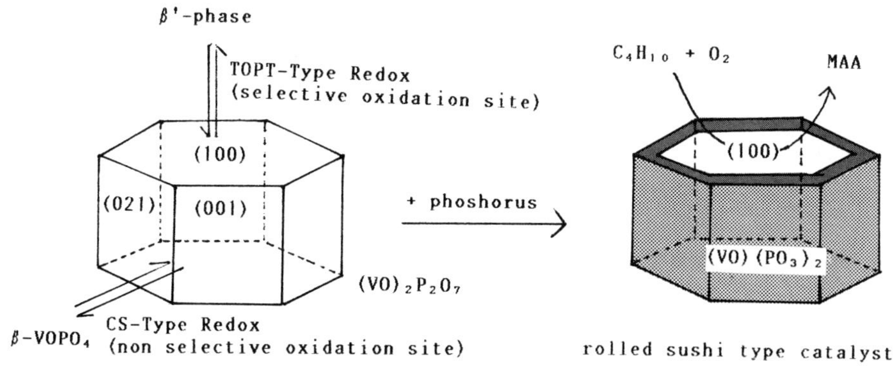

E. BORDES (Univ.technologie de Compiegne, France):
I regret to mention that the methods used to prepare pure α,β,γ forms of $(VO)_2P_2O_7$ are not adequate and I completely agree with Dr.Volta concerning his remarks about morphological effects.

I. MATSUURA (Toyama University, Japan): Although the methods used to prepare the pure α,β,γ forms of $(VO)_2P_2O_7$ may not be perfect, the data shown in Table 1 more than adequately support our preparation method.

G. Centi and F. Trifiro' (Editors), *New Developments in Selective Oxidation*

INVESTIGATION OF ACTIVE AND SELECTIVE OXYGEN IN V-P-O CATALYSTS FOR n-BUTANE CONVERSION TO MALEIC ANHYDRIDE

M. E. Lashier, T. P. Moser and G. L. Schrader

Department of Chemical Engineering, Ames Laboratory-USDOE, Iowa State University
Ames, IA 50011

ABSTRACT

An ^{18}O-enriched β-$VOPO_4$ catalyst was synthesized by the conversion of $(VO)_2P_2O_7$ to β-$VOPO_4$ in the presence of $^{18}O_2$. Characterization by laser Raman spectroscopy indicated the isotopic label was incorporated into specific tetrahedral lattice positions. On-line mass spectrometry of the products of n-butane and 1-butene reaction over the labeled catalyst (in the absence of gas phase oxygen) demonstrated that selective oxidation and combustion occurred at different sites. In situ Raman experiments indicated that the initial high activity of the catalyst was not associated with bulk catalyst reduction; over extended time periods, however, extensive reduction did occur.

INTRODUCTION

Commercial interest in vanadium-phosphorus-oxygen (V-P-O) catalysts has resulted from the high activity and selectivity these materials have for C_4 hydrocarbon oxidation to maleic anhydride. Several studies have linked catalytic activity and selectivity to specific V-P-O phases (refs. 1-3) or to P-to-V ratios used in catalyst formulations (refs. 4-6). The active centers of V-P-O catalysts have been primarily characterized by surface acidity (ref. 7) and texture (refs. 7-9). The active sites responsible for paraffin activation, oxygen incorporation, and complete combustion have yet to be identified. Participation of lattice oxygen during the selective oxidation of hydrocarbons is a general characteristic of many metal oxide catalysts, including V-P-O catalysts (refs. 10-12). Studies utilizing $^{18}O_2$ labeling of gas feeds (refs. 13-17) and ^{18}O enriched bismuth molybdate catalysts with characterization by laser Raman spectroscopy has revealed that specific oxygen sites are responsible for particular reaction steps. Similar identification of the active sites for C_4 hydrocarbon oxidation by V-P-O catalysts has not been reported (ref. 18).

In the present study, examination of the lattice oxygen involved in paraffin activation, oxygen incorporation, and complete combustion has been performed using in situ laser Raman spectroscopy of an ^{18}O-labeled β-$VOPO_4$ catalyst and mass spectroscopy of the reaction products.

EXPERIMENTAL PROCEDURE

Synthesis of ^{18}O-Enriched β-$VOPO_4$

^{18}O-enriched β-$VOPO_4$ was prepared by the solid state reaction of $(VO)_2P_2O_7$ (ref. 1) with $^{18}O_2$. $^{18}O_2$ was obtained from Merck, Sharp and Dohme with an atom enrichment of 97.8%. The synthesis of $(VO)_2P_2O_7$ has been described previously. Powdered $(VO)_2P_2O_7$ (0.50 g) was charged to a 9 mm O.D. Pyrex tube which was evacuated and backfilled with a stoichiometric quantity of $^{18}O_2$ gas. The reaction tube was heated at 550°C for 24 h followed by cooling to 200°C at a rate of 50°C/h.

Laser Raman Spectroscopic Characterization

Laser Raman spectra were obtained using a Spex 1403 laser Raman spectrometer with the 514.3 nm line of a Spectra Physics Model 2020-05 argon ion laser operated at 100 mW at the source.

Reactor Studies

Reactions of n-butane and 1-butene using the ^{18}O-enriched catalyst were performed in a continuous flow microreactor system in the absence of gas phase oxygen. The microreactor was a 1/4" stainless steel tube passivated by calcination in O_2 after treatment with phosphoric acid. 0.3 g of pressed and sieved catalyst (10-20 mesh) was used in each experiment. The feed gas was delivered at 50 cm^3/min with a composition of 2% n-butane or 1-butene (Matheson, instrument grade) in He (Matheson, zero grade). A reduced copper catalyst (BASF) was used to remove residual oxygen.

Mass spectral analysis of the products of n-butane and 1-butene reaction over the ^{18}O-enriched catalyst was performed by 100C precision quadrupole mass analyzer controlled by a PDP 11/23 computer. The mass analyzer was interfaced with the microreactor system by a glass SGE single stage molecular jet separator.

The ^{18}O content of maleic anhydride, CO_2, and H_2O produced by the ^{18}O-enriched β-$VOPO_4$ catalyst was determined for the oxidation of n-butane and 1-butene. The percent ^{18}O content of maleic anhydride was calculated from the specific ion current intensities of maleic anhydride molecules with mass-to-charge ratios (m/e) of 98, 100, 102, and 104. For 1-butene, phthalic anhydride was formed as the catalyst became reduced. Phthalic anhydride also has a 104 m/e peak, and the maleic anhydride data were corrected by monitoring other phthalic anhydride peaks.

The ^{18}O content of carbon dioxide and water were determined in a similar manner. The CO_2 data were corrected by subtracting the minor interferences at m/e 44 from n-butane or 1-butene fragmentation and background CO_2. CO could also be observed, but the data were significantly complicated by interference from background N_2 and $C^{16}O$.

In Situ Laser Raman Spectroscopy

A Spex 1877 Triplemate laser Raman spectrometer was used with 514.3 nm line of a Spectra Physics Model 164 argon ion laser operated with 200 mW at the source. *In situ* laser Raman spectra of the functioning catalyst were obtained using a tubular controlled atmosphere cell (ref. 12). The *in situ* Raman studies did not involve ^{18}O-enriched catalysts.

EXPERIMENTAL RESULTS

Characterization of ^{18}O-Enriched β-$VOPO_4$

The Raman spectrum of ^{18}O-enriched β-$VOPO_4$ prepared by the solid state reaction of $(VO)_2P_2O_7$ with $^{18}O_2$ was compared to the Raman spectrum of similarly prepared β-$VOPO_4$ (ref. 1) using $^{16}O_2$ (Fig. 1). ^{18}O could be detected in the catalyst lattice by a P-^{18}O band at 886 cm^{-1}; the P-^{16}O stretch at 896 cm^{-1} had

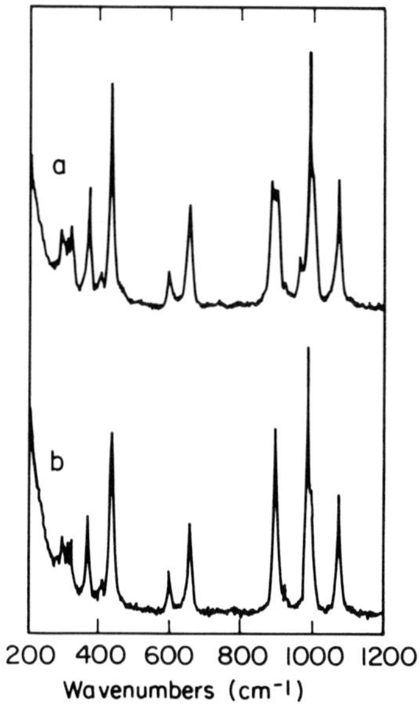

Fig. 1. Raman spectra of
(a) ^{18}O-enriched β-$VOPO_4$ prepared by reaction of $(VO)_2P_2O_7$ with $^{18}O_2$ and
(b) β-$VOPO_4$ prepared by reaction of $(VO)_2P_2O_7$ with $^{16}O_2$

near equal intensity. Another P-^{18}O band was observed at 961 cm^{-1}. The intensity of this band, however, was rather weak (about 5% of the P-^{16}O band at 987 cm^{-1}). A slight broadening of the P-O band at 1072 cm^{-1} was also observed. Unreacted $(VO)_2P_2O_7$ was detected by a very weak band at 923 cm^{-1} in the spectrum of ^{18}O-enriched β-$VOPO_4$. The Raman spectrum of β-$VOPO_4$ prepared by the reaction of $(VO)_2P_2O_7$ with $^{16}O_2$ also indicated the presence of a small amount of unreacted $(VO)_2P_2O_7$.

Mass Spectrometry Studies

^{18}O Incorporation into the products of n-butane oxidation. As the exposure to n-butane proceeded, the conversion decreased from about 5% to nearly 0% (Fig. 2). Significant levels of maleic anhydride were produced for nearly 20 mins. As the catalyst was further exposed to n-butane, furan was detected.

The ^{18}O content of maleic anhydride, CO_2, and H_2O are presented in Fig. 3. The ^{18}O content of the maleic anhydride ranged from an initial level of 11-12% to an ultimate level of about 13-14%. The ^{18}O content of the CO_2 produced was initially at the 7-8% level, dropped to a low of about 6% after about two minutes, and then leveled off at about 6%. The ^{18}O content in the H_2O produced in this reaction varied from an initial level of about 8% to a final level near 12%. The most striking feature is that maleic anhydride was produced containing ^{18}O at nearly twice the level of CO_2, while H_2O contained ^{18}O at an intermediate level.

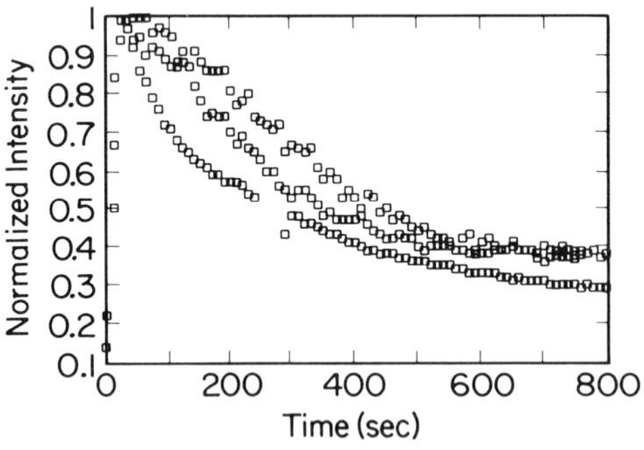

Fig. 2. Maleic anhydride production from 2% n-butane over ^{18}O-enriched β-$VOPO_4$ (500°C): m/e 98 normalized intensity

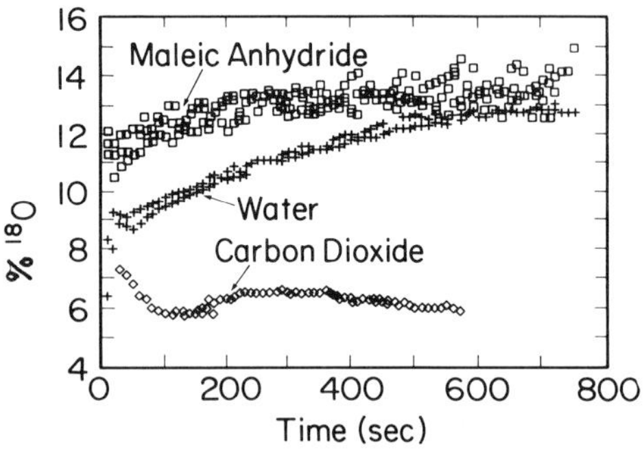

Fig. 3. ^{18}O content of maleic anhydride, CO_2, and H_2O during 2% n-butane oxidation by ^{18}O-enriched β-$VOPO_4$ (500°C)

^{18}O Incorporation into the products of 1-butene oxidation. As the catalyst was exposed to 1-butene, the conversion decreased from nearly 50% to around 0%. The production of maleic anhydride decreased rapidly to a low--but fairly steady--level after about 2 mins (Fig. 4). Furan could be detected throughout the experiment, while phthalic anhydride could be detected at a longer time for exposure to 1-butene.

The levels of ^{18}O in maleic anhydride, CO_2, and H_2O are presented in Fig. 5. ^{18}O levels in maleic anhydride ranged from an initial level of 12-13% to a final value of 14-15%. The ^{18}O level in CO_2 started at about 7-8%, rose to a maximum of about 10% at about 90 sec, and then decreased to the 4-6% level. As with n-butane, the maleic anhydride ^{18}O level and CO_2 ^{18}O level produced from 1-butene were significantly different. The level of ^{18}O incorporation into maleic anhydride for each feed gas was similar with 1-butene producing slightly higher levels. The ^{18}O content in CO_2 from 1-butene behaved somewhat differently in comparison to n-butane. This was probably due to the difference in catalyst selectivity for these feeds. The ^{18}O content of H_2O produced in this reaction was initially about 6% and increased to about 10%.

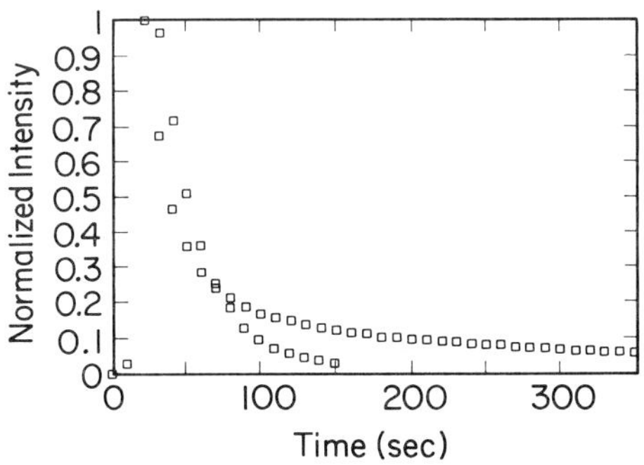

Fig. 4. Maleic anhydride production from 2% 1-butene over ^{18}O-enriched β-$VOPO_4$ (450°C): m/e 98 normalized intensity

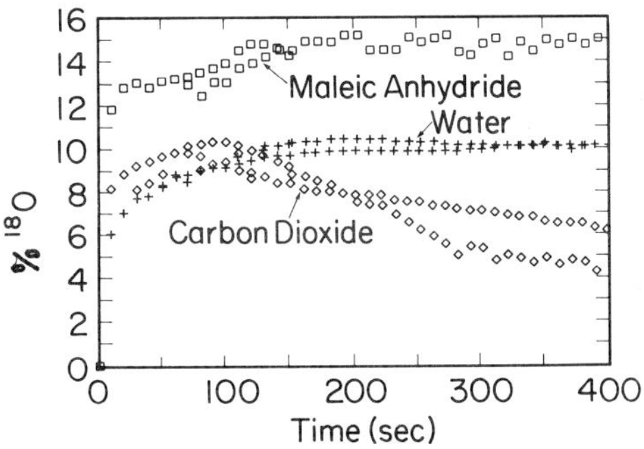

Fig. 5. ^{18}O content of maleic anhydride, CO_2, and H_2O during 2% 1-butene oxidation by ^{18}O-enriched β-$VOPO_4$ (450°C)

IN SITU LASER RAMAN SPECTROSCOPY

Characterization during n-butane oxidation

In situ Raman spectra of β-$VOPO_4$ under conditions of n-butane oxidation at 500°C are presented in Fig. 6. Comparison of the room temperature spectrum and the reaction temperature spectrum illustrates temperature effects: thermal broadening and slight peak shifts occurred, but no bulk thermal reduction was evident. As the reaction of 2% n-butane in He with the catalyst began, no reduction was apparent until nearly 30 mins when a $(VO)_2P_2O_7$ band at 970 cm^{-1} began to appear. During this time period, however, there was a reduction in intensity of the Raman scattering from the β-$VOPO_4$ phase which could be due to reduction or other effects.

Characterization during 1-butene oxidation

In situ Raman spectra of β-$VOPO_4$ under conditions of exposure to 1-butene at 450°C are presented in Fig. 7. As for the previous n-butane studies at 500°C, thermal broadening and an intensity loss occurred upon heating, but these effects were not as severe at 450°C. As the reaction of 2% 1-butene commenced, all peak intensities decreased continuously. $(VO)_2P_2O_7$ was formed after about six minutes, as evidenced by the appearance of a band at 930 cm^{-1}. The intensity of this band increased continuously, accompanied by a decrease in the β-$VOPO_4$ bands.

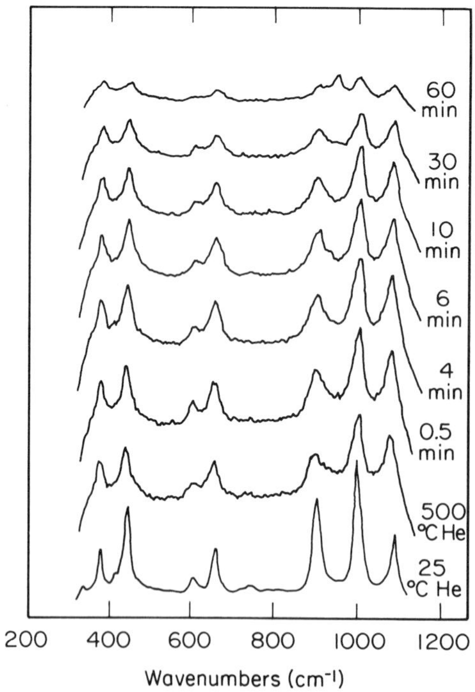

Fig. 6. Raman spectra of β-$VOPO_4$ during n-butane oxidation (500°C)

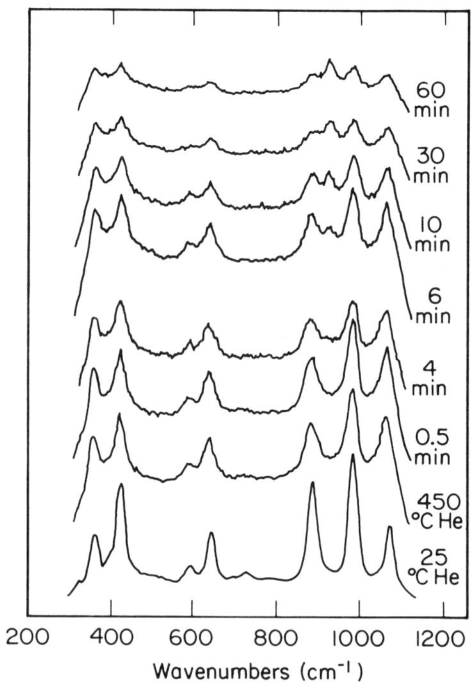

Fig. 7. Raman spectra of β-$VOPO_4$ during 1-butene oxidation (450°C)

DISCUSSION OF RESULTS

The Raman spectra of the isotopically enriched catalysts provided information about the nature of the $(VO)_2P_2O_7$ to β-$VOPO_4$ phase transformation. The solid state reaction of $(VO)_2P_2O_7$ with $^{18}O_2$ produced an ^{18}O-enriched phase according to specific stoichiometry: one mole of $(VO)_2P_2O_7$ reacted with one-half mole of $^{18}O_2$ to form the ^{18}O-enriched β-$VOPO_4$ phase. The Raman results indicated that incorporation of the ^{18}O occurred in specific lattice sites. The Raman spectrum of ^{18}O-enriched β-$VOPO_4$ had an isotopically shifted P-^{18}O band at 886 cm^{-1} which was of nearly equal intensity as the related P-^{16}O band at 896 cm^{-1}. ^{18}O was also incorporated to a much lesser extent at other P-O positions as indicated by a band at 987 cm^{-1}. The incorporation of ^{18}O into the PO_4 groups of β-$VOPO_4$ therefore occurs very specifically; random distribution of ^{18}O in the β-$VOPO_4$ phase clearly was not observed. In addition, complete incorporation of ^{18}O into a limited region or portion of the material was not detected.

$$(VO)_2P_2O_7 + 1/2\ {}^{18}O_2 \longrightarrow 2\beta\text{-}VOPO_{7/2}{}^{18}O_{1/2} \quad (1)$$

The structures of the catalysts suggest that the β-$VOPO_4$ to $(VO)_2P_2O_7$ transformation involves the cooperative movement of VO_6 octahedra to form double octahedral chains characteristic of $(VO)_2P_2O_7$ (ref. 3). Concurrently, pyrophosphate structures are formed from neighboring (above and below) phosphate tetrahedra of β-$VOPO_4$.

Because of the specific incorporation of ^{18}O into the lattice of the catalytically active β-$VOPO_4$ phase, it was possible to relate the production of oxygenated products with the reactivity of the oxygen sites. The oxidation of n-butane by ^{18}O-labeled β-$VOPO_4$ resulted in the preferential incorporation of ^{18}O into maleic anhydride as compared to CO_2 and H_2O. For example, the initial ^{18}O content of maleic anhydride was approximately two times greater than for CO_2 for n-butane reaction at 500°C; similar results were observed for 1-butene at 450°C. According to the Raman spectrum of ^{18}O-labeled β-$VOPO_4$, the P-^{18}O stretch at 886 cm^{-1} and the complementary P-^{16}O stretch and 896 cm^{-1} had relative intensities indicating that approximately 40% of the oxygen associated with this stretching vibration were labeled with ^{18}O. A small degree of ^{18}O incorporation into another P-O lattice position was detected: the intensity of a P-^{18}O band at 961 cm^{-1} was approximately 5% compared to the related P-^{16}O stretch at 987 cm^{-1}.

Due to the stoichiometric nature of the preparation, only 10% of the total oxygen in the catalyst can be ^{18}O. Based on the Raman characterization, all of the ^{18}O is incorporated at P-O positions. For both n-butane and 1-butene feeds, maleic anhydride was produced which contained nearly 13% ^{18}O. To account for this selective incorporation, a "pool" of 13% ^{18}O in the catalyst must exist. If all oxygen associated with P-O bonding were considered to be equivalent, such a pool of 13% ^{18}O would exist. The Raman data would appear to indicate, however, that ^{18}O tends to be associated with two of the three P-O oxygen stretches. It is quite, however, possible that these oxygen positions are structurally more similar at the catalyst surface than in the catalyst bulk.

The ^{18}O levels found in carbon dioxide indicate that total oxidation likely occurs through more than one pathway. Direct combustion of maleic anhydride to carbon dioxide is known to occur over β-$VOPO_4$ (ref. 19). However, if complete oxidation of maleic anhydride occurred randomly at all available oxygen sites, the ^{18}O found in carbon dioxide should be higher than the 6-8% observed. Similarly, if combustion occurred only at unlabeled sites, the ^{18}O content of the carbon dioxide should be lower than the observed values. An additional route (or routes) to complete combustion products must also exist involving utilization of some oxygen from labeled sites. This reaction pathway could proceed by an initial electrophilic attack on the C-C bonds of n-butane or other C_4 hydrocarbon intermediates. The V=O site has been identified as being electrophilic (ref. 20) and therefore is likely to be involved in this non-selective activation. Such "cracking" reactions would produce highly activated C_1-C_3 species which could interact with any available oxygen site to produce carbon dioxide. It is possible that combustion could proceed exclusively on the V=O sites, but the Raman spectra indicate that no ^{18}O is incorporated at these positions. The C_1-C_3 reactive intermediates are likely also to undergo reaction at other oxygen sites, including the ^{18}O-labeled P-O sites (also involved in maleic anhydride production).

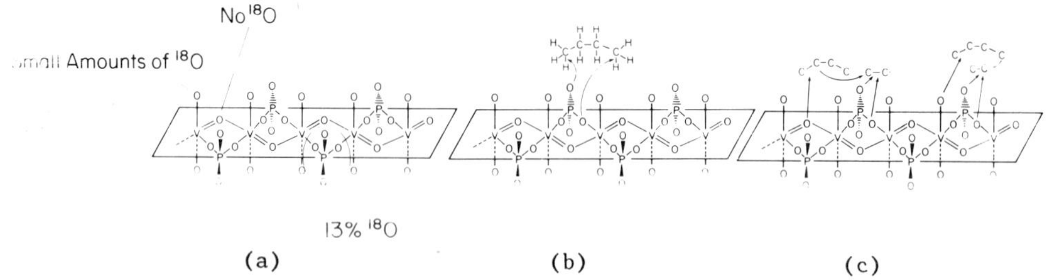

Fig. 8. Plausible reaction surface: (a) location of ^{18}O as determined by laser Raman spectroscopy, (b) activation of n-butane and oxygen insertion to produce maleic anhydride, (c) example of complete combustion of hydrocarbons and maleic anhydride involving several oxygen sites

Shown in Fig. 8(a) is a depiction of the location of the ^{18}O labeled sites as identified by the laser Raman studies. Also shown in Fig. 8(b) is the activation of n-butane and the insertion of oxygen at P-O sites, resulting in the production of maleic anhydride. Combustion of C_4 hydrocarbons and maleic anhydride involving C_1-C_3 reactive intermediates is depicted in Fig. 8(c).

CONCLUSIONS

The incorporation of ^{18}O into maleic anhydride (about 13%) is very similar for both n-butane and 1-butene feeds, although the reaction rates differ significantly. The reaction pathways for the selective oxidation of these species would appear to be similar after the initial activation. The source of these selective oxygen atoms was identified as being associated with P-O structures. CO_2 formation occurs through at least two possible pathways. The ^{18}O levels observed indicate that in addition to the route from the combustion of maleic anhydride, CO_2 is formed from highly reactive species produced by breaking of carbon-carbon bonds by electrophilic V=O species.

ACKNOWLEDGMENT

This work was conducted through the Ames Laboratory which is operated for the U.S. Department of Energy by Iowa State University under contract No. W-7405-ENG-82. Support from the Office of Basic Energy Sciences is acknowledged.

REFERENCES

1 Moser, T.P. and Schrader, G.L., J. Catal., 92 (1985) 216.
2 Morselli, L., Trifiro, F., and Urban, L., J. Catal., 75 (1982) 112.
3 Bordes, E. and Courtine, P., J. Chem. Soc., Chem. Commun., (1985) 294.
4 Wenig, R.W. and Schrader, G.L., Ind. Eng. Chem. Fund., 25 (1986) 612.
5 Garbassi, F., Bart, J., Tassinari, R., Vlaic, G., and Lagarde, P., J. Catal., 98 (1986) 317.
6 Hodnett, B. and Delmon, B., J. Catal., 88 (1984) 43.
7 Busca, G., Centi, G., and Trifiro, F., J. Am. Chem. Soc., 107 (1985) 7758.
8 Busca, G., Cavani, F., Centi, G., and Trifiro, F., J. Catal., 49 (1986) 400.
9 Cavani, F., Centi, G., and Trifiro, F., J. Chem. Soc., Chem. Commun., (1985) 492.
10 Pepera, M., Callahan, J., Desmond, M., Milberger, E., Blum, P., and Bremer, N., J. Amer. Chem. Soc., 107 (1985) 4883.
11 Kruchinin, Yu., Mishchenko, Yu., Nechiporuk, P., and Gel'bshtein, A., Kinet. Katal. (English Translation), 25(2) (1984) 328.
12 Moser, T.P. and Schrader, G.L., J. Catal., 104 (1987) 99.
13 Wragg, R., Ashmore, P., and Hockey, J., J. Catal., 22 (1971) 49.
14 Krenzke, D. and Keulks, G., J. Catal., 61 (1980) 316.
15 Sancier, K., Wentrcek, P., and Wise, H., J. Catal., 39 (1975) 141.
16 Hoefs, E., Monnier, J., and Keulks, G., J. Catal., 57 (1979) 331.
17 Miura, H., Otsubo, T., Shirasaki, T., and Morikawa, Y., J. Catal., 56 (1979) 84.
18 Glaeser, L., Brazdil, J., Hazle, M., Mehicic, M., and Grasselli, R., J. Chem. Soc., Faraday Trans., 1(79) (1985) 2903.
19 Moser, T.P., Wenig, R.W., and Schrader, G.L, Appld. Catal., 34 (1987) 39.
20 Haber, J. and Serwicka, E.M. React. Kinet. Catal. Lett., 35(1-2) (1987) 369.

E. BORDES (Université de Technologie de Compiègue, France): May I tell that I do appreciate your work which contributes greatly to the understanding of the VPO system.

Have you studied by mass spectrometry the distribution of ^{18}O in maleic anhydride? If ^{18}O belongs to a P-O bond, such oxygen would be the bridging oxygen between carbons in maleic anhydride.

G. L. SCHRADER (Iowa State University, U.S.A.): The 54 m/e peak which would provide information about the isotopic enrichment of the terminal maleic anhydride oxygens was obscured by much larger n-butane peaks in that region. However, forthcoming work involving pulse reactor studies indicates that the highly labeled (40% ^{18}O) P-O site is responsible for the formation of strongly adsorbed furan on the catalyst surface (ref. 1).

1 M. E. Lashier, G. L. Schrader, J. Catal., (submitted).

J. VOLTA (Institut de Recherches sur la Catalyse, France): Many authors consider now that the effective phase for mild oxidation of butane to maleic anhydride is vanadium pyrophosphate since this phase is detected by techniques characteristic of the bulk.

Your communication brings a new interesting insight into the role of the $VOPO_4$ phases in the mechanism of butane mild oxidation. You bring proof that the source of selective oxygen atoms are associated with the P-O structures, even on β-$VOPO_4$.

My opinion is that the actual catalytic sites should be associated with the interface between microdomains of $VOPO_4$ phases (short range order) and the $(VO)_2P_2O_7$ phase (long range order), as was proposed by Dr. Bordes. This model was demonstrated in our laboratory using the radial electron distribution of x-rays (ref. 1) and ^{31}P solid state NMR (ref. 2). In this model, the role of the P-O structures around the VO_6 octahedra of the $VOPO_x$ structure is emphasized. Can you comment on this point?

G. L. SCHRADER (Iowa State University, U.S.A.): We believe that active catalysts are likely to involve both V^{4+} and V^{5+} sites. In fact, strong similarities exist for oxygen incorporation for $(VO)_2P_2O_7$ catalysts which have been labeled with ^{18}O (ref. 3). We have recently reported ^{31}P spin-echo-mapping solid state NMR results which confirm the presence of these oxidation states in catalysts which have been exposed to 1.5% n-butane or 1-butene in air at 450°C (ref. 4). The surface structures and arrangement of active phases are more difficult to determine, but, nonetheless, likely include V^{+4} and V^{+5} species.

1 G. Berguet, M. David, J. P. Breyer, J. C. Volta, G. Decquet, Catal. Today, 1 (1987) 37.
2 F. Lefebre, M. David, J. C. Volta, 12th Iberoamerican Symposium on Catalysis, Guanajuato, 1988.
3 M. E. Lashier, G. L. Schrader, J. Catal., (to be submitted).
4 M. E. Lashier, R. D. Walker, J. Li, B. C. Gerstein, G. L. Schrader, J. Catal., submitted.

G. BUSCA (Instituto di Chimica, Italy): The V=O stretching band in the Raman spectrum of β-$VOPO_4$ that you have shown is partially superimposed on a strong VPO band and is only visible as a shoulder at higher frequency. If partial exchange occurs, leading to $V=^{18}O$ groups, a new weak component at lower frequency should appear, superimposed on the VPO band cited above. Can you be sure from the spectrum of partially $^{16}O/^{18}O$ exchanged β-$VOPO_4$ that his component is really absent?

G. L. SCHRADER (Iowa State University, U.S.A.): Spectra obtained at higher resolution indicate that no distinct band appears at a lower frequency; there also is no observed shoulder near the estimated band frequency for $V={}^{18}O$ at about 971 cm^{-1}. In addition, calculations involving the other labeled bands at 886 cm^{-1} and 961 cm^{-1} indicate that there should be no ^{18}O available for incorporation at other positions for a β-$VOPO_4$ phase labeled with 10% ^{18}O. Additionally, FTIR spectra indicate no changes occur in the position of the V=O bands.

R. K. GRASSELLI (Mobil Research and Development Corporation, U.S.A): Based on your $^{18}O_2$ Raman studies, you conclude that the oxygen inserted into butane to produce maleic anhydride stems from nucleophilic P-O or P-O-V moieties. Considering the redox requirements of the reaction and the structural constraint of the $(VO)_2P_2O_7$ catalyst, it appears to me that the oxygen inserting species is much more likely to be the bridged P-O-V oxygen rather than an oxygen derived from P-O surface moieties.

G. L. SCHRADER (Iowa State University, U.S.A.): The notation we have used in our paper stems from the spectroscopist's terminology for of these bands. This "shortened" reference does not seek to imply that P-O bonds are present in the bulk β-$VOPO_4$ structure; clearly, only P-O-V bonding occurs. We believe that this is the site involved in the oxygen insertion mechanism. Whether P-O moieties can exist at the surface is a question that awaits more extensive characterization of V-P-O catalysts by surface sensitive techniques.

G. Centi and F. Trifiro' (Editors), *New Developments in Selective Oxidation*

COMPARATIVE STUDY OF STRUCTURE-SENSITIVE OXIDATIONS OF n-BUTANE AND 1-BUTENE IN MALEIC ANHYDRIDE ON TWO KINDS OF CATALYSTS

E. BORDES

Département de Génie Chimique, Université de Technologie de Compiègne, B.P. 649, 60206 Compiègne Cédex (France).

ABSTRACT

The selective oxidation of n-butane and 1-butene in maleic anhydride is studied on two kinds of catalytic solids, bulk V-P-O phases and multicomponent catalysts made with cobalt molybdate and uranium molybdate supported by MoO_3/TiO_2. The comparison between their catalytic properties in both reactions is made according to several criteria of selectivity used in oxidation catalysis and to the kind of hydrocarbon. It is shown that the structural criterion already used in structure-sensitive reactions accounts for the observed differences and analogies.

INTRODUCTION

Up to now, in the field of the selective oxidation of alcanes, satisfying catalytic performances (high activity and selectivity) are found in the case of n-butane oxidation only, using a V-P-O catalyst (ref. 1). The properties of V-P-O phases, which are also used in the oxidation of butenes, have been extensively studied and reviewed (Refs. 3-5). In the case of butene oxidation, the best catalytic results are correlated with the presence of both (α or) β-$VOPO_4$ and $(VO)_2P_2O_7$, whereas in the case of n-butane $(VO)_2P_2O_7$ is generally found alone (refs. 2-9).

Since the oxidation of n-butenes to MA is kinetically described by a consecutive "rake" mechanism involving butadiene and furan (ref. 10), we looked for a multicomponent catalyst able to make this kind of reaction, starting from n-butane. MoO_3/TiO_2 being active in the second step butadiene-MA (ref. 11), uranyl-based compounds and $CoMoO_4$ known as catalysts for oxidation of olefins (refs. 12, 13) and dehydrogenation of n-butane (ref. 14) respectively, were tried for the first part of the reaction (C_4-butadiene). Molybdates salts were chosen because of their structural compatibility with molybdenum oxides leading to coherent interfaces (refs. 15-18), and were therefore supported on MoO_3/TiO_2. The main catalytic and structural properties of these multicomponent catalysts will be presented and compared with those of extensively studied V-P-O phases in the same reaction, in order to find analogies from which could be drawn selectivity criteria.

EXPERIMENTAL

MoO_3/TiO_2-based catalyst was prepared first, by dissolving ammonium paramolybdate with oxalic acid in H_2O and adding powdered TiO_2 ($7\ m^2.g^{-1}$) (30 mol.% active phase). After heating this slurry with stirring (80°C, 2 hrs), evaporation to dryness followed by drying (110°C, 12 hrs) and calcination in air (500°C, 6 hrs) gave the final support. This powdered solid was dispersed in a cobalt or uranyl nitrate and paramolybdate solution (10 wt% active phase on MoO_3/TiO_2) (refs. 17, 18), which was treated as above, yielding the final catalysts $CoMoO_4$-MoO_3/TiO_2 or $UMoO_6$-MoO_3/TiO_2 (atomic ratio $Mo_{tot.}/U = 10$). The surface area can be increased by use of TiO_2 (H) obtained by hydrolysis of $TiCl_4$ during the solubilization step of ammonium paramolybdate.

n-Butane and 1-butene oxidation experiments were conducted using a continuous-flow, fixed-bed reactor, and effluents analyzed by gas chromatography, as already described (ref. 16). The reactor was operated in an integral mode, in the following conditions : 1 vol.% C_4H_8 (or 1.5% C_4H_{10}) in air, GHSV 4,500 - 36,000 h^{-1}, temperature 553-823 K.

The reactivity of the catalysts was examined by TGA and/or DTA under N_2, O_2 or reactive gas, and samples identified before and after catalytic reaction by X-Ray diffraction (XRD) using a diffractometer (X-Ray reflexion method) and Cu-Kα radiation.

RESULTS

Catalytic properties

As already found with V-P-O in the oxidation of *1-butene* (ref. 16), CO, CO_2, H_2O and MA are detected, together with some butadiene, furan, methacrolein, crotonaldehyde and C_2, C_3 by-products, even at high conversion, on U-Mo-Ti-O catalysts : activity and selectivity of UO_2MoO_4/TiO_2 (Mo/U = 1) and of UO_2MoO_4-$9MoO_3/TiO_2$ (Mo/U = 10) are close from those of bulk β-$VOPO_4$ and of β-$VOPO_4/TiO_2$ respectively (693 K, GHSV = 4,500 hr^{-1}) (Fig. 1). The selectivity in MA shown by Mo/U = 1 increases to $S_{MA} = 49$ mol.% at C = 98 mol.% for Mo/U = 10. Both activity and selectivity are enhanced when TiO_2 is used as a support.

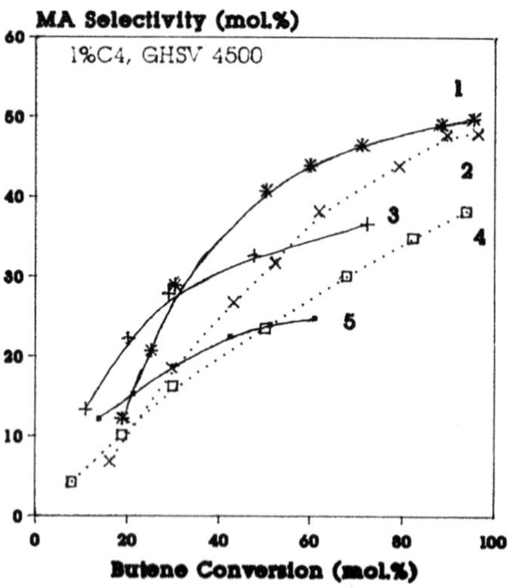

Fig. 1. Selectivity in MA *vs* 1-butene conversion. Curve 1 : UO_2MoO_4-$9MoO_3/TiO_2$; 2 : β-$VOPO_4/TiO_2$; 3 : UO_2MoO_4/TiO_2 ; 4 : β-$VOPO_4$; 5 : UO_2MoO_4-$9MoO_3$.

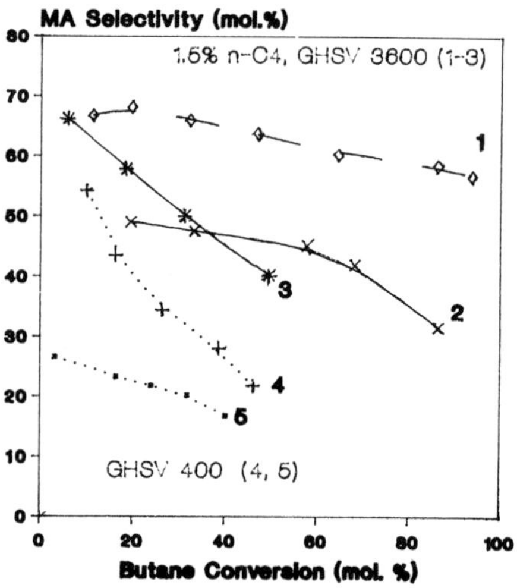

Fig. 2. Selectivity in MA *vs* n-butane conversion. Curve 1 : γ-$(VO)_2P_2O_7$; 2 : $CoMoO_4$-$/MoO_3/TiO_2$ H ; 3 : id., TiO_2 7 $m^2.g^{-1}$; 4 : MoO_3/TiO_2 ; 5 : $CoMoO_4$.

By contrast, very few by-products (mainly acetic acid and CO, CO_2) are detected during the oxidation of n-butane with P/V = 1. High temperature and GHSV are necessary for pure $CoMoO_4$ and MoO_3/TiO_2, which are poorly active and selective (763 K, GHSV = 324 hr^{-1}), as compared with multicomponent $CoMoO_4/MoO_3/TiO_2$ (693 K, GHSV = 3,600 hr^{-1}) (Fig. 2). More interesting performances are obtained with $CoMoO_4/MoO_3/TiO_2$(H) (S_{MA} = 52 mol.%, C = 46 mol.%). By comparison, unsupported γ-$(VO)_2P_2O_7$ prepared by the right method is more active and selective (S_{MA} = 60 mol.%, C = 80 mol.%), and even more if the preparation is improved (refs. 1, 20).

Structural properties and reactivity of multicomponent catalysts

1- Catalysts $CoMoO_4$-MoO_3/TiO_2. MoO_3 and TiO_2, α and β-$CoMoO_4$ (with $\alpha > \beta$) are identified by XRD. Crystal structures of α and β-$CoMoO_4$ essentially differ in the 6- and 4-coordination of molybdenum, for the low and high temperature α and β-forms respectively (10). The large excess of MoO_3 and TiO_2 hinders to see if the proportion $\alpha > \beta$ is, or not, modified after the catalytic reaction. The surface planes are mostly represented by crystal planes (220) of α and of β-$CoMoO_4$ and (020) of MoO_3 as revealed by the reinforced intensity of the respective lines obtained by use of the X-ray reflexion method, at least in first approximation.

Experiments were undertaken by DTA under inert gas (N_2) to see if the reactivity of $CoMoO_4$ is modified by the presence of MoO_3 and/or MoO_3/TiO_2 (Table 1). The temperature T_{all} of the allotropic transition α--->β-$CoMoO_4$ is higher for $CoMoO_4$ supported on MoO_3 or on MoO_3/TiO_2, and lower for $CoMoO_4/TiO_2$ respectively than for pure α-$CoMoO_4$ (495°C) (ref. 14). Therefore, it can be assumed that the presence of MoO_3 and of MoO_3/TiO_2 increases the range of stability of α-$CoMoO_4$ (octahedral Mo) as compared with β, by delaying the transition α--->β. Attempts to relate T_{all} to catalytic properties and surface area are not easy (Table 1) : in the first case, it can be rougly said that the highest T_{all}, the highest activity and yield in MA, and in the second the influence of the chemical nature of the "support" is greater than that of surface area.

TABLE 1
Temperature T_{all} of allotropic transition α--->β-$CoMoO_4$ (endothermic) and related catalytic performances in the oxidation of n-butane

CATALYST	Transition temp.[b] T_{all} (°C)	Surf.Area $m^2.g^{-1}$	Conv.[a] C_4H_{10}	Yield MA	Select. MA
α-$CoMoO_4$	495	3	21.5	04.8	22.4
$CoMoO_4/MoO_3$	507	-	19.0	05.2	17.3
$CoMoO_4$-MoO_3/TiO_2	530	7	45.0	15.5	30.0
$CoMoO_4/TiO_2$	450	7	26.O	10.5	42.5
$CoMoO_4/TiO_2$(H)	500	16	-	-	-
$CoMoO_4$-MoO_3/TiO_2(H)	520	17	53.5	24.8	46.4

a : mol.% ; 1.5 vol.% n-butane in air, 763 K, GHSV = 400 hr^{-1}.
b : $\Delta T = \pm 5$°C

2- Catalysts UO_2MoO_4-$9MoO_3/TiO_2$ anatase. XRD experiments show that UO_2MoO_4, MoO_3 and TiO_2-anatase are present in fresh and used catalysts. TGA of the isothermal reduction of the system were performed in order to detect possible reduced phases. With C_4H_8/N_2 = 0,4 at 420°C, pure UO_2MoO_4 is reduced to $UMoO_5$ and UO_2MoO_4-$9MoO_3$ to $UMo_{10}O_{32}$ or $U_3Mo_{20}O_{64}$ (Fig. 3). In both cases, the reoxidation by O_2/N_2 = 0,4 proceeds faster than the reduction, which account

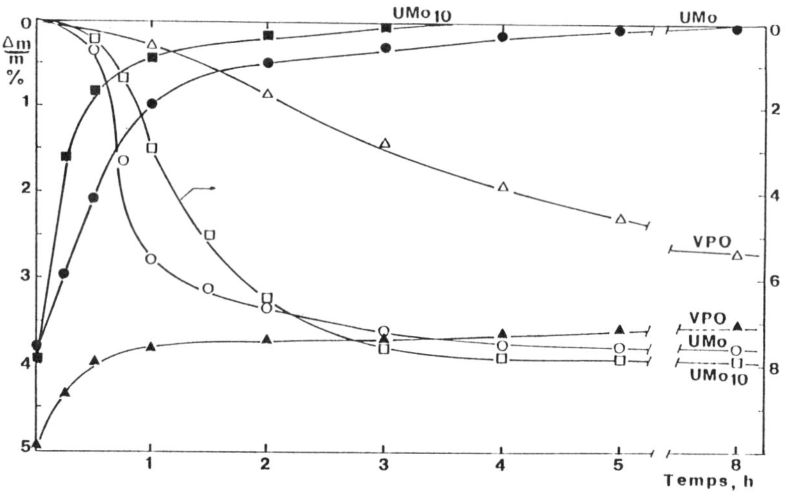

Fig. 3. TGA of reduction of catalysts UO_2MoO_4 and UO_2MoO_4-$9MoO_3$ under butene/N_2, and β-$VOPO_4$ under butene/air (open symbols) ; reoxidation of reduced phases, $UMoO_5$, $U_3Mo_{20}O_{64}$ under O_2/N_2 and $(VO)_2P_2O_7$ under butene/air (full symbols) ; T = 693 K.

for the oxidized state of the solid always found after catalysis. Such reduced phases $UMo_{10}O_{32}$ and $U_3Mo_{20}O_{64}$ were previously identified, and their formation correlated with the use of oxalic acid during preparation, and/or excess of MoO_3 (refs. 16, 17). The optimum stoichiometry Mo/U = 10 corresponding to their possible formation instead of $UMoO_5$ can be related to the best catalytic properties in oxidation of butene (Fig. 1). On Fig. 3 are also reported the isothermal reduction of pure β-$VOPO_4$ by 1 vol.% C_4H_8/air (same as during catalysis), and oxidation of the reduced phase $(VO)_2P_2O_7$ in the same atmosphere. The mean stoichiometry $VPO_{4.7}$ is reached by both phases. By comparison with the preceding U-Mo-O system, the nature of phases which participate in the redox system during catalysis is easier to determine.

DISCUSSION

Recall on V-P-O catalysts

Redox couples and phases present in fresh V-P-O catalysts and at the steady state, according to the reaction, butane-MA or butene-MA, are recalled and gathered with those of multicomponent catalysts deduced from the just described experiments (Table 2).

The selective P/V = 1,0 catalysts in the oxidation of n-butane are constituted almost exclusively by (100) planes of γ-$(VO)_2P_2O_7$ (≈100 % de V^{4+}) (refs. 1-9), obtained by the topotactic dehydration of the precursor $VOHPO_4$·$0.5H_2O$, and able to yield δ or γ-$VOPO_4$ according to temperatu-

TABLE 2

Main characteristic of V-P-O and multicomponent molybdates related to mild oxidation of n-butane and 1-butene in maleic anhydride.

CATALYSTS for :	n-BUTANE	1-BUTENE
V-P-O		
initial phase(s)	γ-$(VO)_2P_2O_7$	$(VO)_2P_2O_7$ + β-$VOPO_4$
redox couple	"γ" V^{5+}/V^{4+}	"β" V^{5+}/V^{4+}
steady state :		
- mean stoichiometry	$VPO_{4.5}$	$VPO_{4.7}$
- microdomains	(100) γ-$(VO)_2P_2O_7$ with surface V^{5+}	microdomains $VOPO_4$ in $(VO)_2P_2O_7$
$MMoO_4$-MoO_3/TiO_2		
initial phases	α,(β)-$CoMoO_4$, MoO_3, TiO_2	UO_2MoO_4, MoO_3, TiO_2
redox couple(s)	(Co^{2+}/Co^{3+}), Mo^{6+}/Mo^{5+}	Mo^{6+}/Mo^{5+}, U^{6+}/U^{4+}
steady state :	α,(β)-$CoMoO_4$, MoO_3, TiO_2 (+ Co^{3+} and Mo^{5+})	UO_2MoO_4, MoO_3, TiO_2 + $UMo_{10}O_{32}$ or $U_3Mo_{20}O_{64}$

re and oxidizing atmosphere (air or oxygen) (refs. 22, 23). Edge-sharing octahedra are found in the crystal structure of both oxidized and reduced forms. These features allow to differentiate the "γ" redox, which is involved in the reaction n-butane-MA, from the "β" redox discussed below. Any other method of preparation leads to less active and selective catalysts (refs. 7, 9).

For the same atomic ratio P/V = 1,0, classical preparations give mainly β-$VOPO_4$ (or eventually α). Reaction with 1-butene leads to the partial reduction of β-$VOPO_4$ (or α) into $(VO)_2P_2O_7$ (Fig. 3). Contrary to the preceding case, such reduction necessitates the sharing of single octahedra found in the structure of β-$VOPO_4$ (or α) by means of crystallographic shear planes (refs. 7, 9, 16). The redox was called "β" because these phases are found in working catalysts. Owing to their structural analogies, interfaces between β-$VOPO_4$ and $(VO)_2P_2O_7$ microdomains are coherent (ref. 15). The validity of such models has been recently demonstrated by ERD (ref. 24).

Now we have to compare the two kinds of catalysts, the catalytic properties of which are very close, and examine them in terms of selectivity criteria, in order to see if analogies are found.

Comparison between catalysts for oxidation of 1-butene

The *first criterion* on selectivity generally used when dealing with *olefin oxidation* is that a π-allylic intermediate is formed on cationic sites with electronic d^0 or d^{10} configuration (refs. 25, 26). In the case of butene, this step is followed by formation of butadiene, itself followed by its desorption or its oxidation in MA, according to the catalyst. This criterion is satisfied by V^{5+}, Mo^{6+}, U^{6+} ions contained in $VOPO_4$, UO_2MoO_4 (tetrahedral Mo), or in Bi_2MoO_6 (octahedral Mo) which however makes butadiene to desorb. Other molybdates such as isostructural β-$MMoO_4$ (M = Co, Mn, Fe) (tetrahedral Mo) are satisfactorily active in butene-butadiene, but far less in butene-MA reaction (refs. 27, 28). Selectivity appears to be rather related to redox V^{5+}/V^{4+} or Mo^{6+}/Mo^{5+},

U^{6+}/U^{4+}, the ionization potential of which is not the same according to the phases (V_2O_5 or $VOPO_4$, MoO_3 or UO_2MoO_4, etc,...) in which they are involved. Unfortunately, the actual values of such potentials are not available and cannot therefore be compared.

In the case of structure-sensitive reactions, catalysts are also characterized by the framework of their active surface planes (refs. 7, 15, 29, 30). From this *structural viewpoint* , analogies between oxidized UO_2MoO_4 and β-$VOPO_4$ forms on the one hand, and reduced $U_3\phi Mo_{20}O_{64}$ and $(VO)_2P_2O_7$ forms on the other, are noticeable. Both UO_2MoO_4 and $VOPO_4$ frameworks show single distorted octahedra $(UO_2)O_4$ or $(VO)O_5$ isolated by MoO_4 or PO_4 groups, with perpendicular U=O and V=O bonds (ref. 31). Therefore, one can think that this kind of surface framework is convenient for the butene molecule, which can be at least oxidatively dehydrogenated on single octahedra. Further, adsorbed butadiene has to find, through several elementary steps, three oxygen atoms on the surface of the catalyst in order to be transformed into MA (and H_2O). Such a drastic diminution of the number of oxygen on the solid surface must be either quickly compensated by gaseous oxygen or accommodated by the structure without collapse. This does not seem to be possible on Bi_2MoO_6 nor to proceed easily on the above-mentionned $MMoO_4$ phases, since butadiene desorbs in both cases. On the contrary, the particular structure of $VOPO_4$ and of $UO_2MoO_4/9MoO_3$ allows their reduction in $(VO)_2P_2O_7$ and in something like $U_3Mo_{20}O_{64}$ respectively, by means of crystallographic shear-plane mechanism (refs. 16, 17). In the crystal structure of $U_3\phi Mo_{20}O_{64}$ (ϕ = cationic vacancy) (ref. 21) are found double layers (Mo_2O_7) of corner-shared ($Mo^{6+}O_6$) octahedra similar to those of (010) MoO_3, which are sandwiched by layers ($U_3\phi Mo_4O_8$) containing hexagonal bipyramids UO_8 sharing edges with ($Mo^{5+}O_6$). Even if this phase does not crystallize as such during catalysis, its structure, which accomodates Mo^{6+}, Mo^{5+} and U^{4+} (and perhaps U^{5+}), gives a model of active sites in the Mo/U = 10 catalyst. Both reduced phases indeed show edge-sharing sites, $[VO^{2+}\text{-}VO^{2+}]$ in $(VO)_2P_2O_7$ and $[UO^{2+}\text{-}MoO^{3+}]$ in $U_3Mo_{20}O_{64}$. Edge-sharing of single octahedra would be therefore consecutive to the insertion of lattice oxygens in butadiene to yield MA. Such a structural analogy could account by itself for the catalytic properties of $VOPO_4$ and UO_2MoO_4-MoO_3/TiO_2 (activity, selectivity, product distribution *vs* conversion, experimental conditions), which are indeed very close.

Comparison between n-butane oxidation catalysts

Owing to the reactivity of butene, its activation and the following oxidation steps seem to proceed on the same active site. In the case of *n-butane* , activation is rate-limiting and very probably happens on another kind of site, namely strong acid sites (ref. 4). Oxygenated cations VO^{2+} and MoO^{3+} (Mo^{5+}), which are known to activate alcanes, are harder acids than Co^{2+} which is borderline (ref. 32). This *acidity criterion* is therefore satisfied by $(VO)_2P_2O_7$. The case of $CoMoO_4$ which is active in the reaction butane-butadiene is more complicated. Hard acid Co^{3+} (or MoO^{3+}) ions should exist on the surface of α or β forms, as been indeed found in the oxidation of acrolein in acrylic acid (ref. 33). On the other hand, high activity and selectivity correspond roughly to high values of T_{all} (Table 1), that is to greater amounts of the α form in which octahedral Mo allows the formation of Mo^{5+} ion known to only accept distorted 6- or 5-coordination (ref. 34). The same argument is valid for MoO_3 in MoO_3/TiO_2, which itself is partly active and selective (C = 45, S_{MA} = 28 mol.%) in MA formation from n-butane (ref. 18).

The special framework of (100)$(VO)_2P_2O_7$ has been assumed by several authors (refs. 3-5, 8, 24, 35) to be the key of its selective properties. This *structural criterion* is also fulfilled by the surface plane (110) of β-$CoMoO_4$, the framework of which is indeed strikingly similar (edge-sharing CoO_6 octahedra linked to MoO_4 tetrahedra) (Fig. 4). Both compounds are able to activate n-butane, but further oxidation to MA, which needs oxidized sites, proceeds on $CoMoO_4$ only when MoO_3/TiO_2 is present (Table 1). When dealing with γ-$(VO)_2P_2O_7$, VO^{2+} in double octahedra are easily oxidized (by air) to VO^{3+} without drastic change of coordination, the more so as these pairs of

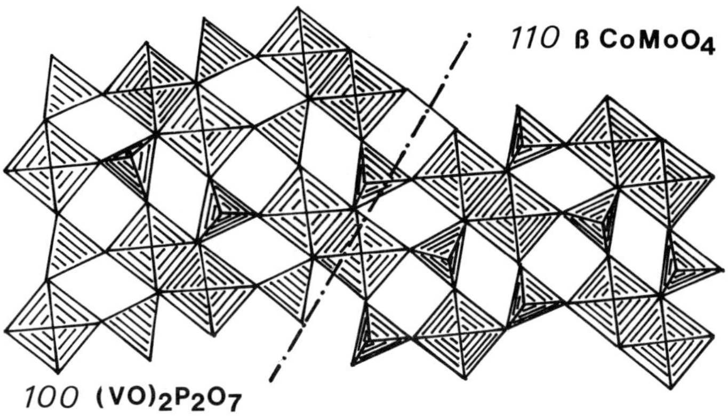

Fig. 4. Common surface framework of (110) β-$CoMoO_4$ and (100) γ-$(VO)_2P_2O_7$, made with edge-sharing octahedra containing Co^{2+} or V^{4+} and MoO_4 or PO_4 tetrahedra respectively.

octahedra are found also in δ,γ-$VOPO_4$ (refs. 7, 9). The molecule is able to be oxidized while reduction to VO^{2+} proceeds again and, as a result, $(VO)_2P_2O_7$ is still detected. On the contrary, three oxygens do not seem to be available on either form of $CoMoO_4$ alone, although the catalytic reactions proceed in a temperature range close from the transition one. Probably the only oxygen that does the transformation α <--> β by its own movement is not available for oxidation. In a recent work dealing with 1-butene oxidation (ref. 36), β-$CoMoO_4$ has been exclusively found and shown not to transform to α, even when an excess of MoO_3 is present. In our work both forms are present, which make difficult to recognize if a specific role is to be attributed to each one. It is probable however that the common 6-coordination of Mo in α-$CoMoO_4$ and MoO_3 allows the transfer of electron and oxygen species to be easier through the coherent interfaces. In the multicomponent catalyst $CoMoO_4/MoO_3$ activated by TiO_2, excess MoO_3 provides active sites and selective redox (as modified by the presence of Co) for the second part of the reaction. Since these phenomena occur at coherent interfaces, but of course with less efficiency than in V-P-O where the same reaction step is made with very small change in the solid, lower performances in butane oxidation result.

CONCLUSION

The whole results and experiments concerning the comparison between *a-priori* different catalysts, which in fact have similar chemical and catalytic properties, outline the importance of the structural criterion, the validity of which was demonstrated many times (refs. 7, 15, 29, 30, 35). This criterion depends also on the methods of preparation and activation and is to be added to acid-base properties, in order to provide new orientations in mild oxidation of hydrocarbons, especially of alcanes.

REFERENCES

1 T.C. Yang , K.K. Rao and I. Der Huang , US Patent 4,392,986 (1987).
2 R.M. Contractor, H.E. Bergna, H.S. Horowitz, C M. Blackstone, U. Chowdhry and A.W. Sleight, Stud. Surf. Sci. Catal., 38 (1988) 645-654.
3 B.K. Hodnett, Catal. Rev.-Sci. Eng., 27 (1985) 373.
4 Papers in "Selective Catalytic Oxidation of C-4 Hydrocarbons to Maleic Anhydride", Catal. Today, 1 (1987).
5 F. Centi and F. Trifiro, Chim. Indust., 68 (1986) 74.
6 M. Ai, J. Catal., 100 (1986) 336.
7 E. Bordes , Catal. Today, 3 (1988) 163-174.
8 M.A. Pepera, J.L. Callahan, M.J. Desmond, E.C. Milberger, P.R. Blum and N.J. Bremer, J. Am. Chem. Soc., 107 (1985) 4883-92.
9 E. Bordes , Catal. Today, 1 (1987) 499-526.
10 M. Ai, P. Boutry and R. Montarnal, Bull. Soc. Chim. Fr., (1970) 2775 and 2783.
11 M. Akimoto and E. Echigoya, Bull. Chem. Soc. Jap., 48 (1975) 3518.
12 V.C. Corberan, A. Corma and G. Kremenic, I&EC, Prod. Res. Dev., 23 (1984) 546.
13 R.K. Grasselli, J.D. Burrington and J.F. Brazdil, J. Chem. Soc., Far. Disc., 72 (1982) 273.
14 J.C. Daumas, Thèse, Paris (1970).
15 P. Courtine, ACS Symp. Series, 279 (1985) 37.
16 E. Bordes and P. Courtine, J. Catal., 57 (1979) 236 ; E. Bordes, Thèse, Compiègne (1979).
17 E. Bordes, S.J. Jung and P. Courtine, Proc. 9th. Ibero-America Symp. on Catalysis, Lisboa (Portugal) (1984), vol.2, 983-991.
18 S.J. Jung, E. Bordes and P. Courtine, "Adsorption and Catalysis on Oxide Surfaces", M. Che and G.C. Bond (Eds), Stud. Surf. Sci. Catal., 21 (1985) 345 ; S.J. Jung, Thèse, Compiègne (1984).
19 E. Bordes, unpublished results.
20 M. David , Thèse, Lyon (1988).
21 V.N. Serezhkin, L.M. Kovba and V.K. Trunov V.K., Sov. Phys. Crystal., 17 (1973) 999.
22 E. Bordes, J.W. Johnson and P. Courtine, J. Sol. State Chem., 55 (1984) 270.
23 E. Bordes, J.W. Johnson, A. Raminosona and P. Courtine, Mater. Sci. Monograph., 28 B (1985) 887.
24 G. Bergeret, M. David, J.P. Broyer, J.C. Volta and G. Hecquet, Catal. Today, 1(1987) 37.
25 R.K. Grasselli and D.D Suresh, J. Catal., 25 (1972) 273.
26 J.E. Germain, Intra Sci. Chem. Rep., 6 (1972) 273.
27 G. Centi and F. Trifiro, Appl. Catal., 12 (1984) 1-21.
28 U. Ozkan and G.L. Schrader, J. Catal., 95 (1985) 120 and 137.
29 J.C. Volta and J.L. Portefaix , Appl. Catal., 18 (1985) 1-32 ; J.M. Tatibouet, J.E. Germain and J.C. Volta, J. Catal., 82 (1983) 240.
30 J. Ziolkowski, J. Catal., 80 (1983) 263.
31 E. Bordes and P. Courtine, Bull. Soc. Chim. Fr., submitted.
32 R.G. Pearson, "Hard and Soft Acids and Bases", Dowden, Hutchinson & Ross Inc., Stroudsburg, Pensylv., USA, (1973).
33 J. Haber, M. Witko and A. Golobiewski, J. Molec. Catal., 3 (1977) 213.
34 M. Che, F. Figueras, M. Forissier, J. McAteer, M. Perrin, J.L. Portefaix, H. Praliaud, Proc. 6th Int. Cong. Catal., 1 (1976) 261.
35 J. Ziolkowski, E. Bordes and P. Courtine, J. Catal., submitted.
36 U. Ozkan and G.L. Schrader, Appl. Catal., 23 (1986) 327.

C.B. MURCHISON (Dow Chemical, USA) : For $CoMoO_4$ catalyst containing no excess MoO_3, is the O of the water formed during the oxydehydrogenation of butane from the lattice or gas phase ?

E. BORDES (Université de Technologie de Compiègne, France) : Lattice oxygen of $CoMoO_4$ is responsible for such reactions, as shown in several other works (see e.g. ref. 14).

B. DELMON (Université Catholique de Louvain, Belgium) :
1) I refer to your catalyst $CoMoO_4/MoO_3/TiO_2$. It is impossible to avoid a redissolution-redeposition of MoO_3 when an MoO_3-containing catalyst is reimpregnated by a second compound. On the other hand, it is well-known that the α <---> β transition of $CoMoO_4$ depends on its stoichiometry. Therefore, I do not believe any argument could be taken from a change of the transition temperature. Another reason for raising doubts is that this transition is strongly nucleation-limited and that nucleation rate is dependent on dispersion. Dispersion of $CoMoO_4$ is not identical in your comparison.
2) If there is a structural fit between $CoMoO_4$ alone and $(VO)_2P_2O_7$ (Fig. 4), this can be proven by direct techniques (high resolution microscopy) or, indirectly, using the techniques of OZKAN and SCHRADER or those we used in our laboratory. What physico-chemical evidence have you of the "epitaxy" you propose ?

E. BORDES (Université de Technologie de Compiègne, France) :
1) I agree with your remarks, in the sense that the error on the temperature at which the transition proceeds is large. However synergetic effects between $CoMoO_4$ and MoO_3 (and TiO_2) do exist, as found also by the authors you mention (ref. 28), and we have to find explanations for that. We demonstrated in several other examples that crystallographic fit between the frameworks can account for such effects.
2) Fig. 4 shows only the similarity between the surface frameworks of $CoMoO_4$ and $(VO)_2P_2O_7$, and nothing else.

R. CHUCK (Lonza A.G., Switzerland) : In Fig. 1 you show that the selectivity of oxidation of 1-butene with V-P-Ti catalyst is higher than V-P catalysts alone.
- Have you an explanation for that ?
- Have you characterized the V-P-Ti catalyst in comparison with the V-P catalyst ?
- If so, are there any clues to structural differences that might account for the superior performance of the former ?

E. BORDES (Université de Technologie de Compiègne, France) :
- TiO_2 (anatase) used as support enhances activity and selectivity (in oxidation reactions) of any vanadium-containing catalyst (and also of MoO_3). It must be first noticed that distorted octahedra MO_6 of nearly the same size, allowing an eventual anchoring, exist in all these compounds. Our interpretation is that the crystallographic misfit between surface patterns of the active phase and the support is low. Interfaces are said to be coherent, which means that the transfer of species is facilitated, but also that metastability occurs owing to tensions and compressions more or less accommodated by the lattices (refs. 9, 15, 16). As a result activity is enhanced. Another point is that the support can "stabilize" a particular reduced oxide, e.g. in the case of V_2O_5/TiO_2 (o-xylene - phthalic anhydride), which results in a better selectivity (*).
- In the case of V-P-Ti, no difference is seen with pure V-P phases (in this case, mainly $\beta VOPO_4$ and few $(VO)_2P_2O_7$), except that the surface of fresh β–$VOPO_4$ is enriched with V^{4+}. Since no mixed V-P-Ti phase exists, the only interpretation should be as above.
(*) : see also A. Véjux et al., J. Sol. State Chem., 23 (1978) 93 ; ibid., 63 (1986) 179.

G. Centi and F. Trifiro' (Editors), *New Developments in Selective Oxidation* 595

IDENTIFICATION OF A CATALYTICALLY ACTIVE COPPER OXYCHLORIDE PHASE FOR THE SYNTHESIS OF MALEIC ANHYDRIDE

M.J. DAVIES[1], D. CHADWICK[2] and J.A. CAIRNS[1]

[1]Applied Chemistry Group, B429, Materials Development Division, Harwell Laboratory, Oxon. OX11 0RA, U.K.

[2]Department of Chemical Engineering and Chemical Technology, Imperial College of Science Technology and Medicine, London, SW7 2BY, U.K.

ABSTRACT

An investigation of the partial oxidation of n-butane to maleic anhydride has identified a new catalytically active material for this conversion, namely copper (II) oxychloride. The formation of this phase from copper (II) trihydroxychloride has been characterised by XRD and TGA. Catalytic activity was measured using a microprocessor controlled microreactor under flow conditions with product stream analysis being performed by a single injection into a gas chromatograph. The catalytic activity was observed to deactivate rapidly. It was concluded, by the use of TGA and XRD, that the deactivation process was due to chlorine loss.

INTRODUCTION

The selective, partial oxidation of n-butane to maleic anhydride is both commercially and scientifically important, and is currently being performed and extensively studied using vanadium phosphorus mixed oxide catalysts (ref. 1). Cuprous oxide is known to be active for the partial oxidation of alkenes to aldehydes (ref. 2), cupric oxide being principally used for the catalytic combustion of carbon monoxide. Cupric chloride is used as a catalyst for the oxychlorination of alkenes eg. ethene to dichloroethene (ref. 3). The objective of this work was to study the catalytic activity of copper oxychloride material for the heterogeneous selective partial oxidation of alkanes.

EXPERIMENTAL

Preparation of the catalyst.

The precursor material, atacamite, was prepared by suspending 25.0 g of Analar grade cupric oxide in 60.0 ml of distilled water by stirring. To this a stoichiometric ratio of hydrochloric acid (15.0 ml of 35% HCl) was added to give the desired copper (II) trihydroxychloride ($CuCl_2.3[Cu(OH)_2]$). The suspension was left stirring at room temperature for 1 hour before being evaporated to dryness in a oven overnight at 80 Celsius. The resultant powder was then gently ground with a pestle and mortar. This method is significantly simpler and quicker than that reported by Oswald et al. (ref. 4).

Five point nitrogen physisorption analysis using a Micromeritics Digisorb 2600 gave a BET surface area of 13.0 m^2/g for the precursor material.

Catalyst characterisation.

X-ray diffraction (XRD) measurements were taken with a Philips 1050 goniometer on a

Philips 1010 X-ray generator using nickel filtered Cu K_{α} radiation and scanning at 2 minutes per degree (2θ).

Thermogravimetric analysis (TGA) was performed using a DuPont 951 thermogravimetric analyser module linked to a DuPont 1090 microcomputer. Temperature profiles were recorded using a linear ramp rate of 10 Celsius/min. with an atmosphere of air flowing at 40 ml/min.

Catalytic activity.

(i) The reactor system. Shown in figure 1, the gases pass through 1 of 3 parallel mass flow controllers, for accurate volumetric metering of the reactant gases, followed by a mixing chamber and a capacitance pressure transducer before being introduced into the top of the reactor. The reactor tube (see Fig. 2) comprises of a silica glass tube with a sintered glass frit approximately half way along its length onto which the catalyst, in powder form, was placed. The reactor tube was resistively heated with thermocouple feedback to a Honeywell digital process controller DCP 7700 which was used to control all reactor conditions. All effluent lines were kept at 180 Celsius to prevent condensation of the products. A heated gas sampling valve, also controlled by the Honeywell DCP, was used to sample the effluent gas stream for analysis by gas chromatography. Cold traps were then used to remove condensibles before the exhaust.

(ii) The analysis of the reactants and products. This was performed by gas chromatography using a single sample taken from the effluent gas stream by the automatic heated gas sampling valve. The GC, a Perkin Elmer Σ1 was fitted with a heated 10 port 2 way valve and two columns, the first was a glass 1 m long by 6 mm diameter porapak QS and the second was a stainless steel 1 m long by 3 mm diameter spherocarb, see Fig. 3. Initially the two columns were held in series at 90 Celsius for 5 minutes, the sample was separated on the porapak column with the air, carbon monoxide and carbon dioxide being eluted as one peak and the organics being retained. The permanent gases were fed onto the spherocarb column and separated. The air and carbon monoxide eluted and were detected using a thermal conductivity detector before the temperature was ramped to 160 Celsius for 7 minutes where the carbon dioxide eluted and was detected. The valve was then switched allowing the species eluting from the porapak to go straight to the detector (a flame ionisation detector) and the oven was ramped to 200 Celsius for elution and detection of the n-butane and maleic anhydride. The detectors were calibrated using certified gas mixtures and the maleic anhydride peak was calibrated using weighed samples of pure maleic anhydride injected onto the columns using a pyroprobe injector.

RESULTS

Catalytic activity.

A premixed gas of 1% n-butane (by volume) in synthetic air was passed through the catalyst bed at a gas hourly space velocity of 1000 hr^{-1}, and the reactor temperature was ramped stepwise from 250 to 500 Celsius in 25 Celsius steps with a sample being taken 15 minutes into

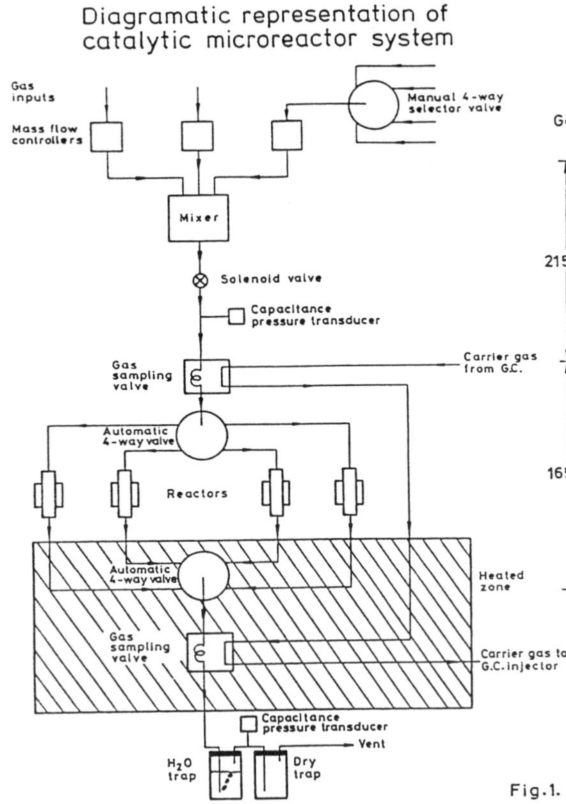

Fig.1.

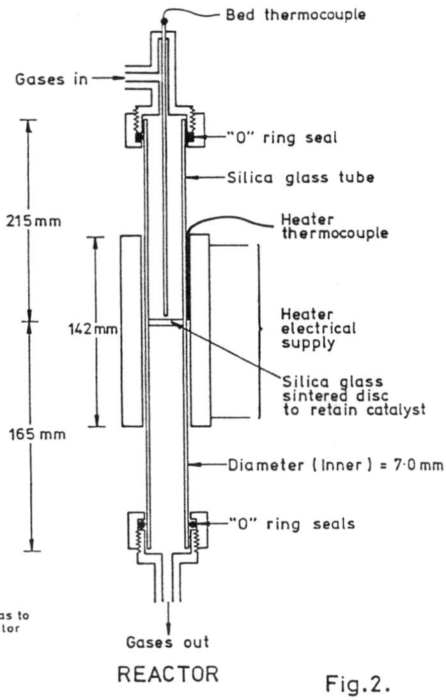

Fig.2.

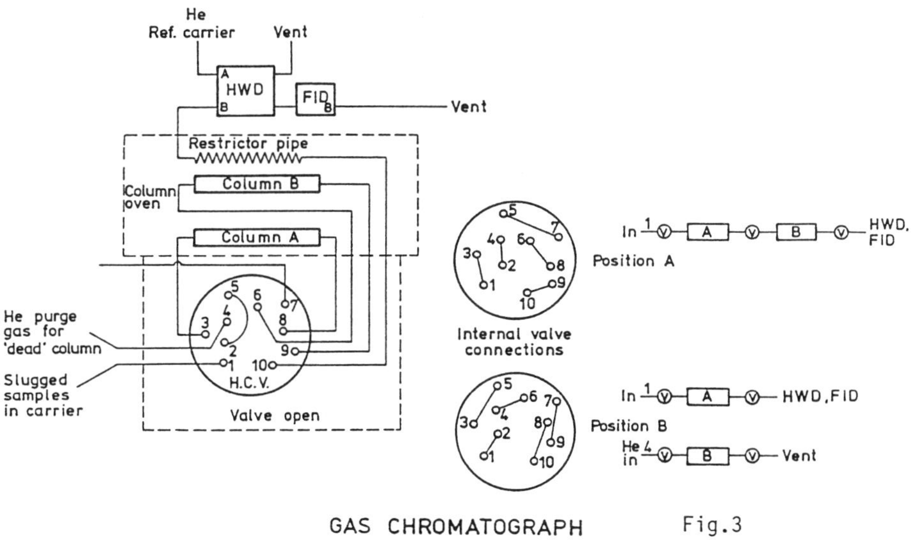

Fig.3

each step. An initial reactant gas sample was taken for analysis at room temperature to act as an internal calibration for the detectors.

Figure 4 shows n-butane conversion and the yields of maleic anhydride and carbon dioxide over the temperature range 250 to 500 Celsius. Figure 5 shows the corresponding variation of maleic anhydride yield and selectivity as a function of temperature. Inspection of the chromatograms obtained showed that other species, either by-products or reaction intermediates, are produced in reasonable quantities over the temperature range that maleic anhydride production was observed. These other products vary together in concentration with temperature following the trend in maleic anhydride yield.

Gas chromatography - mass spectroscopy was used to identify the unknown products by adsorbing a sample of the product gas stream onto a Tenax column. The GC-MS system used a capillary column which was ramped from 20 to 300 Celsius at 6 Celsius/min. Therefore, direct relation of the identified compounds to peaks seen in the on-line GC was not possible. The species identified were HCl, butenes, CH_2Cl_2, $CHCl_3$, CCl_4 and other C_2-C_4 chlorinated organics, often unsaturated, in addition to the known compounds.

Deactivation of the catalyst was examined in the following way. A second run was performed with the n-butane in air feed. This time the temperature was held at 350 Celsius and the samples were taken at two hourly intervals, the first being taken after 15 minutes into the isothermal period. The second sample showed that the maleic anhydride yield had fallen from 1.7% to 0.7% whilst the n-butane conversion had risen from 7% to 9%. The third injection corresponding to 6 hours on stream showed that the catalyst had completely deactivated giving no reaction products other than carbon dioxide and that the n-butane conversion had further risen to 11%. The results of detailed catalyst characterisation (see below) showed that this deactivation was due to chlorine being stripped from the surface of the catalyst in the formation of HCl and the chloro compounds. The resultant cupric oxide, identified in the deactivated catalyst by XRD, being known to only combust n-butane to carbon dioxide. It was also noticed that carbon monoxide was not produced at any temperature which is consistent with cupric oxide being a carbon monoxide oxidation catalyst.

Catalyst characterisation.

Thermogravimetric analysis shows the catalyst, as prepared, to have two sets of decompositions, the first between 250 and 350 Celsius and the second from 450 Celsius to 650 Celsius, see Fig. 6. As the results of the catalytic activity temperature profiles show the activity to increase from 275 Celsius to a peak at 400 Celsius before deactivation occurs, the species present between the two sets of decompositions is most probably the catalytically active species. As the isothermal catalytic activity experiment had shown the material to deactivate, a thermogravimetric run was performed at 350 Celsius in atmospheres of air, argon and the n-butane/air mix. The weight loss observed in both argon and air atmospheres was found to be 0.5 weight% per hour, but

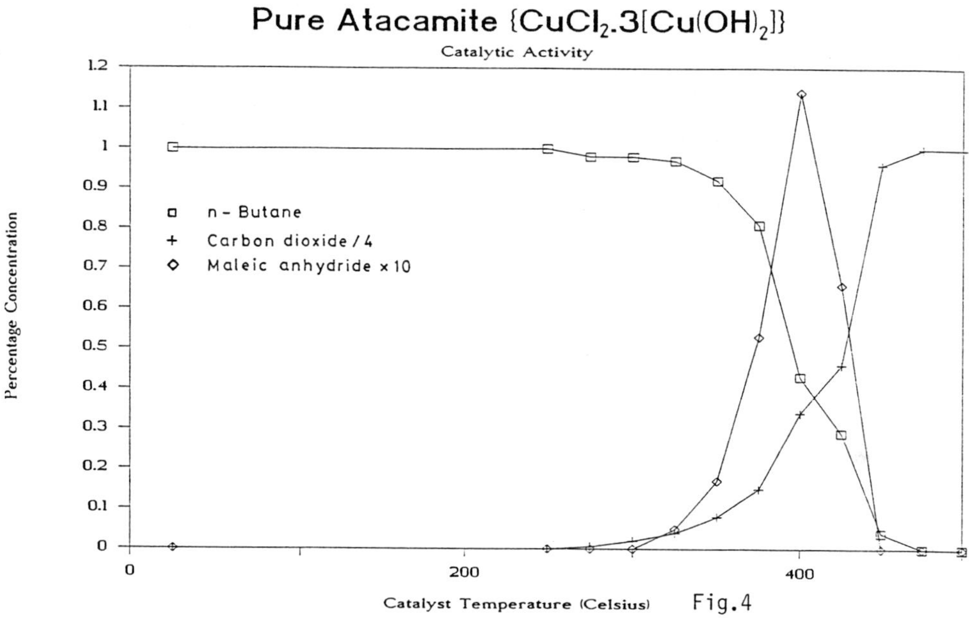

Pure Atacamite {CuCl2.3[Cu(OH)2]}
Catalytic Activity
n - Butane
Carbon dioxide / 4
Maleic anhydride x 10
Percentage Concentration
Catalyst Temperature (Celsius)

Fig.4

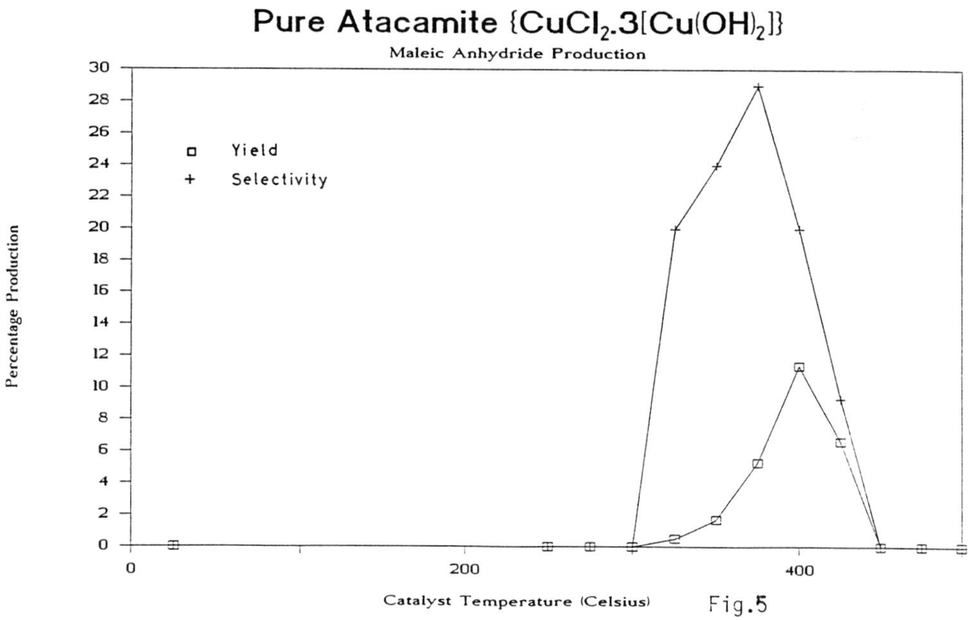

Pure Atacamite {CuCl2.3[Cu(OH)2]}
Maleic Anhydride Production
Yield
Selectivity
Percentage Production
Catalyst Temperature (Celsius)

Fig.5

Thermogram of pure Atacamite in air

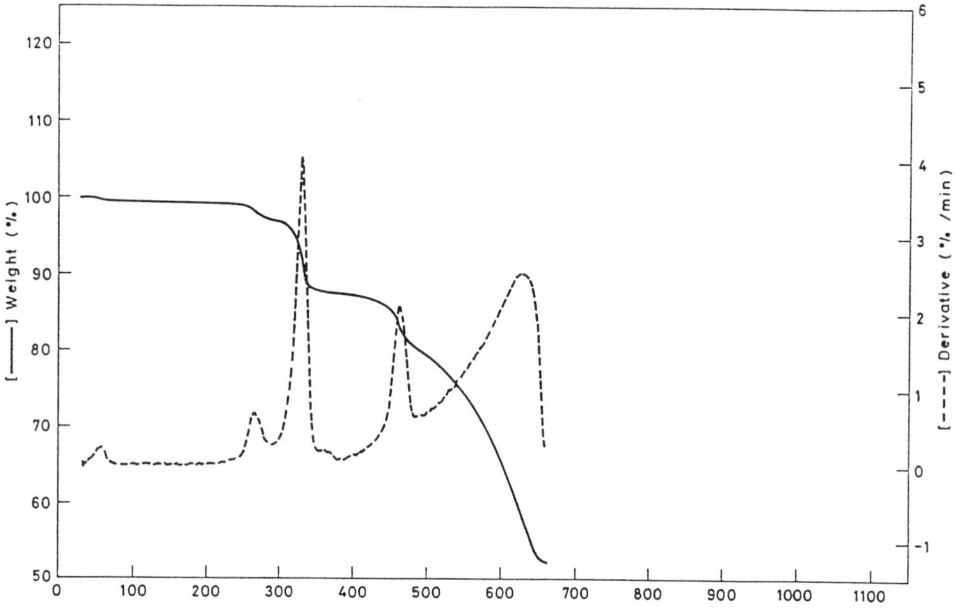

Fig.6

XRD pattern of pure Atacamite calcined at 325 Celsius

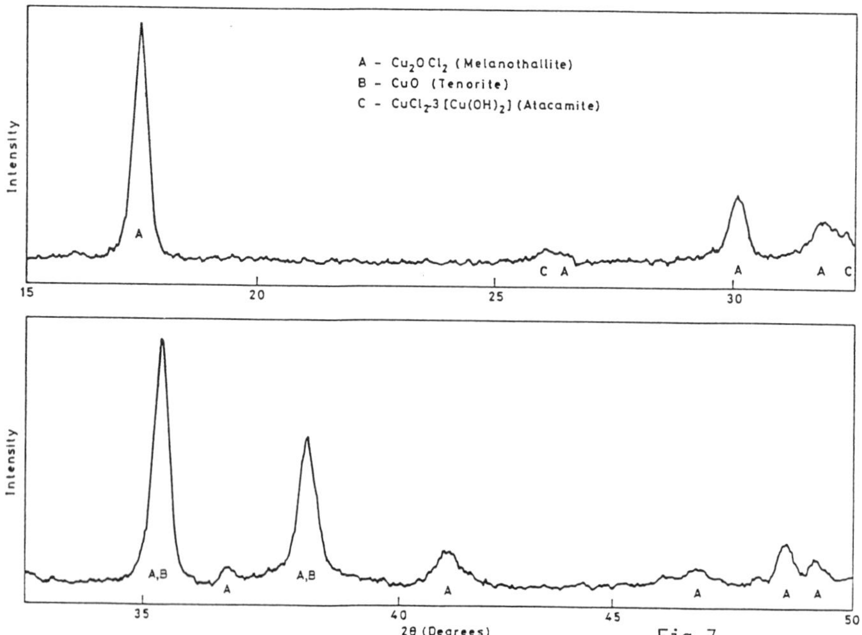

Fig.7

in n-butane/air the rate was found to be enhanced to 1.1 weight% per hour. Since the final decomposition corresponded to the formation of cupric oxide (see below) it was concluded that the decomposition of the active species is by loss of chlorine and that the presence of n-butane accelerated the chlorine removal process by reaction.

XRD analysis of the precursor material, as prepared, confirmed it to be copper (II) trihydroxychloride (atacamite, ref. 4) with small amounts of cupric oxide and cupric chloride present. Taking a sample of the precursor and calcining it at 325 Celsius in air gave a material with the XRD pattern of copper (II) oxychloride (Cu_2OCl_2) known as melanothallite, see Fig. 7. However, when the material was calcined at 500 Celsius the XRD pattern showed the presence of cupric oxide alone. The same result was observed by XRD analysis of the totally deactivated material. The melanothallite would appear, therefore, to be the active material. It was found that when this material was left in moist air, it reverted to atacamite by water adsorption and rehydroxylation.

DISCUSSION

Comparison of the results obtained from the TGA, XRD and catalytic activity experiments show that the active material is copper (II) oxychloride, known as melanothallite. This material is formed by the thermal decomposition of the prepared precursor, copper (II) trihydroxychloride in the atacamite phase. It was also found that melanothallite would revert to atacamite by rehydroxylation from the adsorption of moisture. However, the material as prepared deactivates rapidily under reaction conditions by the loss of chlorine. This has been shown by TGA to be accelerated in the presence of n-butane by reaction to form chlorinated organic products.

Preliminary work using n-butene feeds has shown that the activation energy for n-butane conversion is, as expected, much greater than for n-butenes. The latter also produce maleic anhydride and a range of chlorinated organics although the chloro compounds are not identical to the n-butane case. Chlorine is well-known to facilitate abstraction of hydrogen from hydrocarbons, but it is not clear for the melanothallite whether loss of chlorine from the solid is necessary for the catalyst to function effectively in this manner. As far as the lifetime of the catalyst is concerned, the loss of constitional chlorine results in loss of partial oxidation activity. It may be that the chlorine content is critical and that an active catalyst requires sufficient chlorine to facilitate hydrogen abstraction whilst not being too labile to avoid rapid deactivation.

The importance of oxygen in the feed gas has been investigated in a preliminary study using n-butene in helium over the copper oxychloride catalyst which yielded no maleic anhydride. However, the production of carbon dioxide was still observed initially. After a period of time as the lattice oxygen and chlorine were lost, the hydrocarbon conversion decreased. These results indicate that chemisorbed oxygen species are involved in the formation of the partial oxidation product, maleic anhydride. Although the initial production of carbon dioxide could be explained partly by the presence of a small amount of cupric oxide impurity, it appears that lattice oxygen from the melanothallite may oxidise hydrocarbons to carbon dioxide by a parallel route. In addition to facilitating hydrogen abstraction, the presence of chlorine may limit the lability of the lattice and chemisorbed oxygen species and therefore limit the extent of oxygen insertion into the hydrocarbon molecule and so produce partially oxidised products rather than carbon dioxide.

The ability of the copper oxychloride to partially oxidise n-butane is particularly interesting in the light of the work by Ueda and Thomas (ref. 5.) on layered bismuth oxyhalide catalysts for the partial oxidation of methane. It would appear that oxychlorides may be active catalysts for a range of partial oxidation reactions. The basic material, bismuth oxychloride, was found to deactivate rapidly as observed for copper oxychloride, but other members of the same structural family with bismuth ions replaced by alkali or alkaline earth ions were reported to be much more stable. Whether it is possible to promote or modify the copper oxyhalide catalyst to improve its stability requires further study.

REFERENCES

1 B.K. Hodnett, Catal. Rev. -Sci. Eng., **27(3)** (1985) 373-424.

2 D.J. Hucknall, Selective Oxidation of Hydrocarbons, Academic Press, London, 1974, p.24.

3 R. Vetrivel, K.V. Rao, K. Seshan, K.R. Krishnamurthy and T.S.R. Prasada Rao, Proc. 9th Int. Congr. Catal., Calgary, July 1988, **4**, 1766-1773

4 H.R. Oswald and E.W. Feitknecht, Helvetica Chimica Acta, **47[I](35)** (1964) 272-289.

5 W. Ueda and J.M. Thomas, Proc. 9th Int. Congr. Catal., Calgary, July 1988, **2**, 960-967.

P. RUIZ (Univ. Catholique du Louvain, Belgium): It seems to me that the XRD characterisation of the calcined precursor has identified the active phase. Do you have any results concerning the XRD analysis of the catalyst after the catalytic tests ?

M.J. DAVIES (Harwell Laboratory, U.K.): XRD analysis of the catalyst after complete deactivation showed the presence of copper (II) oxide (tenorite) alone. This confirms the results of the TGA study which showed a weight loss consistent with the formation of cupric oxide.

J. HABER (Inst. of Catalysis and Surface Chemistry, Poland): Your results do not justify a general conclusion that lattice oxygen is responsible for total oxidation, and gas phase oxygen for selective oxidation. Oxidation of butane is a multistep process and from the fact that in the absence of gas phase oxygen no reaction is observed one can only conclude that in the first step of reaction, i.e. the activation of butane, gas phase oxygen plays an important role. However, in the next steps it may be lattice oxygen which is involved in the reaction as indeed it was shown by may authors.

M.J. DAVIES (Harwell Laboratory, U.K.): The experiments in the absence of gaseous oxygen in the reactant gas stream were performed with 1-butene and not n-butane to reduce problems associated with the initial activation of the hydrocarbon. The experiment described is only preliminary and I agree that it does not conclusively establish the respective role of chemisorbed oxygen species and lattice oxygen in partial oxidation. However, the absence of maleic anhydride, in the presence of gaseous oxygen, does indicate that chemisorbed oxygen plays an important role in the production of maleic anhydride and not just in initial activation. As you point out, this does not exclude involvement of lattice oxygen.

The initial production of carbon dioxide in the absence of air at the same concentration as in the presence of gaseous oxygen does suggest that lattice oxygen may be combusting the hydrocarbons by a parallel route. However, we have not claimed that carbon dioxide is produced exclusively in this way.

V. CORTES CORBERAN (Inst. of Catalysis C.S.I.C., Spain): It is a little surprising that no calcination step is used (at least not reported) in the preparation of your catalyst, taking into account that the oxidation of butane usually takes place between 300 - 500°C. By comparing figures 3 and 4, it seems to be clear that loss of chlorine is not only the cause of deactivation but the real cause of selective oxidation products formation. This situation appears to be parallel to the one produced in the promoting effect of alkali halides on catalysts for oxidative dimerisation of methane: the increase of activity and selectivity can be ascribed to the formation of active chlorine species upon decomposition of the alkali halide. Then, melanothallite would be a chlorinating agent rather than an active oxidation catalyst of the n-butane itself. Can the activity of the deactivated catalyst be regenerated? By Cl_2? Can the activity of the working catalysts be maintained by adding small amounts (of the order of ppm) of chlorinated hydrocarbons such as $ClCH_3$?

M.J. DAVIES (Harwell Laboratory, U.K.): The comment regarding the possible use of a calcination step during the preparation of the catalyst is not appropriate. The transformation from atacamite to the active melanothallite occurs rapidly under reaction conditions and is reversed on exposure to the atmosphere due to water adsorption. In these circumstances there would be no point in calcining the catalyst prior to insertion in the reactor.

The point raised concerning the role of active chlorine species and the chlorinating agent are covered adequately by the discussion in the paper where we express the view that an active catalyst may require sufficient chlorine to facilitate hydrogen abstraction whilst not being too labile to avoid rapid deactivation.

Regarding the stabilisation of the catalyst by addition of a chlorine species to the feed, we do believe this to be possible and experiments are in progress to establish the optimum procedure.

G. Centi and F. Trifiro' (Editors), *New Developments in Selective Oxidation*
1990 Elsevier Science Publishers B.V., Amsterdam — Printed in The Netherlands

SELECTIVE OXIDATION OF n-BUTANE TO MALEIC ANHYDRIDE OVER WELL-CHARACTERIZED VANADIUM-PHOSPHORUS MIXED OXIDES

Makoto MISONO, Koichi MIYAMOTO, Katsuyuki TSUJI, Tatsuya GOTO, Noritaka MIZUNO, and Toshio OKUHARA
Department of Synthetic Chemistry, Faculty of Engineering, The University of Tokyo, Hongo, Bunkyo-ku, Tokyo 113, Japan

SUMMARY

Catalytic oxidations of n-butane, n-butene, and butadiene as well as isotopic experiments using O-18 were carried out over three well-characterized crystalline V-P oxides (P/V=1). Conclusions are as follows. (1) The critical step in the selective oxidation of n-butane to maleic anhydride is the first step, that is, the dehydrogenation of butane to butene. (2) Reduction-oxidation only near the surface is involved in the catalysis. (3) The morphology of $(VO)_2P_2O_7$ has phenomenologically a great influence on the catalytic performance at high conversion levels; thick platelet crystals gave better performance than crystals having rose-like shape.

INTRODUCTION

$(VO)_2P_2O_7$ is known to be a good catalyst for the production of maleic anhydride (MA) from n-butane (ref. 1-3). We previously reported the preparation of $(VO)_2P_2O_7$, α-$VOPO_4$ and β-$VOPO_4$ which were pure and crystalline as far as XRD, XPS, and IR were concerned. It was demonstrated by using these catalysts that the $(VO)_2P_2O_7$ catalyst was much superior to the others and its performance was comparable or better than those claimed in patents (ref. 2). We further demonstrated in our preliminary communication that the $(VO)_2P_2O_7$ catalyst was superior in the step of the selective dehydrogenation of n-butane into n-butenes (ref. 4).

In the present study, we compared in more detail the rates and selectivities of the oxidation of butane, butene, and butadiene over the above three catalysts to determine in which step of butane $\longrightarrow$ butenes $\longrightarrow$ butadiene $\longrightarrow$ furan $\longrightarrow$ MA the $(VO)_2P_2O_7$ catalyst is most effective. We also investigated the dynamic behavior of lattice oxygen during the catalytic oxidation of carbon monoxide and butane by using ^{18}O tracer and the influence of the morphology of $(VO)_2P_2O_7$ catalysts prepared by different methods on the catalytic performance.

EXPERIMENTAL

Catalysts

(i) $(VO)_2P_2O_7$ (abbreviated as **P**). Precursor, $VO(HPO_4)\cdot 0.5H_2O$, was prepared

with V_2O_5, 85%H_3PO_4, $NH_2OH \cdot HCl$ and water following the literature as described in the previous paper (ref. 2), and the structure was confirmed by XRD. The precursor was heated in a nitrogen stream (the rate of temperature rise: 10 deg·min^{-1}) up to 550 °C and kept for 3 h at 550 °C. Then it was further calcined in O_2 at 500 °C for 2 h. Washing with water and subsequent drying increased crystallinity. XRD and IR indicated the presence of only $(VO)_2P_2O_7$. Surface area: 10.1 $m^2 \cdot g^{-1}$.
(ii) α-$VOPO_4$ (abbreviated as **A**). The precursor, $(NH_4)_2[(VO)_2C_2O_4(HPO_4)_2] \cdot 5H_2O$ (P-I), which was prepared with V_2O_5, $(COOH)_2 \cdot 2H_2O$, $NH_4H_2PO_4$ and water, was calcined in an O_2 stream for 4 h at 300 °C and for 8 h at 600 °C as in the previous paper (the rate of temperature rise: 5 deg·min^{-1} for 25 - 300 °C) and 2.5 deg·min^{-1} for 300 - 600 °C)(ref. 2). The XRD indicated that the catalyst contained about 94% α-VOPO4 and 6% β-$VOPO_4$. IR agreed with that of α-$VOPO_4$. Surface area: 5.2 $m^2 \cdot g^{-1}$.
(iii) β-$VOPO_4$ (abbreviated as **B**). The precursor (P-I) prepared as above was heated in an O_2 stream with the rate of temperature increase of 10 deg·min^{-1} to 300 °C, kept at 300 °C for 4 h and further calcined in O_2 at 600 °C for 8 h. This was then washed and dried. The last procedure eliminated impurity X_1 phase (cf. ref. 2) and increased the purity. XRD and IR were consistent with pure β-VOPO4. Surface area: 2.6 $m^2 \cdot g^{-1}$.
(iv)$(VO)_2P_2O_7$-2 (abbreviated as **P'**). $VO(HPO_4) \cdot 0.5H_2O$ was prepared from V_2O_5, H_3PO_4, isobutanol and benzyl alcohol. This precursor was calcined in a similar way as catalyst **P**. Surface area: 56 $m^2 \cdot g^{-1}$.

According to XPS and redox titration, vanadium ions of catalysts **A** and **B** were in the +5 state and those of catalyst **P** were in the +4 state both near the surface and in the bulk as in the previous work (ref. 2). The XPS analysis based on the method in the literature (ref. 5) indicated that the surface phosphorus to vanadium ratios of these catalysts were close to unity.

Oxidation reactions

Oxidation of butane was performed at 440 °C using a conventional flow reactor at atmospheric pressure. The feed gas consisted of 1.5 vol% butane, 14 vol% oxygen and nitrogen (balance). Oxidations of 1-butene and butadiene were carried out at the temperature from 360 to 440 °C using the same concentrations of reactant and oxygen as those of the butane oxidation. Prior to the reaction, $(VO)_2P_2O_7$ and $VOPO_4$ were pretreated in N_2 and O_2 flow, respectively, at 500 °C for 1 h. All products were analyzed using an on-line gas chromatograph. Exit lines from the reactors were heated at about 150 °C to prevent the condensation of maleic anhydride (MA). MA, acetic acid, furan, and crotonaldehyde were analyzed by a flame ionization detector with Porapak QS column, of which the temperature was programmed to rise from 80 to 240 °C at the rate of 20 deg·min^{-1}. The amount of acids was confirmed by titration. Oxygen, nitrogen, CO, and CO_2 were analyzed by a thermoconductivity detector.

Isotopic experiments

$CO-{}^{18}O_2$-catalyst (CO: 30 Torr, O_2: 15 Torr, catalyst: 0.5 - 1 g) and $C^{18}O_2$-catalyst (CO_2: 20 Torr, catalyst: 0.2 g) reactions were carried out at 350 and 380 °C in a closed circulation system. The composition and the ^{18}O content of each molecule in the gas phase were analyzed by GC and mass spectrometry, respectively. Butane-$^{18}O_2$-catalyst reactions (butane: 20 Torr, O_2: 80 Torr, catalyst: 0.17 - 1 g) were carried out at 300 - 380 °C in a closed circulation system, to which a specially designed trap kept at 0 °C (for MA) and an ordinary trap (for butane, CO_2, and H_2O) were attached. The other isotopic experiments were performed in a similar way.

Pretreatments of catalysts were evacuation at the reaction temperature for 0.5 h for catalyst **P** and O_2-treatment at the reaction temperature for 0.5 h followed by evacuation for 5 - 10 min for catalysts **A** and **B**.

RESULTS AND DISCUSSION

Oxidations of butane, butene and butadiene over PVO catalysts

It is very probable that butane is initially dehydrogenated to butene and then oxidized to maleic anhydride through butadiene and furan by a scheme shown in Eq. 1 (ref. 4), where CO_x (CO and CO_2) is possibly produced from all reactants. In order to determine in which step of Eq. 1 the $(VO)_2P_2O_7$ catalyst (**P**)is effective, the rates and selectivities of the oxidations of butane, butene, and butadiene were compared over the three catalysts.

$$\begin{array}{ccccccccc} \text{Butane} & \xrightarrow{1} & \text{Butene} & \xrightarrow{2} & \text{Butadiene} & \xrightarrow{3} & \text{Furan} & \xrightarrow{4} & \text{Maleic anhydride} \\ \searrow & & \searrow & & \searrow & & \searrow & & \searrow \\ CO_x & & CO_x & & CO_x & & CO_x & & CO_x \end{array} \qquad (1)$$

Figures 1 and 2 respectively show the selectivities of each product obtained for oxidations of butane and 1-butene as a function of conversion. The products in butane oxidation were only MA and CO_x for all catalysts. The selectivity to MA decreased as the conversion of butane increased for all catalysts. It is clear from Fig. 1 that $(VO)_2P_2O_7$ is prominent for the selective formation of MA from butane (the selectivity was comparable with those of the best data in patents). On the other hand, the selectivity was low over α- (**A**) and β-$VOPO_4$ (**B**).

In the case of the oxidation of 1-butene (Fig. 2), the three catalysts showed no significant differences in the selectivity-conversion correlation of the formations of MA, butadiene, and furan. This is in marked contrast to the result of the butane oxidation shown in Fig. 1. Butadiene was the main product at low conversions, and MA increased as the conversion increased. This trend is consistent with the reaction sequence of Eq. 1.

Table 1 shows some results of oxidations of butane (440 °C), 1-butene, and butadiene (440 and 360 °C)for the three catalysts. For each catalyst, butene and butadiene reacted 10 - 20 times faster than butane. This indicates that the step 1 of Eq. 1 is the rate-determining step for the oxidation of butane. As for the catalytic activity which was normalized to the surface area, **A** and **B** were two to four times more active than **P** for each of the three hydrocarbons.

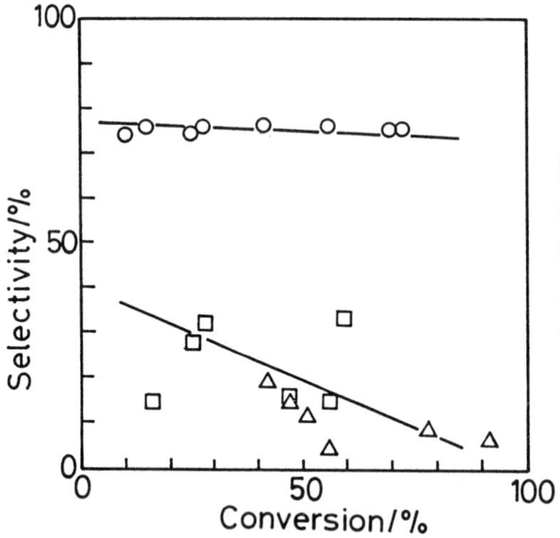

Fig. 1. The selectivity to maleic anhydride as a function of the conversion of butane oxidation at 440 °C.
○: $(VO)_2P_2O_7$, △: α-$VOPO_4$, □: β-$VOPO_4$

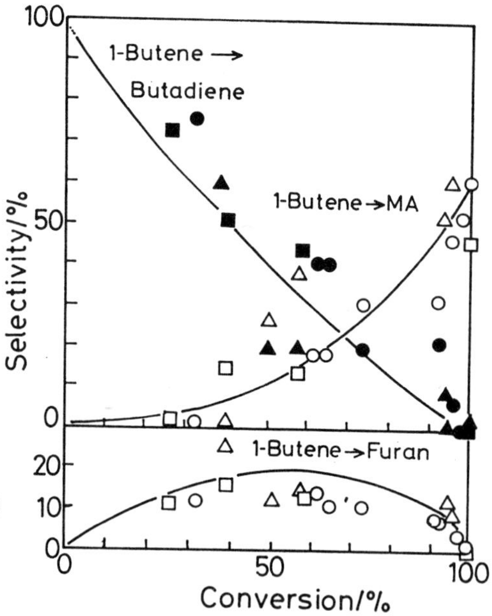

Fig. 2. The selectivity as a function of the conversion of 1-butene oxidation at 360 - 440 °C.
○●: $(VO)_2P_2O_7$, △▲: α-$VOPO_4$, □■: β-$VOPO_4$

TABLE 1. Catalytic Activities and Selectivities in Oxidations of Butane, 1-Butene, and Butadiene over V-P Oxide Catalysts

Catalyst	Act.[a]	Selectivity[b]/%				Act.[a]	Selectivity[b]			
		MA	HC	CO_x	(Conv./%)		MA	HC	CO_x	(Conv./%)
Butane 440 °C										
$(VO)_2P_2O_7$	1	76	0	24	(52)					
α-$VOPO_4$	3	19	0	81	(42)					
β-$VOPO_4$	4	32	0	68	(28)					
1-Butene 440 °C						360 °C				
$(VO)_2P_2O_7$	17	51	19	30	(91)	2	15	69	16	(20)
α-$VOPO_4$	65	59	24	17	(85)	7	28	63	9	(21)
β-$VOPO_4$	35	65	24	11	(87)	3	26	64	10	(26)
Butadiene 440 °C						360 °C				
$(VO)_2P_2O_7$	16	22	35	43	(36)	2	37	35	28	(21)
α-$VOPO_4$	-	-	-	-	-	7	22	41	37	(25)
β-$VOPO_4$	46	29	49	22	(17)	8	19	47	34	(13)

a) Rate of oxidation (averaged values); 10^{-4} $mol \cdot g^{-1} \cdot h^{-1}$, b) Selectivities at the conversions given in parentheses; on the basis of C_4. MA; maleic anhydride, HC; product hydrocarbons other than MA. The reaction was carried out at 1 atm (C_4-hydrocarbon; 1.5 vol%, O_2; 17 vol%).

As for the selectivity, if one considers the selectivity vs. conversion correlations shown in Figs. 1 and 2, it is obvious again in Table 1 that **P** is much superior in the oxidation of butane, but it is comparable with the others in the oxidations of butene and butadiene.

On the basis of these results, it may be concluded that **P** is superior in the selectivity of the step 1 of Eq. 1, that is, the dehydrogenation of butane to butene. Probably, the C-H dissociation of butane proceeds selectively on **P**, while it is not so on the others.

The superiority of **P** must be intrinsic of its structure, since no changes were observed by XRD and XPS for three catalysts after use in the oxidation of butane of the three catalysts (ref. 4). Even after use for butene oxidation, **P** showed no changes either in the structure or in the oxidation state, indicating that the V(IV) ions of **P** in the surface layer are stable and most of the vanadium ions are present in the oxidation state of +4 under the reaction conditions. As reported in the literature (ref. 6), the decreases in the oxidation state of the surface layers were observed by XPS in the cases of butene and butadiene oxidation; from V(V) (518.5 eV) to mostly V(IV) (517.6 eV) for **B** and from V(V) (518.6 eV) slightly to V(IV) (broader peak at 518.3 eV) for **A**. Therefore, the oxidation state of V of **A** and **B** during butene and butadiene oxidation was a little different from that with butane oxidation. So, there is a possibility that butene formed in the oxidation of butane over **A** and **B** was overoxidized due to their higher oxidation state (+5), and that this was the reason of the low selectivity. However, this possibility is not likely, since the three catalysts having different oxidation states showed a very similar selectivity for the oxidations of butene and butadiene.

Effect of the morphology of the $(VO)_2P_2O_7$ crystallites was examined by using catalysts **P** and **P'**. SEM and XRD measurements showed that the catalyst **P** had plate-like morphology, of which thickness in the [020] direction was about 40 nm, and the catalyst **P'** had rose-like shape with about 10 nm thickness. The catalytic activity of the catalyst **P'** was higher due to its higher surface area. However, the catalyst **P** gave higher yields of MA at high conversions (70% selectivity at 80% conversion) than the catalyst **P'**(48% selectivity at 80% conversion), while the difference in the selectivity between the two catalysts was small at low conversions.

Isotopic exchange

(i) $CO-^{18}O_2$-PVO catalysts. No changes in XRD and IR were observed after the reactions for catalysts **P**, **A**, and **B**. However, oxidation of the surface layer of catalyst **P** was demonstrated by the color change, XPS, formation of water-soluble phase (X_1 or X_2 after recrystallization (cf. ref. 2)) and the oxygen uptake during the reaction. Due to this, the rate which was expressed by rate $= k\ p(CO)^{0.8-1}p(O_2)^{0.1}$ increased by repeated runs in the case of catalyst **P**. The rates of these three catalysts were comparable.

An example of the results of tracer experiments is shown in Fig. 3 (catalyst **P** at 350 °C). The ^{18}O content of CO_2 decreased at first. The isotopic distribution in CO_2 approached to the statistical one, the fraction of $C^{16}O^{18}O$ being initially greater than the statistical distribution. No exchange of isotopic oxygen was observed for CO and O_2. The diffusion of ^{18}O into catalyst took place rather slowly (calculated from the ^{18}O balance). The hypothetical surface layers in which ^{18}O % was isotopically in equilibrium with CO_2 in the gas phase (cf. ref. 7) slowly increased from 0.5 layers at 15 min to 3.7 layers at 240 min at 350 °C and from 2.1 (30 min) to 4.5 (120 min) at 380 °C. These results indicate that the CO oxidation proceeds at least in the initial stage by picking up either adsorbed ^{18}O or oxide ion on the surface that was oxidized by $^{18}O_2$. Since the oxygen exchange of the surface was limited to the surface monolayer in the case of CO_2-PVO catalyst reaction (see below), the diffusion of ^{18}O was likely accelerated by the redox cycles of the surface, which took place during the $CO-O_2$ reaction. Similarly the diffusion of ^{18}O into catalyst bulk was slow for catalysts **A** and **B**.

(ii) $C^{18}O_2$-catalyst **P**. After 1 h, $C^{18}O_2$ was isotopically in equilibrium with the oxide ions in the 0.6 surface layers at 350 °C and 1.0 layer at 400 °C. The rate of exchange was several times faster than the rate of CO oxidation.

(iii) Butane-$^{18}O_2$-PVO catalysts. No changes in XRD, XPS, IR, and color were detected after reactions for catalysts **P**, **A**, and **B**. Both of the carbon and oxygen balances were excellent (95 - 100%). Figs. 4a and 4b show the changes of the composition in the gas phase and the ^{18}O contents of each molecule in the reaction over catalyst **P** (the data were obtained by separate runs). The selectivity to MA and conversion were very similar to those obtained by flow experiments. For example, they were 82% and 74%, respectively, after 60 min. All the products, as shown in Fig. 4b, exhibited similar ^{18}O contents and no exchange was observed for O_2. From the above $C^{18}O_2$-**P** experiment, CO_2 may be considered to be isotopically in equilibrium with the top surface. The hypothetical surface layer in equilibrium with CO_2, calculated from the ^{18}O content of CO_2 and the ^{18}O balance, stayed almost constant after 30 min, that is, 1.3 layers at 350 °C and 2.8 layers at 380 °C (Fig. 5). Oxygen exchange only in the surface layer has been indicated by Pepera et al. based on the isotopic analysis of CO_2 (ref. 8). The analysis of the whole products in the present study under the conditions that gave high selectivity to MA confirmed the indication. Results obtained for catalysts **A** and **B** were very similar to those for catalyst **P**.

The ^{18}O distributions in CO_2 were close to the statistical one. The difference in the ^{18}O content between O_a and O_b of MA which was determined by the ^{18}O content of CO and CO_2 formed in mass spectrometer by Eq. 2 was small.

It was observed that non-catalytic reaction of butane over catalyst **P** at 300 °C formed MA very selectively.

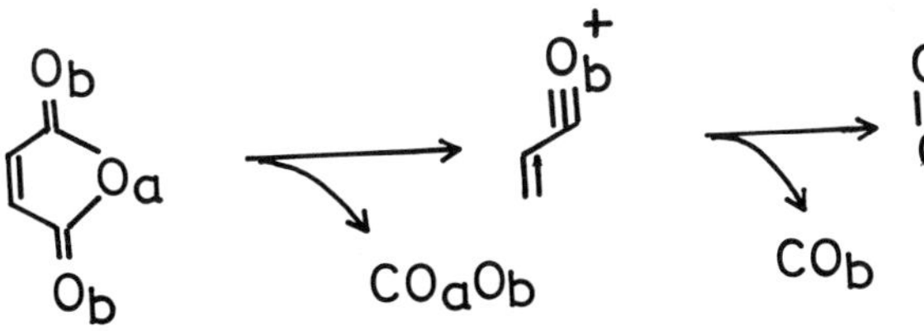

(2)

(iv) MA-$H_2{}^{18}O$-catalyst **P** and MA-catalyst **P**. ^{18}O exchange between gas phase and catalyst was observed for both reactions. Prior to the latter exchange reaction, the catalyst was treated with $C^{18}O_2$ at 440 °C (1.7 surface layers were equilibrated). The rate was in the order of

oxidation of butane > MA-H_2O-**P** exchange > MA-**P** exchange.

(v) Furan (20 Torr)-$^{18}O_2$-catalyst **P** (at 380 °C). Although the carbon balance was rather low (55 - 88%), it was found that no isotopic exchange took place for furan and O_2 and that the ^{18}O content of MA formed was about 60%.

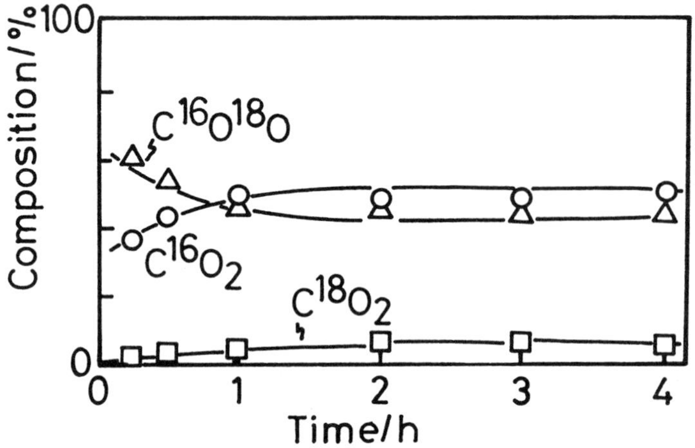

Fig. 3. The isotopic distribution in CO_2 formed by CO + $^{18}O_2$ reaction over $(VO)_2P_2O_7$ (catalyst **P**) at 350 °C.

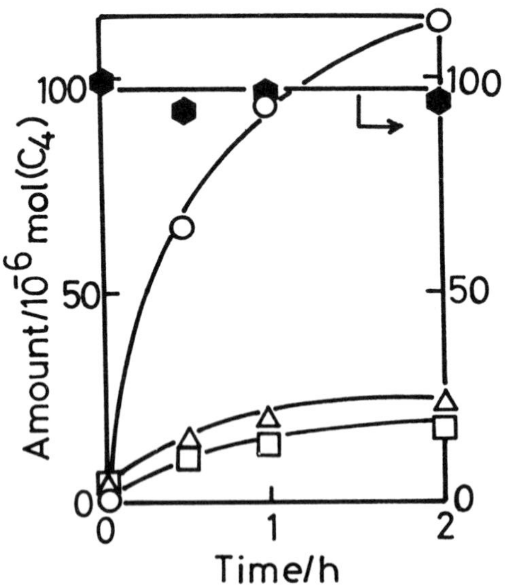

Fig. 4a. Changes of the composition in the gas phase in butane-$^{18}O_2$ reaction over $(VO)_2P_2O_7$ (**P**) at 380 °C.
○:MA, △:CO, □:CO_2

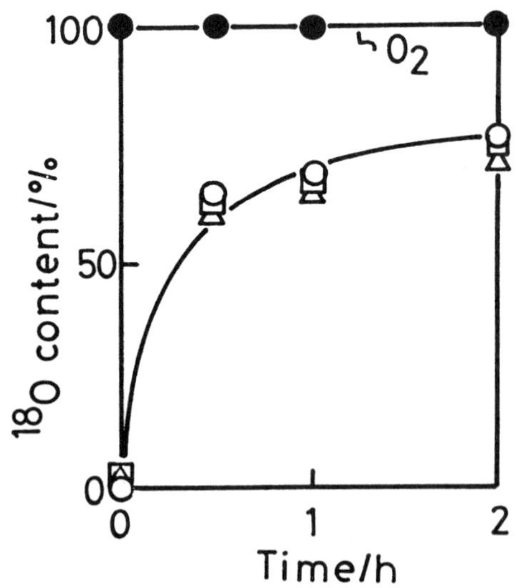

Fig. 4b. Changes in ^{18}O contents of each molecule in butane-$^{18}O_2$ reaction over $(VO)_2P_2O_7$ (**P**) at 380 °C. Marks are the same as in Fig. 4a.

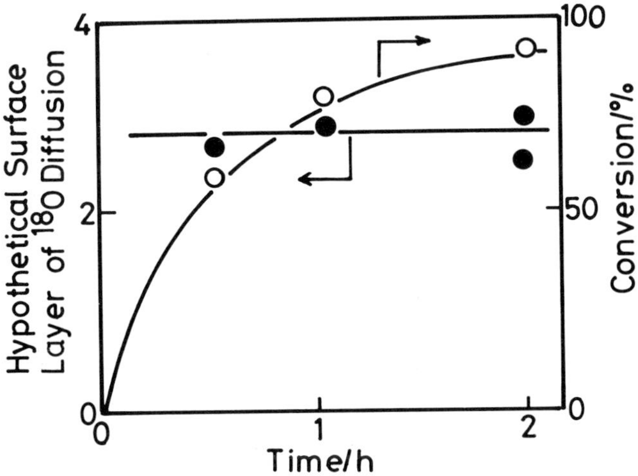

Fig. 5. Changes of the hypothetical surface layer isotopically in equilibrium with CO_2 in the butane-$^{18}O_2$ over $(VO)_2P_2O_7$ (**P**) at 380 °C.

CONCLUSION

(1) $(VO)_2P_2O_7$ was much more selective than α- and β-$VOPO_4$ for MA formation from butane. This superiority of $(VO)_2P_2O_7$ is due to its high selectivity in the dehydrogenation of butane to butene.

(2) Isotopic study using $^{18}O_2$ for butane oxidation over $(VO)_2P_2O_7$ revealed that reduction-oxidation involved in the catalysis occurred only very near the surface.

(3) Among various $(VO)_2P_2O_7$ having different morphology, $(VO)_2P_2O_7$ having plate-like shape showed higher yield of MA at high conversions.

This study was supported by the Grant-in-Aid for Scientific Research on Priority Areas from the Ministry of Education, Science and Culture.

REFERENCES

1 G. Centi and F. Trifiro, Chem. Rev., 88 (1988) 55.
2 T. Shimoda, T. Okuhara and M. Misono, Bull. Chem. Soc. Jpn., 58 (1985) 2163.
3 T.P. Moser and G.L. Schrader, J. Catal., 92 (1985) 216.
4 K. Miyamoto, T. Nitadori, N. Mizuno, T. Okuhara and M. Misono, Chem. Lett., 1988, 303.
5 J.H. Scofield, J. Electron Spectr., 8 (1976) 129.
6 T.P. Moser and G.L. Schrader, J. Catal., 104 (1987) 99.
7 T. Nitadori and M. Misono, J. Catal., 93 (1985) 459.
8 M.A. Pepera, J.L. Callahan, M.J. Desmond, E.C. Milberger, R.P. Blum and N.J. Bremer, J. Am. Chem. Soc., 107 (1985) 4883.

B. Delmon (University of catholique de Louvain): In all selective catalytic oxidations, the supply of oxygen tends to be rate-limiting. The steady state of the near surface layers thus depends on the rate at which the hydrocarbon reacts. Butene reacts with a much higher rate than butane. This implies that the steady state of the surface corresponds to deeper reduction. In order to be really compared, all experiments (including selectivity measurements, mechanistic studies) should correspond to identical surface states (for example by using different O_2/hydrocarbon ratios). I notice, in Table 1, that the most selective for butane is $(VO)_2P_2O_7$ (more reduced), for butene β-$VOPO_4$ (more oxidized). In the time of your experiment, surface slowly tends to the steady rate (this steady state takes a long time to establish in C_4 oxidation). Couldn't we speculate that, with butene, starting from the more oxidized surface of β-$VOPO_4$, that becomes progressively reduced, you reach the same surface state as when starting with $(VO)_2P_2O_7$ in the case of butene? The steady state oxid-reduction level of the catalyst would be the most important factor, rather than other surface characteristics.

M. MISONO (The University of Tokyo, Japan): We also think that the oxidation state of the surface layer at the working conditions is important, and that it depends on the reaction conditions. Nevertheless, our conclusion is that $(VO)_2P_2O_7$ is superior for the selectivity of the first C-H dissociation of butane and there is little difference in the selectivity for the following steps among the catalysts. Your question (as well as comments) can be answered based on the XPS results. No changes in the oxidation state of V was observed by XPS for all catalysts in butane oxidation (at least, up to about 70% conversion). Furthermore, even after the use of butene oxidation, the oxidation state of $(VO)_2P_2O_7$ (V^{4+}) was stable. Thus the results for $(VO)_2P_2O_7$, that is, high selectivivity for both butane and butene oxidations, are considered to be its intrinsic characteristics.

On the other hand, in the case of butene oxidation, V^{5+} of the surface layer of β-$VOPO_4$ changed to V^{4+}, while α-$VOPO_4$ was rather stable and $(VO)_2P_2O_7$ very stable (thus, in reply to your question, it can be said that the oxidations states were different among the three catalysts). In spite of the different oxidation states, the selectivities for the butene oxidation over these three catalysts were similar. Therefore, we can conclude that butane is selectively oxidized only on the surface of $(VO)_2P_2O_7$ and the oxidation of butene takes place with a similar selectivity on these catalysts, regardless of the difference in the oxidation state.

S. L. Kiperman (N. D. Zelinsky Institute of Organic Chemistry, Academic of Science of the USSR): Some short comments on this interesting work. The authors have proposed parallel-consecutive scheme of the reaction. Although they did not study the reaction kinetics, the data on the selectivity allow to examine this scheme. Let us see the changes of the selectivities with conversion. It is seen that the selectivity does not depend on conversion over one catalyst (as the results in Paper C5) but depends with conversion on the other catalysts. If the selectivity does not depend on conversion, it means only parallel scheme. It is possible only very slight contribution of consecutive scheme. In the case of decreasing selectivity the case of consective scheme is most probable. Besides if the selectivity does not depend on conversion, it means not only parallel scheme but in this case the kinetic equations of all these reactions of the process should be the same. This conclusion should be of interest to prove in kinetic investigations. But now is seen that the data on the selectivity contradict to the parallel-consecutive scheme.

M. MISONO (The University of Tokyo, Japan): Over $(VO)_2P_2O_7$, oxidations of butene and butadiene were about 20 times faster than that of butane (Table 1), and oxidation of maleic anhydride was slow. Therefore, in this case, the oxidation of butane is regarded as a parallel reaction from butane to maleic anhydride and CO_x. We are at present more interested in the reason why the $(VO)_2P_2O_7$ is more selective for the butane oxidation than the others and in how the reaction takes place on the surface of these catalysts than the detailed kinetics.

G. Centi and F. Trifiro' (Editors), *New Developments in Selective Oxidation*

PRODUCT DISTRIBUTION IN THE V-P-O CATALYST BED AND AN EFFECT OF REACTION PRODUCTS ON THE CATALYST IN C_4 OXIDATION

V.A.ZAZHIGALOV, V.M.BELOUSOV, A.I.PYATNITSKAYA, G.A.KOMASHKO, Yu.N.MERKUREVA, J.STOCH[1]

Institute of Physical Chemistry, Ukrainian Acad.Sci., 252028 Kiev, (USSR)

[1]Institute of Catalysis and Surface Chemistry, Polish Acad.Sci., 30-239 Kraków, Poland

SUMMARY

Changes in the nature of V-P-Me-O contacts during their activation and work in fixed-bed reactors were studied. Variation in a products composition along the bed were reveled for the catalysts and their precursors. The character of V-P-Me-O catlyst modification during its activation by the reacting mixtures has been shown.

INTRODUCTION

Oxidation of butane to maleic anhydride in an industrial scale can be also realized on installations which are running with benzene, if nothing but a change of the raw material is done. On the other hand, there are many parameters (eg. reagents and products concentrations, temperature profiles etc.) characterizing the work of a contact in fix-bed reactors. Literature data concerning n-butane oxidation in such reactors, are not available. The contacts in the form of $VOHPO_4 \cdot 0.5H_2O$ loaded into reactors, are transformed to an active component $(VO)_2P_2O_7$ during the subsequent activation step. Only few paper [1,2] concern with a dependence of an active phase formation on parameters of a pretreatment or catalytic reaction. The paper concerns these problems.

METHODS

Promoted V-P-O catalysts were prepared from a butanol bath [3]. They were finished in the form of pellets (ϕ=4 mm and l=6-8 mm). The promoter (Me) was added in the form of a phosphate. A precursor of the catalyst was containing $VOHPO_4 \cdot 0.5H_2O + MePO_4$ while the activated catalyst contained $(VO)_2P_2O_7 + MePO_4$. The ratios P/V = 1.08, Me/V < 0.1. The surface area of contacts before and after work are 17-19 m^2/g. The reaction of n-butane oxidation in an air was studied in: (A) an isotermic reactor with the sampling of the gaseous reacting mixture at every 0.25 m of its length. The total bed length was 1.5 m with the temperature difference at the length ΔT = 7-8 K; (B) a flow-type Temkin-Kulkova reactor, 0.2 m length; (C) a differential-flow reactor (4 pellets); (D) an impulse micro-reactor [4].

The X-ray diffractometer DRON-3 was used for a phase analysis. Contents of $(VO)_2P_2O_7$ and β-$VOPO_4$ were determined from the intensity ratio of 0.387 and 0.341 nm lines for these phases and that one of 0.2087 line for α-Al_2O_3 standard [5]. IR spectra were obtained with the FTIR spectrometer Bruker IVS-113 (samples pressed with KBr (1:15), detector Ag-Cd-Fe, resolution 4cm^{-1}).

The surface stste was controlled with the ESCA-3 VG spectrometer using Al $K\alpha_{1,2}$ radiation. The XPS spectra were calibrated against the C 1s line (285.0 eV).

RESULTS AND DISCUSSION

The product distribution at the bed length

Results obtained for an isothermic reactor (A) with the precursor $VOHPO_4$ $0.5H_2O$ and the activated catalyst $(VO)_2P_2O_7$ are shown in Fig. 1a and 1b. In the case of activated catalyst we found (Fig. 1a) that concentrations of products were increasing along the length of the bed, while the CO content exceeded that one of CO_2 up to the end of the bed (Table 1). We have concluded that oxidation of C_4H_{10} to $C_4H_2O_3$ on $(VO)_2P_2O_7$ is parallel with its oxidation to CO. Inside the bed CO is oxidize to CO_2 and near the bed end we found their concentrations equal.

TABLE 1

The ratio CO_2/CO along the length of the bed catalysts

	$(VO)_2P_2O_7$						$VOHPO_4 \cdot 0.5H_2O$					
Bed length (cm)	25	50	75	100	125	150	25	50	75	100	125	150
CO_2/CO	0	0.2	0.4	0.6	0.9	0.95	1.8	2.0	1.8	1.9	1.9	2.0

Behaviour of the precursor was different (Fig. 1b). In this case a new specimen of catalyst have been used for each experi-

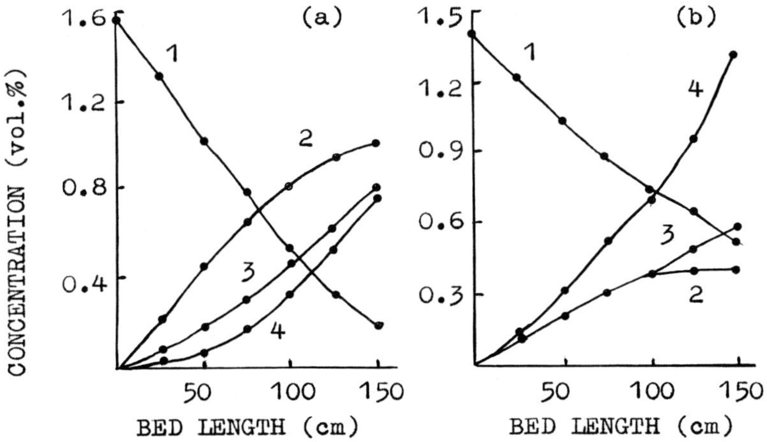

Fig. 1. The concentration profiles of n-butane (1), $C_4H_2O_3$ (2), CO (3), and CO_2 (4) on the catalyst activated (a) and containing precursor (b). T = 688 K, W = 5.1 l/h·g cat.

ment (T, W/F). Duration of experiment is 1 h. It was less active comparing with $(VO)_2P_2O_7$. The $C_4H_2O_3$ content was increasing along the length of bed but near the end it became constant or even little decreasing while the CO_2 content increased considerable. The CO_2 concentration exceeding that one of CO along the length of the bed. Thus, we concluded that $VOHPO_4$ $0.5H_2O$ is active in combustion of maleic anhydride. It necessary to mark of high concentration of the products combustion in the end of the bed length.

<u>The catalyst composition along the bed length</u>

After 400 h work (T = 678 K) of both catalysts the beds were pulled out in 6 parts and then, they were analyzed by X-ray diffraction (XRD) and FTIR spectroscopy.

We found that morphology of "used" and "as received" form of activated catalysts is the same and the surface of all 6 samples of the "used" catalyst contains only $(VO)_2P_2O_7$. BE V $2p_{3/2}$ are 517.6 eV.

In the case of the precursor we found changes of composition along the length of the bed during the reaction course. The β-$VOPO_4$ formation at the end of the bed (after 100 cm bed length - XRD and after 75 cm - FTIR) can be caused by the decrease of the n-butane concentration and high oxygen content or the influence of products. A high oxidation state of vanadia of the surface is evidenced by the BE (V $2p_{3/2}$) value 518.4 eV. The compound β-$VOPO_4$ is inactive in partial oxidation of C_4H_{10} [6,7] while it is active in the total oxydation of $C_4H_2O_3$ to CO and CO_2 [8, 9]. Thus, its presence at the end of the bed is unwanted, because it lowers the quality of catalyst.

It is known [6,10-12] that $VOHPO_4 \cdot 0.5H_2O$ transforms to $(VO)_2P_2O_7$ (T = 653 K) in flowing inert (reaction mixtures) or to β-$VOPO_4$ (T = 673 K) in flowing O_2 (air). The transformation $(VO)_2P_2O_7 \rightarrow VOPO_4$ is possible when $(VO)_2P_2O_7$ is heated in air at T > 873 K [11,12].

In this case observed (Fig. 1b) sufficiently high concentration of n-butane (~0.6 vol.%, $O_2/C_4H_{10} < 40$) at end of the bed (100-150 cm) but concentration of the products combustion is also high. Therefore, the causes of the formation $VOPO_4$ in the stage activation it is not clear.

The influence of n-butane conversion on the catalyst composition

In order to study conditions for the formation of inactive $VOPO_4$ phase from $VOHPO_4 \cdot 0.5H_2O$ during the catalysis, experiments concerned with an effect of n-butane conversion when the reaction products are absent has been performed. Mixtures of different amount of butane with an air (Table 2) was passing at 693 K (τ_c= 0.1 s, X_{C4} = 0) during 48 and 120 h through the reactor (C) filled with the precursor.

TABLE 2

Dependence of the catalyst composition on the n-butane conversion

No	Reaction mixture		Conversion 1)	Relative XRD-line intensities				BE V $2p_{3/2}$
				After 48 h		After 120 h		
	C_4H_{10} in air,	O_2/C_4H_{10}	%	$(VO)_2P_2O_7$ 0.387 nm	$VOPO_4$ 0.341 nm	$(VO)_2P_2O_7$ 0.387 nm	$VOPO_4$ 0.341 nm	surface pellet, eV
1.	1.55	13	0	100	0	100	0	517.8
2.	0.49	43	68	100	0	100	0	517.7
3.	0.32	65	79	100	0	100	0	517.7
4.	0.16	130	90	93	0	85	10	517.8
5.	air		-	0	100	-	-	518.7

1) conventional conversion of butane relative to the mixture (1).

Results (Table 2) show that $VOPO_4$ is not produced when conversion of n-butane is less than 80 % (mixtures 1-3), while it is fairly well formed in air (mixture 5). The low rate of the formation $VOPO_4$ is observed when the concentration of n-butane in air is very little (mixture 4). Or course, the $VOPO_4$ formation from $VOHPO_4 \cdot 0.5H_2O$ during the catalytic oxidation of n-butane also can be caused by an action of the reaction products.

The catalyst activation with different mixtures

The reactor (B) was filled with pellets of the precursor $VOHPO_4 \cdot 0.5H_2O$ and at 693 K mixtures with different compositions (Table 3) were passing (τ_c = 1 s) during 6 and 24 h. After such pretreatment, the catalyst composition (XRD) and its catalytic activity were studied.

The XRD analysis has revealed (Table 3), that 5 types contact compositions can be distinguished:

1. $VOPO_4$ with promoter in the form of a phosphate,
2. promoter in the form of Me/Me_xO_y and X-ray amorphous V-P-O,
3. $(VO)_2P_2O_7$ with $VOPO_4$ and promoter in the form of a phosphate,
4. $(VO)_2P_2O_7$ with promoter in form of a phosphate and Me/Me_xO_y,
5. $(VO)_2P_2O_7$ with promoter in the form of a phosphate.

A catalytic activity of these catalytic systems was measured in reactors of (D) and (B) type. A lower temperature of these tests as compared to that of activation enables one to keep unchanged the state of the surface resulted from pretreatment. In the case of the impulse reactor (D), during every of 10 pulses of reaction mixture (T = 673 K) the rate of butane oxidation and selectivity to $C_4H_2O_3$ changed a little. These facts allowed to begin tests in the flow-type reactor. Results concerning the flow-type reactor (B) (after 10 min work, T = 673 K, τ_c = 2 s) are presented in Fig. 2. The catalyst composition after these tests (duration 30 min) remain unchanged. It is seen, that properties of the systems are different and following:

1 and 2 types (Fig. 2a). Low activity and selectivity caused by the absence of the active pyrophosphate (reduction mixtures 20, 20', 4' [1]) and air - 5, 5');

3 type (Fig. 2b). The selectivity decreases when a relative content of $VOPO_4$ increases. We found that activity and selecti-

---

[1] No' - after 24 h activation

TABLE 3

The catalyst composition after the 6 and 24 hours activation

No	Mixture content	Phase content, % [1]					
		After 6 h			After 24 h		
		A	B	C	A	B	C
1	C_4H_{10}(2.3%)+O_2(8.2%)+He [2]	79	0	0	95	0	0
2	C_4H_{10} (1.7%) + air [3]	60	0	0	78	0	0
3	C_4H_{10} (1.7%) + air [4]	40	9	0	45	11	0
4	C_4H_{10} (100%)	18	0	37	0	0	90
5	Air	0	67	0	0	100	0
6	Ar	47	13	0	70	23	0
7	N_2	48	32	0	55	39	0
8	CO_2 (2%) + He	66	0	0	60	0	8
9	CO (45%) + He	58	0	31	-	-	-
10	CO (2%) + He	44	0	5	38	0	12
11	H_2O (7%) + He	60	0	0	90	0	3
12	$C_4H_2O_3$(0.8%)+O (17%)+He	51	20	0	76	15	0
13	$C_4H_2O_3$ (0.8%) + He	78	0	22	70	0	30
14	C_6H_6 (2%) + air	59	0	0	64	0	0
15	2-C_4H_8 (1.1%) + air	37	3	0	50	16	0
16	1-C_4H_8 (1.1%) + air	57	13	0	68	26	0
17	R-OH (2%) + air	73	0	0	100	0	0
18	C_3H_6(2%) + O_2(10%) +He	44	0	0	67	0	0
19	SO_2 (1.5%) + He	37	0	0	52	0	0
20	H_2 (100%)	0	0	90	0	0	100 [5]

[1] A - $(VO)_2P_2O_7$, B - $VOPO_4$, C - Me/Me_xO_y, moreover catalysts contain $MePO_4$ and X-ray amorphous V-P-O

[2,3,4] conversion of C_4H_{10} = 65, 60 and 75-85 %, respectively

[5] the sample is standard for determination of Me/Me_xO_y containing in other catalysts

vity of catalysts this system do not depend on the contents of $(VO)_2P_2O_7$ (inert content O_2 - 6, 6', 7, 7', reaction mixtures at high conversion - 3, 3', 12, 12', 15, 15', 16, 16');

4 type (Fig. 2c). A relation between catalytic activity or selectivity and the $(VO)_2P_2O_7$ content is undefined, though as the rule, they decrease when the promoter content (in metal-

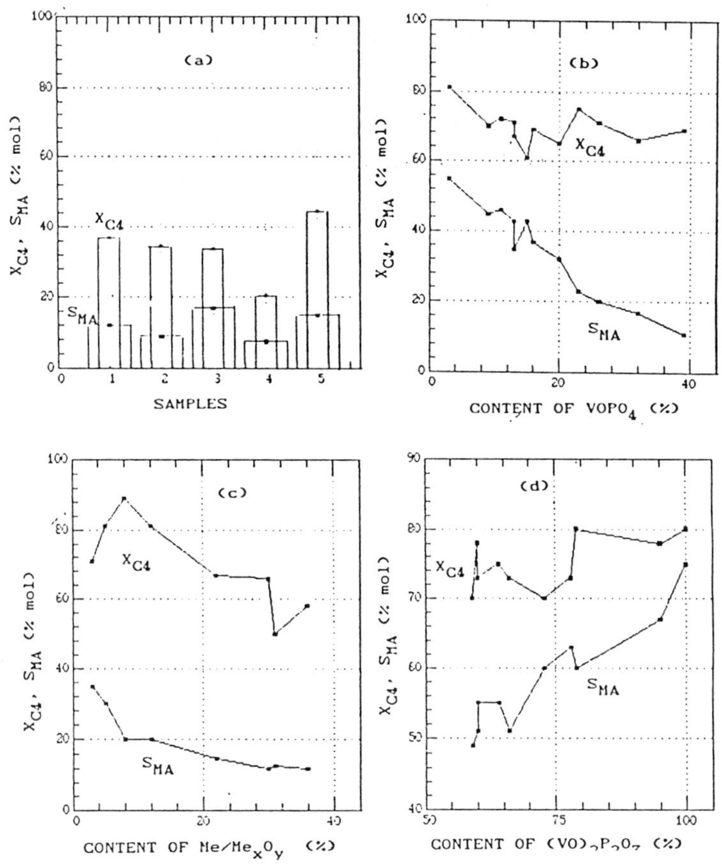

Fig.2. n-Butane conversion and selectivity to maleic anhydride on the catalyst activated with different mixtures (Tab.3) for the phase systems (a)- 1,2; (b)- 3; (c)- 4; (d)- 5. Numbers of the samples in the figure (a) corresponds to the following numbers in Tab.3: 1,2 - 5, 5'; 3,4 - 20, 20'; 5 - 4.

lic or oxide form) increases (mixture content products - 8', 9, 10, 10', 11', 13, 13', 4).

5 type (Fig. 2d). The catalytic activity and selectivity are high. There is a weak dependence between the catalytic activity and the $(VO)_2P_2O_7$ content, while the selectivity increases parallel to the increase of the $(VO)_2P_2O_7$ content (reaction mixtu-

res at conversion $< 65\%$ - 1, 1', 2, 2', 8, 11, 14, 14', 17, 17', 18, 18', 19, 19').

The results obtained show that under long activation by butane oxidation products ($C_4H_2O_3$, CO, CO_2, and H_2O) a reduction of promoter to metal or its oxide takes place. Oxidation of $C_4H_2O_3$ results in formation $VOPO_4$.

Taking into consideration data on products distribution under butane oxidation in the presence precursor (Fig. 1b) we can conclude, that products do act harmfully on the catalyst during the activation period. It one can recomend: catalyst activation is performed under reaction mixture; hydrocarbon conversion has not to exceed 65%.

REFERENCES

1. J.S. Buchanan, J. Apostolakis and S. Sundaresen, Appl. Catal., 19 (1985) 65-75.
2. H. Bosch, A.A. Bruggink and J.R.H. Ross, Appl. Catal., 31 (1987) 323-337.
3. V.A. Zazhigalov, V.M. Belousov, G.A. Komashko, A.I. Pyatnitskaya, T.A. Kriger, O.Ya. Polotnuk and Yu.I. Sorokin, Zhurn. Prikl. Khim., 60 (1987) 865-869.
4. V.A. Zazhigalov, Yu.P. Zaytsev, V.M. Belousov, N. Wyustnek and H.Wolf, React. Kinet. Catal. Lett., 24 (1984) 375-378
5. R.A. Schneider, U.S.Pat. 3,864.280 Chevron Res. Comp., (1975)
6. G.Centi, F.Trifiro, J.R. Ebner, and V.M. Franchetti, Chem. Rev., 88 (1988) 55-80.
7. T. Shimoda, T. Okuhara and M. Misono, Bull. Chem. Soc. Japan, 58 (1985) 2163-2171.
8. V.A. Zazhigalov and Yu.P. Zaytsev, Kataliz i Katalizatory, 20 (1982) 51-54.
9. T.P. Moser, R.W. Wenig and G.L. Schrader, Appl. Catal., 34 (1987) 39-48.
10. G. Busca, F. Cavani, G. Centi and F. Trifiro, J. Catal., 99 (1986) 400-414.
11. E. Bordes, P. Courtine and J.W. Johnson, J. Solid. State Chem., 55 (1984) 270-279.
12. J.W. Johnson, D.C. Johnston and A.J. Jacobson, in B.Delmon, P.Grange (Eds.), Prep.Catal.IV, Elsev., Amsterd.,1987,181-189
13. V.A.Zazhigalov, V.M.Belousov, G.A.Komashko, J.Stoch, Proc. 9-th Intern. Congr. on Catal., Calgary, 1988, 1546-1553.

G. Centi and F. Trifiro' (Editors), *New Developments in Selective Oxidation*

A DYNAMIC MODEL OF THE OXIDATION OF n-BUTANE AND 1-BUTENE ON VARIOUS CRYSTALLINE FACES OF $(VO)_2P_2O_7$

J. ZIÓŁKOWSKI[1], E. BORDES[2] and P. COURTINE[2]

[1]Institute of Catalysis and Surface Chemistry, Polish Academy of Sciences, ul. Niezapominajek, 30-239 Cracow, Poland.
[2]Département de Génie Chimique, Université de Technologie de Compiègne, B.P. 649, 60206 Compiègne Cedex, France.

ABSTRACT

The structure and performance of various crystal faces of $(VO)_2P_2O_7$ in the mild oxidation of n-butane and butene in maleic anhydride are analyzed using the crystallochemical model of active sites (CMAS). The model provides geometric and energetic maps of each face and allows to calculate the energetics of the elementary steps involving adsorption, desorption, as well as movements of oxygen, hydrogen and water along the surface. Surface energy calculations and Curie-Wulff plot show that the crystals of $(VO)_2P_2O_7$ should be bordered by (100), (021) and (001). Application of CMAS to these faces accounts for the favourable specificity of (100) $(VO)_2P_2O_7$ in oxidation of butane.

INTRODUCTION

The concept of catalytic anisotropy (ref. 1) has been introduced to account for the catalytic specificity exhibited by various crystal faces of a given compound with respect to the particular reactions they promote. Many examples in the field of mild oxidation and oxidative dehydrogenation catalysis have recently been provided and theoretical models proposed (refs. 1-9). Practical consequences of this concept are also important, such as the influence of a support when a synergistic effect appears between it and the active phase (ref. 6), or the choice of the method of preparation which influences the morphology of the catalyst. The last case is well illustrated by V-P-O catalysts, the phase $(VO)_2P_2O_7$ being active in oxidation of n-butane in maleic anhydride (MA) and present during the same reaction starting from n-butenes. In the case of butane, the best activity and selectivity are found only when the precursor of the catalyst is $VOHPO_4.0.5H_2O$. This is related to the fact that layered crystals of $(VO)_2P_2O_7$ exhibiting predominantly (100) faces are formed by the topotactic dehydration of the precursor (refs. 10, 11). The main differences between the two reactions lie in the kinetics (direct one-step and rake mechanism for butane and butene, respectively) and the mean oxidation state of vanadium necessitated to activate the molecule (V^{4+} and V^{5+} for butane and butene, respectively). In particular, different surface active species or "clusters" should be responsible for each considered reaction.

According to the well-known Mars and Van Krevelen's mechanism (ref. 12), the mild oxidation of an hydrocarbon -(CH)- proceeds on an oxidized form KO of the catalyst which is consecutively reduced (K). Several steps must be considered, which are (i), the adsorption of -(CH)-on KO, (ii), abstraction of hydrogen and insertion of lattice oxygen, (iii), desorption of water and of the product -(C)O-, and (iv), reoxidation of K by gaseous oxygen. The crystallochemical model of active sites

(CMAS) developed by one of us (refs. 1, 9) allows to calculate the energetics of these steps involving adsorption, desorption, as well as movements of oxygen, hydrogen and water along the surface. It provides also geometric and energetic maps of each exposed face according to the distribution of atoms and the values of the length, the strength and the energy of metal-oxygen bonds. On the other hand, active surface energies E_c (responsible for cleavage and various aspects of the surface reactivity, including catalysis) and Curie-Wulff plots allow to predict the equilibrium shape of crystals (refs. 13), which can be compared with the experimentally observed one. It will be shown in this paper how these models, applied to the case of $(VO)_2P_2O_7$, account for the abovementionned differences between the oxidation of n-butane and butene in MA, and for catalytic anisotropy of this active phase.

METHOD

The crystallochemical model of active sites (CMAS) is based on the bond-length – bond-strength – bond-energy concept (refs. 1, 9). The following formulas (1), (2) give respectively the strength s (vu, valence unit) and the atomization enthalpy e (kcal.mol.$^{-1}$), for M^{n+}-O bonds in an oxide crystal :

$$s = d\ z / (R-R_0) \quad (1)$$

where d, R_0 are empirical parameters related to the ionic radius (Å) of the considered cation, z is its valence (vu), and,

$$e = J\ s \quad (2)$$

where J is the standard molar atomization enthalpy of the simple oxide of the considered cation per valence unit (kcal.mol.$^{-1}$.vu^{-1}) .

The coefficients d, R_0, J, used in this paper have already been presented (refs. 14, 15) ; s and e values have been calculated for bulk $(VO)_2P_2O_7$ (ref. 16) according to cell parameters and coordinates (ref. 17). Bond energy data for C-H, C-C, C-O,... in organic molecules are taken from the literature as listed formerly (ref. 9).

The geometric map of the surface is constructed by cutting the crystal along the considered crystallographic plane in such a way as to break the weakest bonds, to retain the configuration of atoms and bond lengths characteristic for the bulk, and to conserve the stoichiometric composition. As a consequence, some of the surface atoms gain an energy undersaturation, conditioning further interactions with an adsorbing molecule. These atoms form an active multiplet site in the sense proposed by Balandin (ref. 18). The dynamic aspect of the model is involved in such a way that species abstracted from the molecule may move along the surface and interact with surface atoms localized in the vicinity. Therefore, a catalytic reaction is thought to be composed of a number of elementary steps in which usually one bond is broken and another formed, but concerted exo- and endothermal steps are also considered. The algebraic sum of the energies of the broken and formed bonds gives the net energy of the elementary step. The most probable pathway corresponds to the smallest endothermal barriers. The calculated specific enthalpies (in kcal.mol^{-1}) characterize the following processes : adsorption of hydrogen (first atomized) over undersaturated oxygens (e_H), transfer of hydrogen along the surface (q_{Htr}), formation and desorption of water (q_W). In the same way, the energy states e_{Ox} of the surface oxygen on various sites of the considered face, the dissociative adsorption of oxygen q_{Ox} on various pairs of adjacent cationic sites, and the energy of oxygen transfer q_{Otr} between possible vacant sites, are calculated. At last, the same kind of calculation is performed for elementary steps according to n-butane or butene transformation, and the energetically-easiest pathway chosen among them.

RESULTS AND DISCUSSION

1- Oxidation of n-butane on the (100) face of $(VO)_2P_2O_7$

The structure of (100) $(VO)_2P_2O_7$ (ref. 15) can be described by layers alternating along [100] (Fig. 1). The so-called "reduced"-layer contains all V and P cations and all oxygens belonging to

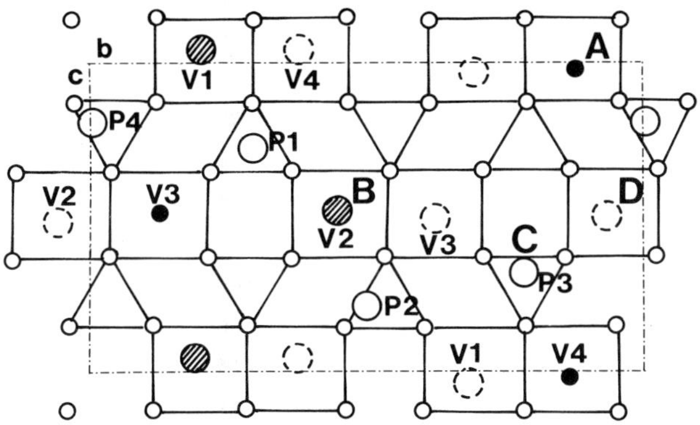

Fig. 1. (100) face of $(VO)_2P_2O_7$. Vanadium and phosphorus are labelled as in ref. 15. Symbols refer to oxygen sites : small open circles : inactive equatorial O ; small black : inactive very strongly bonded O=(V) (site A) ; large hatched : strongly saturated adsorbed O over V (B) ; large open : moderately saturated O adsorbed over P (C) ; large, dotted : weakly adsorbed O over V (D).

the equatorial plane of pairs of edge-sharing octahedra ; the "oxidized" and external layer contains only oxygen adsorbed over vanadium and phosphorus in the former layer, including vanadyl O=V oxygens. Adsorption sites are labelled A, B, C, D (Fig. 1, Table 1), according to their mean bonding energies. Very strongly bonded vanadyl oxygen over V_3 and V_4 and equatorial oxygens are considered as saturated and inactive sites, and excluded from the further considerations. The calculated heats of adsorption e_{Ox} of atomized oxygen (equal to the energies of the bonds broken when the surface is formed) are negative (exothermal), as well as the heats of dissociative adsorption q_{Ox} on adjacent pairs of sites V-V and V-P (Table 1). This means that the surface energy of (100) (ref. 13) is diminished when the surface is oxidized and that nearly all V and P sites are covered with oxygen (and/or hydroxyles). In the case where few oxygen vacancies exist, values of q_{Otr} (Table 1) show that oxygen, trying to move along the surface to replenish them, meets with important energy barriers .

The same kind of calculation is made for the energy states e_H of atomized hydrogens which, during the reaction, are abstracted from the adsorbed molecule and adsorbed over undersaturated

TABLE 1

Energy states $-e_{Ox}$ of surface oxygen, enthalpies of dissociative adsorption $-q_{Ox}$ of O_2 and of oxygen transfer q_{Otr} between adjacent sites on (100) $(VO)_2P_2O_7$ (all energies in kcal.mol^{-1}).

Oxygen sites	Mean $-e_{Ox}$*	Diss. Adsorp. of O_2 over	$-q_{Ox}$ range**	O-transfer from to	q_{Otr} range
A	200	A + D	120 - 150	A ---> D	150 - 170
B	120	B + D	30 - 50	B ---> D	80 - 90
C	87	A + C	160 - 185	A ---> C	130
D	40	B + C	83 - 100	B ---> C	30 - 40
		D + C	0 - 20	C ---> D	40 - 50

* : a low $-e_{Ox}$ means weakly bonded oxygen.
** : ranges are due to crystallographically nonequivalent V, P, O sites.

TABLE 2

Energy states $-e_H$ of hydrogen adsorbed on various sites, enthalpies q_{Htr} of H-transfer between adjacent sites and q_W of desorption of water on (100) $(VO)_2P_2O_7$ (all energies in kcal.mol^{-1}).

H on sites :	Mean $-e_H$	H-transfer from to	q_{Htr} range	Form. of H_2O* from to	q_W range
A	-	-	-	-	-
B	102 - 90	D ---> B	14 - 25	B ---> D	9 - 20
C	113	C ---> B	10 - 25	C ---> D	30 - 40
D	116	D ---> C	0 - 6	B ---> C	58

* : H adsorbed on B moves towards OH site (D) to make water which finally desorbs (from D).

surface oxygens, move along the surface (q_{Htr}) using adjacent oxygens and combine with them to form H_2O which is finally desorbed (q_W) (Table 2). With the help of enthalpies known for the various C-H, C-C, C-O, Me^{n+}-O bonds, energies of elementary and of concerted steps of the catalytic reaction are calculated, and the energetically-easiest pathway chosen among them.

In view of the above calculations, the following mechanism of n-butane oxidation is suggested (to be explained in detail in ref. 16). It is assumed that the molecule of butane is first adsorbed on the surface by means of hydrogen bonds between C-H and undersaturated oxygens. The properties of the suggested active site, represented as a cluster on Fig. 2, are the following :

a- The site is located on the cross of three routes of easy movement of H along the surface. Energies $-e_H$ are high, which facilitates the abstraction of hydrogen from the molecule by neighbouring oxygens. Hydrogens may move away (low q_{Htr}) to form H_2O on other sites, setting free those situated in the vicinity of the molecule which are consequently available in the next step (see point b-). Eight hydrogens can be abstracted in such a manner during the whole reaction.

b- The abstraction of the first two hydrogens from the terminal C1 and C4 carbons is concerted with the formation of two very strong bonds between the molecule and the catalyst. As a result, the

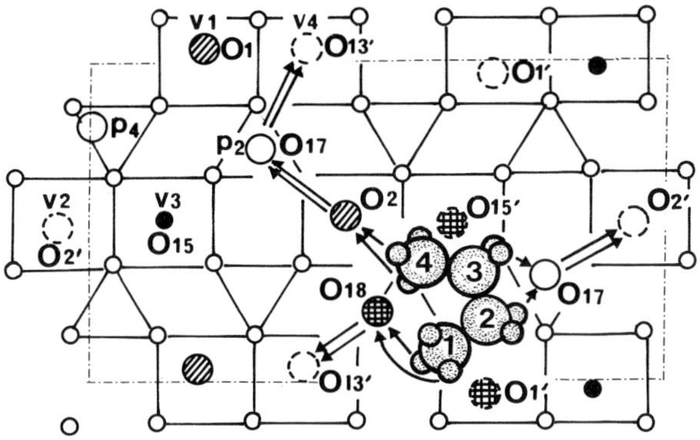

Fig. 2. Suggested adsorption of n-butane on a cluster of (100) $(VO)_2P_2O_7$, on the cross of three routs of movement of hydrogen (some hydrogen transfer are indicated by arrows) : when two H have met the same oxygen, H_2O can evolve (on sites with low q_W) ; cross-hatched circles represent oxygen to be inserted into the final molecule. Other symbols as in Fig. 1.

molecule is strongly anchored for a time long enough for the reaction to be completed. These steps called as "bifurcate activation" can be described by the following equations (see also Fig. 2) :

H_3C1 ---> H_2C1-O-V3 + H-O-P2 (3)

H_3C4 ---> H_2C4-O-V1 + H-O-V2 (4)

(where H_3C1 is a methyl group with carbon labelled C1).

c- The molecule has a direct contact with five active oxygens on sites B, C, D (3 O-V, 2 O-P). Three of them (2 O-V, 1 O-P) make bridges through which hydrogens from all four carbons may move away (as in point a-), in order to let them available. Two O-V (D sites) participate first in the strong anchoring to be finally incorporated with one O-P (C) to form the molecule of maleic anhydride. Low $-e_{Ox}$ facilitate the abstraction of oxygen from the surface to be inserted into the molecule. It is stressed that the bifurcate activation of butane is possible only when each terminal carbon is in the vicinity of two oxygens. The abstraction of the first H is thus immediately compensated by the formation of the C-O_{surf} bond, and the second abstraction of H is made without producing a high local undersaturation on these carbons. All proceeds as if butene-like intermediate did not appear and the oxidation was in fact prior to the formation of any double carbon-carbon bond. This point is very important as it should make the difference with the case of butene oxidation.

d- As a result of the whole process, seven oxygen vacancies are formed on the surface (six D-, one C-type). The mechanism of the catalyst reoxidation involves dissociation of O_2 (q_{Ox}) and some steps consisting in the transfer of oxygen. These steps, along with those of evolution of water, (moderate q_{Otr} and q_W) produce the highest energy barriers and determine *in fine* the rate of the reaction.

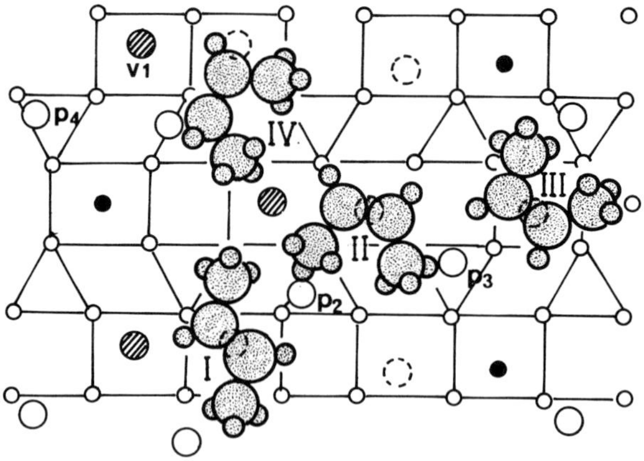

Fig. 3. Four possible adsorption schemes of butene on (100) $(VO)_2P_2O_7$; molecules I-III should be transformed into epoxybutane, crotonaldehyde, dihydrofuran, furan ; molecule IV in acetaldehyde,...

2- Oxidation of butene on the (100) face of $(VO)_2P_2O_7$

Due to the strongly reactive double bond present in butene, it is thought that adsorption through C=C is significantly competitive with respect to H-bonding followed by linking the terminal carbons to the surface, as just proposed for n-butane. As (100) is oxidized, the most probable interaction seems to be that between C=C and weakly bound (highly undersaturated) oxygens 0...(V) with formation of epoxybutanes (Fig. 3, molecules I-III). Other possibility consists in the formation of "peroxy"-type complex resulting in the formation of acetaldehyde (Fig. 3, molecule IV). Molecules (I-III) of butene adsorbed in the above indicated way are in contact with only one highly active and two moderately active oxygens ; anchoring of terminal carbons and insertion of three oxygens are therefore excluded. In fact, carbon oxides or/and oxidized by-products containing two and three carbons should form, the reaction being non selective. Other possibilities involve the case where oxygen vacancies exist on the surface, allowing the formation of classical π-complexes over cations. The number and geometry of active oxygens around such complexes being non convenient, non-selective products should similarly form. Isolated oxygen vacancies are expected on the surface in order for the less oxygenated by-products to be obtained. The reoxidation of the catalyst would thus require a long-range movement of oxygen along the surface, so as to have two adjacent vacancies able to bind O_2 molecules. This is difficult owing to important energy barriers.

3- Oxidation of n-butane on other faces of $(VO)_2P_2O_7$

Active surface energies E_c have been calculated (as in ref. 13) for (100), (021), (001) and (010) and are 1330, 1470, 1610 and 1960 $mJ.m^{-2}$ respectively. Using these values instead of free energies and making Curie-Wulff plot lead to the conclusion that the equilibrium form of a grain should be a hexagonal pillar "standing" on (100) face and bordered by (021) and (001). The area of indicated faces should make 22, 62 and 16 % of the external area, respectively. The frameworks of faces perpendicular to (100) are characterized by single octahedra (instead of pairs of edge-sharing octahedra as in (100)) linked to pyrophosphate groups. On (001) face, pairs of octahedra sharing corners are found along [010] and vanadium is not hidden by oxygen.

The same calculations as above have been therefore performed for these faces. As compared to (100), perpendicular faces have very high energy states $-e_{Ox}$ (Table 3). This means that oxygen is

TABLE 3

Energies e_{Ox}, e_H of O and H states , q_{Otr}, q_{Htr} of transfer of O and H, and q_W of water desorption for various (hkl) faces of $(VO)_2P_2O_7$ (all heats in kcal.mol^{-1})

Planes	$-e_{Ox}$	q_{Otr}	q_W	$-e_H$	q_{Htr}
(100)	V sites : 35 - 45 P sites : 87 - 91*	45	11 - 40	90 - 116	1 - 26
(021)	137 - 142	1 - 3	35 - 40	62 - 67	1 - 3
(001)	91 - 105	0 - 11	78 - 80	109 - 112	0 - 2
(010)	137 - 138	1	39	67	1

* only one P-site is used ; the heat is diminished in the concerted step.

strongly adsorbed and correlatively that its abstraction from the surface to be further inserted in the molecule is strongly hindered. In the case of (001), energy q_W of desorption of water is very high, which corresponds to an additional, high energy barrier limiting the rate. Moreover in the case of (021) and (010), there are conformations in which a bifurcate activation of n-butane is not favourable, and/or finding an additional active oxygen to bridge the terminal carbons is difficult. Consequently, these faces should theoretically give a minor yield of products of oxidation.

CONCLUSIONS

Using CMAS, we have shown that the essential properties of (100) $(VO)_2P_2O_7$ favouring the direct, selective oxidation of n-butane to maleic anhydride are the following :

- The surface is considered as (at least) partly covered by O and OH, which is in agreement with several experimental findings, such as (i), Raman spectroscopy (ref. 19), (ii), activity drastically decreasing with the number of pulses of butane (ref. 20). Such a surface resembles that of oxidized form γ-$VOPO_4$, which is assumed to be present as microdomains at the steady state (refs. 9, 10).
- The size of the active cluster is close from one surface unit cell and sufficient to transform one molecule during one catalytic cycle. $(VO)_2P_2O_7$ is detected by XRD when such clusters are ordered. Conversely, if they are linked in a disordered way, an "amorphous" state of the catalyst can result, which does not hinder the catalytic properties as noticed by several authors studying industrial catalysts.
- The convenient number and geometry of the undersaturated oxygens around adsorbed butane allows a "bifurcate" activation of butane, and an additional active oxygen exists to bridge the terminal carbons (furan-type cycle). Abstraction of hydrogen from hydrocarbon and transport of H from

the molecule along several routes for water to be formed are easy (high $-e_H$, low q_{Htr}), and abstraction of oxygen from the surface to be inserted into the molecule (low $-e_{Ox}$) is possible.

- Although movements of oxygen and evolution of water produce the energy barriers limiting the rate, these barriers are passable because heats q_{Otr} and q_W are moderate.

The less selective oxidation of butene on $(VO)_2P_2O_7$, as compared with butane, is due to the way of adsorption (C=C over undersaturated oxygen of cation) which limits the number of oxygens accessible to an adsorbed molecule. Low activity in the latter reaction results from difficult reoxidation of the surface.If oxygen vacancies over vanadium are present, butene should prefer to be adsorbed so as to give π-allylic intermediate. It must be kept in mind however that activation of butene is better made on V^{5+} (in $VOPO_4$) than on V^{4+}. Application of CMAS to this case will be examined in a further work.

Local site-energetics (Table 3) are the main reason of the worse efficiency of the other faces of $(VO)_2P_2O_7$ in the catalytic transformation of butane. According to the equilibrium shape of crystals obtained by Curie-Wulff plot, the (100) face, which shows the best activity and selectivity in oxidation of butane, makes 22 % of total area only. It is clear that any special way of preparation which contributes to the development of (100) area, such as one consisting in precipitation followed by topotactic decomposition of $VOHPO_4.0.5H_2O$ precursor, enhances strongly the catalytic properties of VPO catalysts. On the contrary, this does not seem to be required in the case of butenes.

REFERENCES

1 J. Ziolkowski, J. Catal., 80 (1983) 263 ; ibid, 84 (1983) 317.
2 J.C. Volta and J.L. Portefaix, Appl. Catal., 18 (1985) 1, and papers quoted therein.
3 J.C. Volta and J.M. Tatibouët, J. Catal., 93 (1985) 467.
4 J.M. Tatibouët and J.E. Germain, J. Catal., 72 (1981) 375.
5 J. Ziolkowski, J. Catal., 81 (1983) 298, 311 ; ibid, 84 (1983) 74.
6 P. Courtine, Am. Chem. Soc. Symp. Series, 279 (1985) 37.
7 A. Andersson, J. Solid State Chem., 42 (1982) 263.
8 J.N. Allison and W.A. Godard, J. Catal., 92 (1985) 127.
9 J. Ziolkowski, J. Catal., 100 (1986) 45.
10 E. Bordes, J.W. Johnson and P. Courtine, J. Solid State Chem., 55 (1984) 270.
11 E. Bordes, Catal. Today, 1 (1987) 499 ; ibid, 3 (1988) 163.
12 J. Mars and D.W. Van Krevelen, Chem. Eng. Sci., Suppl., 3 (1954) 41.
13 J. Ziolkowski, Surf. Sci., 209 (1989) 536.
14 J. Ziolkowski, J. Solid State Chem., 57 (1985) 269 ; ibid, 61 (1986) 343.
15 J. Ziolkowski and L. Dziembaj, J. Solid State Chem., 57 (1985) 291.
16 J. Ziolkowski, E. Bordes and P. Courtine, J. Catal., submitted.
17 Yu.E. Gorbunova and S.A. Linde, Dokl. Akad. Nauk SSSR, 245 (1979) 5864.
18 A.A. Balandin, Adv. Catal., 19 (1969) 1.
19 S.J. Puttock and C.H. Rochester, J. Chem. Soc., Faraday Trans. I, 82 (1986) 2773.
20 M.A. Pepera, J.L. Callahan, M.J. Desmond, E.C. Milberger, P.R. Blum and N.J. Bremer, J. Am. Chem. Soc., 107 (1985) 4883.

G. CENTI (University of Bologna, Italy) : In our experiments on butane oxidation on $(VO)_2P_2O_7$ we have observed a correlation between catalytic behavior and presence of disorder in the layered structure. Furthermore, when disorder is present in the structure of $(VO)_2P_2O_7$ we observe by EXAFS (Faraday Disc., in press) a change in the V-O distances and in particular a shortening of the V=O double bond. Is it possible to include the effect of disorder in your model ?

E. BORDES (Université de Technologie de Compiègne, France) : If you refer to the fact that $(VO)_2P_2O_7$ is often found partly amorphous by X-ray diffraction, such disorder means that short-range order only exists. At the moment, the model shows that a particular cluster of sites arranged as in (100) $(VO)_2P_2O_7$ is necessary for the transformation of one molecule of butane (some sites allowing capture of H and desorption of H_2O and others insertion of O in the molecule and reoxidation). It does not tell anything about the necessity for such clusters to be linked in a special way leading to the crystalline state of $(VO)_2P_2O_7$. The model should allow to study the effect of shortening of the vanadyl bond. This is an interesting suggestion.

G. BUSCA (University of Genova, Italy) : In your model you neglect possible interactions of the hydrocarbon molecules (butane, butene) with surface cationic sites (coordinatively unsaturated vanadium sites). It is known both from homogeneous catalysis and from surface chemistry studies that these interactions can occur, and can activate hydrocarbons towards selective oxidation. Is it possible to check this with your kind of calculations ?

E. BORDES (Université de Technologie de Compiègne, France) : The calculation of interactions has been made between (100) $(VO)_2P_2O_7$ covered by O and/or OH, and butane or butene. It shows in the last case that butene is converted to unselective products. One kind of "defect" on the surface of $(VO)_2P_2O_7$ should be indeed the presence of coordinatively unsaturated vanadium V(IV) ot V(V) sites. On the other hand, such sites are necessary on the surface of $VOPO_4$, in order that time to oxidize selectively butene. We intend to do the same calculations in these cases.

M. HADDAD (Amoco Chemical Comp., USA) : Could you review the structural and energetic reasons for the higher activity and selectivity of the (100) face of vanadyl pyrophosphate ?

E. BORDES (Université de Technologie de Compiègne, France) : The essential properties of (100) face favouring the direct, selective oxidation of butane to maleic anhydride are related with the surface framework.

- The "active site" is a cluster located on the cross of three routes of easy abstraction of H from hydrocarbon (high $-e_H$) and transport of H (low q_{Htr}) along several routes for water to be formed, near oxygen easily abstracted from the surface to be inserted in the molecule (low e_{Ox}) and from which water can desorb (moderate q_W). Vacancies can be refilled by molecular oxygen (moderate q_{Otr}).
- On this cluster, a bifurcate activation of butane is possible, which means that the abstraction of the first two hydrogens from the terminal C1 and C4 carbons is concerted with the formation of two very strong bonds $C\text{-}O_{surf.}$. As a result, the molecule is strongly anchored for the whole reaction to be finished. It has a direct contact with 5 active O. Two O-P and one O-V make bridges through which abstraction of H is possible and two O-V ensure the strong anchoring and are finally inserted. Local site-energetics are the main reasons of the worse efficiency of the other faces of $(VO)_2P_2O_7$ in the catalytic oxidation of butane.

Synthesis of Phthalic and Maleic Anhydrides from n-Pentane: Reactivity of Possible Intermediates and co-Feeding Experiments

G. Centi, J. Lopez Nieto[1], D. Pinelli, F. Trifiro' and F. Ungarelli

Dep. of Industrial Chemistry and Meterials, V.le Risorgimento 4, I-40136 Bologna (Italy)
[1] *On leave from the Institute de Catalisis y Petroleoquimica, Serrano 119, 28006 Madrid (Spain)*

The mechanism of synthesis of phthalic and maleic anhydrides from n-pentane on vanadyl pyrophosphate is analyzed. The research involved studying the reactivity of some of the possible intermediates or of useful probe molecules (C_5 linear and cyclic olefins, 2-methyl furan, n-amyl alcohol, 1-bromopentane, 4 methyl tetrahydrophthalic anhydride and hexahydrophthalic anhydride), co-feeding experiments (pentane or 1-pentene and butadiene, 2 methyl furan or n-butane), and analysis of the catalytic behavior of C_1-C_{10} alkanes. The results indicate the specificity of $(VO)_2P_2O_7$ for the formation of phthalic anhydride from n-pentane which forms by a parallel reaction with respect to maleic anhydride. The reaction of phthalic anhydride synthesis does not involve gas phase Diels-Alder like reactions, but rather (**i**) a surface reaction between an adsorbed diene and on adsorbed O-containing intermediate or (**ii**) a surface template reaction between two adsorbed, possibly cyclic, activated hydrocarbons. The possible mechanisms of formation of phthalic and maleic anhydrides are discussed taking into account the strong inhibition and adsorption effects observed by feeding C_5 dienic molecules.

Introduction

n-Pentane is present in small amounts in natural gas and in the gasoline fraction of petroleoum; C_5 olefins and diolefins are significant by-products in the steam reforming process. However, due to their low cost and the absence of processes for their transformation to more valuable products, their upgrading by selective oxidation seems promising, notwithstanding the general absence of literature indications [1-4]. In a previous paper [5] we have shown that in the oxidation of n-pentane using the most active and selective catalyst for the oxidation of n-butane to maleic anhydride [$(VO)_2P_2O_7$], phthalic anhydride (**PA**) and maleic anhydride (**MA**) form in comparable amounts. The formation of PA, in particular, is interesting both from the fundamental point of view since it is one of the few examples of a selective oxidation reaction with the formation of products with higher C atom numbers than the starting alkane, and from the industrial point of view as an alternative process for the synthesis of phthalic anhydride.

A kinetics investigation of the reaction network in n-pentane oxidation [6] has shown that the surface reactions in the synthesis of PA and MA are not consecutive but parallel with a different dependance on the n-pentane and oxygen concentrations. Higher selectivities to PA can be obtained at the lower temperatures and oxygen concentrations, but MA is always present together with carbon oxides. The kinetics analysis does not give further indications about the surface properties of the catalyst necessary to increase the maximum yield of PA and about the mechanism of formation of PA. In order to gain information on this problem, in this paper we compare **(i)** the behavior of $(VO)_2P_2O_7$ in the oxidation of several compounds that are possible reaction intermediates in the synthesis of the two anhydrides or useful probe molecules and **(ii)** the behavior in the oxidation of mixtures of pentane with activated C_4 and C_5 compounds. These last experiments of co-feeding may clarify the role of small amounts of gas-phase and adsorbed possible intermediates on the competitive pathways of synthesis of MA and PA from n-pentane.

Experimental

Catalyst Preparation. The catalyst used in this study was prepared in organic medium according to well established literature procedures [6-9]. V_2O_5 is refluxed in isobutyl alcohol up to reduction of the vanadium and the blue $[VOHPO_4]_2.H_2O$ precursor complex is precipitated by addition of 99% H_3PO_4, refluxing and distilling to remove solvent. The final catalyst was formed in- situ by running the butane oxidation reaction at 1.5% butane, air and 1000 G.H.S.V. The equilibrated catalyst sample had ca. 750 h continuous on-steram time in the butane oxidation reaction. Analysis of the sample showed the average vanadium oxidation state to be 4.00.

Catalytic Testing. The experimental investigations were conducted using a tubular fixed bed down-flow reactor as previously described [6]. n-Pentane or other liquid reagents were added to the preheated gaseous feed composed of calibrated amounts of the other reagents. The organic feed was vaporized in a chromatographic-like injector. The reagent composition and the reaction products were analyzed using two on-line gas chromatographs, the first one with a flame ionization detector and the second one with a thermoconducibility detector. Further experimental details have been reported elsewhere [5-13]. The catalyst charges used in the 2-5 g range, with total flow at STP conditions usually higher than $60\text{-}80 \times 10^{-3}$ L min^{-1} in order to avoid interphase diffusional limitations. The catalyst was used in the 40-60 mesh range of particle diameter.

Results

Oxidation of Alkanes The behavior of $(VO)_2P_2O_7$ in the oxidation of C_1-C_{10} paraffins is compared in Figure 1. The selectivity is reported at 50% of conversion in the formation of main products. The specific selective behavior of $(VO)_2P_2O_7$ in the oxidation of C_4 and C_5 alkanes and the specificity of the formation of PA from n-pentane is shown. Only traces of PA are observed for higher alkanes.

The effect of the reaction temperature and of space-velocity (W/F, grams of catalysts/moles.h^{-1} of pentane) in n-pentane oxidation on $(VO)_2P_2O_7$ is summarized in Figure 2 **a** and **b**, respectively. The reactions of synthesis of MA and PA are parallel, the latter being favoured by low temperatures of reaction due to the higher rate of consecutive oxidation to carbon oxides in comparison to MA [6].

Oxidation of Activated Hydrocarbons. The oxidation of 1-pentene, pentadiene and

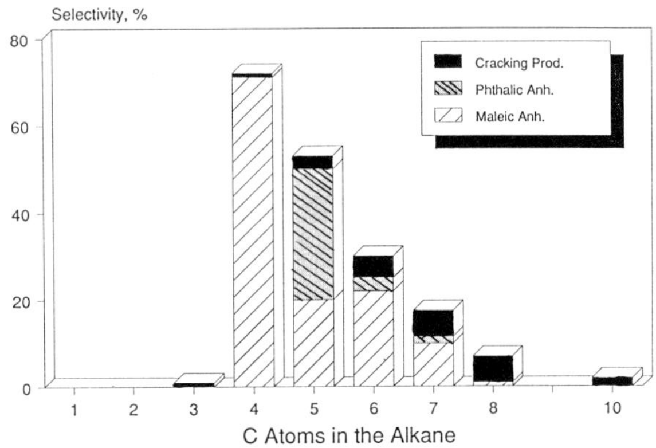

Figure 1 Selectivity to main products (apart from COx) at 50% of conversion (reaction temperatures variable) in the oxidation of C_1-C_{10} linear alkanes on $(VO)_2P_2O_7$. Exp. conditions: 1.5% of hydrocarbon, 20% O_2, W/F = 300 g.h/moles C_5.

cyclopentadiene is compared in Figure 3 as a function of the reaction temperature. Contrary to that expected, all these C_5 olefins (**i**) are much less selective to anhydrides in the entire range of temperatures and (**ii**) give rise to an higher ratio of MA to PA. It should be noted that the formation of cyclopentadiene is observed in the oxidation of the C_5 linear olefins.

In order to further explore the reactivity of other possible intermediates as well as useful probe molecules, the behaviors of 2 methyl furan, n-amyl alcohol, 1 bromopentane, 4 methyl tetrahydrophthalic anhydride (**4Me4HPA**) and hexahydrophthalic anhydride (**6HPA**) are compared in Table I (temperature of 20% conversion and relative selectivities to main products). As observed for C_5 olefins, the oxidation of other C_5 molecules also does not give rise to a selective formation of anhydrides and, in particular, of PA. On the contrary, the catalyst shows a very high selectivity in dehydrogenation/aromatization of the ring of 4Me4HPA forming the corresponding 4MePA or PA and, in the former case, PA by demethylation at higher temperatures.

Experiments of co-Feeding. The oxidation of mixtures of n-pentane with butadiene, 2 methyl furan and n-butane and the oxidation of mixtures of 1- pentene with butadiene is summarized in Table II. In all the co-feeding experiments, an inhibition of the formation of PA is observed, whereas the synthesis of MA is much less affected.

Discussion

Specificity of $(VO)_2P_2O_7$ for n-Pentane Oxidation

The results reported in Figuress 1-3 and Table I show that the formation of PA is a reaction specific for n-pentane. Paraffins with a higher number of carbon atoms are oxidized essentially to carbon oxides with the contemporaneous formation of cracking products (lower olefins) or of small yields of MA. n-Butane is selectively oxidized to MA, while paraffins with less than 4 carbon atoms

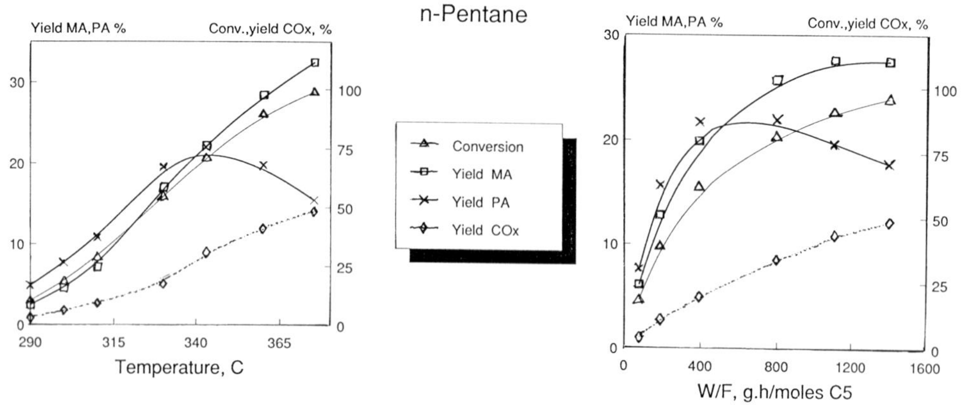

Figure 2 Catalytic behavior of $(VO)_2P_2O_7$ in the oxidation of n-pentane as a function of the reaction temperature (**a**) or of the space-velocity (**b**). Exp. conditions: 2.5% n-pentane, 20% O_2, W/F = 780 g.h/moles C_5 (Fig. 2a), T = 350 C (Fig. 2b).

are oxidized only to carbon oxides. In the case of the oxidation of C_5 activated hydrocarbons the main products also are carbon oxides, but small amounts of the two anhydrides are formed at low conversion together with other intermediates.

This specificity of $(VO)_2P_2O_7$ for the oxidation of n-pentane is not due to a further transformation of MA, but rather to the presence of a parallel surface pathway of synthesis of PA, as indicated by the tests with various space-velocities (Fig. 2b). The decrease in the amount of PA observed with increasing contact time (Fig. 2b) as well as increasing reaction temperature (Fig. 2a) is related to the higher rate of consecutive oxidation of PA with respect to that of MA [6].

Mechanism of Oxidation of n-Pentane

In order to simplify the discussion on the nature of this specific behavior of vanadyl pyrophosphate in the oxidation of n-pentane, we have assumed that the syntheses of MA and PA involves a similar sequence of adsorbed intermediates similar to that proposed for n-butane oxidation to maleic anhydride on vanadyl pyrophosphate [7-9].

Maleic Anhydride. We may start the discussion of the mechanism of MA formation on the basis of the following general reaction pattern, taking into account that the reaction occurs entirely on the surface without desorption of any observable intermediate:

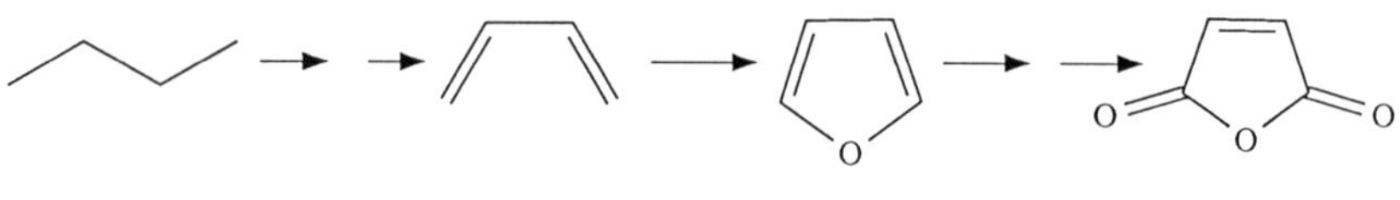

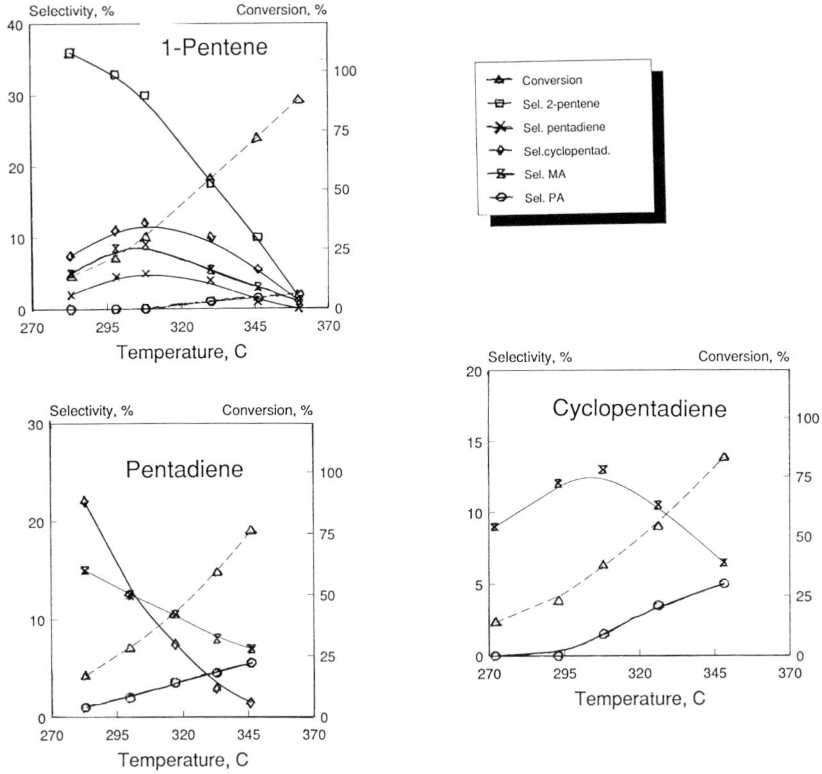

Figure 3 Catalytic behavior of $(VO)_2P_2O_7$ in the oxidation of 1-pentene (**a**), 1,3-pentadiene (**b**) and cyclopentadiene (**c**). Exp. conditions: (**a**) 2.2% 1- pentane, 18% O_2, W/F = 265 g.h/mol C_5; (**b**) 1.3% pentadiene, 11% O_2, W/F = 410 g.h/mol C_5; (**c**) 1.6% cyclopentadiene, 13.5% O_2, W/F = 260 g.h/mol C_5.

Table I shows that in the oxidation of all the hypothesized intermediates of this mechanism (**i**) the selectivities to PA and MA are much lower than the values found in the oxidation of n-pentane and (**ii**) the ratio of MA to PA is higher. Also, in the oxidation of C_4 molecules [10,11] a higher selectivity to MA has been observed at low conversion from n-butane than from 1-butene, butadiene or furan due to self-inhibition of the dienic molecules on the synthesis of MA [12]. However, the maximum yields to MA were comparable due to the different trend of the selectivity to MA vs. the conversion. In the case of n-butane the selectivity to MA decreases with increasing conversion, and in the case of other activated C_4 molecules increases with increasing conversion. Therefore, the higher selectivity of paraffin oxidation seems to be a typical property of the vanadyl pyrophosphate, but this property is more evident in the oxidation of the C_5 fraction. This is related to the stronger adsorption of the C_5 intermediates as compared to corresponding intermediates from C_4 hydrocarbons and therefore more drastic effects of self-inhibition. The results of co-feeding experiments (Table II) are in agreement with these indications, showing that the co-feeding of even small amount of dienic hydrocarbons strongly inhibits the selective behavior of the catalyst. This suggests that the direct oxidation of the intermediates does not provide sufficient evidence about the

Table I Comparison of the behavior of $(VO)_2P_2O_7$ in the oxidation of C_5 molecules: temperature of 20% of conversion (T_{20}) and relative selectivities to main products, apart from carbon oxides. Exp. conditions: W/F = 300 g.h/mol C_5, O_2/C_5 = 20:2.2.

		Selectivity, %							
Reagent	T_{20}	PA	MA	2Pe	Pd	CPd	AcFu	2MeFu	3MePA
n-pentane	330	37	13	-	-	-	-	-	-
1-pentene	303	-	8	33	3	10	-	-	-
1,3 pentadiene	290	2	14	-	-	13	-	-	-
cyclopentadiene	285	-	10	-	-	-	-	-	-
2 methyl furan	264	-	5	-	-	-	39	-	-
n-amyl alcohol	<250	-	-	64*	-	-	-	14	-
1 Br pentane	308	2	4	37*	-	-	-	-	-
4 Me4HPA	320	24	-	-	-	-	-	-	40
6HPA	305	65	5	-	-	-	-	-	-

Note: **MA**: maleic anhydride, **PA**: phthalic anhydride, **2Pe**: 2-pentene, **Pd**: pentadiene, **CPd**: cyclopentadiene, **AcFu**: Furanoic Acid, **2MeFu**: 2 methyl furan, **3MePa**: 3 methyl phthalic anhydride, **4Me4HPA**: 4 methyl tetrahydro phthalic anhydride, **6HPA**: hexahydro phthalic anhydride.

*: 1 pentene + 2 pentene.

participation or not of that intermediate in the mechanism of alkane transformation to anhydrides. In fact, Table I shows that the more oxidized the hydrocarbon, the lower the temperature at which 20% conversion is obtained. This fact together with the strong adsorption properties of $(VO)_2P_2O_7$ could explain the absence of desorbed intermediates in the oxidation of n-pentane. Furthermore, this indicates that the surface concentration of the possible intermediates is very low and thus phenomena of self-inhibition may be absent.

In conclusion, the behavior of the $(VO)_2P_2O_7$ in the oxidation of some of the possible C_5 intermediates is not indicative of the participation or not of these in the mechanism of anhydride synthesis, due to the different surface adsorption and self-inhibition effects.

Assuming therefore that the above mechanism is still valid, the lower selectivity to anhydrides shown in the oxidation of activated C_5 hydrocarbons, may be tentatively attributed to the strong adsorption of these compounds and to the higher probability of side waste reactions due to the higher surface life-times and competition with active oxygens. A further difference with respect to C_4 hydrocarbons is the possibility of consecutive dehydrogenation of pentadiene with cyclization and formation of cyclopentadiene. Further dehydrogenation of butadiene, on the contrary, is much more difficult due to the absence of allylic H atoms. The formation of cyclopentadiene and the high rate of isomerization of olefins is thus an indication of the possibility of additional surface pathways from C_5 hydrocarbons in comparison with corresponding C_4 hydrocarbons.

Also surprising is the very low selectivity in maleic anhydride obtained from 2 methyl furan. In the case of this compound furanoic acid has been observed with a relatively high selectivity at low conversion (Table I), but the consecutive transformation of this acid takes place mainly to carbon oxides and only to a small extend to to MA. The catalyst quickly deactivates in 2 methyl furan

Table II Selectivities to anhydrides at 40% of conversion and relative temperature (T_{40}) in the co-feeding of n-pentane or 1-pentene with dienic C_4 and C_5 molecules. Experimental conditions as in Table I.

Reagents		*Selectivity, %*		T_{40}
		MA	**PA**	
	(pure)	17	30	355
	n-butane (0.9%)	29	28	360
n-pentane	butadiene(0.2%)	29	3	370
(2.2%) +	2 Methyl furan (0.1%)	10	traces	350
1-pentene	*(pure)*	6.2	1	355
(2.0%) +	butadiene (0.2%)	6.9	1.3	330

Note: the value after the reagents indicates their amount (%) in the feeding mixture at the reactor.

oxidation and is not more selective for n-pentane oxidation. Therefore, 2 methyl furan or some products of its oxidation considerably modify the surface reactivity of the catalyst. This effect can explain the low selectivity obtained in MA from this compound which is very near in oxidation state and conformational structure to the final product. The modification of the surface reactivity of $(VO)_2P_2O_7$ is a further indication of the enhanced strong adsorption of these compounds in comparison to corresponding compounds from C_4 hydrocarbons in which no deactivation effects could be observed.

Phthalic Anhydride. In a first approach we may assume that the most probable mechanism of PA formation from the possible reaction intermediates in n- pentane oxidation is a Diels-Alder like reaction between MA and a diene (such as pentadiene) with consecutive dehydrogenation/aromatization and demethylation. Table II shows that PA can be formed very easily from 4 Me4HPA or 6HPA in a range of temperatures comparable to that of the oxidation of n-pentane. Therefore, if a Diels-Alder reaction occurs, the product evolves easily to PA. The addition of butadiene in the gas phase should favour PA synthesis according to this mechanism, but the co-feeding experiments do not provide evidence on this question (Table II). In fact, the addition of butadiene decreases selectively the formation of PA. The increase in the formation of MA is not directly related to butadiene. In fact, the addition of butadiene to 1- pentene has no effect on the formation of the two anhydrides. Probably, butadiene adsorbs selectively on the sites responsible for PA formation inhibiting this synthesis. The inhibition effect of diolefins can also explain the fact that in the oxidation of C_5 activated hydrocarbons PA forms in a very small amount in relation to MA and essentially at high temperature where all the olefins and diolefins are converted.

If the above mechanism is valid, the Diels-Alder reaction must occur in the adsorbed phase between adsorbed intermediates/products and we can exclude gas phase reactions. However, both the reactivity of possible intermediates and the experiments of co-feeding do not provide further evidence about the possible nature of intermediates involved in the reaction. In particular, there are two principal hypotheses for the mechanism: (**i**) surface Diels-Alder reaction between an adsorbed diene (such as pentadiene) and one O-containing adsorbed product (such as maleic anhydride) or (**ii**) a surface template reaction between two adsorbed activated hydrocarbons (such as two cyclopentadiene molecules). It is interesting to note in this respect that the study of the benzene oxidation on $(VO)_2P_2O_7$ [13] has shown the formation of PA with significantly high selectivities, higher than those observed from C_5 pentenes but lower than that observed from n-pentane. This observation may be more in favour of the second hypothesis, but is not conclusive evidence.

Conclusion

The tests of the oxidation of several alkanes or of activated hydrocarbons evidence the specific selectivity of $(VO)_2P_2O_7$ for the formation of phthalic anhydride from n-pentane. High selectivities in the sum of phthalic anhydride and maleic anhydride are observed only from n-pentane, while more oxidized compounds give essentially carbon oxides. The effect may be explained on the basis of the strong adsorption of dienic compounds on the vanadyl pyrophosphate which causes a self-inhibition effect and decreasing selectivity to anhydrides. The effect is much more evident in the oxidation of C_5 molecules than in the corresponding C_4 molecules. The reactions which lead to phthalic anhydride occur only between intermediates in the adsorbed phase and the presence of a gas phase Diels-Alder reaction between maleic anhydride and diolefins can be excluded. However, due to the strong adsorption and self-inhibition of dienic compounds, no conclusive evidence about the mechanism of phthalic anhydride formation from n-pentane can be drawn from the study of the reactivity of the possible intermediates and in co-feeding experiments.

Acknowledgements

This work was sponsored by the Ministero Pubblica Istruzione (60%).

References

[1] N.S. Butt, A. Fish, *J. Catal.*, 5 (1966) 205, 494 and 508.
[2] G. Mattson, D. Sasser, *Oxid. Commun.*, 7 (1984) 333.
[3] T. Seiyama, K. Nita, T. Maehara, N. Yamazoe, Y. Takita, *J. Catal.*, 49 (1977) 164.
[4] D. Honicke, K. Griesbaum, Y. Yang, *Chem.-Ing. Techn.*, 59 (1987) 222.
[5] G. Centi, M. Burattini, F. Trifiro', *Appl. Catal.*, 32 (1987) 353.
[6] G. Centi, J. Lopez Nieto, D. Pinelli, F. Trifiro', *Ind. Eng. Chem. Research*, 28 (1989) 400.
[7] G. Centi, F. Trifiro', J.R. Ebner, V. Franchetti, *Chem. Rev.*, 88 (1988) 55.
[8] G. Busca, G. Centi, J. *Am. Chem. Soc.*, 111 (1989) 46.
[9] G. Centi, F. Trifiro', G. Busca, J.R. Ebner, J.T. Gleaves, in Proceedings, 9th Int. Congress on Catalysis, Calgary 1988, M.J. Phillips and M. Ternan Eds., The Chemical Inst. of Canada Pub., Vol. IV, p. 1538.
[10] F. Cavani, G. Centi, I. Manenti, F. Trifiro', *Ind. Eng. Chem. Prod. Res. Dev.*, 26 (1985) 221.
[11] G. Centi, F. Trifiro', *J. Molec. Catal.*, 35 (1986) 255.
[12] F. Cavani, G. Centi, F. Trifiro', *Ind. Eng. Chem. Prod. Res. Dev.*, 22 (1983) 570.
[13] F. Ungarelli, Thesis, Univ. of Bologna, July 1988.

G. Centi and F. Trifiro' (Editors), *New Developments in Selective Oxidation*

NON-FARADAIC ELECTROCHEMICAL MODIFICATION OF CATALYTIC ACTIVITY: PARTIAL OXIDATION OF C_2H_4 ON Ag AND CH_3OH ON Pt

C.G. VAYENAS, S. BEBELIS and S. NEOPHYTIDES

Institute of Chemical Engineering and High Temperature Chemical Processes, Department of Chemical Engineering, University of Patras, Patras 26110, Greece

SUMMARY

It was found that the catalytic activity and selectivity of Ag and Pt porous catalyst films can be altered dramatically and reversibly by electrochemically supplying or removing oxygen anions O^{2-} to or from the catalyst surface via polarized solid electrolyte cells. Oxygen anions forced electrochemically to adsorb on the catalyst surface during C_2H_4 oxidation on Ag and CH_3OH oxidation on Pt alter the catalyst work function in a predictable way and cause dramatic changes in catalytic activity and selectivity.

INTRODUCTION

Solid electrolytes can be used both to study and to influence catalytic phenomena on metal surfaces. Progress in this area has been reviewed recently (refs. 1,2).

Very recently it has been found that the catalytic activity and selectivity of metal surfaces can be altered dramatically and reversibly in solid electrolyte cells of the type:

$$\text{gaseous reactants, metal catalyst} \mid ZrO_2\ (8\ \text{mol\%}\ Y_2O_3) \mid M,\ O_2 \qquad (1)$$

where the metal M catalyzes the reaction $O_2+4e^- \leftrightarrows 2O^{2-}$ and serves as a means of supplying or removing O^{2-} to or from the catalyst, respectively, through the zirconia solid electrolyte under the influence of an external voltage (refs. 1-10).

In several cases the observed increase Δr in catalytic reaction rate is non-Faradaic, i.e., Δr can be a factor of 10^2 up to $3.\cdot 10^5$ higher than the rate I/2F of O^{2-} supply or removal to of from the catalyst (refs. 1,2,4-10), with a concomitant up to fifty-fold enhancement in catalytic rate (refs. 7,10). This new phenomenon has been described under the acronym NEMCA, i.e., Non-Faradaic Electrochemical Modification of Catalytic Activity (refs. 2,7-10). Table 1 lists the catalytic reactions which have been found already to exhibit the NEMCA effect. The enhancement factor Λ and rate enhancement ratio ϱ parameters shown in Table 1 have been defined (refs. 2,7-10) as:

$$\Lambda = \Delta r/(I/2F) \qquad ; \ \varrho = r/r_0 \qquad (2)$$

where r_0 is the open-circuit, i.e., regular catalytic rate and $r=r_0+\Delta r$ is the NEMCA induced catalytic rate. A reaction is said to exhibit the NEMCA effect when $|\Lambda| \gg 1$.

TABLE 1
List of catalytic reactions found to exhibit the NEMCA effect

	Reactants	Products	Catalyst	T[°C]	Λ range	ϱ range	References
1	$CH_2=CH_2$, O_2	ethylene oxide, CO_2	Ag	320-470	[0,+300]	<3	4,this work
2	propylene, O_2	propylene oxide, CO_2	Ag	320-420	[0,+300]	<2	5
3	$CH_2=CH_2$, O_2	CO_2	Pt	290-450	$[0,+3\cdot10^5]$	<55	7,8,10
4	CO, O_2	CO_2	Pt	300-550	[-500,+500]	<6	6
5	CO, O_2	CO_2	Pd	550-400		<1.5	10
6	CH_3OH, O_2	CO_2,H_2CO	Pt	300-500	$[-3\cdot10^4,+10^4]$	<10	this work,14
7	CH_3OH	H_2CO, CO, CH_4	Pt	400-500	[-10,0]	<3	14
8	CH_3OH	H_2CO, CO, CH_4	Ag	550-750	[-25,0]	<6	9

The main features of the NEMCA effect observed in all previous studies where the catalyst potential V_{WR} (with respect to a reference air electrode) and the exchange current I_0 of the catalyst-solid electrolyte interface were measured (refs. 7-10) are the following:

1. Catalytic rates usually depend exponentially on catalyst potential V_{WR}:

$$\ln(r/r_0) = \alpha F(V_{WR} - V^*_{WR})/RT \tag{3}$$

where α and V^*_{WR} are reaction- and catalyst-specific constants.

2. The magnitude of the enhancement factor Λ can be estimated from: $\Lambda \approx 2Fr_0/I_0$. This implies that high Λ values can be obtained by using low I_0, i.e., highly polarizable catalyst-electrode interfaces.

3. The catalytic rate relaxation time constant τ during galvanostatic transients can be estimated by $\tau = 2FN/I$, where N is the independently measured total catalyst surface area, expressed in surface metal g-atoms. This implies that the catalytic properties of the entire catalyst surface change when the catalyst-solid electrolyte interface is polarized.

The proposed semi-quantitative explanation of the NEMCA effect (refs. 7-10) is based on the changes induced in the average work function eΦ of the gas-exposed catalyst surface resulting from the polarization of the catalyst-solid electrolyte interface and from the spatial uniformity of the Fermi level in the conductive catalyst. This explanation is also corroborated by a very recent study which showed that β''-Al_2O_3, which is a Na^+ conductor, can be used instead of zirconia, to induce the NEMCA effect (ref. 11). Thus the nature of the solid electrolyte plays very little role in the NEMCA effect. In this paper the NEMCA effect is studied for the oxidation of C_2H_4 on Ag and CH_3OH on Pt. The former reaction was the first for which a non-Faradaic rate enhancement was reported (refs. 4,5). However this early study employed a two-electrode configuration and therefore V_{WR} and I_0 could not be measured.

EXPERIMENTAL METHODS

The experimental apparatus has been described in detail previously (refs. 6-10). It utilizes on-line gas chromatography (Perkin-Elmer Sigma 300), mass spectrometry (Balzers QMG 311) and IR spectroscopy (Beckman 864) for continuous analysis of the feed and products of the continuous well-mixed zirconia reactor shown in Fig. 1 and described in detail elsewhere (refs. 6-10).

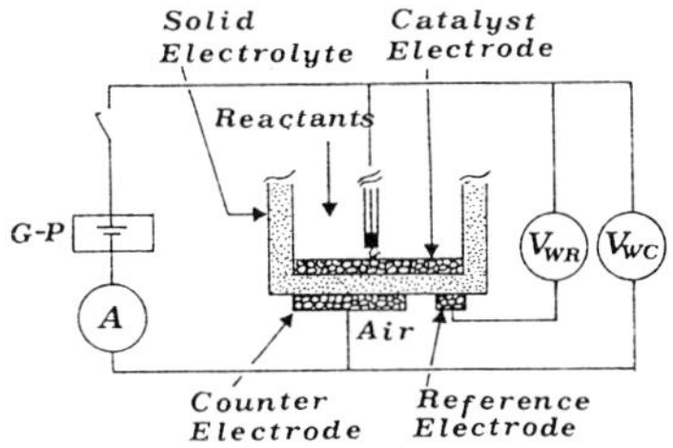

Fig. 1. Reactor configuration

Porous Pt and Ag catalysts and auxiliary electrode films were deposited on the zirconia electrolyte using Engelhard Pt pastes and GC Electronics Ag pastes, followed by calcination at 900-1100 K as described in refs. 4-10, where catalyst characterization details are also presented. The catalyst films have typically thicknesses of 5μm, porosities of 30%, superficial surface areas of $2cm^2$ and true surface areas of the order of 10^2-10^3 cm^2 as measured by surface titration techniques (refs. 5,9).

In the experiments one monitors the open-circuit, i.e., regular catalytic rate r_0 and then uses a galvanostat (AMEL 553) to apply a constant current I between the catalyst and the counter-electrode while monitoring the rate r and the catalyst potential V_{WR} vs the reference air electrode (Fig. 2). A typical example for the case of C_2H_4 oxidation on Pt (ref. 8) is given in Fig. 2. When r and V_{WR} reach a steady-state, then the circuit can be opened (I=0), after which both r and V_{WR} gradually resume their open-circuit values (Fig. 2). As in previous papers (refs. 4-10) we have defined I to be positive when O^{2-} are pumped to the catalyst. The experimental and theoretical

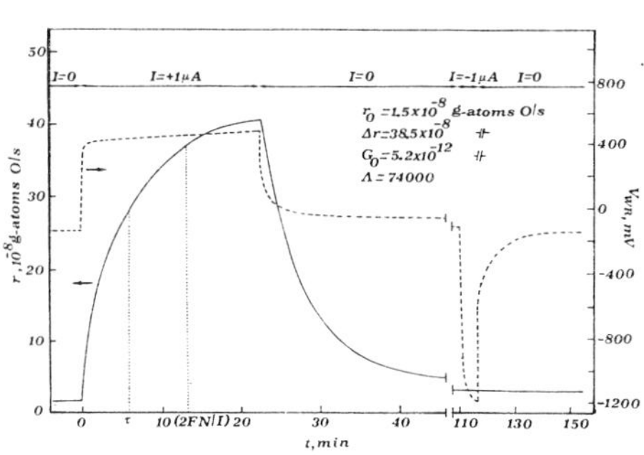

Fig. 2. Typical catalytic rate response to step changes in cell current (galvanostatic transients); comparison of experimental (τ) and computed (2FN/I) relaxation time constants; C_2H_4 oxidation on Pt; T=370°C, P_{O_2}=4.6·10^{-2}bar, P_{ET}=3.6·10^{-3}bar, Reactive oxygen uptake N=4.2·10^{-9} g-atoms.

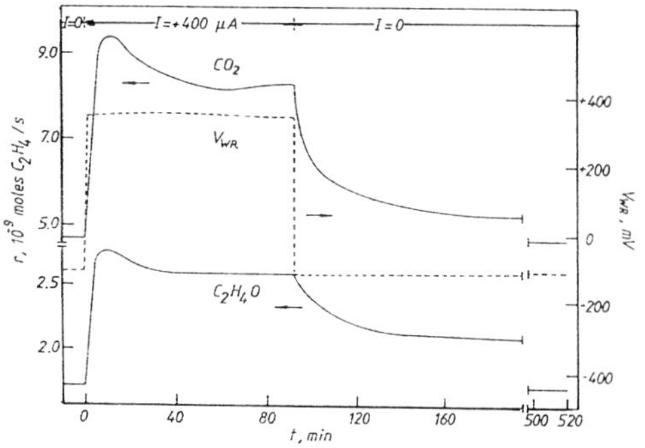

Fig. 3. Typical catalytic rate response to step changes in cell current (galvanostatic transients); C_2H_4 oxidation on Ag; Catalyst RC2; T=435°C, P_{O_2}=3.1·10^{-2}, P_{ET}=2.2·10^{-2} bar.

procedure of measuring the exchange current I_0 of the catalyst-solid electrolyte interface through steady-state current-potential data has been described previously (refs. 6,8,9).

RESULTS

Ethylene Oxidation on Ag

This is one of the most thoroughly studied catalytic systems (e.g. ref.12,13). Under the conditions of this investigation (ref. 14) ethylene epoxidation and oxidation to CO_2 can be treated as two parallel reactions (e.g. ref. 4).

Figure 3 shows a typical galvanostatic transient when a current I=400μA is applied and O^{2-} are pumped to the catalyst at a rate $I/2F=2.07\cdot 10^{-9}$gatoms O/s. The observed steady state increase in the rate of oxygen consumption is $22.5\cdot 10^{-9}$ gatoms O/s, i.e., 11 times larger than I/2F. As also shown in Fig. 3 the effect is quite reversible, i.e., both rates return to their open-circuit values upon current interruption.

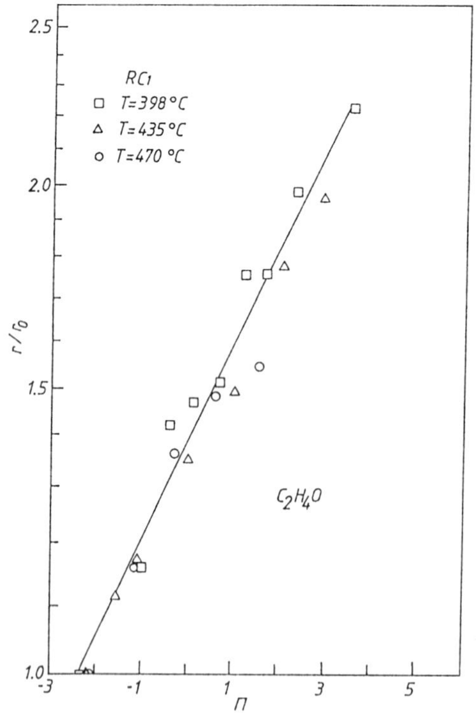

Fig. 4. Effect of catalyst IR-free potential V_{WR} and of $\Pi=FV_{WR}/RT$ on the rate enhancement ratio $\rho=r/r_0$; Catalyst RC1; $P_{O_2}=3.1\cdot 10^{-2}$, $P_{ET}=2.5\cdot 10^{-2}$ bar.

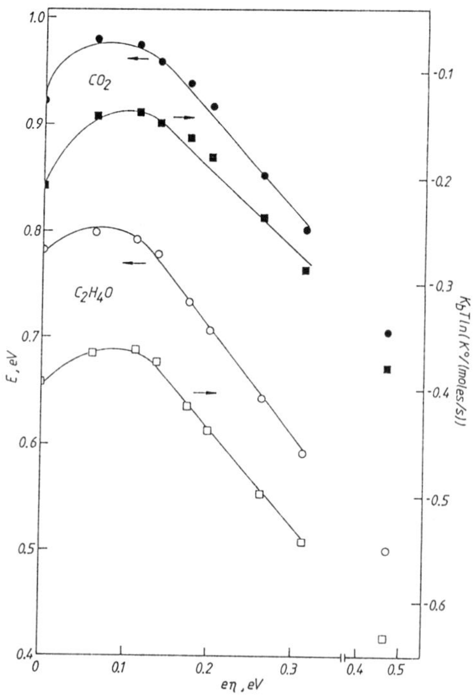

Fig. 5. Effect of overpotential η on the activation energies and preexponential factors of the reactions of production of C_2H_4O and CO_2; Catalyst RC1; $P_{O_2}=3.1\cdot 10^{-2}$, $P_{ET}=2.5\cdot 10^{-2}$bar.

As shown in Fig. 4 the rate of C_2H_4O formation depends exponentially on catalyst potential V_{WR}. Similar is the behaviour for CO_2 formation. Results obtained at different temperatures can be unified by the dimensionless parameter $\Pi = FV_{WR}/RT$ as in previous studies of the NEMCA effect. The constants α and V^*_{WR} of the NEMCA equation (3) take the values α=0.14, V^*_{WR}=-140 mV for ethylene epoxidation. and α=0.16, V^*_{WR}=-120mV for ethylene oxidation to CO_2.

Figure 5 shows the effect of overpotential $\eta = V_{WR} - V^0_{WR}$ (where V^0_{WR} is the open-circuit potential) on the apparent activation energies E_i and preexponential factors k^0_i of the epoxidation (i=1) and deep oxidation (i=2) reactions. After a slight initial increase both E_1 and E_2 decrease substantially with increasing V_{WR} according to:

$$\Delta E_1 = -1.03\ \Delta eV_{WR}, \qquad \Delta E_2 = -0.93\ \Delta eV_{WR} \tag{4}$$

Interestingly, the preexponential factors behave similarly and $k_bT\ln K_i$ decreases linearly with increasing V_{WR} (Fig. 5):

$$k_bT\Delta \ln k^0_1 = -0.93\ \Delta eV_{WR}, \qquad k_bT\Delta \ln K^0_2 = -0.81\ \Delta eV_{WR} \tag{5}$$

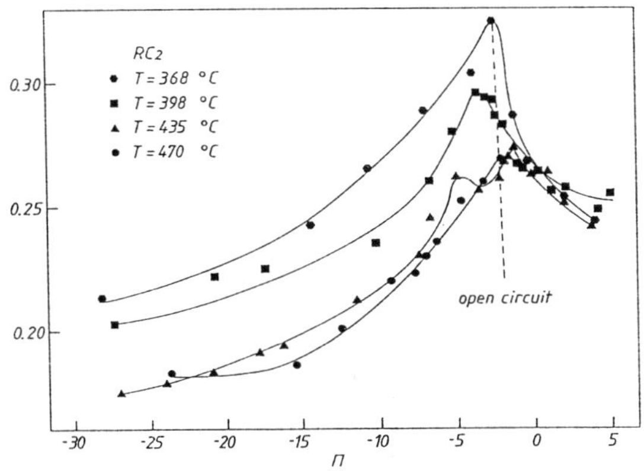

Fig. 6. Steady state effect of catalyst IR-free potential on the selectivity to C_2H_4O; $P_{O_2}=3.1\cdot10^{-2}$, $P_{ET}=2.5\cdot10^{-2}$ bar.

Catalyst potential was found to have a pronounced effect in selectivity as shown in Fig. 6 for a relatively reducing gaseous composition. Similarities and differences with previous results obtained under oxidizing conditions (refs. 4,5) are discussed elsewhere (ref. 14).

Methanol Oxidation on Pt

The reaction was investigated at temperatures 350° to 650°C, CH_3OH partial pressures P_M between $5.\cdot10^{-2}$ and 10^{-2} bar and oxygen partial pressures P_{O_2} between 0.01 and 0.2 bar. Formaldehyde and CO_2 were the only products detected in measurable concentrations. The open-circuit selectivity to H_2CO is of the order of 50% and is practically unaffected by gas residence time over the above conditions for methanol conversions below 30%. Consequently the reactions of H_2CO and CO_2 formation from CH_3OH and O_2 can be considered kinetically as two parallel reactions.

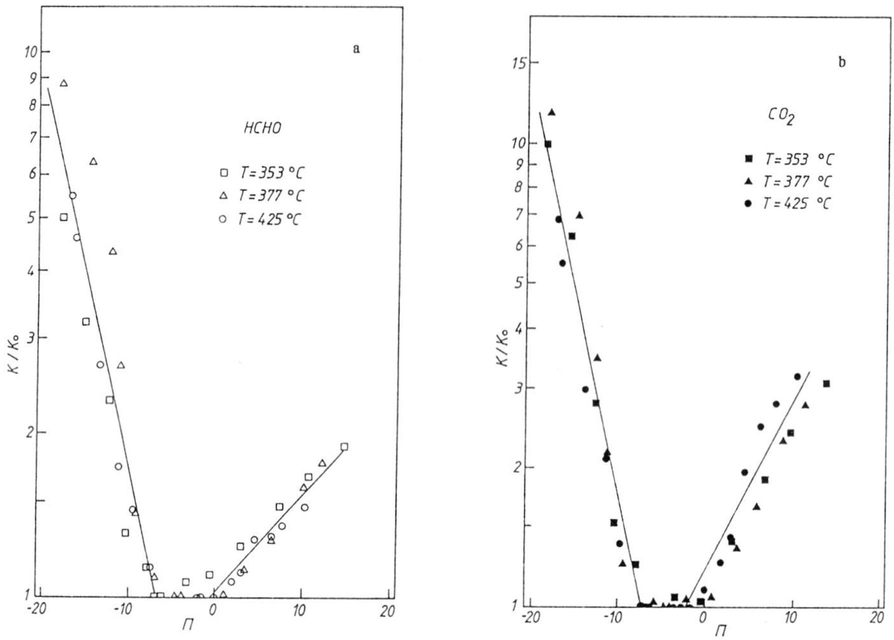

Fig. 7. Effect of catalyst IR-free potential and of $\Pi=FV_{WR}/RT$ on the kinetic constants of the reactions of H_2CO and CO_2 formation.

The open circuit kinetics are rather complex and are described in detail elsewhere (ref. 15). Figures 7a and 7b show the effect of catalyst potential V_{WR} and of the parameter $\Pi=FV_{WR}/RT$ on the kinetic constants of the two reactions. It can be seen that both reactions exhibit both "positive" ($\Lambda \gg 1$) and "negative" ($\Lambda \ll -1$) NEMCA behaviour. The constants α and V^*_{WR} of the NEMCA equation (3) take the values (α=0.043, V^*_{WR}=-30mV) and (α=0.087, V^*_{WR}=-120mV) for H_2CO and CO, formation, respectively, when the current is positive, i.e., for high Π, and the values (α=-0.16, V^*_{WR}=-395mV) and (α=-0.21, V^*_{WR}=-430 mV) for H_2CO and CO_2 formation, respectively, when the current is negative, i.e., for low Π values.

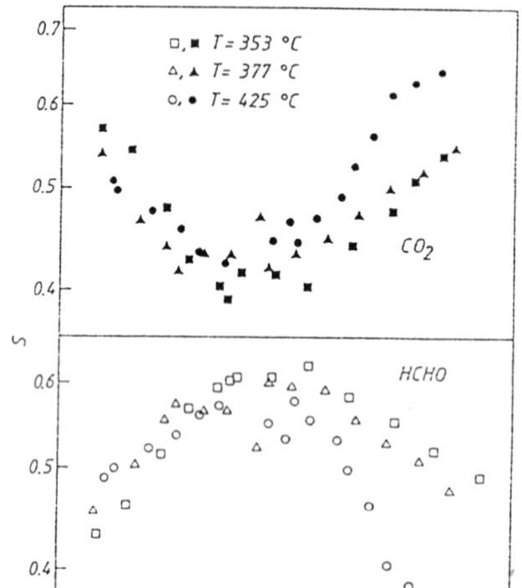

Fig. 8. Effect of the dimensionless potential Π on product selectivities.

Figure 8 shows the effect of Π on product selectivity. The selectivity to H_2CO can be varied deliberately between 35 and 60% and goes through a maximum near the open-circuit potential value.

DISCUSSION

The present results show that both C_2H_4 epoxidation on Ag and CH_3OH oxidation on Pt exhibit the NEMCA effect. The observed exponential rate dependence on catalyst potential and on the dimensionless group $\Pi = FV_{WR}/RT$ conforms to the NEMCA equation (3). Also the magnitude of the observed enhancement factors Λ conform to $\Lambda \approx 2Fr_0/I_0$.

Both reaction systems studied exhibit positive NEMCA behaviour for I>0. This applies both for the partial oxidation products (C_2H_4O, H_2CO) and for CO_2. Also in both cases the selectivity to the partial oxidation product decreases with increasing V_{WR}. This behaviour can be understood as follows: It is known (refs. 7-10) that $\Delta eV_{WR} = \Delta e\Phi$, where $e\Phi$ is the average catalyst surface work function. Therefore increasing V_{WR} corresponds to an increase in $e\Phi$ and to a concomitant decrease in the strengths of chemisorptive bonds (refs. 7-10,16). According to early theoretical consideration of Boudart (ref. 16) it should be $\Delta(-\Delta H_{ad}) = -(n/2)\,\Delta e\Phi$, where n is the number of valence electrons of the adatom taking part in the bonding and $-\Delta H_{ad}$ is the heat of adsorption. Therefore increasing V_{WR} causes a near-linear decrease in the heats of adsorption of the adsorbed species. If the rate limiting step of a catalytic reaction involves cleavage of a metal-adsorbate bond, then the rate of this reaction will increase exponentially with ($-\Delta H_{ad}$) and with V_{WR}, provided the activated complex remains unaffected (ref. 8). This explains both the observed exponential increase in the rates with V_{WR} and also the observed quasilinear decrease in activation energy with eV_{WR} with a slope of order -1 (Fig. 5). It is also likely that the observed decrease in partial oxidation product selectivity at high V_{WR} values is due to the weakening of the metal-oxygen bond and the concomitant enhancement in the mobility of chemisorbed oxygen which might favor rapid combustion of intermediates to CO_2.

In the case of negative currents, V_{WR} and $e\Phi$ decrease and, according to Boudart (ref. 16) chemisorptive bonds become stronger. In the case of C_2H_4O formation this was mainly found to decrease the rate, which can be accounted for by the above discussion. However in the case of CH_3OH oxidation, both H_2CO and CO_2 formation exhibit negative NEMCA behaviour in that region, i.e., both rates are accelerated by decreasing $e\Phi$ and, therefore, by an increase in the strength of the metal-adsorbate bond. Similar behaviour has been observed during CH_3OH dehydrogenation on Ag (refs. 7,9) and has been explained by considering that a strengthening in the metal-adsorbate bond causes a weakening of the inter-adsorbate bond strengths and a concomitant acceleration in the rate of adsorbate decomposition rate (ref. 9). The same explanation can be used here: As V_{WR} and $e\Phi$ decrease, chemisorptive bonds become stronger and hydrogen abstraction from methanol to form H_2CO and chemisorbed CO is accelerated. The latter is rapidly oxidized to CO_2 by chemisorbed oxygen, thus the rate of CO_2 formation is also accelerated. In fact, more negative potentials favor complete hydrogen abstraction from CH_3OH and H_2CO to form CO and therefore selectivity to CO_2 increases, as experimentally observed.

In summary, the rate and selectivity of metal catalyzed partial oxidations can be altered dramatically by controlling the catalyst work function independently from the gas phase via the NEMCA effect. Although with both reactions discussed here maximum partial oxidation product selectivity is near the open-circuit, i.e., regular catalytic, conditions there is no reason to expect that this will turn out to be a general observation.

REFERENCES

1 C.G. Vayenas, Catalytic and electrocatalytic reactions in solid oxide fuel cells, Solid State Ionics 28-30 (1988) 1521-1539.

2 H.-G. Lintz and C.G. Vayenas, Solid ion conductors in heterogeneous catalysis, Angewandte Chemie, in press (1989)

3 T.M. Gür and R.A. Huggins, Methane synthesis over transition metal electrodes in a solid state ionic cell, J. Catalysis 102 (1986) 443-446.

4 M. Stoukides and C.G. Vayenas, The effect of electrochemical oxygen pumping on the rate and selectivity of ethylene oxidation on polycrystalline silver, J. Catalysis 70 (1981) 137-146.

5 M. Stoukides and C.G. Vayenas, Electrochemical modification of the activity and selectivity of silver for light olefin oxidation, 8th Intnl. Congr. on Catalysis, Dechema, Berlin, 1984, pp.827-833.

6 I.V. Yentekakis and C.G. Vayenas, The effect of electrochemical oxygen pumping on the steady-state and oscillatory behavior of CO oxidation on polycrystalline Pt, J. Catalysis 111 (1988) 170-188.

7 C.G. Vayenas, S. Bebelis and S. Neophytides, Non-Faradaic electrochemical modifcation of catalytic activity, J. Phys. Chem. 92 (1988) 5083-5086.

8 S. Bebelis and C.G. Vayenas, Non-Faradaic electrochemical modification of catalytic activity: 1. The case of C_2H_4 oxidation on Pt, J. Catalysis 118 (1989) 125-146.

9 S. Neophytides and C.G. Vayenas, Non-Faradaic electrochemical modification of catalytic activity : 2. The case of CH_3OH dehydrogenation and decomposition on Ag, J. Catalysis 118 (1989) 147-163.

10 C.G. Vayenas, S. Bebelis, S. Neophytides and I.V. Yentekakis, Non-Faradaic electrochemical modification of catalytic activity in solid electrolyte cells, Applied Physics A 49 (1989) 95-103.

11 C.G. Vayenas, S. Bebelis and M. Despotopoulou, NEMCA effect using a β''-Al_2O_3 solid electrolyte, to be submitted to J. Catalysis (1989).

12 X.E. Verykios, F.P. Stein and R.W. Coughlin, The oxidation of ethylene on silver: Adsorption, kinetics, catalyst. Catal. Rev. Sci. Eng. 22(1980) 197-234.

13 R.B. Grant and R.M. Lambert, A single crystal study of the silver-catalyzed selective oxidation and total oxidation of ethylene, J. Catal. 92 (1985) 364-375.

14 S. Bebelis, Ph.D. Thesis, Univ. of Patras, Greece (1989).

15 S. Neophytides and C.G. Vayenas, Non-Faradaic electrochemical modification of catalytic activity: 3. The case of methanol oxidation on Pt, submitted to J. Catalysis (1989).

16 M. Boudart, Heterogeneity of metal surfaces, J. Am. Chem. Soc. 74 (1952) 3556-3561.

B. DELMON (University Catholique de Louvain, Belgium) : There is probably a narrow analogy, or possibly an identity, between the action of spillover oxygen ions in Professor VAYENAS' experiments and that in our experiments with mixtures of oxides which do not contaminate each other. It seems that here too we have a "remote control", namely oxygen ions sent by another part of the system (the "Donnor" part) creates catalytic sites or adjusts catalytic active centers on an Acceptor surface.

It is clear here that we have a case of remote control: one single O^{2-} species permits the reaction of several hundreds of thousands of O_2 molecules. The Y stabilized zirconia and the electric potential constitute the Donnor (instead of α-Sb_2O_4, $BiPO_4$, Bi_2O_3 etc in our case).

The results of Professor VAYENAS seem to confirm our conclusion that the O_{SO} species is O^{2-}.

We have another dramatic example of the crucial role of surface dynamic phenomena (spill-over and reaction of this species with surface for modifying catalytic activity) in catalysis.

How would Professor VAYENAS contemplate the modification or reaction of active centers on the Acceptor, namely Ag or Pt?

C.G. VAYENAS (University of patras, Greece): Indeed there appear to be interesting similarities between NEMCA and the remote control phenomena, caused by oxygen spillover between oxides, studied by Professor DELMON. In a very broad sense one can consider the NEMCA effect as a remote control effect, a very remote one indeed, since some of the metal crystallites (acceptors) changing dramatically their catalytic activity are 50,000 Å away from the solid electrolyte-catalyst-gas three-phase-boundaries (donnor). It would be very interesting to use a zirconia electrolyte to study NEMCA on some of the "acceptor" oxides investigated by Professor DELMON during partial oxidation reactions. This could help identify similarities between the two phenomena.

Regarding the nature of oxygen anions which spill-over the metal surfaces we have studied so far, causing the NEMCA effect, there is strong electrokinetic evidence (Refs. 1,2) that it is O^- produced by partial oxidation of O^{2-} at the three-phase-boundaries. There is also some XPS evidence supporting this (Refs. 1,3), but further spectroscopic investigation is definitely necessary. It may very well be that the charge of spill-over oxygen is surface-specific.

A unique feature of using a solid electrolyte as a "donnor" in Professor DELMON's terminology, and a conductive catalyst film as an "acceptor", is that one can electrochemically both control and measure the reversible changes induced by ion spillover on the electronic properties of the metal catalyst and the concomitant dramatic changes in catalytic activity. Our results (Refs. 1-4) shows a) that ion spillover causes a change in the gas exposed catalyst surface work function $\Delta e\Phi$, measured independently by a Kelvin probe (ref. 4), which equals $e\eta$, where η is the overpotential developed at the catalyst-solid electrolyte interface. b) that catalytic rates depend exponentially on catalyst work function and c) that reaction activation energy usually changes linearly with catalyst work function. These observations can be made at practically constant coverages of adsorbed species (ref. 2) since the coverage of spill-over ions is typically less than 5% (ref. 4).

Therefore our NEMCA results on Pt, Ag and Pd can be explained without invoking the creation of new active centers although up to 6,000% increases in catalytic activity are observed. They seem to indicate a rather continuous variation in the binding energies of chemisorbed species on the already existing catalytic sites with varying catalyst work function.

1 C.G. Vayenas, S. Bebelis, S. Neophytides, **J. Phys. Chem.** 92 (1988) 5083.
2 S. Bebelis, C.G. Vayenas, **J. Catal.** 118 (1989) 125.
3 T. Arakawa, A. Saito, J. Shiokawa, **J. Appl. Surf. Sci.** 16 (1983) 365).
4 C.G. Vayenas S. Bebelis, S. Ladas, submitted for publication (1989).

M. MICHMAN (Hebrew University, Jerusalem, Israel) : Can the open circuit reaction and closed-circuit reaction be mechanistically distinguished? Because the experiments deal both with generation of reaction induced by O^{2-} and surface reactions like the open-circuit catalysis which are also potential dependent. The two reaction routes may not always be quantitatively distinct. Can this affect the quantitative methods described here?

C.G. VAYENAS (University of Patras, Greece) : For the systems where the NEMCA effect has been studied so far (e.g. Refs. 1-4) the electrocatalytic and catalytic reactions are kinetically and mechanistically distinguishable.

The electrocatalytic reaction of O^{2-} deelectronation to O^- takes place at the zirconia-metal-gas three-phase-boundaries (Refs. 1-3). Its rate is I/2F and its mechanism can be studied by examining the dependence of I on the overpotential η (Refs. 2,3).

The catalytic reaction takes place over the entire gas-exposed catalyst surface, which is typically 100-2000 cm^2 in our experiments (Refs. 1-4). Its open-circuit rate r is usually 10^2-10^3 times higher than I/2F values typically employed in NEMCA studies. The NEMCA induced increase in catalytic rate Δr is typically 10^3-10^5 time higher than I/2F. One of the central findings of NEMCA is that r changes exponentially with η. This is because an overpotential η at the catalyst-electrolyte interface causes a change $\Delta e\Phi = e\eta$ to the work function of the gas-exposed catalyst surface. This result, which stems from the spatial uniformity of the Fermi level in the conductive catalyst, has been predicted theoretically (Ref. 1-4) and recently verified experimentally using a Kelvin probe (Ref. 5).

The spillover oxygen anions do not participate in any direct way in the catalytic reaction. The consumption rate of spillover oxygen is, at most, I/2F, i.e., negligible both with respect to r and Δr. However, it is the spillover oxygen ion presence on the catalyst surface which changes the catalyst work function and this induces the observed dramatic changes in catalytic activity.

It is worth considering as an example the case of C_2H_4 oxidation on Pt, the NEMCA behaviour of which has been studied very thoroughly (Ref. 2). It was shown conclusively that NEMCA does not change the form of the catalytic kinetic expression and thus the catalytic reaction mechanism, but rather induces dramatic changes in the kinetic constant values. This, of course, was found to change the rate limiting step but not the reaction mechanism (Ref. 2).

1 C.G. Vayenas, S. Bebelis, S. Neophytides, **J. Phys. Chem.** 92 (1988) 5083.
2 S. Bebelis, C.G. Vayenas, **J. Catal.** 118 (1989) 125.
3 S. Neophytides, C.G. Vayenas, **J. Catal.** 118 (1989) 147.
4 C.G. Vayenas, S. Bebelis, S. Neophytides, I.V. Yentekakis, **Appl. Physics A** 49 (1989) 95.
5 C.G. Vayenas, S. Bebelis, S. Ladas, submitted for publication (1989).

K.OGURA (Yamaguchi University, Japan) : How do you think the charging current is involving in your catalytic reaction? Because you use a constant current technique, I don't think you can neglect the charging current problems.

C.G. VAYENAS (University of Patras, Greece) : There are no charging current "problems". In general, creating a net charge on a metal surface changes its Volta and Galvani potentials but leaves its work function and catalytic properties practically unaffected. In our system the polarizing current changes the (equal and opposite) charges at the metal-solid electrolyte interface but does not create any net charge on the catalytically active gas-exposed catalyst surface, as one can show from simple electrostatic considerations (Ref. 1). The charge of each spillover ion is compensated by an equal and opposite charge in the metal. It is the creation of these surface dipoles which changes the catalyst surface potential and work function and induces the observed dramatic changes in catalytic activity. Incidentally, both galvanostatic and potentiostatic operation lead to the same steady-state rate enhancement when compared at the same catalyst potential.

1 S. Bebelis, C.G. Vayenas, **J. Catal.** 118 (1989) 125.

G. Centi and F. Trifiro' (Editors), *New Developments in Selective Oxidation* 653
© 1990 Elsevier Science Publishers B.V., Amsterdam — Printed in The Netherlands

ELECTROGENERATIVE OXIDATION OF ETHANOL ON PLATINIZED-CARBON ELECTRODES

Huang Zhong-Tao, Ye Daiqi and Pang Xianxing

Department of Chemical Engineering, South-China University of Technology, Guangzhou 510641, P.R.China

SUMMARY

High performance platinized-carbon electrodes are developed for utilization in electrogenerative process. Load current density as high as 600 mA per square centimeter can be achieved for oxygen reducing in 6N sulfuric acid with a good stability. With these electrodes and sulfuric acid electrolyte in electrogenerative cell, ethanol vapor carried by nitrogen gas can be selectively oxidized to give acetaldehyde. Selectivity of acetaldehyde depends on potential of the cell and the feed rate of ethanol vapor and it can be more than 80% in optimized conditions. The initial product of ethanol oxidized on platinized-carbon electrode is acetaldehyde.

INTRODUCTION

The selective oxidation of ethanol to acetaldehyde is one of special interest in areas where biomass-based economics are under development since this product can be used for the synthesis of other basic chemicals such as acetic acid, acetic anhydride, butanol, etc.(ref. 1) With the advancement of fuel cell technology and electrocatalysis stimulated by world energy and raw material situation provide some special opportunities for investigating electrosynthesis of organic chemicals not only without a power source but with concurrent electricity generation(ref. 2, 3). Electrogenerative processes produce desired chemicals while generating electricity. However, chemical synthesis, i.e. producing desired chemicals, is the primary motivation and electricity is the by-product. Apart from energy recovery, there are possibilities of controlling reaction rate as well as selectivity through potential and other electrocatalytic means. For high free energy change and using air(O_2) as one of reactants, the electrogenerative oxidation of ethanol to give acetaldehyde is a very promising electrogenerative process(ref. 4). The electrooxidation of ethanol on platinum has been the subject of numerous investigations with the purpose of mainly to elucidate the mechanism with the double-layer region on platinum(ref. 4-10), but not yet well understood. Most reported oxidations of ethanol to acetadehyde in the literature have been with dissolved subsrate systems, very few devoted to the investigation of vapor phase oxidation on gas porous electrodes, and the recovery of acetadehyde has been pursued in only a few instances(ref. 9, 10). Work reported here is the electrogenerative oxidation of ethanol with air(O_2) on porous platinized-carbon electrodes.

EXPERIMENTAL

Preparation of catalyst and electrodes

Active carbon chosen as catalyst's support is subjected to repeated gravity separation with saturated calcium chloride solution and "Soxhlet" treatment with azeotropic hydrochloric acid, the ash content of so treated carbon is found to be less than 0.03%. Then activated by heating in CO_2 atmosphere at 900℃ for 4 h. The surface area of so activated carbon obtained by BET method is found to be more than 1500 m^2/g. Acetylene black, used as basic material for preparation of porous hydrophobic(i.e. gas-permeable liquid-impermeable) membrane is also subject to pretreat with the similar procedure as active carbon. The BET surface area of gas activated acetylene black is about 500 m^2/g. Catalyst are prepared by depositing platinum from chloroplatinic acid onto the active carbon by sodium formate reduction, platinum loading is 10 wt.%. The subsequent heat treatment is conducted in a quartz tube for 3 h at 300℃ hydrogen atmosphere.

Electrode are manufactured by pressing Ni screen(about 80 mesh), diffusion membrane and catalytic membrane together under about 500 kg/cm^2 pressure for one minute. Diffusion membrane(i.e. porous hydrophobic membrane) is made by following procedure: mix the 60% (in weight) Teflon(PTFE) emulsion, gas activated acetylene black and "OP" emulsifier with the ratio of 10:4:2(in weight), add suitable volume of distilled water until it become dough state, then roll it carefully and repeatedly, a very thin and uniform membrane can be obtained. Its thickness is about 0.5 mm. The resulting film is subjected to repeated soxhlet treatment with acetone for 48 h in order to remove the

"OP" and create the fine pores within the membrane. Catalytic membrane(semi-hydrophobic) is manufactured with same procedure as diffusion film, but the Telflon(PTFE) content is about 20%(in weight) and the thickness is about 0.2mm. The resulting platinum loading of catalytic membrane is found to be about 6-7 mg/cm^2.

Electrochemical cells

The electrochemical performance of electrode are tested in elcetrochemical cell as shown in Figure 1 by reducing O_2(air) at room temperature. The current-potential curves were obtained galvanostatically and are not IR corrected. Each experimental run is repeated with different electrodes of the same type in order to ascertain the reproducibility of the data. Lifetime tests are conducted on the optimized electrodes by loading them upto acceptable load currents over extended durations and the accompanying changes in the electrode potential are recorded.

The electrogenerative cell and system are represented in Figure 2 and could be operated at different temperature. Two porous electrodes with about 5.5 cm^2 of exposed geometric are separated by a 1 cm thick electrolyte chamber containing 6N H_2SO_4 which can be forced to flow during the measurement. The substrate carrier gas N_2 is passed through a gas washing bottle with about 100ml ethanol liquid. A bubble type flow meter is used to measure flow rates by diverting streams before and after the cell. Air(O_2) flow rate at the cathode exceeds required flow rates for complete combustion at the anode. Analyses of the

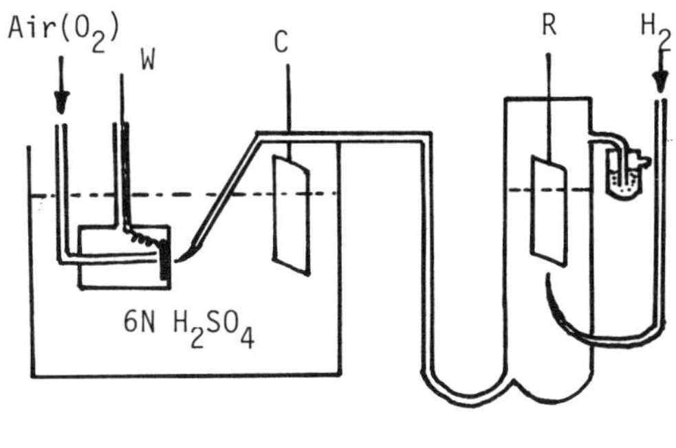

Fig. 1. Electrochemical cell for the evaluting performance of electrode, work in room temperature.

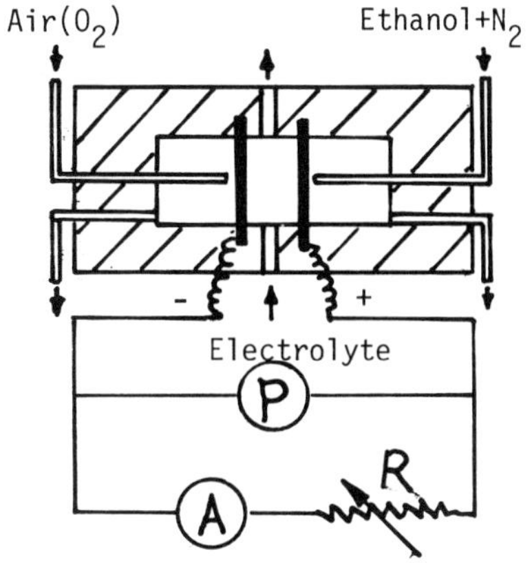

Fig. 2. Electrogenerative cell.

inlet and outlet streams are carried out using a Shimadzu 7A gas chromatograph equipped with a 3m× 3mm Carbowax 1500 column. Product CO_2 is analyzed with a Porapak Q column. The polarization curves reported here are the second ones taken after a period of cell operation. The third polarization experiment gave comparable results. Current and potential values are measured with instruments in the external circuit(see Figure 2) on a reaching a relative stable potential after about ten minutes' operation. The reported cell potential are IR corrected using cell resistance values measured under hydrogen at both electrodes before polarization experiments were initiated.

RESULTS

Electrochemical performances of electrodes

The current-potential curves for O_2 reducing in 6N H_2SO_4 at room temperature are given in Figure 3. Four types of electrodes are evaluated. Two of these four without Pt catalyst loading, and two are not gas activated. Except these differences the manufacture procedure and measure conditions are the same. It clearly shows that the loading of Pt enhances the electrode's performance significantly. However, with the same Pt content, gas activated one gives a much better performance. The platinized-carbon electrodes are made more cost effective by enhancing their catalytic activity through preparation of highly dispersed platinized platinum catalyst though with a relative lower Pt loading. From Figure 3, one can see that having gas

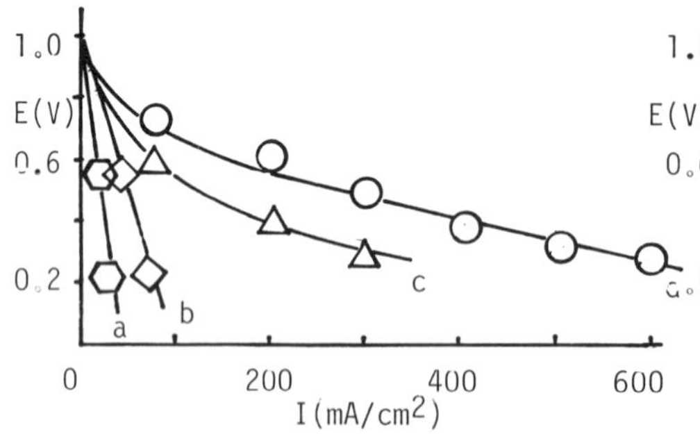

Fig. 3. Current-potential curves for O_2 reducing in 6N H_2SO_4, 20°C
a: untreated carbon; b: gas activated carbon; c: 10% Pt on a; d: 10% Pt on b. E(V) vs rhe.

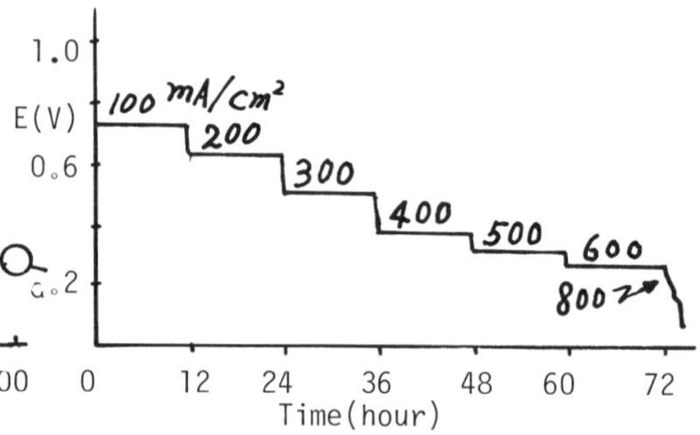

Fig. 4. Lifetime test for 10% Pt on gas activated active carbon electrode, measure conditions are same as Fig. 3.
E(V) vs. rhe.

activated high-surface-area active carbon with well dispersed platinum black deposited on it, the performance of the electrode is the best, it can withstand load current upto 600 mA/cm^2. The current density output for electrooxidation of ethanol is also comparable to the value(50 mA/cm^2) reported in literature though the electrode has a lower Pt catalyst loading. Electron microscopy shows the gas activated one has uniform pore and Pt particle size distribution. While inactivated one has a formation of platinum metal agglomerates which result in the reduction of active surface area and then the poor activity. Lifetime test conducted on the optimized type electrode, as shown in Figure 4, demonstrated the very good stability of these electrodes. Running hydrogen-air(O_2) fuel cell before and after every electrogenerative oxidation of ethanol can give reproducible current and potential output indicates that the electrodes were not affected by the time of experiment, this result agrees well with the lifetime test shown in Figure 4.

<u>Electrogenerative oxidation of ethanol</u>

Selected results of several experiments together with experimental parameters are list in Table 1. For electrogenerative oxidation of ethanol, the measured open circuit potential is about 0.64-0.65V. Using air(O_2) but not pure oxygen as one of the reactants, the measured open circuit potential values (0.64-0.65V) are a little lower than the value (0.67 V) using pure O_2 as reported in literature(ref. 10) but much lower

TABLE 1
Selected results of ethanol oxidation on platinized-carbon electrodes in 6N H_2SO_4

T(℃)	R(Ω)	E(V)	I(mA/cm^2)	P_e(V)^a	F(cc/min)	S(%)^b
		0.41	10	0.49		82.1
20	0.58	0.36	25	0.54	60	75.3
		0.16	45	0.74		71.4
		0.43	20	0.47		76.2
40	0.52	0.32	40	0.58	60	70.7
		0.17	52	0.73		67.8
		0.42	20	0.48		73.5
60	0.48	0.38	40	0.52	60	68.6
		0.16	60	0.74		64.2

a: P_e is the anode potential estimated from the overpotential of air electrode and cell's operating voltage; b: S is the selectivity of of acetaldehyde from gas phase product.

than calculated one(1.05V), this is because at an open circuit a significant amount of cell voltage can be lost initially due to the rest overpotential especially the air electrode. The down of the potential with the first current is due to the activation overpotential. However, the open circuit(rest) overpotential and activation overpotential can be reduced with the effective electrocatalyst at one or both electrodes.

Figure 5 and Figure 6 shown the effects of the increase of substrate's flow rate and cell's temperature on polarization curves. Since reactions take place on the interface between gas-electrocatalyst-electrolyte, the basic challenge is to provide effective transport paths for reactant access, product removal, and current output. Fast flow rates enhanced mass transfer, then the operation performance improved significantly as shown in Figure 5 indicates that the transfer of ethanol molecules to catalyst sites through electrode inner pores is the critical factor determining the whole process rate. Temperature increase is expected to enhance the mass transfer and reaction rate, and cause a decrease of cell's ohmic losses, the current output is increased as shown in Figure 6 and Table 1. For weakly adsorbed reactants, the reaction rate may significantly increase with temperature, however, for reactants adsorbing strongly on the electrocatalyst, temperature increase will probably have a small effect. Since the effect of raising temperature is not so significantly, ethanol should be adsorbed on Pt active sites but not very strong.

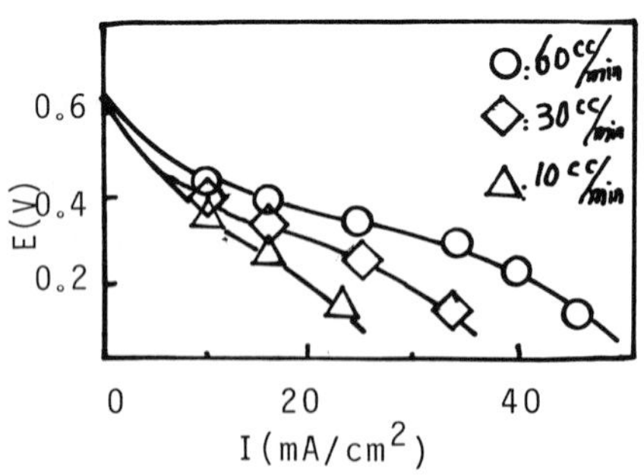

Fig. 5. Effect of flowrate on operation characteristics. (at room temperature)

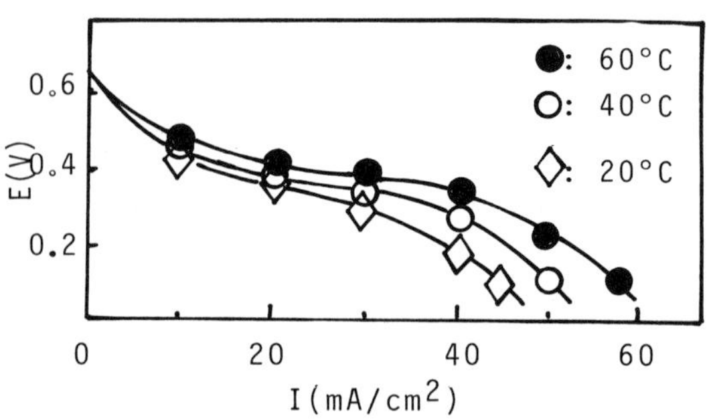

Fig. 6. Effect of temperature on operation characteristics. (at flowrate of 60 cc/min)

The main product of oxidation in gas phase are acetaldehyde and CO_2, no CO has been detected. At the electrolyte side, no significant by-product such as CH_3COOH and diethyl acetal produced. The selectivity of ethanol oxidation to give acetaldehyde is determined by cell's potential, substrate flow rate and reaction temperature: The higher the temperature, the faster the reaction rate but the lower the selectivity because the acetaldehyde is oxidized to CO_2 before it can be desorbed from catalytic sites. The faster the flowrate the faster the mass transfer and the higher the selectivity because fresh ethanol tend to displace and to sweep acetaldehyde from catalytic sites. The lower the cell's voltage, the larger the current output and then the lower the selectivity again because acetaldehyde is oxidized to CO_2 before it is desorbed for the faster reaction rate.

DISCUSSION

It is essential to employ porous electrodes since the electrode current, that is reaction rate, attainable under given conditions is not only a function of the potential, but it also dependent on the surface available for the reaction to occur. However, this lead to the complexity of the process and make it difficult to study the reaction mechanism, since the reactant's access and product's removal through micropores, that is mass transfer or diffusion, are necessary and these will influence the process significantly.

A lot of investigations have been devoted to the study of the ethanol oxidation mechanism. Several authors share the opinion that in acidic solutions and potentials less than E_{rhe} = 0.9 V, ethanol can only be oxidized to acetaldehyde(ref. 5, 6, 8). Snell and Keenan suggested that adsorbed ethanol reacted with the PtOH or PtO species to give acetadehyde(ref. 5, 6). However, Willsau and Heitbaum think that the ethanol is oxidized to acetaldehyde directly at a lower potential by the cleavage of one hydrogen of the α -C-atom and the hydroxyl hydrogen, and at a little higher potential CO_2 produced parallelly(ref. 7). In our experiment, the ethanol anode is operated in the potential region of 0.50-0.80V as listed in Table 1. Considering the ohmic and concentration polarization can develop along the pores, it should work in even lower potential than that values. In these range of potential, platinum is known not to undergo any surface oxidation or oxide film formation: the PtOH or PtO species should not be formed significantly. Based on the fact of ethanol chemisorbed on Pt and oxidation product analyses, for ethanol oxidized to acetaldehyde, a mechanism consistent with the above considerations involves a fast and chemisorption step followed by a electrochemical step,

$$CH_3CH_2OH \text{ --- } \underset{Pt}{CH_3^{*}CHOH} + H^+ + e^- \quad (1)$$

$$\underset{Pt}{CH_3^{*}CHOH} \text{ --- } CH_3CHO + H^+ + e^- \quad (2)$$

Since a fast flow rate improve the selectivity of acetaldehyde, and acetaldehyde has a similar adsorption and oxidation property, it seems it is further oxidized to CO_2:

$$CH_3CHO \overset{Pt}{\text{----}} CH_3CHO_{ads} \quad (3)$$

$$CH_3CHO_{ads} + 3\ H_2O \text{ --- } 2\ CO_2 + 10\ H^+ + 10\ e^- \quad (4)$$

In conclusion, on platinized-carbon electrode, the ethanol is oxidized to acetaldehyde directly, and the acetaldehyde can be further oxidized to CO_2.

ACKNOWLEDGMENT This work was partially supported by National Education Commission of China.

REFERENCES

1. Palsson, B.O., Fathi-Afshar, S., Rudd, D.F. and Lightfoot, E.N., Biomass as a source of chemical feedstocks: an economic evaluation, Science, 213(1981)513.
2. Huang Zhong-Tao and Ye Daiqi, The research and development of electrogenerative processes, Chem. Eng. & Ind. Prog., 1(1987)21.
3. Langer, S.H. and Sakellaropoulos, G.P., Electrogenerative voltameiotic processes, Ind. Eng. Chem. Process. Des. Dev., 18(1979)567.
4. Ye Daiqi and Huang Zhong-Tao, A research on electrocatalytic oxidation of ethanol, Proceedings of 4th National Catalysis Congress, 1-F-10, 1988, Tianjin, P.R.China.
5. Snell, K.D. and Keenan, A.G., Chloride inhibition of ethanol electrooxidation at a platinum electrode in aqueous acid solution, Electrochim. Acta, 26(1981)1339.
6. Snell, K.D. and Keenan, A.G., Effect of anions and pH on ethanol electrooxidation at a platinum electrode, Electrochim. Acta, 27(1982)1683.
7. Willsou, J. and Heitbaum, J., Elementary steps of ethanol oxidation on Pt in sulfuric acid as evidenced by isotope labeling, J. Electrochem. Soc., 194(1985)27.
8. Rightmire, R., Rowland, R.L., Boos, D.L. and Beals, D.L., Ethyl alcohol oxidation at platinum electrodes, J. Electrochem. Soc., 111(1964)242.
9. See Meshbesher, T.M., U.S.Patent No. 4,347,109(Aug. 31, 1982)
10. Langer, S.H., Card, J.C. and Foral, M.J., Electrogenerative and related processes, Pure & Appl. Chem., 58(6)(1986)895.

G. Centi and F. Trifiro' (Editors), *New Developments in Selective Oxidation*
© 1990 Elsevier Science Publishers B.V., Amsterdam — Printed in The Netherlands

OXIDATION OF L-PHENYLALANINE AND RELATED REACTIONS : PREPARATION OF L-DOPA.

M. BLANCHARD, C. BOUCHOULE, G. DJANEYE-BOUNDJOU, P. CANESSON
Laboratoire de Catalyse, Faculté des Sciences, 40 Avenue du Recteur Pineau
86022 POITIERS (FRANCE).

SUMMARY

L-Phenylalanine and phenylethylamine are oxidized at low temperature with Fe^{2+} complexes : L-DOPA and tyrosins, tyramine and Dopamine are respectively obtained. The catalytic cycle is created by the use of electrons provided by the cathode of an electrochemical cell. In order to improve the yield in hydroxy compounds these are continuously evacuated out of the reactor.

INTRODUCTION

Aromatic compounds are oxidized at low temperature on the benzenic ring (1) (2) by dioxygen and a catalyst such as Fe^{2+} complex. As the reaction proceeds the metallic cation is oxidized and a reducing agent must be added in order to use catalytic amounts of metals ; this can be done with organic compounds such as ascorbic acid (Udenfriend's system) but the byproduct of its oxidation, namely oxalic acid, displaces the metal from its ligand and the reaction does not proceed any more. This catalytic system is therefore limited to low concentrations of substrate.

We wish to report the results that we have obtained with a relatively more simple system, including dioxygen, a catalyst and electrons ; these are provided by the cathode of an electrochemical cell (3). The substrate is L-Phenylalanine or phenylethylamine which are selected as a source of valuable products such as tyrosines, tyramines, DOPA or Dopamine.

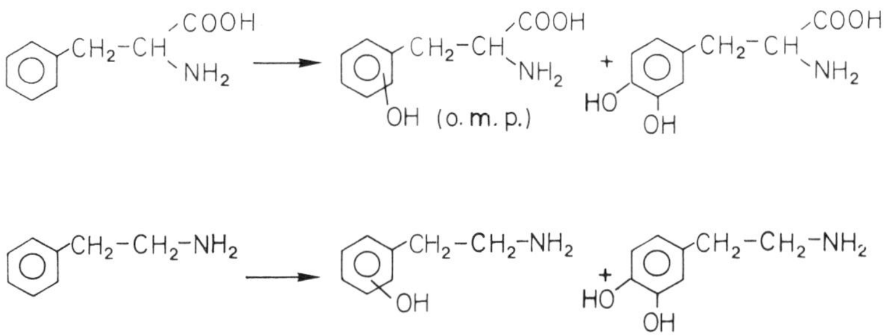

EXPERIMENTAL

The reactions are carried out at 40°C in an electrochemical cell (200 ml) equipped with a cathode made of a ribbon of carbon fiber (45 cm), a platinum anode and a reference electrode (calomel). The average intensity is 90 mA. The aromatic substrate (1.50 mmole) is dissolved in 100 ml of buffer solution (pH = 3) containing 1.50 mmole of a Fe^{2+} complex ; the solution is continuously swept with dioxygen and analysis are performed by HPLC. The column (C18 Rosil) is eluted with a H_3PO_4 0.05M -MeOH solution (1:1) containing 0.005 mole l^{-1} of sodium dodecylhydrogenosulfate. The identification of the products is made by comparison of their elution times with those of known samples.

For the in infrared experiments, the spectra are recorded in water or if necessary in heavy water with a spectrophotometer IRFT, NICOLET MX-10.

RESULTS

a - Oxidation of L-Phenylalanine

In the experimental conditions above mentioned, the oxidation of L-Phenylalanine has been carried out with various Fe^{2+} complexes. The activity and selectivity are reported in the Table 1. The activity is expressed as the amount of substrate consumed per unit of time and the selectivity for the production of a compound i (DOPA or Tyrosin) is the yield of this product versus the amount of substrate consumed. By initial activity we mean the slope of the graph % product vs time measured at zero time.

TABLE 1

Rate and selectivity of the oxidation of L-Phenylalanine pH = 3, T = 40°C, i = 50-90 mA.

Ligand	Initial rate ($mmole.\ h^{-1}$) of oxidation	Initial rate of formation (DOPA) ($mmole.h^{-1}$)	Ratio DOPA/Tyr
Trans 1,2 diamino Cyclohexane N,N,N',N' Tetracetic Acid	0.360	0.08	0.5
1,3 diamino 2-hydroxypropane N,N,N',N' tetracetic acid	0.450	0.18	1.4 - 1.5
Diethylene triamine Pentacetic acid	0.740	0.60	0.4 - 0.6
1,2 diaminopropane N,N,N',N' tetracetic acid	0.690	0.16	0.6 - 0.7
E.D.T.A.	1.080	0.11	0.6 - 0.8

As a first conclusion it appears that all of the catalysts tested are rather active, except may be for the trans 1,2 diamino cyclohexane N,N,N',N' tetracetic acid. In most of the experiments all the substrate is consumed after 4 hours, but the selectivity decreases as soon as the conversion increases because the tyrosins and the DOPA are more sensitive to oxygen : they are converted into polyphenols which cannot be detected by chromatography. As far as the kinetics are concerned it is worthwhile to study the influence of the ratio substrate = catalyst : the results are reported in the Fig. 1 for the Fe^{2+}/EDTA catalyst. Above the ratio substrate : catalyst = 0.3 the rate remains constant : this data would be in agreement with the formation of a substrate - catalyst complex. Moreover the I.R. spectra recorded in water or heavy water show a shift of the 1408 cm^{-1} band to 1386 cm^{-1} when L-Phenylalanine and Fe EDTA are mixed in equal proportions. It would indicate an interaction of the ionised carboxyl group with the catalyst [5].

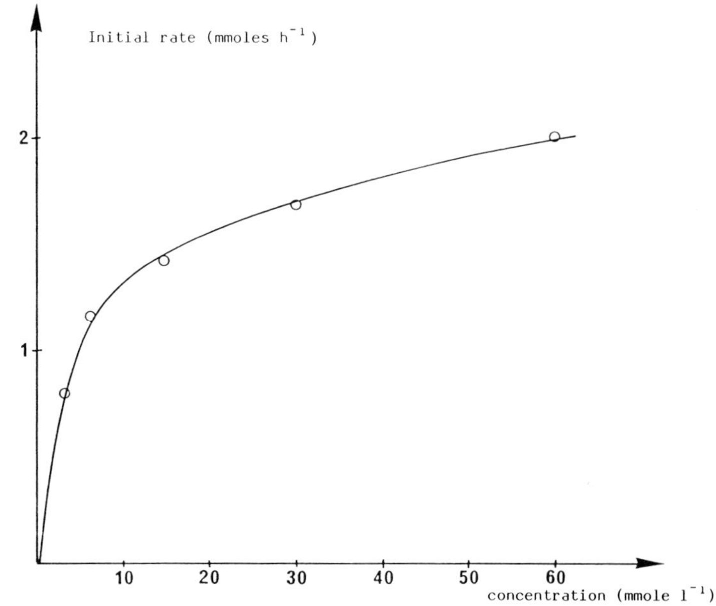

Fig. 1 : Variation of the initial rate of disappearance of substrate (phenylalanine)
[catalyst] = 0.05 mole l^{-1} pH = 3 T = 40°C.

As already mentioned the selectivity of the DOPA production decreases when the conversion increases, even with the best catalyst : 1.3 diamino 2-hydroxypropane N,N,N',N' tetracetic acid. This arises from the oxidation of the DOPA into polyphenols, undetected by chromatography. In order to improve the yield, the experimental apparatus has to be modified in such a way that the required products, hydroxy compounds, are displaced from the reaction medium. We report the results obtained with phenylethylamine which was chosen as a model molecule, and which can lead to the tyramines and dopamine.

b - Oxidation of phenylethylamine.

In similar experimental conditions as those which are used for the phenylalanine oxidation the tyramine yield was 23% for a conversion of 37% of the amine. Moreover the ratio o : m : p was 40 : 28 : 32 (expressed as %). The dopamine yield was 27%.

Here again the yield is affected by the formation of polyphenols arising from the oxidation of the dihydroxyamine. In a second set of experiments the electrochemical cell is connected to a column packed with Amberlite resin (IRA 400) and the solution is pumped at the top of this column : at the same point a 10 N solution of NaOH is continuously added in order to transform the tyramines and the dopamine into their sodium salt. At the bottom of the column the solution is reinjected into the cell after an addition of a 10 N solution of HCl whose amount is adjusted for the neutralization of the excess NaOH.

It is expected from this device that the phenate ions are absorbed by the column and that the unreacted amine is recycled into the reactor. If the recycling rate is high, the residence time of the hydroxy compounds in the reactor is low and a better selectivity must be obtained. At the end of the test the column is eluted with 10 N HCl and the solution is analysed by HPLC for the estimation of the products (as amine salts).

In these conditions, the amount of dopamine could not be estimated precisely because it was partially retained on the column, but for a conversion of 21%, a 33% yield of tyramines was obtained with a o : m : p ratio of 29 : 29 : 40. This has to be compared with 40 : 28 : 32 obtained without the resin.

Work is actually in progress in order to improve the resin but it appears from preliminary results that there is some possibility to increase the selectivity.

REFERENCES

1 S. Udenfriend, T. Clark, J. Axelrod, B.B. Brodie, J. Biol. Chem. 208, 1954, 741.
2 A.A. Akhrem, D.I. Metelitsa, E. Skurdo, Russ. Chem. Rev. 44, 1975, 398.
3 J.M. Maissant, C. Bouchoule, P. Canesson, M. Blanchard, J. Mol. Cat. 9183, 18, 1983, 189.
4 J.F. Pearson, M.A. Slifkin. Spectrochimica Acta 28A, 1972, 2403.

1) MICHAEL MICHMAN (The Hebrew University of Jerusalem)

Is there no loss of Fe^{2+} catalyst by deposition of iron on the cathode ? The factors which would control this are a cathode with low overpotential for hydrogen as well as the type of complex ligand to which Fe^{2+} is bound. Have this been checked ?

MICHEL BLANCHARD (University of Poitiers - France)

There is no loss of iron on the cathode, but it may occur that some of the Fe^{2+} complex is oxidized. This last point has to be checked.

2) MICHAEL MICHMAN (The Hebrew University of Jerusalem)

Table 1 shows various initial rates for individual iron complexes. Is "rate" standing for current or chemical analysis. The various complexes differ in their redox potential and also their interaction at electrode surface and the double layer. Voltammetric analysis could screen the functionality and efficiency of the different catalysts and provide rational informations both on this electrolytic behaviour and the consequent chemical steps. Have any of this been attempted ?

MICHEL BLANCHARD (University of Poitiers - France)

The initial rates reported inthe Table 1 are calculated by using the results of a chemical (chromatography) analysis.
As for the second point, the voltammetric analysis were not carried out but this is a very good suggeestion.

3) J.C. VEDRINE (Institut de Catalyse - Lyon France)

You have used a resin to improve electron transfer and catalytic feature. What kind of resin do you use (acidic, cation exchanged type .....) and how is the iron complex attached to the resin in order to influence the properties.

M. BLANCHARD

The resin was an anionic one (Amberlite IRA 400). It is outside the electrochemical cell ; its purpose is to fix the phenols (tyrosins and L.DOPA) as phenate ions, in order to avoid their oxidation into polyphenols. There is no catalytic effect of this resin.

G. Centi and F. Trifiro' (Editors), *New Developments in Selective Oxidation*

Selective Electrooxidation of 4-(Di-n-propylsulfamyl)toluene to 4-(Di-n-propylsulfamyl)benzoic acid.

M. Michman and M. Weiss, Department of Organic Chemistry, The Hebrew University of Jerusalem, Jerusalem Israel 91904

SUMMARY

4-(Di-n-propylsulfamyl)benzoic acid is an important drug, known under commercial names such as benemid or probenecid [1]. It's preparation under mild conditions by an electrochemical method is described. The method involves cathodic generated oxygen radical-anion which selectively converts the methyl group of 4-(Di-n-propylsulfamyl)toluene to carboxylate. The sulfonamide group which would be reduced under high cathodic potential is not attacked under these conditions. This method circumvents safety problems and the need for corrosive reagents involved in chemical methods of preparation.

INTRODUCTION

Interest in electrogenerated superoxide anion radical $[O_2^-]^*$ and its potential as a reagent is reflected in recent works [2] and reviews [3]. The cathodic reduction of oxygen can be closely followed by electroanalytical methods and $[O_2^-]^*$, in aprotic solvents has shown selective reactivity as an electrogenerated base (EGB) and nucleophile. Low selectivity is often a problem with organic oxidations in general but with electroorganic synthesis in particular when more than one chemical process is activated under a required electrode potential. The introduction of *indirect electrolysis* [4] or of electrogenerated active species [5] has been recognized as a promising remedy to low selectivity.

Some years ago we have described a method for the oxidation of 4-(Di-n-propylsulfamyl)toluene (I) to 4-(Di-n-propylsulfamyl)benzoic acid (II) which is an important drug with uricosuric activity, known under commercial names such as *benemid* or *probenecid* [1]. Oxidation of I was by oxygen or air in alkaline aprotic media [6a] in similarity to reactions described by Russell et. al. [7] where the role of $[O_2^-]^*$ has been suggested but in our case not unequivocally proved.

Most Russell-type reactions of hydrocarbons yield peroxides, yet reactions with electrogenerated $[O_2^-]^*$ reported by Lund and Osa et. al. [2a,b] yielded carboxylic acids from alkyl aromatics . It was our purpose to examine $[O_2^-]^*$ in the oxidation of I as well as test the practicality of the *electrochemical* route as a selective method in this synthesis. The *chemical* procedure requires pressurized oxygen, high concentrations of potassium *tert*-butanolate or KOH in powder form, and suffers from difficulties in initiation, unpredictable induction periods and safety problems regarding the handling of alkaline aprotic solvents, all the more with oxygen. The *electrochemical* introduction of $[O_2^-]^*$ has none of the mentioned difficulties and requires no alkali. Since it is run under mild conditions and low potential the sulfamido group which is sensitive to higher cathodic potential [8], is preserved. The reaction is easy to control and can be adapted to a continuous process [6b]. By comparison, several attempts at the direct anodic oxidation of I on preparative scale failed, and indeed we found no anodic wave for I at potentials as far as 2.5V in DMF.

RESULTS

Oxygen is reduced by standard method [2,3] in dry DMF in an H form cell, divided by fritted glass with a mercury pool cathode and a graphite anode. Cathode potential during voltammetry and preparative electrolysis was kept at -1.1V vs. (Ag| Ag^+ (0.1M) ($AgBF_4$) as reference electrode. Known values for oxygen reduction are E_0 in DMF = -0.60 V vs. NHE [2] and -0.9V vs. SCE [3]. With I in catholyte at 20-25°C, at the end of 24 h, 4-(Di-n-propylsulfamyl)benzoic acid (II) forms in 45-50% yield with practically 100% selectivity (equations 1 and 2) even though the charge was provided in large excess - ca. 8mF for 1.75mM of I converted. The reaction stops short of complete conversion probably because of water formation, which consumes $[O_2^-]^*$. However, the balance of I is recovered and can be used again. TLC and (300MHz)^{1}H-NMR detect only I and II in the product mixture. Surprisingly, in AN the process was ineffective and I was recovered unchanged.

Formally:

$$O_2 + e \longrightarrow [O_2^-]^* \quad (1)$$

$$4\text{-}(CH_3CH_2CH_2)_2NSO_2\text{-}C_6H_4CH_3 + [O_2^-]^* + 1/2\,O_2 \longrightarrow 4\text{-}(CH_3CH_2CH_2)_2NSO_2\text{-}C_6H_4CO_2H + H_2O \quad (2)$$

Substitution of the mercury pool electrode (nominal surface 4.50 cm^2) with a small platinum cathode (nominal surface 0.16 cm^2) and bubbling oxygen in DMF yielded exclusively II, (0.63mM,17%) after 12mF, hence it is possible to use a Pt cathode as well.

A typical CV trace for reduction of $[O_2^-]^*$ in DMF at 20°C (Fig. 1 broken line), shows $E_{1/2}$ at -0.8V and I_p red of 0.75mA (320 μA mm^{-2}) at 0.98V. Intensity of the reverse current I_p ox, is about 0.65 times that of I_p red. When oxygen is reduced in presence of I, I_p ox.has only about 0.29 the intensity of I_p red. (Fig. 1 bold line). Slowing the sweep rate, from 0.1 to 0.05 Vs^{-1} shows a clearer CV curve and a ratio I_p ox / I_p red = 0.35 as expected for the extended time allowed for reaction of $[O_2^-]^*$. Reported values for the lifetime of are $t_{1/2}$ of ca.40 min. at ambient temperature [3a] which is long for the time scale of this experiment so that the decrease in value of I_p ox due to the presence of I definitely implies reaction between $[O_2^-]^*$ and I. In AN $I_p$red of O_2 is only 2.18 mA mm^{-2} (0.05 mA as compared to 0.75 mA in DMF) under same conditions. The ratio of I_p ox / I_p red is 0.75 in the absence of I, and 0.6 when I is present. Hence $[O_2^-]^*$ reacts with I in AN and it is rather the slow reduction of oxygen that explains failure to obtain II in AN. The

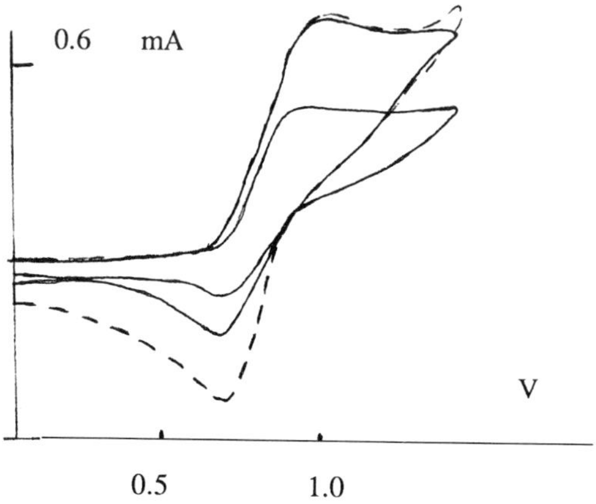

Fig. 1. CV for the reduction of O_2 in dimethylformamide (DMF) 0.1M tetraethylammonium perchlorate (TEAP) as electrolyte. O_2 (- - - - - -); O_2 with 4-(Di-n-propylsulfamyl)toluene (———). Large cycle: sweep rate 100 mV s^{-1}; Small cycle: sweep rate 50 mV s^{-1};

reason for the slow reduction is not clear. We find no indication in the literature [3,9] that reduction of oxygen should be *a priori* slower in AN than DMF.

DISCUSSION

The method has proved especially suitable for weak organic acids [2] for which a direct anodic oxidation [10] is often ineffective such as anthracene, alkyl-pyridines or toluene derivatives 4-$XC_6H_4CH_3$ where X has a high free-energy substituent constant. A pKa limit of 24 for a hydrocarbon was defined as a prerequisite for reactivity towards $[O_2^-]^*$ in aprotic solvents [3] however several toluenes which fit these criteria were found unreactive.

Although in the oxidation of toluene derivatives alcohols and aldehydes are plausible intermediates, the full conversion to carboxy acids is typical in the reactions with $[O_2^-]^*$ [2a,b] and the mechanism for this last point is only partially understood. Electrochemical analysis clearly shows $[O_2^-]^*$ to be the initial reactant consumed by I. If oxygen were a mere electron-transfer mediator the value of I_p red of O_2 would have increased by additions of I and this is not the case. The decrease in I_p ox when I is added shows that $[O_2^-]^*$ is actually consumed. After formation of a hydroperoxide $[O_2^-]^*$ may further deprotonate it at the benzylic position with consequent cleavage of the oxygen- oxygen bond to give the araldehyde. This is a familiar reaction of a base with peroxides bearing α-hydrogen [11]:

$$B: + ArCH_2OOH \;\text{--->}\; [ArCHOOH]^- + BH^+$$

$$[ArCHOOH]^- \;\text{--->}\; ArCHO + OH^-$$

$$BH^+ + OH^- \;\text{--->}\; H_2O + B:$$

B: may be $[O_2^-]^*$ as well as OH^- at later stages of the reaction. The nature of consequent steps cannot be deduced from CV analysis. Aromatic aldehydes might react by hydrogen radical abstraction with $[O_2^-]^*$ [2a] or O_2H^* [12,13] or alternatively by a Cannizzaro type reaction with $[O_2^-]^*$ or O_2H^- as a base [13, 2f] to yield the carboxylic acid. The reaction mixture is alkaline anyway since water the other product, reacts with $[O_2^-]^*$ to yield OH^-. Practically, for easy recycle of materials it is an advantage that intermediates do not survive the procedure.

EXPERIMENTAL

Chemicals

Dimethylformamide (DMF) was dried over 3A molecular sieve (Merck) for 48h. Acetonitrile (AN) spectroscopic grade was distilled over P_2O_5 and stored over 3A molecular sieve. Tetraethylammonium perchlorate (TEAP) and Tetrabutylammonium bromide (Fluka) were used as supplied.

4-Di-N-(n-propyl)toluenesulfonamide (I)

Di-n-propylamine (20.7 ml, 0.15 mol.) was added dropwise, with stirring to a solution of toluenesulfonylchloride (9.5 g, 0.05 mol.) in benzene at room temperature, refluxed for 3 h and left overnight. TLC ($CHCl_3$ - petroleum ether 4:1), showed complete consumption of toluenesulfonylchloride. The gelatinous precipitate was washed (aq. HCl 10%, water) and the organic phase dried over $MgSO_4$. Removal of the solvent yielded a colourless liquid which crystallized after few hours to yield I (11.8g).

^1H NMR of I (300MHz in $CDCl_3$) δ: 0.8 (6H, triplet); 1.46 (4H, sextet); 2.32 (3H, singlet); 3.0 (4H,triplet); 7.22 (2H, doublet); 7.6 (2H, doublet);

Cyclic voltammetry

CV was performed with a Princeton Applied Research B62 Model scanning potentiometer with a Tokogawa Hokeeshin Model 3025 X-Y recorder. CV was carried out in DMF or AN saturated with oxygen with TEAP 0.1M, in a dry H cell with a fritted glass separator. Working electrode (cathode) was a Pt wire tip (0.02 cm^2) and a graphite rod (0.2cm^2, 10% porosity) was the anode. Reference electrode was (Ag| $AgBF_4$ | 0.1M Bu_4NClO_4 in DMF or AN).

Constant potential electrolysis

CPE was carried out in the same cell as above except for using a mercury pool cathode (4.50 cm^2) with a Pt wire connection and a graphite anode (0.63 cm^2). The reference was held at a distance of 0.5cm from the pool which was agitated by a stream of dry O_2. Electrolysis was carried out at 20°C, in 40 ml DMF or AN solutions of 0.1M TEAP in both compartments and with (0.9g, 3.7 mM) of I in catholyte. Cathode potential was adjusted at -1.0 to -1.1V and about 8 mF passed. The current, 10-11mA stable for the first 4-5 h, eventually drops to 4.5 mA towards the end of 24 h. (During reaction, the catholyte solution attains a yellow colour. This colouration does not cross the barrier). Catholyte is then poured into an excess of 10% $NaHCO_3$ solution and residual I extracted with CH_2Cl_2. The aqueous solution is acidified with 5% HCl to precipitate II (yield 45-50%) m.p. 193-4° (lit 194 [1]). No other products were detected in the aqueous or organic extract.

IR of II in nujol shows (cm^{-1}):3400, 1700, 1160, 1620, 1590, 1440-1400, 1360, 1340-1300, 1190-1170, 1000, 880, 800.

^1H NMR of II (300MHz in $CDCl_3$) d: 0.813 (6H, triplet); 1.47(4H, sextet); 3.1 (4H, triplet); 7.87 (2H, doublet); 8.25 (2H, doublet);

Further tests, acetonitrile as solvent;

When DMF is replaced by AN and the graphite anode replaced by Pt no product is obtained. Most of the starting material I was recovered - 0.77g out of 0.9g. There is also no evidence for products which could originate in AN. With a cathode potential of -1.1 to -1.2V the current was 2.4-2.5mA. At the end of 24 h cathode potential was -0.6 and current 2.0mA. A separate reaction was carried out in the presence of 18-crown-6 (3.5 mM) to enhance reactivity of $[O_2^-]^*$. No product was obtained.

CONCLUSION

In the method described here electrogenerated radical-anion of oxygen is an effective substitute for direct anodic oxidation and oxidation with oxygen or other chemical reagents. In principle organic *electrochemical* oxidation accommodates mild conditions and fits in well with a continuous process. This compares favourably in safety and environment control to chemical oxidation reagents of which many require high pressures and temperatures, are batch processes, imbued with hazards and involve expensive disposal procedures. The use of a mediator or electrogenerated reagent could provide a measure of selectivity not attainable by direct electrolysis and also facilitate work at lower potentials namely, that efficient currents can be obtained with smaller energy loads.

REFERENCES

1. Merck Index, Merck & Co Publ. 10th' edition, (M. Windholz ed.), (1983) 1116
2. a. H. Sagae, M. Fujihira H. Lund and T. Osa, *Heterocycles* 13 (1979) 321.

b. H. Sagae, M. Fujihira, T. Osa and H. Lund, *Chemistry Letters,* (1977) 793.
c. S. Mitchio and M.M. Baizer, *J. Org. Chem.,* 48 (1983) 9331.
d. M. Sugawara and M.M. Baizer, *Tetrahedron Lett.*, 24 (1983) 2223.
e. R.R. Mehta, V.L. Pardini and J.H.P.Utley, *J.Chem Soc. Perkin Trans.* I, 4 (1982) 2921.
f. J.M. Saveant and S.K. Binh, *J. Org. Chem.* 42 (1977) 1242.
g. Y.J. Page and J. Simonet, *Electrochim. Acta,* 23 (1978) 445.
h. R.C. Hallcher, R.D. Goodin and M.M. Baizer, U.S. Pat. 4293393, (1981)
i. C. Degrand, B. Gautheron, M. Bikrani, F. Gasquez and P.-L. Compagnon, *J. Organometallic Chem.* 273 (1984) 319.
j. M. Tezuka, Y. Ohkatsu and T. Osa, Bull. Chem. Soc. Jap. 48 (1975) 1471.
k. R. Rastogi, G. Dixit and K. Zutshi, *Electrochimica Acta,* 28 (1983) 129.
l. M. Gareil, J. Pinson and J.M. Saveant, Nouveou Journal de Chimie, 5 (1981) 311.

3. a. D.T. Sawyer and J. S. Valentine *Acc. Chem. Res.*, 14 (1981) 393.
b. E. Lee-Ruff, *Chem Soc. Rev.* 6 (1977) 195.
c. T. Shono, *Electroorganic Chemistry as a New Tool in Organic Synthesis* , Springer Verlag, Berlin , 1984 159
d. J. Simonet in *Organic Electrochemistry,* (M.M. Baizer and H. Lund Eds.), Marcel Dekker 1983 862.
e. C.P. Andrieux, P. Hapiot and J.M. Saveant, J. Amer. Chem. Soc. 109 (1987) 3768.
f. A. A. Frimer, *The Chemistry of Functional Groups, Peroxides* (S. Patai Ed.), John Wiley & Sons 1983 429.
4. E. Steckhan in *Topics in Current Chemistry, Electrochemistry* I, (E. Steckhan Ed.), Springer Verlag, Berlin, 1987 1.
5. J.H.P.Utley in *Topics in Current Chemistry, Electrochemistry* I, (E. Steckhan Ed.), Springer Verlag, Berlin, (1987) 133.
6. a. Y. Migron and M. Michman, Israeli Pat. 52814, 1979; b. M. Michman and M. Weiss, Israeli Pat. Application 089366 1989.
7. G. A. Russell and A.G. Bemis, J. Amer. Chem. Soc. **83** (1966) 5491.
G. Russell, A.G. Bemis, E. J. Geels, E.G. Janzen and A. J. Mote *Oxidation of Organic Compounds, Advances in Chem. Ser,* (R.E. Gould ed.) A.C.S. Pub., 1968, 174;
8. L. Horner and H. Lund, *Organic Electrochemistry,* (M.M. Baizer and H. Lund Eds.), Marcel Dekker, 1983, 747.
9. P.S. Jain and S. Lal, *Electrochimica Acta* , 27 (1982) 759.
10. I. Nishiguchi qnd T. Hirashima, *J.Org. Chem.*, 50 (1985) 539.
11. N. Kornblum and H.E. De LaMar, *J. Amer. Chem. Soc.* 73 (1951) 880.
12. M.J. Gibian, D.T. Sawyer, T. Ungermann, R. Tangpoonpholvivat and M. M. Morrison, *J. Amer. Chem. Soc.* 101 (1979) 640.
13. W.T. Monte, M.M. Baizer and R.D. Little, *J.Org. Chem.*, 48 (1983) 803.

G. Centi and F. Trifiro' (Editors), *New Developments in Selective Oxidation*
© 1990 Elsevier Science Publishers B.V., Amsterdam — Printed in The Netherlands

PHOTOCATALYTIC OXIDATIONS AT ROOM TEMPERATURE IN VARIOUS MEDIA

J.M. HERRMANN, H. COURBON, J. DISDIER, M.N. MOZZANEGA and P. PICHAT

URA au CNRS Photocatalyse, Catalyse et Environnement ;

Ecole Centrale de Lyon, BP 163, 69131 Ecully Cédex (France)

SUMMARY

Various inorganic compounds, such as CO, SO_3^{2-}, I^-, Br^- and NH_3, are photocatalytically oxidized at room temperature. The photocatalytic oxidation of hydrocarbons yields aldehydes and ketones, either in the gas phase or in the neat-liquid phase with a selectivity that depends on the molecular structure. For instance, cyclohexane is mainly transformed into cyclohexanone, and alkyltoluenes are selectively converted into the corresponding alkyltolualdehydes. Among various semiconductor oxides (TiO_2, ZrO_2, ZnO, CeO_2, SnO_2, Sb_2O_4, V_2O_5), titania was generally the most active, whereas V_2O_5 was totally inactive. As expected, for a given oxide, the activity depends on the particular specimen. From photoconductivity and oxygen isotope exchange measurements, it was inferred that the photoactivated oxygen species is atomic in gas phase reactions. In aqueous medium, the photoproduced holes react with OH^- groups thus forming strongly oxidizing OH° radicals as shown by the primary products of the oxidation of monochlorophenols. These oxidations could be of interest for the synthesis of fine chemicals or for the removal of pollutants.

INTRODUCTION

Oxidations by oxygen (or air) of a variety of inorganic and organic compounds can be photocatalyzed at room temperature by oxide semiconductors. Reviews on this topic have been published (refs. 1-4). Photons whose energy is equal to or greater than the band gap of the oxide can excite an electron from the valence band into the conduction band, thus leaving a positive hole in the valence band. In the presence of an electrophilic compound, such as oxygen, the surface is covered with negatively charged adsorbed species. Consequently, the photoproduced holes are attracted to the surface, where they are able to react with oxidizable (i.e. electron-donating) species.

Examples of photocatalytic oxidations of organic or inorganic compounds in various media (gas, liquid or aqueous solution) are presented here.

METHODS

Photoreactors

In our laboratory, for gas-phase reactions, two types of reactors were used according to the nature of the reaction and the analysis required. A static photoreactor connected to a grease-free vacuum-line including a quadrupole gas analyzer, was employed for oxygen isotope exchange, and oxidations by NO. In the other cases, a differential flow-photoreactor with a fixed-bed of catalyst and on-line gas chromatography analysis was used. Reactions in the liquid phase were carried out in a static reactor where the catalyst was kept in suspension by oxygen bubbling and/or a magnetic stirrer.

Optical system

UV light was provided by a Philips mercury lamp (HPK 125 W) equipped with a circulating-water cuvette to remove IR beams and convenient filters were used to determine the light-energy dependence of the photocatalytic reactions and/or to avoid photochemical side-reactions. The light-beams were admitted into the reactors through flat optical windows.

RESULTS AND DISCUSSION

Photocatalytic oxidations over TiO_2

1. Oxidation of inorganic compounds

(i) Gas-phase reactions. CO is oxidized into CO_2 by oxygen. The reaction can proceed over a carefully dehydroxylated surface (calcination at 500°C in oxygen for 15h, evacuation to 10^{-6} Torr and cooling in O_2). This demonstrates that OH groups are not necessary for gas-phase photocatalytic oxidations. This reaction can contribute to the progressive elimination of CO in the atmosphere, where particles of semiconductor oxides are present ; for example, ca. 1% TiO_2 (ref. 5)in weight has been found in fly ash samples from coal-fire power plants.

Ammonia is photocatalytically oxidized into N_2(80%) and N_2O (20%). A kinetic study showed that the reaction rate r follows a Langmuir-Hinshelwood mechanism with two kinds of sites (ref.6).

$$r = \frac{k\, Ko_2^n\, K_{NH_3}\, Po_2^n\, P_{NH_3}}{(1 + Ko_2^n\, Po_2^n)\,(1 + K_{NH_3}\, P_{NH_3})}$$

The linear transforms $r^{-1} = f(P_{02}^{-n})$ and $r^{-1} = f(P_{NH3}^{-1})$ shows that the same molecular oxygen species (n=1) is involved in the formation of N_2 and N_2O , after the initial attack by activated atomic oxygen. Since the main product is nitrogen, this reaction can also be considered as contributing to the removal of pollutants.

(ii) Aqueous-phase reactions. The photocatalytic oxidation of halide ions was studied with aerated aqueous suspensions (ref. 7 and 8) . Iodide ions are oxidized with a conversion rate 80 times greater than that of Br^- ions, whereas Cl^- ions withstand oxidation. The reaction rate r follows a Langmuir-Hinshelwood mechanism : $r = k K [X^-] / (1 + K[X^-])$. By analogy with gas-phase reactions, it was tentatively suggested that the reaction proceeds via the neutralization of adsorbed O^- ions by photogenerated holes, thus producing activated oxygen species which oxidize adsorbed halide ions into hypohalite ions (ref. 7). The latter ions are stable in basic medium, whereas in acidic medium hypohalite ions are converted into halogen with the excess of halide ions

$$IO^- + I^- + 2H^+ \rightarrow I_2 + H_2O$$

On the other hand, it was concluded that the initial oxidation step produces X atoms in anhydrous acetonitrile (ref. 9).

Other anions are oxidized in aqueous solution. For instance, SO_3^{2-} and HSO_3^- ions are transformed into sulfate ions. In photographic effluents, whereas silver ions are recovered as metal by photo-assisted deposition on titania, the thiosulfate ions are simultaneously oxidized into sulfate ions as measured by ion chromatography.

2. Oxidation of organic compounds

A large variety of organic compounds can be oxidized by heterogeneous photocatalysis.

Alkanes. Various alkanes are oxidized in gas phase using a differential flow-photoreactor with a fixed-bed of catalyst : under such conditions, with low conversions ($\leq$ 2.5%), high selectivities (55-95%) in aldehydes and /or ketones are obtained. For instance, the oxidation of isobutane yields 90% acetone plus CO_2 (ref. 10).

Alkanes can also be oxidized in the liquid phase at room temperature. For instance, cyclohexane is transformed into cyclohexanone (83%), cyclohexanol (5%) and CO_2(12%) (ref. 10).

Propene. Propene is photocatalytically oxidized into acetone, acrolein, propene oxide, ethanal and CO_2. By decreasing the conversion, somewhat parallel increases in ethanal and acrolein percentages are observed, whereas the percentage of acetone is less affected and that of propene oxide is markedly enhanced. This latter increase and the fact that the photocatalytic oxidation of propene oxide under the same conditions, yields all the other oxidation products of propene suggests that this compound could be the initial product arising from the attack of adsorbed propene by an activated and dissociated oxygen species (ref. 12).

Alcohols. Alcohols are photocatalytically oxidized into their corresponding aldehyde or ketone either in the gas phase or in the liquid phase.

The liquid-phase oxidation of 2-propanol, which yields pincipally acetone, has been chosen as a test reaction to control the photostability of pigmentary titania. (ref. 13). We have compared, with this rather short test (~ 1.5h), a series of industrial pigments.

Alkyltoluenes. For the photocatalytic oxidation of several alkyltoluenes in gas phase, the selectivity in the corresponding alkyltolualdehyde was very high (ref. 14)

$$R - C_6H_4 - CH_3 + 02 \rightarrow R - C_6H_4 - CHO + H_2O$$

with R = CH_3 , C_2H_5, i - C_3H_7, t - C_4H_9, OCH_3, Cl.

This shows that the methyl group is preferentially attacked with respect to the other substituents and that the aromatic ring is stable under these conditions. These oxidations are not classical, since, usually, the secondary or tertiary hydrogen atoms are preferentially eliminated to give rise to hydroperoxides.

For 4-tertbutyloluene, the oxidation was also carried out in the liquid phase, in a static photoreactor. The methyl group is oxidized, whereas the t-butyl group withstands oxidation. However, the alkylbenzaldehyde is photochemically transformed into the corresponding acid in the presence of O_2, although with a slow rate. This limits the degree of conversion in a static reactor to preserve the selectivity to 4-tertbutylbenzaldehyde. Considering the initial quantum yield (0.2), this type of selective oxidation could be of interest in the synthesis of fine chemicals (ref. 15).

Oxalic acid. Oxalic acid is oxidized into carbon dioxide in aerated aqueous suspensions of TiO_2 . The reaction follows the Langmuir-Hinshelwood mechanism between chemisorbed acid and oxygen. The reaction rate is maximum at about pH ~ 2.3, i.e. when the acid is present as $HC_2O_4^-$ ions (ref.16). This could be of interest to destroy this pollutant in waste water.

<u>Comparison of TiO_2 with other semiconductor oxides</u>

In the case of the oxidation of propene, various oxides have been tested and the following activities have been found (ref. 12) :

$$TiO_2 > ZrO_2 > CeO_2 > ZnO > Sb_2O_4 > SnO_2 > WO_3 >> V_2O_5 = 0$$

In the oxygen isotope exchange (ref. 17b), the activity pattern was similar :

$$TiO_2 > ZrO_2 > ZnO > SnO_2 >> V_2O_5 = 0$$

Note that these orders have only a relative meaning, since various factors, in particular the texture, can modify the value of the photocatalytic rate for a given oxide.

<u>Photoactivated oxygen species</u>

The photoactivated oxygen species have tentatively been identified by various methods in our laboratory.

(i) <u>Photoconductivity measurements</u>. When the semiconductor oxides are submitted to illumination under vacuum, they generally become photoconductors. Adsorption of oxygen consumes electrons and therefore the photoconductance decreases. In the log-log plot $\sigma = f(P_{O_2})$ (fig. 1), the slope - 1 relative to TiO_2 is indicative of the presence of O_2^- adsorbed species that control the electron transfer between the excited semiconductor and the gas phase :

$$O_2(g) \rightleftarrows O_2(ads),$$

$$O_2(ads) + e^- \rightleftarrows O_2^- (ads)$$

For TiO_2 at low pressures and for other oxides, the slope - 1/2 can be interpreted as indicating that O^- species control the electron equilibrium between the solid and O_2 :

$$O_2(g) \rightleftarrows 2O(ads),$$

$$O(ads) + e^- \rightleftarrows O^-(ads).$$

V_2O_5, which is photocatalytically inactive in oxidations, is not photoconducting, either in vacuo or in an oxygen atmosphere.

In situ simultaneous measurements of the photoconductance σ and the photocatalytic activity A of titania during the oxidation of isobutane give the relationships :

$$\sigma = k\, P_{O_2}^{-1}\, P_i^{O}$$

$$A = k_A\, P_{O_2}^{O}\, P_i^{0.35}$$

The independence of σ with respect to isobutane pressure P_i shows that this reactant is not in electronic interaction with the solid. The fractional kinetic order 0.35 shows that isobutane reacts in the adsorbed phase, since this value corresponds to the apparent order of adsorption determined by thermogravimetry in the pressure range investigated. The variations in σ as $P_{O_2}^{-1}$ show that gaseous oxygen is in equilibrium with O_2^- species (see fig.1), whereas O^- species are adsorbed at saturation. Since A is independent of P_{O_2}, it is inferred that active oxygen is associated with O^- species.

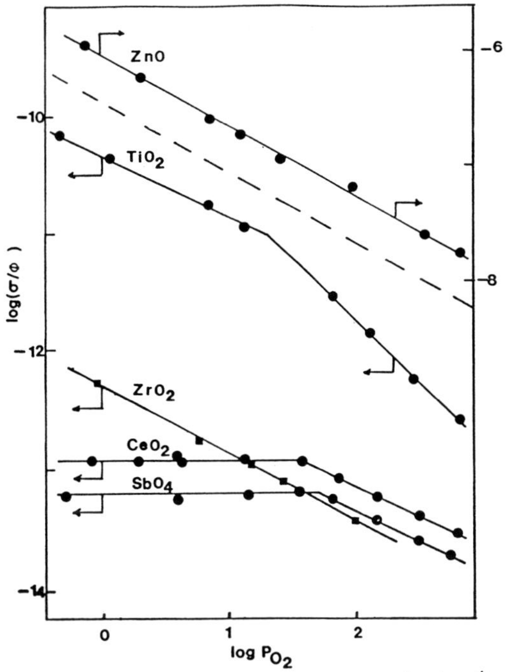

Fig. 1 Variations fin the photoconductance σ (in ohm^{-1}) of various oxides at equilibrium per mW of radiant flux as a function or the oxygen pressure (in Pa).

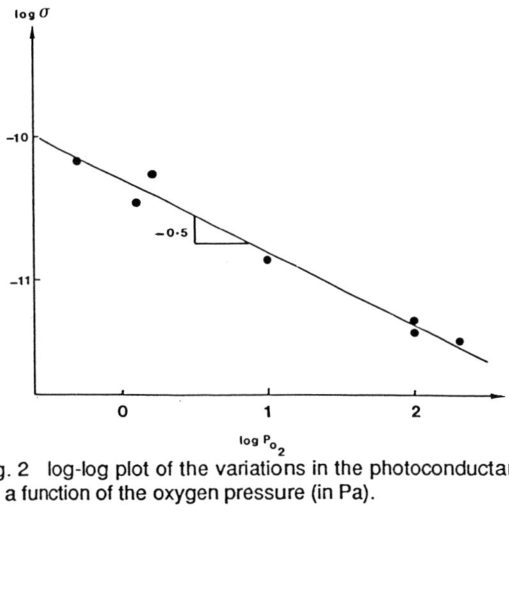

Fig. 2 log-log plot of the variations in the photoconductance σ (in ohm^{-1}) of 5wt%Pt/TiO_2 at equilibrium as a function of the oxygen pressure (in Pa).

(ii) Oxygen Isotopic exchange. The kinetics of oxygen isotope exchange perfectly follows the model developed by Boreskov (ref.18) usually called R' mechanism (ref. 19) that concerns the exchange of one adsorbed surface oxygen atom O_S at a time for each gaseous O_2 molecule.

$$^{18}O_2 + {}^{16}O_S \rightarrow {}^{18}O - {}^{16}O + {}^{18}O_S$$

When this exchange is carried out in the presence of isobutane, the oxidation of isobutane proceeds first and the exchange reaction starts only when this oxidation is achieved (ref. 17a). It was inferred that both reactions include a common step that involves dissociated oxygen species, in agreement with the simultaneous measurements of the photoconductance and the photocatalytic activity described in the preceding section.

(iii) Use of NO as an oxidant

NO was used as a source of atomic oxygen to substantiate the preceding conclusions. An inverse dependence of the photoconductance of TiO_2 on the pressure of NO is observed (ref. 20) which corresponds to the consumption of one electron by one molecule of NO. However, other data are required to discriminate between the formation of NO^- or NO_2^- (where the second oxygen atom is supplied by the surface). The N-O bond is weakened by the presence of an additional electron in the antibonding π^* orbital. However, the absence of decomposition of NO in the dark at room temperature demonstrates that photoproduced holes are needed for this decomposition which yields N_2 and N_2O (ref. 20). If an easily oxidizable substrate, such as butanols, is added, the corresponding molecules containing carbonyl groups are formed and, moreover, the isotopic exchange of $N^{18}O$ with TiO_2 is suppressed (ref. 21). It is concluded that NO provides oxygen atoms to the surface, which corroborates the importance of dissociated oxygen species in the photocatalytic oxidations with gaseous oxygen.

(iv) Influence of metal deposits or cation doping

The deposition of a noble metal, such as platinum, on titania creates electron traps as shown by photoconductance measurements.

$$e^- + Pt \rightleftarrows e^-_{Pt}$$

The higher the metal loading, the smaller the photoconductance σ of TiO_2. On the other hand, the isotherm log σ = f(log P_{O_2}) of a 5wt% Pt/TiO_2 sample shows that σ varies as $P_{O_2}^{-1/2}$ (see Fig. 2). This means that titania is partially depleted of electrons and O^- species control the electron transfer between the solid and the gas phase. Since Pt/TiO_2 samples are active in photocatalytic oxidations, this is also in favour of associating active oxygen with O^- species.

n-type or p-type doping of TiO_2 obtained with pentavalent (Sb^{5+} and Nb^{5+}) or trivalent (Ga^{3+}, Cr^{3+}) ions, respectively, creates donor or acceptor centres which behave as recombination centres of the charge carriers and are detrimental to the photocatalytic activity as shown, for example, by the oxidation of cyclohexane into cyclohexanone (ref. 11).

Influence of OH groups in aqueous medium

In aqueous medium, the surface of titania is fully hydroxylated and the OH^- groups can capture photoproduced holes

$$OH^- + p^+ \rightarrow OH^\circ$$

forming hydroxyl radicals which are strong oxidants. For example, in the case of the photocatalytic degradation of monochlorophenols in diluted aqueous solutions (ref. 22), parahydroxylation has been observed. The polyhydroxylated aromatic ring is degradated to CO_2 in many subsequent steps. Therefore, in aerated aqueous phase, the aromatic ring can easily be oxidized in contrast to what happens in the gas phase where it is quite stable. Since chlorophenols, as well as other toxic halogenated molecules, can be rapidly dehalogenated and eventually mineralized (however with a much slower rate), intensive current research concerns the usage of heterogeneous catalysis to decontaminate water.

CONCLUSION

Heterogeneous photocatalysis constitutes another method of oxidizing organic and inorganic molecules. The conditions are attractive : room temperature, air, inexpensive catalysts. The selectivity can be different of that of other methods, which could be of interest in synthesis, although research is still needed to improve it in numerous cases. In aqueous medium, OH° radicals are generated, so that heterogeneous photocatalysis appears as a potential means of decontaminating water. Finally, the knowledge of gas-phase photocatalysis can be useful to understand the role of aerosol particles of semiconductor oxides in the chemical transformations that occur in the atmosphere.

REFERENCES

1 M.A. Fox, Acc. Chem. Res. 16 (1983) 314 ; Top. Org. Electrochem. 1 (1986) 177.
2 P. Pichat, in : M. Schiavello (Ed.), Photoelectrochemistry Photocatalysis and Photoreactors, Reidel, Dordrecht, 1985, p. 425.
3 S.J. Teichner and M. Formenti, ibid, p. 457.
4 M. Anpo, Res. Chem. Intermediates, 11 (1989) 67.
5 G.L. Fisher, D.P.Y. Chang and M. Brummer, Science, 192 (1976) 553.
6 H. Mozzanega, J.M. Herrmann and P. Pichat, J. Phys. Chem., 83 (1979) 2251.
7 J.M. Herrmann and P. Pichat, J. Chem. Soc. Faraday Trans 1, 76 (1980) 1138.
8 P.R. Harvey and R. Rudham, J. Chem. Soc. Faraday Trans 1, 84 (1988) 4181.
9 M.A. Fox and T.L. Pettit, J. Org. Chem., 50 (1985) 5013.
10 J.M. Herrmann, J. Disdier, M.N. Mozzanega and P. Pichat, J. Catal., 60 (1979) 369.
11 Unpublished results.
12 P. Pichat, J.M. Herrmann, J. Disdier and M.N. Mozzanega, J. Phys. Chem., 83 (1979) 3122.
13 G. Irick, Jr., J. Appl. Polym. Sci., 16 (1972) 2387.
14 M.N. Mozzanega, J.M. Herrmann and P. Pichat, Tetrahedron Lett., 34 (1977) 2965.
15 P. Pichat, J. Disdier, J.M. Herrmann and P. Vaudano, New J. Chem., 10 (1986) 545.
16 J.M. Herrmann, M.N. Mozzanega and P. Pichat, J. Photochem., 22 (1983) 333.
17 (a) H. Courbon, M. Formenti and P. Pichat, J. Phys. Chem., 81 (1977) 550 ; (b) H. Courbon, P. Pichat, Compt. Rend. Acad. Sci., Ser. C., 285 (1977) 171.
18 G.K. Boreskov, Adv. Catal., 15 (1964) 285.
19 J. Novakova, Catal. Rev. 4 (1970) 77.
20 P. Pichat, H. Courbon, J. Disdier, M.N. Mozzanega and J.M. Herrmann, Proc. 7th Int. Cong. Catal. Tokyo, T. Seiyama and K. Tanabe (Ed.) Elsevier, Amsterdam, Part B,1981, p. 1498.
21 H. Courbon and P. Pichat, J. Chem. Soc. Faraday Trans. 1, 80 (1984) 3175.
22 G.H. Al Sayyed, J.C. D'Oliveira and P. Pichat (to be published) ; P. Pichat, in : M. Schiavello (Ed.), Photocatalysis and Environment, Kluwer Acad. publ. : Dordrecht, Vol. 237, 1988, p. 399.

J. VEDRINE (Institut de Recherches sur la Catalyse, France) : You have mentioned that under photons 0_2^- or 0^- are formed (and evidenced for $Ti0_2$ by ESR technique). In a catalytic process lattice oxygen ions are usually involved in the reaction. In photocatalysis can you tell if active 0 species are adsorbed species arising from the gas phase oxygen activated by photons or from lattice ions, eventually activated by photons.

Jean-Marie HERRMANN (CNRS, Ecole Centrale de Lyon, France) : This is right that two oxygen species 0_2^- and 0^- have been evidenced at the surface of titania under illumination by in situ photoconductivity measurements. They can be defined as **ionosorbed** species, since they result from the simultaneous adsorption of oxygen from the gas phase and the creation of photo-induced electrons by photons, whose energy hν is ≥ E_G (E_G being the band gap energy of the semiconductor oxide). All the photocatalytic oxidation reactions mentioned have been carried out at room temperature. In the case of conventional heterogeneous catalysts, in the dark, when temperature is progressively increased, the weakly adsorbed species progressively desorb in the following order : 0_2^- , 0^- and 0^{2-}. At relatively high temperature $\theta \geq$ 400°C), lattice oxygen anions 0^{2-} are directly in interaction with the gas phase. In reaction conditions, they can give a redox process between the reactant and the surface of the solid. This process is generally illustrated by the Mars and Van Krevelen mechanism (Chem. Eng. Science ; Spec. Sup., (1954), 9, 41).

Heterogeneous photocatalytic reactions occur at room temperature and lattice oxygen cannot be thermally activated. However, UV-illumination by promoting electrons from the valence band into the conduction band weakens the Ti-0 bonds and makes surface lattice oxygen species more labile. Accordingly, the atomic oxygen which initiates the photocatalytic oxidation process could originate from such coordinatively unsaturated surface species as well as from ionosorbed 0^- entities.

G. PAJONK (Université Claude Bernard, (Lyon I), France) : You showed us that NH_3 and 0_2 as well as NO react photocatalytically in your experimental conditions so I wonder if it should be possible to take advantage of these reactivities in order to produce ammoxidation reactions and even nitroxidation ones.

Jean-Marie HERRMANN (CNRS, Ecole Centrale de Lyon, France) : Ammonia and hydrocarbons can undergo a mild oxidation in our photocatalytic conditions. The study of photo-assisted ammoxidation reactions by using a mixture of a hydrocarbon, ammonia and oxygen could be envisaged. However, this could be obtained only if the hydrocarbon and ammonia have similar reactivities with respect to photoactivated oxygen.

A photocatalytic nitroxidation based on NO would suppose that the oxygen atom oxidatively dehydrogenates a hydrocarbon molecule, whereas the N atom simultaneously incorporates into this molecule to form the corresponding nitrile. This would mimic what happens in your system (paper E2). However in our case we found that photoactivated N0 is almost unreactive with respect to isobutane chosen as a model of alkanes.

J.M. BREGEAULT (Université Pierre et Marie Curie, Paris, France) : Can you give the complete balance for one of the reactions cited, for instance :

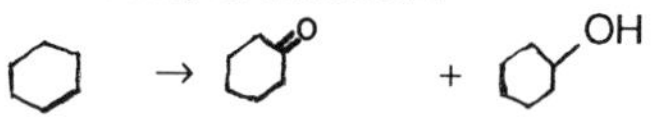

+ $C0_2$

Can you give the number of moles of ketones produced per unit of time and per photon received ?
Did you try the oxidation of cyclohexanone separately ?
Did you observe the formation of dicarboxylic acids ?

Jean-Marie HERRMANN (CNRS, Ecole Centrale de Lyon) : In all the reactions described, the mass balance was systematically determined and for organic reactions, the carbon balance was tentatively established. This is easy to do in a differential flow photoreactor. In a static liquid-phase photoreactor, it is more difficult. For instance, with liquid cyclohexane, we obtained mainly cyclohexanone (83%), with 5% cyclohexanol and 12% $C0_2$ as indicated in the text. $C0_2$ was measured by GC-analysis of the gas phase, whereas cyclohexane, cyclohexanone and cyclohexanol were determined by GC-analysis of the liquid phase. For small conversions, since cyclohexane is in large excess, it is difficult to determine with accuracy the exact number of C_6H_{12} molecules converted.

For all the reactions and especially for the oxidation of cyclohexane, we have determined the quantum yield, which is defined as the ratio of the reaction rate (in molecules s^{-1}) to the photonic flux (in photons s^{-1}). It is generally of the order of a few percents. It was found equal to 10% for cyclohexanone formation.

We have not tried the photocatalytic oxidation of cyclohexanone in the liquid phase, since we already know that for alkanes mild oxidation yields aldehydes and/or ketones as final products. This was confirmed by the liquid phase oxidation of cyclohexanol into cyclohexanone.

Concerning the formation of dicarboxylic acids, we have not observed such products in our reaction medium. Since $C0_2$ is also formed, this implies the existence of an oxidative ring opening with probably, as you suggest, the formation of a dicarboxylic acid. Two explanations can be given for the non-observation of such dicarboxylic acids : (i) a lack of occuracy in our detection method or (ii) a non-desorption of these acids in the organic phase. In this latter case, the intermediate acids would remain strongly adsorbed at the surface of titania where they would undergo the final oxidative degradation to carbon dioxide.

G. Centi and F. Trifiro' (Editors), *New Developments in Selective Oxidation*

DYNAMIC STUDIES OF PHOTOCATALYTIC OXIDATION OF C_3H_6 WITH O_2 ON VANADIUM OXIDE SUPPORTED ON POROUS VYCOR GLASS

M. ANPO,*[1] T. SUZUKI,[1] Y. YAMADA,[1] Y. OTSUJI,[1] E. GIAMELLO,[2] and M. CHE[3]

[1]Department of Applied Chemistry, College of Engineering, University of Osaka Prefecture, Mozu-Umemach, Sakai, Osaka 591, Japan

[2]Dipartimento di Chimica Inorganica, Chimica Fisica e Chimica dei Materiali, Università di Torino, Via P. Giuria 9, 10125 Torino, Italy

[3]Laboratoire de Réactivité de Surface et Structure, Université P. et M. Curie, UA1106, CNRS, 4 Place Jussieu, Tour 54, 75252 Paris, Cedex 05, France

ABSTRACT

The photocatalytic oxidation of C_3H_6 with O_2 on highly dispersed supported vanadium oxides has been studied by dynamic photoluminescence, ESR and analysis of the reaction products. These studies indicate that both O_2 and C_3H_6 are activated through interaction with the charge transfer excited complexes of vanadyl species, $(V^{4+}-O^-)^*$ of the catalyst. The reaction of these activated species results in the oxidation of C_3H_6 to produce CH_3CHO, C_2H_5CHO, and $CH_2=CH-CHO$.

INTRODUCTION

Recent dynamic studies of the excited state of highly dispersed supported vanadium oxides and of their photocatalytic activity for the isomerization of butenes have indicated that the photo-reactivities of the catalyst are directly associated with the activation of surface vanadyl groups (V = O double bond).[1-3] Vanadium oxides supported on silica[4] or Vycor glass[1] have been known as photocatalysts for the partial oxidation of alkenes, though the detailed mechanism is not understood. Therefore, in the present work the photocatalytic oxidation of C_3H_6 with O_2 on vanadium oxide supported on porous Vycor glass (V/PVG oxide) is investigated by applying dynamic photoluminescence, ESR and analysis of the reaction products.

EXPERIMENTAL

Vanadium oxides supported on porous Vycor glass (transparent 1 mm thick sheet) (Corning code: 7930, BET surface area: 160 m^2/g, major composition: SiO_2 ≒ 96%, B_2O_3 ≒ 3%) were prepared by impregnation of the support with an aqueous solution of NH_4VO_3. Prior to the experiments, the catalysts were first degassed at 773 K for 5 hs, then heated in oxygen (about 100 Torr) at 773 K for 3 hs, and then finally evacuated at 550 K for 2 hs. Dynamic photoluminescence measurements were carried out at 298 K (or 77 K) with a Shimadzu RF-501 spectrofluorophotometer and an apparatus for the lifetime measurements consisting of a N_2 laser, a monochromator, lenses, a photomultiplier, and a storage scope.[3] ESR spectra were measured at 77 K with a Jeol (JES-ME-1) spectrometer (X-band). Photoreactions were carried out at 298 K using a Toshiba SHL-100UV mercury lamp ($\lambda > 280$ nm). Quantum yields were determined for the photons absorbed by vanadium oxides by using potassium ferrioxalate actinometry. Analysis of the reaction products was made by gas-chromatography.

RESULTS AND DISCUSSION

1. Excited state of supported vanadium oxide catalyst and its reaction with C_3H_6 or O_2 molecules

Figure 1 shows the typical photoluminescence spectra of highly dispersed V/PVG oxide catalyst obtained at 298 K (1) with the corresponding excitation (a) and UV absorption spectrum (b). The absorption at around 300-340 nm and the photoluminescence at around 450-550 nm can be attributed to the charge transfer processes,[1-5] i. e., an electron transfer from O^{2-} ion to V^{5+} ion and a reverse radiative decay process

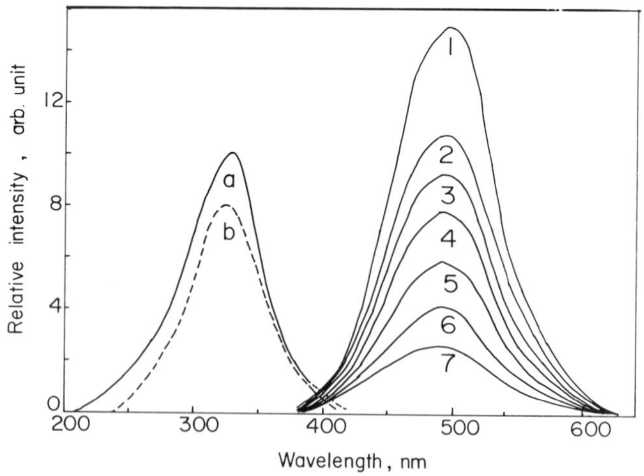

Fig. 1. Phosphorescence spectra of V/PVG oxide (0.018 V wt%) at 298 K in the absence (1) and presence of O_2 (2-7), excitation spectrum (a), and absorption spectrum (b). (pressure of O_2 (in Torr), 2: 0.0053, 3: 0.009, 4: 0.012, 5: 0.02, 6: 0.063), 7: 15.0).

from the charge transfer excited triplet state of vanadyl species (i. e., phosphorescence), respectively. As described previously,[1-3,5] only the vanadyl species located in tetrahedral coordination can act as emitting sites.

Figure 1 also shows that the phosphorescence from the excited triplet state of vanadyl species is easily quenched by adding O_2. The addition of C_3H_6 also led to the quenching of phosphorescence, its extent depending on the added C_3H_6 pressure. The lifetimes, measured in parallel with the quenching, were found to become shorter by increasing the pressure of O_2 or C_3H_6. Thus, the results clearly indicate that the quenching of phosphorescence occurs through the interaction of O_2 or C_3H_6 with the charge transfer excited triplet state of vanadyl species (i. e., dynamic quenching but not static quenching) to enhance the radiationless deactivation. The Stern-Volmer plots,[3] i. e., Qo/Q values against the pressure of added O_2 and C_3H_6 were found to be linear functions of their pressure, where Qo and Q are the yields (intensities) of the phosphorescence of V/PVG oxide in the absence and presence of quencher molecules, respectively. From the slopes of the Stern-Volmer plots it was found that the reactivity of O_2 with the charge transfer excited triplet state at 298 K (k_q = 34.1 x 10^4 $Torr^{-1}$) is much higher than that of C_3H_6 (k_q = 5.8 x 10^4 $Torr^{-1}$).

Figure 2 (a) shows the ESR spectrum obtained by UV-irradiation of V/PVG oxide at 300 K in the presence of O_2. The spectrum scarcely

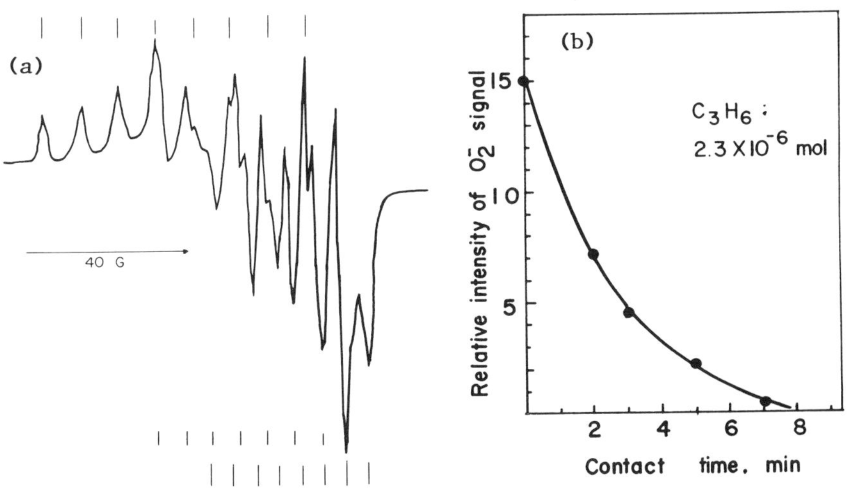

Fig. 2. ESR spectrum of photo-fromed O_2^- anion radicals on V/PVG oxide (a) and kinetic curve of the destruction of the signal by the reaction with C_3H_6 at 300 K for 30 min (b). (ESR spectra were recorded at 77 K.)

changed by the addition of CO onto the catalyst at 300 K, indicating that neither O^- nor O_3^- species are involved in the spectrum since both anion radicals are known to react easily with CO molecules on oxides.[6) The signal (g_{zz} = 2.0235, g_{yy} = 2.0110, and g_{xx} = 2.0035) which increased in intensity with increasing UV-irradiation time is assigned to the photo-formed O_2^- species adsorbed on V^{5+} ions on the basis of the g_{zz} value of the superhyperfine splitting due to the interaction of the unpaired electron with 100% naturally abundant ^{51}V isotope (I = 7/2).[6) The O_2^- species was thermally stable at least up to 323 K. It is likely that the photo-uptake of O_2 on V/PVG oxide is closely associated to the photo-formation of these thermally stable O_2^- anion radicals.

On the other hand, as reported previously,[1,7) UV-irradiation of V/PVG oxide at 77 K in the presence of C_2H_4, the more stable of the olefins studied, led to the appearance of an ESR spectrum which consists of five lines with an intensity ratio of about 1:4:6:4:1 and a hyperfine splitting of a_H = 21.5 G having a g value of 2.0025. The signal has been assigned to a bridged type π complex.[1,7) These radicals were stable at 77 K, in contrast to what was observed on supported Mo oxides where the complex breaks into Mo= CH_2 and HCHO, the former inducing ethylene homologation (or olefin metathesis).[1, 7) It is likely that the photo-adsorption of C_2H_4 at 298 K is associated with the formation of these species on the catalyst. The photo-adsorption of C_3H_6 was also observed, though its amount was much smaller than that of the O_2 photo-adsorption.

Thus, these results obtained by dynamic photoluminescence and ESR studies clearly indicate that the interaction of O_2 or C_3H_6 with the charge transfer excited triplet state results in the activation of O_2 (formation of O_2^- anion species) and C_3H_6 (formation of bridged type complex), the activation of O_2 being much faster than that of C_3H_6.

2. Photocatalytic oxidation of C_3H_6

As shown in Table 1, UV-irradiation of V/PVG oxide in the presence of a mixture of C_3H_6 and O_2 at 298 K leads to the photo-oxidation reaction of C_3H_6 (experiment; d, e). Under such photooxidation conditions, the photo-induced uptake of O_2 was enhanced as compared with that in O_2 alone, its extent increasing with the pressure of C_3H_6. After photooxidation for 30 min, the temperature of the catalyst was raised stepwise (2 ^{o}C/min) and then the desorption products were analyzed. As shown in Fig. 3 and Table 1 (d, e), desorption of the major oxygen-containing products, i. e, CH_3CHO, C_2H_5CHO, and CH_2=CHCHO occurs at around 373 K. At higher temperatures, hydrocarbons such as CH_4 and C_2H_4, as well as CO and CO_2 (minor products) were desorbed.[8)]

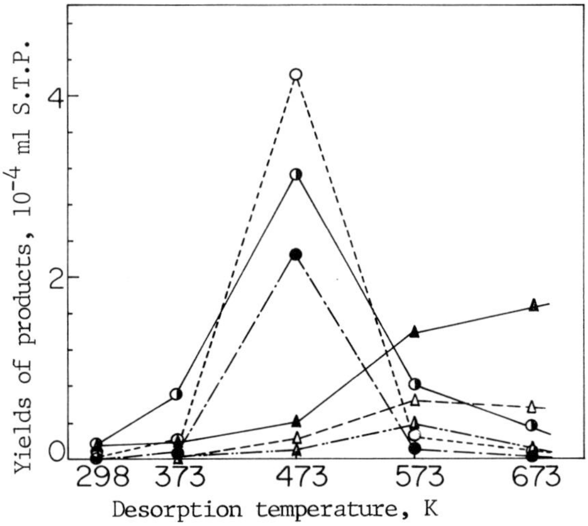

Fig. 3. Desorption profiles of photooxidation products on V/PVG oxide (0.031 V wt%) (experiment: d). (O: C_2H_5CHO, ◐: CH_3CHO, ●: CH_2=CHCHO, ▲: CH_4, Δ: C_2H_4, ◭: C_4H_8, UV-irradiation of V/PVG oxide was carried out at 298 K for 30 min in the presence of C_3H_6 (0.26 Torr) and O_2 (0.65 Torr)).

TABLE 1

Yields of the reaction products of C_3H_6 over V/PVG oxides under various reaction conditions at 298 K.

	10^{-4} ml S.T.P.		
	CH_3CHO	C_2H_5CHO	CH_2=CHCHO
Dark reaction of C_3H_6 and O_2[a)]	0	0	0
UV; in O_2 alone, and then introduced C_3H_6[b)]	0.81	0.42	0.058
UV; in C_3H_6 alone[c)]	0.54	0	0
UV; in C_3H_6 and O_2[d)]	5.36	4.95	2.42
UV; in C_3H_6 and O_2[e)]	20.4	12.8	7.01
Quantum yields[f)]	0.00275	0.0017	0.0009

a): thermal reaction of C_3H_6 and O_2 over V/PVG oxide (0.031 V wt%) at 298 K for 1 h.

b): UV irradiation of V/PVG oxide (0.031 V wt%) was carried out in the presence of O_2 alone, and then C_3H_6 (0.26 Torr) was introduced onto the catalyst and kept for 1 h at 298 K.

c): UV irradiation of V/PVG oxide (0.031 V wt%) was carried out in the presence of C_3H_6 alone (0.026 Torr) at 298 K for 1 h.

d): UV irradiation of V/PVG oxide (0.031 V wt%) was carried out in the presence of C_3H_6 (0.26 Torr) and O_3 (0.65 Torr) for 1 h at 298 K.

e): UV irradiation of V/PVG oxide (0.143 V wt%) was carried out in the presence of C_3H_6 (0.46 Torr) and O_2 (0.33 Torr) for 1 h at 298 K.

f): Quantum yields were determined at 330 nm excitation at 298 K.

On the other hand, UV-irradiation of V/PVG oxide in the presence of C_3H_6 alone led to the uptake of C_3H_6 (experiment; c). After the photo-adsorption of C_3H_6, the temperature of the catalyst was raised stepwise (2 ^oC/min) and the desorption products were analyzed. As shown in Table 1 (c), CH_3CHO is found to be the only oxygen-containing product with much lower yield than that in experiment (d, e). The formation of 1-C_4H_8 and 2-C_4H_8 was also observed, their yields being much lower than those in the photooxidation. Table 1 also shows that the selectivity of CH_2=CHCHO formation becomes higher on the catalyst having a higher V content (experiments d and e).

Taking into account the thermal stability of O_2^- anion radicals on V/PVG oxide, their reactivity of O_2^- toward C_3H_6 was investigated as follows: After formation of O_2^- by UV-irradiation of V/PVG oxide catalyst in the presence of O_2 alone, C_3H_6 was introduced onto the catalyst without evacuation of O_2 (experiment b). As shown in Fig. 2, O_2^- anion radicals react with C_3H_6 molecules at 300 K, their concentration gradually decreasing with reaction time. After sufficient contact of C_3H_6 with O_2^-, the temperature of the catalyst was raised stepwise (2 ^oC/min). The products are summarised in Table 1 (experiment b). It is seen that exactly the same oxygen-containing products, i. e., CH_3CHO, C_2H_5CHO, and CH_2=CHCHO are desorbed, though their yields are much lower than those of the photooxidation (experiment d) by less than one order of magnitude. The O_2^- species adsorbed on V^{5+} ions did not react with C_3H_6 at 77 and 198 K; reacted only above 273 K.

3. Mechanism of photooxidation of C_3H_6

On the basis that the O^- and/or O_3^- radical species are not observed and that the reaction of O_2^- species with C_3H_6 leads to the formation of the same products as those of the photooxidation, it may be concluded that the photooxidation of C_3H_6 with O_2 is closely associated with the presence of O_2^- anion radicals and that activation of C_3H_6 occurs through its interaction with the charge transfer excited state of vanadyl species of the catalyst.

As reported previously,[3)] UV-irradiation of V/PVG oxide in the presence of 2-C_4H_8 leads to the trans$\leftrightarrow$cis and 2$\rightarrow$1 isomerization. The C = C double bond of butene is opened on interaction with the charge transfer excited complex, suggesting that recombination of the photo-formed electron (V^{4+}) and hole (O^-) is required for photocatalytic isomerization to occur. It was found that the photocatalytic isomerization was markedly inhibited by O_2 and that the photooxidation of butene took place in place of the isomerization. These results suggest that olefin molecule can interact with a photo-formed hole center (O^-) of the excited vanadyl species but no longer reacts with trapped electron center (V^{4+}) within the short lifetime of the complex of about 10^{-4} sec., because, as mentioned above, O_2 rapidly reacts with V^{4+} center to form the V^{5+}-----O_2^- species.[9] As a result, as shown in the following reaction scheme, O_2^- anion radicals are produced on V^{5+} ions (step II) while the propene cation radical species are expected to be formed on O^{2-} ions (step III). Simultaneously, V^{4+} ions and O^- ions go back to the ground state ($V^{5+} = O^{2-}$) vanadyl species. When O_2^- anion radicals and propene cation radicals react, the oxidation of C_3H_6 is expected to proceed (step IV) in a manner similar to the photo-oxygenation of organic compounds in which the reaction of the radical cations of substrates produced via photo-induced electron transfer and O_2^- anion radicals has been proposed.[10] Thus, the results clearly indicate that the propene cation radicals react with O_2^-. However, it seems difficult to exclude the contribution of the reaction with unactivated O_2 molecules to the oxidation of propene. The qantitative comparison between the rates of the propene cation radicals with O_2^- and O_2 is a further aim of the present study.

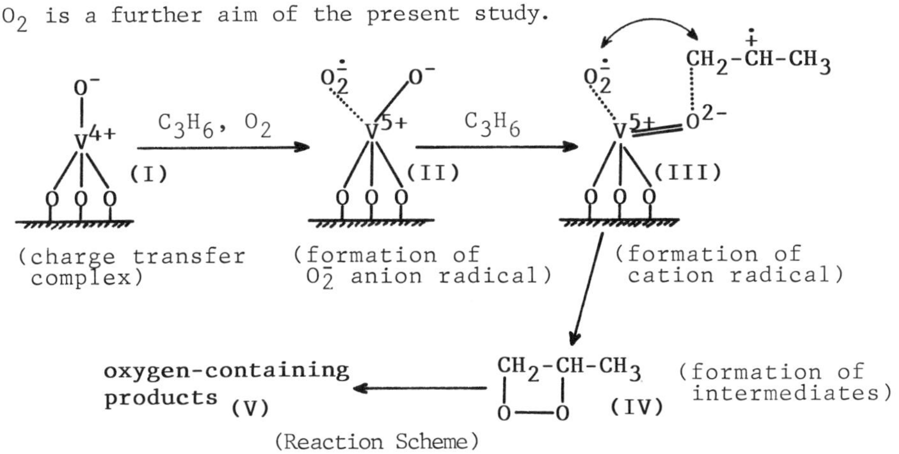

(Reaction Scheme)

Recently, Yoshida et al.[4] have studied the photooxidation of C_3H_6 on V/SiO_2 oxide and proposed the mechanism in which the excited lattice oxygen atom (O^- in $(V^{4+}-O^-)^*$ complex) is the active species to attack C_3H_6 which results in the formation of π-ally intermediates, and the π-ally species reacts with O_2 to form oxygen-containing products. However, the results obtained in the present dynamic photoluminescence and ESR studies indicate that the activation of O_2 occurs on the V^{4+} site of the excited vanadyl species and simultaneously the activation of

C_3H_6 takes place on the O^- site.[11] As shown in Table 1, with V/PVG oxide catalysts with higher V contents the selectivity of CH_2=CHCHO formation became higher. This might be consistent with the mechanism proposed by Yoshida et al.[4] since contribution of lattice O^- free radical species, which separate from the pair state, V^{4+}—O^-, becomes important with the catalyst with higher V contents.[1]

CONCLUSION

Both O_2 and C_3H_6 were activated through the interaction with the charge transfer excited triplet of the vanadyl species, $(V^{4+}—O^-)^*$, of V-oxide supported on Vycor glass to form O_2^- and C_3H_6 cation radicals, respectively. It was found that the reaction of these photo-activated species led to the formation of CH_3CHO, C_2H_5CHO, and CH_2=CH-CHO.

ACKNOWLEDGEMENTS

M. Anpo would like to thank the Universite P. et M. Curie for an appointment as Professor Associe in 1988. Thanks are due to The Ministry of Education of Japan (Grant-in-Aid for Scientific Research No. 62550595 and Grant-in-Aid for Special Project Research No. 61223022).

REFERENCES

1. M. Anpo and Y. Kubokawa, Res. Chem. Intermedi., **8**, 105 (1987), and in "Adsorption and Catalysis on Oxide Surfaces", eds. M. Che and G. C. Bond, (Elsevier) (Amsterdam), 127 (1985).
2. M. Anpo, I. Tanahashi, and Y. Kubokawa, J. Phys. Chem., **84**, 3440 (1980), ibid., **86**, 1 (1982).
3. M. Anpo, T. Suzuki, Y. Yamada, and M. Che, Proc. 9th I. C. C. (Calgary), **4**, 1513 (1988), and M. Anpo, M. Sunamoto, and M. Che, J. Phys. Chem., **93**, 1187-1189 (1989).
4. S. Yoshida, T. Tanaka, M. Okada, and T. Funabiki, J. Chem. Soc., Faraday Trans. I, **80**, 119 (1984), T. Tanaka, M. Ooe, T. Funabiki, and S. Yoshida, ibid., **82**, 35 (1986), and H. Kobayashi, M. Yamaguchi, T. Tanaka, and S. Yoshida, ibid., **81**, 1513 (1985).
5. V. B. Kazansky, Proc. 6th Intern. Congr. Catal., (London), 50 (1976).
6. M. Che and A. J. Tench, Adv. Catal., **32**, 2 (1983).
7. M. Anpo and Y. Kubokawa, J. Catal., **75**, 204 (1982), and M. Anpo, Y. Kubokawa, T. Fujii, and S. Suzuki, Chem. Express, **1**, 41 (1986).
8. Desorption patterns of all products indicate that the product distribution is not affected by thermal desorption up to 673 K.
9. Without propene O^- deactivates to O^{2-} ion within the short lifetime and only O_2^- species is observed. This is the case of Fig. 2.
10. for example, Proc. Intern. Symp. Activation of Dioxygen and Homogeneous Catalytic Oxidation, (Elsevier) (Amsterdam) (1988).
11. M. Che and A. J. Tench, Adv. Catal., **31**, 77 (1982).

J.-M. BREGEAUCT (Université P. et M. Curie, France): As your mechanism is in favor of O_2^- as an intermediate activated species, I suggest that you try the cleavage of ketones which are cleaved at room temperature in phase transfer catalysis by $K^+O_2^-$. It could be proof of the formation of this species.

E. GIAMELLO (Universita' di Torino, Italy) and M. ANPO (University of Osaka Prefecture, Japan): Thank you very much for your comment on the O_2^- species. It would be useful to make sure our mechnaism. However, the reactivities of O_2^- species with akenes leading to the formation of oxygen-containing compounds such as aldehydes on various oxides involving supported vanadium oxides are rather well established facts (ref. 1).

1. M. Che and A. J. Tench, **Adv. Catal.**, 32, (1983) 1.

J. HERRMANN (Ecole Centrale de Lyon, France): I would like to make two comments. First, you are using highly dispersed vanadium oxide, whereas, in our previous study (published paper), we have found bulk V_2O_5 quite photoinactive. The dispersion state can be an important factor in the photoreactivity.

Second, since you have to make a thermodesorption to collect reaction products, I think it would be better to define your reaction as "surface photo-assisted reaction" rather than a photocatalytic one. This is in line with the discussion which I had with Prof. J. HARBER at the end of the proceeding paper.

E. GIAMELLO (Universita' di Torino, Italy) and M. ANPO (University of Osaka Prefecture, Japan): Thank you very much for your comments. Bulk V_2O_5 catalysts did not exhibit any photoreactivity, being in agreement with your results. Whereas, as published by M. ANPO et al. (ref. 2), highly dispersed supported vanadium oxides exhibit the high photocatalytic reactivities for various reactions. Recently, M. ANPO et al. (ref. 3) have clearly exhibited that such photocatalytic activities of supported vanadium oxides are closely associated with the reactivities of the charge transfer excited triplet state of tetrahedrally coordinated vanadium oxide species.

Of course, we agree the comment about the concept of photocatalysis. However, as we have mentioned in the text, all reaction products were produced only under UV irradiation, but all oxygen-containing products easily adsorb onto Vycor glass after the cease of UV irradiation. If we did the UV irradiation in the flow system, such adsorption would be avoided. However, our interesting point is to see the primary processes of the photocatalytic reactions in the closed system. We would like to say that the reaction are really photocatalytic one, as mentioned in our previous paper (ref. 4).

2. M. Anpo, I. Tanahashi, Y. Kubokawa, **J. Phys. Chem.**, 84 (1980) 3440; **ibid.**, 86 (1982) 1; M. Anpo, Y. Kubokawa, **Rev. Chem. Intermedi.**, 8 (1987) 105; M. Anpo, T. Suzuki, Y. Yamada, M. Che, **Proc. Int. Congr. Catal., 9th (Calgary)** 4 (1988) 1513; Y. Kubokawa, M. Anpo, "**Adsorption and Catalysis on Oxide Surfaces**", Eds. M. Che, G. C. Bond, Elsevier (Amsterdam) p. 127 (1985).

3. M. Anpo, M. Sunamoto, M. Che, **J. Phys. Chem.**, 93 (1989) 1187; M. Anpo, M. Sunamoto, T. Fujii, H. Patterson, M. Che, **Res. Chem. Intermedi.**, 11 (1989) 245.

4. Y. Kubokawa, M. Anpo, C. Yun, **Proc. Int. Congr. Catal., 7th (Tokyo)** (1980) 1170.

G. Centi and F. Trifiro' (Editors), *New Developments in Selective Oxidation*

ON PRINCIPLES OF CATALYST CHOICE FOR SELECTIVE OXIDATION

G.I.GOLODETS

The L.V.Pisarzhevskii Institute of Physical Chemistry, Academy of Sciences of the Ukrainian SSR, Prospekt Nauki 31, Kiev, USSR

ABSTRACT

Elementary steps of heterogeneous catalytic processes of oxidation have been classified on the basis of redox and acid - base interactions of reactants with catalyst. General types of mechanisms have been distinguished. Main factors (energetic, structural, etc.) determining catalytic properties of metal oxides for each type of the mechanism are examined.

INTRODUCTION

A progress in the selective oxidation requires an intensive development of its theory. The proposed paper represents the results in this field obtained by the author with his coworkers. In this connection, some remarks should be made. (1) As a consequence of the selected aim, we shall discuss here mainly our own data (not thinking, of course, that they are better than those of many other authors which are summarized, for instance, in Refs. /1,2/). (2) We shall analyse not only our recent results but also some earlier ones to exhibit a whole system of our views. (3) Nevertheless our approach is far of completeness, we hope it to be worthy of discussion.

CLASSIFICATION OF ELEMENTARY STEPS AND MECHANISMS OF THE OXIDATION PROCESSES OVER OXIDE CATALYSTS

It was earlier shown /3/ that in the oxidation catalysis not only redox but also acid-base interactions of reagents R (or products RO_n) with catalysts are essential. Therefore, it is natural to classify elementary steps using this principle. Typical redox steps are the reduction of surface oxides, $M^{m+}O^{2-}$, with R and the reoxidation of their reduced form, $M^{(m-1)+}\square^{-}$ ($\square^{-}$ is an oxygen vacancy) by O_2. Acid - base interactions are assumed in a wide sense involving the formation of complexes with the Lewis or Brönsted active sites, salt-like compounds, π- complexes, etc.

A comprehensive system of such a classification is given in Ref./2/. Using it, one can construct the majority of known mechanisms of selective and deep oxidation /1/. It can be illustrated by the methanol oxidation /2/:

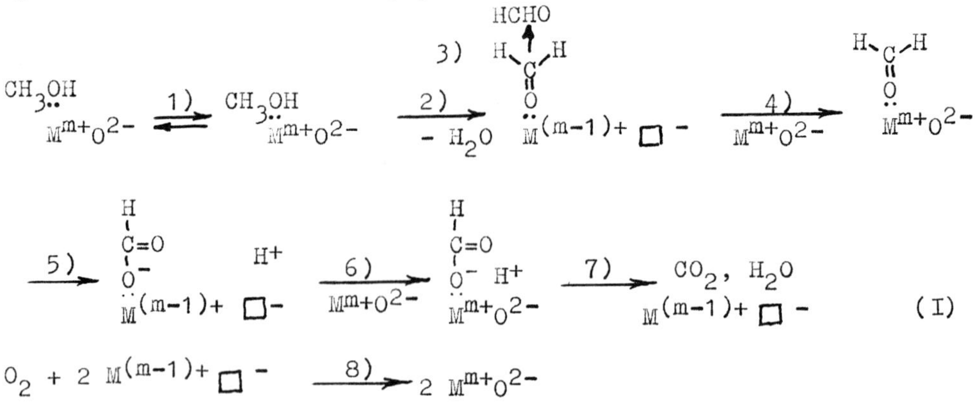

(I)

In this scheme step 8) is purely redox one; steps 1),3) are purely acid-base stages; steps 2),5),7) are "mixed" ones involving the both types of interaction; steps 4),6) include the migration of oxygen or organic intermediates.

The mechanism of a catalytic reaction on a given catalyst can change significantly with temperature (Table).

At moderate temperatures which are typical of industrial catalysis, mechanisms of alternating surface reduction-reoxidation involving O^{2-} species predominate. These reaction pathways (Type I) have been studied far better than other ones. For example, the Type I mechanism of the o-xylene oxidation over V-oxide catalysts has been proved by the ESR method "in situ" /4/, by comparison of kinetics of separate steps and the overall reaction /5/, by the method of competing reactions /6/. Similar evidences have been obtained for selective oxidation of ammonia at $\sim$150-350°C /7/.

At low temperatures, when endothermal desorption of the intermediates (RO, etc.) is especially retarded, the mechanisms with complex reoxidation of a surface (Type IIa) become advantageous; in this case the formation of final products occurs simultaneously with exothermal surface reoxidation /8/. In catalysis over diluted layers of supported metal ions, when the reduction of O_2 to 2 O^{2-} is inhibited, the mechanisms involving reactive adsorbed $\dot{O}^-$ species are profitable /9/ (Type IIb).

At elevated temperatures the desorption of radicals or atoms initiating the gas-phase reaction (heterogeneous-homogeneous catalysis /10/) becomes possible. Such mechanisms (Type III) are typical, for instance, of the CH_4 oxidative coupling /11, 12/.

TABLE

Classification of typical mechanisms of oxidation

t, °C	Mechanism	Scheme
900–700	Type III. Heterogeneous-homogeneous radical-chain mechanisms	$R + M^{(m-1)+}\dot{O}_2^- \rightarrow RO_2 + M^{(m-1)+}\square^-$ $R\dot{O}_2 \rightarrow R'CHO + OH^{\cdot}$ $R'CHO + O_2 \rightarrow RCO^{\cdot} + H\dot{O}_2 \ldots$
500–300	Type I. Mechanisms of alternating surface reduction-reoxidation	$R + M^{m+}O^{2-} \rightarrow M^{(m-1)+}(RO)^- \rightarrow$ $\rightarrow RO + M^{(m-1)+}\square^-$ $O_2 + 2\,M^{(m-1)+}\square^- \rightarrow 2\,M^{m+}O^{2-}$
300–100	Type IIa. Mechanisms with complex reoxidation of surface	$R + M^{m+}O^{2-} \rightarrow M^{(m-1)+}(RO)^-$ $O_2 + M^{(m-1)+}(RO)^- \rightarrow RO_2 + M^{m+}O^{2-}$
100	Type IIb. Mechanisms involving adsorbed anion-radicals of oxygen	$R + M^{(m-1)+}\dot{O}^- \rightarrow RO + M^{(m-1)+}\square^-$

ON PRINCIPLES OF CATALYST SELECTION FOR VARIOUS TYPES OF MECHANISM

Mechanisms of Type I.

Since a dominating intermediate is O^{2-}, catalytic properties should depend on its bond energy expressed as heat, Q_s, of the process $O_2 + 2\,M^{(m-1)+}\square^- \rightarrow 2\,M^{m+}O^{2-}$. In simple cases like the H_2 oxidation an exact relationship between Q_s and specific catalytic activity, r, is observed (ln r decreases with growing Q_s since M-O bonds are broken in a slow step) /1,13/. Since in the formation of deep oxidation products more number of M-O bonds are broken than in partial oxidation, selectivity towards mild oxidation increases with Q_s /7/.

However, in the majority of reactions typical deviations appear in the region of high Q_s values (see Fig. 1a).

This is observed in the oxidation of aromatics, olefins, alcohols, acrolein, etc. /1/. The oxides exhibiting "elevated" activity contain Ti^{4+}, V^{5+}, Nb^{5+}, Mo^{6+}, W^{6+}, i.e. ions with electron configuration of d^0 which are strong Lewis acids. For such catalysts,

$$\ln r = \ln r(Q_s) + \ln r(Q_A),$$

where $\ln r(Q_s)$ is a contribution determined by Q_s (i.e. by the energy of redox processes like $M^{m+} + e \rightarrow M^{(m-1)}$) while $\ln r(Q_A)$

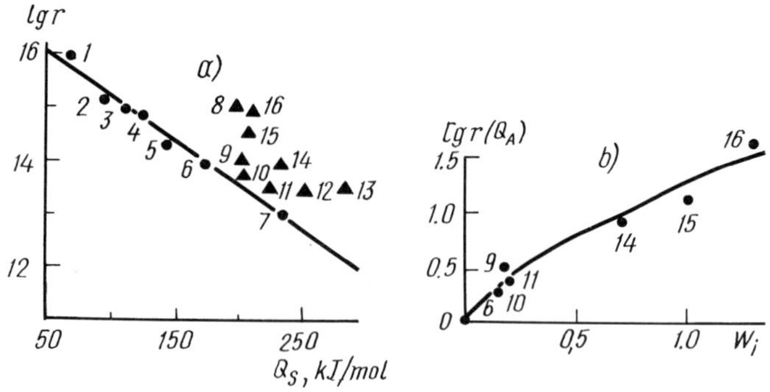

Fig. 1. Plots of lg r - Q_s(a) and lg r(Q_A) - acidity (b) for the CH_3OH oxidation: 1-Co_3O_4, 2-MnO_2, 3-NiO, 4-Cr_2O_3, 5-Fe_2O_3, 6-$CuMoO_4$, 7-MoO_3, 8-V_2O_5, 9-$ZnMoO_4$, 10-$CoMoO_4$, 11-$NiMoO_4$, 12-TiO_2, 13-$Bi_2(MoO_4)_3$, 14-$MnMoO_4$, 15-$Cr_2(MoO_4)_3$, 16-$Fe_2(MoO_4)_3$ /2/

is a fraction of activity depending on the energy of acid-base interactions, Q_A. Using mechanisms like (I), one can divide an influence of these two factors /2/.

The values of ln r(Q_s) correspond to points on the straight line of ln r - Q_s (Fig. 1a). The values of ln r(Q_A), equalling to vertical deviations from the line, can be expressed as

$$\ln r(Q_A) = \frac{1}{RT}\left[(1-\alpha)Q_{CH_3OH} + \alpha Q_{HCHO}\right] + \text{const},$$

where Q_i are adsorption heats of CH_3OH and HCHO; α is a transfer coefficient in the Brönsted-Temkin relationship. Since CH_3OH and HCHO react here as the Lewis bases, the values of Q_i should grow with surface acidity (the rates W_i of butene-1 isomerization can serve as a relative measure of acidity /14/). The fact that the observed values of Δ = ln r(Q_A) increase regularly with W_i (Fig. 1b) confirms our assumptions. In analysis of selectivity, one should also consider stability of salt-like complexes (formiates, etc.) determining the probability of deep oxidation /2/.

A consistent application of such an approach to various reactions of selective oxidation leads to formulation of rather general concept of combined influence of energy of the redox and acid-base surface interactions on the catalytic properties of oxidation catalysts /2,3/. It allows predicting new catalysts (for instance, in the methanol oxidation /15/) and new catalytic reactions (in particular, the oxidation of acetone into acetic acid /16/). At the same time, even within the mechanisms of Type I catalytic properties are determined not only by energy factors but also by other characteristics.

As to electronic structure, the behaviour of ions of transition and non-transition metals in many cases is quite different /17, 18/. In the oxidative coupling, specific properties are exhibited by the oxides of p-elements. On the other hand, in deep oxidation, the oxides of d- and p-elements behave as a common group of catalysts /2/.

In many cases, the geometric (structural) characteristics of active surface are of primary importance. Let us consider two aspects of the problem.

When the formation of key intermediate requires multi-point adsorption, a distance between active sites influences selectivity. So, in the low-temperature oxidation of ammonia into N_2O, the two-point adsorption of HNO-species is necessary. An increase in the distance between active centres causes low probability of existence of the complexes and, as a result, N_2 becomes a dominating product /7/.

With the mechanism like (I), for the formation of the least oxidized product, RO, a single active site, $M^{m+}O^{2-}$, is sufficient. But for the more oxidized product, RO_2, double "clusters" $...M^{m+}O^{2-}M^{m+}O^{2-}...$ are necessary. Hence, one should expect that the selectivity towards mild oxidation will increase when the concentration of the "clusters" (determined by the concentration of an active component, C_M, supported on inert carrier) is decreased. The experimental data on the o-xylene oxidation over V_2O_5/Al_2O_3 (Fig. 2) quantitatively confirm this conclusion /19/.

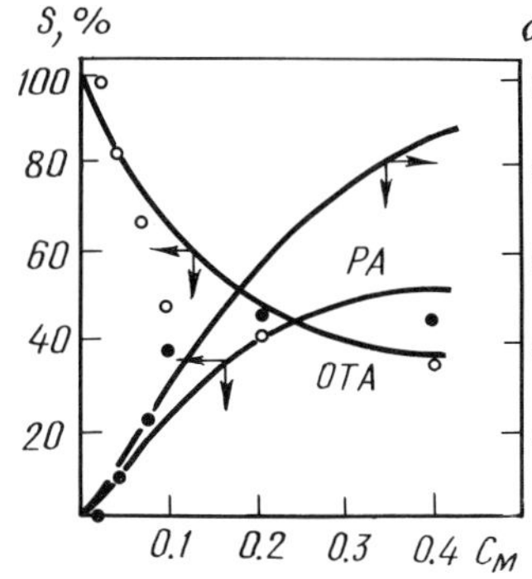

Fig. 2. Plot of ω (fraction of V-ions united in the "clusters") and selectivities S towards o-toluic aldehyde (OTA) and phthalic anhydride (PA) vs surface concentration of V (C_M) /19/

Mechanisms of Type IIa

They usually coexist with the mechanisms of Type I. The following rather general regularity was deduced /20/: mild oxidation products appear only via mechanism I while the more oxidized products are formed via the both mechanisms, I and IIa. For instance, benzaldehyde from toluene is formed only via mechanism I, while maleic anhydride and CO_2 are produced via mechanism IIa (CO_2 also via mechanism I). The same is observed in the oxidation of derivatives of toluene, in the CH_4 oxidation, etc.

Hence, the following typical scheme of conversion of hydrocarbon, R, into mild oxidation products, RO (for example, aldehydes) and into more oxidized products, RO_2 (anhydrides of organic acids, CO_2) can be proposed:

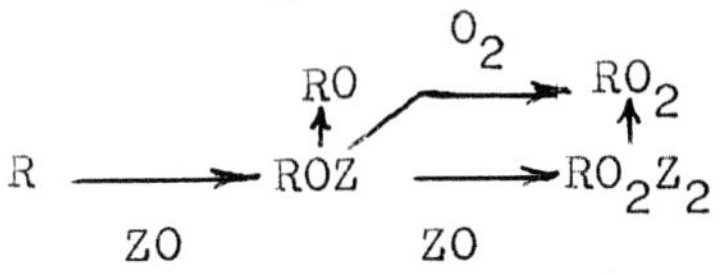

(II)

Complex ROZ is an adsorbed aldehyde bound to $M^{(m-1)+}$ by means of unshared electron pair of the C=O group; RO_2Z_2 is a carboxylate or carbonate complex. Applying the Brönsted-Temkin relationship to the rate constants, one can deduce the equation for selectivity /21/:

$$S_{RO} = \left[1 + A \exp\left(\frac{\alpha}{RT} Q_{RO_2}\right) \exp\left(-\frac{\alpha}{2RT} Q_s\right) + B \exp\left(\frac{\alpha}{2RT} Q_s\right)\right]^{-1},$$

where Q_{RO_2} is heat of the RO_2 adsorption; A and B are constant (at a given composition of the reacting mixture). If in the series of compared catalysts the values of Q_{RO_2} change much less than those of Q_s, we obtain (for competing mechanisms I, IIa):

$$S_{RO} = \left[1 + A' \exp\left(-\frac{\alpha}{2RT} Q_s\right) + B \exp\left(\frac{\alpha}{2RT} Q_s\right)\right]^{-1}.$$

Hence it follows that on increasing Q_s selectivity should pass through a maximum. This is really observed in the oxidation of CH_4 or toluene (Fig. 3) and in the naphthalene oxidation /22/.

At small values of Q_s, the mechanism of Type I predominates, and selectivity grows with Q_s; at high values of Q_s the mechanism of Type IIa prevails and selectivity decreases with an increase in Q_s.

In a more strict examination one should also take into account the values of Q_A, i.e. energies of acid-base interaction.

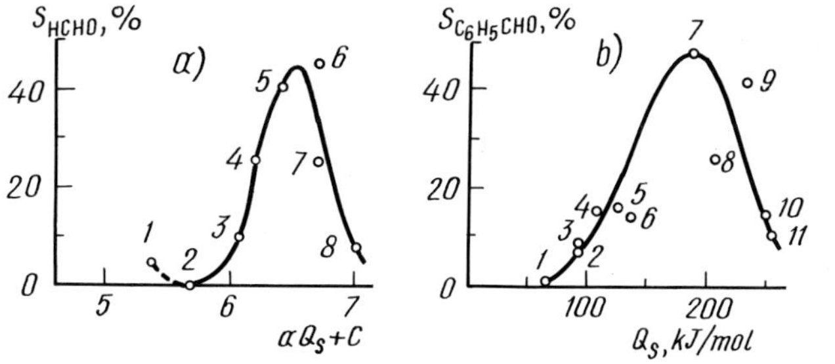

Fig. 3. Dependence between Q_s and selectivity: (a) in the CH_4 oxidation on phosphates (1-Cr, 2-Mn, 3-Ni, 4-Fe, 5-B, 6-Al, 7-Mg, 8-Co); (b) in the toluene oxidation on oxides (1-Co, 2-Cu, 3-Mn, 4-Ni, 5-Cr, 6-Fe, 7-V, 8-Mo, 9-W, 10-Ti, 11-Nb) /21/

Mechanisms of Type IIb

There are few evidences in favour of these mechanisms /9/. The partial oxidation of methane over diluted layers of supported metal oxides is likely to belong to such reaction schemes /23/.In this case the catalytic activity in total process decreases with an increase in the energy bond of metal ions with $\dot{O}^-$ (Q_s'); selectivity towards mild oxidation (into methanol and formaldehyde) increases with growing Q_s' values /24/.

Mechanisms of Type III

These mechanisms have been studied insufficiently. In Ref. /25/ the relationship between probability of heterogeneous-homogeneous mechanism and the chemical nature of oxide catalysts have been clarified for the model reaction of hydrogen oxidation. It was shown that in the presence of active heterogeneous catalysts (the oxides of Co, Cu, V, etc.) the lower ignition limit, P_1, of the detonating gas increases in comparison with the P_1 value in a quartz vessel (i.e. in the presence of SiO_2, see Fig. 4).

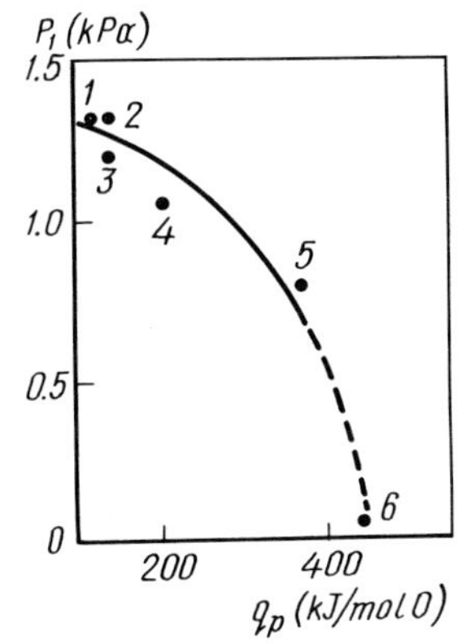

Fig. 4. Comparison of values of the 1st ignition limit of detonating gas and the values of q_p for metal oxides:
1-V_2O_5, 2-Co_3O_4, 3-CuO, 4-MoO_3, 5-TiO_2, 6-SiO_2 /25/

Since in the course of reaction the catalysts were reduced, it was supposed that an intensive chain termination takes place by way of the irreversible reduction of M_mO_n with H-atoms (which are known to be the main intermediates in the radical-chain combustion of H_2). Then one should expect P_1 to be greater (heterogeneous-homogeneous mechanism to be less probable) when reducibility of the oxide increases. Really, as one can see in Fig.4, the P_1 value increases with a decrease in the heat of dissociation, q_p, of metal oxides (i.e. with increasing heats of reduction). The same was observed for the oxidation of methane, carbon monoxide and some other processes, so that the above mentioned relationship is likely to be rather widespread.

REFERENCES

1 G.I. Golodets, Heterogeneous Catalytic Reactions Involving Molecular Oxygen, Elsevier, Amsterdam, 1983
2 G.I. Golodets, Problemy Kinetiki i Kataliza, 19(1985)28
3 G.I. Golodets, Doklady Akad. Nauk SSSR, 184(1969)1334
4 V.M. Vorotyntsev, Yu.I. Pyatnitsky, G.I. Golodets, Teor.Eksper. Khim., 12(1976)488
5 Yu.I. Pyatnitsky, V.M. Vorotyntsev, G.I. Golodets, Kataliz i Katalizatory, 12(1974)28
6 Yu.I. Pyatnitsky, T.G. Skorbilina, Kinet.i Katal.,21(1980)451
7 N.I. Ilchenko, Uspekhi Khimii, 45(1976)2168
8 V.D. Sokolovsky, Teor.Probl.Kataliza,Novosibirsk, 1977, p.33
9 V.B. Kazansky, Problemy Khim.Kinet., Nauka, Moscow, 1979,p.232
10 M.V. Polyakov, Kataliz i Katalizatory, 1(1965)35
11 M.Yu.Synev, V.N.Korchak,O.V.Krylov,Kinet.i Kataliz.,28(1987) 1376
12 N.I.Ilchenko,L.N.Raevskaya,A.I. Bostan, G.I. Golodets, Teor. Eksper. Khim., 24(1988)638
13 G.K.Boreskov,Heterogeneous Catalysis (in Russ.),Nauka,Moscow, 1986
14 T.G.Alkhazov, K.Yu.Adzhamov, E.A.Mamedov, V.P.Vislovsky, Kinet. i Kataliz., 20(1979)118
15 G.I. Golodets, Teor. Eksper. Khim., 18(1982)37
16 G.I.Golodets, V.V. Borovik, V.M. Vorotyntsev, Teor.Eksper. Khim., 22(1986)252
17 O.V. Krylov, V.F.Kiselev, Adsorption and Catalysis on Transition Metals and Their Oxides(in Russ.),Khimia,Moscow,1981,p.288
18 J. Haber, M. Sochacka, B. Grzybovska, A.Golzbiewskii, J. Molec. Catal., 1(1975)35
19 Yu.I. Pyatnitsky,G.I.Golodets,React.Kinet.Catal.Lett.,5(1976) 345
20 G.I.Golodets, Yu.I.Pyatnitsky,L.N.Raevskaya,Kinet.i Katal., 25(1984)571
21 G.I.Golodets, Kinet. i Katal., 28(1987)837
22 G.I.Golodets, Teor. Eksper.Khim., 1(1965)756
23 N.I.Ilchenko, A.I.Bostan,L.Yu.Dolgikh, G.I.Golodets,Teor.Eksp. Khim., 23(1987)641
24 N.I.Ilchenko, A.I.Bostan, G.I.Golodets, Teor.Eksper.Khim., 24(1988)455
25 G.I.Golodets, React.Kinet.Catal. Lett., 28(1985)131

G. Centi and F. Trifiro' (Editors), *New Developments in Selective Oxidation* 701

MUTUAL ORIENTATION OF OXYGEN AND HYDROCARBON MOLECULES AS FACTOR DETERMINING THE REACTION PATHWAY

M. WITKO, J. HABER and E. BROCLAWIK

Institute of Catalysis and Surface Chemistry, Polish Academy of Sciences, ul. Niezapominajek, 30 239 Kraków (Poland)

SUMMARY

Quantum chemical calculations of simple potential energy maps for system composed of toluene and oxygen moieties are presented. Reaction pathways following from energy profiles for interaction between reagents in different mutual orientations are discussed.

INTRODUCTION

In the studies of chemical reactions two fundamental questions must be answered before a general theoretical description of their mechanism might be attempted:

- which is the reactivity of the given molecule towards other molecules of the system and which products may be formed as the result of their reactions;
- which is the molecular description of reactions in terms of the shifts of nuclei and electrons.

Quantum chemistry provides the theoretical basis for the empirical rules of molecular reactivity and supplies quantitative data for the selection of possible reaction mechanisms. The analysis of molecular electronic structure and its changes along different reaction pathways enables information to be obtained concerning the probability of the required transformations and conclusions to be drawn as to which mechanism constructed on the basis of phenomenological studies corresponds to physical reality.

Basing on the Born-Oppenheimer approximation, fundamental in quantum chemical considerations, the concept of the potential energy hypersurface for molecular motions may be introduced. The minima on such a hypersurface correspond to stable systems, i.e. to reactants and products of the investigated chemical reaction network. As the chemical systems interesting from the point of view of catalysis are relatively large, semiempirical computational methods must usually be applied, among them only MNDO and its schemes, which correctly predict molecular energies and geometries, are the most appropriate. Even in approximate approach the computation of the reaction pathway is a very time consuming procedure, because at each value of the reaction coordinate (for each separation between the reacting moieties) a full optimization of geometry must be carried out. Therefore further simplifications of the model must be adopted in order to make full description feasible.

Analysis of experimental data suggest that the form of the transition state is predetermined already at the preliminary stage of the reaction, the choice of the reaction pathway being decided by the form of the potential energy hypersurface even at large distance between the reacting molecules. Assuming thus that the long range interaction energies between molecules in different mutual orientations are approximately proportional to the activation barriers encountered on their approach in these orientations, the calculations of the total energy of the transition complex along different reaction pathways may be replaced by the estimation of the reaction energies along these pathways at certain chosen distance between the reacting molecules i.e. at certain chosen value of the reaction coordinate.

In our previous study we have shown that products of the reaction between benzene and molecular oxygen depends on the direction of the approach of the oxygen molecule to benzene ring. An interesting question arises as to what extent the presence of the methyl group at the benzene ring influences the reaction pathways between benzene and oxygen. On the other hand it is well known that it is difficult to achieve high selectivity in heterogeneous oxidation of toluene to benzaldehyde and considerable amounts of CO_2 are always formed. An additional question may be raised as to which is the role of molecular oxygen in favouring total oxidation of toluene in contrast to the case of benzene, in which maleic anhydride may be obtained with high selectivity. Thus the reaction between toluene and oxygen moieties was chosen for the present study. It was hoped that ample experimental data and detailed calculations for some reaction pathways concerning the catalytic oxidation of aromatic hydrocarbons would enable the confrontation of the presented model with reality.

METHODS AND SYSTEMS

Quantum chemical calculations for the systems under consideration were carried out by means of MNDO method with geometry optimization procedure [1].

The reaction may be considered as proceeding between following reacting species:

- nonactivated toluene and oxygen molecules;
- toluene activated by abstraction of hydrogen and oxygen molecule;
- nonactivated toluene and oxygen atom;
- toluene activated by abstraction of hydrogen and oxygen atom.

Both molecular and atomic oxygen may carry different numbers of electrons, moreover in the course of the catalytic reaction the interaction with the catalyst surface entails transfer of electrons between reactants and the catalyst. Therefore the total number of electrons in the system is not apriori known. Thus, in order to account for this transfer, the calculations were performed for different numbers of electrons in the system, their distribution between the reactants resulting from the interactions. Thus, the system containing e.g. two excess electrons may be considered as the reaction between neutral toluene and O^{2-} or toluene with -1 charge and O^-.

As discussed in the introduction conclusions concerning the most probable reaction pathways, which follow from the calculations of potential energy hypersurface, may be drawn already from the picture obtained by the calculation of potential energies at certain chosen distance between the molecules. Toluene and oxygen may approach each other from all possible directions. Most probable reaction pathways may be however selected by considering several limiting cases:

- the trajectory around the toluene molecule in the plane passing through the molecular axis perpendicularly to the ring. Above the ring the trajectory is a straight line parallel to the molecular axis at the distance of 2.0 Å, and then curves around the methyl group at the distance of 2.6 Å (Fig. 1);
- the trajectory in plane of the aromatic ring, oxygen molecule approaching toluene from different directions with molecular axis perpendicular to this plane or in plane. The calculations were carried out for oxygen molecule (atom) moving around the aromatic ring at the distance of 4.0 Å from the center of the ring, and around the methyl group at the distance of 2.6 Å from the carbon atom of this group;
- the trajectory around toluene molecule in the plane perpendicular to the molecular axis crossecting the aromatic ring.

RESULTS AND DISCUSSION

Attack of molecular oxygen

Figures 1 and 2 show the energy profiles against the coordinate ξ measuring the position of oxygen on the trajectory for activated and nonactivated toluene, the charge on the system being 0 and -1. Results illustrated in Fig. 1 indicate that energy minima appear over the aromatic ring in all studied cases. For nonactivated toluene this minimum is located over the centre of the ring, whereas in the case of activated toluene it is shifted slightly away from the methyl group. When toluene is activated, a second minimum appears over the bond between the ring and the methylene group. The minimum at the end of each curve corresponds to the in plane attack illustrated in Fig.2. Results drawn in this figure indicate that for the activated toluene this minimum is the deepest one, whereas for nonactivated toluene the attack at the methyl group encounters similar barrier as those on the ring.

Detailed quantum chemical calculations were carried out for entire reaction pathways from the directions, corresponding to the discussed minima of the potential energy. The side attack of oxygen onto the ring may be the model of the oxidation of benzene described in paper [2]. It was found that in such a case the most possible is the perpendicular attack of molecular oxygen onto the ring, leadind to degradation of the molecule, which proceeds with low activation barrier.For the attack of oxygen from above the ring plane, parallelly to the diagonal, the results suggest that the system must pass through relatively high energy barrier corresponding to the stretching of O - O bond and evolves further towards to the metastable transition state with bridging oxygen, rearranging eventually to maleic anhydride [3]. Basing on our present results we can conclude that the oxidation of aromatic ring should proceed along similar pathways for benzene and its alkyl-substituted derivatives.

The attack of oxygen onto the methylene group may be the model of the oxidation of activated o-xylene [4]. In this paper the reaction pathway for the attack of perpendicularly oriented molecule onto CH_2 group was calculated, which lead to the o-xylene peroxide. Our present calculations show, that this process should be the easiest to proceed in the case of activated toluene.

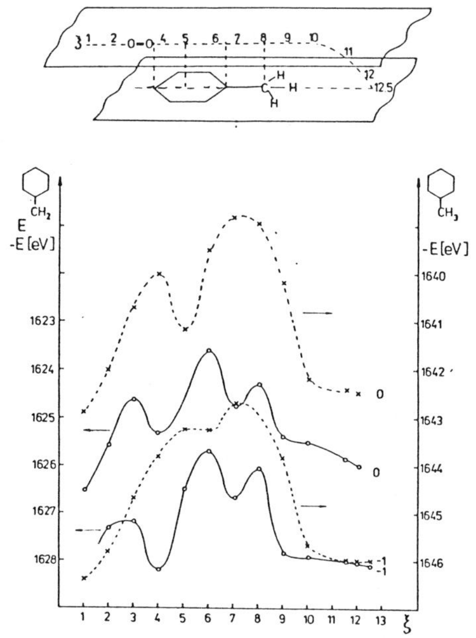

Fig. 1. Energy profiles for the movement of oxygen molecule at the distance d=2.0 Å above the plane of toluene ring paralelly to the molecular axis. (--- curves for nonactivated toluene, —— curves for activated toluene, charge on the system is given by curves)

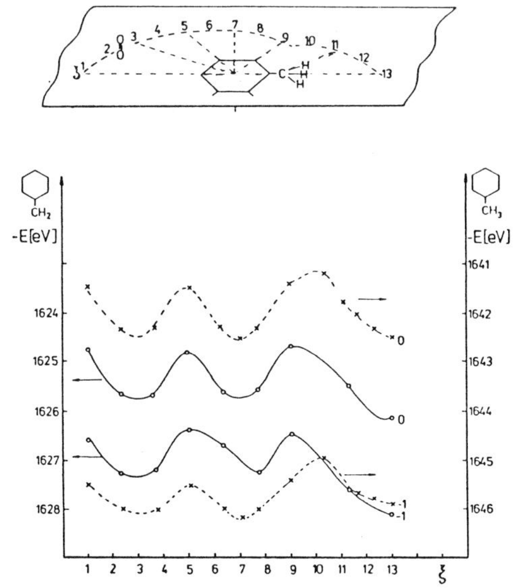

Fig. 2. Energy profiles for the movement of perpendicularly oriented oxygen molecule around toluene in the plane of the ring. (curve captions the same as in Fig. 1.)

Attack of atomic oxygen

From similar calculations of energy profiles of subsystems (toluene or activated toluene and oxygen atom) one can conclude that in the case of nonactivated toluene the most probable direction of the attack is that perpendicular to the ring plane and pointing towards ring C - H bonds for all numbers of electrons. The energy minima for the in plane attack on the C = C bonds of the ring as well as on the methyl group were less pronounced.

The situation for the activated toluene is different and depends on the number of electrons in the system. Both 0 and -1 excess charges lead approximately to the neutral oxygen and toluene with 0 and -1 charge. For these cases it is the C atom from CH_2 group which is the most susceptible to the attack. Shallow local minima appear on the energy profiles in the directions of hydrogen atoms which may correspond to hydrogen bonds.

A possibility should be also considered of the attack of oxygen atom from above the plane on the π system directly towards the C = C bonds, which would probably lead to the degradation of the ring. This attack provides the most probable reaction pathway for -2 excess charge on the system, corresponding approximately to O^- ion and $C_6H_5CH_2^{-1}$. Apparently the negative charge accumulating on the methylene group makes the attack of oxygen ion more difficult.

The results for the attack of atomic oxygen are in agreement with the conclusions obtained from the detailed study of the reaction pathways for the oxidation of o-xylene [4]. Indeed, the attack of atomic oxygen onto CH_2 group of activated o-xylene from above the molecular plane leads to the tolualdehyde type product.

The examples discussed in the present paper seem to confirm the assumption that essential information about the shape of reaction pathways is already given by the energy of interaction of reagents at the distance corresponding to early stages of the reaction. The comparative analysis of these interactions for different mutual orientations of reacting molecules may help in the difficult task of choosing possible reaction pathways.

REFERENCES

1 M.J.S. Dewar and W. Thiel, J. Am. Chem. Soc., 99 (1977) 4899.
2 E. Broclawik, J. Haber and M. Witko, J. Mol. Catal., 26 (1984) 249-258.
3 M. Witko, E. Broclawik and J. Haber, J. Mol. Catal., 35 (1986) 179-189.
4. M. Witko, E. Broclawik and J. Haber, J. Mol. Catal., 45 (1988) 183-191.

G. Centi and F. Trifiro' (Editors), *New Developments in Selective Oxidation*

CHARACTERIZING OXYGEN ADSPECIES FOR ETHYLENE EPOXIDATION OVER SILVER WITH THE TAP REACTOR SYSTEM

N.C. RIGAS, J.T. GLEAVES and P.L. MILLS

The Fred Gasche Laboratory, Department of Chemical Engineering, Washington University, 1 Brookings Drive, Box 1198, St. Louis, MO 63130-4899 (U.S.A.)

SUMMARY

The nature of the active oxygen species in the selective oxidation of ethylene to ethylene oxide over a silver metal catalyst has been examined using transient pulse experiments with the TAP (Temporal Analysis of Products) reactor system. Through the application of multipulse and pump-probe experiments using $^{16}O_2$, $^{18}O_2$ and C_2D_4 as reactant molecules, and by monitoring the transient formation of $C_2D_4{}^{16}O$, $C_2D_4{}^{18}O$, $C^{16}O_2$, $C^{16}O^{18}O$, and $C^{18}O_2$, it is possible to distinguish between oxygen adsorbed on the surface versus subsurface or lattice oxygen. The results suggest that a migration of lattice oxygen to both selective and nonselective catalytic sites occurs. Comparison of transient response curves for the various isotopic forms of carbon dioxide and ethylene oxide suggest that more than one type of catalytic site is involved in CO_2 production.

INTRODUCTION

Determining the nature of the active oxygen species in heterogeneous catalysts for selective vapor phase oxidation of small paraffinic and olefinic molecules is an essential part of unraveling the mechanism for these systems. This type of knowledge has significant potential for providing insights into improving the performance of existing catalytic processes, or designing new ones.

Transient response experiments based upon the TAP reactor system (refs. 1-2) have been employed to characterize the active oxygen species for butane oxidation to maleic anhydride over $(VO)_2P_2O_7$ catalysts (refs. 2-6), propene oxidation to acrolein over Bi_2MoO_6 catalysts (ref. 1), methanol ammoxidation to hydrogen cyanide over $Mn_2P_2O_7$ and $Fe_2(MoO_4)_3$ catalysts (ref. 7), and ethylene oxidation to ethylene oxide over a silver metal catalyst (ref. 8). In the PVO system, for example, it has been shown that various surface atomic oxygen species with different surface lifetimes and reactivities are present, depending upon the nature of the catalyst and the reaction conditions (refs. 3-5). Surface lattice oxygen is responsible for allyl oxidation and ring insertion, whereas labile oxygen is involved in other mechanistic steps. Both the selective and nonselective pathways are clearly influenced by the presence of these different types of oxygen species. Further discussion on this subject for the remaining systems is given in the above cited references to which the interested reader is referred for details.

Ethylene epoxidation over unsupported silver metal catalysts has been used as a model reaction for mechanistic studies of the commercial catalytic system (refs. 9-11). More recent results have suggested that both atomic surface oxygen and subsurface oxygen are necessary for catalytic epoxidation (refs. 8, 11). Recent TAP studies on ethylene epoxidation (ref. 8) have lead to several key insights. It has been shown that adsorbed molecular oxygen, whether formed from the gas phase or through the recombination of adsorbed atomic oxygen, does not participate in ethylene epoxidation. Experiments using $^{18}O_2$ support the previous assertions (refs. 9-10) that both surface and lattice oxygen play a key role in the epoxidation reaction.

The primary objective of this work is to examine the role of surface and lattice oxygen in ethylene epoxidation through more detailed transient TAP experiments using isotopic ethylene and oxygen. This differs from previous TAP reactor studies (ref. 8) in that the oxygen transfer between the surface and lattice is inferred using various TAP transient pulse formats that lead to the formation of key isotope bearing products.

EXPERIMENTAL

Materials

A nonporous granular silver metal catalyst (Johnson-Mathey) was used as received for all TAP experiments. It was sieved to give particles having diameters between 300 and 600 μm to yield an average diameter of 400 μm. This latter value yields a specific external surface area of 1.43×10^3 cm^2/g.

All reactant gases were obtained from Matheson and also used as received without any further treatment. The particular ones used here and their minimum purities include C_2D_4 (99%), $^{16}O_2$ (99.98%), $^{18}O_2$ (97%), and CO (99.98%).

TAP Reactor System

All reaction experiments were performed in the TAP (Temporal Analysis of Products) reactor system. A detailed description is given elsewhere (refs. 1-2) so only the key features pertinent to this study will be given here. Referring to Fig. 1, the silver catalyst is maintained within a 6.4 mm i.d. microreactor whose overall length is 38.2 mm. The catalyst was packed to a depth of 12.4 mm and was preceded by a 16 mm layer of glass beads where the bead diameter was 1 mm. The layer of glass beads at the reactor inlet provides a zone where the inlet gas pulse can expand from the continuum regime to the Knudsen or molecular flow regime before entering the catalyst bed. The reactor is attached to two high speed pulse valves and a continuous flow valve through a zero volume manifold. The reactant and product pulses that exit the reactor are detected using a quadrupole mass spectrometer which can either scan a predefined range of m/e values, or sense the temporal variations of a given m/e value. The reactor, valve assembly, and QMS are located within a differentially pumped vacuum system which minimizes the distortion of the pulse as it travels through the system as a molecular beam. Average pressures in the reactor chamber are ca. 10^{-6} to 10^{-7} torr when pulse intensities between 10^{15} to 10^{18} molecules/pulse are used. When the reactor is removed, a hydrogen pulse can typically be characterized by a mean residence time of ca. 1.75 milliseconds and a width at half-height of 200 microseconds.

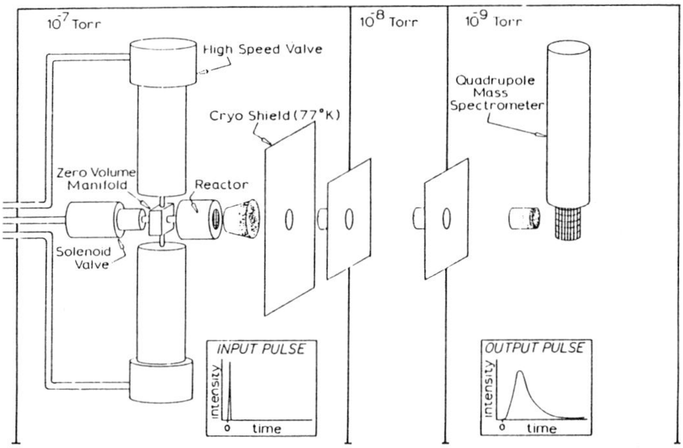

Fig. 1. Basic components of the TAP reactor system showing the fixed-bed microreactor, continuous flow and high speed pulse valves, quadrupole mass spectrometer detector system, and differentially pumped high vacuum system.

Transient TAP system experiments are defined according to various methods used to actuate the pulse valves. These include the single pulse, multipulse, and pump-probe pulse formats. The results given below are based upon two different pulse formats, namely, the pump-probe and multipulse formats. In the pump-probe experiment (Fig. 2), the two pulse valves are sequentially actuated at user-defined values for the pulse rate (e.g., 1 pulse per second), the interval between firing of the pulse valves (e.g., 0.25 s), and the pulsing duration (e.g., 60 s). Intermediates formed on the catalyst after actuating the first pulse valve and introducing the pump molecules can be trapped by the probe molecules introduced by actuating the second pulse valve. By varying the time interval between the pump and probe molecules, it is possible to monitor the transient response of a particular m/e value that characterizes a given desorbed species. Although the intermediates on the catalyst surface may not necessarily desorb for a variety of reasons, the probe molecule may react with it to form a species which can desorb and be detected. Both the nature of surface on intermediates and their relative lifetimes can be studied in this fashion. The results given later employ O_2 as the pump molecule and C_2D_4 as the probe molecule.

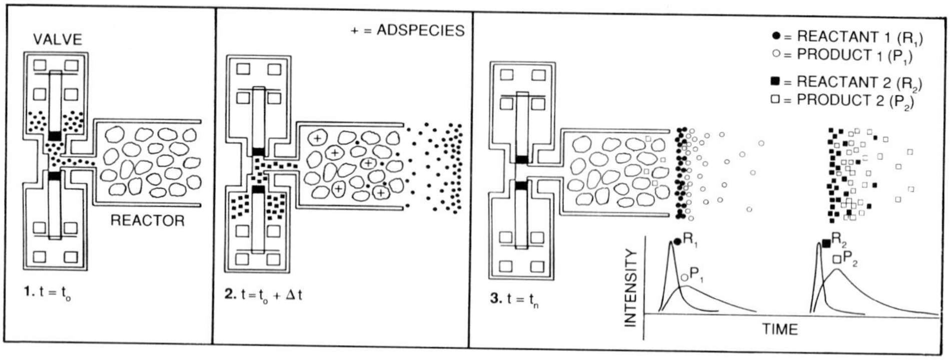

Fig. 2 Schematic of the TAP reactor system pump-probe experiment.

The multipulse experiment is a variation of the pump-probe experiment in which only one of the valves is utilized. A pulse train of reactant gas is introduced to the microreactor at a user-defined pulse rate (e.g., 1 pulse per second) over an assumed time interval (e.g., 60 seconds) while monitoring the transient response of a given m/e value. This allows one to examine the relative increase or decrease of one or more reaction products under a desired set of conditions. A common application in our previous selective oxidation work cited above involves multipulse experiments with O_2 while monitoring the transient formation of CO, CO_2, and H_2O. This provides a very precise method for titrating or dosing the catalyst since the pulse characteristics can be carefully controlled.

RESULTS AND DISCUSSION

Silver Catalyst Activation

Various catalyst activation procedures were compared at the onset of this work to ensure that reproducible transient experiments could be obtained. Multipulse experiments conducted on fresh catalyst using $^{16}O_2$ at 1 pps and T = 290 C showed that the transient peaks associated with CO_2 production would increase to a maximum and then asymptotically decrease to a small peak having a constant height. These results indicated that carbon or carbon-bearing species were being removed from the catalyst. It was also shown that a minor carbon-containing impurity in the $^{16}O_2$ was responsible for the small asymptotic peaks evident in the data.

Pump-probe experiments using a pump-probe interval of 0.25 s with $^{18}O_2$

and C_2D_4 as the pump and probe molecules, respectively, did not yield any deuterated ethylene oxide $C_2D_4{}^{16}O$ as a transient product at m/e = 48. By reducing the catalyst with CO at 500 C, Fig. 3 shows that C_2D_4O could be obtained as a product. Both the conversion of C_2D_4 and selectivity of C_2D_4O increase with the degree of CO reduction, which is indirectly measured here by the CO exposure time.

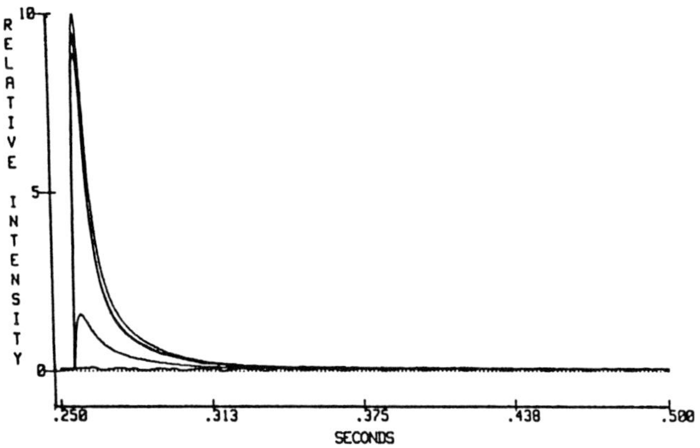

Fig. 3. Transient responses of C_2D_4O obtained from pump-probe experiments at various degrees of CO reduction. Pump molecule = O_2 pulsed at t = 0, probe molecule = C_2D_4 pulsed at t = 0.25 s.

The above procedure of reducing the silver surface with CO has been previously studied (ref. 12). The results in Fig. 3 support the finding that reduction of the silver catalyst with CO is necessary for activation and to produce ethylene oxide as a product. By contrast, activation of the silver with oxygen only does not produce ethylene oxide. Additional details are provided elsewhere (ref. 13).

Multipulse Results

Multipulse experiments were conducted to investigate the exchange of atomic oxygen between the lattice and surface. The silver catalyst was first exposed to a continuous flow of ca. 2 sccm ${}^{16}O_2$ for one hour at 290 C which presumably saturates the lattice and surface. A pulse train of C_2D_4 was introduced to the microreactor at the rate of 2 pps, and the transient responses of $C^{16}O_2$ (m/e = 48) were monitored. Results for both species are compared in Fig. 4. The overall reactions occurring in this case would be

$$C_2D_4 + 1/2\ O_2 \rightarrow C_2D_4O \quad (1)$$

$$C_2D_4 + 3\ O_2 \rightarrow 2\ CO_2 + 2\ D_2O \quad (2)$$

$$C_2D_4O + 11/2\ O_2 \rightarrow 2\ CO_2 + 2\ D_2O \quad (3)$$

where O_2 is understood to be $^{16}O_2$. Elementary steps associated with each of these are not given, as are the overall reactions that describe CO formation. Examination of Fig. 4 shows that the peak areas for CO_2 are noticeably greater than the corresponding ones for C_2D_4O. This suggests that formation of CO_2 is more facile than C_2D_4O under these conditions. In addition, it illustrates how the multipulse experiment can be used to titrate a particular component, which is CO_2 in this case, from the catalyst.

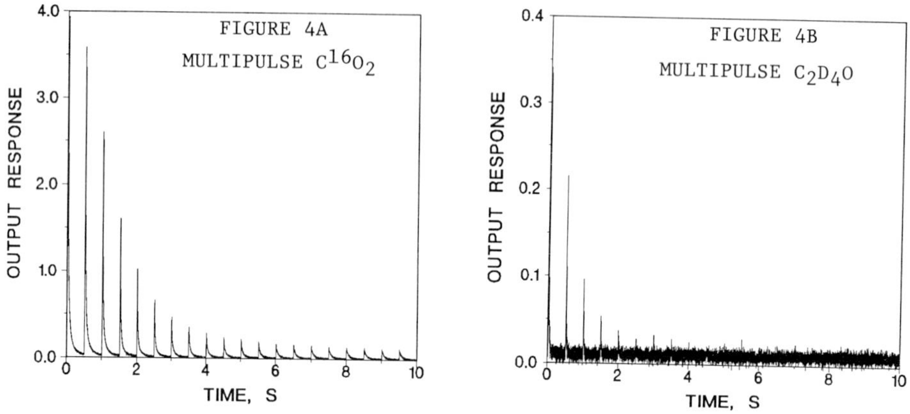

Fig. 4. Transient responses of $C^{16}O_2$ (Fig. 4a) and C_2D_4O (Fig. 4b) obtained from multiphase experiments in which C_2D_4 is pulsed at 2 pps. Conditions: T = 290 C, 0.3 g of silver catalyst, 10^{17} molecules/pulse of O_2 and C_2D_4.

Pump-Probe Results

Additional information about the oxygen exchange was obtained by using a sequence of continuous flow, multipulse, and pump-probe experiments. Initially, the activated silver catalyst was exposed to a continuous flow of $^{16}O_2$ for ca. 30 minutes at 2 sccm and 290 C to saturate the surface and lattice with ^{16}O, or other species containing ^{16}O. Next, C_2D_4 was pulsed at 2 pps for a duration of 60 s in a multipulse experiment while monitoring the transient responses of $C^{16}O_2$ (m/e = 44) and $C_2D_4{}^{16}O$ (m/e = 48). Results similar to those given above in Fig. 4 were obtained which indicated that ^{16}O, or a species containing it, was being titrated from the catalyst surface. Immediately following this titration, a pump-probe experiment in which $^{16}O_2$ and C_2D_4 were used as the pump and probe molecules, respectively, was carried out. The primary transient products identified were $C_2D_4{}^{16}O$ and $C^{16}O_2$. Figure 5a shows the transient response of O_2 and C_2D_4, while the transient response of $C_2D_4{}^{16}O$ is given in Fig. 5b.

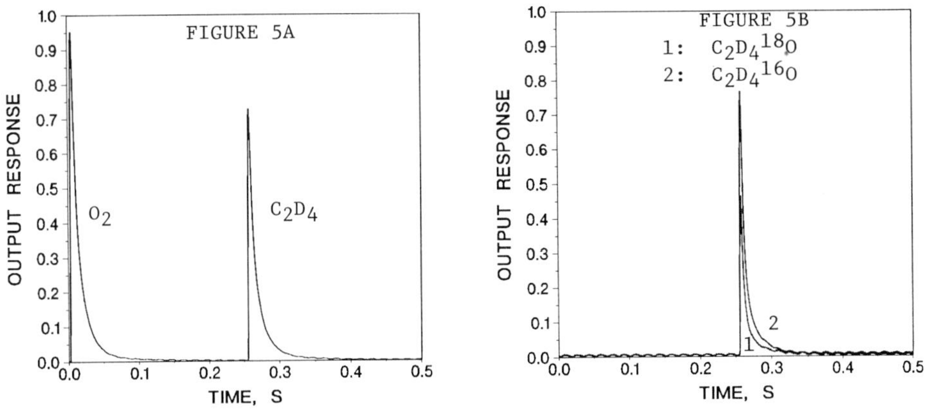

Fig. 5. Transient response of O_2 (pump) and C_2D_4 (probe) molecules (Fig. 5a), and either $C_2D_4{}^{16}O$ or $C_2D_4{}^{18}O$ (Fig. 5b). Conditions: same as Fig. 4.

The above procedure was repeated, except $^{18}O_2$ was substituted for $^{16}O_2$ in the pump-probe portion of the experimental sequence. Possible products associated with the transient responses include $C_2D_4{}^{16}O$, $C_2D_4{}^{18}O$, $C^{16}O^{18}O$, $C^{18}O_2$ and $C^{16}O_2$. The primary source of ^{16}O is the lattice since adspecies containing ^{16}O had been reduced to undetectable levels with the multipulse experiment. In Fig. 5b, the pump-probe transient response associated with $C_2D_4{}^{18}O$ (m/e = 50) is given for comparison to $C_2D_4{}^{16}O$ obtained in the previous pump-probe experiments in which $^{16}O_2$ was used as the pump molecule. It is worth noting that the $C_2D_4{}^{18}O$ peak area is ca. 60% of the $C_2D_4{}^{16}O$ peak area. Since it is unlikely that the differences in peak areas are due to kinetic isotope effects between $^{16}O_2$ and $^{18}O_2$, the difference is ascribed here to some $C_2D_4{}^{16}O$ being produced by ^{16}O in the lattice when $^{18}O_2$ is being pulsed in as the pump molecule. The overlap in the m/e values for $C_2D_4{}^{16}O$ and $C^{18}O_2$ prevents a direct confirmation of this result. Other experiments involving carbon isotopic labelled species have been suggested to confirm this.

The transient responses of $C^{16}O_2$, $C^{16}O^{18}O_2$ and $C^{18}O_2$ obtained in the above pump-probe experiments using $^{18}O_2$ as the pump molecule are shown in Fig. 6a. Since the only source of ^{16}O is in the lattice, the formation of $C^{16}O_2$ and $C^{16}O^{18}O$ requires migration of ^{16}O from the lattice to the catalytic sites. The only source of ^{18}O is that introduced in the gas pulse, hence the formation of $C^{18}O_2$ and $C^{16}O^{18}O$ must likely involve ^{18}O adsorbed on the surface. Inspection of the peaks in Fig. 6a shows that the peak areas follow the order $C^{16}O^{18}O > C^{18}O_2 > C^{16}O_2$. It is interesting to note that the smallest peak area is associated with the species containing only lattice oxygen. This is consistent with the expectation that the characteristic time for migration would be greatest for oxygen transport from the lattice to a surface active site as compared to migration on the surface only to an active site.

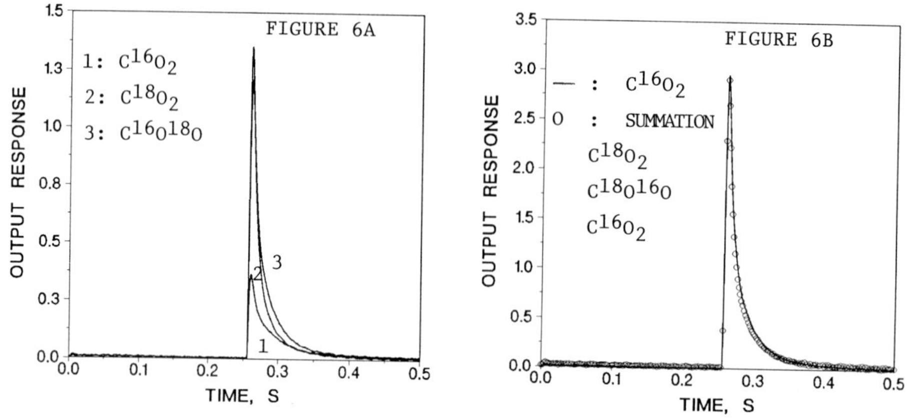

Fig. 6. Transient responses of $C^{16}O_2$, $C^{16}O^{18}O$, and $C^{18}O_2$ obtained in pump-probe experiments with $^{18}O_2$ (pump) and C_2D_4 (probe) molecules (Fig. 6a). Comparison of the $C^{16}O_2$ transient response obtained with $^{16}O_2$ (pump) and C_2D_4 (probe) to the sum of the transient responses of $C^{16}O_2$, $C^{16}O^{18}O$, and $C^{18}O_2$ (Fig. 6b). Conditions: same as Fig. 4.

Fig. 6b shows a comparison between the $C^{16}O_2$ transient response obtained when $^{16}O_2$ is the pump molecule and C_2D_4 is the probe molecule, to the transient response obtained by summation of $C^{16}O_2$, $C^{16}O^{18}O$, and $C^{18}O_2$ where $^{18}O_2$ was used as the pump molecule and C_2D_4 as the probe molecule with ^{16}O present in the lattice. As suggested above, formation of CO_2 involves at least three different reaction pathways. If the relative QMS response to each isotopic form is the same, then the sum of the responses CO_2 isotopes should be equal to the CO_2 response obtained in the absence of the isotopes. As seen by Fig. 6b, nearly perfect agreement is obtained between the sum of the transient responses of the isotopes (denoted by points), to the transient response obtained without isotopes.

CONCLUSIONS

The possible modes of oxygen activation and pathways for oxygen incorporation into reaction products of ethylene epoxidation over a silver metal catalyst have been examined using multipulse and pump-probe TAP reactor system experiments. Formation of CO_2, which is an undesired nonselective product, can occur from surface adsorbed atomic oxygen, lattice or subsurface atomic oxygen, or through atomic oxygen supplied from both the surface and the lattice. The latter case implies the formation of a surface adsorbed CO intermediate which combines with lattice oxygen, while the remaining ones suggest formation of CO_2 occurs through one or more elementary steps where either surface adsorbed oxygen or lattice oxygen are involved. Comparison of relative amounts of isotopic CO_2 indicates that formation of CO_2 from lattice oxygen is the slowest process when compared to the remaining two.

Formation of C_2D_4O from gas phase O_2 and C_2D_4 also involves both surface adsorbed and lattice oxygen. As in the case of CO_2, precise specification of the intermediates and elementary steps is not possible. However, both CO_2 and C_2D_4O appear to compete for both types of oxygen sources at different net values for the local reaction rate. Additional transient experiments involving carbon-labeled species are necessary to discriminate between the reactant sources and possible reaction pathways.

REFERENCES

1 J.T. Gleaves, J.R. Ebner and T.C. Kuechler, Temporal analysis of products (TAP). A unique catalyst evaluation system with submillisecond time resolution, Catal. Rev.-Sci. Eng., 30(1) (1988) 49-116.

2 J.T. Gleaves, J.R. Ebner and P.L. Mills, A novel catalyst evaluation system for temporal analysis of reaction products with submillisecond time resolution, in: J.W. Ward (Ed.), Studies in Surface Science and Catalysis-Catalysis 1987, Elsevier, Amsterdam, 1988, pp. 633-644.

3 J.R. Ebner and J.T. Gleaves, TAP studies of oxygen activation on vanadium phosphorus oxide catalysts, paper presented at the Tenth North American Meeting of the Catalysis Society, San Diego, California, May 17-22, 1987.

4 J.R. Ebner and J.T. Gleaves, The activation of oxygen by metal phosphorus oxides-The vanadium phosphorus oxide catalyst, in: A.E. Martell and D.T. Sawyer (Eds.), Oxygen Complexes and Oxygen Activation by Transition Metals, Plenum, New York, 1988, pp. 273-292.

5 G. Centi, F. Trifiro, G. Busca, J.R. Ebner and J.T. Gleaves, Selective oxidation pathways at the vanadyl pyrophosphate surface in light paraffin conversion, in: M.J. Phillips and J. Ternan (Eds.), Catalysis: Theory to Practice, Proc. of the 9th Int. Congr. on Catalysis - Vol. 4. Oxide Catalysts and Catalyst Development, Chemical Institute of Canada, Ottawa, Canada, 1988, pp. 1538-1545.

6 J.R. Ebner, J.T. Gleaves and P.L. Mills, Transient analysis of reaction products (TAP). A new device for high-speed pulsed reactant studies on heterogeneous systems with application to butane oxidation, poster presented at the Tenth International Symposium on Chemical Reaction Engineering (ISCRE 10), Basle, Switzerland, August 28-31, 1988.

7 J.R. Ebner, J.T. Gleaves, T.C. Kuechler and T.P. Li, Ammoxidation of methanol to hydrogen cyanide. Binary oxide catalysts and mechanistic aspects, in: D.R. Fahey (Ed.), Industrial Chemicals via C_1 Processes, ACS Symp. Ser. Vol. 328, American Chemical Society, Washington, D.C., 1987, pp. 189-205.

8 A.G. Sault, J.T. Gleaves, J.R. Ebner and R.J. Madix, Ethylene oxidation on silver powder: A TAP reactor study, J. Catalysis (submitted for publication), 1989.

9 R.A. Van Santen and H.P.C.E. Kuipers, The mechanism of ethylene epoxidation, Adv. Catalysis, 35 (1987) 265-321.

10 R.A. Van Santen and C.P.M. deGroot, The mechanism of ethylene epoxidation, J. Catalysis, 98 (1986), 530-539.

11 R.P. Grant and R.M. Lambert, The mechanism of the silver-catalyzed heterogeneous epoxidation of ethylene, J. Chem. Soc., Chem. Commun., 1983, pp. 662-663.

12 A.W. Czanderna, The effect of cyclic oxygen adsorption and reduction on a silver surface, J. Phys. Chem., 70(7) (1966) 2120-2125.
13 N.C. Rigas, J.T. Gleaves and P.L. Mills, Some observations on activation of silver catalysts for ethylene epoxidation using the TAP reactor system, Catalysis Lett. (in preparation), 1989.

G. Centi and F. Trifiro' (Editors), *New Developments in Selective Oxidation* 717

ETHYLENE OXIDATION OVER α-ALUMINA SUPPORTED SILVER-GOLD CATALYSTS

Herrera Rafael[1], Varma Arvind[2], and Martínez Enrico[3]

[1] Universidad Nacional Autónoma de México, Departamento de Ingeniería Química, Avenida 3, # 167, San Pedro de los Pinos, México D.F. 03800 (Mexico).

[2] University of Notre Dame, Chemical Engineering Department, Notre Dame IN-46556 (U.S.A).

[3] Universidad Autónoma Metropolitana-Iztapalapa, División de Ciencias Básicas e Ingeniería, Apartado Postal 55-534, México D.F. 09340 (México).

SUMMARY

The effect of Au in the oxidation of ethylene over α-alumina supported Ag-Au catalysts was investigated. Catalysts were prepared by sequential impregnation and by coimpregnation, and tested by means of x-ray diffraction, oxygen chemisorption, and ethylene oxidation. Results indicated that the catalysts were α-alumina supported particles of Ag and Au with crystallite size of 380-660 Å, with different amounts of Ag on the surface. Reaction rates and catalytic properties were expressed in terms of the available surface Ag providing a link for correlating all catalysts. It was found that the addition of Au to Ag (>20 at%) produces a bimetallic specie that, in comparison with a pure Ag sample, exhibited lower oxygen chemisorption capacity, and higher or lower ability for ethylene epoxidation, depending upon the relative amount of Ag and Au in the catalyst. The promoter effect of Au in the epoxidation of ethylene could be explained in terms of electronic interactions of the bimetallic catalytic surface and the gas phase components.

INTRODUCTION

The oxidation of ethylene over Ag is the only commertial method to produce ethylene oxide, a key compound in the petrochemical industry [1]; it has been represented by the following reaction scheme: $2C_2H_4 + O_2 \rightarrow 2C_2H_4O$; $C_2H_4 + 3O_2 \rightarrow 2CO_2 + 2H_2O$; and $2C_2H_4O + 5O_2 \rightarrow 4CO_2 + 4H_2O$; the preferential occurrence of these reactions depends upon the reactor operating conditions and the catalyst composition and history. Despite the enormous effort spent in studying this reaction, there are fundamental questions that still have not been clearly explained, such as the uniqueness of Ag to epoxidize ethylene with high yields, and the way in which promoters work. It is known that Ag catalyzes ethylene oxidation only when it is, or has been previously, exposed to an oxygen-containing atmosphere; it is generally accepted that the interation between oxygen and Ag produces chemisorbed oxygen, in both dissociative and non-dissociative forms, and subsurface oxygen [2, 3]. It has explained [1-7] that, in comparison with pure Ag samples, substances such as Cl_2 Mg, Cd or Au change the chemisorption of oxygen on the silver-modified catalysts, enhancing ethylene oxide formation. Based on the complete miscibility of Au and Ag, and the inertness of Au for both chemisorption of oxygen and oxidation of ethylene, researchers [5-7] have used Au as an inert diluent of Ag. Results for unsupported [6, 7], and supported Ag-Au catalysts [5, 7] are contradictory. In the former case [6], an intermediate Ag/Au atomic ratio which favors ethylene oxide production was observed; while in the others it was found that relatively small amounts of Au (<10%), produced a sharp decrease in the selectivity towards ethylene oxide [7], and more recently [5] it was found that small amounts of Au increased ethylene oxide and carbon dioxide production; however, selectivity remains approximately constant up to 10% Au and drop at higher Au contents. Therefore, it is worthwhile to study the role of Au in the oxidation of ethylene over supported Ag-Au catalysts. There are indications [5-10] that Ag, Au, and Ag-Au bimetallic species all have different geometric and electronic properties. The importance of the geometric factor in the oxidation of ethylene was demonstrated by establishing a relationship between the lattice parameter of unsupported Ag-Au samples and their overall activity [7]. The observed promoter effect of Au in the epoxidation of ethylene was explained by considering that, following a substitution mechanism, some Au atoms are introduced in the Ag structure (FCC) producing a structure with a different lattice parameter, and therefore different catalytic properties. The electronic interaction between Ag and Au has been demonstrated by several experiments [8,9]. XPS analysis of Ag-Au alloys [8] showed electron transfer from Ag to Au at the *s* level, and a simultaneous compensatory effect at the *d* level. The net depletion of the electronic charge of Ag in the presence of Au, shows that Au is more electronegative than Ag. The electronic properties of Ag-containing samples have also been studied through measurements of the electron work function, Ψ, [9]. Measurements of Ψ for Ag-Au alloys showed that by increasing the content of Au, Ψ increases from the value for Ag to a higher value for Au, following a nonlinear relationship. Moreover, exposure of the alloy to O_2 produces a positive increase in the work function, $\Delta\Psi$, showing a maximum for samples containing about 5 at% Au. These results were explained considering an interaction between the alloy and oxygen, which results in the adsorption and diffusion of oxygen on and into the alloy; the diffusion of oxygen into the bulk is thought to reduce the surface potential of the oxygen-exposed sample, thus increasing the $\Delta\Psi$, and therefore reducing its surface electron density. On the other hand, in order to improve the catalytic properties of Ag-based catalysts for ethylene oxide production, different substances have been added to the reaction system, either during the course of catalyst preparation, and/or with the reactant stream [10]. Using the non-modified Ag catalyst as a reference, it was established that the modification of Ag with more electronegative substances, such as Cl_2, S, Mo, or F produces an increase in the work function, as well as an increase in the selective epoxidation of ethylene with a simultaneous decrease in the overall activity for ethylene oxidation. The opposite effect was observed with the addition of substances less electronegative than Ag, such as K, Ca, Sr, or Ba; in all these cases, both the electron work function and the selectivity suffered a reduction, while the overall activity was increased. These effects are explained in terms of an electronic interaction between the modifier and Ag. Since the geometric and electronic characteristics of a given material are closely related to each other, changes in the lattice parameter of Ag produced by the adition of Au, should also imply changes in the electronic characteristics of the modified sample. Therefore, the catalytic behavior of bimetalic Ag-Au particles is expected to be the result of the geometric and electronic interaction of these two metals.

RESULTS

Catalyst Preparation. Four types of catalysts were prepared by impregnation of the support with solutions of Ag and/or Au: i) two types of Ag catalysts, and ii) two types of Ag-Au catalysts, one by sequential impregnation and other by co-impregnation [11]. Ag Catalysts. Type I Ag catalyst was prepared by wet impregnation of α-alumina in a Ag lactate solution. Type II Ag catalyst was also prepared by wet impregnation; a Ag cyanide/ ethylene-diamine solution was used. The last thermal treatment for the Ag catalysts corresponded to the decomposition-annealing-activation process for the bimetallic catalysts. Ag-Au Catalysts. Type I Ag-Au catalysts were prepared by sequential impregnation following a method that involves impregnation of an insoluble Ag salt (AgO-lactic acid) with an aqueous solution of Au (tetrachloroauric acid), followed by an annealing process. Type II Ag-Au catalysts were prepared by impregnation of the support with a solution containing the appropiate concentrations of salts of Ag and Au. The metallic bulk composition of the catalysts was determined by atomic absorption spectroscopy. Analysis of the samples by X-ray diffraction allowed estimation of the average crystallite size for the metallic particles, and gave an indication of the mixing of Ag and Au, forming bimetallic crystallites; a qualitative agreement with data for lattice parameter of Ag-Au alloys was observed, showing a minimum at ca. 60 at%. All samples exhibited large crystallites, in the range 380-660 Å, showing no dependence between the catalyst composition and crystallite size, therefore, the effect of crystallite size could be ignored when comparing the catalytic behavior of these catalysts in the oxidation of ethylene.

Oxygen Chemisorption. Oxygen chemisorption and kinetic experiments were conducted *in-situ* with a fixed bed reactor that could be operated either in the pulse mode or as a fixed-bed with external recycle. Based on published [12], and experimental [11] data, a procedure was adopted to determine surface Ag via oxygen chemisorption. Briefly, the procedure consists of two main steps: i) sample pretreatment (cycles: 2h, O_2, 200°C/ 0.5h, Ar, 200°C/ 1h, H_2, 200°C); and ii) saturation of the sample with oxygen (Ar: 60 cc/min, pulse volume: 60 µl of O_2, 200°C). Figure 1 shows the results for the characterization of all catalysts by oxygen chemisorption; from these results, the following observations can be made: i) both types of Ag catalysts exhibited similar capacity for oxygen chemisorption in both stages: fresh and used; ii) the capacity for oxygen chemisorption of the catalyst decreases as the content of Au increases; iii) fresh type II samples with less than about 30 at% Au exhibited practically the same capacity for oxygen chemisorption as the unpromoted Ag sample; iv) the use of the catalysts in ethylene oxidation reduces their capacity for oxygen chemisorption; this effect is more pronounced in the cases of type I and type II Ag catalysts, as well as in all type II Ag-Au catalysts. These results were explained in terms of the catalysts preparation procedures [11]. The amount of oxygen chemisorbed of the used samples was translated into surface Ag to define the reaction rates on the basis of available surface Ag.

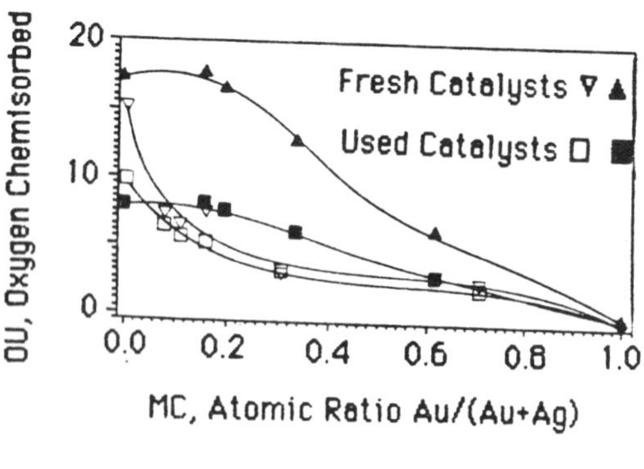

Figure 1. Oxygen chemisorption capacity, OU, of catalysts type I (filled in figures) and type II (unfilled figures) as a function of their metallic bulk composition, MC. OU is given in: mole O_2 chem. $\times 10^{-7}$/g catal.; and MC is given in terms of the atomic ratio: Au/(Au+Ag). Conditions: 200°C, and 60 cc/min.

<u>Ethylene Oxidation</u> Analysis of reaction products was performed by gas chromatography (Varian 920; porapak-Q column). An Infrared Analyzer (Beckman, 865) was used continuously to monitor the production of CO_2. In agreement with previous reports [13] it was observed that the overall activity and selectivity for ethylene oxide changed relatively fast during the first few hours of reaction, then a slow change occured until a constant value was reached. Consequently, in order to obtain catalysts with constant catalytic properties, all samples were subjected to a stabilization process first, and then tested under operating conditions shown in Table 1. Neither the support, nor the Au/α-Al_2O_3 sample showed measurable conversion in the range of 240-300°C.

Table 1. *In-situ* stabilization and reaction conditions

Mass of catalyst: ~1g; external flow rate: ~0.15 l/min; recycle ratio: ~180
Stabilization conditions:
Feed composition: 6.0 mole% C_2H_4, balance O_2; 300°C; ~40 h
Reaction conditions:
Feed composition: 0.5-5.0 mole% C_2H_4, balance O_2; 240-270°C

Examples of the correlation between the composition of a catalyst and its catalytic properties are shown in Figure 2. From these results, it can be concluded that the presence of relatively small amounts of Au (<20 at.%) increase the activity towards ethylene epoxidation, while the activity for the total combustion remains constant or decreases; the combined effect of these individual changes is an increase in the selectivity towards ethylene oxide production. However, the magnitude of that increment seems to depend, also, upon the method used to prepare the catalyst. Type I catalysts showed higher selectivity than type II catalysts, such difference could have been due to an involuntary chlorine promotion, because type I catalysts were prepared from tetrachloroauric solution. On the other hand, both types of high-Au catalysts, i.e. those containing more than 50 at% Au, exhibited relatively high overall activity and low selectivity in comparison with the low-Au catalysts

DISCUSSION.

Catalyst behavior is determined by the method of preparation, its composition, history and testing conditions. In this work, the effect of catalyst composition was isolated from other factors by preparing each set of samples by the same method, and by treating and testing them under practically the same conditions. Results indicated that, in comparison with the pure Ag sample, bimetallic samples exhibited lower capacity for oxygen chemisorption [Fig.1], and higher or lower ability to epoxidize ethylene, depending upon the relative amount of the two metals on the catalytic surface [Fig. 2]; maximum selectivity was observed with catalysts containing about 15 at% Au [Fig. 2d]; that resulted from the combined effect of R_1 increase and R_2 decrease [Fig. 2c].

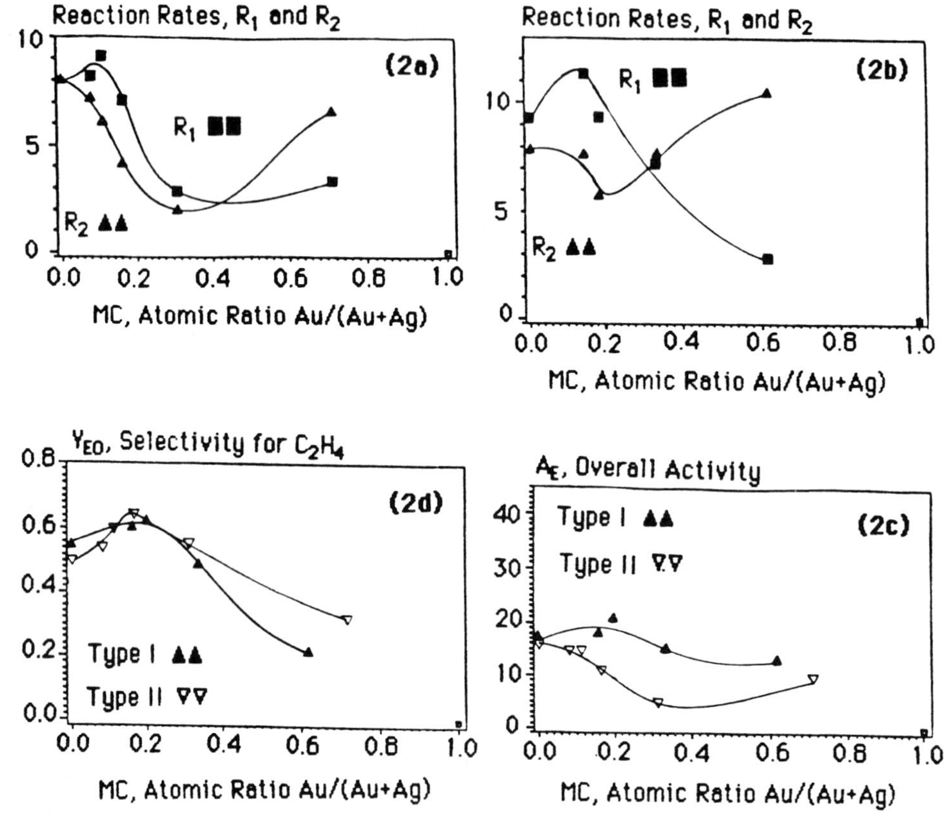

Figure 2. Catalytic properties of type I and type II catalysts as a function of their metallic bulk composition. The rates of conversion of C_2H_4 to C_2H_4O and CO_2, R_1 and R_2 respectively, are given in mole C_2H_4 reacted/h g-surface Ag. **(2a)** Type I catalysts. **(2b)** Type II catalysts. **(2c)** Overall activity, A_E;is given in mole C_2H_4 reacted/h g-surface Ag. **(2d)** Selectivity towards C_2H_4O, Y_{EO}; is given in mole C_2H_4O produced/mole C_2H_4 reacted; metallic bulk composition is given in terms of the atomic ratio Au/(Au+Ag). Reaction conditions: 1.5 mole% C_2H_4, balance O_2; temperature: 240°C.

According to previous considerations it could be expected that an interaction between Au and Ag determines the geometric and electronic characteristics of the Ag-Au catalysts and, consequently, threir capacity for oxygen chemisorption, and ability for the epoxidation of ethylene. To explain these results in terms of the electronic exchange between gas phase components and the catalytic surface, the following considerations are made: i) the characteristics of the surface oxide layer are determined by the relative amount of Ag and Au in the sample, since the other determining factors, such as history and testing conditions, were the same for all samples; ii) electron density at the surface oxide layer, characteristic of the geometric-electronic Ag-Au interaction, determines the chemisorptive and catalytic properties; iii) the molecular and atomic oxygen species are responsible for the epoxidation and total combustion of ethylene, respectively; iv) the total combustion of ethylene oxide is produced by the reaction between adsorbed ethylene oxide and adsorbed atomic oxygen. Based on the previous considerations, it is proposed that the addition of relatively small amount of Au to Ag (~15 at%) produces bimetallic species that, in comparison with pure Ag, have lower electron density at the surface, thus decreasing its capacity for chemisorption of electron aceptor species; however, the ratio $O_2^{\delta-}/O^{\delta-}$ is increased, due to the higher amount of electrons required for the dissociative chemisorption; adsorption of ethylene oxide and ethylene are also decreased, because ethylene oxide is an electron acceptor molecule, while ethylene is thougt to be adsorbed on electropositive centers created during the dissociative adsorption of oxygen [14]. These changes in the catalytic surface could produce a decrease in the combustion of ethylene and ethylene oxide; however the epoxidation of ethylene could be increased, and consequently the selectivity towards ethylene oxide production is increased.

This reaction model does not explain the behavior of the high-Au samples, which exhibit low capacity for oxygen chemisorption, and relatively high overall activity coupled with low selectivity as compared to those of low-Au catalysts. Such behavior resembles that of well-dispersed Ag samples, which exhibit high activity and low selectivity [1,15]. There are no direct measurements of the amount of surface Au; however, it is possible that the addition of large amounts of this metal (> 50 at% Au), could isolate Ag atoms in small groups, which could in fact behave like small Ag crystallites.

REFERENCES

1 Verykios, X. E., Stein, F. P., and Coughlin, R. W., "Oxidation of Ethylene over Silver: Adsorption, Kinetics, Catalyst", Catal. Rev.- Sci. Eng., **22**(2), 197-234 (1980).

2 Kilty, P. A., Rol, N. C., and Sachtler, W. M. H., "Identification of Oxygen Complexes Adsorbed on Silver and Their Function in the Catalytic Oxidation of Ethylene", Catal. Proc.,Int. Congr., 5th, Hightower, J. W., North-Holland, Amsterdam, Neth, **64** , 929-944(1973).

3 Grant, R.B., and Lambert, R.M., "Mechanism of the Silver-Catalyzed Heterogeneous Epoxidation of Ethylene", J. Chem. Soc. Commun., **58**, 662(1983).

4 Brandt K. B., and Verykios X.E., "The oxidation of Ethylene over Silver-Base Alloy Catalysts. II Silver-Palladium Alloys", J. Catal., **91**, 36(1985).

5 Noreddine Toreis, and Verykios X.E., "The oxidation of Ethylene over Silver-Base Alloy Catalysts. 3 Silver-Gold Alloys", J. Catal., **108**, 161(1987).

6 Flank, W. H., and Beachell, H. C., "The Geometric Factor in Ethylene Oxidation over Gold-Silver Alloy Catalysts", J. Catal., **8** , 316-325 (1967).

7 Genen, P. V., Boss, H. J., and Pott, G. T., "A Study of the Vapor-Phase Epoxidation of Propylene and Ethylene on Silver and Silver-Gold Alloy Catalysts", J. Catal., **77** , 499-510 (1982).

8 Watson, R. E., Hudis, J., and Perlman, M. L., "Charge Flow and d Compensation in Gold Alloys", Phys. Rev. B, **4**(12), 4139-4143 (1971).

9 Fain, S. C., and McDavid J. M., "Work-Function Variation with Alloy Composition: Ag-Au", Phys. Rev. B, **9**(12), 5099-5107 (1974).

10 Margolis, L. Ya., Enikeev, E. Kh., Isaev, O. V., Krylova, A. V., and Kushnerov, M. Ya., "Modification of Catalysts for the Oxidation of Hydrocarbons", Kin. Catal. **3**(2) , 153-159 (1962).

11 Herrera, R., "Effect of Gold in the Oxidation of Ethylene over α-Alumina Supported Silver-Gold Catalysts", Doctoral Disertation, Chemical Engineering Department, University of Notre Dame (1987).

12 Seyedmonir, S. R., Strohmayer, D. E., Geoffroy, G. L., Vannice, A. M., Young, H. W., and Linowski, J. W., "Characterization of Supported Silver Catalysts. I. Adsorption Of O_2, H_2, N_2O, and the H_2-Titration of Adsorbed Oxygen on Well-Dispersed Ag on TiO_2", J. Catal., **87** , 424-436 (1984).

13 Ayame, A., Shibuya, Y., Yoshida, T., and Kano, H., "Oxidation of Ethylene and Surface Residues on a Silver Catalyst", Int. Chem. Eng., **14**(3), 577-588 (1974).

14 Beran, S., Jiru, P., Wichterlova, B., and Zahradnik, R., "A Molecular Orbital Study of the Catalytic Oxidation of Ethylene and Propylene on Silver", Proc. Int. Congr. Catal. 6th, Bond, G. C., Wells, P. B., and Tompkins, F. C. (Editors), Chem. Soc., Letchworth, Engl., 324-335 (1978).

15 Wu, J. C., and Harriot, P., "The Effect of Crystallite Size on the Activity and Selectivity of Silver Catalysts", J. Catal., **39** , 395-402 (1975).

G. Centi and F. Trifiro' (Editors), *New Developments in Selective Oxidation*
1990 Elsevier Science Publishers B.V., Amsterdam — Printed in The Netherlands

ETHYLENE EPOXIDATION ON SILVER-BASED ALLOY CATALYSTS

N. TOREIS[1] AND X.E. VERYKIOS[2]

[1]Ecole Mohammadia d' Ingenieurs, BP 765 Agdal, Rabat (Morocco)
[2]Institute of Chemical Engineering and High Temperature Chemical Processes
Department of Chemical Engineering, University of Patras, GR26110 Patras (Greece)

ABSTRACT

Ethylene epoxidation and combustion is investigated over Ag-Cd, Ag-Zn and Ag-Au alloy catalysts. Surface composition of alloy particles is determined by selective chemisorption techniques and ESCA and found to be significantly different than bulk composition. Turnover frequencies and in certain cases activation energies of the two reaction routes are found to be significantly affected by alloying. The mode of oxygen adsorption was also found to be affected by alloying Ag with Au. These results are used to formulate a mechanism for the two reaction routes.

INTRODUCTION

Because of its industrial and scientific importance, the selective oxidation of ethylene over Ag catalysts has received great attention. Of particular interest is the uniqueness of Ag in its ability to promote the epoxidation reaction so as to obtain significantly high yields of ethylene oxide. Fundamental as well as practical aspects of this catalytic system have been reviewed frequently (1,2).

Catalysis by bimetallic or alloy catalysts is also of great fundamental and practical interest. Many bimetallic systems have been shown to be superior to their monometallic counterparts, exhibiting higher activity, selectivity and stability in processes of industrial importance (3). In studies of fundamental concepts, alloy catalysts have been used to investigate the influence of the geometric and electronic structure of surfaces on chemisorptive and kinetic parameters.

In the present study, the effects of alloying silver with a second metal on kinetic parameters is ethylene epoxidation and combustion were investigated. The alloying metals selected are Cd and Zn whose electronegativity is lower than that of Ag, and Au whose electronegativity is higher. Other criteria applied in the selection include bulk composition range in which monophasic alloys can be achieved, expected surface composition range, and thermodynamics of alloy formation.

METHODS

Supported Ag-Cd and Ag-Zn alloy catalysts were prepared by impregnation of α-Al_2O_3 with mixed silver nitrate and the nitrate of the alloying metal, drying at 110°C and reduction in flowing hydrogen at 200-400°C for 24h. Silver cyanide and gold cyanide were used in the preparation of Ag-Au catalysts which were calcined in air at 250°C for 24h. All catalysts were examined with respect to alloying achieved by X-ray diffraction. Surface composition of alloy particles was determined as a function of bulk composition by selective chemisorptive titrations and by ESCA. Specific reaction rates and kinetic parameters of ethylene epoxidation and combustion were determined in the temperature range of 200-250°C, at a pressure of 15 atm. The feed composition consisted of 3.0% ethylene, 3.3% oxygen, balance nitrogen. The oxygen concentration was kept at low levels so as to avoid phase separation of the alloy catalysts during reaction. In all cases, conversion was maintained at low levels (less than 8%) so as to treat the reactor as differential one. Thus, initial rates were measured. Details of these procedures have been reported elsewhere (4-6).

RESULTS AND DISCUSSION

Bulk characterization of Alloy Particles

Precautions were taken in this study to form monophasic alloys by preparing catalysts with second metal content below the solubility limit. All alloys were subjected to X-ray diffraction analysis before and after exposure to reaction conditions. Lattice constants were estimated and are shown on Fig. 1 for the Ag-Au and Ag-Zn alloys as a function of bulk composition. These results compare favorably with those reported in the literature (7), which are also shown on the same figure. The small differences can probably be attributed to structural differences between small supported alloy particles and bulk alloys. The X-ray diffraction analysis confirmed that alloy formation had been achieved and that neither phase shift nor separation to individual components had occured upon reaction.

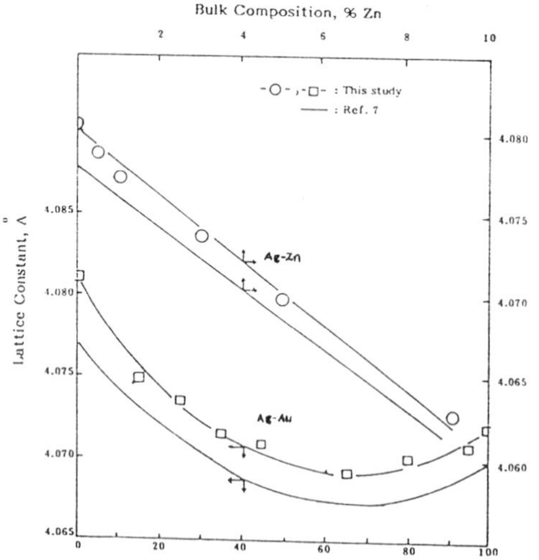

Fig. 1. Variation of lattice constants of Ag-Au and Ag-Zn alloys with bulk composition

In order to further characterize Ag-Au alloy catalysts, extended X-ray absorption fine structure (EXAFS) analysis was performed on alloys of composition 80 at %Ag -20 at % of the alloying metal at 80K. Similar analysis, under identical conditions was performed on a silver foil which was used as a standard. Extended fine structure beyond the silver K-edge was measured. The associated phase and amplitude-corrected Fourier transforms of the EXAFS functions are shown in Fig. 2.

The silver EXAFS results of the Ag-Zn catalyst reveal that both X(K) function and its corrected Fourier transform are essentially identical to those of Ag in the silver foil. This indicates that the

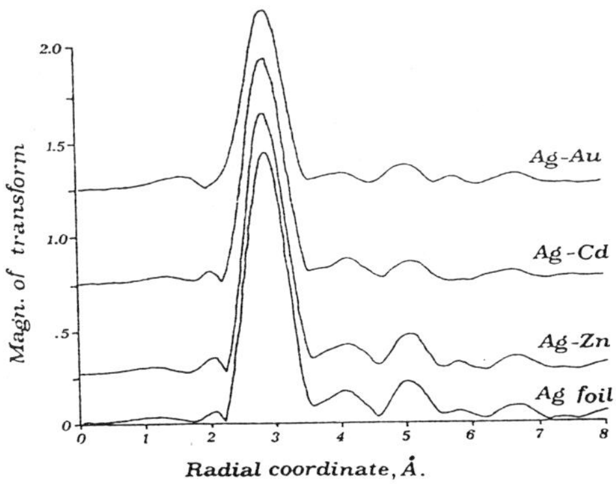

Fig. 2. Phase and amplitude-corrected Fourier transforms of EXAFS spectra.

environment about a silver atom in Ag-Zn alloy particles is, on the average, not different from that is the reference foil. Furthermore, matching of the Fourier transform functions occurs at even higher radial distances, indicating that even higher shells are identical. The average radial distance between Ag atoms and their nearest neighbors was also determined and found to be 2.89Å, equal to that of the standard Ag foil. These results are consistent with a view in which a Ag-Zn cluster consists of a central core of primarily silver atoms with zinc atoms present primarily at the surface of the cluster. In contrast to the Ag-Zn alloy system, silver EXAFS spectra of Ag-Cd and Ag-Au alloy particles are significantly different from those of the Ag foil, indicating that the environment about a silver atom in these alloys is different from that in the foil. In the Fourier transform plots, the amplitude of the Ag-Cd and Ag-Au system is smaller than that of the Ag foil in the main peak and subsequent ones. A tendency for centroids to shift to a higher radial coordinate is also apparent. It is also observed that zero nodes have the tendency to shift to lower values of K and the difference is progressively greater as K increases. The amplitudes are smaller than the ones corresponding to Ag foil. With the silver foil as a reference, the nearest neighbor distance of silver in the Ag-Cd and Ag-Au alloys was determined to be 2.91 Å which is higher than that of the foil by 0.02 Å. The average radial distance between Ag and its nearest neighbors in the nth shell were also calculated. The ratios of the average radial distance in the nth shell, r_n, to that of the first shell, r_1, was computed and found to be in good agreement with the relationship: $r_n/r_1=n^{1/2}$, which is specific at an FCC structure. Thus, it can be concluded that in all alloy particles the FCC structure is conserved even at higher shells.

Surface Characterization

It is well known that surface composition of equilibrated alloys can vary drastically from that of the bulk. In the case of Ag-Cd and Ag-Zn alloys, determination of surface composition was achieved by selective chemisorption of O_2 and H_2, and in the case of Ag-Au by ESCA analysis. ESCA analysis was also performed on other alloys to verify surface compositions obtained by other techniques.

Surface composition of alloy particles as a function of bulk composition is shown in Fig. 3. In the cases of Ag-Zn and Ag-Cd the surface is significantly enriched with the alloying metal. In the cases of Ag-Au the opposite is observed. This is primarily due to

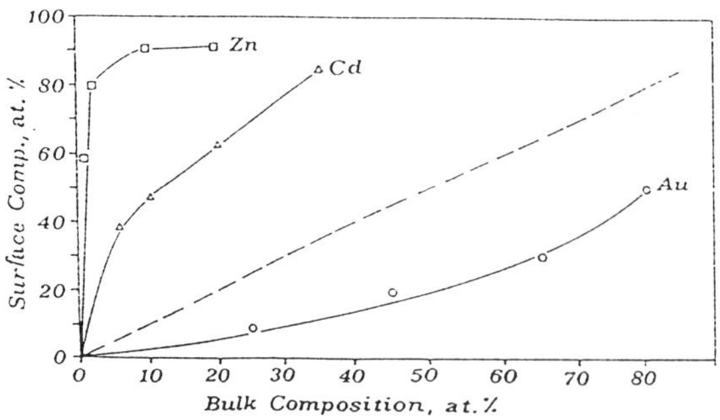

Fig. 3. Surface composition of Ag-based alloy catalysts as a function of bulk composition.

a thermodynamic factor and to the phenomenon of chemisorption induced surface enrichment. Thermodynamically, the component with the lower heat of sublimation tends to segregate at the surface. This component is Ag in the case of Ag-Au and the alloying component in the other two alloys. If the alloy is at equilibrium with a gas atmosphere, the component which forms the strongest chemisorption bonds with the gas tends to segregate at the surface. In some cases these two factors point towards the same direction while in other cases they point in opposite directions. Thus, when Ag-Au alloys are exposed to oxygen (as under reaction conditions) the surface tends to be enriched with Ag since it forms stronger chemisorption bonds with oxygen. The surface of Ag-Cd and Ag-Zn alloys under same conditions tends to be enriched in Cd or Zn since they form stronger chemisorption bonds with oxygen than Ag does.

Effects of Alloying on kinetic Parameters

Specific rates per exposed silver atom (turnover numbers) of ethylene epoxidation and combustion reactions were determined in an isothermal reactor operating in the differential mode, utilizing the surface compositions reported in the previous section. Preliminary experiments conducted with supported Zn, Cd and Au catalysts indicated that these metals do not exhibit any catalytic activity under the conditions employed in this study. Normalized turnover numbers (turnover number obtained with alloy catalyst over turnover number obtained with pure silver catalyst) of ethylene epoxidation are shown in Fig. 4, as a function of alloying metal content (in atom%) of the catalyst. It is apparent

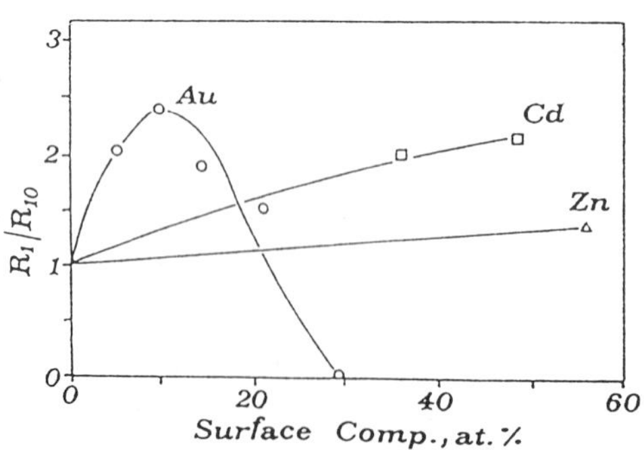

Fig. 4. Influence of surface composition on specific activity of Ag for ethylene epoxidation.

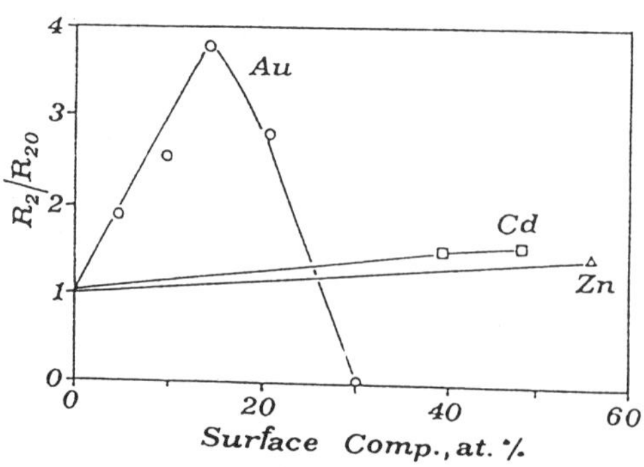

Fig. 5. Influence of surface composition on specific activity of Ag for ethylene combustion.

that epoxidation activity of Ag in Ag-Cd alloys increases with increasing Cd content of the surface. In the case of Ag-Zn, epoxidation activity increases only very slightly with alloying, while in the case of Ag-Au epoxidation activity goes through amaximum at a surface containing approximately 10% Au and it drops to zero when the surface contains 30%Au or more.Similar variations in normalized turnover numbers of ethylene combustion are shown in Fig. 5. Cd and Zn seem to influence slightly the activity of Ag while, in the case of Au activity goes through a maximum at approximately 15% Au in the surface and again drops to zero when the surface contains 30% Au or more.Variation of activation of epoxidation and combustion with surface composition is shown in Fig. 6. Alloying with Zn, was found not to influence significantly the activation energy of either reaction. However, alloying with Cd was found to reduce the activation energy of both reactions while alloying with Au was found not to affect activation energy of combustion but to influence the activation energy of epoxidation in the manner shown in Fig. 6.

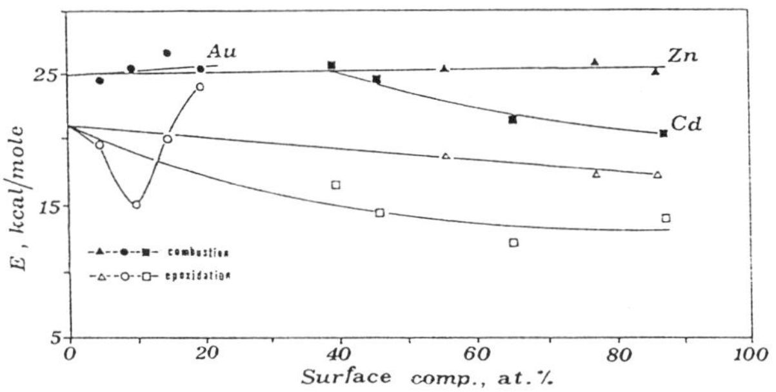

Fig. 6. Influence of surface composition on apparent activation energies of ethylene epoxidation and combustion.

Effects of Alloying on Oxygen Adsorption

TPD of oxygen adsorbed on Ag-Au alloy particles was conducted to determine if the presence of gold on the surface affects the adsorption/desorption behavior of O_2 on Ag. TPD spectra obtained with alloy catalysts of various compositions are shown in Fig. 7. Two oxygen peaks are observed over alloys with low surface Au content (0-14%) while only a single peak is observed over alloys with higher gold content. The positions of peak maxima is found to be a function of composition of the alloy particles. Both, the low-temperature and high-temperature peak maxima decrease with increasing gold content. The maximum of the low-temperature peak is found to vary from 170°C at 0% gold to 115°C at 14%Au at the surface. Similarly, the maximum of the high-temperature peak varies from 300 to 265°C, within the same composition range. In cases in which only a single peak is present, its maximum shifts to higher temperatures as Au content increases. Based on results of other investigators (8-11), both of these peaks can be assigned to atomic oxygen species.

It is obvious from Fig. 7 that the coverage of the H-T adsorbed atomicoxygen species decreases while that of the L-T species increases with increasing goldcontent. Dissociative adsorption of oxygen on Ag has been proposed to proceed on sites of four adjacent silver atoms (12). Recent work (9) has also shown that multiatom sites are

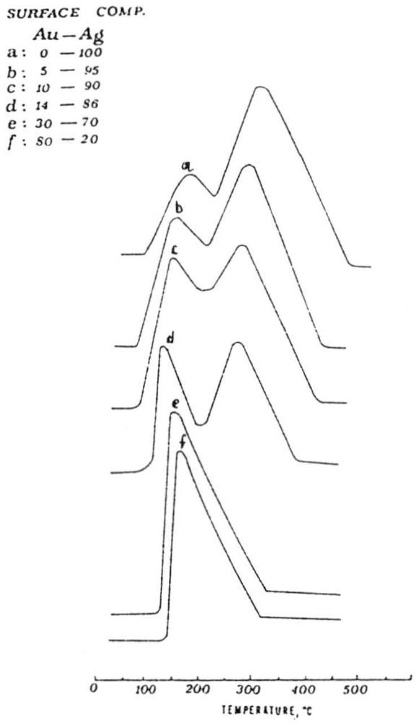

Fig. 7. TPD spectra of oxygen adsorbed on Ag-Au alloy catalysts

required for dissociative adsorption. The decrease of the H-T peak with increasing Au content can be explained by the following reasoning: Au acts as a diluent, destroying a number of four adjacent Ag atom sites, thus hindering the dissociativen adsorption of oxygen on such sites. As a result, not only the surface population of multiatom site adsorbed species decreases but also the population of species which require one site increases. This is precisely what is observed in the present study. After all four adjacent silver atom sites are destroyed (30% Au or higher) then only one species dominates the surface, the species which requires adsorption sites composed of a fewer number of atoms.

Proposed Mechanism of Ethylene Oxidation

Based on the results presented earlier, the following mechanism of ethylene oxidation can be formulated: Oxygen adsorbs on Ag surfaces in two modes: a strongly adsorbed, multicoordinated, monatomic species and a weakly adsorbed monocoordinated species which is also atomic. The presence of strongly adsorbed species is necessary for catalytic action. Ethylene adsorbs on electropositive silver sites created by the multicoordinated adsorption of oxygen, and it reacts with the weakly adsorbed monocoordinated oxygen to form an intermediate complex which either branches to ethylene oxide or to total combustion products. This branching is affected by the presence or absence of adsorbed species in neighboring silver atoms and by the electronic configuration of silver atoms.

Results of the present study can be explained in terms of the mechanism outlined above. At low gold surface contents, turnover frequencies of ethylene epoxidation and combustion increase with increasing gold content. This is accompanied by a similar increase in the amount of adsorbed monocoordinated atomic oxygen. The fact that a direct correlation between these two phenomena exists indicates that this oxygen species is the one which directly participates in all reactions. The fact that the multicoordinated oxygen species follows the exact opposite trend indicates that this species does not participate directly in the reactions. Nevertheless, this species is responsible for generation of electropositive silver sites for ethylene adsorption. Although ethylene adsorption decreases with increasing Au content, its effect is not immediately felt because this adsorption step is not the slowest one in the sequence.

These considerations are valid as long as there is sufficient ethylene on the surface. When the concentration of gold is sufficiently high so as to reduce multicoordinated adsorption of oxygen and subsequent adsorption of ethylene, turnover numbers begin to decrease and finally fall to zero at sufficiently high surface gold content. At that level (≥30%), no ethylene can adsorb on the surface and thus no reaction occurs.

In addition to these geometric effects, an electronic effect is also operable. Near-neighbor electronic interactions between Ag and Au atoms result in electron transfer from Ag to Au, rendering surface silver atoms electropositive. The fact that the activation energy of epoxidation decreases with increasing gold content at low gold concentrations implies that the transformation of the intermediate complex involves electron transfer from the complex to Ag atoms. This is satisfied by the breaking of an Ag-O bond. An electropositive silver surface facilitates such a step which manifests itself as lower activation energy since this is the rate-controlling step. At higher gold contents the activation energy of epoxidation begins to increase as the reduced population of ethylene oxide on the surface is felt. Thus, the rate-controlling step shifts from the transformation of the intermediate complex to ethylene oxide, to the formation of the intermediate complex. As a result, activation energy of epoxidation gradually increases with increasing gold content and approaches that of combustion. Activation energy of combustion is not affected by electronic near-neighbor interactions between Ag and Au atoms since the rate controlling step for this route is the formation of the intermediate complex. This conclusion is derived from the observation that at high Au content, activation energy of epoxidation approaches that of combustion. In this region, the formation of the intermediate complex controls both reaction routes.

There is yet another very important factor which must be considered. This is related to hindrance effects of species adsorbed on atoms neighboring the adsorbed intermediate complex. The intermediate complex results in ethylene oxide if it is sterically hindered so that abstraction of hydrogen from the ethylene molecule cannot occur easily. This consideration explains why the maximum in epoxidation turnover number occurs at a lower gold content than that of combustion. Since Au does not adsorb oxygen, such a hindrance is not present and the intermediate complex decomposes to CO_2 and H_2O at the expense of ethylene oxide.

Results obtained over the other alloy catalysts can also be explained in terms of the mechanism outlined above. Results of Ag-Zn alloy catalysts are explained on the basis of structural information obtained by EXAFS analysis. In the case of Ag-Cd alloys, turnover frequency of epoxidation was found to rapidly increase with increasing Cd content, while that of combustion to increase at a lower rate. This behavior is similar to the one observed over Ag-Au alloy catalysts of low gold content. The increase in turnover numbers is due to increased concentration of monocoordinated atomically adsorbed oxygen. The increase of epoxidation activity is higher than that of combustion due to a steric hindrance effect. This effect comes about from oxygen species adsorbed not only on Ag but also on Cd atoms, neighboring the intermediate complex. At higher Cd contents, in contrast to the case of Au, turnover numbers continue to increase because, unlike Au, Cd adsorbs oxygen and, in cooperation with Ag atoms, it provides sites for the multicoordinated dissociative adsorption of oxygen. Thus, the adsorption of ethylene on the alloy surface is not hindered. For this reason, rates per surface Ag atom continue to increase, even at very high Cd contents. The different effects of Cd and Au on turnover frequencies is due to the fact that Cd provides sites for the multicoordinated adsorption of oxygen and simultaneously provides the hindrance effect which is necessary for selective oxidation.

The most controversial aspect of the mechanism of ethylene oxidation catalysis is the nature of the adsorbed oxygen species which participates in the reaction. TPD results of the present study strongly indicate that weakly adsorbed monocoordinated atomic oxygen participates in both reactions. This weakly adsorbed oxygen has not been detected in ultrahigh vacuum single crystal experiments probably due to the drastically different conditions under which oxygen is adsorbed. The present work offers experimental evidence of the existence of a weakly adsorbed atomic oxygen species on Ag surfaces which participates directly in ethylene epoxidation and combustion chemistry.

REFERENCES

1 X.E. Verykios, F.P. Stein and R.W. Coughlin, The oxidation of ethylene on silver: adsorption, kinetics, catalyst, Catal. Rev. Sci. Eng. 22 (1980) 197-234.
2 W.M.H. Sachtler, C. Backx, and R.A. van Santen, On the mechanism of ethylene epoxidation, Catal. Rev. Sci. Eng., 23 (1981) 127-149.
3 V. Ponec, Surface composition and catalysis on alloys, Surf. Sci., 80 (1979) 352-365.
4 B.K. Bonin, and X.E. Verykios, The oxidation of ethylene over silver-based alloy catalysts. 2. Silver-Cadmium Alloys, J. Catal., 91 (1985) 36-43.
5 N. Toreis and X.E. Verykios, The oxidation of ethylene over silver-based alloy catalysts. 3. Silver-Gold Alloys, J. Catal., 108 (1987) 161-174.
6 N. Toreis, X.E. Verykios, S.M. Khalid, G. Runker and Z.R. Korszum, The Oxidation of Ethylene over Silver-based alloy catalysts. 4. Silver-zinc alloys, J. Catal., 109 (1988) 143-155.
7 W.B. Pearson, Handbook of Lattice Spacings and Sructure of Metals, Vol. 4, Pergamon, New York, 1958.
8 C. Backx, C.P.M. de Groot, P. Biloen, and W.M.H. Sachtler, Interaction of O_2, CO_2, CO, C_2H_4 and C_2H_4O with Ag(110), Surf. Sci., 128 (1983) 81-103
9 C.T. Campbell, and M.T. Paffett, The interaction of O_2, CO and CO_2 with Ag(110), Surf. Sci., 143 (1984) 517-535.
10 C.T. Campbell, Atomic and molecular oxygen adsorption on Ag(111), Surf. Sci., 157 (1985) 43-60.
11 R.B. Grant, and R.M. Lambert, Basic studies of the oxygen surface chemistry of silver: chemisorbed atomic and molecular species, Surf. Sci., 146 (1984) 256-268.
12 P.A. Kilty, and W.M.H. Sachtler, Mechanism of selective oxidation of ethylene to ethylene oxide, Catal. Rev. Sci. Eng., 10 (1974) 1-16.

G. Centi and F. Trifiro' (Editors), *New Developments in Selective Oxidation*

ETHYLENE OXIDATION OVER HYDROPHOBIC THIN LAYER CATALYSTS

F. FRUSTERI[1], A. IANNIBELLO[2], A. PARMALIANA[3], A. CANNIZZARO[1] and N. GIORDANO[1]

[1]CNR-TAE Institute, 98126 S. Lucia - Messina (Italy)

[2]Department of Chemistry, University of Calabria, 87030 Arcavacata di Rende, Cosenza (Italy)

[3]Department of Industrial Chemistry, University of Messina, C.P. 29, 98010 S. Agata, Messina (Italy)

SUMMARY

The oxidation of ethylene to acetaldehyde over noble metal (Pt and Pd) hydrophobic thin layer catalysts, in a three phase system, at mild conditions was investigated. Fe^{+3}/Fe^{+2} redox couple in sulfuric acid solution (pH=1) has been used as oxygen carrier. The reaction was studied in the temperature range 353-393 K. Kinetic measurements by operating in a batch mode have been performed. A novel approach to attempt selective oxidation reactions of hydrocarbons is pointed out.

INTRODUCTION

Partial oxidation of hydrocarbons to obtain chemicals is one of the most challenging field for catalysis research. Success in this area could lead to major innovations in the production of chemical feedstocks by a single-step processes. The renewed interest currently focused on heterogenization of the Wacker reaction could be ascribed to this context "(refs. 1-2)". In fact there are some engineering concerns that must be overcome in the conventional liquid-phase Wacker process: i) the separation of catalyst solution from the product; ii) the severe corrosion problems associated with the catalyst solution "(ref. 2)". Liquid - solid and gas-liquid-solid catalytic reactions are generally controlled by mass-transfer limitations. To reduce resistances of interphase and intraparticle mass transfer in an aqueous reaction system, "water-repellent" or "hydrophobic" catalysts have been designed and investigated "(ref. 3-4)". Recently Hatziantonlou et al. "(ref. 5)" have claimed the peculiar features of thin catalytic plates in enhancing the mass transfer of gas and liquid in three phase reactions. Partial selective oxidation of hydrocarbons can be achieved, both in homogeneous "(ref. 6)" and heterogeneous "(ref. 7)" reaction system, by using redox couple as oxygen carrier.

Therefore the aim of this work is to link together these concepts proposing an original approach for the hydrocarbons selective oxidation under mild conditions. A three phase reaction system has been designed and operated for the evaluation of noble metal hydrophobic thin layer catalysts mediated by Fe^{+3}/Fe^{+2} redox couple acting as oxygen carrier. Ethylene selective oxidation to acetaldehyde has been taken as a reaction model.

EXPERIMENTAL

Thin layer catalyst

Thin layer catalysts were prepared according to the scheme reported in Fig. 1. Carbon supported noble metal catalysts (10 wt% Pt/C; 10 wt% Pd/C) were dispersed ultrasonically in H_2O. By adding teflon and isopropanol, a paste is formed. This is spread on a carbon paper (Stackpole PC 206) wet proofed with a FEP solution. The catalyst plate so obtained was pressed, dried in air at 373 K and activated at 573 K in N_2 flow. Thin layer catalysts with different hydrophobic-hydrophilic characteristics are obtained by varying teflon content in the range 20-60 wt%. BET surface area of thin layer catalyst (PTFE = 30%) was 134 m^2g^{-1} (Pore volume = 0.16 cm^3g^{-1}). The diameter of the catalyst plates used for ethylene oxidation was 80 mm (thickness - 0.6 mm; W = 2.2 g). Noble metal loading and PTFE content of the thin layer catalysts prepared are listed in Table 1.

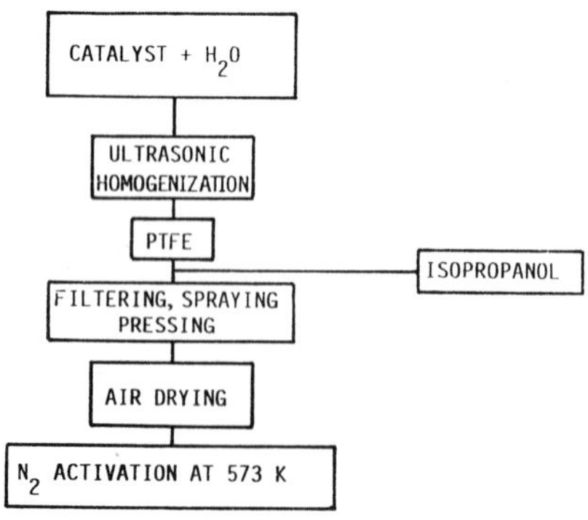

Fig. 1 - Thin layer catalyst preparation

TABLE 1

List of thin layer catalysts

Sample	Noble metal	Metal loading (wt%)	PTFE (wt%)
E20	Pt	2.05	20
E30	Pt	2.2	30
E40	Pt	2.43	40
E50	Pt	2.47	50
E60	Pt	2.36	60
E40-A	Pd	1.27	40

Apparatus and procedure

The thin layer catalyst reactor (Cell Reactor) used in this study is schematically represented in Fig. 2. The thin layer is mounted between two stainless steel plates provided with turbulence nets. Care was taken to maintain a pressure difference of 3 KPa ($\Delta P = P_g - P_l$ = 3 KPa) between the gas and liquid sides of the catalyst plate in order to control: i) the filling up of the pores in the hydrophobic catalytic layer with gas; ii) the liquid and gases leakages; iii) the cracks of catalyst plate.

A diagram of the experimental equipment is shown in Fig. 3. It is provided with liquid and gas recycle pumps which allow operations in batch, semi-batch and continuous mode. Catalytic measurements have been performed in the range 353-393 K with an ethylene - nitrogen mixture ($P_{C_2H_4}/P_{N_2}$ = 5) at 120 KPa. The liquid phase was a sulfuric acid solution of Fe^{+3} 1 M (pH=1).

The evolution of catalytic activity has been followed by: i) continuous potentiometric measurement; ii) volumetric titration of Fe^{+2} formed; iii) gas-chromatographic analysis of liquid and gas phases. GC analysis was carried out by a two column system (A - Porapak QS, l= 2.5 m, i.d.= 2 mm; B - Molecular Sieve 5A, l= 3 m, i.d.= 2 mm) operated at 353 K, using a TC detector, for separation and detection of N_2, O_2, CO_2, C_2H_4 and H_2O. A Carbopack B-3% SP1500 column (l= 2.5 m; i.d. = 2 mm) operated at 363 K and a flame ionization detector for oxygenated products were used.

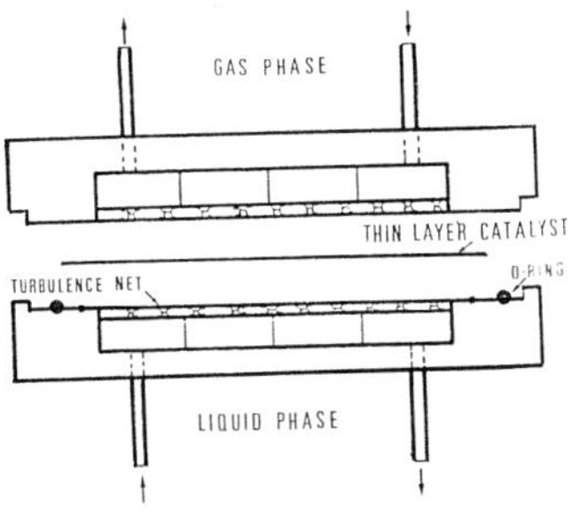

Fig. 2 - Cell reactor

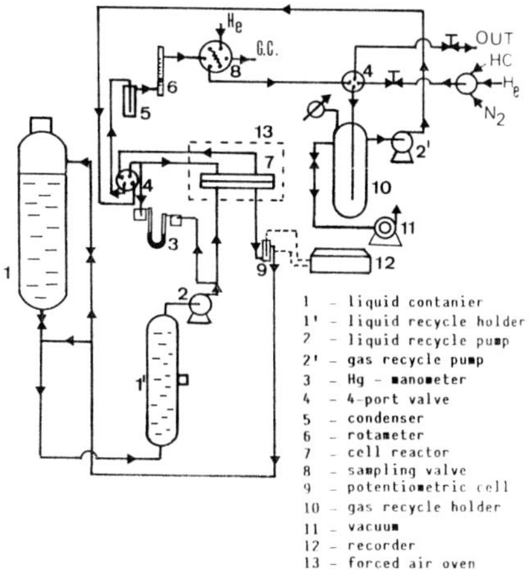

Fig. 3 - Experimental equipment

RESULTS AND DISCUSSION

Preliminary runs were performed to assess the reliability of our reaction system. The results of a typical run, expressed in terms of C_2H_4 molar conversion (as measured by GC analysis), $[Fe^{+2}]$ formed and potential change ($\Delta E = E_t - E_o$ where E_t and E_o are respectively the cell potential at reaction time t and $t_o = 0$) versus the reaction time are reported in Fig. 4. These data indicate that the monitored change in the potential well correlates with ethylene conversion. Therefore potential measurement can be used as a tool for istantaneous kinetic measurement.

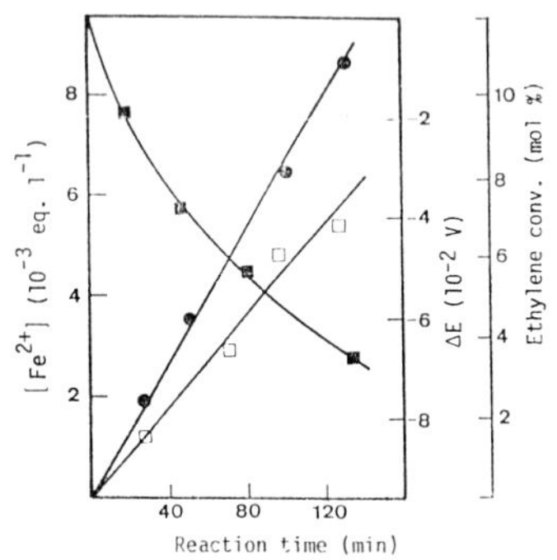

Fig.4 - Ethylene conversion versus reaction time on Pt thin layer catalyst (E30 sample) at 373 K.
● $[Fe^{+2}]$ (10^{-3} eq. l^{-1})
■ $\Delta E = E_t - E_o$ potential change (10^{-2} V)
□ C_2H_4 conversion (mol %)

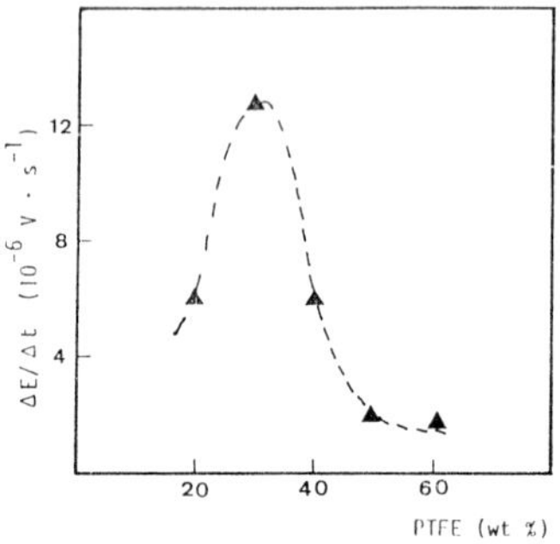

Fig.5 - Ethylene oxidation. Influence of PTFE content on catalytic activity of thin layer catalysts. Batch reactor. T = 373K. Catalytic activity is expressed as potential change rate, $\Delta E/\Delta t$

Effect of PTFE content

As generally accepted rules on the effects of PTFE content on activity of hydrophobic catalysts are lacking ("ref. 3-5)", we have made further attempts to evaluate its influence on the performance of thin layer catalysts in our specific reaction model. The volcano shaped relationship between the catalytic activity, expressed as potential change rate ($\Delta E/\Delta t$, $V \cdot s^{-1}$) and PTFE loading reported in Fig. 5 indicates the existence of an optimum PTFE content corresponding about to 30wt% on thin layer catalysts. This trend can be explained by inferring that PTFE till to 30 wt% exerts a positive enhancement of mass transfer of the gas phase due to high water repellency, while at higher PTFE content (> 30 wt%) a detrimental effect prevails due to reduced liquid - solid contacting effectiveness. To better understand these effects further investigations, mostly devoted to catalysts characterization, are required.

Kinetic measurements

Kinetic measurements were carried out with the Pt hydrophobic thin layer catalyst (sample E30), operating in a batch reactor. Each run was carried out using a fresh catalytic plate and under standard operating conditions in the temperature range 353-393 K. Ethylene conversion versus reaction time, at different reaction temperatures, is reported in Fig.6. A pseudohomogeneous kinetic analysis leads to consider an initial apparent reaction rate of zero order in $[C_2H_4]$. Activity and selectivity data are summarized in Table 2.

TABLE 2

Catalytic activity of Pt and Pd hydrophobic thin layer catalysts in ethylene oxidation

CAT.	NOBLE METAL	T (K)	CONV.[a] (mol %)	S_{CH_3CHO} (mol %)	S_{CO_2} (mol %)	RATE 10^{-8} mol/g_{cat} · s	YIELD g_{CH_3CHO}/Kg_{cat} · h
E30	Pt	353	2.2	72	28	4.6	5.2
E30	Pt	373	4.1	67	33	8.6	9.1
E30	Pt	393	7.6	56	44	15.9	14.2
E40A	Pd	373	7.2	98	2	15.0	23.2

[a]Conversion at 120 min.

For comparison the feature of Pd thin layer catalyst (sample E40A) is also given. As expected, Pd behaves more selective to CH_3CHO than Pt catalysts. However our data indicate that reaction temperature significantly affects the CH_3CHO selectivity on Pt.

As shown in Fig. 7, the temperature dependence of the rates of ethylene oxidation were found to fit the Arrhenius plot, with an apparent activation energy, Ea, of 8.8$\pm$0.5 Kcal/mol. Further studies are planned to have more detailed kinetic analysis.

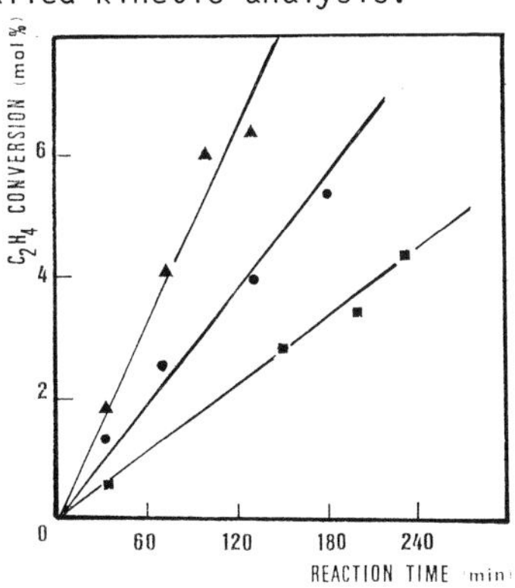

Fig. 6 - Ethylene conversion vs. reaction time at different reaction temperatures. Cat E30. T : (■) 353K; (●) 373K; (▲)393 K.

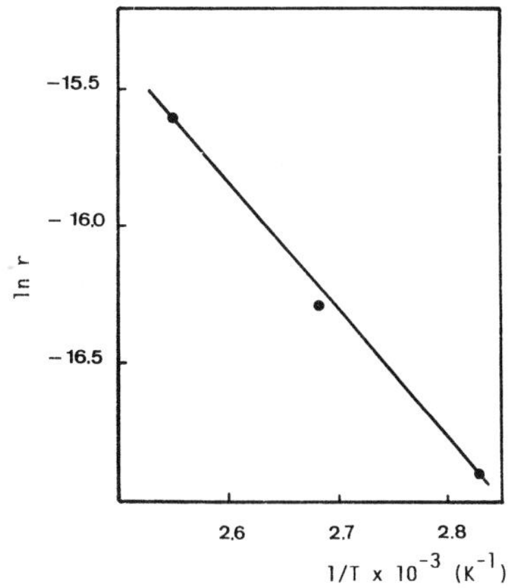

Fig. 7 - Arrhenius correlation for C_2H_4 oxidation on Pt thin layer catalysts.

Final remarks

The feasibility of a thin layer reactor has been successfully demonstrated in ethylene oxidation to acetaldehyde on noble metal hydrophobic thin layer catalysts. The suitability of Fe^{+3}/Fe^{+2} redox couple as oxygen carrier has been assessed. It is noticeable that the three phase reaction system outlined here allows an independent control of the flow regime of gas and liquid phases.

Finally the overall results indicate that the proposed experimental approach could open new perspectives in hydrocarbon selective oxidation reactions at mild conditions.

REFERENCES

1 I. Sine Shaw, J.S. Dranoff and J.B. Butt, Ind. Eng. Chem. Res. 27 (1988) 935-942.
2 V. Rao and R. Datta, J. Catal., 114 (1988) 377-387.
3 S. Goto and M. Morita, Chem. Eng. Comm., 60 (1987) 253-269.
4 S. Matsuda, T. Mori, S. Takeuchi, A. Kato and F. Nakajima, J. Catal., 79 (1983) 264-270.
5 V. Hatziantonlou, B. Andersson, T. Larsson, N.H. Schöön, L. Carlsson, S. Schwarz and K.B. Wideen, Ind. Eng. Chem. Process Des.Dev., 25 (1986) 143-150.
6 L.N. Arzamaskova, A.V. Romanenko and Yu. I. Ermakov, Kin. Kat., 22 (1981) 1438-1445.
7 G. Konig, German Patent n. DE3101024 A1 (1982).

Question: (Prof. S.Coluccia Dip.Chimica-Univ. Torino)
Do you think the apparatus described in the paper might be adapted for methane activation?

Answer:
Yes! We are working on methane activation by thin layer catalysts in three phase system and our preliminar results appear to be promising!

G. Centi and F. Trifiro' (Editors), *New Developments in Selective Oxidation*

The Mechanism of Alkene Epoxidation and Epoxide Conversion on Single Crystal Silver Surfaces

S. HAWKER , C. MUKOID , J.P.S. BADYAL , R.M. LAMBERT

Department of Chemistry , University of Cambridge , Lensfield Rd., Cambridge CB2 1EW (England).

ABSTRACT

Primary chemistry (alkene to epoxide) and the secondary chemistry (epoxide to CO_2 + H_2O) have been investigated on Ag (111) and (331) single crystal surfaces. The use of higher alkenes as model compounds has shown that epoxidation is indeed due to chemisorbed oxygen atoms, and that this species acts as an electrophile towards the C=C bond. The mechanistic model proposed is confirmed by examining the opposite effects on reaction selectivity which are induced by potassium and chlorine promotion. Temperature programmed reaction studies (TPR) and ultra violet photoelectron spectroscopy (UPS) data obtained with ethene epoxide and its further oxidation products suggest that isomerisation to the aldehyde is the rate determining step in the overall conversion of epoxide to CO_2 + H_2O.

1. INTRODUCTION

The molecular mechanism and control of reaction selectivity in the heterogeneously catalysed epoxidation of ethene continue to attract much academic and technological interest. Considerable controversy (ref. 1) has centred on the nature of the oxygen species responsible for partial oxidation of the alkene; observations on well-defined single crystal model catalyst systems should in principle provide a means of resolving this issue. Indeed our single crystal data have clearly indicated that it is the atomic oxygen species which is responsible for all the chemical activity involved - both in the primary and secondary chemistry (ref. 2,3). The difficulties encountered in working with ethene itself under low pressure conditions may be circumvented by using higher alkenes which bind more strongly to the silver surface.

We report here on the selective oxidation of two higher alkenes on Ag(111) , and on the modifying effects induced by the technologically important promoters chlorine and potassium. The particular alkenes investigated were chosen in order to test certain mechanistic proposals ; firstly, that it is the atomic state which is the active oxygen species , and secondly that the absence of γ-hydrogens in the alkene is necessary for high selectivity (ref. 4). The secondary chemistry was examined by studying ethene epoxide and acetaldehyde , since, unlike ethene, the reactive behaviour of these species *can* be investigated under conditions of ultra high vacuum. The object here was to characterise reaction intermediates and to identify the rate determining step which controls the further conversion of ethene epoxide - a process which plays a significant role in limiting overall reaction selectivity towards epoxide formation.

2. EXPERIMENTAL

Experiments were carried out in two different vacuum chambers details of which have been reported before (ref. 5). The crystal surface was cleaned by Ar^+ etching (330 eV; 7.10^{-2} Am^{-2}; 650 K) until no impurities were detectable by Auger spectroscopy; final traces of carbon were removed by repeated oxygen adsorption/desorption cycles. Oxygen dosing was carried out by means of a calibrated quartz capillary array, while hydrocarbon exposures were performed by backfilling the chamber *via* a leak valve. Chlorine was dosed from a solid state electrochemical source (ref. 6) and potassium from a thermal evaporation source.

3. RESULTS

3.1. Primary Chemistry : The Epoxidation of Higher Alkenes

As already noted , ethene oxidation cannot be studied under UHV conditions : the alkene adsorbs at low temperature (100K) but desorbs at ~150K (E_d = 38 kJ mol^{-1}) without undergoing reaction with preadsorbed oxygen. In order to increase the barrier to desorption , thereby increasing reaction probability , two substituted ethene compounds were used , styrene (ethenyl benzene) and t-butyl ethylene (3,3-dimethyl-1-butene , DMB). In addition, the choice of these particular alkenes allowed us to test further an important aspect of the reaction mechanism which we proposed in an earlier publication (ref. 2). According to this view, epoxidation activity should increase with nucleophilicity of the alkene π-bond.

Styrene desorbs molecularly from the clean Ag surface at 305K; *in the presence of preadsorbed atomic oxygen alone it reacts to give styrene epoxide* , which desorbs at 500K accompanied by combustion products and some unreacted alkene (fig.1). Unambiguous identification of styrene epoxide is provided by fragment ion spectra of the correct relative intensity at 39 amu and 65 amu (ref. 7) which mirror the 91 amu spectrum shown in fig.1; the 104 amu ion fragment corresponds to unreacted styrene. On a chlorine precovered surface the alkene desorbs at higher temperature (345K) and the extent of oxidative reaction is less , but the selectivity towards the epoxide product is substantially increased (fig. 2) . In marked contrast to this , the reaction of styrene with preadsorbed atomic oxygen on a potassium promoted surface (fig.3) shows that the selective oxidation is completely suppressed and only combustion products are observed.

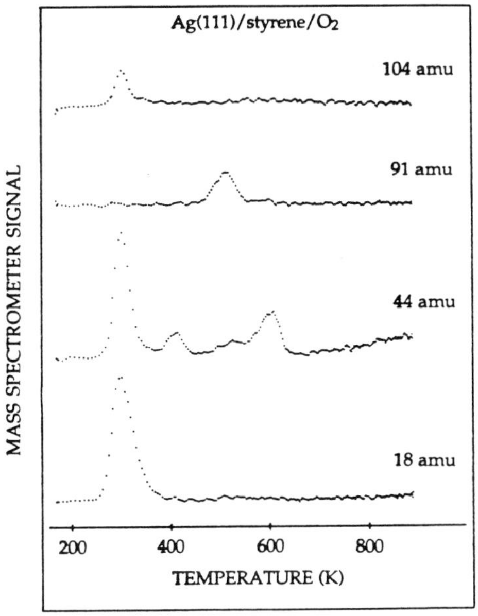

Figure 1. Reaction products from styrene + atomic oxygen overlayers on Ag(111): 1L styrene/236K + 4800L O_2/236K.

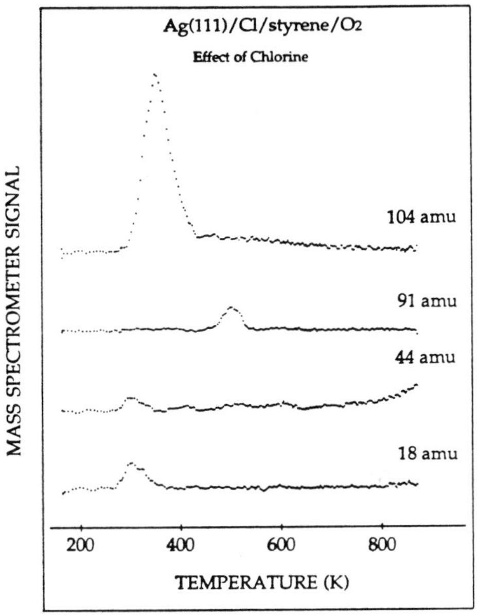

Figure 2. Reaction products from styrene + atomic oxygen overlayers in the presence of preadsorbed Cl on Ag(111): 1L styrene/236K + 4800L O_2/236K.

In the case of 3,3-dimethyl-1-butene, desorption from clean Ag(111) occurs before any detectable reaction with preadsorbed atomic oxygen can take place ; the same behaviour was observed with a potassium-promoted surface. However, preadsorbed chlorine (θ_{Cl} = 0.14ML) increases the alkene desorption energy, resulting in a highly selective oxidation with the formation of epoxide and a small amount of CO_2 and H_2O (fig. 4). Once again, unambiguous identification of the epoxide is provided by a fragment ion spectrum of the correct relative intensity, in this case at 58 amu (ref. 8), which mirrors the 43 amu spectrum shown in fig.4; the 41 amu ion fragment corresponds to unreacted DMB.

3.2. Secondary chemistry : Conversion of ethene epoxide

The adsorption and reactivity of ethene epoxide (EO) is dependent upon the adsorption temperature and the choice of crystallographic plane. On the atomically flat Ag(111) surface there is no reaction observable with preadsorbed atomic oxygen irrespective of the adsorption temperature or the presence of coadsorbed potassium. However , on the stepped Ag(331) surface the behaviour is different; once again there is no reaction when EO is adsorbed at low temperature , however adsorption at 300K results in the formation of combustion products (fig.5). Acetaldehyde (AA) , which is well known as the isomerisation product of EO (ref. 9,10) , reacts readily on Ag(111) even at low temperature giving deep oxidation products as seen in fig. 6 .

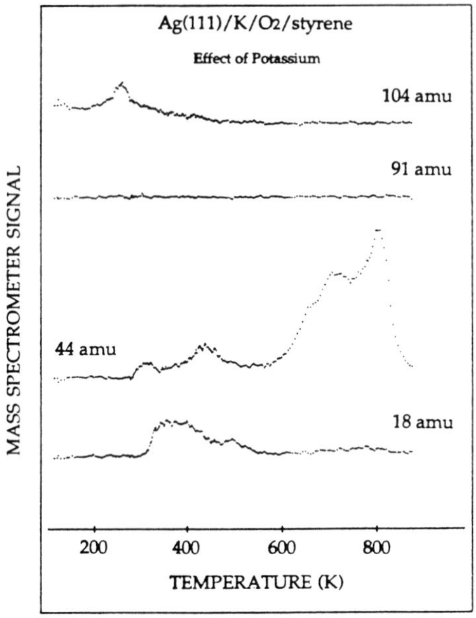

Figure 3. Reaction products from styrene + atomic oxygen overlayers in the presence of preadsorbed K on Ag(111): 8L O_2/290K + 1L styrene/105K.

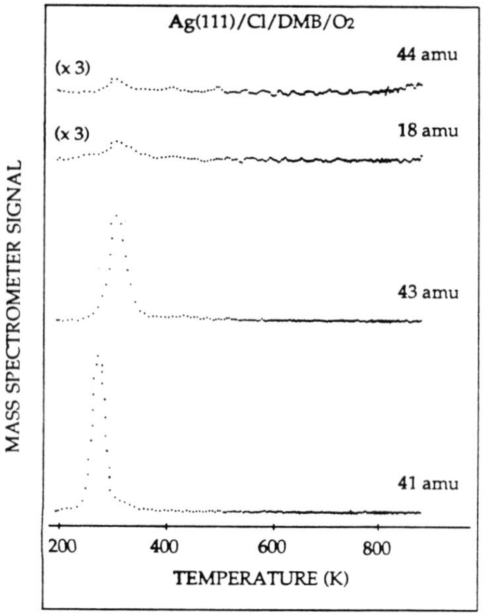

Figure 4. Reaction products from DMB + atomic oxygen overlayers in the presence of preadsorbed Cl on Ag(111): 1L DMB/195K + 4800L O_2/225K.

The temperature dependence of the HeI ultra violet photoelectron spectra of the Ag(111)/K/oxygen/EO system (fig.7) following low temperature adsorption of the epoxide reveal that the latter desorbs without detectable isomerisation or oxidation over the temperature range 105 - 300K in agreement with the TPR data. However, similar observations on the Ag(111)/K/oxygen/acetaldehyde system (fig. 8) show that reaction products are formed and that these desorb from the surface by 540 K, in line with the temperature for the CO_2 peak in the TPR spectrum.

In these experiments, preadsorbed potassium (θ_K = 0.16 ML) was used to catalyse the adsorption of oxygen by Ag(111); this procedure increases the oxygen sticking probability by a very large factor (~ 10^5) making it practical to carry out experiments with relatively modest oxygen exposures. It is therefore important to note that the presence of alkali does not induce the chemistry which is reported here. On the contrary, we have shown that alkalis actually inhibit the further conversion of EO (ref. 3), so that the observed behaviour reflects an intrinsic property of the epoxide/Ag system.

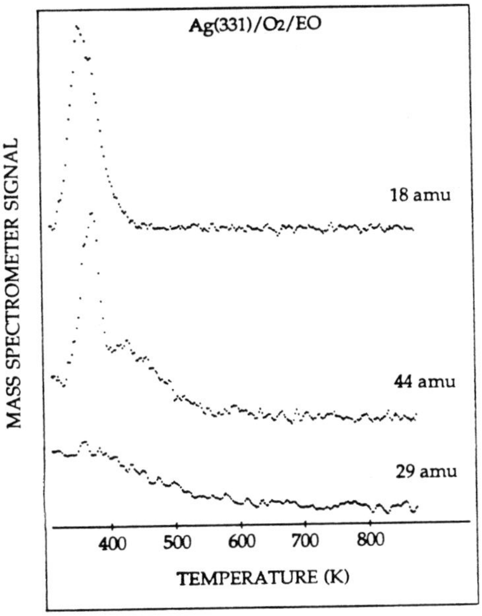

Figure 5. Reaction products from EO + atomic oxygen overlayers on Ag(331): 1800L O_2/320K + 1000L EO/320K.

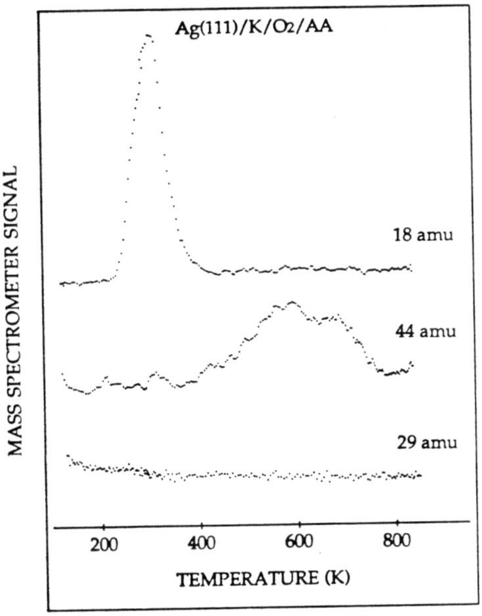

Figure 6. Reaction products from AA + atomic oxygen overlayers in the presence of preadsorbed K on Ag(111): 3L O_2/300K + 2L AA/115K.

4. DISCUSSION

The substitution of a phenyl group for a hydrogen atom in the ethene molecule leads to the styrene system , in which the π electrons are delocalised over the whole of the molecule. This gives rise to a much enhanced desorption enthalpy (E_d = 53.5 kJ mol^{-1}) which in turn makes it possible to carry out reaction measurements under UHV conditions: *the present results show unambiguously that both styrene and DMB are efficiently epoxidised by atomic oxygen.*

The apparent uniqueness of ethene itself in displaying usefully high selectivity towards epoxide formation may be associated with the absence of γ-hydrogens: alkenes which contain such γ-hydrogens are susceptible to hydrogen abstraction due to the stability of the allyl radical which is formed i.e.

$$R\text{-}CH_2\text{-}CH{=}CH_2 + O\,(a) \rightarrow R\text{-}CH{\cdots}CH{\cdots}CH_2\,(a) + OH\,(a)$$

Once such hydrogen abstraction takes place, epoxide formation is effectively blocked and stepwise degradation of the organic species leading ultimately to carbon dioxide and water is the inevitable result. Absence of γ-hydrogens should therefore result in an increased propensity towards epoxide formation, and indeed this has been elegantly demonstrated by Roberts and Madix (ref. 11) who used norbornene to show that epoxidation of higher alkenes of this type can be carried out successfully under ultra high vacuum conditions. The alkenes used in the present work also belong to this class. Our results indicate that the absence of γ-hydrogens is indeed a crucial feature if epoxide is to be obtained in high yield. Moreover, they demonstrate yet again that atomic oxygen is in fact the crucial surface species which is responsible for *all* the observed catalytic chemistry - i.e. both partial and complete oxidation of the alkenes. It therefore appears that such 'atomic epoxidation' is a general feature of selective alkene oxidations with silver catalysts.

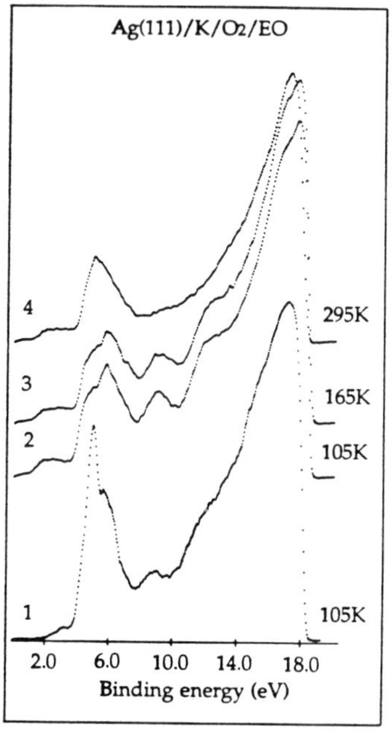

Figure 7. HeI UP temperature dependence spectra of EO + atomic oxygen in the presence of preadsorbed K on Ag(111): 2L O_2/300K + 2L EO/105K. Spectrum 1: EO on clean Ag(111).

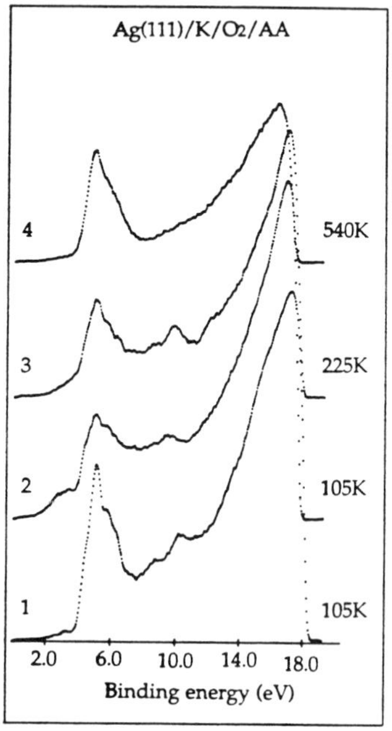

Figure 8. HeI UP temperature dependence spectra of AA + atomic oxygen in the presence of preadsorbed K on Ag(111): 2L O_2/300K + 2L AA/105K. Spectrum 1: AA on clean Ag(111).

The observed effects of potassium and chlorine are in good accord with the mechanistic proposals which we advanced in an earlier publication (ref. 12) for the case of ethene itself . For the alkenes used in our experiments we may calculate an approximate selectivity towards epoxide formation under the conditions of these experiments as follows

$$\text{SELECTIVITY} = \frac{A_E \cdot S_E}{(A_E \cdot S_E) + 1/n(A_{44} \cdot S_{44})} \times 100\ \%$$

where A_E = area of the epoxide peak ; A_{44} = area of CO_2 peak
S_E , S_{44} = mass spectrometer sensitivities for epoxide and CO_2 respectively
n = number of carbon atoms in an alkene molecule

In the case of styrene this procedure yields a figure of 60% for the unpromoted surface; on a chlorine promoted surface (fig. 2) selectivity increases to > 90%, while on the potassium pretreated surface the selectivity is zero. The opposing effects of chlorine and potassium may be explained in terms of their electronegativities. Charge transfer occurs from the metal to adsorbed chlorine, thus reducing the negative charge residing on adsorbed oxygen atoms and making them more likely to participate in electrophilic attack on the alkene , resulting in epoxidation (ref. 2). The reverse is true for potassium : this promoter leads to the formation of strongly basic oxygen species which abstract hydrogen from the alkene , leading ultimately to combustion. A related electronic effect is also very apparent in the desorption enthalpy for the alkene : π donation to the metal is increased by the presence of chlorine so E_d increases (64.5 kJ mol^{-1}) ; consistent with this , a marked effect in the opposite direction is produced by potassium (E_d = 46.5 kJ mol^{-1}).

By comparison with styrene , DMB is less strongly adsorbed because it does not possess an extended π system. Fig. 4 shows that in the presence of chlorine the oxidation to the epoxide proceeds with almost 100% selectivity . A possible reason for the greater selectivity exhibited by DMB may be that neither the parent alkene nor the product are as tightly bound to the surface as in the case of styrene. This should enhance the rate of product desorption relative to that of competing reactions.

The temperature programmed oxidation of EO on both Ag(331) and Ag(111) does not occur following low temperature adsorption, because the molecule is only weakly adsorbed, so that desorption occurs before it receives sufficient energy to open the epoxide ring. Ring opening is the first step in the isomerisation to acetaldehyde (ref. 13,14) and once the isomerisation has taken place, then oxidation to CO_2 and H_2O is rapid (ref. 3). On the (331) face, room temperature adsorption of EO using much larger gas exposures *does* lead to the uptake of a species which is capable of further reaction. This relatively low probability process is presumably due to an activated step which results in the direct adsorption of a species other than EO itself , which is the precursor to the subsequent deep oxidation reaction. It is interesting to note that the (111) face does not behave correspondingly - room temperature dosing with EO does not lead to detectable uptake and oxidation of any organic species. It thus appears possible that in contrast with the primary chemistry, the secondary chemistry of the system may exhibit a degree of structure sensitivity - in particular, these findings suggest that small Ag particles may be more effective than large ones in the combustion of EO.

The UPS data for the (111) face are in accord with the TPR results. They confirm that adsorbed epoxide is unreactive toward atomic oxygen at temperatures up to the desorption temperature; at the same time it is found that acetaldehyde is readily oxidised under similar conditions (fig. 6). This suggests that isomerisation to the aldehyde is the rate determining step in the further conversion of the epoxide. Earlier work on single crystal Ag (111) has shown that alkalis inhibit the isomerisation of EO on the otherwise clean metal surface. Subsequent microreactor /single crystal work by Tan et al (ref. 15) demonstrated that alkalis are also effective in suppressing the burning of EO under oxidising conditions (i.e. in the presence of gaseous oxygen). These are significant observations because they demonstrate that purely metal-catalysed chemistry can lead to appreciable loss of epoxide - even in the absence of a contribution from the alumina support (ref. 16,17). The present work establishes by direct observation that the rate determining step in this process is the Ag - catalysed isomerisation of the epoxide to the aldehyde - previously this was no more than an inference - it also suggests that Ag particle size effects may be of significance in this secondary chemistry.

REFERENCES

1. R.A. van Santen and H.P.C.E. Kuipers , Adv. Cat. Vol.35 , Ch. 4 , (Academic Press Inc. , London 1987).
2. R.B. Grant and R.M. Lambert , J. Cat. 92 (1985) 364.
3. R.B. Grant and R.M. Lambert , J. Cat. 93 (1985) 92.
4. S. Hawker , C. Mukoid , J.P.S. Badyal , R.M. Lambert , Surface Science in press.
5. R.A. Marbrow, R.M. Lambert, Surf. Sci. 61 (1977) 329.
6. N.D. Spencer , P.J. Goddard , P.W. Davies , M. Kitson , R.M. Lambert , J. Vac. Sci. Tech. A1 (1983) 1554.
7. A. Cornu , R. Massot , "Compilation of mass spectral data" , 2nd edition. (HEYDEN , London- Philadelphia-Rheine 1979).
8. "Eight peak index of mass spectra" , 3rd edition. (The Mass Spectrometer Data Centre , The Royal Society of Chemistry , Nottingham University , UK. 1983).
9. R.E. Kenson and M. Lapkin , J. Phys. Chem. 74 (1970) 1493.
10. Y. Ide , T. Takagi , T. Keii , Nippon Kagaku Zasshi 86 (1965) 1249.
11. J.T. Roberts and R.J. Madix , J. Am. Chem. Soc. 110 (1988) 8540.
12. R.B. Grant , C.A.J. Harbach , R.M. Lambert , S.A. Tan , J. Chem. Soc., Faraday Trans. 1, 1987, 83, 2035.
13. E.L. Force and A.T. Bell , J. Cat. 40 (1975) 356.
14. E.L. Force and A.T. Bell , J. Cat. 38 (1975) 440.
15. S.A. Tan , R.B. Grant , R.M. Lambert , J. Cat. 106 (1987) 54.
16. A. Orzechowski and K.E. MacCormack , Canad. J. Chem. 32 (1954) 432.
17. J.W. Woodward , R.G.Lindgren , W.H. Corcoran , J. Cat. 25 (1972) 292.

G. Centi and F. Trifiro' (Editors), *New Developments in Selective Oxidation*

ACTIVE SITES IN PROPYLENE MILD OXIDATION AS STUDIED ON NEW [100] ORIENTED MoO_3 CATALYSTS

M. ABON, B.MINGOT, J. MASSARDIER and J.C. VOLTA
Institut de Recherches sur la Catalyse, C.N.R.S.
2, avenue Albert Einstein, 69626 Villeurbanne cédex France

SUMMARY

The reducibility and the acidic properties of MoO_3 [100] and MoO_3 [010] catalysts have been investigated in relation with their activity in the propylene oxidation reaction. It appears that a correlation exists between some acid sites -evidenced by TPD of pyridine and dehydration of isopropanol- and the selectivity for propylene mild oxidation to acrolein.

INTRODUCTION

In previous studies (ref.1-3), it has been shown that olefin oxidation is a structure-sensitive reaction on graphite-supported α-MoO_3 catalysts ; further mechanistic studies (ref.4,5), involving the comparative oxidation of propylene and allyl iodide and also propylene oxidation by $^{18}O_2$, led us to infer that both the initial hydrogen abstraction leading to allylic intermediates and the subsequent step of oxygen insertion were favoured on the (100) faces.

However the role of the various surface sites found on the basal (010) face or the side faces of α-MoO_3 is still largely debated in theoretical (ref.6,7) and experimental (ref.8) studies. Even the fundamental question of crystal-face specificity in the oxidation of propylene on MoO_3 has been questioned in a recent paper of Oyama (ref.9).

Most recent experimental works (ref.8-10) relevant to this question rely on the study of unsupported vapour-grown α-MoO_3 in an oxygen stream (ref.11) giving after sieving or (and) grinding thin crystallites of different sizes. As outlined by Oyama (ref.9), the structural differences between such samples may be exceedingly small, the crystallites always exposing predominantly the most stable (010) face. Furthermore, grinding or even sieving is expected to induce a more or less important density of various structural defects.

Another way to induce changes in the crystal habit of α-MoO_3 is to use a suitable support inactive in propylene oxidation. This was the case for graphite. However the presence of a support seriously hampers the characterization of the active phase by most physical surface-sensitive techniques.

We therefore tried to devise a new preparation yielding unsupported α-MoO_3 catalysts exposing on their outer surface the (100) side faces.

The preparation and characterization of [100] oriented MoO_3 catalysts are only briefly exposed in so far as the present work is mainly devoted to a comparative study of the reducibility and the acidic properties of these [100] MoO_3 catalysts with respect to a more conventional vapour grown [010] MoO_3 catalyst.

PREPARATION AND CHARACTERIZATION OF [100] ORIENTED MoO_3 CATALYSTS (ref.12)

Total oxidation of metallic molybdenum sheets (25 m in thickness) in the temperature range 773-973K leads to a pure α-MoO_3 phase as checked by XRD and XPS. The obtained MoO_3 platelets display a preferential [100] orientation as shown by XRD in Fig 1a whereas vapour-grown MoO_3 crystallites are [010] oriented (Fig 1b). This peculiar orientation, which could be imposed by the oxidation mechanism of Mo, has been also checked at a single crystal scale, by electron microdiffraction and HREM.

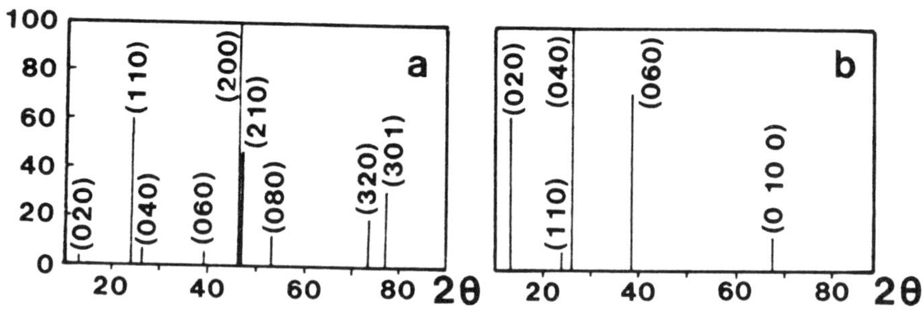

Fig.1. X-Ray diffraction diagram of (a) MoO_3 [100] 943 K and (b) MoO_3 [010].

Scanning Electron Microscopy (SEM) has shown that the expected (100) surface planes are actually truncated to (110), (120) or (130) faces, as represented in the scheme of Fig.2. These (1k0) faces, likely more stable that (100) faces, can be viewed as stepped-surfaces composed of (100) terraces and normal (010) steps. It is worth adding that the oxide crystals are thicker than vapour-grown crystals in the [010] direction, with therefore more-developped side faces and especially (1k0) faces when the temperature of preparation is increased up to about 940K, leading to a non-homothetic growth of the oxide crystals.

REDUCIBILITY OF [100] AND [010] MoO_3 CATALYSTS AS STUDIED BY XPS

It is usually considered that propylene mild oxidation involves a redox mechanism (13) and it is then important to compare the relative reducibility of (1k0) and (010) faces. When both [100] or [010] MoO_3 samples are first analysed under vacuum prior to any thermal treatment, the Mo 3d doublet is typical of

Mo^{VI} with a binding energy of 232.6 eV for the Mo 3d 5/2 level. The oxygen O_1s peak is fairly symmetrical on both [010] or [100] samples with no marked shoulder at higher binding energy (+ 1.7 eV) indicative of the presence of OH species (ref.14). Within the sensitivity of the technique, these observations suggest the absence of a large density of OH groups, in agreement with IR results of Vergnon et al (ref.15). When the samples are heated in vacuum (for 30 mn at each temperature up to 723K) the evolution of the Mo 3d doublet is

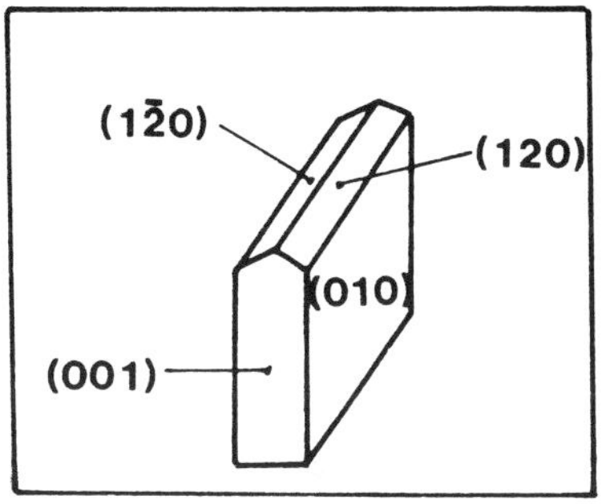

Fig.2. Schematic drawing of a MoO_3 [100] crystallite

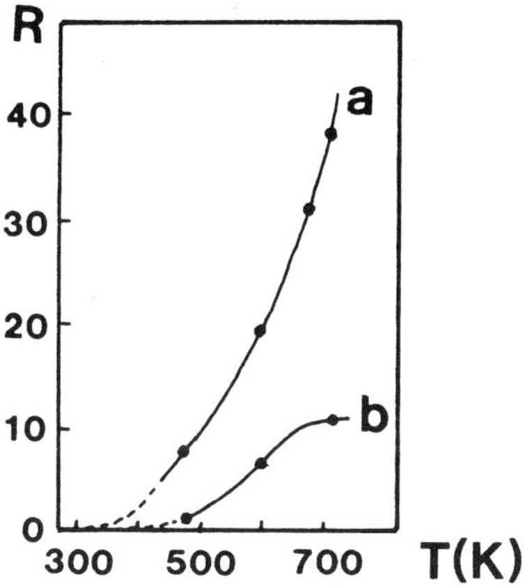

Fig.3. Relative amount R of Mo^V measured by XPS on (a) MoO_3 [010] and (b) MoO_3 [100] 943K vs the temperature.

indicative of a partial and progressive reduction to Mo^V (binding energy of 231,3 eV). As shown in Fig.3, the thermal reduction to Mo^V is much more effective on the basal (010) faces than on (1k0) faces, especially at 673K and above.

Fig.4 shows that the reduction is more pronounced when a sample is heated for 30 mn at 673K under a low pressure of propylene (0.4 Torr) instead of a similar treatment under vacuum.

These observations show that (Mo=O) groups typical of the (010) faces (ref.1,2,16)can be easily reduced at least to the Mo^V state. Under the conditions of the reaction of propylene oxidation at 673K, due to the consumption of lattice oxygens, a large number of surface oxygen vacancies can be postulated, especially on the (010) faces, with the presence of Mo^V in addition to Mo^{VI} ions.

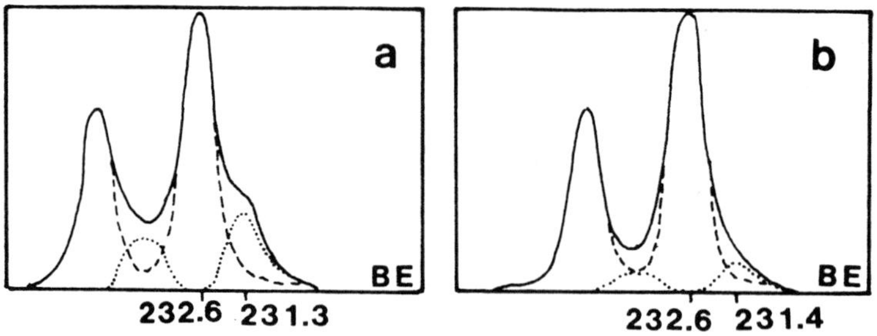

Fig.4. XPS Mo 3d level of a catalyst MoO_3 [100] 943K after heating at 673K (a) under propylene (b) under vacuum.

STUDY OF THE ACIDIC SITES

The mild oxidation of propylene would require Lewis acid sites created by oxygen vacancies where the initial activation to allylic species occurs (ref.1-3). As shown in the structural model of Fig.5, there are coordinately unsaturated Mo^{VI} on the (100) face, such sites being able to activate propylene to allyl intermediates (ref.17). It may be noticed that Lewis acid sites do not preexists on the basal (010) faces, they must be created by partial reduction under the catalytic conditions.

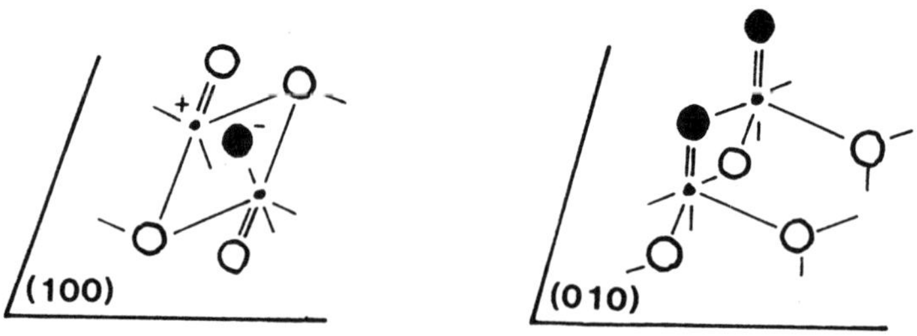

Fig.5. Structural model of MoO_3 (100) and (010) faces (black circles are surface oxygens pointing outward)

Temperature-Programmed Desorption (TPD) of basic probe-molecules

(i) Ammonia

The MoO_3 samples, first heated at 673K under oxygen and then evacuated under vacuum, have been exposed to NH_3 (100 Torr) at 300 K. TPD spectra recorded by mass spectrometry, after pumping down to the 10^{-8} Torr range, are represented in Fig.6. A fairly similar low-temperature peak is observed on both samples whereas

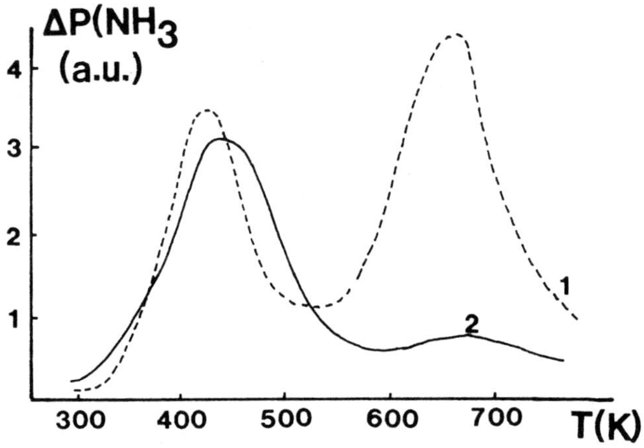

Fig.6. NH_3 TPD spectra on (1) MoO_3 [010], (2) MoO_3 [100] 809K.

a high temperature peak is only important on MoO_3 [010]. It is frequently postulated that NH_3 may reduce the surface of oxides (18) and the high-temperature peak found on the MoO_3 (010) sample could be indicative of strong acid sites created by reduction during the TPD process, in agreement with the easier reducibility of the (010) faces as shown by XPS experiments.

(ii) Pyridine

In order to avoid a possible surface reduction by NH_3, further experiments have been made using pyridine which is known to probe the acidic sites of oxidized molybdenum (ref.11,19). The samples have been exposed to pyridine at 273K (5 Torr). TPD spectra were recorded by mass spectrometry after pumping down to the 10^{-8} Torr range. It has been checked that no thermal cracking of pyridine occurs during TPD. Only molecular desorption (m/e = 79) takes place.

Different TPD spectra have been obtained when adsorption is done either on a fresh sample (Fig.7-1) or a sample already used in the propylene oxidation test

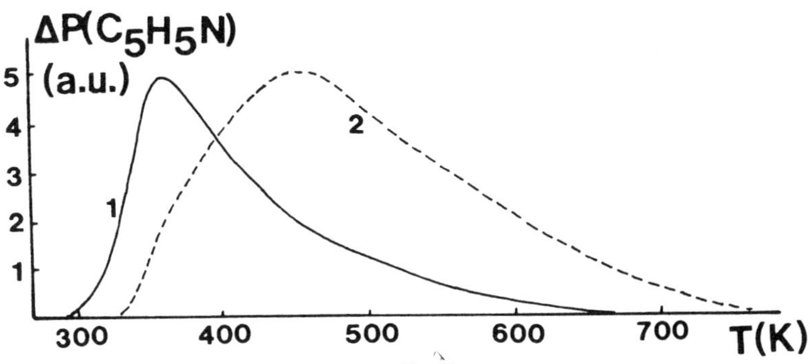

Fig.7. Pyridine TPD spectra on MoO_3 [100] 943K (1) fresh (2) already used.

at 673K (Fig.7-2). Taking in account both the shift of the peak and the surface under the peak, it may be concluded that the surface reduction (which likely occurs during the catalytic reaction) brings about an increase of acidity.

Therefore further comparative experiments have been carried out only on MoO_3 catalysts already used in the propylene oxidation test. TPD spectra in Fig.8 show distinct low and high temperature peaks. The MoO_3 [100] sample prepared at the highest temperature (943 K) gives a very large and asymetric peak which can be deconvoluted as shown in the figure.

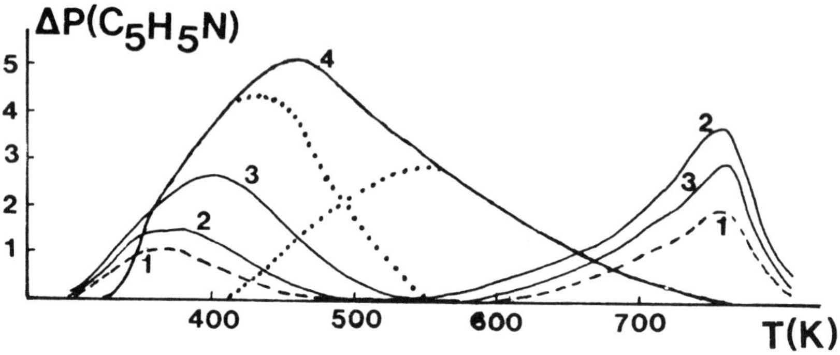

Fig.8. Pyridine TPD spectra on (1) MoO_3 [010] and MoO_3 [100] prepared at (2) 809K, (3) 893 K and (4) 943 K.

If one focuses on the low temperature peak which may be tentatively attributed to acidic Lewis sites of medium strength, the number and the strength of these sites increase from the MoO_3 [010] to the MoO_3 [100] samples prepared at increasing oxidation temperature.

As shown in Fig.9, a fairly good correlation is obtained by plotting the area under the pyridine low temperature peak and the selectivity in acrolein of the four studied MoO_3 catalysts in the propylene oxidation reaction (ref.20). On the basis of these observations, it may be infered that these medium-strength acidic sites, preferentially located on (1k0) faces, would be active in the

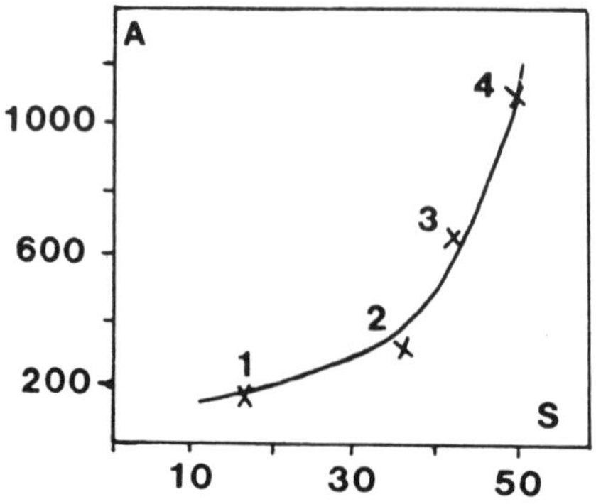

Fig.9. Amount A of adsorbed pyridine (see text) vs the selectivity S in acrolein for the (1)(2)(3)(4) MoO_3 catalysts already defined in Fi.8.

mild oxydation of propylene to acrolein. The high temperature peak correponds to stronger acidic sites which could be related to the total oxidation of propylene. Further experiments should be done to ascertain this assumption.

Isopropanol conversion reaction

This reaction was studied at low conversion between 403 and 453 K in a flow differential micro reactor (C_3H_7OH : O_2 : N_2/13 : 150 : 600). The main products were propylene (dehydration reaction) and acetone (dehydrogenation reaction). It is known that dehydration occurs on acid sites whereas dehydrogenation is related to redox and /or basic sites (ref.21). Results are summarized in Table 1. The three MoO_3 [100] catalysts are more selective with respect to propylene than the MoO_3 [010] sample but there is no clear relation between this selectivity and the acidic properties evidenced by the TPD of pyridine. The redox properties have also to be considered in the production of acetone. However it is noteworthy that the activity in propylene production increases monotonically from MoO_3 [010] to MoO_3 [100] 943 K as their acidic properties do (Fig.9).

TABLE 1

Results of the isopropanol conversion at 160°C on the different MoO_3 catalysts.

Samples	Selectivity Propylene	Selectivity Acetone	Rate of propylene Production ($x 10^{-7}$) ($mol.m^{-2}.s^{-1}$)
MoO_3 [100] 809K	94	3	7,6
MoO_3 [100] 893K	88	10	8,5
MoO_3 [100] 943K	82	16	10,6
MoO_3 [010]	63	37	3,1

CONCLUSIONS

MoO_3 [100] catalysts prepared by molybdenum oxidation have been compared to a vapour-grown MoO_3 (010) catalyst with respect to their surface reducibility and acidic properties. The basal MoO_3 (010) faces appear to be more easily thermally reduced giving rise to strong redox sites likely active in propylene total oxidation. The adsorption of ammonia and pyridine reveals the presence of a spectrum of acid sites, the number and the strength of which vary with the orientation and the temperature of preparation of the samples. Moreover the population of such sites appears to be developed during the catalytic process. Low or medium strength acid sites present on the (100) faces would be active in the mild oxidation of propylene to acrolein, as suggested by the adsorption of pyridine and the dehydration reaction of isopropanol.

ACKNOWLEDGMENTS

The authors thank Drs Tatibouet and Abou-Akar for kindly providing us with vapour-grown MoO_3 (010) catalyst and Drs Floquet and Bertrand for the SEM characterization of [100] oriented MoO_3 catalysts.

REFERENCES

1 J.C. Volta and J.M. Tatibouet, J. Catal., 93 (1985) 467-470.
2 J.C. Volta, J.M. Tatibouet, C. Phichitkul and J.E. Germain, Dechema (Ed), Proc. 8th Intern. Congress Catalysis, IV, Berlin 1984, pp 451-461.
3 J.C. Védrine, G. Coudurier, M. Forissier, and J.C. Volta, Catalysis Today, 1 (1987) 261-280.
4 A. Guerrero-Ruiz, M. Abon, J. Massardier and J.C. Volta, J. Chem. Soc., Chem. Commun (1987) 1031-1033.
5 A. Guerrero-Ruiz, J. Massardier, D. Duprez, M. Abon and J.C. Volta, in : M.J. Phillips and M. Terman (Ed), Proc. 9th Intern. Congress Catalysis, Calgary 1988, pp 1601-1608.
6 J. Ziolkowski, J. Catal. 80 (1983) 263-273.
7 J. Ziolkowski, J. Catal. 100 (1986) 45-58.
8 K. Brückman, R. Grabowski, J. Haber, A. Mazurkiewicz, J. Sloczynski and T. Wiltowski, J. Catal. 104 (1987) 71-79.
9 S.T. Oyama, Bull. Chem. Soc. Japan, 61 (7), (1988) 2585-2594.
10 A. Abou Akar, Thesis n°87-537, Univ. Claude Bernard, Lyon I, 1987.
11 J.M. Tatibouet, C.R. Acad. Sci. Paris, 297 (1983) Serie II 703-708.
12 B. Mingot, N. Floquet, O. Bertrand, M. Treilleux, J.J. Heizmann, J. Massardier and M. Abon, J. Catal., in press.
13 P. Mars and D.W. Van Krevelen, Chem. Eng. Sc., 3 (1954) 41.
14 M.W. Roberts and R. St. C. Smart, J. Chem. Soc, Faraday Trans. 1, 80, (1984), 2957-2968.
15 P. Vergnon, D. Bianchi, R. Benali Chaoui and G. Coudurier, J. Chim. Phys. 77 (1980) 1043-1049.
16 J.E. Germain, in M. Che and G.C. Bond (Ed), Adsorption and catalysis on oxide surfaces, ELsevier, Amsterdam, 1984 pp 355-368.
17 Doe Bok Kang and Euk Suk Lee, Bull. Korean Chem. Soc. 8, 1987, 482-484.
18 A. Miyamoto, T. Ui and Y. Murakami, J. Catal., 80, (1983), 106-113.
19 R.M. Henry, B.W. Walker and P.C. Stair, Surf. Sci., 155, (1985) 732-750.
20 B. Mingot, J. Massardier, N. Floquet, O. Bertrand, J.C. Volta and M. Abon, J. Catalysis, submitted for publication.
21 M. El Jamal, M. Forissier, G. Coudurier and J.C. Védrine, in : M.J. Phillips and M. Ternan (Ed), Proc. 9th Intern. Congress Catalysis, Calgary 1988, IV, 1988, pp 1617-1623.

Prof. J. HABER (Institute of Catalysis and Surface Chemistry, Polish Academy of Sciences, Krakow, Poland).

I would like to make two comments. In your ammonia TPD graphs there is a peak at 700K which you assign to strong acidity. However, ammonia may be used to measure acidity only in conditions in which it is reversibly adsorbed. It is well known that above 650K ammonia is oxidized by MoO_3, the latter being reduced with the generation of shear planes. Thus the conclusion about the existence of strong acidity may not be valid.
Your data show that the redox properties are mainly located at the (010) plane, however you correlate the yield of acrolein with the acid properties of the (100) plane. Why formation of acrolein which is a redox process does not correlate with redox properties ?

M. ABON (Institut de Recherches sur la Catalyse, Villeurbanne France).

1) It is true that the temperature of the second TPD peak of ammonia seems fairly high (about 670 K). However the molecular desorption of ammonia was observed by mass spectroscopy. It may be added that the ammonia residual pressure was reduced in the 10^{-8} Torr range at room temperature before starting the thermal desorption, in order to minimize the possibility of reaction between ammonia and the oxide.

2) The thermal partial reduction of the (010) planes appears to be some what easier than on the (100) planes on the basis of XPS measurements. However it turns out that the sites active in the formation of acrolein must have not only redox but also appropriate acidic and structural properties. The active faces would not be (100) but rather (120) faces (see the next and ref. 1).

(1) B. MINGOT, N. FLOQUET, O. BERTRAND, M. TREILLEUX, J.J. HEIZMANN, J. MASSARDIER and M. ABON.
J. Catal. 118 (1989) 424-435.

Profs. L.T. WENG, P. RUIZ and B. DELMON. (Unité de Catalyse et Chimie des Materiaux divisé, Place Croix du Sud, 1, 1348, Louvain-la-Neuve Belgium).

Comment : We have indirect data, strongly supporting the emphasis on the role of acidic sites in propylene oxidation. This comes from parallel studies in an oxygen aided dehydration [1] and oxidation of isobutene to methacrolein [2] on the same catalysts (MoO_3+α-Sb_2O_4). In the first reaction, Zhou showed a direct proportionality between catalytic activity and number of acidic sites. The activity of the catalyst in the second reaction varies in a parallel way with that in the first reaction. This strongly suggests that the acidic sites play the central role in oxidation.
It is indeed paradoxial that the crucial parameter is not the oxidation-reduction properties of the catalyst. We could speculate that the existence of acidic sites provides entry ports for oxygen, thus allowing an easy reoxidations of the reduced sites. More precisely, the existence of the "acid site involving sites", or entry ports, would make that the replenishment of the catalyst with oxygen is much less rate limiting than other wise.

Question : We have a remark concerning the nature of acidic sites (NH_3 TPD, around 450°K). Belokopytov, on one hand, and Groff, on the other hand, through IR studies of adsorbed pyridine, demonstrated that the acidic sites are Brönsted sites.

What are your arguments for Lewis sites ?

[1] B. Zhoa, Ph. D. Thesis, U.C.L. (1988)
[2] L.T. Weng et al. 9th Int. Cong. Catal., Calgary, P.1609 (1988).

M. ABON (Institut de Recherches sur la Catalyse, Villeurbanne France).

It is well known that MoO_3 has only fairly weak acidic properties. This oxide is not able to dissociate water (ref. 1 and 2) and Bianchi et al (ref. 2) only observed weak and strong Lewis acidic sites by IR studies. The adsorption of ammonia or pyridine on Mo Lewis sites has been also claimed by Alsdorf et al (ref. 3) and Tatibouet (ref. 4).

(1) P. Vergnon, D. Bianchi, R. Ben Ali Chaoui and G. Coudurier, J. Chim. Phys. 77 (1980) 1043-1049.
(2) D. Bianchi, J.L. Bernard, M. Camelot, R. Ben Ali Chaoui and S.J. Teichner, Bull. Soc. Chim. Fr., 7-8 (1980) 275-280.
(3) E. Alsdorf, W. Hanke, K.H. Schnabel and E. Schrieir, J. Catal. 98 (1986) 82-87.
(4) J.M. Tabibouet, C.R. Acad. Sci. Paris, 297 (1983) 703-708.

Prof. V. CORTES CORBERAN (Instituto Catalysis y Petroleoquimica Serrano, 119, 28006 Madrid, Spain).

1) No data on the activity of the different samples for propylene oxidation are given in this paper to substantiate the conclusion that "low or medium strength acid sites present on the (100) faces are active in mild oxidation". Alcohol dehydration indicates the presence of acid sites on the samples, that could activate propylene, but not neccesarily to selective oxidation products. As propylene is formed almost quantitatively from isopropanol on some of the samples already at low temperature, it can be infered that raising the temperature (provided the medium acid centers were not changed, as implicitly hypothesized by the authors) to the range of allylic oxidation should give place to acrolein formation, according to authors' conclusions. Have the authors tried such a possibility ?

2) As (100) faces are actually not present excepting in terraces with (010) steps, have the authors considered the possibility that active centers in Mo [100] samples could be in the edges of steps instead (100) plane ?

M. ABON (Institut de Recherches sur la Catalyse, Villeurbanne France).

1) It is true that the catalyc properties of [100] oriented MoO_3 are not given in the present paper, with the exception of some indication in their selectivity for acrolein formation in the propene oxidation reaction (see Fig. 9). These catalytic properties will be detailed in a forthcoming paper in preparation (to be submitted to J. Catal.).

The conversion of isopropanol has been studied at 160°C and this temperature is indeed much too low to allow the allylic oxidation of propylene. We did not try to raise the temperature in so far as we were interested in the primary products, indicative of the acidity.

2) The definition of edges on oxides is difficult at an atomic level and coordinatively unsaturated Mo on the (100) terraces of (120) faces might well be considered also as edge sites.

G. Centi and F. Trifiro' (Editors), *New Developments in Selective Oxidation*

EFFECTS OF ADDED SnO_2, WO_3, MoO_3 AND α-Sb_2O_4 ON THE CATALYTIC PROPERTIES OF Bi_2MoO_6 IN SELECTIVE OXIDATION OF ISOBUTENE TO METHACROLEIN

L.T. WENG, E. SHAM*, B. DOUMAIN, P. RUIZ and B. DELMON

Unité de Catalyse et Chimie des Matériaux Divisés, Université Catholique de Louvain
Place Croix du Sud 1, 1348 Louvain-la-Neuve (Belgium)

*Fac. de Cs. Tecnologicas, Universidad Nacional de Salta, Buenos Aires 177, 4400 SALTA, RA

SUMMARY

Bi_2MoO_6 alone is moderately active and selective in the oxidation of isobutene to methacrolein. Mechanically admixed α-Sb_2O_4 improved selectivity while SnO_2, WO_3 or MoO_3 enhanced activity and selectivity. Characterization of the mixed samples by XRD, BET, CTEM, AEM, XPS and ISS ruled out mutual contamination for the mixtures with SnO_2, WO_3 or α-Sb_2O_4. The observed effects are explained by a creation of selective catalytic sites thanks to spillover oxygen (remote control mechanism).

INTRODUCTION

Since the discovery of multicomponent Bi-molybdate (MCM) catalysts, many attempts have been made to elucidate the working mechanism of these catalysts [1-6]. These attempts included studies of the structure, composition, promoting effect of the added elements, mutual cooperation between phases etc. It is now usually accepted that the MCM catalysts have the "layer" structure, i.e. that the surface layer is constituted of Bi-molybdates : Bi_2MoO_6 (γ) and $Bi_2Mo_3O_{12}$ (α) (probably with MoO_3[6]), followed by a second layer composed of Fe-molybdate and the core of M^{II}-molybdates [1]. Many models have been advanced but no one is universally accepted [4-6].

Since Bi-molybdates are the main active phases in MCM, one interesting approach would be to look at the possible cooperation between phases in Bi-molybdates, for instance of γ with the α phase or with MoO_3. If some positive cooperation is detected, this would provide a new point of view to understand the working mechanism of MCM catalysts. This has been indeed the subject of many studies [7-9]. Carson et al. have shown that the mechanical mixture of α and γ phases is more active and selective in the selective oxidation of propene than either of the starting phases. They explained this synergistic effect by the appropriate addition of the useful properties of each phase [7-8]. Batist has claimed that γ-Bi_2MoO_6 can be improved by adding a small amount of MoO_3 and he explained this effect by a partial coverage of "Bi_2O_2" layer of γ-Bi_2MoO_6 by MoO_3 [9]. A similar effect of MoO_3 on γ-Bi_2MoO_6 was also observed by Matsuura et al.[10] and their explanation was the formation of an active surface layer with a composition corresponding to Bi/Mo≈1. El Jamal et al.[11] recently studied the cooperation between phases in Bi-molybdates in detail and they concluded that the elimination of the "impurity" (Bi) on the surface of Bi_2MoO_6 was the essential origin of the cooperation.

On the other hand, experiments with mechanical mixtures of two simple oxides, e.g. MoO_3+α-Sb_2O_4 [12-13], or SnO_2+α-Sb_2O_4 [14-15], conducted in our laboratory have demonstrated that the catalytic synergy between these oxides could be explained by a remote control mechanism. In remote control, an oxide such as MoO_3 or SnO_2 carries the necessary functions for selective oxidation, while the other, α-Sb_2O_4, though inactive when alone, activates molecular oxygen into mobile species (spillover oxygen) which migrate onto the first oxide and react with the surface to improve its catalytic properties. The migration of oxygen species from α-Sb_2O_4 to MoO_3 has been demonstrated using the ^{18}O technique [16]. We were led to distinguish between two classes of oxides; (i) acceptors (A) such as MoO_3 or SnO_2 and (ii) donors (D) such as α-Sb_2O_4 or $BiPO_4$ [17-18].

We thus speculated that a compound oxide, e.g. γ-Bi_2MoO_6, although possessing the necessary functions for selective oxidation, might not possess the optimal balance of D and A functions. In this case, its activity and/or selectivity could be increased by mixing it with another oxide of the opposite class, namely A in order to take full advantage of the D properties, or viceversa. In such a situation, this would justify the necessity to have several phases in efficient MCM catalysts, even when **compound** oxides are present. The proposed communication will examine this point. We studied the effect of an acceptor phase or a donor phase on the catalytic properties of Bi_2MoO_6. Although MoO_3 is a good acceptor phase, it gives complicated solid state reactions with Bi_2MoO_6; this can certainly complicate the explanation. Moreover, the effect of MoO_3 on the catalytic properties of Bi_2MoO_6 have been studied in literature as mentioned above. These were the reasons why we chose, in addition, two other **new** acceptor phases : SnO_2 and WO_3, the effects of which have never been mentioned in literature. α-Sb_2O_4 was selected as added donor phase.

EXPERIMENTAL

Catalyst preparation and characterization

α-Sb_2O_4, MoO_3 and WO_3 were obtained by calcination of Sb_2O_3, $(NH_3)_4Mo_7O_{24}\cdot 4H_2O$, and WO_3 in air at 500°C for 20h. SnO_2 was prepared by precipitation of stannous hydroxide from $SnCl_2\cdot 2H_2O$, followed by washing to eliminate Cl^-, drying and finally calcining in air at 600°C for 8h.

Bi_2MoO_6 was prepared by the citrate method [19]. An anqueous solution of Bi^{3+} and Mo^{6+} with an atomic ratio (Bi/Mo=2) was prepared from commercial products $Bi(NO_3)_3\cdot 5H_2O$ and $(NH_3)_4Mo_7O_{24}\cdot 4H_2O$ (Merck, p.a.). After addition of a small amount of HNO_3, citric acid was added in such a manner that the number of moles of citric acid was equal to the total number of moles of Bi and Mo ions together. The solution thus obtained was evaporated under reduced pressure in a rotary evaporator, first at 60°C until the liquid became viscous; thereafter the temperature was subsequently raised to 90°C until the liquid became solid. The solid was finally decomposed at 300°C for 16h and calcined at 500°C for 18h.

The mechanical mixtures were obtained by dispersing the separately prepared oxides in n-pentane for 10 minutes, evaporating the solvent and drying at 80°C overnight.

The composition of the mechanical mixture was defined by the mass ratio:

$$R_m = \frac{\text{Mass } Bi_2MoO_6}{\text{Mass } Bi_2MoO_6 + \text{Mass } MO_X} \quad (1)$$

where MO_X refers to SnO_2, WO_3, MoO_3 or α-Sb_2O_4. Hereafter, the mechanical mixtures will be designated as $M_M{}^Y$, where M refers to MO_X while Y refers to the concentration of Bi_2MoO_6. For instance, $M_{Sn}{}^{75}$ corresponds to the mechanical mixture Bi_2MoO_6+SnO_2 with 75 wt% Bi_2MoO_6.

Both fresh and used samples were characterized using XRD, BET surface measurements, Electron Microscopy (CTEM and microanalysis), XPS and ISS. In XPS, we calculated the M/(Bi+Mo) (in which M refers to Sb, Sn or W) and Bi/Mo ratios by using the sensitivity factors determined by Wagner et al.[20]. In the ISS measurements, the M/(Bi+Mo) and Bi/Mo ratios represent the intensity ratios. Contamination of one phase by the other can be revealed by the first ratio while the surface modification of Bi_2MoO_6 can be detected by the latter.

Catalytic activity measurement

The catalytic test was selective oxidation of isobutene to methacrolein, carried out in a fixed-bed reactor system working under atmospheric pressure [15]. The reactor was an U-shape Pyrex tube of a diameter of 8mm, into which a small tube of 4mm diameter was inserted for loading a thermocouple. The catalysts were screened and the fractions of 500-800 μm were used. The catalyst (300 mg) was diluted with small glass spheres (500 μm) in order to occupy a fixed height of 1.5 cm in the reactor. The standard reaction conditions were: $i\text{-}C_4H_8/O_2/N_2 = 1/2/7$, total feed = 60 ml/min., temperatures = 420-460°C.

The unreacted reactants and products were analyzed by "on-stream" gas chromatograph.

A catalytic synergy was often observed. It was calculated using the following formula:

$$\text{Synergy} = \frac{Y_{AB} - (R_m\, Y_A + (1-R_m)\, Y_B)}{R_m\, Y_A + (1-R_m)\, Y_B} \qquad (2)$$

where Y_{AB}, Y_A and Y_B represent the methacrolein yields of the mechanical mixture AB, oxides A and B respectively.

RESULTS

Physico-chemical characterization

1. BET surface area and XRD

Table 1 summarizes the BET surface areas and XRD phases for some pure oxides and mechanical mixtures.

The surface areas of the pure oxides did not change during the catalytic reaction. The surface area of the mechanical mixture is the simple sum of those of pure oxides for $Bi_2MoO_6+SnO_2$ and $Bi_2MoO_6+Sb_2O_4$ systems. It was not modified after the catalytic reaction.

Table 1 : BET surface areas and XRD phases for the mechanical mixtures

Samples	Fresh		Used	
	BET surface ($m^2 \cdot g^{-1}$)	XRD phase	BET surface ($m^2 \cdot g^{-1}$)	XRD phase
Bi_2MoO_6	1.92	$\gamma\text{-}Bi_2MoO_6$	1.95	$\gamma\text{-}Bi_2MoO_6$
M_{Sn}^{50}	8.68	$\gamma\text{-}Bi_2MoO_6 + SnO_2$	8.59	$\gamma\text{-}Bi_2MoO_6 + SnO_2$
SnO_2	13.80	SnO_2	13.16	SnO_2
M_W^{50}		$\gamma\text{-}Bi_2MoO_6+WO_3$		$\gamma\text{-}Bi_2MoO_6+WO_3$
WO_3	3.00	WO_3	3.10	WO_3
M_{Mo}^{50}		$\gamma\text{-}Bi_2MoO_6+MoO_3$		$\gamma\text{-}Bi_2MoO_6+MoO_3$
M_{Sb}^{50}	1.98	$\gamma\text{-}Bi_2MoO_6+\alpha\text{-}Sb_2O_4$	2.00	$\gamma\text{-}Bi_2MoO_6+\alpha\text{-}Sb_2O_4$
Sb_2O_4	2.00	$\alpha\text{-}Sb_2O_4$	2.00	$\alpha\text{-}Sb_2O_4$

For all mechanical mixtures prepared, the XRD patterns are the simple addition of those of the starting oxides. No new peak was observed. The patterns did not change after the catalytic reaction.

2. Electron microscopy

Figure 1-a presents a typical CTEM micrograph for M_{Sn}^{50} after the catalytic reaction. The bigger particles correspond to Bi_2MoO_6 and the smaller ones to SnO_2. Compared with pure oxides and the fresh mechanical mixture, the size of both particles remained unchanged. The corresponding EDS-STEM microanalysis spectra are reported in Fig.1-b. The identification numbers correspond to those in the CTEM picture. Only the Sn signal is obtained when the analysis is made on SnO_2 particles and only the Bi and Mo signals appear on Bi_2MoO_6 particles. No mutual contamination between the two phases was detected with EDS. Similar results have been obtained for Bi_2MoO_6+WO_3 mixtures.

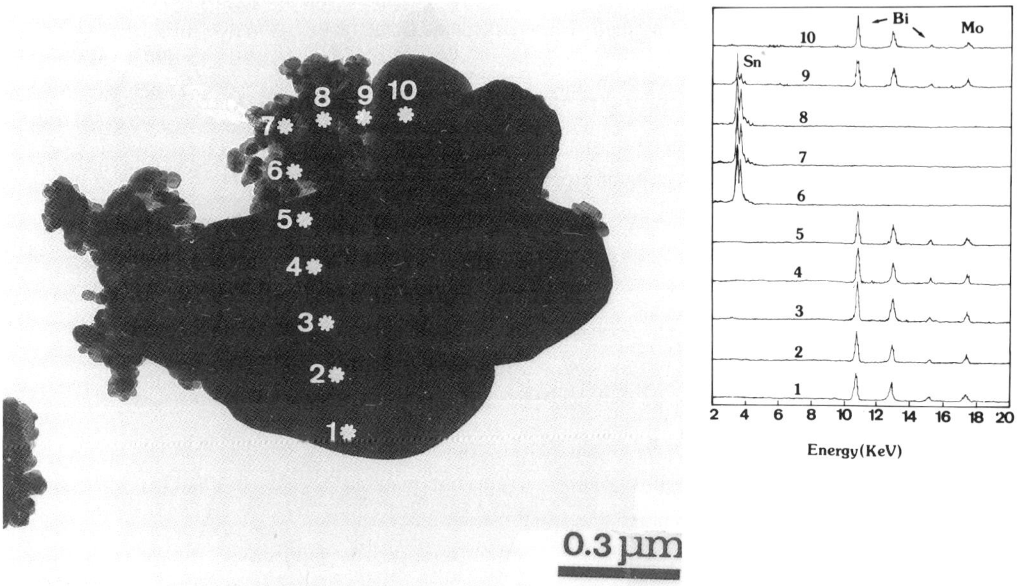

Figure 1 : a) CTEM micrograph and b) the corresponding EDS-STEM microanalysis patterns for M_{Sn}^{50} after the catalytic reaction

3. XPS and ISS

Table 2 reports the M/(Mo+Bi) and Bi/Mo ratios determined by XPS and ISS for Bi_2MoO_6 and its mechanical mixtures with SnO_2, WO_3 and Sb_2O_4 respectively. Both the M/(Mo+Bi) and Bi/Mo ratios are the same before and after the catalytic reaction. No mutual contamination is detected. The Bi/Mo ratios for the mixtures are equal to that of pure Bi_2MoO_6, indicating that the surface of Bi_2MoO_6 is not modified either during the preparation of the mixture or during the catalytic reaction.

Table 2 : XPS and ISS results for γ-Bi_2MoO_6 and its mechanical mixtures with SnO_2,WO_3 and α-Sb_2O_4 mixtures

Samples	XPS		ISS	
	M/(Mo+Bi)[a]	Bi/Mo	M/(Mo+Bi)	Bi/Mo
γ-Bi_2MoO_6		1.26		33.30
M_{Sn}^{75}	1.170	1.21		
M_{Sn}^{75} (after reaction)[b]	1.158	1.20		
M_W^{75}	0.791	1.29		
M_W^{75} (after reaction)[b]	0.830	1.30		
M^{50}_{Sb}	0.265	1.28	0.165	35.71
M^{50}_{Sb} (after reaction)	0.275	1.26	0.165	35.71

a : M refers to Sn, W or Sb; b : the sample was calcined at 400°C for 20h in order to eliminate deposited coke.

Catalytic activity

1. Pure oxides

The catalytic activity results of pure oxides at 440°C and 460°C are reported in Table 3.

Table 3: The catalytic activity and selectivity of pure oxides at 440°C and 460°C

Samples	440°C			460°C		
	Conversion (%)	Yield (%)	Selectivity (%)	Conversion (%)	Yield (%)	Selectivity (%)
Bi_2MoO_6	25.06	16.66	66.48	33.05	23.97	72.53
SnO_2	43.85	0.55	1.30	45.00	0	0
WO_3	10.82	1.92	17.65	15.14	2.64	17.44
MoO_3	11.07	4.50	40.65	15.71	6.47	41.18
Sb_2O_4		no detectable			no detectable	

Pure Bi_2MoO_6 is active and selective and its selectivity increases with reaction temperature. Compared with Bi_2MoO_6, much less methacrolein was produced for pure single oxides, MoO_3, WO_3, SnO_2 and Sb_2O_4. MoO_3 and WO_3 have almost the same conversion, but the former exhibits a higher selectivity to methacrolein. SnO_2 is very active and poorly selective. Pure Sb_2O_4 is inactive.

2. Mechanical mixtures

Figures 2 a and b correspond to Bi_2MoO_6+SnO_2 system. They show the methacrolein yield and selectivity as a function of the composition R_m at three temperatures. A conspicuous synergy for methacrolein yield is observed for all compositions. The maximum synergy is observed around R_m=0.75. The synergistic effect is due partially to an improvement in selectivity (see the selectivity curves in Fig. 2b), and partially to an increase of the conversion at the higher temperature. Similar figures have been obtained for Bi_2MoO_6+WO_3 and Bi_2MoO_6+MoO_3.

In order to compare the effects on the properties of Bi_2MoO_6, Table 4 summarizes the catalytic activity, selectivity and calculated synergy for the best mechanical mixture (R_m=0.75) of these systems obtained at two temperatures. The catalytic activity results with Bi_2MoO_6+Sb_2O_4 mixture are also reported.

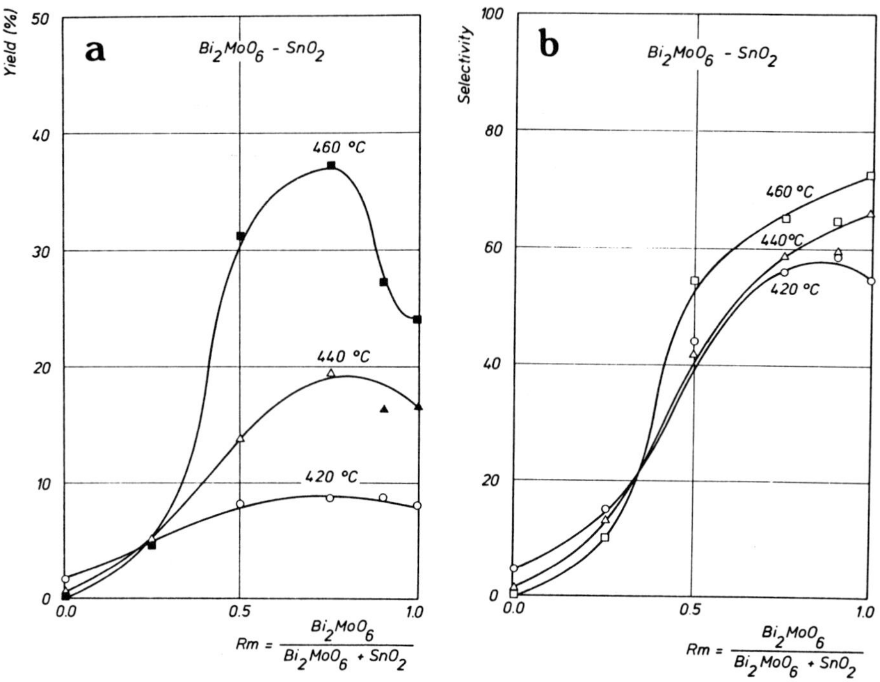

Figure 2 : a) methacrolein yield and b) selectivity to methacrolein as a function of mixture composition R_m for Bi_2MoO_6-SnO_2 system

Table 4 : Catalytic activity, selectivity and calculated synergy with (2) for 4 systems of mechanical mixtures with R_m=0.75 at 440°C and 460°C respectively

Samples	440°C			460°C		
	Yield (%)	Selectivity (%)	Synergy (%)	Yield (%)	Selectivity (%)	Synergy (%)
M_{Sn}^{75}	19.42	58.81	53.73	37.04	64.66	106.04
M_W^{75}	14.87	60.00	14.61	24.00	69.60	28.77
M_{Mo}^{75}	14.27	66.94	17.45	21.00	70.00	26.37
M_{Sb}^{75}	12.44	80.54	-0.44	17.71	81.56	-1.49

The table shows that there exists a strong cooperation between Bi_2MoO_6 and SnO_2, WO_3 and MoO_3 for the methacrolein yield. The intensity of the cooperation, as indicated by the synergy, increases with reaction temperature (by a factor of 2). The selectivity of these mechanical mixtures is smaller than that of pure Bi_2MoO_6 but higher than that which should be obtained if no catalytic synergy existed, i.e. the simple sum of the selectivities of pure oxides. The catalytic synergy follows the order : $SnO_2 > WO_3 \approx MoO_3$.

For the mechanical mixtures Bi_2MoO_6+Sb_2O_4, almost no catalytic synergy is observed for the methacrolein yield. However, the selectivity to methacrolein is enhanced (by ~10%). Similar results have been obtained for the other compositions of mechanical mixtures.

DISCUSSION

The catalytic activity results show that the catalytic properties of Bi_2MoO_6 can be improved by simply mixing it with acceptor phases such as SnO_2, WO_3 or MoO_3 or a donor phase such as α-Sb_2O_4. The former improve the catalytic activity and selectivity while the latter only enhances selectivity. In order to ensure that the remote control mechanism explains the observed results, it is necessary to first examine if solid state interactions between the two phases could take place.

Possible solid state interaction between two phases

Let us first look at the mechanical mixtures of Bi_2MoO_6 with SnO_2, WO_3 and α-Sb_2O_4. The XRD measurements showed that there is no indication of the formation of new phases either in fresh or in used mixtures. The BET surface area of both oxides (Bi_2MoO_6 and added phase) is not changed either during the mixture preparation or during the catalytic reaction. This is in agreement with the observations by electron microscopy, namely that the particle size of both oxides in the mixtures remains practically unchanged. Electron microanalysis indicated that the particles of both oxides remain pure, i.e. that no mutual contamination takes place. This is consistent with XPS results, showing that the M/(Bi+Mo) and Bi/Mo ratios remain unchanged for the fresh and used mixtures and the latter ratio is the same as that of pure Bi_2MoO_6. These results are further confirmed by the results obtained by ISS (the technique more sensitive to surface) for the Bi_2MoO_6+α-Sb_2O_4 system. Taking into account all these results, we can conclude that the mechanical mixtures of Bi_2MoO_6 with SnO_2, WO_3 or α-Sb_2O_4 are constituted of two separate (non-contaminated) oxides phases. In this case, the only possible explanation for the observed results is the occurrence of a remote control mechanism.

The situation of the mechanical mixtures of Bi_2MoO_6 with MoO_3 is more complicated, because it has been proposed by several authors [10-11] that a solid state reaction takes place between Bi_2MoO_6 and MoO_3. They showed that the surface of Bi_2MoO_6 could be changed to β-$Bi_2Mo_2O_9$ when mixed with a small quantity of MoO_3 and calcined at temperatures higher than 480°C and to α-$Bi_2Mo_3O_{12}$ when a great amount of MoO_3 was used [11]. Therefore, the formation of new phases (α or β phase) on the surface cannot be totally ruled out, although the XRD measurements did not reveal any new phase. However, the fact that the effect of MoO_3 on the catalytic properties of Bi_2MoO_6 is very similar to that of WO_3 or SnO_2 in this study may suggest that the remote control mechanism could also operate in the Bi_2MoO_6+MoO_3 system. The exact origin of the cooperation between Bi_2MoO_6 and MoO_3 would necessitate further investigation.

Explanation of the catalytic results based on the remote control mechanism

We shall discuss the effects of added phases on the catalytic properties of Bi_2MoO_6 in two parts : i) the case where the added phase is an acceptor and ii) the case where the added phase is a donor.

The first case is that of Bi_2MoO_6 mixed mechanically with SnO_2, WO_3 or MoO_3. MoO_3 has been shown to be a typical acceptor [12,13]. Chemical analogy leads us to attribute the same role to WO_3. SnO_2, although possessing a slight donor character, is mainly an acceptor, as shown by its behaviors when mixed with α-Sb_2O_4 [14,15]. It is thus quite logical to think that Bi_2MoO_6, apart from possessing itself all the functions necessary for oxidation, plays an additional role as a D, and cooperates with the A phases SnO_2, WO_3 or MoO_3 by activating them. The high mobility of oxygen ions in Bi_2MoO_6 has been shown in several works [21-22]. The interpretation of our results is that Bi_2MoO_6 produces spillover oxygen and this mobile oxygen species migrates to the surface of A (SnO_2, WO_3 or MoO_3) to create and/or regenerate catalytic centers. The increase of catalytic activity reflects the creation of total catalytic centers and the increase of selectivity to methacrolein reflects to the increase of selective ones.

The synergistic effect is higher with SnO_2 than with WO_3 or MoO_3. According to the remote control mechanism, the synergistic effect depends strongly on the number of contacts between A and D. The latter depends greatly on the particle size of the two phases (or surface area). In the present case, the D phase is the same; the synergistic effect is thus directly related to the A phase. The surface area of SnO_2 is about 6 times greater than that of MoO_3 or WO_3. It can be expected that the number of contacts between Bi_2MoO_6 and SnO_2 is greater than that between Bi_2MoO_6 and WO_3 or MoO_3, hence explains the higher synergy.

The fact that the catalytic synergy (Table 4) increases with reaction temperature is due to the fact that the increase of temperature would facilitate the migration of spillover oxygen.

Let us now look at the second case where Bi_2MoO_6 is mixed with α-Sb_2O_4. In this case, Bi_2MoO_6 is an acceptor while α-Sb_2O_4 is a donor. The improvement of selectivity by α-Sb_2O_4 is related to its ability to act as a very strong donor D. It is striking that Bi_2MoO_6, which has been considered as a model catalyst in selective oxidation, can have its selectivity improved by a new contact with a donor.

CONCLUSIONS

The present work shows that Bi_2MoO_6 is moderately active and selective in the oxidation of isobutene to methacrolein. Its catalytic properties can be improved by the added phases such as SnO_2, WO_3, MoO_3 (acceptor phase) or α-Sb_2O_4 (donor phase). The acceptor phase increases its activity and selectivity, while the donor can only improve its selectivity. This shows that the optimal balance of the D and A functions in a catalyst is not necessarily obtained in a compound oxide and that the addition of other phases may be beneficial.

REFERENCES

1 M.W.J. Wolfs and J.H.C. van Hooff, in B. Delmon, P.A. Jacobs and G. Poncelet (Eds.) "Preparation of Catalysts", Stud. Surf. Sci. Catal., Vol. 1, Elsevier, Amsterdam (1976) 161
2 I. Matsuura and M.W.J. Wolfs, J. Catal., 37 (1975) 174
3 T. S. Prasada Rao and P.G. Menon, J. Catal., 37 (1978) 64
4 Y. Moro-oka, W. Ueda and T. Ikawa, 187th ACS meeting, April 8-13 (1984) St Louis, USA
5 M. Sh. Zurmukhtashvili, Yu. V. Maksimov, M.Yu. Kutyrev, L.Ya. Margolis, D.D.Shaskkin and O.V. Krylov, Kinet. Katal. 25, No. 4 (1984) 955-961
6 I. Matsuura, in T. Inui (Ed.) "Successful design of catalysts", Stud. Surf. Sci. Catal. Vol. 44, Elsevier, Amsterdam (1989) 111
7 D. Carson, G. Coudrier, M. Forssier and J.C. Vedrine, J. Chem. Soc. Faraday Trans. I, 79 (1983) 1921
8 D. Carson, M. Forssier and J.C. Vedrine, J. Chem. Soc. Faraday Trans. I, 80 (1983) 1017
9 P.A. Batist, J. Chem. Biotechnol., 29 (1979) 451
10 I. Matsuura, R. Shuit and K. Hirakwa, J. Catal., 63 (1980) 152
11 M.El. Jamal, M.D. Forissier, G. Coudurier and J.C. Vedrine, 9th Congr. on Catal., Vol. 4, Calgary, Canada (1988) 1617
12 L.T. Weng, B. Zhou, B. Yasse, B. Doumain, P. Ruiz and B. Delmon, Ibid. p. 1609.
13 P. Ruiz, B. Zhou, M. Remy, T. Machej, F. Aoun, B. Doumain and B. Delmon, Catalysis Today, 1 (1987) 181
14 L.T. Weng, P. Ruiz and B. Delmon, 2nd Int Conf. Spillover, June 12-16, Leipzig, D. R. G.(1989)
15 L.T. Weng, N. Spitaels, B. Yasse, J. Ladrière, P. Ruiz and B. Delmon, to be published.
16 F.Y. Qiu, L.T. Weng, P. Ruiz and B. Delmon, Appl. Catal., 47, (1989) 115-123
17 L.T. Weng, P. Ruiz, B. Delmon and D. Duprez, J. Mol. Catal., 52 (1989) 349-360
18 S. Ceckiewcz and B. Delmon, Bull. Soc. Chim. Belg., 93, No.3, (1984) 163
19 Ph. Courty, H. Ajot, Ch. Marcilley and B. Delmon, Powder Technol., 7 (1973) 21
20 C.D. Wagner, L.E. Davis, H.V. Zeller, P.A. Taylor, R.H. Raymond and L.H. Gale, Surf. Inter. Anal., 3, No.5 (1981) 21
21 G.W. Keulks and T. Matsuzaki, in Che and Bond (Eds) "Adsorption and Catal. on Oxides Surf.", Stud. Surf. Sci. Catal, Vol 21, Elsevier, Amsterdam (1985) 297,
22 J.T. Gleaves, J.R. Ebner and T.C. Kuechler, Catal. Rev. Sci. Eng., 30(1) (1988) 49

I. OSIPOVA (Institute of Catalysis, Novosibirsk, USSR) : 1) Sb_2O_4 may act as a catalyst of oxidation of isobutene. Did you observe its activity and if it was so, what is the additive activity of these mixtures ?
2) You observe the dependence of catalytic activity on the acidic centers concentration of catalysts ? Did you investigate the influence of the basic properties of the catalyst on its activity ?
3) What is the influence of the particle size of mixed oxides on the reaction rate ? Is there optimum in the dependence of catalytic activity on the size of particles of mixed oxides ?

L. T. WENG (Université Catholique de Louvain, Belgium) : 1) As shown in our paper, we did not observe any activity of pure Sb_2O_4 in the selective oxidation of isobutene. But when it was mixed with Bi_2MoO_6, the selectivity to methacrolein of the latter was greatly enhanced, although almost no improvement on activity was observed. We explained these results by the improvement of the catalytic properties of Bi_2MoO_6 by the reaction of its surface with spillover oxygen emitted by Sb_2O_4.

2) The dependence of catalytic activity of oxidation on the acidic sites concentration was indeed observed in our laboratory, but it was not presented in the present paper. In fact, we have shown that there exists a linear proportionality between the Brönsted acidic sites concentration measured by TPD of NH_3 and catalytic activity in the oxygen-aided dehydration of N-ethyl formamide to propinitrile on MoO_3-Sb_2O_4 mixtures catalysts [1]. On the other hand, the catalytic activities (precisely methacrolein yields) of these catalysts in the oxidation of isobutene vary in a parallel way as those in the above dehydration [2]. Therefore, if we correlate the methacrolein yields with the Brönsted sites concentration, we can find a similar linear proportionality. Although we did not measure the acidic sites concentration of the mechanical mixtures presented in this paper, we may expect the existence of such dependence in these cases.

It is interesting to note that the paper by Abon et al. (Paper A3) in this symposium found a similar dependence in the selective oxidation of propene on MoO_3.

It seems necessary to add here that the acidic sites observed on MoO_3 surface are not the strong ones (corresponding to $\approx$ 200°C in TPD NH_3) . We speculate thus that these sites would not contribute to the total oxidation (due to strong adsorption of isobutene or propene or difficult desorption of selective oxidation products), but rather to the activation of molecular oxygen, i.e. that these sites are used as the "entry" of oxygen ions for the reduced sites on MoO_3 and keeps it in an optimum oxidation state.

We did not investigate the influence of basic properties of the catalyst on this activity. The reason is that almost no basic sites are found on MoO_3 surface.

3) Indeed, the particle size of both oxides in mechanical mixtures can influence the catalytic activity in a critical way. According to the remote control mechanism, the catalytic synergy is due to the migration of oxygen species from the Donor (producing spillover oxygen, e.g. Sb_2O_4) to the Acceptor (possessing all functions for oxidation, e.g. MoO_3). While the migration of oxygen species depends on the "contacts" between two phases. Therefore, when the particle size of one or both of two phases is changed, the number of contacts will be changed too, and consequently the catalytic activity will be changed. Theoretically, when the particle size of two oxides is comparable, the maximum of the contacts between two phases will be observed at the point (50/50). Now if we change the particle size of one of oxides, the maximum will be changed in a manner that the maximum is obtained at a point where the surface area of both oxides is balanced. The influence of particle size on catalytic activity has been shown in our previous papers [3, 4].

1. B. Zhou, S. Cieckiewcz and B. Delmon, J. Phys. Chem., 91 (1987) 5061
2. P. Ruiz, B. Zhou, M. Remy, T. Machej, F. Aoun, B. Doumain and B. Delmon, Catalysis Today, 1 (1987) 181
3. B. Delmon and P. Ruiz, React. Kinet. Catal. Lett., 35 (1987) 303
4. L. T. Weng, P. Ruiz and B. Delmon, Proc 2nd Inter. Conf. Spillover (H. K. Steinberg), University of Leipzig, GDR (1989) 39

G. Centi and F. Trifiro' (Editors), *New Developments in Selective Oxidation* 767

MoO_3 ON ANATASE AND RUTILE LOW-SURFACE-AREA TiO_2: CHARACTERIZATION AND OXIDATION OF o-XYLENE AND BUTENE-1

E. FILIPEK[1], B. GRZYBOWSKA[1], E. SERWICKA[1], Y. BARBAUX[2], J.P. BONNELLE[2] and J. GRIMBLOT[2]

[1]Institute of Catalysis and Surface Chemistry, Polish Academy of Sciences, 30-239 Kraków (Poland)

[2]Laboratoire de Catalyse, Université des Sciences et Techniques de Lille, 59655 Villeneuve d'Ascq (France)

SUMMARY

Catalysts containing 0.2 - 4 wt.% MoO_3 on anatase (AN) and rutile (RT) titania were characterized by ESR, ISS, surface potential (SP) techniques, acidity probe reactions (decomposition of isopropanol and butene-1 izomerization) and tested in butene-1 and o-xylene oxidation. On both AN and RT a MoO_x species of low acidity and high dehydrogenating properties are observed for MoO_3 content up to monolayer coverage. Above monolayer content other species of higher acidity and less strongly bound to the support appear, corresponding most probably to highly dispersed MoO_3. Their activity in the oxidation reaction is higher than that of monolayer content catalysts and of unsupported MoO_3. Results are compared with those for V_2O_5/TiO_2 system.

INTRODUCTION

It has been shown previously (ref. 1) that acid-base properties of V_2O_5/TiO_2-AN catalysts, measured by isopropanol (iPrOH) decomposition, are modified at the monolayer (mnl) content of V_2O_5, at which the optimal performance of this system in o-xylene oxidation has been observed. The iPrOH decomposition was proposed (ref. 2) as a probe for type of dispersion of V_2O_5 on TiO_2 : dehydrogenation to acetone increasing up to a maximum at a mnl content of vanadia, and suppression of dehydration to propene were suggested as an indicator of a mnl dispersion, whereas rapid increase in the propene formation as an indicator of appearance of a tridimensional oxide phase. It seemed of interest to examine properties of MoO_3 deposited on low surface area AN and RT. The MoO_3/TiO_2 system has been recently a subject of numerous studies (refs. 3-11), most of which, however, have been concerned with structural characterization of MoO_3 on high-surface area AN titania, and moreover only few with oxidation reactions on this system (refs. 9-13).

EXPERIMENTAL

Preparation of catalysts

The Mo/AN and Mo/RT catalysts were prepared by impregnation of the support with ammonium paramolybdate solution at pH=6, evaporation, drying at 120°C and calcination in flowing air at 500°C for 5 h. The Mo content, checked by atomic spectroscopy, varied between 0.2 and 4 wt.% MoO_3, which corresponded to 0.2-5 theoretical monolayers of MoO_3, calculated with the assumption that ϕ (MoO_3) = 15.4 $Å^2$ (ref. 14). No crystalline MoO_3 was found in the studied samples by XRD technique, though mechanical mixture 4 wt.% MoO_3 + TiO_2 gave clear diffraction pattern of MoO_3. AN (7.0 m^2/g) and RT (5.4 m^2/g) titania were product of Chemical Works, Police, Poland. Spectral analysis showed P, Al, Si, V, Fe impurities ($\leq$ 0.01%) in both supports; with the XPS technique P and K were seen on AN, and K, P and Al on RT surface. RT samples contained ca.7% of AN, no RT was found by XRD method in AN. Specific surface areas of the Mo/TiO_2 samples were equal within 0.5 m^2 to those of the pure supports.

Methods

(i) iPrOH decomposition and butene-1 isomerization were measured at 200°C with a pulse method, using dried helium as a carrier gas. 0.5 g of the sample and 0.5 μl and 0.5 ml pulses of iPrOH and butene,respectively, were used. AN and RT supports were only slightly active under the adopted conditions (conversion <5%). The conversion and yield of products did not change with number of pulses for Mo/AN samples and decreased for Mo/RT preparations.

(ii) ESR spectra of the samples after iPrOH test (15 pulses at 200°C) were recorded at 77 and 293 K with a SE/X Technical University Wrocław spectrometer.

(iii) ISS spectra were taken with a Leybold-Heraus LHS-10 spectrometer: $^4He^+$ ions with incident energies E_o= 0.5,1 and 2 keV were succesively used.

(iv) Surface potential (SP) was measured with a vibrating condenser method under a flow of Ar/O_2 mixture [p (O_2) = 0.05 atm] in an apparatus described in ref. 15. The reported values, V are relative to graphite reference electrode, the increase in V indicating that the surface becomes more negatively charged.

(v) Oxidation of butene-1 (1.2 vol% in air) and o-xylene (1.6 vol% in air) were tested in a fixed bed flow apparatus, the contact time being 6 and 1 s respectively.

RESULTS

Decomposition of isopropanol

Fig. 1 shows total conversion of iPrOH and amounts of acetone and propene for the first pulse, as a function of MoO_3 content for Mo/AN and Mo/RT catalysts. After treating the sample containing 4% MoO_3 on AN with concentrated ammonia, which removes the soluble Mo species decreasing MoO_3 content to ~ 0.7%, high amounts of acetone ($0.48 * 10^{-6}$ mol/m^2) and negligible of propene are observed.

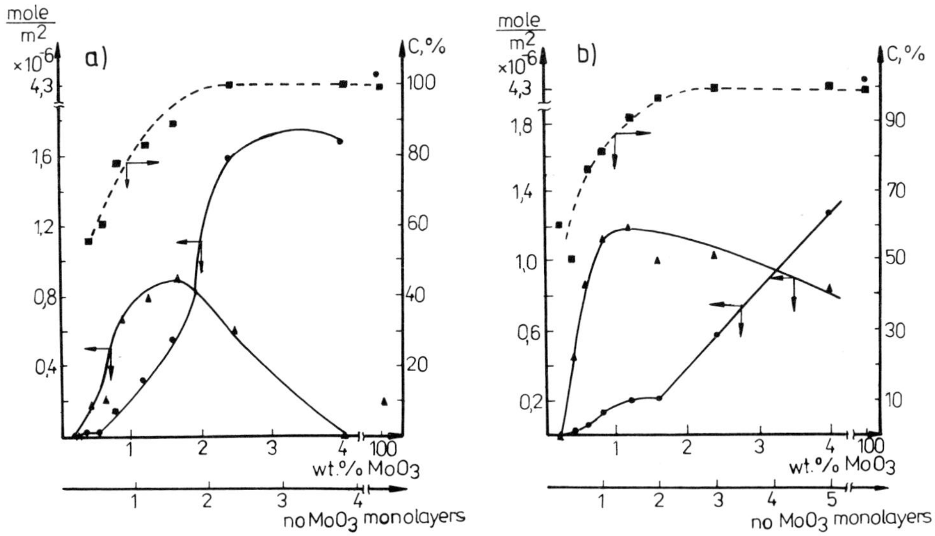

Fig. 1. Isopropanol decomposition on Mo/An (a) and Mo/RT (b) catalysts as a function of MoO_3 content; ■ - iPrOH conversion, ● - propene, ▲ - acetone.

Isomerization of butene-1

The results of butene-1 isomerization for selected samples from the two regions of the Mo content are given in Table 1.

TABLE 1

Butene-1 isomerization on MoO_3/TiO_2 catalysts

Catalyst wt.% MoO_3	Bu-1 conv. %	trans Bu-2	cis Bu-2
		10^6 mol/m^2	
0.83/AN	12	0.22	0.12
4.2/AN	88	2.42	1.80
0.84/RT	<1	0.0	0.0
4.2/RT	22	0.6	0.7

ESR spectra

ESR spectra of the catalysts after the iPrOH test reveal two types of paramagnetic centres both on Mo/AN and Mo/RT samples (Fig. 2). One at g_{av}= 1.930, denoted further A, visible at room temperature, is present as the only ESR active species in samples with low MoO_3 content (below 1 mnl on RT and 1.5 on AN).

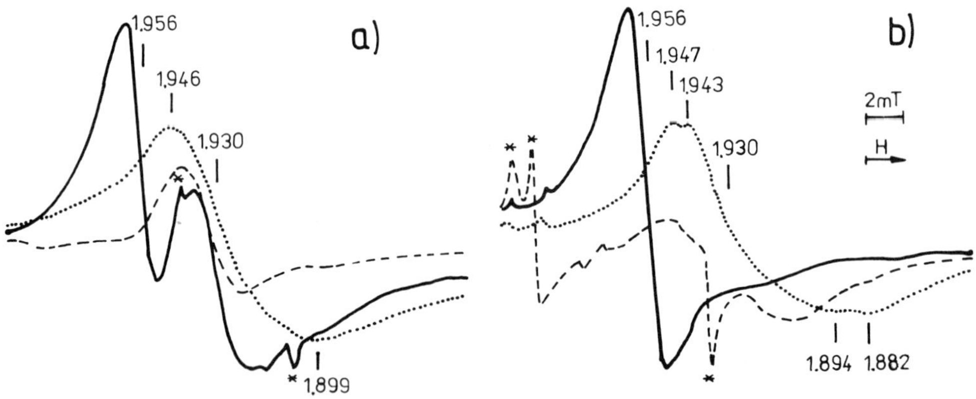

Fig. 2. ESR spectra of Mo/RT (a) and Mo/AN (b) catalysts.---; 0.8% MoO_3 recording at 293 K, —— ; 4.2% MoO_3 recording at 77 K, ...; samples 4.2% washed with ammonia after 5 min treatment with 10 Tr of iPrOH at 200°C, recording at 77 K. * signals present in the pure supports.

The other, at g_{av}= 1.956, visible only at 77 K, denoted B, appears in addition to the former at higher concentrations. Both are present already in the freshly calcined samples, albeit with lower intensities. Their ESR parameters fall in the range observed for Mo^{5+} species in oxide matrices (ref. 16). After washing with ammonia signal B disappears.

Interaction with isopropanol vapour of the washed samples resulted in an enhancement of signal A (Fig. 2 a b dotted lines). Higher concentration of the A centres resulted in line broadening, nevertheless the absence of signal B allowed more accurate determination of the ESR parameters. Mo/RT sample displayed an asymmetric absorption at $g_{\perp} = 1.946$, $g_{\parallel} = 1.899$. On the Mo/AN sample signal A consisted in fact of two overlapping absorptions of similar parameters and temperature dependence, at $g_{\perp} = 1.947$, $g_{\parallel} = 1.894$, and $g_{\perp} = 1.943$, $g_{\parallel} = 1.882$ respectively.

ESR characteristics of signal A resembles that found for molybdenyl species in strongly distorted, coordinatively unsaturated octahedra (ref. 17). Signal B may be tentatively assigned to quasi crystalline MoO_3 (ref. 16).

ISS measurements

The results of ISS measurements on Mo/AN samples are shown in Fig.3 and 4. Two different regions of the Mo content can be observed (Fig.3).

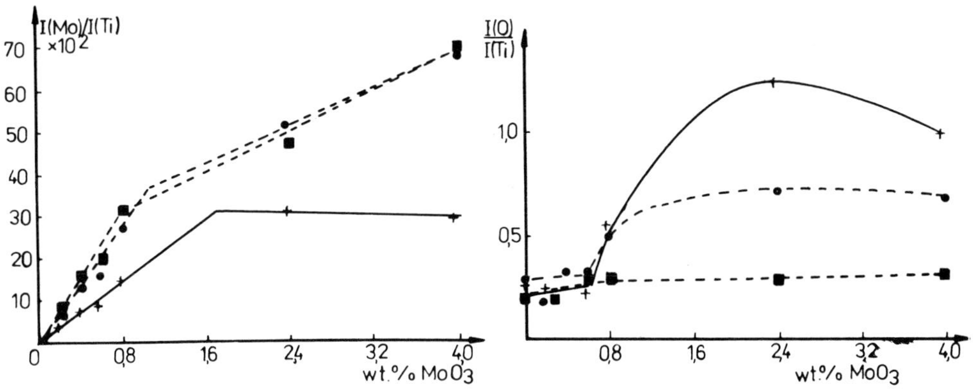

Fig. 3. Relative ISS intensities in Mo/AN catalysts as a function of MoO_3 content; E_o = 0.5 keV (+), 1 keV (●), 2 keV (■).

In the first of them up to ca. 1 mnl of MoO_3 the I_{Mo}/I_{Ti} ratio increases practically linearly with the Mo content indicating gradual covering of TiO_2 with Mo species and then remains constant with the further increase of Mo loading for E_o= 0.5 keV, or it increases further but more slowly for E_o= 1 and 2 keV. At the same time the I_o/I_{Ti} ratio is practically constant irrespectively of the incident beam energy in the first concentration range. In the second region above 1 mnl this ratio increases distinctly for low energy (0.5 eV): its changes with the Mo content become levelled off as the energy increases to 2 eV. These results indicate that the species in the second region are more densely covered with oxygen than those in the first region. Fig. 4 presents the depth profile of Mo at 1 keV for samples of different Mo content. The rate of Mo removal, estimated by the value $\Delta I = I_{10min} - I_{100min}$ is lower for samples in the mnl region of concentration as compared with this rate for samples above the mnl coverage, which indicates that the Mo species in the submnl range are more strongly bound.

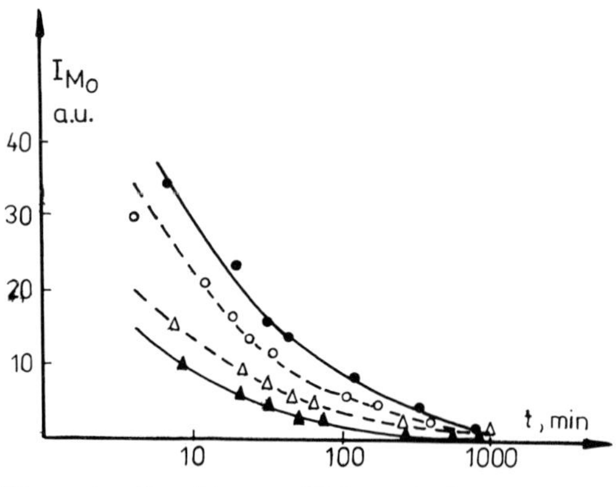

Fig. 4. Depth profile of Mo at E_o = 1 keV for Mo/AN catalysts of different Mo content. ▲ - 0.42 % MoO_3 (8), Δ - 0.62% MoO_3 (10.8), o - 2.49% MoO_3 (20.2), ● - 4.15% MoO_3 (20.4). Numbers in brackets give $\Delta I = I_{10} - I_{100min}$.

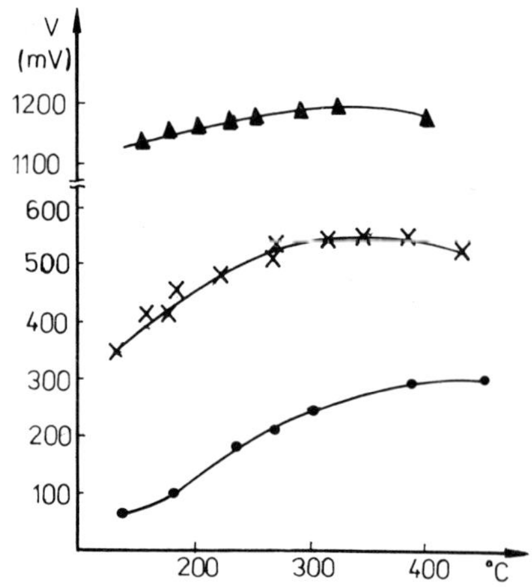

Fig. 5. Variations of surface potential, V with temperature. ● - AN-TiO_2, x - 1.25 % MoO_3/AN, ▲ - 0.8 % V_2O_5/AN.

Surface potential measurements

Fig. 5 presents variations of SP with temperature for pure AN, for the sample 1.25 % MoO_3/AN and for V_2O_5/AN sample of the mnl content prepared by impregnation (ref. 1).The course of the changes for Mo/AN sample resembles that of the pure support, the values of V at 400°C being intermediate between that of AN and of pure MoO_3 (1.4 V). This result indicates that only a fraction of TiO_2 surface is covered with MoO_3. By analogy with other systems (ref. 18) the increase in V can by ascribed to conversion of adsorbed oxygen species from O_2^- to more charged O^- and O^{2-}. The properties of AN are masked for V_2O_5/AN sample, which indicates better coverage with vanadium.

<u>Oxidation of butene-1 and o-xylene</u>

The data for selected samples from the two regions of Mo content are presented in Table 2 in terms of total hydrocarbon conversions and selectivities to main products. Oxidation of both hydrocarbons show similar trends for the studied catalysts: a) the samples of the Mo content close to 1 mnl are less active than those containing several monolayers, b) activity of Mo/AN samples is higher than that of the Mo/RT samples, this effect being more marked at higher Mo content and in butene-1 oxidation, c) selectivities to the anhydrides (maleic and phthalic) at low Mo content and at comparable conversions are lower for supported samples than for pure MoO_3.

TABLE 2
Oxidation of butene-1 and o-xylene on MoO_3/TiO_2 system.

Catalyst, wt.% MoO_3/TiO_2	Butene oxidn. (400°C) conv. %	selectivity*, % BD	MA	$CO+CO_2$	O-xylene oxdn. (410°C) conv. %	selectivity, % TA	PA	$CO+CO_2$
0.6/AN	--	--	--	--	15.8	73.4	<0.1	26.6
0.8/AN	13.0	28.0	2.4	47.0	--	--	--	--
1.2/AN	--	--	--	--	15.7	77.0	0.6	22.3
4.0/AN	96.0	10.5	50.5	20.0	97.0	45.2	11.0	38.4
0.8/RT	8.0	25.0	6.0	37.6	--	--	--	--
1.2/RT	--	--	--	--	15.2	79.0	<0.1	20.6
4.0/RT	17.4	31.0	11.6	29.4	36.2	67.7	4.1	28.2
100(MoO_3)	7.2	26.0	10.0	32.7	14.0	77.5	1.0	22.5

BD-butadiene,MA-maleic anhydride,TA-tolualdehyde,PA-phtalic anhydride.*-other products: acetaldehyde (4-8%), acetic acid (3-9%), crotonaldehyde (3-5%), crotonic acid (2-5%), furan (2-4%), butyraldehyde (1-2%), butyric acid (2-4%), acrolein (1-3%). acrylic acid (2-4%).

DISCUSSION

The results described above show two regions of the MoO_3 concentration on TiO_2, in which different Mo-O species are present. These species, present on both AN and RT titania, exhibit different behaviour in acidity probe reactions and different activity in oxidation of o-xylene and butene-1. They are characterized by different ESR Mo^{5+} signals and different strength of interaction with the support, as indicated by the ISS measurements and solubility with ammonia. The presence of the Ti signal in ISS spectra in the whole range of MoO_3 studied and the SP data indicate moreover, that they do not cover completely the support even at Mo content exceeding the theoretical mnl coverage. In the first region corresponding to submnl coverage the suppression of acidic properties with respect to bulk MoO_3 is observed as shown by low yields of propene in iPrOH dehydration and low extent of butene-1 isomerization. At the same time the rate of acetone formation increases linearly indicating gradual covering of TiO_2 with the species of high dehydrogenating properties up to ca 1 mnl content of MoO_3, and then decreases. The nature of the species present in this region cannot be given unanimously on the basis of the present results: the ESR signal suggests the presence of reduced molybdenyl species. Mo=O species were in fact observed by Busca at low loading of Mo on MoO_3/high area AN (ref. 7). Above the mnl content additional new species appear which exhibit properties close to those of MoO_3 (high dehydrating rate of isopropanol, ESR signal), though no crystalline MoO_3 is observed. These species can be of the form of amorphous or disordered MoO_3 as suggested in refs. 4,8. The increase in Mo/Ti ratio at higher energy in ISS spectra with the Mo content in this region indicate that they cover a part of unocupied TiO_2 surface. At the same time the decrease of acetone formation at higher Mo content and the fact that after removing these species by ammonia the high amount of acetone is observed, suggest that a fraction of the paracrystalline MoO_3 covers also the mnl species, in agreement with the model proposed by Bond (ref. 4). The catalytic activity in oxidation reactions of the species in the supramnl region is higher than that of those of the mnl Mo content and higher than that of bulk MoO_3. At higher conversions in the supramnl region the acidic products (maleic and phthalic anhydrides) appear in oxidation of butene-1 and o-xylene respectively. Higher activity of the AN samples as compared with the RT ones can be related to the higher amount of the MoO_3-like species on the surface of the catalysts, suggested by the higher amounts of propene in iPrOH decomposition on Mo/AN samples. Comparison of the data for MoO_3/TiO_2 catalysts with previous ones for V_2O_5/TiO_2 system (ref. 1) show some differences. Firstly the enhancement of dehydrogenating properties in isopropanol decomposition and decrease of acidity with respect to pure V_2O_5 was observed for V_2O_5/TiO_2 only for the AN modification, whereas similar behaviour in this reaction has been shown for MoO_3 deposited on both AN and RT TiO_2. This difference can be due to different RT supports used in V/RT and Mo/RT samples, RT used for vanadia deposition containing some Si on the surface and no K, whereas that in the present study had no Si but K and moreover contained some AN. In fact similar catalytic properties of V_2O_5 deposited on AN and RT were reported by Trifiro and coworkers (ref. 19). Limiting the comparison of the Mo/TiO_2 and V_2O_5/TiO_2 catalysts to AN support (the same in the both studies) we observe that the catalytic activity in oxidation of o-xylene is not ammeliorated with respect to bulk oxide in the mnl region of Mo content on AN, in contrast to V/AN system, in spite of the similar modification of the acid-base properties for both V and Mo on AN. One of the tentative explanation of this fact could be that VO_x units at mnl content cover better the AN surface forming chains, whereas MoO_x units are isolated. Better coverage of TiO_2 in case of vanadia deposition as compared with molybdena was in fact observed by Bond (ref. 4) and confirmed in this paper by SP studies. One can envisage then better transport of oxygen species and of charge in the case of V/AN catalysts and hence higher activity as compared with Mo/AN at the mnl coverage.

Acknowledgements

The help of Mr J. Janas and Mr W. Janiszewski in performing oxidations of butene-1 and o-xylene and of Mr R Dula in recording ESR spectra is gratefully acknowledged.

REFERENCES

1 I. Gąsior, M. Gąsior and B. Grzybowska, Appl. Catal., 10 (1984) 87-100.

2 B. Grzybowska-Świerkosz, Catalysis by Acids and Bases, eds., B. Imelik, C. Naccache, G. Coudurier, Y. Ben Taarit, J.C. Vedrine, Studies in Surface Science and Catalysis, 20 (1985) 45-56.

3 K.Y.S. Ng and E. Gulari, J. Catal., 92 (1985) 340-354.

4 G.C. Bond, S. Flamerz and L. Van Wijk, Catal. Today, 1 (1987) 229-243.

5 M.I. Zaki, B. Vielhaber and H. Knozinger, J. Phys. Chem., 90 (1986) 3176-3183.

6 K. Segawa, Du Soung Kim, Y. Kurusu and I.E. Wachs, Proc. 9th Int. Congr. on Catalysis, 4 (1988) 1960-1967. eds. M.J. Phillips and M. Ternan.

7 G. Ramis, G. Busca and V. Lorenzelli, Z.Phys. Chem. N.F., 153 (1987) 189-200.

8 T. Machej, B. Doumain, B. Yasse and B. Delmon, J. Chem. Soc. Faraday Trans I, 84 (1988) 3905-3916.

9 T. Ono, Y. Nakagawa, H. Miyata and Y. Kubokawa, Bull. Chem. Soc. Jap., 57 (1984) 1205-1210.

10 A.J. van Hengstum Thesis, Twente University of Technology, Enschede, The Netherlands (1984) p.35.

11 Y.C. Liu, G.L. Griffin, S.S. Chan and I.E. Wachs, J. Catal., 94 (1985) 108-119.

12 M. Akimoto and E. Echigoya, J. Catal., 31 (1973) 278-286.

13 M. Ai, Bull. Chem. Soc. Jap., 49(1976) 278-286.

14 T. Fransen, P.C. van Berge and P. Mars, Prep. of Catalysts eds. B. Delmon, P.A. Jacobs and G. Poncelet, Elsevier 1976 405-416.

15 Y. Barbaux, J.P. Bonnelle and J.P. Beaufils, J. Chem. Phys., 73 (1976) 25-31.

16 E. Serwicka, J. Solid State Chem., 51 (1984) 300-306 and the references therein.

17 G. Busca and L. Marchetti, J. Chem. Res. (S), (1986) 174-175.

18 Y. Barbaux, A. Elamrani and J.P. Bonelle, Catal. Today, 1 (1987) 147-156 and references therein.

19 F. Cavani, G. Centi, E. Foresti and F. Trifiro, J. Chem. Soc. Faraday Trans.I, 84 (1988) 237-254.

D.SCHOLL (Alusuisse AG, Neuhausen, Switzerland)

1. What is the transition temperature anatase-rutile?
2. What is the detection method and "real purity" of final catalyst?
3. Rutile or anatase in oxidation reactions under view of selectivity?
4. Laboratory equipment and reactor design.

B. GRZYBOWSKA (Institute of Catalysis, Polish Academy of Sciences, Kraków, Poland): 1/ The transition temperature for the studied MoO_3/anatase catalysts was 550-580°C depending on MoO_3 concentration. Generally the transition temperature for V_2O_5 and MoO_3 on anatase catalyst depends on concentration of supported oxide, presence of impurities (e.g. potassium lowers it considerably), specific surface area of titania and atmosphere of calcination (oxidizing or reducing). 2/ Presence of rutile in final MoO_3/anatase catalysts was checked by XRD method 3/ The effect of the titania modification on selectivity of titania supported catalysts in oxidation reactions is still a matter of discussion: the available data indicate that it may depend on: a/ purity of titania, b/ nature of the oxidized molecule c/ nature of the dispersed oxide. Higher selectivities to phthalic anhydride in o-xylene oxidation on anatase supported vanadia (ref. 1,2) are questioned by other authors (ref.3,4) who claim the same selectivities for anatase and rutile supported catalysts and ascribe the earlier observed differences to the presence of impurities such as K and P on anatase and Al and Si on rutile. It should be however noted that the selectivities values given in refs. 3 and 4 ($\leqslant$ 70%) are lower than those reported in other papers for V_2O_5/anatase catalysts (76-80%). On the other hand for the same catalysts the effect of the titania modification may be different for different molecules: for instance oxidation of methoxytoluene is less effected by the type of TiO_2 than oxidation of toluene or p-tertbutyltoluene (ref.5) or of o-xylene (ref. 1), or methanol oxidation to formaldehyde depends more strongly on crystal structure of titania than oxidation of o-xylene (ref. 3). 4/ Catalytic measurements were performed in a flow system the reactor (150 x 15 mm) being connected on-line with gas chromatographs (ref.6).

1, M.Gąsior, B.Grzybowska, Appl. Catal. 10 (1984) 87.
2. Y,Saleh, I.E.Wachs, S.S.Chan, C.C.Cheresich, ACS Symposium Div. Petroleum Chem. 31 (1986).
3. A.J. van Hengstum, J.G. van Ommen, H.Bosh, P.J.Gellings, Proc. 8th Int. Congr. Catalysis, Berlin 4 (1984) 297-307.
4. F.Cavani, G.Centi, E.Foresti, F.Trifiro, J.Chem.Soc. Faraday Trans. I, 84 (1988) 237-284.
5. M.Czerwenka, B.Grzybowska, M.Gąsior, Bull. Polish Acad. Sci. Chemistry, 35 (1987) 353-363.
6. M.Gąsior, B.Grzybowska, Bull. Acad. Polon Scic ser. sci. chim. 27, 1979 (835-841).

J. VEDRINE (Institut de Catalyse, CNRS, Villeurbanne, France): I have appreciated very much your way to compare catalytic properties of isolated molybdenum-oxygen species, monolayer molybdate and bulk MoO_3 crystallities on TiO_2 support, which clearly show that they are drastically different. Comparison with V_2O_5 and the use of electric type feature of oxygen species by surface potential measurements are interesting. One conclusion is that V_2O_5 spreads better on the surface than MoO_3 which seems reasonable. Do you have better proof of it from ISS and XPS data not given for V_2O_5 in your paper.

You emphasize the influence of K or Si impurity on the TiO_2 surface rather than that of the difference in structure of anatase against rutile. This is rather original conclusion and at variance with Bordes and Courtine conclusion on V_2O_5/TiO_2 due to structural fitting feature between the active phase and the support. Could you comment on this?

B.GRZYBOWSKA (Institute of Catalysis, Polish Academy of Sciences, Kraków, Poland): We have not performed the ISS measurements on the V_2O_5/TiO_2 system mainly because of the resolution problems: with helium as an ion source it is not possible with the spectrometer we used to separate V and Ti signals as their atomic masses are too close. The XPS data for these catalysts (ref.1) show, however, a good agreement between the thickness of a vanadia layer determined by XPS and the value of this thickness calculated assuming full coverage of titania with V_2O_5.

The origin of the difference between vanadia catalysts deposited on anatase or rutile titania is still not understood and even questioned by some authors (ref. 19 of the paper). The role of impurities on titania surface has been suggested to explain this difference (ref. 2), though not proved experimentally as yet. The conclusion of Bordes and Courtine is not the only one to explain the modification of vanadia properties in contact with anatase titania: other authors (ref.3,4) ascribe this modification to formation of a monolayer of V-O polyedra on anatase surface, this monolayer having different properties than bulk V_2O_5. According to the results obtained in our Institute (ref.5) both effects i.e. formation of a monolayer at low concentrations of vanadia and morphology of vanadia phase at higher concentrations determine the properties of vanadia-titania catalysts.

1. J.Mendialdua, Y.Barbaux, L.Gengembre, J.P.Bonnelle, B.Grzybowska, M.Gąsior, Bull. Polish Acad. Sci. Chemistry, 35 (1987) 213-220.
2. A.J. van Hengstum, J.G. van Omen, H.Bosh and P.J.Gellings Appl. Catal. 8 (1983) 369-382.
3. G.C.Bond, J.Catal., 116 (1989) 531-539, and references therein.
4. M.Gąsior, I.Gąsior, B.Grzybowska, Appl. Catal., 10 (1984) 87-100.
5. M.Gąsior, B.Grzybowska, Vanadia catalysts for processes of oxidation of aromatic hydrocarbons, Polish Scientific Publishers Warsaw-Cracow, 1984, p. 133-159.

G. Centi and F. Trifiro' (Editors), *New Developments in Selective Oxidation*
1990 Elsevier Science Publishers B.V., Amsterdam — Printed in The Netherlands

THE ROLE OF THE DISTRIBUTION OF DOUBLE BONDED OXYGEN ON SURFACE OF V-Mo-O CATALYSTS IN SELECTIVE OXIDATION OF BENZENE TO MALEIC ANHYDRIDE

M. Najbar[1], W. Wal[2], J. Chrząszcz

[1]Faculty of Chemistry, Jagiellonian University, Karasia 3, Kraków, Poland

[2]Institute of Industrial Chemistry, Rydygiera 8, Warszawa, Poland

SUMMARY

The oxidation of the benzene by air at temperature range 641-683 K over V-Mo-O catalysts was investigated. O_2^- ions adsorbed at the catalyst surface were considered to be responsible for selective benzene oxidation to maleic anhydride (MA) and atomic oxygen ions as causing total benzene oxidation.

The distribution of doubly bonded oxygen ions on catalyst surfaces was suggested to be a main factor determining the species of adsorbed oxygen and thus a selectivity of benzene oxidation to MA.

INTRODUCTION

(001) V_2O_5 and (010) MoO_3 planes, containing doubly bonded oxygen, have been widely discussed in literature as active in benzene oxidation to MA. Grussenmeyer (ref. 1) have postulated that the strength of a double bond in solid solution of MoO_3 in V_2O_5, being function of MoO_3 concentration, is a factor controlling their activity in benzene oxidation to MA. Grussenmeyer (ref. 1) has also suggested that (010) MoO_3 plane, showing usually very poor activity in benzene oxidation to MA, become active if some Mo-cations were replaced by vanadium ones, playing the role of the sites of adsorption. The previous investigations of this laboratory (ref. 2) showed that the epitaxial layers of MoO_3, formed on the best developed faces of the crystals of V-Mo-O system during the reduction, revealed particularly high selectivity in benzene oxidation to MA. Waugh (refs. 3-4) showed undebatably that oxygen adsorbed in the molecular form on the surface of vanadia is responsible for selective benzene oxidation to MA.

The aim of this paper was to elucidate a role of the double bonded oxygen distribution on the surface of V-Mo-O catalysts in the formation of O_2^- ions and, thus, in the selective benzene oxidation.

EXPERIMENTAL

Supported catalysts were obtained by evaporation of aqueous

solutions of ammonium methavanadate, molibdic acid and oxalic acid on a sintered low surface area corrundum support. The dry catalysts were calcined at 590-610 K in air atmosphere. The BET surface areas of all investigated catalysts were very close. Activity and selectivity of the catalysts were studied using a flow reactor of the diameter of 21 mm. The length of the catalyst bed was 30 cm. Benzene was oxidized in mixture containing ca 40 g benzene /1 N m^3 air at catalyst load of 120 g benzene (catalyst liter)$^{-1}$(hour)$^{-1}$.

The reaction was performed within the range of temperature 641-683 K. The stationary state in benzene oxidation was achieved not later then after two hours of a catalyst use. The content of benzene in the effluent gases was determined by gas chromatography and the concentration of MA by the titration with NaOH of its water solution. The phases present in the catalysts were determined by X-ray powder diffraction using DRON-3 diffractometer with copper filtered (Ni) radiation (λ = 1.5418 Å). The degree of the vanadium reduction ($V^{4+}/V^{4+}+V^{5+}$) by manganometric titration assuming that the reduction results in V^{4+} ions formation.

RESULTS AND DISCUSSION

The characterization of the catalysts is presented in Table.

TABLE

Characterization of the catalysts

MoO_3- content [mole %]	Phases	Degree of V reduction [%]	C_6H_6- conversion [%]	Selectivity at 673K [%]
0	V_2O_5	42	94	27
10	- -	21	96	54
25	- -	21	91	35
35	- -	22	97	25
45	V_2O_5+I.C.	36	91	35
60	I.C.	61	89	34
85	I.C. + MoO_3	64	32	68
95	MoO_3	75	10	75
97	- -	15	19	69
100	- -	-	41	1

It is interesting to notice that catalyst with 95 mole % MoO_3 and the highest degree of vanadium reduction, determined by X-ray analysis as MoO_3 phase, shows the lowest activity in benzene oxidation and the highest selectivity to MA while the catalysts with the structure of V_2O_5 and rather low degree of vanadium reduction show the highest activity in the benzene oxidation. Their selectivity depends strongly on the molybdena content and achieves the highest value for the catalyst containing 10 mole % MoO_3.

In Fig 1a changes of maleic anhydride yield with temperature are presented for particular catalysts. As is easy to notice, the values of maleic anhydride yield for the catalysts containing 0-35 mole % MoO_3 pass through maximum within the investigated range of temperatures. On the other hand, the MA yield of the catalysts with molybdena content 45-100 mole % MoO_3 continuously increases with temperature. To explain the maximum on the curve "α_{MA} - T" it should be assumed that increase of the rate of MA formation over particular active centers with temperature is accompanied by the processes leading to the decrease of the number of these centers, occurring with higher energy of activation then the process of the MA formation. The continuous increase of the of maleic anhydride yield with temperature shows that the diminishing of the number of the active centers, if any, is not big enough to influence the direction of the changes of the value of α_{MA}. In the range of temperatures 653-683 K the sharpest increase of the MA yield with temperature was observed for the catalyst with 95 and 97 mole % MoO_3, containing only MoO_3 phase. These increase of α_{MA} may be treated as caused mainly by the increase of the rate of selective benzene oxidation. The less sharp increase of α_{MA}, within above mentioned range of temperatures, for the catalysts containing 85, 60 and 45 mole % of MoO_3 suggests diminishing with temperature of the number of the centers responsible for the selective benzene oxidation .

The dependence of the benzene conversion and of selectivity to MA at 673 K on catalyst composition is illustrated in Fig. 1b. As it is seen, the selectivity of the pure V_2O_5 catalyst is relatively low and it increases rapidly when 10 mole % MoO_3 is added. Further addition of MoO_3 causes decrease of the selectivity and then for the catalysts containing above 35 mole % MoO_3 its repeated increase. Maximum is attained for the catalyst with 95 mole % MoO_3. Further increase of molybdenum content leads to the decrease of the selectivity. The value of the selectivity of the catalyst containing 100 % MoO_3 is close to zero. It is interesting to notice that selectivity of the catalysts containing more then 45 mole % MoO_3 changes with the molybdenum content in opposite direction then the activity.

It seems that observed dependence of the catalytic properties on the temperature and catalyst composition can be explained by the evolution of the structures of the V-Mo-O system occurring

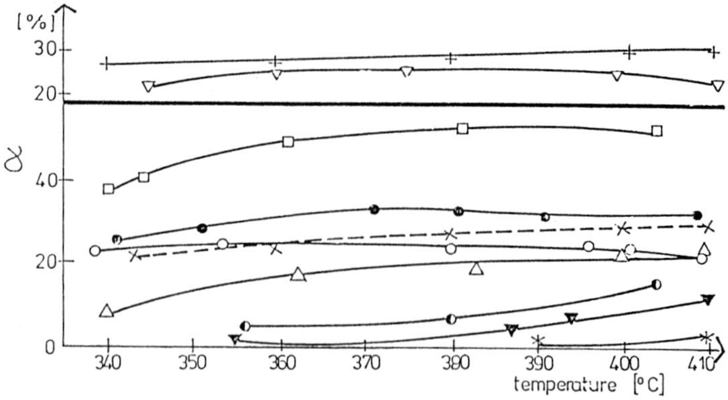

Fig. 1a. Dependece of the MA yield on temperature ; ▽ -100 % V_2O_5, □ -10 mole % MoO_3, ● -25 mole % MoO_3, ○ -35 mole % MoO_3, ✕ -45 mole % MoO_3, + -60 mole % MoO_3, △ -85 mole % MoO_3, ▼-95 mole % MoO_3, ◐ -97 mole % MoO_3, ✱ -100 % MoO_3.

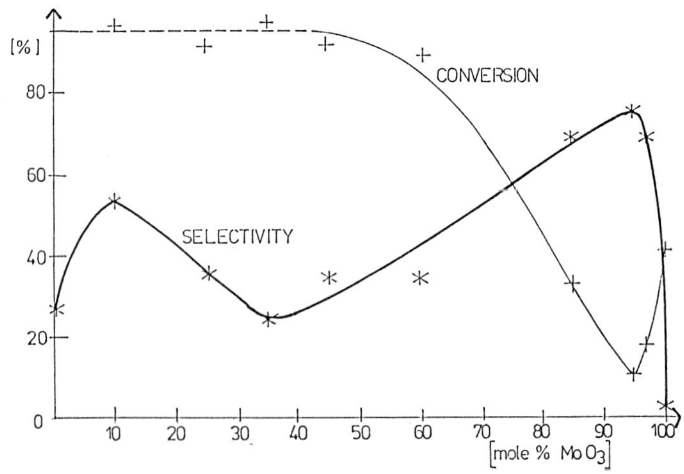

Fig. 1b. Dependence of the benzene conversion and selectivity to maleic anhydride at 673 K on the catalyst composition.

with the increase of the MoO_3 content. It should be remembered that following phases could exist in V-Mo-O catalysts in the conditions of the benzene oxidation: orthorhombic V_2O_5 and solid solution of MoO_3 in V_2O_5 (s.s.) (ref. 6), monoclinic solid solution $(Mo_{0.7}V_{0.3})O_5$ (ref.6) orthorhombic I.C. (ref. 5), and orthorhombic MoO_3 (ref. 7).

In agreement with Andersson's considerations (ref. 8), the oxygen atoms on (100) and (010) planes of V_2O_5 are useless in benzene oxidation as they occur in the form of OH groups and (001) plane containing double bonded oxygen atoms and free adsorption sites should be considered as active in this process. In Fig. 2 the structures of V_2O_5 and solid solution of MoO_3 in V_2O_5, of $(Mo_{0.3}V_{0.7})_2O_5$, of V_2MoO_8 as well as of MoO_3 have been shown. The double bond was marked by showing the distortion of the metal atoms from the centers of octahedra toward (001) surfaces of V_2O_5, of $(Mo_{0.3}V_{0.7})O_5$, as well as of V_2MoO_8 and toward (010) surface of MoO_3. As can be noticed, in all discussed structures the metal atoms in pairs of octahedra sharing edges are displaced in reverse directions. The difference between orthorhombic V_2O_5 or s.s. and monoclinic $(Mo_{0.3}V_{0.7})_2O_5$ is connected with the distortions of the metal atoms within ReO_3-type slabs. In V_2O_5 structure all metal atoms within these slabs are displaced in the same direction while in $(Mo_{0.3}V_{0.7})_2O_5$ they are shifted pair wise in opposite directions. Thus, on (001) V_2O_5 surface within ReO_3 slabs either all metal atoms play the role of the adsorption sites or all are blocked by the doubly bonded oxygen atoms. On the contrary, within ReO_3 slabs on (001) $(Mo_{0.3}V_{0.7})_2O_5$ surface the single rows of the adsorption sites are adjusted to the single rows of the metal atoms blocked by doubly bonded oxygen atoms. In V_2MoO_8 structure the slabs of ReO_3-type have thickness of three octahedra but interconnection of the slabs remains the same. The displacement of the metal atoms within the middle octahedra of each slab goes in both directions. In effect, there are equal amounts of the rows of the single and of the double adsorption sites on (001) I.C. surface. In MoO_3 structure all metal atoms on (010) surface are blocked by doubly bonded oxygen atoms and strongly reducing conditions would be necessary to create adsorption sites. The rows of the pairs of adsorption sites form more comfortable conditions for the dissociative oxygen adsorption then rows of the single adsorption sites. Geometrical limitations for the dissociative adsorption on single row sites cause increase of the participation

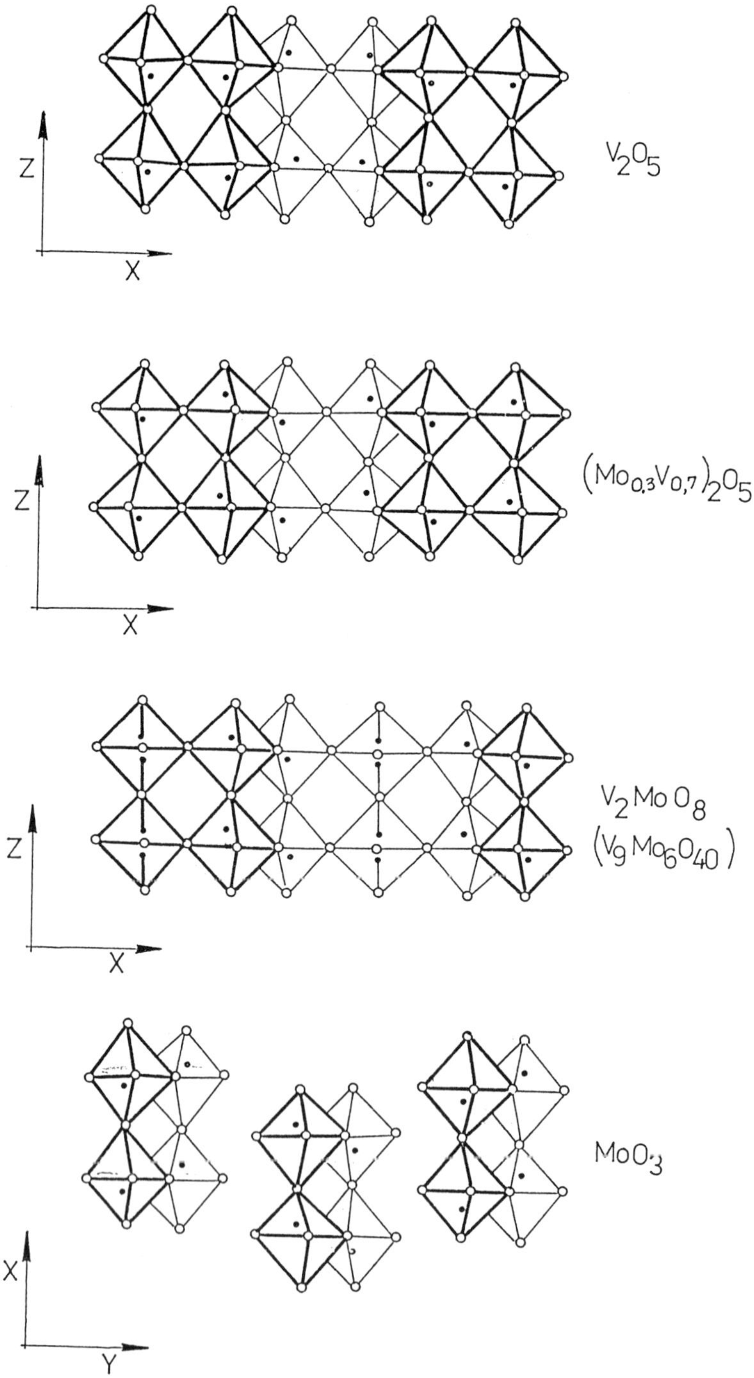

Fig. 2. Evolution of structures with the increase of Mo-content

of molecular species in oxygen adsorption. Therefore, the highest selectivity would be expected for $(Mo_{0.3}V_{0.7})_2O_5$, lower for V_2MoO_8 and the lowest for V_2O_5. The selective benzene oxidation would not be expected over stoichiometric MoO_3. The experimental dependence of the selectivity on molybdenum content (Fig. 1b) showing maximum for 10 mole % of MoO_3 is not contradictory to above considerations because catalyst crystals undergoing some reduction in the described conditions (ref. 8) become covered by epitaxial layers of the phases enriched in Mo (refs. 2, 10). In such a way on the (001) s.s. plane the epitaxial layer of $(Mo_{0.3}V_{0.7})_2O_5$ exposing (001) surface, on (001) $(Mo_{0.3}V_{0.7})_2O_5$ epitaxial layer of V_2MoO_8 exposing (001) surface, and on (001) V_2MoO_8 epitaxial layer of MoO_3 exposing (010) surface are formed (refs. 2, 10) as initial result of the phase segregation. The last step of the segregation in V-Mo-O crystals is the formation of the MoO_3 epitaxial layers preventing further reduction (ref. 10). The lower molybdena content the slower is formation of MoO_3 layers in determined reducing condition. Therefore, catalysts containing up to 35 mole % MoO_3 undergo continuous changes with temperature, resulting in the decrease of the amount of the sites responsible for maleic anhydride formation, while the catalyst with higher molybdenum content become stable at lower temperatures of the investigated range. The reduction of the crystals of pure vanadia causes diminishing of the number of the active sites due to decrease of the surface concentration of the doubly bonded oxygen atoms. Simultaneous increase of the selectivity and decrease of benzene conversion with the molybdenum content, observed for the catalysts containing more then 45 mole % MoO_3, can be explain by the increasing participation of (010) MoO_3 plane in the formation of the active catalyst surface. In agreement with Grussenmeyer (ref. 1) V-atoms incorporated in (010) MoO_3 surface play a role of adsorption sites (ref. 1). Well dispersed V-sites able to adsorb oxygen in the form of molecular ions are considered as responsible for benzene oxidation to MA. Although the solubility of V_2O_5 in MoO_3 is rather low (ref. 11) the concentration of vanadium on (010) MoO_3 surface may be quite high (ref. 1). This is because structural limitations of the solubility in the bulk of MoO_3 crystals are not valid at the surface (ref. 1). In the investigated range of temperatures MA is practically not formed over pure molybdena (Fig. 1a). This means that the creation of its substantial amount needs the presence of the V-adsorption sites. The observed activity of pure MoO_3 catalyst in nonselective benzene oxidation seems to be connected with double bonded oxygen at (010) MoO_3 surface. This oxygen is more labile in MoO_3 then in V_2O_5 because double bond in MoO_3 (1.67 Å) (ref. 7) is weaker then in V_2O_5 (1.58 Å) (ref. 6). The addition of 3 mole % vanadia causes formation of a solid solution of V_2O_5 in MoO_3. A low degree of reduction of V (Table) suggests, in agreement with Grussenmeyer (ref. 1), that vanadium is incorporated mainly into the bulk of MoO_3 crystals with formation of oxygen bridges between the layers of octahedra sharing the edges. Such bridging does not exist, of course, at (010) MoO_3 surface which can contain five coordinated V^{4+} ions. The well dispersed V^{4+} ions on the (010) MoO_3 surface of crystals containing 3 mole % V_2O_5 are responsible for substantial increase of α_{MA}. Simultaneous decrease of the amount of the double bonded oxygen atoms on (010) MoO_3 surface and increase of the strength of double bond (ref. 12) causes decrease of the total activity. Further addition of vanadia up to 5 mole % MoO_3 leading to the sharp increase of the degree of reduction of the catalyst suggests a big increase of the total amount of the V-sites. The observed decrease of the maleic anhydride yield suggests diminishing of the amount of the V-sites able to oxygen adsorption in the molecular form.. It seems, therefore, that surface concentration of V-sites is high enough to enable atomic adsorption of the large part of oxygen. The observed decrease of benzene conversion is connected mainly with the increase of the double bond strength.

REFERENCES

1 J. Grussenmeyer, Thesis, Lyon, 1978.
2. M. Najbar, Proc. 8 th Int. Congress Catal., Berlin, 1984, pp. 323-332.
3 J. Lucas, D. Vandervell and K.C. Waugh, J. Chem. Soc., Faraday Trans. 1, 77 (1981) 15, 77 (1981) 31.
4 R.W. Petts and K.C. Waugh, J. Chem. Soc., Faraday Trans. 1, 78 (1982) 803-815.
5 H.A. Eick and L. Kihlborg, Acta Chem. Scand., 20 (1966) 1658-1666.
6 L. Kihlborg, Acta Chem. Scand., 21 (1967) 2496-2502.
7 L. Kihlborg, Arkiv Kemi, 21 (1963) 357-364.
8 A. Andersson, J. Solid. State Chem., 42 (1982) 263-275.
9 A. Bielański, M. Najbar, J. Chrząszcz, W. Wal, in B. Delmon and G. Froment (Eds), Studies in Surface Science and Catalysis, Elsevier, Amsterdam, 1980, pp. 127-140.
10 M. Najbar, J. Chem. Soc., Faraday Trans. 1, 82 (1986) 1673-1680.
11 R. Khulbe, R. Mann and A. Manoogien, J. Chem. Phys., 60 (1974) 12.
12 W. Wal, to be published

J. VEDRINE (Institut de Recherches Catalyse-Villeurbanne, France):
You have shown nicely how the distribution of Mo and V (therefore of M=O bonds) influences catalytic properties particularly selectivity. Part of your assignment involves O_2^- type species identified by ESR. This conclusion is rather surprising to me since such species have been observed to result in total oxidation. Did you use ^{17}O (for ESR) and ^{18}O labelling to bring more proofs to your statements and could you give us the variation of ESR O_2^- signal intensity versus Mo-content (since you only gave V^{4+} signal intensity changes)?. O^- should also be detected by ESR and identify using ^{17}O labeling. Did you observed such ESR signal?
Moreover, doubled bond M=O species should give total oxidation while M-O-Mo species may result in selective or mild oxidation . Could you comment on this ?

M. NAJBAR (Jagiellonian University, Kraków, Poland) : Waugh has shown that O_2^- form is responsible for selective benzene oxidation on V_2O_5-MoO_3 catalysts (citation 5-7 in presented paper).
We have only notice that the distribution of the double bonded oxygen on the basal faces of the crystals of V_2O_5-MoO_3 catalysts determine their selectivity in the same way in which the concentration of O_2^- species is expected to be determine (for the geometrical reason). We dont use ESR method in our investigation.

G. Centi and F. Trifiro' (Editors), *New Developments in Selective Oxidation*

Effect of Promoters on Activity and Selectivity of Benzene Oxidation on V_2O_5 Catalysts: Study by Well-Defined V_2O_5 Catalysts

Atsushi Satsuma, Michiatsu Nakata, Shun-ichi Iwasaki, Tadashi Hattori and Yuichi Murakami

Department of Synthetic Chemistry, Faculty of Engineering, Nagoya University, Furo-cho, Chikusa-ku, Nagoya 464, Japan.

ABSTRACT

The effects of promoters on the selective oxidation of benzene over V_2O_5 catalysts were investigated with a novel approach, i.e., the use of the model catalysts; 010 catalysts selectively exposing the (010) plane of V_2O_5 crystal, and monolayer V_2O_5/MoO_3 and V_2O_5/WO_3 catalysts. The improvements in the activity and selectivity with the addition of promoters to V_2O_5 catalysts were attributed to the specific crystallographic planes or the specific types of active sites.

INTRODUCTION

V_2O_5 is one of the most important catalysts in the industry, especially for the selective oxidation of benzene, butane and o-xylene. In order to improve the activity and selectivity, MoO_3, P_2O_5, WO_3 and SnO_2 are frequently added to V_2O_5 catalysts as the promoters. However, the difficulty arises in the study of promoted V_2O_5 catalysts because of the complexity of its structure. The exposure of the various crystallographic planes is one of the factors for the complexity. In the case of metal catalysts, the different crystal planes show the different catalytic properties because of the presence of geometrically different sites, such as, kink, step and terrace. Further, the activity and selectivity of the catalysts may depend on the types of active sites, i.e., the configuration of vanadium and promoter ions in the surface. Although the active sites on unpromoted V_2O_5 catalysts are the surface V=O species on (010) plane (A) shown in Fig. 1, conventionally "mixed" V_2O_5 catalysts have the active sites on both of the (010) and the other planes.(ref. 1) Further, various types of active sites may exist on mixed catalysts as shown in Fig. 1: (B) the surface V=O species interacted with promoter ions in lower layer, (C) the surface V=O species interacted with neighbor promoter ions in the surface, and (D) the surface Mo=O, W=O and Sn=O species activated by synergistic effect of vanadium ions. The presence of the last type of active sites have been confirmed previously.(ref. 1)

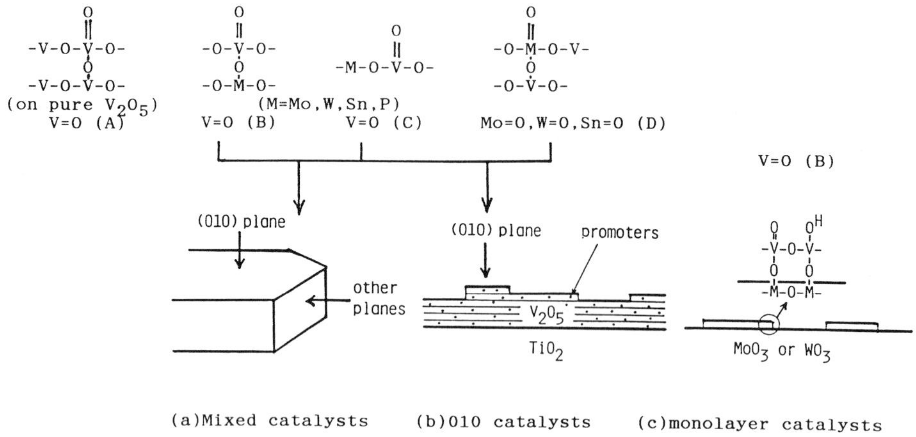

Fig. 1. Structure and active sites of promoted V_2O_5 catalysts.

In order to clarify the effect of promoters on V_2O_5 catalysts, we have prepared two types of model catalysts; 010 catalysts to clarify the effect on a specific plane and monolayer catalysts to clarify the effect on a specific type of active sites.

EXPERIMENTAL SECTION

Preparation of Catalysts.

010 catalysts were prepared by supporting V_2O_5 on TiO_2 support so that the (010) plane of V_2O_5 can be selectively exposed and the structure of V_2O_5 can remain unchanged with the addition of promoters. The content of V_2O_5 is 10 mol%. An oxalic acid solution of NH_4VO_3 and TiO_2 powders (Nippon Aerosil, P-25) were mixed, and evaporated with stirring. The precursor was dried at 393 K overnight, calcined at 673 K for 3 h in air, and calcined again at 873 K for 10h in O_2. Promoter oxides were added to thus obtained 010 catalyst by the incipient wetness method by using an aqueous solution of H_3PO_4, an oxalic acid solution of $(NH_4)_{10}W_{12}O_{41}\cdot 5H_2O$, an aqueous solution of $(NH_4)_6Mo_7O_{24}\cdot 4H_2O$ and a methanol solution of $SnCl_2\cdot 2H_2O$, respectively. The content of promoter ion was 10 atom% of the supported ions (W/(V+W)=0.1, for example). Then the catalysts were dried at 393 K overnight, and calcined at 823 K for 3 h in O_2.

Monolayer catalysts were prepared by CLD (Chemical Liquid-phase Deposition) method which is the little modification of the method proposed by Bond et al.(ref. 2) 10 g of MoO_3 or WO_3 was dried in a flask with flowing dry N_2 at 393 K for 18 h. Then, predetermined amount of $VOCl_3$ was dissolved in 50 ml of toluene which had been dried with Molecular Sieve 3A for 3 days, and the solution was added to the flask. The amount of added $VOCl_3$ corresponds to the theoretical amount for the monolayer coverage calculated from the density of vanadium ions on the (010) plane of V_2O_5 crystal (15.78 $\mu mol/m^2$) and BET surface area of MoO_3 (2.5 m^2/g) and WO_3 (7.6 m^2/g). These operations were carried out in a dry-box filled with dry N_2. The flask was set in a thermostatted oil-bath, and was heated under reflux for 5 h. Then, the solid was filtered and washed with toluene. The catalyst was dried at 393 K for 1h, hydrated at 473 K for 3h in flowing O_2 containing water vapor of saturated pressure at room temperature, and calcined in flowing dry O_2 for 1h at 473 K.

Mixed catalysts were prepared by evaporating mixed solution. An oxalic acid solution of NH_4VO_3 and an aqueous solution containing promoter ions were mixed in various compositions and evaporated. Then the precursors were calcined in flowing O_2 at 773 K for 3h.(ref. 1) All of the catalysts were pressed and sieved in the range of 28-48 mesh.

Oxidation of Benzene.

Benzene oxidation was carried out in a conventional continuous-flow apparatus at atmospheric pressure. The catalyst (0.1 g) dispersed in 2.5 g of fused Al_2O_3 powders was placed in a Pyrex glass tube (10mm i.d.). At the center of the catalyst bed, a Pyrex glass tube (4mm o.d.) was installed as a sheath of the thermocouple. Reactions were carried out under the following conditions: reaction temperature was 603-723 K, W/F was 0.4 $g\cdot hr\cdot mol^{-1}$, partial pressure of benzene and O_2 were 2.23 and 20.3 kPa, respectively, with N_2 balance. The following products were detected gas-chromatographically; maleic acid anhydride (MA), little amount of benzoquinone (BQ) and carbon oxides (CO and CO_2 indicated as COx). Turn-over frequencies (TF) of products, e.g. TF(MA) for MA, total TF of all products (TF(all)), and selectivity to MA (S(MA)) were calculated as follows;

$$TF(MA) = \frac{\text{Reaction Rate of MA } (mol\cdot s^{-1}\cdot g^{-1})}{\text{Surface Concentration of Redox Sites } (mol\cdot g^{-1})} \quad (s^{-1}) \tag{1}$$

$$TF(all) = \frac{TF(COx) - 2\cdot TF(MA)}{6} + TF(MA) + TF(BQ) \quad (s^{-1}) \tag{2}$$

$$S(MA) = \frac{TF(MA)}{TF(all)} \times 100 \quad (\%) \tag{3}$$

RESULTS AND DISCUSSION

010 and Monolayer Catalysts as Model Catalysts.

010 Catalysts: The surface area of V_2O_5 (010) plane (S_{010}) was determined by NARP ($NO-NH_3$ Rectangular Pulse) technique.(ref. 3) S_{010} (7.8 m^2/g) was close to the BET surface area (9.5 m^2/g), indicating that the 010 catalyst have sufficiently high exposure of V_2O_5 (010) plane for the study on the catalytic properties of the (010) plane of V_2O_5, i.e., the effect of the other planes can be neglected in the first approximation. It was confirmed by the BAT (Benzaldehyde-Ammonia Titration) method (ref. 4) that TiO_2 surface is not exposed. Therefore, the catalytic properties of the 010 catalyst can be regarded as the indication of those of the (010) plane of V_2O_5 within the reasonable error.

When the promoters were added to the catalyst, any changes were not observed in the XRD patterns, IR spectra, and UV spectra, indicating that the structure of the 010 catalysts remained unchanged. Further, it was confirmed by the depth profiles obtained by XPS and SIMS that the added promoter ions are dispersed uniformly in V_2O_5 layers. Therefore, the promoted 010 catalysts can be depicted as Fig. 1b, and they can be a good model catalysts for the study of the promoting effects on V_2O_5 (010) plane.

Monolayer Catalysts: The number of layers of monolayer catalysts was confirmed by the NARP technique. In the NARP experiment, the concentration profile of N_2 from pure V_2O_5 consists of two parts, i.e., the initial sharp N_2 and the tailing N_2.(ref. 3) However, the results obtained on the V_2O_5/MoO_3 and V_2O_5/WO_3 catalysts were very similar to that of the monolayer V_2O_5/TiO_2 catalysts prepared by Inomata et al.(ref. 3); the profiles consist only of the initial sharp N_2, and has no tailing part. Therefore, it can be concluded that monolayer V_2O_5 was formed on MoO_3 and WO_3 supports. The number of the surface V=O species were determined from the amount of initial sharp N_2 produced at the calcination temperature of 673 K.

When monolayer of V_2O_5 covers MoO_3 or WO_3 support, the chemical bonds are formed between all the vanadium ions in the top layer and the cations of support oxides in the under layer. The surface V=O species on the monolayer V_2O_5 catalysts should act as the active sites which arc affected by the promoter ions. Thus, the monolayer V_2O_5 catalysts on these oxides can be good model catalysts having the specific type of active sites, i.e., the surface V=O species (B) in Fig. 1c.

Activity and Selectivity in Benzene Oxidation.

Fig. 2 shows the selectivity to MA (S(MA)) over mixed, 010 and monolayer catalysts. As for the mixed catalysts (open symbols), S(MA) increased with the addition of P_2O_5, MoO_3 and WO_3. Only SnO_2 decreased S(MA). In the case of the 010 catalysts (closed symbols), S(MA) also increased with the addition of MoO_3 in the same way as the mixed catalysts, but it was not varied with the addition of the other promoters.

Figs. 3 and 4 show the turn-over frequency (TF) of MA and COx (CO and CO_2) on mixed catalysts and 010 catalysts, in order to examine the effect of the promoters on the activity of individual active sites. TF(MA) and TF(COx) can be discussed independently from one another, since the consecutive oxidations

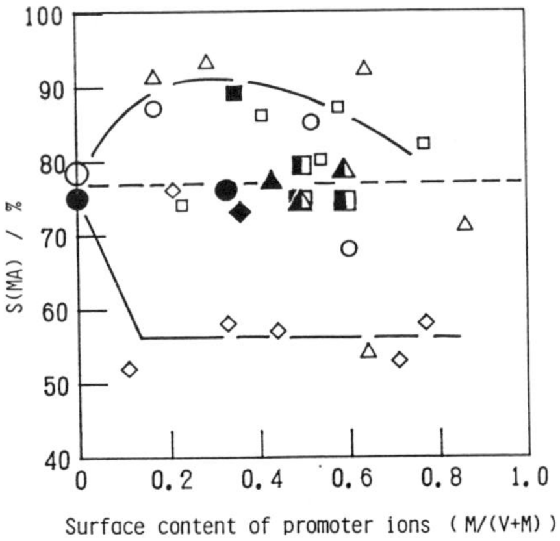

Fig. 2. Selectivity to MA (S(MA)) over promoted V_2O_5 catalysts:
Promoters; (○) P_2O_5, (□) MoO_3, (△) WO_3, (◇) SnO_2.
Symbols; (opened) mixed catalysts; (closed) 010 catalysts; (half closed) monolayer catalysts.

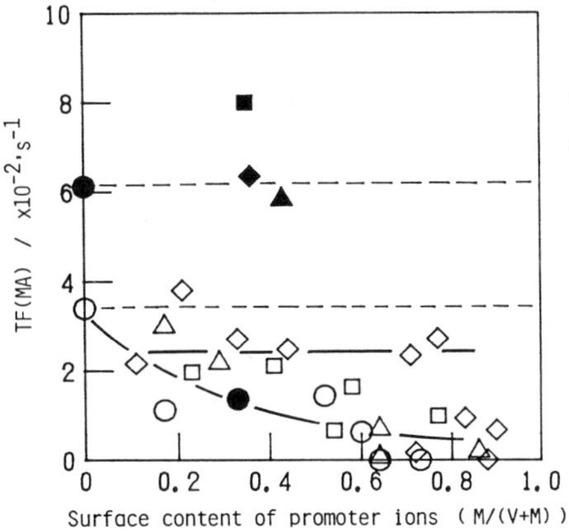

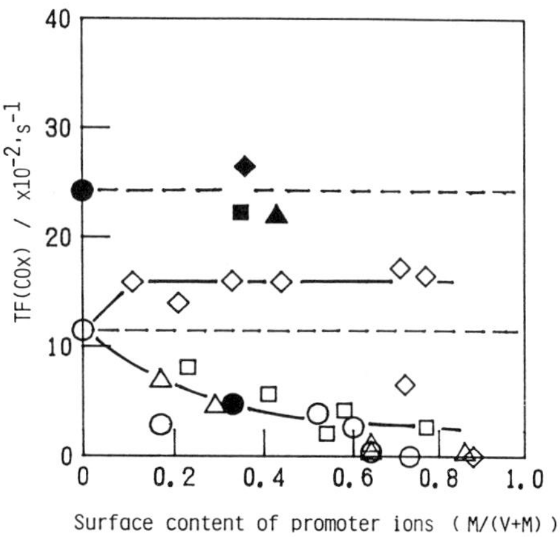

Fig. 3(Left). Turn-over frequency of MA (TF(MA)) over promoted V_2O_5 catalysts. **Fig. 4(Right).** Turn-over frequency of CO and CO_2 (TF(COx)) over promoted V_2O_5 catalysts: Promoters; (○) P_2O_5, (□) MoO_3, (△) WO_3, (◇) SnO_2. Symbols; (opened) mixed catalysts; (closed) 010 catalysts.

can be neglected in the present reaction condition. As for the mixed catalysts (open symbols), TF(MA) was decreased by the addition of all the promoters. TF(COx) also decreased with the addition of P_2O_5, MoO_3 and WO_3. Only in the case of SnO_2, TF(COx) increased.

As for the unpromoted 010 catalysts, both TF(MA) and TF(COx) were higher than those of unpromoted V_2O_5, which may be due to the effect of TiO_2 supports. It follows that only the relative changes of TFs with the addition of the promoters should be discussed hereafter. While TF(MA) increased with the addition of MoO_3 (closed square in Fig. 3), TF(COx) did not change (Fig. 4).

This is the reason why S(MA) increased on the (010) plane of V_2O_5 with the promoting effects of MoO_3. (Fig. 2) Although both TF(MA) and TF(COx) decreased with the addition of P_2O_5, S(MA) was the same as that of the unpromoted 010 catalyst. The effects of WO_3 and SnO_2 on the 010 catalyst were different from the mixed catalysts, that is, neither the selectivity (Fig. 2) nor the activity (Figs. 3 and 4) were varied on the 010 catalysts.

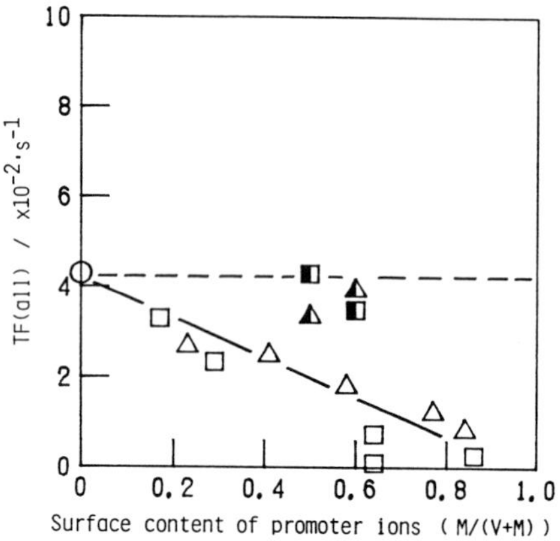

Fig. 5. Turn-over frequency of total products (TF(all)) over promoted V_2O_5 catalysts:
Promoters; (□) MoO_3, (△) WO_3.
Symbols; (opened) mixed catalysts; (half closed) monolayer catalysts.

As shown the half closed symbols in Figs. 2 and 5, both S(MA) and TFs on monolayer catalysts were the same as those of unpromoted V_2O_5. It suggests that the selectivity and activity of the surface V=O species (B) is not different from those of the surface V=O species (A) on unpromoted V_2O_5 catalysts, or, in other words, the promoters in the lower layer have no effect on the catalytic properties of the surface V=O species.

Promotion Effect on Specific Sites over Specific Planes.

Table I summarizes the above-mentioned results. The last row (TOTAL) indicates the results on the mixed catalysts. The selectivity is increased with the addition of MoO_3, WO_3 and P_2O_5 in the mixed catalysts, but it is decreased with the addition of SnO_2. The activity in terms of TF is decreased with the addition of MoO_3, WO_3 and P_2O_5, but it dose not changed with SnO_2.

The results on the monolayer V_2O_5/MoO_3 catalyst (Figs. 2 and 5) indicate that MoO_3 has no effect on the surface V=O species (B). Or, in other words, the changes in the activity and selectivity with the addition of MoO_3 can be attributed to the promotion of the surface V=O species (C) or the surface Mo=O species. Figs. 3 and 5 indicate that MoO_3 increased the selectivity on the (010) plane, and Fig. 2 indicates that the increase in the selectivity on the mixed catalyst is very close to that on the 010 catalyst. It follows that MoO_3 has a promoting effect on the selectivity of the surface V=O (C) species or Mo=O species over the (010) plane, as indicated by "+" mark in Table I.

Similar discussion on the activity leads to the conclusion that MoO_3 decreases the activity of the surface V=O (C) species or Mo=O on the other planes, which cause the decrease in the activity of mixed V_2O_5-MoO_3 catalyst.

Table I: Effect of Promoters on Selectivity and Turn-over Frequency of Benzene Oxidation over V_2O_5 Catalysts.

<table>
<tr><td colspan="2"></td><td colspan="2">V_2O_5-MoO_3</td><td colspan="2">V_2O_5-WO_3</td><td colspan="2">V_2O_5-P_2O_5</td><td colspan="2">V_2O_5-SnO_2</td></tr>
<tr><td>planes[a]</td><td>sites[b]</td><td>S(MA)</td><td>TF</td><td>S(MA)</td><td>TF</td><td>S(MA)</td><td>TF</td><td>S(MA)</td><td>TF</td></tr>
<tr><td rowspan="2">(010) plane</td><td>V=O (B)</td><td>0</td><td>0</td><td>0</td><td>0</td><td rowspan="2">0</td><td rowspan="2">–</td><td rowspan="2">0</td><td rowspan="2">0</td></tr>
<tr><td>V=O (C) and Mo=O or W=O</td><td>+</td><td>+</td><td>0</td><td>0</td></tr>
<tr><td rowspan="2">other planes</td><td>V=O (B)</td><td>0</td><td>0</td><td>0</td><td>0</td><td rowspan="2">+</td><td rowspan="2">–</td><td rowspan="2">–</td><td rowspan="2">0</td></tr>
<tr><td>V=O (C) and Mo=O or W=O</td><td>0</td><td>–</td><td>+</td><td>–</td></tr>
<tr><td colspan="2">TOTAL</td><td>+</td><td>–</td><td>+</td><td>–</td><td>+</td><td>–</td><td>–</td><td>0</td></tr>
</table>

+, increase; -, decrease; 0, not varied; in S(MA) or TF relative to the surface V=O (A) species estimated on the basis of the results on (a) 010 catalysts and (b) monolayer catalysts.

The effect of the other promoters were also attributed to the specific planes and the specific types of the active sites. According to the Table I, the promoting effect on the activity and selectivity can be classified into three types. The addition of the promoters leads to the following changes in the active sites compared with the unpromoted V_2O_5 catalysts;

(1) increase in both the selectivity and the activity.
  MoO_3: V=O (C) and Mo=O species on (010) plane.
(2) increase in the selectivity and decrease in the activity.
  WO_3 : V=O (C) and W=O species on the other planes.
  P_2O_5: V=O species on the other planes.
(3) decrease only in the selectivity.
  SnO_2: V=O and Sn=O species on the other planes.

Mechanism of Promotion Effect

These differences can be accounted for by the electronegativity of each ions and the charge transfer in the promoted active sites. Electronegativity of each ions is as follows;

ions	Sn^{4+}	<	V^{5+}	<	W^{6+}	<	P^{5+}	<	Mo^{6+}
χ_i	16.2		17.6		22.1		23.1		23.4

The addition of more electronegative ions (i.e., W^{6+}, P^{5+} and Mo^{6+}) should result in the decrease of the electron density of the surface V=O bond. Then, the V=O bonds are weakened, and the activity of the V=O species decreases in the step of the hydrogen-abstraction from benzene, which should lead to the decrease of TFs. As in the case of WO_3 and P_2O_5, the addition of these promoters thus improve the selectivity of V_2O_5 catalysts, when the degree of the decrease in TF(COx) is smaller than that in TF(MA). On the other hand, the addition of less electronegative Sn^{4+} promotes the combustion of reactant over the active sites.

In the charge transfer between vanadium and promoter ions, the V-O-M (M=Mo, W) bond length should be the one of the most important factors for promoting the surface V=O species. The bond length of V-O-M in the V=O (C) species is 3.78 Å, while that in V=O (B) is 4.35 Å. The shorter bond length in the former may be the reason why the promoting effect on the V=O (C) species is more effective than that in the V=O (B).

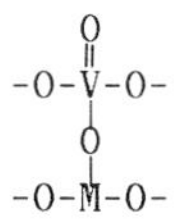

V=O (B)
V-O-M length: 4.35 Å
not effective

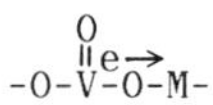

V=O (C)
V-O-M length: 3.78 Å
effective

(M=Mo,W)

In conclusion, the effect of promoters can be separated and classified by using well-defined model catalysts, indicating such approach will be a great help for the deep understanding of catalysis as the scientific research. Such deep understanding would give the useful informations for the catalyst design. For example, the addition of MoO_3 increases both the activity and selectivity on V_2O_5 (010) plane, as shown in Table I. This suggests the selective exposure of the (010) plane of V_2O_5, which may be achieved by the supporting V_2O_5-MoO_3 catalyst, is effective for the selective oxidation. Actually, V_2O_5-MoO_3 catalysts are frequently supported on TiO_2 and so on for the industrial use for the benzene oxidation.(ref. 5)

REFERENCES

1 Satsuma, A.; Hattori, A.; Mizutani, K.; Furuta, A.; Miyamoto, A.; Hattori, T.; Murakami, Y. J. Phys. Chem., **1988**, 92, 2275; **1988**, 92, 6052; **1989**, 93, 1484; Okada, F.; Satsuma, A.; Furuta, A.; Miyamoto, A.; Hattori, T.; Murakami, Y. submitted.
2 Bond, G.C.; Brüchman, K. Faraday Disc. Chem. Soc., **1981**, 72, 235.
3 Miyamoto, A.; Yamazaki, Y.; Inomata, M.; Murakami, Y. J. Phys. Chem. **1981**, 85, 2366; Inomata, M.; Miyamoto, A.; Murakami, Y. Ibid. **1981**, 85, 2372.
4 Niwa, M.; Matsuoka, Y.; Murakami, Y. J. Phys. Chem., **1987**, 91, 4519.
5 U.S. Pat., 3,005,831(1961); Jpn. Kokai, 135(1963); U.S. Pat., 3,221,671(1965); U.S. Pat., 3,417,108(1968).

G. Centi and F. Trifiro' (Editors), *New Developments in Selective Oxidation*
© 1990 Elsevier Science Publishers B.V., Amsterdam — Printed in The Netherlands

COOPERATION BETWEEN PHASES IN MIXED SnSbO SELECTIVE OXIDATION CATALYSTS

L.T. WENG, P. PATRONO*, E. SHAM, P. RUIZ and B. DELMON

Unité de Catalyse et Chimie des Matériaux Divisés, Université Catholique de Louvain
Place Croix du Sud 1, 1348 Louvain-la-Neuve (Belgium)
*IMAI-CNR Aria, Ricerca, Via Salaria, CP 10 Monterolonctoscalo, Rome (Italy)

SUMMARY

A synergy is observed in the oxidation of isobutene to methacrolein when a Sn-rich SnSbO oxide calcined at high temperature is mechanically mixed with an Sb-rich SnSbO oxide calcined at the same temperature (or if mixed with pure Sb_2O_4). The catalysts were characterized by XRD, BET surface area measurements, Electron microscopy and XPS and it was found that i) the Sn-rich samples are either a pure Sb^{5+}-SnO_2 solid solution or a solid solution with a small quantity of segregated antimony oxide; ii) the Sb-rich samples contain α-Sb_2O_4 with a trace of solid solution and iii) mechanical mixtures are constituted of two separate phases : solid solution and α-Sb_2O_4. The observed synergy is explained by a remote control mechanism; α-Sb_2O_4 controls the number of active sites on the surface of Sb^{5+}-SnO_2 solid solution.

INTRODUCTION

Mixed SnSbO oxides have received much attention in scientific research, since they constitute the basis of efficient commercial oxidation catalysts [1]. Although there exist some divergence regarding the nature of the active site, the solubility of Sb^{5+} in SnO_2, the role of segregated antimony oxide etc..., it is usually agreed that the best catalysts (both in activity and selectivity) are those calcined at high temperatures. Such catalysts contain two phases: a solid solution of Sb^{5+} dissolved in SnO_2 (Sb^{5+}-SnO_2) and an antimony oxide (α or β-Sb_2O_4). Many hypotheses have been put forward to explain the roles played by these two phases, but no general agreement has been reached [2-4].

Our study with mechanical mixtures of SnO_2 and α-Sb_2O_4 [5] had shown that the cooperation between the oxides can be explained by a remote control mechanism; α-Sb_2O_4, itself not active in allylic oxidation, produces spillover oxygen. This mobile species, by reacting with the surface of SnO_2, brings about the formation of selective oxidation sites : α-Sb_2O_4 controls, from a distance, the active sites of SnO_2.

In order to investigate whether the same explanation could be given for the mixtures of $Sb_xSn_{1-x}O_2$ and α-Sb_2O_4 (especially with respect to the role of α-Sb_2O_4), we studied mechanical mixtures containing the $Sb_xSn_{1-x}O_2$ mixed oxides (with different Sb contents) and α-Sb_2O_4. A gentle mechanical mixing procedure is used in order to avoid (or to minimize) mutual contamination. The results obtained with these catalysts will constitute the first part of the present communication.

One could argue that some mutual contamination explains the observed synergy. In order to examine this possibility, we studied mechanical mixtures constituted of two mixed $Sb_xSn_{1-x}O_2$ oxides: one rich in Sn and the other in Sb. In such a case, each phase has already been contaminated. If some cooperation still manifests itself, the only possible explanation is remote control. This is dealt with in another section of the present paper.

EXPERIMENTAL

Catalyst preparation and characterization

α-Sb_2O_4 (2 $m^2 \cdot g^{-1}$) was prepared by calcination of Sb_2O_3 at 500°C for 20h. It will be referred to as Sb_2O_4(I).

Mixed $Sb_xSn_{1-x}O_2$ oxides with different Sb contents (O,1,5,90,95,99,100 at%) were prepared by conventional coprecipitation [2] followed by calcination at 900°C for 16h. Hereafter, the corresponding samples will be designated as S^x, where x refers to Sb content.

Mechanical mixtures of two oxides in equal proportion (50/50 on a weight basis) were obtained by dispersing and mixing the respective powders in n-pentane for 10 minutes, evaporating the solvent and drying at 80°C overnight. The mixture was not subjected to further calcination.

Fresh and used samples were characterized using XRD, BET surface area measurements, Electron Microscopy (CTEM, SEM, EDS-STEM) and XPS techniques. For XPS, we calculated the surface concentration of Sb (Sb/(Sb+Sn)) using the sensitivity factors given by Wagner et al. [6].

Catalytic selective oxidation of isobutene

Selective oxidation of isobutene to methacrolein was carried out in a conventional fixed bed reactor system working at atmospheric pressure [5]. The reaction conditions were as follows : $C_4H_8/O_2/N_2$ (diluting gas) = 1/2/7; total flow: 60 ml/min.; reaction temperatures: 420 - 460°C; catalyst: 300mg.

A catalytic synergy was often observed. It was calculated using the following formula :

$$\text{synergy} = \frac{Y_{AB} - 0.5\,(Y_A + Y_B)}{0.5\,(Y_A + Y_B)} \qquad (1)$$

where Y_{AB}, Y_A and Y_B are the yields obtained with 300 mg of the mechanical mixture AB, oxides A or B, respectively.

RESULTS

Physico-chemical characterization

a) Mixed $Sb_xSn_{1-x}O_2$ oxides

Table 1 reports on the crystallographic phases detected by XRD, the BET surface areas and the surface concentration of Sb as determined by XPS, for the various mixed oxides.

Table 1 : XRD phases, surface areas and surface Sb concentration of $Sb_xSn_{1-x}O_2$ oxides

Sb content (%)	XRD phases	BET surface ($m^2.g^{-1}$)	Sb/(Sb+Sn)XPS
1	SnO_2	8.49[a]	7.6
5	SnO_2	14.48[a]	21.4
90	α-Sb_2O_4+SnO_2	1.40[b]	72.6
95	α-Sb_2O_4+SnO_2	1.00[b]	79.7
99	α-Sb_2O_4	0.60[b]	96.1
100	α-Sb_2O_4	0.60[b]	100.0

a : measured by adsorption of N_2 at 77 °K; b : measured by adsorption of He at 77 °K

As in the literature, we observe an enrichment in Sb at the surface for the Sn-rich samples by XPS. For the Sb-rich samples, however, one observes an enrichment of Sn on the surface with respect to bulk concentration.

Electron microscopy measurements carried out on the Sn-rich samples (S^1 and S^5) show that the particles have a size of ca. 30 nm on average; the Sb/Sn ratios detected by microanalysis are almost constant from particle to particle for the same sample. For the samples rich in Sb, two types of particles are present, namely particles similar to those in Sn-rich samples (namely Sb^{5+}-SnO_2 solid solution) and pure Sb_2O_4. The particle size of the solid solution is almost the same for the three samples but the relative number of particles depends on the nominal Sn content. The particle size of Sb_2O_4 in these samples is also comparable, but it is slightly smaller than in S^{100}. Figure 1 presents a typical SEM micrograph and the corresponding EDS-STEM microanalysis patterns for S^{90}. The particle size of the Sb^{5+}-SnO_2 solid solution is much smaller than that of pure Sb_2O_4.

b) Mechanical mixtures

The X-ray diffraction patterns of mechanical mixtures are the simple addition of those of the starting oxides. Only the peaks characteristic of SnO_2 or α-Sb_2O_4 were observed. The diffraction patterns did not change after the catalytic reaction.

Electron microscopy measurements indicate that the particles in the mechanical mixtures have the same size as those in the starting oxides. The particle size did not change during the catalytic reaction. Figure 2 presents a typical CTEM micrograph of S^5 + Sb_2O_4(I) when fresh. Electron microanalysis of this sample indicates that there was no mutual contamination between Sb_2O_4 and the solid solution. The same conclusion is valid for the other mechanical mixtures (between Sn-rich oxides and Sb-rich oxides). No difference was observed for the samples after the catalytic reaction.

Table 2 reports on the Sb surface concentration of the mechanical mixtures containing S^5 before and after the catalytic reaction. In reference [5], we have shown that coke can be deposited preferentially on the SnO_2 surface in the mechanical mixtures of SnO_2-α-Sb_2O_4 after catalytic reaction. This complicates the interpretation of the XPS results, because the Sb/(Sb+Sn) ratio as determined by XPS did not correspond to the real composition near the surface (the reason being that deposited coke hides a fraction of the photoelectrons emitted by SnO_2). One way to obtain the real surface composition is to calcine the catalyst samples collected after reaction in air under mild conditions, in order to eliminate the coke formed. For this reason, we have also calcined our used samples in air at 400°C for 20h. The results in the table show that Sb/(Sb+Sn) ratios remain almost constant before and after catalytic reaction if proper precautions are taken to eliminate deposits.

Table 2 : XPS results for the mechanical mixtures containing S^5

Samples	Sb/(Sn+Sb) by XPS (%) before test	after test and regeneration*
$S^5 + S^{90}$	33.1	37.0
$S^5 + S^{95}$	35.0	39.0
$S^5 + S^{99}$	36.8	40.0
$S^5 + S^{100}$	34.1	37.5
S^5 + Sb_2O_4(I)	41.8	45.3

*regeneration refers to a calcination of the sample at 400°C for 20h.

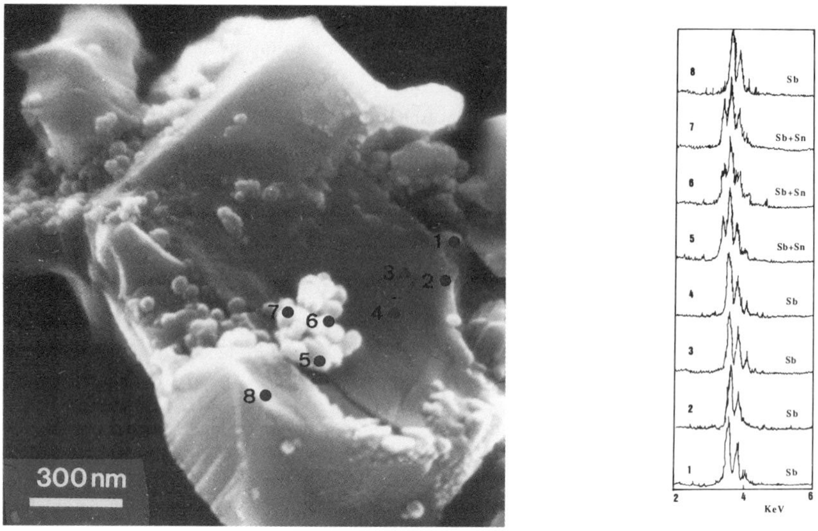

Figure 1 : SEM micrograph and the corresponding EDS-STEM microanalysis patterns for S^{90}

Figure 2 : CTEM micrograph for mechanical mixture S^5 + $Sb_2O_4(I)$

Catalytic activity

a) Pure mixed $Sb_xSn_{1-x}O_2$ oxides

The catalytic activity and selectivity of pure $Sb_xSn_{1-x}O_2$ oxides are reported in Table 3. For Sn-rich oxides (S^1, S^5), the methacrolein yield and selectivity at a given temperature increase with the Sb content. The methacrolein selectivity decreases with reaction temperature, especially for S^1.

Compared with the Sn-rich samples, Sb-rich samples exhibit a much lower activity, but possess a higher selectivity. The total catalytic activity decreases with the increase of Sb content, S^{99} and S^{100} being almost inert.

We also calculated the intrinsic activity, namely the yield/surface area ratio. It increases with the increase of the Sb content for all samples having a detectable activity.

Table 3 : Catalytic activity and selectivity for pure $Sb_xSn_{1-x}O_2$ mixed oxides at 420 and 460°C and their intrinsic activity

Samples	420°C			460°C		
	Yield (%)	Selectivity (%)	Yield/Surface (%)	Yield (%)	Selectivity (%)	Yield/Surface (%)
S^1	1.52	18.81	0.18	2.45	9.39	0.29
S^5	7.01	39.72	0.48	14.30	34.80	0.99
S^{90}	1.19	42.92	0.85	2.57	47.68	1.82
S^{95}	0.98	44.95	0.98	1.98	48.53	1.98
S^{95}		no detectable			no detectable	
S^{100}		no detectable			no detectable	

b) Mechanical mixtures

The catalytic activity results for the mechanical mixtures between Sn-rich oxides on the one hand, and Sb-rich oxides (or Sb_2O_4) on the other hand, at 420°C and 460°C, are reported in Table 4. The catalytic synergies for methacrolein yield calculated according to formula (1) are also presented in the table.

The table shows that there exists a noticeable, and sometimes very marked, synergy when methacrolein yield or selectivity are considered. This is true for all mixtures. The magnitude of the catalytic synergy increases with reaction temperature. With the same Sn-rich mixed oxide, the synergistic effect increases when the other phase contains more Sb. If the same Sb-rich mixed oxide is considered, the synergistic effect increases when the other phase contains more Sn. The largest synergistic effects are observed for the mixtures containing Sb_2O_4(I). It is interesting to note that although S^{100} and Sb_2O_4(I) have the same structure, namely α-Sb_2O_4, the latter brings about a higher synergistic effect whichever the Sn-rich mixed oxide is associated with it.

DISCUSSION

Our results show that a strong cooperative effect exists between Sn-rich and Sb-rich $Sb_xSn_{1-x}O_2$ mixed oxides in the selective oxidation of isobutene. When a catalytic cooperation between two phases is observed, it may be explained in several ways: (i) the formation of a new phase or a solid solution, (ii) surface contamination of one phase by elements coming from the other, e.g. the vapor transfer of Sb from an Sb-rich phase to an Sb-lean phase; (iii) bifunctional catalysis, and (iv) a remote control mechanism, etc. In the case of selective oxidation of isobutene, a bifunctional mechanism has to be ruled out : the intermediate, namely the allylic species is too bulky and too strongly adsorbed to be able to migrate from

Table 4 : Catalytic activity results for the mechanical mixtures between Sn-rich oxides and Sb-rich oxides at 420°C and 460°C and the catalytic synergies calculated with (1)

Samples	420°C Yield (%)	420°C Selectivity (%)	420°C Synergy (%)	460°C Yield (%)	460°C Selectivity (%)	460°C Synergy (%)
S^1+S^{90}	1.67	28.40	12.20	4.58	34.86	82.47
S^1+S^{95}	1.63	37.39	30.40	3.76	29.01	52.14
S^1+S^{99}	1.51	32.13	98.87	3.70	34.39	202.00
S^1+S^{100}	1.20	34.01	57.89	2.50	33.36	122.22
$S^1+Sb_2O_4$(I)	1.61	36.26	111.80	3.81	37.43	211.00
S^5+S^{90}	4.62	56.83	12.68	11.26	60.15	33.41
S^5+S^{95}	4.47	43.57	12.03	10.86	54.93	33.42
S^5+S^{99}	4.80	54.61	37.14	11.43	54.98	59.86
S^5+S^{100}	4.64	55.77	32.57	9.53	50.72	33.29
$S^5+Sb_2O_4$(I)	6.74	51.45	93.14	12.90	52.27	80.42

one phase to the other for further transformation. In what follows, we shall examine the three other possibilities.

Investigation of the possible formation of a new compound

In this section, we shall examine whether some chemical interaction between the starting oxides could take place and explain the catalytic results.

The Sn-rich mixed oxides have been extensively studied in the literature [2-4]. Our results with XRD, electron microscopy and XPS are similar to those reported previously [7-11]. Taking into account all this information, we can conclude that S^1 is a solid solution of Sb^{5+} dissolved into SnO_2 with an enrichment in Sb at the surface (Sb^{5+}-SnO_2). For the samples similar to S^5, the presence of segregated antimony oxide has been deduced from careful Mössbauer studies [3] and XPS measurements [11]. It therefore seems likely that S^5 may contain a small quantity of antimony oxide segregated at the surface, in addition to the Sb^{5+}-SnO_2 solid solution.

Much less attention has been paid in the literature to Sb-rich samples. Our XRD measurements show that not only α-Sb_2O_4 but also traces of SnO_2 were observed for S^{90} and S^{95}. Observations of all Sb-rich samples (S^{90}, S^{95} and S^{99}) by electron microscopy reveal the existence of two kinds of particles : pure α-Sb_2O_4 and a solid solution of Sb in SnO_2. This implies that the formation of a solid solution of Sn in α-Sb_2O_4 or surface contamination of α-Sb_2O_4 by Sn is thermodynamically unfavorable. This conclusion is in complete agreement with those we arrived at previously for impregnated catalysts [12]. In one such study, we had impregnated the surface of α-Sb_2O_4 with a small amount of Sn^{4+} (or Sn^{2+}) ions in order to facilitate the mutual contamination; but the characterization results showed that Sn^{4+} (or Sn^{2+}) ions tended to crystallize to SnO_2 either during thermal calcination or during catalytic reaction, with, as a result, the formation of a two-phase catalyst containing α-Sb_2O_4 and SnO_2 [12]. The fact that we observed an apparent enrichment of Sn at the surface by XPS can be easily explained by the fact that the particle size of α-Sb_2O_4 is much greater than that of the Sb^{5+}-SnO_2 solid solution, as shown by electron microscopy.

In conclusion, two situations should be distinguished for the $Sb_xSn_{1-x}O_2$ mixed oxides calcined at high temperatures : one containing only pure solid solution Sb^{5+}-SnO_2 with an enrichment of Sb at the surface (S^1) and the other containing two phases : a Sb^{5+}-SnO_2 solid solution with a segregated antimony oxide Sb_2O_4.

Let us now look at what happened when we mechanically mixed a Sn-rich and an Sb-rich mixed oxide. The XRD measurements did not reveal any new phases. This excludes the first possibility. But one may argue the mutual contamination between two starting phases during reaction condition, e.g. the vapor transfer of Sb. But this is difficult to be accepted because : i) Electron microscopy showed that two kinds of particles, Sb^{5+}-SnO_2 and α-Sb_2O_4, coexisted and no detectable contamination was observed between two phases; ii) If the transfer of Sb from Sb-rich phase to Sb-lean phase exists, the surface Sb concentration should increase after catalytic reaction, this is against the XPS results; iii) If the transfer of Sb is the origin of cooperation, the magnitude of synergetic effects between Sn-rich samples with Sb_2O_4(I) and S^{100} should be the same, but this is not in agreement with our results. All these suggest that two separated phases coexist in all mechanical mixtures : a Sb^{5+}-SnO_2 and a pure α-Sb_2O_4. Consequently, we must conclude that the only possible explanation for the observed synergy is the existence of a remote control mechanism.

Explanation of the catalytic activity results based on the remote control mechanism

As indicated in the introduction, the remote control mechanism assumes two different roles for the two phases, one being to carry the catalytic centers for oxidation (called an acceptor) and the other to produce spillover oxygen (called a donor). In our case, α-Sb_2O_4 is absolutely inactive; it can therefore only play the role of a donor. Sn-rich samples are active even without α-Sb_2O_4; they carry active centers and they can play the role of an acceptor. A complicating factor is that, for the Sb-rich samples S^{90}-S^{99}, two phases (α-Sb_2O_4 and Sb^{5+}-SnO_2) are present : the acceptor action will be exerted both by the Sn-rich sample and by the solid solution present in Sb-rich samples. In all cases, we can conclude that the Sb^{5+}-SnO_2 solid solution is the acceptor and α-Sb_2O_4 is the donor.

According to the remote control mechanism, the role of spillover oxygen produced by the donor is to react with the surface of the acceptor to create new selective sites and/or regenerate the sites which get deactivated during catalytic reaction (e.g. due to deep reduction or coke deposition). If this is true, the effect of α-Sb_2O_4 in our mechanical mixtures would be to improve the selectivity to methacrolein. This is exactly what we observed in the catalytic activity results (Table 4).

The effect of reaction temperature on methacrolein selectivity is quite different for pure Sn-rich samples and their mechanical mixtures with Sb-rich or pure α-Sb_2O_4 (Tables 3 and 4). For pure Sn-rich samples, the increase of reaction temperature decreases the methacrolein selectivity. This indicates that reaction temperature favours total oxidation. For mechanical mixtures, except the above factor, another parameter should be taken into account, namely that the increase of reaction temperature would favour the production of spillover oxygen and its migration. This is favorable for methacrolein selectivity. The fact that the methacrolein selectivity for mechanical mixtures remains almost constant with reaction temperature suggests that the influence of above two parameters, acting in opposite directions in magnitude, is comparable.

The fact that, taking the same acceptor, the catalytic synergistic effect increases with the increase of Sb content is quite logical because the higher the Sb content, the larger the amount of α-Sb_2O_4 present, and, consequently the more spillover oxygen is available. With the same donor, the synergistic effect is higher for S^1 than for S^5. This is due to the fact that S^5 may already contain a small quantity of segregated antimony oxide at the surface; the latter acts as a donor, and consequently less additional donor is needed.

The comparison of the catalytic synergies observed with mechanical mixtures containing Sb_2O_4(I) and those containing S^{100} (Table 4) shows the influence of another important parameter, namely surface area. With the same acceptor, the increase of the surface area of the donor should bring about two effects: i) more surface accessible to gaseous oxygen (or more mobile oxygen can be produced) and ii) more contacts with acceptor. These two effects would favour the production of spillover oxygen. Therefore, with the same acceptor, higher catalytic synergy should be observed with the donor with the higher surface area. This is what we observed in Table 4 (bearing in mind the fact that Sb_2O_4(I) and S^{100} have the surface areas of 2 m^2/g and 0.6 m^2/g respectively).

Finally, let us look at the activity of pure mixed $Sb_xSn_{1-x}O_2$ oxides. If the remote control mechanism operates, the intrinsic activity of the Sb^{5+}-SnO_2 solid solution in Sb-rich oxides should be higher than that in Sn-rich region because the first can benefit from the presence of α-Sb_2O_4. This is exactly what is shown in Table 3, especially by the selectivity results.

In conclusion, the catalytic activity can be satisfactorily explained by the remote control mechanism.

OUTLOOK

As indicated in the introduction, many hypotheses have been put forward to explain the roles played by the Sb^{5+}-SnO_2 solid solution and antimony oxide in $Sb_xSn_{1-x}O_2$ mixed oxides. The diverging views are essentially due to the difficulties encountered in the characterization of these catalysts. The approach used in the present study was different, namely studying mechanical mixtures of a Sn-rich oxide with pure α-Sb_2O_4 or Sb-rich oxides. The results demonstrate that the solid solution benefits from the presence of α-Sb_2O_4 even when the two phases are only in "physical" contact. This, in association with the results presented previously [13-16] for the systems containing α-Sb_2O_4, allows us to attribute with certitude the role of α-Sb_2O_4 as that of a donor of spillover oxygen. The roles played by the phases present in the $Sb_xSn_{1-x}O_2$ system appear clearly in the context of the remote control mechanism. The pure solid solution is an acceptor while pure Sb_2O_4 is a donor. When the Sb content increases, the donor ability increases while the acceptor properties decrease.

REFERENCES

1 J.R. Bethell and D.J. Hadley, U.S. Patent, 3,094,565 (1963)
2 F.J. Berry, Adv. Catal., 30, (1980) 97
3 J.C. Volta, P. Bussière, C. Coudrier, J.M. Herrmann and J.C. Vedrine, Appl. Catal., 16, (1985) 315
4 B. Viswanathan and S. Chokkalingam, Surf. Techol., 26, (1984) 231-344
5 L.T. Weng, N. Spitaels, B. Yasse, B. Ladrière, P. Ruiz and B. Delmon, to be published
6 C.D. Wagner, L.E. Davis, H.V. Zeller, P.A. Taylor, R.H. Raymond and L.H. Gale, Surf. Inter. Anal. 3, No. 5 (1981) 21
7 Y.M. Cross and D.R. Pyke, J. Catal., 58, (1979) 61-67
8 Y. Boudeville, F. Figueras, M. Forissier, J.L. Portefaix and J.C Vedrine, J. Catal., 58, (1979) 52-60
9 D.R. Pyke and R. Reid and J.D. Tilley, J.C.S. Faraday I, 76, (1980) 1174-1182
10 J.C. Volta, B. Benaichouba, I. Mutin and J.C. Vedrine, Appl. Catal., 8, (1983) 215-233
11 B. Viswanathan, S. Chokkalingam, TK. Varadarajan and S. Badrinarayanan, Surf. Coating Technol., 28, (1986) 201-206
12 L.T. Weng, B. Yasse, B. Ladrière, P. Ruiz and B. Delmon, to be published
13 P. Ruiz, B. Zhou, M. Remy, T Machej, B. Yasse, F. Aoun, B. Doumain and B. Delmon, Catalysis Today, 1, (1987) 181
14 L.T. Weng, B. Zhou, B. Yasse, B. Doumain, P. Ruiz and B. Delmon, 9th Int. Congr. Catal., Calgary, Cadana, Vol.4, (1988) 1609
15 F.Y. Qiu, L.T. Weng, P. Ruiz and B. Delmon, Appl. Catal., 47, (1989) 115-123
16 L.T. Weng, P. Ruiz, B. Delmon and D. Duprez, J. Mol. Catal., 52 (1989) 349-360

J. C. VEDRINE (Institut de Recherche sur la Catalyse, CNRS, France) : I would like to propose two other explanations for your results, although I have no objection to your own explanation.

In our old work (your references 3, 8 and 10), we have shown that Sb_2O_4 or Sb_6O_{13} species laying on the SbSnO solid solution was the active phase and not the solid solution because of nearly (vicinal) $Sb^{3+}Sb^{5+}$ ions favourable to propene oxidation to acrolein. Large Sb_2O_4 particles have low external surface and therefore their properties were negligible.

For SMSI type study on metal on oxide support the authors favour the overlapping of the metal by the oxide after high temperature treatment. This was reversible when contacting the catalyst with air. Thus XPS and ISS techniques were not conclusive.

Don't you think that such a phenomenon does occur, i.e. Sb_2O_4 was partly covering your catalyst in a resemble manner and in a bidimensional type layer. Therefore for all catalysts you will get enhanced properties.

L.T. WENG (Université Catholique de Louvain, Belgium) : The work you mention (refs 3, 8 and 10 of our paper) do not conclude exactly what your comments suggest. Your work suggested that the Sb_2O_4 or Sb_6O_{13} particles, because of a strong interaction with the SbSnO solid solution, would lead to the preferential development of faces with special Sb-Sb arrangements (like in (001) of α-Sb_2O_4) containing either only Sb^{3+} or only Sb^{5+}, which would be the active and selective sites. You suggested that the catalytic active and selective sites in propene oxidation were associated with these ions, and you said that the solid solution was not the main responsible for catalytic activity.

Actually, our results show :

(i) that pure antimony oxide (α-Sb_2O_4) is absolutely inactive, this makes difficult to accepte that one specific face of α-Sb_2O_4 would be extremely active;

(ii) that the solid solution is active;

(iii) that the solid solution tends to segregate antimony oxide as a separate phase due to high temperature calcination, as shown by our electron microscopy study [1] or by the results from literature: this suggests that there is no strong interaction between SbSnO solid solution and Sb_2O_4;

(iv) that antimony oxide in simple mechanical contact with the solid solution (no strong interaction) strongly increases the activity and particularly the selectivity of the solid solution.

If we consider now your second explanation, SMSI, it is incompatible with the above mentioned results : the antimony oxide and the solid solution have no tendency to enter into SMSI association, neither in the fresh catalyst nor in the used catalyst. You certainly remember the results of your group (ref. 3) and those of Viswanathan et al.(ref. 11), which show that, far from having a tendency to mutual contamination, the phases tend to further segregate during the catalytic reaction.

Concerning both of your suggestions, we have an experiment in which we tried to impregnate Sb^{3+}/Sb^{5+} ions over the surface of SnO_2 to form a monolayer of Sb_2O_4 on SnO_2. The characterization results showed that this artificially created contamination decreased during catalytic work [2]. This experiment also strongly suggest the absence of any strong interaction between SbSnO solid solution and Sb_2O_4.

1. P. Patrono, L. T. Weng, E. Sham, P. Ruiz and B. Delmon, XI Simposio Iberoamericano de Catalisis, Guanajuato, Mexico, p.991 (1988)
2. L. T. Weng, N. Spitaels, B Yasse, J. Ladrière, P. Ruiz and B. Delmon, XI Simposio Iberoamericano de Catalisis, Guanajuato, Mexico, p.929 (1988)

G. M. PAJONK (Université Claude Bernard Lyon 1, France) : While listening to your interesting paper, I could not refrain from making analysis with the studies we have made in Lyon for H_2 spillover [1, 2].

I suggest you to activate your "acceptor" as usual and then look at your so activated "acceptor" and perform oxidation reaction as well. Perhaps entirely new type(s) of sites would have been created in this way. At least this procedure would be able to add a contribution to the reality of oxygen spilling species.

1. W. C. Conner, G. M. Pajonk and S. J. Teichner, Adv. Catal., 34 (1986) 1
2. G. M. Pajonk, Proc. 2th Inter. Conf. on Spillover (K. H. Steinberg, Ed.), University of Leipzig, GDR (1989) 1

L. T. WENG (Université Catholique de Louvain, Belgium) : Your beautiful experiments with your experimental system, namely first 'irrigating' the Acceptor phase with spillover species from a Donor, then removing the Donor and measuring the activity of the so activated Acceptor, necessitates lengthy pretreatment and give rise to treated Acceptor with modest activity, which deactivate during experiment. The remote control mechanism, although similar, corresponds to a continuous process during the catalytic reaction in which the oxygen species emitted by the Donor steadily creates the new active and selective centers and/or regenerates the centers which could get deactivated during catalytic reaction; these activated centers work continuously. The magnitude of the synergies observed in our cases (increases by a factor of 2 ($\Delta Y/Y$ = 200%) in the present case, a factor of 1.5 in Sb_2O_4-MoO_3 and 10 for Sb_2O_4-SnO_2) corresponds to a quite spectacular increase. The corresponding activity remains at this high level for the whole course of the experiment. The created sites (on MoO_3 or SbSnO solid solution) are the same as those existing on the Acceptor phase in the absence of Donor (Brönsted sites in the case of MoO_3 [1]) but become more numerous, let us mention, in addition, that the spillover of oxygen has been demonstrated in a specially designed experiment (ref. 16 of our paper).

1. B. Zhou, Ph.D Thesis, Université Catholique de Louvain 1988

J. OTAMIRI (University of Lund, Sweden) : What is the difference of the spillover oxygen species and an oxygen species obtained from molecular oxygen adsorption and activation at the reaction site ?

L. T. WENG (Université Catholique de Louvain, Belgium) : We should indeed distinguish between the roles played (i) by spillover oxygen, produced by the Donor, which creates catalytic sites on the Acceptor phase and (ii) reactant molecular oxygen, which is activated on the reaction sites on the Acceptor phase and is used to oxidize the hydrocarbon. The Acceptor phase possesses all functions necessary for oxidation, namely abstraction of α-H, insertion of oxygen and reoxidation of reduced sites by molecular oxygen. The largest part of molecular oxygen is used directly by the Acceptor phase for reaction ((2)as shown in following schema). The role of the Donor is to dissociate a small part of oxygen to spillover oxygen species and these species migrate to the surface of the Acceptor to improve its catalytic properties, e.g. create new selective sites and/or regenerate the sites which become deactivated during the reaction (2).

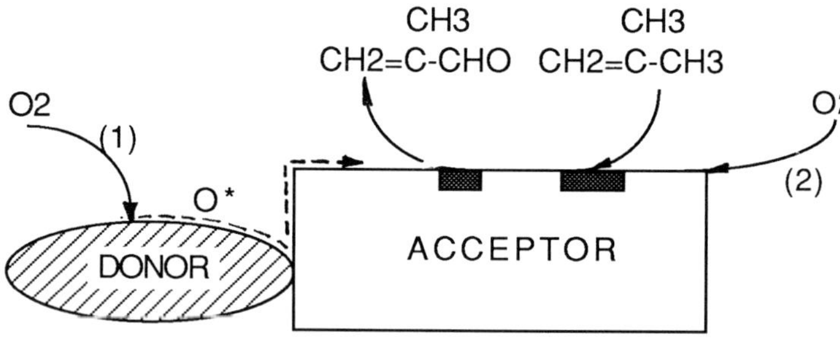

REMOTE CONTROL MECHANISM

G. Centi and F. Trifiro' (Editors), *New Developments in Selective Oxidation*
1990 Elsevier Science Publishers B.V., Amsterdam — Printed in The Netherlands

IR SPECTROSCOPIC IDENTIFICATION OF ADSORBED SURFACE SPECIES ON OXIDATION CATALYSTS EXPOSED TO PROPENE AND ETHENE IN AIR

M.J.PIRES[1], N.T.DO[2], M.BAERNS[2] and M.F.PORTELA[1]

[1] Grupo de Estudos de Catálise Heterogénea, Centro de Processos Químicos (INIC), Universidade Técnica de Lisboa, Instituto Superior Técnico, Aven. Rovisco Pais, 1096 Lisboa Codex (Portugal)

[2] Lehrstuhl fur Technische Chemie, Ruhr-Universität Bochum, Postfach 102148, D-4630 Bochum (West Germany)

SUMMARY

IR spectroscopic studies have shown different surface complexes on three catalysts, i.e. $Bi_2O_3.MoO_3$, γ-Al_2O_3 supported thallium oxide and γ-Al_2O_3 supported metallic silver when exposed under continuous flow to a mixture of C_3H_6 or C_2H_4 and air. Spectra recorded after desorption in air were also different. These results can be related to the different catalytic behaviours observed in previous studies.

INTRODUCTION

Selectivity in hydrocarbon oxidation is associated with the structure and the energy differences between the surface intermediates formed by the hydrocarbon and oxygen and the catalyst.

Infrared spectroscopy is a powerful technique for the study of adsorbed species and a helping tool for the understanding of the mechanisms in complex catalytic reactions. For these reactions we may assume that they involve several chemisorption forms at the catalyst surface, leading to different products.

The aim of this work is the elucidation of the nature of adsorbed propene species on the surfaces of silver, bismuth-molybdate and thallium based catalysts, for which previous studies (refs.1, 2) have shown different activities (epoxidation, allylic and total oxidations).

EXPERIMENTAL

Catalysts

The low temperature $Bi_2O_3.MoO_3$ pure phase was prepared by a reproducible coprecipitation technique (ref.3). X-ray diffraction and Raman spectroscopy did not show traces of impurities. The IR spectrum recorded with the catalyst in air at 303 K was identical to the ones reported in literature (refs. 4,5). XPS confirmed the expected Bi/Mo atomic ratio on the surface. The BET surface area was 2.1 m^2 g^{-1}.

Thallium(III) oxide catalyst was prepared from thallium(III) nitrate by addition of nitric acid (pH = 1.5) and ammonium hydroxide (pH = 8). The precipitate was dried 2 h at 353 K and activated under air at 573 K (4 h).

The silver supported in γ-alumina catalyst was prepared by dry impregnation technique, with silver nitrate, followed by drying and further reduction with formaldehyde (8 h at 393 K).

<u>IR equipment and method of investigation</u>

For measuring infrared spectra a double beam IR spectrophotometer (Perkin-Elmer, model 580A) attached to a minicomputer (Dietz, model 621 x 2/Mulheim) was used. Two cells built after Gallei and Stolz (ref. 6) and described by Ramstetter (ref. 7) were incorporated into the spectrometer. Adsorbate spectra were obtained by compensating the overall spectrum by subtraction of the spectra of the gasphase and the clean catalyst from it. A detailed description of the method was given by Baerns and Ramstetter (refs. 8,9).

<u>Experimental conditions of irs measurements</u>

To obtain disks of good quality all catalysts except Bi_2O_3,MoO_3 were mixed with γ-Al_2O_3 as support before pressing the disk. BET surface areas after this pretreatment were 140 and 156 m^2 g^{-1} for silver and thallium oxide catalysts. Each catalyst was left overnight prior to the experiment under air stream at temperatures lower than 720 K and was tested in continuous flow adsorption runs with C_3H_6-air mixture followed by desorption runs in air at the same temperature. The gases were dried and purified before contact with the catalyst. Adsorbate formation lasted for several hours. Spectra corresponding to different adsorbed amounts were recorded as function of time. Experimental conditions were empirically established by previous tests, including checking of the inertness of alumina for the used experimental conditions.

With Bi_2O_3.MoO_3 catalyst no adsorption was observed at temperatures above 473 K; thallium oxide and silver required higher olefin concentration in order to observe any adsorbates.

RESULTS

Olefin species adsorbed on the catalysts were identified by

their ir spectra after adsorption and after subsequent desorption.

$Bi_2O_3.MoO_3$

A) Adsorption and desorption of C_3H_6 in the presence of air

Fig.1 presents the characteristic bands observed in the range 4000 - 800 cm^{-1}.

In the region near 3000 cm^{-1}, corresponding to C-H stretching vibrations, bands above 3000 cm^{-1} suggest that the hydrocarbon fragment is olefinic, i.e., propene is absorbed on the surface without breaking the double bond; this is confirmed by the presence of ir bands due to out-of-plane deformation vibrations C-H at 990 and 910 cm^{-1}. The band at 3450 cm^{-1} developped after 15 h at 303 K can be reasonably assigned to an OH frequency; hence it appears that dissociation accompanies propene adsorption. At 373 K a different spectrum was recorded in this region: bands at 3450, 3080 and 2860 cm^{-1} disappear while the intensities of other bands at 3100, 2960, 2935 and 2885 cm^{-1} are reduced by broadening.

In the range of 1600-1200 cm^{-1} the bands 1475 and 1445 cm^{-1} can be assigned to C-H deformation vibrations, but there are other unusual bands: 1665, 1655, 1640, 1560, 1545, 1510 and 1340 cm^{-1}.Considering the double-bond stretching vibration band at 1652 cm^{-1} for gaseous propene one could interpret the values around 1545 cm^{-1} as a band shifted due to the interaction of the double bond with the surface. The presence of the 1655 cm^{-1} band at the same time would indicate a partial interaction of π-bonding. The same effect of broadening is shown in the spectrum at 373 K for the 1650, 1530, 1400 and 1340 cm^{-1} bands.

During desorption of C_3H_6 from a Bi/Mo oxide surface at 303 K the following observations were made:

1. Olefinic C-H band 3080 cm^{-1}, 990 and 910 cm^{-1} bands disappeared.
2. Bands at 1665-1640 cm^{-1} were also removed after 3 h in air.

These bands are possibly due to a more weakly bound form of propylene. A longer desorption time did not essentially change the surface except that a band developped at 1465 cm^{-1} and that the two other bands at 1540 and 1560 cm^{-1} disappeared.

The effect of cleaning of the surface by desorption at 373 K during 3 h in air was also observed but a new thin band was detected at 1510 cm^{-1}.

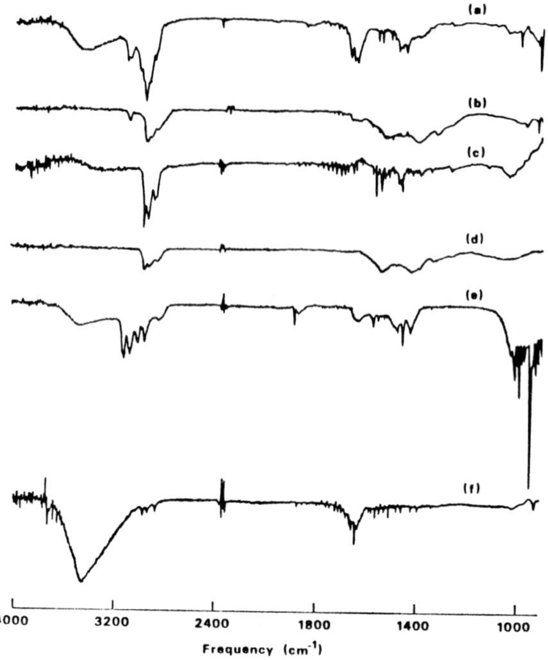

Fig.1. Compensated IR spectra of $Bi_2O_3.MoO_3$ after adsorption of 15% C_3H_6 in air (a) 303 K (15 h), (b) 373 K (20 h) and desorption in air (c) 303 K (14 h), (d) 373 K (3 h); after adsorption of 40% C_2H_4 in air, (e) 303 K (10 h) and desorption in air, (f) 303 K (15 min).

B) Adsorption and desorption of C_2H_4 in the presence of air

The same disk of $Bi_2O_3.MoO_3$, after being heated at 673 K during 1.5 h under air, was tested in an adsorption run with ethene at 303 K, followed by desorption at the same temperature.

The adsorption spectrum after 1 h showed nothing. But after 10 h there was a development of bands, resulting in a rather complex spectrum. By comparing these results with the characteristic bands of gaseous ethene (see Fig.1) we verify the presence of all these bands and other little ones at 1660, 1625, 1565 and 1510 cm^{-1} already detected during propene adsorption.

Fifteen minutes under desorption conditions were enough to change the catalyst surface which presented mostly a development of the bands of adsorbed water at 3450 cm^{-1} and 1640 cm^{-1}.

Thallium oxide/γ-Al_2O_3

The results obtained over the supported thallium oxide are in Fig.2. The spectra were recorded in the range 4000-600 cm^{-1} but below 1000 cm^{-1} the catalyst was not transparent. Adsorption and desorption spectra of C_3H_6 on the supported thallium oxide show no bands in the frequency range 4000-2000 cm^{-1}.

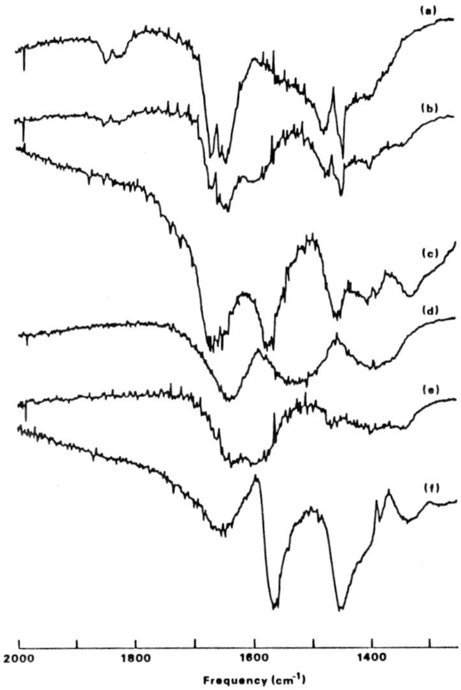

Fig.2. Compensated IR spectra of thallium oxide/γ-alumina after adsorption of 30% C_3H_6 in air (a) 303 K (24 h), (b) 373 K (12 h), (c) 573 K (8.5 h) and desorption in air, (d) 303 K (4 h), (e) 373 K (4 h), (f) 573 K (8).

Adsorption spectra at 303 K show bands at 1985, 1845 and 1825 cm^{-1} which decrease with temperature and disappear at 573 K. The double bond stretch vibration band (1670-1640 cm^{-1}) was always observed. But some changes occur in the region near 1600 cm^{-1}:

1. A broad band arises around 1595 cm^{-1} at 373 K.
2. Adsorption at 573 K during 1.5 h leads to a better definition of the broad band at 1595 cm^{-1} into 1580, 1575 and 1565 cm^{-1} and to a new band at 1335 cm^{-1}.

Spectra recorded after desorption runs show a cleaner surface at 303 and 373 K. But at 573 K after 8 h in air the spectrum is not very different from the one recorded after adsorption run at the same temperature except the relative intensities of bands around 1655 and 1575 cm^{-1}. Bands under desorption conditions become better defined and seem to increase with time.

Silver/γ-Al_2O_3

This catalyst was studied in the range 4000-800 cm^{-1}.

After 1.5 h and 4 h at 303 K under a mixture of 15% C_3H_6 in air the catalyst did not present measurable adsorption. In Fig.3 spectra show only a few adsorbed species under 30% C_3H_6 in air and indicate fast and complete desorption at both temperatures.

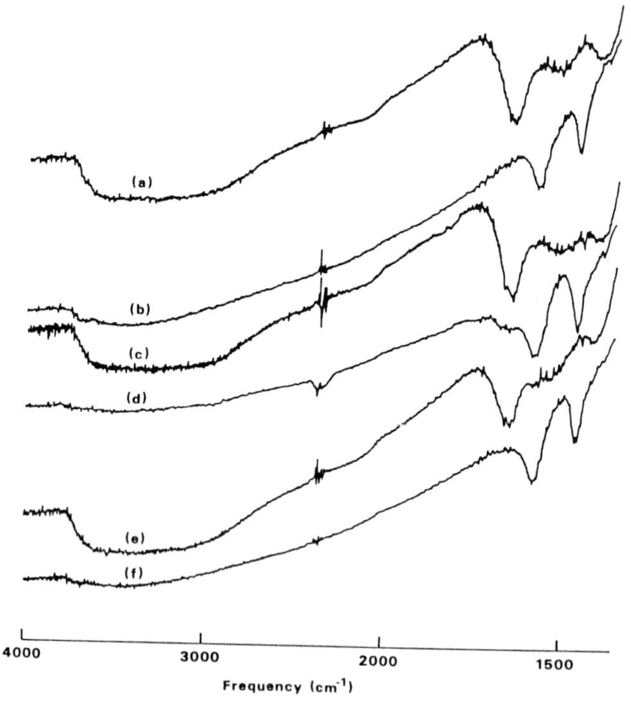

Fig.3. IR spectra of silver/γ-alumina in air (a) 303 K, (b) 623 K; after adsorption of 30% C_3H_6 in air (c) 303 K (1.5 h), (d) 623 K (9 h) and desorption in air (e) 303 K (3 h), (f) 623 K (15 min).

DISCUSSION AND CONCLUSIONS

According to literature there are two adsorbed forms of C_3H_6 at the surface of catalytic oxides. Davydov and Budneva (ref.10) refered a reversible weakly bounded form, precursor of π-allylic and σ-allylic complexes and a irreversible one which undergoes dissociation on desorption. The latter will be responsible of oxidated complexes, carbonate and carboxylate type and π-complexes. Gerei et al. (ref.11) observed on the total oxidation catalysts an adsorbed propylene mainly in an irreversible form with evidence for double bond scission. In the mild oxidation catalysts the double bond is preserved though somewhat perturbed; the result being a π-complex, weakly bound and precursor of the π-allylic complex. Both complexes are also mentioned by Dent and Kokes (ref.12). Force and Bell (refs. 13,14) refer the formation of carbonated species over metallic silver; the band 2180 cm^{-1} can be assigned to a carbon-metallic structure Ag-CO.

Table 1 presents the characteristic bands of the different adsorbed species, possible intermediates in total and mild oxidation of C_3H_6, according to literature. The comparison with the bands observed in our work leads to the following conclusions:

1. The $Bi_2O_3.MoO_3$ catalyst seems to present π-allylic complexes as well as precursors of total oxidation even at low temperatures. This result agrees with our previous studies (ref.1). The structures of irreversible form leading to the total oxidation could be

TABLE 1

IR bands of C_3H_6 adsorbed species according to literature

		Davydov and Budneva (ref.10)	Gerei et al. (ref.11)	Dent and Kokes (ref.12)
Reversible Form	π-allylic complex	1440 1350	no π-allylic complex	1545
	σ-allylic complex	1600 1580		
	weakly bound π-complex		1626 1430 1364	1620
Irreversible Form	strongly bound π-complex	1510 1410		
	Formate	1560 1370	1575 1390	
	Acetate	1560 1450	1575 1440 1410	
	Carbonate	1620-1660 1320	1730 1654 1610 1550	

explained by the desorption spectra: bands at 1540-1560, 1465 and 1385 cm^{-1} can be assigned to a formate (bands at 1560 and 1370

cm^{-1}) as well as an acetate structure (bands at 1540-1560 and 1450 cm^{-1}). The CH stretching vibration of both complexes absorb at 2870 and 2950 cm^{-1}.

2. The surface complexes on Ag^{o}/γ-Al_2O_3 seem to be carbonated species, precursors of total oxidation in agreement with the low epoxidation selectivities of this catalyst. Bands of adsorbed propene on Ag^{o} at 303 K at 1840, 1820, 1655 and 1440 cm^{-1} correspond to those of unadsorbed propene (bands at 1830, 1810, 1630-1660 and 1444-1476 cm^{-1}). Physisorbed propene desorbs by purging with air at 303 K and was not observed by adsorption at 623 K. The bands at 1650 and 2180 cm^{-1} observed at 623 K can be assigned to a carbonate and a Ag-CO structure.

3. γ-Al_2O_3 supported thallium oxide also shows precursors of total oxidation, whose presence becames more evident at higher temperature. Propene adsorbs on thallium oxide at low temperature (303 K) as on Ag. The bands at 1565-1580 and 1335 cm^{-1} observed at high temperature (573 K) can be assigned to a formate structure.

4. Ag^{o}/γ-Al_2O_3 catalyst presents a clean surface 15 minutes after desorption in air, unlike $Bi_2O_3.MoO_3$. This fact may indicate different binding energies for the surface complexes on both catalysts. Thallium oxide presents a different behaviour under desorption conditions at 573 K characterized by a better definition and development of the bands.

The assignment of the bands around 1650 cm^{-1} is rather difficult. They are present in the adsorbate spectra of the three catalysts. This is the region of the double bond stretch vibration but it is also assigned to some carbonate structures.

AKNOWLEDGEMENT

The experimental work has been supported by a grant of the Volkswagen Foundation.

REFERENCES

1 M.F. Portela, M.M. Oliveira, M.J. Pires, F.M.S. Lemos and L. Ferreira in: Proceedings of the 8th International Congress on Catalysis, Verlag Chemie, Berlin, 1984, II 533.
2 M.F. Portela, C. Henriques, M.J. Pires, L. Ferreira and M. Baerns, Catalysis Today, 1 (1987) 101.
3 M.J. Pires, M.F. Portela, M. Oliveira, A. Saraiva and T. Miranda in: Proceedings of the 7th Iberoamerican Symposium on Catalysis, La Plata, Argentina, 1980, 189.
4 F. Trifiro, H. Hoser and R.D. Scarle, J. Catal., 25 (1972) 12.
5 P.A. Batist, A.H.W.M. Kindern, Y. Leeuwenburg, F.A.M.G. Metz and G.C.A. Schuit, J. Catal., 12 (1968) 45.

6 E. Gallei and H. Stolz, Appl. Spectrosc., 28 (1974) 430.
7 A. Ramstetter, Doctoral Dissertation, Ruhr-Universitat Bochum, 1983.
8 A. Ramstetter and M. Baerns, Ber. Bunsenges. Phys. Chem., 86 (1982) 1156.
9 A. Ramstetter and M. Baerns, J. Catal., 109 (1988) 303.
10 A.A. Davydov and A.A. Budneva, Kinet. Katal., 15 (1974) 1557.
11 S.V. Gerei, E.V. Rozhkova and Y.B. Gorokhovatsky, J. Catal., 28 (1973) 1.
12 A.L. Dent and R.J. Kokes, J. Amer. Chem. Soc., 92 (1970) 6718.
13 E.L. Force and A.T. Bell, J. Catal., 3 (1975) 440.
14 E.L. Force and A.T. Bell, J. Catal., 40 (1975) 356.

G. Centi and F. Trifiro' (Editors), *New Developments in Selective Oxidation*
1990 Elsevier Science Publishers B.V., Amsterdam — Printed in The Netherlands

OXIDATION OF PROPENE AND TOLUENE ON CoO-MgO SOLID SOLUTIONS: A SPECTROSCOPIC STUDY.

E. GIAMELLO, E. GARRONE, S. COLUCCIA, G. SPOTO and A. ZECCHINA.

Dipartimento di Chimica Inorganica, Chimica Fisica e Chimica dei Materiali, Università di Torino, Via P. Giuria 7, 10125 Torino Italy.

ABSTRACT

IR evidence shows that the oxidation of propene and toluene with dioxygen on 5% CoO-MgO solid solution basically follows the same course as on MgO, although at much higher rate. Reaction products are acetates and formates, and benzoates respectively. EPR studies indicate that surface Co ions are able to coordinate dioxygen in a fashion close to oxygen natural carriers. Active species are superoxo radicals formed on Co ions but spilled over Mg cations. A detailed reaction scheme is proposed.

INTRODUCTION

Diluted solid solutions of CoO in MgO are model systems for the study of the basic features of adsorption and catalysis on transition metal ions (ref. 1). Co^{2+} ions dispersed in the bulk of MgO are in the octahedral coordination typical of rocksalt structure. The coordination number of surface ions ranges instead between 5 and 3 (refs. 2-4): 5-coordinate ions are found at the planar (100) faces, 4-coordinate at steps or edges and 3-coordinate at vertexes and kinks of cubic microcrystals, respectively. Surface Co ions show an interesting coordination chemistry, in particular as far as dioxygen is concerned. Five-coordinate Co^{2+} ions are able to bind reversibly O_2 forming superoxo adducts (refs. 5-8): the solid solution behaves as a heterogeneous analogue of the homogeneous oxygen carriers.

The ability to form surface superoxo complexes, by mere O_2 adsorption, makes the CoO-MgO system of particular interest in the study of catalytic oxidation of hydrocarbons in that the superoxide ion is an intermediate in many oxidation reactions, e.g., those leading to the cleavage of the double bonds (ref. 9).

The MgO matrix itself is not inert: several hydrocarbons undergo oxidation at the MgO surface, though at low rate, following a complex mechanism involving the heterolytic dissociation of a C-H bond at the basic O^{2-} sites and the superoxide formation via electron transfer from the resulting carbanion to the O_2 molecule (refs. 10-11).

CoO-MgO catalysts offers both functionalities, Co^{2+} ions activating the O_2 molecule and O^{2-} ions activating the hydrocarbon. For this reason we have studied, by coupling infrared and EPR spectroscopies, the mechanism of interaction of molecular oxygen and propene or toluene at the surface of CoO-MgO.

EXPERIMENTAL.

CoO-MgO solid solutions have been prepared by decomposition of mixed hydroxides following a method thoroughly described elsewhere (ref. 2). Surface area of the samples varies between 200 and 120 m^2g^{-1} in the range of composition 1-5% Co molar fraction. Samples were outgassed under vacuum at 1073 K in situ, then contacted with oxygen or with a mixture of O_2 and the hydrocarbon at a given temperature. IR spectra were taken by means of a PE 580B spectrometer equipped with a data station both at room temperature and at a nominal temperature of 77 K. EPR spectra have been recorded usually at 77 K on a Varian E109 machine using Varian Pitch (g = 2.0028) for g values calibration.

RESULTS AND DISCUSSION.

a) Spectroscopic features of oxygen adducts on CoO-MgO.

Oxygen adsorption at 77 K on CoO-MgO solid solutions gives rise to both EPR and IR signals. The EPR spectrum, recorded at low oxygen pressure in order to avoid line broadening, already reported (ref. 7), is due to two slightly different paramagnetic centres having an apparently axial g tensor and a hyperfine structure caused by the ^{59}Co nucleus (I=7/2). The two centres are Cobalt superoxo adducts characterised by a bent, "end-on" structure with a significant degree of electron transfer, so to be roughly describable as $Co^{3+}O_2^-$ (ref. 8).

Owing to their bent structure, the oxygen species are IR active. Actually, several absorption bands are seen in the region 1050-1200 cm^{-1}, whose intensity is pressure dependent (Figure 1). At low oxygen pressure (i.e. in the circumstances needed for EPR spectra recording), a band is seen at 1132 cm^{-1} together with some shoulders at lower frequency.

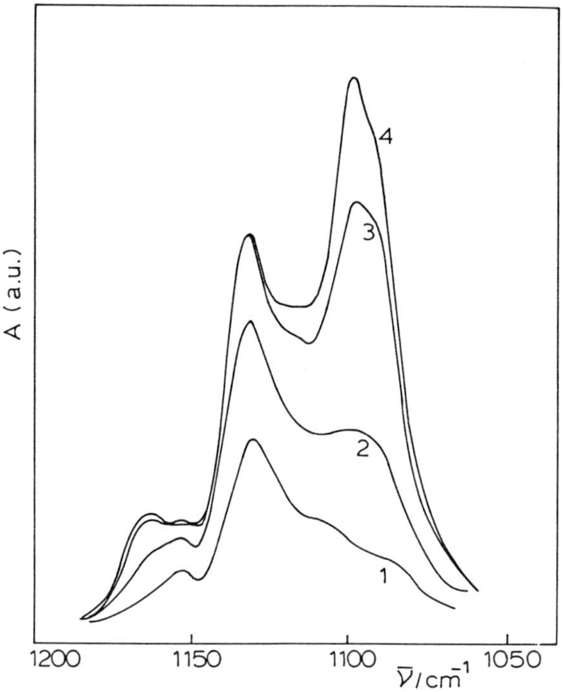

Figure 1. IR spectra of dioxygen adsorbed on CoO-MgO 5% at 77 K (Absorbance vs wavenumber). Oxygen pressure less than 0.1 torr in curves 1 and 2; 4.0 torr in curve 3; 12 torr in curve 4.

Homogeneous Co-based oxygen carriers invariably show a square pyramidal coordination of the central ion: 5-coordinate Co^{2+} ions at the (100) faces exhibit the same geometrical arrengment, and are likely to be involved in O_2 binding. The striking variety of alike species is amenable to structural effects (ref. 6). As to the 4- and 3-coordinate Co^{2+} ions, it is probable that, for the same reason, they do not form stable O_2 adducts, in contrast with the general observation that the lower the coordination of a surface ion, the higher the activity in adsorption.

When oxygen is first adsorbed at room temperature under a mild pressure, the spectrum recorded at 77 K (Figure 2) is due to the superposition of a structured signal due to a cobalt superoxo adduct, whose features are labelled A in the Figure, with that of a superoxide O^{2-} ion (indicated with C) adsorbed onto a Mg^{2+} ion at the surface. Species A disappears by pumping at room temperature, whereas species C is stable. Species A is restored by oxygen readsorption: such a reversibility indicates that a fraction of the Co^{2+} ions behaves as oxygen carriers. Note that species A does not correspond, however, to any of the species formed by adsorption at 77 K.

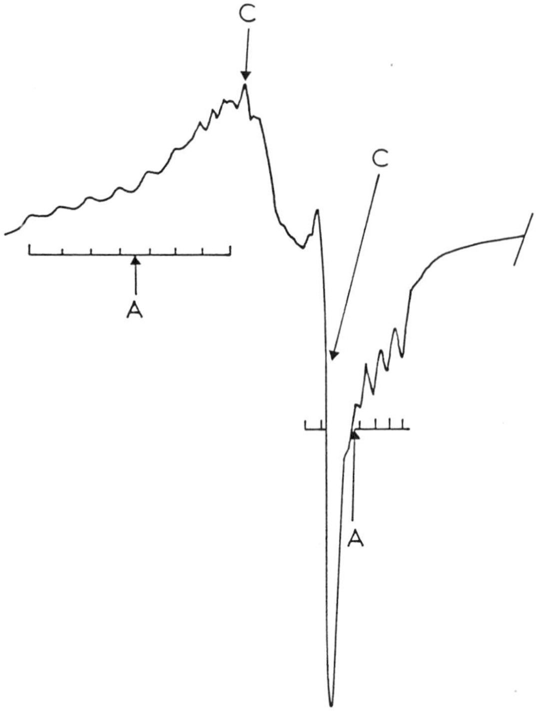

Figure 2. EPR spectrum of superoxo species formed at the surface of CoO-MgO 5% after room temperature adsorption of oxygen (pressure less than 0.1 torr). A = $Co^{3+}O_2^-$; C = O_2^- on Mg^{2+}.

The process leading to the spectrum in Fig. 2 probably is:

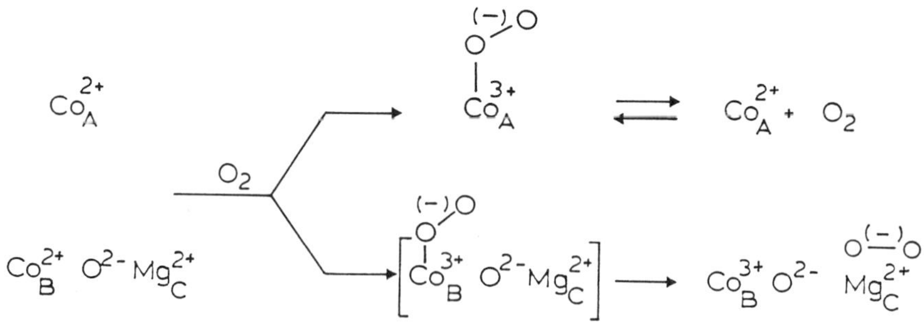

O_2^- sitting on Mg^{2+} is formed by spillover of the superoxo species from a second family of cobalt ions ($Co^{2+}{}_B$) which are irreversibly oxidized by interaction. As to the nature of the B family of Co ions, it is feasible to assume that these are the low-coordinate species, unable to coordinate stably dioxygen. The driving force for the oxidation of B ions is, on the one hand, the gain in Madelung energy of the Mg^{2+} ion, brought about by O_2^- coordination, and, on the other hand, the stabilization of Co ions in low coordination when in the trivalent state, because of the definite preference of Co^{3+} for tetrahedral coordination. All this probably renders the oxidation of $Co^{2+}{}_B$ ions thermodynamically favoured. Some activation energy seems however to be required.

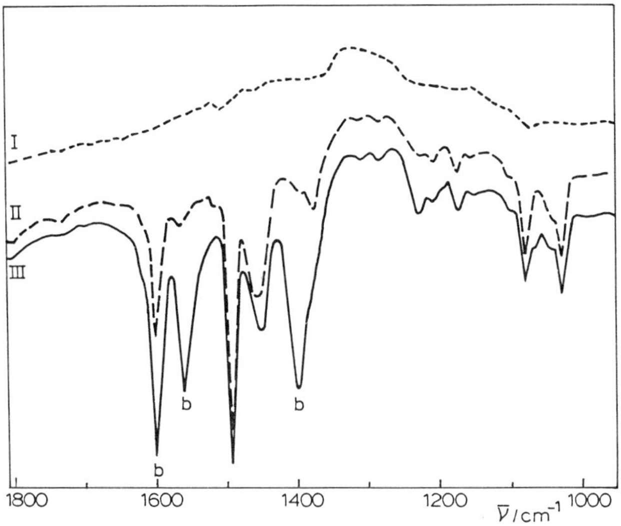

Figure 3. IR spectra concerning the interaction toluene/dioxygen. (Transmission vs wavenumbers). Curve I: background; Curve II: 15 torr toluene; Curve III: 15 torr O_2 added, after 2.5 hours contact and evacuation.

b) Oxidation of toluene and propene.

IR spectra in Figure 3 illustrate the reaction between toluene and oxygen at room temperature. Bands due to reaction products are readily assigned to adsorbed benzoates anions. A similar experiment concerning the coadsorption of propene and oxygen (spectra not reported) leads to the formation of adsorbed acetates and formates, thus clearly showing the cleavage of the double bond.

Oxidation products are in both cases the same observed when the reactions are carried out at the surface of MgO. The reaction rate, however, is higher in the case of the CoO-MgO solid solutions by a factor of at least 10.

In order to elucidate the role of the transition metal ion, the effect of toluene admission on a sample of CoO-MgO previously contacted with oxygen at room temperature and kept at 77 K, has been followed in EPR (Figure 4). The starting spectrum is similar to that reported in Figure 2, and clearly shows the features of both $Co^{3+}O_2^-$ and $Mg^{2+}O_2^-$. Admission of a small toluene pressure at 77 K causes the interconversion of the former species into the latter, as witnessed by the disappearence of the hyperfine features of the Cobalt centre and by the constancy of the overall intensity as results from spin counting. Warming up of the sample to room temperature causes the rapid disappearence of the EPR spectrum due to the onset of the oxidation process that consumes the superoxide ion.

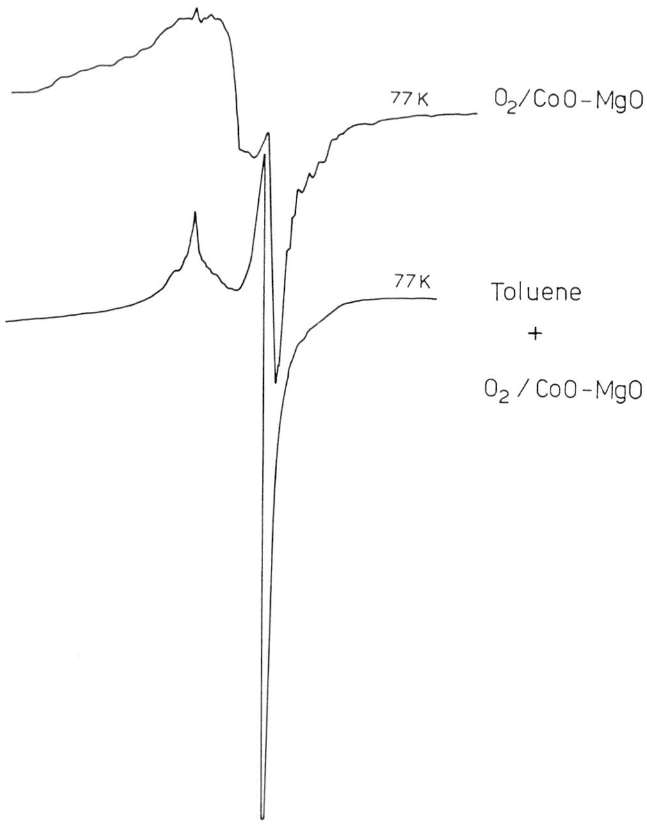

Figure 4. EPR spectra at 77 K concerning the interaction of oxygen and toluene on CoO-MgO 5%. Oxygen pressure less than 0.1 torr; toluene pressure 25 torr.

The above experiment allows to establish the basic patterns of toluene (and propene) oxidation on CoO-MgO. Contrary to what expected, the superoxo adducts bound to cobalt are not the active intermediates as such, as they spillover onto Mg^{2+} in the presence of the hydrocarbon. The role of the cobalt seems instead to be to coordinate, probably dissociatively, the hydrocarbon molecule in cooperation with a basic O^{2-} anion. Further steps of reaction are the formation of the neutral radical species HO_2, extremely active (ref. 10), and the yield of a peroxo species XO_2H, as shown in the following scheme:

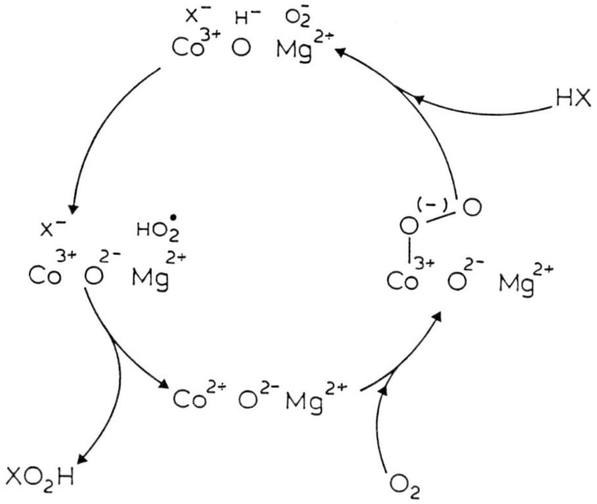

The fate of this latter is probably different for propene and toluene. In the former case, the allylperoxide $CH_2{=}CH{-}CH_2OOH$ decomposes into acetic and formic aldehydes: in the latter, benzylperoxide $C_6H_5CH_2OOH$ dehydrates to benzoic aldehyde. All kinds of aldehydes are then oxidized to acids following a mechanism already proposed, envisaging the abstraction of a hydride species (as in the related Cannizzaro disproportionation) and its oxidation to water (ref. 10):

$$R{-}CHO + O^{2-}{}_{cus} \longrightarrow R{-}COO^- + H^-$$

$$2\ H^- + O_2 \longrightarrow 2\ OH^-$$

Both reaction products (benzoates species and water) are strongly held by the surface at room temperature and gradually inhibit the catalytic cycle with the eventual loss of any activity of the sample.

REFERENCES

1 A. P. Hagan, M. G. Lofthouse, F. S. Stone and M. Trevethan, in B. Delmon, P. Grange, P. Jacob and G. Poncelet (eds), Preparation of Catalyst II: Scientific Bases for the Preparation of Heterogeneous Catalysts, Elsevier, Amsterdam, 1979, p. 417.
2 A. Zecchina, G. Spoto, S. Coluccia and E. Guglielminotti, J. Phys. Chem., **88** (1984) 2575.
3 A. Zecchina, G. Spoto, E. Borello and E. Giamello, J. Phys. Chem., **88** (1984) 2582.
4 E. Giamello, E. Garrone, E. Guglielminotti and A. Zecchina, J. Mol. Catal., **24** (1984) 59.
5 A. Zecchina, G.Spoto and S. Coluccia, J. Mol. Catal., **14** (1982) 351.
6 E. Giamello, Z. Sojka, M. Che, and A. Zecchina, J. Phys. Chem., **90** (1986) 6084.
7 Z. Sojka, E. Giamello, M. Che, A. Zecchina and K. Dyrek, J. Phys. Chem., **92** (1988) 1541.
8 M. Che, K. Dyrek, E.Giamello and Z. Sojka, Zeit. Phys. Chem. Neue Folge,**152** (1987) 397.
9 J. Haber, Proc. 8th Int. Congr. Catal., Verlag Chemie, Weinheim, Berlin, 1984, vol. 1, p. 85.
10 E. Garrone, E. Giamello, S. Coluccia, G. Spoto and A. Zecchina, in M. J. Phillips and M. Ternan (Eds), Catalysis: Theory to Practice. Proc. 9th Int. Congr. Catal.,The Chemical Institute of Canada, 1988, vol. 4, p.1577.
11 E. Garrone and E. Giamello, unpublished data.

G. Centi and F. Trifiro' (Editors), *New Developments in Selective Oxidation* 825
© 1990 Elsevier Science Publishers B.V., Amsterdam — Printed in The Netherlands

FT-IR STUDY OF SELECTIVE OXIDATION INTERMEDIATES OF BENZENE ON THE SURFACE OF VANADIA-TITANIA "MONOLAYER" CATALYSTS

Guido Busca*, Gianguido Ramis and Vincenzo Lorenzelli
Istituto di Chimica, Facoltà di Ingegneria, Università, P.le Kennedy, 16129 Genova (Italy).

SUMMARY.

The adsorption of benzene and the evolution of the adsorbed species upon heat treatment in oxygen over vanadia-titania have been studied by FT-IR spectroscopy. For comparison analogous experiments have been carried out by adsorbing phenol and p-benzoquinone. It has been demonstrated that benzene reacts with the clean surface producing different phenol and quinone species, some of which elvolve to maleic anhydride while other produce carboxylate species and carbon oxides by reaction in oxygen. Some modifications on the reaction network proposed in the literature for the heterogeneously-catalyzed oxidation of benzene have been suggested.

INTRODUCTION

The selective oxidation of benzene represented for many years, and still is in western Europe, the main industrial process for the synthesis of maleic anhydride (1). The catalysts used are constituted by vanadium-molybdenum mixed oxides (V:Mo > 2:1) supported on low-surface-area inert carriers such as corundum-type alumina (2,3). Titania-supported vanadia-molybdena catalysts (4,5) have also been patented and their behaviour has been investigated.

Between pure oxides vanadia is reported to be the most selective catalyst for this reaction (6). It has been demonstrated that it works in a slightly reduced state (7). Mixing with small amounts of molybdena improves its selectivity (7) while supporting on titania strongly increases its activity (8). Vanadia supported on titania is a well characterized selective oxidation catalyst (9) representing the basic formulation of the industrial catalysts for the phtalic anhydride synthesis by o-xylene oxidation.

The present paper reports the results of our attempts to identify surface intermediates

---

* to whom correspondence should be addressed

that may be involved in the selective and unselective benzene oxidation pathways on vanadia-titania using FT-IR spectroscopy.

EXPERIMENTAL

The TiO_2 support was P25 (90% anatase from XRD, 53 m^2/g) from Degussa (Hanau, West Germany). Vanadia-Titania was prepared by dry impregnation with boiling water solution of ammonium metavanadate (Carlo Erba, Milano), followed by drying and calcination in air at 720 K for 3 h. The loaded amount was 10% as V_2O_5 by weight, slightly higher than that needed to complete the theoretical monolayer (9). The catalyst surface area was 48 m^2/g.

The IR spectra were recorded by a Nicolet MX1 Fourier Transform instrument, connected to conventional evacuation/gas-manipulation ramp and IR cell (NaCl windows). The powder samples were pressed into self-supporting disks of appropriate thicknesses and activated in the IR cell by evacuation at 673 K before adsorption experiments.

RESULTS AND DISCUSSION

Fig. 1,a shows the spectrum of the surface species arising from the irreversible adsorption of benzene at r.t. on vanadia-titania. The comparison with the spectra of benzene, both liquid and adsorbed as such on pure titania (10), evidences that reaction occurred, producing several strongly adsorbed species. Pure and adsorbed benzene in fact only show one strong IR active band in the region 1700-1100 cm^{-1} (1479 cm^{-1} for pure benzene, poorly shifted upon adsorption). In Fig. 1,a instead we observe at least four sharp bands in the region 1620-1570 cm^{-1}, and four more bands in the region 1500-1440 cm^{-1}. These regions are those where the 8, a and b, and 19, a and b, ring vibrations of substituted benzenes fall

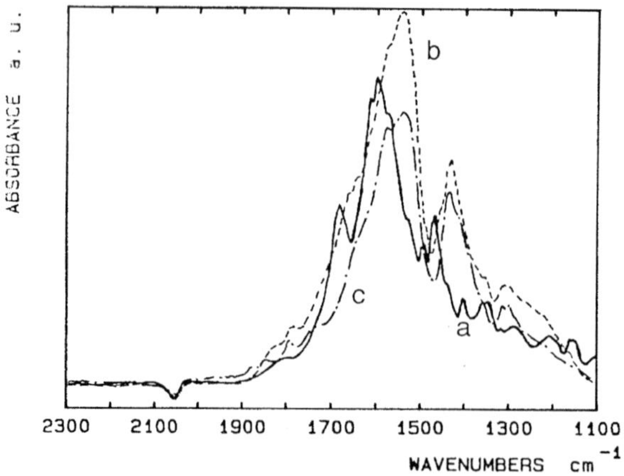

Fig. 1. FT-IR spectra of the adsorbed species arising from the adsorption of benzene on vanadia-titania at r. t. (a) and after successive contact with O_2 200 torr at 150 °C (b) and 200 °C (c).

respectively (notation of Wilson (11)). Moreover, we observe also bands in the 1900-1630 cm^{-1} region, reasonably due to the C=O stretching vibrations of carbonyl compounds. According to the conditions of our experiment (adsorption at r.t. on the clean surface without gas-phase oxygen), substituents on the benzene ring in these surface species would be oxygen atoms arising from the surface. The formation of C-V or C-Ti bonds, as well as of polymerization products (polyphenyls), seems unlikely on chemical bases, and would not justify the observed spectra. If comparison is made with the spectrum of phenol adsorbed in the same conditions (Fig. 2,a) we note that all main bands are present even after benzene adsorption, although very weak. Bands at 1590 (ring vibration ν8a), 1575 (ν8b), 1488 (ν19a), 1482 (ν19b), 1225-1210 (νC-O), 1165 (ν9a), and 1155 cm^{-1} (ν9b) can be then

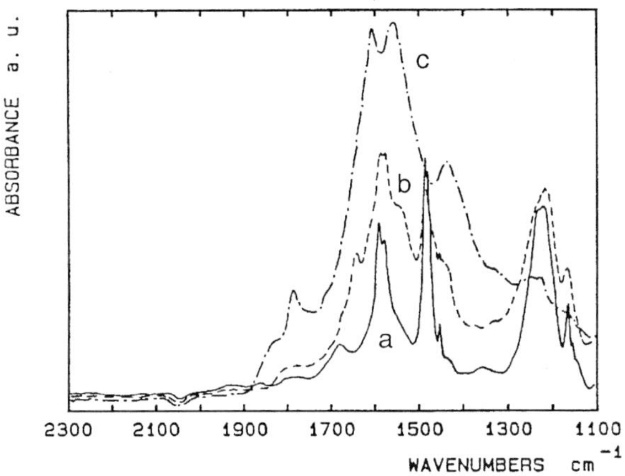

Fig. 2. FT-IR spectra of the adsorbed species arising from the adsorption of phenol on vanadia-titania at r. t. (a) and after successive contact with O_2 200 torr at 100 °C (b) and 200 °C (c).

assigned to surface phenate species (12). The presence of all these bands, although very weak, after benzene adsorption indicates that substitution of an hydrogen by an oxide or hydroxide species occurred.

If p-benzoquinone is adsorbed at r.t. on our catalyst (Fig. 3,a) bands very similar to those of the pure compound (13) are observed . However by heat treatment both in the absence and in the presence of oxygen (Fig. 3) it is clear that a second more stable adsorbed species is present, showing rather strong bands at 1620 and 1605 cm^{-1} (probably ring vibrations) and weaker bands at 1662 and 1640 cm^{-1} (νCO), as well as a characteristic weak triplet at 1352, 1320 and 1290 cm^{-1}. All these bands are also observed in the spectrum of benzene transformation products, even before contact with oxygen, but become more evident after heating the adsorbed species in oxygen atmosphere (Fig. 1, b-d). We conclude then that an adsorbed form of p-benzoquinone is also formed by benzene oxidation on the surface of vanadia-titania. The oxidation of phenate ions to p-benzoquinone species by vanadates, as do for exemple chromates in solution, is not surprising, and does not need necessarily dioxygen to be present.

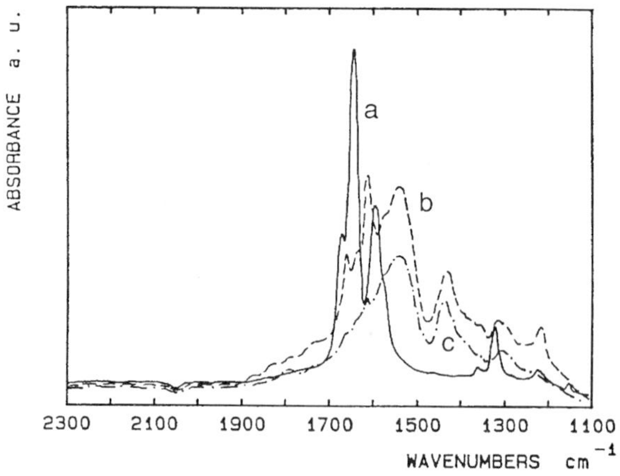

Fig. 3. FT-IR spectra of the adsorbed species arising from the adsorption of p-benzoquinone on vanadia-titania at r. t. (a) and after successive contact with O_2 200 torr at 150 °C (b) and 200 °C (c).

A relatively broad band is also observed near 1680 cm^{-1} after both benzene and phenol adsorption at r.t. (Fig. 1,a and 2,a). This absorption is almost certainly due to a C=O stretching vibration, possibly corresponding to another quinone species. Nevertheless, the spectra of adsorbed p-benzoquinone do not show this band (Fig. 3). The band at 1680 cm^{-1}, relatively strong, seems related to a band at 1400 cm^{-1} and both may be assigned tentatively to adsorbed o-benzoquinone (14). According to the known behaviour of o-quinone, this species is very labile and disappears upon heating in oxygen.

Other bands at 1611, 1598, 1470, 1462 cm^{-1}, well evident when starting from benzene but also present when phenol is adsorbed, evidence the presence of other monosubstituted or bisubstituted aromatic compounds. They can be assigned to an adsorbed form of o-dihydroxy benzene on the basis of the similarity of the observed bands (15).

Heat treatment of adsorbed benzene species in the presence of gas-phase oxygen (Fig. 1, b-c) shows the progressive growth of two relatively broad absorptions at 1540 and 1330 cm^{-1}. Bands having the same frequencies and shapes are also produced by oxidation of butenes and butadiene and by adsorption of maleic anhydride on the same surface (16), and can be assigned to surface maleate species. These bands grow by treatment in oxygen at 100 and 150 °C but decrease at 200 °C, probably because in these conditions the adsorbed species start to decompose giving carbon oxides. In this temperature range bands are also detected at 1850 and 1790 cm^{-1}, with a third component near 1880 cm^{-1}. These bands are very similar to the most intense ones of maleic anhydride (symmetric and asymmetric C=O stretching, respectively (17)) and are very strong when maleic anhydride is directly adsorbed on vanadia-titania (16). The spectrum of the adsorbed species arising from oxidation of adsorbed benzoquinone (Fig. 3, b-c) looks exactly the same to that observed after oxidation of benzene.

In the spectra recorded after oxygen treatment of adsorbed phenol (Fig.2, b-c) the bands due to maleic anhydride appear much stronger than after benzene and quinone oxidation, and grow together with those of maleate species even at 200 °C. The bands of p-quinone are well evident after treatment at 100-150 °C. A band near 1610 cm^{-1} grows by heat treatment of adsorbed phenol in oxygen 200 torr between 150 and 200 °C and might tentatively be assigned to polyphenols produced by the oxidative coupling of phenate species, according to their very high concentration in our experimental conditions.

CONCLUSIONS

The most clear result of the present investigation is that several oxidized species are formed on the surface of vanadia-titania by benzene adsorption in the absence of gaseous or adsorbed dioxygen. Other species are formed by successive contact with oxygen gas at increasing temperatures. For many of these species only tentative assignments can be given. However, the presence of phenate ions and, at least after contact with oxygen, of p-quinone species looks unambiguous. Maleic anhydride and carboxylate species, very probably mainly maleates, are also finally detected. These results parallel those reported in the literature concerning catalytic benzene oxidation over vanadia-titania (8), titania-supported vanadia-molybdena (4) and pure vanadia (6), that reported maleic anhydride, p-benzoquinone and traces of phenol to be the products besides carbon oxides. The formation of several adsorbed phenols and quinones during benzene oxidation over V_2O_5-MoO_3/TiO_2 has also been previously suggested (4).

Even the aromatic C-H bonds, like the more reactive benzylic C-H bonds (18), can then react already at r.t. with surface vanadate species. This reaction is expected to occur faster

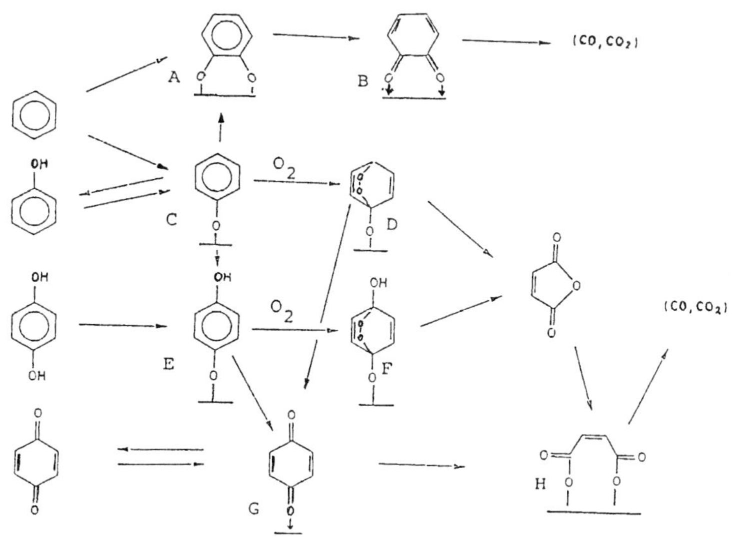

Scheme 1. Proposed reaction network of benzene oxidation on vanadia-titania.

at higher temperatures, as in reaction conditions. This suggests a modification of the mechanism proposed in the literature for the heterogeneously catalyzed oxidation of benzene (scheme 1). Waugh and co-workers (2) proposed as the first step an 1,4-electrophilic reaction of an adsorbed superoxo-species with an adsorbed benzene molecule, that would produce hydroquinone. This mechanism has been suggested on the basis of the results of TPD experiments and justified on theoretical bases. Successively, Haber and co-workers (19) confirmed on quantum chemical grounds that dioxygen can react with unperturbed benzene only in the state of O_2^- superoxide species. To match the hypotheses of their theoretical study (dioxygen approaching parallel to an unperturbed although adsorbed benzene molecule) these authors speculated about the possibility of a vertical adsorption mode of benzene on V^{5+} sites and of dioxygen on a near V^{4+} site. It is however reasonable to think that this type of reactivity is easier on surface phenate species, that have higher electron density on the aromatic ring, and stand almost certainly vertical on the surface. From our data it seems then reasonable to propose alternatively that benzene can first be activated as a phenate species (C in scheme 1) and later undergo electrophilic attack. According to this, phenol has been reported to be oxidized very rapidly to maleic anhydride (20). On vanadia-silica, if an oxidizing agent different from O_2 (for instance N_2O) is used, phenol may be recovered in relevant yields (21). It is then possible that adsorbed phenates, react rapidly whith dioxygen (adsorbed or gaseous) producing an intermediate (D) that later evolves towards maleic anhydride, possibly giving also p-quinone (G) as a side product. Another difference between scheme 1 and that reported in ref. 2 is that hydroquinone (or its adsorbed form) is not necessarily intermediate in benzene oxidation, although it may also be oxidized to maleic anhydride via a parallel probably faster way (E and F) (2).

The production of p-quinone and its further oxidation to maleate species (H) and carbon oxides, as well as a double substitution on the benzene ring in ortho positions, giving first adsorbed o-dihydroxy-benzene (pyrocatecholate species A), later adsorbed o-benzoquinone (B) and finally carbon oxides, would both represent parallel nonselective oxidation routes. The selectivity to the desired product maleic anhydride can also be lowered by its successive oxidation, probably involving again maleates (H) as intermediates. It may be of interest to remark that while in the case of benzene oxidation neither 1,2- nor 1,4-benzoquinone are thougth to be intermediates in the selective pathway (2), in the case of naphtalene oxidation it has been reported that both 1,2- and 1,4naphtoquinone are intermediates in phtalic anhydride synthesis (22).

On the basis of previous data (16,18) it is reasonable to propose coordinatively unsaturated vanadyl species, probably VO^{2+} that have the character of both Lewis sites and radical centers, as the active sites for benzene activation. The strong bond of phenate species on the relatively strongly acidic (9) surface of vanadia-titania and their tendency to undergo further oxidation prevent their desorption as phenol and, eventually, the production of this molecule by selective oxidation of benzene with O_2.

ACKNOWLEDGEMENTS

This work has been supported by Italian Government (MPI 40%). The technical collaboration of Dr. G. Oliveri is gratefully acknowledged.

REFERENCES

1. G. Chinchen, P. Davies and R.J. Simpson, in "Catalysis, Science and Technology" (J.R. Anderson and M. Boudart eds.), Springer Verlag, Berlin, Vol. 8, 1987, p.1.
2. J. Lucas, D. Vandervell and K.C. Waugh, J. Chem. Soc. Faraday Trans. I, 77 (1981) 15 and 31 ; R.W. Petts and K.C. Waugh, J. Chem. Soc. Faraday Trans. I, 78 (1982) 803.
3. M. Najbar, A. Bielanski, J. Camra, E. Bielanska, W. Wal, J. Chrzaszcz and W. Ormaniec, in "Preparation of Catalysts IV" (B. Delmon, P. Grange, P.A. Jacobs and G. Poncelet eds.), Elsevier, Amsterdam, 1987, p. 217.
4. E. Fiolitakis, M. Schmid, H. Hofman and P.L. Silveston, Can. J. Chem. Eng. 61 (1983) 703.
5. O. Rodriguez Dominguez and C. Laguerie, Bull. Soc. Chim. France I - (1983) 155.
6. J.E. Germain and R. Laugier, Bull. Soc. Chim. France (1972) 2910.
7. A. Bielanski, J. Piwowarczyk and J. Pozniczek, J. Catal. 113 (1988) 334.
8. A. Miyamoto, K, Mori, M. Inomata and Y. Murakami, Proc. 8th ICC, Berlin, 1984, Vol. IV, p. 285.
9. G. Busca, G. Centi, L. Marchetti and F. Trifirò, Langmuir 2 (1986) 568 ; G. Busca, Langmuir, 2, (1986) 577.
10. G. Busca, L. Marchetti, T. Zerlia, A. Girelli, M. Sorlino and V. Lorenzelli, Proc. 8th ICC, Berlin, 1984, Vol. III, p. 299.
11. E.B. Wilson, Phys. Rev. 45 (1934) 706.
12. M.F. Berny and R. Perrin, Bull. Soc. Chim. France(1967) 1013; B.Q. Xu, T. Yamaguchi and K. Tanabe, Mater. Chem. Phys. 19 (1988) 291 ; J.C. Evans, Spectrochim. Acta 16 (1960) 1383 .
13. E. Charney and E.D. Becker, J. Chem. Phys. 42, (1965) 910.
14. W. Otting and G. Staiger, Chem. Ber. 88 (1955) 828.
15. H.W. Wilson, Spectrochim. Acta 30A (1974) 2141.
16. G. Busca, G. Ramis and V. Lorenzelli, J. Mol. Catal., in press.
17. P. Mirone and P. Chiorboli, Spectrochim. Acta 18 (1962) 1425.
18. G. Busca, F. Cavani and F. Trifirò, J. Catal. 106 (1987) 471.
19. E. Broclawik, J. Haber and M. Witko, J. Mol. Catal. 26 (1984) 249.
20. J.E. Germain, Catalytic Conversion of Hydrocarbons, Academic Press, New York, 1969, p. 258.
21. M. Iwamoto, J. Hirata, K. Matsukami and S. Kagawa, J. Phys. Chem. 87 (1983) 903.
22. M.S. Wainwright, H. Ali, A.J. Bridgewater and R.P. Chaplin, J. Mol. Catal. 38 (1986) 383

G. Centi and F. Trifiro' (Editors), *New Developments in Selective Oxidation* 833
1990 Elsevier Science Publishers B.V., Amsterdam — Printed in The Netherlands

PROPOSAL FOR ACTIVE SITES OF IRON PHOSPHATES IN ISOBUTYRIC OXIDATIVE DEHYDROGENATION REACTION

Jean-Marc M. MILLET [1], Jacques C. VEDRINE [1] and Gérard HECQUET [2]

[1] Institut de Recherche sur la Catalyse, CNRS associé à l'UCB, Lyon I, 2 avenue A. Einstein, 69626 Villeurbanne (France)

[2] Centre de Recherche du Nord, Orkem, BP 57, 62670 Mazingarbe (France)

ABSTRACT

Iron phosphate based catalysts are claimed to be active and selective for the oxidative dehydrogenation of isobutyric acid to methacrylic acid. In order to progress in the knowledge of these catalysts we have synthesized, studied as catalyst and characterized well defined phases in the ternary phase diagram Fe_2O_3 - FeO - P_2O_5 . The characterization by RGN and XRD of the samples before and after catalysis shows that the starting phases were transformed under the conditions of reaction and that a new phase was formed containing both ferric and ferrous ions in the same framework which should favor, through easier electron transfer, a redox type mechanism. This phase which has also been identified in industrial catalyst could be responsible for the high performances of the latter.

INTRODUCTION

Iron phosphate catalysts are known to be active in isobutyric acid oxidative dehydrogenation to methacrylic acid (ref.1-2). Except alkali metal, which can be present in non negligible amount, the industrial catalysts contain as major elements iron, phosphorus and oxygen with a P/Fe ratio close to one. We have shown previously that, after catalysis, both ferrous and ferric cations are present in the catalysts (ref.3). It is also interesting to note that vanadium phosphate catalysts, used for butane oxidation to maleic anhydride, also contain metallic cations in two valence states (ref.4-5). The presence of both cations in large quantities favor a mechanism of Mars and Van Krevelen type (ref.6).

In order to understand the details of this redox mechanism our effort has been focused on the characterization of the active phase in iron phosphates based catalysts.
In the present work different phases have been chosen within the Fe_2O_3 - FeO - P_2O_5 ternary phase system with a P/Fe ratio close to one. Their catalytic properties have been studied and various physical techniques, before and after catalytic reaction has occured, have been used for characterizing the nature of the different phases.

EXPERIMENTAL

Three iron phosphates have been prepared in the laboratory by the following procedures : $FePO_4$ was prepared by mixing aqueous solutions of $FeNO_3.9H_2O$ and H_3PO_4. The solution was evapored to dryness three times and the residue was subsequently heated to 723 K. $Fe_2P_2O_7$ was obtained by reduction of the latter at 1023 K in a N_2-H_2-H_2O gas mixture. Finally $Fe_7(PO_4)_6$ was formed by reduction in a similar gas mixture of a precipitate obtained with ferric nitrate and di-ammonium hydrogen-phosphate (ref.7).Fe/P content ratio of the solids was checked by chemical analysis using atomic absorption technique and found to be close to the theoretical stoichiometry. B.E.T. area measurements were preformed using a volumetric apparatus by adsorption of nitrogen at liquid nitrogen temperature.

All samples were examined at room temperature by X-ray diffraction using a Siemens D500 diffractometer with CuKα radiation and by Mössbauer spectroscopy using a conventional constant acceleration spectrometer with a 2 GBq ^{57}Co/Rh source. Isomer shifts, given with respect to αFe, and quadrupolar splittings are determined within the precision of 0.02 $mm.s^{-1}$.The accuracy for hyperfine field calculations was 2 kOe.
For the two techniques, peak positions of the spectra were determined by computer fits using a least-squares minimization assuming Lorentzian line shapes.

Oxidative dehydrogenation of isobutyric acid (IBA) to methacrylic acid was carried out at 653 K in a flow differential microreactor containing 50 to 100 mg of catalyst. The total flow rate was 1 $cm^3.s^{-1}$ with IBA : H_2O : O_2 : N_2 = 5.94 : 77.2 : 4.32 : 15.4 expressed in kPa. Catalytic properties were determined in steady state conditions.

RESULTS AND DISCUSSION

It was observed by Mössbauer spectroscopy and X-ray diffraction (ref.3) that $FePO_4$ underwent complete transformation during the catalytic reaction. The $Fe_7(PO_4)_6$ phase was then identified by the two techniques as the major component while a new phase in small amounts was also detected. In the case of $Fe_2P_2O_7$ its caracterization after catalysis showed that the pyrophosphate was only partially transformed and that this transformation was related to the appearance of the same new phase.

As the catalytic efficiency of the phase $Fe_7(PO_4)_6$ alone do not match, under steady state conditions, that of $FePO_4$ and $Fe_2P_2O_7$ which present comparable properties (table 1) we can assume that the presence of the new phase is important to obtain efficient catalysts. Moreover this phase was also observed in the industrial catalyst.

TABLE 1

Catalytic properties of the compounds tested at 653 K.

starting compound	Selectivity CO_2	Propene	Acetone %	MAA[1]	Rate of formation of MAA[1] $mol.s^{-1}.m^{-2}.10^{8}$
$FePO_4$	1	8	25	66	107
$Fe_2P_2O_7$	3	14	16	67	70
$Fe_7(PO_4)_6$	9	7	52	32	184

[1]methacrylic acid

In order to isolate and to study this new phase we have conducted an experiment which consisted of following the oxidation of $Fe_2P_2O_7$ at 723 K under various oxygen partial pressures. The results obtained are presented in table 2 and were given in ref 3.

First it can be observed that at low oxygen partial pressure the oxidation of $Fe_2P_2O_7$ did not yield $FePO_4$ although the reduction of $FePO_4$ under various hydrogen partial pressures always gave $Fe_2P_2O_7$. For partial oxygen pressures less than 27-28 kPa two phases were observed by X-ray diffraction and Mössbauer spectroscopy to be present namely $Fe_2P_2O_7$ and the new phase (A2,A3,A4). The new phase is characterized by the presence of both ferric and ferrous ions with Mössbauer parameters as follows :

$$Fe^{3+} \quad \delta = 0.47 \pm 0.05 \text{ mm.s}^{-1} \qquad \Delta = 0.68 \pm 0.02 \text{ mm.s}^{-1}$$
$$Fe^{2+} \quad \delta = 1.20 \pm 0.05 \text{ mm.s}^{-1} \qquad \Delta = 2.73 \pm 0.02 \text{ mm.s}^{-1}$$

For higher partial oxygen pressures of oxygen the new phase and Fe_2O_3 were detected (A5,A6). As the oxygen partial pressure increased, the amount of Fe_2O_3 detected increased . At 723 K it seemed that the equilibrium between the two phases was relatively stable since after 48 hours at this temperature with a 51.4 kPa oxygen pressure no other phase was detected. When the temperature was raised further iron orthophosphate $FePO_4$ appeared detrimentally to the new phase and to Fe_2O_3 which have therefore reacted together to form $FePO_4$ which is the only phase observed at 893 K.

These new results lead us to consider that the new phase should present a P/Fe ratio equal to 1.33 and a Fe^{3+}/Fe^{2+} ratio equal to 2 which correspond to the stoichiometry $Fe_3(P_2O_7)_2$.The partial phase equilibria of the system at 723 K that can be drawn from our experimental results are presented on figure 1.

TABLE 2

Mössbauer parameters computed from the spectra, recorded at 295 K of $Fe_2P_2O_7$ heated at 723 K for 5 hours in various oxygen partial pressures (0,12.3,22.6,25.7,28.8,51.4 kPa).
δ : isomeric shift / Fer α, Δ : quadrupolar splitting, H : internal magnetic field

oxygen partial pressure kPa	sample	δ mm.s^{-1}	Δ mm.s^{-1}	H kOe	relative intensity %	phase assignement
0	A1	1.23	2.37		73	$Fe_2P_2O_7$
		1.20	2.56		27	
12.3	A2	1.23	2.36		66	$Fe_2P_2O_7$
		1.21	2.56		31	
		0.42	0.67		3	
22.6	A3	1.23	2.34		63	$Fe_2P_2O_7$
		1.21	2.55		24	
		0.52	0.72		9	new phase
		1.19	2.75		4	
25.7	A4	1.22	2.39		42	$Fe_2P_2O_7$
		1.21	2.52		15	
		0.47	0.68		26	new phase
		1.12	2.75		16	
28.	A5	0.44	0.67		53	new phase
		1.21	2.70		34	
		0.36	-0.23	511	13	Fe_2O_3[1]
51.4 [2]	A6	0.49	0.65		51	new phase
		1.27	2.75		28	
		0.39	-0.22	512	21	Fe_2O_3[1]

[1] Fe_2O_3 : haematite
[2] sample calcined 48 hours

We have not detected the phase Fe_2O_3 in A2,A3 and A4 samples as we should but the quantity of this phase was very low and probably too low for the accuracy of both X-ray diffraction and Mössbauer spectroscopy. However the light brownish color of the solids led us suppose that iron oxide was present in all of them. According to the phase equilibria samples A5 and A6 should present the $Fe_2P_2O_7$ phase. It is obvious that for A6 sample the quantity of ferrous

FIGURE 1

Loop in the phase diagram Fe_2O_3 - FeO - P_2O_5 close to the samples under study and showing the subsolidus phase relations observed at 723 K.
o definite phases
• other compositions studied (A1-A6)

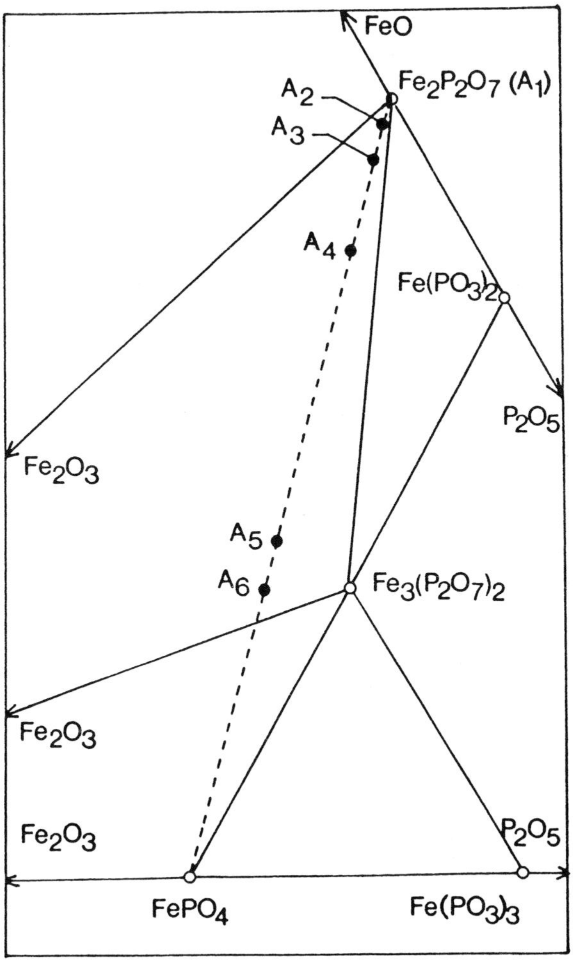

pyrophosphate is very small and cannot be detected but for sample A5 we should see it (table 3.a). In the presence of the new phase it is difficult to identify $Fe_2P_2O_7$ as their characteristic XRD peaks are common. However changes in relative intensity for some diffraction peaks in the A5 pattern indicate the presence of the pyrophosphate.

Assuming the presence of this phase we have attempted new fits of the Mössbauer spectrum of the A5 sample including one and two extra ferrous doublets. Only the addition of one extra doublet improve the fit of the spectrum but we can see that the isomer shift of this new site is approximately the same as those of the sites of $Fe_2P_2O_7$ and its quadrupolar splitting is intermediate between the ones of the sites of $Fe_2P_2O_7$ (table 3.b). It is interesting to note that the relative spectral areas of the two sites of the new phase is now equal to 2 as it should be. Another fit has been then performed where two doublets with the parameters of $Fe_2P_2O_7$ and the spectral areas corresponding to the doublet of the precedent fit were imposed. It appears to be rather acceptable (table 3.c).

TABLE 3

Mössbauer parameters computed from the spectrum of the sample A5, obtained from sucessive fits.a : original fit, b : a + one ferrous doublet, c : a+ two ferrous doublets with imposed $Fe_2P_2O_7$ parameters and the relative intensity of the extra ferrous doublet of fit b.
δ : isomeric shift / Fer α, Δ : quadrupolar splitting, H : internal magnetic field

fit	δ	Δ	H	relative intensity	phase assignement
	mm.s^{-1}		kOe	%	
a	0.44 1.21	0.67 2.70		53 34	new phase
	0.36	-0.23	511	13	Fe_2O_3[1]
b	0.44 1.21	0.67 2.75		53 27	new phase
	0.36	-0.23	511	13	Fe_2O_3[1]
	1.17	2.47		7	$Fe_2P_2O_7$?
c	0.44 1.21	0.68 2.70		53 27	new phase
	0.36	-0.23	511	13	Fe_2O_3[1]
	1.22 1.20	2.56 2.37		5 2	$Fe_2P_2O_7$

[1] Fe_2O_3 : haematite

All attempts to grow single cristals for determination of the unit cell have failed. However the X-ray diffraction data obtained for the phase from the pattern of A6 sample have been taken to calculate the unit cell using a method proposed by De Wolff and programmed by Visser (ref.7).

This program calculates a number of unit cells which are tested. A precision of 0.03 ° 2θ is required to pass the test (ref.8). The simplest unit cell which accounts for the data appears to be monoclinic with parameters :

$$a = 1.2640(6)\ \text{nm} \quad b = 0.8353(4)\ \text{nm} \quad c = 0.9742(5)\ \text{nm} \quad \beta = 108.25(3)\,°$$

The comparison of experimental and calculated data is presented in table 4.

Note that the actual industrial catalyst contains alkali-metal pyrophosphate (1) which crystallized in the same system with cell parameters close to some of the new phase. It may therefore be suggested that the addition of dopes or of another phase like the alkali-metal pyrophosphate most probably favors crystallization or stabilisation of the new phase.The cell volume of the new phase, 0.9776 nm^3, is approximatively eight times that of $Fe_2P_2O_7$ (ref.10). It may thus be suggested that the structure contains iron vacancies with would following formula $Fe^{3+}{}_2Fe^{2+}{}_1\square_1(P_2O_7)_2$.

The presence in the same framework of ferrous and ferric ions should favor the electron transfer during catalytic reaction.

TABLE 4

Proposed indexation for the X-ray powder pattern of the new phase

hkl	2θ° obs	2θ° calc	I / I0
1,0,0	7.35	7.358	5
0,1,1	14.27	14.273	10
2,0,0	14.74	14.747	5
1,1,1	17.42	17.407	25
1,0,-2	18.24	18.249	5
0,0,2	19.17	19.170	5
2,0,-2	20.22	20.218	5
1,2,0	22.52	22.520	5
3,1,-1	23.78	23.780	20
2,2,0	25.98	25.967	5
1,2,-2	28.14	28.154	15
0,2,2	28.76	28.773	10
1,1,-3	29.53	29.532	100
4,0,0	29.72	29.745	65
3,2,-1	30.23	30.221	10
1,2,2	31.24	31.250	15
3,2,-2	32.56	32.586	5
2,2,2	35.24	35.232	10
3,0,-4	38.34	38.334	5

CONCLUSIONS

The results described in this paper clearly show that :

(i) like vanadium phosphates, iron phosphates are evolving under catalytic reaction conditions. This evolution is characterized by a structural rearrangement which in the case of iron phosphates leads to the formation of a new phase.

(ii) this new phase, shown to be present in all performant catalysts, is the clue of the redox mechanism. Its characterization by Mössbauer spectroscopy and X-ray diffraction showed that it is composed of ferric and ferrous ions in a ratio equal to 2 and that it crystallizes in the monoclinic system with following lattice parameters:

a = 1.2640(6) nm b = 0.8353(4) nm c = 0.9742(5) nm β = 108.25(3) °

Subsolidus compatibility relations, observed between phases in the course of a study of the oxidation of $Fe_2P_2O_7$, suggest that this new phase present a P/Fe equal to 1.33 and its stoichiometry can be formulated as $Fe_3(P_2O_7)_2$ with the presence of iron vacancies.

(iii) the presence in the new phase of ferric and ferrous ions should favor a rapid and reversible electron tranfert during the catalytic act.

ACKNOWLEDGMENT

The authors are gratefull to Dr. Mentzen for his support in the interpretation of XDR pattern and computer work.

REFERENCES

1. E. Cavatera et al., assigned to Montedison, US Patent 3 948959 (1976)
2. C. Daniel et al. assigned to Ashland Oil Inc., US Patent 4 298755 (1981)
3. J.M.M. Millet, C. Virely, M. Forissier, P. Bussière and J.C. Vedrine, Hyperfine Interactions, 46 (1989) 619-628
4. E. Bordes and P. Courtine, J. Catal., 57 (1979) 236-252
5. G. Centi, G. Fornasari and F. Trifiro, J. Catal., 89 (1984) 44-51
6. C. Virely, M. Forissier (to be published)
7. A. Modaressi, Thesis Nancy (1982)
8. P.M. De Wolff, J. Appl. Crystallogr., 1 (1968) 108-113
9. J.W. Visser, J. Appl. Crystallogr., 2 (1968) 89-95
10. T. Stefanidis and A.G. Nord, Z. Kristallogr., 159 (1982) 255-264

Prof. C. CENTI Dipartimento di Chimica Industriale e dei Materiali
V. Le Risorgimento 4
40136 Bologna (ITALY)

In your lecture you comment about the analogies of iron-phosphates with vanadium-phosphate catalysts. It was shown during the meeting that on the surface at V-P-O catalyst there is an excess of P, with a P : V ratio around 2.0 ?
You believe that a similar situation on the Fe-phosphate catalysts, taking into account that the new phase you indicated has a P : Fa ratio of about 1.3 ?

J.C. VEDRINE (Institut de Recherches sur la Catalyse, Villeurbanne).

XPS experiments have been carried out on an industrial-type and on several iron phosphates with chemical Fe/P ratio of one. In all cases the P/Fe ratio found by XPS was more than one (in the 1.3 to 1.7 ratio range). However it is difficult to make precise XPS determination since the strongest XPS peaks correspond to 2p levels where secondary electron emission sharply enhance the background. Moreover calculations involve the use of electron cross section values taken from table, as Scofield's one, which may be discussed in the absence of standard (really standard on the surface).

Prof. HADDAD (Amoco Chemical Company)
P.O. Box 400
Naperville Illinois 60566
U.S.A.

Have you been able to prepare a pure sample of the new phase $Fe_3\ (P_2O_7)_2$ and if so have you tested it catalytically ?

J.C. VEDRINE (Institut de Recherches sur la Catalyse, Villeurbanne).

We have not yet been able to prepare a pure $Fe_3\ (P_2O_7)_2$ sample but we obviously work very hard on its preparation and progresses have be gained. We do hope for future by using dopes as alkaline cations able to favor similar phases such as $AFe_2P_2O_-$

G. Centi and F. Trifiro' (Editors), *New Developments in Selective Oxidation*

HETEROPOLYACID CATALYSTS FOR OXYDEHYDROGENATION OF ISOBUTYRIC ACID: KINETICS AND DEACTIVATION

O. WATZENBERGER, Th. HAEBERLE, D. T. LYNCH[1] , G. EMIG

Institut für Technische Chemie I, Universität Erlangen-Nürnberg, D–8520 Erlangen, F.R.G.

[1]Department of Chemical Engineering, University of Alberta, Edmonton, Canada T6G 2G6

ABSTRACT

Based on the experimentally observed reaction scheme, a Mars–van Krevelen type kinetic model was derived and the kinetic parameters for different types of catalysts determined. Details of the fundamental mechanisms responsible for the deactivation processes (redox processes with Mo volatilization) have been elucidated. The effects of catalyst composition and reactor operating conditions on kinetics and deactivation have been determined. A technique for stabilizing the activity and selectivity of the catalyst is presented.

INTRODUCTION

In recent years increasing attention has been focussed on the use of heteropolyacid (HPA) catalysts for the oxydehydrogenation of isobutyric acid (IBA), $(CH_3)_2CHCOOH$, to form methacrylic acid (MAA), $CH_2{=}C(CH_3)COOH$, which is an important intermediate in the production of methylmethacrylate esters (ref. 1). Selectivities of over 70% can be achieved when HPA catalysts with the overall composition $M_yH_{3+x-y}Mo_{12-x}V_xPO_{40}$ (where $x = 0, 1, 2, \ldots$, and M = Li^+, Na^+, K^+, Cs^+, etc.) are used (ref. 2). There is significant commercial interest in producing MAA via this route because of the lack of environmental problems, unlike the current acetone cyanohydrin process where ammonium bisulfate is produced as a by–product. However, the large–scale industrial use of HPA catalysts for MAA production has not occurred due to relatively strong deactivation effects associated with these catalysts.

EXPERIMENTAL

Heteropolyacids of the overall composition $H_{3+x}Mo_{12-x}V_xPO_{40}$, prepared by methods described by Kürzinger (ref. 3), were used as catalysts. Further details of the chemistry and properties of these compounds have been reported by Pope (ref. 4) and Tsigdinos (ref. 5). The oxydehydrogenation was carried out in a conventional continuous flow fixed-bed reactor, 30 cm in length with a 10 mm inner diameter. As shown in Fig. 1, the gas flow was metered by mass flow controllers and a pulsation–free double reciprocating pump was used to feed the liquid IBA into the evaporator. Product stream composition was de-

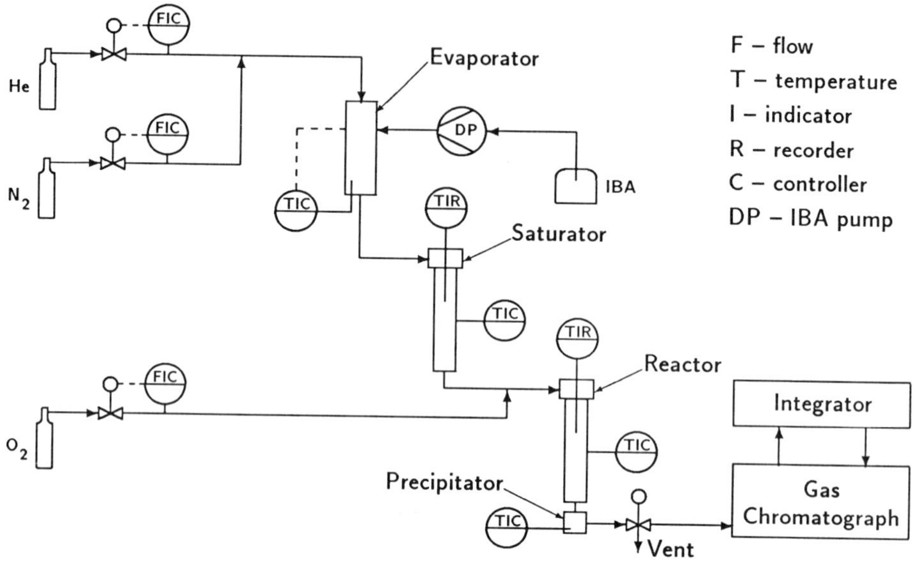

Fig. 1. Schematic of experimental fixed–bed reactor system with Mo feed saturator

termined using a gas chromatograph which was controlled by a microprocessor. TCD and FID detectors were used for the total analysis of the product gas stream (ref. 6). The item labelled as "saturator" in Fig. 1 was used for the deactivation experiments and will be described later. For the analysis of catalyst deactivation a modified integral reactor (multibasket reactor) with a special segmented inner device was used, as shown in Fig. 2. Small screens are used to contain the catalyst in each segment and the segments are screwed together. The inner (open) diameter of the segments is 10 mm, thus, the flow conditions are similar to that in the standard integral reactor. The six reactor segments (baskets) can be separated after an experimental run, and for each segment the catalyst activity can be separately measured in the standard integral reactor followed by a composition analysis of the catalyst using XFS. Additional details of the multibasket reactor have been presented by Haeberle and Emig (ref. 6).

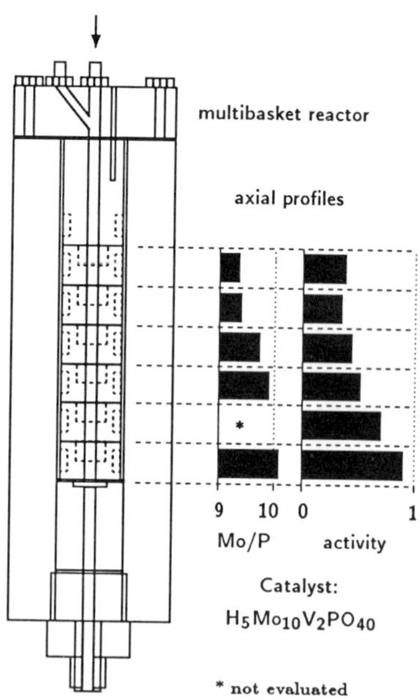

Fig. 2. Multibasket reactor showing axial variation of catalyst activity and Mo/P ratio

RESULTS

Kinetics

From an examination of various experimental conditions, the reactions shown in Fig. 3 have been found to occur during the heterogeneously catalyzed oxydehydrogenation of IBA. For a reaction temperature of less than 600 K consecutive reactions and the total oxidation of isobutyric acid can be neglected. Therefore the reaction scheme can be restricted to the description of three parallel reactions; the main reaction to MAA and the side reactions to acetone and CO_2, and to propene and CO. The reactions to MAA and acetone which proceed through the consumption of lattice oxygen can be well described using Mars–van Krevelen type rate equations (Eq. 1 and Eq. 2), whereas the propene formation, which is related to the acidic properties of the catalyst (strong acid sites are apparently responsible for propene formation), can be represented by a power–law rate expression (Eq. 3) (ref. 6).

$$r_1 = \frac{k_1\, k_2\, P_{O_2} P_{IBA}}{k_1\, P_{O_2} + k_2\, P_{IBA}} \qquad \text{(MAA formation)} \tag{1}$$

$$r_2 = \frac{k_3\, k_4\, P_{O_2} P_{IBA}}{k_3\, P_{O_2} + k_4\, P_{IBA}} \qquad \text{(acetone formation)} \tag{2}$$

$$r_3 = k_5\, P_{IBA}^{0.7}\, P_{O_2}^{0.2} \qquad \text{(propene formation)} \tag{3}$$

The terms k_1 and k_3 are rate constants for the reoxidation of the catalyst sites responsible for the formation of MAA and acetone, respectively, whereas, the terms k_2 and k_4 are rate constants for the reactions by which the adsorbed IBA species are converted into MAA or acetone, respectively. The estimates of the parameters for this intrinsic kinetic model are listed in Table 1. Further information on the methods used in these kinetic investigations and more details regarding the kinetic behaviour of these catalysts have been previously described (ref. 6).

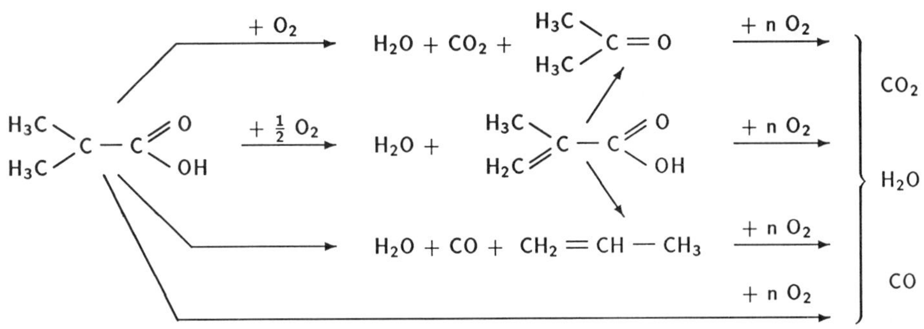

Fig. 3. Overall reaction scheme with main and side reactions

TABLE 1
Kinetic constants for MoO_3 and heteropolyacid catalysts

Component / Parameter	MoO_3	$H_4Mo_{11}V_1PO_{40}$	$H_5Mo_{10}V_2PO_{40}$	$H_6Mo_9V_3PO_{40}$
k_1	5.38	1.12	3.76	3.15
k_2	0.27	94.0	27.9	37.3
k_3	1.33	0.39	0.83	0.60
k_4	1.77	24.6	17.5	21.4
k_5	1.36	5.09	6.25	5.27

rate constant k_i $[\times 10^{-10}\ mol/g_{cat}/s/Pa^n]$ at 573 K

It was found that the reoxidation of the catalyst, which is the rate limiting step, is enhanced when the vanadium content in the HPA is increased. It reaches a plateau value (see k_1 in Table 1) at the overall composition $H_5Mo_{10}V_2PO_{40}$. This kinetic model was also used to describe the catalytic behaviour of molybdenum trioxide, and the kinetic parameter estimates for MoO_3 are also given in Table 1. While the reoxidation steps proceed more easily for pure molybdenum trioxide than for any of the HPA catalysts (see k_1 and k_3 in Table 1), it is unfortunately also seen that MoO_3 possesses very few active sites for MAA formation (or for acetone formation) relative to the HPA catalysts (see k_2 and k_4 in Table 1). This leads to the speculation that it is the specific atomic distances between the molybdenum (and/or vanadium) atoms in the HPA catalysts that produce oxygen which is active for MAA formation. The incorporation of one vanadium atom in the Mo–HPA may optimize this oxygen activation. Calculations, using these kinetic parameters, showed that at a low oxygen to IBA ratio the selectivity and conversion increases with vanadium content, whereas at a high oxygen concentration the reduction step is limiting. Therefore the maximum in yield can be reached using the V_1–acid.

Deactivation

Large–scale industrial use of HPA catalysts for MAA production has not occurred due to relatively strong deactivation effects associated with these catalysts. The deactivation behaviour of the different HPA catalysts, determined in an integral reactor, is in some respects comparable to the decrease in thermal stability, as measured by DTA, which results from an increase in vanadium content (an increase in vanadium content leads to a decrease in the temperature at which a HPA decomposes to the bulk oxides). A typical experimental run is shown in Fig. 4, where effluent partial pressures of the main product species from a fixed–bed integral reactor are shown as a function of time (ref. 7).

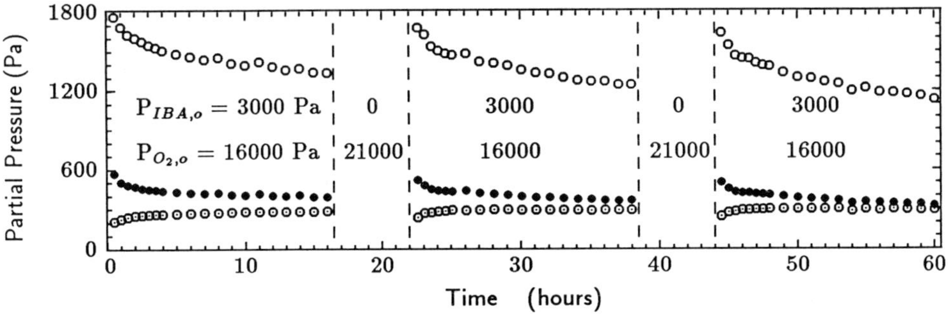

Fig. 4. Integral Reactor Exit Partial Pressures (o MAA, • Acetone, ⊙ Propene) (2 g $H_6Mo_9V_3PO_{40}$ catalyst, T = 590 K, total feed rate 1.5 mol/h)

For the initial 16 h period, it is seen in Fig. 4 that the catalytic activities for MAA and acetone production continuously decreased whereas that for propene formation increased. For comparison, shown in Fig. 5 are the decreases in activity with time on stream for several molybdenum containing HPA type catalysts of various compositions obtained from literature data. The deactivation behaviour is similar to the curves in Fig. 4 indicating that it is likely due to the same fundamental mechanism.

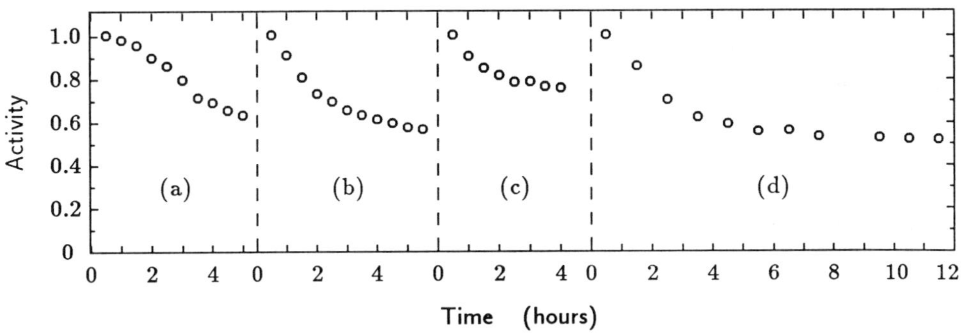

Fig. 5. Decrease of activity with time on stream. (a) $H_3Mo_{12}PO_{40}$, (b) $Cu_3(Mo_{12}PO_{40})_2$, (c) $H_3Mo_{10}W_2PO_{40}$, Oxydehydrogenation of isobutyric acid, Akimoto, et al. (refs. 8–10); (d) $BiMo_{12}PO_{40}$ + 2 VO^{2+}, Partial oxidation of n-butane, Ai (ref. 11).

Shown in the middle and last sections of Fig. 4 are the effects of reoxidizing the catalyst with air at 590 K. It is seen that only a certain fraction of the catalytic activity could be restored, and that irreversible deactivation had occurred. The fraction of the activity which could be restored can be related to the oxidation state of the catalyst, however, the irreversible deactivation has been found to be related to the destruction of the Keggin structure of the HPA catalyst. If decomposition of the Keggin units (the primary structure of the HPA) leads to metal oxides, especially to molybdenum oxides, then the formation of MAA should decrease whereas the formation of propene should increase due to the catalytic properties of molybdenum oxide. This is what is seen to occur in Fig. 4.

Experimental evidence indicates that the destruction of the HPA Keggin structure is linked to a loss of Mo from the catalyst (ref. 6). This can be seen in Fig. 2 by comparing the axial profiles of activity and Mo/P ratio determined in the multibasket reactor where, after the deactivation experiment, the catalyst bed was divided into six different samples, each of which was analyzed separately. The Mo/P ratio was measured by XFS analysis. These results indicate that the loss of activity is due to a loss of Mo and it is seen that the activity decreases markedly even when the Mo loss is very small. The comparatively greater Mo loss in the inlet section, indicates a parallel deactivation mechanism, caused by the reactants IBA and/or O_2.

The Mo must be transported out of the reactor via the gas phase. Most of the molybdenum which was lost could be found in the outlet of the reactor in the condensed liquid. Deeper insight into the fundamental mechanism responsible for the irreversible deactivation can be gained from Table 2, in which are given the partial pressures of various volatile molybdenum containing compounds based on both thermodynamic calculations and experimental Mo loss measurements. The experimental values were determined by flowing a particular “solvent” (e.g. IBA, MAA, etc. in an inert gas carrier) at a, constant flow rate (1.5 mol/h total feed rate) over a measured quantity of catalyst for a fixed period of time. From the amount of Mo (determined gravimetrically) in the condensed liquid product, the partial pressure of a volatile molybdenum containing compound which is needed to account for this loss of molybdenum from the catalyst was calculated.

TABLE 2
Mo partial pressures at reactor exit based on thermodynamic calculations and experimental molybdenum loss observations

	P_{Mo}, Pa
HPA catalyst under reaction conditions (this work)	0.077 – 0.13
MoO_3 sublimation, calculated (ref. 12)	$6\times10^{-9} - 7\times10^{-7}$
$Mo(CO)_6$ formation, calculated (ref. 13)	10^{-25}
$MoO_2(OH)_2$ formation (refs. 14-15)†	$5\times10^{-5} - 2\times10^{-4}$
Volatilization from Fe/Mo catalyst with methanol (ref. 16)†	0.056
Volatilization from HPA with methanol (this work)†	0.067
Volatilization from HPA with acetone (this work)†	0.008
Volatilization from HPA with isobutyric acid (this work)†	0.06 – 0.16
Volatilization from HPA with methacrylic acid (this work)†	0.14 – 0.17

†T = 590 K, P_{total} = 101.3 kPa, P_x = 1 kPa (x = MAA, IBA, acetone, methanol or water)

From the values in Table 2 it can be clearly seen that under these reaction conditions the transport of Mo out of the catalyst, and out of the reactor, can only proceed via the formation of a gaseous MoO_x–acid (acid means IBA and/or MAA) compound. The degree of formation of this species is in the same range as Popov (ref. 16) reported for the volatilization of Mo with methanol from a Mo/Fe catalyst. The Mo pressures calculated from the experimental Mo/P profiles for several HPA catalysts investigated in the multi-basket reactor are in good agreement with the value determined from the volatilization experiments.

The following reaction between molybdenum within the catalyst and acid from the gas phase can be postulated to occur:

$$y \text{ Acid (gas)} + MoO_x \text{ (catalyst)} \rightleftharpoons MoO_xAcid_y \text{ (catalyst)} \rightleftharpoons MoO_xAcid_y \text{ (gas)}$$

The deactivation occurs preferentially in the inlet section of the reactor where nearly no Mo in the gas phase is present, whereas in the outlet zone no Mo–loss can be observed, probably due to the gas phase being saturated with the Mo–acid component. Thus, this indicates that deactivation could possibly be avoided by saturating the feed stream with a volatile molybdenum containing compound prior to the feed entering the reactor. This was experimentally accomplished by passing the IBA–containing feed through a fixed bed of MoO_3 (denominated as "saturator" in Fig. 1) prior to the feed entering the main reactor which contains the HPA catalyst. The saturator contained 24 g of MoO_3 in the form of 0.5 to 2.5 mm particles. The temperature of this vessel was held constant at 250°C. The effect of using the saturator is shown in Fig. 6. For a time on stream of 16 h, essentially no deactivation was observed compared to a MAA yield loss of 25% which occurred when presaturation with Mo was not employed. From Fig. 6 it is seen that it is possible to eliminate both the linear activity decrease due to the loss of volatile Mo as well as the exponential activity decrease due to the change in the overall oxidation state of the catalyst at the beginning of the deactivation period. Thus, use of the saturator prevents both the initial reduction of the catalyst as well as the loss of molybdenum from the catalyst.

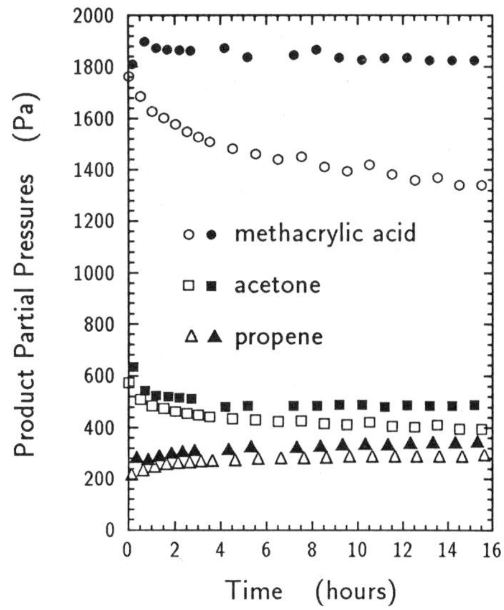

Fig. 6. Effect of Molybdenum in the Feed (open symbols – without Mo; filled symbols – with Mo)

SUMMARY

The catalytic activity of Mo–HPA catalysts can be explained by their redox behavior and can be described using Mars–van Krevelen type kinetic equations. The reoxidation of the catalyst is the rate–limiting step. Successive substitution of Mo by V improves the reoxidation power, reaching a maximum at the overall composition of $H_5Mo_{10}V_2PO_{40}$. At the beginning of a reaction period a certain oxidation state of the HPA catalyst is eventually reached depending on the ratio of reoxidation rate to reduction rate. This results in a marked decrease of the initial activity. This decrease in activity can be reversed by reoxidation in air. The irreversible, nearly linear deactivation with time on stream is due to the formation of a volatile Mo–acid compound causing Mo loss from the catalyst. In the latter part of the reactor, an equilibrium apparently exists between the Mo–acid compound in the gas phase and the Mo–acid compound on the catalyst. The HPA catalyst seems to be protected by the decomposition products of this volatile MoO_x compound. The MoO_x component covering the surface also improves the reoxidizing potential.

Acknowledgement—The sponsorship of part of this work by the Alexander von Humboldt–Stiftung (research fellowship for D.T.L.) is gratefully acknowledged.

REFERENCES

1. K. Kürzinger, G. Emig, H. Hofmann, *in* Proc. 8th Int. Congr. Catal., Vol. **V**, Berlin, July 2–6, 1984, Verlag Chemie, Weinheim, 1984, pp. 499–507.
2. M. Misono, Catal. Rev.–Sci. Eng., **29** (1987) 269–321.
3. K. Kürzinger, Ph. D. Dissertation, Universität Erlangen–Nürnberg, 1983.
4. M. Th. Pope, Heteropoly and Isopoly Oxometalates, Springer, Berlin, 1983.
5. G. A. Tsigdinos, Top. Curr. Chem., **76** (1978) 1–64.
6. Th. Haeberle, G. Emig, Chem. Eng. Technol., **11** (1988) 392–402.
7. O. Watzenberger, H. Brünner, D. T. Lynch, G. Emig, *in* Abstracts 38th Can. Chem. Eng. Conf., Edmonton, October 2–5, 1988, p. 14.
8. M. Akimoto, Y. Tsuchida, K. Sato, E. Echigoya, J. Catal., **72** (1981) 83–94.
9. M. Akimoto, K. Shima, H. Ikeda, E. Echigoya, J. Catal., **86** (1984) 173–186.
10. M. Akimoto, H. Ikeda, A. Okabe, E. Echigoya, J. Catal., **89** (1984) 196–208.
11. M. Ai, *in* Proc. 8th Int. Congr. Catal., Vol. **V**, Berlin, July 2–6, 1984, Verlag Chemie, Weinheim, 1984, pp. 475–486.
12. O. Kubaschewski, E. L. Evans, Metallurgical Thermochemistry, Pergamon Press, New York, 1958.
13. G. Pilcher, M. J. Ware, D. A. Pittam, J. Less-Common Met., **42** (1975) 223-228.
14. I. Nicolau, A. Aguiló, P. B. DeGroot, *in* Proc. 4th Int. Conf. Chem. Uses of Molybdenum, Climax Molybdenum Co., Ann Arbor, 1982, pp. 234–240.
15. J. Buiten, J. Catal., **10** (1968) 188–199.
16. B. I. Popov, V. N. Bibin, G. K. Boreskov, Kinet. Catal., **17** (1976) 322–327.

M. Misono (University of Tokyo, Japan):

- Your idea of presaturating is interesting. Salts of heteropoly acids are often more stable then the acid form. Did you examine the effect of salt formation on the deactivation?
- You assumed two independent active sites for the formation of methacrylic acid and acetone. Could you suggest those sites on heteropoly catalyst substantially?

O. Watzenberger (University of Erlangen, FRG):

- Kinetic investigations using a $Cs_2H_3Mo_{10}V_2PO_{40}$ catalyst (*ref. 1*) indicate that the salt is more stable then the acid. It was found that the degree of deactivation strongly depends on the oxygen to isobutyric acid ratio. In a first deactivation experiment, similar to those reported in Fig. 4 (except for $m_{Cat} = 0.5g, p_{IBA,o} = 1500Pa$) qualitatively a similar deactivation behaviour as shown in Fig. 4 and Fig. 5 was found.
- There are two reasons that confirm the existence of two different sites for the formation of methacrylic acid and acetone, respectively.
  - The methacrylic acid and the acetone formation show different deactivation behaviour (*ref. 2*)
  - Bulk MoO_3 is mainly active for propene and acetone formation while, for example, a $Cs_2H_3Mo_{10}V_2PO_{40}$ catalyst can be very selective for methacrylic acid formation (*ref. 1, Table 2*).

  If there is an active site for the acetone and/or propene formation on the heteropoly compound, it should have the same structure as on the bulk MoO_3. MoO_3 and other intermediate decomposition products can be expected on the surface of a heteropoly acid catalyst after calcination (*ref. 3*).

1. Haeberle, Th., and Emig, G., *Chem. Eng. Technol.* **11**, 392 (1988).
2. Watzenberger, O., Lynch D. T., and Emig, G., submitted for publication in *J. Catal.*
3. Black J. B., Clayden, N. J., Gai, P. L., Scott, J. D., Serwicka, E. M., Goodenough, J. B., *J. Catal* **106**, 1 (1987)

G. Centi (University of Bologna, Italy)

- In table I you have utilized the k_1 and k_2 parameters of the rate equation (*1*) to evaluate the oxidation and reoxidation properties of the catalyst. Usually these parameters are very correlated and it is not possible to obtain a good evaluation. May you give some details about the correlation matrix of these parameters?

O. Watzenberger (University of Erlangen, FRG):

- We did not calculate the correlation matrix because we did not expect any strong correlation between these parameters. This can be demonstrated in Figure A where the sensitivity of the parameters k_1 and k_2 is shown depending on the inlet concentration of oxygen. It can be seen that the model is very sensitive on the reoxidation parameter k_1 at low oxygen concentrations whereas, at higher oxygen concentrations the system becomes very sensitive on the reduction parameter k_2. Both parameters are not likely to be very much correlated to parameters for the acetone and propene formation k_3, k_4 and k_5 because, the gas phase concentration does not depend very much on the by-product formation due to the relative high selectivity.

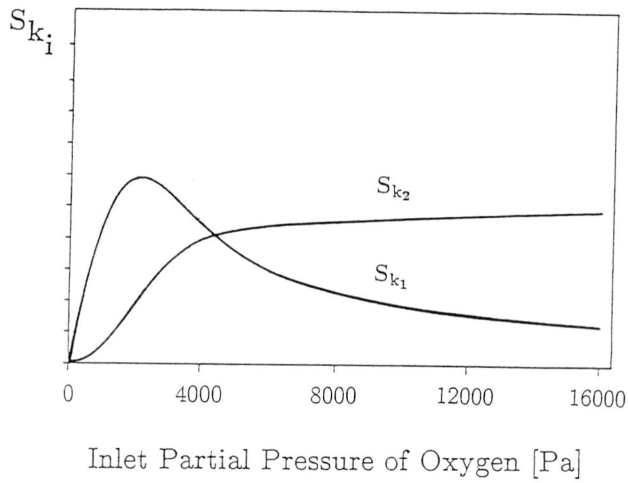

Figure A: Sensivity S of the parameters k_1 and k_2 as function of the oxygen concentration (*ref. 4*).

4. Haeberle, Th., Ph. D. Dissertation, Universität Erlangen-Nürnberg, 1987.

J. M. Bregeault (University P. et M. Curie, Paris, France)

- Do you prepare "$PMo_{11}V_1$", "$PMo_{10}V_2$", "PMo_9V_3" without Na^+ ?
- Can you characterize your precursor by ^{31}P NMR spectroscopy?, V NMR spectroscopy?
- What about the phase diagram of "$PMo_{11}V_1$", "$PMo_{10}V_2$", "PMo_9V_3" with increasing temperature?

O. Watzenberger (University of Erlangen, FRG):

- These catalysts were prepared by boiling stoichiometric amounts of molybdenum trioxide, vanadium pentoxide and phosphoric acid in water. After vacuum filtration of the solution through a 7 μm quartz filter the solid catalysts were obtained by recrystallization at room temperature, followed by drying and calcination at 340°C for 5 hours. The solids were then crushed and sieved. XFS analysis showed that the solids produced by this proceduce have the correct Mo/P and V/P ratios of 11/1 and 1/1, 10/1 and 2/1 or 9/1 and 3/1, respectively.
- From ^{31}P NMR measurements it was found that the catalysts are not pure substances, instead the V_2 catalyst contains a certain fraction of V_1 and V_3 based compounds. Similarly, the V_3 catalyst contains HPA molecules with other numbers of substituted vanadium atoms. We did not investigate the catalyst using V NMR spectroscopy.
- To our knowledge, there exists only a phase diagram for aqueous solutions which can not be applied for higher temperatures.

G. Centi and F. Trifiro' (Editors), *New Developments in Selective Oxidation*

A STUDY OF THE CATALYTIC OXYDEHYDROGENATION OF ETHYLBENZENE USING NITROBENZENE

R. Fiedorow[1], S. Beszterda[1], W. Przystajko[1] and I.G. Dalla Lana[2]

[1] Faculty of Chemistry, A. Mickiewicz University, 60-780 Poznan, Poland

[2] Department of Chemical Engineering, University of Alberta, Edmonton, Canada, T6G 2G6

SUMMARY

During the oxydehydrogenation of ethylbenzene with nitrobenzene on either alumina or magnesium sulphate surfaces, carbonaceous deposits which exhibit dehydrogenating activity are formed. The activity of these coke catalysts decays with time because growth of this non-porous coke limits access of reactant molecules to the residual external surface of the parent (alumina) catalyst. The $MgSO_4$-based cokes were also tested using model reactions. Paramagnetic centers in the coke (unpaired electrons localized on carbon atoms) and phenolic-type hydroxyl groups are the likely active sites.

INTRODUCTION

Typically, cokes deposited onto catalyst surfaces exhibit a deactivating influence, being, in themselves, catalytically inactive or unsuitable. However, the oxydehydrogenation (ODH) of ethylbenzene using oxygen as a hydrogen acceptor is known to be coke-catalyzed (refs. 1-7). Compared to non-oxidative dehydrogenations, ODH is exothermic and proceeds at lower temperatures with higher conversions. In this study, nitrobenzene is investigated as an oxidizing agent for this coke-catalyzed reaction of ethylbenzene to form styrene and aniline.

EXPERIMENTAL DETAILS

Aluminas, pure or impregnated with different loadings of NaOH or H_3BO_3, were prepared as described previously (ref. 6). Their impregnation loadings, wt %, are labelled 0.5 S, 1.0 S, . . .; whereas the orthoboric-acid impregnated alumina is 0.05 B (atomic ratio, B/Al = 0.05).

To characterize the cokes produced during these experiments, access to coke free of its coke-generating matrix would be preferable. To dissolve high-temperature alumina requires highly aggressive solvents such as 40 % hot KOH (refs. 8, 9) or concentrated hot HCl (ref. 10) with incomplete removal of alumina (ref. 10) and possible alteration of the coke properties. For this reason, magnesium sulphate ($9\ m^2g^{-1}$) (precalcined at 500° C for 2 h) was used as a water-soluble (at 40° C, continuous stirring) coke-generating matrix, producing as much as 8.1 wt % of coke after 7 h onstream. This magnesium sulphate contains acid centers (refs. 11, 12), but its usefulness is diminished by the instability of the acid properties of $MgSO_4$ at the oxydehydrogenation temperatures used. Thus, the mechanism of coke production on $MgSO_4$ may differ from that on alumina.

Dissolution of cokes in organic solvents was limited, only 3 wt % being extracted with pyridine (the most efficient organic solvent tested) using a Soxhlet apparatus for 70 h. IR spectra of coke extracts were recorded on a Perkin-Elmer Model 580 spectrophotometer by evaporating a drop of extract placed on a NaCl plate.

Elemental analyses of coke were obtained with a Perkin-Elmer Model 240 Analyzer; surface areas were measured by N_2 adsorption at liquid N_2 temperature in an OMNISORP-360 Sorptometer. XRD spectra were recorded on a Phillips X-ray diffractometer, with Cu tube and graphite monochromator. A Bruker Model CXP-200 provided ^{13}C MAS NMR spectra and a SE/X-2542 (RADIOPAN Poznan) ESR spectra of coke-covered samples.

Catalytic activities for the ODH of ethylbenzene were obtained using a flow reactor loaded with 0.25 g of catalyst (0.25-0.50 mm particles) and into which a 0.008 cm^3 min^{-1} equimolar stream of nitrobenzene and ethylbenzene was introduced. With impregnated samples of 0.25 g, additional experiments involved the injection of 0.5 microliter pulses of the feed into a He flow of 37.5 cm^3 min^{-1}. A reaction temperature of 450° C was used throughout.

The $MgSO_4$-generated water-insoluble coke (0.1g samples) was also tested using the pulse technique with several model reactions (see ref. 6 and list in table 3), including dehydrogenation of cumene at 370° C.

RESULTS AND INTERPRETATION

The development of catalytic activity for ODH is shown by plotting conversion to styrene (St) and to aniline (An) as a function of time-on-stream in figure 1. The accompanying measured properties are shown in table 1 for the coke produced on the non-impregnated alumina. It is apparent that the catalytic activity develops to a maximum value and then gradually diminishes with prolonged time-on-stream. This could correspond to formation of coke, its expansion over the surface and interior of pores eventually leading to blocking of access to the pores. The decreases in surface area and accessible pore volume parallel this trend.

Table 1. Changes in surface area, coke content and C/N ratio of coke-covered alumina after different time-on-stream

Time-on-stream, min.	Surface area, m^2/g	Total pore volume, cm^3/g	Coke content, wt %	C/N atomic ratio
0	155	0.53	0.0	---
10	158	0.46	10.6	10.1
25	142	0.29	12.3	10.2
45	135	0.27	16.5	10.6
105	107	0.18	22.1	13.1
165	80	0.13	28.0	14.2
225	73	non det'd	29.4	15.3
420	11	0.12	29.8	16.1

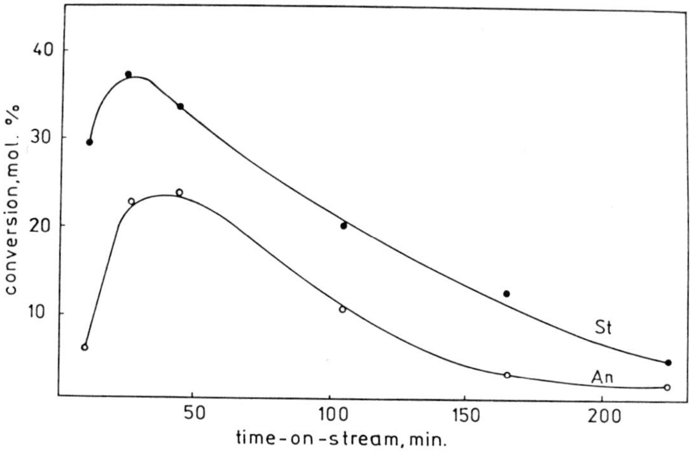

Figure 1. Conversion of ethylbenzene to styrene (St) and aniline (An) for varying times-on-stream.

Earlier (ref. 6), when using air instead of nitrobenzene for the ODH of ethylbenzene, the styrene yield stabilized and coke surface area did not change significantly after about 7h on stream. It was believed that the combustion of fragments of coke to form CO and CO_2 in large amounts (8.2%) created a pore structure within the carbonaceous deposit. The use of milder oxidizing agents (nitrobenzene or sulphur dioxide) yielded one and zero percent of carbon oxides, respectively, with eventual decline in activity occurring for both oxidizing agents. This coke seems to be "non-porous".

Table 1 shows that C/N ratio also increases with the amount of coke being formed, corresponding to increased carbonization of the coke deposit. The ^{13}C MAS NMR spectra for these coke-catalyst samples show increasing amounts of quaternary carbon atoms accompanied by fewer protonated carbon atoms, indicative of increased condensation of aromatic hydrocarbons in the coke. This observation is confirmed by the elemental analyses; the C/H ratio rises from 1.6 after ten minutes to a maximum value of 2.1.

Impregnation of alumina with NaOH or H_3BO_3

The influence of impregnation of γ-alumina with either NaOH or H_3BO_3 upon the ODH of ethylbenzene by nitrobenzene was tested using the pulse reactor. The results in figure 2 show a plot of styrene conversion versus number of pulses. Activity was totally suppressed for additions of 6 and 10 wt % NaOH. Since NaOH poisons acidic sites on alumina (ref. 13), the decreasing catalytic activity with increasing NaOH content suggests involvement of acid sites in generating active coke. However, the remaining catalytic activity, even with 3 wt % NaOH, indicates that some weak acid

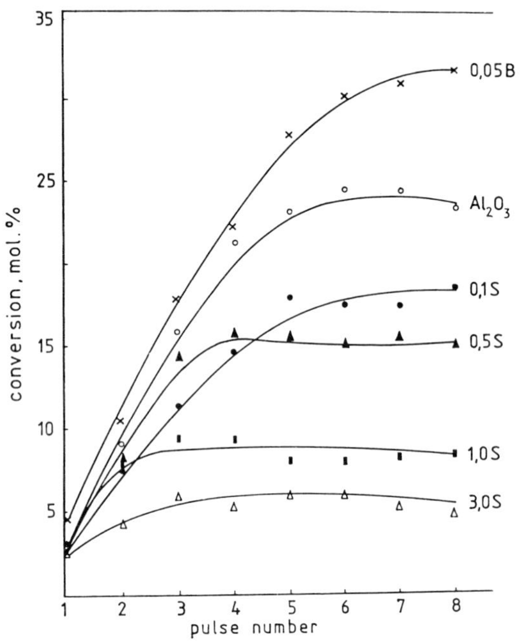

Figure 2. Effect of impregnation of alumina with NaOH and H_3BO_3 on conversion to styrene as measured by pulse technique.

centers ($H_0 \geq 6.8$) residual on the surface of alumina likely take part in formation of coke.

Impregnation with boric acid clearly increases the catalytic activity for ODH above that of alumina; the result of changes in the distribution of acid strength and promotion of acidity (refs. 11, 14).

ESR measurements on catalysts

ESR measurements on samples (225 min on-stream) reveal that the largest number of paramagnetic centers, 3.40 (10^{17}) spins/g of coke-covered sample, formed on coke-covered H_3BO_3-impregnated alumina. NaOH additions to alumina result in lowered spins/g. The alumina shows growing numbers of spins/g, starting from 1.22 (10^{17}) after 10 minutes and reaching a maximum value, 2.36 (10^{17}), after 225 min. In earlier work on ODH of ethylbenzene with air (ref. 6), the number of unpaired electrons correlated with yield of styrene. In this work, the concentration of unpaired electrons appears to correlate only with the coke content. These two studies suggest that the paramagnetic centers (probably in the form of quinone-type complexes with radical character, ref. 2) relate to coke activity. If the accessible surface area of the coke diminishes, some of these active centers become submerged within the coke and cannot participate.

<u>Comparisons between alumina- and $MgSO_4$-based cokes</u>

XRD patterns of the two cokes show differences in their ordering: alumina-coke appears to be completely amorphous whereas $MgSO_4$-cokes exhibit both amorphous and crystalline character. The amorphous structure is characterized by a very broad maximum at $2\Theta = 20°$; whereas the crystalline structure exhibits maxima at 37, 43.5, 50, 53, 74 and 82°. Diffraction line-broadening for peaks at $2\Theta = 20°$ (amorphous) and 43.5° (crystalline) suggest average coke particle sizes of 1.3 and 11 nm, respectively. Thus, the amorphous coke particles are an order of magnitude finer-sized than those of the crystalline coke.

Comparison of the IR-based chemical composition of the two pyridine-extracted soluble portions of the coke were similar. Figure 3 shows the IR spectra obtained. This may not be significant as a comparison because the extraction may have been selective; however, their common chemical composition is of interest. In both spectra, intensive bands at 2900-3000 and at 1460 and 1380 cm^{-1} indicate aliphatic structural fragments. Small bands between 1500 and 1600 and smaller ones between 3000 and 3100 cm^{-1} indicate aromatic compounds. Other bands suggest, at near 1720 cm^{-1} carbonyl groups

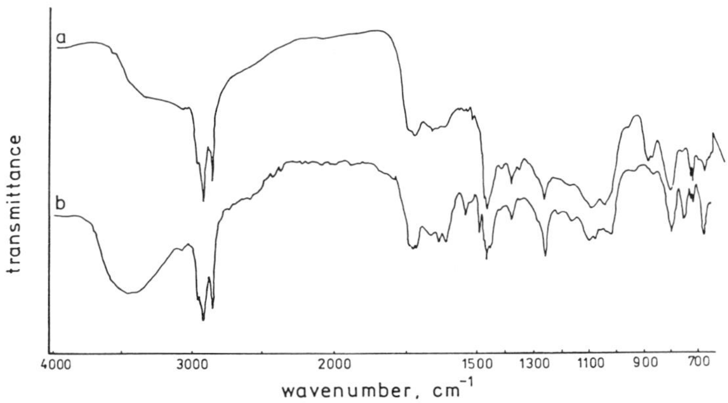

Figure 3. IR spectra of pyridine-extracted matter from coke deposited on alumina (a) and on magnesium sulphate (b).

of ketonic or carboxylic origin, at about 1650 cm^{-1} quinone structures, and between 1040 and 1260 cm^{-1} stretching vibrations of CO groups in alcohols, phenols or ethers. By diluting the non-IR-transparent coked alumina with KBr, 1:100 by weight, a semblance of the complete IR spectra for the carbonaceous deposit on alumina was obtained. Here, bands were observed at 3470 cm^{-1} (hydroxyl groups from alumina or phenolic groups in coke), 3060 and 3030 cm^{-1} (ν (CH) aromatic), 1600 cm^{-1} (ν (C=C) aromatic), 1520 -1450 cm^{-1} (ν (C-C) aromatic), and very weak bands at 2920 and 2840, 1470-1450, and 1375 cm^{-1} (aliphatic structural fragments).

The coke separated from $MgSO_4$ (8.4 wt % soluble in pyridine) exhibited higher C/H and C/N atomic ratios, an empirical formula, $C_{23.9}H_{10.9}O_{2.1}N$, and 120 $m^2 g^{-1}$ surface area (possibly from many very small not necessarily porous particles: electron micrographs show many fine particles of 0.20 micrometers).

Table 2. Catalytic activity of unsupported coke

Reaction	Composition of reaction products, mol %		
Dehydrogenation of cyclohexene	benzene		cyclohexene
	8.2		91.8
Dehydrogenation of cumene	α - methylstyrene		cumene
	3.8		96.2
Decomposition of 2-propanol	propylene	acetone	2-propanol
	72.9	9.5	17.6
Isomerization of 1-butene	inactive		

Note: The above data were obtained for the first pulse of a reactant, but no significant change in activity was observed after several pulses.

Catalytic ativities for model reactions

The coke separated from $MgSO_4$ was tested for catalytic character by using a number of well-studied model reactions. Table 2 lists these model reactions and summarizes the results of these tests.

With cyclohexene, only dehydrogenation to benzene occurred (without skeletal isomerization or hydrogen disproportionation) indicative of the absence of strong acid sites. Failure to isomerize 1-butene also is indicative of the absence of acid or base centers of strength sufficient to catalyze double bond migration in olefins. The considerable dehydration activity observed with 2-propanol relates to weak acid sites, most likely phenolic hydroxyl groups. The simultaneous dehydrogenation to acetone (along with cyclohexene to benzene) shows the coke to possess considerable dehydrogenation activity. The dehydrogenation of cumene to α-methyl styrene implies that the active sites involve a free radical mechanism (15); cumene cracking via an ionic mechanism should produce benzene and propylene, both of which were absent. These results also suggest that strong acid sites are absent. This isolated coke also catalyzed the ODH of ethylbenzene either by air or nitrobenzene. In the former ODH, the activity of the coke decreased monotonically to zero after ten hours on-stream because of the occurrence of combustion of coke in the absence of coke-generating matrix.

CONCLUSIONS

Catalytically active coke was generated during the ODH of ethylbenzene by nitrobenzene. The weak acid centers on alumina appear to be primarily responsible for generating this active coke, whose activity for ODH relates to the presence of paramagnetic centers on the surface of the coke (surface accessible to ethylbenzene). Deactivation of this coke is likely caused by its non-porous physical form, which eventually grows in extent thereby sealing off the pore structure of the parent alumina.

Further more, one observes from this work:

1. Unsupported coke was active for several additional dehydrogenation reactions. Weak acid centers on the coke surface made it also active for the dehydration of alcohol.
2. The active coke formed during ODH of ethylbenzene by nitrobenzene is a mixture of aliphatic and aromatic compounds responsible for catalyzing the redox reactions studied, e.g. quinonic carbonyl groups belonging to paramagnetic complexes existing within the carbonaceous material are probably the ones involved in catalyzing ODH.
3. The coke is largely insoluble in either polar or non-polar solvents.
4. Impregnation of alumina with orthoboric acid enhances the catalytic activity of the coke for ODH of ethylbenzene beyond that exhibited by the coke formed on non-impregnated alumina.

REFERENCES

1 T.G. Alkhazov and A.E. Lisovski, Oxidative Dehydrogenation of Hydrocarbons, Khimiya Publishing House, Moscow, 1980, Chapter 4.
2 A. Schraut, G. Emig and H.-G. Sockel, Appl. Catal. 29 (1987) 311-326.
3 A. Schraut, G. Emig and H. Hofmann, J. Catal. 112 (1989) 221-228.
4 G.E. Vrieland, J. Catal. 111 (1988) 1-13.
5 G.E. Vrieland, J. Catal. 111 (1988) 14-22.
6 R. Fiedorow, W. Przystajko, M. Sopa and I.G. Dalla Lana, J. Catal. 68 (1981) 33-41.
7 R. Fiedorow, W. Kania, K. Nowińska, M. Sopa and M. Wojciechowska, Bull. Acad. Pol. Sci., Ser. Sci. Chim. 27 (1978) 641-649.
8 R. Fiedorow, S. Kowalak and M. Sopa, to be submitted for publication.
9 T.G. Alkhazov and A.E. Lisovskii, Kinet. Catal. (Eng. transl.) 17 (1976) 375-379.
10 R. Fiedorow, W. Gut, W. Przystajko and M. Sopa, in Wybrane Zagadnien z Fizykochemii Ukladow Homo-i Heterogennych, UAM Seria Chemia 43 (1982) 59-75.
11 K. Tanabe, Solid Acids and Bases, Kodansha Tokyo - Academic Press, New York, 1970, Chapter 4.5.2.
12 K. Tanabe and T. Takeshita, Adv. Catal. 17 (1967) 315-349.
13 R. Fiedorow and I.G. Dalla Lana, J. Phys. Chem. 84 (1980) 2779-2782.
14 W. Kania, K. Nowińska, M. Sopa and M. Wojciechowska, Roczn. Chem. 51 (1977) 1787-1789.
15 S.E. Tung and E. McIninch, J. Catal. 4 (1965) 586-593.

G. EMIG (University of Karlsruhe, Germany): Is it possible with nitrobenzene system, as we know for O_2, to oxidize the coked catalyst separately and react then ethylbenzene to styrene ?

R. FIEDOROW (A. Mickiewicz University, Poland): The treatment of coke of nitrobenzene origin with 175 ethylbenzene pulses has resulted in the decrease in the conversion to styrene. When such coke was again contacted with nitrobenzene and this was followed by an injection of a pulse of ethylbenzene, a considerably higher conversion to styrene (compared to that after 175 ethylbenzene pulses) was observed. The coke of nitrobenzene origin is a very rich reservoir of oxygen-containing groups, which could not be completely exhausted (judging by the activity for styrene formation) even after one-hour treatment with hydrogen at 450°C.

G. EMIG: Did you observe also species derived from nitrobenzene other than aniline, e.g. partially reduced species ?

R. FIEDOROW: We have detected only a secondary reaction product, namely diphenylamine which was formed by the condensation of aniline on Lewis acid centres of coke-free fragments of alumina surface. Its presence was proved by the appearance of a green colour during a qualitative test with HCl and $FeCl_3$. As far as partial reduction products are concerned, we did not perform their determination because they either decompose at temperatures below our reaction temperature (e.g. azoxybenzene) or have boiling points too high (e.g. phenylhydrazine, diphenylhydrazine) to leave analytical column at the temperature used in our G.C. measurements, i.e. 90°C.

G. Centi and F. Trifiro' (Editors), *New Developments in Selective Oxidation*

THE ADSORPTION OF ALIPHATIC NITROCOMPOUNDS ON OXIDES INVESTIGATED BY FT-IR SPECTROSCOPY

P.A.J.M. ANGEVAARE, E.J. GROOTENDORST, A.P. ZUUR and V. PONEC

Gorlaeus Laboratories, Leiden University, P.O.Box 9502, 2300 RA Leiden, The Netherlands.

ABSTRACT

In this study FT-IR spectroscopy was used to examine the adsorption and surface reactions of nitromethane on three oxides: SiO_2, Al_2O_3 and α-Mn_3O_4. In order to study the products of the surface reactions the gasphase was analysed by mass spectrometry. Upon adsorption at 300 K on SiO_2 only physisorption took place. On Al_2O_3 and α-Mn_3O_4 at 300 K physisorption as well as dissociative adsorption (nitronate ions were formed) and surface oxidation (leading to formate species) took place. At higher temperatures formation of (iso)cyanate surface species is observed by IR spectroscopy. In the gasphase the presence of NO and formaldehyde at lower temperatures and CO_2 at higher temperatures were detected.

INTRODUCTION

An important way of producing valuable compounds on industrial scale is selective oxidation of organic molecules[1,2,3]. In most cases of heterogeneous catalytic oxidation the so called "Mars and van Krevelen" mechanism[4] operates. The essential feature of this mechanism is that a constituent of the catalyst appears in the product molecules. A molecule that has to be oxidized extracts oxygen atoms out of the oxide, which serves as catalyst. The oxygen vacancies that arise, are subsequently filled up by oxygen, usually from molecular oxygen. The essentially same mechanism operates in chlorination/dechlorination[5] processes catalyzed by solid chlorides, in hydrodesulphurization[6] by sulphides, etc. This shows that the mechanism is quite general and it is surprising how little is known about the use of the same principle in selective reductions. If one creates an oxygen vacancy in an oxide (catalyst) by a common reductor, one can use the reoxidation of the catalyst for the selective reduction of a molecule containing oxygen. Patents[7,8] are known, which suggest a heterogeneous catalytic reduction of nitro- to nitrosobenzene, or of carbonic acids to aldehydes by the use of a suitable oxide, but next to nothing is known about the detailed mechanism of such reductions. In this study an attempt is being made to produce information on this point.

In this paper the interaction of nitromethane with surfaces of oxides has been studied. The intention is to gain information leading to a better understanding of the reduction of nitro- to nitrosobenzene, which is catalyzed by α-Mn_3O_4[9]. Three oxides have been used: SiO_2, Al_2O_3 and α-Mn_3O_4 and the main technique applied is FT-IR spectroscopy. Additional information on surface reactions of nitromethane has been obtained by a temperature programmed surface reaction on α-Mn_3O_4, monitored by a mass spectrometer.

METHODS

The silica and alumina used in this study were Aerosil 200 and Al_2O_3-C (Degussa, West-Germany), respectively. α-Mn_3O_4 was prepared by thermal decomposition of manganese(II) hydroxide in air at 390 K. The hydroxide was precipitated from manganese(II) nitrate ($Mn(NO_3)_2.4H_2O$, Fluka, Switzerland) solution with ammonia. The nitromethane adsorbates, CH_3NO_2 and CD_3NO_2, are over 99% pure and from Aldrich Chemie, West-Germany.

Self-supporting pellets of the various oxides were obtained by pressing about 50 mg of the oxide at pressures of 1.10^7 Pa. Pellets were placed in an all-metal transmission cell[10] mounted with CaF_2-windows, which are transparent in infrared down to 1000 cm^{-1}. The cell is connected with a conventional gas manipulation/evacuation system (during evacuation of a sample generally a pressure in the range 10^{-4}-10^{-5} Pa is achieved). The pellet can be heated and cooled while being in the beam and the oxidation, reduction and adsorption procedures are also performed in situ.

Silica and alumina were activated by evacuation at 600 K (generally for 15 hrs) to remove adsorbed gases. α-Mn_3O_4 was evacuated for 1 hr at 575 K, which step was followed by an oxidation of 1 hr at 575 K in O_2 (99.998% Messer Griesheim, West-Germany) and a reduction in H_2 (99.9990% Messer Griesheim, West-Germany) at 575 K for 1 hr. This procedure was followed by evacuation at 575 K and cooling to the standard (beam)temperature before adsorption of the individual compounds.

The infrared spectra were recorded by an evacuable FT-IR spectrometer, Bruker IFS-

113v, equipped with a liquid nitrogen cooled MCT (mercury cadmium telluride) detector.

Before adsorption of one of the compounds an infrared spectrum of the oxide is recorded at the temperature of adsorption. This background spectrum, transformed in the absorbance mode, is automatically substracted from the absorbance spectrum obtained after adsorption.

Surface reactions were monitored by introducing the catalyst (α-Mn_3O_4) in a closed vessel connected to a mass spectrometer (Vacuum Generators, MM8-80S). After nitromethane was admitted, the temperature was raised in steps of 25 K. The temperature was kept constant for 15 minutes before recording a spectrum. After recording the spectrum the reaction vessel was evacuated before raising the temperature by 25 K. To distinguish the (m/e) peak of NO from that of formaldehyde, fully deuterated nitromethane was used.

RESULTS

Experiments with SiO_2 (under the used conditions fully hydroxylated oxide surface) showed at 300 K a weak adsorption (physisorption) of nitromethane by formation of hydrogen bonds to isolated Si-OH groups at the surface. This interaction is revealed by a decrease in the IR-absorption band corresponding to the isolated silanol groups (3741 cm^{-1}), an increase of hydrogen-bond bound OH groups (a broad band around 3600 cm^{-1}), and by the appearance of bands of almost unperturbed nitromethane. The spectrum is dominated by the 1565 cm^{-1} band of $\upsilon_a(NO_2)$. See for more details table 1. Evacuation at 300 K showed complete desorption of nitromethane.

With Al_2O_3 at 300 K, physisorption is also detectable. This physisorbed CH_3NO_2, however, reacts further by splitting off a H-atom. The H-atom is used to form additional Al-OH groups (3550 cm^{-1}) at the surface. At the same time IR bands which are assignable to nitronate ions ($H_2C{=}NO_2^-$) appear (see table 2). Support for the idea of nitronate formation is gained from adsorption of CD_3NO_2. The shifts of the IR bands obtained after using CD_3NO_2 instead of CH_3NO_2 are alike to the shifts observed comparing the spectra of the sodium salts of these nitrocompounds[12].

In a following step the nitronate ions become oxidized at the surface. This reaction leads to the formation of surface formate species ($HCOO^-$), as is revealed by IR bands at 1600 cm^{-1} ($\upsilon_a(CO_2)$) and at 1390-1320 cm^{-1} (δ(CH) and $\upsilon_s(CO_2)$).

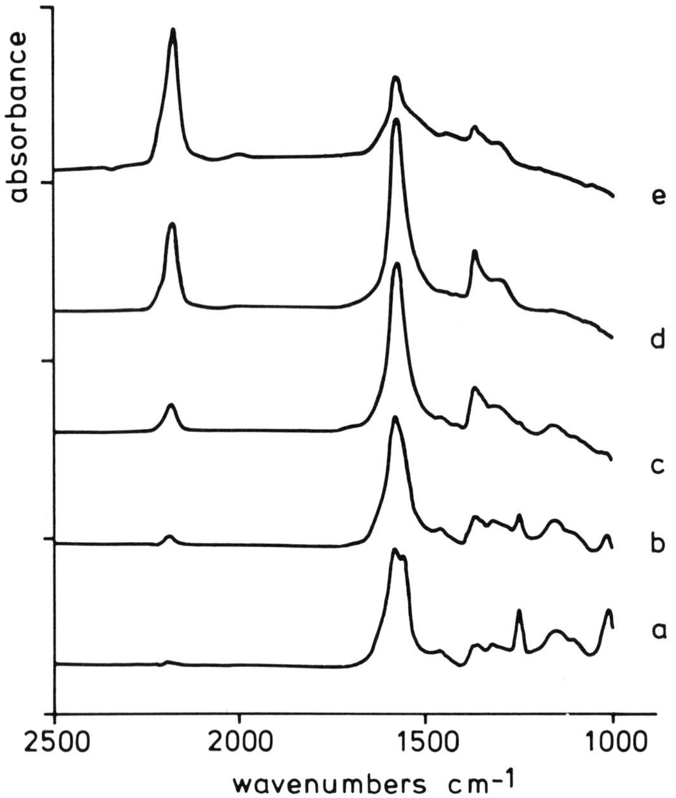

Figure 1. IR spectra obtained after adsorption of CH_3NO_2 on α-Mn_3O_4, followed by evacuation at 300 K (a). After a subsequent heating at 350 K (b), heating at 425 K (c), heating at 500 K (d), and heating at 575 K (e).

On α-Mn_3O_4 at 300 K, nitromethane (figure 1a) also forms nitronate ions, which is shown by the corresponding IR bands (see table 2). Subsequently the reaction to formate species is seen, with corresponding IR bands found at 1580 and 1375-1355 cm^{-1}, respectively.

Table 1: Physisorbed nitromethane: observed bands and their assignment, region 1800-1000 cm^{-1}, T = 300 K.

Gasphase band (cm^{-1})[11]	CH_3NO_2/SiO_2 (cm^{-1})	CH_3NO_2/Al_2O_3 (cm^{-1})	assignment[11]
1582 vs	1565 vs	1560 vs	$\upsilon_a(NO_2)$
1449 w	1429 m	1427 m	$\delta_a(CH_3)$
1413 w	1403 s	1402 s	$\delta_s(CH_3)$
1384 vs	1380 s	1375 s	$\upsilon_s(NO_2)$
1097 m	no	1103 m	$r(CH_3)$

vs = very strong, s = strong, m = medium, w = weak.
no = not observable.

Figure 1 shows also the change of the infrared spectrum caused by the increase of temperature from 300 to 575 K. In the range 300-500 K the peaks, which are characteristic of the nitronate ion disappear. In the same temperature range, the peaks characteristic of formate species become more pronounced. In the temperature range 350-575 K the spectra show a peak (2185 cm^{-1}), with an increasing intensity, and this absorption can be assigned to (iso)cyanate species[13,14].

Table 2: Observed bands and assignments of surface nitronate species, T = 300 K.

Sodium salt $[CH_2NO_2Na]$ (cm^{-1})[12]	$CH_2NO_2^-/Al_2O_3$ (cm^{-1})	$CH_2NO_2^-/\alpha\text{-}Mn_3O_4$ (cm^{-1})	assignment[12]
1582 s	1627 vs	1560 s	$\upsilon(C{=}N)$
1272 vs } 1261 vs }	1281 vs } 1257 vs }	1252 m	$\upsilon_a(NO_2)$
1032 vs } 1018 vs }	1045 s	1012 m	$r(CH_2)$
983 s	no	no	$\upsilon_s(NO_2)$

vs = very strong, s = strong, m = medium, w = weak.
no = not observable.

The results of the temperature programmed surface reaction of CD_3NO_2 are shown in figure 2. At lower temperatures (300-400 K) the main products detected in the gasphase are NO (m/e=30) and deuterated formaldehyde ($D_2C{=}O$; m/e=32). At higher temperatures (400-525 K) mainly CO_2 is detected.

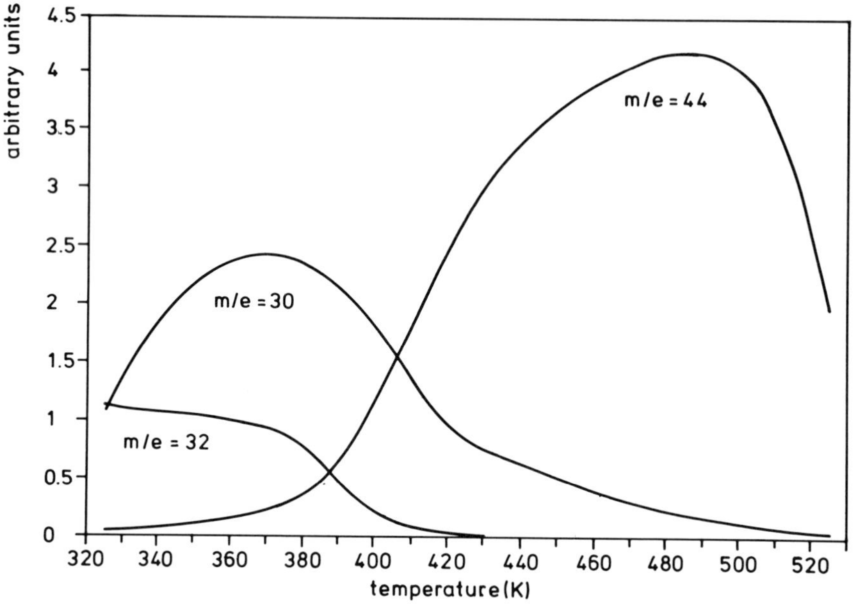

Figure 2. Temperature programmed surface reaction of CD_3NO_2 on α-Mn_3O_4, followed by mass spectrometry (m/e=30: NO; m/e=32: $D_2C{=}O$; m/e=44: CO_2).

DISCUSSION

Obviously, the interaction of nitromethane with an oxidic surface starts in all cases by physical adsorption. On Al_2O_3 and α-Mn_3O_4 this molecule, stabilized in its form of nitronic acid, splits of a H-atom to form a nitronate ion. The two mentioned oxides facilitate the deprotonizing step by exposing oxygen ions and cations; the former ones accept the H-atom and the latter ones stabilize the nitronate ions. The slight difference in the positions of the IR bands for the nitronate ions on alumina and α-Mn_3O_4 reveals the influence of the cations on this position.

Literature concerning homogeneous oxidation of nitromethane by $KMnO_4$[15] offers an explanation for the other above mentioned observations. The CH_2= group is converted into formaldehyde, which either desorbs into the gasphase (figure 2) or stays in the adsorbed state forming a formate species (figure 1). Most likely this oxidation to formaldehyde is accomplished with lattice oxygen. The oxygen vacancies created in this way, can then reduce the "NO_2" species to NO, which is observed after its desorption (figure 2). The possibility of NO to readsorb and react further at the surface cannot be excluded, but since (iso)cyanate species are also formed when the catalyst is heated under evacuation (with NO pumped out of the reaction vessel), a direct conversion of the nitronate ion into (iso)cyanate species seems a more likely step to occur.

No nitrosomethane has been detected by the simple arrangement used in this study. This is in agreement with the practice of organic chemistry, that such a reduction is difficult. Our experiments seem to point out that this difficulty is caused by the easy migration of the hydrogen atom along the molecule and the very fast formation of nitronate species. Once this pathway is opened, other reactions than reduction to the nitroso compound become more probable. In absence of a hydrogen atom on the α-carbon atom, there will be no nitronate formed in the adsorbed state. Currently, we investigate whether this absence increases the probability to form nitroso compounds. With nitrobenzene the reduction to nitrosobenzene is indeed quite selective (around 80%)[9].

REFERENCES

1. B. Delmon and P. Ruiz, Catal. Today, 1 (1987) 1.
2. K. van der Wiele and P.J. van den Berg, in C.H. Bamford and C.F.H. Tipper (Eds.) Chemical Kinetics, Elsevier, Vol. 20, p. 123.
3. J. Haber, Proc. 8th Int. Congr. Catal., Berlin, Vol. I (1984) 85.
4. P. Mars and D.W. van Krevelen, Chem. Eng. Sc., 3 (1954) 41.
5. F. Wattimena and W.M.H. Sachtler, Proc. 7th Int. Congr. Catal., Tokyo, Vol. B (1980) 816.
6. B.C. Gates, J.R. Katzer, and G.C.A. Schuit, Chemistry of catalytic processes, McGraw-Hill Book Company, New York, 1979, Ch. 5, p. 424.
7. H.G. Zengel and M. Bergfeld, Ger. Offen 2939692 (1981).
8. N.L. Holy, A.P.Gelbein, and R.Hansen, U.S. Patent 4585900 (1986).
9. T.L.F. Favre, P.J. Seijsener, P.J. Kooyman, A. Maltha, A.P. Zuur, and V. Ponec, Catal. Lett., 1 (1988) 457.

10. H.A.C.M. Hendrickx, Ph.D.Thesis 1988, Leiden University, The Netherlands.
11. A.J. Wells and E.B. Wilson Jr., J. Chem. Phys., 9 (1941) 314.
12. M.J. Brookes and N. Jonathan, J. Chem. Soc.(A), (1968) 1529.
13. S. Rosen and D. Swern, Anal. Chem., 38 (1966) 1392.
14. J. Rasko and F. Solymosi, J. Catal., 71 (1981) 219.
15. E. Breuer, in S. Patai (Ed.) The chemistry of Amino, Nitroso, and Nitro Compounds and their Derivatives, John Wiley & Sons Ltd., New York, 1982, Ch. 13.

G. Centi and F. Trifiro' (Editors), *New Developments in Selective Oxidation*
© 1990 Elsevier Science Publishers B.V., Amsterdam — Printed in The Netherlands

THE MODEL OF ACTIVE CENTERS FOR PARTIAL OXIDATION IN THE STRUCTURE OF HETEROPOLY COMPOUNDS

M.YU.KUTYREV, I.N.STAROVEROVA, N.Z.THIÊP, O.V.KRYLOV.
Institute of Chemical physics, USSR Ac.Sci. Moscow 117334, ul.Kossygina 4.

SUMMARY

The paper deals with possible use of 12-Mo-P-Heteropoly compounds (HPC) as a matrix for introduction of various modifying ions: V,Cu,Cr,Fe,Zn,K in the state controlled by physical methods: EPR, UV and Mössbauer spectroscopy. By varying the preparation techniques it appeared possible to obtain samples containing these ions either in the cationic or anionic sublattice of HPC crys tallines, The ions coordination in cationic position was of the tetragonally distorted octahedra and in the anionic one - close to the regular octahedra.The catalytic reactions revealed the effect of the active centers obtained on the rate and route of the processes. The oxidation of methacrolein to methacrylic acid was found to occur over ions in anionic position only. Oxidative dehydrogenation occurs mainly over ions in cationic position. The selective oxidation of n-butane and propene into maleic anhydride and acrolein requires the formation of pair centers.

INTRODUCTION

Two basic factors: the composition and functions of active centers (a.c.) and the relevant reaction mechanism have to be known in order to understand the nature of heterogeneous catalysts and to find new active systems. The reactions of the selective oxidation can be considered according to three main types (ref. 1):i) oxidation as insertion of oxygen atoms into the substrate (e.g. transformation of aldehydes into acids); ii) oxidative splitting of the C-H bonds in organic molecules resulting in unsaturated products formation (oxidative dehydrogenation), iii) complex processes involved combinations of both steps (e.g. alkane or olefine oxidation into oxygen-containing products).

It is of great interest to compare thecomposition and properties of a.c. with their catalytic activities in reactions of the types mentioned above.

Earlier research provides ground for the choice of modified 12-Mo-P hetero poly compounds (HPC) as one of the best objects for such comparison (ref.2-5): 1) their cristalline structure is described in detail (ref.6) and is specified by the low dencity of the elementary cell (fig.1) which permits the addition of different elements without any lattice changes,

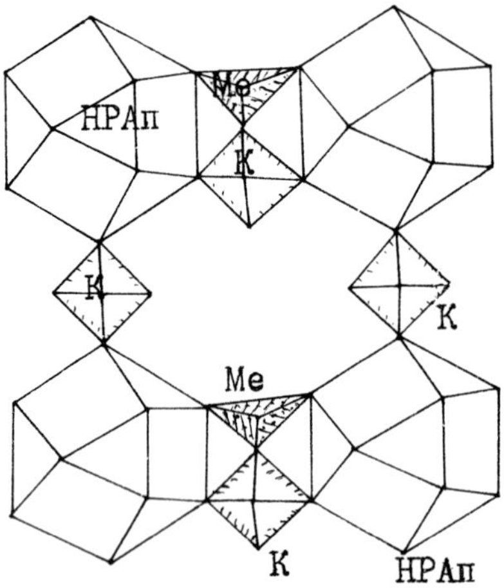

Fig. 1. Fragment of a heteropoly compound structure with added ions.

2) the introduced ions can become located either in the heteropoly anion (HPAn – – Keggin cell) substituting molybdenum ions, or in the cation sublattice between the HPAn (ref.5,6);

3) the initial HPC is active virtually in all known reactions of the selective oxidation and acid-base catalysis (ref.2,3).

The aim of the presented research was to obtain well-specified single-phase catalysts of the 12-Mo-P HPC structure containing added ions of transition metals in different positions and to find the regularities of their influence on catalytic activities in some selective oxidation processes.

METHODS

Catalysts of the composition $K_1P_1Mo_{12-X}Me_XO_{40}$ (where Me stands for V,Cu,Cr, Fe) with "X" varying from 0 to 8 for V, and was 1.0 for Fe,Cu,Cr were prepared by "solid phase" synthesis (ref.4,5). Potassium ions were added to enhance the stability of these systems (ref.2,5). Zn ions were also also introduced into certain samples. These catalysts are designated as HPCMe.

Catalysts containing the same elements incorporated by dissolution of the relevant oxide or salts in the 12-Mo-P heteropoly acid in acidic medium were produced too. These systems were designated hereafter as HPA+Me.

The IR spectra were recorded by UR-20W apparatus, the X-ray patterns – by a DRON-3, the EPR spectra – by "Bruker", the Mössbauer spectra – by an apparatus with the ^{57}Co source in Cr (the samples were additionally enriched with ^{57}Fe to 10%), the UV spectra – by "Pye Unicam".

The catalysts activities were tested in the selective oxidation of propene, methacrolein, n-butane and oxidative dehydrogenation of diethylbenzene.

EXPERIMENTAL

The catalysts structure according to the X-Ray diffraction results was that of potassium salt of the 12-Mo-P heteropoly acid (ref.2,7). The existance of the Keggin unit structure was confirmed by the IR-spectra (ref.2,6).

The DTA curves are also characteristic for HPC (ref.2,4).

The EPR spectra of HPC+Cu and HPC+V displayed individual signals with parameters corresponding Cu^{2+} and $|V=O^{2+}|$ ions coordinated as tetragonally distorted octahedron (ref.5,8) (table 1). Taking into account the preparation method, the salt of corresponding addition had to be formed, i.e. these ions are located in the cationic positions between the HPAn (fig.1).

TABLE 1
Parameters of EPR spectra for vanadyl and copper(2+) ions in the samples HPC

Samples studied	EPR spectrum parameters				Ascription
	A_I(G)	g_I	A_{II}(G)	g_{II}	
HPA+V	77.5	1.974	198.8	1.936	Distorted octahedron
HPCV	56.1	1.976	164.0	1.940	Close to regular octahedron
HPA+Cu	24.0	2.061	125.0	2.315	Tetr.distorted octahedron
HPCCu	16.0	2.066	91.0	2.325	Slightly dist. octahedron

Signals with lesser splitting of HFS were detected in HPCV and HPCCu spectra, which is evidence of a greater symmetry of their environment, i.e. coordination approaches a regular octahedron characteristic for ions substituting Mo^{6+} in the Keggin cell (ref.5,9–11). (see Annex).

The simultaneously presence of two signals can be explained as follows: due to the partially replacement of Mo^{6+} ions in HPAn by ions of lower charge (5+ or 2+) increases the anion negative charge. In water solution and in hydrated form of HPC at temperatures lower 250° this excess charge may be compensated by protons. After the finally calcination temperature (380°) the samples were remarcable dehydrated. To stabilise the totally electroneutrality some introduced ions becomes located in the cationic sublattice.

Thus an additional cation can be introduced to shift the equilibrium between the concentration of modifyed ions in different positions to the anion side. Only the involvement of this ion into the anionic position must be impossible. Zinc was found (ref.5) to be most effective "compensating ion. It does not change the crystalline structure, its effect on the catalytic activity in most cases was low, and it gives no EPR signals. The addition of calculated amounts of Zn(2+) ions (in the range $Zn_{0.5\div2.0}$) permitted to obtain pure anionic forms of vanadyl and copper(2+).

The X-ray diffraction data, UV, IR and EPR spectra of prepared standart samples are given in Annex. The most useful was the possibility to distinguish two forms of Cu and V ions based on the position of the 1-st parallel component position in the EPR spectrum.

The Mössbauer spectra of the Fe-containing catalysts also displayed two signals depending on the catalysts preparation techniques, one was ascribed to octahedrally, and the other to tetrahedrally (or strong distorted octahedrally) coordinated Fe^{3+} ions (ref.12),(fig.2). The parameters for cationic form are IS=0.64; QS=1.02, and for anionic one IS=0.67, QS=0.36÷0.52. Similar to the EPR spectra the big difference in the quadrupole shift can be caused by the greater distortion of ions arrangement.

The presence of two forms of Fe^{3+} can be confirmed by UV-Vis spectra - the catalysts displayed well resolved maxima at wavenumber 700 and 750 nm corresponding to the tetrahedral and octahedral coordinated ions respectively (d-d - transfer, ref.13).

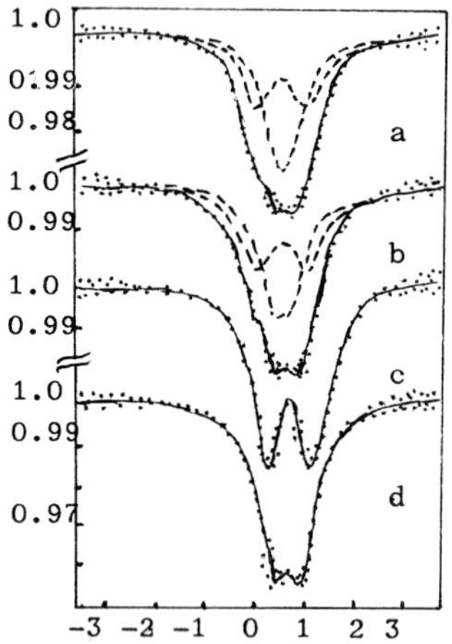

Figure 2. Mössbauer specrta of catalysts: HPCFe(a),HPCFeCr(b) HPA+Fe(c), HPCFe(+Zn)(d)
-- - - decomposition

As stated above, an octahedral coordination is observed when the additive is introduced into HPAn instead of Mo, and the tetrahedral one corresponds to the cation position with ion surrounded by four Keggin cells. (fig.1).

Simultaneous addition of iron (X=1) and chromium(0.5) results in the decrease of anionic Fe^{3+} form concentration (from 56 to 42%). This seems to be due to competition between the Fe^{3+} and Cr^{3+} ions with close radii value for the anionic position.

Conversely, the addition of compensating Zn ions transfers all iron ions to the position inside HPAn (fig.2d).

As result of this research, it became possible to vary the state of metal ions introduced into HPC matrix. These samples were testsd in several model oxidation reactions.

The methacrolein oxidation into methacrylic acid process parameters over certain HPC catalysts are shown in the Table 2.

The species introduced into the cationic position have virtually no effect on the reaction indexes, whereas their appearance in the anion position improves in the case of vanadium and chromium the selectivity of acid formation. However in the presence of copper and iron the selectivity decreases.

TABLE 2
Methacrolein conversion and the selectivity of methacrylic acid formation as a function of the catalysts composition and preparation method (max.acid yield).

Samples	t^oC	MA,conv.%	MAA sel.%	Additives	positions
HPC (K-salt)	290	71	43	-	-
HPA+V	295	70	47	$V=O^{2+}$	- cation 100%
HPCV	295	71	64	$V=O^{2+}$	- cation/anion 50%/50%
HPA+Cr	290	73	45	not analised	
HPCCr	290	71	56	"	"
HPA+Cu	290	72	41	Cu(2+)	- cation 100%
HPCCu	290	72	32	Cu(2+)	- cation 95%, anion 5%
HPCCu(+Zn)	290	74	23	Cu(2+)	- anion 100%
HPA+Fe	285	75	46	Fe(3+)	- cation 100%
HPCFe	280	81	15	Fe(3+)	- cation 65%, anion 35%
HPCFe(Zn)	280	83	12	Fe(3+)	- anion 100%

It has been suggested earlier (ref.2,14) that the aldehydes oxidation proceeds on the active centers fdrmed from octahedrally coordinated V^{5+} or Mo^{6+} ions. The result of the present research confirm this suggestion - well formed octahedra are locate_ in the anionic positions of HPC.

The oxidative dehydrogenation of diethyl benzene is catalysed in industry by the modified iron oxide system (ref.1). The modelling of similar active centers in the structure of HPC (table 3)* showed in contrast to the oxygen insertion that the oxidative dehydrogenation is sensitive to the properties of the cation sublattice only and is independent on the HPA composition.

TABLE 3
Oxidative dehydrogenation of diethyl benzene over Fe-containing catalysts

Samples	t^o	DEB conv%	Divinylbenzene sel.%	Additives positions
HPA(K-salt)	350	9	24	-
HPA+Fe,Cr	330	27	73	Fe^{3+} - pure cation
HPCFeCr(+Zn)	350	12	23	Fe^{3+} - pure anion

Oxidation of n-butane into maleic anhydride (MAA) is one of the most structure sensitive reactions of partial hydrocarbons oxidation, its rate being high only over the vanadyl pyrophosphate catalyst (ref.15). It was reported that a certain activity was displayed also by V-containing HPC. (ref.16).

To find out the nature of active centers in this reaction use was made of KPMo(12-X)V(X) O(40) samples with X-value varying from 0 to 8.0 (table 4).

*The results were obtained in the cooperation with professor E.Rico and collaborators in Instituto Polytechnico National, Mexico D.C., Mexico.

TABLE 4

n-Butane oxidation over V-containing HPC (conditions given for the maximal anhydride yield)

"X"-vanadium content	t °C	C_4H_{10} conv.%	selectivities% MAA	-enes	diene	Ratio V_c/V_a	notes
0	350	28.5	0	2.0	1.7	- -	
0,125	330	6.5	0	12.0	1,4	0	pure anion
0.25	318	19.0	37.0	2.7	3.7	1.2	
0.5	329	30.0	14.0	1.5	4.5	11.0	
1.0	342	45.3	43.8	0.7	3.4	2.3	
1.25	334	24.1	16.3	4.3	4.1	0.5	
2.0	346	45.3	26.9	4.2	7.0	0.6	
5.0	335	37.2	15.1	19.0	8.8	-	no resolution
1.0	356	11.9	0	6.4	3.9	∞	HPA+V - pure cation
0.3	364	8.0	0	11.7	4.1	0	HPCV(+Zn)-pure anion

No traces of vanadyl pyrophosphate structure were found in the catalysts either before or after the reaction,i.e. only vanadium containing HPC structure is responsible for the activity in MAA formation. After the decomposition of this structure by heating at 500° the catalysts became inactive.

According to data given in the table 4 two forms of vanadyl ions,cationic and anionic, must be simultaneously present in the catalyst to obtain the maleic anhydride. The ratio of the concentration of these forms was estimated from the intensity of the first parallel component of the EPR signals. The best catalytic results were obtained on the samples with close concentrations of both vanadyl forms (table 4).

By adding of small amounts of the compensating Zn^{2+} ions(0.03÷0.20) into the samle with constant total vanadium concentration (X=1.0) it appeared possible to obtain a different ratio between the quantities of the two vanadyl forms (fig.3). The maximal selectivity of MAA formation was reached just with $\{V=O^{2+}_{cat.}\}/\{V=O^{2+}_{an.}\}$ close to unity.

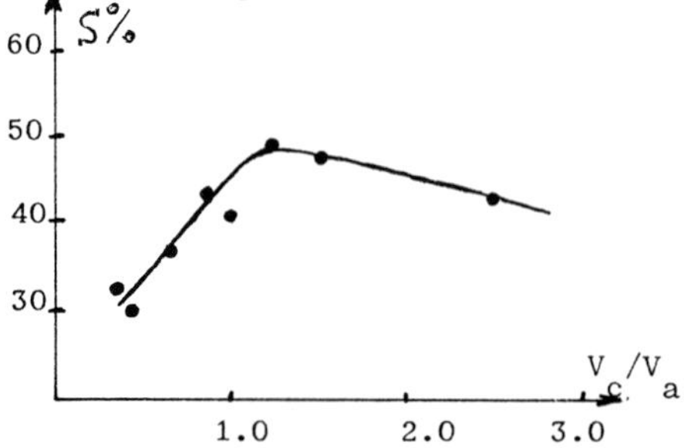

Fig.3.Selectivity of MAA formation as a function of the ratio of cationic to anionic VO^{2+} concentration

The high specific activity of$(VO)_2P_2O_7$ in butane oxidation is ascribed to the sence of a.c. consisting of two V^{5+} ions in a qudratic pyramid coordination occupying nonequivalent positions on the catalyst surface (ref.15).

Similarly in HPC the vanadium ions are present in different positions. Activation of the hydrocarbon by C-H bond rupture seems to occur over distorted octahedrally (or tetrahedrally) coordinated vanadyl, and the regular octahedrally coordinated one acts as an oxygen source. These conclusions are in god agreement with regularities foud above for simple reactions.

The formation of pair centers in the $(VO)_2P_2O_7$ structure is caused by the specific properties of its crystal cell and is extremely important for the high selectivity of the MAA formation (ref.15).

In generally the presence of two V ions types in HPC does not mean the formation of pair centers - in the case of their statistic distribution the probability of direct interaction would be small.

However certain chemical forces may cause the attraction between different vanadium forms. As mentioned above vanadium in cation sublattice compensates the excess anion charge arising on the replacement of Mo^{6+} by V^{5+}, i.e. the interaction of cationic form with modified Keggin cell is more probable than with the unmodified one. This can be the reason for quite high selectivity of HPC catalysts, which is comparable with that of vanadium pyrophosphate.

The formation of a pair a.c. seems to be a more important factor for butane oxidation route than the ions coordination as such. This follows from the difference in local surrounding of pyrophosphate and HPC vanadium containing a.c.

The reaction of propene selective oxidation was studied on the Fe-containing HPC that modelled the famous multicomponent catalysts (ref.1).

The position of iron ions has revealed a fundamental effect on the propene transformation route (table 5). The sample containing pure cationic Fe^{3+} gives acetone. According to (ref.17) propene oxidation to acetone seems to consist of two main steps: 1) olefine dehydrogenation with breaking of 2 C-H bonds; 2)the susequent fast hydratation. The first step is rate limiting and, in accordance with examples given above, can occur on tetrahedrally coordinated Fe^{3+} ions.

TABLE 5 *
Propene oxidation over Fe-containing catalysts at 355° and 365^o(last sample)

Samples	C_3H_6 conv.%	Selectivities % into acetone	acrolein	Addition positions
HPA+Fe	35	93	-	Fe^{3+} - 100% cation
HPCFe	7	-	83	Fe^{3+} -40%cation+60%anion
HPCFe(+Zn)	20	52	38	Fe^{3+} -100% anion+Zn^{2+} -cation

*Results from A.A.Firsova - Inst.of Chem.Physics, USSR Ac.Sci., Moscow.

The two other catalysts show activities in acrolein formation which may be ascribed to the presence of pair centers $Fe_{cat.} - Fe_{an.}$ in the case of HPCFe or $Zn_{cat.} - Fe_{an.}$ in the HPCFe(+Zn). (Zn oxide is known as a low active hydrogen abstraction catalyst (ref.18).

CONCLUSIONS

The additional transition metal ions introduced into the 12-Mo-P heteropoly compound depended from the preparation technique can became located either in the cationic or anionic sublattice. The coordination of these ions is that of a tetragonally distorted octahedra or of a slightly distorted octahedra consequently.

The composition of cationic sublattice effects on the processes of hydrogen abstraction; ions in anionic position displays their activity in the oxygen incorporation into the reacting molecules.

The potentialities of a.c. model in the matrix of HPC are demonstrated. It appears possible to obtain both individual and pair centers of a fixed composi-- tion and coordination state of the ions added. The application of this concept makes possible the preparation of new complex partial oxidation catalysts.

REFERENCES

1.J.E.Germain in"Chemical and Physical Aspects of Catalytic Oxidation" ed.CNRS, Paris,1979,p.207.
2.M.Misono, Catal.Revs.,Sci.Eng.,29 (1987) 269-321.
3.I.V.Kozhevnikov,K.N.Matveev, Applied Catal., 5 (1983) 135.
4.M.Yu.Kutyrev and I.N.Staroverova "Preparation of Catalysts" Louvan-la-Neuve prep. C 3. 1.
5.I.N.Staroverova,B.V.Rozentuller,M.Yu.Kutyrev,Kin.i Cat.,27(1986) №3,698.
6.E.N.Yurchenko,J.Mol.Structure,60 (1980) 325.
7.ASTM 9-403, 9-412.
8.M.Akimoto,K.Shima,E.Echigoya,J.Chem.Soc.Far.Tr.1,79 (1983) 2467-74.
9.C.Sanches,J.Liverage,J.Lunay,J.Am.Chem.Soc.,104 (1982) 3194-99.
11.N.G.Maksimov,V.F.Anufrinko,Dokl.Ac.Nauk SSSR,228 №6 (1976) 1391-96.
10.R.Fricke,H.-G.Jerschkevitz and G.Ohlmann,J.Chem.Soc.Far.Tr.1, 82 (1986) 3479.
12.In "Mössbauer effect data index,Ed.J.G.Stievens,IFI Plenum,N.Y.,1972.
13.F.Ballhausen, Optical Electron Spectroscopy of Transitient Metal Ions,N.Y.1971.
14.O.S.Morozova e.a.,Izv.Ac.Sci.USSR,Ser.Chim.,3 (1977) 527.
15.G.Centi,F.Trifro,J.B.Ebner,V.M.Franchetti,Chem.Revs.,88 (1988) 55-80.
16.M.Ai,J.of Catal., 85 (1984) 324-330.
17.M.Yu.Kutyrev,e.a. Proc. 8th Cogr.on Catal.,W.Berlin,(1984) v.IV,463-472.
18. A.L.Dent,R.J.Kokes, J.Amer.Chem.Soc.,92 (1970) 1092.

ANNEX

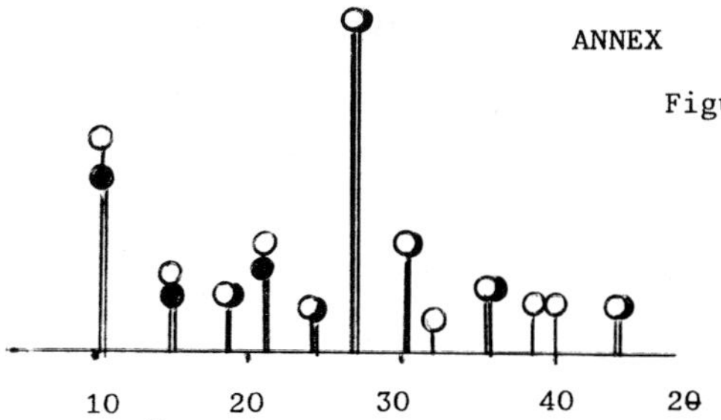

Figure 1 X-Ray diffraction patterns of catalysts $K_1P_1Mo_{12-X}Me_XO_{40}$ prepared by solid phase synthesys (o) and using $H_3PMo_{12}O_{40}$ HPA (•).

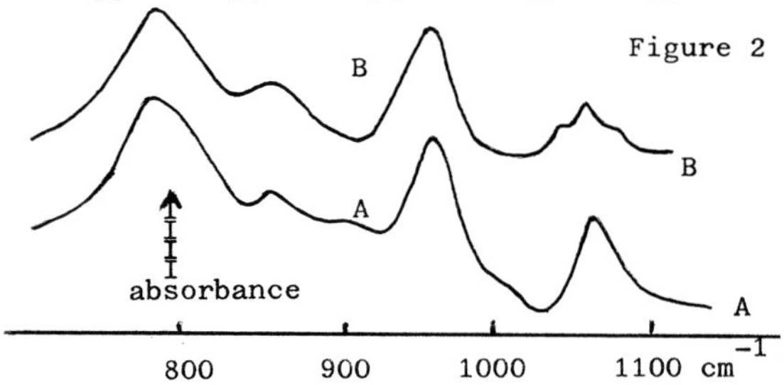

Figure 2 IR-spectra of catalysts: $KPMo_{12}O_{40}$ (A) $KPMo_{11}V_1O_{40}$ (B).

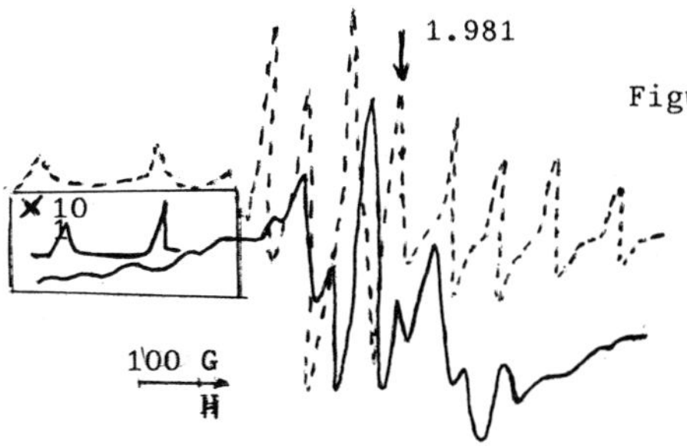

Figure 3 EPR spectra of V-containing catalysts: $KPMo_{12}O_{40}+V_1$ (——) $KPMo_{11}V_1O_{40}$ (- - -) both recorded at 77K

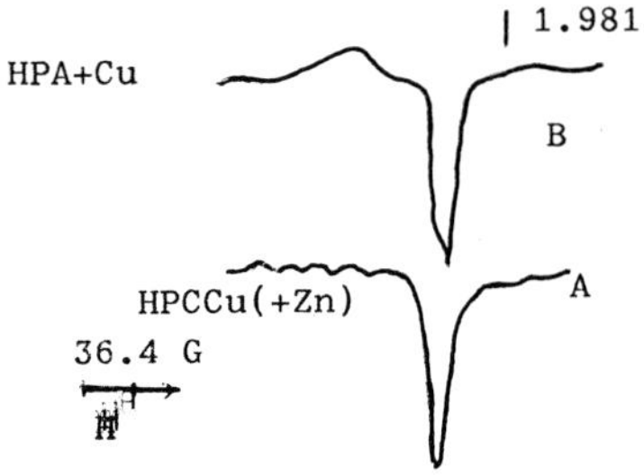

Figure 4 EPR spectra of Cu-containing catalysts: $KPMo_{12}O_{40}+Cu$ (A); $KPMo_{11}Cu_1O_{40}+Zn_2$ (B). both recorded at 77K

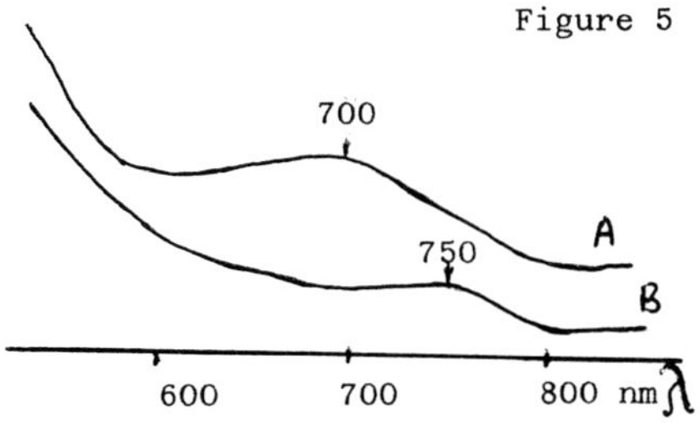

Figure 5 Diffusion reflection spectra of Fe-containing catalysts: $KPMo_{12}O_{40}+Fe_1$ (A) $KPMo_{11}Fe_1O_{40}$(+Zn).(B)

M.MISONO[1] : I agree that the introduction of other transitient elements in both or either of cationic and anionic sites is an important method for the catalysts design based on heteropoly compounds.
However, the introduction at definite site and the assignement are rather difficul You assigned tetrahedral site to the cationic site to the octahedral one, but the correlation may not be so straightforward due to the variable nature of the secondary structure, cationic site may become both tetrahedral and octahedral.
Do you have any direct evidence to prove what metal ion is actually in Keggin unit Probable, the change of IR spectrum due to Keggin unit is helpful.

M.Kutyrev[2] : Of course the coordination of ions introduced in both sites is far from being regular octahedral or tetrahedral. It seems to characterise the cationi position as a tetragonally distorted octahedron. The presence of a new ion inside the Keggin unit may be detected by means of various methods: i) The symmetry of Keggin unit posessing an introduced ion does beeh changed-that results in the shifting of the central IPO_4I octahedron and splitting of P-O-Me bands in IR-spec (shown in the Annex). The other physical methods permit only to distinguish different coordination state.

J.KIWI[3] : You have made some anionic & cationic substitutions in Keggin structure related polyhetero acids and obtained some values for efficiances in various catalytic processes. Have you been able to correlate these observations with the red properties of these, Cu, Fe substituted Keggin structures?
A.LA GINESTRA[4]: What about the influence of copper ions inserted in your compounds Did you find, as like as in other ion exchangers, an apparent increase in the acidic strengh? Which are the products obtained in the investigated reactions?

M.KUTYREV[2]: Unfortunately the redox and acid-base properties of substituted systems were not investigated, but we belive that the difference between modified and initial samples is very significant.
The copper containing systems we tested in different reactions are usually active in total oxidation.

D.MILLET[5] : How do you explain that the iron ions which are in the octahedral posit in HPAn present an quadrupolar splitting lower than the other iron ions which have an tetrahedral coordination?

M.KUTYREV[2]: The ascription was made in accordance to many publications, e.g. Garten R.L.,Delgass W.N.and Boudart M. J.Catal.,18 (1970) 90.(Fe^{3+}-cation tetrahedrally coordinated in the zeolites)
Carbucicchio M. and Trifiro F., J.Catal.,43 (1975) 77. (Fe^{3+}- octahedrally coordinated ions in iron molybdate).
The general tendency in sygnals change is the increasing of quadrupol splitting by the distortion of coordination polyhedra.

T.C.VEDRINE[5]: How did you quantify the relative number of vanadium ions in cationic and/or anionic state?
Depending on the elements substituted in the HP compounds and cation or anion state the stability of the catalyst under catalytic conditions i.e. 400°C with H_2O present may be different. Are you able to test your samples stability. If not are your correlations done in the starting or aged catalysts?

M.KUTYREV[2]: We were able to provide only the estimation of V^{4+} ions concentrations based on the intensivities of the 1-th parallel component in the EPR spectra (see Annex). The ratios beteween these intensivities retain unchanged for each sample both bevore and after catalytic reaction.
The samples syntesied were thermally stable up to the 430-450°C. The first reason was the introduction of potassium ions (as mentioned in "METHODS"), the second reason was the presence of low valence ions in HPAn.(experimentally observed). Our correlations were done for starting catalysts, but no remarkable changes in spectroscopic data were find after the reaction.

1- The University of Tokyo, Japan
2- Institute of Chemical Physics, USSR AC.Sci.,Moscow, USSR.
3- EPFL, IPC II, Lausanne, Switzerland
4- University of Rome Italy.
5- Institute de Catalyse, CNRS, Villeurbanne, France.

Author Index

Abadjieva N., 287
Abon M., 747
Ai M., 257
Aliev S.M., 437
Andersson A., 275
Anelli P.L., 63
Angevaare P.A.J.M., 861
Anpo M., 683
Anshits A.G., 483
Aparicio L.A., 417

Badyal J.P.S., 739
Baerns M., 247, 807
Bagnasco G., 327
Banfi S., 63
Barbaux Y., 767
Baronetti G.T., 373
Bartsch S., 353
Bazhan O.V., 469
Bebelis S., 643
Beck I.E., 213
Belousov V.M., 239, 617
Belyaev V.D., 469
Beszterda S., 853
Bielsa R., 295
Biermann J.J.P., 381
Blanchard M., 661
Bonnelle J.P., 767
Bordes E., 625, 585
Boronin A.I., 89
Bouchoule C., 661
Bovin J.-O., 275
Bregeault J.-M., 205
Bressan M., 119
Broclawik E., 701
Budi F., 537
Busca G., 305, 825
Buyevskaya O.V., 437

Cairns J.A., 595
Canesson P., 661
Cannizzaro A., 733
Castro A.A., 373
Centi G., 515, 635
Cesana A., 43
Chadwick D., 595
Chaki S., 335
Che M., 683
Chrzaszcz J., 779
Ciambelli P., 327
Clerici M., 33
Coluccia S., 817
Contractor R., 553

Courbon H., 675
Courtine P., 625
Cristiani C., 305
Cvengrosova Z.. 169

Dalla Lana I.G., 853
Davies M.J., 595
Davydov A.A., 437
Deller K., 159
Delmon B., 757, 797
Despeyroux B.M., 159
Dimitriev Y., 287
Disdier J., 675
Djaneye-Boundjou G., 661
Do N.T., 247, 807
Dolphin D., 109
Doumain B., 757
Dumesic J.A., 417

Ebner J., 553
El Ali B., 205
Elmi A.S., 305
Emig G., 843
Esposito A., 33

Feringa B.L., 177
Ferrini C., 53
Fiedorow R., 853
Fierro J.L.G., 295
Filipek E., 767
Forlani O., 417
Forzatti P., 305
Frusteri F., 733
Fumagalli C., 537

Gargano M., 139
Garrone E., 817
Geerts J.W.M.H., 343
Gervasutti P., 43
Giamello E., 683, 817
Giordano N., 733
Girerd J.-J., 97
Gleaves J.T., 707
Golodets G.I., 693
Goor G., 71
Goto T., 605
Grasselli R.K., 515
Grimblot J., 737
Grinenko S.B., 239
Grootendorst E.J., 841
Grzybowska B., 767
Gusevskaya E.V., 213

Haber J., ... 701
Habersberger K., ... 317
Haeberle Th., ... 843
Hansen S., ... 275
Hattori T., ... 789
Hawker S., ... 739
Hayakawa T., ... 133
Hecquet G., ... 833
Herrera R., ... 717
Herrmann J.M., ... 675
Hodnett B.K., ... 459
Hofmann H., ... 353
Hronec M., ... 169
Huang Zhong-Tao, ... 653

Iannibello A., ... 733
Ilavsky J., ... 169
Ito S., ... 125
Iwanejko R., ... 195
Iwasaki S., ... 789

James B.R., ... 109
Jiru P., ... 317

Kaddouri A., ... 365
Kadushin A.A., ... 447
Kalinkin A.V., ... 527
Kalthoff R., ... 247
Keulks G.W., ... 427
Kieffer R., ... 365
Kiennemann A., ... 365
Kiers N.H., ... 177
Kim, Y. Ch. 491
Klissurski D., ... 287
Komashko G.A., ... 617
Korf S.J., ... 381
Kotter M., ... 267
Kouwenhoven H.W., ... 53
Kremenic G., ... 295
Krylov O.V., ... 447, 477, 869
Kunai A., ... 125
Kutyrev M. Yu., ... 869
Kuznetsova N.I., ... 89

La Ginestra A., ... 327
Lambert R.M., ... 739
Lashier M.E., ... 573
Lazzari E.J., ... 373
Leofanti G., ... 43
Li D.X., ... 267
Likholobov V.A., ... 89, 213
Lisitsyn A.S., ... 89
Lopez Nieto J.M., ... 295, 635
Lorenzelli V., ... 825
Lupieri M., ... 417
Lynch D.T., ... 843

MacGiolla Coda E., ... 459
Maddox P.J., ... 453
Mamedov A. Kh., ... 477
Mano V., ... 185
Mantegazza M., ... 43
Martin C., ... 205
Martin J., ... 205
Martinez E., ... 717
Maspero F., ... 33
Massardier J., ... 747
Matsuura I., ... 563
McCarty J.G., ... 405
McEwen A.B., ... 405
Menage S., ... 97
Mercier J., ... 205
Merkureva Yu. N., ... 617
Michman M., ... 667
Migita C.T., ... 335
Millet J.-M.M., ... 833
Mills P.L., ... 707
Mimoun H., ... 97
Min Yu., ... 427
Minchev L., ... 287
Mingot B., ... 747
Misono M., ... 605
Miyamoto K., ... 605
Mizuno N., ... 605
Mlodnicka T., ... 195
Mohammedi O., ... 205
Montanari F., ... 63
Moro-Oka Y., ... 491
Morvillo A., ... 119
Moser T.P., ... 573
Mozzanega M.N., ... 675
Mukoid C., ... 739
Mummey M., ... 553
Murakami Y., ... 789

Najbar M., ... 779
Nakata M., ... 789
Neophytides S., ... 643
Neri C., ... 33

Ogura K., ... 335
Okuhara T., ... 605
Orita H., ... 133
Osipova Z.G., ... 527
Otsuji Y., ... 683
Otamiri J.C., ... 275

Padovan M., ... 43
Pajonk G.M., ... 229
Pang Xianxing, ... 653

Parmaliana A., 733
Parmon V.N., 469
Patane' E., 515
Patrono P., 797
Peldszus E., 159
Pesheva J., 287
Petrini G., 43
Piccoli V., 417
Pichat P., 675
Pickering I.J., 453
Pinelli D., 635
Pinna F., 81
Pires M.J., 807
Plate S.E., 447
Poix P., 365
Poltowicz J., 195
Ponec V., 861
Portela M.F., 807
Prescher G., 71
Przystajko W., 833
Pyatnitskaya A.I., 617

Quici S., 63
Quinlan M.A., 405

Rajapakse N., 109
Ramis G., 825
Ravasio N., 139
Rehspringer J.L., 365
Rekoske J.A., 417
Riekert, L., 267
Rigas N.C., 707
Roffia P., 43
Romano U., 33
Roos J.A., 381
Ross J.R.H., 381, 505
Rossi M., 139
Rossini S., 417
Ruiz P., 757, 797
Russo G., 327

Sakharov A.M., 221
Sanfilippo D., 417
Sasaki K., 125
Satsuma A., 789
Saussine L., 97
Scelza O.A., 373
Schmidt M., 71
Schrader G.L., 573
Schuchardt U., 185
Serwicka E., 737
Seshan K., 505
Sham E., **757, 797**
Shashkin D.P., 477
Sheldon R.A., 1
Shimizu M., 133
Shiryaev P.A., 477
Shishkina N.N., 483
Skibida I.P., 221
Smits R.H.H., 505
Sobyanin V.A., 469
Sokolovskii V.D., 437, 527
Spinicci R., 393
Spoto G., 817
Staroverova I.N., 869
Stefani G., 537
Stepanov A.G., 213
Stoch J., 617
Strukul G., 81
Suciu, G.D. 537
Suzuki T., 683
Swaan H.M., 505

Takehira K., 133
Thiep N.Z., 869
Thomas J.M., 453
Tonti S., 43
Toreis N., 725
Trautmann S., 247
Trevino A.A., 417
Trifiro' F., 515, 635
Tronconi E., 305
Tsuji K., 605
Tuleja J., 169
Tulenin Yu.P., 447
Turco M., 327
Tvaruzkova Z., 317

Ueda W., 491
Ungarelli F., 635
Ushkov S.B., 527

Van Bekkum H., 147
Van Dam H.E., 147
Van der Wiele K., 343
Van Kasteren J.M.N., 343
Van Leeuwen P.W.N.M., 177
Van Ommen J.G., 381, 505
Varma A., 717
Vayenas C.G., 643
Vedrine J.C., 833
Vereshchagin S.N., 483
Verykios X.E., 725
Vinke P., 147
Volta J.C., 747

Wal W., 779
Watanabe Y., 133
Watzenberger O., 843
Weiss M., 667

Weng L.T., ... 757, 797
Witko M., ... 701

Yamada T., ... 335
Yamada Y., ... 683
Yamazaki M., ... 563
Ye Daiqi, ... 653

Zamaraev K.I., ... 213
Zanardo A., ... 81
Zazhigalov V.A., ... 617
Zecchina A., ... 817
Ziolkowski J., ... 625
Zuur A.P., ... 861

Subject Index

Acetaldehyde, ... 653
Acid-base Properties, ... 257, 693
Acrolein, ... 491
Activation of Catalyst, ... 393, 617
Active Sites, ... 437, 833, 869
Activity, ... 779
Adsorbed Intermediates, ... 807, 825
Adspecies, ... 707
Aerial Oxidation, ... 125
Ag-doped Bi-V-molybdate, ... 491
Alcohol Oxidation, ... 97, 169
Aldehydes, ... 177
Alkali- and Alkaline Earth Vanadate, ... 505
Alkali-doped Metal Oxides, ... 353
Alkane Oxidation, ... 119, 185, 505
Alkene Epoxidation, ... 739
1-Alkenes, ... 177
Alloy Catalysts, ... 725
Alpha-Methylstyrene Oxidation, ... 257
Alumina, ... 853
Ammoxidation of Propane, ... 491, 515
Ammoxidation, ... 275
Ammoximation, ... 43
Aromatic Compounds, partial oxidation of .. 257
Axial Ligands, covalently anchored, ... 63

Benzene Oxidation, ... 125, 789, 825
Bi-dopant, ... 159
Bi_2MoO_6, ... 757
Bifunctional Catalysts, ... 81
Bimetallic Catalysts, ... 725
Biomimetic, ... 221
Bond Energy of Surface Oxygen, ... 693
n-Butane Oxidation, ... 537, 553, 563
.... 573, 585, 595, 625
Butene Oxidation, ... 767
1-Butene Oxidation, ... 585

C_4 Oxidation, ... 617
Calcium Oxide, ... 393
CaO Addition to MgO, ... 365
Carbonyl Groups, ... 97
Catalytic Anisotropy, ... 625, 747
Catalyst Development for Alkane Oxidation, 515
Catalyst Work Function, ... 643
Catalytic Testing, ... 247
Charge Transfer Excited Triplet, ... 683
Classification of Catalysts, ... 693
Clusters as Active Sites, ... 693
Coke, ... 853
CoO-MgO Solid Solution, ... 817
Cooperation between Phase, ... 797
Copper Catalysts, ... 139
Copper Oxychloride, ... 595
Copper Trihydroxychloride, ... 595
Copper-Amine Catalysts, ... 133
Crystallochemical Model, of Active Sites, . 625
Cu(I) , Cu(II) Catalysis, ... 221
Cu(II)-chloride/Amine Hydrochloride/O_2, ... 133
Cyclic Voltammetry, ... 667
Cyclohexane Oxidation, ... 185, 195
Cyclohexanonoxime, ... 43
Cyclohexene Epoxide, ... 89

Deactivation, ... 169, 343, 843
Dehydrogenation with CO_2, ... 477
Dimerization of Methane, ... 437
Dioxygen, ... 205, 265
DOPA, ... 661
Dopant, ... 459
Dynamic Model of Catalysis, ... 625
Dynamics of Photoxidation, ... 683

Effect of Gas Composition, ... 381
Effect of Process Conditions, ... 381
Effect of Products, ... 617
Effect of Promoters, ... 789
Ehylene Oxidation, ... 733
Electrocatalysis, ... 469
Electrochemical Modification of Selectivity, 643
Electrolysis, ... 667
Electroxidation, ... 653
Energy Profiles, ... 701
Enzymatic Oxidation, ... 221
Epoxidation, ... 63, 71, 89, 707, 739
Epoxidation Catalysts, ... 63
Epoxide Conversion, ... 739
ESR, ... 817
Ethane, ... 477, 483
Ethane Dehydrogenation, ... 405
Ethanol, ... 653
Ethene, ... 807
Ethers, ... 119
Ethylendiamine, ... 335
Ethylene, ... 213, 707, 717
Ethylene Epoxidation, ... 643
Ethylene Oxidation, ... 725

Fe-Mn Oxide Catalyst, ... 477
Fine-Chemical Production, ... 1
Fluidized Bed Reactor, ... 537
Formaldehyde, ... 287, 459
FTIR Characterization, ... 825, 861
Fuel Cell, ... 469

Gas-solid Reaction, ... 453

GIF System, ... 185
Glucaric /Gluconic Acid Oxidation, ... 159
Gold, ... 717

H_2/O_2 Mixture, ... 89
H_2O_2 Alkene Epoxidations, ... 63
Heterogeneous Catalyst, ... 125
Heterolytic Activation, ... 437
Heteropolyacid Catalysts, ... 265, 843
Higher Aldehydes, ... 317
Homogeneous Oxidation, ... 661
HOCl Alkene Epoxidation, ... 63
o-Hydroxylation, ... 139
Hydrocarbon Oxidation, ... 169, 701
Hydrocarbon Oxydehydrogenation, ... 853
Hydrocarbon, ... 229
Hydrochinone, ... 33
Hydrogen Peroxide, ... 43, 71, 81
Hydrophobic Catalysts, ... 733
Hydroxyl Radical, ... 125
5-Hydroxymethylfurfural, ... 147
Hypochloride, ... 119

Identification of Adsorbed Species, ... 807
In-situ FTIR, ... 247
In-situ Raman, ... 573
In-situ Studies, ... 447, 453, 477
Intermediates, ... 707
IR Spectroscopy, ... 807
Iron Phosphate, ... 833
Isobutene Oxidation, ... 757
Isobutyric Acid, ... 833
Isotopes, ... 707
Isotopic Studies (^{18}O), ... 573

Keto Acids, synthesis of 265
Kinetic, ... 185, 239, 247, 393
Kinetics of Methane Oxidation, ... 405

Li-doped MgO, ... 343, 353, 373
Li-Ni-O Catalysts, ... 453
$LnLiO_2$ Catalysts, ... 365

Maleic Anhydride from n-Pentane,.............. 635
Maleic Anhydride, .537, 553, 563, 573, 595, 605
Mechanism of Methane Oxidation, ... 427
Mechanism, ... 213
Metal/Substrate Interaction, ... 147
Metallic Molybdate Catalyst, ... 585
Metallic Copper, ... 139
Metalloporphyrins, ... 195
Methacrolein, ... 727
Methacrylic Acid, ... 843
Methane, ... 335, 343, 353, 365, 381, 393, 417
.... 427, 437, 453, 469

Methane Oligomerization, ... 373
Methane Oxidation to Formaldehyde, ... 459

Methanol Oxidation, ... 287, 305, 643
Methanol Oxidative Transformation, ... 317
Methyl Formate, ... 305
Methylstirene, partial oxidation of 257
Methylamine, ... 335
MgO. 373
Mild Oxidation, ... 747
Mn(III) oxo Carboxylato Complexes, ... 97
Mn(II) tetraarylporphyrins, ... 63
$Mo-SiO_2$, ... 459
Model Catalysts, ... 789
Molecular O_2 oxidation, ... 139
Molecular Oxygen Ions, ... 779
Molecular Oxygen, ... 133
Monolayer V_2O_5, ... 789
MoO_3 on TiO_2, ... 767
MoO_3, ... 459, 747, 757
Multicomponent Mo-catalysts, ... 585

Naphtols, ... 139
Nickel-Lead Oxides, ... 229
Nitriles, ... 229, 239
Nitro and Nitrato Complexes, ... 213
Nitrocompound, ... 861
Nitromethane, ... 861
Nitroxidation, ... 229, 239
NO 483, 675

^{18}O-Labelled Catalyst, ... 573
Olefin, ... 63, 71, 81
Oxidation of Alcohol Group, ... 147
Oxidation Mechanisms, ... 573
Oxidative Cleavage, ... 265
Oxidative Conversion, ... 483
Oxidative Coupling, ... 343, 365, 381
..... 393, 417, 427, 453, 469
Oxidative Dehydrogenation of Propane, 505, 515
Oxidative Dehydrogenation, ... 833, 869
Oxides, ... 437
Oxometalates, ... 205
Oxoporphyrinato Mo(V) , W(V) complex, ... 71
Oxovanadium Complexes, ... 265
Oxydehydrogenation, ... 843
Oxyesterification, ... 305
Oxygen, ... 177
Oxygen Activation, ... 139
Oxygen Adsorption, ... 725
Oxygen Isotope Exchange, ... 675
Oxygen-containing Functional Groups, ... 853
Oxyhalides, ... 595

P-Mo Heteropoly Compounds, ... 869

Palladium Nitro, ... 177
Palladium, ... 81, 159, 169, 213
Paramagnetic Centers, ... 853
n-Pentane Oxidation, ... 635
Perovskite Catalysts, ... 405
Phenol Adsorption and Oxidation, ... 825
Phenol Hydroxylation, ... 53
Phenol Oxidation, ... 139
Phenol Synthesis, ... 125
Phenol, ... 33, 109, 139
Phenylacrolein, ... 257
l-Phenylalanine, ... 661
Phorphyrins, ... 63
Photocatalysis, ... 675
Photocatalytic Oxidation, ... 683
Photochemistry, ... 335
Photoconductivity, ... 675
Phthalic Anhydride from n-Pentane, ... 635
Phthalic Anhydride, from o-xylene, ... 267
Platinized-carbon, ... 653
Platinum, ... 81, 89, 147, 159
Polycyclic Aromatic Oxidation, ... 247
Porous Electrode, ... 653
Porphyrins, ... 109
Propane Ammoxidation, ... 515
Propane Oxidation with SO_2, ... 527
Propane Oxidation, ... 491, 505
2-Propanol, ... 109
Propene, ... 683, 747, 807, 817
Propylene Oxidation, ... 295
4-(di-n-Propylsulfamyl)benzoic acid, ... 667
4-(di-n-Propylsulfamyl)toluene, ... 667

Quantum Chemical Calculation, ... 701

Radical, ... 335
Rare Earth Molybdate, ... 295
Reactor, Structured Fixed Bex, ... 267
Redox Kinetics, ... 823
Redox Operation in Riser, ... 553
Redox Properties, ... 693
Role of Double-Bond, ... 779
Ruthenium, ... 109, 119

Sb_2O_4, ... 757
Selective Oxidation, ... 1
Selective Reduction, ... 861
Selectivity, ... 147, 779
Selectivity Criteria, ... 585
SERS, ... 447
Shape Selectivity, ... 53
Silver, ... 707, 717, 725
Single Crystal Silver Surfaces, ... 739
SnSbO Mixed Oxides, ... 767
Solid Electrolyte Aided Modification of Work Function, ... 697
Spetroscopic Characterization, ... 295
Stability of Redox Chemistry, ... 553
Structured Fixed Bed Reactor, ... 267
Structure-sensitive, ... 717
Structure-sensitivite Reactions, ... 585
Styrene, ... 327, 739
Superconductivity, ... 287
Superoxide Radical Anion, ... 667
Supported La-oxides, ... 373
Supported Noble Metal Catalysts, ... 717
Supported Vanadium Oxide, ... 683
Surface Reaction, ... 861
Surface Species, ... 447

Tellurium Oxid containing Catalysts, ... 257
TGA Redox Studies, ... 553
Thin Layer Catalysts, ... 733
Thioethers, ... 109
Three Phase Reactor, ... 733
Ti-Silicalite, ... 33, 43, 53
Tin-germanium Phosphates, ... 327
TiO_2, ... 675
Toluene, ... 275, 817
TPR, ... 295
Transient Response, ... 707
Trimethyl-p-benzoquinone, ... 133
Trimethylphenol, ... 133
Turnover Number, ... 185
Two-phase Catalysts, ... 63

V-Antimonate based Catalysts, ... 515
V-Ti Oxides, ... 305, 825
V_2O_5 Catalysts, ... 789
Vanadium Oxide, ... 779
Vanadium Phosphorus Oxide, 563, 573, 605, 617
Vanadyl Pyrophosphate, 537, 553, 563, 625, 635
$VOPO_4$, ... 573

Well Characterized VPO, ... 605

o-Xylene Oxidation, ... 267,767

$YBa_2Cu_3O_{6+x}$ ($0<x<1$), ... 275

Zeolite-oxide Catalysts, ... 317
Zeolites, ... 43, 53, 483
ZrO_2 Catalyst, ... 343
ZSM-5, ... 483

STUDIES IN SURFACE SCIENCE AND CATALYSIS

Volume 1 **Preparation of Catalysts I.** Scientific Bases for the Preparation of Heterogeneous Catalysts. Proceedings of the First International Symposium, Brussels, October 14–17, 1975
edited by **B. Delmon, P.A. Jacobs and G. Poncelet**

Volume 2 **The Control of the Reactivity of Solids.** A Critical Survey of the Factors that Influence the Reactivity of Solids, with Special Emphasis on the Control of the Chemical Processes in Relation to Practical Applications
by **V.V. Boldyrev, M. Bulens and B. Delmon**

Volume 3 **Preparation of Catalysts II.** Scientific Bases for the Preparation of Heterogeneous Catalysts. Proceedings of the Second International Symposium, Louvain-la-Neuve, September 4–7, 1978
edited by **B. Delmon, P. Grange, P. Jacobs and G. Poncelet**

Volume 4 **Growth and Properties of Metal Clusters.** Applications to Catalysis and the Photographic Process. Proceedings of the 32nd International Meeting of the Société de Chimie Physique, Villeurbanne, September 24–28, 1979
edited by **J. Bourdon**

Volume 5 **Catalysis by Zeolites.** Proceedings of an International Symposium, Ecully (Lyon), September 9–11, 1980
edited by **B. Imelik, C. Naccache, Y. Ben Taarit, J.C. Vedrine, G. Coudurier and H. Praliaud**

Volume 6 **Catalyst Deactivation.** Proceedings of an International Symposium, Antwerp, October 13–15, 1980
edited by **B. Delmon and G.F. Froment**

Volume 7 **New Horizons in Catalysis.** Proceedings of the 7th International Congress on Catalysis, Tokyo, June 30–July 4, 1980. Parts A and B
edited by **T. Seiyama and K. Tanabe**

Volume 8 **Catalysis by Supported Complexes**
by **Yu.I. Yermakov, B.N. Kuznetsov and V.A. Zakharov**

Volume 9 **Physics of Solid Surfaces.** Proceedings of a Symposium, Bechyňe, September 29–October 3, 1980
edited by **M. Láznička**

Volume 10 **Adsorption at the Gas–Solid and Liquid–Solid Interface.** Proceedings of an International Symposium, Aix-en-Provence, September 21–23, 1981
edited by **J. Rouquerol and K.S.W. Sing**

Volume 11 **Metal-Support and Metal-Additive Effects in Catalysis.** Proceedings of an International Symposium, Ecully (Lyon), September 14–16, 1982
edited by **B. Imelik, C. Naccache, G. Coudurier, H. Praliaud, P. Meriaudeau, P. Gallezot, G.A. Martin and J.C. Vedrine**

Volume 12 **Metal Microstructures in Zeolites.** Preparation – Properties – Applications. Proceedings of a Workshop, Bremen, September 22–24, 1982
edited by **P.A. Jacobs, N.I. Jaeger, P. Jirů and G. Schulz-Ekloff**

Volume 13 **Adsorption on Metal Surfaces.** An Integrated Approach
edited by **J. Bénard**

Volume 14 **Vibrations at Surfaces.** Proceedings of the Third International Conference, Asilomar, CA, September 1–4, 1982
edited by **C.R. Brundle and H. Morawitz**

Volume 15 **Heterogeneous Catalytic Reactions Involving Molecular Oxygen**
by **G.I. Golodets**

Volume 16 **Preparation of Catalysts III.** Scientific Bases for the Preparation of Heterogeneous Catalysts. Proceedings of the Third International Symposium, Louvain-la-Neuve, September 6–9, 1982
edited by **G. Poncelet, P. Grange and P.A. Jacobs**

Volume 17 **Spillover of Adsorbed Species.** Proceedings of an International Symposium, Lyon-Villeurbanne, September 12–16, 1983
edited by **G.M. Pajonk, S.J. Teichner and J.E. Germain**

Volume 18 **Structure and Reactivity of Modified Zeolites.** Proceedings of an International Conference, Prague, July 9–13, 1984
edited by **P.A. Jacobs, N.I. Jaeger, P. Jirů, V.B. Kazansky and G. Schulz-Ekloff**

Volume 19 **Catalysis on the Energy Scene.** Proceedings of the 9th Canadian Symposium on Catalysis, Quebec, P.Q., September 30–October 3, 1984
edited by **S. Kaliaguine and A. Mahay**

Volume 20 **Catalysis by Acids and Bases.** Proceedings of an International Symposium, Villeurbanne (Lyon), September 25–27, 1984
edited by **B. Imelik, C. Naccache, G. Coudurier, Y. Ben Taarit and J.C. Vedrine**

Volume 21 **Adsorption and Catalysis on Oxide Surfaces.** Proceedings of a Symposium, Uxbridge, June 28–29, 1984
edited by **M. Che and G.C. Bond**

Volume 22 **Unsteady Processes in Catalytic Reactors**
by **Yu.Sh. Matros**

Volume 23 **Physics of Solid Surfaces 1984**
edited by **J. Koukal**

Volume 24 **Zeolites: Synthesis, Structure, Technology and Application.** Proceedings of an International Symposium, Portorož-Portorose, September 3–8, 1984
edited by **B. Držaj, S. Hočevar and S. Pejovnik**

Volume 25 **Catalytic Polymerization of Olefins.** Proceedings of the International Symposium on Future Aspects of Olefin Polymerization, Tokyo, July 4–6, 1985
edited by **T. Keii and K. Soga**

Volume 26 **Vibrations at Surfaces 1985.** Proceedings of the Fourth International Conference, Bowness-on-Windermere, September 15–19, 1985
edited by **D.A. King, N.V. Richardson and S. Holloway**

Volume 27 **Catalytic Hydrogenation**
edited by **L. Červený**

Volume 28 **New Developments in Zeolite Science and Technology.** Proceedings of the 7th International Zeolite Conference, Tokyo, August 17–22, 1986
edited by **Y. Murakami, A. Iijima and J.W. Ward**

Volume 29 **Metal Clusters in Catalysis**
edited by **B.C. Gates, L. Guczi and H. Knözinger**

Volume 30 **Catalysis and Automotive Pollution Control.** Proceedings of the First International Symposium, Brussels, September 8–11, 1986
edited by **A. Crucq and A. Frennet**

Volume 31 **Preparation of Catalysts IV.** Scientific Bases for the Preparation of Heterogeneous Catalysts. Proceedings of the Fourth International Symposium, Louvain-la-Neuve, September 1–4, 1986
edited by **B. Delmon, P. Grange, P.A. Jacobs and G. Poncelet**

Volume 32 **Thin Metal Films and Gas Chemisorption**
edited by **P. Wissmann**

Volume 33 **Synthesis of High-silica Aluminosilicate Zeolites**
by **P.A. Jacobs and J.A. Martens**

Volume 34 **Catalyst Deactivation 1987**. Proceedings of the 4th International Symposium, Antwerp, September 29–October 1, 1987
edited by **B. Delmon and G.F. Froment**

Volume 35 **Keynotes in Energy-Related Catalysis**
edited by **S. Kaliaguine**

Volume 36 **Methane Conversion**. Proceedings of a Symposium on the Production of Fuels and Chemicals from Natural Gas, Auckland, April 27–30, 1987
edited by **D.M. Bibby, C.D. Chang, R.F. Howe and S. Yurchak**

Volume 37 **Innovation in Zeolite Materials Science.** Proceedings of an International Symposium, Nieuwpoort, September 13–17, 1987
edited by **P.J. Grobet, W.J. Mortier, E.F. Vansant and G. Schulz-Ekloff**

Volume 38 **Catalysis 1987**. Proceedings of the 10th North American Meeting of the Catalysis Society, San Diego, CA, May 17–22, 1987
edited by **J.W. Ward**

Volume 39 **Characterization of Porous Solids**. Proceedings of the IUPAC Symposium (COPS I), Bad Soden a. Ts., April 26–29, 1987
edited by **K.K. Unger, J. Rouquerol, K.S.W. Sing and H. Kral**

Volume 40 **Physics of Solid Surfaces 1987.** Proceedings of the Fourth Symposium on Surface Physics, Bechyne Castle, September 7–11, 1987
edited by **J. Koukal**

Volume 41 **Heterogeneous Catalysis and Fine Chemicals.** Proceedings of an International Symposium, Poitiers, March 15–17, 1988
edited by **M. Guisnet, J. Barrault, C. Bouchoule, D. Duprez, C. Montassier and G. Pérot**

Volume 42 **Laboratory Studies of Heterogeneous Catalytic Processes**
by **E.G. Christoffel**, revised and edited by **Z. Paál**

Volume 43 **Catalytic Processes under Unsteady-State Conditions**
by **Yu. Sh. Matros**

Volume 44 **Successful Design of Catalysts.** Future Requirements and Development. Proceedings of the Worldwide Catalysis Seminars, July, 1988, on the Occasion of the 30th Anniversary of the Catalysis Society of Japan
edited by **T. Inui**

Volume 45 **Transition Metal Oxides**. Surface Chemistry and Catalysis
by **H.H. Kung**

Volume 46 **Zeolites as Catalysts, Sorbents and Detergent Builders.** Applications and Innovations. Proceedings of an International Symposium, Würzburg, September 4–8, 1988
edited by **H.G. Karge and J. Weitkamp**

Volume 47 **Photochemistry on Solid Surfaces**
edited by **M. Anpo and T. Matsuura**

Volume 48 **Structure and Reactivity of Surfaces.** Proceedings of a European Conference, Trieste, September 13–16, 1988
edited by **C. Morterra, A. Zecchina and G. Costa**

Volume 49 **Zeolites: Facts, Figures, Future.** Proceedings of the 8th International Zeolite Conference, Amsterdam, July 10–14, 1989. Parts A and B
edited by **P.A. Jacobs and R.A. van Santen**

Volume 50 **Hydrotreating Catalysts.** Preparation, Characterization and Performance. Proceedings of the Annual International AIChE Meeting, Washington, DC, November 27–December 2, 1988
edited by **M.L. Occelli and R.G. Anthony**

Volume 51 **New Solid Acids and Bases.** Their Catalytic Properties
by **K. Tanabe, M. Misono, Y. Ono and H. Hattori**

Volume 52 **Recent Advances in Zeolite Science.** Proceedings of the 1989 Meeting of the British Zeolite Association, Cambridge, April 17–19, 1989
edited by **J. Klinowski and P.J. Barrie**

Volume 53 **Catalyst in Petroleum Refining 1989.** Proceedings of the First International Conference on Catalysts in Petroleum Refining, Kuwait, March 5–8, 1989
edited by **D.L. Trimm, S. Akashah, M. Absi-Halabi and A. Bishara**

Volume 54 **Future Opportunities in Catalytic and Separation Technology**
edited by **M. Misono, Y. Moro-oka and S. Kimura**

Volume 55 **New Developments in Selective Oxidation.** Proceedings of an International Symposium, Rimini, Italy, September 18–22, 1989
edited by **G. Centi and F. Trifiro'**